THE

MATHEMATICAL PAPERS OF

ISAAC NEWTON

VOLUME I

1664-1666

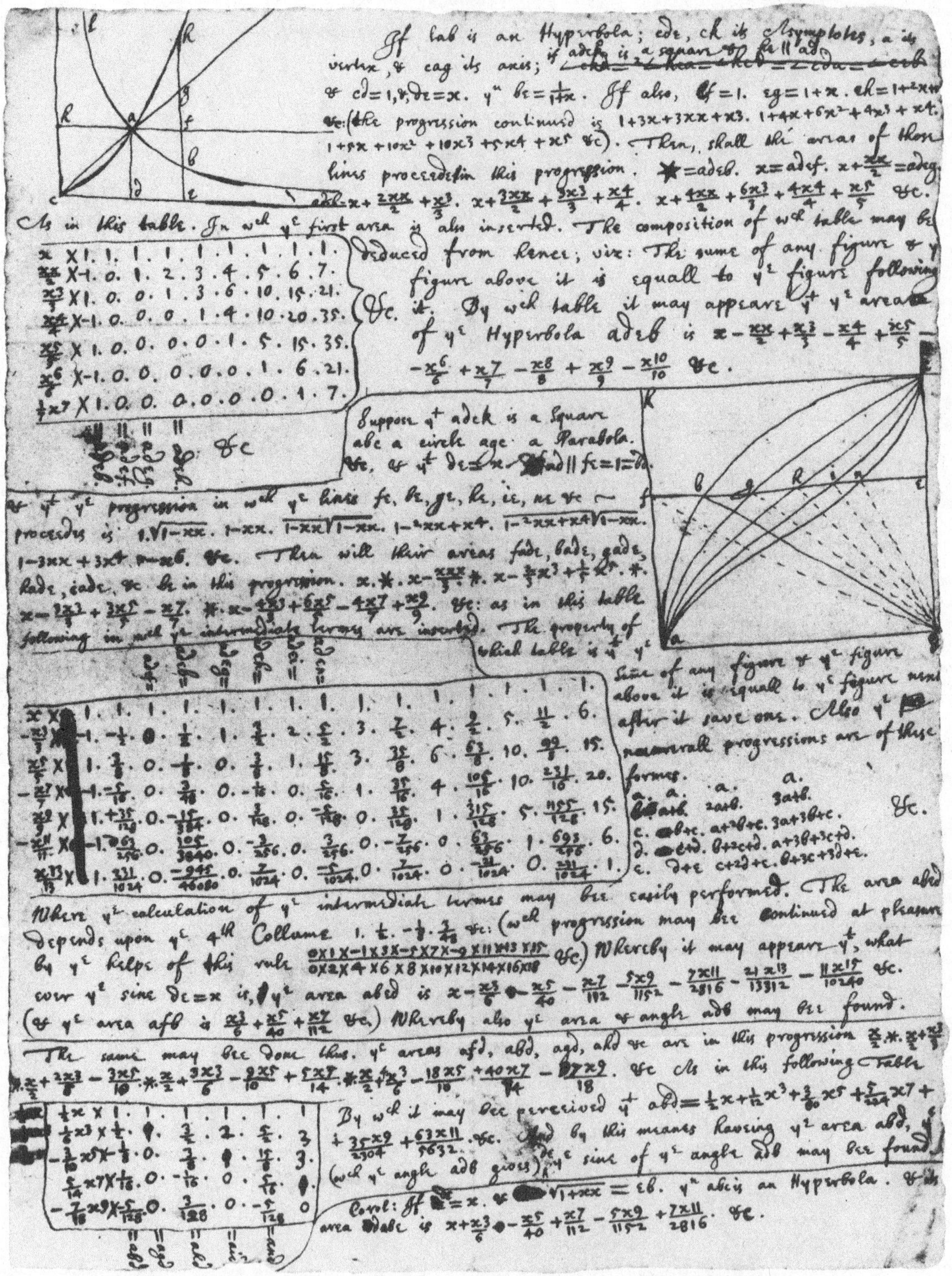

Binomial researches: interpolation of the extended Pascal triangle (I, 3, §4.1).

THE MATHEMATICAL PAPERS OF ISAAC NEWTON

VOLUME I

1664-1666

EDITED BY

D. T. WHITESIDE

WITH THE ASSISTANCE IN PUBLICATION OF

M. A. HOSKIN

CAMBRIDGE

AT THE UNIVERSITY PRESS

1967

CAMBRIDGE UNIVERSITY PRESS
Cambridge, New York, Melbourne, Madrid, Cape Town, Singapore, São Paulo

Cambridge University Press
The Edinburgh Building, Cambridge CB2 8RU, UK

Published in the United States of America by Cambridge University Press, New York

www.cambridge.org
Information on this title: www.cambridge.org/9780521058179

First published 1967
This digitally printed version 2008

A catalogue record for this publication is available from the British Library

Library of Congress Catalogue Card Number: 65–11203

ISBN 978-0-521-05817-9 hardback
ISBN 978-0-521-04595-7 paperback
ISBN 978-0-521-72054-0 paperback set (8 volumes)

TO RUTH

PREFACE

The depth and complexity of Isaac Newton's scientific writings—to make no mention of the involved psychological make-up of the man concealed beneath the public *persona*—have for almost three centuries defied the combined efforts of his various editors to encompass them in an orderly, definitive form. At the time of his death in 1727 he had already, largely through the publication of the several editions of his *Philosophiæ Naturalis Principia Mathematica* and *Opticks*, secured the dominating position in intellectual history which he has retained to the present day through many upsets and changes in critical enthusiasm. His contributions to mathematical thought have always been less immediately obvious and in their details little understood. Until the last decade, indeed, his public claim to mathematical pre-eminence rested on a handful of published tracts, mostly brief, never (with the possible exception of the *Arithmetica Universalis*) widely read and not wholly representative of his achievement as we may now know it from his private papers. No one, not even Newton himself, has ever made a complete gathering of the original texts on which such an assessment must be based. That I have sought to accomplish in this edition, the primary aim of which is to make available an adequate reproduction of those of Newton's mathematical papers now known to exist, with sufficient commentary to clarify for the modern reader the particularities of seventeenth-century idiom and to illuminate the contemporary significance of the text discussed. Incidentally, many texts of main importance to Newtonian scholarship are here printed for the first time, but I have sought to resist the temptation to establish an empty priority in publication by including trivia or pieces not directly pertinent to my theme.

Though I have long accepted the traditional high estimate of Newton's mathematical ability, my interest in the details of his achievement dates from the autumn of 1957 when, engaged in a general study of the sources of mathematical development in the seventeenth century, I first began to survey the Cambridge University Library holdings of his papers. I was immediately impressed by the depth and luxuriance of his unexplored manuscript researches. After a somewhat nervous, disorganized beginning I slowly took the measure of the particular problems which their study and editing for publication presented and in the spring of 1959 incorporated my preliminary findings in a research thesis.* During the following summer I made the acquaintance of Professor H. W. Turnbull who, in conversation and by correspondence, gave me the benefit of his lifetime's study of mathematical history and was a timely

* ULC. Ph.D. 3878, published subsequently as 'Patterns of Mathematical Thought in the later Seventeenth Century' (*Archive for History of Exact Sciences*, **1** (1961): 179–388).

steadying influence in curbing my wilder hypotheses. He himself had come to study Newton in depth only a few years before, on retirement from university teaching—too late, as he remarked, for him to attempt the comprehensive edition of Newton's mathematical papers which he now relinquished to me. Not my least debt to him is that he significantly widened the breadth of my documentary knowledge by making available to me a rich private store of Newton's papers. (Once owned by William Jones and deriving ultimately from the estate of John Collins, these fill the main gaps in the public holdings of Newtonian manuscript. It was through the agency of another friend, Sir Harold Hartley, that permission to reproduce them in the present work was later obtained.) Since that time I have been afforded opportunity to devote undivided attention to the preparation for publication of the extant corpus of Newton's mathematical papers. Of the eight volumes in which, it is intended, those papers will be reproduced, this is the first to appear. The remainder will be published in chronological sequence at appropriate intervals.

It remains to acknowledge the many institutions and individuals who have helped in the preparation of the present work. This could never have been undertaken without either generous financial support over many years or access to the various stores of Newtonian manuscripts. For monetary assistance during the period of preparation I have been indebted in turn to the former Ministry of Education, the Trustees of the Leverhulme Research Awards, the Royal Society and most recently to the former Department of Scientific and Industrial Research, whose help has culminated in a substantial research grant. For allowing reproduction of source material in their archives I thank collectively the Syndics of the University Library, and the Librarians of King's College, Trinity College and the Fitzwilliam Museum, Cambridge; of the Bodleian, Oxford; of the British Museum and the Royal Society, London; of the Niedersächsische Landesbibliothek, Hanover; and of the Pierpont Morgan Library, New York. The efficiency, courtesy and tolerance of their staffs I gratefully acknowledge and would express my especial debt to Mr P. J. Gautrey of Cambridge University Library. To a private owner of manuscript, also, my thanks for permission to reproduce his Newtonian papers.

Individually, I have been helped by many, not all of whom there is space to mention. For his technical advice and valuable help I would like to single out Mr Adolf Prag of Westminster School, London, who has read this first volume many times in manuscript and proof, contributing a number of necessary emendations to my editing of it. Again, I am greatly indebted to the interest of many senior friends, particularly Professor I. Bernard Cohen of Harvard University and Professor Sir William Hodge, former Physical Secretary of the Royal Society. To Sir Harold Hartley, whose encouragements have been manifold and unceasing, and to the late Professor Herbert W. Turnbull I express once more my particular gratitude. Above all, I must acknowledge a fundamental debt to Dr Michael A. Hoskin, for so many years my friend, tutor, mentor and collaborator, who has from the first been closely involved in the preparation of this edition. For its deficiencies, omissions and imperfections I alone stand responsible, but it is not too much to say that were it not for his patient wisdom and care I would long ago have been discouraged by the difficulties and disappointments which have arisen over the years.

To the Syndics and Secretaries of the Cambridge University Press I owe a final word of appreciation for the meticulous pains taken by their staff during each of the stages through which my submitted manuscript has passed in attaining the elegance and excellence of its present printed form.

D.T.W.

1 January 1966

EDITORIAL NOTE

The great majority of Newton's papers, carefully selected and ordered by him for posterity, passed at the time of his death into the hands of John Conduitt and subsequently into the possession of the Portsmouth family. The cumulative result of the well-intentioned efforts of the several scholars who have, during the past two centuries and a half, been permitted to divide and rearrange those 'Portsmouth' papers has been effectively to destroy all traces of Newton's own ordering of them. No longer can we know with certainty what he himself intended when he gathered his writings together in the last months of his life, burning a great part of his personal correspondence and, we may suspect, certain inferior technical papers he was unwilling to communicate to his successors. The point need not be laboured that even after the 1888 cataloguing committee had performed its task (diligently, if not with a full measure of insight) the Portsmouth manuscripts remained in a state of confusion which recent editors have found difficult to resolve. In our own preliminary examination of Newton's mathematical development as recorded in those papers neither the committee's scheme of classification nor the previous listings of Pellet and Horsley, nor indeed the traditional account of that development (evolved over the years by Newton's several biographers largely from his own later observations and Conduitt's collection of anecdote), were found to be of real help. In consequence, I felt obliged to evolve an independent chronology of that development solely, in the first instance at least, on the basis of available contemporary documents. The resulting scheme, the product of many years' study of manuscript sources, is the major hypothesis in the interpretative structure here imposed on Newton's papers. Let us dare to hope that the editorial techniques employed in filling out that scheme have approached asymptotically to the truth.

The primary intention of this edition is the publication of an accurate text of all Newton's significant mathematical manuscripts.(1) In achieving that aim

(1) Our definition of Newton's 'significant' mathematical work does not comprise the entirety of his calculating sheets or of his annotation of the work of others. In this first volume, for example, the notes made by him in his copy of Barrow's *Euclid* (Trinity College, NQ. 16.201) and the crude, uninteresting algebraic passages in a small quarto book (ULC. Add. 3995, otherwise containing extracts in a non-Newtonian hand from Quintus Curtius together with miscellaneous theological memoranda) are both omitted. In assessing what is significant the degree of excitement aroused on initially encountering individual documents has possibly been a determining influence: conversely, over-familiarity with several of Newton's most fundamental papers may have bred an undeserved complacency which, we may hope, does not obtrude itself in the commentary.

An 'accurate' text is, of course, a literary fiction: such a text is perhaps that which best accords at any given time with current, contingent scholarly and typographical criteria of

the first concern has been to establish the text of those papers, making use of modern photo-copying techniques where these proved convenient but in all doubtful cases reverting to the original autograph manuscripts. The majority of these were found to be in excellent condition and easily legible, but the remainder were less tractable, many having suffered advanced deterioration in paper quality from age and ill-use, while others had at various times been subjected to scorching by fire, staining by damp and other soiling. Where complete disintegration had not occurred, restoration of Newton's text was made with the aid of the magnifying glass and, in extreme cases, ultra-violet light: words and phrases then remaining incomplete were subsequently restored on the basis of existing fragments and surrounding context. Progressively a first version of the manuscript was transcribed (on loose sheets for easier reordering) according to a convention of coloured inks established *ad hoc* for the purpose.(2)

Having then an accurate transcription of pertinent manuscript, we entered on the secondary stage of shaping that relatively amorphous mass into the edited version here reproduced. At this level the mathematical significance of

---

manuscript reproduction and can at best hope to set a standard for future scholarship. These criteria in turn are imbedded in our capacity to comprehend a sequence of irregularly patterned ink-marks absorbed not always legibly into a thickness of paper, and upon them we raise a complex structure of hypothesis which combines a not completely consistent mass of external documentary knowledge with our interpretative assessment of authorship and dating of handwriting, speed and sequence of composition and, not least, the significance of cancellations. At a more sophisticated level we dare to restore fragmented phrases and amend syntactical inconsistencies and then seek to place each piece of text as a component in a higher-order structure of explanation, guessing at the previous existence of documents needed to fill gaps in our scheme, revising first interpretations as our knowledge of context widens and continuously checking our understanding against simplifying canons of logical consistency and historical reasonableness. Ineluctably, each stage of the editing process imposes its compromises on the naïve ideal of unaltered facsimile reproduction of the original manuscript filled out with pertinent objective comment. In framing the pattern of the present edition we have been profoundly aware of these fundamental problems of critical scholarship.

(2) We may add that in the years it took to prepare an intimate knowledge, in part intuitive, was acquired of the various changes undergone by Newton's writing style between early youth and old age and that we came ultimately to trust our ability to fix the date of composition of an unexamined portion of autograph manuscript accurately by sight to within half a dozen years (and sometimes even more narrowly still). That skill has been employed to good purpose, we hope, in determining bounding dates for certain sets of papers whose composition history is wholly unknown. In the event, the resulting chronology of Newton's mathematical development has proved both credible and self-consistent: occasionally, indeed, additional documentary evidence much later came to light which has amply confirmed (but never refuted) its accuracy. (An example in point is the mass of Newton's geometrical research, to be reproduced in our seventh volume, which was evidently intended for a mysterious 'Liber Geometriæ' which had left no trace in known Newtonian literature. Failing all else, its date of composition was set tentatively by eye as mid-1693 (± 2 years). Subsequently, the researches of H. W. Turnbull brought to light an unpublished memorandum of David Gregory which fully confirmed the conjecture.)

the text is paramount and it has been our worry to make that meaning as evident as possible to the modern reader while remaining close to the manuscript original. It will be obvious that facsimile reproduction even of figures has, not a little regretfully, been rejected. The main reasons for so doing are that the physical condition of the original is not infrequently unsuitable for such reproduction(3) and, more generally, that Newton's text is on occasion too involuted and overwritten to be readily comprehensible except in an unravelled version. No less clearly, any other manner of reproducing a historical text is justifiable only in so far as we transpose it following a consciously framed set of conventions. Those chosen here for the most part conform to accepted scholarly standards and we need not detail their nature. It has been our ideal to use appropriate visual equivalents everywhere, preserving contractions, superscripts, subscripts and the like in the verbal no less than in the mathematical text,(4) accurately redrawing all figures and setting them in a place in the printed text consonant with their position in Newton's manuscript: non-trivial cancellations are indicated either in the commentary or, in the case of longer passages, by reproducing them as uncancelled text but setting against them in the left-hand margin a double vertical bar. In one or two instances, not finding it possible uniquely to determine Newton's spatial layout of his manuscript page we have had to compromise but as far as can be ascertained this has nowhere resulted in an intolerable simplification of the original text.(5) Inevitably, however, despite all effort the reproduced version does not wholly reflect the visual ambience of the autograph source. So, therefore, that the reader may be able both to check the accuracy of our reproduction in particular cases and to encounter, vicariously though it may be, Newton's mathematical

(3) This will be evident from several of the photographic illustrations: of the original manuscripts reproduced in Plates II and III of the present volume, for example, the page-bottoms are badly damp-stained, that of the latter being in addition fragmentary.

(4) Thus we print 'w^{ch}', 'y^t', 'ꝑtis' and '${}_2A$' rather than 'which' (or the misbegotten 'wch'), 'that' (or 'yt'), 'partis' (or 'ptis') and '$2A$'.

(5) In illustration, we have not been able to decide whether, in reproducing the early algebraic texts, to set coefficients of algebraic variables as superscript or on the line. Newton himself is not consistent in this matter, sometimes writing '$2x^3$', at others '${}^2x^3$' but mostly penning a mean between the two which could be taken for either and which it seems impossible to reproduce in a fixed-level typesetting. The notation '${}^2x^3$' was established in print by Frans van Schooten (compare Florian Cajori, *A History of Mathematical Notations*, **2** (Chicago, 1928): 351, §307), but he likewise makes inconsistent use of it and we may guess that his printer, Elzevier, decided arbitrarily on the level of each coefficient as he came to it in the copy. Whether Newton here consciously followed Schooten's practice would appear impossible to adjudge: his favourite use of it, indeed, is in such examples as '${}^2\not{3}x^3$', where his written superscript '2' may have been intended by him to be in line with his cancelled '3'. Rather than decide each occurrence arbitrarily on our assessment of the coefficient's horizontal position we have chosen to print all coefficients in line with the variable they denominate (following Newton's own invariant custom in his later years).

arguments in their original dress, we have inserted photocopies of typical leaves of manuscript.

In our commentary we have sought to help the modern reader to understand, with a minimum of unnecessary strain, the substance and idiom of the papers reproduced, and to that end we make use of historical introductions, technical footnotes and English paraphrase. In the prefaces and appendices we have sought to present items of historical and biographical interest which illustrate and aid the appreciation of Newton's text but which are not so well known as to seem superfluous, seeking to create—in the pattern of our own image of him—a true picture of the character and surroundings of the man who wrote the papers now edited. The notes(6) are more narrowly technical: there we explain points of idiom and mathematical usage, recast Newton's sequence of argument in modern notation (amending it where necessary) and refer to pertinent secondary works. (It should be remembered that during his adult years Newton had access to several of the foremost libraries of his day and that little of the weighty mathematical heritage of the period can be disregarded as irrelevant to his developing knowledge and genius.) In addition, since his manner of expression is at times—and particularly in his early papers—excessively abrupt, it has on occasion been thought appropriate to add a paraphrase, either in footnote or following on his text. Where the original manuscript is in Latin we have considered it obligatory to juxtapose an accurate English translation (usually, when the Latin piece is extensive and continuously written, on the facing page though shorter phrases, including cancelled variants, are rendered in footnote). Little variant revisions of papers already translated on previous pages and non-Newtonian appendices are not usually given in English as well. At all times our transliterative ideal has been to render Newton's prose word for word in modern equivalent rather than attempt the spurious archaism of seventeenth-century phrasing or remould the balanced clauses of the original's syntactical structure into an elegantly paraphrased contemporary form.(7)

In collective identity, lastly, of details in this edition it has been felt more appropriate and immediately useful to introduce each volume with an analytical table of contents and manuscript locations rather than to add to it the now usual elaborate concluding indexes of names and topics. Except in the

(6) These we place at the foot of the page to which they pertain wherever possible (though printing difficulties on occasion dictate that they be advanced or retarded), while corresponding places in the text are identified by superscript numbers in round brackets. We may note that, as with verbal text, square-bracketed numbers are invariably our insertions in (or emendations of) Newton's text.

(7) Newton, of course, worked easily and naturally—from about 1668 at least—in the thought-forms of classical (scholastic) Latin and only rarely made a preliminary English draft which would serve as an unexceptionable model for translation.

forewords, footnotes and appendices names of individuals occur but rarely and we choose, in older style, to outline the continuous development of mathematical structure in each volume rather than pinpoint relatively unimportant discrete points of historical detail. For the time being a short author index in each volume serves the latter purpose, while in the final one an elaborate, systematically cross-listed general index will, it is hoped, be a sufficient pointer to the macrocosm of individuals who, along with their works, have entered these pages.

GENERAL INTRODUCTION

Before Halley edited the *Principia* for him in the summer of 1687 no significant portion of Newton's extensive mathematical researches had been printed,(1) though, as we shall see in future volumes, Newton in the early 1670's had (with John Collins' earnest support) fought hard to have both his annotations on Kinckhuysen's Dutch *Algebra* and an extensive fluxional tract of his own appear in the bookshops. Thwarted by the unsaleability of technical works at a time of acute depression in the book trade following the Great Fire of London, he ultimately relinquished hope of publication. (The tract on fluxions was, in consequence, to reach the public only after his death while the Kinckhuysen commentary disappeared into centuries' long oblivion.) We may suspect that the not wholly relevant mathematical sections which he packed into the opening pages of his *Principia* were at once his outlet for a long frustrated talent and his conscious revenge on a world of mathematical smatterers and practitioners which had hitherto scorned him. His first independent mathematical publications, his 'Two Treatises of the Species and Magnitude of Curvilinear Figures' (namely, the *Enumeratio Linearum tertij Ordinis* and *Tractatus de Quadratura Curvarum*), were sneaked by him into print in 1704 in appendix to his *Opticks*, while their English versions were to appear even more obscurely half

(1) We except some improvements which Newton added to the printed version of Isaac Barrow's scientific lectures (*Lectiones XVIII Cantabrigiæ in Scholis publicis habitæ; in Quibus Opticorum Phænomenωn Genuinæ Rationes investigantur, ac exponuntur. Annexæ sunt Lectiones aliquot Geometricæ*, London, 1670) and a few excerpts from his mathematical correspondence with Leibniz in 1676 which John Wallis published, without Newton's express permission, in his *A Treatise of Algebra, both Historical and Practical.... With some Additional Treatises* (London, 1685). By comparing the text of Barrow's 'Optical Lectures' with his own subsequent *Lectiones Opticæ* (ULC. Dd. 9.67, published in full in their original Latin at London in 1729, though an English version of the first, theoretical part had appeared the previous year as *Optical Lectures Read in the Publick Schools of the University of Cambridge, Anno Domini, 1669*) we see that Newton's help was virtually restricted to two passages (*Lectio* XIII, §XXVI: 94 and *Lectio* XIV: 103–4) giving improved variants of Barrow's own theorems and for which the latter thanks an unnamed 'amicus' (though Newton is cited by name in the preceding *Epistola ad Lectorem* as 'collega noster (peregregiæ vir indolis ac insignis peritiæ) [qui] exemplar revisit, aliqua corrigenda monens, sed & de suo nonnulla penu suggerens, quæ nostris alicubi cum laude innexa cernes'). Equally, his assistance with the following *Lectiones Geometricæ* was apparently restricted to the section (at the end of *Lectio* X) on James Gregory's analytical tangent-method: as he wrote to Chamberlayne in June 1714, 'a paper of mine [the *De Analysi*] gave occasion to Dr Barrow to shew me his method of Tangents before he inserted it into his 10th Ge[o]metrical Lecture. For I am that friend [the "amicus" (p. 80) on whose advice the section was inserted] wch he there mentions' (ULC. Add. 3968.30: 441^{r}; cf. L. T. More, *Isaac Newton: a Biography*, New York, 1934 (reissued 1962): 185, note 35). Wallis restricted his quotation (*Algebra*, chs. LXXXV, XCI–XCIII: 318–20, 330–8 and XCIV–XCV: 338–47 respectively) to passages from Newton's letters of 13 June and 24 October 1676, the famous 'epistolæ prior et posterior', relating to his invention of the general binomial expansion and his extraction of the roots of 'affected' equations.

a dozen years later as articles on 'CURVES' and 'QUADRATURE' in the second volume of John Harris' *Lexicon Technicum*. The *Arithmetica Universalis: sive De Compositione et Resolutione Liber*, published at Cambridge in 1707 by William Whiston from his deposited Lucasian lectures of thirty years before, was scorned by Newton till he found time and patience to issue his own improved edition of the work in 1722: Raphson's English version of the book he had no authority to supervise. The short pieces *De Analysi per Æquationes Numero Terminorum Infinitas* and *Methodus Differentialis* he allowed William Jones in 1711 to print (along with reproductions of his tracts on quadratures and cubics) as the *Analysis per Quantitatum Series, Fluxiones, ac Differentias; cum Enumeratione Linearum Tertii Ordinis*. That, apart from a few cautious references, made at various times during his priority dispute with the Leibnizians, to his manuscripts on fluxional analysis, was his total public contribution to mathematics.(2)

As we now know Newton's creative mathematical researches continued with little break between the opening of his last undergraduate year at Cambridge, say from the early summer of 1664,(3) till he left university life behind him more than thirty years afterward in early 1696 for the bustle of London and the office of Warden of the Mint. Even then his inspiration flickered spasmodically on for another two decades, blazing up from time to time in answer to challenges from Leibniz and Johann Bernoulli and in support of his own claim to priority in systematizing the calculus, and dying out finally only in the very last years of a long life. During his thirty years of active mathematical research there issued from his pen a continuous stream of preliminary worksheets, revised drafts and final versions carefully polished for intended deposit in public archive or for the printed page. Through various circumstances (which we will examine in their due place) a large portion of those mathematical papers remained hidden from his contemporaries throughout his life and when he died these autograph manuscripts became the only source of knowledge, not always accessible, never closely studied and rarely published even in fragment, of a significant area of Newton's achievement in exact science.

At no period in history have all these papers been gathered together, nor are

(2) It is not our intention to go deeply into bibliographical aspects of Newton's mathematical work or systematically to assess its impact on his contemporaries and successors, though we will discuss some important neglected features of it (in our final volume) on the basis of original manuscript and printed sources. When in due course we come to reproduce the autograph drafts of his published works we will indicate for each the major points of difference between manuscript and published versions and also sketch its printing history. For the moment the reader is referred to our introductions to the recent collected reprint of English renderings of those works (*The Mathematical Works of Isaac Newton*, ed. D. T. Whiteside, 2 vols., New York, 1964–66).

(3) See the Introduction to Part 1 of the first volume, and compare D. T. Whiteside, 'Isaac Newton: Birth of a Mathematician', *Notes and Records of the Royal Society of London*, **19** (1964): 53–62.

they now ever likely to be. As a young man Newton had sought consciously to make his name in the intellectual world by communicating certain (not always the severest and most advanced) of his manuscripts to influential acquaintances and those papers were not in every case returned intact. Through Isaac Barrow in July 1669, for example, he made known his *De Analysi* first to John Collins and subsequently to other interested parties in England and France; a little after, he allowed Collins both to examine a wide range of his mathematical and optical papers and to copy certain portions of them, if not to abstract the Newtonian fragments which were found with those copies after his death. Later, Newton's motive changed more to one of disinterested generosity and the natural desire to reveal to others what he had discovered. Throughout his life he remained willing to permit a number of his fellows to view his private papers and, indeed, at times gave away items from them to a favoured few. On one occasion or another John Craige, David Gregory, John Raphson, Edmond Halley, Fatio de Duillier, William Jones, John Keill, Henry Pemberton and possibly John Wallis, Abraham de Moivre and Nicholas Bernoulli were allowed varying degrees of freedom to inspect his store of manuscript, and Gregory and Jones in particular became possessed thereby of certain autograph fragments of considerable interest (the former, apparently, so that he might incorporate them into the extensive commentary on the *Principia* which he was preparing in the early 1690's, the latter that he might publish them in an abortive intended sequel to his collection of Newton's shorter mathematical tracts, the *Analysis per Quantitatum Series, Fluxiones ac Differentias*). These exceptions apart, however, the bulk of Newton's mathematical papers—essentially all but some important early researches in calculus and geometry—remained intact in his hands till his death, along with his other personal, scientific, religious and administrative drafts and correspondence. The subsequent history of this main corpus of Newton's autograph legacy has never been recorded in detail. Since it explains in large part why no serious effort to publish a definitive edition of Newton's contribution to any branch of intellectual history has yet been made, we may here examine it.

Newton died intestate on 20 March 1726/7(4) and without immediate family. Immediately there arose among the surviving relatives a quarrel of no

(4) A 'Munday', as his contemporary biographer Stukeley records (A. Hastings White, *Memoirs of Sir Isaac Newton's Life by William Stukeley, M.D., F.R.S. 1752. Being some Account of his Family and chiefly of the Junior Part of his Life* (London, 1936): 83; compare *The London Gazette*, no. 6569 for 1–4 April 1727). The dates which Newton himself set on his papers are likewise invariably Julian. For those who wish to find the weekday corresponding to any given date in that calendar we might note that the d-th day of the m-th month of year y is the k-th day of the week where $d+[\frac{1}{5}(13m+3)]+[\frac{5}{4}y] \equiv k \pmod 7$, providing always that we take January and February as belonging to the previous year: thus 21 May 1665 (O.S.) is the first weekday

mean dimensions over the division of his estate, built by him in his lifetime into a considerable fortune. John Newton of Colsterworth, great grandson of Isaac's uncle Robert, was his heir-at-law, but the main dissension occured between the offspring of his mother's second marriage (to the Reverend Barnabus Smith) and above all between Newton Smith and Thomas Pilkington, eldest sons of his step-brother Benjamin and step-sister Mary, and John Conduitt, acting for his wife, Catherine, daughter of Newton's step-sister Hannah. The issue was further complicated by Newton's position at his death as Master of the Mint: in that appointment he was personally accountable for all outstanding debts at the Mint and his estate now became liable to seizure until the account was settled. Only Conduitt of all the relatives was wealthy enough to stand bond for Newton's 'debts' and it was in virtue of the considerable expense and trouble to which he had put himself that he later claimed all Newton's unpublished papers. As he wrote:[5]

When Sr Isaac Newton died he was indebted to the Crown £34,330. Mr John Conduitt was the only person who could without difficulty make up & pass that account & till that was done none of the relations could have any benefit from Sr I. Newtons effects. Besides the account of £34,330 Sr I. Newtons Adm^{rs}[6] were to pass a trial of all the monies coined by Sr I. Newton from the 24 August 1724 to the time of his death, & unless he were acquitted by a Jury of Goldsmiths at the trial of the Pix before my Ld Chan^{r} & the Privy Council, his estate was subject to fine & ransom...wch could not be got over...by any of the relations but Mr Conduitt.

After several contests & disputes in the Prerogative Court about the Administration & the disposal of the Manuscripts the following agreement was made... & the same was made an act of the Court of the second sessions in Easter term 1727 & signed by Mr Conduitt, & Mr Tho. Pilkington & Mr Benjamin Smith, the two Adm^{rs}[7]. What relates to the Manuscripts is as follows. Viz—

That the papers & Manuscripts be first perused by the parties & such as are treatises be afterwards examined by Dr Pellett & printed if thought proper by him & the Adm^{rs} & sold to the best advantage, & in the meantime be loged with Mrs Conduitt. That

---

and so (perhaps surprisingly when we recall Newton's religious severity) a Sunday. Dates in the Gregorian calendar (introduced into England in 1752) are $\left[\frac{3c}{4}-2\right]$ days ahead of Julian ones, where c is the century: that is, 10 in the 17th century (up to 29 February 1699/1700) and 11 thereafter (up to 1800).

(5) King's College, Keynes MS 127^{A}.5: 'An Account of the right & title John Conduitt his heirs & assigns have to the Manuscripts of Sr Isaac Newton & of the papers touching the same.' Since Conduitt notes that Newton's *Prophecies* 'are printed', the account was written up in or after 1733 (compare note (13)).

(6) The Administers appointed by the Prerogative Court of Canterbury (on 18 April 1727) were Thomas Pilkington, Benjamin Smith and Catherine Conduitt (Keynes MS 127^{A}.2).

(7) Conduitt's copies of this agreement and the court order (made in Doctor's Commons on 3 April) confirming it are now Keynes MS 127^{A}.1/3. Clarification of his temporary guardianship of the manuscripts was made in a further agreement on 27 April (Keynes MS 127^{A}.6).

Mr Conduitt be oblig'd to pass the account with the Crown or indemnify the deceased's estate from the same.... In consideration of wch he shall have such papers wch shall not be thought proper to be printed giving bond that if ever he shall publish any or make any advantage thereof to be accountable for the same....(8)

All the relations perused all the papers & Manuscripts at Sr I. Newton's dwelling house & took Catalogues of them. The appraisers Comins & Ward examined them & thought only the Chronology & Prophecies fitt to be appraised & sett the value of £250 upon the Chronology, & no value upon the Prophecies they being imperfect....(9)

Dr Pellett examined them all & certified under his hand what were proper to be printed, what not, & what were to be reconsidered wch certificate is attested by Thomas Pilkington one of the two Adm^{rs} & a copy of the schedule of the Manuscripts & the said certificate is lodged in the prerogative court, being annexed to the bond of £2000 wch I gave in pursuance of the agreements made in Court to be answerable to the relations for all profit I should make by selling or otherwise disposing of the said Manuscripts.(10)

Dr Pellet certified that none of the Manuscripts were fitt to be printed except N°. 80 being the Chronology, wch was sold for £350 & printed.(11)

(8) Conduitt preserved the certificates from the Deputy Auditor, J. Oakley, that he did in fact indemnify the estate from the Crown's demands on 19 May 1727 and that he settled the Mint account on 17 January 1728/9 (Keynes MS 127^A.8/9).

(9) This agrees with the 'inventary' of Newton's house made under the Court's Commission of Appraisement of 18 April 1727 by Valens Comyns and Thomas Ward together with Thomas Money, William Carr and Fletcher Gyles. A 'short chronicle from the first memory of things in Europe, being an introduction to the Chronology of ye Antients, containing twelve pages in folio' and 'the Chronology of the Antients in five chapters, containing ninety pages in folio' are 'both valued att £250. 0. 0', while 'the history of the prophecies in ten chapters containing 88 pages in folio and also part of the eleventh chapter containing two pages folio unfinisht' is not priced. The inventory notes also 'Manuscripts in a box sealed up at the house of John Conduitt Esq.' and 'several boxes at Sir Isaac Newton's house wherein are contained many loose papers and Letters, relating to the office of the Mint, the manuscripts above mentioned and his Mathematical Works already published and likewise two small parcels of papers of the same kind in the box att the house of John Conduitt Esq'. See R. de Villamil, *Newton: the Man* (London, [1931]): 49–61, especially 54–5.

(10) Conduitt's copy of his bond (given in late May) and his preliminary draft of it are now Keynes MS 127^A.2/7. Pellet's inventory of the manuscripts, made in three stages on 20, 22 and 26 May, was published (perhaps from the deposited original) by Charles Hutton in his *A Mathematical and Philosophical Dictionary...In Two Volumes*, London, 1795 (2, 148–57: Article 'Newton (Sir Isaac)', especially 154–7). Conduitt's copy of the list, which does not agree with Hutton's in every detail, is preserved in King's College (Keynes MS 127^A.4). Though the fact does not seem to have been officially recorded Pellet made at least a partial reassessment in the following autumn, for in the frontispieces to two of Newton's early scientific notebooks (ULC. Add. 3996, 4000) he has signed his name to 'Sep. 25 1727 Not fit to be printed'.

(11) This manuscript—actually two, 'twelve half sheets in folio' and 'ninety two half sheets in folio' (whose originals, bound in red morocco and labelled 'John Conduitt', are now ULC. Add. 3988 and 3987)—was published by the Strand bookseller J. Tonson as *The Chronology of Ancient Kingdoms Amended. To which is Prefix'd, A Short Chronicle from the First Memory of Things in Europe, to the Conquest of Persia by Alexander the Great*, London, 1728.

N°. 33. A Mathematical tract de Motu Corporum [Liber Secundus] wch was sold to Tonson for £31–10 & printed under the title of De Mundi Systemate.(12)

N°. 81. The prophecies wch are printed.(13)

N°. 38. Paradoxical questions concerning Athanasius.(14)

N°. 61. an Imperfect Mathematical tract.(15)

Which three last are not judged by Dr Pellet absolutely to be printed but only to be reconsidered as appears by his certificate.

All the other Manuscripts belong to me & are mine not only by way of custody as acting Adm^r in right of my wife, but in property on account of a valuable consideration I gave for them, viz—passing the accounts of Sr I. Newton & the trial of the Pix, wch were points of the greatest consequence to the relations & were not effectuated by me without great trouble & expence, wch I paid out of my own pocket over & above the fees allowed in my account & have not charged to the Adm^n, but tho^u these manuscripts are my property I & my heirs & assigns must account to the relations & their representatives for their share of all profit that shall be made by selling or otherwise disposing of the same, reserving always one eighth share of the profit for themselves.(16)

It seems abundantly clear that the sole interest of Newton's relatives—John and Catherine Conduitt apart—in his estate was a financial one. Thomas Pellet's thick clumsy annotations 'Not fit to be printed', which appear on the

(12) This tract of 'Fifty six half-sheets in folio de motu corporum the greater part not in S^r I's hand' (Pellet)—whose original, bound and lettered uniformly with the preceding and following works, is preserved as ULC. Add. 3990—was published jointly by Tonson, J. Osborn and T. Longman as *De Mundi Systemate Liber Isaaci Newtoni*, London, 1728. In 1964 William Dawson's in their Catalogue 130 (*Cambridge Science and Medicine, 1551–1958*) listed for sale Conduitt's presentation copy of this to William Stukeley: a little confusingly Stukeley himself has noted on its title page 'edidit Johēs Conduit' but we should understand that Conduitt was merely reponsible for its publication (owning the manuscript) and perhaps corrected its proofs, not that he edited the work in any creative sense. An unfinished revision of the piece, which Newton deposited as five of his 1687 lectures (beginning Thursday, 29 September) in Cambridge University Library, is now ULC. Dd. 4.18, and in that form it was copied by Roger Cotes in 1700 (Trinity College, Cambridge, R. 16.39). The work is, of course, Newton's first version (composed perhaps in mid-1685) of Book 3 of his published *Principia*.

(13) These 'forty half sheets in folio' (now ULC. Add. 3989) appeared as *Observations upon the Prophecies of Daniel, and the Apocalypse of St John. In Two Parts*, London, 1733 (printed by 'J. Darby and T. Browne in Bartholomew Close').

(14) This still unpublished manuscript, 'Thirty one half sheets in folio' (now Keynes MS 10 in King's College, Cambridge) has been discussed, with extracts, by David Brewster (*Memoirs of the Life, Writings and Discoveries of Sir Isaac Newton*, Edinburgh, 1855: **2**: 342–6), H. McLachlan (*Sir Isaac Newton: Theological Manuscripts*, Liverpool, 1950: 102–3) and F. E. Manuel (*Isaac Newton, Historian*, Cambridge, 1963: 158–9).

(15) This, listed by Pellet as 'One hundred & forty four quarter sheets & fifty half sheets in folio being loose mathematical papers', very probably combined the unfinished 1666 English and 1671 Latin fluxional tracts (now ULC. Add. 3958.3: 48^v–63^v and 3960.14 respectively) which Horsley was to find packaged together in 1777. See also note (17).

(16) We have not discovered whether this last provision, presumably still legally effective, was implemented when the Portsmouth family in 1936 (note (57)) sold that portion of Newton's papers then remaining in their possession.

flysheets of Newton's surviving notebooks and other manuscripts, may now seem at once pitiful and ludicrous, but we should remember his brief: he was engaged by the executors of the estate to go through the papers singling out those which could be turned into ready cash by a quick sale to the printer—and no publisher at that time would purchase any but a finished, ordered tract. Presumably Pellet, a Fellow of the Royal Society in a period when Fellowship was no unimpeachable mark of scientific ability, was competent enough in his restricted role as printer's skivvy and it was not his fault but rather that of Newton's pennypinching relations that no one better able to appreciate the riches of the unpublished mathematical papers was appointed to the task.(17) We may be exceedingly grateful in retrospect that Conduitt, acting ultimately out of the purest of motives (that of honouring a great man into whose family he had married), was able and willing to manœuvre the papers into his own control. Had he failed, the consequences for future Newtonian scholarship would have been disastrous.

Pellet's list(18) is no more informative than we might reasonably expect. Apart from the four manuscripts singled out by Conduitt, which may be accurately identified, it is not easy to trace corresponding items in his and later catalogues. We have already(19) tentatively identified No. 61 ('One hundred & forty four quarter sheets & fifty half sheets in folio being loose mathematical papers [in part] to be reconsider'd') as the 1666 and 1671 fluxional tracts which Horsley was to find parcelled together in 1777, while No. 4 ('Trigonometria, about five sheets') is probably part at least of Newton's 'Trigonometria succinctè proposita et nova methodo demonstrata a S^{to} Joanne Hareo Arm^{r}'.(20) No. 63 ('A Folio Cōmonplace book part in Sr Isaac's hand') is doubtless his Waste Book, and we may be sure that his mathematical notebooks(21) are

(17) In this connection we may note that when Henry Pemberton sought, shortly after Newton's death, to publish the 1671 fluxional tract he was forestalled because 'the owners of the copy asked more money than the booksellers cared to advance'. (See James Wilson's preface to his edition of Pemberton's *Course of Chemistry*, London, 1771: xvi.)

(18) Conduitt described it (Keynes MS 127^{A}.4) as 'A List of the papers & manuscripts belonging to Sr Isaac Newton K^{t} deceased taken by the relations upon perusing & examining the same', Hutton (*A Mathematical and Philosophical Dictionary*, 2: 155) more narrowly as 'A Catalogue of Sir Isaac Newton's Manuscripts and Papers, as annexed to a Bond, given by Mr. Conduit, to the Administrators of Sir Isaac: by which he obliges himself to account for any profit he shall make by publishing any of the papers. Dr. Pellet, by agreement of the executors, entered into Acts of the Prerogative Court, being appointed to peruse all the papers and judge which were proper for the press'. (Hutton may have had his information second-hand through Samuel Horsley rather than directly from the deposited Bond.) It is divided into three sections (Nos. 1–20, 21–40 and 41–82) which Pellet examined on 20, 22 and 26 May 1727 respectively. See also note (10) above.

(19) Note (15) above.

(20) ULC. Add. 3959.4: to be reproduced in volume IV.

(21) That now in the Fitzwilliam Museum, Cambridge, and ULC. Add. 4000.

included in No. 82 ('Five small bound books in duodecimo [the greatest part] not in Sir Isaac's hand being rough calculations'), but the other mathematical entries are wholly vague.(22) Conduitt himself seems never to have attempted further definition of the anonymous, conglomerate mass of Newtonian manuscript in his possession, but during the remaining decade of his life interested himself almost entirely in preparing a non-technical biography of Newton, never completed and indeed hardly begun.(23) His wife Catherine, a little before her death in January 1739, toyed with those of her uncle's papers relating to religion and chronology and in 1737 added a codicil to her will directing her executor to

lay all the Tracts relating to Divinity before Dr Sykes...in hopes he will prepare them for the press....all of them I ordain shall be printed and published, so as they be done with care and exactness: and whatever proffit may arise from the same, my dear

(22) No. 24 lists 'Three hundred & fifty three half sheets in folio & fifty seven in small Quarto being foul & loose papers relating to figures & mathematicks', No. 44: 'Four hundred & ninety five half sheets in folio being loose & foul papers relating to Calculat^{ns} & Mathematicks'; Nos. 48, 57, 61, 70 and 77 together comprise 619 'half sheets' and 215 'quarter sheets' in-folio 'being loose mathematical papers', while No. 76 is described scarcely more precisely as 'Forty half sheets being loose papers foul & dirty relating to Calculations'.

(23) Conduitt did little more than make a reasoned, not very critical compilation of the many anecdotes for which he canvassed among Newton's acquaintances or which he pencilled down in his little green notebooks from his reading of published literature. His collection (now for the most part gathered in Keynes MS 130–137 in King's College, Cambridge) is the source for most of the usual stories about Newton the man, some accurate and invaluable, others mythical and worthless: we shall draw upon it from time to time in this edition. Conduitt, too, made extensive use of his gathered material in preparing the *Memoir* (Keynes MS 129.1–3) which he sent to Fontenelle in the early summer of 1727 for incorporation in the latter's widely popular 'Éloge de M. Neuton' (*Histoire de l'Academie Royale des Sciences de l'Année M. DCCXXVIII*, Paris, 1728: 151–72), but otherwise his rich store of anecdote remained unused till Edmund Turnor inserted portions of it in his *Collections for the History of the Town and Soke of Grantham* (note (36) below). Brewster in the two volumes of his *Memoirs* (note (14)) was the first to exploit the material systematically.

We may note that Roger Cotes' executor Robert Smith lent Conduitt his fair copies of 'about 20 or 30 [actually 38] letters written by Sr Isaac to Mr Cotes during the printing of the 2^d Edition of the Principia. Mr Conduit borrowed them of me when he collected materials for writing Sr Isaac's Life, & promised to return them, and with them to send me those which Mr Cotes wrote to Sr Isaac upon the same Philosophical Subjects, but forgot his promise' (Smith to William Hanbury, 16 March 1757 = Trinity MS. R. 16. 38: 413^r). In return for his trouble, and no doubt ever worried at forfeiture of his bond to Newton's relatives, Conduitt in February 1733 loaned Smith a single letter (that of Cotes to Newton, 29 April 1715) and was quick to secure its speedy return. Smith's copies (of the originals, now in Trinity College, Cambridge, which were later published by J. Edleston in his *Correspondence of Sir Isaac Newton and Professor Cotes*, London, 1850) were sold by the Portsmouth family in 1936 and acquired by J. M. Keynes (now King's College, Cambridge, Keynes MS 110). Edleston, who could never gain access to the Portsmouth papers, rather misleadingly suggested (Preface: xvii) that these were lost originals but in fact Newton's papers (ULC. Add. 3983) contain only 35 letters of Cotes to Newton, a few wholly or partially unknown to Edleston but most accurately published by him from Cotes' little variant drafts and minutes.

Mr. Conduitt has given a bond of £2000, to be responsible to the seven nearest of kin to Sir Is. Newton. Therefore the papers must be carefully kept, that no copys may be taken and printed, and Dr. Sykes desired to peruse them here, otherways if any accident comes to them the penalty of the Bond will be levy'd.(24)

Evidently Conduitt's bond remained for many years in danger of being surrendered, and we may imagine the care which he took to ensure that no unauthorized person had access to Newton's manuscripts, especially when publishers were becoming increasingly ready to print pirated versions of any of his papers, previously published or no.(25) In the outcome no new work of Isaac Newton's appeared in the stationers' catalogues until Samuel Horsley printed the original Latin version of the 1671 fluxional tract fifty years later.(26)

In the middle years of the eighteenth century, as people who had known Newton personally began to die off at an increasing rate, it became more and more difficult to obtain accurate information relating to the content of his papers and indeed for the few who tried to seek them out their very whereabouts came to be clothed in mystery. John Conduitt died in May 1737, his wife Catherine not quite two years later, and on her death the Newton papers passed (without recorded objection from the other surviving relatives) to her daughter Catherine. When in 1740 the latter married John Wallop, styled Viscount Lymington after his father was created the first Earl of Portsmouth in 1743, the papers disappeared unobtrusively from public knowledge into the possession of the Portsmouth family, where they survived virtually untouched for a century.(27) Already in the middle 1750's when Roger Cotes' cousin and

(24) Quoted from Brewster's *Memoirs* (note (14)) **2**: 341, note 3, where he reproduces a copy of the codicil sent him by Jeffery Ekins (then, as we shall see, in possession of the papers finally taken away by Sykes for examination in 1755).

(25) Already in Newton's lifetime Nicolas Fréret had printed an unauthorized French version of the Chronology abstract (*Abrégé de la Chronologie de M. le chevalier Isaac Newton fait par lui-même, & traduit sur le manuscrit Anglois*, Paris, 1725). Other minor religious papers of Newton's appeared without authority in 1737 and 1754. In 1736 John Colson, using a transcript from the original prepared by William Jones in Newton's lifetime, issued his English version of the 1671 tract as *The Method of Fluxions and Infinite Series; with its Application to the Geometry of Curve-lines* (London, 1736, translated into French by Buffon in 1740 and back into Latin by Castiglione in 1744), and this itself was pirated anonymously the following year (by Wilson?) as *A Treatise of the Method of Fluxions and Infinite Series, With its Application to the Geometry of Curve-lines* (London, 1737; reissued 1738).

(26) Horsley set it in print initially from a copy of Jones' transcript of the work made by Cavendish, using the former's conjectured title of *Artis Analyticæ Specimina sive Geometria Analytica.* In press, however, he came in October 1777 upon Newton's incomplete autograph original of the piece and made appropriate revisions. See note (34).

(27) However, shortly before his death in November 1756 the latitudinarian pamphleteer Dr Arthur Ashley Sykes, implementing the 1737 codicil to Catherine Conduitt's will, compiled a 'digest' of Newton's religious and chronological papers (cf. William Hanbury to Robert Smith, 11 April 1757 = Trinity MS. R. 16. 38: 422r). Soon after, in November 1755, a selected portion of the theological papers were forwarded to London for his detailed appraisal: Lord

executor Robert Smith searched for missing items of the former's correspondence with Newton and for the latter's 'common place book' (presumably the Waste Book) he could for a time find no one who knew where the papers were then located. His friend William Hanbury, who conducted the greater part of his search for him, reported that

Mr Wilcocks...who has carefully perused [the papers in the hands of the Rev[d] M[r] Ekins] assures me there is nothing of the nature of a commonplace book among them, & M[r] Ekins tells me he never saw or heard of such a piece. L[d] Macclesfield, to w[m] the library of M[r] Jones was left, said to me last winter that ke knew nothing of it but that he would when he went into the country carefully examine the Manuscripts [in his library but now] he declares there is no such thing in his collection.... Henry Stevens...says the book you mean was called the Green book, after w[ch] great enquiry was made long since, but no trace of it could ever be discovered.(28)

Only weeks later, after further inquiries made to Ekins, did he learn through Dr Habberdon that some of Newton's papers had passed to Lord Portsmouth. At last, however, an intermediary, Philip Barton, was able to write to him:

Lord Portsmouth...says that all the papers he has, w[ch] relate to S[r] Isaac Newton are in the Country at his seat [Hurstbourne Park] in Hampshire, that they were lately

Portsmouth's list drawn up on 12 November (King's College, Cambridge, Keynes MS 127[A].4: 'An Acc[t] of Sir Isaac Newton's papers—sent to the Rev[d] D[r] Sikes to London') details eleven items relating to chronology, biblical criticism and church history. Years later Lady Urania Portsmouth noted on its verso: 'Found Feb[y]: 1770. This Memo: preserv'd by U: Portsmouth as she knows not whether the Papers specify'd were return'd by D[r] Sykes.' In fact they never were but after Sykes' death passed subsequently into the hands of his friend Jeffery Ekins, in whose family they remained till in the nineteenth century they were presented—legally or no—to New College, Oxford. (They are now on permanent deposit in the Bodleian, listed New College MS 361.) As Ekins wrote to Joseph Wilcox on 27 March 1757, 'M[rs] Conduit had once an intention of publishing the Theological tracts of S[r] Isaac's and added a Codicil to her will for that purpose desireing that they might further be revised by D[r] Sykes. Some few of these were found in the hands of her Executor after his decease & are at present in my custody. None of them are perfect excepte one little tract relating to y[e] controverted text 1. John 5, 7 &c which surreptitiously got into print about two or three yeares ago [in 1754, in fact]. The rest seem to be very inconsiderable except[g]...that the manuscript of S[r] Isaac's Chronology is amongst these papers & is pretty fairly wrote' (quoted from Hanbury's copy in Trinity MS. R. 16.38: 416[r]–417[r]). These 'Ekins' papers, a few scraps of calculation excepted, contain almost nothing relevant to an assessment of Newton's mathematical achievement. (In his covering letter to Robert Smith on 8 March 1757 (Trinity MS R. 16.38: 410[r]) Hanbury confirms that it was Sykes who was Catherine Conduitt's executor while Ekins, executor of her daughter's will, merely 'transacted the affairs of M[rs] Conduit'. In his *Memoirs* (note (14)) **2**: 342 Brewster is somewhat in error here, but has been followed in his presumption by all later editors of Newton's theological works.)

(28) Hanbury to Smith, 8 March 1757 (Trinity MS R. 16.38: 410[r]–412[r]). Stevens perhaps mistook the commonplace book (described by Smith as a 'great' book, one 'in w[ch] the late S[r] Isaac Newton used to enter such observations as he made in the course of his reading under proper Heades to w[ch] he might the more easily refer on occasion') for the small green notebooks (King's College, Keynes MS 130.6) in which Conduitt jotted down his Newtonian memoranda, or perhaps for one (or several) of Newton's early scientific notebooks.

under the Perusal of a Gentleman[29] who died before y[t] Perusal was finished, y[t] they were very voluminous & it will be a matter of much time & trouble to examine them. At the same time his Lordship says that they shall be examined & if either M[r] Cotes's or any other Letters will be of Use to the Learned World He is ready to communicate them in a proper manner.[30]

Evidently no one remembered—or at least felt bound by—Conduitt's bond any more, and Lord Portsmouth was obviously amenable to having portions of his Newton manuscripts examined by competent scholars. Smith himself did not seize his opportunity. Both he and Portsmouth were then old men and we may guess that if he ever viewed the papers he quickly realized that he no longer had time or stamina sufficient for their thorough examination. But their new resting-place was now known with certainty to the English learned world and no serious student of Newton's achievement could in future claim ignorance of their existence.

As it happened the Reverend (later Bishop) Samuel Horsley was the first scholar to profit by the relaxing of the secrecy imposed by the conditions of Conduitt's bond, though he too found considerable difficulty in gaining access to the Portsmouth papers. We presume that the new Earl was not so willing as his grandfather to have a commoner invade the privacy of his country estate. Only briefly during the autumn of 1777 when it was too late for Horsley to edit any new mathematical pieces for the five-volume collected edition of Newton's works he was then preparing was he allowed down to Hurstbourne to study them.[31] Helped by William Mann Godschall he began a preliminary

(29) Dr Sykes: see note (27).

(30) Barton to Hanbury, 25 March 1757 (Trinity MS. R. 16.38: 415[r]).

(31) Horsley published proposals in 1776 for his 'complete' edition, later to appear as *Isaaci Newtoni Opera quæ exstant Omnia. Commentariis illustrabat Samuel Horsley, LL.D., R.S.S.* (5 vols., London, 1779–85, reprinted in photo-offset Stuttgart-Bad Cannstatt, 1964), but publication of the work was at first delayed through the serious illness of his wife (who at length died in August 1777) and because of the difficulty of finding a draughtsman competent enough to fashion the complex wood-cuts Newton's text demanded: fortunately, for the figures of the shorter mathematical pieces he was able to use the blocks which had been cut in 1711 for their edition in Jones' *Analysis*. (See John Nichols, *Literary Anecdotes of the Eighteenth Century; comprizing Biographical Memoirs of William Bowyer, Printer, ...and many of his Learned Friends* (London, 1812), **4**: 5–677, especially Horsley's letters to Bowyer of 6 July 1776 and 23 June, 20 July, 17 September and 2 November 1777, where he discusses various points of printing style in his projected edition. On 17 September 1777, in particular, he wrote from his (formerly Oughtred's) rectory at Aldbury in Surrey that 'I hope that you will some time tomorrow receive my copy of the *Arithmetica Universalis*, which is to be the first tract in my first volume.... You will set M[r] Gilbert about the figures immediately, and let the printing go on with as much expedition as possible. I would wish that my subscribers should have their first volume before next Midsummer; it will be a very large one....I intend to go into Hampshire in about ten days, to visit a repository of manuscripts Sir Isaac Newton left behind him, to which I have with great difficulty procured access. I may perhaps stay there ten days...—you will never have copy to wait for; as the whole of the first volume is actually finished, and a great part of

'catalogue' of the papers on 15 October and preparation of this continued through the following day.(32) The resulting list is the first careful record we have of Newton's manuscripts and is invaluable as a check on the not wholly productive efforts of later cataloguers who, unlike Horsley himself, were allowed freedom to resort the papers. We should remember, too, that the paper slips in his hand (assessing the fitness for printing of various groups of manuscript) which he interpolated were possibly no more than personal memoranda and not necessarily the pompous *ex cathedra* pronouncements which they may now appear.(33) Horsley's ill fortune was that he was not always competent to assess the importance and genius of the technical papers in science and mathematics which he sifted through, and that he had so few days in which to broach their detailed study. Some of the vagueness of Pellet's inventory reappears in his appraisals (we read once more of 'foul Copies of Mathematical Works' and 'loose Sheets of Calculations') and yet again the crucially significant Waste Book is noted merely as 'a Commonplace Book' without indication of its content but such notebooks as 'a Duodecimo Book composed of Extracts from different Writers perhaps with some Notes of Sir Isaac Newtons interspersed' (ULC. Add. 3996) and 'Another Duodecimo Book containing extracts out of Wallis & Oughtred, some easy Quadratures, Notes about Telescopes & Music' (Add. 4000) are now accurately described. Apart from 'Fragments of pieces upon Fluxions. . .never Published. a Great many loose & foul Papers' (Add. 3960 and 3962: passim), among the calculus

another.' In fact, Horsley's preliminary list (note (32)) and his subsequent cataloguing slips, still preserved with Newton's manuscripts, show that he studied the 'repository' only during the period 15–26 October 1777: already on 2 November, when he wrote to Bowyer again, he was back at Aldbury, hard at work seeing his first volume through the press.) Apart from the replies to two challenge-problems (originally published anonymously in the *Philosophical Transactions*), which were delayed till Volume 4, all the mathematical tracts published by Horsley are contained in the first volume (which finally appeared in mid-1779).

(32) The list—in Godschall's hand—is now, together with its drafts, in King's College, Cambridge (Keynes MS 127A.4: 'Catalogue taken of Sr: Isaac Newtons M: S: S: Octr: 15th: & 16th: in the Year 1777. By Wm: Mann Godschall Esqr: & the Revd: Dr: Horsley'). J. M. Keynes has written that Newton's papers 'on esoteric and religious matters' were then in a box 'packed up when he finally left Cambridge in 1696' and that when Horsley 'was asked to inspect the box with a view to publication [he] saw the contents with horror and slammed the lid' ('Newton, the Man', *The Royal Society Newton Tercentenary Celebrations, 15–19 July 1946*, Cambridge, 1947: 27–34, especially 27, 30 and 31). This is surely apocryphal and in part, no doubt, Keynes' misreading of Horsley's catalogue. The latter, in fact, found the papers not at all carefully divided by subject but somewhat randomly tied up as parcels in the drawers of a 'Cabinet' (which he called 'A') and three 'Bureau[x]' (called by him 'B', 'C' and 'D'): the papers on religion and alchemy were concentrated but not exhaustively confined in one of the latter, more probably (and as we now understand it) a writing desk with sets of drawers than a compartmented travelling chest.

(33) In so far as these pertain to the mathematical papers they will be reproduced at appropriate points in the text.

manuscripts he noted 'Geometria Analytica with the English Tract written in 1666' (Add. 3960. 14 and Add. 3958. 3: 48^v–63^v respectively) and 'a Paper on the quadrature of trinomial Curves with some Fragments relating to Fluxions' (Add. 3962.6). In the geometrical papers he found 'Fragments relating to the antient Problem de Loco Solido' and 'of a Work entitled Geometriæ Libri tres, & a restitution of Euclids Porisms which perhaps made only a part of these three Books' together with 'Scraps of the Lines of the 3^d Order' (Add. 3963: passim and Add. 3961 respectively); among the algebraic, 'Scraps' and 'a fair Copy of part of the Arithmet. Univers.' (Add. 3963.9/3964.2 and Add. 3993) and 'an elementary treatise of Trigonometry' together with 'A Compend^m of Trigonometry by one Hare' (Add. 3959.2–5). The following week, between the 20th and 26th of October, Horsley made a closer examination of portions of the mathematical manuscripts and also studied certain of the theological papers along with letters relating to optics and the fluxions priority dispute. But with his first volume already in press in London and anxiously awaited by his subscribers there was little he could now do to publish new mathematical material, much of which required long years of patient editing before its content could be made comprehensible and fitted into the pattern of Newton's creative development. In the outcome he suitably improved certain texts already scheduled for publication by collating them with the manuscript (notably emending the text of the 'Geometria Analytica', which he had previously derived from two inferior copies, one indeed derivative from the other[(34)]) and later published substantial selections from Newton's correspondence with Oldenburgh in his fourth volume, but otherwise introduced no major changes into the scheme of his edition after his visit to the Portsmouth's Hampshire seat.

As far as we know, Horsley never renewed his brief acquaintance with the treasures of Hurstbourne Park but came more and more to be claimed by the demands of church and university politics. His fellow scholars, innocent in their ignorance, were not stirred by the opportunities for further research which he himself had neglected fully to implement. Indeed, Horsley's bulky Latin edition, not incompetent if neither outstanding nor comprehensive, seemed then to fill all reasonable demands for accurate knowledge of the source-works on which Newton's vast scientific prestige was factually based. There were few

(34) See note (26) above and Horsley's *Opera Omnia*, **1** (1779): 389–518, especially 390. Except for suppressing an irrelevant 'Auctore Isaaco Newtono, Equite Aurato' Horsley retained William Jones' title for the piece (whose manuscript original now—as then?—lacks its title-page but was always referred to obliquely by Newton himself as 'my 1671 tract') together with his anachronistic substitution of dot-notation for the original's literal fluxions (because the cost of resetting the type was too great?). Some errors of transcription which had arisen in Jones' and Cavendish's copies were corrected and a few lacunae filled. The division of the tract (in Horsley's version) into chapters and paragraphs is uniquely his.

enough in England who could begin to appreciate the subtleties of his *Principia* and his published mathematical texts, let alone who had the intellectual hunger to savour and devour unfamiliar material and the ability to digest it. For almost a century the technical portion of Lord Portsmouth's papers were to lie unopened in their packets, their location known(35) but their importance not admitted. They were not again broached till historians began to recognize the need for an accurate control of sources and to fill gaps in their knowledge not by extrapolation but through manuscript study.

Little enough was done for a long time to examine and publish the non-technical papers. In 1806 the Lincolnshire antiquary Edmund Turnor, who had recently purchased Newton's traditional birthplace (the little 'manor house' at Woolsthorpe), printed some of the biographical material collected by Conduitt.(36) A quarter of a century later James Henry Monk, engaged in revising his biography of the Trinity authoritarian Bentley in the early 1830's, approached the Portsmouth family for permission to study the letters which Cotes wrote to Newton during the production of the second edition of the *Principia*, and was later able to acknowledge that he had seen them.(37) He found at Hurstbourne a situation not wholly conducive to calm scholarship: the third Earl, John Charles Wallop, had some years before become mentally deranged and the family estates were administered by trustees, notably the Earl's brother, Newton Fellowes. Fortunately for Monk, Fellowes' son, Henry, had become personally interested in the papers and through him Monk was able to satisfy

(35) In his widely read *Mathematical and Philosophical Dictionary* (note (10)) Charles Hutton, seemingly for the first time in print and probably with Horsley's aid, pinpointed the location of the theological portion at least of Newton's papers: 'It is astonishing what care and industry Sir Isaac had employed about the papers relating to Chronology, Church History, &c: as, on examining the papers themselves, which are in the possession of the family of the Earl of Portsmouth, it appears that many of them are copies over and over again, often with little or no variation; the whole number being upwards of 4000 sheets in folio...; besides the bound books &c in [Pellet's] catalogue' (**2**: 157).

(36) In his *Collections for the History of the Town and Soke of Grantham. Containing Authentic Memoirs of Sir Isaac Newton, now first published from the Original MSS. in the possession of the Earl of Portsmouth* (London, 1806). Apart from printing (on p. 162, n. 2) Charles Montague's letter to Newton of 19 March 1695/6, Turnor reproduced (pp. 158–67) Conduitt's original English version of the 'Memoir' he sent to Fontenelle in 1727; (pp. 172–3) 'A remarkable and curious conversation between Sir Isaac Newton and Mr. Conduitt'; (pp. 174–80) the greater part of William Stukeley's letter of 26 June 1727 to Richard Mead; and (pp. 181–6) Rutty's 'Extracts from the Journal Books of the Royal Society, relating to Sir Isaac Newton'. (These are now all in King's College, Cambridge, Keynes MSS 129.1, 130.11, 136 and 128 respectively.) As Turnor notes, Gough had printed an inferior partial transcript of Stukeley's letter (derived from its author?) in the *Gentleman's Magazine* for November 1772 (pp. 520–2).

(37) J. H. Monk, Life of *Richard Bentley* (London, $_2$1833): **1**: 230, n. 47: 'Some letters which are, properly speaking, part of this series, are among the collection of Sir Isaac Newton's papers, belonging to the Earl of Portsmouth, at Hartsbourne House (sic), Hampshire, where they were obligingly shown to me by the Hon. H. Fellowes.'

his curiosity. Less successful was the astronomical historian, Francis Baily, then busy editing Flamsteed's papers for publication, who also at this time was allowed to 'inspect the large and valuable collection of Newton MSS' at Hurstbourne in search of the latter's Newtonian correspondence, but though their custodian 'H. Fellowes...was so obliging as to afford me every facility for that purpose...I [have not] been able to throw any light on the special object of my inquiries'.(38) It was through Henry Fellowes, too, that David Brewster was four years afterwards permitted access to much of the Portsmouth material when he came to revise his short, popular *Life of Sir Isaac Newton* (London, 1831) into the two bulky volumes of his *Memoirs of the Life, Writings and Discoveries of Sir Isaac Newton* (Edinburgh, 1855), which for more than a century have remained the standard biographical account.

Contrary to some modern opinions,(39) Brewster was apparently never allowed unrestricted access to the Portsmouth manuscripts but limited almost wholly to the selection of biographically pertinent material, letters and papers selected for him by Fellowes. Though usually adept at cloaking his lack of complete information in his text, he makes the matter evident in his prefatory remarks:

> Mr. Henry Arthur Fellowes...met me in June 1837, at Hurstbourne Park, to assist me in examining, and making extracts from the large mass of papers which Sir Isaac had left behind him. In this examination our attention was particularly directed to such letters and papers as were calculated to throw light upon his early and academical life, and with the assistance of Mr. Fellowes, who copied for me several important documents, I was enabled to collect many valuable materials unknown to preceding biographers.... in so far as Mr. H. A. Fellowes and I could make an abstract of these and other manuscripts during a week's visit at Hurstbourne Park, I have availed myself of them in composing the first volume of this work, which was printed before the papers themselves came into my hands....

(38) *An Account of the Revd. John Flamsteed, the first Astronomer-Royal; compiled from his own Manuscripts, and other Authentic Documents, never before published* (London, 1835): xx. As Brewster remarked in his *Memoirs* (**2**: 161; compare **1**: xi–xii) Baily had not looked very hard, for 'the letters of Flamsteed to Newton...had been carefully preserved'.

(39) L. T. More, for example, following the 1888 cataloguers of the Portsmouth papers, has suggested that Brewster 'made a very considerable use of Conduitt's manuscripts and of abstracts from Newton's correspondence, and some use of the mathematical notes and papers [but] used his discretion in extracting and in omitting many important documents which seemed to him not advantageous to Newton's reputation' (*Isaac Newton, A Biography*, Chicago, 1934 (reissued New York, 1962): Preface: ix). In his *Memoirs* there is no clear instance of Brewster's having studied any of the mathematical papers: his accounts (*Memoirs*, **1**: 21–5; and **2**: 10–17) of Newton's early mathematical development, apparently based on his examination of the original documents in the Portsmouth papers, are in fact little variant reproductions and summaries of material gathered by Conduitt (notably King's College, Keynes MS 130.4 and ULC. Add. 3960.2) or of items in the Macclesfield collection copied for him by S. P. Rigaud. We will discuss this point in greater detail in the final volume when we examine the immediate impact of Newton's researches upon his contemporaries.

Before I began the second volume, which contains the history of the Fluxionary controversy, and the Life of Newton subsequent to the publication of the first edition of the *Principia*, I had the good fortune to obtain from the Earl of Portsmouth, through the kindness of Lord Brougham, the collection of manuscripts and correspondence which the late Mr. H. A. Fellowes had examined and arranged as peculiarly fitted to throw light on the Life and Discoveries of Sir Isaac. In these manuscripts I found much new information respecting the history of the *Principia*, which, though it might have been more appropriately placed in the first volume, I have introduced into those chapters of the second which relate to the period when the other editions of the *Principia* were published.(40)

Because of Brewster's otherwise reprehensible habit of inscribing the documents entrusted to his care in his fine spidery handwriting, we may indeed confirm that he had access to a tolerably complete selection from Newton's correspondence, to the biographical material gathered by Conduitt and to the papers relating to the fluxion priority dispute, but there is no indication that he made a wider study of original manuscript. Composed under such limitations his biography is, for its period, a minor masterpiece of shrewd insight into technical questions he could many times not examine on the basis of Newton's autograph papers. Historically it was equally important, together with the minutely documented critical essays of his fellow scholars Rigaud and Edleston,(41) in reviving interest in the accurate documentation of Newton's life and scientific achievement. Together, they were in particular to encourage two prominent Cambridge mathematicians, John Couch Adams and George Stokes, to travel down to Hurstbourne Park in late July 1872, the official delegates of Newton's university entrusted with the commission to examine the entire corpus of the Portsmouth papers in detail and to report back.

The circumstances which provoked this delegation are not clear.(42) However, having been approached through the Duke of Devonshire, Lord Ports-

(40) *Memoirs*, **1**: Preface: vii–viii, x–xi. Henry Fellowes died in 1847 before his father succeeded as fourth Earl. A great many of his transcripts, some annotated in Brewster's hand, are now in ULC. Add. 4007.

(41) Stephen Peter Rigaud, *Historical Essay on the First Publication of Sir Isaac Newton's Principia*, Oxford, 1838 (whose appendix reproduced several important Newtonian documents and letters for the first time); Joseph Edleston, *Correspondence of Sir Isaac Newton and Professor Cotes*, London, 1850 (with 'Notes, Synoptical View of the Philosopher's Life and a Variety of Details illustrative of his History' and 'An Appendix containing other Unpublished Letters and Papers by Newton').

(42) L. T. More (*Isaac Newton* (note (39)): Preface: ix; and 4, n. 6) suggested that Lord Portsmouth was encouraged to approach Cambridge University after a fire in his house which destroyed one of the Conduitt manuscripts. The appropriately baptised fifth Earl, Isaac Newton Wallop, was (if otherwise known to his contemporaries as a sporting figure) intensely proud of his family's connection with his namesake and throughout showed the utmost generosity in allowing free access to his Newton papers and in ultimately presenting the technical portion to Cambridge University as a gift.

mouth wrote to the Cambridge Vice-Chancellor on 23 July that he was 'happy to receive the Deputation...elected to look over the Newton papers and hopes they will accept his hospitality while engaged in their research'. Upon their arrival Adams and Stokes quickly realized that a brief visit was totally inadequate for their purpose, and so with Portsmouth's approval they returned to Cambridge a few days later, bearing with them the entire collection of Newtonian manuscript for a thorough, unhurried examination. In his covering note of 2 August Lord Portsmouth specified the terms on which he had loaned them:

Professor Adams & Stokes have looked over the Newton papers & I have handed over to them (for the University) the Newton manuscripts and two *copies* of the Principia 1st & 2d Editions corrected by Newton. There are a Number of Fragments relating to Mathematics &c which Professors Adams & Stokes think should be carefully investigated to see whether they relate to, or are rough drafts of his works. These I am willing to *lend* but in the event of their proving to be the calculations from which he made his deductions in his works, I am willing to make them over to the University. I also *lend* some very interesting letters from Eminent Men to Newton in order that any letters which may not have *been published* may be published if they throw light on scientific questions....all the *letters* as well as memoranda Books &c relating to personal matters I *lend* only, wishing to retain them as Heirlooms with other *personal* Property of his. I am sure that Professors Stokes & Adams will scrupulously carry out my wishes. I wish to advance the interests of science by placing these Papers at the service of the University, but I would rather cut my hand off than sever my connection with Newton which is the proudest Boast of my Family.(43)

In a following letter to the Vice-Chancellor a few days later Lady Portsmouth further clarified the transaction:

The reason [why the papers were taken by Professors Stokes & Adams to Cambridge there to be sorted] was that from the state of confusion in wh[ich] all Sir Isaac Newton's papers were, such sorting & dividing at Hurstbourn would have required so indefinitely long a time that it was deemed more advisable to remove them to Cambridge where the investigation could be carried on with thoroughness & at leisure. Sir Isaac's papers appear never before now to have been carefully arranged or examined by any competent person so that in bundles marked 'worthless' Ld Pth understood that the Professors thought they had found calculations of considerable interest & importance.(44)

In retrospect Lord Portsmouth must have been highly pleased at his decision temporarily to relinquish custody of the documents rather than continue to extend his hospitality to Adams and Stokes, for during that thorough, leisurely examination the papers were retained in Cambridge for sixteen long years.

The two members of the original deputation were competent to assess the more technical (especially mathematical and mechanical) papers but when it

(43) ULC. Add. 2588.494/495.
(44) ULC. Add. 2588.496.

came to evaluating the mass of personal, alchemical, chronological and theological manuscript also thrust on them to classify Adams and Stokes found themselves somewhat at a loss. To assist them in their task they were joined, in the official University cataloguing syndicate appointed on 6 November 1872, by H. R. Luard (who devoted his principal efforts to the non-scientific and personal papers) and G. D. Liveing (who concerned himself more narrowly with the chemical and alchemical manuscripts). Evidently they felt it their duty to retain some record of the more important items, particularly correspondence, scheduled to be returned to Hurstbourne, for both Adams and Luard spent many hours carefully if not always accurately transcribing a selection of these.[45] In the preface to the long-awaited printed catalogue of the Portsmouth papers which appeared finally in 1888[46] its compilers noted that their examination, classification and final division 'has proved a lengthy and laborious business, as many of the papers were found to be in great confusion—mathematical notes being often inserted in the middle of theological treatises, and even numbered leaves of MSS having got out of order. Moreover, a large portion of the collection has been grievously damaged by fire and damp. The correspondence, however, is in a very fair condition throughout, and had been arranged [by Fellowes] in an orderly manner'. For better or worse their classification has been retained virtually intact to the present day in the portion of the papers retained by the University. Now for the first time Newton's manuscripts were conveniently subdivided into subject categories under the principal headings of 'Mathematics' (including *Principia* documents and papers on astronomy, optics, hydrostatics, sound and heat), 'Chemistry' (including alchemy), 'Chronology', 'History', 'Theology', 'Letters' and 'Books' together with some ancillary sections relating in particular to Newton's

(45) Their transcripts, particularly Luard's and to some extent Adams', are now scattered in the miscellany of copies preserved in ULC. Add. 4007. In addition, copies of Brewster's *Memoirs* (note (14): now ULC. Adv. c. 76.1/2), most of whose reproduced letters and documents were 'carefully collated with their originals' by Adams and Luard, and of Edleston's *Correspondence* (note (41): originally Add. 4010, now Adv. c. 76.3), with its reproduced text of many of Cotes' draft letters to Newton collated by Adams with the originals received by Newton (now ULC. Add. 3983), were placed with the papers subsequently retained by the University. Their efforts have not been wholly in vain: in the first volume, for example, we have had to make use of Luard's copy in reproducing extracts from a memorandum given in November 1727 by De Moivre to Conduitt since the original (returned to Hurstbourne and sold in 1936 to an undisclosed private owner) has now disappeared.

(46) *A Catalogue of the Portsmouth Collection of Books and Papers written by or belonging to Sir Isaac Newton, the Scientific Part of which has been presented by the Earl of Portsmouth to the University of Cambridge, drawn up by the Syndicate appointed the 6th November 1872*, Cambridge, 1888. The indicated division between papers kept and those returned to Portsmouth is not accurate, but a corrected copy is now retained with the 'Portsmouth Collection' in Cambridge University Library (ULC. Add. 3958–4007, though recently certain of the printed volumes—3991, 3992, 3994, 3999 and 4001—have been reclassified as 'Adversaria').

family and Conduitt's intended biographical study. In the Preface and its lengthy Appendix(47) as well as in the sections on 'Mathematics' Adams' influence is predominant: the sections on the *Principia* and the lunar theory (given a separate status not really warranted by the profusion or importance of the corresponding documents) are acceptably divided, those on pure mathematics much less satisfactorily so, and it is abundantly clear that none of the cataloguers were authorities in the latter field.(48) Of the other sections Liveing's classification of the chemical papers is extremely competent and Luard's listing of the letters thorough. Perhaps the most unhappy effect of the syndicate's hard work was that virtually all traces of Newton's original ordering of his papers, hitherto preserved more or less intact, were removed. Most, indeed, of Horsley's hurriedly scrawled observations (of some historical interest and not always uninformative) were removed from the manuscripts in which they had been interposed a century before and gathered rather uselessly in a separate folder (now in Add. 4005).

With all its failings and weaknesses, however, the catalogue should have been an eye-opening stimulus to late nineteenth-century Newtonian scholarship. Inexplicably its contemporary impact was almost nil and Lord Portsmouth's portion of the papers was returned to him without comment. No member of the cataloguing syndicate implemented the official report with an enlightening secondary study, historical or biographical, of any of the documents he had pondered over so long and for the better part of fifty years little use was made of the Portsmouth papers, either in Cambridge or back in Hurstbourne. The lone significant exception to this general apathy was the Cambridge amateur historian of mathematics, Walter William Rouse Ball, and even he ceased his active interest in 1893 when appointed tutor at Trinity—Cambridge's gain but the world's loss. Having in December 1890 read to the London Mathematical

(47) About half the preface (five out of twelve pages) is devoted to a brief discussion of three topics, Lunar Theory, the Theory of Atmospheric Refraction and the Solid of Least Resistance, cherished by Adams while the Appendix (pp. xxi–xxx), which reproduces 'a few extracts from the Newton papers on some of the subjects which have been referred to', is devoted wholly to exemplifying those three aspects.

(48) The collection of geometrical fragments (Add. 3963) remains especially disordered: no attempt has been made to establish any sequence, logical or chronological, in them and one or two closely germane groups of papers are now split between this and other manuscript parcels. The sections on 'Fluxions' and 'Quadrature of Curves' (Add. 3960, 3962) are quite arbitrarily divided and choked with irrelevant, fragmentary transcripts by Jones and Wilson. Documents relating to Newton's bulkiest published mathematical piece, the *Arithmetica Universalis*, have not been accorded their separate division but are strewn almost at random over Add. 3959, 3962, 3963, 3964 and 3993. In general, indications of autograph manuscript are not everywhere accurate while, here as elsewhere in the non-mathematical portions, several of the groupings are not felicitous: in particular, the unjustifiable juxtaposition of two interpolation manuscripts in Add. 3964.5 (as the 'Regula Differentiarum &c') was to cause Duncan Fraser considerable confusion, never by him resolved, when he edited them in 1918.

Society a lengthy study, based principally on unpublished manuscript in the Portsmouth Collection, of Newton's classifications of cubic curves,(49) to the same Society he communicated in May 1892 a brief critique of a Newtonian fragment on central forces(50) and in the following year produced his authoritative essay on the genesis of the *Principia*:(51) much later he published the text of a minor paper in which Newton, among other things, discoursed on the place of mathematics in a university education.(52) For the rest, shortly before the First War a German scholar, Alexander Witting, studied the fluxional manuscripts in Cambridge and prepared a preliminary report,(53) but the ensuing outbreak of hostilities prevented him from continuing his research and he never returned to it. A little afterward Duncan C. Fraser, a professional actuary who had interested himself in the history of interpolation, gathered all he knew of Newton's researches in that topic in a series of periodical articles, making some use of unpublished material in Cambridge University Library.(54)

(49) 'On Newton's Classification of Cubic Curves', *Proceedings of the London Mathematical Society*, **22** (1891): 104–43. This efficient essay reproduced (pp. 132–40, 140–4) extracts from unpublished autographs in ULC. Add. 3961.4/1. We will examine it more closely in the seventh volume.

(50) 'A Newtonian Fragment relating to Centripetal Forces', *Proceedings of the London Mathematical Society*, **23** (1892): 226–31. Ball there reproduced the significant portion of ULC. Add. 3965.2, reconstructing the theoretical basis which underlies the unproved assertions of the text itself.

(51) *An Essay on Newton's 'Principia'*, London, 1893. Apart from publishing in appendix the first tolerably complete version of Newton's correspondence with Hooke and Halley during the period 1679–86 and giving in his text an able summary of the *Principia's* technical content, on pp. 33–56 he contributed a discussion of the major variants between the various manuscripts of *De Motu Corporum* in the Portsmouth papers (ULC. Add. 3965.7, first printed by S. P. Rigaud in his 1838 *Essay* (note (41)) from the inferior Royal Society version) and on pp. 116–20 added the essence of Newton's English manuscript (Add. 3965.1) of the 'Locke' proof of *Principia's* Book 1, Prop. XI, previously available only in Whiston's incomplete Latin translation in his *Prælectiones Physico-Mathematicæ* (Cambridge, $_1$1710) and the Locke version printed by King in his *Life of Locke* (London, $_2$1830: 389–400).

(52) 'Isaac Newton on University Studies', *Cambridge Review* for 21 October 1909, reprinted in Ball's *Cambridge Papers* (London, 1918): Part II: 244–51. The original autograph is ULC. Add. 4005: 14^r–15^r.

(53) 'Zur Frage der Erfindung des Algorithmus der Newtonschen Fluxionsrechnung', *Bibliotheca Mathematica*, $_3$**12** (1911–12): 56–60. Witting first quoted [James Wilson's] introductory comment to William Jones' somewhat reordered transcript (Add. 3960.1) of Newton's October 1666 tract and went on to exemplify its contents, stressing Problems 7, 8, 16 and 17. In sequel he described the fluxional notations used in an unpublished chapter (Add. 3960.4) of Newton's 1671 tract, the discussion of tangents by limit-motion considerations in an October 1665 draft (Add. 3958.2), an example of Newton's early technique of differentiation by substitution in the autograph (Add. 3958.3: 48^v–63^v, wrongly identified as 'von anderer Hand') of the 1666 tract, and finally noted Newton's first algorithmic fluxional formulations in a mid-1665 manuscript (Add. 3960.12).

(54) Fraser, in particular, made a none too accurate transcript (with English translation) of the 'Regula Differentiarum' papers (ULC. Add. 3964.5) in his 'An unpublished Manu-

Only when Louis Trenchard More at the time of the bicentenary of Newton's death began to gather material for a new biography(55) were the wider resources of the Portsmouth Collection drawn upon. With the exception, however, of a few non-technical excerpts from Flamsteed's correspondence with Newton and some phrases quoted in his notes from unpublished manuscript, More made no effective use of the scientific portion of the papers deposited in Cambridge but, in imitation of his predecessor Brewster, confined himself to the correspondence and alchemical, theological and biographical documents at Hurstbourne. In consequence his judgements on the mathematical and scientific aspects of Newton's genius are conventional and shallow when not actually misleading.

The year 1936 is a turning point in the history of Newtonian scholarship. On 13 and 14 July the whole of the Newton papers returned in 1888 to Hurstbourne Park 'where they would be carefully preserved'(56) were put up for public auction in London. At the end of the second day for a mere £9030. 10*s.* 0*d.* an estimated three million words of Newton's autograph manuscript(57) was scattered literally to the farthest corners of the world. Through the valiant efforts of Keynes and others a significant portion came subsequently to rest at various locations in Cambridge and in London, but certain important scientific and biographical items together with a few letters have (temporarily at least) vanished from public knowledge.(58) The one positive result of the sale has, paradoxically, been to arouse interest in Newton's still unpublished papers, an interest reinforced by the approach of the tercentenary of his birth in 1942 (though the main celebrations were deferred for four years by war). Since then active Newtonian research has grown swiftly into a minor industry, its pace further accelerated since 1959 by the publication of the first volumes of the Royal Society's edition of Newton's correspondence under the able editorship of Herbert Westren Turnbull.(59) Most recently

script by Sir Isaac Newton', *Journal of the Institute of Actuaries*, **58** (1927): 53–95, especially 75–84 (reproduced in the same pagination in his *Newton's Interpolation Formulas*, London, 1927).

(55) Published subsequently as his *Isaac Newton* (note (39)). For his convenience Lady Portsmouth sent the Hurstbourne papers to the British Museum so that 'I might examine and use [them] at my leisure' (Preface: viii). More seems to have had no clear idea of the significance of the 1888 division or that the Portsmouth family had long before relinquished ownership of the Cambridge portion of the papers.

(56) *Catalogue of the Portsmouth Collection* (note (45)): x.

(57) John Taylor's 'conservative estimate' in his foreword to Sotheby's *Catalogue of the Newton Papers Sold by Order of The Viscount Lymington* (London, 1936).

(58) Compare A. N. L. Munby, 'The Keynes Collection of the Works of Sir Isaac Newton at King's College, Cambridge', *Notes and Records of the Royal Society of London*, **10** (1952): 40–50. The annotated copy of Sotheby's sale catalogue (note (57)) in King's College Library lists the present location of many of the papers sold.

(59) *The Correspondence of Isaac Newton* (Cambridge, 3 vols. 1959–61, continuing). As Turnbull noted in the preface (p. xxix) to his first volume, 'the search for letters among the

A. Rupert Hall and Marie Boas Hall have together published a selection of texts from the Newton papers in Cambridge,(60) while I. Bernard Cohen,(61) John W. Herivel(62) and Richard S. Westfall(63) have further editions of the unpublished dynamical and physical manuscript in press. For the present, unfortunately, much of this current effort remains uncoordinated and a tendency to duplication—even triplication—of research is pronounced. In the meanwhile, the definitive collected edition of Newton's intellectual achievement is an unapproached ideal. May this present edition be a small step toward that long-overdue monument to a man who in so many areas of human thought himself took a giant's leap.

---

collections of manuscripts in Newton's handwriting has brought to light many mathematical or physical notes...and a few of the shorter pieces among them have been included here'. Apart from the mathematical content of the letters (which we will summarize at appropriate places in the present volumes), the *Correspondence* indeed reproduces extracts from a variety of Newton's early fluxional and dynamical papers along with (in volume 3) contemporary memoranda of David Gregory which throw a great deal of light onto the chronological sequence of Newton's mathematical development in his middle years. Recent summaries of current Newtonian research, rendered already obsolescent by the present spate of publication of manuscript investigations, are presented in I. B. Cohen, 'Newton in the Light of Recent Scholarship', *Isis*, **51** (1960): 489–514 and in D. T. Whiteside, 'The Expanding World of Newtonian Research', *History of Science*, **1** (1962): 16–29.

(60) *Unpublished Scientific Papers of Isaac Newton. A Selection from the Portsmouth Collection in the University Library, Cambridge*, Cambridge, 1962. Some necessary criticism of their title and the mathematical portion of their commentary is made in D. T. Whiteside, 'Scientific Papers of Newton', *History of Science*, **2** (1963): 125–30.

(61) Cohen has been working principally for the last half dozen years, partially with the collaboration of the late Alexandre Koyré, on a detailed variorum edition of the *Principia*. Volume 1 (Latin text) of his provisionally entitled 'Critical Edition of Newton's *Principia*' (see the *Year Book of the American Philosophical Society*, Philadelphia, 1960: 516–20) is now in press at Cambridge and the edition is scheduled to be completed with an English translation and several volumes of commentary based on manuscript findings. We may note that a not quite complete collection of Koyré's *Newtonian Studies* (London, 1965) reproduces some discussion, in part not previously published, of documents in the Portsmouth papers.

(62) *The Background to Newton's Principia: A Study of Newton's Dynamical Researches in the Years 1664–84. Based on Original Manuscripts from the Portsmouth Collection in the Library of the University of Cambridge*, Oxford, 1966.

(63) Provisionally entitled *Force in Newton's Physics* (London, no date given) Westfall's study will trace the growth of Newton's concept of force on the basis of unpublished manuscript.

FOREWORD TO VOLUME I

This first volume contains the texts, until now almost wholly unpublished, which must provide the documentary basis for any accurate study of the flowering of Newton's mathematical genius during his late undergraduate and early graduate years at Trinity College, Cambridge. Some further insight into the early history of his fluxional discoveries may be gained from Newton's own comments about them when half a century afterwards he sought to justify his priority in their invention to the world, while the edited summary of his researches which he transmitted to Leibniz in 1676 throws additional light on his formulation of the binomial expansion: these we will examine in later volumes.

The impact of these first creative mathematical outbursts on his contemporaries was virtually nil. It is true that Newton permitted a restricted circulation of portions of his early papers among immediate acquaintances at various times in his later life and that certain of his fluxional manuscripts were copied in his lifetime by John Collins, William Jones and James Wilson, but their versions (inferior transcripts which in no way clarify the meaning of Newton's original texts) were never widely known and soon forgotten. We possess, unfortunately, no true letter of his written before 1669 and his later correspondence till the time of the fluxion priority squabble with the Leibnizians sheds little light on his earliest creative years. Through the centuries, therefore, Newton's first mathematical researches have remained intangible, known but dimly through his own capriciously chosen revelations and the distorted, inadequate secondary accounts furnished by the editors of his few published mathematical pieces. With his original texts before us we can now know that Newton's capacity for self-criticism and impartial evaluation, never strong and emasculated in his middle age by a recurrent overriding impulse to secrecy, here proved largely inadequate to the task of defining and illuminating the riches of his first researches. But more important still, these autograph manuscripts, now reproduced as accurately as possible, allow an hitherto unobtainable exact knowledge of the details of his rise to mathematical maturity and must be fundamental in any reconstruction and estimation of that growth.

The first section collects the significant portion of Newton's annotations of the first mathematical texts he read, notes which from the beginning reveal his genius and cannot be clearly divided from his earliest independent research. As we shall see, the works he read were few but of the highest quality and of considerable difficulty even for the mature mathematicians of his day. After a quick passage to enlightenment through his own untutored reading, he was soon inspired to explore the fruitful amalgam of Cartesian geometry and the calculus of fluxional increase which was to be the keystone of his future mathematical method. The manuscripts in which he formed his outlook and recorded his earliest discoveries in that field make up the second section. A following one complements this main avenue of research by reproducing some minor offshoots in algebra and trigonometry together with some miscellaneous matter not better placed earlier in the volume. In a short appendix, finally, we have a selection of Newton's researches at this time in geometrical optics: these are inserted in illustration of the ease and fertility with which he was able to apply his pure techniques in appropriate areas of applied science.

ANALYTICAL TABLE OF CONTENTS

PART I
THE FIRST MATHEMATICAL ANNOTATIONS
(1664–1665)

PART II
RESEARCHES IN ANALYTICAL GEOMETRY AND CALCULUS
(1664–1666)

PART III
MISCELLANEOUS EARLY MATHEMATICAL RESEARCHES
(1664–1666)

APPENDIX
EARLY NOTES ON GEOMETRICAL OPTICS
(1664–1666)

LIST OF PLATES

PART 1

THE FIRST MATHEMATICAL ANNOTATIONS (1664-1665)

INTRODUCTION

We know nothing of Newton's early mathematical training, but it seems certain that some time in his boyhood, possibly as a child in the little day-schools in Skillington and Stoke, but more probably at his Grantham grammar school, he became familiar with the elementary rules of manipulation which were the basis of school arithmetics of his day.(1) In the little evidence we have of his youthful characteristics(2) there are no signs of any pronounced facility in

(1) The primary schools of Newton's day, the maid or 'petty' schools, taught the bare rudiments of reading, writing and reckoning. As for the country grammar schools of the period, Ben Jonson's comment that they taught 'small Latine and lesse Greeke' was not outdated, though from the beginning of the century a growing number of schools were beginning to teach the elements of practical computation. Geometrical studies were, in all but a few exceptional cases, confined to the universities. (See Foster Watson: *The Beginnings of the Teaching of Modern Subjects in England* (London, 1909): ch. VIII, 'The teaching of arithmetic', 288–331. Further details on the organization and curricula of grammar schools may be had from Watson's *The English Grammar Schools to 1660* (Cambridge, 1908), and from T. W. Baldwin's *William Shakspere's Small Latine & Lesse Greeke* (Urbana, Illinois, 1944).

(2) Of a large fund of contemporary anecdote, partially inconsistent and in many respects not substantiated, which refers to Newton's boyhood, nothing speaks of any early technical mastery of mathematics. Almost all the well-known stories of Newton's Lincolnshire days derive from the assorted collection of scribbled notes, memoranda and letters which John Conduitt, the husband of Newton's niece, Catherine Barton, began to gather in 1727 immediately after Newton's death in preparation for an authoritative biography. This *Life* was seemingly never completed nor have its existing drafts been printed in a collected form, though Fontenelle drew extensively on Conduitt's knowledge in writing his *Eloge de Monsieur le Chevalier Newton* (Paris, 1728), and both David Brewster and L. T. More made liberal use of the drafts in composing their standard biographies. (See their respective *Memoirs of the Life, Writings and Discoveries of Sir Isaac Newton* (Edinburgh, 1855): chs. I, II; and *Isaac Newton, a biography* (New York, 1934): chs. I, II.) Conduitt's papers formed section V of the sale at Sotheby's in July 1936 of that portion of the Portsmouth Collection of Newton papers which had been returned to the family in 1888. Of these lots Lord Keynes eventually managed to acquire all but No. 218 (the de Moivre memorandum) and they are now deposited in the library of King's College, Cambridge (Keynes MSS 129–37). (A. N. L. Munby describes these manuscripts generally in his 'The Keynes Collection of the works of Sir Isaac Newton at King's College, Cambridge', *Notes and Records of the Royal Society*, **10** (1952): 40–50, especially 47. They are listed in the *Catalogue of the Newton Papers sold by Order of the Viscount Lymington*, Sotheby and Co., 1936: 53–60.) The important letter of William Stukeley to Richard Mead, dated Grantham, 26 June 1727 (Keynes MS 136, partially printed by Edmund Turnor in his *Collections for the History of the Town and Soke of Grantham* (London, 1806): 174–80) formed the basis for Stukeley's own *Memoirs of Sir Isaac Newton* (1752) (ed. A. Hastings White, London, 1936). Several interpretative essays have appeared in recent years which deal with this material, adding a certain amount that is new. Two of Newton's early notebooks, one in the Pierpont Morgan Library, New York, the other now in Trinity College, Cambridge, give valuable insights into his mechanical facility and his formal education in the classical languages, while some years ago several geometrical figures, arguably Newton's, were found scratched on the walls of his Woolsthorpe manor-house. G. L. Huxley has summarized these new findings in 'Newton's Boyhood Interests', *Harvard Library Bulletin* **13** (1959): 348–54. For the general details of Newton's family background

computation, and indeed if he ever was an infant prodigy his later life reveals clearly that his arithmetical gifts soon degenerated to the level of no more than average competence.(3) While he was always prepared to face a long and tedious calculation with a dogged perseverance which speaks well for Stokes, his Grantham schoolmaster, we should never overstress Newton's arithmetical abilities. A better argument could be raised, as Stukeley in his *Life* has done,(4) in support of innate geometrical powers which showed themselves in Newton's boyhood expertness in the fashioning of models, sundials and water-wheels and we might even attribute to him in youth a certain artistic deftness. This, no doubt, was the source of much of the adult Newton's capabilities as an experimental scientist, but his easy power to comprehend the complex geometrical diagram must be seen as an offshoot of his intellectual powers rather than as any development of youthful manual dexterity. We may readily agree that Newton revealed a little of his innate gifts as a boy, but the intellectual explosion of his powers was not touched off till his university days.

Newton entered Trinity College, Cambridge, in the summer of 1661 and we may suppose that he endured the standard curriculum of the period, one still heavily slanted towards scholasticism.(5) We know a little, indeed, of his first leanings towards academic science from a small pocket-book(6) which contains a sequence of early annotations, made perhaps in 1662 or early 1663, of his reading in Aristotle and such Aristotelian texts as the *Physiologiæ peripateticæ libri sex, cum commentariis* of Johannes Magirus.(7) These notes, together with our knowledge of the contemporary system of education, allow us to

C. W. Foster's 'Sir Isaac Newton's Family' (*Associated Architectural Societies' Reports and Papers*, **39** (1928); *Reports and Papers of the Architectural and Archaeological Society of the County of Lincoln*: 1–62) is indispensable.

(3) The mistakes which occur in his lengthy calculations of hyperbola areas (3, §§3/5 below) are damning evidence against the myth of Newton's prodigious facility in numerical calculation. Lengthy calculations appear frequently in his later scientific papers and not infrequently they too contain small numerical errors.

(4) *Memoirs* (note (2) above): 54–5.

(5) A summary of the prevailing Cambridge educational system may be read in W. T. Costello's *The Scholastic Curriculum at Early Seventeenth-Century Cambridge* (Harvard, 1958). A more general and not entirely consistent viewpoint is presented by M. H. Curtis in his *Oxford and Cambridge in Transition 1558–1642* (Oxford, 1959).

(6) Briefly referred to by Conduitt as one of several 'paper books' in his jottings on the contents of early Newton papers (Keynes MS 130.12), but more fully and accurately described by A. R. Hall in 'Sir Isaac Newton's Note-book, 1661–65', *The Cambridge Historical Journal*, **9** (1948): 239–50.

(7) Probably in the 1642 Cambridge edition.

(8) Curtis, in his *Oxford and Cambridge in Transition* (note (5) above): 257–8, quotes an instructive passage (ULC. MS Baker 37: $163^r/163^v$), written in the late 1660's by Roger North, which throws an interesting sidelight on the stir which Descartes' ideas were then beginning to provoke in 'the brisk part of the university'.

conclude that Newton's first two undergraduate years were taken up in conventional studies. However, new things were stirring in Cambridge[8] and some time in 1663 (or so the writing suggests) Newton began a new section in his notebook, the *Qu[æ]stiones quædam Philosoph[i]cæ*, in which he started to collect a miscellany of scientific jottings, scholastic in tone at first but soon with references to the new philosophies expounded in Descartes' *Principia Philosophiæ* and the works of Boyle and Hobbes. A little later, perhaps about the middle of 1664, Newton began to enter separately his own notes and observations on the planetary positions, and it is clear that he had deeply immersed himself in the work of Galileo and was familiar with Thomas Street's *Astronomia Carolina* and the observations of Jeremy Horrox, presumably from Hevelius' *Mercurius in sole visus* of 1662. By 1664, then, his reading had attained a level at which a fair measure of contemporary mathematical knowledge was required for its full appreciation. In fact, Newton composed his first mathematical essays in the late summer of this same year, 1664, and we may fairly settle on that point of time as origin in his exploration of the subtleties of advanced mathematics.

Inevitably, then, Newton's attention was directed to higher mathematics at an early stage in his scientific development. His first baptism was, if we are to believe de Moivre, somewhat unusual:[9]

In 63 [Newton] being at Sturbridge fair[10] bought a book of Astrology, out of a curiosity to see what there was in it. Read in it till he came to a figure of the heavens which he could not understand for want of being acquainted with Trigonometry.

Bought a book of Trigonometry,[11] but was not able to understand the Demonstrations.

---

(9) The following series of recollections, presumably passed on at second hand from Newton's own mouth, are quoted from Conduitt's *Memorandum relating to Sr Isaac Newton given me by Mr Demoivre in Novr 1727*. (The original, sold in 1936 at Sotheby's, is now in private possession in New York and the text is transcribed from a late-nineteenth-century copy in ULC. Add. 4007: 706^r–707^r.) With some variants and interpolations the memorandum was repeated by Conduitt in his account of Newton's life at Cambridge (Keynes MS 130.4), from which Brewster gave the extracts which appear in his *Memoirs* (note (2) above) 1: 21–2, 24. Conduitt's narrative interpolates several well-known stories whose truthfulness is dubious but which now appear in biographies as factual, and is interesting in itself. Since it has never been printed, a relevant extract is here set in Appendix 1 below.

(10) The annual international trade fair held in midsummer on the expanses of Stourbridge meadows, then immediately to the east of the town. See W. W. R. Ball, *A History of the Study of Mathematics at Cambridge* (Cambridge, 1889): 223–4, 250.

(11) Newton's library at his death contained several early works on trigonometry. The 1636 edition of Gunter's collected works (Trinity. NQ. 9.160) we know he did not buy till 1667 (when an undergraduate notebook now in the Fitzwilliam Museum, Cambridge, lists its purchase for five shillings). The texts of Newton's copies of Norwood's *Doctrine of Triangles* (London, 1645) (Trinity. NQ. 9.33) and Seth Ward's *Idea Trigonometriæ* (Oxford, 1654) (Trinity, NQ. 8.71) are both devoid of notes. However, in his copy (Trinity. NQ. 16.183) of Oughtred's *Trigonometria* (London, 1657), Newton has added two illustrative diagrams in an

Got Euclid(12) to fit himself for understanding the ground of Trigonometry.

Read only the titles of the propositions, which he found so easy to understand that he wondered how any body would amuse themselves to write any demonstrations of them. Began to change his mind when he read that Parallelograms upon the same base & between the same Parallels are equal, & that other proposition that in a right angled Triangle the square of the Hypothenuse is equal to the squares of the two other sides.(13)

Began again to read Euclid with more attention than he had done before & went through it.

Read Oughtreds [Clavis] which he understood tho not entirely, he having some difficulties about what the Author called Scala secundi & tertii gradus, relating to the solution of quadratick [&] Cubick Equations.(14) Took Descartes's Geometry in hand, tho he had been told it would be very difficult, read some ten pages in it, then stopt, began again, went a little farther than the first time, stopt again, went back again to the beginning, read on till by degrees he made himself master of the whole, to that degree that he understood Descartes's Geometry better than he had done Euclid.

Read Euclid again & then Descartes's Geometry for a second time. Read next Dr Wallis's Arithmetica Infinitorum, & on the occasion of a certain interpolation for the quadrature of the circle, found that admirable Theorem for raising a Binomial to a power given. But before that time, a little after reading Descartes Geometry, wrote many things concerning the vertices Axes [&] diameters of curves, which afterwards gave rise to that excellent tract de Curvis secundi generis.(15)

In 65 & 66 began to find the method of Fluxions, and writt several curious problems relating to that method bearing that date which were seen by me above 25 year ago.(16)

---

early hand at the top of p. [2]. Oughtred's text, with its weight of unusual symbolism, would have been formidable indeed for a beginner in mathematics. (For details of Newton's library see note (22) below.)

(12) Newton's library contains an anonymous early edition, *Euclidis Elementorum Libri XV, Græcè & Latinè. Quibus, cùm ad omnem Mathematicæ scientiæ partem, tùm ad quamlibet Geometriæ tractationem, facilis comparatur aditus* (Paris, 1573) (Trinity. NQ. 16.60), whose foreign origin would suggest that it is probably the one here referred to. The work limits itself to an enunciation of the propositions, in both Greek and Latin, without proof or commentary apart from diagrams. Newton's better-known 1655 Latin edition of Barrow's *Euclidis Elementorum Libri XV. breviter demonstrati* (Trinity. NQ. 16.201) does indeed have copious annotations in his hand, but reveals on Newton's part a deep study of the text which is hardly consistent with the hasty consultation implied here. It is feasible that, after a first glance through the propositions as stated in the earlier text (which is unmarked), Newton went on to study their proofs in Barrow's edition, already at that time in Cambridge established as a university text. (Compare note (28) below.)

(13) Propositions 35 and 47 of Euclid's first book. The latter is, of course, the celebrated 'theorem of Pythagoras'. When Newton studied it in Barrow's edition, certainly, he found it none too straightforward, for over the printed figure in his copy he has inked in several construction lines.

(14) Apparently a reference to Oughtred's solution of the general quadratic equation in the *Clavis Mathematicæ* (Oxford, 1652): Cap. XVI, 'De Æquatione & De quæstionibus per Æquationem solvendis': 52–3, §9, 'Æquationum in quibus sunt tres species æqualiter in ordine scalæ ascendentes constitutio' (that is, equations of the form $x^{2p}+\alpha x^p+\beta=0$). (Oughtred finds the two roots x_1^p, x_2^p from the symmetric evaluations $x_1^p+x_2^p=-\alpha$, $x_1^p x_2^p=\beta$: so that

But whatever the truth of Newton's first readings in Euclid and Descartes, there can be no doubt that he studied the works of both minutely, deriving particular inspiration from the latter's *Geometrie*.(17) As for the works of Oughtred and Wallis—and those of François Viète, Frans van Schooten and others not here mentioned—we have, in confirmation of de Moivre's account, not only the direct evidence of Newton's extant annotations of them (reproduced below) but his own independent testimony repeated on several occasions in little differing form. Thus, having occasion to turn back in his notebooks to his annotations out of John Wallis' work, he entered on a facing page:

July 4th 1699. By consulting an accompt of my expenses at Cambridge in the years 1663 & 1664(18) I find that in ye year 1664 a little before Christmas I being then senior Sophister,(19) I bought Schooten's Miscellanies & Cartes's Geometry (having read this

$x_1^p - x_2^p = \pm\sqrt{(\alpha^2 - 4\beta)}$ and the familiar algorithm for the roots, $\frac{1}{2}(-\alpha \pm \sqrt{(\alpha^2-4\beta)})$, follows.) Oughtred's exposition is indeed more than a little obscure, but in the appended *De Æquationum Affectarum Resolutione* he refuses to treat quadratics by numerical methods, referring them back to his preceding resolution with the words: 'Æquationum Quadraticarum, omniumque in quibus sunt tres species in ordine scalæ æqualiter adscendentes Analysin supersedebo: quia in cap. XVI, Sect. 9 *Clavis* modus facilior tradita est, quàm per genèralem hanc methodum præstari poterit: Et ad Exempla Æquationum aliter affectarum progrediar'. The reference to cubic equations is mysterious, but Newton (if these are his reported words) may confuse the present resolution of quadratics with the logarithmic machine for the numerical resolution of general algebraic equations, whose basic idea he took also from Oughtred. (See **3**, 3, §1 below.)

(15) This surely refers to Newton's earliest researches in analytical geometry (ULC. Add. 4004: 8r ff., reproduced in **2**, 1 below), whose main object is to find general analytical methods for locating the vertices, axes and diameters of curves. The lessons gained in the use of coordinate transforms were indeed applied about 1668 in his first, inadequate enumeration of cubics or *curvæ secundi generis*. (Newton's text will be reproduced in the next volume of this edition.) Very much altered and abridged, the *Enumeratio linearum curvarum tertii ordinis* was finally published as the first of the Latin appendices to the first edition of Newton's *Opticks* (London, 1704): 139–62.

(16) The 'several curious problems' written in 1665 and 1666 were probably those gathered in the October 1666 tract (**2**, 7 below), but it is possible that de Moivre saw a selection of the original papers, reproduced in **2**, 2–6, which Newton composed between December 1664 and May 1666.

(17) A well-known story, deriving ultimately from John Conduitt's manuscript *Life* (note (2) above), records that Newton's first reaction to Descartes' *Geometrie* was to point out a multitude of errors in it and to assert that it was not geometry at all. The foundation for this legend (or, as we believe, myth) is examined in Appendix 1, note (11) below. (There can be no doubt that Newton read the *Geometrie* in Schooten's second Latin edition. All his explicit references to it, in particular, are quoted by page from the latter.)

(18) This 'accompt' of expenses for the years 1663 and 1664 is lost. Presumably it carried on the expenses listed by Newton in an early notebook now in Trinity College, Cambridge, and was itself continued by the notebook now in the Fitzwilliam Museum. (Compare Brewster's *Memoirs* (note (2) above), **1**, 23, note 2.)

(19) That is, senior undergraduate. (For the function of the sophister in the seventeenth-century Cambridge scholastic system, consult pp. 14 ff. and pp. 88–9 of the works quoted in note 5 by Costello and Curtis respectively.) Newton in fact graduated B.A. in January 1664/5. (See note (25) below.)

Geometry & Oughtreds Clavis above half a year before) & borrowed Wallis's works and by consequence made these Annotations out of Schooten & Wallis in winter between the years 1664 & 1665. At wch time I found the method of Infinite series. And in summer 1665 being forced from Cambridge by the Plague(20) I computed ye area of ye Hyperbola at Boothby in Lincolnshire(21) to two & fifty figures by the same method.

Is. Newton(22)

But an exact chronology of Newton's first steps in mathematics is not, after all, so very important. What is vital to our appreciation of Newton's developing mathematical intelligence is to grasp clearly its strength and facilities and to examine its play in his mathematical creations. Though our view of that development must be a somewhat dubious amalgam of contemporary account

(20) A note made on 23 May 1665 ('R^d 10^{li}... whereof I gave my Tutor 5^{li}') in the pocket-book in the Fitzwilliam Museum confirms that Newton was still in Cambridge on that date. However, since Newton did not claim the extra $6\frac{1}{2}$ weeks commons paid to those who stayed in residence till Trinity College was dismissed on 8 August (J. Edleston, *Correspondence of Sir Isaac Newton and Professor Cotes..., with Notes, Synoptical View of the Philosopher's Life, and a variety of details illustrative of his History* (London, 1850): xlii, note 8), we may conclude that Newton went down some time in June. The belief is widespread that Newton did not return to Cambridge till early in 1667. In fact the Fitzwilliam notebook records journeys made to Cambridge on 20 March 1666 and on 22 April 1667. Since Newton was paid 13 weeks' commons for 1666 (Edleston: *loc. cit.*) he would seem to have left Cambridge again in the middle of June of that year. It appears, therefore, that Newton was in Cambridge between September 1664 and June 1665, again between 20 March and mid-June 1666 and from 22 April 1667. If we compare these dates with those of his early dated papers, the conclusion seems inescapable that Newton wrote most of them while in Cambridge (and hence with full access to university and college libraries) rather than in seclusion in Lincolnshire.

(21) Almost certainly the parish of Boothby Pagnell, about three miles to the north-east of Woolsthorpe. As Stukeley reminds us (*Memoirs* (note (2) above): 51–2), the rector of 'Boothby Pannel' at the time was Dr Babington, 'senior fellow of Trinity College; a person of learning and worth'. Newton, away from the books available to him in Cambridge, could very well have profited by the riches of Babington's library. More importantly perhaps for Newton at that time, Babington was brother-in-law to the apothecary Clark with whom Newton lodged during his schooldays in Grantham, and uncle to the pretty young girl, Miss Storey, who played with Newton when he was a schoolboy and for whom 'Sir Isaac entertain'd a passion... when they grew up' (Stukeley's *Memoirs*: 46). Newton's mother and Babington's sister were close friends and it may have been Babington rather than Aiscough who secured for Newton his sizar's place at Trinity in 1661.

(22) ULC. Add. 4000: 14^v, transcribed by Brewster in his *Memoirs* (note (2)), **1**: 23–4, and more correctly by A. R. Hall in his *Sir Isaac Newton's Notebook, 1661–65* (note (6) above): 240. In Brewster's version there is an important mistranscription (of 'clean over' for 'above') which considerably alters the time-scale of Newton's remark. Brewster probably made his copy not from the original but from Conduitt's transcription (King's College. Keynes MS 130.4), and it is possible that he misread Conduitt's 'clavis above' as 'clean over' and then added his own 'Clavis' in attempted clarification. (Brewster prints his 'Clavis' in italics as though to indicate an insertion.) A second misreading of 'such' for 'w^{ch}' is insignificant, but probably has a similar explanation.

For equivalent accounts see ULC. Add. 3968.41: 76^r ('In the winter between the years 1664 & 1665 upon reading Dr Wallis's Arithmetica Infinitorum & trying to interpole his

and the individual impressions we draw from the books Newton read(23) and the papers he wrote, its basic pattern is clear. According to Conduitt Newton once told Richard Bentley that 'all his merit was patient thought',(24) and this trait of stubborn perseverance allied to an acute perception of mathematical structure, trained by years of assiduous reading, and a formidable power of mental concentration (well described if not comprehended by Humphrey Newton at the time when he was composing his *Principia*(25)) carried Newton through most of his mathematical difficulties. His handwriting, delicately fine yet firm and powerful, is typical of the man. His essential loneliness of spirit, undeniable if not to be overstated, his inability to collaborate and even the difficulties he found in communicating his thoughts to his contemporaries are revealing aspects of the mental isolation he found necessary for the effective

progressions for squaring the circle, I found out first an infinite series for squaring the circle & then another infinite series for squaring the Hyperbola & soon after') and the draft (ULC. Add. 3968.41: 85^r) of a letter to Desmaizeaux, which, since its first printing in the preface (p. xviii) of the 1888 *Catalogue of the Portsmouth Collection of Books and Papers written by or belonging to Sir Isaac Newton*, has taken on a modern independent existence as the 'Portsmouth Draft Memorandum'.

(23) In July 1727, a little after Newton's death, his books were purchased by John Huggins, warden of the Fleet Prison, and sent to his son Charles, rector of Chinnor (near Oxford) where they eventually passed into the possession of Huggins' successor, James Musgrave. When Musgrave's son became owner of Barnsley Park in Gloucestershire in the late eighteenth century, the books were removed there, catalogued and then forgotten till 1920 when many of the more important items were sold in bundles by public auction. The remaining 860 books, rather less than half the original total, were finally bought by the Pilgrim Trust and presented to Trinity College in 1943. Two listings of Newton's library exist, neither completely accurate: the first was made at its sale on 20 July 1727 to Huggins, the second (rather fuller) by James Musgrave about 1760. (See R. de Villamil, *Newton: the man* (London, [1931]): 2–7, 62 ff.) Luckily most of the mathematical works were not sold in the 1920 sale (the main exceptions being such folio volumes as Vlacq's *Trigonometria Artificialis* and the three volumes of Wallis' *Opera omnia mathematica*) and such priceless items as a complete set of James Gregory's printed works are to be found in the Trinity collection. Of those sold in 1920 several important books remain in England in public collections, particularly the annotated copies of Barrow's *Euclid* and of Newton's own *Arithmetica Universalis* (London, $_1$1707) and *Opticks* (first Latin edition, 1706). Newton was never a systematic margin-scribbler and the large majority of his books, some apparently well-thumbed, remain virtually unmarked. Occasionally, however, as with the autograph scrap inserted in his copy of Schooten's *Exercitationes Mathematicæ* (reproduced as 1, §1 below), it is possible to find something of value.

(24) King's College. Keynes MS 130.5, quoted by Conduitt as though he had the anecdote by word of mouth from Bentley. Newton's remark is, in fact, drawn from his letter to Bentley of 10 December 1692, where, in discussing the philosophical basis of *Principia*, he commented: 'When I wrote my treatise about our Systeme I had an eye upon such Principles as might work with considering men for the beleife of a Deity & nothing can rejoyce me more then to find it usefull for that purpose. But if I have done ye publick any service this way 'tis due to nothing but industry & a patient thought' (*Correspondence of Isaac Newton* **3** (1961): 233).

(25) In his two letters to John Conduitt of 17 January and 14 February 1727/8 (King's College. Keynes MS 135, printed by Brewster in his *Memoirs* (note (2) above), **2**: 91–5 and 95–8 respectively).

prosecution of his research. Beyond reasonable doubt he was self-taught in mathematics, deriving his factual knowledge from the books he bought or borrowed, with little or no outside help. De Moivre's account confirms it and who anyway was there in Cambridge to teach him? (The pleasant story of Barrow's tutorial guidance is ill-founded and there was no formal instruction in higher mathematics to be had in the Cambridge of Newton's undergraduate years.[26]) Very quickly he advanced to the frontier of existing knowledge and cut a broad path into new mathematical country. Soon he began to prepare a series of papers on analytical geometry and calculus which went far beyond anything known, and the initial impetus of that first headlong advance was to bear him easily through the rest of his mathematical days. The price he paid for the depth of his penetration was the eternal warring within him of a natural instinct of pride in his findings coupled with the wish to publish them to the world, in conflict with his conscious knowledge of how far he had outranged his fellow mathematicians and his refusal to face their misunderstanding. The one impulse made him ever willing to show his unpublished papers to those whose ability he respected,[27] the other hindered their public communication to the world (with a few notable exceptions), and the ensuing deadlock has led to the intolerable present situation in which many of his crucially important early papers have lain, unknown and unsuspected, in darkness for more than two centuries.

Inevitably we are led back to examine those papers and they must remain our fundamental source of knowledge of Newton's mathematical development. In this first part of the present edition of Newton's mathematical work are reproduced his annotations, as he set them down in his notebooks, of the standard mathematical treatises of his youth—of Schooten, Viète, Wallis,

(26) The tradition of Barrow's early influence on Newton was already accepted by Conduitt at the time of Newton's death. (See Appendix 1.) His authority seems to have been Stukeley, who in his letter to Mead of 26 June 1727 wrote: 'The famous Dr Barrow...was Sir Isaac's tutor. If he did not take a byass in favor of mathematical studies from him, at least he confirm'd it thereby....His tutor...conceiv'd the highest opinion and early prognostic of his excellence; would frequently say that truly he himself knew something of the mathematics, still he reckon'd himself but a child in comparison of his pupil Newton'. On to this account Conduitt himself grafted the anecdote (not in itself implausible) of Barrow's examining Newton in Euclid and finding him wanting. However, Barrow was not Newton's undergraduate tutor when Newton took his B.A. in January 1664/5 (a position filled by Benjamin Pulleyn: see Edleston (note (20) above): xli, note 4, and J. and J. A. Venn's *Alumni Cantabrigienses*, Part 1, Vol. 3 (Cambridge, 1924): 406). Whether Barrow's own mathematical maturity was sufficient at this time for him to be potentially helpful to Newton is an open question, but it is clear that his early Lucasian lectures (those for 1664–66, printed in 1685 as *Lectiones Mathematicæ XXIII in quibus Principia Matheseos generalia exponuntur...habitæ Cantabrigiæ 1664–1666*) were too elementary and philosophical in approach to appeal to a young man whose mind was already afire with new ideas in mathematics. (Their quality, indeed, is an unflattering comment on the prevailing intellectual level of scientific instruction in the university.

Descartes, Oughtred, Huygens, de Witt and others. They give us a vivid picture of Newton's likes and dislikes, of what he thought significant and what he passed lightly over in his reading. Like all true notes these annotations are not mere inferior, copied images but have their own life, revealing a young mathematician at work stretching his mind, shaping what he read and recording his own impressions in comment upon it. Sometimes indeed, and especially with the Wallis notes, there is no true dividing line between the summarized impact of the original and the following wave of new ideas which became a piece of original research. Elsewhere, as in his annotations on Schooten, Newton seems more obviously to have been taking factual notes of what interested him in his text. In the case of Descartes' *Geometrie*, whose decisive influence on Newton is clear in his early research-papers, explicit notes are few and those almost wholly refer to the appendices added by Schooten in his Latin editions. We may perhaps frame the loose criterion that the more clear-cut the annotation, the less fiery its ultimate effect on Newton's development. Of the individual variations in this general pattern, however, detailed discussion will be found in the summaries and footnotes added to the text reproduced below. Here it remains to evaluate the importance of these annotations as a necessary first stage in Newton's growth to mathematical maturity.

A mathematician needs an adequate notation, a competent knowledge of mathematical structure and the nature of axiomatic proof, an excellent grasp of the hard core of existing mathematics and some sense of promising lines for future advance. By and large Newton took his arithmetical symbolisms from

See M. H. Curtis' *Oxford and Cambridge in Transition* (note (6) above): 227–60: ch. IX, 'The Universities and the Advancement of Learning', and Costello's *Scholastic Curriculum* (note (6) above): ch. III, 'The Undergraduate Sciences', especially 102–4, 'Mathematics'.) About 1713, in the midst of his priority dispute with Leibniz, Newton wrote down an off-hand note that 'The same year [1665] I got some light into the method of moments & fluctions. And its probable that Dr Barrows Lectures might put me upon considering the generation of figures by motion, tho I not now remember it.' (ULC. Add. 3968.41: 84v). It is interesting to have irrefutable evidence that Newton attended Barrow's lectures, but there is here nothing of the cordial intimacy of later anecdote, and not till 1669, when Newton communicated to Barrow his *De Analysi*, do we have formal evidence of personal friendship. Later still, apparently, when Barrow left Cambridge for London the friendship—if indeed it was ever more than a professional relationship—lapsed again. (In 1670 Barrow gave Newton a presentation copy (Trinity. NQ. 16.181) of his *Lectiones XVIII...in quibus Opticorum Phænomenωn genuinæ Rationes investigantur ac exponuntur* with the dedication 'Isaaco Newtono Reverendus Author hunc dono dedit. July 7th 1670', but it cost Newton 3/6d to buy his copy (Trinity. NQ. 16.193, ironically a presentation copy to 'Edm: Matthews'!) of Barrow's 1675 *Archimedis Opera: Apollonii Pergæi Conicorum libri IIII....*)

(27) Apart from Barrow, who probably saw only the *De Analysi*, Collins, John Craig, Fatio de Duillier, Raphson, Halley, de Moivre, David Gregory and William Jones (but not, significantly, John Wallis) are known to have seen substantial portions of Newton's mathematical papers during his lifetime.

Oughtred and his algebraical from Descartes, and on to them where necessary, particularly in developing his calculus methods, he grafted new modifications of his own. Traditional forms of proof he learnt from his repeated reading of Barrow's *Euclid*(28) and these, coupled with the elementary scholastic logic he learnt at grammar school and in his early years at Cambridge,(29) gave him an adequate training in simple deductive procedures. More complex patterns of deduction, particularly the method of exhaustion, he seems never to have liked.(30) Use of the free algebraic variable he acquired from the distinction, hazarded by Viète and more simply presented by Oughtred,(31) between logical restrictions given numerically (*in numeris*) and as freely variable (*in speciebus*) and from their confident exploitation by Descartes who, by a suitable interpretation of geometrical entities, expanded the variable into a new system of geometrical structure. Subtler usages, especially the manipulation of tied

(28) *Euclidis Elementorum Libri XV. breviter demonstrati, Operâ Is. Barrow Cantabrigiensis* (Cambridge, 1655), to which is added *Euclidis Data succinctè demonstrata: Unâ cum Emendationibus quibusdam & Additionibus ad Elementa.* Newton's copy (Trinity. NQ. 16.201) is described briefly by H. Zeitlinger in his *A Newton Bibliography* = *Isaac Newton, 1642–1727* (ed. W. J. Greenstreet) (London, 1927): 148–70, especially 169–70. (The 'few notes in a different and later handwriting' mentioned by Zeitlinger are also Newton's.) The annotations reveal two levels of study on Newton's part: a first reading impatient of minor points of detail, and a later exhaustive examination whose notes in part correct the earlier, more superficial ones. It is important to observe that he gave most of his attention to the 'arithmetical' Books (II, V, VII and X) of the *Elements*, while Books III, IV, VI, VIII, IX, XII and the pseudo-XIV contain only trivial corrections to Barrow's marginal references. In Book I Newton has slightly altered the proof of Prop. 7, added three extra construction lines to the diagram of I, 47 and made trivial additions to I, 15 and I, 16. Book XI is blank apart from a cancelled alternative proof of XI, 4 (printed in photocopy on p. 170 of Zeitlinger's article). Book XIII also is blank except that to Barrow's appendix *Ex Herigonio* (which adds some numerical details, taken from Hérigone's 1642 *Cursus Mathematicus*, on the five regular solids) Newton contributes some further amplifications, probably taken from Oughtred's *De Solidis Regularibus Tractatus* (= pp. 23–41 of his *Elementi decimi Euclidis declaratio*, printed with the *Clavis Mathematicæ* in editions from 1647). Book XV is likewise clean except for a final addition, 'Haud secus icosaëdrũ in dodecaedro describes'. The notes on Books II, V, VII and X are wholly explanatory and almost always brief. (The lengthy addition to X, 18, printed by Zeitlinger on his p. 168, is unique.) In the main Newton has used Barrow's symbolism as listed in the *Notarum explicatio* (itself based closely on Oughtred's notation in the *Clavis*) to render the meaning of Euclid's verbal exposition immediately obvious, though he is quick to spot duplication (as at X, 14, which he notes to be 'Eadem cum priori'). The appended *Data* has been read fairly closely, and though Newton has entered no notes on it he has corrected two wrong marginal references.

(29) From Conduitt (see Appendix 1) derives the tradition that Newton read Saunderson's 'logick' (*Logicæ artis compendium*), a popular university text of the period often used as an introduction to the rigours of Aristotle's text. As is usual with Conduitt's anecdotes the details are overloaded, but we may easily believe that Newton did read the work some time during his early years at Trinity. The evidence of Newton's own undergraduate notes in ULC. Add. 3996 (note (6) above) with its sets of annotations under such headings as *Ex Aristotelis Stagiritæ Peripateticorum principis Organo Definitionum, sumarumq; sententiarum recollectio* is more direct evidence of Newton's early schooling in the niceties of Aristotelian logic.

variables, Newton learnt more slowly. His reading ranged widely over a number of topics: simple number theory (in Schooten), traditional and elementary modern pure geometry (in Euclid, Viète, Schooten and de Witt), topics in algebra and the theory of equations (in Viète, Oughtred, Schooten and Wallis), in elementary analytical geometry (in Descartes and possibly Wallis, with some applications in Schooten) and in the twin calculus procedures of method of tangents (in Descartes) and quadratures (in Wallis). A strong bias to the analytical side of mathematics is noticeable and Newton's own early researches show a similar emphasis. We miss, too, a few important names: Napier, Briggs(32) and Harriot;(33) Desargues, Pascal and Fermat;(34)

(30) In a few early papers Newton did, in fact, use variants of the exhaustion proof, but the logical structure of these papers, first proofs that tangent- and quadrature-methods are strictly inverse procedures, is taken over from Heuraet's general rectification procedure as printed by Schooten in his second Latin edition of Descartes' *Geometrie*. (See **2**, 5, §1, especially note (51).) Newton took no part in the refinement of the exhaustion-method, which in the hands of Torricelli, Pascal, Huygens, and James Gregory became a near equivalent of the Riemann integral. (Compare D. T. Whiteside, 'Patterns of Mathematical Thought in the later seventeenth century', *Archive for History of Exact Sciences* **1** (1961): ch. IX, 'Calculus 2. The method of proof by exhaustion'.)

(31) Viète, in so far as we may interpret the illusive presentation he gave in his *In Artem Analyticen Isagoge*, *Ad Logisticen Speciosam Notæ priores* and *De Æquationum Recognitione et Emendatione Tractatus Duo*, had difficulty in separating the logical and the extralogical in the process of algebraic analysis (*Zetetice*). However, his distinction between *Logistice numerosa* and *Logistice speciosa* is explicit: 'Logistice numerosa est quæ per numeros, Speciosa quæ per species seu rerum formas exhibetur, ut[-]pote per Alphabetica elementa' (*In Artem Analyticen Isagoge:* Cap. IV, 'De præceptis Logistices speciosæ' = *Opera Mathematica* (ed. Schooten) (Leyden, 1646): 4). In other words, and as Schooten amplifies the remark in his commentary, *Logistice numerosa* relates roughly to the logical structure of numerical procedures, while *Logistice speciosa* uses the free variable to explore algebraic structure. Oughtred in his *Clavis* cast away the Greek elements in Viète's presentation and settled for a strict distinction between magnitudes denoted 'vel numeris...vel etiam speciebus' (*Clavis Mathematicæ* (Oxford, $_3$1652: 3)), devoting much of his work to expounding the elementary techniques of 'Speciosa hæc Arithmetica' through a notation for variables and logical symbols based on Viète's.

(32) Several decades ago Duncan C. Fraser, in an examination of Newton's work in finite differences first printed as a series of articles in the *Journal of the Institute of Actuaries* and later collected as *Newton's interpolation formulas* (London, 1927), put forward the hypothesis that Newton drew his ideas on subtabulation from a reading of the introductory matter in Briggs' *Arithmetica Logarithmica* (London, 1624). In particular from an analysis of two manuscripts collectively labelled *Regula differentiarum* (ULC. Add. 3964.7) he concluded (*Newton's Interpolation Formulas*: 58) that 'no one who compares the directions given in the MS with the example given by Briggs can doubt where Newton found his starting point [though] Newton never mentions Briggs'. However, the evidence Fraser gives in support of his claim is flimsy and circumstantial, while Briggs' name seems to occur nowhere in any of Newton's unpublished papers. Further, Newton's early researches in the binomial expansion and the resolution of 'affected' equations show no trace of Briggs' own methods as developed in the *Arithmetica Logarithmica* and *Trigonometria Britannica* (Gouda, 1633), and while it is true that certain portions of the latter were inserted by Vlacq in his *Trigonometria Artificialis* of the same year, a copy of which was in Newton's library (de Villamil, *Newton: the man* (note (23) above): 69), these insertions were restricted to the unexciting second book of Briggs' introduction. As

Apollonius, Archimedes and indeed any Greek geometer but Euclid (and he represented only by his *Elements* and *Data* in Barrow's translation): Stevin, Girard and even Kepler (though Newton was soon to be familiar with his optical work); Cavalieri, finally, and Torricelli.(35) Of all that Newton did read, however, it is not difficult to isolate Descartes and Wallis as the great formative influences. In them Newton rightly seized on the two fields of analytical geometry and calculus, making them the twin foundation of his mature work in analysis, whose nature will be analysed in detail in future volumes.

One final remark. Newton's annotations do not form a first stage in his development which can be set off as the period of his apprenticeship to the mathematical discipline. His first researches—those in the transformation of co-ordinate systems, in the binomial theorem and Wallisian interpolation and in the geometrical optics of white light—were indeed pursued while he was in the thick of taking notes on his reading, and the two aspects are essentially interdependent. But already by mid-1665, one short crowded year after his

---

for Napier, there is nothing to show that Newton read either his *Logarithmorum Canonis Descriptio* (Edinburgh, 1614) or the deeper *Mirifici ipsius Canonis Constructio* (Edinburgh, 1619) where the theory of the logarithm was first expounded in its Napierian form. Almost certainly Newton learnt the techniques of logarithmic manipulations elsewhere, perhaps from the tract *De Æquationum Affectarum Resolutione* which Oughtred appended to his *Clavis Mathematicæ* (editions from 1647).

(33) On the flyleaf of an undergraduate notebook (ULC. Add. 3996: 31^r) Newton entered, about the end of 1664 if we date the writing correctly, the cryptic phrase 'Harriots geometry or Algebra. Or Praxis artis Analiticæ'. No other reference to Harriot appears in Newton's early mathematical papers, and it seems likely that this is merely a bibliographical reference to be checked, taken perhaps from Wallis' *Mathesis Universalis* (*Dedicatio:* [v]) as printed in his *Operum Mathematicorum Pars Prior* (Oxford, 1657). Beyond doubt Harriot's work was not studied in detail by Newton who formed his algebraical ideas from a reading of Viète, Oughtred and, supremely, Descartes.

(34) Though Newton read parts at least of Wallis' *Commercium Epistolicum* (Oxford, 1658), in his annotations on the work (3, §3 below) there is no mention of Fermat's name or of his letters there published. (Indeed he may not have had access to an issue of Wallis' work which included the important letter of Fermat to Kenelm Digby sent on to Wallis on 13 June 1658: see Appendix 2.) Great effort, again, has been spent in trying to trace a connection between Newton's and Fermat's method of tangents. In fact Newton in an early paper expresses his debt to de Beaune alone for drawing his attention to the advantages of working with the subtangent rather than with the subnormal. (See **2**, 4, §3, especially note (33).) Later, it is true, he was willing to admit that he 'met with the method of Fermat in ye second book of Schooten's Commentaries' (ULC. Add. 3968.19: 290^v), that is, in Schooten's second Latin edition of Descartes' *Geometrie* (**1**: *Commentarii in Librum II*, 0: 253–5). The question will be discussed in greater detail in the introduction to Part 2 below.

(35) Wallis wrote in preface to his *Arithmetica Infinitorum* (printed in his *Operum Mathematicorum Pars altera* (Oxford, 1656)): 'Exeunte anno 1650 incidi in Torricellii scripta mathematica [that is, *Opera Geometrica.* Florence, 1644], (quæ ut per alia negotia licuit, anno sequenti evolvi), ubi inter alia Cavallerii Geometriam indivisibilium exponit. Cavallerium ipsum nec ad manum habui, et apud bibliopolas aliquoties frustra quæsivi'. There is nothing which

first beginnings, the urge to learn from the work of others was largely abated. The indication of his rapid rise to mathematical maturity is telling. It was time for him to go his own way in earnest and thereafter, though he continued to draw in detail on the ideas of others, Newton took his real inspiration from the workings of his own fertile mind.

APPENDIX 1

JOHN CONDUITT'S ACCOUNT OF NEWTON'S LIFE AND WORK AT CAMBRIDGE

Extract from the original in King's College, Cambridge(1)

Mr Aiscough(2) had given Sr Isaac before he sett out for Cambridge Sanderson's logick & told him that was the first book his tutor would read to him, this Sr I. read over by himself & when he came to hear his tutor's lectures upon it found he knew more of it than his tutour, who finding him so forward told him he was going to read Kepler's Opticks to some gentlemen commoners & that he might come to those lectures. Sr I. immediately read it at home & when his tutour gave him notice of the lectures he told him he had already read that book throu.(3) He bought a book of Judicial(4) Astrology out of curiosity (wch Hobbes calls the mother of all Philosophy. Human nature—p. 112(5)) to see

reveals any direct knowledge on Newton's part of the works of either Cavalieri or Torricelli, and he seems to have been content to take Wallis' account of indivisibles as an adequate rendering of the new 'Geometria indivisibilibus continuorum...promota' (as Cavalieri's 1635 source-work entitled the theory).

(1) Keynes MS 130.4: 1–8 (Conduitt's pagination). This portion is reproduced both for its intrinsic interest and to show the liberties which Conduitt took with de Moivre's memorandum in drafting what was to become the main source for later accounts of Newton's introduction to mathematics.

(2) Newton's uncle, William, on his mother's side. Relations between the two at this time seem to have been good. Thus, among other notes entered on a loose vellum sheet perhaps in the autumn of 1665 Newton added an abrupt phrase as a reminder 'of wt hath past twixt my unkle ayscough & mee' (ULC. Add. 3958.2: 45v).

(3) This anecdote is in neither de Moivre's nor Stukeley's accounts and we may reasonably doubt its truth.

(4) Conduitt's interpolation in de Moivre's account.

(5) The quotation is, for what it is worth, from the second edition, 'augmented and much corrected by the Author's own hand', *Humane Nature: Or, the fundamental Elements of Policy, being a discovery of the Faculties, Acts and Passions of the Soul of Man, from their original causes* (London, 1651): 112: 'And from this Passion of Admiration and Curiosity, have arisen not onely the invention of Names, but also supposition of such Causes of all things as they thought might produce them. And from this beginning is derived all Philosophy; as Astronomy from the admiration of the course of Heaven, Naturall Philosophy from the strange effects of the Elements and other Bodies.'

what there was in that science & read in it till he came to a figure of the heavens wch he could not understand for want of being acquainted with Trigonometry, & to understand the ground of that bought an English Euclid with an Index of all the problems at the end of it & only turned to two or three wch he thought necessary for his purpose & read nothing but the titles of them finding them so easy & self evident that he wondered any body would be at the pains of writing a demonstration of them & laid Euclid aside as a trifling book, & was soon convinced of the vanity & emptiness of the pretended science of Judicial astrology.(6)

About Midsummer 1664 he read Oughtred's Clavis wch he understood tho^{u} not entirely he having some difficulties about what the author calls scala secundi et tertii gradus relating to the solution of Quadratick [&] cubic Æquations. (The opinion he had of Oughtred's Clavis appears by the following memorandum found among his papers in his own writing & signed with his name viz—Mr Oughtred's Clavis being one of the best as well as one of the first Essays for reviving the Art of Geometrical Resolution & Compositon—I agree with the Oxford professors that a correct edition thereof to make

(6) Probably Conduitt's own fabrication rather than inserted anecdote.

(7) The original (ULC. Add. 3965.6: 35^{v}) was scrawled hurriedly and with some cancellation at the foot of a page of drawings of scientific instruments and on the back of part of a draft revision of his *Principia* begun about 1693. (It is reproduced as Plate IV, facing p. 176, in A. R. and M. B. Hall's *Unpublished Scientific Papers of Isaac Newton* (Cambridge, 1962).) Conduitt's first copy (Keynes MS 130.12) is a more accurate transcription of Newton's 'Mr Oughtred Clavis being one of ye best as well as one of ye first Essays for reviving ye Art of Geometricall Resolution & Composition I agree wth ye Oxford Professors that a correct edition thereof to make it more usefull & bring it into more hands will be both for ye honour of o^{r} nation & advantage of Mathematicks. Is. N.' The Savilian Professors at the time were John Wallis (of Geometry) and David Gregory (of Astronomy), with both of whom Newton was in correspondence. A fifth Latin edition of the *Clavis* (unaltered from the third) was published at Oxford in 1693, but more probably the work to which Newton was giving his support was the new English translation, printed at London in 1694 but not particularly free from error, which Halley recommended in his preface 'to all Beginners in the Analytical Art, especially to such, who tho they may be Ignorant of the Latin tongue, may yet be desirous to inform themselves in Geometry'.

(8) Descartes' 'Epistles' were first published by Clerselier in his *Lettres de M. Descartes* 3 vols. (Paris, 1657–67), with the bulk of the mathematically interesting letters inserted in Tome III (1667), 'Où il répond à plusieurs difficultez qui luy ont esté proposées sur la Dioptrique, la Geometrie, & sur plusieurs autres sujets'. (A second edition, with Latin and French translations added to French and Latin originals, appeared in six volumes at Paris, 1724/1725.) Conduitt seems to refer to Descartes' letter to de Beaune on 20 February 1639 (Clerselier **3**: Letter 71: 409–16, especially 410 = *Œuvres des Descartes* (ed. Adam and Tannery), **2** (1898): CLVI: 510–19, especially 511–12), where he wrote: '...i'auois preueu que certaines gens, qui se vantent de sçauoir tout, n'eussent pas manqué de dire que ie n'auois rien écrit qu'ils n'ayent sceu auparauant, si ie me fusse rendu assez intelligible pour eux;...Outre que ce que i'ay obmis ne nuit à personne; car pour les autres, il leur sera plus profitable de tascher à l'inuenter d'eux mesmes, que de le trouuer dans vn Liure. Et pour moy, ie ne crains pas que ceux qui s'y entendent m'imputent aucune de ces obmissions; car i'ay par tout

it more usefull & bring it in to more hands will be both for the honor of our nation and advantage of Mathematicks.(7)) He then young as he was took in hand Des-Cartes's Geometry (that book wch Descartes in his Epistles with a sort of defiance says is so difficult to understand(8)). (He began with the most crabbed studies & books, like a high spirited horse who must be first broke in plowed grounds & the roughest & steepest ways, or could otherwise be kept within no bounds.(9)) When he had read two or three pages & could understand no farther he being too reserved or modest to trouble any person to instruct him(10) begain again & got over three or four more till he came to another difficult place, & then began again & advanced farther & continued so doing till he not only made himself master of the whole without having the least light or instruction from any body(9), but discovered the errors of Descartes, as appears by the original book wch he read at that time [&] is still in being & marked in many places in his own hand writing with these words *Error—Error non est Geom.* (Mem^m Mr Professor Smith told me he had seen the book[:] it will be proper to mark the passages & shew they are errours.)(11)

eu soin de mettre le plus difficile, & de laisser seulement le plus aisé.' Descartes made many similar remarks in his correspondence. On 31 March 1638, for example, he wrote to Mersenne: '...ie mets dans la question de Pappus tout ce qu'il faut sçauoir de plus pour les entendre. Mais le bon est touchant cete question de Pappus, que ie n'en ay mis que la construction & la demonstration entiere, sans en mettre tout l'analyse, laquelle ils s'imaginent que i'ay mise seule, en quoy ils tesmoignent qu'ils y entendent bien peu. Mais ce qui les trompe, c'est que i'en fais la construction, comme les Architectes font les bastimens, en prescrivant seulement tout ce qu'il faut faire, & laissant le trauail des mains aux charpentiers & aux masons.' (Clerselier, **3**: Letter 69: 395–6 = *Œuvres*, **2**: CXIX: 83.) To Schooten, then engaged in preparing the commentary to his Latin translation of the *Geometrie*, he sent a similar passage for insertion: 'Ejusdem brevitatis studio nulla etiam hîc mentio sit oppositarum Hyperbolarum, non quod ab Authore ignorentur.... Sed notandum est illum faciliora ferè semper in hac Geometria neglexisse, nihil autem ex difficilioribus, inter ea quæ tractanda suscepit, omisisse.' (Clerselier, **3**: Letter 82: 471 = *Œuvres*, **2**: CLXII: 577, inserted with slight alterations in Schooten's *Commentarii in Librum II*, *E* = *Geometria*, $_1$1649: 197/$_2$1659: 225.)

(9) Conduitt's interpolations, both dubious and highflung.

(10) Conduitt's interpolation in de Moivre's account.

(11) David Brewster wrote in 1855: 'Newton's copy of Descartes' Geometry I have seen among the family papers. It is marked in many places with his own hand, *Error, Error non est Geom.*' (*Memoirs...of Sir Isaac Newton*, **1**: 22, note 1.) Brewster's note is clearly based on Conduitt's present account, yet is at odds with the latter's affirmation that he had the story from 'Mr Professor Smith' (the well-known Master of Trinity and editor of Roger Cotes' posthumous *Harmonia Mensurarum*) and the implication that the copy of Descartes' *Geometrie* in question was not among the family papers in Conduitt's possession. We must surely believe Conduitt's version, but need not thereby convict Brewster of consciously cheating. It would be easy for him to have misread his notes on Conduitt's manuscript and ultimately to have convinced himself that he had seen the book. We may add that the 1888 *Catalogue of the Portsmouth Collection* enters (at section VII, 6) 'A copy of Schooten's edition of Des Cartes' Geometry, Lugd. [Batav.] 1649, with a few notes in Newton's hand'. The assertion is false. The listed work (ULC. Adv.d. 39.1) has indeed some minor corrections but the hand is not Newton's. It is in any case beyond doubt that Newton read the *Geometrie* in its second Latin

Soon after he stood to be a Scholar of the House & Dr Barrow examined him in Euclid wch he knew so little of that Dr Barrow conceived a very indifferent opinion of him. The D^r never asked him about Descartes's Geometry not imagining that any one could be master of that book without first reading Euclid & Sr Isaac was too modest to mention it himself so that he was not made Scholar of the House(12) till the year following. Upon this Sr I. read Euclid over again & began to change his opinion of him when he read that Parallelograms upon the same base & between the same parallels are equal & that other proposition that in a right angled triangle the square of the Hypothonuse is equal to the squares of the two other sides(13) (& in his latter days he spoke with regret of his mistake at the beginning of his Mathematical studies in applying himself to the works of Descartes & other Algebraic writers before he had considered the Elements of Euclid with that attention wch so excellent a writer deserved—Pemberton in the preface.(14))

About Xmas 1664 He read Schooten's Miscellanies & Dr Wallis's Arithmetica Infinitorum & on the occasion of a certain interpolation for the quadrature of the circle found that admirable Theorem for raising a binomial to a power given, but before that time a little after reading Des Cartes's Geometry wrote many things concerning the vortices(15) Axes [&] Diameters of curves wch afterwards gave rise to that excellent tract de Curvis secundi generis & made several notes & remarks on Schooten's Miscellanies wch are still in being, that was his usual method in all the books he read.

edition (Leyden, 1659/61) with its appended tracts by Hudde, Heuraet and de Witt. An extensive search through the Wren Library at Trinity College and the University Library has turned up no copy which can be identified with that claimed to have been seen by Smith, himself hardly familiar enough with Newton's hand to be able to distinguish it with certainty. If Newton did, in fact, enter an 'Error, non est Geom.' in Descartes' work we may justly convict him of an excess of undergraduate callowness and incomprehension. But the story conflicts with the unchallangeable manuscript evidence of Descartes' overwhelming influence on Newton's early mathematical development, and may for that reason be safely disregarded.

(12) That is, of Trinity College.

(13) A neat example of Conduitt's grafting of anecdote on to de Moivre's account.

(14) Conduitt refers to Henry Pemberton's *View of Sir Isaac Newton's Philosophy* (London, 1728): Preface: [iii]: 'Sir Isaac Newton has several times particularly recommended to me Huygen's stile and manner. He thought him the most elegant of any mathematical writer of modern times, and the most just imitator of the ancients. Of their taste, and form of demonstration Sir Isaac always professed himself a great admirer: I have heard him even censure himself for not following them yet more closely than he did; and speak with regret of his mistake at the beginning of his mathematical studies, in applying himself to the works of Des Cartes and other algebraic writers, before he had considered the elements of Euclide with that attention which so excellent a writer deserves.' Here Newton was doing himself down a little too strongly. His copy of Barrow's *Euclid* (see note (28), page 12 above) shows that he went through the geometrical argument carefully, correcting several mistakes in proposition references.

(15) Conduitt intends 'Vertices' of course!

(In the winter between the years 1664 & 1665 he found the method of infinite series, & in summer 1665 being forced from Cambridge by the plague computed the area of the Hyperbola at Boothby in Lincolnshire to two & fifty figures by the same method—Mem^{m}. This is writt in a pocket book in Sr I.'s own hand writing.(16))

APPENDIX 2
BIBLIOGRAPHICAL NOTE ON THE WORKS ANNOTATED BY NEWTON

As far as we have been able to discover, Newton in the years 1664 and 1665 made detailed notes upon six standard mathematical texts of the period, three edited or written by Frans van Schooten, one by William Oughtred and two by John Wallis. For convenience of reference we collect here some pertinent facts relating to their composition and printing history.

Having written it between the autumn of 1635 and the beginning of 1637, René Descartes first published his *Geometrie* in June of the latter year as the third of the scientific tracts appended, ostensibly so at least, as practical applications of his *Methode...pour chercher la verité dans les sciences.*(1) From the first he was little satisfied with its form and argument and quickly encouraged the writing of commentaries clarifying its obscurities and developing its approach.(2) Already in the summer of 1638 he had sanctioned the anonymous *Introduction à la Geometrie* (perhaps by Haestrecht),(3) but more importantly he

(16) The transcription of this passage (quoted in the preceding introduction from Newton's original) is taken from Conduitt's primary copy (Keynes MS 130.12), where he noted that 'In one of Sr I.N.'s paper books near some annotations of Dr Wallis's Arithmetica Infinitorum There is the following Mem^{m} in Sr Isaac's hand writing—July 4. 1699...by the same method. Isaac Newton. Mem^{m} in the same book [ULC. Add. 4000] a good deal upon chances & Musick.' Conduitt probably made this latter copy from the original, but it is possible (if we interpret correctly the dubious ordering on pp. 8 and 9 of Conduitt's Keynes MS 130.4) that he had it at second hand from William Jones, who himself had taken, along with a transcription (now in private possession) of the whole of Newton's annotations of Wallis, a 'Cop. ver.' of the passage.

(1) [René Descartes], *Discours de la Methode pour bien conduire sa raison, & chercher la verité dans les sciences. Plus La Dioptrique. Les Meteores. Et la Geometrie. Qui sont des essais de cete Methode*, 4° (Leyden, 1637): ($_1$1–$_1$78) *Discours de la Methode*; ($_2$1–$_2$153) *La Dioptrique*; ($_2$155–$_2$294) *Les Meteores*; ($_2$295–$_2$413) *La Geometrie*; (30 pp. unpaginated) *Table*.

(2) See Gerard Milhaud's introduction to his first printing of de Beaune's *Notes brèves* in their original French in [Ch. Adam and G. Milhaud], *Descartes, Correspondance* 3 (Paris, 1941): Appendice II: 352–67, especially 360–1.

(3) Printed for the first time (in modernized French) on pp. 328–52 of the work cited in the preceding note (2).

welcomed, as a perceptive exposition of the more elementary aspects of his work, the *Notes brèves sur la geometrie de Mr. D.C.* which Florimond de Beaune composed in the winter of 1638/39.[4] Ten years later Schooten translated both the *Geometrie* and de Beaune's notes into Latin and published them in 1649 with an informative commentary of his own,[5] many of whose details were derived directly from Descartes' own criticisms made in his correspondence with Schooten. A decade later still, nine years after Descartes' death, Schooten revised his Latin edition, doubling its bulk with letters and tracts by Johann Hudde, Henrik van Heuraet and Jan de Witt together with two further tracts of his own authorship and one of de Beaune's (as posthumously edited by Erasmus Bartholin). This final compendium,[6] overwhelming the hundred small pages of Descartes' original text with more than eight hundred pages of commentary, is the work which Newton borrowed and annotated in the summer and autumn of 1664 (though the copy he bought for personal use the following winter may have been a 1649 edition).[7] To distinguish it from

(4) Printed (in modernized French) on pp. 368–401 of the work cited in note (2).

(5) *Geometria à Renato des Cartes Anno 1637 Gallicè edita; nunc autem Cum Notis Florimondi de Beaune, In Curia Blæsensi Consiliarii Regii, In linguam Latinam versa, & commentariis illustrata, Operâ atque Studio Francisci à Schooten, Leydensis, in Academiâ Lugduno-Batavâ, Matheseos Professoris, Belgicè docentis*, Leyden, 1649 (4°): (Unpaginated) *Epistola Dedicatoria, /Ad Lectorem,/ Index Eorum, quæ in hac Geometria continentur;* (1–18/19–74/75–118) *Geometriæ Liber primus/ secundus/ tertius;* (119–61) *Florimondi de Beaune In Geometriam Renati des Cartes Notæ Breves;* (162–294) *Francisci à Schooten In Geometriam Renati des Cartes Commentarii;* (295–336) *Additamentum.*

(6) *Geometria, à Renato des Cartes Anno 1637 Gallicè edita; postea autem Unà cum Notis Florimondi de Beaune, In Curia Blesensi Consiliarii Regii, Gallice conscriptis in Latinam linguam versa, & Commentariis illustrata, Operâ atque studio Francisci à Schooten, in Acad. Lugd. Batava Matheseos Professoris. Nunc demum ab eodem diligenter recognita, locupletioribus Commentariis instructa, multisque egregiis accessionibus, tam ad uberiorem explicationem, quàm ad ampliandam hujus Geometriæ excellentiam facientibus exornata,* (2 parts) Amsterdam, 1659/61 (4°). [*Pars prima*]: (unpaginated) *Epistola Dedicatoria/Præfatio ad Lectorem;* (1–16/17–66/67–106) *Renati des Cartes Geometriæ Liber primus/ secundus/tertius;* (107–42) *Florimondi de Beaune in Geometriam Renati des Cartes Notæ Breves;* (143–344) *Francisci à Schooten in Geometriam Renati des Cartes Commentarii;* (345–68) [*ejusdem*] *Appendix, de Cubicarum Æquationum Resolutione;* (369–400) *Additamentum,* [*in quo continetur solutio artificiosissima difficilis cujusdam Problematis; & Generalis Regula de extrahendis quibuscunque Radicibus Binomiis*]; (401–506/507–16) [*Johannis Huddenii*] *Epistolæ duæ, quarum altera de Æquationum Reductione, altera de Maximis et Minimis agit;* (517–20) *Henrici van Heuraet Epistola de Transmutatione Curvarum Linearum in Rectas. Pars secunda:* (1–48) *Principia Matheseos Universalis, sev Introductio ad Geometriæ Methodum Renati des Cartes, Conscripta ab Er. Bartholino, Casp. Fil. Editio tertia, priore correctior* [first published at Leyden, 1651]; (49–152) *De Æquationum Natura, Constitutione, & Limitibus Opuscula Duo. Incepta à Florimondo de Beaune...; Absoluta vero, & post mortem ejus edita ab Erasmio Bartholino, Medicinæ & Mathematum in Regia Academia Hafniensi Professore publico;* (153–340) *Johannis de Witt Elementa Curvarum, Edita Operâ Francisci à Schooten...;* (341–420) *Francisci à Schooten, Leidensis,... Tractatus de Concinnandis Demonstrationibus Geometricis ex Calculo Algebraïco. In lucem editus à Petro à Schooten, Francisci Fratre.* (The tracts in the second part have individual title-pages. An unchanged reissue of the whole was made at Amsterdam in 1683.)

Descartes' *Geometrie* and for convenience of reference we shall universally dub this second Latin edition the *Geometria.*

Earlier, in 1646, conscious of the high importance of François Viète's scattered works on algebra, geometry and analysis which had been published singly over the period 1579–1615 and already become exceedingly rare, Schooten had gathered most of them,(8) suitably edited with commentary and a valuable discussion by Alexander Anderson of Viète's researches in angular sections,(9) in his *Francisci Vietæ Opera Mathematica.*(10) The work, having been quickly established as an indispensable gathering of mathematical source-material, was soon widely circulated and twenty years later Newton had a choice among several copies in Cambridge libraries from which to make his notes.(11) Schooten had also published in 1646 a very readable piece of geometrical research *De Organica Conicarum Sectionum in Plano Descriptione,*(12) and

(7) In the Portsmouth Collection in the University Library, Cambridge, there is a copy (ULC. Adv.d.39.1) of the 1649 edition which seems to have been in Newton's possession at his death. Despite the affirmation of the *Catalogue* (Cambridge, 1888: 47 = vII, 6) its brief manuscript annotations are not in Newton's hand.

(8) The most important exception is Viète's work on trigonometry, *Canon Mathematicus, Sev Ad Triangula Cum Adpendicibus* (Paris, 1579), with its appendix *Francisci Vietæi Vniversalium Inspectionum Ad Canonem Mathematicum Liber Singularis.* (These, it appears, were already sufficiently rare for Schooten not to be able to obtain a copy.)

(9) [Alexander Anderson]: *Ad Angularium Sectionum Analyticen Theoremata καθολικώτερα. A Francisco Vieta...primum excogitata, et absque ulla Demonstratione ad nos transmissa, iam tandem Demonstrationibus confirmata...studio Alexandri Andersoni,* Paris, 1615.

(10) *Francisci Vietæ Opera Mathematica, In unum Volumen congesta, ac recognita, Operâ atque studio Francisci à Schooten Leydensis, Matheseos Professoris Lugduni Batavorum,* Leyden, 1646 (folio). (Unpaginated) [*Epistola Dedicatoria*],/*Ad Lectorem,*/*Francisci Vietæ Vita Ex Iac. Augusti Thuani Historiarum Libro CXXIX;* (1–12) *In Artem Analyticen Isagoge;* (13–41) *Ad Logisticen Speciosam Notæ Priores;* (42–81) *Zeteticorum Liber Primus/Secundus/Tertius/Quartus/Quintus;* (82–158/159–61) *de Æquationum Recognitione et Emendatione Tractatus Duo./Appendix ab Alexandro Andersono Operi subnexa* (from his 1615 edition at Paris); (162–228) *de Numerosa Potestatum Purarum, atque Adfectarum Ad Exegesin Resolutione Tractatus;* (229–39) *Effectionum Geometricarum Canonica Recensio;* (240–57) *Supplementum Geometriæ;* (258–74/275–85) *Pseudo-Mesolabum/& alia quædam Adiuncta Capitula;* (286–304) *Ad Angulares Sectiones Theoremata ΚΑΘΟΛΙΚΩΤΕΡΑ, Demonstrata per Alexandrum Andersonum;* (305–24) *Ad Problema, Quod Omnibus Mathematicis Totivs Orbis construendum proposuit Adrianus Romanus;* (325–38/339–42/343–6) *Apollonius Gallus. Sev, Exsuscitata Apollonii Pergæi ΠΕΡῚ 'ΕΠΑΦΩΝ Geometria./Appendicula I. De Problematis quorum Geometricam Constructionem se nescire ait Regiomontanus./Appendicula II. De Problematis quorum Factionem Geometricam non tradunt Astronomi, itaque infeliciter resolvunt;* (347–435) *Variorum de Rebus Mathematicis Responsorum, Liber* vIII: (436–46) *Munimen adversus Nova Cyclometrica, seu, Anti ΠΕΛΕΚΥΣ;* (447–508) *Relatio Kalendarii vere Gregoriani;* (509–39) *Canones In Kalendarium Gregorianum. Canon I. De Cyclo Decennovennali Aurei Numeri;* (540–4) *Adversus Christophorum Clavium Expostulatio;* (545–54) *Francisci à Schooten Notæ.*

(11) Newton appears never to have bought a copy of Schooten's edition, but to have made his notes wholly from library copies. (Viète's name does not appear in the list of authors of the books in Newton's library at his death.)

(12) *Francisci à Schooten de Organica Conicarum Sectionum in Plano Descriptione Tractatus....Cui subnexa est Appendix, de Cubicarum Æquationum Resolutione. Leyden, 1646.*

this in 1657 he incorporated, along with two sets of mathematical problems and two further tracts on pure geometry, in his *Exercitationum Mathematicarum Libri Quinque.*(13) (Its citation by Newton as the 'Miscellanies' refers strictly only to the fifth book, on the first half of which he made elaborate notes.) In appendix to these *Exercitationes* (which title we give it for brevity) and in his own Latin translation from the original Dutch, Schooten printed a small tract of Christiaan Huygens' on elementary probability, *De Ratiociniis in Ludo Aleæ.*(14)

In contrast, William Oughtred, employed about 1628 as tutor to an English nobleman's son, had composed his *Clavis Mathematicæ* as an elementary introduction to arithmetic and algebra for his pupil's use.(15) From the time of its small first edition in 1631(16) its qualities of clean workmanlike notation allied to concise exposition of basic methods quickly made it a widely read text-book. Later in 1647 Oughtred revised the original text, translating it into English, slightly shortening it but adding several new brief mathematical appendices. Further appendages appeared in the second and third Latin editions of 1648 and 1652, and the latter,(17) reprinted without change in 1667, became the standard edition of the *Clavis* and the one apparently studied by Newton.(18)

(13) *Francisci à Schooten Exercitationum Mathematicarum Libri Quinque* (Leyden, 1657) (4°). (Unpaginated) *Ad Lectorem;* (1–112) *Liber Primus, Continens Propositionum Arithmeticarum et Geometricarum Centuriam;* (113–90) *Liber* II. *De Constructione Problematum Simplicium Geometricorum, seu Quæ solvi possunt, ducendo tantùm rectas lineas;* (191–292) *Liber* III, *Continens Apollonii Pergæi Loca Plana restituta;* (293–368) *Liber* IV, *sive de Organica Conicarum Sectionum in Plano Descriptione Tractatus. Geometris, Opticis; præsertim verò Gnomonicis & Mechanicis Utilis;* (369–516) *Liber* V, *Continens Sectiones trigintas Miscellaneas;* (517–34) [Appendix] *De Ratiociniis in Ludo Aleæ.* Newton's copy of this work is now in Trinity College, Cambridge (NQ. 16.184).

(14) See the previous note. Huygens' original Dutch rendering, *Van Rekeningh in Spelen van Geluck*, is printed (with an interleaved French translation) in *Œuvres complètes de Christiaan Huygens* **14** (La Haye, 1920): 50–91.

(15) See Florian Cajori, *William Oughtred, a Great Seventeenth-Century Teacher of Mathematics* (Chicago and London, 1916): ch. II, especially 17–46.

(16) *Arithmeticæ in Numeris et Speciebus Institutio: Quæ tum Logisticæ, tvm Analyticæ, atqve adeo totivs Mathematicæ, quasi Clavis est* (London, 1631).

(17) *Guilelmi Oughtred Ætonensis...Clavis mathematicæ denvo limita, sive potius fabricata. Cum aliis quibusdam ejusdem commentationibus....Editio tertia auctior & emendatior* (Oxford, 1652) (8°). Main text: (unpaginated) *Ad Lectorem;* (1–109) *Clavis Mathematicæ denuo limata;* (110–51) *De Æquationum Affectarum Resolutione in Numeris.* Minor tracts (individually paginated): *Elementi Decimi Euclidis Declaratio; De Solidis Regularibus Tractatus; De Anatocismo, sive Usura Composita; Regula Falsæ Positionis; Theorematum in Libris Archimedis de Sphæra et Cylindro Declaratio; Horologia Scioterica in Plano, geometricè delineandi Modus.*

(18) There is a curious fragmented copy of the text of the *Clavis* in this edition which is preserved in Newton's library (Trinity. NQ. 8.59). The volume in which it is contained consists otherwise of handwritten copies (not in Newton's hand) of elementary mathematical texts. Presumably it was composed by an older contemporary of Newton's at Trinity and then passed on to him after graduation. After a preliminary two pages (in Latin transcript with some Greek) relating to *Archimed. de Sphæra et Cylindro*, there follow 39 manuscript pages, separately paginated, which discuss the 'Aurea Regula in Fractionibus', 'Rationum Numeratio', 'Numeri Figurati' and finally the definition of the conics. Then come pp. 1–108 of the

John Wallis, finally, who had become Savilian Professor of Geometry at Oxford in 1649,[19] published as his first *Opera Mathematica* in 1656/7[20] a fat and rather diffuse two-part collection of his early mathematical lectures, commentaries and researches, of which his *Arithmetica Infinitorum* (composed partially in collaboration with Brouncker) was to become particularly well known and read, not least by Newton. Around the same time and again with Brouncker's collaboration he became engaged in an active mathematical correspondence, carried on for the most part through Kenelm Digby, with the French mathematicians Fermat and Frénicle. These letters, at first restricted to specialized topics in number theory but gradually broadened in theme to include general topics in analysis and mutual criticism of mathematical techniques and attitudes, were presented by Wallis in abridged form in his 1658 *Commercium Epistolicum*,[21] a work clearly published with the main inten-

1652 edition of the *Clavis Mathematicæ*, with a handwritten copy of the final p. 109 inserted immediately after and complemented by 3 pp. of elementary notes (likewise not in Newton's hand) *In Clavem*. The volume is completed by a closely written, continuously paged set of notes: (1–62) *E P.R. Geometria* (a Latin summary, with some additional material, of the geometrical propositions in the *Elements*); (63–79) *Ex Archimede, de Sphæra & Cylindro* (in Greek); (81–115) [*Notæ*] *In P.R. Geometriam;* (116–45) *Paradigmata Clavi inservientia;* ([146]–[156]) *Archimed. de Sphæra et Cylindro* (in Latin); ([157]–[160]) [Notæ] *In P.R. Geometr.;* ([160]) *De Sphæra et Cylindro;* ([161]–[164]) *In Clavem.*

(19) See J. F. Scott, *The Mathematical Work of John Wallis* (London, 1938); and also A. Prag, 'John Wallis, 1616–1703. Zur Ideengeschichte der Mathematik im 17. Jahrhundert', *Quellen und Studien zur Geschichte der Mathematik. Abt. B, Studien*, **1** (1930): 381–412.

(20) *Johannis Wallisii, SS.Th.D. Geometriæ Professoris Saviliani in Celeberrimâ Academia Oxoniensi; Operum Mathematicorum Pars Prima/Altera*, Oxford 1657/1656 (4°). *Pars prima* (1657): (18 pp. unpaginated) *Johannis Wallis, Geometriæ Professoris Saviliani, Oratio Inauguralis: In Auditorio Geometrico, Oxonii, habita; ultimo die Mensis Octobris, Anno Æræ Christianæ 1649;* ([i–viii]+(1–398)) *Mathesis Universalis: Sive, Arithmeticum Opus Integrum, tum Philologice, tum Mathematice traditum; Arithmeticam tum Numerosam, tum Speciosam sive Symbolicam complectens, sive Calculum Geometricum; tum etiam Rationum Proportionumve traditionem; Logarithmorum item Doctrinam; aliaque quæ Capitum Syllabus indicabit; Adversus Marci Meibomii De Proportionibus Dialogum, Tractatus Elencticus* (comprising ($_1$1–$_1$50) *Dedicatio* [*sive Epistola; qua de Paraboloide Cubicali; & Æquationibus Cubicis agitur*]; ($_2$1–$_2$62) *De Proportionibus, Dialogi; A Marco Meibomii Conscripti; Refutatio;* (2 pp. unpaginated) *M. Mersenni locus notatur*). *Pars altera* (1656): ([i–ii]+(1–51)) *De Angulo Contactus et Semicirculi Disquisitio Geometrica;* ([i–v]+(1–108)) *De Sectionibus Conicis, Nova Methodo Expositis Tractatus;* ([i–xii]+[i–ii]+(1–298)) *Arithmetica Infinitorum, sive Nova Methodus Inquirendi in Curvilineorum Quadraturam, aliaq; difficiliora Matheseos Problemata;* ([i–iii]+(1–9)) *Eclipsis Solaris Oxonii Visæ Anno Æræ Christianæ 1654. 2° Die Mensis Augusti, Stilo Veteri, Observatio. Observatore Johanne Wallisio.*

(21) *Commercium Epistolicum, De Quæstionibus quibusdam Mathematicis nuper habitum Inter Nobilissimos Viros D. Gulielmum Vicecomitem Brouncker, Anglum. D. Kenelmum Digby, item Equitem Anglum. D. Fermatium, in suprema Tholosatum Curia Iudicem Primarium. D. Freniclum, Nobilem Parisium. Una cum D. Joh. Wallis Geomet: Profess: Oxonii. D. Franc: a Schooten, Math: Prof: Lugduni Batavorum; Aliisque. Edidit Johannes Wallis,* Oxford, 1658 (4°): (2 pp. unpaginated) *Dedicatio* [to Digby]; (1–182) *Commercium Epistolicum;* (183–90) *Appendix ad Literas præcedentes.* (The appendix is not present in all copies, and in particular it is lacking in the copy in the University Library, Cambridge (M. 6.26) which Newton may have consulted.)

tion of adding to Wallis' mathematical prestige but nevertheless containing several interesting ideas which there appeared in print for the first time. Newton's annotations show that he read parts of it at least with keen interest. (Note further that, though he made no explicit notes on the work, it is probable that Newton also read Wallis' 1659 *Tractatus Duo*,(22) which presented his researches on the cycloid, cissoid and other geometrical figures and is today often to be found bound up with the *Opera Mathematica* and the *Commercium Epistolicum*.)

It is, in conclusion, interesting to recall an incident which shows clearly the faith Newton had in these texts as a propaedeutic. In 1691 when Richard Bentley, then intent on mastering Newton's newly published *Philosophiæ Naturalis Principia Mathematica*, anxiously applied to its author for a suitable introductory course of study, he was favoured with a set of 'Directions' which included:

'Next after Euclid's Elements the Elements of ye Conic sections are to be understood. And for this end you may read...the first part of ye Elementa Curvarum of John de Witt....

For Algebra read first Barth[ol]in's introduction & then peruse such Problems as you will find scattered up & down in ye Commentaries on Cartes's Geometry & other Al[geb]raical writings of Francis Schooten. I do not mean y^t^ you should read over all these Commentaries, but only ye solutions of such Problems as you will here & there meet with. You may meet with De Witt's Elementa curvarum & Bartholin's introduction bound up together wth Cartes's Geometry & Schooten's commentaries.(23)

(22) *Johannis Wallisii...Tractatus Duo. Prior, De Cycloide et corporibus inde genitis. Posterior, Epistolaris; In qua agitur De Cissoide, et Corporibus inde genitis: et De Curvarum, tum Linearum Εὐθύνσει, tum Superficierum Πλατυσμῷ.* Oxford, 1659. (The latter part, pp. 75–121, is presented as an open letter to Huygens but it was never sent as such. Compare Wallis' letter to Huygens of 24 November 1659 = *Œuvres complètes de Christiaan Huygens* **2** (La Haye, 1889): 518–20.)

(23) 'Directions from Mr Newton by his own hand', printed by J. Edleston in his *Correspondence of Sir Isaac Newton and Professor Cotes* (Cambridge, 1850): 274 = *Correspondence of Isaac Newton* **3** (Cambridge, 1961): 155.

1

ANNOTATIONS FROM OUGHTRED, DESCARTES, SCHOOTEN AND HUYGENS

§1. MATHEMATICAL NOTES IN THE FITZWILLIAM POCKET-BOOK

[Late 1664?]

From the original[1] in the Fitzwilliam Museum, Cambridge

[1] *Of right angled triangles*[2]

I. Any two leggs given to find y^e other.

$$\left.\begin{array}{l} 1.\ bq+cq=hq. \\ 2.\ r\!:\!hq-bq\!:=c. \\ 3.\ r\!:\!hq-cq[:]=b. \end{array}\right\} \begin{array}{l}\text{Eucl}\\ \text{lib. 1. pr: 47.}^{(3)}\end{array}$$

(1) This small notebook, 4·9 in. × 2·8 in. of 118 leaves (with entries on about 50 pp.) and bound in calf, is described by Brewster, *Memoirs* 1: 31–3. The long list of expenses (which begins in 1665 with the entry, 'R^d 10^{li} May 23^d whereof I gave my Tutor 5^{li}' and ends in the autumn of 1668) is a later addition at the reverse end of the notebook and is partially written in over the numbered blank sheets of a *Canones sinuum.* The original inside cover is blank except that Brewster has added a classifying number '(5)' in his resorting of the Portsmouth papers in 1855. The flyleaf bears Thomas Pellet's adjudication, 'Sep. 25 1727/Not fit to be printed/T Pellet', entered when he roughly sorted the Portsmouth papers immediately after Newton's death. On the reverse inside cover Newton has entered, probably when he first purchased the book, 'Isaac Newton/pret 8^d'. Newton's first entry seems to have been a page of dictionary entries in Hebrew and English, headed 'Nova cubi Hæbræi Tabella.'. The notebook is not paginated except for Newton's own numbering of the pages he intended to allot to his unfinished sine table, and the headings in square brackets are introduced to break the text up suitably into natural divisions.

(2) Newton has cancelled the Latin title, 'De Triangulis rectangulis'.

(3) The celebrated 'Theorem of Pythagoras' relating the sides of right triangles. Newton has entered this reference in cancellation of the previous 'Ought. c. 18 Theo: 14', that is, Oughtred[, *Clavis*]: Cap. xviii, [Penus Analytica: 63–74;] Theorem 14 [:68] (and also Theorem 15: 68–69).

II. (4) y^e^ $b.\ c.$ & h given to find p.

1. $\frac{b\times c}{h}=p$. Euc[l]id 6.8.(5)

III. $c.\ h.\ p$ given to find dsh.

1. $H-2r:bq-pq:=dsh$.

IIII. $b.\ p.\ h.$ given to find dsh.

1. $2r:bq-pq:-h=dsh$.

V. $b.\ c.\ h$ given to find dsh.

1. $H-2r:cq-Q:\frac{b\times c}{h}:=dsh$.

2. $2r:bq-Q:\frac{b\times c}{h}:+h=dsh$.

VI. $b.\ c$ or $b.\ h$ or $h.\ c$ given to find p:

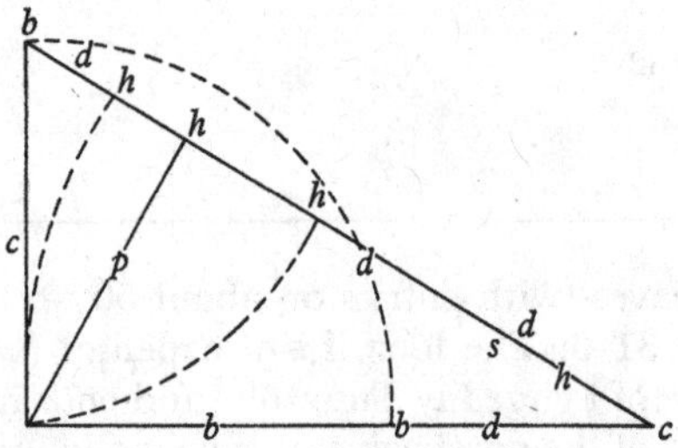

h=hypotenusa.
b=basis.
c=Cathetus.
p=perpendicular.
hdc=diff: hypot: & Cath[:]
bdc=diff: basis & cathet:
bdh=difference basis & hyp[:]
dsh=diff: seg: hypoten:
sh=segment: hypoten:
bh=lesse seg[:] hyp[:]
ch=greater seg: hypot:

1. $\frac{b\times c}{r:bq+cq:}=p$.

2. $\frac{b\times r:hq-bq:}{h}=p$.

3. $\frac{c\times r:hq-cq:}{h}=p$.

(4) This was originally added to I as '4. $\frac{b\times c}{h}=p$.', but immediately cancelled and made a separate section.

(5) Euclid, *Elements*: Book VI, Prop. 8. (The usual interpretation of this in contemporary texts, particularly Barrow's *Euclid*, would be Book VIII, Prop. 6.)

VII. $b.\ h.$ or $c.\ h.$ or $b.\ c$ given to find dsh.

1. $[2bh - h = dsh.]$

[2. $h - 2ch = dsh.$]

[3. $\frac{bq - cq}{r : bq + cq :} = dsh.$](6)

Explanation. Newton uses Pythagoras' theorem and the properties of similar triangles to derive relations between line-segments on his figure. In treatment and notation Newton models himself closely upon Cap. XIV, *Penus Analytica* of Oughtred's *Clavis*, but his theorems are obvious when expressed in modern notation.

I. 1. $b^2 + c^2 = h^2$.

2. $\sqrt{h^2 - b^2} = c$.

3. $\sqrt{h^2 - c^2} = b$.

II. 1. $\frac{bc}{p} = h$.

III. 1. $h - 2\sqrt{b^2 - p^2} = (dsh)$.

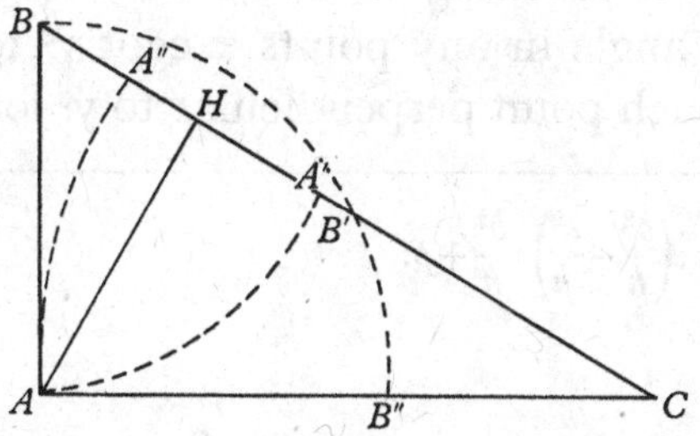

$h =$ hypotenusa $= BC$,

$b =$ basis $= AC$,

$c =$ Cathetus $= AB$,

$p =$ perpendicular $= AH$,

$hdc = BC - AB = CA'$,

$bdc = AC - AB = B''C$,

$bdh = BC - AC = BA''$,

$dsh = CH - BH = CB'$,

$sh = BH$ or CH,

$bh = BH$,

$ch = CH$.

IIII. 1. $2\sqrt{b^2 - p^2} - h = (dsh)$.

V. 1. $h - 2\sqrt{c^2 - \left(\frac{bc}{h}\right)^2} = (dsh)$.

2. $2\sqrt{b^2 - \left(\frac{bc}{h}\right)^2} + h = (dsh)$.

(6) Newton broke off and did not complete the entry, which is restored on the lines of his earlier theorems.

VI. 1. $\frac{bc}{\sqrt{b^2+c^2}}=p$.

2. $\frac{b\sqrt{h^2-b^2}}{h}=p$.

3. $\frac{c\sqrt{h^2-c^2}}{h}=p$.

VII. 1. ______________________

[2] *Theorem* 1.

As y^e^ difference twixt y^e^ base & cath[:] (in rectang: triang:) is to y^e^ greater side:: so is y^e^ difference of y^e^ segm̄ of y^e^ base; to y^e^ greater segm^nt^ of y^e^ base & perpendicular.(7)

Theorem 2.

As y^e^ difference twixt y^e^ base & cathetus to y^e^ less side::so ye diff[:] of ye segm^ts^ of ye base to ye lesse segment of ye base & perpendicular.(8)

Theorē 3^d^.

base−Cathetus:hypotenusa::greate[r] seḡ: base−less seg[:] base:base+ Cathetus.(9)

Theor. 4.

If w^th^in a circle be described an Ellipsis touching y^e^ Circle in 2 opposite points [&] if y^e^ Diameter cut it at right angle in any points except y^e^ touch point[,] y^n^ a line drawne f[ro]m either touch point perpendicular to y^e^ former

(7) In the notation of the first section, $b-c:b::\left(\frac{b^2}{h}-\frac{c^2}{h}\right):\frac{b^2}{h}+p$.

(8) $b-c:c::\left(\frac{b^2}{h}-\frac{c^2}{h}\right):\frac{c^2}{h}+p$.

(9) $b-c:h::\left(\frac{b^2}{h}-\frac{c^2}{h}\right):b+c$.

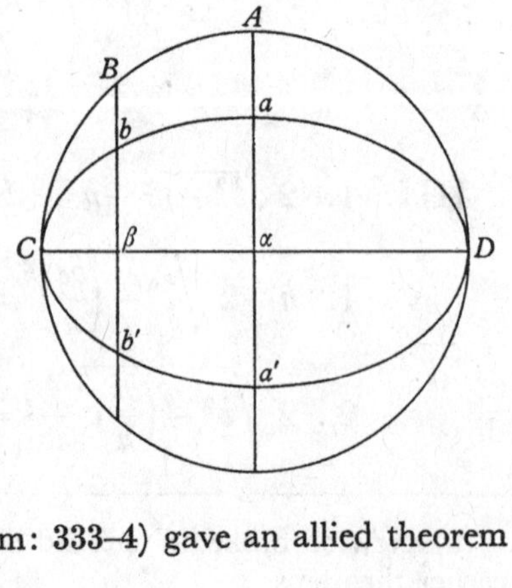

(10) Read 'circle'.

(11) Newton has drawn no diagram with his text, though without one it is a little obscure. The inscribed ellipse $aCa'D$ meets the circle $ABCD$ in two diametrically opposite 'touch points', C, D, such that, where $Aa\alpha$ is the normal 'semidiameter' and $Bb\beta$ an arbitrary ordinate parallel to it, $B\beta : b\beta = A\alpha : a\alpha$.

This construction for the ellipse from the generating circumscribed circle was widely known in the early seventeenth century. Implicitly it occurs in Oughtred's *Clavis*: Cap. XIX: Prob. XXIII: Scholium: 103, while Schooten in his *Exercitationes* (Liber IV: Cap. VI: Scholium: 333–4) gave an allied theorem for the hyperbola.

diameter will bisect it & being produced will cut y[e] ⊙[(10)] in y[e] other touch point & all y[e] lines drawne twixt y[e] ⊙ & y[t] line parallell to y[t] diameter shall be di[vi]ded by y[e] Ellipsis so as one segment shall bee to y[e] other as y[e] segments of y[e] semidiameter are to one another. they being divided by y[e] same Ellip:[(11)]

let ab bee equall to 10[0] ꝑts.

$eb=157$[,] 079 = Periph: [ac] &

p[e]riph − Rad : Rad :: Rad : db.

$db=175$, 1938394.[(12)]

$de=18$, 1142067.[(13)]

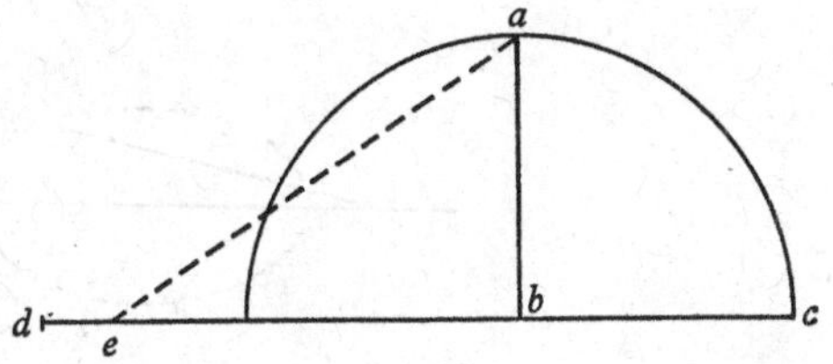

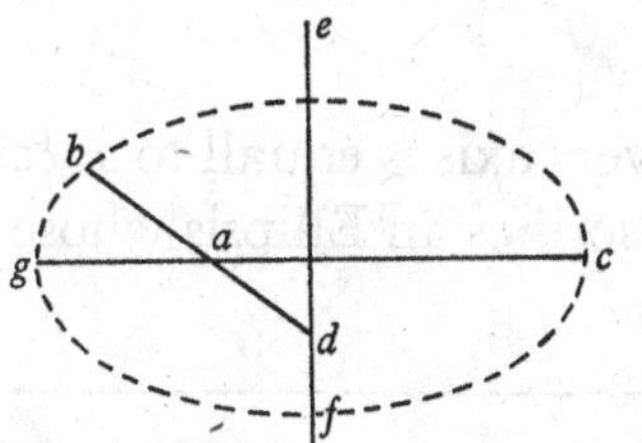

[3] *To describe an ellipsis.*

[a] Let fe & gc be two lines [&] ef make righ[t]-angles wth gc. let a point be taken in bd as at a & let y[t] point move along y[e] line gc. & d y[e] one end of y[e] line db move on y[e] line ef & y[e] other end b shall describe y[e] Ellipsis $gbcf$.[(14)]

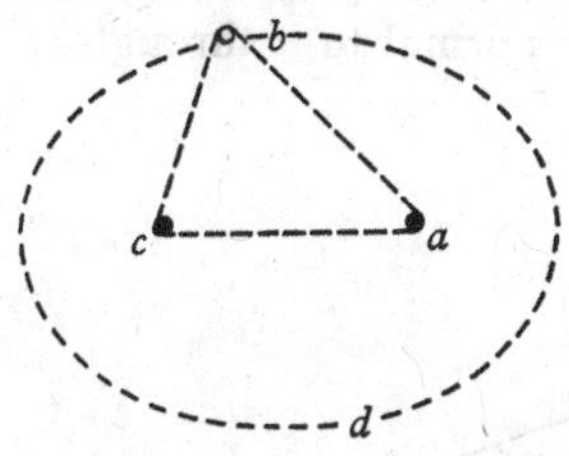

[b] Let c & a b[e] two fixed points about w[ch] let a loose cord be put haveing both ends tyed together. as is signified by y[e] 3 lines cb. ba. ac. strech it out wth another point as b. & keeping it so streched out draw y[e] point b about & it shall describe y[e] Ellipsis bd.[(15)]

Chartesij Dioptr[:][(16)]

---

(12) $\frac{(100)^2}{57{\cdot}079}$.

(13) We have not been able to trace the source of this note, and perhaps it is Newton's own. It is, anyway, an immediate corollary of the approximation

$$eb = \tfrac{1}{2}\pi \times 100 = 157{\cdot}079 \quad (\text{or } \pi = 3{\cdot}14159).$$

(14) The well-known trammel construction, noted by Newton from Schooten's *Exercitationes*, Liber IV: Cap. IV: 323–5.

(15) The 'gardener's' construction of the ellipse, probably noted by Newton from Schooten's *Exercitationes*, Liber IV: Cap. IV: 325–6, or Cap. IX: 345–6.

(16) This is added in a slightly later hand, and the reference is to Descartes' use of the construction in his *Dioptrique*. Newton's knowledge of French was weak and almost certainly he read Descartes' work in one of its Latin editions. (The first was *Renati des Cartes Specimina Philosophiæ: seu Dissertatio de Methodo rectè regendæ rationis & veritatis in scientiis investigandæ: Dioptrice, et Meteora. Ex Gallico translata, & ab Auctore perlecta, variisque in locis emendata*, Amsterdam, 1650: see especially Cap. VII, *De figuris quas pellucida corpora requirunt, ad detorquendos refractione radios omnibus modis visioni inservientibus, II, Quid sit Ellipsis, & quomodo sit describenda.*: 142–3.) Probably Newton added the reference to the note he had already made from Schooten.

[*c*] Let yᵉ line *ae* be infinitely extended [&] in it take yᵉ point *o* about [wᶜʰ] yᵉ line *oc* shall turne[.] at yᵉ point *c* in *oc* let yᵉ point *c* in yᵉ line *ab* be fastened & let *a* yᵉ end of yᵉ line *ab* move on yᵉ line *ae* & *oc* turning round, each point of yᵉ line

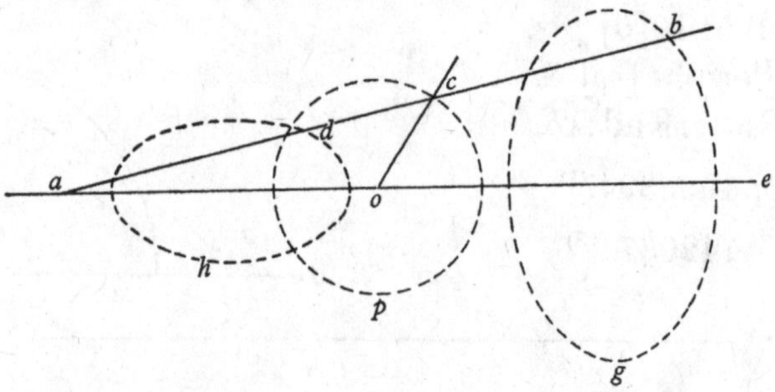

ab betwixt *a* [&] *c* will describe an Ellipsis whose transvers axis is equall to *oc* & parallell to *ae* but each point on yᵉ other side [of] *c* describes an Ellipsis whose righ[t] axis is equall to *oc* & parallel to *ae*.(17)

(17) An incorrect generalization by Newton of Schooten's ellipse-construction in [*d*] below. The locus of *d* will be, in general, one of the two non-elliptical ovals of an irreducible quartic whose Cartesian curve is symmetric both about *aoe* and the normal to it through *o*.

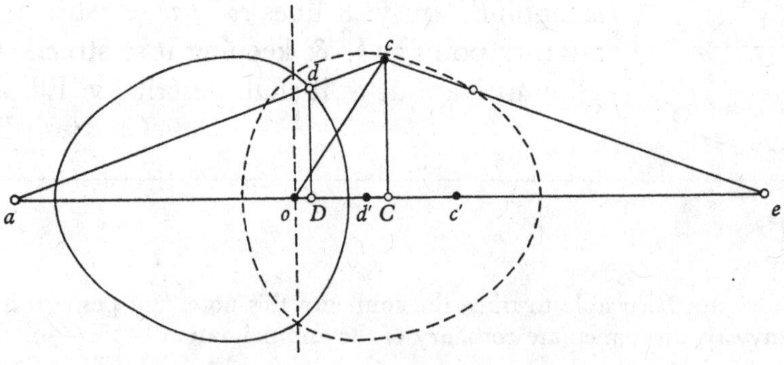

In proof, take c', d' in *ae* with $od' = oc - cd$ and $oc' = oc$; take also, to define *d*, co-ordinate line-lengths $d'D = x$, $Dd = y$ and the constants $cd = c'd' = k$, $da = l$, $oc = oc' = m$. Then, where $\lambda = (l+k)/l$, $aD = \surd(l^2-y^2)$ and $oC = \surd(m^2-\lambda^2y^2)$, so that

$$(k/l)\,\surd(l^2-y^2) = DC = (Dd'+d'c')-(oc'\mp oC)$$
$$= x+k-m\pm\surd(m^2-\lambda^2y^2),$$

a quartic. Only in the particular case (Schooten's) where $ac = oc$ ($k+l = m$, $\lambda = m/l$) does the locus reduce to the conic-pair

$$\left.\begin{cases} x-l = (2k+l)/l\,\surd(l^2-y^2) & \text{(an ellipse)} \\ x-l = \surd(l^2-y^2) & \text{(a circle)} \end{cases}\right\}.$$

(18) Read 'one'. The two 'other's are a straight translation of the Latin 'alius...alius', idiomatically translated as 'one...other'.

[*d*] Extend *de* both ways[.] take y^{e} lines *ca* & *ab* equall to one another [&] fasten[d] together at one end as at *a*. set y^{e} other[18] end of *ca* at y^{e} point *c* in *db*. & let y^{e} other end of *ab* slide on *db*. y^{n} take a point in *ab* as *o* & turne *ac* about & it shall describe y^{e} ellipsis *dgoe*.

Shooten in lib. 2^{d} Cartesij Geometriæ.[19]

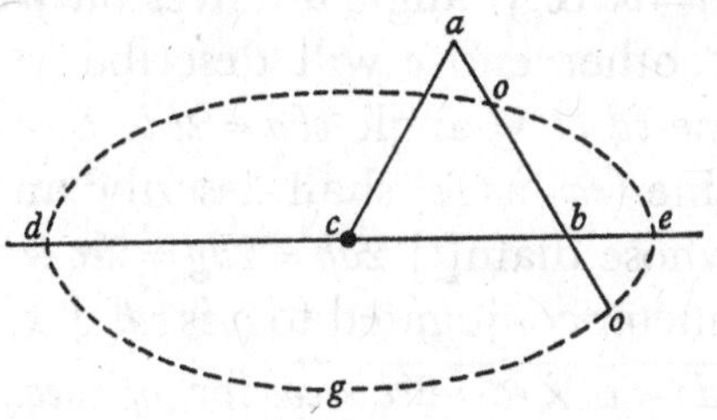

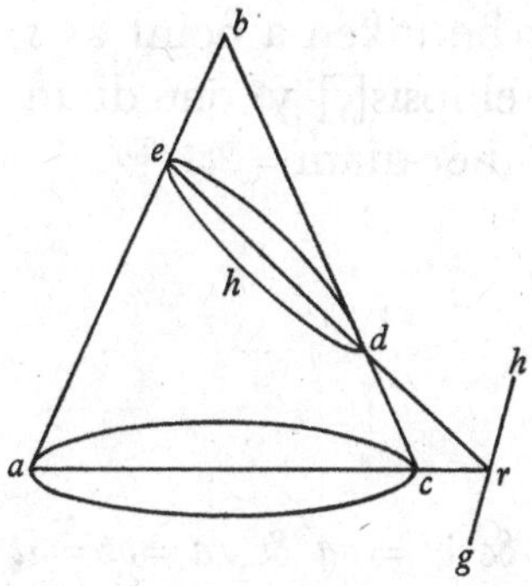

[*e*] Cut y^{e} cone *abc* so y$^{[t]}$ y^{e} diam[:] of y^{e} section *ed* produced cute[20] y^{e} base of y^{e} triangle *ac* produced wthout y^{e} cone as at *r* & makes right angles wth *gh* y^{e} base of y^{e} sectiō.[21]

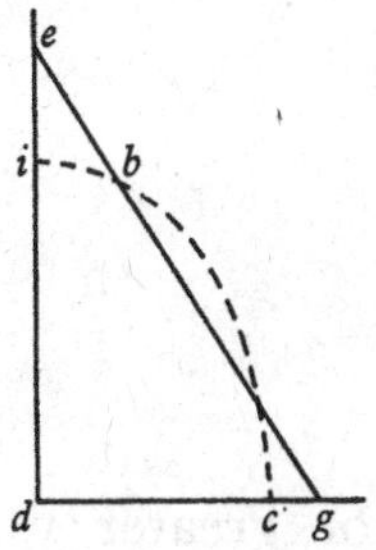

[*f*] If *eg* be moved twixt y^{e} lines *ed* & *gd*[,] a point in it as (*b*) shall describe an ellipsis whose semi-axis *ad* is equall to *b*[*g*][22] & semiaxis $dc = eb$.[23]

---

(19) Newton's reference is to Schooten's analytical proof in his *Commentarii in Librum II*, *A* = *Geometria*: 172–4, but the note itself seems taken from Schooten's synthetic treatment in his *Exercitationes*, Liber IV: Cap. II: 309–11 (theory); Cap. IV: 321–3 (construction).

It is the particular case of the preceding construction with $ca = ab$: the quartic locus then reduces to the conic-pair, circle (o') × ellipse (o), where o' is taken in *ca* with $co' = bo$ ($o'a = oa$).

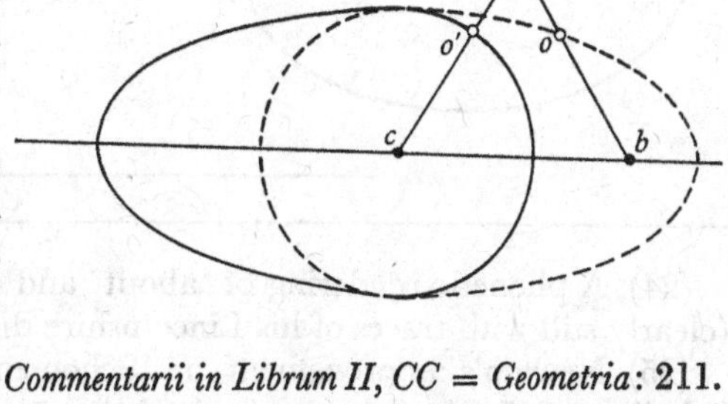

(20) Read 'cut'.

(21) Newton's simplification from Schooten's *Commentarii in Librum II*, *CC* = *Geometria*: 211.

(22) Newton wrote '*bd*'.

(23) Compare Schooten's *Exercitationes*, Liber IV: Cap. III: 314–17; Cap. IV: 323–5, especially 325. The condition that *ad*, *dc* be semiaxes limits the angle $\widehat{adc}$ to being right. (Newton's own drawing, incorrectly, has *eg* tangent to the ellipse at *b*.)

[*g*] If *dc* revolve abute(24) y^e center *d.* & to y^e other end *b* be fastend a triangle *bca* & $db=ba=bc$ & y^e angle *a* moves on y^e line *ad*[,] y^e other end *c* will describe y^e streigh[t] line *cd* & y^e angle $cba=2cda$ & a point in y^e line (*ca*) as (*e*) shall describe an ellipsis *ehg* whose diam[:] $2dh=2dg=2ec$ & y^e other diameter conjugated to it is *od*[.] & $od=\sqrt{4db\times db-ec\times ec-2ec\times ea}$ for $op=ec$. $oq=ea$. $dp=2db$.

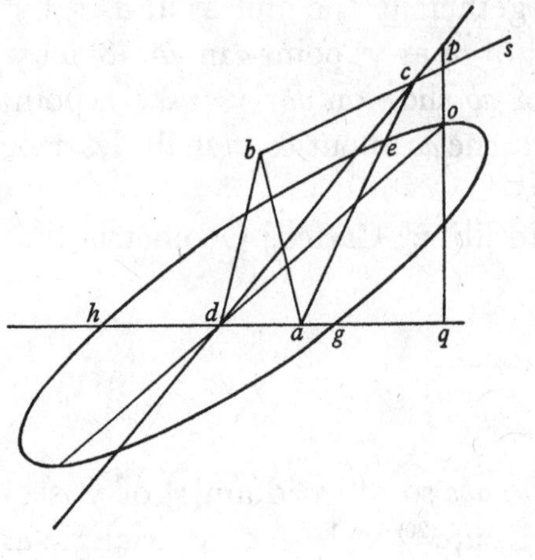

& if in y^e line *bc* be taken a point as *s*, it shall describe an ellipsis[,] y^e one diam: being $2db+2bs$, y^e other diam $=2cs$.(25)

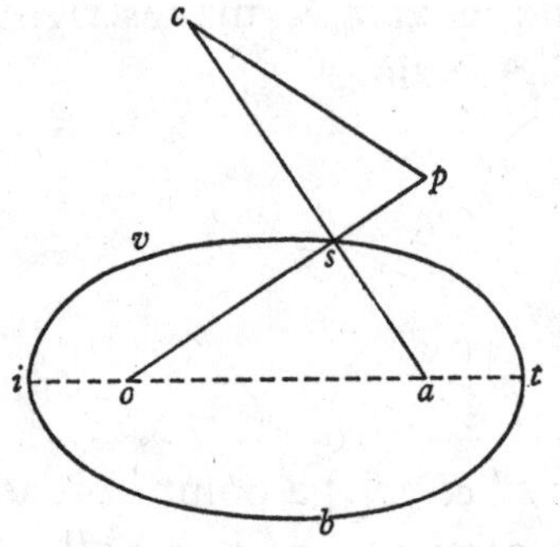

[*h*] If *o* & *a* be y^e foci & $cp=oa$ & $ca=op=it$ theire section in *s* shall describe an ellipsis.(26)

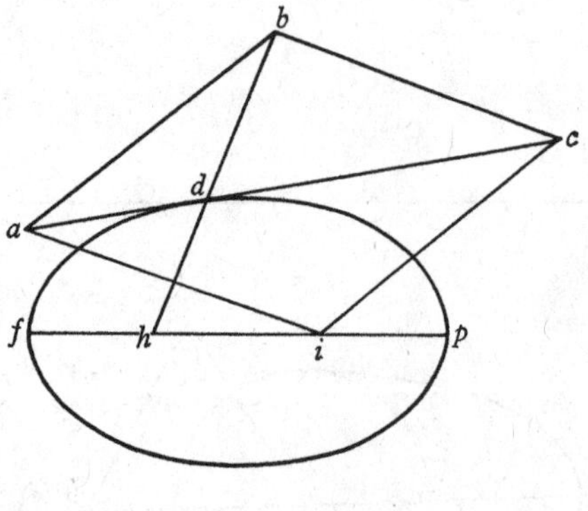

[*i*] If $ab=bc=ci=ai=if$ or greater y^n (*if*) & $bh=fp$ & *ac* bisects y^e angles *bai. bci.* y^n if *bh* turne round y^e intersections of *bh* & *ac* shall describe an Ellipsis. & *h* & *i* are y^e foci.(27)

(24) A phonetic rendering of 'about' and an interesting commentary on Newton's speech (clearly still with traces of his Lincolnshire dialect).

(25) Newton's improvement on Schooten's *Exercitationes*, Liber IV: Cap. III, especially Scholium: 317–21. Schooten proved that:

(*a*) The locus of *c* is the fixed line *cd*. (The point *b* is the circumcentre of the variable triangle *adc*, in which *ac* is a chord of fixed length; or $a\hat{b}c$ is constant, with $a\hat{d}c$ (subtended by the chord *ac* in the circumcircle) $=\frac{1}{2}\times a\hat{b}c$.)

[4] *To describe a Parabola.*

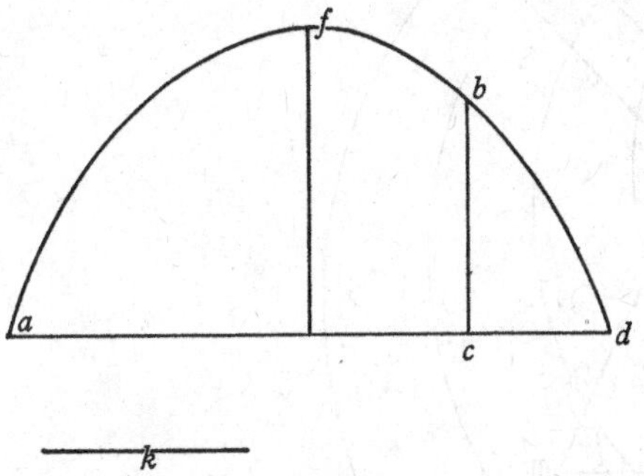

[*a*] Let *bc* fall perpendicular on *ad* & let *c* y[e] one end thereof move[(28)] uppon *ad* a given line[.] & if $bc \times k$ a given line, be equall to $ac \times cd$. y[n] shall *b* y[e] other end of *bc* describe y[e] Parabola *afd*.[(29)]

[*b*] Draw *ah* perpendic[:] to *ap*. & [*v*]*b* from *ah* parallell to *ap*. divid *bh* into equall ꝑts as *bcdefgh*. & divide *ap* into parts equall to y[e] former as *iklmnop*. draw lines cros[(30)] to each part of y[e] lines *ah* & *ap* as *ib*. *kc*. *ld*. *me*. *nf*. &c [&] wth half of each line descri[be] a circle as *b*[*q*]*i* w[th] [y[e] radius] $\frac{1}{2}$ *ib*. from *bv* in y[e] poin[ts] cut by y[e] diameters of y[e] circle[s] draw lines perpendicula[r] to y[e] diamet[r[s]] untill they reach y[e] circle from whose diameter they are drawne as y[e] lines *pw*,

(*b*) The locus of *e* can be reduced to [*d*] above, and so to an ellipse. (Take λ in *be* with $b\lambda = ba = bd$. It follows that the locus of λ is a fixed line $d\lambda$, and the reduction is immediate.)

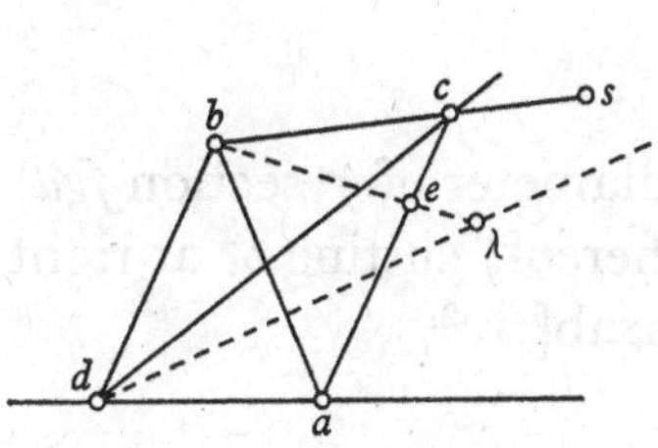

(*c*) Since the locus of *c* is the fixed line *dc*, the locus of *s* reduces straightforwardly to [*d*], and so is an ellipse.

To this Newton adds some results which define the ellipse-locus with greater precision. Clearly, the semi-diameter *do* conjugate to *hdg* is fixed by finding *o*, the point on the locus (*e*) farthest from *da*; or by taking *ced* normal (at *q*) to *da*. In this position *dp* is the diameter of the (fixed) circumcircle of the variable triangle *adc* (here coincident with *qdp*), and so

$$dp = 2bd = 2ba = 2bc.$$

Finally,
$$od^2 = oq^2 + dq^2 = ea^2 + pd^2 - pq^2$$
$$= 4bd^2 - [(ce+ea)^2 - ea^2] = 4bd^2 - ce(ce+2ea).$$

(26) Schooten's *Exercitationes*, Liber IV: Cap. VII: 339–41. The triangles *osa*, *csp* are congruent, or $os + sa = op = it$, constant; so that the locus reduces to the gardener's construction, [*b*].

(27) Schooten's *Exercitationes*, Liber IV: Cap. VII; *idem aliter:* 341–3. The locus-point *d* is the meet of *ac*, normal bisector of *bi*, with *hb*; $db = di$, therefore, so that $hd + di = hb = fp$, constant, and the locus again reduces to [*b*]. (In Newton's sketch the line *adc* is drawn, incorrectly, not tangent to the ellipse at *d*.)

(28) Newton has cancelled the word 'perpendicular' here.

(29) Schooten's *Commentarii in Librum II, A = Geometria:* 175. (The property has a long tradition and was known to both Apollonius and Archimedes.)

(30) Read 'across'.

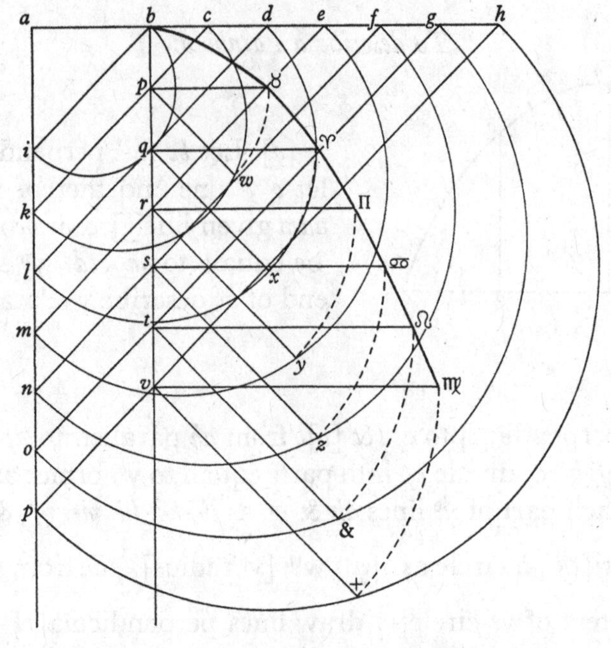

qx, *ry*, *sz*, *t*&, *v*+. Erect those lines perpendicular to y[e] line *bv* as *p*♉, *q*♈, *r*Π, *s*♋, *t*♌, *v*♍. & by y[e] end of those lines draw a line & it shall be a parabola. as *b*♉ ♈ Π ♋ ♌[♍].(31)

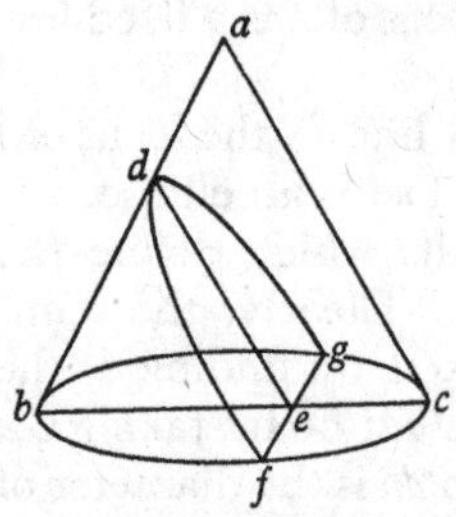

[*c*] If *abc* be a cone. *de* (y[e] diameter of y[e] section *fgd*) parallell to *ac*: & *fg* (y[e] base thereof) cutting *bc* at right angles y[n] is y[e] section *dfg* a Parab[:](32)

(31) Newton had drawn a preliminary diagram in which the circles are all quadrants of centre *a*, and then cancelled. This seems to suggest that the construction is Newton's: a conclusion reinforced by its relative crudity of structure in comparison with the elegances of Schooten's expositions. However, some allied constructions for the hyperbola will be found in de Witt's *Elementa Curvarum*: Liber I: Cap. IV: 231–8.

The construction itself requires that $ai = ab$, as well as $bc = cd = \ldots = ik = kl \ldots$. It follows that $ib = kp = lq = \ldots = \sqrt{2}ab$, and $pc = \sqrt{2}pb$, $qd = \sqrt{2}qb$, ...; and so at a general point ♉

$$(p♉)^2 = (pw)^2 = kp \times pc = \sqrt{2}ab \times \sqrt{2}pb$$
$$= pb \times 2ab.$$

That is, *b* ♉ ♈ Π ♋ ♌ ♍ is a parabola of latus rectum $2ab$. (Note that Newton's diagram has two points *p* and that in it the various circle arcs are but crudely drawn. The semicircle on diameter *bi*, here correctly drawn tangent to *ld* at *q*, is made by him to pass very nearly through *r* and was denoted in the manuscript by '*bri*'.

(32) Simplified from Schooten's *Commentarii in Librum II*, *CC* = *Geometria*: 207.

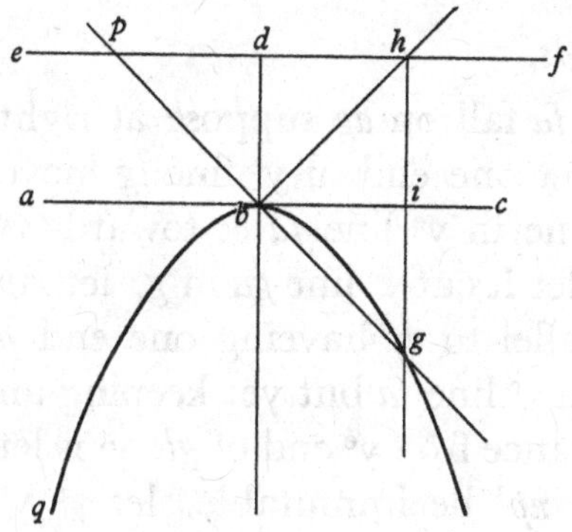

[d] Make *db* perpendicular to *ef*[&] on yᵉ center *b* let yᵉ right angled figure *pbgh* turne. let *gh* move perpendicularly on *ef* ever intersecting *ef* & *bh* in one point[.] yⁿ *pbgh* moveing round yᵉ intersections made twixt *pg. gh* describe yᵉ parabola *qbg*.[33]

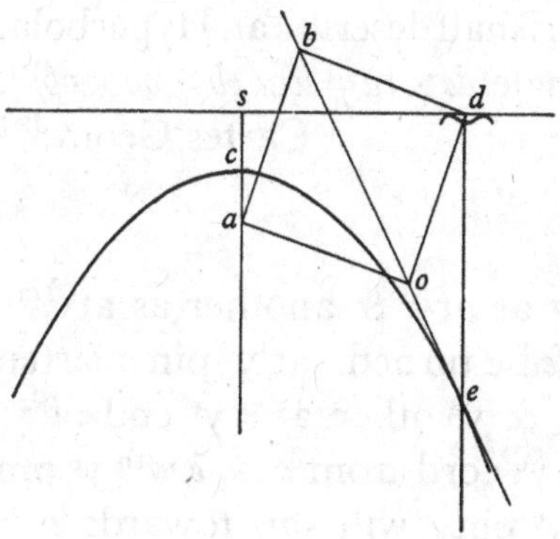

[e] If *ab*=*bd*=*do*=*ao* is greater then *ac* & *ac*=*cs* yᵉ corner (*a*) fastend to yᵉ focus (*a*). & yᵉ line *de* fastened to yᵉ corner *d* & moveing perpendicularly on *sd*[.] & yᵉ line *boe* crossing yᵉ corners *b* & *o*. yⁿ yᵉ line[s] *boe* & *de* at theire intersections shall describe a Parab & yᵉ line *boe* always toucheth yᵉ Parabola in (*e*) &c.[34]

[f] If (*d*) be yᵉ focus[,] *od*=*oe*[,] yᵉ ruler *fc*=to yᵉ thred *fad* & [yᵉ] thred [be] fastened to yᵉ ruler at *f* & to yᵉ focus *d* & yᵉ ruler move perpendic̄ to *ce* & parallell to *de*. yⁿ yᵉ parting of yᵉ thred from yᵉ ruler as at (*a*) shall describe a Parabola.[35]

---

(33) The first, particular case of de Witt's *Elementa Curvarum*: Th. 1: 162–4. (Newton later transcribed the complete theorem—see below.) The proof of the construction follows by the similarity of the triangles *bdh*, *big*, since *bd*:*dh* (or *bi*) = *bi*:*ig*, or (*bi*)² = *db*×*ig*.

(34) Schooten's *Exercitationes*, Liber IV: Cap. XIII: 356–9. Since *boe* is the normal bisector of *ad* (where *a* is the focus and *d* on the directrix), the focus-directrix defining property, *ae*=*de*, is an immediate consequence. Further, since *boe* bisects *aêd*, *boe* is tangent at *e*.

(35) Schooten's *Exercitationes*, Liber IV: Cap. XIV: 360–1. The construction is a simple variant on [*e*], and akin to the 'gardener's' method for drawing the ellipse.

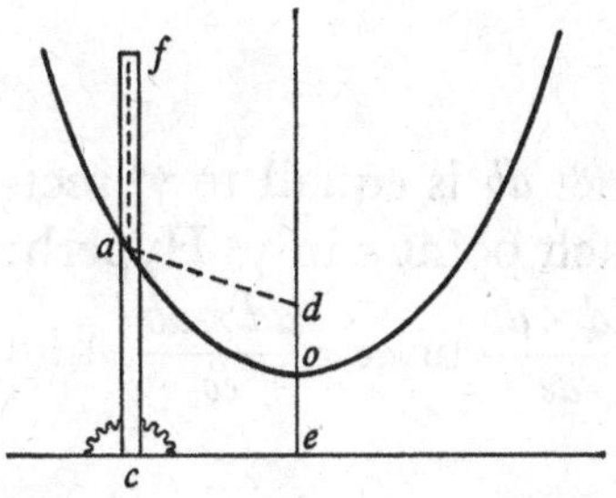

[5] *To describe an Hyperbole.*

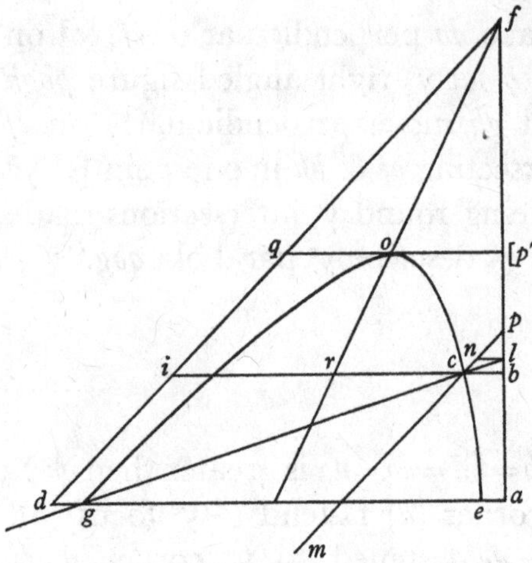

[*a*] Let *fa* fall on *ag* suppose at right angles[.] let one end of y^e line *lg* move up & downe in y^e line *fa* & towards y^e other end let it cut y^e line *ga* in *g*. let *mp* keepe parallel to *df* haveing one end *p* moveing in y^e line *fa* but yet keeping an equall distance frō *l* y^e end of *gl*. y^t is let y^e triangle *npl* be immutable. let y^n y^e lines *mp* & *gl* thus move to & fro & theire intersections shall describe an Hyperbola. & y^e rectangle $de\times ea = ic\times cb = qo\times op^{[\prime]}$. Cartes Geom:(36)

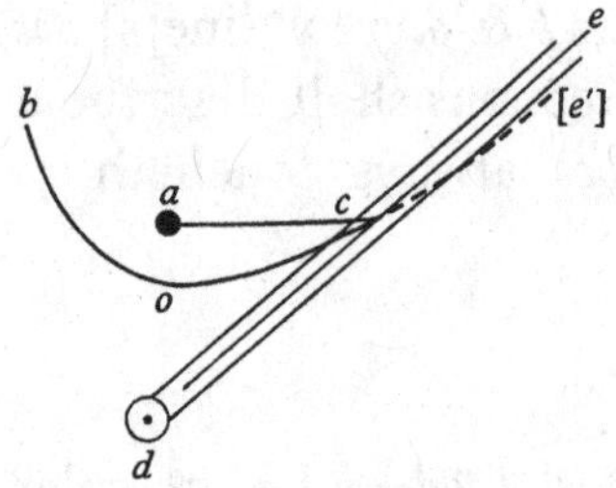

[*b*] Fasten a pegg as at *a* & another as at *b*(37) upon wch let y^e line *de* be turned. at y^e pin *a* fasten one end of a chord & y^e other at *e* y^e end of y^e line *de*. y^n streching y^e cord from *a* & *e* wth y^e pin *c* turne *de* about & y^e pin *c* will slip towards *e* & describe $\frac{1}{2}$ y^e Hyper: *oce*[′].(38)

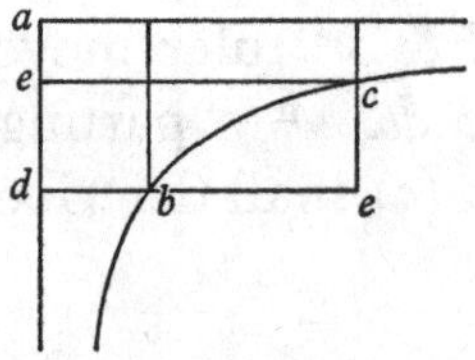

[*c*] the rectangle twixt *ad* & *db* is equall to y^e rectangle twixt *ae* & *ec*. so y^t each point *c* in y^e Hyperb: *bc* is found by makeing $ec = \dfrac{ad\times db}{ae}$ or $ae = \dfrac{ad\times db}{ec}$. also $be\times ce = be\times da - db\times ec$.(39)

(36) This is Descartes' well-known construction of the hyperbola by moving angles, first published in Book 2 of his *Geometrie* (= *Geometria*: 21–3). Descartes himself gave only an analytical proof of the construction and did not consider the 'opposite' branch of the hyperbola, also yielded by the method. (See also Schooten's *Commentarii in Librum II, H* = *Geometria*: 236–8.) Schooten, in his *Exercitationes*, Liber IV: Cap. VI: 331–2, was the first to give a synthetic proof, resolving the locus by reducing the construction, as here, to the property $ic\times cb = qo\times op'$. (Note that Newton has added the reference to 'Cartes Geom:' at a later date, which seems to show that he entered the annotation from Schooten; also, that by *p* he designates two separate points, of which one is taken in the text as *p′*.)

(37) Read '*d*'.

(38) Schooten's *Exercitationes*, Liber IV: Cap. VII: *idem aliter, datis axibus*: 337–8. The construction fixes the defining condition $dc - ac = do - ao$, constant, where *a* and *d* are the foci.

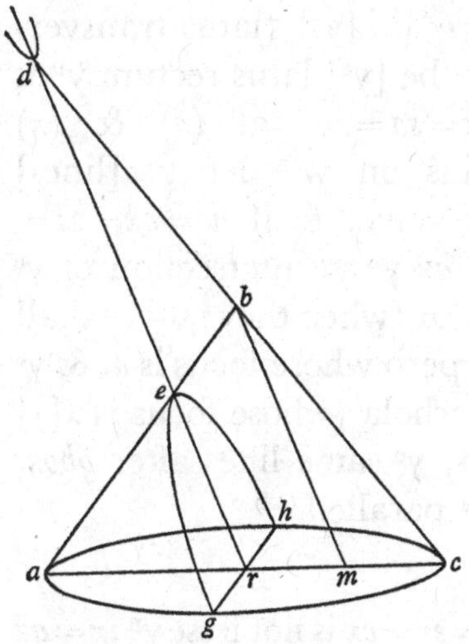

[*d*] Cut y^e cone *abc* so y^t y^e dia[me]ter of y^e section *er* produced cuteth one side of y^e Cone *bc* produced as at *d.* y^e base thereof *gh* cutteth *ac* y^e base of y^e triang: *abc* a[t] right angles.(40)

[*e*] If (*of*) touch y^e Hyperb: & (*as*) be its transverse diam: & (*gb*) keepe parallel to (*eo*) & (*cag*) a[l]ways pass through (*a*) y^e vertex of y^e Hyperb. & (*bc*) be always in y^e line (*fh*) fastend to (*gb*) & equall to $fd = de = \frac{fh}{4}$. y^n y^e lines (*agc*) & (*gb*) moveing by theire intersection shall describe an Hyperbola whose asymtotes are *oe*, *fe*; *eh*, *eb* & *wx* is a right line conjugate to y^e transverse diameter (*as*). viz: it is y^e right diameter.(41)

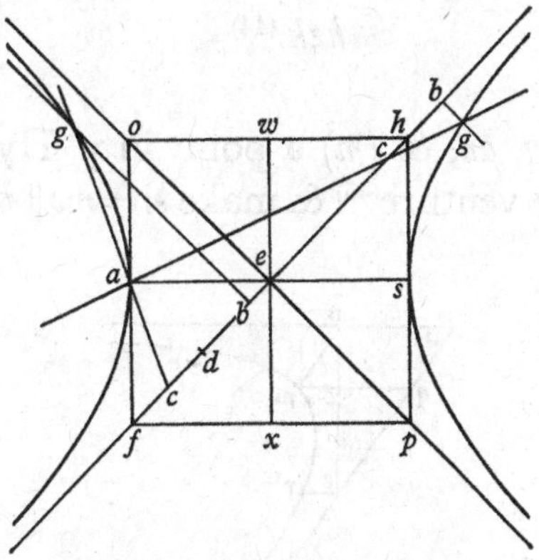

[(] tis y^e same with y^e first.)

(39) Schooten's *Commentarii in Librum II = Geometria*: 175–6. This familiar property of the hyperbola was, a little later, to be the basis of Newton's researches in the theory of the logarithm.

(40) Newton's simplification from Schooten's *Commentarii in Librum II, CCC = Geometria*: 209.

(41) Schooten's *Exercitationes*, Liber IV: Cap. VI: 331–2. As Newton notes it is identical with [*a*] above. The asymptote '*eb*' should presumably be '*ep*'.

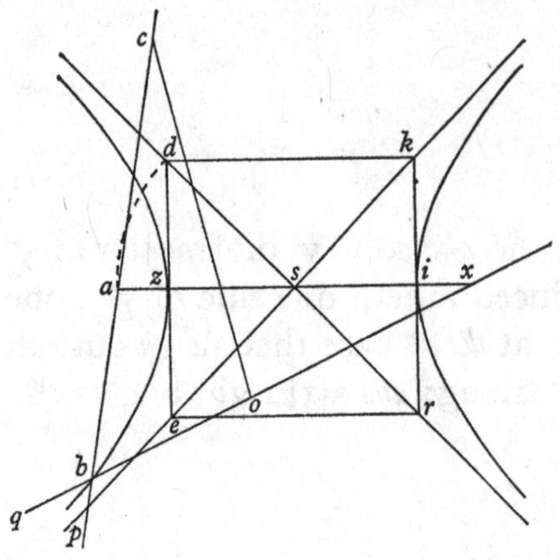

[*f*] If $dk=er$ be [y^{e}] (latus transversum) & $de=kr$ be [y^{e}] latus rectum y^{n} is $sd=sr=se=sk=sa=sx$. at (a) & (x) fa[s]ten 2 pins on wch let y^{e} [lines] ($acbp$, $xobq$) revolve, & if $ac=ox=zi=dk=er$, & $co=ax$ y^{n} y^{e} intersection of y^{e} lines $cabp$, & $qbox$ (when they move) shall describe a Hyperb whose focus is a, & y^{e} opposite Hyperbola (whose focus is x[)] is described by y^{e} same lines after $qbox$, esk & $cabp$ are parallell.[42]

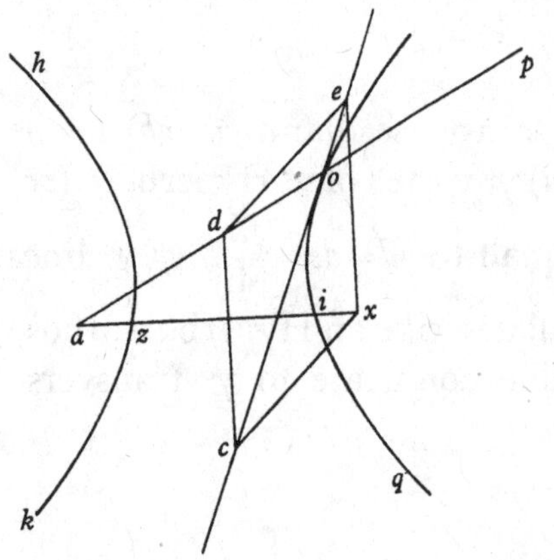

[*g*] If $de=dc=ex=cx$ is not lesse y^{n} $ix=az$ & 2 of theire ends loose pind[43] together at (e) & 2 at (c) on wch 2 corners lyes y^{e} line (coe)[,] two of theire ends are loosely pinnd on y^{e} focus (x) [&] y^{e} last two are pind on y^{e} line (adp) at (d) soe y^{t} y^{e} ruler adp being pinnd to y^{e} focus (a), $ad=zi$[.] y^{n} y^{e} intersections of y^{e} lines (adp, coe) describe y^{e} Hyperbola oiq. & after they are parallell they shall describe y^{e} opposite Hyperbola hzk.[44]

[*h*] The Asymptotes aq, an, & (m) a point in y^{e} Hyperbola draw $mq \parallel an$. & $mn \parallel aq$. Then draw en at a venture[45] & make $er=mc \parallel er$ & r shall bee a point in y^{e} Hyperbola.[46]

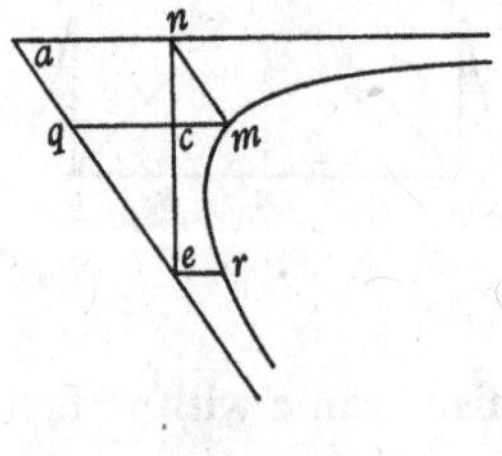

(42) Schooten's *Exercitationes*, Liber IV: Cap. IX: 344–6 (and *scholium*: 347–8). The construction yields the defining condition $bx-ba$ (or bo) $= ox =$ transverse axis zi, the focal property of the hyperbola. (Note that Newton, cramped for room in his notebook, drew a badly misshapen figure which is here redrawn more accurately.)

(43) Read 'loose-pinned'.

(44) Schooten's *Exercitationes*, Liber IV: Cap. IX: 348–52. Since coe is the normal bisector of dx, $do = ox$, and the construction reduces to the focal property $ao-ox = ad = zi$. Further,

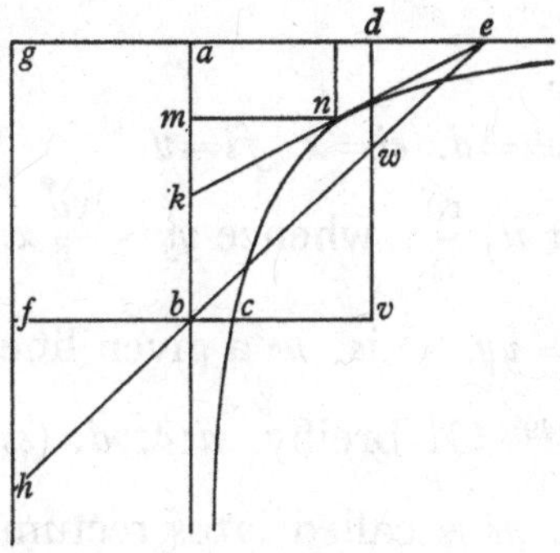

[*i*] If y^e^ position of y^e^ Asymptotes (*ad*) & (*ab*) bee given & any pointe as (*c*) in y^e^ Hyperbola. then draw $vcbf \parallel ad$. $vd \parallel ab \parallel fg$ making $bf = bv = 4bc$. Then at a venter draw *bewh*, through y^e^ point *b*. & make $ak = fh = vw$. Or $dw = bk$. & from y^e^ point *k* draw *ke*, wch shall touch y^e^ Hyperpola in *n*, if $kn = ne$.(47)

[*j*] The foci (*a*, *d*) & (*c*) a point in one Hy[p]erbo: given to describe them.

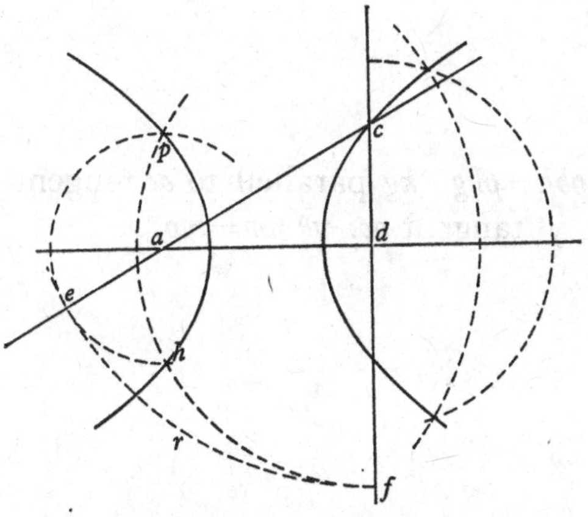

Draw *ac*, *cd*, frō the given point *c* to the foci, y^n^ upō the center *c* wth any radius *ce* describe y^e^ circle *erf*. soe y^t^ $ec = cf$. y^n^ wth y^e^ Rad *ae* & *df* upon y^e^ centers *a* & *d* describe y^e^ circles *hep*[,] *fhp*. their points of intersection *p*, *h* shall be in y^e^ hyperbola. The intermediate distance twixt divers points thus found may bee completed by y^e^ helpe of tang^nt^ lines or circles or a steady hand.(48)

---

coe, since it bisects $a\hat{o}z$, must be tangent at *o* and is so drawn here, though in Newton's sketch it is conspicuously not so.

(45) That is, arbitrarily.

(46) This is apparently Newton's own deduction from [*c*]. The result follows from the hyperbola property $an \times nm = ae \times er$ since, by the similarity of the triangles *ean*, *nmc*, $ae : an = nm : mc$.

(47) Probably Newton's adaptation from Schooten's *Commentarii in Librum II*, *CCC* = *Geometria*: 219–20, though many details are Newton's own. Since the tangent *kne* included between the asymptotes *ab*, *ad* is bisected by the contact-point *n*, immediately $ak = 2am$ and $ae = 2mn$; so that $ab \times bc = am \times mn = \frac{1}{4}ak$ (or $vw = fh$) $\times ae$, or $fh : 4bc = ab : ae = fh : fb$, since the triangles *bae*, *hfb* are similar.

(48) This simple construction is apparently original with Newton, and reduces immediately to the focus property: for *pd* (or *df*) $-$ *pa* (or *ae*) $= (cf - cd) - (ce - ca) = ca - cd$, constant, since $cf = ce$. Newton's last remark is probably his first reference to interpolating a continuous curve between given points on it by drawing the tangents at these points (here found by bisecting *apd*) or by freely joining up the set of constructed points with circle arcs or arbitrary portions of smooth curves drawn in with 'a steady hand'.

[6]

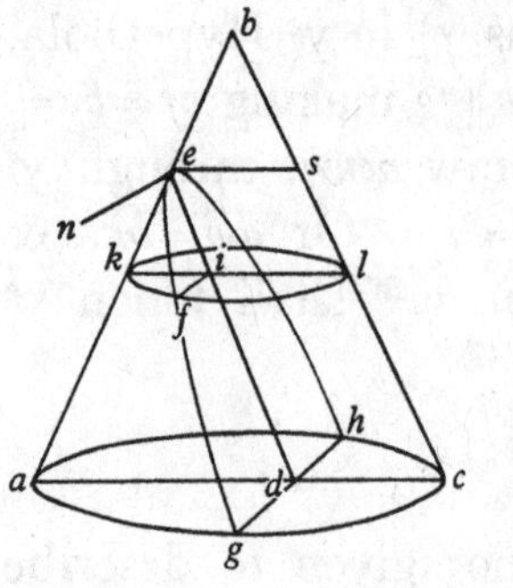

The properties of y^e Parabola.

[*a*] $ab=a.\ bc=b.\ ac=c.\ eb=d.\ ei=x.\ fi=y.$

$b:c::x:(ik)\,\frac{cx}{b}.\ a:c::d:(es,\text{ or }il)\,\frac{cd}{a}.$ whence $yy=\frac{ccd}{ab}x.$

$ab:cc::d:(en)\,\frac{ccd}{ab}.\ \frac{ccd}{ab}=r.\ rx=yy.$ y^t is. *ne* a given line multipl[y]ing $ei=if$ square.[49] Or breifly. $a:c::d:(es$ or $il)\,\frac{cd}{a}.\ b:c::\frac{cd}{a}:(en)\,\frac{ccd}{ab}=r.$ *ne* is called latus rectum of[50] Apollon[ius] & Parameter by Mydorgius. *gh* is its base [&] *ed* its Diameter.[51]

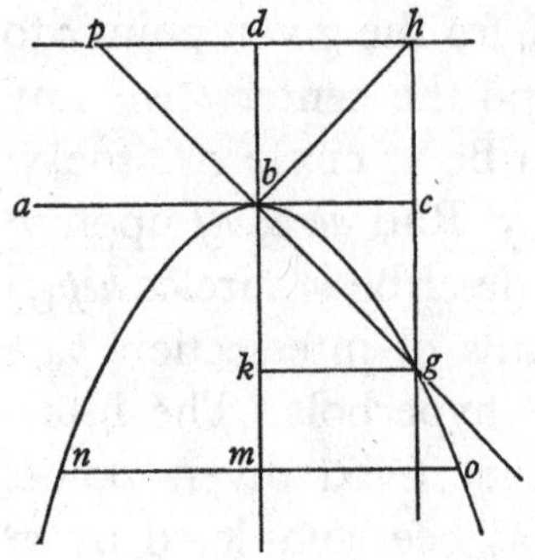

[*b*] ang[:]$pbh=phg$. *kg* parallell to *ac* tangent. *no* parallell to y^e tangent *ac*. y^n $nm=mo$.

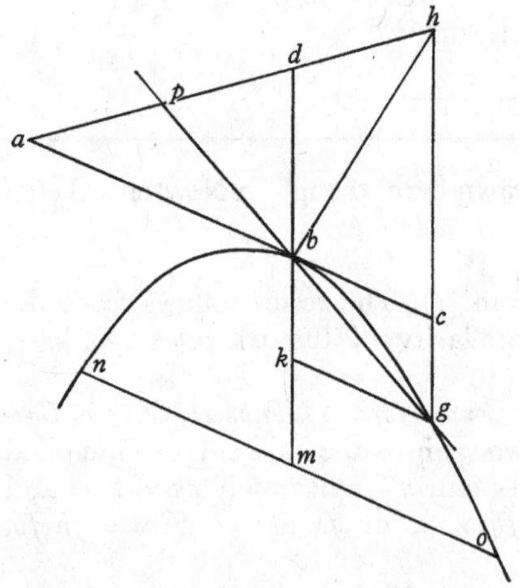

(2) $db\times bk=kg\times kg.$
$kg\times kg:nm\times nm::db\times bk:db\times bm::$
$bk:bm.$[52]

(49) Where *ne* is constant, $ne\times ei=(if)^2$.
(50) Read 'by'.
(51) A basic proposition in conics taken over unaltered, along with the reference to Apollonius and Mydorge, from Schooten's *Commentarii in Librum II*, *CCC* = *Geometria*: 207–8. (In the accompanying diagram a first, wrongly drawn line *fi* has been omitted.)
(52) The generalization of the construction already noted by Newton (in section 4*d*) from de Witt's *Elementa Curvarum*: Th. 1: 162–4. Here *ph* is no longer parallel to *ac* but, as before,

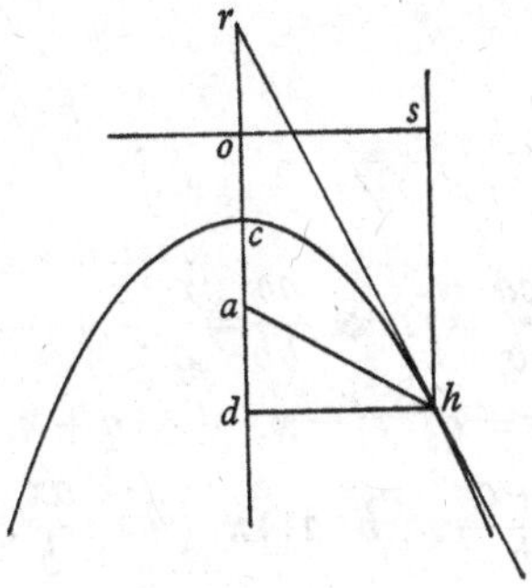

[*c*] $a = $ foco. $ac = \frac{1}{4}$ lateris recti. $ac = oc$. $ah = do$, sit (sh) Parallela ad (dr). & (rh) contingat Parab̄: in h. & (dh) perpend: ad (dr). erit ang: $ahr = rhs$.(53)

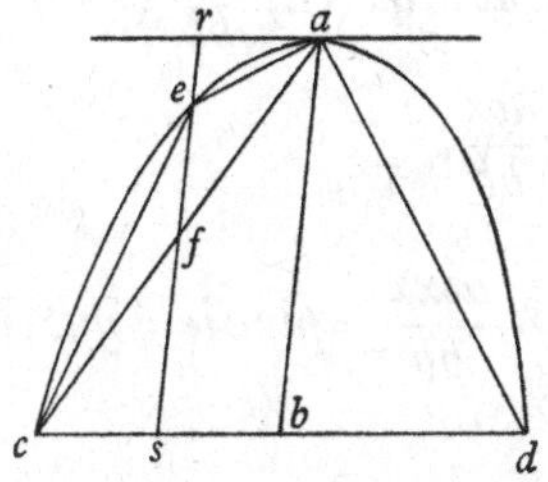

[*d*] If $cs = sb$ & sr parallell to ab yⁿ yᵉ triang $cea:cab::1:4$. & so it may be saide infinitely. If ab & cd are ordinately applyed yᵉ Parabola *ceadb* is to yᵉ triangle *cda* as eight to six.(54) & $rf \times rf = rs \times re$. or, $re:rf::rf:rs$.(55)

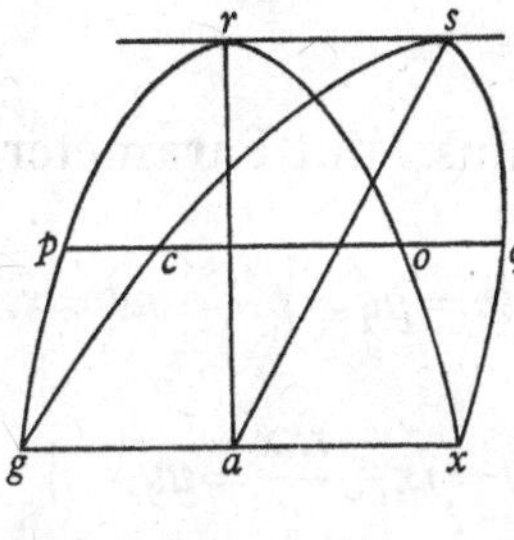

[*e*] If rs is parallell to gx yⁿ are yᵉ 2 segments of Parabolas [(]$gproxa = gcsqxa$) equall & $po = cq$. & if $ga = ax$ then yᵉ diameters ar. as cut yᵉ line rs in its touch points.(56)

---

the essence of the proof is that the triangles *bdh*, *bcg* are similar, with $a\hat{b}d = p\hat{b}h = p\hat{h}c = a\hat{d}b$ (or $ad = ab$, $dh = bc$). Then $bd:dh$ (or bc) $= bc:cg$ (or bk), and so db (constant) $\times bk = bc^2 = kg^2$.

(53) 'a = focus. $ac = \frac{1}{4}$ of latus rectum. $ac = oc$. $ah = do$. let sh be parallel to dr, rh tangent to the parabola in h, and dh perpendicular to dr. then $a\hat{h}r = r\hat{h}s$.'

The proposition is adapted from Schooten's *Exercitationes*, Liber IV: Lemma: 355–6, where the parabola property follows from $ah = do$ by

$$dh^2 = ah^2 \text{ (or } od^2) - ad^2 = oa^2 + 2oa \times ad = 4ca \times cd.$$

(54) Newton originally wrote 'five'!

(55) A familiar Archimedean proposition given an original slant by Schooten in his *Exercitationes*, Liber IV: [general] Scholium: 361–8. The pole-polar property that re, rf, rs are in continued proportion (I: 361–2) is fundamental in the proof.

(56) Adapted from Schooten's *Exercitationes*, Liber IV: Cap. III, Scholium: 316–17 (where the theorem is given for ellipse segments).

[7] *The properties of yᵉ Hyperbola.*

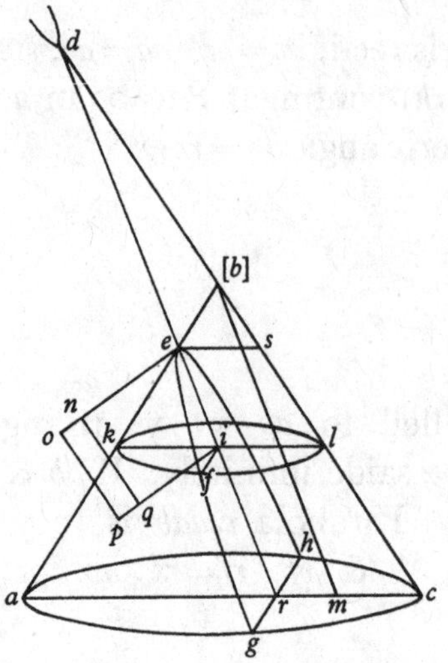

$[a]\ rx+\frac{acxx}{bb}=yy.$

☞ $rx+\frac{r}{q}xx=yy.$ for $\frac{acq}{bb}=r.$ & $\frac{ac}{bb}=\frac{r}{q}.$

$am=a.\ mb=b.\ mc=c.\ de=q.\ ei=x.\ di=q+x.$

$fi=y.\ b:c::q+x:(il)\ \frac{cq+cx}{b}.\ \ b:a::x:(ik)\ \frac{ax}{b}.$

$il\times ik=yy=\frac{cqax+caxx}{bb}.\ bb:ac::q:(en)\ \frac{acq}{bb}\ (=r).$

$bb:ac::q:\frac{acq}{bb}::x:(qp$ or $on)\ \frac{acx}{bb}.$

Whenc $\frac{acqx+acx[x]}{bb}=rx+\frac{acxx}{bb}=pi\times ie=yy.$

More breifly thus.

$b:c::q[:]\ (es)\ \frac{cq}{b}.\ \ b:a::\frac{cq}{b}:\frac{acq}{bb}\ (=r).$

de is called latus transversum & *en* latus rectum by Apollonius. but Parameter by Mydordgius.(57)

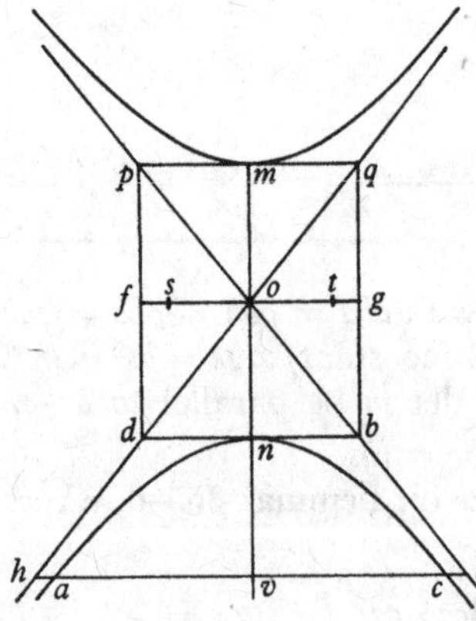

$[b]\ mn=pd=bq=q.\quad fg=db=pq=p.\quad nv=x.$
$av=y.\ ha=ck=b.\ st=r.$

(1) $q:r::qx+xx:yy.$ & $yy=rx+\frac{rxx}{q}=yy.$

(2) $2by+bb=\frac{1}{4}pp.$(58)

(3) $q:p::p:r.$

(4) $q:r::qq:pp.$

(5) $yy:qx+xx::qq:pp.$(59)

$pq=fg=db=$ axi secundo & recto. & diam. rectæ $pd=mn=qb=$ axi primo, transverso & lateri sive diametro transversæ. $st=r=$ Lateri recto.(60)

(57) Noted, along with the references to Apollonius and Mydorge, from Schooten's *Commentarii in Librum II*, *CCC* = *Geometria*: 209–10.

(58) Since $ha\times ak = hc\times ck = dn\times nb$.

(59) Theorems (1) and (3) are standard properties of the hyperbola, from which theorems (4) and (5) are immediately deducible.

(60) '$pq = fg$ = second & right axis. & the right diameters $pd=mn=qb$ are equal to the first, transverse & lateral axis or transverse diameter. $st = r$ = Latus rectum.'

[*c*] If $xt=p$. $sr=q$: $r+$Param̄ & [area] $iry=a$. $eno[r]=b$. $in=y$. $en=z$. $rn=x$. Then if $p=q=r$ as in (*a*)[61]: (*a*) is y^e simplest of all Hyperbola's, & y^n, $yy=xx+qx$. & if (*q*) is y^e same in both (*a* & *b*) & ($xt=p$) is propper[62] to (*b*) then $yy:zz::qq:pp$. & therefore Hyperbolas are to one another as theire rig[h]t axis[63] are supposeing theire transverse axes equall. viz

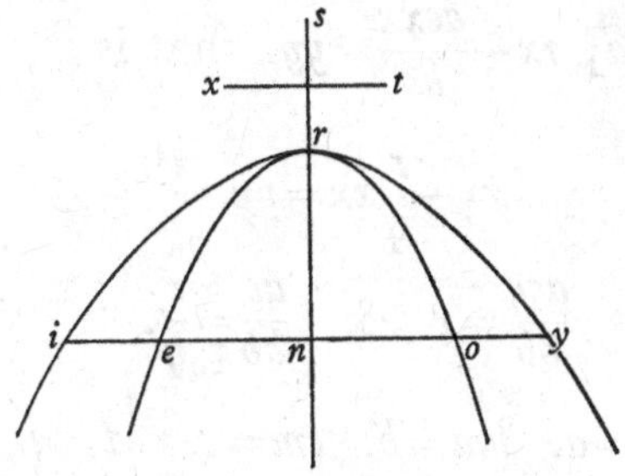

$$iryeon:eron::in:en::[q]^{(64)}:p.$$

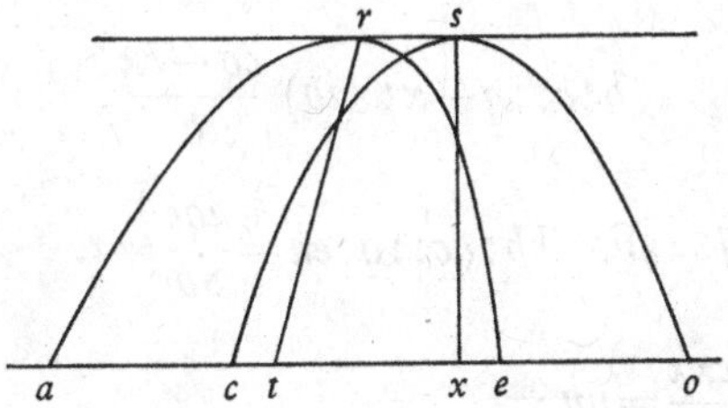

therefore if (*rs*) is parallell to *ao*, & $ae=co$. y^n ($arextc=csoext.$) & if

$$at=te=cx=xo.$$

tr & *xs* (cutting *rs* in y^e touch points) are ordinately applyed to y^e Diameters & bisect y^e Hyperbolas.[65]

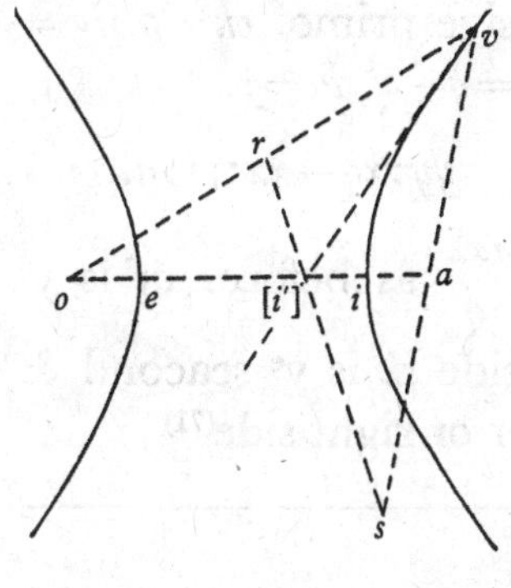

[*d*] If (*o* & *a*) are y^e foci & (*v*) a point in one of y^e Hyberb:s. then $av+ei=ov$. & if $as=ei=or$. y^n $vs=vo$. & $rs=oa$. & ($i^{[\prime]}v$) bisecting y^e angle ($ri^{[\prime]}a$.) it shall touch y^e Hyperb in *v*.[66]

(61) That is, the parabola *iry*.

(62) Read 'belongs'.

(63) Read 'axes'.

(64) Newton wrote '*p*' in error.

(65) Newton's analytical rendering of Schooten's *Exercitationes*, Liber IV: Cap. VI, Scholium: 333–5.

(66) The gist of Schooten's *Exercitationes*, Liber IV: Cap. IX: 344–6, and Scholium: 347–8. The proposition reduces to the focus-property of the hyperbola that $ov-av = or = ei$, constant.

[8] *The Properties of yᵉ Ellipsis.*

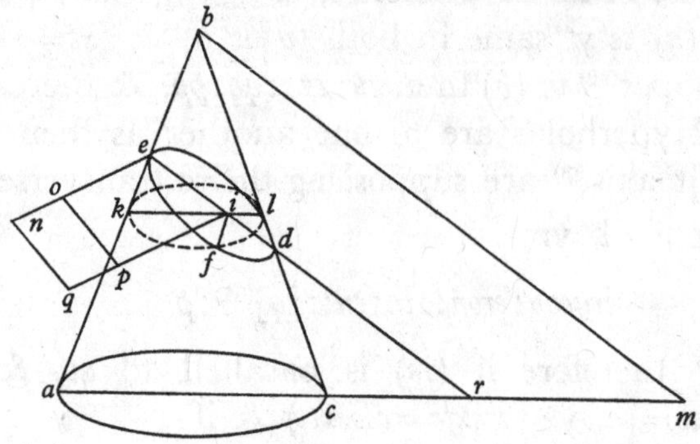

[*a*] $rx-\frac{acxx}{bb}=yy$. that is

☞ $rx-\frac{r}{q}xx=yy$.

for $\frac{acq}{bb}=r$ & $\frac{ac}{bb}=\frac{r}{q}$.

$am=a$. $bm=b$. $cm=c$. $ed=q$. $ei=x$. $id=q-x$. $fi=y$. $en=r$.

$b:c::q-x:$ (*ib*) $\frac{cq-cx}{b}$.

$b:a::x:$ (*ik*) $\frac{ax}{b}$. $ki\times il=\frac{cqax-acxx}{bb}=fi\times fi=yy$. $bb:ac::q:en=\frac{aqc}{bb}=r$.

$bb:ac::x:on=\frac{acx}{bb}$. wherefore $rx-(on\times x)=$(67) $\frac{acxx}{bb}=yy$.(68)

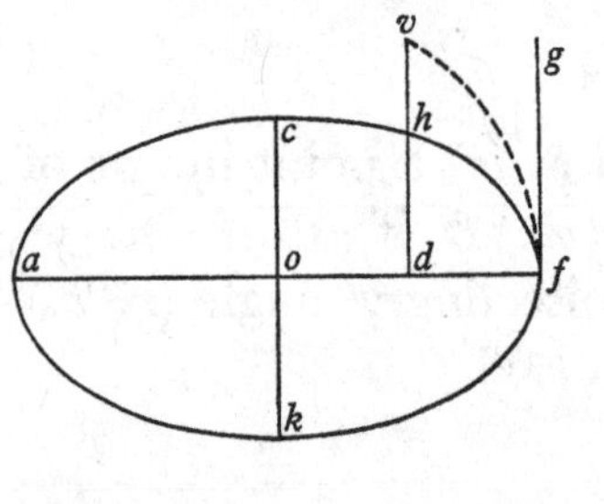

[*b*] $Af=q=$axi trans *v*: sive primo. $ck=p$. $fg=r=$lateri recto.(69) $ad=x$. $df=q-x$. $dh=y$.

(1) $q:p::p:r$. (2) $yy:xq-xx::r:q$.

therefō(70) [(]3[)] $yy=rx-\frac{rxx}{q}$ as before. *af* is yᵉ first & transverse axis or side *ck* is yᵉ seacond & right axis *fg* is yᵉ Parameter or right side(71)

(67) Read '$(on\times x=)$'.

(68) Transcribed from Schooten's *Commentarii in Librum II*, *CCC* = *Geometria*: 211–12.

(69) '$af=q=$ transverse or first axis... $fg=r=$latus rectum.'

(70) Read 'therefore'.

(71) Newton's reworking of [*a*]. As his cancellation indicates parts (1) and (3) were originally combined and placed after (2).

(72) Read '*oekl*'.

(73) 'Take $pn=qn$. $ne=no$. then will segment *oekl* be to segment *cbd* as *cbd* to $gbh'cd::fh:ab::(afbhcd)$ ellipse: $(ah'bg)$ circle.'

The theorem is adapted from Schooten's *Exercitationes*, Liber IV: Cap. II and Scholium: 309–11, 312–13, which in essence proves that, where λ, μ are the respective meets *fp*, *ncq* and *h'cdg*, *plqb*, then

$$pl:p\mu = ol \text{ (or } c\mu):h'\mu = \text{ellipse } (afbh):\text{circle } (ah'bg)$$
$$= fp \text{ (or } cq):pb \text{ (or } \lambda c) = \mu q:p\mu,$$

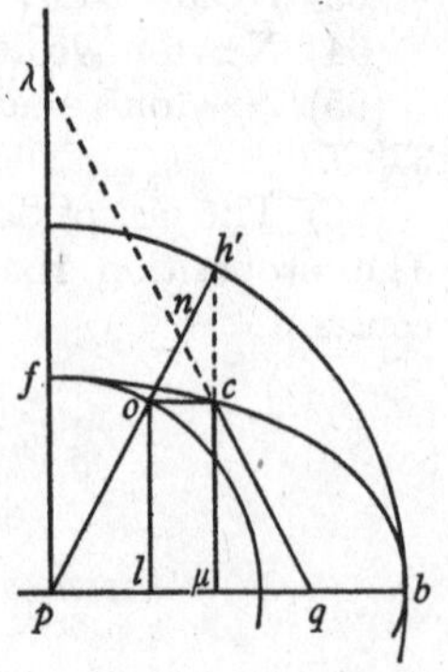

so that $pl=\mu q$. It follows that *p*, *o*, *n*, *h'* are collinear (and so similarly are *p*, *k*, *g*). (Note that Newton has two points *h* in his diagram, one of which is here transcribed *h'*.)

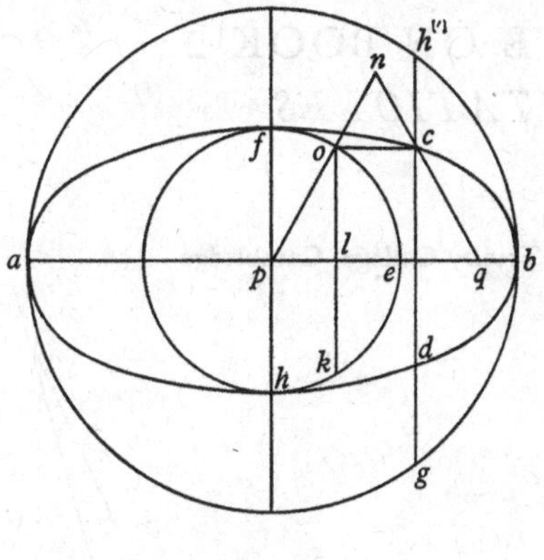

[*c*] sit $p[n] = qn$. $nc = no$. erit segmentum *oetl*(72) ad segm̄ *cbd*, ut *cbd* ad *gbh*['] *cd*::*fh*:*ab*:: (*afbhcd*) elipsis: (*ah*[']*bg*) circulum.(73)

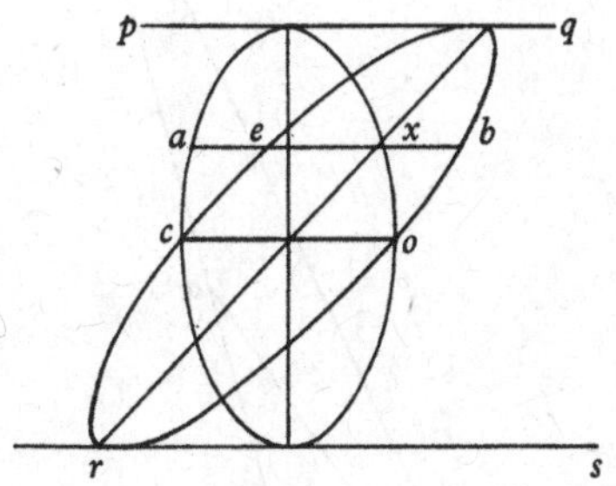

[*d*] If y^e lines (*pq*, *rs*) are parallell & *co* y^e common axis of both y^e Ellipses y^n are y^e 2 Ellipses equall to one another, for $ax = be$. y^e conjugated diam: cut y^e touch points of *pq*, *rs*. & parallells to these are also conjugated.(74)

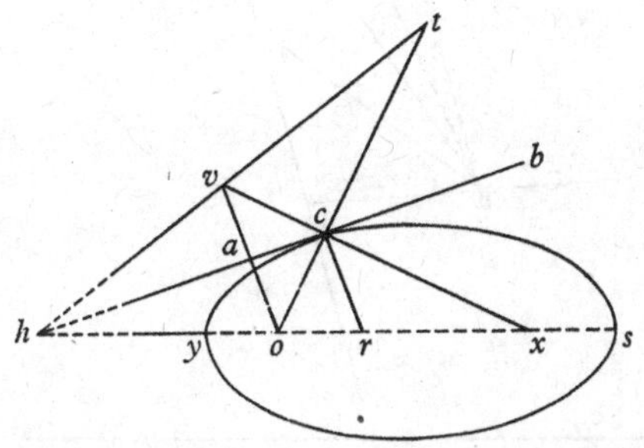

[*e*] If *ab* touch an Ellipsis & (*o*) & (*x*) be y^e foci y^n y^e angle $aco = bcx$. & if (*ocx*) be bisected by (*cr*) y^n $acr = bcr =$ right angle.

If $xv = ot = ys$. & *vo* bisected in *a* then $vac = oac =$ to a right angle.

If also $vt = ox$. & *vt* & *xo* be produced till they meete in *h*. y^e angle *vho* shall be bisected by y^e line *acb*.(75)

---

(74) Noted from Schooten's *Exercitationes*, Liber IV: Cap. III, Scholium: 316–17.

(75) Newton's transcription from Schooten's *Exercitationes*, Liber IV: Cap. VIII: 339–41, whose foundation is the focal property of the ellipse that $oc + cx = ys$, constant. Tangent *ab* bisects $o\hat{c}x$ externally, while from $vx = ot = ys$ $(= oc + cx)$, or $vc = oc$, $et = cx$, it follows that the triangles *ocx*, *vct* are congruent and symmetric round the line *ab*.

§2. A MISCELLANEOUS NOTE ON BOOK 2 OF SCHOOTEN'S *EXERCITATIONES*

[Late 1664?]

From the original pasted in Newton's library copy(1) in Trinity College, Cambridge

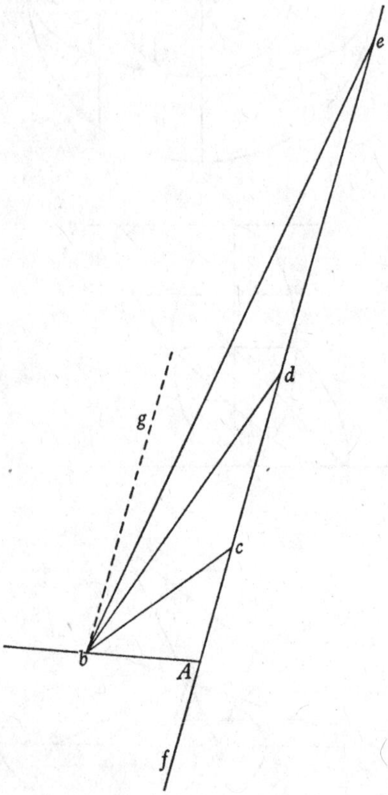

If $ab=ac$, & $bc=cd$, & $bd=de$ &c.

Then is $\angle baf=2bcf=4bdf=8bef$. &c. & $\angle Abc=2cbd=4dbe$ &c. Or if $bg \parallel Af$. then is $gbA=2gbc=4gbd=8gbe$. &c.(2)

(1) Trinity. NQ. 16. 184. The text of the *Exercitationes* itself is clean, but the insert is pasted in between pp. 178/9—that is, in the *Appendix Simplicium Problematum* at the end of *Liber* II.

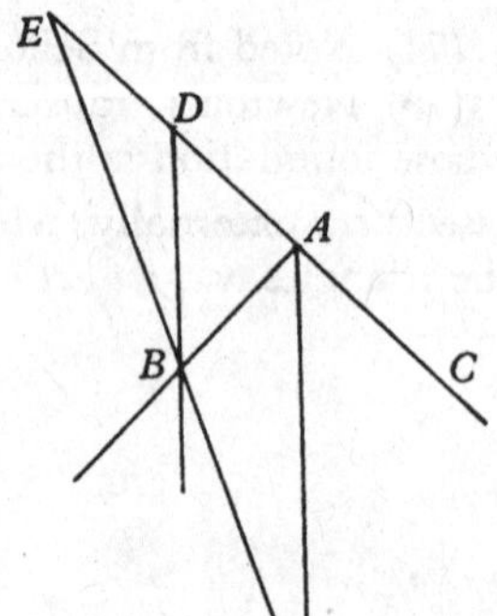

(2) This seems to be Newton's improvement of a construction by 'simple geometry' (that is, by using a graduated straight edge) which Schooten gave in *Liber* II, *De Constructione Problematum Simplicium Geometricorum, seu Quæ solvi possunt, ducendo tantum rectas lineas.: Problema I, Datum angulum rectilineum BAC bifariam secare.*

Schooten's construction finds the bisector AF of $B\hat{A}C$ by taking $AB = AD = DE$ and F on EB with $EB = BF$. On the same basis Newton would bisect his $b\hat{A}f$ by constructing $bg \parallel fA$ and finding the bisector bc of $g\hat{b}A = b\hat{A}f$ by constructing $bA = Ac$. It is in this form, indeed, that a few years later in an (unpublished) tract on constructing equations now in the University Library, Cambridge (Add. 3963.9), Newton entered the proposition (on 72v) as his *Prob 16*: *Datum angulum BAC bisecare idq continuò.*

§3. NEWTON'S NOTES ON THE FIRST HALF OF BOOK 5 OF SCHOOTEN'S *EXERCITATIONES*

[Late 1664?]

From the original in a pocket book(1) in the University Library, Cambridge

Propositiones Geometriæ ex Schootenij Sectionibus miscellaneis.(2)

Sectio 2ma(3)

To know how many changes 6 Bells, *abcdef.* or how [many] divers conju[n]ctions ye 7 Planets can make ♄ ♃ ♂ ☉ ♀ ☿ ☽.(4) or how many divisors *abcde*(5) hath, or how many divers compositions ye 24 leters(6) can make &c the examples following show.

1. *a*. 1.

2. *b*. *ab*. 3.

4. *c*. *cb*. *cab*. *ac*. 7.

8. *d*. *da*. *db*. *dab*. *dc*. *dac*. *dcb*. *dcab*. 15.

16. *e*. *ea*. *eb*. *eab*. *ec*. *eac*. *ecb*. *ecab*. *ed*. *eda*. *edb*. *edab*. *edc*. *edac*. *edcb*. *edcab*. 31.

32. *f*. *fa*. *fb*. *fab*. *fc*. *fcb*. *fac*. *fcab*. *fd*. *fda*. *fdb*. *fdab*. &c. 63.

64. *g*. *ga*. *gb*. *gab*. *gc*. *gcb*. *gac*. *gcab*. *gd*. *gda*. *gdb*. &c. 127.(7)

wch shows yt in 7 letters 127 elections may be made[.] yt 7 Planets may be conjoyned 120(8) divers ways. yt *abcdefg* hath 128 divisors for an unite is one of ȳ & $1\times2\times3\times4\times5\times6.=720$. are ye number of changes in six bells.(9)

---

(1) Add. 4000: 12r–14r. In private possession there exists a contemporary copy (in John Collins' hand) which later passed into William Jones' library.

(2) 'Propositions of geometry out of Schooten's miscellaneous sections.'

(3) *Exercitationes*: 373–80: *Sectio 1, Ratio inveniendi electiones omnes, quæ fieri possunt, datâ multitudine rerum.*

(4) The conventional astronomical symbols for Saturn, Jupiter, Mars, Sun, Venus, Mercury and Moon respectively.

(5) Read '$a\times b\times c\times d\times e$'.

(6) Newton notes this from the original written in Latin, which lacks *w* and *z*.

(7) The left-hand numbers enumerate the number of combinations written out at each level, the right-hand ones the total number of combinations in this and all higher levels.

(8) Read '127'.

(9) Two simple problems in combinations and permutations:

(*a*) Total number of combinations of *n* objects taken one, two, three, ... (up to) *n* at a time is $\sum_{1\leqslant i\leqslant n}\binom{n}{i} = 2^n-1$ (which Schooten proves by total enumeration successively for $n = 1, 2, 3, \ldots$). The additive 'zero' is not counted by Schooten since for him a null-set is not a real combination, but the multiplicative 'zero' (Schooten's 'unite') is allowed.

(*b*) Total number of permutations of *n* objects is $n!$.

Sec 2(10)

[1] *To know how many things & of wt sort they are wch may be chosen 15 ways.*

$15+1=16.\ \frac{16}{2}=8.\ \frac{8}{2}=4.\ \frac{4}{2}=2.\ \frac{2}{2}=1.$ & $2-1.\ 2-1.\ 2-1.\ 2-1.\ =1.\ 1.\ 1.\ 1.\ =4.$ [so] that 4 things all unequall may be varyed 15 ways. also $\frac{16}{4}=4.\ \frac{4}{2}=2.\ \frac{2}{2}=1.\ 4-1.\ 2-1.\ 2-1.\ =5.$ & 5 things whereof 3 are equall viz: *a. a. a. b. c.* [may be varyed 15 ways.] & $\frac{16}{4}=4.\ \frac{4}{4}=1.\ 4-1.\ +4-1\ =3+3=6.$ & 6 thin[g]s whereof 3 & 3 are eaquall as *a. a. a. b. b. b.* may be varyed 15 ways. & $\frac{16}{8}=2.\ \frac{2}{2}=1.\ 8-1.\ 2-1=8=7+1.$ & 8 things whereof 7 are =(11) may be varyed 15 ways as *a. a. a. a. a. a. a. b.* & $\frac{16}{16}=1.\ 16-1=15.$ wherefore 15 alike things &c as a^{15}.(12)

2 Wt things vary 23 ways. $23+1=24$ & 24 admitts a 7 fold divison(13) therefore ye multitude of things sought may be 7 fold.(14)

[3] but sinc 43 is a primary(15) number (viz wch cannot bee divided) $42+1=43.\ \frac{43}{43}=1.\ 43-1=42.$ therefore onely 42 like things can be varyed 42 ways as a^{42}.(16)

To know how many divers ways things, whereof some of ym are equall, may bee ordered. as of *a. b. b. c. c. c. d. d.* doe thus

a b b c c c d d

$$\frac{1}{1}\times\frac{2\times 3}{1\times 2}\times\frac{4\times 5\times 6}{1\times 2\times 3}\times\frac{7\times 8}{1\times 2}=112^{(17)}$$ the number of changes, in order.

(10) *Exercitationes*: 380–3: *Ratio inveniendi multitudinem rerum, earumque inter se habitudinem, quæ datis vicibus eligi possunt.*

(11) Read 'equall'.

(12) Read '*a. a. a. a. a. a. a. a. a. a. a. a. a. a. a.*'. In this note from *Exercitationes*: 380–1: 1*mum* *Exemplum* Newton follows Schooten's application of the general rule for combinations of objects not all different, using the factorizations

$$(15+1=)\ 16 = 2\times 2\times 2\times 2 = 4\times 2\times 2 = 4\times 4 = 8\times 2 = 16\times 1$$

to derive the sets of $(2-1)+(2-1)+(2-1)+(2-1)$, $(4-1)+(2-1)+(2-1)$,

$$(4-1)+(4-1),\quad (8-1)+(2-1)\quad\text{and}\quad (16-1)$$

objects whose elements may be combined in 15 ways. (The factorization is exhaustive in positive integers, and so no other combinations are possible.)

(13) $24\times 1 = 2\times 12 = 2\times 2\times 6 = 2\times 2\times 2\times 3 = 2\times 4\times 3 = 4\times 6 = 3\times 8$.

To know how many elections may bee made doe thus

$$a \quad bb \quad ccc \quad dd \quad eeee \qquad\qquad a \qquad bb \qquad ccc \qquad dd \qquad eeee$$

$$2\times3\times4\times3\times5=360=2\times\frac{2\times3}{2}\times\frac{2\times3\times4}{2\times3}\times\frac{2\times3}{2}\times\frac{2\times3\times4\times5}{2\times3\times4}.$$

therefore there are $359=360-1$ elections in $abbc^3dde^4$.[18]

Sec 3[19]

Every quantity hath one divisor more [y^n][20] it hath aliquote ꝑts (y^t is ꝑts of whole numbers).[21] Now to find a quantity haveing a given multitude of divisors or aliquote ꝑts: suppose its aliq: ꝑts must be 15. $15+1=16$ & soe by y^e former section[22] $abcd$. a^3bc. a^3b^3. a^7b. a^{15}. may be varyed 15 ways therefore they shall have 15 aliquote ꝑts & 16 divisors.[23] but since onely 42 like things (as a^{42}) can be varyed 42 ways therefore oenely[24] a^{42} hath 42 aliquote ꝑts & 43 divisors. &c.

Sec 4[25]

To find y^e least numbers haveing a given multitude of divisors & aliquote ꝑts instead of soe many letters in y^e former sec:[26] put soe many least primary numbers & take y^e least result from y^m. as from y^e former example: $abcd$. a^3bc. a^3b^3. a^7b. a^{15}. that is 2. 3. 5. 7. or 2. 2. 2. 3. 5. &c.[27] now $2\times3\times5\times7=210$. & $2\times2\times2\times3\times5=120$. &c therefore $2\times3\times5\times7=210$.[28] is y^e least number having 16 divisors.

Sec: 5[29] conteines a table of Primary numbers.

---

(14) 2um *Exemplum:* 381–2.　　　　　(15) Read 'prime'.

(16) 3um *Exemplum:* 382–3.　　　　　(17) Read '1680'.

(18) These last two remarks are Newton's own addition, clarifying Schooten's numerical examples. He finds respectively that:

(*a*) The number of distinct permutations of 8 things, made up of sets of $1+2+3+2$ identical objects, is $\dfrac{8!}{1!\,2!\,3!\,2!}$.

(*b*) The total number of combinations of 12 objects, made up of sets of $1+2+3+2+4$ identical objects, is $(1+1)(2+1)(3+1)(2+1)(4+1)-1=359$ (omitting the null-set).

(19) *Exercitationes*: 383–7: *Ratio inveniendi quantitates, datam habentes partium aliquotarum aut divisorum multitudinem.*

(20) Newton wrote 'y^t'.

(21) Newton has cancelled his first version, 'greater y^n an unite'.

(22) Section 2.　　　　　　　　　　　(23) Adding the unit.

(24) Read 'onely'.

(25) *Exercitationes*: 387–92: *Ratio inveniendi minimos numeros, datum habentes partium aliquotarum aut divisorum multitudinem.*

(26) Read 'sections'.　　　　　　　　(27) Taking $a=2$, $b=3$, $c=5$, $d=7$.

(28) Read '$2^3\times3\times5=120$'!

(29) *Exercitationes*: 393–403: *Syllabus numerorum primorum, qui continentur in decem prioribus chiliadibus.*

Sec 6[30]

To find progressions constituteing rectangular triangles wth sides rationall y^e examples following shew.

take two numbers as 1. 2. y^n $1\times2=2$. since y^e product is eaven double it viz: $2\times2=4$. & 4 is y^e numerator. y^n $1+2=3$ & since 3 is od multiply it by y^e difference of y^e termes: $1\times3=3$ & 3 is y^e denominator. & y^e first terme $\frac{4}{3}$. y^n since (1) y^e difference of y^e termes is od multiply it by 4. $4\times1=4$ & $4\times$per 2 majorē terminum.[31] $4\times2=8$. $8+4$ (the former numerator) $=12=$ numerator 2^d. y^n 3 (y^e former denom:) added to 2 (y^e double square of y^e diff: of y^e termes because y^e square (1) is odd) $=5$ y^e 2^d denominator.[32]

I ad anoth^r example. take 1. 3. y^n $1\times3=3=1^{st}$ numerator. y^n $1+3=4$ & since 4 is eaven $\frac{4\times2}{2}$ (diff: of y^e termes) $=4$ & y^e first denom: is 4. y^e first terme $\frac{3}{4}$. y^n becaus y^e diff: of y^e termes 2 is eaven[33] $2\times2=4$ & $4\times3=12$ & $12+3=15$. y^n $2\times2=4$. $4+4=8$. & $\frac{15}{8}$ y^e 2^d terme.[34]

& now termes may be had by Arithmeticall proportion. thus. $\frac{4}{3}.\frac{12}{5}$ [&c] or $1\frac{1}{3}.2\frac{2}{5}.3\frac{3}{7}.4\frac{4}{9}.5\frac{5}{11}.6\frac{6}{13}.7\frac{7}{15}.8\frac{8}{17}.9\frac{9}{19}.10\frac{10}{21}$. &c[35] & $\frac{3}{4}.\frac{15}{8}$ [&c] or $\frac{3}{4}.1\frac{7}{8}$. $2\frac{11}{12}.3\frac{15}{16}.4\frac{19}{20}.5\frac{23}{24}.6\frac{27}{28}.7\frac{31}{32}.8\frac{35}{36}$. &c.[36] thus may other progressions be

---

(30) *Exercitationes*: 404–6: *De Progressionibus, rectangula triangula constituentibus, quorum latera sint rationalia.*

(31) 'By the greater terme.' Newton, a little carelessly, both symbolizes Schooten's 'per' by '×' and keeps the preposition in his annotation.

(32) Newton, following Schooten, constructs successively $\frac{2mn}{m^2-n^2}$, $\frac{2(2m-n)\,m}{(2m-n)^2-m^2}$, for the particular values $m=2$, $n=1$.

(33) Newton wrote 'is eaven 2', but the transposition is made to make the phrase intelligible and avoid a clash with the following calculation.

(34) Newton constructs, for $m=3$, $n=1$, the fractions $\frac{mn}{\frac{1}{2}(m^2-n^2)}$, $\frac{(2m-n)\,m}{\frac{1}{2}[(2m-n)^2-m^2]}$. (Since $m+n$ is here supposed even, then $m-n$ will be even and so $\frac{1}{2}(m^2-n^2)$ and

$$\tfrac{1}{2}[(2m-n)^2-m^2]=\tfrac{1}{2}(m-n)\,(3m-n)$$

are both integral.)

(35) Where $m=r+1$, $n=r$ (or $2m-n=r+2$), then

$$f(r)=\frac{2mn}{m^2-n^2}=r+\frac{r}{2r+1} \quad\text{and}\quad \frac{2(2m-n)\,m}{(2m-n)^2-m^2}=f(r+1).$$

Newton tabulates successively $f(r)$, $r=1, 2, 3, \ldots$.

obteined. For y^e^ Use take y^e^ numerator for one leg & y^e^ denom for another & y^e^ Hypoten: will be rationall.(37) as in $2\frac{2}{5}$ or $\frac{12}{5}$. $\sqrt{144+25}=\sqrt{169}=13$. & in this $1\frac{7}{8}$ or $\frac{15}{8}$. $\sqrt{225+64}=17$.

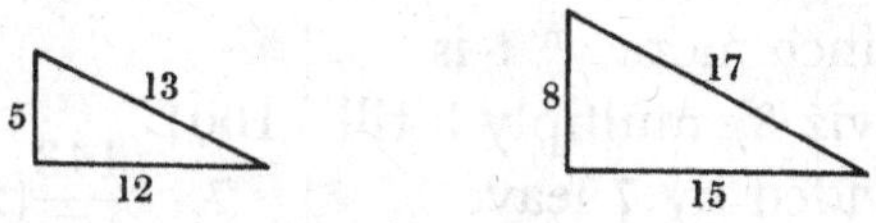

If y^e^ suposed numbers be 2. 5. y^n^ $2\times5=10$. $10+10=20$. & $2+5=7$. $3\times7=21$. so y^t^ $\frac{20}{21}$. y^n^ $4\times3=12$. $12\times5=60$. $60+20=80$. & $3\times3=9$. 9 doubl[e]d=18. $18+21=39$. & y^e^ 2 first termes $\frac{20}{21}.\frac{80}{39}$ or $2\frac{2}{39}$.(38) Againe, if y^e^ numbers be 3. 4. $3\times4=12$. $12\times2=24$. & $3+4=7$. $1\times7=7$. therefore $\frac{24}{7}$. y^n^ $4\times1=4$. $4\times4=16$. $16+24=40$. & $1\times1=1$. $2\times1=2$. $7+2=9$. therefore $\frac{40}{9}$ is y^e^ 2^d^(39) & y^e^ progres may be continued, as $\frac{20}{21}.2\frac{2}{39}.3\frac{5}{57}.4\frac{8}{75}.5\frac{11}{93}$. [&c] & $3\frac{3}{7}.4\frac{4}{9}.5\frac{5}{11}.6\frac{6}{13}$. &c(40)

(36) Where $m = 2s+1$, $n = 2s-1$ (or $2m-n = 2s+3$),

$$g(s) = \frac{mn}{\frac{1}{2}(m^2-n^2)} = s-\frac{1}{4s} \quad\text{and}\quad \frac{(2m-n)\,m}{\frac{1}{2}[(2m-n)^2-m^2]} = g(s+1),$$

Newton tabulates $g(s)$ for $s = 1, 2, 3, \ldots$.

(37) This follows immediately, since $(m^2+n^2, 2mn, m^2-n^2)$ and $((2m-n)^2+m^2, 2(2m-n)m, (2m-n)^2-m^2)$ are Pythagorean triads.

(38) Newton calculates $\dfrac{2mn}{m^2-n^2}, \dfrac{2(2m-n)\,m}{(2m-n)^2-m^2}$ respectively with $m = 5$, $n = 2$.

(39) The particular values of $\dfrac{2mn}{m^2-n^2}, \dfrac{2(2m-n)\,m}{(2m-n)^2-m^2}$ with $m = 4$, $n = 3$.

(40) The second of these sequences, $f(r) = r+\dfrac{r}{2r+1}$, $r = 3, 4, 5, \ldots$, has already been derived. The former, $h(t) = t+\dfrac{3t-4}{3(6t+1)}$, $t = 1, 2, 3, \ldots$, follows by taking $m = 3t+2$, $n = 3t-1$ (or $2m-n = 3t+5$), so that

$$\frac{2mn}{m^2-n^2} = h(t) \quad\text{and}\quad \frac{2(2m-n)\,m}{(2m-n)^2-m^2} = h(t+1).$$

Sec 7(41)

To find a [number] wch divided by 7 leaves 2. by 11 leaves 1. by 13 leaves 9. the least common divisor of 7. 11. 13 is $7\times 11\times 13=1001$. divide 1001 twice by each & consider y^{e} remainder of y^{e} seacond division thus.

1 Since more y^{n} 1 is left (viz 3) multiply 3 till it divided by 7 leave 1. $\frac{5\times 3}{7}=2\frac{1}{7}$ therefore $5\times 143=715$ y^{e} multiplier.	$\frac{1001}{7}\left(\frac{143}{7}\left(20\frac{3}{7}\right.\right.$.
2 Since more y^{n} 1 is left (viz: 3) $\frac{3\times 4}{11}=1\frac{1}{11}$ therefore $4\times 91=364$ y^{e} multipl:	$\frac{1001}{11}\left(\frac{91}{11}\left(8\frac{3}{11}\right.\right.$.
3 If but 1 had beene left 77 had beene divisor but now $\frac{12\times 12}{11}=13\frac{1}{11}$. therfore $12\times 77=924$ is [y^{e}] multiplyer.	$\frac{1001}{13}\left(\frac{77}{13}\left(5\frac{12}{13}\right.\right.$.

now the number sough[t] is thus found.

Divisor.	Reliq:(42)		Multip:
7.	2	×	715 = 1430.
11.	1	×	364 = 364.
13.	9	×	924 = 8316.
		The Sum̄e	10110.

Lastly divide by y^{e} least com: divis:(43) $\frac{10110}{1001}\left(10\frac{100}{1001}\right.$. wherefore 100 y^{e} number left is y^{e} number sought.(44)

(41) *Exercitationes*: 407–10: *De modo inveniendi numeros, qui per datos divisi certos post divisionem relinquunt.*

(42) Read 'reliquum'.

(43) Read 'common divisor'.

Sec: 8.(45)

Touching y^e^ Method of weights suppose a man have weig[ht]s of 1. 2. 4. 8. 16. 32 pounds &c. by y^m^ all intermediate pounds may be thus weighed

1.	2.	3.	4.	5.	6.	7.	8.	9.	10.
1.	2.	1+2.	4.	1+4.	2+4.	1+2+4.	8.	8+1.	8+2.

11.	12.	13.	14
8+1+2.	8+4.	8+4+1.	8+4+2

&c or if his w[e]ights be 1. 3. 9. 27. 81. [&c] all weights will be supplyed thus.

1.	2.	3.	4.	5.	6.	7.	8.	9.
1.	3−1.	3.	3+1.	9−1−3.	9−3.	9+1−3.	9−1.	9.

10.	11.	12.	13.	14
9+1.	9+3−1.	9+3.	9+3+1.	27−9−3−1

[&c] Note y^t^ weight[s] marked w^th^ − signifie ye w[e]igh[ts] to be put in y^e^ opposite ballance.(46)

Sec. 9.(47)

To find numeri amicabiles that is 2 numbers whose aliquote ꝑts are mutually equall to theire wholes. take this Des-Cartes his rule.(48)

---

(44) The details of the argument are clearer when transposed in modern notation. Thus, since

$$\left\{\begin{array}{l} 11\times 13 = 143 \equiv 3 \pmod 7 \\ 13\times 7 = 91 \equiv 3 \pmod{11} \\ 7\times 11 = 77 \equiv 12 \pmod{13} \end{array}\right\}, \quad \left\{\begin{array}{l} 5\times 143 = 715 \equiv 1 \pmod 7 \\ 4\times 91 = 364 \equiv 1 \pmod{11} \\ 12\times 77 = 924 \equiv 1 \pmod{13} \end{array}\right\},$$

so that

$$\left\{\begin{array}{l} 2\times 715 = 1{,}430 \equiv 2 \pmod 7 \\ 1\times 364 = 364 \equiv 1 \pmod{11} \\ 9\times 924 = \underline{8{,}316} \equiv 9 \pmod{13} \\ \qquad\qquad\quad 10{,}110 \end{array}\right\},$$

with $10{,}110 \equiv 100 \pmod{7\times 11\times 13}$, or finally $100 \equiv 2 \pmod 7$, $\equiv 1 \pmod{11}$, $\equiv 9 \pmod{13}$. Newton has cancelled an earlier example (not completed) which clearly solves the allied problem of finding the least integer x which satisfies $x \equiv p \pmod 2$, $\equiv q \pmod 3$, $\equiv r \pmod 5$, $\equiv s \pmod 7$, though Newton did not specify particular values for the integers p, q, r, s.

(45) *Exercitationes*: 410–19: *Praxis ponderandi.*

(46) Newton has cancelled 'scale'. The notes in this section relate to the famous problem of weights, first satisfactorily resolved by C. G. Bachet (Sieur de Méziriac) in his *Problemes plaisans et delectables, qui se font par les nombres*, Lyon, 1624: Appx, Probl. V: 215 ff.

(47) *Exercitationes*: 419–26: *Ratio inveniendi numeros amicabiles, hoc est, duos numeros, quorum partes aliquotæ, cum adduntur, eos ipsos vice versâ componunt.*

(48) Schooten gave, on p. 423 of his *Exercitationes*, the reason for his attribution of the rule to Descartes: 'Cæterum, ut iis, qui Algebræ ignari sunt, ratio quoque amicabiles numeros inveniendi constet, in medium adducturus sum Regulam, quam olim ab illustri Viro Renato

If (2) or any other number produced out of 2 as 2×2. $2\times 2\times 2$. &c (viz[:] 2. 4. 8. 16. 32 &c) bee such a number y^t 1 taken out of it[s] triple there rests a primary number, & y^t if 1 taken from it[s] sextuple there rests a primary number, & if 1 taken from its square octodecuple[49] a primary number rests: y^n multiply this last prime number by y^e assumed number doubled & y^e product is one amicable number & y^e aliquote ꝑts of it make y^e other.[50]

Example. if 2 be taken. $2\times 3-1=5$ numero primario primo. $2\times 6-1=11$ numero primario scdo. $2\times 2\times 18-1=71$ numero primario tertio. $4\times 71=284$, one amicable number, & y^e 2 former prime numbers $\times$ one another & y^e product $\times$ 4 y^e double of y^e assumed number viz $5\times 11=55$. $55\times 4=220$. Thus from 8. & 64 &c may be deduced amicable numbers.[51]

Sec 10[52]

To find triangles whose sides, segments of theire bases, & Perpendiculars are expressible by rationall numbers.

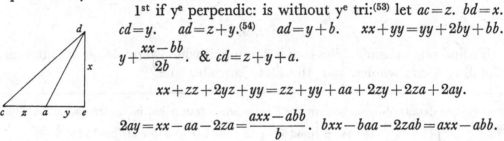

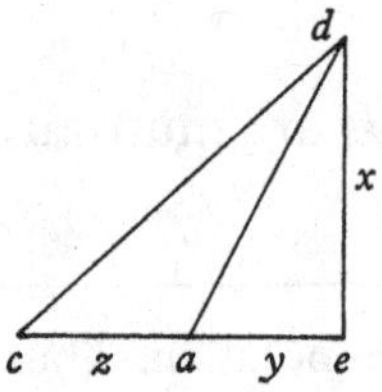

1st if y^e perpendic: is without y^e tri:[53] let $ac=z$. $bd=x$. $cd=y$. $ad=z+y$.[54] $ad=y+b$. $xx+yy=yy+2by+bb$. $y+\dfrac{xx-bb}{2b}$. & $cd=z+y+a$.

$$xx+zz+2yz+yy=zz+yy+aa+2zy+2za+2ay.$$

$$2ay=xx-aa-2za=\frac{axx-abb}{b}.\quad bxx-baa-2zab=axx-abb.$$

$\dfrac{bxx-baa-axx+abb}{2ab}=z$. puting any numbers for a, b, & x; y & z may be found.

---

des-Cartes didici, qui Algebræ peritiâ doctissimorum consensu non modò summas in hisce Disciplinis difficultates superare novit; sed etiam, quicquam demum circa illas ab humano ingenio cognosci potuit determinare.' How and when Descartes communicated his rule for amicals to Schooten we do not know, but he had already sent it to Mersenne on 31 March 1638 in reply to a challenge from Fermat, communicated by Mersenne, to formulate such a rule. (See Descartes' *Œuvres* (ed. Adam and Tannery), **2** (Paris, 1898): 93–4.) Later Schooten passed Descartes' rule on to Huygens, who wrote to Claude Mylon on 15 March 1656 asking to be sent other similar rules (Huygens' *Œuvres*, **1** (La Haye, 1888): 391) and in reply was sent a rule equivalent to Descartes' together with the information that it was Frénicle's invention. (See Huygens' letter to Schooten of 20 April 1656 = Huygens' *Œuvres* **1**: 405.) Fermat, however, who has provoked Descartes to frame his rule, seems its first European inventor. The study of amical pairs themselves, however, has a much longer history, dating back at least to Iamblichus' treatment of them in his commentary on Nicomachus' *Arithmetic.* In particular, the amical pair (220, 284) was widely known among Arabic mathematicians, and since Descartes' rule is a simple generalization of their structure we should not be surprised to find that F. Woepcke claims priority in its invention for Thâbit ibn Qurra in the eleventh century (in his 'Notice sur une théorie ajoutée par Thâbit ben Corrah à l'arithmétique spéculative des Grecs', *Journal asiatique* (October/November 1852): 420–9).

(49) Read 'eighteen times'.

then ad(55) $=z+y=\dfrac{xx+bb}{2b}$. $cd=z+y+a=\dfrac{xx+aa}{2a}$. wch reduced to y^{e} common denominator $2ab$; & y^{t} cast away $cd=bxx+baa$. $ad=axx+abb$. $de=2abx$. $ae=axx-abb$. $ce=bxx-baa$. $ac=bxx-axx+abb-aab$.

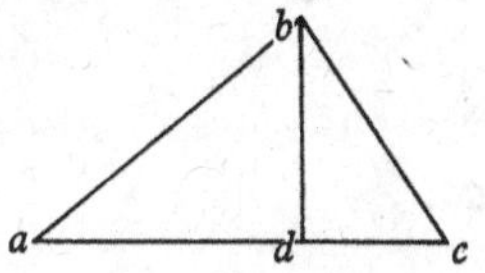

In like manner if y^{e} perpendicular fall wthin side. $ab=bxx+baa$. $bd=2abx$. $ad=bxx-baa$. $dc=axx-abb$. $bc=axx+abb$. $ac=bxx+axx-abb-baa$.

Also by y^{e} conjunction & disju[n]ction of 2 triangles it may be found y^{t} $ab=bbx+axx$. $ad=bbx-aax$.

$$ac=bbx-aax-axx+abb.\ bc=axx+abb.\ db=2abx.\ dc=axx-abb.$$

For if $bd=x$. $dc=\dfrac{xx-bb}{2b}$. $bc=\dfrac{xx+bb}{2b}$. that is $bd=2bx$. $dc=xx-bb$. $bc=xx+bb$. Likewise $bd=2ab$. $ad=bb-aa$. $ab=bb+aa$. $2abx$ y^{e} least quantity divisible by $2bx$ & $2ab$, being divided by y^{m}, leaves a & x wch must multiply y^{e} bases & hypotenusas. If y^{e} Perpendic: fall wthout y^{e} legs [it] may be thus exprest.

$$cd=acc+ayy. \qquad da=yyc+aac.$$

$$ca=acc-ayy+cyy-aac.^{(56)} \qquad ae=yyc=aac.$$

$$ce=acc-ayy. \qquad ad^{(57)}=2acy.^{(58)}$$

(50) Descartes' rule yields the amical pair $[2^{r+1}(3\times2^{r}-1)\,(6\times2^{r}-1),\ 2^{r+1}(18\times2^{2r}-1)]$, where $3\times2^{r}-1$, $6\times2^{r}-1$ and $18\times2^{2r}-1$ are each prime integers.

(51) The pair (220, 284) is the particular case when $r=1(2^{r}=2)$. When $r=3$, $6(2^{r}=8, 64)$, the amical pairs which result are respectively (17,296, 18,416) and (9,363,584, 9,437,056).

(52) *Exercitationes*: 426–30: *De modo inveniendi triangula, quorum singula latera, segmenta basis & perpendicularis exprimuntur per numeros rationales absolutos.*

(53) Read 'triangle'. In this first case Newton considers the situation where the perpendicular de is external to the triangle acd.

(54) Read '$de=x$. $ae=y$. $ce=z+y$'. (Newton confuses the points on his figure with those of that following.)

(55) Read 'ce'.

(56) Read '$acc-ayy-cyy+aac$'.

(57) Read 'ed'.

(58) From the parametrizations

$$2abx = a(x^2+b^2)-a(x^2-b^2) = b(x^2+a^2)-b(x^2-a^2) = x(b^2+a^2)-x(b^2-a^2)$$

and

$$2acy = c(y^2+a^2)-c(y^2-a^2) = a(c^2+y^2)-a(c^2-y^2)$$

are derived Pythagorean triads with a common number: as for example

$$(a(x^2+b^2),\ 2abx,\ a(x^2-b^2)) \quad\text{and}\quad (b(x^2+a^2),\ 2abx,\ b(x^2-a^2)).$$

Sec 11(59)

To make y^t two such tri: be of y^e same base & altitude. Suppose an equation twixt y^e bases & perpendiculars of y^e 2 last tri: as $2abx = 2acy$. $x = \frac{cy}{b}$. $xx = \frac{ccyy}{bb}$. [&] $bbx - aax - axx + abb = acc - ayy + yyc - aac$.

$$\text{or} \quad \frac{bbcy - aacy}{b} - \frac{aacyy}{bb} + abb = acc - ayy + yyc - aac.$$

$$\& \; yy = \frac{\begin{matrix} +b^3cy & +aabbc \\ -aabc & +ab^4 \\ & -abbcc \end{matrix}}{bbc + acc - abb}. \text{ suppose } aabbc + ab^4 = abbcc.^{(60)} \text{ or } a = c - \frac{bb}{c}.$$

let $c = 3$ greater y^n $b = 2$. $a = \frac{5}{3}$. $y = \frac{22}{61}$. $x = \frac{33}{61}$. & consequently

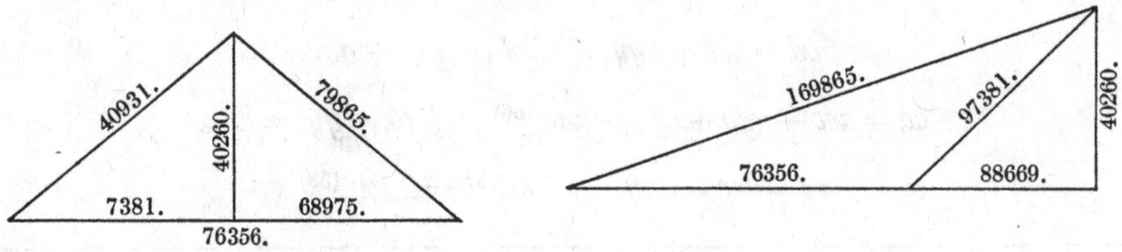

(61)

Sec 14(62) differs not from Cap. 19: prob 18 Oughtred.(63)

Sec: 15(64) *Of Polygons or multangular numbers.*

The sum̄e of all y^e tearmes in an arithmet: progres: increasing from an unite by 1 comp[o]seth triangles. by 2, composes □s(65). by 3, composes pentangles. by 4, hexang: &c. as 1. 2. 3. 4. 5. 6. compos y^e triangles

(59) *Exercitationes*: 430–2: *Modus inveniendi duo triangula ejusdem basis & altitudinis, quorum singula latera, basis segmenta & perpendiculares exprimuntur per numeros rationales integros.*

(60) This reduces the value of yy to $\frac{(b^3c - aabc)\,y}{bbc + acc - abb}$, or $y = \frac{bc(b^2 - a^2)}{b^2c + ac^2 - ab^2}$.

(61) Newton does not annotate sections XII/XIII = *Exercitationes*: 432–6.

(62) *Exercitationes*: 436–42: *De Progressionibus Arithmeticis.*

(63) Taken literally, this refers to *Clavis*, $_3$1652: 94–5: *Ad datam rectam lineam...parallelogrammum adplicare, deficiens figura parallelogramma, quæ similis alteri parallelogrammo...dato.* But presumably Newton intends 'prob. IV Oughtred', a reference to the preceding sixth problem (denoted in error there by IV instead of VI) of *Clavis*: 78–80: *Problematum circa Progressionem Arithmeticam solutio in viginti Propositionibus.*

(64) *Exercitationes*: 442–6: *De numeris Multangulis seu Polygonalibus.*

(65) Read 'squares'.

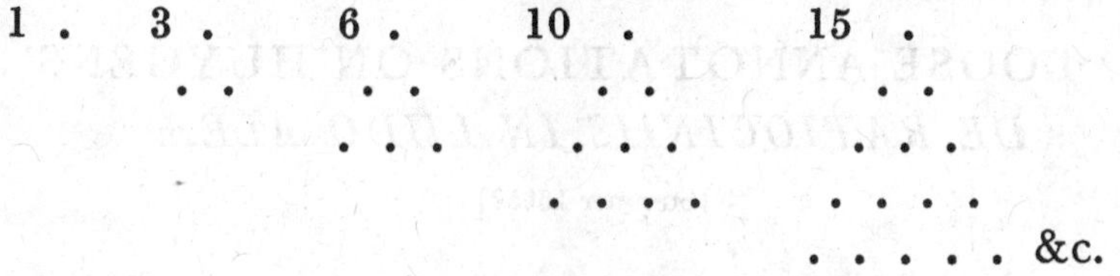

likewise 1. 3. 5. 7. 9. compose

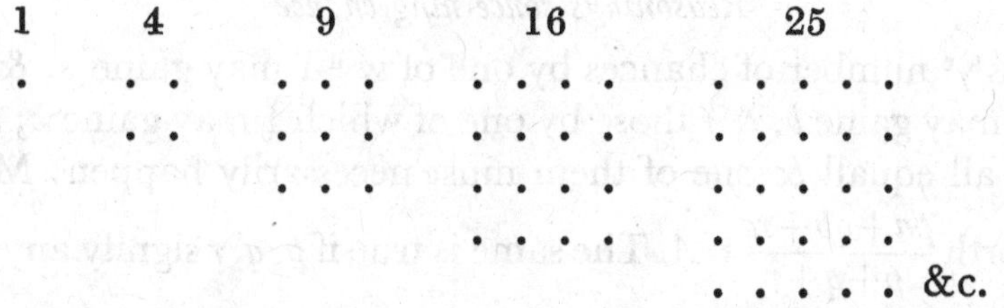

so 1. 4. 7. 10. 13 compose y^e^ quintangles 1. 5. 12. 22. 35. 51. 70. &c. If $a=1=$ y^e^ first terme. y^e^ excess of y^e^ progression $=x$. y^e^ sum̄e of y^e^ termes $=z=$ to y^e^ polygon. y^e^ multitude of y^e^ termes $=t=$ to y^e^ side of y^e^ Polygon. Suppos t given to find z. $z=\dfrac{1tt+1t}{2}$ or $z=\dfrac{tt+t}{2}$ in trigons. $z=tt$ in 4^gons^. $z=\dfrac{3tt-t}{2}$ in 5^gons^. $z=2tt-t$ in 6^gō^. $z=\dfrac{5tt-3t}{2}$ in 7^gons^. $z=3tt-2t$ in 8^gons^. $z=\dfrac{7tt-5t}{2}$ in 9^gons^. &c.

& z given t is found thus. $t=\dfrac{-1+\sqrt{1+8z}}{2}$ in tri. $t=\dfrac{\sqrt{0+16z}}{4}$ in 4^gō^. $t=\dfrac{+1+\sqrt{1+24z}}{6}$ in 5^gons^. $t=\dfrac{+2+\sqrt{4+32z}}{8}$ in 6^gons^. &c. As y^e^ side 12 of a tri given. y^e^ tri[:] $=z=\dfrac{12\times 12+12}{2}=78$ &c. & if $z=21$ be octangled.

$$t=\frac{+4+\sqrt{16+48z}}{12}=\frac{4+\sqrt{16+48\times 21}}{12}.\ t=\frac{4+\sqrt{1024}}{12}=3.^{(66)}$$

(66) In general, an m-gon has $z=\frac{1}{2}((m-2)t^2-(m-4)t)$ and conversely

$$t=\frac{(m-4)+\sqrt{\{(m-4)^2+8(m-2)z\}}}{2(m-2)}.$$

Newton, here at least, has not annotated the remaining 15 sections of Schooten's Liber v. As we shall see later (compare **2**, 7: note (151) below), there is good reason to believe that Newton did, in fact, study them thoroughly.

§4. LOOSE ANNOTATIONS ON HUYGENS' *DE RATIOCINIIS IN LUDO ALEÆ*

[Summer 1665?]

From the original(1) in the University Library, Cambridge

Reasonings concerning chance

1. (2)If p is y^e^ number of chances by one of wch I may gaine a, & q those by one of wch I may gaine b, & r those by one of which I may gaine c; soe y^t^ those chances are all equall & one of them must necessarily happen: My hopes or chance is worth $\frac{pa+qb+rc}{p+q+r}=A$. The same is true if p, q, r signify any proportion of chances for a, b, c.(3)

2. If I bargaine for more y^n^ one chance (viz: y^t^ after I have taken y^e^ gaines by my first chance, from the stake $a+b+c$; I will venter(4) another chance at y^e^ remaining stake, &c) my second lott is worth $A-\frac{AA}{a+b+c}=B$. My third lot is worth $A\frac{-AA-AB}{a+b+c}=C$. My Fourth lot is worth $A\frac{-AA-AB-AC}{a+b+c}=D$. My Fift lot is worth $A\frac{-AA-AB-AC-AD}{a+b+c}=E$. My sixt lot is worth

$$A-A\times\frac{A+B+C+D+E}{a+b+c}.\ \&c.^{(5)}$$

As if 6 men (1. 2. 3. 4. 5. 6.) cast a die so y^t^ he gaines a who throws a cise(6) first: since there is but one chance(7) to gaine a & 5 to gaine nothing at each cast, I make $b=0=c=r$. $p=1$ & $q=5$. Therefore by the [rule] The first mans lot is worth $\frac{a}{6}$. The 2^ds^ is worth $\frac{a}{6}-\frac{a}{36}=\frac{5a}{36}$. The Thirds is worth $\frac{5a}{36}-\frac{5a}{216}=\frac{25a}{216}$. The 4^ths^ is $\frac{25a}{216}-\frac{25a}{1296}=\frac{125a}{1296}$. The fifts lot is worth $\frac{125a}{1296}-\frac{125a}{7776}=\frac{625a}{7776}$. The

(1) Add. 4000: 90^r^–92^r^. Newton nowhere mentions Huygens' name explicitly, but it seems beyond reasonable doubt that the following notes were based on his reading of the *De Ratiociniis in Ludo Aleæ* as appended to Schooten's *Exercitationes*. In particular, Newton's title is a close translation of Huygens'. (See Appendix 2 above for bibliographical details.)

(2) Newton has cancelled a first draft: 'If the equall chances a, b, c, &c are such y^t^ one of y^m^ must necessarily happen, & y^t^ if one of y^e^ chances a happen I gaine p thereby, or q by one of y^e^ chan[c]es b, or r by one of y^e^ chances c. My chance or expectation is worth $\frac{pa+qb+rc}{a+b+c}$'.

(3) A slight generalization of Huygens' Prop. III (= *Exercitationes*: 523–4), which gives the expectation $\frac{pa+qb}{p+q}$ for two 'gaines' a and b, with respective 'chances' p and q.

sixts lot is $\frac{625a}{7776}-\frac{625a}{46656}=\frac{3125a}{46656}$. &c. Soe y[t] their lots are as 7776: 6480: 5400: 4500: 3750: 3125.[(8)]

Soe y[t] if I cast a die two or more times tis 1 to 5 y[t] I cast a cise at y[e] first cast & 11 to 25 y[t] I throw it at two casts, & 91 to 125 y[t] I cast it at thrice, & 671 to 625 y[t] I cast it once in 4 trialls, & 4651 to 3125 y[t] I cast it once in 5 times, &c.[(9)]

3. If I bargaine to cast severall sorts of lots successively at y[e] same stake y[e] valor of each lot is thus found viz: The first prop: gives y[e] valor of y[e] first lot: w[ch] valor being detracted[(10)] from y[e] stake, y[e] remainder is y[e] stake of y[e] 2[d] lot w[ch] therefore may bee also found by y[e] first prop: &c.

As if I gaine a by throwing 12 at y[e] first cast, or 11 at y[e] 2[d] or 10 at y[e] 3[d] &c wth two dice. Since at y[e] first cast there is but one chance for a (viz 12) & 35 for nothing Therefore its valor is $\frac{a}{36}$ (by Prop 1). & y[e] stake for y[e] 2[d] cast is $a-\frac{a}{36}=\frac{35a}{36}$. Now since there are two chances for it

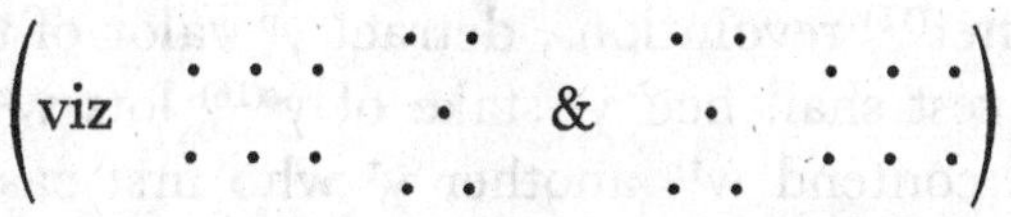

& 34 for 0 at y[e] 2[d] cast therefore its valor is $\frac{2\times 35a}{36\times 36}=\frac{35a}{648}$. & y[e] stake for y[e] 3[d] lot[(11)] is $\frac{595a}{648}$. for w[ch] there are 3 chances

& 33 for nothing. Therefore its valor is $\frac{595a}{7776}$.[(12)]

---

(4) Read 'venture'.

(5) Newton's explicit formulation of what is given verbally by Huygens or assumed in his numerical examples.

(6) A rendering of the French 'six'. The more common variant is 'sice'—compare 'cinque' for a five-spot.

(7) Newton has cancelled 'point'.

(8) That is, $6^5:6^4\times 5:6^3\times 5^2:6^2\times 5^3:6\times 5^4:5^5$.

(9) The chances are, respectively,

$$\tfrac{1}{6},\quad \tfrac{1}{6}+\tfrac{5}{36},\quad \tfrac{1}{6}+\tfrac{5}{36}+\tfrac{25}{216},\quad \tfrac{1}{6}+\tfrac{5}{36}+\tfrac{25}{216}+\tfrac{125}{1296},\quad \tfrac{1}{6}+\tfrac{5}{36}+\tfrac{25}{216}+\tfrac{125}{1296}+\tfrac{625}{7776}.$$

Newton here has adapted the results of Huygens' Prop. X = *Exercitationes*: 530.

(10) Read 'subtracted'.

(11) Newton has cancelled 'chance'.

(12) Newton uses Huygens' ideas but the details are his own.

4. If I bargaine wth one or two more to cast lots in order untill one of us by an assigned lott shall win y^e stake *a*: Since y^e chances may succede infinitely I onely consider ye first revolution of them. The valor of each mans whole expectation being in such proportion one to another as ye valors of their lots in one revolution. & y^e valor of each mans first lot to y^e valor of his whole expectation as y^e summe of y^e valors of their first lots to y^e stake *a*.

As if I contend wth another y^t who first throws 12 wth 2 dice shall have *a*, I haveing y^e dice.[(13)] My first lot is worth $\frac{a}{36}$ (by prop 1), The 2^d his first lot is worth $\frac{35a}{36\times 36}$. And $\frac{a}{36}:\frac{35a}{36\times 36}::36:35::$ my expectation : to his. for y^e two first lots make one revolution because I have y^e same lot If I throw a 2^d time y^t I had at y^e first. Therefore $\left(36+35=71:a::36:\frac{36a}{71}\right)$ $\frac{36a}{71}$ is my interest in y^e stake.

If o^r[(14)] bargaine bee soe y^t there is some lott at y^e beginning of o^r play wch returnes not in y^e after[(15)] revolutions, detract y^e valor of those irregular lotts from y^e stake & y^e rest shall bee y^e stake of y^e[(16)] lots wch follow & revolve successively. As if I contend wth another y^t who first casts 11 must have *a*, onely I have y^e first cast for 12. My first lot is worth $\frac{a}{36}$. & y^e stake for o^r after throws is $\frac{35a}{36}$. his first lot being $\frac{35a}{684}$. & my next lot $\frac{595a}{11664}$. soe y^t his share in y^e stake $\frac{35a}{36}$ is to mine as $\frac{35a}{648}:\frac{595a}{11664}::18:17$. Soe y^t my share in it is $\frac{17a}{36}$. To wch adding y^e valor of my first lot viz: $\frac{a}{36}$, y^e summe is $\frac{18a}{36}=\frac{a}{2}$, my interest in y^e stake *a* at y^e begining.[(17)]

5. If y^e Proportion of the chances for any stake bee irrationall the interest in the stake may bee found after y^e same manner. As if y^e Radij *ab*, *ac*, divide y^e horizontal circle *bcd* into two pts *abec* & *abdc* in such proportion as 2 to $\sqrt{5}$. And if a ball falling perpendicularly upon y^e center *a* doth tumble into y^e portion *abec*[(18)] I winn (*a*): but if into y^e other portion, I win *b*. my hopes is worth $\frac{2a+b\sqrt{5}}{2+\sqrt{5}}$.

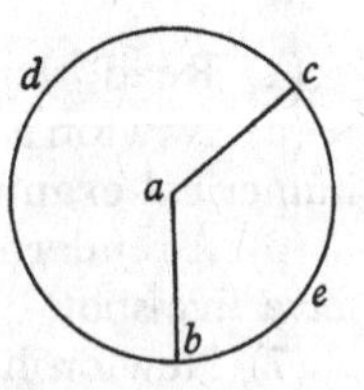

(13) And so first throw.
(14) Read 'our'.
(15) Newton has cancelled 'next'.
(16) Newton has cancelled 'regular'.
(17) This section has its base firmly in ideas derived from Huygens' tract, but the details are wholly original with Newton.
(18) Newton seems to have in mind a primitive form of roulette.

Soe if a die bee not a Regular body but a Parallelipipedon or otherwise unequall sided, it may bee found how much one cast is more easily gotten then another.(19)

☞ 6. Soe y^t ye facility of ye chances & ye stake belonging to each chance being knowne ye worth of the lott may bee ever found by ye precedent precepts. And if they bee not both immediatly known(20) they must bee sought before ye valor of ye lott can bee found.

As if(21) I want two games at Irish(22) & my adversary three to win a, & I would know(23) my interest in y^e stake (a.) my first lot can gaine me nothing but y^e advantage of another lot, & therefore to know its vallue I must first find y^e value of y^t other lot &c. First therefore if wee each wanted one lot to win a o^{r}(24) interest in it would bee equall viz my lot worth $\frac{a}{2}$. 2dly If I want one game & my adversary two, & I gaine y^e next game y^n I gaine a but if I loose it I onely gaine an equall lot for a at y^e next game wch is worth $\frac{1}{2}a$, Therefore my interest in y^e stake is $\frac{a+\frac{1}{2}a}{2}=\frac{3a}{4}$. 3dly If I want one game & my adversary three & I gaine y^e next game I get a; but if I loose it, then I want one game & my Adversary but two, y^t is I get $\frac{3a}{4}$: Therefore (there being one chance for a & one for $\frac{3a}{4}$.) my interest in y^e stake is $\frac{a+\frac{3a}{4}}{2}=\frac{7a}{8}$. 4^{thly} If I want 2 games & my adversary 3; & I win I get $\frac{7a}{8}$. but if I loose I get $\frac{1}{2}a$ for o^r chances will then bee equall; Therefore my interest in y^e stake is $\frac{11a}{16}$. Soe if(25) I want 1 game & my adversary 4 my interest in a is $\frac{15a}{16}$. If(26) I want two & hee 4, it is $\frac{13a}{16}$. If I want 3 & hee 4 it is $\frac{21a}{32}$. If I 1 & hee 5 it is $\frac{31a}{32}$. If I 2 & hee 5 it is $\frac{57a}{64}$. If I 3 & hee 5 it is

(19) The whole of this section seems original with Newton. In the last paragraph he clearly opts for a frequency theory of probability—that is, the absolute probabilities are not given *a priori* but are to be determined as the asymptotic limit of the numerical probabilities observed over a succession of occurrences of a state.

(20) 'found' is cancelled.

(21) Huygens' Prop. VI = *Exercitationes*: 526, slightly adapted.

(22) A game of dice, popular at the time and much like backgammon. Charles Cotton in his *The Compleat Gamester* (London, 1674: ch. XXVII, 'Of Irish') relates that 'Irish is an ingenious game, and requires a great deal of skill to play it well, especially the After-game'.

(23) Newton has cancelled 'y^e value of'.

(24) Read 'our'.

(25) The second part of Huygens' Prop. V = *Exercitationes*: 526.

(26) Huygens' Prop. VII = *Exercitationes*: 526.

$\frac{99}{128}a$. If I 4 & hee 5, it is $\frac{163}{256}a$.(27) (The like may bee done if 3 or more play together. as if(28) one wants one game, another 3 [&] a third 4: Their lots are as 616:82:31.(29) &c. As also if their lots bee of divers sorts.)(30)

By this meanes also some of y^e^ precedent questions may bee resolved. as if I have two throws for a cise to win a, wth one die; If I have missed my first lot alredy I have at my second cast five chances for nothing & one for a. therefore y^t^ cast is worth $\frac{a}{6}$. Soe y^t^ in my first cast I had five chances for $\frac{1}{6}a$ & one for a, wch therefore (wth my 2^d^ cast) is worth $\frac{11}{36}a$. That is tis 11 to 25 y^t^ I cast a cise once in two throws. as before.

By this meanes also my lot may bee known if I am to draw 4 cards of severall sorts out of 40 cards, 10 of each sort.

Or if out of two white & 3 black stones I am blindfold to chose a white & [a] black one.(31)

(27) These are Newton's results, in amplification of Huygens'.

(28) Huygens' Prop. IX = *Exercitationes*: 527–9.

(29) The tenth entry in the table of Prop. IX.

(30) For historical details on this famous *problème des partis* consult Huygens' *Œuvres complètes*, **14** (1920): 21–5.

(31) These last three paragraphs are original with Newton.

It is, perhaps, not unfair to point out that Newton, for all his excellent theoretical knowledge and practical interest, was not markedly successful in games of chance: an entry in the Fitzwilliam notebook in 1667 records 'Lost at cards twice—0. 15. 0'. He seems eventually to have learnt his lesson, for Conduitt, shortly after Newton's death in March 1727, wrote in one of his little green notebooks (Keynes MS 130.6 (2): [5^v^]) that he 'had not played at cards in 40 years'. At games of skill Newton fared much better, or so Conduitt reassures us. In an immediately preceding entry he reports, on the word of his wife and Newton's niece, Catherine, that 'When he was young and first at university he played at drafts & if any gave him first move [he was] sure to beat them'. (Instead of 'drafts' Conduitt first wrote 'chess' but then cancelled it.)

2

ANNOTATIONS FROM VIÈTE & OUGHTRED

[Late 1664?]

§1. NOTES ON VIÈTE'S *DE NUMEROSA POTESTATUM AD EXEGESIN RESOLUTIONE*(1)

From the original(2) in the University Library, Cambridge

[1] *Of ye Extraction of Pure Square. Cubick. Squaresquare & squarecubick rootes &c.*(3)

Let ye number whose roote is to bee extracted bee pointed makeing ye first point under ye unite & comprizeing soe many numbers under each point as ye number hath dimensions. as if ye number be square-cube tis thus pointed

57086352410802.

Then out of ye figures of ye first point next ye left hand extract ye greatest roote proper to ye power of ye number & set yt downe in ye Quotient wch is ye fir[s]t side & is called A. (as ye roote quintuplicate of 5708 is (5), & (5) quintuplicate is 3125.) yn takeing yt roote duely multiplied out of ye number (as 3125 out of 5708) wth ye rest of ye numbers to ye next point. seeke ye seacond side wch is found by divideing yt number by another number made out of ye first side (wch is called ye Divisor) & this second side I name E. (thus by divideing 258363524 by $5Aqq+10Ac+10Aq+5A$ after such a maner yt

$$5AqqE+10AcEq+10AqEc+4AEqq+Eqc$$

(1) Viète first published the tract as *De numerosa Potestatum ad Exegesim Resolutione: Ex Opere restitutæ Mathematicæ Analyseos, seu Algebra nova* (Paris, 1600). In Schooten's collected edition (*Francisci Vietæ Opera Mathematica, in unum Volumen congesta, ac recognita* (Leyden, 1646): 163–228) it is retitled *De numerosa Potestatum purarum, atque adfectarum ad Exegesin Resolutione Tractatus*. Oughtred, in editions of his *Clavis Mathematicæ* from 1647 (and in particular in the third Latin edition of 1652 which Newton read) included a revised summary of the work, the first part being introduced into the body of the *Clavis* itself (= *Clavis*: 42–5: Cap. XIV, *Sequitur ANALYSIS: quæ est eductio radicis ex numerosa potestate data.*) with the second added as an appendix (*Clavis*: 110–43/144–51: *De Æquationum Affectarum Resolutione in Numeris.* with added *Notæ*). Apparently, Newton has annotated both Viète and Oughtred together, taking the structure of his argument from the former, but making wide use of Oughtred's simplified notations.

(2) Add. 4000: 2^r–3^r, 4^r–6^r.

(3) *Vieta*: 163–72: *De numerosa Potestatum Purarum Resolutione.*

may be conteined in y^e number y^e product of y^t division shall be

$E=$ [ )][4]

The extraction of y^e sqare roote.[5]

The square to be resolved	29	16	(54 The Product.
The square of y^e first side to	25		be taken away.
The rest of y^e sqare to be	4	16	resolved.
The divisor for finding y^e seacond	1	0	side.[6] wch is y^e first side doubled.
The rectangle by 2*A* & *E*	4	0	} to be substracted.
The square of *E*		16	
The sum̃e of y^e rectangles	4	16	to be subducted.
	0	00	The remainder.

The extraction of y^e cube roote.[7]

The cube to be resolved	157	464	(54
The cube to be subducted	125		whose roote is $A=5$
The remainder for y^e finding	32	464	of *E*
The divisors for y^e finding	7	5	3*Aq*
of (*E*) y^e seacond side		15	3*A*
The sume of y^e divisors	7	65	
Sollids to be substracted	30	0	3*AqE*
	2	40	3*AEq*
		64	*Ec*
The sume of those	32	464	sollids
The remainder	00	000.	

(4) Adapted from Viète's introduction = *Vieta*: 163–4. Viète finds the real root of $N = x^p = (A+E)^p$: where $A = [N^{\frac{1}{p}}]$, $N-A^p = (A+E)^p - A^p$ and this he takes approximately as $E \times ((A+1)^p - A^p - 1)$, naming $(A+1)^p - A^p - 1$ the 'Divisor'. The first approximation $E \approx \dfrac{N-A^p}{(A+1)^p - A^p - 1}$ follows immediately and successive repetition of the operation gives narrowing bounds to the root.

The idea of 'pointing' the figures on their underside is taken over from Oughtred, along with the modified cossic notation (*A*, *E* for algebraic variables with their second, third, fourth, fifth, ... powers represented by adjoining *l*, *q*, *c*, *qq*, *qc*, ...).

(5) *Vieta*: 166: *Paradigma analyseos quadrati puri.*

(6) Newton wrote 'sidie' here in error.

(7) *Vieta*: 167–8: *Paradigma analyseos cubi puri.*

(8) *Vieta*: 169: *Paradigma analyseos quadrato-quadrati puri.*

The extraction of y^e square square roote.(8)

The square-square	33	1776	(24
The square-squ: to be subdūc:	16		$=Aqq$
Remainder	17	1776	
Divisors for finding y^e seacond side E	3	2	$4Ac$
		24	$6Aq$
		8	$4A$
Theire sum̄e	3	448.	
Squ-squares to be subducted	12	8	$4AcE$
	3	84	$6AqEq$
		512	$4AEc$
		256	Eqq
Theire Sum̄e	17	1776	

The Extraction of y^e Square-Cube roote.(9)

The squ: cube to be resolved	79	62624	(24
Substract	32		Aqc
Remaind^r	47	62624	
Divisors	8	0	$5\,Aqq$
		80	$10Ac$
		40	$10Aq$
		10	$5A$
The Sume of y^e divisors	8	8410	
Plano-sollids to be substracted	32	0	$5AqqE$
	12	80	$10AcEq$
	2	560	$10AqEc$
		2560	$5AEqq$
		1024	$Eqc.$
Theire Sum̄e	47	62624	
Remainder	00	00000	

Note y^t y^e 3^d 4th 5th & other figures are found by y^e same manner y^t y^e seacond figure is found. Onely makeing all y^e figures found to stand for A y^e first side & y^e figure sought for E or y^e 2^d side.

And if [y^e] roote(10) is found inexpressible in whole numbers y^n [proceed] adding ciphers & pointing them from y^e Unite towards y^e right=hand as was before explained & soe holde on y^e worke in decimalls.

(9) *Vieta*: 170–1: *Paradigma analyseos quadrato-cubi puri.*

(10) Newton has cancelled 'y^e number propounded'.

As for y^{e} Divisors they are easily found by y^{e} 2^{d} Table of Powers from a Binomiall roote.[11]

If y^{e} number bee of 6. 7. 8. 9. 10 &c dimensions The roote may be extracted after ye same manner.

[2] *Of ye Extraction of Rootes in Affected powers.*[12]

The manner of ye extraction of rootes in pure & affected powers is verry much alike especially when ye affected powers are decently prepared, y^{t} is, when theire affections are not over large[13] & those altogether either affirmative or negative, & ye power affirmative, affirmations & negations so mixt y^{t} there be noe ambiguity & all fractions & Asymmetry taken away.[14]

All ye figures in ye coeff[ici]ents & affected power[s] are to be pointed (after ye manner before e[x]plained in ye Analisis of pure powers) according to ye degree of theire dimensions & the work onely differs from y^{t} in pure powers in y^{t} ye coefficients enter into y^{e} divisors.

Let y^{e} first side be called *A*. y^{e} 2^{d} be called *E*. y^{e} Roote of y^{e} equation *L*. y^{e} cöefficients *B*. *Cq*. *Dc*. *Fqq*. *Gqc*. *Hcc* &c. y^{e} Power *P*. *Pq*. *Pc*. *Pqq* &c & y^{e} Operation follows.[15]

The analysis of Cubick Equations.

The equation supposed $Lc * + 30L = 14356197$. $Lc + CqL = Pc$.[16]

The square coëfficient		3	0	
The cube affected to be	14	356	197	(243
Sollids to be substracted	8			$=Ac$
		6	0	$=ACq$
Theire sum̄e	8	006	0	[17]
Rests	6	350	197	for finding y^{e} 2^{d} side.
The extraction of y^{e} seacond side				
Coëfficient			30	or superior divisor.
The rest of y^{e} cube to be	6	350	197	resolved
The inferior divisors	1	2		$3Aq$
		6		
Their sum̄e	1	260	30	[18]
Sollids to be subtracted	4	8		$=3AqE$
		96		$=3AEq$
		64		$= Ec$
		1	20	$=ECq$
Their sum̄e	5	825	20[19]	

(11) Compare Oughtred's *Clavis*: 45: 'Si numerus propositus non sit verus sui generis figuratus, sed peracta Analysi aliquid restet: punctationes...pro suo genere, quot opus erit statuendæ sunt: & continuanda Analysis post lineam separatricem.' The '2^{d} Table of Powers'

[The extraction of ye 3d	side]		
The superior part of ye divisor		30	or ye square coefficient
The remainder for finding	524	997	ye third side(20)
The inferior part of ye divisor	172	8	$3Aq$ that is $3 \times 24 \times 24$
		72	$3A$ or 3×24
The sume of ye divisors	173	550(21)	
Sollids to be taken away	518	4	$3AqE$
	6	48	$3AEq$
		27	Ec
		90	Ecq
Theire sume	524	997(22)	(23)
Remaines	000	000	

is Oughtred's 'tabula potestatum ascendentium in scala à radice binomia: quæ POSTERIOR vocetur' (= *Clavis*: 37).

(12) *Vieta*: 173–228: *De numerosa Potestatum Adfectarum Resolutione.*

(13) That is, when the number of terms in the equation is reduced to a minimum.

(14) Almost exactly a translation of *Vieta*: 173: 'Numerosam resolutionem potestatum purarum imitatur proxime resolutio adfectarum potestatum, præsertim cum potestates adfectæ decenter præparatæ fuerint. Tunc autem decenter præparari intelliguntur, cum parcissime fuerint adfectionibus obrutæ, iisque omnino adfirmatis aut negatis omnino, ita tamen ut potestas adfirmata sit, non etiam ab homogenea vel homogeneis gradu insignitis avellatur, ac denique mixtim ita negatis & adfirmatis, ut non insit ambiguitas.'

(15) As Newton (following Viète) says, everything carries over smoothly. The extended problem is now to resolve $N = x^p$ 'affected' with 'homogeneous powers', that is,

$$N = x^p + a_1 x^{p-1} + a_2 x^{p-2} + \ldots + a_{n-1} x, \quad \text{or} \quad N = f(x) = f(A+E),$$

where A is a near approximation to a 'true' root $x = A+E$. Viète takes as his divisor

$$g(A) = f(A+1) - f(A) - 1$$

(where the unit is omitted probably because it is too small to take into account). He then takes approximately $E \times g(A) \approx N$ [or $f(A+E)$] $- f(A)$, so that $E \approx \dfrac{N-f(A)}{g(A)}$ is a narrower bound to E.

Note further that the successively nearer approximations to $N, f(A_i)$, where $A_i = A + \sum_i (E_i)$ and $E = \lim_{i\to\infty} \sum_i (E_i)$, are not calculated outright each time, but that $f(A_i) - f(A_{i-1})$ is first expanded as a polynomial in A_{i-1} and E_i with $f(A_i) = f(A_{i-1} + E_i)$ calculated as

$$f(A_{i-1}) + [f(A_i) - f(A_{i-1})].$$

(Correspondingly, at each stage E_i is taken approximately as $[N - f(A_{i-1})]/g(A_{i-1})$.)

(16) *Vieta*: 177–8: *Paradigma* (for $x^3 + C^2 x = N$).

(17) $f(200)$, where $f(x) = x^3 + 30x$.

(18) $10 \times g(200)$, where the 'divisor' $g(200) = f(201) - f(200) - 1$.

(19) The first approximation E_1 (to E) is taken $E_1 \approx [N - f(200)]/g(200)$, or $E_1 = 40$, with $f(240) - f(200) = 5825200$.

(20) $[N - f(200)] - [f(240) - f(200)] = N - f(240)$.

(21) $g(240)$. It follows that $E_2 \approx [N - f(240)]/g(240)$, or $E_2 = 3$.

(22) $f(243) - f(240)$.

(23) $0 = [N - f(240)] - [f(243) - f(240)]$, or $N = f(243)$.

But yᵉ Coëfficient may be greater yⁿ yᵉ Power soe yᵗ it cannot be substracted from it wᶜʰ argues yᵗ yᵉ Cube more propperly affects yⁿ is affected. In this case yᵉ coëfficient must descend towards yᵉ unite soe many points untill it may be substracted, & soe many points as yᵉ coëfficient is devolved soe many pricks must be blotted out towards yᵉ left hand in yᵉ power affected. As yᵉ example shews.

$Lc+95400L=1819459.$[24]

	9	540	0	Coefficiens
			· · ·	
	1	819	459	The Power
	·	·	·	
Since 9 is greater yⁿ 1 I make a devolution thus.				
		954	00	The Coefficient
			·	
The Quote (19	1	819	459	The affected power
	[·]	·	·	
Sollids to be substracted		954	00	*ACq*
		1		*Ac*
Sum̃a		955	00	substrahenda
Divisorū superior pars		95	400	Coefficiens Planum
			·	
		864	459	Potestas reliqua
Divisorū pars inferior			3	*3Aq*
			3	*3A*
Divisorū sum̃a		95	730	
Sollida ablativa		858	600	*ECq*
		2	7	*3AqE*
		2	43	*3AEq*
			729	*Ec*
Eorū summa		864	459	
Restat		000	000[25]	

To place yᵉ unite of yᵉ coefficient in its right place in respect of yᵉ power make so many pricks above as there are under yᵉ power beginning at yᵉ unit. & if yᵉ coefficient be one dimension lesse yⁿ yᵉ power make a prick on every figure. if 2 dimensions les[26] yⁿ on every other figure. if 3 dimensions lesse make it one each third figure &c.

If there be many coefficients in yᵉ equation each must be placed according to this rule.

(24) $f(x) = x^3+C^2x = N$ with $C^2 = 95{,}400$, $N = 1{,}819{,}459$ (= *Vieta*: 179: *Paradigma*).

(25) $A = 10$, $E_1 = 9$ with $1{,}819{,}459 = f(A+E_1)$. (Note that Newton begins to relapse into Viète's original Latin phrasings.)

Some times y^e^ coeff[ic]ient is under a negative sine(27) as $Lc-10L=13584$.(28) & y^e^ Analysis is as follows.

Coëfficiens planum	−	10	sublaterale(29)
Cubus resolvendus	+13	584	(24
Solida ablativa	{+ 8		*Ac*
	−	20	*ACq*
Suma	+ 7	80	
Restat	+ 5 [.]	784	resolvendum
Divisorū ꝑs superior		−10	coëfficiens planum
Divisorū ꝑs inferior	{+ 1	2	$+3AA$
	+	6	$+3A$
Suma divisorū	+ 1	25	
Solida ablativa	{+ 4	8	$3AAE$
	+	96	$3AEE$
	+	64	EEE
		−40	EEC (30)
Eorū suma	+ 5	784	

But sometimes y^e^ square coëfficient hath more paires of figures y^n^ y^e^ cube to be analysed hath threes, & y^n^ præfixing so many ciphers to y^e^ cube as figures are wanting, y^e^ first side will not much differ from y^e^ square roote of y^e^ coefficient. as $Lc-116620L=352947$.(31)

	−11	662	0	Coefficiens planū
Cubus resolvendus	00	352	947	(343
Sollida Ablativa	{+27			*Ac*
	−34	986		*ACq*
Restat auferendū	− 7	986		
Reliquum resolvendi	+ 8	338	947	Cubi(32)

(26) Read 'less'.

(27) Read 'signe'.

(28) $f(x) = x^3+C^2x = N$ with $C^2 = -10$, $N = 13,584$ (= *Vieta*: 198–9: *Paradigma*).

(29) 'The square sublateral coefficient', that is C^2, the coefficient of x.

(30) $A = 20$, $E_1 = 4$ with $13,584 = f(A+E_1)$.

(31) $f(x) = x^3-116,620x = 352,947$ (= *Vieta*: 199–200: *Paradigma*).

(32) Newton here reached the bottom of a page in his notebook and repeats this entry two lines further on (on his own next page).

Divisorū ꝑs superior	− 1	166	20	Coeff: planum
Reliquū resolvendi cubi	+ 8	338	947	negative affecti
Divisorū ꝑs inferior	+ 2	7		$3AA$
	+	9		$3A$
Sumā Divisorum	+ 1	623	8.	$3AA+3A+Cq$
Sollida ablativa	+10	8		$3AAE$
	+ 1	44		$3AEE$
	+	64		EEE
	− 4	664	8	CCE
Eorum summa	+ 7	639	2	
Restat Resolvend[ū]	+	699	747	pro 3° latere
Divisorū ꝑs superior	−	116	620	CC
Divisorū ꝑs inferior	+	346	8	$3AA$
		+1	02	$3A$
Eorum summa		231	200	$=3AA+3A+CC$
Sollida Ablativa	+ 1	040	4	$3AAE$
	+	9	18	$3AEE$
	+		27	EEE
		−349	860	ECC
Eorum summa		+699	747	(33)

Sometimes though there be as many 2 figures in y[e] coefficient as 3 figures in y[e] Cube affected yet y[e] coëfficient may be so greate as to deceive an u[n]wary Analist. As in this $Lc-6400L=153000$.(34) where y[e] roote of 64 is 8 w[ch] cubed is 512 w[ch] added to 153 makes 665[,] thē(35) whose [cube] roote y[e] number immediately greater is 9 w[ch] is y[e] first sid$=A$.

But if y[e] coefficient had beene affirmative, y[n] not y[e] aggregate of y[e] facts but y[e] difference must be taken as in this. $Lc+64L=1024$.(36) Since y[e] roote of 64 is 8. w[ch] cubed is 512. & $1024-512=512$. y[e] roote of w[ch] is $8=A$. The like is observable in equations of higher powers.(37)

If y[e] Cube be Affected wth a negative sine as $13{,}104L-Lc=155{,}520$.(38) Then y[e] equation is expressible of 2 rootes: whereof y[e] square of one is lesse &

(33) $A = 300$, $E_1 = 40(A_1 = 340)$, $E_2 = 3$ with $352{,}947 = f(343)$.
(34) $x^3-6400x = 153{,}000$.
(35) Read 'then' (Newton's form of the modern 'than').
(36) $x^3+64x = 1024$.
(37) A rough translation of *Vieta*: 200–1 = *De numerosa Potestatum adfectarum Resolutione:* Prob. XI (conclusion).
(38) $13{,}104x-x^3 = 155{,}520$ (*Vieta*: 214–15: *Paradigma*).

y^e square of y^e other is greater then $\frac{13104}{3}$. & therefore one roote is lesse y^e other greater then $\frac{155520}{13104}$. & in this equation $27755L - Lqq = 217944$[39] are two rootes whereof one is greater y^e other lesse then $\frac{217944}{27795}$.

☞ Suppose in y^e former cubick equation y^e lesse roote be 12. y^n $\frac{155520}{12} = 12960$. or else $13104 - 12 \times 12 = 12960$. & $Lq + 12L = 12960$. where $L = 108$ is y^e greater roote.[40]

And in y^e latter equation if y^e greater roote be 27. & $\frac{217944}{27} = 8072$, C[41]. or $-27 \times 27 \times 27 + 27755 = 8072$. $27 \times 27 = 729$.[42]

If there be 4 cubes continu[a]lly proportionall whose greate extreame is $27c$[43] $= 19683$. & y^e aggregate of y^e 3 rest is 8072 & Lc y^e lesse extreame, therefore $Lc + 27Lq + 729L = 8072$. y^e roote of wch is 8 y^e other roote of y^e equation.[44]

☞ Or haveing one roote of an equation y^e equation may be lessoned[45] by division thus. $13104l - lc = 155520$. or $l^3 - 13104l + 155520 = 0$. & one roote is 12. therefore divide this equation by l–12 & y^e Quote is an equation conte[in]-ing y^e other roote viz: $lq + 12l = 12960$.[46]

---

(39) $27{,}755x - x^4 = 217{,}944$ (*Vieta*: 219–20: *Paradigma*).

(40) $0 = x^3 - 13{,}104x + 155{,}520 = (x-12)(x^2 + 12x - 12{,}960)$, where

$$0 = x^2 + 12x - 12{,}960 = (x-108)(x+120).$$

(41) In the reduced cubic $Lc + 27Lq + 729L = C$.

(42) Newton calculates the coefficients of the cubic which results when the factor $L - 27 = 0$ is taken out: specifically

$$0 = x^4 - 27{,}755x + 217{,}944 = (x-27)(x^3 + 27x^2 + 729x - 8072).$$

(43) Read '27 cubed'.

(44) $0 = x^3 + 27x^2 + 729x - 8072 = (x-8)(x^2 + 35x + 1009)$, where $x^2 + 35x + 1009 = 0$ has the complex ('imaginary') roots $x = \frac{1}{2}(-35 \pm \sqrt{-2811})$.

(45) Read 'lessened'.

(46) $0 = x^3 - 13{,}104x + 155{,}520 = (x-12)(x^2 + 12x - 12{,}960)$, as before. Note Newton's sudden jump from Oughtred's capital L (inspired from Viète's cossic notation) to a Cartesian lower-case l, and, similarly, his preference for the Cartesian power-notation l^3 over the older lc (Lc).

§2. NOTES ON VIÈTE'S GEOMETRICAL PROPOSITIONS[1]

Propositiones Geometricæ Franc: Vietæ[2]

[1][3]

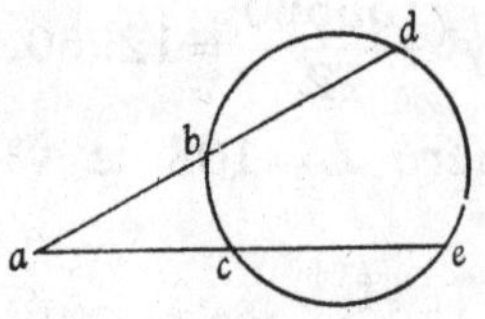

prop 1

$ab:ac::ce:bd.$[4]

prop 2

& if $ab:ac::ac:bd.$ then $ac:ab::ab:ce.$[5]

prop 3

If $ab\times ac=bd\times ce.$ then $bd:ac::ac:ab::ab:ce$ ∻[6]

prop 3['].

To find two me[a]ne proportionalls twixt *BC* & *IK*.[7]

On y^e^ center *a* wth y^e^ Rad *ai* describe y^e^ circle *ibck*. inscribe $bc=cd$. draw *da* through y^e^ center & *bg* parallel to it. draw *hk* through *A* soe y^t^

$gh=ab(=ai).$

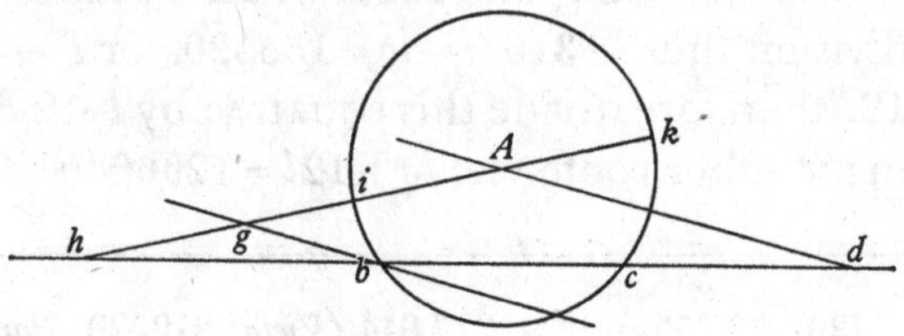

& [y^n^] $ik:hb::hb:hi::hi:bc.$ ∻[8]

To find two meane proportionalls twixt *BC* & *IK*.

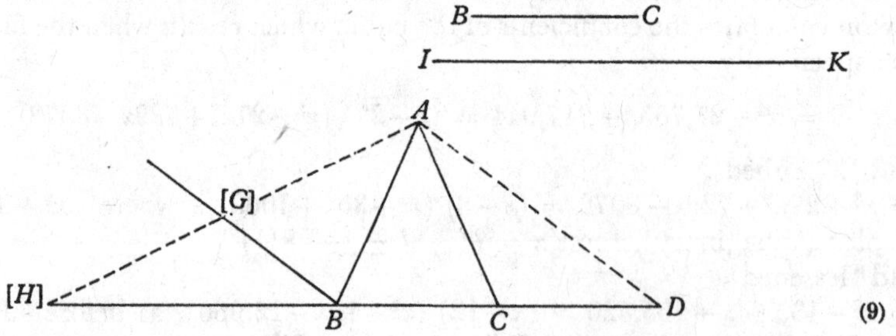

(9)

(1) Add. 4000: 8^r^–11^v^. This second part is a set of notes on miscellaneous geometrical and trigonometrical theorems in Schooten's *Vieta*.

(2) 'Geometrical propositions of François Viète.'

(3) Annotations on *Vieta*: 240–57 = *Supplementum Geometriæ* (reprinted by Schooten from Viète's *Supplementum Geometriæ. Ex Opere restitutæ Mathematicæ Analyseos, seu Algebra nova* (Tours, 1593)).

(4) Read '*ae*:*ad*', mistranscribed from *Vieta*: 240–1: Prop. I (where the points are lettered differently).

Prop: 4

If $ad = db = cb$. y[n] y[e] Angle *cbe* is tripple to y[e] Angle *abd*.(10)

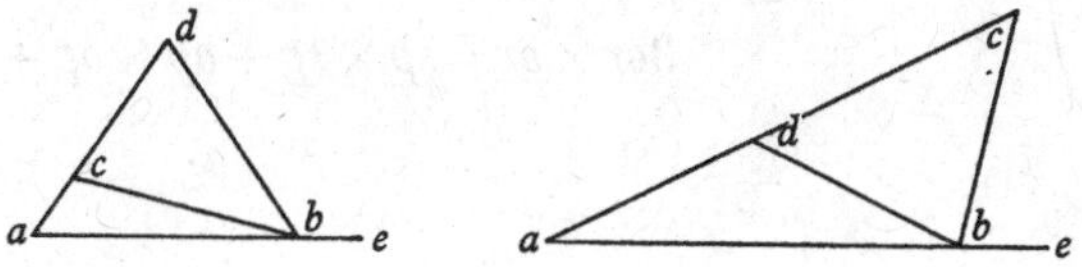

Prop 5

If $ab = bd =$ Rad.(11) [y[n]] 3 Ang: $bad = cde$.(12)

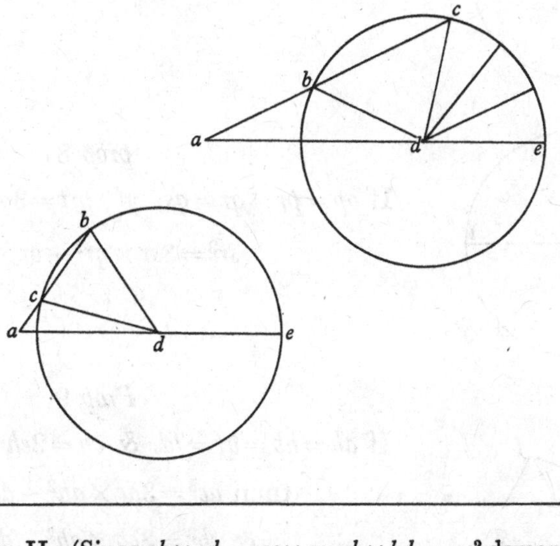

---

(5) *Vieta*: 241: Prop. II. (Since $ab \times ad = ac \times ae$, $ab \times bd = ac^2$ demands that $ad : bd = ae : ac$, or $ab : bd = ce : ac$, and so $ab^2 : ab \times bd = ac \times ce : ac^2$.)

(6) *Vieta*: 242: Prop. III/IV, the converse of Prop. 2 (with the implicit condition that b, c, e, d are concyclic). The proof is immediate, since $ab \times ad = ac \times ae$, or

$$ab \times ad \times bd = ac(ac \times bd + bd \times ce) = ac^2(bd + ab),$$

so that $ab \times bd = ac^2$. (The symbol '$\because\!\!\because$' for continued proportion is taken from Oughtred.)

(7) Read '*bc* & *ik*'. (Newton relapses temporarily into Viète's capitals.)

(8) *Vieta*: 242–3: Prop. V. In proof,

$$hi \text{ (or } ga) + hb = hg \text{ (or } ia) \times bd \text{ (or } 2bc) = (2ia \text{ or) } ik \times bc,$$

so that the conditions of Prop. 3 are satisfied.

(9) This Newton added on the facing page (in a slightly later hand) and then cancelled. Viète's construction is simplified to the finding of H in BC such that $GH = BC$, where G, H are collinear with A, D is in BC with $BC = CD$, and GB is parallel to AD with $AH = IK$. It follows that HB is the first of the two means between IK and BC.

(10) *Vieta*: 245: Prop. VIII. ($\widehat{abd} = \widehat{bad}$, $\widehat{cdb} = \widehat{abd} + \widehat{bad} = \widehat{dcb}$ with $\widehat{cbe} = \widehat{dcb} + \widehat{dab}$.)

(11) Read 'Radius' (that is $bd = cd$).

(12) *Vieta*: 245–6: Prop. IX (identical in structure with the preceding Prop. 4).

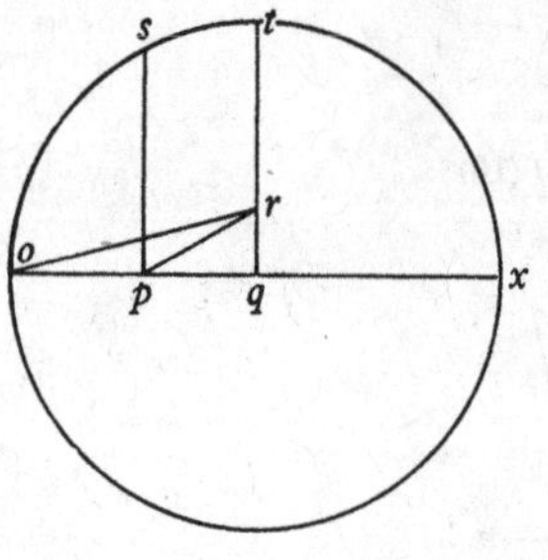

Prop 6

If $3rpq = spq$: recto.[13] that is If $2qr = pr$. then

$$3or \times or = sp \times sp + op \times op + px \times px.^{(14)}$$

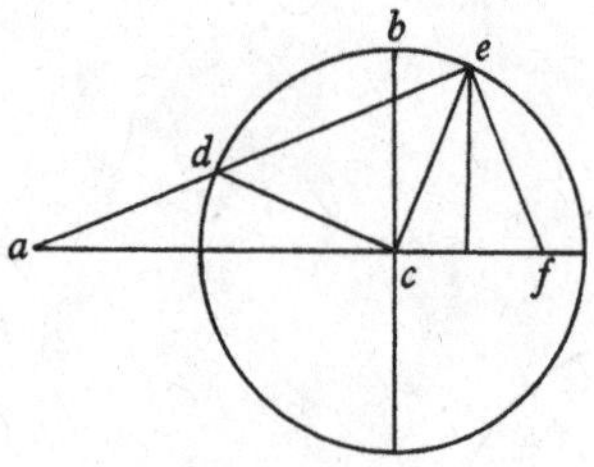

Prop 7

If $ad = dc = ce = ef$. then $ecf = efc = 3dac = 3dca$.

& $[ac]^3 = 3[ac] \times ad^2 + cf \times ad^2$.

$z^3 = aaz + b^3$.[15]

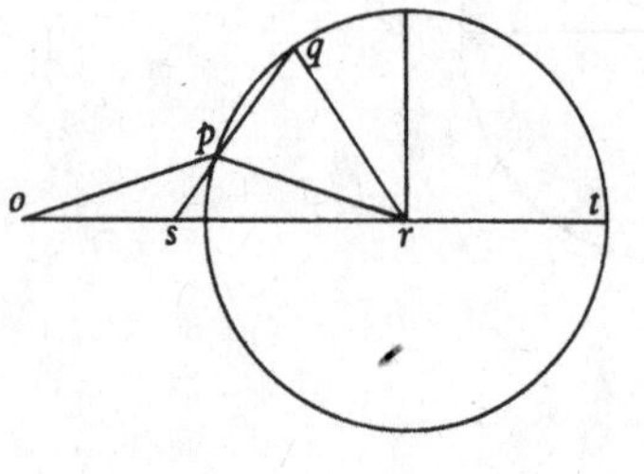

prop 8

If $op = pr = qr = qs$. y^n^, $prt = 3qsr$ &c and

$sr^3 = 3sr \times qr^2 - or \times qr^2$.

$z^3 = aaz - b^3$[16]

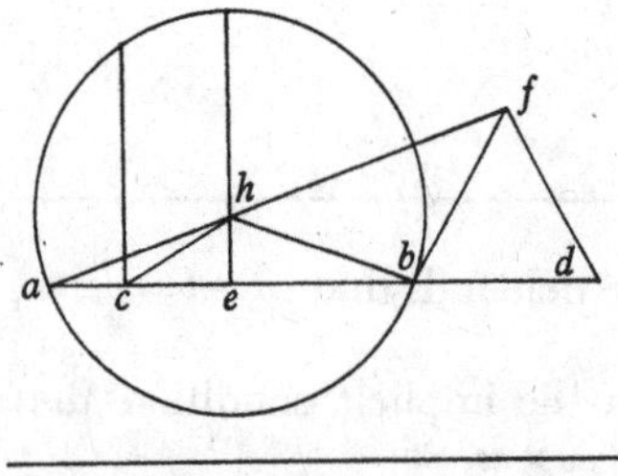

Prop 9

If $ah = hb = bf = fd$. & $ch = 2eh$ or $ceh = 3hce$.

then $ac^3 = 3ac \times ah^2 - db \times ah^2$.
& $cb^3 = 3cb \times ah^2 - db \times ah^2$ }

$z^3 = aaz - b^3$.[17]

(13) Read 'a right angle'.

(14) *Vieta*: 248: Prop. XV. In proof note that

$$sp^2 \text{ (or } oq^2 - pq^2) + op^2 \text{ (or } (oq - pq)^2) + px^2 \text{ (or } (oq + pq)^2) = 3oq^2 + pq^2 \text{ (or } 3qr^2) = 3or^2.$$

(15) *Vieta*: 248–9: Prop. XVI. In proof,

$$\frac{\frac{1}{2}ac}{ad} = \frac{\frac{1}{2}(ad+ae)}{ac} = \frac{ac + \frac{1}{2}cf}{ae} = \cos \widehat{dac} = \frac{\frac{3}{2}ac + \frac{1}{2}cf}{ad+ae},$$

or
$$ac^2 = ad(ad+ae) = \frac{ad}{ac} \times ad(3ac+cf).$$

In the concluding squared off line Newton, following Viète, adds the analytical representation: where $ad = a$, $ac = z$ and $b^3 = a^2 \times cf$ (with, say, $\widehat{dac} = \frac{1}{2}\theta$ and $CS(\theta) = 2\cos\frac{1}{2}\theta$, the complementary subtense), then $ac/ad = CS(\theta)$, $cf/ad = CS(3\theta)$ and so Viète's result is

$$[CS(\theta)]^3 = 3CS(\theta) + CS(3\theta),$$

where the magnitude of $3\theta = \widehat{ecf}$ is supposed given and the trisection $\widehat{eaf}$ is to be found.

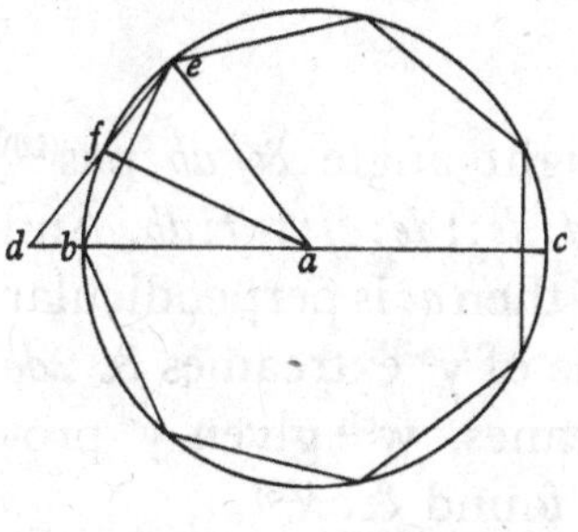

prop 10

If $de = ea$ & $db : da :: ab \times ab : dc \times dc$. y[n] be is a side of a 7 equall sided & angled figure. or $7eab = 4$ right angles.(18)

(16) *Vieta*: 250–1: Prop. XVII. Here

$$\cos \widehat{qsr} = \frac{\frac{1}{2}sr}{qs} = \frac{qs-\frac{1}{2}qp}{sr} = \frac{sr-\frac{1}{2}or}{qs-qp} = \frac{\frac{3}{2}sr-\frac{1}{2}or}{2qs-qp},$$

or

$$sr^2 = qs(2qs-qp) = \frac{qs}{sr}(3sr-or).$$

In the analytical equivalent, $sr = z$, $qs = qr = pr = a$, and so (where $\widehat{qsr} = \frac{1}{2}\theta$)

$$\frac{\frac{1}{2}sr}{qs} = CS(\theta), \quad -\frac{\frac{1}{2}or}{op \text{ (or } qs)} = CS(3\theta)$$

and $[CS(\theta)]^3 = 3CS(\theta)+CS(3\theta)$, as before (where $\widehat{opr} = 3\theta$, taken obtuse, is to be trisected).

(17) *Vieta*: 251: Prop. XVIII. For

$$3ah^2 = ab^2 - ac\times cb \quad \text{(by Prop. 6)}$$

and

$$ab^3 = (3ab+bd)\,ah^2 \quad \text{(by Prop. 7)},$$

so that

$$ab^3+ac^3 = (ab+ac)\,3ah^2 = 3ac\times ah^2+ab^3-bd\times ah^2,$$

and

$$ab^3+cb^3 = (ab+cb)\,3ah^2 = 3cb\times ah^2+ab^3-bd\times ah^2.$$

(18) *Vieta*: 255–6: Prop. XXIV. Viète's proof shows that fa bisects $\widehat{eab}$. (The condition is that

$$\frac{da\times ba}{dc^2} = \frac{db}{ab \text{ (or } ae = de)} = \frac{df}{dc}, \quad\text{or}\quad \frac{df}{ab \text{ (or } de)} = \frac{da}{dc},$$

so that

$$\frac{df}{fe} = \frac{da}{ac\text{(or } ae)}.)$$

It follows that $\widehat{fea} = \widehat{efa} = \widehat{eda}$ (or $\widehat{ead} = 2\widehat{fad}) + \widehat{fad} = 3\widehat{fad} = 3\widehat{eaf} = \frac{3}{7}\pi$, or $ead = \frac{2}{7}\pi$, so that eb is the side of the inscribed heptagon.

Analytically (and perhaps more revealingly), taking $db = y$, $ba = 1$, $be = x$, the subtense $S(\frac{2}{7}\pi)$, we have immediately

$$\frac{da}{ea} = y+1 = 2\cos(\tfrac{2}{7}\pi) = 2-x^2;$$

further,

$$7x-14x^3+7x^5-x^7 = S(7\times\tfrac{2}{7}\pi) = 0;$$

so that, eliminating $x = 0$ and substituting $x^2 = 1-y$, $y(y+2)^2 = y+1$.

[2](19)

prop 10[']

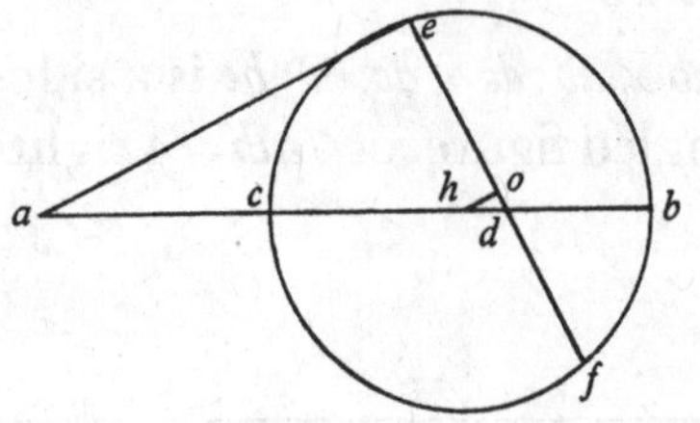

If $ac=ef$. & aef a right angle & ab pas(20) through y^e center then $cd:de::de:df::df:db$. And if $cd:de::de:df::df:db$.(21) then ae is perpendicular to ef. $2hd$ is y^e difference of y^e extreames & $2do$ is y^e difference of y^e meanes. wch given y^e proportionall lines may be found &c.(22)

prop 11

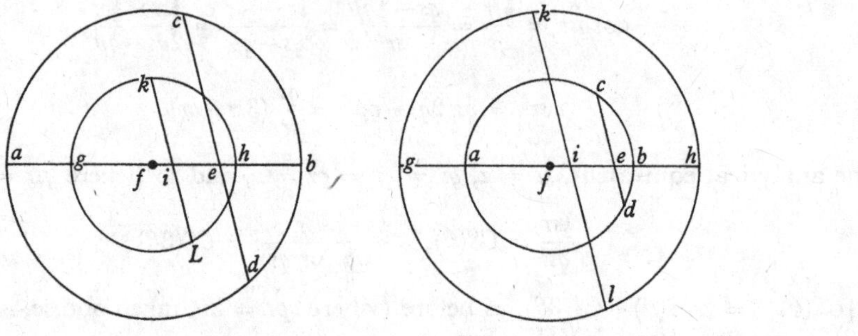

Pseudomesolabium wherby To find 2 meane proportionalls. If,

$$ae:ec::ec:ed::ed:eb.$$

[&] they be inscribed in y^e circle $acbd$ y^e diam: being $ae+eb$. [&] If twixt gi & ih two meane proportionalls are sought. on y^e same center f wth y^e Rad: $\dfrac{gi+ih}{2}$ describe $gkhl$ & inscribe a line kl parallell to cd cutting ab in y^e point i &

$$gi:ki::ki:il::il:ih.^{(23)}$$

Examine it.(24)

(19) Annotations from Viète's *Pseudo-Mesolabum* (= *Vieta*: 258–74) together with the *Adjuncta Capitula* (275–85).

(20) Read 'pass'.

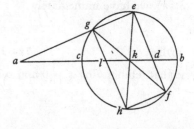

(21) The condition that 'ab pas through y^e center' is still understood.

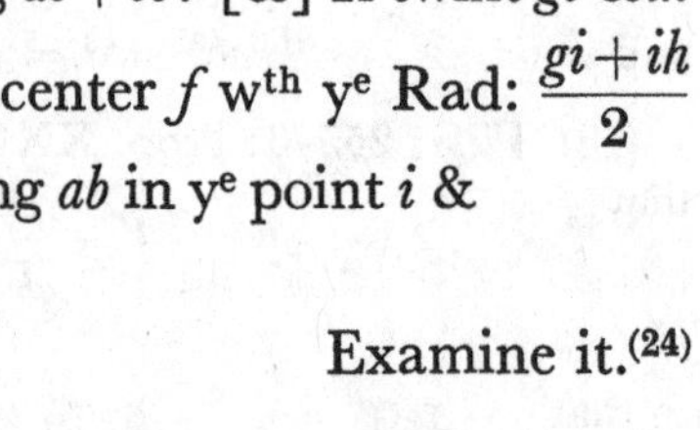

(22) *Vieta*: 263/265–6: Prop. IX/XII. In proof Viète takes g to be the second meet of ae with the circle. Then, since $\widehat{aef}$ is right, gf is a diameter, and so, drawing the second diameter ekh, $fegh$ will be a rectangle. Finally, ad (or $ac+cd$) : al (or $ac+db$) $= ed:gl$ (or df), so that

$$(ef+cd):(ef+db) = ed:df = (cd+df):(ed+db),$$

with $cd:ed = df:db = (cd+df):(ed+db)$, or $cd:ed = ed:df = df:db$.

(23) *Vieta*: 274: ψευδοπρόβλημα.

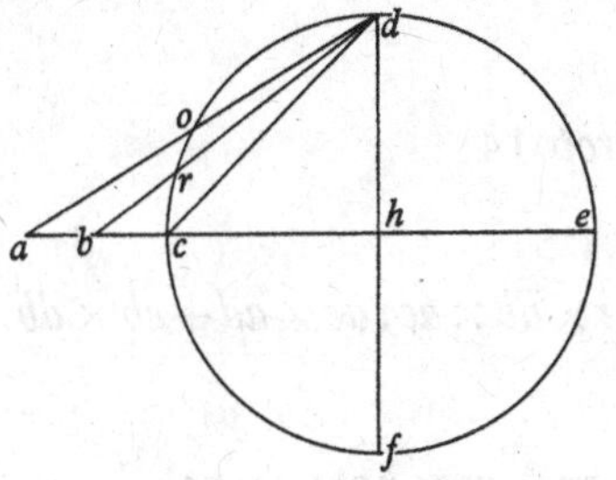

prop 12.

If $do = dh$ & ac bisected in b & bd bee drawne rd is $\mathrm{y^e}$ side of a pentagon $\mathrm{w^{ch}}$ may be inscribed in *defcro*.[25]

prop 13.

If rd be $\mathrm{y^e}$ side of a $\left\{\begin{matrix}\text{octogon}\\ \text{decagon}\end{matrix}\right\}$ & pd $\mathrm{y^e}$ side of an $\left\{\begin{matrix}\text{hexagon}\\ \text{octogon}\end{matrix}\right\}$ [&] $\mathrm{y^e}$ arch rp divided in o, [$\mathrm{y^n}$] od will be the side of an $\left\{\begin{matrix}\text{heptagon}\\ \text{enneagon}^{(26)}\end{matrix}\right\}$ to be inscribed in $\mathrm{y^e}$ circle *ord* & $\mathrm{y^e}$ arch RP is rightly[27] divided by Bisecting $\mathrm{y^e}$ Line ac.[28]

Examine it.

prop 12. 13 & I think 11 are trew onely mechanically.[29]

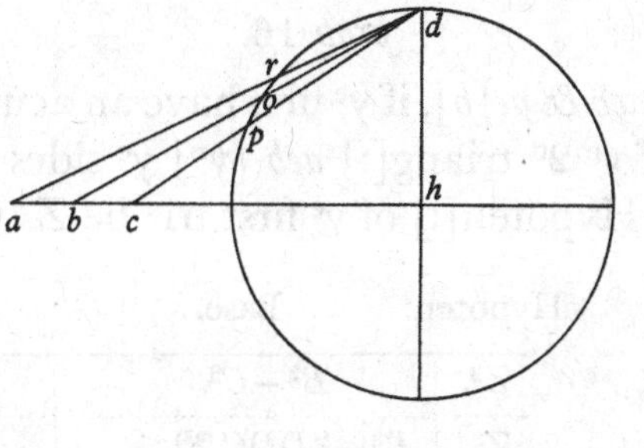

---

(24) Apparently Newton never did, nor would he have found anything amiss if he had (since this proposition is merely a corollary of the preceding, where a crucial point in the argument has been passed over in silence). It is not in fact possible, with implicit restriction to a straight-edge and compasses construction, to inscribe $ef = ac$ (prop: 10) in the circle so that $a\hat{e}f$ is a right angle.

(25) *Vieta*: 283: *Adj. Capitula*: Caput II.

(26) A regular polygon of nine sides (Viète's word).

(27) 'Correctly.'

(28) *Vieta*: 283–4: Caput II.

(29) Newton added this last sentence, possibly at a later date, in a small gap on the left of his figures. His choice of the word 'mechanically' is curious. All three propositions are, indeed, 'trew onely mechanically' in the Euclidean sense that their construction is impossible by straight-edge and compasses. However, though all the propositions are constructible by cubic curves (or, equivalently, by insertions of two means or angle trisections), only Viète's construction in prop: 11 is exact while those of prop: 12 and prop: 13 (which use only a graduated ruler and a fixed circle) are only approximate.

[3] *Of Angular sections*(30)

Prop 14

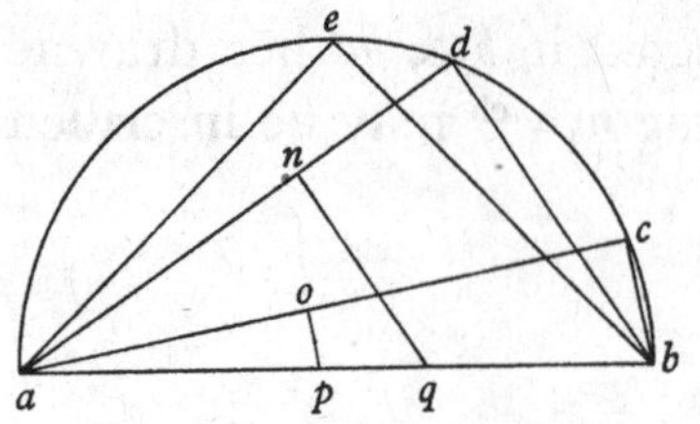

If *ead*=*cab*. Then

$ab : ab^2 :: cb : eb \times ad - ae \times db :: ac : ae \times ad + eb \times db.$

Or,

$ap : aq \times ab :: op : eb \times an - ae \times nq :: ao : ae \times an + eb \times nq.$

but y[e] angles *anq*, *aop* are right ones & $ean = oap = eab - dab$.(31)

prop 15

If y[e] angle $cab + dab = eab$.(32) or $naq + oaq = eab$. & *anq*, *aop* are right angles then $Ab : ab^2 :: Eb : ad \times bc + ac \times db :: ea : ad \times ac - db \times cb$. Or the triang: unequall.(33) $ab : ap \times aq :: eb : an \times op + ao \times nq :: ea : an \times ao - nq \times op$.(34)

prop 16.

In 2 rectang: triang: *acb* & *ae*[*b*], if y[e] first have an acute angle *cab* submultiple to y[e] acute angle *eab* of y[e] 2[d] triang[:] *aeb* [y[n]] y[e] sides of y[e] seacond have this proportion. suppose y[e] Hypoten[:] of y[e] first tri: be Z. y[e] base *b*. y[e] Cathetus *c*.

If y[e] acute angle of y[e] seacond Triangle be to y[e] acute angle of y[e] first triangle in a proportion		Hypoten:	Base.	Perpendicular
	Duple,	Z^2.	$B^2 - C^2$.	$2BC$.
	Triple,	Z^3.	$B^3 - 3DDC$.(35)	$3BBC - C^3$.
	Quadruple,	Z^4.	$B^4 - 6B^2C^2 + C^4$.	$4B^3C - 4BC^3$.
	Quintuple,	Z^5.	$B^5 - 10B^3C^2 + 5BC^4$.	$5B^4C - 10B^2C^3 + C^5$.(36)

(30) Annotations from *Vieta*: 287–304: *Ad Angulares Sectiones Theoremata ΚΑΘΟΛΙΚΩΤΕΡΑ, demonstrata per Alexandrum Andersonum* (reprinted by Schooten from Anderson's commentary, *Ad Angularium Sectionum Analyticen Theoremata καθολικώτερα. A Francisco Vieta...primum excogitata, et absque ulla Demonstratione ad nos transmissa, iam tandem Demonstrationibus confirmata* (Paris, 1615)).

(31) *Vieta*: 287–8: Theorema I. Where $\widehat{eab} = \alpha$, $\widehat{eac} = \widehat{dab} = \beta$; then

$$\frac{cb}{ab}\,[= \sin(\alpha-\beta)] = \frac{eb}{ab}\times\frac{ae}{ab}-\frac{ad}{ab}\times\frac{db}{ab}\,[= \sin\alpha\cos\beta - \cos\alpha\sin\beta],$$

and $$\frac{ac}{ab}\,[= \cos(\alpha-\beta)] = \frac{ae}{ab}\times\frac{ad}{ab}+\frac{eb}{ab}\times\frac{db}{ab}\,[= \cos\alpha\cos\beta + \sin\alpha\sin\beta].$$

(32) That is, $\widehat{ead} = \widehat{cab}$, as before.

(33) A mysterious addition to Viète's text, probably the product of Newton's incomplete comprehension. (The truth of the theorem continues to demand $\widehat{ead} = \widehat{cab}$.)

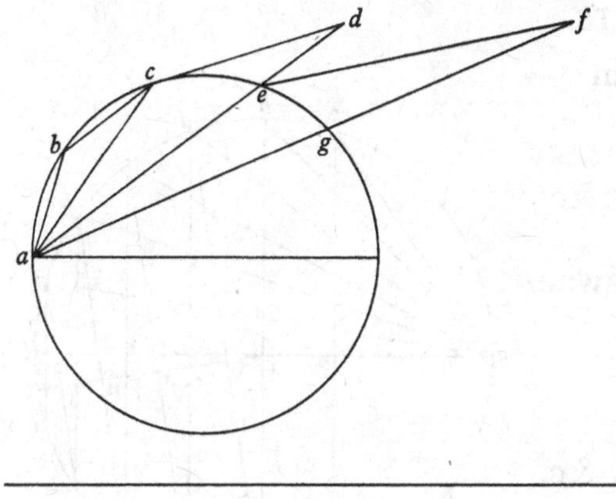

Prop 17.

If $ab=bc=ce=eg$ &c: & $ac=cd$. & $ae=ef$. &c then $ab:ac::ac:ad::ae:af$ &c. & $ed=ab$. & $ac=gf$. &c.(37) Nam Δcde & cba, efg & eac & c: = &sim.(38)

---

(34) *Vieta*: 288: Theorema II. Where $d\hat{a}b=\alpha$, $e\hat{a}d=c\hat{a}b=\beta$, then

$$\frac{eb}{ab}\,[=\sin(\alpha+\beta)]=\frac{ad}{ab}\times\frac{bc}{ab}+\frac{ac}{ab}\times\frac{db}{ab}\,[=\cos\alpha\sin\beta+\cos\beta\sin\alpha],$$

and
$$\frac{ea}{ab}\,[=\cos(\alpha+\beta)]=\frac{ad}{ab}\times\frac{ac}{ab}-\frac{db}{ab}\times\frac{cb}{ab}\,[=\cos\alpha\cos\beta-\sin\alpha\sin\beta].$$

(35) Read '$3BBC$'.

(36) *Vieta*: 289–90: Theorema III. Where $c\hat{a}b=\theta$ and $e\hat{a}b=n\theta$, $n=2,3,4,\ldots$, with

$$Z:B:C=1:\cos\theta:\sin\theta,$$

$$\text{then}\left\{\begin{array}{lllll} Z^2: & B^2-C^2 & : 2BC & =1: & \cos 2\theta: \sin 2\theta. \\ Z^3: & B^3-3B^2C & : 3B^2C-C^3 & =1: & \cos 3\theta: \sin 3\theta. \\ Z^4: & B^4-6B^2C^2+C^4 & : 4B^3C-4BC^3 & =1: & \cos 4\theta: \sin 4\theta. \\ Z^5: & B^5-10B^3C^2+5BC^4: & 5B^4C-10B^2C^3+C^5 & =1: & \cos 5\theta: \sin 5\theta. \end{array}\right\}$$

Viète probably derived this scheme by successive applications of the preceding Prop: 14 and Prop: 15 (the addition theorems for sine/cosine). Quickly, as Newton noted below, he gave a more direct approach.

Newton's text has no diagram, but its restoration is not troublesome.

(37) *Vieta*: 291: Theorema IV.

(38) 'For the triangles cde & cab, efg & eac, &c are equall & similar' (that is, congruent).

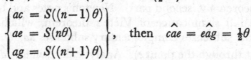

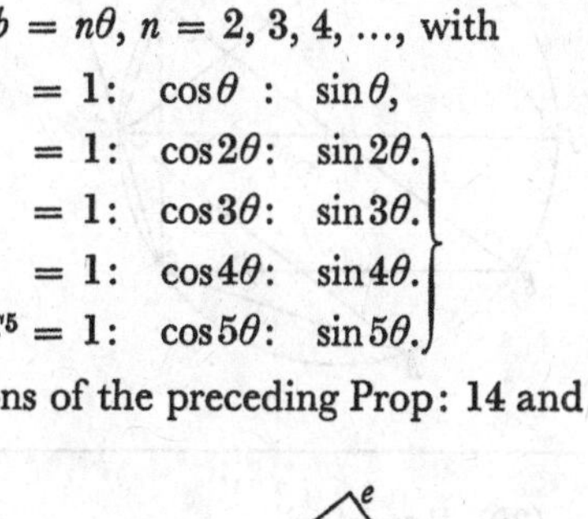

This important theorem is the foundation for Viète's tables of $S(n\theta)=2\sin(\frac{1}{2}n\theta)$ and $CS(n\theta)=2\cos(\frac{1}{2}n\theta)$ below. In proof,

where
$$\left\{\begin{array}{l} ac=S((n-1)\,\theta) \\ ae=S(n\theta) \\ ag=S((n+1)\,\theta)\end{array}\right\},\quad \text{then } c\hat{a}e=e\hat{a}g=\tfrac{1}{2}\theta$$

and $af/ae=2\cos\frac{1}{2}\theta$, so that $ac+ag=S((n-1)\,\theta)+S((n+1)\,\theta)=2\cos\frac{1}{2}\theta\times S(n\theta)=af$; or $ac=gf$.

Since $S(n\theta)=CS(\pi-n\theta)=CS(m\theta)$, say, a second analytical model is deducible from this theorem: namely,

where
$$\left\{\begin{array}{l} ac=CS((m+1)\,\theta) \\ ae=CS(m\theta) \\ ag=CS((m-1)\,\theta)\end{array}\right\},\quad \text{so that } c\hat{a}e=e\hat{a}g=\tfrac{1}{2}\theta$$

and $af/ae=2\cos\frac{1}{2}\theta$, then again $ac+ag=CS((m+1)\,\theta)+CS((m-1)\,\theta)=2\cos\frac{1}{2}\theta\times CS(m\theta)$.

Prop. 18.

If $bd=dg=gh=hk=pq=pw$(39) &c Then

$al:ak::pe:pc::ed:do::pd:pc+do::$

$rg:gs::rq:qo::qg:qo+gs::$ &c.

& if $\left(\frac{lf}{2}=l\zeta\right)$ from ζ to y^e center [c] be drawne $c\zeta$ then

$al:ak::di:dv::iq:qx::dq:dv+qx::$

$gz:gx::wz:wn$ &c.

& Ergo

$ac:ak::ab[:]\ a\delta+ad[::]\ ad:ab+ag:[:]\ ag:$

$ad+ah:[:]\ ah:ag+ak$ &c.(40)

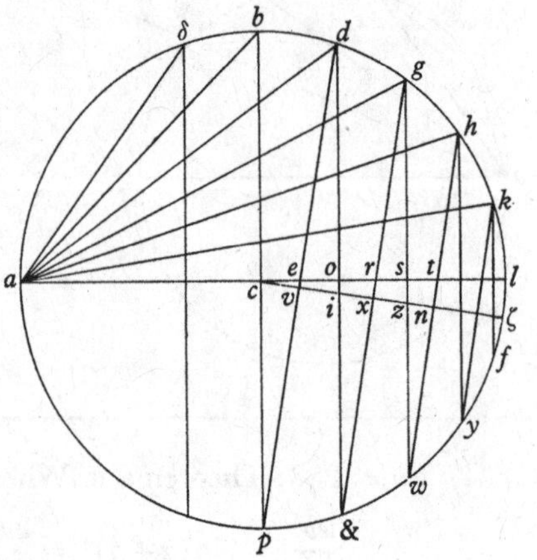

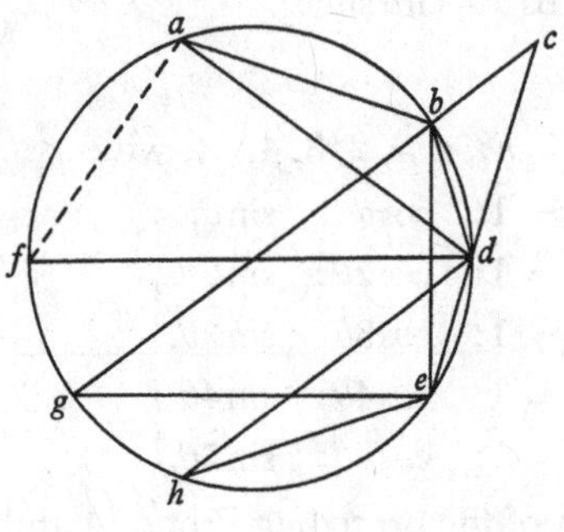

Prop 19

If $fa=ab=be=eh$ &c. & $af+ab+be+eh$ are greater y^n y^e semiperiphery.(41) & dh is y^e greatest, db y^e least line drawn from d to these points a, b, e, h. y^n rad: dh(42)$::db:da-de$.(43)

(39) Read 'qw'.

(40) *Vieta*: 293: Theorema v (first part). The last sentence restates the preceding proposition.

(41) Newton has cancelled

'then, Rad: $ge(=ec)\ (=ad)::bd(=bc):cd(=ad-de)::de:(dh-bd)$'.

(42) Read 'ah', which Newton first wrote and then cancelled!

(43) *Vieta*: 293–4: Theorema v (second part). Newton's notes and his incorrect emendation seem not to reveal the full significance of Viète's theorem, which is the complement of Prop: 17 above (where the subtenses/complementary subtenses da, db, de, dh/eb, ed, eh straddle the diameter fd and that through the point e). Analytically, where $a\hat{d}b = b\hat{d}c = e\hat{d}h = \theta$ and $db = S(n\theta)$, then $de = S(2\pi-(n+1)\theta) = -S((n+1)\theta)$, so that

$$da-de = S((n-1)\theta)+S((n+1)\theta) = 2\cos\tfrac{1}{2}\theta\times S(n\theta),$$

as before, or $(da-de):db::fd\times\cos\frac{1}{2}\theta:\text{radius } \frac{1}{2}fd$.

In Alexander Anderson's proof, c is taken in ed with $bc = be$. Then it follows that

$$b\hat{c}d = b\hat{d}c = b\hat{g}e \quad (\text{subtended by } be) = a\hat{d}b \quad (\text{subtended by } ab = be) = \tfrac{1}{2}\theta;$$

so that $\triangle abd \equiv \triangle ebc$, or $ce = da$ and $da-de = dc$ (with $dc/db = 2\cos\frac{1}{2}\theta$, Viète's 'least chord').

prop 20

Out of y^e 18th & 19th Prop:

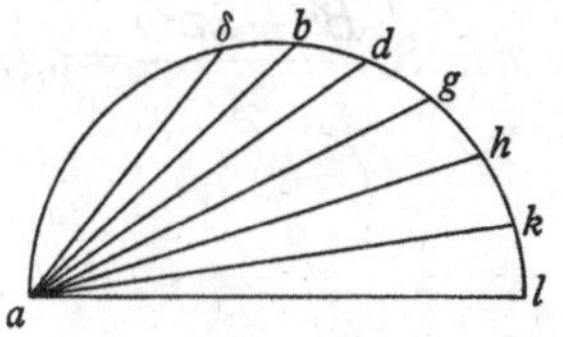

To divide An angle into any number of pts in y^e figure of y^e 18th prop:(44) $al = \text{diam} = 2z$. ak is y^e greatest of y^e inscribed lines $= B$: now

$$z:B::B:ah+2z.^{(45)}$$

therfore $bb = ah$ in $z+2z^2$. & $\dfrac{bb-2zz}{z} = ah$. And $z:B::\dfrac{b^2-2z^2}{z}:b+ag$. therfore $\dfrac{b^3-2zzb}{zz} - b = ag$ or $\dfrac{b^3-3zzb}{zz} = ag$. Likewise $\dfrac{B^4-4zzbb+2z^4}{z^3} = ad$. & $\dfrac{B^5-5zzB^3+5z^4B}{z^4} = ab$.

$$\frac{B^6-6zzB^4+9z^4BB-2z^6}{zzzzz^{(46)}} = a\delta.$$

$$\frac{B^7-7zzb^5+14z^4b^3-7z^6b}{z^6} = \text{to a seaventh line.}$$

$$\frac{B^8-8zzb^6+20z^4b^4-16z^6bb+2z^8}{z^7} = \text{to an eight line.}$$

$$\frac{B^9-9zzb^7+27z^4b^5-30z^6b^3+9z^8b}{z^8} = \text{a nineth line.}$$

$$\frac{B^{10}-10zzb^8+35z^4b^6-50z^6b^4+25z^8bb-2z^{10}}{z^9} = \text{[a] tenth \&[c].}^{(47)}$$

(44) The accompanying figure does not occur in Newton's text, but has been adapted from that of Prop: 18 and is here inserted for convenience.

(45) That is, *al*.

(46) Read 'z^5'.

(47) *Vieta*: 294–5: Theorema VI. Prop: 17 and Prop: 19 together yield the recursive rule, $CS(n+1(\theta)) = xCS(n\theta) - CS((n-1)\theta)$, where $ak = x = B/Z = 2\cos\frac{1}{2}\theta$, and $CS(n\theta)$ is the complementary subtense of $n\theta$ (or $CS(n\theta) = 2\cos(\frac{1}{2}n\theta)$). This rule Viète used to tabulate successively, where z is taken as unity for simplicity,

$$ah = CS(2\theta) = xCS(\theta) - CS(0) = x^2-2,$$

$$ag = CS(3\theta) = xCS(2\theta) - CS(\theta) = x^3-3x,$$

....

Prop 21

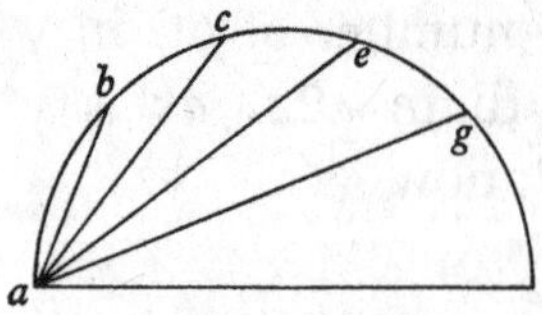

out of y^e^ 17th Theor:(48) in y^e^ figure whereof if ab y^e^ least inscribed line $=z$. & ac y^e^ next line bee B. then $z:B::B:z+ae$. & $\frac{bb-zz}{z}=ae$. & $\frac{B^3-2z^2b}{zz}=ag$. & $\frac{B^4-3z^2bb+z^4}{z^3}=$ to a fift line.

$\frac{B^5-4zzb^3+3z^4b}{z^4}=$ a sixt. & $\frac{B^6-5zzb^4+6z^4bb-z^6}{[z^5]}=$ a seaventh.

$\frac{B^7-6zzb^5+10z^4b^3-4z^6b.}{z^6}=$ to an eight line.

$\frac{B^8-7zzb^6+15z^4b^4-10z^6bb+z^8}{z^7}=$ to a nineth line.

$\frac{B^9-8zzb^7+21z^4b^5-20z^6b^3+5z^8b}{z^8}=$ to a tenth line.

$\frac{B^{10}-9zzb^8+28z^4b^6-35z^6b^4+15z^8bb-z^{10}}{z^9}=$ eleventh.(49)

Prop 22

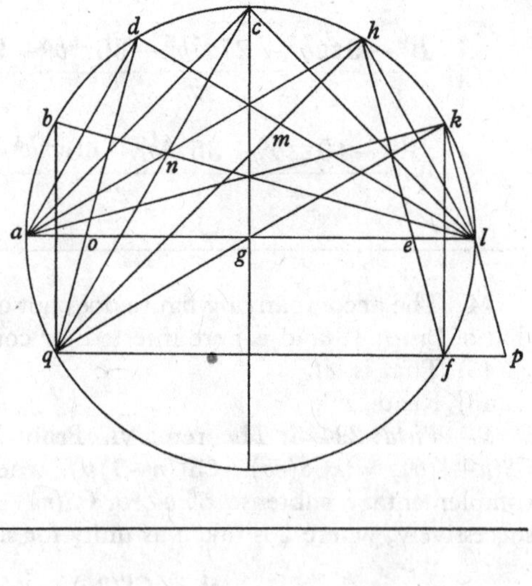

If $aq=ab=bd=dc=ch=hk=kl=lf$. Then Gk rad $:kl::kl:el\ (=al-ah)::hl:hm\ (=qh-qd=ak-ac)::dl:do\ (=qd-qa=ac-ab)::lc:cn(=qc-qb=ah-ad$ &c). Soe that y^e^ Periph: [being] divided into any number of pts $gl:lk::lk:al-ah::lh:ak-ac::lc:ah-ad::dl:ac-ab$ &c. & $gl:lk::ah:lc-lk::ac:ld-lh::ad:lb-lc::ab:al-ah$ &c.(50) hence

(48) And 19th.

(49) *Vieta*: 296–7: Theorema VII. Here $S(\theta)=z$, $S(2\theta)=b$ (or B) and the recursive rule yields $S(n+1(\theta))=yS(n\theta)-S((n-1)\theta)$, where $y=b/z=CS(\theta)$. The tabulation then proceeds: $ae=z(y^2-1)$, $ag=z(y^3-2y)$, ...

(50) *Vieta*: 297–8: Theorema VIII. Where $\hat{bad}=\hat{dac}=\ldots=\hat{hak}=\hat{kal}=\frac{1}{2}\theta$, then

$$2\sin\tfrac{1}{2}\theta=\frac{S((n+1)\theta)-S((n-1)\theta)}{CS(n\theta)},$$

or

$$S((n+1)\theta)=2\sin\tfrac{1}{2}\theta\times CS(n\theta)+S((n-1)\theta),$$

Prop 23.

In y^e^ former scheame

If $al=2x=$ hypotenusa. $kl=b$. $x:b::b:2x-ah$. & $\dfrac{-bb+2xx}{x}=ah$.

$x:b::\dfrac{-bb+2xx}{x}:\dfrac{2bxx-b^3}{xx}$ $(=lc-b)$ [&] therefore $\dfrac{3bxx-b^3}{x^3}=lc$.

& $\dfrac{2x^4-4bbxx+b^4}{x^3}=ad$ y^e^ base of y^e^ 4th triang:

& $\dfrac{5bx^4-5b^3xx+b^5}{x^4}=$ y^e^ perpendicular (bl) of y^e^ 5^t^ tri:

& $\dfrac{2x^6-9x^4bb+6xxb^4-b^6}{x^5}=$ base of y^e^ 6^t^ triang.

$\dfrac{7x^6b-14x^4b^3+7x^2b^5-b^7}{x^6}=$ perpendic[:] of y^e^ 7th tri

$\dfrac{2x^8-16x^6bb+20x^4b^4-8xxb^6+b^8}{x^7}=$ base of y^e^ 8th tri:

$\dfrac{9x^8b-30x^6b^3+27x^4b^5-9xxb^7+b^9}{x^8}=$ perp: of y^e^ 9th tri:(51)

---

and
$$2\sin\tfrac{1}{2}\theta = \frac{CS((n-1)\theta)-CS((n+1)\theta)}{S(n\theta)},$$
or
$$CS((n+1)\theta) = -2\sin\tfrac{1}{2}\theta S(n\theta)+CS((n-1)\theta).$$

(51) *Vieta*: 298–300: Theorema IX. Viète uses the preceding *Theorema* to tabulate
$$S((2n-1)\theta) \text{ and } CS(2n\theta)$$
alternately, thus:
$$[CS(0) = 2]$$
$$S(\theta) = b/x = s, \text{ say},$$
$$CS(2\theta) = -s(s)+2 = 2-s^2,$$
$$S(3\theta) = s(2-s^2)+s = 3s-s^3,$$
$$\cdots\cdots\cdots\cdots$$
The recursive scheme is $\left\{\begin{aligned} S((2n+1)\theta) &= sCS(2n\theta)+S((2n-1)\theta).\\ CS((2n+2)\theta) &= -sS((2n+1)\theta)+CS(2n\theta).\end{aligned}\right\}$

Prop 24

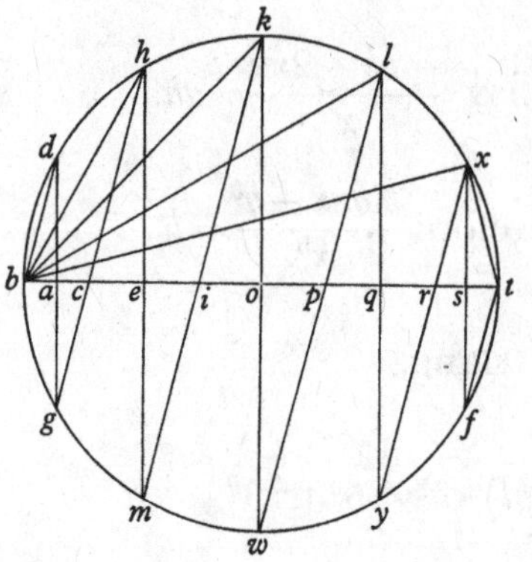

If $bd = dh = hk = kl = bg$ &c: then $bh = gd$ & $bk = gh$ & $bl = hm$ &c: & then $xt:bx[::ba:da]::ac:ag::ce:eh::ei:em::io:ok::op:ow::pq:ql::qr:qy::rs:sx::st:sf::ba:ad$. therefore $xt:bx::bt:dg+hm+kw+ly+xf$.(52) againe $xt:bt::ab:bd::ac:cg::ce:ch::ei:im$ &c. Therefore $xt:bt::bt:bd+gh+mk+wl+yx+ft$. & since, as $xt:bt::bx:dg+hm+kw+ly+xf$. Therefore $xt:bt::bx+bt:bd+dg+gh+hm+mk+kw+wl+ly+yx+xf+ft$. And $xt:bt::bx+bt+xt$: to all y^e^ perpendicular & transverse line[s]$+bt$. that is $(bd)\ xt:bt::xt+bt+bx:2bd+2bh+2bk+2bl+2bx+2bt$.(53)

[4](54)

Prop 24[']

If in y^e^ circle *cfgh* be inscribed y^e^ helix(55) *bedc* & *ac* touch it in y^e^ point *c* then $ab =$ to y^e^ circumference.(56)

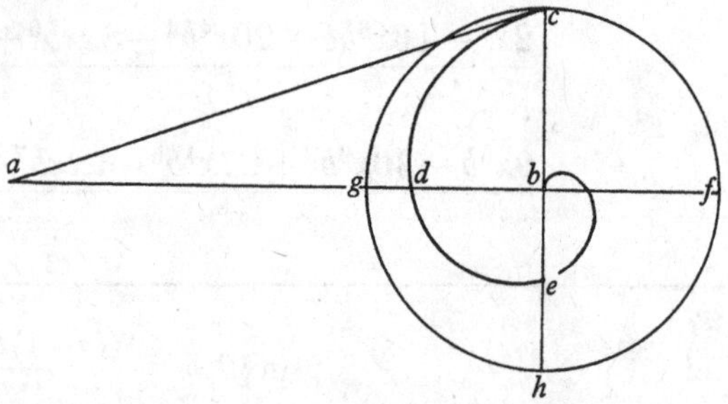

(52) This is, in fact, Prop. 20 of the first book of Archimedes' *On the Sphere and Cylinder*. It does not seem likely that Newton was aware of the connection at this time, for there is no record of his having read any of Archimedes' work in the original as an undergraduate.

(53) *Vieta*: 300: Theorema x. Where $d\hat{b}h = h\hat{b}k = \ldots = l\hat{b}x = x\hat{b}t = \frac{1}{2}\theta$, $bt = 2$ and $n\theta = \pi$, Viète's result is that

$$S(\theta) \times \sum_{1 \leqslant i \leqslant n} [2 \times S(i\theta)] = 2 \sum_{1 \leqslant i \leqslant n} [\cos(i-1)\tfrac{1}{2}\theta - \cos(i+1)\tfrac{1}{2}\theta]$$
$$= 2[\cos(0) + \cos(\tfrac{1}{2}\theta) - \cos(\tfrac{1}{2}\pi) - \cos(\tfrac{1}{2}\pi + \tfrac{1}{2}\theta)]$$
$$= 2 + CS(\theta) + S(\theta).$$

(54) Annotations on *Vieta*: 347–435: *Variorum de Rebus mathematicis Responsorum Liber VIII* (reprinted by Schooten from Viète's work of the same title which was published at Tours in 1593).

(55) In modern terminology, an Archimedean spiral.

(56) *Vieta*: 355–6: Caput VI, Prop. I, *Usus volutarum in dimensione circuli*. For proof Viète refers back to Prop. 18 of Archimedes' *On Spirals*. Newton has foreshortened *ab*.

(57) *Vieta*: 357: Caput VI, Prop. VI. Where $rq = 2$, $\widehat{rp} = \theta$, then $rp = S(\theta)$ and

$$po = 2\theta - S(2\theta),$$

so that
$$(rapc) = \tfrac{1}{2}(\theta - S(\theta)CS(\theta)) = \tfrac{1}{4}(po \times \tfrac{1}{2}rq).$$

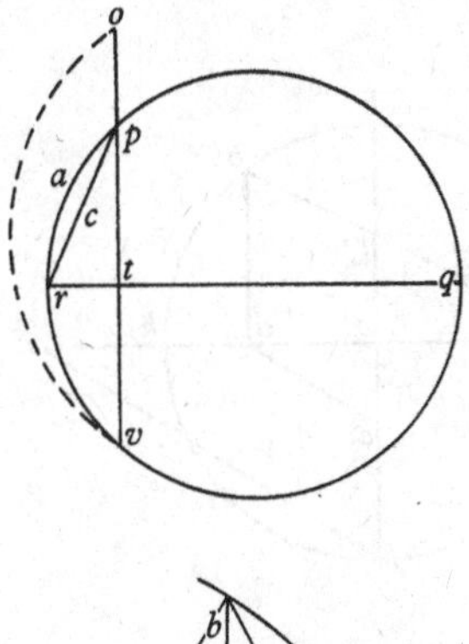

Prop 25

If *apcr* be les y^n halfe y^e circle. & $vt=tp$. & $vo=$ to *vrap*: then $\frac{rq\times po}{2}=4$ times y^e section *rapc*.(57)

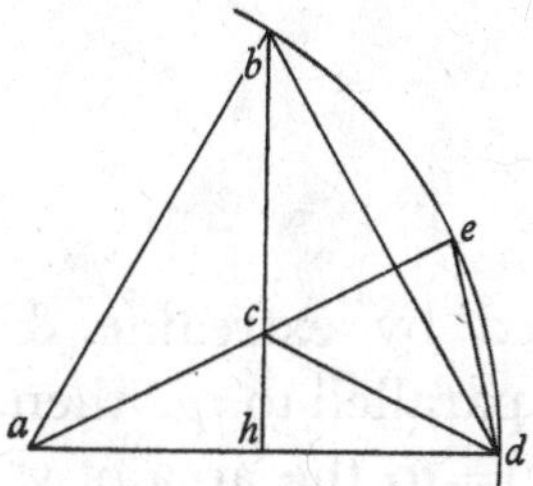

Prop 26

If $ab=bd=ad$ & *bh* perpendicular to *ad* frō y^e angle(58) *b*. $ce=ed$. y^n $aed=ade=3dae$. & *ed* is y^e side of a heptagon.(59)

Prop 27.

If a line be cut by extreame & meane proportion y^e lesse segment almost is to y^e whole line as y^e diameter is to 5 times y^e Periphery divided by 6.(60)

Prop 28

Si secetur linea per extremam & mediam proportionem erit proximè, ut tota linea plus minori segmento ad bis totam lineam, ita quæ potest quadrato sesquialterum semidiametri ad latus quadrati circulo equalis.

linea secta sit 100,000. minus segmentū(61) 38,197. Semidiamet^r 100,000, quæ potest quadrato sesquialterum semidiametri(62) paulo major est quàm 122,474. Radix Peripheriæ,(63) 177,245.(64)

---

(58) Read 'corner' (Latin *angulus*).

(59) *Vieta*: 360: Caput VII, *Ad descriptionem heptagoni propositam à F.F.C.*, Scholium, Problema II. The proof is immediate since $\widehat{ade} = \widehat{adc}$ (or $\widehat{dae}$) $+\widehat{cde}$ (or $\widehat{dce} = 2\widehat{dae}$). (Note that the construction of the equilateral triangle *abd* is largely irrelevant, since Viète uses it merely to construct *bh*, the perpendicular bisector of *ad*.)

(60) *Vieta*: 392: Caput xv, *Geometrica κύκλου μέτρησις, bene proxima veræ*, Prop. I. The approximation is $\frac{5}{6}\pi \approx \frac{2}{3-\sqrt{5}}$ (correct to 4*D*).

(61) $10^5\times\frac{3-\sqrt{5}}{2}$. (62) $10^5\times\sqrt{\frac{3}{2}}$. (63) $10^5\times\sqrt{\pi}$.

(64) *Vieta*: 392: Caput xv, Prop. II. 'If a line be cut in extreme and mean proportion, then approximately as the whole line together with the lesser segment to twice the whole line, so will be the number whose square is half as much again as the radius to the side of the square equal (in area) to the circle.

'Let the sectioned line be 100,000. the lesser segment 38,197. the radius 100,000. the

Prop 28[(')]

If $er = rh = or$. & $ao = fc =$ to y^e^ side of a decagon;[(65)] & fn parallell to cd[,] y^n^ en shall be almost equall to y^e^ fourth ꝑte of a circle. for af[(66)] is divided in extreame & meane propor[:] in y^e^ point c. & $ec:ef::ef:\frac{5}{12}$ Perim̄ $hbkfa::hr:\frac{5}{24}$ Perim$::\frac{6}{10}\,ef(=de)$[:] $\frac{1}{4}$ Perim: by y^e^ 27th prop: & $ec:ef::ed:en$ $(=\frac{1}{4}$ Perim[:]).[(67)]

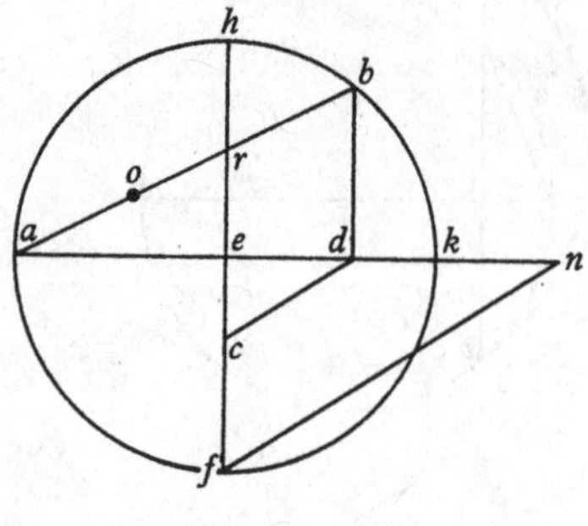

Prop: 29.

If $os = 2cp$. & co is divided by extreame & meane proportion in r. & od parallell to rp. then db is y^e^ side of a square [$bdeg$] = to the area of y^e^ circle. for by y^e^ 28th prop:[(68)] As br (= to line + less segm̄) : bo (= twice y^e^ line) :: bp

$$\left(=\sqrt{\frac{3}{2}}\text{ of y}^\text{e}\text{ square of y}^\text{e}\text{ semidiameter}\right):$$

bd (= to y^e^ roote of a square equall to y^e^ area of a[(69)] circle[)]].[(70)]

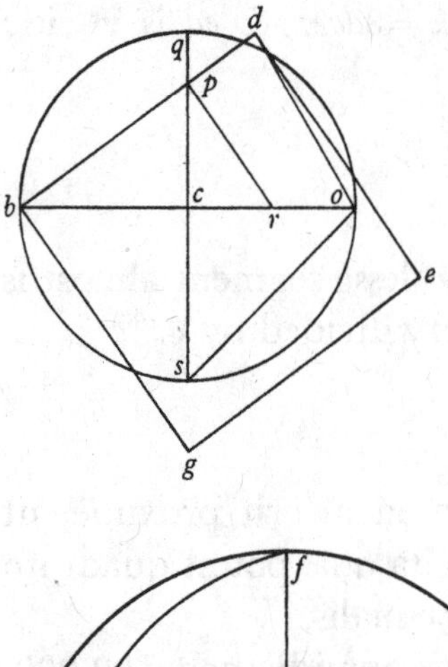

Prop 30

If y^e^ line dc touch y^e^ helix[(71)] in y^e^ line ag. & y^e^ line hf toucheth y^e^ beginning of it in y^e^ center a & $4ac = af$. then[(72)] $2ad$ shall bee equall to perim: asr. & ac being y^e^ Diam̄: y^e^ area of y^e^ triang[:] acd = to y^e^ area of y^e^ circle asr.[(73)]

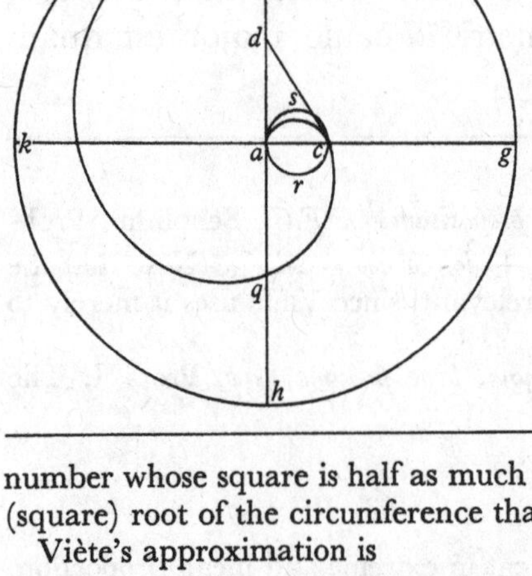

number whose square is half as much again as the radius is a little greater than 122,474, the (square) root of the circumference than 177,245.'

Viète's approximation is

$$\frac{\frac{1}{2}(5-\sqrt{5})}{2} \approx \frac{\sqrt{\frac{3}{2}}}{\sqrt{\pi}}, \quad\text{or}\quad \frac{138{,}197+}{200{,}000} \approx \frac{122{,}474+}{177{,}245+}.$$

(65) That is, $\dfrac{\sqrt{(5)}-1}{2}\times ae$.

(66) Read 'ef'.

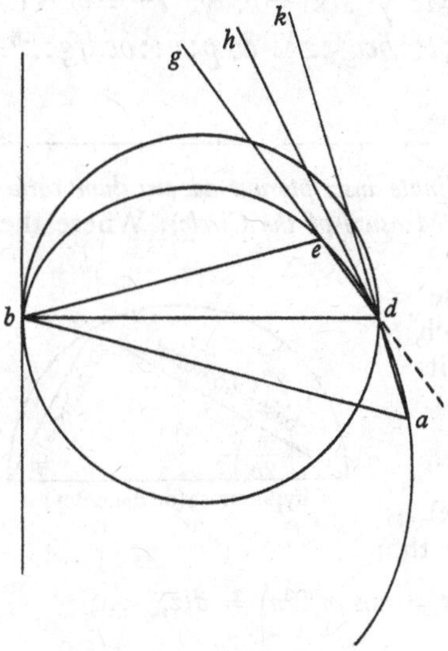

Prop 31

If *bed* be a square of one revolution of an helix & y[e] angle *dbe*=*dba*. & through y[e] points *a*, *d*, in y[e] helix be drawne y[e] line *adk*. & through y[e] points *e*[,] *d* in y[e] Helix be drawne *edg*. & y[e] angle *kdg* bisected by *dh*; then *dh* shall almost touch y[e] helix in *d*. & it shall be soe much y[e] nigher a touch line by how much y[e] angles *ebd*[,] *dba* are lesser.(74)

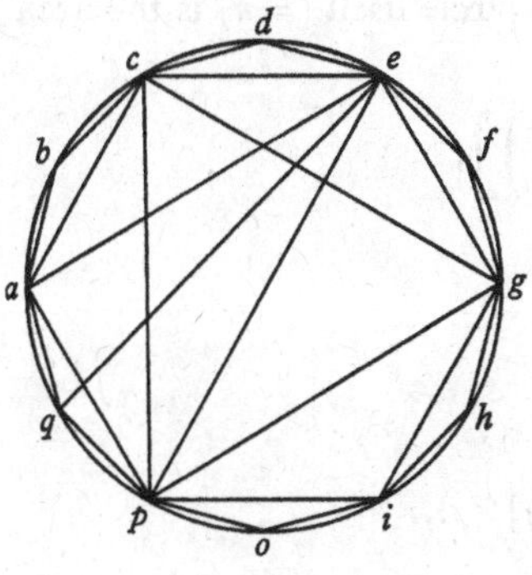

Prop 32

If many Polygons be inscribed in a circle[,] y[e] number of theire sides increaseing in a double proportion. & theire apotomies,(75) or y[e] base of a tri: whose cathetus is a leg of y[e] Polygon & hypotenusa is y[e] Diam̄ (as y[e] apotome of y[e] Polygō *cgp* is *ce*. of *pacegi* is *ae* &c) if y[e] Apotome of y[e] sides of y[e] first Polygō be called *b*. of y[e] 2[d]=*c*. of y[e] 3[d]=*d*. of y[e] 4[th]=*f*. of y[e] 5[t]=*g*. of y[e] sixt=*h*. & y[e] diameter be *z*. And y[e] first Polygon be=*p*. y[e]

(67) *Vieta*: 392–3: Caput xv, Propositio III.

(68) Prop. 28′.

(69) Read 'the'.

(70) *Vieta*: 393: Caput xv, Propositio IV. In the accompanying figure

$$cp = \tfrac{1}{2}os = \frac{1}{\sqrt{2}}\times co \quad\text{and}\quad bp = \sqrt{(1+\tfrac{1}{2})}\times co.$$

(71) Archimedean spiral.

(72) Much as in Prop: 24′.

(73) *Vieta*: 393: Caput xvi, *De altera methodo quadrandi circulum per angulum angulo, qui fit à tangente helicem, & diametro circuli, propemodum æqualem*, Propositio I. In the diagram, circle $(asr) = \frac{1}{4}\pi\times ac^2 = \frac{1}{4}\pi ac$ (or ad) $\times \frac{1}{2}ac = \triangle(acd)$.

(74) *Vieta*: 395–6: Caput xvi, Propositio II. As J. E. Hofmann stresses in his 'François Viète und die Archimedische Spirale' (*Archiv der Mathematik* 5 (1954): 138–45), the important thing in this theorem is that it constructs the tangent *hd* to the 'helix' as the limiting position of *ged*/*kda* as both the points *e* and *a* pass into *d*. (The convexity of the spiral shows that the tangent must lie in $\widehat{gdk}$, whose vanishing is equivalent to finding the tangent *hd* as the limit-chord *ed* (or *ad*) as the points *e* (or *a*) and *d* coincide.) For 'square', read 'quadrant'.

(75) Read 'apotomes'.

$2^d = q$. y^e $3^d = r = abcdefghiopq$. y^e fourth $= s$. y^e $5^t = t$. y^e sixt $= v$. y^e $7^{th} = w$ &c. then $p:q::b:z$. & $p:r::bc:zz$. & $p:s::bcd:zzz$. & $p:t::bcdf:z^4$. & $p:v::bcdfg:z^5$. & $p:w::bcdfgh:z^6$ &c.(76)

(76) *Vieta*: 398–400: Caput XVIII, *Polygonorum circulo ordinate inscriptorum ad circulum ratio*, Propositio I/Propositio II (which summarize Archimedes' *Measure of the Circle*). Where the regular inscribed polygon has $2n$ sides and the diameter $z = 2$, then its 'leg' or 'cathetus' is $2\sin(\pi/2n) = S(\pi/n)$ and its 'base' or 'apotome' is $2\cos(\pi/2n) = CS(\pi/n)$. Therefore, since each succeeding polygon doubles its number of sides but halves its subtended angle $\pi/2n$, the successive 'apotomies' are

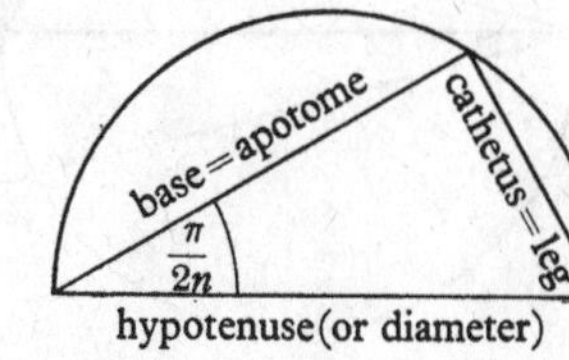

$$b = 2\cos(\pi/2n), \quad c = 2^2\cos(\pi/2^2n), \quad d = 2^3\cos(\pi/2^3n), \quad \ldots,$$

and the areas of the corresponding polygons are $p = n\sin(\pi/n)$, $q = 2n\sin(\pi/2n)$, $r = 2^2n\sin(\pi/2^2n)$, ... It follows therefore that

$$p/q = \cos(\pi/2n) = b/z, \quad q/r = \cos(\pi/2^2n) = c/z, \quad r/s = \cos(\pi/2^3n) = d/z, \quad \ldots,$$

and so $p:q = b:z$, $p:r = bc:z^2$, $p:s = bcd:z^3$, ...

For some reason Newton has not noted Viète's well-known *Corollarium* (= *Vieta*: 400) which expands $2/\pi$ as an infinite product. Specifically, since the circle itself ($=\pi$) is the area of the regular inscribed polygon of an infinite number of sides,

$$\frac{p}{\text{circle}} = \frac{n\sin(\pi/n)}{\pi} = \lim_{\nu\to\infty} \prod_{1\leqslant i\leqslant \nu} \left[\cos\left(\frac{\pi}{2^i n}\right)\right];$$

and when $n = 2$ (or p is a square)

$$\frac{2}{\pi} = \lim_{\nu\to\infty} \prod_{1\leqslant i\leqslant \nu} \left[\cos\left(\frac{\pi}{2^{i+1}}\right)\right],$$

which (correcting Viète's own formulation) we may write

$$2/\pi = \sqrt{\tfrac{1}{2}} \times \sqrt{(\tfrac{1}{2}+\tfrac{1}{2}\sqrt{\tfrac{1}{2}})} \times \sqrt{[\tfrac{1}{2}+\tfrac{1}{2}\sqrt{(\tfrac{1}{2}+\tfrac{1}{2}\sqrt{\tfrac{1}{2}})}]} \times \ldots.$$

3

ANNOTATIONS FROM WALLIS

§1. AN EARLY NOTE ON INDIVISIBLES

[1664?][(1)]

From the original in the University Library, Cambridge[(2)]

Of Quantity.

As finite lines added in an infinite number to finite lines, make an infinite line: so points added twixt points infinitely, are equivalent to a finite line.

All superficies beare the same proportion to a line yet one superficies may bee greater y^n another (ye same may be said of bodys in respect of surfaces) wch happens by reason y^t a surface is infinit[(3)] in respect of a line, so though all infinite extensions beare y^e same proportion to a finite one yet one infinite extension may be greater y^n another soe one angle of contact may exceed another, yet they are all equal when compared to a rectilinear angle viz wch is infinitely greater. Thus $\frac{2}{0}$ is double to $\frac{1}{0}$ & $\frac{0}{1}$ is double to $\frac{0}{2}$, for multiply y^e 2 first & divide y^e 2^{ds} by 0, & there results $\frac{2}{1}:\frac{1}{1}$ & $\frac{1}{1}:\frac{1}{2}$. yet if $\frac{2}{0}$ & $\frac{1}{0}$ have respect to 1 they beare y^e same relation to it[,] y^t is $1:\frac{2}{0}::1:\frac{1}{0}$. & ought therefore to bee considered equall in respect of a unite.[(4)]

(1) Extracted from Newton's *Qu[a]estiones quædam Philosoph[i]cæ*, a miscellany of scientific jottings entered about 1664 (or so we may judge from the writing) in a small pocket-book partially described by A. R. Hall in 'Sir Isaac Newton's Note-book, 1661–65', *Cambridge Historical Journal*, **9**, 2 (1948): 239–50. As Hall remarks on p. 242, the earlier parts of the *Quæstiones* seem 'to be transcripts, more or less direct, from his reading'. In this note Newton names no source for the thoughts he has written up, but the statements on indivisibles and reference to the angle of contact argue strongly that Newton was already familiar with Wallis' *Operum Mathematicorum Pars Altera* (Oxford, 1656), which contains in particular the tracts *De Angulo Contactus & Semicirculi Disquisitio Geometrica* and *Arithmetica Infinitorum*. It is, however, just possible that Newton had first read Thomas Hobbes' belligerent and myopic commentary, *Examinatio et Emendatio Mathematicæ Hodiernæ, Qualis explicatur in libris Johannis Wallisii Geometriæ Professoris Saviliani in Academia Oxoniensi. Distributa in sex Dialogos* (London, 1660). In attacking Wallis' ideas, often on wholly inadequate grounds, Hobbes' *Dialogus Quintus* [*De Angulo Contactus, de Sectionibus Coni, & Arithmetica Infinitorum*] succeeds in conveying a surface-impression of Wallis' thought and it would have been natural for Newton, once his interest was aroused, to pass immediately on to the original.

(2) Add. 3996: 89^r (=Newton's p. 5). (3) Newton first wrote 'infinitly'.

(4) Newton raises one of the fundamental paradoxes of indivisibles. Where '0' is an indefinitely small 'unit' indivisible, the proportion $\frac{2}{0}:\frac{1}{0}::2:1$ holds, though in the limit as 0

The angle of contact is to another angle, as a point to a line, for y[e] crookednes in one circle amounts to 4 right angles & y[t] crookednesse may bee conceived to consist of an infinite number of angles of contact, as a line doth of infinite points. As y[e] point *a* to y[e] line *ab* so y[e] line *ac* to y[e] pgr[(5)] *abcd*::pgr[(5)] *dbef*: y[e] paralelipipedon *bg*.[(6)]

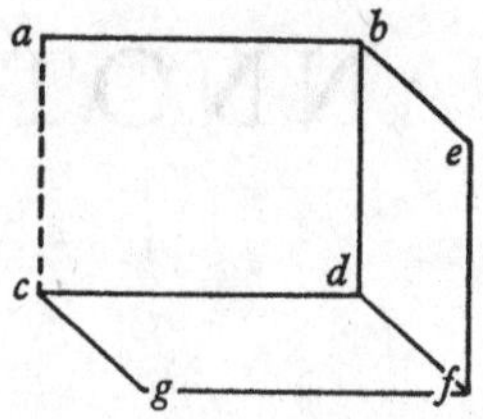

Tis indefinite (y[t] is undetermined) how greate a sphære[(7)] may be made[,] how greate a number may be recconed, how far matter is divisible, how much time or extension wee can fansy[,] but all y[e] Extension y[t] is, Eternity, $\frac{a}{0}$ are infinite. $\frac{a}{0}$ exceeds all number & is soe greate y[t] there can bee noe greater, but (finite) number is called indefinite in respect of a greater.[(8)]

---

passes into absolute zero both $\frac{2}{0}$ and $\frac{1}{0}$ are simple (denumerable) infinities and so 'ought... to bee considered equall...'. On the style of Wallis' *De Angulo Contactus* and *Arithmetica Infinitorum* Newton considers two mathematical models of the paradox: the angle of contact (or horn angle) as the indivisible of a finite angle, and the surface-element as the indivisible of a surface.

(5) Read 'parallelogram'.

(6) Newton adds the familiar model of the geometrical point as the indivisible of a line-segment, and contrasts it with that of the contact angle. The notion of the 'crookednes' of a circle being '4 right angles' is here not wholly clear, but Newton seems to have taken it from Wallis' *De Angulo Contactus*: Cap. XII, *Argumentum quartum*: 40–5, whose argument demands the limit-equation of a circle with a regular *n*-sided polygon whose sides increase indefinitely in number but decrease correspondingly in magnitude. Since the total change in direction in traversing the whole perimeter of the polygon (or the sum of its exterior angles) is 4 right angles, the change in direction from one side to the next (or the magnitude of the exterior angle at each corner) is $4/n$ right angles. In the limit as *n* becomes indefinitely large and the polygon becomes a circle the total curvature remains constant at 4 right angles but each exterior angle becomes an angle of contact (between the circle and the tangent at that point) of unit-indivisible magnitude $4 \times (1/n) = 4 \times 0$ right angles. Since, finally, there is an (equal) angle of contact at each point of the circle perimeter, there are as many angles of contact in the total 'crookednes' of the circle as points in the line-segment equal in length to that perimeter.

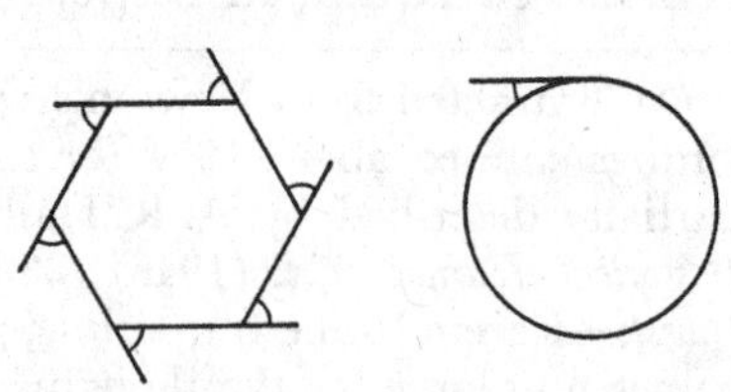

(7) Newton first began a word 'sur[ ]', presumably 'surface', but then cancelled it.

(8) This last paragraph, squashed in at the foot of the page and in a different ink, is clearly a later addition and probably Newton's own thought. Scholastic (Aristotelian) influence, encountered by Newton in his early undergraduate years during formal instruction and his required reading, is here very marked both in the concepts introduced and their expression.

Elsewhere in the *Quæstiones* Newton's mathematical notes are even more strongly scholastic in tone. Compare, for example, the following:

§2. EARLY NOTES AND RESEARCHES ON INDIVISIBLES

[Early(?) 1664][1]

From the original in a pocket-book in the University Library, Cambridge[2]

$166\frac{3}{4}$ January

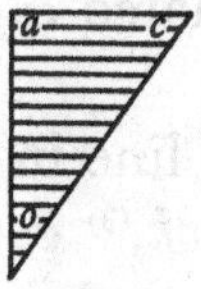

All y^e^ parallell lines wch can be understoode to bee drawne uppon any superficies are equivalent to it, as all y^e^ lines drawne from (*ao*) to (*co*) may be used in stead of y^e^ superficies (*aco*.)

If all ye parallell lines drawne uppon any superficies be multiplied

---

[Newton's p. 8]

Of Place

Extension is related to places, as time to days. yeares &c. Place is ye principium individuationis of streight lines & of equall & like figures....

[Newton's p. 10]

Of Motion

...There are so many parts in a line as there can stand Mathematicall points in a row w^th^out touching (i.e. falling into) one another in it & soe many degrees of motion along y^t^ li[n]e as there can be stops & stays....

[Newton's p. 87]

Of Quantity

If Extension is indefinite onely in greatness & not infinite y^n^ a point is but indefinitely little & yet we cannot comprehend any thing lesse. To say y^t^ extension is but indefinite...because we cannot perceive its limits, is as much as to say God is but indefinitely perfect because wee cannot apprehend his whole perfection.

(1) Newton's notes on what, by the mid-seventeenth century, had already come to be known as *the* 'method of indivisibles' (that is, the mathematical method of reasoning about the quadratures of areas and volumes through examination of their 'indivisible' elements, the line and surface respectively). A first general if not wholly systematic account of the method had been printed thirty years before by Cavalieri as his *Geometria Indivisibilibus Continuorum nova quâdam Ratione promota* (Bologna, ~1~1635) but had been quickly developed both by Cavalieri himself and by his pupil Torricelli. (See D. T. Whiteside, 'Patterns of Mathematical Thought in the later Seventeenth Century', *Archive for History of Exact Sciences*, **1** (1961): 179–388, especially VIII, *Calculus. 1. Indivisibles and the arithmetick of infinites.*) Wallis, in his tracts *De Sectionibus Conicis* and *Arithmetica Infinitorum*, began to popularize the method in England from 1656 and it seems beyond doubt that he is Newton's inspiration here. Ironically, Newton himself was to be a leader in the movement towards algorithmic calculus procedures which made indivisible theories obsolete by the end of the century. Till, however, his own ideas developed (under the overriding influence of Descartes' *Geometria*, as we shall see) he made use of indivisibles on several occasions, but not always wisely as his attempt to square the hyperbola reveals.

Note that the date (January 1663/4) here set at the head of the text is, in fact, separated from it by a considerable blank and may not refer strictly to these notes. However, Newton's writing in this date differs little from that of the text, which we may therefore assume to have been composed in the spring or early summer of 1664.

(2) Add. 4000: 82^r^, 83^r^–84^r^.

by another line they produce a Sollid like y^t wch re[s]ults(3) from ye superficies drawne into y^e [s]ame(4) line. as if either all ye lines in ye superficies (*oac*) or if ye superficies *oac* be drawne into ye line (*b*) they both produce ye same [s]ollid(5) (*d*). Whence All ye parallell superficies wch can bee understoode to bee in any sollid are equivalent to y^t Sollid.(6) And If all ye lines in any triangle, wch are parallell to one of ye sides, be squared there results a Pyramid.(7) if those in a square, there results a cube. If those in a crookelined figure there resu[l]ts a sollid wth 4 sides terminated & bended according to y^e fasshion of y^e crookelined figure.(8)

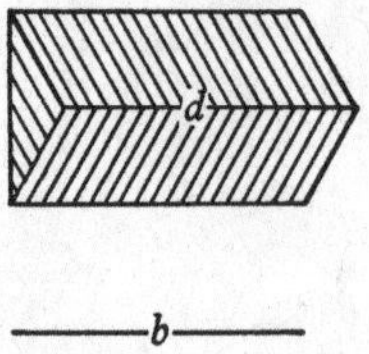

If each line in one superficies bee drawne into each correspondent line in another superfies as in *aebk*,(9) & *omnc* if *ae* × *dh*. *bk* × *cn*. *qv* × *wx* &c. they produce a sollid whos opposite sides are fashioned by one of ye superfic[ies] as ye Sollide *fpsrg*. where all y^e lines drawne from *fr* to *ps* are equall to all the correspondent lines drawne from *ow* to *mx*. & those drawne from *fg* to *fr* are equall to y^e correspondent lines drawne from *qz* to *vz*.(10)

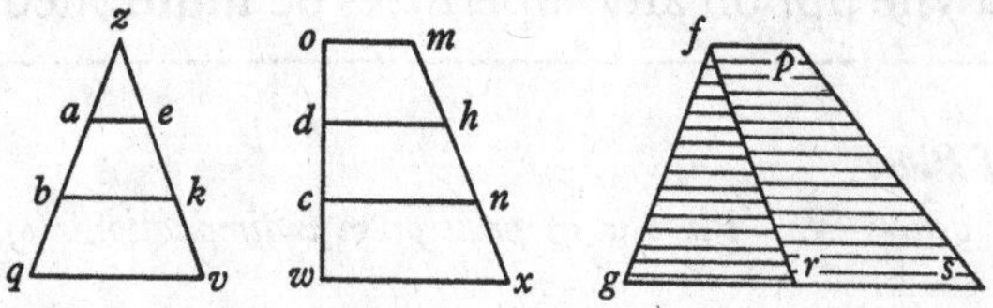

(3) Newton wrote 'relults'.

(4) The text has 'lame'.

(5) Newton wrote 'lollid'.

(6) Compare Wallis' *De Sectionibus Conicis*: 26–7: Prop. XI, *De Cuneo Parabolico.*

(7) *De Sectionibus Conicis*: 14–15: Prop. VI, *De Pyramide.*

(8) Newton generalizes, to an arbitrary 'crookelined figure', Wallis' definition of a conical *Pyramidoides* in his *De Sectionibus Conicis* (of which Prop. IX: 23–4, Prop. XV: 36 and Prop. XIX: 42, define respectively the *Pyramidoides Parabolicum, Ellipticum* and *Hyperbolicum*). In a conical Pyramid all sections through the axis are similar symmetric conics while all sections parallel to the base are squares, and Newton's generalization merely replaces conics by arbitrary symmetric curves.

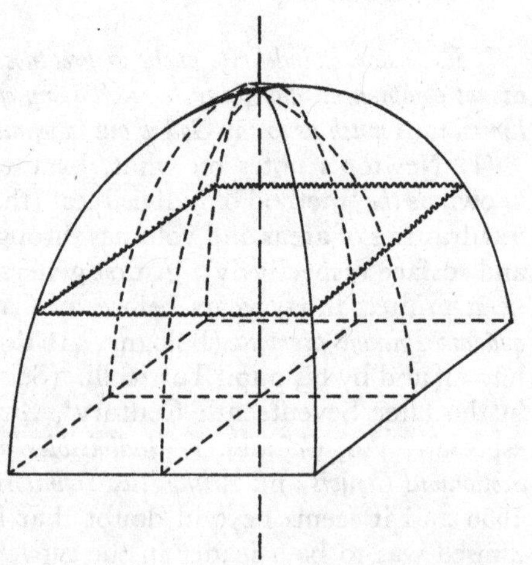

(9) Newton's diagram requires '*zbk*'.

(10) This generalization of the wedge formed by 'multiplying' a given curve into a constant line-length (or, more strictly, rectangle of constant width) was first formulated by Grégoire de Saint-Vincent in his *Opus Geometricum Quadraturæ Circuli et Sectionum Coni* (Antwerp, 1647), whose Liber VII, *De Ductu Plani in Planum*: 703–864, details the method with numerous examples. Wallis, having read Grégoire's work, was very much impressed with its power, introducing it in Prop. LXXV: 60, of his *Arithmetica Infinitorum* and returning to it in many of the following propositions. Since there is no evidence that Newton himself read the *Opus Geometricum*, we may assume that this *ductus plani in planum* method is yet one more annotation from Wallis.

To square y^e Parabola.

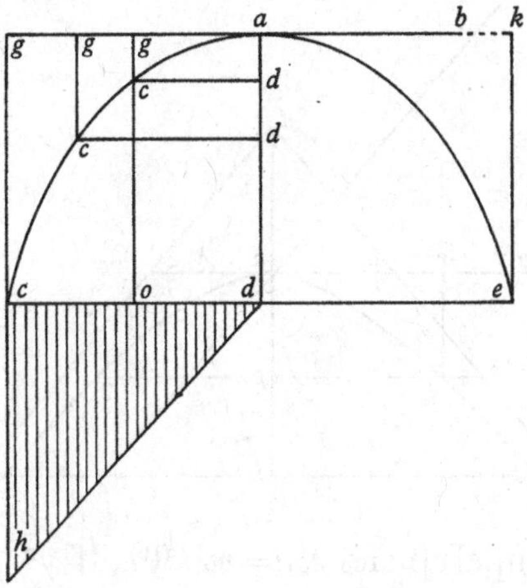

In y^e Parabola *cae* suppose y^e Parameter $ab=r$. $ad=y$. $dc=x$. & $ry=xx$ or $\frac{xx}{r}=y$.(11) Now suppose y^e lines called x doe increase in arithmeticall proportion [y^n] all y^e x's taken together make y^e superficies *dch* wch is halfe a square. let every line drawne from *cd* to *hd* be square & they produce a Pyramid equall to every $xx=\frac{x^3}{3}$.(12) wch if divided by r there remains $\frac{x^3}{3r}=\frac{yx}{3}$ equall to every $\frac{xx}{r}$ [y^t is] equall to every (y) or all y^e lines drawne from *ag* to *accc* equall to y^e superficies *agc* [y^t is] equall to a 3^d pte of y^e superficies *adcg* & y^e superficies $acd=\frac{2yx}{3}$.

Otherwise. suppose $ce=b$. $co=x$. $co=y$. & $ry=bx-xx$.(13) y^e lines x increaseing in arithmeticall proportion every x is equall to 4 times y^e superficies $cdh=\frac{bb}{2}$ wch drawne into b produceth y^e sollid $\frac{b^3}{2}$. but if every x be squarered they produce a pyramid equall to $\frac{b^3}{3}$. wherefore every $bx-xx=\frac{b^3}{6}$ [is] equall to every ry(14) equall to y^e superficies *adcc* drawne into r & $\frac{b^3}{6r}=$ to *cadc* as before.(15)

(11) The defining equation for the parabola.

(12) *Arithmetica Infinitorum*: Prop. XXII: 17. Wallis himself gives an arithmetical proof strictly equivalent to $\int_0^x x^2.dx = \frac{1}{3}x^3$, but Newton here seems to wish to assume the truth of this result from the allied theorem in pure geometry (traditionally ascribed to Democritus but first satisfactorily proved in Euclid's *Elements*) that the volume of a pyramid is one-third that of a prism with equal height and base-area.

(13) Newton finds, where $cd = b$, a new equation for his parabola by now taking $x \to b-x$.

(14) Since $\int_0^b ry.dx = \int_0^b (bx-x^2).dx = b\int_0^b x.dx - \int_0^b x^2.dx.$

(15) Two neat proofs by indivisible methods of a classical Greek result (first proved by Archimedes), yielding the immediate corollary that the general parabolic segment contained between an arbitrary chord and its conjugate diameter is exactly quadrable. Wallis, from whom it seems (note (12)) Newton took the result, himself very probably learnt the result from his reading of Torricelli's *Opera Geometrica* (Florence, 1644). The latter, among a score of proofs of the theorem composed in exemplification of various types of quadrature methods and printed as his *De Dimensione Parabolæ*, gave this present method of comparison with the volume of a pyramid as his Lemma XX: 57–8.

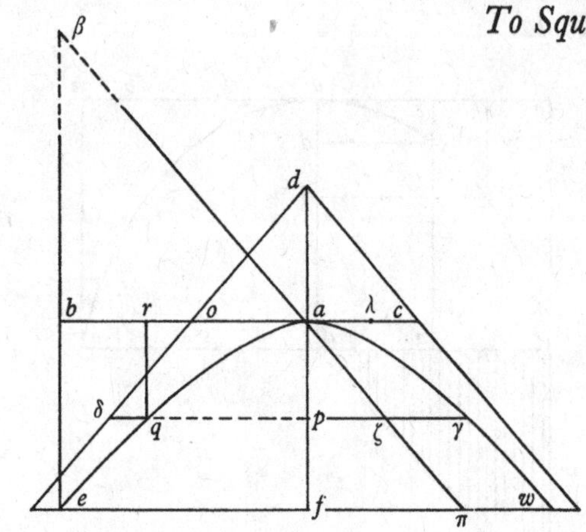

To Square ye Hyperbola.

In ye Hyperbola *eqaw.* suppose $ef=a$. $fa=b$. $ap=rq=y$. $a\lambda=d$. $pq=ar=x$.

$$ad=q={}^5oa={}^5ac={}^5d.^{(16)}$$

& $da+ar:ar::ar:rq.$(17) $xx=dy+yx$. In wch equation every x taken together is equall to ye triangle $a\beta b$ equall to $\frac{aa}{2}$ & every xx taken together is a pyramid $=\frac{a^3}{3}$.(18)

Every y taken together is equall to ye supe[r]ficies $eba=mkt$.(19) If yn $gh=lm=ns=a\lambda=d$. every dy is equall to ye solid *nglmhs*. If ye angle *mhk* is a right one & if $mh=gl=ba=ef=a$ that is if ye triangle $mhk=ab\beta$. every yx will be equall to ye sollid *mhstk*. Joyne these two sollids together as in $lmtng=\frac{a^3}{3}$.(20)

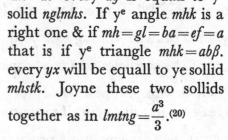

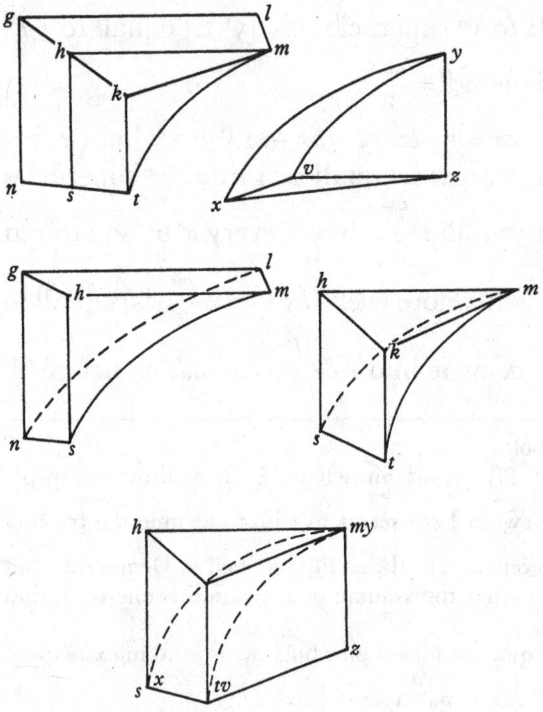

$$p\delta=pd=\frac{q+y}{5}.\ \&\ \frac{qy+yy}{5}=xx.^{(21)}$$

Againe suppose every x taken together to be equall to ye superficies *aef*(22), ye line $q\gamma$(23) squared is $4xx$. every $4xx$ composeth a sollid like (II ♉ ♌) an eighth ꝑte whereof (wch is equall to every $\frac{xx}{2}$) being like II ♉ ♓$o=xyzv$; xv will be equall to

II ♓ $=ef=a=st=xz=o$♓ $=hm$.

& $vz=\text{II}o=a\beta=km$. whence ye

(16) Newton first wrote '$ad=q=oa=ac=d$' and then added the numerical coefficients as shown. What he intended is not clear. (The second hyperbola below, defined by $5x^2=y(q+y)$, has transverse axis $ad=q=2\sqrt{5}d$.) The entry '$a\lambda=d$' in the preceding line (with λ clearly different from d in Newton's accompanying figure) is likewise mysterious, though perhaps intended for the $\frac{1}{4}$-parameter ($\frac{1}{2}d$) of the second hyperbola.

(17) A rare defining 'symptom' for the hyperbola *aqe* touched by *ca* in *a* and having *do* for

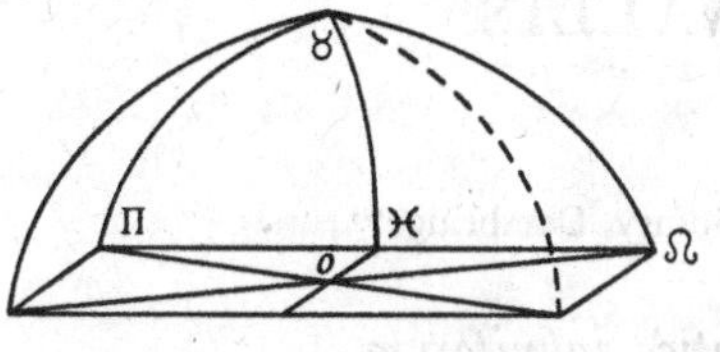

co[n]vexe superficies *xyv* of y^e^ figure *xyvz* will fitly joyne wth y^e^ concave superficies *mst* of y^e^ figure *shmtk*. If every $x^{(24)}$ is equall to y^e^ superficies *aef*, every $y^{(25)}$ shall be equall to y^e^ triangle $af\pi = \frac{bb}{2}$. every $yy^{(26)} = \frac{bbb}{3}$. every $qy = \frac{qbb}{2}$. & therefor y^e^ Sollid $yxzv^{(27)} = \frac{b^3}{30} + \frac{qbb}{20} = \frac{2b^3+3qbb}{60}$. Joyne y^e^ Sollid *shmkt* to *yxvz* & there resulte[t]h $shmzth = \frac{aab}{2}^{(28)}$. from wch againe substract $xvzy = \frac{2b^3+3qbb}{60}$ & there remains y^e^ sollid $mhstk = \frac{30aab - 2b^3 - 3qbb}{60}$. wch substract from y^e^ sollid $ntlmg = \frac{a^3}{3}$ & there remaines $nglmhs = \frac{20a^3+2b^3+3qbb-30aab}{60}$. wch being divided by $d = \sqrt{\frac{5rr}{4}}$ there remaines $\frac{40a^3+4b^3+6qbb-60aab}{r\sqrt{5}\,[\times 60]}$ = to y^e^ superficies *abe*.(29)

---

one of its asymptotes. The branch *aγw* in Newton's figure is not, of course, its continuation beyond *a*. (See note (29).)

(18) Analytically, $\int_0^a x.dx = \frac{1}{2}a^2$ and $\int_0^a x^2.dx = \frac{1}{3}a^3$.

(19) Read '*mhs*', which is $\int_0^a y.dx$ since the bounds are $\begin{cases} x = a \\ y = b \end{cases}$ and $\begin{cases} x = 0 \\ y = 0 \end{cases}$.

(20) Since $y(d+x) = x^2$, it follows that $\int_0^a y(d+x).dx = \int_0^a x^2.dx$.

(21) Newton added this at a later stage on the previous page, and it is inserted in its present position following his instruction. Using the 'symptom' *pd*:*ar*::*ar*:*aq*, he now defines a new hyperbola *aqe* by $5x^2 = y(q+y)$. The point *d* is no longer its centre but on the opposite branch of the curve. The branch *aγw*, however, now becomes the true continuation of *eqa*.

(22) That is, $\int_0^b x.dy$. (23) Equal to $2qp$. (24) $\int_0^b x.dy$. (25) $\int_0^b y.dy$.

(26) $\int_0^b y^2.dy$. (27) $\int_0^b \frac{1}{2}x^2.dy = \int_0^b \frac{1}{10}(y^2+qy).dy$.

(28) That is, the wedge $hs \times \Delta mhk = \frac{1}{2}a^2$.

(29) This is a marvellous tangle, but it is clear that, where $f(x, y) = 0$ is the defining equation of some simple (Apollonian) hyperbola, Newton seeks on the lines of his previous investigation of the parabola to evaluate one of $\int_0^a y.dx$ or $\int_0^b x.dy$ rationally in terms of *a* and *b*.

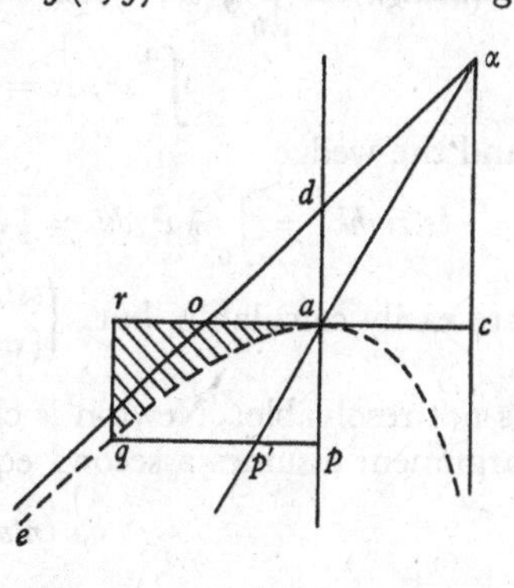

His first choice of equation is the deceivingly simple $x^2 = y(d+x)$, with $qp = x$, $ap = y$, $ad = d$ and $af = b$ (so that $a^2 = b(d+a)$). (This hyperbola, of latus rectum $2d/\sqrt{5}$ and transverse axis $2\sqrt{5}d$, has its centre at α and asymptotes αo and αc, where $\alpha \equiv (-d, -2d)$ and $c \equiv (-d, 0)$. We may easily show that *oa* is tangent at *a*. In Newton's figure the branch *eqa* alone is drawn but its smooth continuity

§3. ANNOTATIONS ON WALLIS

[1664/5][1]

From the original in a pocket-book in the University Library, Cambridge[2]

[1] *Annotations out of Dr. Wallis his Arithmetica infinitorum*

1 A primanary series of quant[it]ys is arithmetically proportionall, as 0, 1, 2, 3, 4. & its index[3] is 1.

Secundanary series are[4] those whose rootes are arithmetically proportionall; as, 0, 1, 4, 9, 16. & its index is 2.

Tertianary, quartanary, quintanary[5] series of quantitys are those whose cube, squaresquare, squarecube rootes are Arithmetically Proportionall as 0, 1, 8, 27, 64./ 0, 1, 16, 81, 156.[6]/ 0, 1, 32, 243, 624.[7] &c. Their indices being 3, 4, 5 &c.[8]

---

with $a\gamma w$, its mirror-image in *daf*, misleadingly suggests that the whole arc $eqa\gamma w$ belongs to a single hyperbola with asymptotes *do* and *dc*.) The area (*abe*), or $\int_0^a y\,.dx$ in the analytical equivalent, Newton now attempts to evaluate by a solid construction. Using his first diagram as basis he builds a set of interlocking volumes whose essential details are shown compactly in our figure. (The crudity of Newton's own illustrations is preserved in the textual diagrams.)

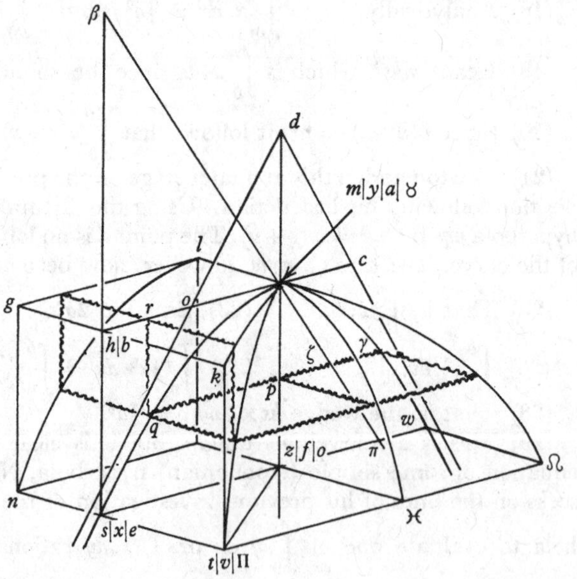

Following Newton's argument, the area (*abe*) sought is to be derived from the volume (*nsmlgh*) $= d \times (abe)$, where

$$(ntmlgk)-(nsmlgh) = (stmhk) = (stzmhk)-(stzm).$$

Here both

$$(ntmlgk) = \int_0^a y(d+x)\,.dx = \int_0^a x^2\,.dx = \tfrac{1}{3}a^3$$

and the wedge

$$(stzmhk) = \int_0^b \tfrac{1}{2}a^2\,.dy = \tfrac{1}{2}a^2 b$$

are easily calculable, but $\begin{Bmatrix}(stzm)\\(xvzy)\end{Bmatrix} = \int_0^b \tfrac{1}{2}x^2\,.dy = \int_0^b \tfrac{1}{2}y(d+x)\,.dy$

is not resolvable. Newton is clearly disappointed but in the hope of successfully revising his argument assumes a second equation for his hyperbola, $5x^2 = y(q+y)$. As before,

$$(ntmlgk)-(nsmlgh) = (stzmhk)-(stzm),$$

2 Primary[,] secundanary, tertianary series &c are said to bee reciprocally proportionall (y^t is to y^e same series increasing) which continually decrease as. $\frac{1}{0}, \frac{1}{1}, \frac{1}{2}, \frac{1}{3}, \frac{1}{4}$. $\frac{1}{0}, \frac{1}{1}, \frac{1}{4}, \frac{1}{9}, \frac{1}{16}$. $\frac{1}{0}, \frac{1}{1}, \frac{1}{8}, \frac{1}{27}, \frac{1}{64}$. Their indices being negative as $-1, -2, -3$.(9)

3 Subsecundanary, subtertianary, series &c: are those whose squares, cubes, &c are arithmetically proportionall, as $\sqrt{0}$, $\sqrt{1}$, $\sqrt{2}$, $\sqrt{3}$. $\sqrt{c:0}$, $\sqrt{c:1}$, $\sqrt{c:2}$, $\sqrt{c:3}$ &c. Theire indices being $\frac{1}{2}, \frac{1}{3}$, &c.(10)

4 The indices of compound or mixt of rationall & irrationall series, by multiplying or dividing y^e indices of y^e simple series may bee found. as in a

---

where $(nsmlgh) = d \times \int_0^a y.dx$, the wedge $(stzmhk) = \frac{1}{2}a^2b$ and (or at least Newton continues to take it so!) $(ntmlgk) = \frac{1}{3}a^3$, but now $(stzm) = \int_0^b \frac{1}{2}x^2.dy$ is equal to

$$\int_0^b \tfrac{1}{10}y(q+y).dy = \tfrac{1}{60}(3b^2q+2b^3),$$

which is rational. Since this revised argument is not cancelled Newton would appear to have persuaded himself that he has succeeded in squaring a general hyperbolic segment. Unfortunately, he has overlooked the crucial point that, since now $5x^2 = y(q+y)$,

$$(ntmlgk) \left(\text{or } \int_0^a y(d+x).dx\right)$$

is no longer $\frac{1}{3}a^3$ and indeed cannot be rationally evaluated. (In fact, the rational quadrature of any segment of a central conic is in general impossible and no revision of Newton's argument can achieve his aim.)

(1) The date is fixed by the autobiographical entry made by Newton on the facing page (14^v) which has already been quoted in full in the introduction: '...I find that in ye year 1664 a little before Christmas I...borrowed Wallis' works & by consequence made these Annotations...in winter between the years 1664 & 1665. At wch time I found the method of Infinite series. And in summer 1665...I computed ye area of ye Hyperbola...to two & fifty figures by the same method.' There is in existence (in private possession) an incomplete contemporary copy of these annotations in the handwriting of John Collins. Neither the copy nor the original seems to have had a wide circulation, though the former passed into William Jones' hands at the turn of the century, when it was checked against the original and a *copia vera* of the autobiographical passage appended in Jones' hand.

(2) Add. 4000: 15^r-17^v. It is partially (ff. 20^r–21^r) entered over some early notes on analytical geometry headed *Epitome Geometriæ* (2, 1, §2). Newton himself has added a pagination 1–19 on the top right corners of ff. 15^r–24^r.

(3) That is, the exponent i of the general term x^i, $x = 0, 1, 2, \ldots$.

(4) Newton first wrote 'A Secundanary series is...'.

(5) Wallis' Latin is perhaps better rendered 'quaternary' and 'quintenary' respectively.

(6) Read '256'. (7) Read '1024'.

(8) Newton restates Prop. XLIV: 35 of Wallis' *Arithmetica Infinitorum.*

(9) A summary of Prop. LXXXVII: 67–8. In the manuscript this paragraph is written below the next, but is here transposed to agree with Newton's numbering.

(10) Prop. LIV: 53. Compare also Prop. LVIII: 46–7.

subsecundanary progression cubed $\sqrt{0}$, $\sqrt{1}$, $\sqrt{8}$, $\sqrt{27}$, $\sqrt{64}$ y^e^ index is $\frac{1}{2}\times 3=\frac{3}{2}$. So in y^e^ cube rootes of a secundanary progression, $\sqrt{c:0}$, $\sqrt{c:1}$, $\sqrt{c:4}$, $\sqrt{c:9}$ &c. y^e^ index is $\frac{1}{3}\times 2=\frac{2}{3}$. so in irrational reciprocall progressions $\sqrt{qq:\frac{1}{0}}$, $\sqrt{qq:\frac{1}{1}}$, $\sqrt{qq:\frac{1}{2}}$, $\sqrt{qq:\frac{1}{3}}$, &c y^e^ index is $-1\times\frac{1}{4}=\frac{-1}{4}$.(11)

Now suppose y^e^ line *ac* be divided into an infinite number of equall pts *ad*, *de*, *ef*, *fg* &c, from each of wch are drawne parallells *nd*[,] *pe*[,] *qf* &c: wch increase continually in some of y^e^ foregoing progressions or in some progresion compounded of y^m^, all those lines may be taken for ye surface *bqnac*, & to know w^t^ proportion that superficies(12) hath to ye superficies *ambc*[,] y^t^ is w^t^ proportion all those lines have to soe ma[n]y equall to y^e^ greatest of y^m^, I say as y^e^ index of y^e^ progression increased by an unite is to an unite soe is y^e^ square(13) *abcm* to y^e^ area of y^e^ crooked line [*bqnac*].(14) As if *abc* is a parabola y^e^ lines *nd*, *pe*, *qf*, &c: are a subsecundanary series (for $y=\sqrt{rx}$) whos index is $\frac{1}{2}$ wch added to an unite is $1+\frac{1}{2}=\frac{3}{2}$. Therefore $\frac{3}{2}:1::3:2$ so is y^e^ square *ambc* to y^e^ area of y^e^ Parab. (y^e^ names of y^e^ lines are (*ad*), *ae*, *af* &c$=x$. *dn*, *pe*, *qf* &c:$=y$. $ac=p$. $bc=q$.)(15) The case is y^e^ same if *abc* bee supposed a sollid, as suppose a Parabolicall conoides. y^n^ since y^e^ nature of it is $rx=yy$. yy designes y^e^ squares *nd*, *pe*, *qf* &c: all wch taken together are equivalent to y^e^ sollid. & those □s(16) increase in y^e^ same proportion w^ch^ *rx*, or *x* doth. y^t^ is they are a primanary series whose index is 1. to w^ch^ (according to y^e^ rule) I ad an unite & tis 2. Therefor 1:2:: soe are all y^e^ □^s^ of y^e^ Primary series to soe many □s equall to y^e^ greatest of y^t^ series, & soe is y^e^ conoides to a cilinder of y^e^ same altitude.(17)

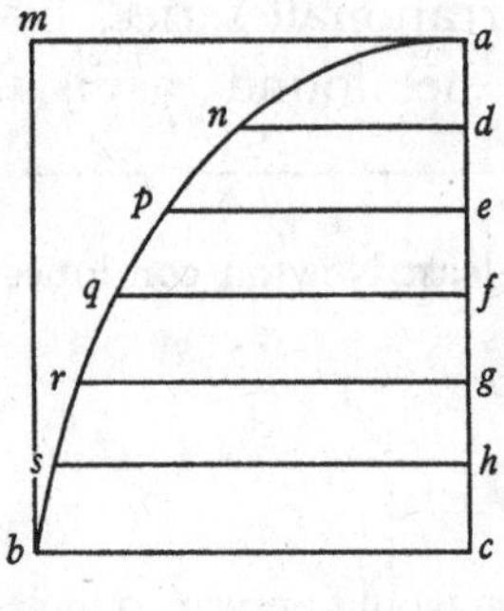

(11) Prop. CV: 79–81.
(12) Newton has cancelled 'all those lines'.
(13) Read 'parallelogram'.
(14) That is, $\int_0^p p^n dx : \int_0^p x^n . dx = (n+1):1$, where $ad=x$, $nd=x^n$, say, and $ac=p$. Wallis gave many treatments of this basic theorem of the method of indivisibles, but dealt with it most generally in Prop. CLXXXII, *Lemma*: 146–61 of his *Arithmetica Infinitorum*.
(15) The gist of Prop. LXIV: 52–3 and Prop. LXV: 53. Note that the accompanying diagram is Newton's, though based on Wallis' in a general way.
(16) Read 'squares'.
(17) A summary of Prop. LXVII–LXXII: 54–8.

Also if a superficies be compounded of 2 or more of these series, Their area is as easily found; as if y[e] nature of y[e] line bee $y=aa-xx$, or

$$y=a^4-2aaxx+x^4 \quad \text{or} \quad y=a^6-3a^4xx+3aax^4-x^6. \text{ \&c.}$$

Their areas will bee to y[e] parallelograms about them as 2 to 3, as 8 to 15, as 48 to 105 &c.[18] but if I put in y[e] intermediate termes in these last named lines their order will bee $y=\sqrt{aa-xx}$, $y=aa-xx$, $y=\overline{aa-xx}\sqrt{aa-xx}$,

$$y=a^4-2aaxx+x^4, \quad y=\overline{a^4-2aaxx+x^4}\sqrt{aa-xx}. \quad y=a^6-3a^4xx+3aax^4-x^6;$$

&c: & since these lines observe a geometricall progression their areas must observe some kind of progression. of w[ch] every other terme is given viz

$$1.\ \square.\ \frac{2}{3}.\ *\ .\ \frac{8}{15}.\ *\ .\ \frac{48}{105}.\ *\ .\ \frac{384}{945}.\ *\ .\ \frac{3840}{10395}.\ [\&c]^{(19)}$$

Twixt w[ch] termes if y[e] intermediate termes $\square$.* can bee found y[e] 2[d] $\square$ will give y[e] area of y[e] line $y=\sqrt{aa-xx}$, y[e] circle.[20] Soe likewise in this progression of lines $y=1$. $y=\sqrt{ax-xx}$. $y=ax-xx$. $y=\overline{ax-xx}\sqrt{ax-xx}$. $y=aaxx-2ax^3+x^4$. &c: y[e] progression of their areas is $1:\square:\frac{1}{6}:*:\frac{1}{30}:*:\frac{1}{140}:*:\frac{1}{630}$: &c. y[e] 2[d] terme if it can bee found givs y[e] area of y[e] $\odot$[21] for as its denominator to its numerator so is y[e] $\square$[22] of y[e] diameter to y[e] area of a semicircle.[23] If this last progresion

(18) Prop. CXI–CXVI: 85–8. Here

$$\int_0^a (a^2-x^2).dx : \int_0^a a^2.dx = 2:3, \quad \int_0^a (a^2-x^2)^2.dx : \int_0^a a^4.dx = 8:15$$

and

$$\int_0^a (a^2-x^2)^3.dx : \int_0^a a^6.dx = 48:105.$$

(19) Newton uses Wallis' reasoning *ex lege continuitatis*, or from a structural pattern. (Compare Prop. CXVIII–CXXI: 88–92; Prop. CXXXII: 104–6; and Prop. CLXIX: 136–7.) In Newton's adaptation a function, say $f(r)$, is defined, for integral values of r, as

$$\frac{\int_0^a (a^2-x^2)^r.dx}{\int_0^a a^{2r}.dx}\left[=\int_0^1 (1-x^2)^r.dx\right],$$

and the problem is posed of interpolating the sequence at $\frac{1}{2}$-intervals. Arranging $f(r)$ in a row, $r = 0, 1, 2, \ldots$, Newton (following Wallis in Prop. CLXXXIV: 161–3) sets $f(\frac{1}{2}) = \square$, the 'square' to be found, and leaves the other $\frac{1}{2}$-intervals blank. (Note that here $\square = f(\frac{1}{2}) = \frac{1}{4}\pi$ is not strictly Wallis' symbol but its reciprocal.)

(20) That is, as $\square a^2$. (21) Read 'circle'. (22) Read 'square'.

(23) Prop. CLXV: 128. This function, say $g(s)$,

$$\frac{\int_0^a (ax-x^2).dx}{\int_0^a a^{2s}.dx}\left[=\int_0^1 (x-x^2).^s dx\right]$$

bee multiplyed by y^e^ respective termes in y^e^ progress 1. 2. 3. 4 &c it may bee diminished, y^e^ res[ul]t[24] being $1.\ 2\square.\ \frac{1}{2}.\ 4*.\ \frac{1}{6}.\ 6*.\ \frac{1}{20}.\ 8*.\ \frac{1}{70}$[. &c]. soe y^t^ in this progression $1,\ b,\ \frac{1}{2},\ c,\ \frac{1}{6},\ d,\ \frac{1}{20},\ e,\ \frac{1}{70},f$, &c: if b can be found y^n^ y^e^ □ of y^e^ diameter to y^e^ area of y^e^ circle is as y^e^ denominator of b to its numerator.[25] Likewise y^e^ 1st series of areas may be diminished by multiplying each terme by its correspondent terme in this progression 1, 2, 3, 4, 5, 6, &c: & it will become $1,\ a,\ 2,\ b,\ \frac{8}{3},\ c,\ \frac{48}{15},\ d,\ \frac{384}{105},\ e,\ \frac{3840}{945}$. &c. In wch if a can bee found y^n^ as y^e^ denom[:] of a to its num:[26] so y^e^ □[27] of y^e^ Radius to a semicircle, y^t^ [is] making y^e^ radius$=q$. $2aq=\odot$.[28] The same kinds of changes may bee performed by any other progressions, as by division by y^e^ geometricall progression 1, 2, 4, 8, 16, &[c] y^e^ first series of areas becomes, $1,\ g,\ \frac{1}{6},\ h,\ \frac{1}{30},\ k,\ \frac{1}{140},\ l,\ \frac{1}{630}$, &c viz y^e^ same wth y^e^ 2^d^ series. Also these changes may be done by addition or substraction of mutuall termes in 2 proportions. Soe y^t^ y^e^ most convenient way may be chosen, wherby to reduce any series of proportions to y^e^ most conv[eni]ent forme.[29]

---

is set out by Newton at $\frac{1}{2}$-intervals, with integral values of the argument, $s = 0, 1, 2, \ldots$, entered numerically, $g(\frac{1}{2}) = \frac{1}{8}\pi$ entered as □ and the remaining spaces left blank. (Note that for integral values of s, $g(s)$ may be evaluated as the continued product $1 \times \frac{1}{2\times 3} \times \frac{2}{2\times 5} \times \frac{3}{2 \times 7} \times \ldots$.) In a theorem (Prop. CXCII: 193) added when the *Arithmetica Infinitorum* was in the press Wallis set the progression in geometrical form: where $VT = s$ and

$$TC = g'(s) = \left(\int_0^1 (x-x^2)^s . dx\right)^{-1}$$

define the curve VC, we can 'square' the circle $y^2 = x - x^2$ if we can find the ordinate

$$g'(\tfrac{1}{2}) = \frac{1}{\square} = \frac{8}{\pi}.$$

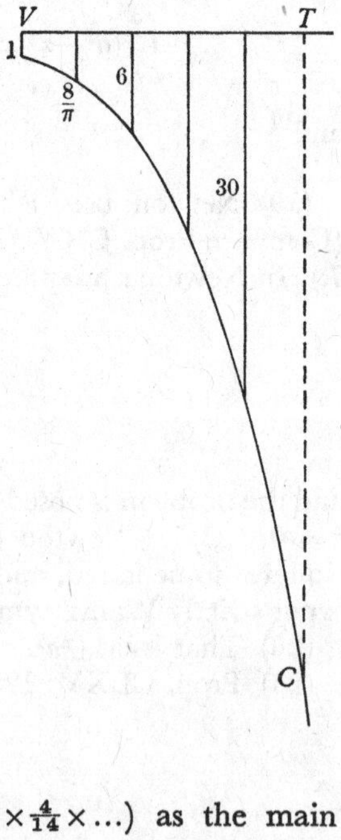

(Compare Wallis' letter to Oughtred set in appendix to the *Dedicatio* and dated *è Typographeo Oxoniensi postridie Paschatis* [*1656*].)

(24) Newton wrote 'reslut'.

(25) Newton's gloss on Prop. CXXXI: 103–4, where the progression,

$$h(s) = \int_0^1 (2s+1)\ (x-x^2) . dx,$$

appears (tabulated only for integral values, $h(s) = 1 \times \frac{1}{2} \times \frac{2}{6} \times \frac{3}{10} \times \frac{4}{14} \times \ldots$) as the main diagonal of the accompanying table.

Now if it be propounded to find these middle termes, It will be convenient to find how the given proportion may bee deduced from an Arithmeticall, Geometricall, or some other familiar proportion, viz whose meane termes may be found; as this progression 1. $\frac{2}{3}$. $\frac{8}{15}$. $\frac{48}{105}$ deduceth its originall from this $A\times\frac{0\times2\times4\times6\times8}{1\times3\times5\times7\times9}$ &c in wch A is an infinine(30) number $=\frac{1}{0}$.(31)

It will also be convenient to find what relatiō all y^e^ other meanes have to y^e^ first soe y^t^ if y^e^ first bee had all y^e^ other[s] may be deduced thence. As in this case suppose y^e^ 1st meane to bee a. The progression will be $\frac{1}{2}a$: 1: a: $\frac{3}{2}$: $\frac{4a}{3}$: $\frac{15}{8}$: $\frac{8a}{5}$: $\frac{105}{48}$: $\frac{64a}{35}$: $\frac{945}{384}$: $\frac{640a}{315}$, deducing its originall from $A\times\frac{0\times2\times4\times6\times8\times10}{1\times3\times5\times7\times9\times11}$(32) & from this $A\left(=\frac{1}{2}\right)\times\frac{2a\times4a\times6a\times8a\times10a}{1\ \times3\ \times5\ \times7\ \times9}$. &c.(33) (note y^t^ y^e^ proportions of these meane termes to one another, or to (a), are found by finding y^e^ proportion of y^e^ circle $y=\sqrt{aa-xx}$ to y^e^ line $y=aa-xx\sqrt{aa-xx}$ &c).(34)

---

(26) Read 'numerator'.

(27) Read 'square'.

(28) Read '[area of the] circle'.

(29) Newton, stimulated by his reading of Wallis, lets his mind roam free.

(30) Read 'infinite'.

(31) Based on Prop. CLXXXVII: 167–9, where Wallis analyses the table accompanying his Prop. CLXXXIV.

(32) Read '$A(=1)\times\frac{3\times5\times7\times9}{2\times4\times6\times8}$ &c'.

(33) Read '$A(=\frac{1}{2}a)\times\frac{2\times4\times6\times8\times10}{1\times3\times5\times7\times9}$ &c'.

(34) Newton reduces his former function

$$f(r) = \frac{\int_0^a (a^2-x^2)^r.dx}{\int_0^a a^{2r}.dx} \quad \left(\text{or more simply } f(r) = \int_0^1 (1-x^2)^r.dx\right)$$

to strict Wallisian form by tabulating its reciprocal

$$F(r) = \frac{1}{f(r)}, \quad r = -\tfrac{1}{2},\ 0,\ \tfrac{1}{2},\ 1,\ 1\tfrac{1}{2},\ \ldots$$

In particular, $a = F(\frac{1}{2}) = 4/\pi$ is the number which Wallis tabulated by □ in Prop. LXXXVIII and succeeding propositions of *Arithmetica Infinitorum*. (Note that

$$F(r) = \frac{\Gamma(r+\frac{3}{2})}{\Gamma(\frac{3}{2})\times\Gamma(r+1)},$$

and compare D. T. Whiteside, 'Patterns of Mathematical Thought in the Later Seventeenth Century', *Archive for History of Exact Sciences*, **1** (1961): 179–388, especially IV, 'Concept of function. 2. Interpolation': 237–40.)

In this case to find y^e quantity a;[35] Naming y^e termes in y^e progress:

$$\begin{array}{cccccccc} \frac{1}{2}a: & 1: & a: & \frac{3}{2}: & \frac{4a}{3}: & \frac{15}{8}: & \frac{8a}{5}: & \frac{35}{16}: \\ b & c & d & e & f & g & h & k \end{array}$$

1st observe y^t $\frac{d}{b}=2.\ \frac{e}{c}=\frac{3}{2}.\ \frac{f}{d}=\frac{4}{3}.\ \frac{g}{e}=\frac{5}{4}.\ \frac{h}{f}[=]\frac{6}{5}$ &c y^e proportions still decreasing & therefore[36] y^t in $\frac{c}{b}.\ \frac{d}{c}.\ \frac{e}{d}.\ \frac{f}{e}.\ \frac{g}{f}.\ \frac{h}{g}.\ \frac{k}{h}$ &c: y^e latter terme is lesse y^n y^e former: & therefore

$$\frac{dd}{cc}\text{ is }\left\{\begin{array}{l}\text{less y}^{\text{n}}\ \frac{d}{c}\times\frac{c}{b}=\frac{d}{b}=2.\\[4pt] \text{greater y}^{\text{n}}\ \frac{d}{c}\times\frac{e}{d}=\frac{e}{c}=\frac{3}{2}.\end{array}\right.$$

or

$$a=d^{(37)}\text{ is }\left\{\begin{array}{l}\text{less y}^{\text{n}}\ 1\times\sqrt{2}=\sqrt{1+\frac{1}{1}}.\\[4pt] \text{greater y}^{\text{n}}\ 1\times\sqrt{\frac{3}{2}}=\sqrt{1+\frac{1}{2}}.\end{array}\right.$$

Also

$$\frac{ff}{ee}\text{ is }\left\{\begin{array}{l}\text{lesse y}^{\text{n}}\ \frac{f}{e}\times\frac{e}{d}=\frac{f}{d}=\frac{4}{3}\\[4pt] \text{greater y}\ \frac{f}{e}+\frac{g}{f}=\frac{g}{e}=\frac{5}{4}\end{array}\right\}=Q:\frac{8a}{9}=\frac{2\times4a}{3\times3}.^{(38)}$$

---

(35) Newton has cancelled 'it may be considered y^t $\frac{a}{\frac{1}{2}a}=2.\ \frac{3}{1\times2}=\frac{3}{2}.\ \frac{4a}{3\times a}=\frac{4}{3}$'.

(36) Appealing to Wallis' continuity principle.

(37) That is, d/c since $c=1$. (38) Read '$\left(\frac{8a}{9}\right)^2=\left(\frac{2\times4a}{3\times3}\right)^2$'.

(39) Newton has cancelled '$3\times3\sqrt{\frac{4}{3}}$'.

(40) A lucid summary of the first part (pp. 178–9) of Wallis' lengthy Prop. CLXXXXI: 178–93. The function $F(r)=\left(\int_0^1(1-x^2)^r.dx\right)^{-1}$ is tabulated at $\frac{1}{2}$-intervals, with

$$b=F(-\tfrac{1}{2})=2/\pi,\quad c=F(0)=1,\quad d=F(\tfrac{1}{2})=a=4/\pi,\quad e=F(1)=\tfrac{3}{2},\quad \ldots.$$

Since $F(r)=\frac{\Gamma(r+\frac{3}{2})}{\Gamma(\frac{3}{2})\times\Gamma(r+1)}$, it follows that $F(r)=\prod_{1\leqslant i\leqslant r}\left(\frac{2i+1}{2i}\right)$ and

$$F(r+\tfrac{1}{2})=\prod_{1\leqslant i\leqslant r}\left(\frac{2i+2}{2i+1}\right)\times a,\quad r=0,\ 1,\ 2,\ 3,\ \ldots.$$

From the (Schwarz) inequalities

$$\frac{F(r)}{F(r-1)}>\frac{F(r+1)}{F(r)}\quad\text{and}\quad\frac{F(r+1)}{F(r)}>\frac{F(r+2)}{F(r+1)},$$

Therefore[39] a is $\left\{\begin{array}{l}\text{lesse y}^{\text{n}}\ \dfrac{3\times 3}{2\times 4}\sqrt{\dfrac{4}{3}}.\\[2ex] \text{greater y}^{\text{n}}\ \dfrac{3\times 3}{2\times 4}\sqrt{\dfrac{5}{4}}.\end{array}\right.$ And so by y^e^ same reasoning,

$$a \text{ is } \left\{\begin{array}{l}\text{less y}^{\text{n}}\ \dfrac{9\times 25\times 49\times 81\times 121\times 169}{2\times 16\times 36\times 64\times 100\times 144\times 14}\sqrt{\dfrac{14}{13}}.\\[2ex] \text{greater y}^{\text{n}}\ \dfrac{3\times 3\times 5\times 5\times 7\times 7\times 9\times\ 9\times 11\times 11\times 13\times 13}{2\times 4\times 4\times 6\times 6\times 8\times 8\times 10\times 10\times 12\times 12\times 14}\sqrt{\dfrac{15}{14}}.\end{array}\right. \quad \text{\&c.}^{(40)}$$

Thus Wallis doth it, but it may bee done thus.

$$a \text{ is } \left\{\begin{array}{l}\text{greater y}^{\text{n}}\ 1.\\[2ex] \text{less then } \dfrac{3}{2}.\end{array}\right. \quad \left|\ \dfrac{4a}{3} \text{ is } \left\{\begin{array}{l}\text{greater then } \dfrac{3}{2}.\\[2ex] \text{lesse then } \dfrac{15}{8}.\end{array}\right.\right. \quad \text{Therefore}$$

$$a \text{ is } \left\{\begin{array}{l}\text{greater y}^{\text{n}}\ \dfrac{3\times 3}{2\times 4}.\\[2ex] \text{lesse then } \dfrac{3\times 3\times 5}{2\times 4\times 4}.\end{array}\right. \quad \dfrac{8a}{5} \text{ is } \left\{\begin{array}{l}\text{greater y}^{\text{n}}\ \dfrac{15}{8}.\\[2ex] \text{lesse y}^{\text{n}}\ \dfrac{35}{16}.\end{array}\right. \quad \text{y}^{\text{t}} \text{ is}$$

$$a \text{ is } \left\{\begin{array}{l}\text{greater then } \dfrac{3\times 3\times 5\times 5}{2\times 4\times 4\times 6} = \dfrac{3\times 5\times 5}{2\times 4\times 4\times 2}.\\[2ex] \text{lesse then } \dfrac{3\times 3\times 5\times 5\times 7}{2\times 4\times 4\times 6\times 6} = \dfrac{5\times 5\times 7}{2\times 4\times 4\times 4}.\end{array}\right. \quad \text{\&c.}$$

By y^e^ same reasoning

$$a \text{ is } \left\{\begin{array}{l}\text{greater y}^{\text{n}}\ \dfrac{3}{2}\times\dfrac{3}{4}\times\dfrac{5}{4}\times\dfrac{5}{6}\times\dfrac{7}{6}\times\dfrac{7}{8}\times\dfrac{9}{8}\times\dfrac{9}{10} = \dfrac{5\times 49\times 81}{2\times 16\times 4\times 64\times 2}.\\[2ex] \text{lesse y}^{\text{n}}\ \dfrac{3}{2}\times\dfrac{3}{4}\times\dfrac{5}{4}\times\dfrac{5}{6}\times\dfrac{7}{6}\times\dfrac{7}{8}\times\dfrac{9}{8}\times\dfrac{9}{10}\times\dfrac{11}{10} = \dfrac{49\times 81\times 11}{2\times 16\times 4\times 64\times 4}.\end{array}\right.$$

---

checked to hold in the tabulated scheme for low values of r, Wallis 'induced' without further analysis the tighter inequalities

$$\frac{F(r)}{F(r-\frac{1}{2})} > \frac{F(r+\frac{1}{2})}{F(r)} \quad \text{or} \quad F(r) > \sqrt{[F(r-\tfrac{1}{2})\times F(r+\tfrac{1}{2})]}$$

and correspondingly $F(r+\frac{1}{2}) > \sqrt{[F(r)\times F(r+1)]}$, $r = 1, 2, 3, \ldots$. Substituting and combining, Wallis derived the inequality

$$\prod_{1\leqslant i\leqslant r}\left(\frac{(2i+1)^2}{2i(2i+2)}\right)\sqrt{\frac{2r+3}{2r+2}} < a\left(\text{or } \frac{4}{\pi}\right) < \prod_{1\leqslant i\leqslant r}\left(\frac{(2i+1)^2}{2i(2i+2)}\right)\sqrt{\frac{2r+2}{2r+1}},$$

here given by Newton for $r = 0$, 1 and 6.

$$\text{Or}\quad a \text{ is}\begin{cases}\text{greater } \bar{y}\ \dfrac{11\times169\times225\times289\times369^{(41)}\times441}{2\times16\times4\times64\times4\times144\times4\times256\times4\times400\times2}.\\[2ex] \text{lesse y}^{\text{n}}\ \dfrac{169\times225\times289\times369^{(41)}\times441\times23}{2\times16\times4\times64\times4\times144\times4\times256\times4\times400\times4}.^{(42)}\end{cases}$$

Note y[t] *a* is greater y[n] $\frac{1}{2}$ these two summes.(43)

---

[2] [*The binomial theorem invented in Wallisian style.*]

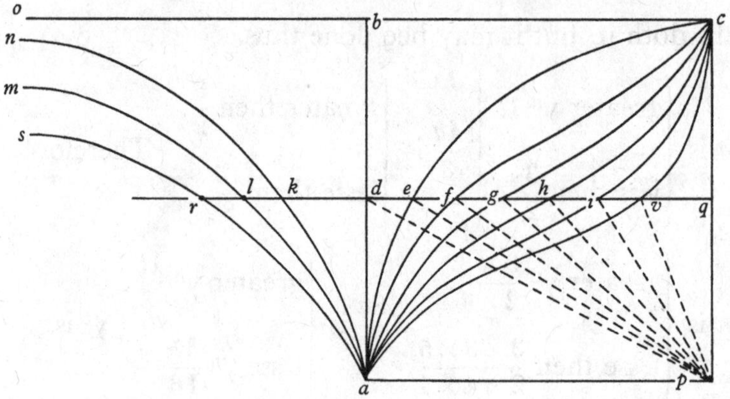

Having y[e] signe(44) of any angle to find y[e] angle or to find y[e] content of any segm[nt] of a circle.

Suppose y[e] circle to be *aec*. its semi=diameter $ap=pc=1$. y[e] given sine $pq=x$, viz: y[e] signe of y[e] angle *epa*. y[e] segment sought *eapq*. *abcp* the □(45) of its Radius. & y[t], *qi*: *qh*: *qg*: *qf*: *qe*: *qd*: *qk*: *ql*: *qr* &c are continually proportionall. Then is $eq=\sqrt{1-xx}$. $fq=1-xx$. $gq=\overline{1-xx}$ in $\sqrt{1-xx}$. $hq=1-2xx+x^4$.

$$iq=\overline{1-2x^2+x^4\sqrt{1-xx}}.\quad dq=1.\quad kq=\frac{1}{\sqrt{1-xx}}.$$

$$lq=\frac{1}{1-xx}.\quad rq=\frac{1}{1-xx \text{ in } \sqrt{1-xx}}.\ \&\text{c.}$$

& since all the ordinately applyed lines in these figures *abcq*, *aecq*, *afcq*, *agcq* &c

---

(41) Read '361'.

(42) Newton weakens Wallis' inequality but gains in simplicity of argument. Noting that $F(r) < F(r+1) < F(r+2)$ in the tabulated scheme, Newton 'induces' by an appeal to Wallis' law of continuity the tighter inequality $F(r) < F(r+\frac{1}{2}) < F(r+1)$. Substituting for these their continued-product expansions (note (40)) and dividing by $\prod_{1\leqslant i\leqslant r}\left(\frac{2i+2}{2i+1}\right)$, we have

$$\prod_{1\leqslant i\leqslant r}\left(\frac{(2i+1)^2}{2i(2i+2)}\right) < a < \prod_{1\leqslant i\leqslant r}\left(\frac{(2i+1)^2}{2i(2i+2)}\right)\frac{2r+3}{2r+2},$$

are geometrically proportionall their areas *adqp*, *aeqp*, *afqp*, *agqp*, *ahpq*(46) &c will observe some proportion amongst one another.(47) To find wch proportion, 1st $adqp = 1 \times x = x$. 2dly *afc* is a parab:(48) therefore $afqp^{(49)} = x - \frac{x^3}{3}$. also since tis $gh = 1 - 2xx + x^4$, therefore $ahqp^{(50)} = x - \frac{2}{3}x^3 + \frac{1}{5}x^5$. Also $vq = 1 - 3xx + 3x^4 - x^6$,

---

Newton's result and 'Wallis'' theorem. This Newton proceeds to simplify, using the reductions:

$$\prod_{1 \leqslant i \leqslant r} \left(\frac{(2i+1)^2}{2i(2i+2)} \right) = \begin{cases} = \dfrac{(r+1) \times \prod_{\frac{1}{2}r+1 \leqslant i \leqslant r} ((2i+1)^2)}{\prod_{1 \leqslant i \leqslant \frac{1}{2}r} ((8i)^2)} & \text{for } r \text{ even (given by Newton for } r = 2, 4, 10), \\[2ex] = \dfrac{\prod_{\frac{1}{2}(r+1) \leqslant i \leqslant r} ((2i+1)^2)}{\prod_{1 \leqslant i \leqslant \frac{1}{2}(r-1)} ((8i)^2) \times 4(r+1)} & \text{for } r \text{ odd (given by Newton for } r = 1\text{); and} \end{cases}$$

$$\prod_{1 \leqslant i \leqslant r} \left(\frac{(2i+1)^2}{2i(2i+2)} \right) \frac{2r+3}{2r+2} = \begin{cases} = \dfrac{(2r+3) \times \prod_{\frac{1}{2}r+1 \leqslant i \leqslant r} ((2i+1)^2)}{2 \times \prod_{1 \leqslant i \leqslant \frac{1}{2}r} ((8i)^2)} & \text{for } r \text{ even (Newton: } r = 2, 4, 10), \\[2ex] = \dfrac{(2r+3) \times \prod_{\frac{1}{2}(r+1) \leqslant i \leqslant r} ((2i+1)^2)}{2 \times \prod_{1 \leqslant i \leqslant \frac{1}{2}(r-1)} ((8i)^2) \times (2(r+1))^2} & \text{for } r \text{ odd (Newton: } r = 1). \end{cases}$$

The primary theorem was discovered, independently and earlier (about 1659), by Pietro Mengoli, but his researches appeared in print only a decade later as his *Circolo* (Bologna, 1672). In particular, Newton's present argument may be found on Mengoli's pp. 26–8. (Mengoli does not appear to have known Wallis' work, but wrote (p. 1): 'Cerceni, sino da giovinetto, il Problema della quadratura del circolo, il più desiderato di tutti nella geometria, come dissi nella prefatione del mio libro delle *Quadrature Arithmetiche* l'anno 1650. Lo trovai, dopo haver serrato il libro de gli *Elementi della geometria speciosa*, che stampai l'anno 1659, cioè l'anno sequente 1660: differendo di conferirlo al mondo, con gli altri *Elementi di Speciosa*, che hò manuscritti dopo, e sino al numero di quattro.')

(43) This last remark of Newton's, not to be found in Wallis, was probably an induction from the numerical cases listed, and suggests that Newton had calculated out several of the continued products as a check on their convergence to $a = 4/\pi$. It may well be that he used this theorem, not too difficult to prove, as a means of shortening his calculations. When, as Wallis did in his Prop. CXCIII: 194, we construct the geometrical curve

$$y = F(x) = \left(\int_0^1 (1-t^2)^x . dt \right)^{-1}$$

with respect to Cartesian co-ordinates $VT = x$, $TC = y$, we see at once visually that it is everywhere convex, so that $F(x) > \frac{1}{2}(F(x+k) + F(x-k))$. This, when $k = \frac{1}{2}$ and x is integral, may have been the source of Newton's insight.

(44) Read '[trigonometrical] sine'.

(45) Read 'square'.

(46) Read more correctly '*ahqp*'.

(47) By Wallis' continuity principle.

(48) Since its Cartesian equation is $y = 1 - x^2$. (Note that in the diagram its representing curve *afc* should be drawn wholly convex upwards.)

(49) $\int_0^x (1-t^2) . dt$.

(50) $\int_0^x (1-t^2)^2 . dt$.

therefore $avqp$(51) $=x-x^3+\frac{3}{5}x^5-\frac{1}{7}x^7$. & by the same proceeding y^e^ proportion may bee still continued after this manner

$$x:\quad x-\frac{1}{3}x^3:\quad x-\frac{2}{3}x^3+\frac{1}{5}x^5:\quad x-\frac{3}{3}x^3+\frac{3}{5}x^5-\frac{1}{7}x^7:$$

$$x-\frac{4}{3}x^3+\frac{6}{5}x^5-\frac{4}{7}x^7+\frac{1}{9}x^9:\quad x-\frac{5}{3}x^3+\frac{10}{5}x^5-\frac{10}{7}x^7+\frac{5}{9}x^9+\frac{1}{11}x^{11}:$$

$$x-\frac{6}{3}x^3+\frac{15}{5}x^5-\frac{20}{7}x^7+\frac{15}{9}x^9-\frac{6}{11}x^{11}+\frac{1}{13}x^{13}:$$

$$x-\frac{7}{3}x^3+\frac{21}{5}x^5-\frac{35}{7}x^7-\frac{35}{9}x^9-\frac{21}{11}x^{11}+\frac{7}{13}x^{13}-\frac{1}{15}x^{15}.\ \&c.^{(52)}$$

And if y^e^ meane termes be inserted it will bee

$$x:\quad x-^{(53)}:\quad x-\frac{1}{3}x^3:\quad x-\frac{3}{6}x^3+\qquad:\quad x-\frac{2}{3}x^3+\frac{1}{5}x^5:\quad x-\frac{5}{6}x^3+\frac{2}{5}x^5-^{(54)}$$

The first letters x run in this progression 1. 1. 1. 1. 1. &c. y^e^ 2^d^ x^3 in this $\frac{-1}{3}.\frac{0}{3}.\frac{1}{3}.\frac{2}{3}.\frac{3}{3}.\frac{4}{3}.\frac{5}{3}$ &c. y^e^ 3^d^ x^5 in this 6. 3. 1. 0. $0+1=1$. $1+2=3$. $3+3=6$. $6+4=10$. $10+5=15$. y^e^ 4th x^7 [in] this(55)

---

(51) $\int_0^x (1-t^2)^3.dt$.

(52) Newton tabulates, for $r = 0, 1, 2, 3, \ldots$, the 2-valued function

$$\phi(r, x) = \int_0^x (1-t^2)^r.dt$$

(where no restriction is placed on x). It is important to notice that the upper bound x of the integral is freely variable (in contrast to Wallis' numerically bounded function

$$f(r) = \int_0^1 (1-t^2)^r.dt$$

in note (19) above) and this is the crux of Newton's breakthrough. The various powers of x order the numerical coefficients and reveal for the first time the binomial character of the sequence where hitherto, at Wallis' hands, it had lain shrouded in numerical complexity.

(53) In a cancelled addition Newton filled this out as '$x:x-\frac{1}{6}x^3:x-\frac{1}{3}x^3:$'.

(54) Newton attempts to insert the $\frac{1}{2}$-intervals in his tabulated sequence, $\phi(r, x)$, $r = 0, \frac{1}{2}, 1, 1\frac{1}{2}, 2, \ldots$. The last entry, $\phi(2\frac{1}{2}, x)$, is incorrect in its last term and should read

$$\text{`}x-\tfrac{5}{6}x^3+\left[\tfrac{1}{5}\binom{\frac{5}{2}}{2}=\right]\tfrac{3}{8}x^5\text{'}.$$

The inference is that Newton, in analogy with the terms in x^3, first tried to interpolate the terms in x^5 as though they were in simple arithmetical proportion. Newton, however, corrects himself immediately.

1st $+x$	$\times 1.$	1.	1.	1.	1.	1.	1.	1.	1.	1.	1.
2d. $-\frac{x^3}{3}$	$\times 0.$	$0+1=1.$	$1+1=2.$	$2+1=3.$	$3+1=4.$	$4+1=5.$	6.	7.	8.	9.	10.
3d. $+\frac{x^5}{5}$	$\times 0.$	$0+0=0.$	$0+1=1.$	$1+2=3.$	$3+3=6.$	$6+4=10.$	15.	21.	28.	36.	45.
4th. $-\frac{x^7}{7}$	$\times 0.$	$0+0=0.$	$0+0=0.$	$0+1=1.$	$1+3=4.$	$4+6=10.$	20.	35.	56.	84.	120.
5. $+\frac{x^9}{9}$	$\times 0.$	$0+0=0.$	$0+0=0.$	$0+0=0.$	$0+1=1.$	$1+4=5.$	15.	35.	70.	126.	210.
6. $-\frac{x^{11}}{11}$	$\times 0.$	$0+0=0.$	$0+0=0.$	$0+0=0.$	$0+0=0.$	$0+1=1.$	6.	21.	56.	126.	252.
	1st. *	2d. *	3d. *	4th. *	5t. *	6t.	1.	7.	28.	84.	210.
						*	7th.	1.	8.	36.	120.
							*	8th.	1.	9.	45.
								*	9th.	1.	10.
									*	10th.	1.
										*	11th.[56]

Now if the meane termes in these progressions can bee calculated y^e^ first of y^m^ gives y^e^ area *aeqp*.[57] Which is thus done

1st. $+x$	$\times 1.$	1.	1.	1.	1.	1.	1.	1.	1.	1.	1.	1.	1.	
2d. $-\frac{x^3}{3}$	$\times 0.$	$\frac{1}{2}.$	1.	$\frac{3}{2}.$	2.	$\frac{5}{2}.$	3.	$\frac{7}{2}.$	4.	$\frac{9}{2}.$	5.	$\frac{11}{2}.$	6.	
3d. $+\frac{1}{5}x^5$	$\times 0.$	$-\frac{1}{8}.$	0.	$\frac{3}{8}.$	1.	$\frac{15}{8}.$	3.	$\frac{35}{8}.$	6.	$\frac{63}{8}.$	10.	$\frac{99}{8}.$	15.	
4. $-\frac{1}{7}x^7$	$\times 0.$	$+\frac{1}{16}.$	0.	$\frac{1}{-16}.$	0.	$\frac{5}{16}.$	1.	$\frac{35}{16}.$	4.	$\frac{105}{16}.$	10.	$\frac{231}{16}.$	20.	
5. $+\frac{1}{9}x^9$	$\times 0.$	$-\frac{3}{128}.$	0.	$\frac{3}{128}.$	0.	$\frac{-5}{128}.$	0.	$\frac{35}{128}.$	1.	$\frac{315}{128}.$	5.	$\frac{1155}{128}.$	15.	
6. $-\frac{1}{11}x^{11}$	$\times 0.$	$\frac{7}{256}.$	0.	$\frac{-3}{256}.$	0.	$\frac{3}{256}.$	0.	$\frac{-7}{256}.$	0.	$\frac{63}{256}.$	1.	$\frac{693}{256}.$	6.	
7. $\frac{1}{13}x^{13}$	$\times 0.$	$\frac{-21}{1024}.$	0.	$\frac{7}{1024}.$	0.	$\frac{-5}{1024}.$	0.	$\frac{7}{1024}.$	0.	$\frac{-21}{1024}.$	0.	$\frac{231}{1024}.$	1.	$\frac{3003}{1024}.$

(55) Newton shows the binomial character of the coefficients $\binom{r}{i}$ of the terms

$$(-1)^i\frac{x^{2i+1}}{2i+1},\quad i = 0,\ 1,\ 2,\ \ldots,$$

in the expansion of $\phi(r, x)$. Straightaway, leaving his first scheme uncompleted, he reconstructs the tabulation in a more visually immediate form.

(56) This is the familiar 'Pascal' triangle of binomial coefficients appropriately filled out with zeros. The entry on row x/column y is $\binom{y-1}{x-1}$, and the composition rule follows by $\binom{y-1}{x-1} = \binom{y-2}{x-1}+\binom{y-2}{x-2}$.

(57) $\phi(\frac{1}{2}, x) = \int_0^x (1-t^2)^{\frac{1}{2}}.dt.$

Soe y^t^ $1\times x-\frac{1}{2}\times\frac{1}{3}\times x^3-\frac{1}{8}\times\frac{1}{5}x^5-\frac{1}{16}\times\frac{1}{7}x^7-\frac{5}{128}\times\frac{1}{9}x^9$ &c: is y^e^ area *apqe*.
y^t^ is $\frac{0}{0}x-\frac{0}{0}\times\frac{1}{2}\times\frac{1}{3}x^3-\frac{0}{0}\times\frac{1}{2}\times\frac{1}{4}\times\frac{1}{5}x^5-\frac{0}{0}\times\frac{1}{2}\times\frac{1}{4}\times\frac{3}{6}\times\frac{1}{7}x^7-\frac{1\times3\times5x^9}{2\times4\times6\times8\times9}$ &c:
The progression may be deduced from hence

$$\frac{0\times1\times-1\times3\times-5\times\ 7\times-9\times11}{0\times2\times\ 4\times6\times\ 8\times10\times\ 12\times14}.\ \&c^{(58)}$$

Soe y^t^ if y^e^ given sine bee $pq=be=x$. & if y^e^ Radius $pc=1$. Then is y^e^ superficies

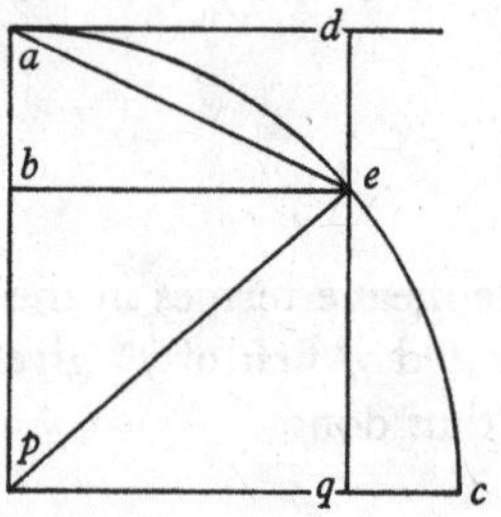

$$ape^{(59)}=x-x\sqrt{1-xx}^{(60)}-\frac{1}{6}x^3-\frac{1}{40}x^5-\frac{x^7}{112}-\frac{5x^9}{1152}-\frac{7x^{11}}{2816}$$

&c:[61] And y^e^ area

$$ade^{(62)}=\frac{x^3}{6}+\frac{x^5}{40}+\frac{x^7}{112}+\frac{5x^9}{1152}+\frac{7x^{11}}{2816}+\frac{21x^{13}}{13312}$$
$$+\frac{11x^{15}}{10240}+\frac{429x^{17}}{557056}+\frac{715x^{19}}{1245184}+\frac{2431x^{21}}{5505024}\ \&c.$$

By wch meanes y^e^ angle *ape* is easily found for *aecpa*: $\angle apc=90^{[gr]}::ape:\angle ape$.[63]

The same may bee thus done.

$$adp=\frac{x}{2}.^{(64)}\ \text{or}\ 2adp=x.\ 2afp^{(65)}=x+\frac{x^3}{3}.$$

$2ahp^{(66)}=x+\frac{2x^3}{3}-\frac{3x^5}{5}$. And $2avp^{(67)}=x+x^3-\frac{9x^5}{5}+\frac{5}{7}x^7$. &c. as in this order[68]

$$x.\ x+\frac{1}{3}x^3.\ x+\frac{2}{3}x^3-\frac{3}{5}x^5.\ x+x^3-\frac{9}{5}x^5+\frac{5x^7}{7}.\ x+\frac{4x^3}{3}-\frac{18x^5}{5}+\frac{20x^7}{7}-\frac{7}{9}x^9.$$

$$x+\frac{5}{3}x^3-\frac{30}{5}x^5+\frac{50x^7}{7}-\frac{35x^9}{9}+\frac{9x^{11}}{11}.\ x+\frac{6x^3}{3}-\frac{45x^5}{5}+\frac{100}{7}x^7-\frac{105x^9}{9}+\frac{54x^{11}}{11}-\frac{11x^{13}}{13}.$$

$$x+\frac{7x^3}{3}-\frac{63x^5}{5}+\frac{175x^7}{7}-\frac{245x^9}{9}+\frac{189}{11}x^{11}-\frac{77x^{13}}{13}+\frac{13x^{15}}{15}.\ \&c.$$

---

(58) That is $\binom{\frac{1}{2}}{i}$, $i = 0, 1, 2, \ldots$, where with a typically Wallisian flourish $\binom{\frac{1}{2}}{0} = 1$ is tabulated as $\frac{\text{‘}0\text{’}}{0}$.

(59) $\frac{1}{2}\sin^{-1}x$.

(60) Read ‘$\frac{1}{2}x\sqrt{1-xx}$’.

(61) That is, $(apqe)-\triangle peq = \int_0^x (1-t^2)^{\frac{1}{2}}.dt-\frac{1}{2}x\sqrt{(1-x^2)}$.

(62) $(apqd)-(apqe) = x-\int_0^x (1-t^2)^{\frac{1}{2}}.dt$.

Which progresions wth their intermediate termes may bee thus exhibited.

		$2\times adpa =$	$2\times aep =$	$2afp =$	$2agp =$	$2ahp =$	$2aip =$	$2avp =$	
$+x$	in	1.	1.	1.	1.	1.	1.	1.	
$+\frac{x^3}{3}$	in	0.	$\frac{1}{2}$.	1.	$\frac{3}{2}$.	2.	$\frac{5}{2}$.	3.	
$\frac{-3}{5}x^5$	in	0.	$\frac{-1}{8}$.	0.	$\frac{3}{8}$.	1.	$\frac{15}{8}$.	3.	
$+\frac{5}{7}x^7$	in	0.	$\frac{1}{16}$.	0.	$\frac{-1}{16}$.	0.	$\frac{5}{16}$.	1.	
$\frac{-7}{9}x^9$	in	0.	$\frac{-5}{128}$.	0.	$\frac{3}{128}$.	0.	$\frac{-5}{128}$.	0.	
$+\frac{9}{11}x^{11}$	in	0.	$\frac{7}{256}$.	0.	$\frac{-3}{256}$.	0.	$\frac{3}{256}$.	0.	&c.(69)

(63) Though Newton does not calculate here the coefficients of the power series for $\sin^{-1}x$, it follows immediately, expanding $\frac{1}{2}x\sqrt{(1-x^2)}$ by the binomial theorem, that

$$\sin^{-1}x = x+\tfrac{1}{6}x^3+\tfrac{3}{40}x^5+\ldots.$$

It is interesting to speculate why Newton deferred introducing this expansion for a few paragraphs. Perhaps Newton was still not completely clear about the validity of the general binomial expansion, which, we note, he had hitherto treated only when it was bounded under an integral sign as $\phi(r, x) = \int_0^x (1-t^2)^r.dt$.

(64) Since $\Delta adp = \frac{1}{2}(ap\times pq)$. (Note that Newton now refers back to his first diagram.)

(65) $2[(afqp)-\Delta fqp] = 2\left[\int_0^x (1-t^2).dt-\frac{1}{2}x(1-x^2)\right]$.

(66) $2\left[\int_0^x (1-t^2)^2.dt-\frac{1}{2}x(1-x^2)^2\right]$.

(67) $2\left[\int_0^x (1-t^2)^3.dt-\frac{1}{2}x(1-x^2)^3\right]$.

(68) Newton tabulates

$$\phi'(r, x) = 2\left[\int_0^x (1-t^2)^r.dt-\tfrac{1}{2}x(1-x^2)^r\right] \quad (r = 0, 1, 2, \ldots),$$

in powers of x^{2i+1}, $i = 0, 1, 2, \ldots$, and then interpolates $\phi'(r, x)$, $r = \frac{1}{2}, \frac{3}{2}, \frac{5}{2}, \ldots$, on the lines of his previous array.

(69) The binomial form of the array is correct. In proof,

$$\phi'(r, x) = 2\int_0^x (1-t^2)^r.dt-x(1-x^2)^r = \int_0^x [(1-t^2)^r+2rt^2(1-t^2)^{r-1}].dt$$

$$= \int_0^x \sum_{0\leqslant i\leqslant\infty}\left[(-1)^i\binom{r}{i}t^{2i}+(-1)^i\,2r\binom{r-1}{i}t^{2i+2}\right].dt$$

$$= \sum_{0\leqslant i\leqslant\infty}\left[(-1)^{i-1}\binom{r}{i}\int_0^x (2i-1)\,t^{2i}.dt\right], \quad \text{since } r\binom{r-1}{i-1} = i\binom{r}{i}.$$

By wch it may appeare y^t if $pe=1$. $pq=x$. y^n

$$aep^{(70)}=\frac{1}{2}x+\frac{x^3}{12}+\frac{3x^5}{80}+\frac{5x^7}{224}+\frac{35x^9}{2304}\ \&c.$$

And y^e area *aep* given gives y^e angle *ape* for $apce : \angle apc = 90^d :: ape : \angle ape$. Likewise y^e angle *ape* given its signe(71) may bee found hereby.(72)

$$\text{Note y}^t\ \left[\frac{x}{2}\times\right]\sqrt{1-xx}=\frac{x}{2}-\frac{2x^3}{8}-\frac{5x^5}{80}-\frac{7x^7}{224}-\frac{45x^9}{2304}-\frac{77x^{11}}{5632}\ \&c.\ \text{that is}$$

$$\left[\frac{x}{2}\times\right]\sqrt{1-xx}=\frac{x}{2}-\frac{x^3}{4}-\frac{x^5}{16}-\frac{x^7}{32}-\frac{5x^9}{256}-\frac{7x^{11}}{512}-\frac{21x^{13}}{2048}\ \&c.$$

According to this progression

$$\frac{1}{2}\times\frac{1}{2}\times\frac{1}{4}\times\frac{3\times5\times\ 7\times\ 9\times11\times13\times15\times17}{6\times8\times10\times12\times14\times16\times18\times20}\ \&c.^{(73)}$$

$$\text{Note also y}^t\text{ y}^e\text{ segment } ae^{(74)}=\frac{x^3}{12}+\frac{3x^5}{80}+\frac{5x^7}{224}+\frac{35x^9}{2304}\ \&c.\ [\&\ \text{y}^t]$$

$$aep^{(75)}=\frac{x}{2}+\frac{x^3}{12}+\frac{3x^5}{80}+\frac{5x^7}{224}+\frac{35x^{9(76)}}{3304}+\frac{63x^{11}}{5632}+\frac{231x^{13}}{26624}+\frac{143x^{15}}{20480}+\frac{6435x^{17}}{1114112}$$
$$+\frac{12155x^{19}}{2490368}+\frac{46189x^{21}}{11010048}\ [\&c].^{(77)}$$

---

(70) $\phi'(\frac{1}{2}, x) = \frac{1}{2}\sin^{-1}x$. Newton does not mention (and perhaps does not yet realize) that

$$\phi'(\tfrac{1}{2}, x) = \int_0^x (1-t^2)^{-\frac{1}{2}}.dt = \phi(-\tfrac{1}{2}, x).$$

(71) Read 'sine'.

(72) There seems no easy way of deriving $\sin x = \int_0^x \cos t.dt$ as an interpolation on Wallisian lines in an array of tabulated instances, $r = 0, 1, 2, \ldots$, of some function $\psi(r, x)$. Perhaps Newton refers vaguely to a derivation by iterating, say, $\sin x = \int_0^x \sqrt{[1-(\sin t)^2]}.dt$ or even $\sin x = \int_0^x \left[1-\int_0^t \sin s.ds\right].dt$, but more probably he intends the derivation of

$$x = \sin y = \sum_i (a_i y^i)$$

by simple reversion of the series

$$y = \sin^{-1}x = \sum_{0\leqslant i\leqslant\infty}\left[\frac{1^2\times3^2\times\ldots\times(2i-1)^2}{(2i+1)!}x^{2i+1}\right].$$

If $pq = a$. $qd = x$. $pc = 1 = pb$. $db = \sqrt{1 - aa - 2ax - xx}$. y^n^ y^e^ areas of y^e^ lines(78) [are] in this progression. (supposeing also $1 - aa = b$.) viz:

$$1.\ \sqrt{b - 2ax - xx}.\ b - 2ax - xx.\ \overline{b - 2ax - xx}|^{\frac{3}{2}}.$$
$$bb - 4abx \quad - 2bxx + 4ax^3 + x^4.\ *.$$
$$+ 4aa$$
$$b^3 - 6abbx \quad - 3bbxx + 8abx^3 \quad + 3bx^4 - 6ax^5 - x^6.\ \&c.^{(79)}$$
$$+ 12aabx^2 - 8a^3x^3 - 12aax^4$$
$$+ 4abx^3$$

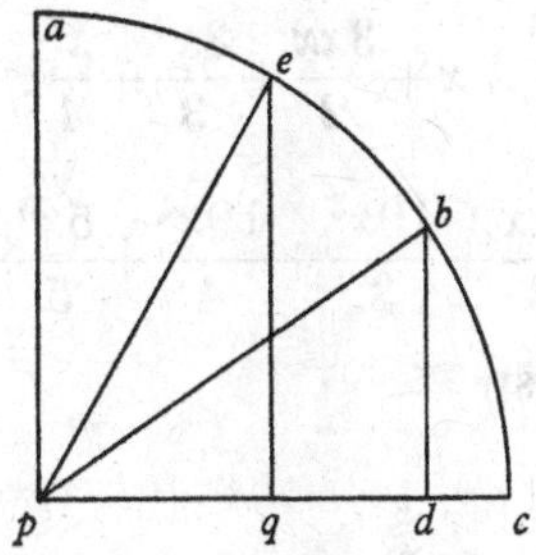

(73) That is, the expansion as a continued product (correct but for omitted signs) of

$$\binom{\frac{1}{2}}{i}, \quad i = 1,\ 2,\ 3,\ \ldots.$$

Using this binomial coefficient for each i, Newton expands $\frac{1}{2}x\sqrt{(1-x^2)}$ in all generality as

$$\sum_{0 \leqslant i \leqslant \infty} \left[\binom{\frac{1}{2}}{i} (-x^2)^i\right] \times \frac{x}{2}.$$

(74) That is, $(aep) - \triangle aep$.

(75) $\phi'(\frac{1}{2}, x) = \int_0^x (1-t^2)^{\frac{1}{2}}.dt - \frac{1}{2}x\sqrt{(1-x^2)}$.

(76) Read $\dfrac{\text{`}+35x^9\text{'}}{2304}$.

(77) In calculating the term coefficients in this power-series for $\frac{1}{2}\sin^{-1}x$, Newton seems a little intoxicated with his new-found numerical facility.

(78) Newton still has in mind his first diagram, where the 'lines' have ordinates

$$(1 - a^2 - 2ax - x^2)^r, \quad r = 0,\ \tfrac{1}{2},\ 1,\ 1\tfrac{1}{2},\ \ldots,$$

and wishes to evaluate the 2-valued function $\int_0^x (b - 2at - t^2)^r.dt$, where $b = 1 - a^2$.

(79) Here, having strained to its limit Wallis' technique of interpolation by seeking structural patterns in tabulated instances, Newton wisely abandons his investigation. In any case, there is no gain in generality since by the linear transform $t \to t - a$

$$\int_0^x (b - 2at - t^2)^r.dt = \int_a^{x+a} (1-t^2)^r.dt = \phi(r, x+a) - \phi(r, a).$$

[3] [*The logarithmic series*]

To Square the Hyperbola

So if *nadm* is an Hyperbola. & $cp=1=pa$. $pq=x$. qd, qe, qf, qg &c$=y$. &

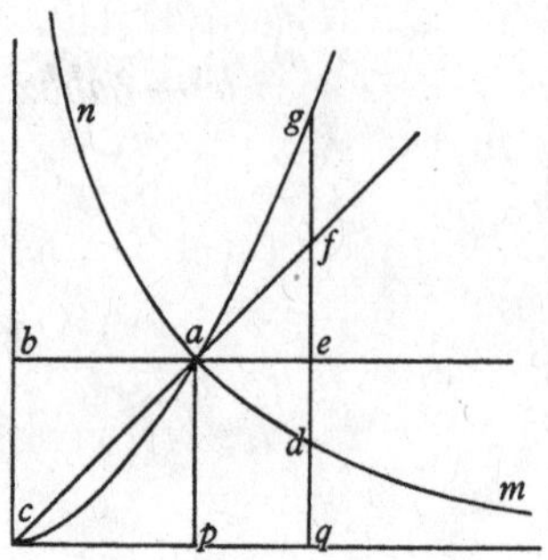

$\frac{1}{1+x}=y=dq$. $1=y=eq$. $1+x=y=qf$.

$1+2x+xx=y=qg$. $1+3x+3xx+x^3$.

$1+4x+6x^2+4x^3+x^4$. &c. There(80) squares are

$*$. x. $x+\frac{xx}{2}$. $x+\frac{2xx}{2}+\frac{x^3}{3}$. $x+\frac{3xx}{2}+\frac{3x^3}{3}+\frac{x^4}{4}$.

$x+\frac{4xx}{2}+\frac{6x^3}{3}+\frac{4x^4}{4}+\frac{x^5}{5}$. $x+\frac{5xx}{2}+\frac{10x^3}{3}+\frac{10x^4}{4}+\frac{5x^5}{5}+\frac{x^6}{6}$.

&c. As in y^e following table.(81)

	$apqd=$	$apqe=$	$apqf=$	$apqg=$					
$x\times$	1.	1.	1.	1.	1.	1.	1.	1.	1.
$\frac{x^2}{2}\times$	$-1.$	0.	1.	2.	3.	4.	5.	6.	7.
$\frac{x^3}{3}\times$	1.	0.	0.	1.	3.	6.	10.	15.	21.
$\frac{x^4}{4}\times$	$-1.$	0.	0.	0.	1.	4.	10.	20.	35.
$\frac{x^5}{5}\times$	1.	0.	0.	0.	0.	1.	5.	15.	35.
$\frac{x^6}{6}\times$	$-1.$	0.	0.	0.	0.	0.	1.	6.	21.
$\frac{x^7}{7}\times$	1.	0.	0.	0.	0.	0.	0.	1.	7.

(80) Read 'their'.

(81) Newton tabulates $\int_0^x (1+t)^r.dt$, $r=0, 1, 2, \ldots,$

in powers of x and then interpolates $\int_0^x (1+t)^{-1}.dt$,

the area under the hyperbola $y(1+x)=1$ (or *nam* in Newton's figure). Clearly the array is one of (positive) binomial coefficients and Newton has but to go one column further to the left:

$$\int_0^x (1+t)^r.dt = \sum_{0\leqslant i\leqslant\infty}\left[\binom{r}{i}\times\int_0^x t^i.dt\right].$$

(82) Newton wrote '$\frac{x^8}{x}$'.

By whose first terme is represented y^e^ square of y^e^ Hyperbola, viz: y^t^ it is

$$x-\frac{x^2}{2}+\frac{x^3}{3}-\frac{x^4}{4}+\frac{x^5}{5}-\frac{x^6}{6}+\frac{x^7}{7}-\frac{x^8}{[8]}{}^{(82)}+\frac{x^9}{9}-\frac{x^{10}}{10}.\ \&c.^{(83)}$$

As if $x=\frac{1}{10}$. or $cq=\frac{11}{10}=1, 1$. y^n^ is

$$dapq=\frac{1}{10}-\frac{1}{200}+\frac{1}{3000}-\frac{1}{40000}+\frac{1}{500000}-\frac{1}{6000000}.\ \&c.$$

y^t^ is

$apqd=[x=]$	0,	10000.	00000.	00000.	00000.	00000.	000	[...]
$\frac{1}{3}x^3=$	+0,	00033.	33333.	33333.	33333.	33333.	333	[...]
$\frac{1}{5}x^5=$	+0,	00000.	20000.	00000.	00000.	00000.	000	[...]
$\frac{1}{7}x^7=$	+0,	00000.	00142.	85714.	28571.	42857.	142	[...]
$\frac{1}{9}x^9=$	+0,	00000.	00001.	11111.	11111.	11111.	111	[...]
$\frac{1}{11}x^{11}=$	+0,	00000.	00000.	00909.	09090.	90909.	090	[...]
$\frac{1}{13}x^{13}=$	+0,	00000.	00000.	00007.	69230.	76923.	076	[...]
$\frac{1}{15}x^{15}=$	+0,	00000.	00000.	00000.	06666.	66666.	666	[...]
$\left[\frac{1}{17}x^{17}=\right]$					0058.	82352.	941	[...]
$\left[\frac{1}{19}x^{19}=\right]$					0000.	52631.	578	[...]
$\left[\frac{1}{21}x^{21}=\right]$					0000.	00476.	190	[...]
$\left[\frac{1}{23}x^{23}=\right]$					0000.	00004.	347	[...]
$\left[\frac{1}{25}x^{25}=\right]$					0000.	00000.	040	[...]
Summa	0,	10033.	53477.	31075.	58063.	57265.	520	[...](84)

(83) $\int_0^x (1+t)^{-1}.dt = \sum_{0\leqslant i\leqslant\infty}\left[\binom{-1}{i}\int_0^x t^i.dt\right] = \sum_{0\leqslant i\leqslant\infty}\left[(-1)^i\frac{x^{i+1}}{i+1}\right]$.

(84) Newton calculates the partial sum $(x+\frac{1}{3}x^3+\frac{1}{5}x^5+\ldots)$, $x = 0.1$, giving the terms up to $\frac{1}{41\times 10^{41}}$ to 68D, but two mistakes vitiate all his hard work. First, the term

$$\frac{1}{19\times 10^{19}} = 10^{-20}\times 0.\ 52631\ \ 57894\ \ 73684\ldots$$

is miscalculated. (Newton's decimal development is wrong from 28D onwards, though Newton recognizes his error, crossing out the first wrong figure '9' at the 28th place and

−0,	00500.	00000.	00000.	00000.	00000.	00000.	00000.	00000.	00000.	00000.	0
	2.	50000.	00000.	00000.	00000.	00000.	00000.	00000.	00000.	00000.	0
		1666.	66666.	66666.	66666.	66666.	66666.	66666.	66666.	66666.	6
		12.	50000.	00000.	00000.	00000.	00000.	00000.	00000.	00000.	0
			10000.	00000.	00000.	00000.	00000.	00000.	00000.	00000.	0
			83.	33333.	33333.	33333.	33333.	33333.	33333.	33333.	3
				71428.	57142.	85714.	28571.	42857.	14285.	71428.	5
				630.	55555.	55555.	55555.	55555.	55555.	55555.	5
					5045.	45454.	54545.	45454.	54545.	45454.	5
						41666.	66666.	66666.	66666.	66666.	6
						384.	61538.	46153.	84615.	38461.	5
						3.	57142.	85714.	28571.	42857.	1
							3364.	58333.	33333.	33333.	3
								29411.	76470.	58823.	5[85]
−0,	00502.	51679.	26750.	72059.	17744.	28779.	27385.	30147.	14044.	12586	

	−277.	77777.	77777.	7
cui addendum	2.	63157.	89421.	0
		02523.	80952.	4[86]
			22727.	3
			217.	4
			2.	1
that is	−280.	43459.	71098.[87]	

And so y^e summes will bee

+0,	10033,	53477.	31075.	58063.	57265.	520	[...]
−0.	00502.	51679.	26750.	72059.	17144.[88]	287	[...]
0.	09530.[89]	01798.	04324.	86004.	40121.[90]	232	[...]

wch is y^e quantity of y^e area *adpq*, If $cpab = 1$. & $cp = ab = 10\ pq$. & $qde \| ap \| bc = ap$.[91]

substituting a correct '8'. He makes no attempt here, however, to calculate the true development.) Secondly, the term $\frac{1}{39 \times 10^{39}}$ is calculated correctly only to 44D, having its period $\overline{256410}$ incorrectly extended to $\overline{2564010}$. (Newton recognizes this mistake also, but here merely goes along the line crossing out superfluous zeros.) Since Newton revised this calculation immediately on a loose sheet (see §5 below) only the correct figures are printed in the text. (The appearance of the original manuscript page may be seen in Plate I. The rather strange arrangement of the figures suggests that Newton first intended to take his calculation only to the edge of the page (that is, to 16D), but then continued it by adding on new rows below.) Note finally Newton's use of the comma as a decimal point (later to become his standard usage), while the point groups the figures by fives.

(85) Note that the 8th, 9th and 13th rows are respectively

$$\left(\frac{1}{16 \times 10^{16}} + \frac{1}{18 \times 10^{18}}\right), \quad \left(\frac{1}{20 \times 10^{20}} + \frac{1}{22 \times 10^{22}}\right) \quad \text{and} \quad \left(\frac{1}{30 \times 10^{30}} + \frac{1}{32 \times 10^{32}}\right).$$

(86) Note that this 3rd row doubles $\left(\frac{1}{40 \times 10^{40}} + \frac{1}{42 \times 10^{42}}\right)$.

(87) Newton's second partial sum, $-(\frac{1}{2}x^2 + \frac{1}{4}x^4 + \ldots + \frac{1}{48}x^{48})$, $x = 0{\cdot}1$, taken in the first place to 'two & fifty figures', that is 51D, but then rounded off (correctly) to 50D.

In like manner if I make $x = \frac{1}{100} = pq$. The opperacon followeth.

$\left[x+\frac{1}{3}x^3 = \right]$0, 01000. 03333. 33333. 33333. 33333. 33333. 33333. 33333. 33333. 3

$+\frac{1}{5}x^5+\frac{1}{7}x^7 \quad = \quad$. 20001. 42857. 14285. 71428. 57142. 85714. 28571. 4

$\frac{1}{9}x^9+\frac{1}{11}x^{11} \quad = \quad$ 11. 11202. 02020. 20202. 02020. 20202. 0

$\frac{1}{13}x^{13}$ ===== 769. 23076. 92307. 69230. 7

$\left[\frac{1}{15}x^{15}\right]$ ==== 6666. 66666. 66666. 6

$\left[\frac{1}{17}x^{17}\right]$ == 58823. 52941. 1

$\left[\frac{1}{19}x^{19}\right]$ = 5. 26315. 7

$\left[\frac{1}{21}x^{21} = \right]$ 47. 6

+0, 01000. 03333. 53334. 76201. 58821. 07551. 40422. 38870. 97309.

$\left[-\frac{1}{2}x^2-\frac{1}{4}x^4-\frac{1}{6}x^6 = \right]$ −0, 00005, 00025, 00166, 66666. 66666. 66666. 66666. 66666. 66666. 6

$-\frac{1}{8}x^8-\frac{1}{10}x^{10}-\frac{1}{12}x^{12} =$ −1250. 10000. 83333. 33333. 33333. 33333. 3

$-\frac{1}{14}x^{14} =$ ————————— −7. 14285. 71428. 57142. 8

$-\frac{1}{16}x^{16}-\frac{1}{18}x^{18}[=]$ ————— −62. 50555. 55555. 5

$-\frac{1}{20}x^{20} =$ ————— 5000. 4

−0, 00005. 00025. 00166. 67916. 76667. 50007. 14348. 21984. 17699.

+0, 00995. 03308. 53168. 08284. 82153. 57544. 26074. 16886. 79610.

Which is y^e quantity of y^e area $apqd$ if $100p[q] = cp$. and $abcp = 1$.(92)

---

(88) Read '17744'. (89) Read '09531'. (90) Read '39521'.

(91) $(adpq) = \int_1^{1\cdot 1} (1+t)^{-1}.dt$, that is log (1·1), though Newton does not introduce the notion explicitly here. Newton in his original takes the calculation to 46D, but the first partial sum (correct only to 28D) imposes a limit to his accuracy. In fact, as we have noted, Newton's result is also vitiated by two small numerical slips, one of addition and one of subtraction.

(92) The similar calculation of

$$\int_1^{1\cdot 01} (1+t)^{-1}.dt = \log(1\cdot 01),$$

using the partial sums

$$(x+\tfrac{1}{3}x^3+\ldots+\tfrac{1}{21}x^{21})-(\tfrac{1}{2}x^2+\tfrac{1}{4}x^4+\ldots+\tfrac{1}{22}x^{22}), \quad x = 0\cdot 01.$$

After widespread revision (not shown in our text) Newton finally reaches a result completely correct to 46D.

[4] [*Annotations on Wallis*' Commercium Epistolicum.]

Dr Wallis in a letter to Sr Kenelme Digby(93) promiseth ye squareing of ye Hyperbola by finding a meane proportion twixt 1, & $\frac{5}{6}$ in the progression $1, \frac{5}{6}, \frac{31}{30}, \frac{209}{140}, \frac{1471}{630}, \frac{10625}{2772}$ &c.(94).

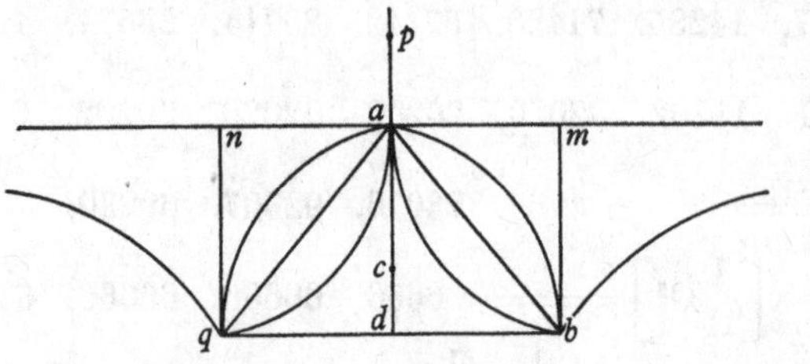

Dr Wallis in a letter to Sr Kenelme Digby(95) teacheth how to find ye center of gravity in divers lines [&] first when their position is as in this figure.(96)

Suppose *ad* ye Axis, *a* their vertex Then saying, as 1 to ye index of ye line increased by an unite (vide pag 2dā(97)) so *cd* to *ca*[,] Then *c* is their center of gravity.(98)

(93) *Commercium Epistolicum:* Epistola XVI [21 November 1657]: 33–51, especially 51: 'Sed alteram [seriem] volo huic consimilem...'.

(94) That Newton adds no explanatory comment seems to indicate that the progression remained a puzzle to him: indeed, a whole page immediately following this insertion has been left blank, presumably for future clarification. The progression is, in fact,

$$g'(s) = \int_0^1 (x+x^2)^s . dx, \quad (s = 0, 1, 2, \ldots),$$

where $g'(\frac{1}{2})$, the 'meane proportion twixt $[g'(0) =]$ 1 & $[g'(1) =] \frac{5}{6}$', is the area (taken between the bounds $x = 0$ and $x = 1$) under the hyperbola $y^2 = x+x^2$. The linear transform $2x+1 = t$ evaluates this area,

$$\int_0^1 (x+x^2)^{\frac{1}{2}} . dx, \quad \text{as} \quad \frac{1}{4}\int_1^3 (t^2-1)^{\frac{1}{2}} . dt = \tfrac{1}{8}[\log(3-2\sqrt{2})+6\sqrt{2}],$$

but there seems no way of deriving this by a Wallisian interpolation scheme or indeed by any process which does not calculate the logarithm directly. Wallis himself in Prop. CLXV: 128 of his *Arithmetica Infinitorum*, annotated by Newton above (see note (23)), had produced the allied sequence $g(s) = \int_0^1 (x-x^2)^s . dx$ with the assertion that the evaluation of $g(\frac{1}{2}) = \frac{1}{8}\pi$ squares the semicircle of diameter 1, but could proceed no further.

Thirty years later Leibniz read this passage in the *Commercium Epistolicum*, and in the late autumn of 1696 wrote to Wallis requesting an explanation of the progression. In his letter of reply dated 21 November 1696 [= *Leibnizens mathematische Schriften* (ed. C. I. Gerhardt) 1, 4 (Halle, 1859): 5–10] Wallis wrote (pp. 5–6): 'Promissum illud meum quod memoras in Commercio Epistolico a me factum (illud, credo, vis quod sub finem Epistolæ XVI habetur) nimirum: exposita serie numerorum $1, \frac{5}{6}, \frac{31}{30}, \frac{209}{140}, \frac{1471}{630}, \frac{10625}{2772}$ &c. si terminum inter 1 et $\frac{5}{6}$ intermedium, seriei congruum, exhiberit Fermatius, exhibiturum me Hyperbolæ Quadraturam. Id ego jam tum præstitueram. Est enim hæc series eadem ipsa quæ habetur Prop. 161 Arithmeticæ Infinitorum; unde colligitur quadratura Prop. 165. Ad quam nihil deest aliud

The Demonstra$\overline{\text{co}}$n(99)

Let p bee y^e^ index of y^e^ series according to wch y^e^ ordinately aplyed lines (parallel to db) increase, y^n^ $1:p+1::$ area of y^e^ line [qab] to $nmbq$.(100) y^e^ distances of those ordinate lines(101) from y^e^ vertex a are equall to y^e^ intercepted diameters & therefore a primanary series (whos index is 1), & since supposing a y^e^

---

quam exhibitio numeri intermedii inter 1 et $\frac{5}{6}$ in illa serie, qui ita respiciat Ordinatas in Hyperbola ut $\frac{5}{6}$ respicit earum Quadraturam. Sicut enim ope seriei Prop. 133, nempe $1, \frac{1}{6}, \frac{1}{30}, \frac{1}{140}, \frac{1}{630}, \frac{1}{2772}$ &c colligitur Circuli Quadratura Prop. 135 ex intermedio numero inter 1 et $\frac{5}{6}$ in illa serie (suntque iidem denominatoris numeri utriusque seriei). Potestque numerus ille approximando pluribus modis exhiberi (quod et a pluribus factum est) sed accurate, credo, (quod quærebatur) numero finito non posse juxta receptam adhuc aliquam notationis formam. Pariter ut ope seriei $1, \frac{2}{3}, \frac{8}{15}, \frac{48}{105}$ &c Prop. 118 colligitur Quadratura Circuli Prop. 121, ex numero intermedio inter 1 et $\frac{2}{3}$, sic ope seriei $1, \frac{4}{3}, \frac{28}{15}, \frac{288}{105}$ &c Prop. 158 colligenda est Quadratura Hyperbolæ, ex interposito numero medio inter 1 et $\frac{4}{3}$ (quæ Hyperbolam exteriorem spectat)....' (Wallis' second hyperbola progression is

$$H(r) = \int_0^1 (1+x^2)^r.dx, \quad r = 0, \; 1, \; 2, \; \ldots,$$

with $H(\frac{1}{2})$ the area under the hyperbola $y^2 = 1+x^2$, that is, $\frac{1}{2}[\sqrt{2}+\log(1+\sqrt{2})]$.)

Two centuries later, finally, Paul Tannery, engaged in editing the *Commercium* for his Fermat edition, published Wallis' progression in 1896 as a challenge to the mathematical world, but drew no reply till G. Vacca in 1903 published the correct formation rule in the form $1, \frac{1}{2}+\frac{1}{3}, \frac{1}{4}+\frac{2}{5}+\frac{1}{6}, \frac{1}{5}+\frac{3}{6}+\frac{3}{7}+\frac{1}{8}, \ldots$, successively. (See *Œuvres de Fermat* (ed. Paul Tannery et Charles Henry), 4 (Paris, 1912): *Notes mathématiques*, XIX: 211–12: *Un problème de Wallis.*)

(95) Epistola XVI [21 November 1657], especially 44–50.

(96) In his letter to Digby, Wallis gave only a general discussion of the proposition, but when he published the *Commercium* in the following year added the proof noted by Newton in a short appended tract on pp. 52–6. (The text of this note in Newton's original is placed after his notes on Wallis' *Adversus M. Meibomii de Proportionibus Dialogum*, but, since it so clearly goes with the previous note, it has been inserted here.)

(97) Newton refers back to his own p. 2 (=f. 15^v^) where, at the beginning of his notes on Wallis' *Arithmetica Infinitorum*, he enters the definition of the index i of the sequence x^i, $x = 0, 1, 2, \ldots$. (Compare notes (3) and (14) above.)

(98) Newton annotates the statement of the problem ('Propositio hæc est.') which opens the appendix.

(99) *Commercium*: 52–3: *Sequitur Demonstratio.*

(100) That is, where the x-axis is ad with ordinates y drawn parallel to qdb with $ad = a$, $y = x^p$ is the defining equation of the 'line' qab, so that

$$(qabq):(nambq) = 2\int_0^a x^p.dx:2\int_0^a a^p.dx = 1:(p+1).$$

(101) Ordinates y drawn parallel to qdb are distant from nam by x (measured along ad).

center(102) of y^e^ ballance y^e^ whole weight of y^e^ surface or figure is composed of its magnitude and distance from y^e^ center & therefore y^e^ index of all its moments or whole weight(103) is $p+1$, viz: y^e^ aggregate of y^e^ other two. Therefore as all its moments (or y^e^ weight of the figure in its site in respect of y^e^ center a) are to soe many of y^e^ greatest (or to y^e^ weight of y^e^ ▭(104) $nmbq$ hung on y^e^ point d) soe is 1, to $p+2$,(105) and if $ap:ad::1:p+2$, then $nmbq$ hung on y^e^ point q(106) shall counterballance y^e^ figure [qab] in its site &c therefore if $ac:cd::p+1:1$, c shall be y^e^ center of gravity of these figures.(107)

Also as the figure is now put extending infinitely towards δ if

$$-2p+1:-p+1::am:ac.$$

m being y^e^ center of $qnbd$ y^n^ c shall bee y^e^ center of gravity of y^e^ whole figure $qn\delta bd$.

Demonstration.

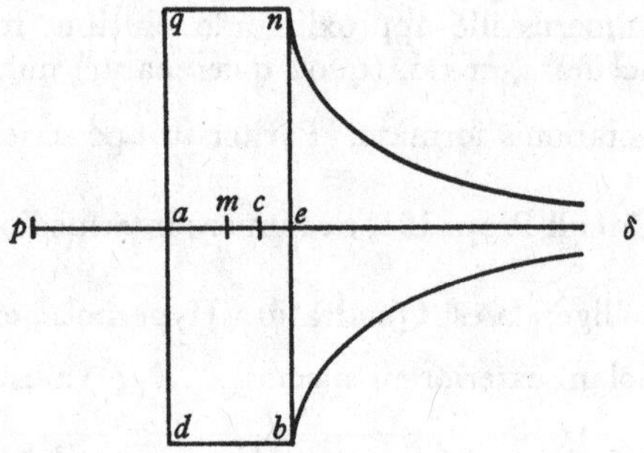

since y^e^ lines parallell to $a\delta$ increase in series reciprocally proportionall their index is $-p$(108) & since y^e^ halfes of those lines increas in y^e^ same proportion their index is $-p$. whose extremitys or middle points of y^e^ whole lines (suposing a y^e^ center of y^e^ ballance) are theire centers of gravity, their distances from a being proportionall to y^e^ lines whose centers they are(109) & consequently their index is $-p$. & since all y^e^ moments(110) (or whole weight of y^e^ figure) increase in a proportion compounded of y^e^ proportion[s] of y^e^ magnitudes & distances of y^e^ lines from y^e^ center a, they will be in a duplicate proportion(111) of y^e^ lines

(102) That is, fulcrum.

(103) The moment round the fulcrum a, $x \times x^p = x^{p+1}$.

(104) Read 'rectangle' (= Newton's 'parallelogram').

(105) Since $\int_0^a x^{p+1}.dx : \int_0^a a^{p+1}.dx = 1:[(p+1)+1]$.

(106) Read 'p'.

(107) For $(qabq)$ 'in its site' is balanced round the fulcrum a by $(nmbq)$ hung at p, where $ap:ad = 1:(p+2)$, and so by $(qabq)$ itself at c where $ap:ac = (qabq):(nmbq) = 1:(p+1)$; that is, where $ac:ad = (p+1):(p+2)$.

(108) That is, where the x-axis is aq with ordinates y taken parallel to $a\delta$ and $aq = a$, the defining equation of $n\delta$ is now $y = x^{-p}$.

(109) In fact, their distance from a is $\frac{1}{2}y$ (or $\frac{1}{2}x^{-p}$).

(110) $\int_0^a \frac{1}{2}y \times y.dx$. (111) $y \times \frac{1}{2}y = \frac{1}{2}y^2$. (112) $\frac{1}{2}y^2 = \frac{1}{2}x^{-2p}$.

(113) That is, $\int_0^a x^{-p}.dx : \int_0^a a^{-p}.dx = 1:[(-p)+1]$.

(114) Read 'parallelogam'.

magnitudes[,] that is a reciprocall series whose index is $-2p$.[112] Therefore y^e figure $[qn\delta bd]$ is to y^e inscribed parallelogram $[qnbd]$ as 1 to $1-p$.[113] & all its moments or whole weight in this its site to the weight of y^e pgr[114] as 1 to $1-2p$.[115] Therefore if, $am:ap::1-2p:1$, the paralelogrā hanging on y^e point p shall counterballanc[e] y^e whole figure in its site &c: whence y^e point c may be found easily, viz $am:ac::1-2p:1-p$.[116]

---

[5] *The resolution of cubick equations out of Dr Wallis in his dedication before Meibomius confuted.*[117]

suppose $x = \mho a \mho e$. y^n $x^3 = \mho a^3 \mho 3aae \mho 3aee \mho e^3$. or $x^3 = +3aex \mho a^3 \mho e^3$. that is making $a^3+e^3=q$. $+3ae=p$. y^n $x^3=+px \mho q$.

Againe suppose $x=a-e$.[118] y^n $x^3=a^3-3aae+3aee-e^3$. y^t is making $a^3-e^3= \mho q$. & $3ae=p$, y^n $x^3=-px \mho q$.[119]

Then in the first of these $p=3ae$. or $\dfrac{p}{3e}=a$. or $\dfrac{p^3}{27e^3}=a^3=q-e^3$. Therefore $e^6=qe^3-\dfrac{p^3}{27}$. & $e^3=\dfrac{1}{2}q \mho \sqrt{\dfrac{1}{4}qq-\dfrac{p^3}{27}}$. & by y^e same reason $a^3=\dfrac{1}{2}q \Omega \sqrt{\dfrac{1}{4}qq-\dfrac{p^3}{27}}$. where y^e irrationall quantitys have divers signes otherwise $a^3+e^3=q$ would bee false. Soe that

$$x= \mho a \mho e = \mho \sqrt{c{:}\ \frac{1}{2}q \Omega \sqrt{\frac{1}{4}qq-\frac{1}{27}p^3}} \mho \sqrt{c{:}\ \frac{1}{2}q \mho \sqrt{\frac{1}{4}qq-\frac{1}{27}p^3}}.$$

is a rule for resolving y^e equation $x^3 * -p[x] \Omega q=0$, when it hath but one

---

(115) That is $\displaystyle\int_0^a \tfrac{1}{2}x^{-2p}.dx : \int_0^a \tfrac{1}{2}a^{-2p}.dx = 1 : [(-2p)+1]$.

(116) Since $ap:ac = (qn\delta bd):(qnbd) = 1:(1-p)$. The second part of this note is an English rendering, virtually word for word, of *Commercium*: 54–5. Wallis' results may be found more easily by equating the total moment of the respective figures round their centres of gravity to zero (a method formulated in all generality by Torricelli and Pascal but not published in 1657). Thus, where $ad = a = \dfrac{1}{k}\times ac$, $0 = \displaystyle\int_0^a (x-ka)\,x^p.dx$ and $k = \dfrac{p+1}{p+2}$; and, where $ae = a^{-p} = \dfrac{1}{l}\times ac$, $0 = \displaystyle\int_0^a (\tfrac{1}{2}x^{-p}-la^{-p})\,x^{-p}.dx$ and $l = \tfrac{1}{2}\times\dfrac{1-p}{1-2p}$.

(117) Notes on Wallis' *Adversus Marci Meibomii de Proportionibus Dialogum Tractatus Elencticus: Dedicatio* (to William Brouncker, separately paginated 1–50).

(118) Newton has cancelled '...suppose $y = \mho b \Omega c$...'. (There is no loss of generality in taking $x = a-e$ instead of $x = \mho a \Omega e$, since by definition a and e must have opposite signs attached.)

(119) Newton uses the signs $\mho / \Omega$ for $\pm/\mp$ (taken over from Liber v of Schooten's *Exercitationes Mathematicæ*) to shorten Wallis' wordy account. Note that the constants a, e, p, q must all be taken positive (a convention firmly established among algebraists since the later sixteenth century), and that the signs $\mho / \Omega$ generalize their range to include negative constants. Here, where Wallis had to consider the cases of $x^3 = +px+q$ and $x^3 = +px-q$ separately, Newton compresses them as the single $x^3 = +px \pm q$; and similarly for $x^3 = -px \pm q$.

roote y^t^ is when it may be generated according to the supposition $x=\text{♉}\,a\,\text{♉}\,e$. &c. By y^e^ same reason $x^3{*}+px\,\text{☊}\,q\,[=0]$ may be resolved by this rule

$$x=a-e=\sqrt{c:\frac{1}{2}q\,\text{♉}\sqrt{\frac{1}{4}qq+\frac{1}{27}p^3}}-\sqrt{c:\frac{1}{2}q\,\text{☊}\sqrt{\frac{1}{4}qq+\frac{1}{27}p^3}}.\text{[120]}$$

But here observe y^t^ Dr Wallis would Argue y^t^ since in the first of these two cases sometimes (viz when y^e^ equation hath 3 reall rootes) y^e^ rule faileth[121] as if it were impossible for y^e^ equation to have rootes when yet it hath, therefore y^e^ fault is in Algebra. & therefore when Analysis leads us to an impossibility wee ought not to conclude y^e^ thing absolutely impossible, untill wee have tryed all y^e^ ways y^t^ may bee.[122]

But let me answer y^t^ y^e^ fault is not in y^e^ Analysis in this example, but in his opperation. for when y^e^ equation $x^3{*}+px\,\text{♉}\,q=0$. hath 3 roots hee supposeth it to have but one roote viz[:] $x=\text{♉}\,a\,\text{♉}\,e$. but since y^e^ equation cannot be then generated according to y^t^ supposition it is impossible it should be re[s]olved[123] by it.[124]

In like manner hee sayeth[125] y^t^ Algebra representeth a thing possible when tis not so as in this example,[126] in y^e^ $\triangle abc$, make $ab=1$. $bc=2$. $ac=4$. Then to find $dc=x$, worke thus, $ad=4-x$. $bd\times bd=1-16+8x-x^2=4-x^2$. therefore $8x=19$. or $x=\frac{19}{8}$.[127] In wch opperacon all things proceede as possible though they are not soe for ac is greater y^n^ $ab+bc$.[128]

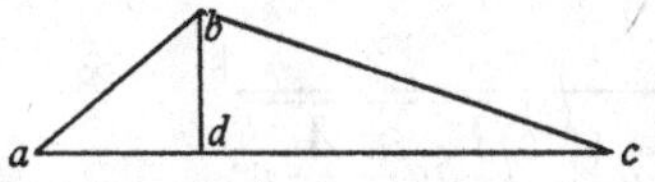

(120) A lucid summary of *Dedicatio*: 17–26, especially 18–20, where Wallis (as he acknowledges explicitly on p. 22) develops 'Cardan's' rule for resolving the reduced cubic

$$x^3\pm px\pm q=0.$$

Note that, as well as the symbols ♉ / ☊ (note (119)), Newton introduces Cartesian notation for Wallis' cossic exposition (derived, as the algebraic treatment itself, by way of Oughtred's *Clavis* from Viète's *De Æquationum Recognitione et Emendatione*, Paris, 1615 [=*Opera* (ed. F. Schooten), Leyden, 1646: 82–158]).

(121) The 'impossible' case when $\frac{1}{4}q^2-\frac{1}{27}p^3<0$.

(122) *Dedicatio*: 24–6.

(123) Newton wrote 'relolved'.

(124) Newton's valid reply to Wallis' criticism is essentially that it is the assumptions made which are false and not, as Wallis would have it, the algebraic structure by which the 'impossible' solution is deduced from the assumptions. Specifically, Wallis has assumed the existence of a real root x of the form $\pm(a+b)$ and deduced by algebraic operations that, even when $\frac{1}{4}q^2<\frac{1}{27}p^3$, a and e must be $\sqrt[3]{[\frac{1}{2}q\pm\sqrt{(\frac{1}{4}q^2-\frac{1}{27}p^3)}]}$ ('impossible' since the root of the negative quantity $\frac{1}{4}q^2-\frac{1}{27}p^3$ is demanded). Newton later came to realize that, in fact, the assumption even in this 'irreducible' case of a root of form $x=\pm(a+e)$ was not intolerable, and when in 1671 he wrote his notes on Kinckhuysen's *Algebra* he included a chapter on the algebraic reduction of this 'irreducible' case. (See his *In Algebram Gerardi Kinckhuysen*

Yet I answer y^t if y^e opperation & conclusion be compared together y^e absurdity will appeare. for in y^e equation $bd \times bd = 4 - xx = 4 - \frac{361}{64} = \frac{256-361}{64}$ or $bd \times bd = \frac{-105}{8}$.(129) but it is impossible y^t a □(130) number should be negative.(131)

Thus $x = \sqrt{-b}$ is impossible. square it & tis $xx = -b$. Againe, & tis $x^4 = bb$. Extract y^e roote & tis $xx = b$, or $x = \sqrt{b}$. wch is possible. The reason of this event(132) is y^t $x^4 - bb = 0$ hath two possible rootes viz[:] $x = \sqrt{b}$. $x = -\sqrt{b}$. & two impossible viz: $x = \sqrt{-b}$. $x = -\sqrt{-b}$.

Thus y^e valors of $x^8 - a^8 = 0$ are $x = a$, $-a$, $\sqrt{-aa}$, $-\sqrt{-aa}$, $\sqrt{4{:}-a^4}$, $-\sqrt{4{:}-a^4}$, $\sqrt{-\sqrt{-a^4}}$, $-\sqrt{-\sqrt{-a^4}}$.(133)

Observationes [=ULC. Add. 3959.1]: *Caput quintum, De solutione Æquationum*, where he considers the cubics $x^3 = 21x+20$ and $x^3 = 15x-4$. In the former, since

$$10 \pm 9\sqrt{-3} = (-2 \pm \sqrt{-3})^3 = \left(\frac{5 \pm \sqrt{-3}}{2}\right)^3 = \left(\frac{-1 \pm \sqrt{-3}}{2}\right)^3,$$

the Cardan solution $x = \sqrt[3]{(10+9\sqrt{-3})} + \sqrt[3]{(10-9\sqrt{-3})}$ yields three real roots $x = -4$, 5 and -1. Since $(-2 \pm \sqrt{-1})^3 = -2 \pm 11\sqrt{-1}$, the resolution of the latter as

$$x = \sqrt[3]{(-2+11\sqrt{-1})} + \sqrt[3]{(-2-11\sqrt{-1})}$$

yields a real root $x = -4$ and the two other roots may easily be found to be real, that is, $2+\sqrt{3}$ and $2-\sqrt{3}$. We will discuss this more fully when we reproduce the text of the *Observationes* in our second volume.)

(125) *Dedicatio*: 17–18.

(126) *Dedicatio*: 18.

(127) Newton has substituted his own algebraic proof for Wallis' geometrical version which finds

$$cd = \frac{bc^2 + ca^2 - ab^2}{2ac} = \frac{4+16-1}{8}$$

immediately by the extended theorem of Pythagoras (the geometrical model of the cosine rule). Note too that Newton's diagram transposes the points a and d, and b and c in Wallis' figure.

(128) This is the crucial point. The algebraic structure is again not at fault, but rather it is the application of it which is erroneous. Since $ab + bc = 3 < 4 = ac$, the points a, b and c cannot be measured by a Euclidean (or Riemannian) metric and so the triangle is not representable in Euclidean space. In particular, in the metric in which the triangle abc can be placed Pythagoras' theorem cannot hold and so both Newton's and Wallis' proofs are invalid.

(129) Read '$\frac{-105}{64}$'.

(130) Read 'square'.

(131) Newton gives the argument that, since bd is $\frac{1}{8}\sqrt{-105}$, not all line-lengths associated with the triangle are constructible in a real Euclidean space. However, this proof depends on the assumption that Pythagoras' theorem holds (which it here cannot).

(132) Newton has cancelled 'proceeding'.

(133) Since $x^8 - a^8 = (x-a)(x+a)(x^2+a^2)(x^4+a^4)$.

§4. FURTHER DEVELOPMENT OF THE BINOMIAL EXPANSION(1)

[Autumn 1665?]

From the original in the University Library, Cambridge(2)

[1] If *lab* is an Hyperbola; *cde*, *ck* its Asymptotes, *a* its vertex, & *cag* its axis; if *adck* is a square & $he \parallel ad$.(3) & $cd=1$, &, $de=x$. y^{n} $be=\frac{1}{1+x}$. If also, $ef=1$. $eg=1+x$. $eh=1+2x+x[x]$. &c: (the progression continued is $1+3x+3xx+x^3$. $1+4x+6x^2+4x^3+x^4$. $1+5x+10x^2+10x^3+5x^4+x^5$. &c).

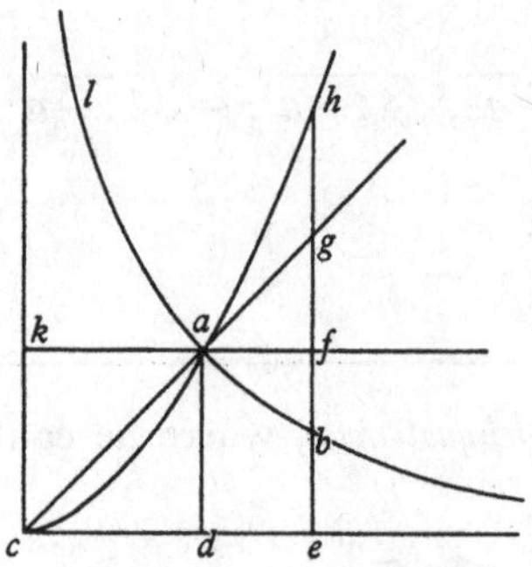

Then, shall the areas of those lines proceede in this progression. $*=adeb$. $x=adef$. $x+\frac{xx}{2}=adeg$. $adeh=x+\frac{2xx}{2}+\frac{x^3}{3}$. $x+\frac{3xx}{2}+\frac{3x^3}{3}+\frac{x^4}{4}$. $x+\frac{4xx}{2}+\frac{6x^3}{3}+\frac{4x^4}{4}+\frac{x^5}{5}$. &c. As in this table

x	×	1.	1.	1.	1.	1.	1.	1.	1.	1.	
$\frac{xx}{2}$	×	−1.	0.	1.	2.	3.	4.	5.	6.	7.	
$\frac{x^3}{3}$	×	1.	0.	0.	1.	3.	6.	10.	15.	21.	
$\frac{x^4}{4}$	×	−1.	0.	0.	0.	1.	4.	10.	20.	35.	&c
$\frac{x^5}{5}$	×	1.	0.	0.	0.	0.	1.	5.	15.	35.	
$\frac{x^6}{6}$	×	−1.	0.	0.	0.	0.	0.	1.	6.	21.	
$\frac{1}{7}x^7$	×	1.	0.	0.	0.	0.	0.	0.	1.	7.	
		$=adeb$.	$=adef$.	$=adeg$.	$=adeh$.		&c.				

In wch y^{e} first area is also inserted. The composition of wch table may be deduced from hence, viz: The sume of any figure & y^{e} figure above it is equall to y^{e} figure following it.(4) By wch table it may appeare y^{t} y^{e} area of y^{e} Hyperbola *adeb* is

$$x-\frac{xx}{2}+\frac{x^3}{3}-\frac{x^4}{4}+\frac{x^5}{5}-\frac{x^6}{6}+\frac{x^7}{7}-\frac{x^8}{8}+\frac{x^9}{9}-\frac{x^{10}}{10}\ \&c.^{(5)}$$

(1) Newton improves on his first researches into the binomial expansion in §3.2/3 above [=ULC. Add. 4000: 18^{r} ff.] and sets them out finally as a set of four propositions. Since the logarithmic series appears as an integral part of the text the date of composition, by Newton's own testimony, is not earlier than summer 1665 and it seems probable that Newton ordered his ideas in their present form when he had had some time to think them over. However, the early autumn of 1665 seems a firm post-date for composition, since parts of the present text

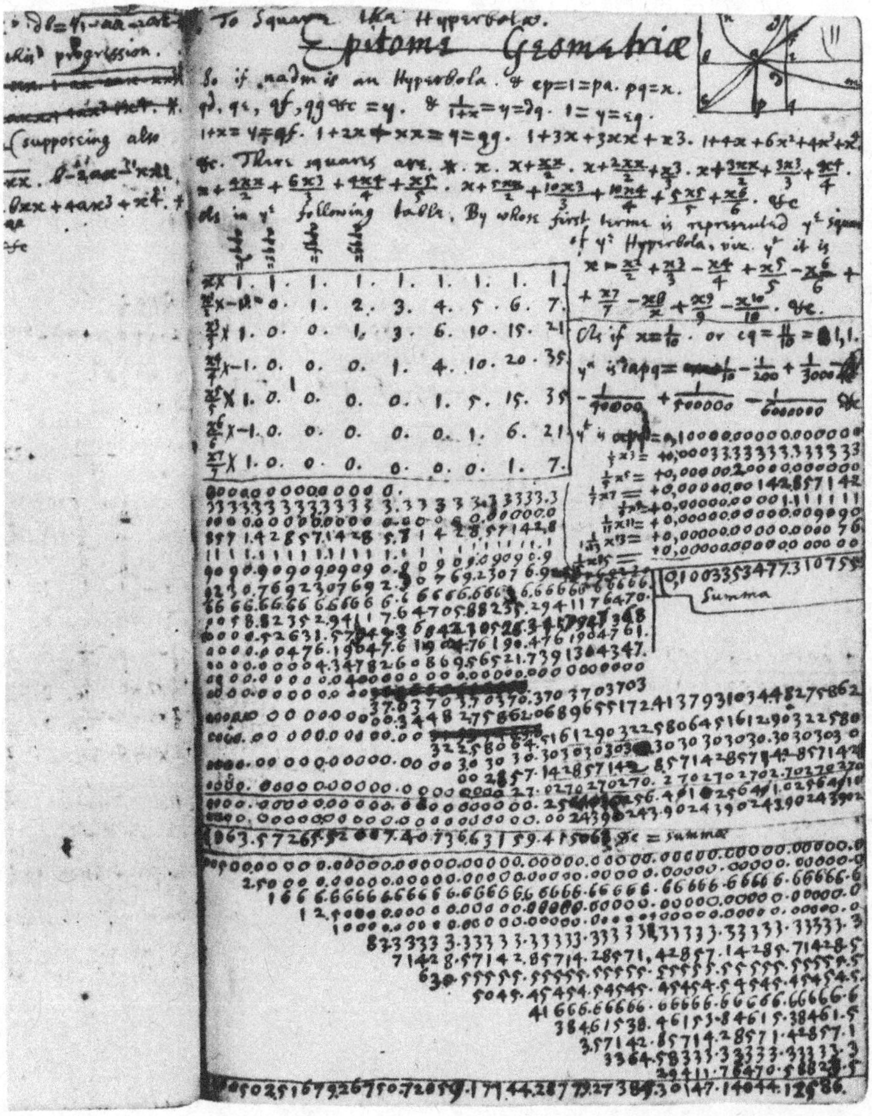

Plate I. Binomial researches: abortive calculation of log (1.1) (I, 3, §3.3).

Suppose y[t] *adck* is a Square[,] *abc* a circle[,] *agc* a Parabola. &c. & y[t] $de=x$. $ad \parallel fe=1=bd$. & y[t] y[e] progression in w[ch] y[e] lines *fe*, *be*, *ge*, *he*, *ie*, *ne* &c proceedes is 1. $\sqrt{1-xx}$. $1-xx$. $\overline{1-xx}\sqrt{1-xx}$. $1-2xx+x^4$. $\overline{1-2xx+x^4}\sqrt{1-xx}$. $1-3xx+3x^4-x^6$. &c. Then will their areas *fade*, *bade*, *gade*, *hade*, *iade*, &c be in this progression. x. *. $x-\frac{xxx}{3}$. *. $x-\frac{2}{3}x^3+\frac{1}{5}x^5$. *. $x-\frac{3x^3}{3}+\frac{3x^5}{5}-\frac{x^7}{7}$. *. $x-\frac{4x^3}{3}+\frac{6x^5}{5}-\frac{4x^7}{7}+\frac{x^9}{9}$. &c: as in this table following in w[ch] y[e] intermediate termes are inserted.

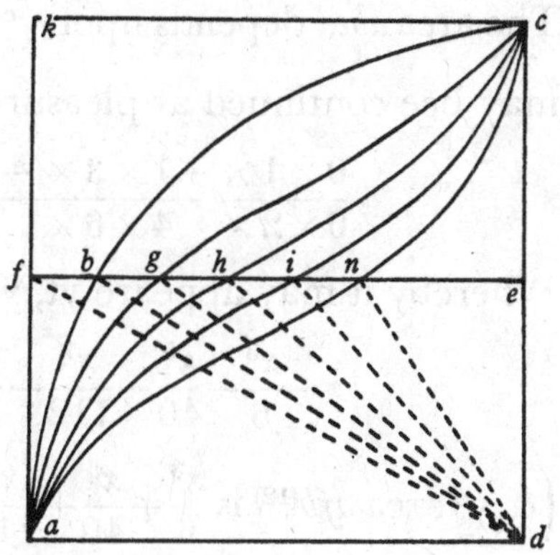

			adef =	*adeb* =	*adeg* =	*adeh* =	*adei* =	*aden* =								
x ×	1.	1.	1.	1.	1.	1.	1.	1.	1.	1.	1.	1.	1.	1.	1.	
$\frac{-x^3}{3}$ ×	−1.	$\frac{-1}{2}$.	0.	$\frac{1}{2}$.	1.	$\frac{3}{2}$.	2.	$\frac{5}{2}$.	3.	$\frac{7}{2}$.	4.	$\frac{9}{2}$.	5.	$\frac{11}{2}$.	6.	
$\frac{x^5}{5}$ ×	1.	$\frac{3}{8}$.	0.	$-\frac{1}{8}$.	0.	$\frac{3}{8}$.	1.	$\frac{15}{8}$.	3.	$\frac{35}{8}$.	6.	$\frac{63}{8}$.	10.	$\frac{99}{8}$.	15.	
$\frac{-x^7}{7}$ ×	−1.	$\frac{-5}{16}$.	0.	$\frac{3}{48}$.	0.	$\frac{-1}{16}$.	0.	$\frac{5}{16}$.	1.	$\frac{35}{16}$.	4.	$\frac{105}{16}$.	10.	$\frac{231}{16}$.	20.	
$\frac{x^9}{9}$ ×	1.	$\frac{+35}{128}$.	0.	$\frac{-15}{384}$.	0.	$\frac{3}{128}$.	0.	$\frac{-5}{128}$.	0.	$\frac{35}{128}$.	1.	$\frac{315}{128}$.	5.	$\frac{1155}{128}$.	15.	[&c]
$\frac{-x^{11}}{11}$ ×	−1.	$\frac{-63}{256}$.	0.	$\frac{105}{3840}$.	0.	$\frac{-3}{256}$.	0.	$\frac{3}{256}$.	0.	$\frac{-7}{256}$.	0.	$\frac{63}{256}$.	1.	$\frac{693}{256}$.	6.	
$\frac{x^{13}}{13}$ ×	1.	$\frac{231}{1024}$.	0.	$\frac{-945}{46080}$.	0.	$\frac{7}{1024}$.	0.	$\frac{-5}{1024}$.	0.	$\frac{7}{1024}$.	0.	$\frac{-21}{1024}$.	0.	$\frac{231}{1024}$.	1.	

The property of w[ch] table is y[t] y[e] Sum̄e of any figure & y[e] figure above it is equall to y[e] figure next after it save one.(6) Also y[e] numerall progressions are of these formes.

a.	a.	a.	a.	
b.	$a+b$.	$2a+b$	$3a+b$.	
c.	$b+c$.	$a+2b+c$.	$3a+3b+c$.	&c.
d.	$c+d$.	$b+2c+d$.	$a+3b+3c+d$.	
e.	$d+e$.	$c+2d+e$.	$b+3c+3d+e$.	

are incorporated in Newton's first thoughts on algorithmic integration in Add. 4000. (See **2**, 5, §1 below.)

(2) Add. 3958.3: 72r, 70r–71r.

(3) Newton has cancelled '$\angle cka=2\angle kca=\angle kcd=\angle cda=\angle ceb$'.

(4) The familiar composition rule in a triangular array of binomial coefficients.

(5) Newton redrafts §3.3 above.

(6) Newton redrafts §3.2 above.

Where[by] y[e] calculation of y[e] intermediate termes may bee easily performed.(7)
The area *abed* depends upon y[e] 4th Collume(8) $1.\ \frac{1}{2}.\ -\frac{1}{8}.\ \frac{3}{48}$ &c: (w[ch] progression may bee continued at pleasure by y[e] helpe of this rule

$$\frac{0\times 1\times -1\times 3\times -5\times\ 7\times -9\times 11\times -13\times 15}{0\times 2\times\quad 4\times 6\times\quad 8\times 10\times\ 12\times 14\times\quad 16\times 18}\quad \&\text{c.}^{(9)})$$

Whereby it may appeare yt, what ever y[e] sine(10) $de = x$ is, y[e] area *abed* is

$$x - \frac{x^3}{6} - \frac{x^5}{40} - \frac{x^7}{112} - \frac{5x^9}{1152} - \frac{7x^{11}}{2816} - \frac{21x^{13}}{13312} - \frac{11x^{15}}{10240}\quad \&\text{c.}^{(11)}$$

$\left(\text{\& y}^{\text{e}}\text{ area } afb^{(12)} \text{ is } \frac{x^3}{6} + \frac{x^5}{40} + \frac{x^7}{112}\ \&\text{c.}\right)$. Whereby also y[e] area & angle *adb* may bee found.(13)

The same may bee done thus. y[e] areas *afd, abd, agd, ahd* &c(14) are in this progression $\frac{x}{2}.\ *.\ \frac{x}{2} + \frac{x^3}{3}.^{(15)}\ *.\ \frac{x}{2} + \frac{2x^3}{6} - \frac{3x^5}{10}.\ *.\ \frac{x}{2} + \frac{3x^3}{6} - \frac{9x^5}{10} + \frac{5x^7}{14}.\ *.$

$$\frac{x}{2} + \frac{4x^3}{6} - \frac{18x^5}{10} + \frac{40x^7}{14} - \frac{7x^9}{18}.\quad \&\text{c.}$$

(7) For the first time Newton introduces a table of finite-differences. Where $a = (0, 0)$, $b = (1, 0)$, $c = (2, 0)$, $d = (3, 0)$, $e = (4, 0)$ and general point (s, k) is the binomial entry $\binom{r+k}{s}$, we may rewrite Newton's array as a unit-lattice (s, k), $s = 0, 1, 2, 3, 4$; $k = 0, 1, 2, 3$. (Newton will extend the lattice both below and to right and left, but not above since $(s, k) = 0$ for $s < 0$.) Since the binomial entries obey the composition rule

$$\binom{r+k+1}{s+1} - \binom{r+k}{s} = \binom{r+k}{s+1},$$

we can define the difference-scheme with unit-intervals,

$$\Delta^1(s, k) \equiv (s+1, k+1) - (s, k) = (s+1, k),$$

or more formally

$$\begin{cases} \Delta^0(s, k) = (s, k), \quad \text{and} \\ \Delta^{i+1}(s, k) = \Delta^1(\Delta^i(s, k)) = \Delta^i(s-1, k) = (s-(i+1), k), \quad i = 0, 1, 2, 3, \ldots \text{ successively,} \end{cases}$$

It follows that

$$\begin{aligned}(s, k) &= (1+\Delta^1)\,(s-1, k-1) = (1+\Delta^1)^k\,(s-k, 0) \\ &= (1+\Delta^1)^k\,\Delta^{s-k}(0, 0) = \sum_{0\leqslant i\leqslant k}\left[\binom{k}{i}\,\Delta^{i+s-k}(0, 0)\right] \\ &= \sum_{k-s\leqslant i\leqslant k}\left[\binom{k}{i}\,\Delta^{i+s-k}(0, 0)\right], \text{ since } \Delta^{i+s-k}(0, 0) = (i+s-k, 0) = 0 \text{ for } i < k-s, \\ &= \sum_{0\leqslant i\leqslant s}\left[\binom{k}{i}\,\Delta^{s-i}(0, 0)\right] = \sum_{0\leqslant i\leqslant s}\left[\binom{k}{i}\,(s-i, 0)\right],\end{aligned}$$

which is Newton's scheme.

The 'calculation of y[e] intermediate termes' by a difference table will be discussed in detail below (note (35)).

As in this following Table.

$\frac{1}{2}x$	$\times[1.]$	1.	1.	1.	1.	1.	1.	
$\frac{1}{6}x^3$	$\times[0.]$	$\frac{1}{2}$.	1.	$\frac{3}{2}$.	2.	$\frac{5}{2}$.	3.	
$-\frac{3}{10}x^5$	$\times[0.]$	$-\frac{1}{8}$.	0.	$\frac{3}{8}$.	1.	$\frac{15}{8}$.	3.	[&c]
$\frac{5}{14}x^7$	$\times[0.]$	$\frac{1}{16}$.	0.	$-\frac{1}{16}$.	0.	$\frac{5}{16}$.	1.	
$-\frac{7}{18}x^9$	$\times[0.]$	$\frac{-5}{128}$.	0.	$\frac{3}{128}$.	0.	$\frac{-5}{128}$.	0.	(16)
	[afd =]	abd =	agd =	ahd =	aid =	and =		

By w[ch] it may bee perceived y[t]

$$abd = \frac{1}{2}x + \frac{1}{12}x^3 + \frac{3}{80}x^5 + \frac{5}{224}x^7 + \frac{35x^9}{2304} + \frac{63x^{11}}{5632}. \quad \&c.$$

And by this meanes haveing y[e] area *abd*, (w[ch] y[e] angle *adb* gives) *de* y[e] sine of y[e] angle *adb* may bee found.(17)

(8) Read 'column'.

(9) Read $`\frac{0\times 1\times -1\times -3\times -5\times -7\times -9\times -11\times -13\times -15}{0\times 2\times 4\times 6\times 8\times 10\times 12\times 14\times 16\times 18}$ &c', that is, $\binom{\frac{1}{2}}{i}$, $i = 0, 1, 2, 3, \ldots$, successively $\left(\text{where } \binom{\frac{1}{2}}{0} = 1 \text{ is entered in Wallisian manner as } \frac{0}{0}\right)$.

(10) Read 'signe'.

(11) That is, $\int_0^x (1-t^2)^{\frac{1}{2}}.dt = \int_0^x \sum_{0\leqslant i\leqslant\infty} \left[\binom{\frac{1}{2}}{i}(-t^2)^i\right].dt$

$$= \sum_{0\leqslant i\leqslant\infty} \left[\binom{\frac{1}{2}}{i}\int_0^x (-t^2)^i.dt\right].$$

(12) $(afed)-(abed) = x - \int_0^x (1-t^2)^{\frac{1}{2}}.dx.$

(13) $\widehat{adb}$ (or $\sin^{-1}x$) $= 2(adb) = 2[(abed)-\triangle bed] = 2\int_0^x (1-t^2)^{\frac{1}{2}}.dt - x\sqrt{(1-x^2)}$.

(14) $\int_0^x (1-t^2)^s.dt - \frac{1}{2}x(1-x^2)^s$, $(s = 0, \frac{1}{2}, 1, \frac{3}{2}, \ldots)$ successively. (15) Read $`\frac{x^3}{6}'$.

(16) $\int_0^x (1-t^2)^s.dt - \frac{1}{2}x(1-x^2)^s = \sum_{0\leqslant i\leqslant\infty}\left[\binom{s}{i}\left(\int_0^x (-t^2)^i.dt - \frac{1}{2}x(-x^2)^i\right)\right]$

$$= \sum_{0\leqslant i\leqslant\infty}\left[(-1)^{i+1}\binom{s}{i}\frac{2i-1}{2(2i+1)}x^{2i+1}\right].$$

(17) $2(abd) = \widehat{adb} = y$, say, $= \sin^{-1}\left(\frac{de}{bd}\right) = \sin^{-1}x$, so that

$$de = x = bd\times\sin[2(abd)] = \sin y.$$

Newton's power-series expansion of $\sin^{-1}x$ allows him to express y as an infinite series in x, and probably $x = \sin y$ is to be expressed as an infinite series in powers of y by reversing the previous series $y = \sin^{-1}x$ (as he was to indicate to Leibniz in his letter of 24 October 1676: see *Correspondence of Isaac Newton*, **2** (1960): 128–9).

Corol: If $de=x$. & $\sqrt{1+xx}=eb$. y^n *abc* is an Hyperbola. & its area *dabe* is $x+\frac{x^3}{6}-\frac{x^5}{40}+\frac{x^7}{112}-\frac{5x^9}{1152}+\frac{7x^{11}}{2816}$. &c.(18)

---

[2](19)

Prop 1. Suppose $ab=x$. $bd=y\perp ab$. And y^t y^e nature of y^e line *addc* is such y^t y^e valor of y is rationall & consists of no fractions in whose denominator x is, or else wholy of such fractions in whose denominators x is, but not of divers dimensions: If I then multiply y^e valor of y by x, & divide each terme of y^t valor by so many units as it hath dimensions in y^t terme; y^e result shall signifie y^e area *abd* of y^e afforesaide line *addc*.(20)

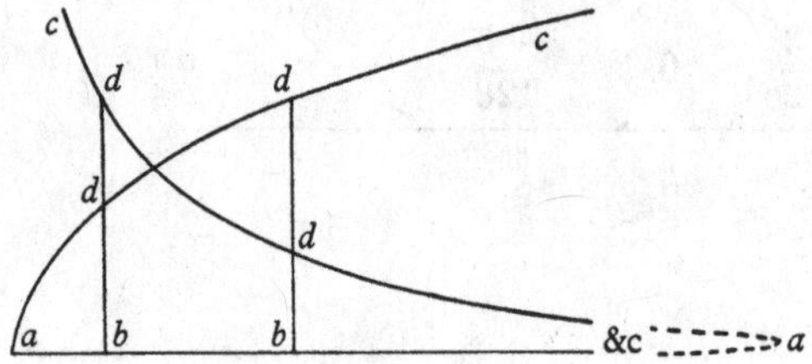

As for example. If $y=1$, or $y=x$, or $y=xx$, or x^3, or x^4 &c y^n y^e area *abd* is $\frac{1\times x}{1}$, or $\frac{x\times x}{2}$, or $\frac{xx\times x}{3}$, or $\frac{x^3\times x}{4}$, or $\frac{x^4\times x}{5}$ &c: so if $y=\frac{a}{b}$, or $y=\frac{ax}{b}$, or $\frac{axx}{b}$ &c: y^n y^e area *abd* is $\frac{ax}{b}$, $\frac{axx}{2b}$, $\frac{ax^3}{3b}$ &c. In like manner if $y=\frac{a}{xx}$, or $y=\frac{a}{x^3}$, or $y=\frac{a}{x^4}$ &c

---

(18) The area $(dabe) = \int_0^x (1+t^2)^{\frac{1}{2}}.dt = \frac{1}{2}(\log(x+\sqrt{\{1+x^2\}})+x\sqrt{\{1+x^2\}})$ under the hyperbola $y^2 = 1+x^2$ is the first interpolated value ($s = \frac{1}{2}$) in the sequence

$$\int_0^x (1+t^2)^s.dt, \quad s = 0,\ 1,\ 2,\ \ldots.$$

Its evaluation is hardly a simple corollary of the preceding but may be carried through in a like way. Thus

$$\int_0^x (1+t^2)^s.dt = \sum_{0\leqslant i\leqslant\infty}\left[\binom{s}{i}\int_0^x (t^2)^i.dt\right] = \sum_{0\leqslant i\leqslant\infty}\left[\binom{s}{i}\frac{x^{2i+1}}{2i+1}\right],$$

so that
$$(dabe) = \sum_{0\leqslant i\leqslant\infty}\left[\binom{\frac{1}{2}}{i}\frac{x^{2i+1}}{2i+1}\right].$$

(19) Newton revises his previous researches as a scheme in four propositions (with provision, unimplemented, for a fifth). This section is described and partially published in D. T. Whiteside, 'Newton's Discovery of the General Binomial Theorem', *Mathematical Gazette*, **45** (1961): 175–80.

(20) Where $bd = \sum_i (a_i x^i)$, $(abd) = \int_0^k (bd).dx = \sum_i \left[\frac{x}{i+1}a_i x^i\right]_0^k$.

(21) Newton in fact, as we have seen above (3. §3), took this general theorem from Wallis.

(22) True since $(abd) = \int_0^k at^{-1}.dt = a[\log(x)]_0^k = \infty$. Four years later Newton added this corollary, unchanged, to the opening page of his tract *De Analysi*. It seems likely that

y^n^ $\frac{a}{-1x}=\frac{-a}{x}$, or $\frac{a}{-2xx}$. or $\frac{a}{-3x^3}$, &c: is y^e^ area *abd* (as by others(21) demonstrated). $\left[\text{so if } \frac{a}{x}=y.\ \text{y}^{\text{n}}\ \frac{a}{0}=\frac{ax^0}{0}\right.$ is y^e^ area *abd*; viz: tis infinite.(22)]. Lastly if $y=1+x+axx-bx^3-\frac{x^4}{d}+\frac{cx^{15}}{e+f}$ &c. y^e^ area *abd* is

$$x+\frac{1}{2}xx+\frac{ax^3}{3}-\frac{bx^4}{4}-\frac{x^5}{5d}+\frac{cx^{16}}{16e+16f}\quad \&\text{c.}$$

For y^e^ area *abd* is compounded of those areas wch are related to & generated by those quantitys of wch y^e^ valor of *y* is compounded. & w^t^ these areas are appeare by y^e^ former example. (Note y^t^ Parabolicall & Hyperbolicall (i:e: (in respect of *bd*,) affirmative & Negative) areas (thus considered) cannot compound any 3^d^ area because they are not on y^e^ same side of y^e^ line *bd*.(23))

This best appeares by multiplying both parts by $b+x$.(25)

Prop. 2. $\frac{aa}{b+x}=\frac{aa}{b}-\frac{aax}{bb}+\frac{aaxx}{b^3}-\frac{aax^3}{b^4}+\frac{aax^4}{b^5}-\frac{aax^5}{b^6}+\frac{aax^6}{b^7}$ &c. For these termes $\frac{aa}{b+x}$. *aa*. *aab*+*aax*. *aabb*+2*aabx*+*aaxx*. $aab^3+3aabbx+3aabxx+aax^3$, [are] $\div\!\!\div$(24), &c. which termes may bee thus ordered———

$\frac{aa}{b+x}$.	*aa*.	*aab*.	*aabb*.	aab^3.	aab^4.	
		aax.	2*aabx*.	3*aabbx*.	$4aab^3x$.	
			aaxx.	3*aabxx*.	6*aabbxx*.	&c:
				aax^3.	$4aabx^3$.	
					aax^4.	

he worried much over this case $i=-1$, which does not apparently fit into the general Wallis theorem though, worked as a problem in hyperbola-area,

$$\int_0^{x_1} at^{-1}.dt-\int_0^{x_2} at^{-1}.dt = a\log\left(\frac{x_1}{x_2}\right).$$

The subtle reconciliation of the two (compare 2, 7: note (12) below) afforded by defining

$$\int_{x_2}^{x_1} at^{-1}.dt \quad\text{as}\quad \lim_{\epsilon\to 0}\left(\int_{x_2}^{x_1} at^{-1+\epsilon}.dt\right) = \lim_{\epsilon\to 0}\left[\frac{at^{\epsilon}}{\epsilon}\right]_{x_2}^{x_1},$$

with $t^{\epsilon}=e^{\epsilon\log(t)}=1+\epsilon\log(t)+O(\epsilon^2)$, appears beyond Newton's present understanding.

(23) Newton has not yet reached the convention that positive and negative areas may, though on opposite sides of *bd*, be compounded as the third area which is their difference.

(24) Read 'continuously proportional' (Oughtred's symbol).

(25) This remark, added here as an afterthought but referring to the proposition as a whole, was apparently inserted immediately after Newton made the division on a loose vellum sheet of calculations (ULC. Add. 3958.2: especially 29^r^), taking the quotient to two terms. In proof

$$\sum_{0\leqslant i\leqslant n}\left[\frac{a^2}{b}\left(-\frac{x}{b}\right)^i\right](b+x) = a^2+(-1)^n a^2\left(\frac{x}{b}\right)^{n+1},$$

so that $\frac{a^2}{b+x}=\sum_{0\leqslant i\leqslant\infty}\left[\frac{a^2}{b}\left(-\frac{x}{b}\right)^i\right]$, where $x<b$ and so $\lim_{n\to\infty}\left[\left(\frac{x}{b}\right)^{n+1}\right]=0$.

Or by supplying y^e vacant places

$$
\begin{array}{lllll}
\frac{aa}{b+x}. & 1\times aa. & 1\times aab. & 1\times aabb. & 1\times aab^3. \quad 1\times aab^4. \\[2ex]
& 0\times\frac{aax}{b}. & 1\times aax. & 2\times aabx. & 3\times aabbx. \quad 4\times aab^3x. \\[2ex]
& 0\times\frac{aaxx}{bb}. & 0\times\frac{aax^2}{b}. & 1\times aaxx. & 3\times aabxx. \quad 6\times aabbxx. \quad [\&c] \\[2ex]
& 0\times\frac{aax^3}{b^3}. & 0\times\frac{aax^3}{bb}. & 0\times\frac{aax^3}{b}. & 1\times aax^3. \quad 4\times aabx^3. \\[2ex]
& 0\times\frac{aax^4}{b^4}. & 0\times\frac{aax^4}{b^3}. & 0\times\frac{aax^4}{bb}. & 0\times\frac{aax^4}{b}. \quad 1\times aax^4.
\end{array}
$$
(26)

Now to reduce y^e first terme $\frac{aa}{b+x}$ to y^e same forme wth y^e rest, I consider in what progressions y^e numbers prefixed to these termes proceede. & find y^m to bee such y^t any number added to y^e number above it is equall to y^e number following it. Whence any termes may bee found wch are wanting, as in y^e annexed Table.(27)

1.	1.	1.	1.	1.	1.	1.	1.	1.	1.	1.	1.	
−4.	−3.	−2.	−1.	0.	1.	2.	3.	4.	5.	6.	7.	
10.	6.	3.	1.	0.	0.	1.	3.	6.	10.	15.	21.	
−20.	−10.	−4.	−1.	0.	0.	0.	1.	4.	10.	20.	35.	&c.
35.	15.	5.	1.	0.	0.	0.	0.	1.	5.	15.	35.	
−56.	−21.	−6.	−1.	0.	0.	0.	0.	0.	1.	6.	21.	
84.	28.	7.	1.	0.	0.	0.	0.	0.	0.	1.	7.	

Also any terme to wch these numbers are prefixed, being multiplyed by b produceth y^e following litterall terme. Or y^e higher terme multiplyed by $\frac{x}{b}$ produceth y^e lower terme. As in y^e following table.

(26) As before, Newton sets out the binomial coefficients $\binom{k}{s} \equiv (s, k)$ as a square array, $s, k = 0, 1, 2, 3, \ldots$, successively; and then, finding the combination rule

$$(s, k) = (s, k-1)+(s-1, k-1)$$

to hold for all (s, k), k positive, he extends the array backwards to the left (that is, for negative values of k) on the suppositions that this combination rule continues to hold and (implicitly) that $(0, k) = 1$ for all k.

(27) With the restriction, necessary for convergence, that x (taken by Newton as ranging only over the positive interval $[0, \infty]$) be less than b.

(28) Where $x > b$.

(29) Where $x < b$.

$\frac{aa}{\overline{b+x})^3}$.	$\frac{aa}{\overline{b+x})^2}$.	$\frac{aa}{b+x}$.	aa.	$aa\times\overline{b+x}$.	$a^2\times\overline{b+x})^2$.	$a^2\times\overline{b+x})^3$.	$a^2\times\overline{b+x})^4$.	
$\Vert$	$\Vert$	$\Vert$	$\Vert$	$\Vert$	$\Vert$	$\Vert$	$\Vert$	
$+\frac{aa}{b^3}$.	$+\frac{aa}{bb}$.	$+\frac{aa}{b}$.	$+aa$.	$+aab$	$+aabb$.	$+aab^3$.	$+aab^4$.	
$-\frac{3aax}{b^4}$.	$-\frac{2aax}{b^3}$.	$-\frac{aax}{bb}$.	$\frac{0\times aax}{b}$.	$+aax$.	$+2aabx$.	$+3aabbx$.	$+4aab^3x$.	
$+\frac{6aax^2}{b^5}$.	$+\frac{3aax^2}{b^4}$.	$+\frac{aaxx}{b^3}$.	$\frac{0\times aax^2}{b^2}$.	$\frac{0\times aaxx}{b^2}$.	$+aaxx$.	$+3aabxx$.	$+6aabbx^2$.	
$-\frac{10aax^3}{b^6}$.	$-\frac{4aax^3}{b^5}$.	$-\frac{aax^3}{b^4}$.	$\frac{0\times aax^3}{b^3}$.	$\frac{0\times aax^3}{bb}$.	$\frac{0\times aax^3}{b}$.	$+aax^3$.	$+4aabx^3$.	&c.
$+\frac{15aax^4}{b^7}$.	$+\frac{5aax^4}{b^6}$.	$+\frac{aax^4}{b^5}$.	$\frac{0\times aax^4}{b^4}$.	$\frac{0\times aax^4}{b^3}$.	$\frac{0\times aax^4}{bb}$.	$\frac{0\times aax^4}{b}$.	$+aax^4$.	
&c.	&c.	&c.	&c.	&c.	&c.	&c.	&c.	

Whence it appeares y^t $\frac{aa}{b+x}=\frac{aa}{b}-\frac{aax}{bb}+\frac{aaxx}{b^3}-\frac{aax^3}{b^4}$ &c. Or y^t $\frac{aa}{b-x}=\frac{aa}{b}+\frac{aax}{bb}+\frac{aaxx}{b^3}+\frac{aax^3}{b^4}$ &c.(27) Or y^t $\frac{aa}{b+x}=\frac{aa}{x}-\frac{aab}{xx}+\frac{aabb}{x^3}-\frac{aab^3}{x^4}$ &c.(28) Or y^t $\frac{aa}{bb+2bx+xx}=\frac{aa}{bb}-\frac{2a^2x}{b^3}+\frac{3aaxx}{b^4}$ &c.(29)

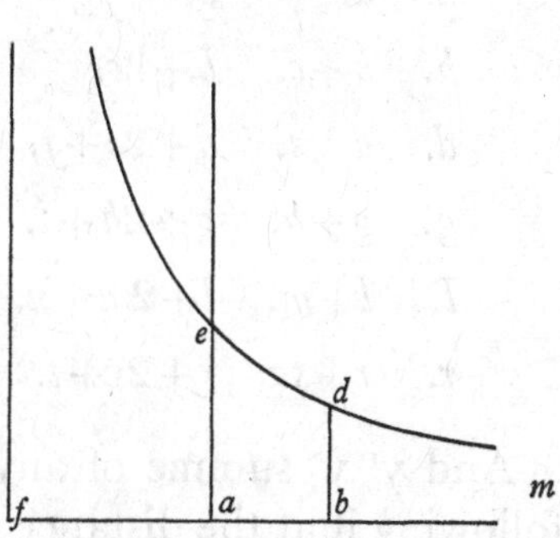

Prop: 3^d. If $ab=x$. $y=db\perp ab\perp ae$. $fa=b$. & $\frac{aa}{b+x}=y=$(prop 2) $\frac{aa}{b}-\frac{aax}{bb}+\frac{aaxx}{b^3}-\frac{aax^3}{b^4}+\frac{aax^4}{b^5}$ &c. Then (by prop 1), $\frac{aax}{b}-\frac{aaxx}{2bb}+\frac{aax^3}{3b^3}-\frac{aax^4}{4b^4}+\frac{aax^5}{5b^5}-\frac{aax^6}{6b^6}$ &c: is *abde*, y^e area of y^e Hyperbola.(30) So if $\frac{aa}{b-x}=y$. &c: In like manner if

$$\frac{aa}{bb+2bx+xx}=y=(\text{prop }2)\ \frac{aa}{bb}-\frac{2aax}{b^3}+\frac{3aaxx}{b^4}-\frac{4aax^3}{b^5}\ \&c.$$

Then (by prop 1) $\frac{aax}{bb}-\frac{aaxx}{bb}+\frac{aax^3}{b^4}-\frac{aax^4}{b^5}$ &c$=$(prop 2) $\frac{aa}{b}-\frac{aa}{b+x}$. is y^e area

(30)
$$\int_0^x \frac{a^2}{b}\left(1+\frac{t}{b}\right)^{-1}.dt=\sum_{0\leqslant i\leqslant\infty}\left[\binom{-1}{i}\frac{a^2}{i+1}\left(\frac{x}{b}\right)^{i+1}\right]$$
$$=a^2\log\left(\frac{b+x}{b}\right),\quad \text{where}\quad x\leqslant b.$$

(It is possible that Newton is aware of this last restriction and seeks to convey it in his accompanying figure by setting $ab<fa$.)

abde.(31) (wch may also thus appeare viz: if $fb=b+x=z$. y^n^ $\frac{aa}{bb+2bx+xx}=\frac{aa}{zz}=y$.

Therefore (prop 1) y^e^ area $dbm=\frac{aa}{z}=\frac{aa}{b+x}$. & $eam=\frac{aa}{b}$. so y^t^ $eadb=\frac{aa}{b}-\frac{aa}{b+x}$(32)).
And so of y^e^ rest. As if

$$\frac{x^3}{aa+bx+x^2}=y=(\text{prop }2)\ \frac{x^3}{aa}\ \frac{-bx^4-x^5}{a^4}\ \frac{+bbx^5+2bx^6+x^7}{a^6}$$

$$\frac{-bbbx^6-3bbx^7-3bx^8-x^9}{a^8}\quad \&c.$$

The area *abde* is

$$\frac{4x^4}{aa}\ \frac{-5bx^5-6x^6}{a^4}\ \frac{+6bbx^6+14bx^7+8x^8}{a^6}\ \frac{-7b^3x^7-24bbx^8-27bx^9-10x^{10}}{a^8}\quad \&c.^{(33)}$$

Prop 4th.(34) To find two or three intermediate termes in y^e^ above mentioned table of numerall progressions, I observe y^t^ those progressions are of this nature viz

$a.$	$a.$	$a.$	$a.$	$a.$	
$b.$	$b+c.$	$b+2c.$	$b+3c.$	$b+4c.$	
$d.$	$d+e.$	$d+2e+f.$	$d+3e+3f.$	$d+4e+6f.$	
$g.$	$g+h.$	$g+2h+i.$	$g+3h+3i+k.$	$g+4h+6i+4k.$	[&c.]
$l.$	$l+m.$	$l+2m+n.$	$l+3m+3n+p.$	$l+4m+6n+4p+q.$	
$r.$	$r+s.$	$r+2s+t.$	$r+3s+3t+v.$	$r+4s+6t+4v+w.$(35)	

And y^t^ y^e^ summe of any terme & y^e^ terme above it is equall to y^e^ terme following it at the distance of y^e^ termes in y^e^ s^d^ numerall table.(36)

---

(31) $$\int_0^x \frac{a^2}{b^2}\left(1+\frac{t}{b}\right)^{-2}.dt = \sum_{0\leqslant i\leqslant\infty}\left[\binom{-2}{i}\frac{a^2}{(i+1)\,b}\left(\frac{x}{b}\right)^{i+1}\right]$$
$$= -\sum_{1\leqslant i\leqslant\infty}\left[(-1)^i\frac{a^2}{b}\left(\frac{x}{b}\right)^i\right], \text{ since } \binom{-2}{i} = (-1)^i(i+1)$$
$$= \frac{a^2}{b}-\frac{a^2}{b+x} = \left[-\frac{a^2}{b+t}\right]_0^x.$$

(32) $$(eadb) = (eam)-(dbm) = \int_b^\infty \frac{a^2}{u^2}.du-\int_z^\infty \frac{a^2}{u^2}.du$$
$$= \int_b^z \frac{a^2}{u^2}.du = \left[-\frac{a^2}{u}\right]_b^z.$$

(33) Through carelessness Newton uses the untenable integration theorem

$$\int_0^x a_i t^i.dt = (i+1)\,a_i x^{i+1}.$$

Suppose I would find y^e^ meane termes in y^e^ 3^d^ progression

$$\begin{matrix} 3. & *. & 1. & *. & 0. & *. & 0. \\ d-4e+10f. & d-3e+6f. & d-2e+3f. & d-e+f. & d. & d+e. & d+2e+f. \end{matrix}$$

$$\begin{matrix} *. & 1. & *. & 3. \\ d+3e+3f. & d+4e+6f. & d+5e+10f. & d+6e+15f. \end{matrix}$$

I compare y^e^ termes of y^t^ progression & of y^e^ correspondent litterall progression & find $d=0=2e+f$. $4e+6f=1$. subduct $4e+2f=0$, or $12e+6f=0$ from $4e+6f=1$. & y^e^ rest is $4f=1$. Or $-8e=1$. wch termes being found viz $d=0$. $e=\frac{-1}{8}$. $f=\frac{1}{4}$. y^e^ progression must be

$$3.\quad \frac{9}{8}.^{(37)}\quad 1.\quad \frac{3}{8}.\quad 0.\quad \frac{-1}{8}.\quad 0.\quad \frac{3}{8}.\quad 1.\quad \frac{15}{8}.\quad 3.\quad \&c.^{(38)}$$

---

(34) Newton has cancelled a first draft which reads:

Prop 4th. If $\frac{a}{b}x^n = y^m$. n & m being numbers y^t^ signifie y^e^ dimensions of x & y: Then is $\frac{mx}{m+n}\sqrt{m:\frac{a}{b}x^n} = \frac{mxy}{m+n}$, The area *abde* if n is affirmative, or *mbd* if n is nega[ti]ve of y^e^ line *edm*.

Prop: 5^t^. If $y=ax^m+bx^n$. y^n^ $\frac{x\times ax^m}{m+1}+\frac{xbx^n}{n+1}=abde$.

(35) Newton proceeds to subtabulate terms at $1/p$ intervals in each row (s, k), k free, in his array $(s, k) \equiv \binom{k}{s}$. Newton had already, as we have seen (note (7)), introduced a difference-scheme where the interpolation was to be made at unit-intervals, and this he generalizes. Where now the binomial coefficients $(s, \lambda+k) \equiv \binom{\lambda+k}{s}$ are entered as a square array and $a = (0, \lambda)$, $b = (1, \lambda)$, $d = (2, \lambda)$, $g = (3, \lambda)$, $l = (4, \lambda)$, $r = (5, \lambda)$, Newton seeks, on the analogy of his previous results, to generate (for each row s) terms $(s, \lambda+k)$ interpolated at $1/p$-intervals by $(s, \lambda+k) = \sum_{0\leqslant i\leqslant s}\left[\binom{kp}{i}c_{s,i}\right]$. When $p = 1$, clearly,

$$(s, \lambda+k) = \sum_{0\leqslant i\leqslant s}\left[\binom{k}{i}c_{s,i}\right], \quad \text{or} \quad c_{s,i} = (s-i, \lambda)$$

in this case. For other values of p, however, though always $c_{s,0} = (s, \lambda)$, there will be no simple way of determining the $c_{s,i}$ and they will have to be calculated anew for each s. (Newton's assumption of the form of his interpolated terms (s, k) is strictly equivalent to positing its identity with a polynomial of degree s whose coefficients may readily be determined. Indeed, as we shall see, Newton quickly concluded that the latter, as an assumption, seemed more plausible than the former and thereafter discarded his treatment by differences.)

(36) The basic composition rule $(s, k) = (s, k-1)+(s-1, k-1)$ is now generalized to cover all rational k. The 'distance of y^e^ termes' is, of course, the unit interval.

(37) Read $\frac{\text{‘}15\text{’}}{8}$.

(38) Newton bisects the intervals of $\binom{k}{2} \equiv (2, k)$, finding (on setting $p = 2$, $\lambda = 0$)

$$d = c_{2,0} = (2, 0) = 0, \quad e = c_{2,1} = -\tfrac{1}{8}, \quad f = c_{2,2} = \tfrac{1}{4}$$

Hence may be deduced this table viz

1.	1.	1.	1.	1.	1.	1.	1.	1.	1.	1.	1.	1.	1.	1.	1.	1.
−3.	$\frac{-5}{2}$.	−2.	$-\frac{3}{2}$.	−1.	$-\frac{1}{2}$.	0.	$\frac{1}{2}$.	1.	$\frac{3}{2}$.	2.	$\frac{5}{2}$.	3.	$\frac{7}{2}$.	4.	$\frac{9}{2}$.	5.
6.	$\frac{35}{8}$.	3.	$\frac{15}{8}$.	1.	$\frac{3}{8}$.	0.	$-\frac{1}{8}$.	0.	$\frac{3}{8}$.	1.	$\frac{15}{8}$.	3.	$\frac{35}{8}$.	6.	$\frac{63}{8}$.	10.
−10.	$\frac{-105}{16}$.	−4.	$\frac{-35}{16}$.	−1.	$-\frac{5}{16}$.	0.	$\frac{1}{16}$.	0.	$-\frac{1}{16}$.	0.	$\frac{5}{16}$.	1.	$\frac{35}{16}$.	4.	$\frac{105}{16}$.	10.
15.	$\frac{1155}{128}$.	5.	$\frac{315}{128}$.	1.	$\frac{35}{128}$.	0.	$\frac{-5}{128}$.	0.	$\frac{3}{128}$.	0.	$-\frac{5}{128}$.	0.	$\frac{35}{128}$.	1.	$\frac{315}{16}$(39)	5.
−21.	$\frac{-3003}{256}$.	−6.	$\frac{-693}{256}$.	−1.	$\frac{-63}{256}$.	0.	$\frac{7}{256}$.	0.	$\frac{-3}{256}$.	0.	$\frac{3}{256}$.	0.	$\frac{-7}{256}$.	0.	$\frac{63}{256}$.	1.
+28.	$\frac{15015}{1024}$.	7.	$\frac{3003}{1024}$.	1.	$\frac{231}{1024}$.	0.	$\frac{-21}{1024}$.	0.	$\frac{7}{1024}$.	0.	$\frac{-5}{1024}$.	0.	$\frac{7}{1024}$.	0.	$\frac{-21}{1024}$.	0.
&c	&c:	&c:	&c:	&c:	&c:	&c:	&c:	&c.	&c.	&c.	&c.	&c.	&c.	&c.	&c.	&c.(40)

Note y[t] y[e] progression $1.\ \frac{1}{2}.\ \frac{-1}{8}.\ \frac{1}{16}.\ \frac{-5}{128}.\ \frac{7}{256}$ &c:(41) may bee deduced from hence $\frac{1\times1\times-1\times-3\times-5\times-7\times-9\times-11}{1\times2\times\ \ 4\times\ \ 6\times\ \ 8\times\ 10\times\ 12\times\ \ 14}$ &c: & one intermediate terme given y[e] rest are easily deduced thence.

In like manner if I would find two meanes twixt every terme of y[t] numerall progrēss(42) I compare y[e] numerall & correspondent litterall progressions, suppose in y[e] 3[d] progression.

$$\begin{matrix} 1. & *. & *. & 0. & *. & *. & 0. \\ d-3e+6f. & d-2e+3f. & d-e+f. & d. & d+e. & d+2e+f. & d+3e+3f. \end{matrix}$$

$$\begin{matrix} *. & *. & 1. & \\ d+4e+6f. & d+5e+10f. & d+6e+15f. & \&c. \end{matrix}$$

And find y[t] $d=0=3e+3f$. & $6f-3e=1$. To w[ch] adding $3e+3f=0$. Or $-6f-6e=0$. y[e] result is $9f=1$. or $-9e=1$. So y[t] y[e] progression must bee

---

and so evaluating $(2, k) = \sum_{0\leqslant i\leqslant 2}\left[\binom{2k}{i}c_{2,i}\right]$. In polynomial form,

$$(2, k) = -\frac{1}{8}\binom{2k}{1}+\frac{1}{4}\binom{2k}{2} = \tfrac{1}{2}k(k-1).$$

(39) Read $\frac{‘315’}{128}$.

(40) Newton tabulates (s, k), $s = 0, 1, 2, \ldots, 6$; $k = -3, -\frac{5}{2}, -2, \ldots, \frac{9}{2}, 5$.

(41) $(s, \frac{1}{2}) = \binom{\frac{1}{2}}{s}$, $s = 0, 1, 2, \ldots$.

(42) Read ‘progression’.

$1.\ \frac{5}{9}.\ \frac{2}{9}.\ 0.\ \frac{-1}{9}.\ \frac{-1}{9}.\ 0.\ \frac{2}{9}.\ \frac{5}{9}.\ 1.\ \frac{14}{9}.\ \frac{20}{9}.\ 3.$ &c.(43) Hence may bee composed this Table

1.	1.	1.	1.	1.	1.	1.	1.	1.	1.	1.	1.	1.
0.	$\frac{1}{3}$.	$\frac{2}{3}$.	1.	$\frac{4}{3}$.	$\frac{5}{3}$.	2.	$\frac{7}{3}$.	$\frac{8}{3}$.	3.	$\frac{10}{3}$.	$\frac{11}{3}$.	4.
0.	$\frac{-1}{9}$.	$\frac{-1}{9}$.	0.	$\frac{2}{9}$.	$\frac{5}{9}$.	1.	$\frac{14}{9}$.	$\frac{20}{9}$.	3.	$\frac{35}{9}$.	$\frac{44}{9}$.	6. [&c.]
0.	$\frac{5}{81}$.	$\frac{4}{81}$.	0.	$\frac{-4}{81}$.	$\frac{-5}{81}$.	0.	$\frac{14}{81}$.	$\frac{40}{81}$.	1.	$\frac{140}{81}$.	$\frac{220}{81}$.	4.
0.	$\frac{-10}{243}$.	$\frac{-7}{243}$.	0.	$\frac{+5}{243}$.	$\frac{+5}{243}$.	0.	$\frac{-7}{243}$.	$\frac{-10}{243}$.	0.	$\frac{25}{243}$.	$\frac{110}{243}$.	1.
0.	$\frac{22}{729}$.	$\frac{14}{729}$.	0.	$\frac{-8}{729}$.	$\frac{-7}{729}$.	0.	$\frac{7}{729}$.	$\frac{8}{729}$.	0.	$\frac{-14}{729}$.	$\frac{-22}{729}$.	0.(44)

Note y[t] this progression viz $\frac{1\times 1\times -2\times -5\times -8\times -11\times -14}{1\times 3\times\ \ 6\times\ \ 9\times\ 12\times\ \ 15\times\ \ 18}$ &c gives y[e] second terme(45) $1.\ +\frac{1}{3}.\ -\frac{1}{9}.\ \frac{+5}{81}.\ \frac{10}{243}.\ \frac{+22}{729}$ &c.(46) & this

$$\frac{1\times 2\times -1\times -4\times -7\times -10\times -13}{1\times 3\times\ \ 6\times\ \ 9\times\ 12\times\ \ 15\times\ \ 18}\ \ \&\text{c}$$

gives y[e] third terme.(47) Also this progression

$$\frac{1\times 4\times 1\times -2\times -5\times -8\times -11\times -14}{1\times 3\times 6\times\ \ 9\times\ 12\times\ 15\times\ \ 18\times\ \ 21}\ \ \&\text{c.}$$

gives y[e] 4th.(48)

AND IN GENERALL if y[e] 2[d] quantity of any terme is $\frac{x}{y}$,(49) this progression gives all y[e] rest viz

$$\frac{1\times x\times \overline{x-y}\times \overline{x-2y}\times \overline{x-3y}\times \overline{x-4y}\times \overline{x-5y}\times \overline{x-6y}}{1\times y\times\ \ 2y\times\ \ 3y\times\ \ 4y\times\ \ 5y\times\ \ 6y\times\ \ 7y}\ \ \&\text{c.}^{(50)}$$

(43) Newton trisects the intervals of $(2, k)$, finding $d = c_{2,0} = (2, 0) = 0$, $e = c_{2,1} = -\frac{1}{9}$, $f = c_{2,2} = \frac{1}{9}$ and so evaluating $(2, k) = \sum_{0\leqslant i\leqslant 2}\left[\binom{3k}{i}c_{2,i}\right]$. In corresponding polynomial form $(2, k) = -\frac{1}{9}\binom{3k}{1}+\frac{1}{9}\binom{3k}{2} = \frac{1}{2}k(k-1)$, as before. (Since $d \leftrightarrow 0$, $\lambda = 0$.)

(44) Newton tabulates (s, k), $s = 0, 1, 2, \ldots, 5$; $k = 0, \frac{1}{3}, \frac{2}{3}, \ldots, \frac{11}{3}, 4$.

(45) Read 'column'.

(46) That is, $(s, \frac{1}{3})$, $s = 0, 1, 2, 3, \ldots$.

(47) $(s, \frac{2}{3})$, $s = 0, 1, 2, 3, \ldots$.

(48) $(s, \frac{4}{3})$, $s = 0, 1, 2, 3, \ldots$.

(49) That is, $(1, x/y)$ or the general fractional binomial coefficient $\binom{x/y}{1}$.

(50) $(s, x/y)$, $s = 0, 1, 2, 3, \ldots$.

Note also y[t]	1. 1. 1. 1. 1. 1.	may bee	$1=y.$
any of these	0. 1. 2. 3. 4. 5.	designed by	$x=y.$
progressions	0. 0. 1. 3. 6. 10.	Geometricall	$xx-x=2y.$
wth their	0. 0. 0. 1. 4. 10. &c.	lines ——	$x^3-3xx+2x=6y.$
intermediate	0. 0. 0. 0. 1. 5.		$x^4-6x^3+11xx-6x=24y.$
termes ——	0. 0. 0. 0. 0. 1.		$x^5-10x^4+35x^3-50x^2+24x=120y.$
	0. 0. 0. 0. 0. 0.		$x^6-15x^5+85x^4-225x^3+274xx-120x=720y.$

In w[ch] x signifieth y[e] distance of any terme from y[e] first 1.0.0.0.0.[0.] & y is the quantity of y[t] terme.(51)

Prop: 5[t].(52)

§5 IMPROVED CALCULATION OF HYPERBOLA-AREAS

[Autumn 1665?](1)

From the original in the University Library, Cambridge(2)

If $ea \parallel vb \parallel dc \perp ac \parallel ev=vb=a$. & $bc=x$. & $dc=y=\dfrac{aa}{b+x}$. Then is vdd an Hyperbola(3) &c: And if $\dfrac{aa}{b+x}$ bee divided as in decimall fractions y[e] product is

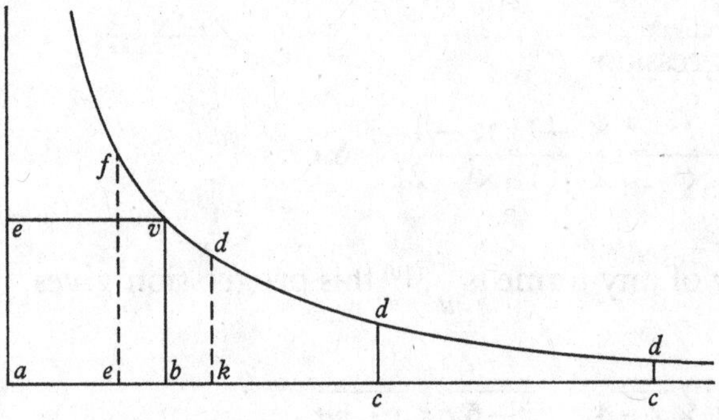

$$\frac{aa}{a+x}=y=a-x+\frac{xx}{a}-\frac{x^3}{aa}+\frac{x^4}{a^3}-\frac{x^5}{a^4}+\frac{x^6}{a^5}-\frac{x^7}{a^6}+\frac{x^8}{a^7}-\frac{x^9}{a^8}+\frac{x^{10}}{a^9}-\frac{x^{11}}{a^{10}}+\frac{x^{12}}{a^{11}}-\frac{x^{13}}{a^{12}+a^{11}x}\ \&c.^{(4)}$$

(51) Newton expands $y=(s,x)$ as the polynomial $\sum_{0\leqslant i\leqslant s}[S_s^i x^i] = s!y$, where the S_s^i are the constants known (since James Stirling tabulated them for $s=0, 1, 2, \ldots, 9$ in his *Methodus Differentialis* (London, 1730): 11) as Stirling's Numbers of the first kind. (See Charles Jordan, *Calculus of Finite Differences* (New York, $_2$1950): ch. IV: 142–68.)

(52) The manuscript breaks off here with only the proposition-head entered. A gap follows till the end of the page (71[r]) and the whole of its verso side is blank.

(1) Newton corrects and expands his logarithmic calculations in 3, §3 above. In turn, when Newton eleven years later came to write his *epistola posterior* to Leibniz (October 1676), these calculations were revised and reordered in two leaves of his mathematical Waste Book (ULC. Add. 4004: 80[r]–81[v]). As a preliminary certain corrections were entered on the present text, and these will duly be recorded in the footnotes. The point lettered 'k' in the figure is perhaps 'c' preceded by an irrelevant vertical stroke.

Which valor of y being each terme thereof multiplyed by x & divided by y[e] number of its dimensions:(5) The product will bee y[e] area *vbcd*.(6) viz[:]

$$vbcd = ax - \frac{xx}{2} + \frac{x^3}{3a} - \frac{x^4}{4a^2} + \frac{x^5}{5a^3} - \frac{x^6}{6a^4} + \frac{x^7}{7a^5} - \frac{x^8}{8a^6} + \frac{x^9}{9a^7} - \frac{x^{10}}{10a^8} + \frac{x^{11}}{11a^9} - \frac{x^{12}}{12a^{10}} \ \&c.$$

As for example. If $a=1$. & $x=0,1$. The calculation is as Followeth.

0,10000,00000,00000,00000,00000,00000,00000,00000,00000,00000,00000 $=ax$.

+0,100 33,33333,33333,33333,33333,33333,33333,33333,33333,33333,33333 $= \frac{x^3}{3a}$.

33, 20000,00000,00000,00000,00000,00000,00000,00000,00000,00000 $= \frac{x^5}{5a^3}$.

53 142,85714,28571,42857,14285,71428,57142,85714,28571,42857 $= \frac{x^7}{7a^5}$.

47 1,11111,11111,11111,11111,11111,11111,11111,11111,11111 $= \frac{x^9}{9a^7}$.

7,31 909,09090,90909,09090,90909,09090,90909,09090,90909 $= \frac{x^{11}}{11a^9}$.

07 7,69230,76923,07692,30769,23076,92307,69230,76923 $= \frac{x^{13}}{13a^{11}}$.

5,5 6666,66666,66666,66666,66666,66666,66666,66666 $= \frac{x^{15}}{15a^{13}}$.

80 58,82352,94117,64705,88235,29411,76470,58823 $= \frac{x^{17}}{17a^{15}}$.

63, 52631,57894,73684,21052,63157,89473,64821 $= \frac{x^{19}}{19a^{17}}$.

57 476,19047,61904,76190,47619,04761,90476 $= \frac{x^{21}}{21a^{19}}$.

26 4,34782,60869,56521,73913,04347,82608 &c.

5,5 4000,00000,00000,00000,00000,00000

20 37,03703,70370,37037,03703,70370

60, 34482,75862,06896,55172,41379

(2) Add. 3958.4: 78[v]/79[r], 78[r]/79[v], 77[v]/80[r]. The original (folded in two) has parted at the folds and the four halves of its two sheets have in recent times been separately paginated in error.

(3) That is, $y(b+x) = a^2$. (Note that Newton's accompanying diagram has two points e.)

(4) As in 3, §4.2: Prop. 2 above when $b = a$, but retaining the error term.

(5) That is, integrated by the fundamental theorem $\int_0^x a_i t^i . dt = \frac{a_i x^{i+1}}{i+1}$.

(6) $\int_0^x \frac{a^2}{a+t} . dt = a^2 \log\left(\frac{a+x}{a}\right) = a^2 . \sum_{0 \leqslant i \leqslant \infty} \left[(-1)^i \frac{1}{i+1} \left(\frac{x}{a}\right)^{i+1}\right]$, with the implicit restriction that $x \in [0, a]$.

11|322,58064,51612,90322,58064

89|3,03030,30303,03030,30303

4,5|2857,14285,71428,57142

26|27,02702,70270,27027

33,|25641,02564,10256

62|243,90243,90243

86|2,32558,13953

9,1|2222,22222

45|21,27659

95,|20408

91|196

43|1

0 = summe.(7)

(7) The partial sum $a^2 . \sum_{1\leqslant i\leqslant 27}\left[\frac{1}{2i-1}\left(\frac{x}{a}\right)^{2i-1}\right]$, $a = 1$, $x = 0\cdot1$, calculated to 55D. A mistake in the addition occurs at the 31st and 32nd decimal places which, as Newton corrected the manuscript a decade later, should read '03' (and not '11'). Further, in 1676 Newton extended the period of several of his rows, presumably to round off the final sum: specifically he adds the figures

$\left[\text{row 15, } \frac{1}{29}(0\cdot1)^{29}\right]$ '31034,48'

$\left[\text{row 16, } \frac{1}{31}(0\cdot1)^{31}\right]$ '56129 03225'

$\left[\text{row 22, } \frac{1}{43}(0\cdot1)^{43}\right]$ '48837, 20930 23255 813 &c.'

$\left[\text{row 24, } \frac{1}{47}(0\cdot1)^{47}\right]$ '57446 &c'

$\left[\text{row 25, } \frac{1}{49}(0\cdot1)^{49}\right]$ '16326 &c'

$\left[\text{row 26, } \frac{1}{51}(0\cdot1)^{51}\right]$ '07843 &c'

$\left[\text{row 27, } \frac{1}{53}(0\cdot1)^{53}\right]$ '88679 &c'.

Finally, Newton corrected the last three figures '430' of his total, extending it by two further figures to '358,63'.

0,00500,00000,00000,00000,00000,00000,00000,00000,00000,00000,00000, $=\frac{1}{2}xx$.

0,0050 | 2,50000,00000,00000,00000,00000,00000,00000,00000,00000, $= \frac{x^4}{4aa}$.

2,5 | 1666,66666,66666,66666,66666,66666,66666,66666,66666,66666, $=\frac{x^6}{6a^4}$.

16 | 12,50000,00000,00000,00000,00000,00000,00000,00000, $= \frac{x^8}{8a^6}$.

79, | 10000,00000,00000,00000,00000,00000,00000,00000,00000, $=\frac{x^{10}}{10a^8}$.

267 | 83,33333,33333,33333,33333,33333,33333,33333,33333, $= \frac{x^{12}}{12a^{10}}$.

50, | 71428,57142,85714,28571,42857,14285,71428,57142, $=\frac{x^{14}}{14a^{12}}$.

72 | 625,00000,00000,00000,00000,00000,00000,00000, $= \frac{x^{16}}{16a^{14}}$.

05 | 5,55555,55555,55555,55555,55555,55555,55555, $=\frac{x^{18}}{18a^{16}}$.

9,1 | 5000,00000,00000,00000,00000,00000,00000, &c.

77 | 45,45454,54545,45454,54545,45454,54545,

44, | 41666,66666,66666,66666,66666,66666,

28 | 384,61538,46153,84615,38461,53846,

77 | 3,57142,85714,28571,42857,14285,

9,2 | 3333,33333,33333,33333,33333,

73 | 31,25000,00000,00000,00000,

85, | 29411,76470,58823,52941,

30 | 277,77777,77777,77777,

42 | 2,63157,89473,68421,

7,5 | 2500,00000,00000,

75 | 23,80952,38095,

03, | 22727,27272,

83 | 217,39134,

73 | 2,08333,

The Summe of these two summes is equall to y^e^ area *befv*, supposeing $ae = 0,9.$[9]

And their Difference is equall to y^e^ area *bcdv*, supposeing $ac = 1,1.$[10] viz:

1,4|2000,

93|19,

63, = Summe.[8]

befv = 0,10536,05156,57826,30122,75009,80839,39279,83061,20372,98327,40793.

If $ae = 0,9.$ or $eb = -0,1.$

bcdv = 0,09531,01798,04324,86004,39521,23280,84509,22206,05365,30864,42067.

If $ac = 1,1.$ or $bc = 0,1.$[11]

In like manner if $x = 0,01.$ & $a = 1.$ The calculation is as followeth.

0,01000,03333,53333,33333,33333,33333,33333,33333,33333,33333,33333 $= ax + \frac{x^3}{3a} + \frac{x^5}{5a^3}.$

0,01000,03333,5333|1,42857,14285,71428,57142,85714,28571,42857,14285 $= \frac{x^7}{7a^5}.$

4,762|11,11202,02020,20202,02020,20202,02020,20202 $= \frac{x^9}{9a^7} + \frac{x^{11}}{11a^9}.$

01,58821,07|769,23076,92307,69230,76923,07692 $= \frac{x^{13}}{13a^{11}}.$

551,4|6666,66666,66666,66666,66666

0422,|58823,52941,17647,05882 &c.

3887|5,26315,78947,36842

0,973|47,61904,76190

08,80|434,78260

734,4|4000

3352 = sum̄e.[12]

0,00005,00025,00166,67916,76667,50007,14285,71428,57142,85714,28571 $= \frac{xx}{2} + \frac{x^4}{4aa} + \frac{x^6}{6a^4} + \frac{x^8}{8a^6} + \frac{x^{10}}{10a^8} + \frac{x^{12}}{12a^{10}} + \frac{x^{14}}{14a^{12}}.$

---

(8) The partial sum $a^2 . \sum_{1 \leqslant i \leqslant 26} \left[\frac{1}{2i}\left(\frac{x}{a}\right)^{2i}\right]$, $a = 1$, $x = 0{\cdot}1$, calculated to 55D. In 1676 Newton corrected the last figure '3', extending it to '6,80'.

(9) $\int_{-0\cdot1}^{0} \frac{1}{1+t} . dt = \log\left(\frac{1}{0\cdot9}\right) \approx \sum_{1 \leqslant i \leqslant 27} \left[\frac{1}{2i-1}(0{\cdot}1)^{2i-1}\right] + \sum_{1 \leqslant i \leqslant 26} \left[\frac{1}{2i}(0{\cdot}1)^{2i}\right].$

0,00005,00025,00166,67916,76667,50007,143 | 62,50555,60556,01010,10101 $= \dfrac{x^{16}}{16a^{14}} + \dfrac{x^{18}}{18a^{16}}$.&c.

48,21984,17698,8672 | 4,16666

8,553 | 38

76 = summe.(13)

The summe of these two summes is equall to y^e^ area *befv*, supposeing *ae* = 0,99. And their Difference to *bcdv*, if *ac* = 1,01. viz:

befv = 0,01005,03358,53501,44118,35488,57558,54770,60855,15007,67462,98736. If *ae* = 0,99.

bcdv = 0,00995,03308,53168,08284,82153,57544,26074,16887,29609,94005,87984. If *ac* = 1,01.(14)

Soe if $x = 0{,}001$. Then the Calculation will bee.

0,00100,00003,33333,53333,34761,90476,19047,61904,76190,47619.04761,9 $= ax + \dfrac{x^3}{3a} + \dfrac{x^5}{5a^3}$ &c.

0,00100,00003,33333,53333,34761,90 | 111,11120,20202,02020,20202,02020,2

587,30167,82107, | 76923,07692,30769,2

55133,8 | 6666,66666,6

2180,04 | 588,2

806,1 = summe

---

(10) $\displaystyle\int_0^{0\cdot1} \frac{1}{1+t}.dt = \log\left(\frac{1\cdot1}{1}\right) \approx \sum_{1\leqslant i\leqslant 27}\left[\frac{1}{2i-1}(0\cdot1)^{2i-1}\right] - \sum_{1\leqslant i\leqslant 26}\left[\frac{1}{2i}(0\cdot1)^{2i}\right]$.

(11) In accordance with the emended partial sums Newton has corrected the 32nd decimal place of (*befv*), '9', by writing in '−8' over it and modified its two last figures '93' to '25,43'; and, similarly, has altered and augmented the last four figures, '2067', of (*bcdv*) to '1991,83' while decreasing its 32nd decimal place by '−8'.

(12) The partial sum $\displaystyle\sum_{1\leqslant i\leqslant 13}\left[\frac{1}{2i-1}(0\cdot01)^{2i-1}\right]$, calculated to 55D. In 1676 Newton increased the array to 57D and so rounded off the last figure more accurately as '6,17'. (Note that the last row is $\frac{1}{25}(0\cdot01)^{25} + \frac{1}{27}(0\cdot01)^{27}$.)

(13) The partial sum $\displaystyle\sum_{1\leqslant i\leqslant 13}\left[\frac{1}{2i}(0\cdot01)^{2i}\right]$, calculated to 55D. Newton in 1676 augmented the last figure to '7,55'. (The third row is $\displaystyle\sum_{8\leqslant i\leqslant 11}\left[\frac{1}{2i}(0\cdot01)^{2i}\right]$.)

(14) $\displaystyle(befv) = \int_{-0\cdot01}^{0} \frac{1}{1+t}.dt = \log\left(\frac{1}{0\cdot99}\right)$ and $\displaystyle(bcdv) = \int_0^{0\cdot01}\frac{1}{1+t}.dt = \log\left(\frac{1\cdot01}{1}\right)$ are the respective sum and difference of the two partial sums (notes (12) and (13)). As before, corrections of the last two figures were added in 1676 to correspond with the improved partial sums.

0,00000,05000,00250,00016,66667,91666,76666,67500,00071,42863,39286 $=$summe of $\frac{xx}{2}+\frac{x^4}{4aa}$ &c.

Therefore $\begin{cases} befv=0{,}00100{,}05003{,}33583{,}53350{,}01429{,}82254{,}06834{,}49607{,}55205{,}25043{,}44092, \text{ If } ae=0{,}999. \\ bcdv=0{,}00099{,}95003{,}33083{,}53316{,}68093{,}98920{,}53501{,}14607{,}55062{,}39316{,}65520, \text{ If } ac=1{,}001. \end{cases}$(15)

And if $x=0{,}0001$. Then

0,00010,00000,00333,33333,53333,33347,61904,77301,58731,06782,10755$=ax+\frac{x^3}{3a}$ &c.

0,00000,00050,00000,02500,00001,66666,66791,66666,76666,66675,00000$=\frac{xx}{2}+\frac{x^4}{4aa}$ &c.

Therefore $\begin{cases} befv=0{,}00010{,}00050{,}00333{,}35833{,}53335{,}00014{,}28696{,}43968{,}35397{,}73457{,}10755. \\ bcdv=0{,}00009{,}99950{,}00333{,}30833{,}53331{,}66680{,}95113{,}10634{,}82064{,}40107{,}10755. \end{cases}$ If $\begin{cases} ae=0{,}9999. \\ ac=1{,}0001. \end{cases}$(16)

If $a=1$ & $x=0{,}2$. The calculation is as followeth[....](17)

---

(15) Having calculated the partial sums

$$\alpha = \sum_{1\leqslant i\leqslant 9}\left[\frac{1}{2i-1}(0{\cdot}001)^{2i-1}\right] \quad\text{and}\quad \beta = \sum_{1\leqslant i\leqslant 9}\left[\frac{1}{2i}(0{\cdot}001)^{2i}\right],$$

Newton evaluates $\quad (befv) = \int_{-0\cdot001}^{0}\frac{1}{1+t}.dt = \log\left(\frac{1}{0{\cdot}999}\right) \approx \alpha+\beta,$

and $\quad (bcdv) = \int_{0}^{0\cdot001}\frac{1}{1+t}.dt = \log\left(\frac{1{\cdot}001}{1}\right) \approx \alpha-\beta.$

Through some confusion over the placing of his commas Newton calculated an extra decimal place in the partial sums, taking them to 56D, but drops it in his final result. Two further places were later added in correction.

(16) The partial sums $\sum_{1\leqslant i\leqslant 7}\left[\frac{1}{2i-1}(0{\cdot}0001)^{2i-1}\right]$ and $\sum_{1\leqslant i\leqslant 7}\left[\frac{1}{2i}(0{\cdot}0001)^{2i}\right]$ are used to evaluate

$$(befv) = \int_{-0\cdot0001}^{0}\frac{1}{1+t}.dt = \log\left(\frac{1}{0{\cdot}9999}\right) \quad\text{and}\quad (bcdv) = \int_{0}^{0\cdot0001}\frac{1}{1+t}.dt = \log\left(\frac{1{\cdot}0001}{1}\right).$$

(17) Newton's calculations are vitiated in several places and their reproduction can be of no use. Formally, Newton attempts to calculate the partial sums

$$\alpha = \sum_{1\leqslant i\leqslant 38}\left[\frac{1}{2i-1}(0{\cdot}2)^{2i-1}\right] \quad\text{and}\quad \beta = \sum_{1\leqslant i\leqslant 38}\left[\frac{1}{2i}(0{\cdot}2)^{2i}\right],$$

and so evaluate $\quad (befv) = \int_{-0\cdot2}^{0}\frac{1}{1+t}.dt = \log\left(\frac{1}{0{\cdot}8}\right) \approx \alpha+\beta$

and $\quad (bcdv) = \int_{0}^{0\cdot2}\frac{1}{1+t}.dt = \log\left(\frac{1{\cdot}2}{1}\right) \approx \alpha-\beta.$

In Newton's array of figures the rows $\frac{1}{55}(0{\cdot}2)^{55}$, $\frac{1}{42}(0{\cdot}2)^{42}$, $\frac{1}{46}(0{\cdot}2)^{46}$, $\frac{1}{48}(0{\cdot}2)^{48}$ and $\frac{1}{72}(0{\cdot}2)^{72}$ are wrongly reckoned, and conversely a correct expansion of $\frac{1}{23}(0{\cdot}2)^{23}$ is cancelled for an

[If $x = 0,02$ the calculation will bee]

0,02000,26673,06849,58071,70371,83954,64639,04807,62053,27455,98440,93

0,02000,26666,66666,66666,66666,66666,66666,66666,66666,66666,66666,66
6,40182,85714,28571,42857,14285,71428,57142,85714,28571,42
5688,88888,88888,88888,88888,88888,88888,88888,88
1,86181,81818,18181,81818,18181,81818,18181,81
63,01538,46153,84615,38461,53846,15384,61
2184,53333,33333,33333,33333,33333,33
77101,17647,05882,35294,11764,70
27,59410,52631,57894,73684,21
998,64380,95238,09523,80
36469,86086,95652,173
13,42177,28000,00
497,10269,62
18512,79
6,92(18)

0,00020,00400,10669,86769,10081,17069,54599,73717,71328,11118,23899,702

0,00020,00400,10666,66666,66666,66666,66666,66666,66666,66666,66666,666
3,20102,43413,33333,33333,33333,33333,33333,33333,333
1,17028,57142,85714,28571,42857,14285,714
40,97456,87984,35555,55555,55555,555
19,06501,81818,18181,818
699,05066,66666,666
25811,10153,846
9,58698,057
357,913
134(19)

---

erroneous one. When Newton checked the manuscript in 1676 he corrected all these errors, not in the text itself but alongside, before redrafting the calculation on Add. 4004: 81v, where revised values for the two areas are given—specifically

$(befv)$ = 0,22314,35513,14209,75576,62950,90309,83450,33746,01085,54800,72136,

and $(bcdv)$ = 0,18232,15567,93954,62621,17180,25154,51463,31973,89337,91448,69839.

(18) The first partial sum $\sum_{1 \leqslant i \leqslant 16} \left[\frac{1}{2i-1}(0{\cdot}02)^{2i-1}\right]$. The twelfth row, that is $\frac{1}{23}(0{\cdot}02)^{23}$, should be '36472,20869,56521,73' and was so corrected by Newton in 1676. It follows that the sum has to be increased by 2,34782,60869,56.

(19) The second partial sum $\sum_{1 \leqslant i \leqslant 16} \left[\frac{1}{2i}(0{\cdot}02)^{2i}\right]$.

If $ae=0{,}98$; or $-eb=0{,}01$[20]; yⁿ

0,02020,27073,17519,44840,80453,01024,19238,78525,33381,38574,22340,63 $=befv$.

[If] $ac=1{,}02$; or $+bc=0{,}01$[20]; yⁿ

0,01980,26272,96179,71302,60290,66885,10039,31089,90725,16337,74541,23 $=bcdv$.[21]

(20) Read '0,02'.

(21) Rounding off to 57D the 58D of the second partial sum, Newton evaluates

$$(befv) = \int_{-0\cdot 02}^{0} \frac{1}{1+t}.dt = \log\left(\frac{1}{0\cdot 98}\right) \approx \sum_{0\leqslant i\leqslant 16}\left[\frac{1}{2i-1}(0\cdot 02)^{2i-1}+\frac{1}{2i}(0\cdot 02)^{2i}\right]$$

and $$(bcdv) = \int_{0}^{0\cdot 02} \frac{1}{1+t}.dt = \log\left(\frac{1\cdot 02}{1}\right) \approx \sum_{0\leqslant i\leqslant 16}\left[\frac{1}{2i-1}(0\cdot 02)^{2i-1}-\frac{1}{2i}(0\cdot 02)^{2i}\right].$$

Newton's error in calculating $\frac{1}{23}(0\cdot 02)^{23}$ in the first partial sum (note (18)) results, as he acknowledged in his 1676 corrections, in both areas being too little by 2,34782,60869,56.

PART 2

RESEARCHES IN ANALYTICAL GEOMETRY AND CALCULUS (1664-1666)

INTRODUCTION

Having set down the evidence of Newton's youthful debt to contemporary mathematical knowledge, we now reproduce the written record of his first researches in the interlocking structures of Cartesian co-ordinate geometry and infinitesimal analysis. In these papers Newton laid the foundations of his mature work in mathematics, revealing for the first time the true magnitude of his genius.

His probings in analytical geometry (here(1) confined to the consideration of a few general co-ordinate-systems, a deepening discussion of axis-transforms, and some not wholly successful attempts at curve-tracing) were not to strike a rich, exploitable vein of discovery till some three years afterwards when (as we shall see in the next volume) they formed the framework for his first detailed analysis of cubics. Unfamiliar to Newton's contemporaries(2) and unsuspected by later scholars they have remained unknown till the present day. In contrast, Newton's early researches in calculus,(3) overwhelming in their bulk, have been in the glare of publicity for more than two centuries and, though their details are still for the most part unpublished, their general content has long been known from Newton's own pronouncements about them. These statements, born in old age of the emotional blindness and prejudice of his priority dispute with the Leibnizians and empty of the fresh, more generous spirit of his early papers and letters, are inevitably both partial and inconsistent.(4) Writing half a century afterward, Newton could indeed recall from memory but little of the period of his earliest mathematical creativeness and was himself forced to use his old papers as a major source for his claims. It is clear, in consequence, that future evaluation of his mathematical thought must be firmly based in a sound textual appreciation of these manuscripts, and that we may introduce details from Newton's secondary account only hesitantly and after due consideration.

As in so many other fields—and though we should not underestimate the impact of Wallis' work upon him—in the creation of his calculus of fluxions

(1) See section 1 below.

(2) De Moivre, who in the memorandum he gave to John Conduitt in November 1727 showed some awareness of their content, is an apparent exception. (See the introduction to Part 1, especially note (15).)

(3) Printed in sections 2–7 below.

(4) We will present a selection of the more significant of these unpublished fragments in our index-volume. For the present we think it valuable to add (in Appendix 1) a pastiche of the chronology which Newton wished to set on his early manuscripts—one on which all previous scholarly judgements have had to rest. Though there is no space here to give its biases careful consideration, we indicate in accompanying footnotes the main inadequacies of Newton's loosely consistent pattern of discovery and in Appendix 2 attempt a schematic reconstruction of the complex growth of Newton's ideas on calculus up to October 1666.

Newton's chief mentor was Descartes. From the latter's *Geometrie*, a veritable mathematical bible in Schooten's richly annotated second Latin edition, he drew a continuous inspiration over the two years from the summer of 1664. Beginning with the Cartesian technique for constructing the subnormal to an algebraic curve,(5) Newton swiftly soaked up the algorithmic facility of Hudde's rule *de maximis et minimis* but quickly came to appreciate its hidden riches, applying it to the construction of the subtangent and of the circle of curvature at a general point on an algebraic curve and ultimately formulating a differentiation procedure founded on the concept of an indefinitely small, vanishing increment.(6) On that basis and little afterwards he was able to set down the standard differential algorithms in the generality with which they were to be expounded by Leibniz two decades later.(7) Along with this a parallel stream of researches, built on Wallis' work in the theory of the algebraic integral(8) and on Heuraet's general rectification procedure,(9) culminated about the same time (mid-1665) in a limited mastery of the quadrature problem and in geometrical insight into the inverse problem of tangents. In the summer and early autumn of that year, away from books in Lincolnshire and with time for unhurried thought, Newton recast the theoretical basis of his new-found calculus techniques, rejecting the concept of the indefinitely small increment in favour of that of the fluxion, a finite instantaneous speed defined with regard to an independent dimension of time and on the geometrical model of the line-segment.(10) Soon after, in the autumn of 1665, he was led to restudy the tangent-

(5) See section 2 below.

(6) These themes are explored in detail in section 4. Newton's differential procedure, embedded at first in his treatment of subnormal, subtangent and curvature problems but explicit by May 1665, essentially found the derivative of an algebraic function, say $f(x)$, as

$$\lim_{o\to\text{zero}}\left[\frac{1}{o}\left(f(x+o)-f(x)\right)\right]$$

with $f(x+o)$ developed in Fermatian style as $f(x)+of'(x)+O(o^2)$, where o denotes the limit-increment (dx) of x. (It appears that Newton first employed the o-notation in a problem in geometrical optics, but soon took it over into his treatment of curvature. See the concluding appendix on geometrical optics 1, §1, especially note (21).) This is, of course, the modern definition of $f'(x)$ as $\lim_{dx\to\text{zero}}\left[\frac{f(x+dx)-f(x)}{(x+dx)-x}\right]$. On one occasion (5, §5.1) Newton used the notation $\dot{p}$ for dp/dx (where p is a function of x), but in general he retained a dot notation for the partial derivatives which he found it necessary to introduce into his discussion of tangents and curvature. In a transitional notation used on 20 May 1665 (4, §3.1 below) he found the total derivatives of a function $\sum_i (p_i y^i) = 0$ (where the p_i are functions of x alone) in terms of the homogenized derivatives $\dot{p}_i = x\frac{dp_i}{dx}$ and $\ddot{p}_i = x^2\frac{d^2p_i}{dx^2}$. On the following day (4, §3.2) he introduced the more general notation $\dot{\mathfrak{X}} = xf_x$, $\dot{\mathfrak{X}} = yf_y$, $\ddot{\mathfrak{X}} = x^2f_{xx}$, $\ddot{\mathfrak{X}} = xyf_{xy}$ and $\ddot{\mathfrak{X}} = y^2f_{yy}$, where $\mathfrak{X}$ denotes the general algebraic function of two variables $f(x, y)$. With slight modification he retained the usage in his October 1666 tract. (At the time of the fluxion

problem by the Robervallian method of combining limit-motions of a point defined in a suitable co-ordinate-system. After an initial crisis in his construction of the quadratrix-tangent(11) he was able correctly to generalize the method, giving in May 1666(12) a comprehensive treatment of tangents by limit-motion analysis and extending its area of application to include the construction of inflexion points. In the autumn of 1666, lastly, and as a not unintended finale, almost all these researches were ordered and condensed in a short, unfinished and till recently unpublished work(13) to which he gave no title but which, following his own practice in later reference, we may name the 'October 1666' tract.

In those two years a mathematician was born: a man, certainly, still capable of profound error but with a depth of mathematical genius which by late 1666 had made him the peer of Huygens and James Gregory(14) and probably the

priority dispute, for example in the appendix he wrote for Raphson's *History of Fluxions* (London, 1715), Newton made vague reference to this dot notation, attempting thereby to backdate the invention of his standard notation for fluxions—for which there is no manuscript evidence before 1691—back to 1665.) With regard to integration Newton made fairly frequent, but not invariable, use from mid-1665 of the notation $\Box P$ (for $\int P.dx$) but we should not press Newton's priority there too hard. Already in 1659 in his *Geometria Speciosa*, which Newton seems never to have read, Pietro Mengoli had used O (for 'Omnes/Omnia') in the same significance.

(7) See 5, §2 below. Leibniz published his version of the algorithms (which he had found in 1676) in his celebrated but unreadable 'Nova methodus pro maximis et minimis, itemque tangentibus, quæ nec fractas, irrationales quantitates moratur, & singulare pro illis calculi genus, per G.G.L.', *Acta Eruditorum* (October 1684): 467–73.

(8) See 1, 3, §3 above.

(9) Compare 5, §1, especially note (51).

(10) In modern language the fluxion of $f(x)$ is the instantaneous 'speed' df/dt, where t is an independent variable of time. (See 5, §4 below.) In Newton's standard early notation the fluxions of x, y and z are represented by the letters p, q and r: in other words, $p = dx/dt$, $q = dy/dt$ and $r = dz/dt$. Taking, further, an arbitrary increment o (or dt) of the 'time' t, we may then represent corresponding increments of x, y and z by op, oq and or. It then follows, in the limit as o vanishes, that $q/p = dy/dx$ and $r/p = dz/dx$ (where the increment o is absorbed into the limit-ratios). In particular, Newton often chooses x itself for the independent variable of time, so that o is now the increment of x (with oq and or corresponding increments of y and z) and $p = 1$, and therefore correspondingly $q = dy/dx$ and $r = dz/dx$.

(11) See 6, §1, especially note (4). (12) 6, §4 below.

(13) Reproduced as section 7 below.

(14) It would be instructive to make a detailed comparative study of these two highly ingenious men and Newton. In 1666 with his *Vera Circuli et Hyperbolæ Quadratura* (Padua, 1667), *Geometriæ Pars Universalis* (Padua, 1668) and *Exercitationes Geometricæ* (London, 1668) still unpublished, James Gregory was still relatively unknown in the mathematical world—indeed, it is true to say that his mathematical eminence has only widely been recognized in the present century with the publication of the *James Gregory Tercentenary Memorial Volume* (ed. H. W. Turnbull, London, 1939) and more recently of Christoph J. Scriba's *James Gregorys frühe Schriften zur Infinitesimalrechnung* (Giessen, 1957). Christiaan Huygens was a

superior of his other contemporaries. His only earnest regret must have been that he had yet found no outlet for communicating his achievement to others. The papers printed in the following pages throb with energy and imagination but yet convey the claustrophobic air of a man completely wrapped up in him self, whose only real contact with the external world was through his books. That was to change somewhat in years to come, but it was Newton's continuing tragedy that he was never to find a collaborator of his own mental stature.

APPENDIX 1

NEWTON'S CHRONOLOGY OF HIS EARLY RESEARCHES IN CALCULUS

[1664–1666](1)

Excerpts from original drafts in the University Library, Cambridge, and the Macclesfield Collection

[A](2) Archimedes in drawing tangents to spirals(3) gave an instance of... the method of fluxions.... Fermat gave another instance of it in determining the greatest & least quant[it]ies & applying the determination to the drawing of Tangents to Curves.(4) This method was published by Herigon in his Cursus

---

scientific giant in his own time, yet many of his more interesting later mathematical researches remained unpublished till half a century ago (when they were printed in his *Œuvres complètes*). We now know that Huygens and Newton, pursuing research into the same problems of tangents, curvature and inflexion points and alike inspired by a reading of Descartes' *Geometrie* in Schooten's Latin version, but neither aware of the other's existence, made many identical discoveries—a fact which we will have occasion to emphasize several times in future pages.

(1) The loosely consistent chronology which follows was never amplified by Newton into an autobiographical study comparable with Leibniz' *Historia et Origo Calculi Differentialis*, but is extracted from the mass of autograph fragments written over the decade (1711–*c.* 1720) of Newton's fluxion priority squabbles. In constructing our patchwork of quotation we have aimed to retain the significant passages in Newton's account while preserving a certain continuity in their presentation, and in so doing have glossed over several of its inconsistencies. (Newton's picture, we must remember, was of events already half a century old and his considered intention in drawing it was to establish priority of invention rather than to conduct an objective self-analysis.) Nevertheless, comparison of the text of his early papers, reproduced below, with his own later appreciation of their content should prove interesting.

(2) ULC. Add. 3968.19: 290^v. On the preceding f. 289^r the fragment is substantially incorporated in a draft preface for his (unpublished) *Extracts of ye MS Papers of Mr John Collins concerning some late improvements of Algebra.*

(3) In Props. 12–20 of his tract *On Spirals.* (See E. J. Dijksterhuis, *Archimedes* (Copenhagen, 1956): ch. VIII: 264–85, especially 264–74.) The justice of this acute observation is argued by C. B. Boyer in his *Concepts of the calculus* (New York, 1939): 57–9. It seems likely that Newton has in mind Viète's commentary on the work in his *Variorum de Rebus mathematicis Responsorum Liber VIII* (=Schooten's *Vieta*: 347–435, Newton's notes on which are reproduced in **1**, **2**, §2.4 above).

A.C. 1631,(5) & a specimen of it was inserted by Schoten in his Commentaries on the 2d book of Cartes's Geometry.(6)... Mr Newton affirms that in ye year 1664 & ye winter following upon reading the Geometry of Des Cartes wth his Commentators he met with the method of Fermat in ye second book of Schooten's Commentaries & not long after(7) applied it to abstracted æquations in the manner described in the first Proposition of his book of Quadrature(8) but did not use ye language of Fluxions from the beginning.

(4) Newton noted elsewhere (ULC. Add. 3968.30: 441r, reproduced by L. T. More in his *Isaac Newton* (New York, 1934): 185, n. 35) that he 'had the hint of this method from Fermats way of drawing Tangents'. In modification of More's accompanying claim that this was 'direct evidence...of Newton's indebtedness to Fermat', it is clear from the present fragment that Newton had no direct knowledge of Fermat's work but knew the outlines of his tangent-method perhaps from Hérigone's first brief printed description (note (5)) or more probably from Schooten's commentary. No adequate account of Fermat's method *de maximis et minimis*—or indeed of the similar tangent-method which Descartes was provoked to formulate in the spring of 1638—appeared till 1667 when Clerselier printed all he could gather of the correspondence relating to Descartes' part in their 'querelle' (in the third volume of his *Lettres de M. Descartes*). Fermat's fundamental tract, *Methodus ad Disquirendam Maximam et Minimam* (the 'Escrit' sent by Mersenne to Descartes early in January 1638) was printed only in 1679 in his posthumous *Varia Opera*. (See Paul Tannery's and Charles Henry's *Œuvres de Fermat*, **1**: 133-79, especially 133-6; and **2**: Letters xxv–xxxvi.)

(5) Newton has his date badly wrong and it would appear that he had never seen Hérigone's work, taking the reference from Schooten (note (6) below). Hérigone, in fact, added his note on Fermat's tangent-method at Mersenne's suggestion in the 1642 *Supplementum* to his *Cursus Mathematicus, nova, brevi et clara Methodo demonstratus* (5 vols., Paris, 1634–7), issuing it with the main text of the *Cursus* only in his second, revised edition (6 vols., Paris, 1644). His account (*Supplementum Cursus Mathematici: Supplementum Algebræ*: 59–69: Prop. XXXVI, *De maximis et minimis. Des maximes et minimes.*) is little more than a summary, transformed into his highly idiosyncratic notation, of Fermat's *Methodus* (note (4) above), closely following its division with a main section on algebraic maxima and minima and the application to tangents set in appendix. The French version of Hérigone's parallel French and Latin text concludes with the attribution: 'Par la mesme methode on trouvera aussi les tangentes de toutes sortes de lignes courbes, en des points donnez en icelles. ... & cette methode ne manque iamais: ce que son inventeur asseure, qui est Monsieur Fermat, Conseiller au Parlement de Toulouse, excellent Geometre, & qui ne cede à aucun en l'Art Analytique: lequel a aussi tres bien restitué tous les lieux plans d'Apollonius Pergæus, que nous les avons veu en cette ville manuscripts entre les mains de plusieurs, en suite desquels se trouve aussi du mesme autheur, une Isagoge aux lieux plans & solides.'

(6) Schooten's 'specimen' (*Commentarii in Librum II, O = Geometria*: 253–5) applied the method to the problem of constructing the normal at a general point on the conchoid. Having described in previous pages two useful approaches, he proposed a third way of investigation 'beneficio Methodi de Maximis & Minimis, cujus Author est Vir Clarissimus D. de Fermat, in Parlamento Tolosano Consiliarius, quam Herigonius in supplemento Cursus sui Mathematici exemplis aliquot illustravit, atque ibidem etiam ad inveniendas tangentes adhibere docuit.' (*Geometria:* 253).

(7) Newton first wrote 'before the end of ye next winter & the summer following'.

(8) That is, the *Tractatus de Quadratura Curvarum*, composed in its final form in late 1691 but printed only in 1704 as the second appendix (pp. 163–211) to Newton's *Opticks: Or, a Treatise of the Reflexions, Refractions, Inflexions and Colours of Light. Also Two Treatises of the Species and*

[B][9] Dr Barrow then read his Lectures about[10] motion & that might put me upon taking these things into consideration.... In the winter between the years 1664 & 1665 I had a method of Tangents like that of Hudden[11] Gregory & Slusius[12] & a method of finding the crokedness of Curves at any given point,[13] & by considering how to interpole certain series of Dr Wallis I found the Rule...for reducing any power or dignity of any Binomium into an approximating Series, & in the following Spring, before the Plague, wch invaded us that summer, forced me from Cambridge, I found how to do the same thing by continual division & extraction of roots....[14] And soon after I extended the Method to the extraction of the roots of affected Equations in species.[15] And from all this I learnt how to deduce the Ordinates or Abscissas of Curvilinear figures from their Areas or A[r]cs given, as well as the Areas & Arcs from the Abscissas & Ordinates.[16] Thus far I proceeded before the plague forced me from Cambridge. And in a paper dated 13 Nov. 1665[17] I find the direct Method of first fluxions set down with examples & a Demonstration.

Magnitude of Curvilinear Figures. Prop. I. Prob. I, set out on pp. 172–5 of the *Opticks*, expounds the Fermatian method of finding the derivative of a general algebraic function under the familiar enunciation, 'Data æquatione quotcunq; fluentes quantitates involvente, invenire fluxiones'. In particular, Newton in his *Demonstratio* and *Explicatio* considers a function, say $f(x, y, z)$, of three variables, finding its derivative

$$\lim_{o\to \text{zero}} \left(\frac{1}{o}\,[f(x+o\dot{x}, y+o\dot{y}, z+o\dot{z})-f(x, y, z)]\right)$$

from the expansion

$$f(x+o\dot{x}, y+o\dot{y}, z+o\dot{z}) = f+o(\dot{x}f_x+\dot{y}f_y+\dot{z}f_z)+O(o^2).$$

(9) ULC. Add. 3968.5: 21r.

(10) Newton has cancelled 'the generation of figures by'. The implications of this offhand comment are discussed in 5, §4, note (4) below.

(11) In elaboration of his algorithm *de maximis et minimis*, sent to Schooten on 27 January 1658 and subsequently printed in Schooten's *Geometria* (1 (1659): 507–16), Hudde expounded his tangent-rule in a second letter to Schooten on 21 November 1659, which remained publicly unknown till, at the time of the fluxion priority dispute, it was printed in the *Journal Literaire* in counterblast to an extract from Newton's letter of 10 December 1672 to Collins which has appeared in a previous issue, and where Newton asserted that Hudde's method was 'limited to æquations wch are free from surd quantities'. (See 'Extrait d'une Lettre de M. Newton à Mr. Collins le 10. de Décembre 1672', *Journal Literaire*, **1**, 3 (May/June 1713): 213–14, and 'Extrait d'une Lettre du feu M. Hudde, à M. van Schooten, Professeur en Mathématiques à Leyde. Du 21. de Novembre 1659', *Journal Literaire*, **1**, 4 (July/August 1713): 460–4. Compare also *Correspondence of Isaac Newton*, **1** (1959): 247–8.)

(12) René-François de Sluse appears to have come upon the tangent-rule some time in the late 1650's, but first asserted his priority only in 1671 in a letter to Oldenburg. (See 'Slusius... his short and easie method of drawing tangents to all geometrical curves without any labour of calculation', *Philosophical Transactions*, **7** (1672): 5143–7. Compare also L. Rosenfeld, 'René-François de Sluse et le problème des tangentes', *Isis*, **10** (1928): 416–34.) With regard to James Gregory's independent discovery of the rule the evidence is less clear. However, on 8 November 1672 Collins wrote to him: 'Dr Wallis in one of the [Philosophical] Transactions published some Methods of Tangents on which Hugens in a Letter made his Animadversions, as I take it to this purpose, that the first Method is Fermats as is extant in Herigon.... That

[C][18] ...there follows the application of this method to the drawing of tangents, by finding the determination of the motion of any point wch describes the curve: & also to the finding the radius of curvity of any curve at any point, by making the perpendicular to the curve move upon it at right angles, & finding that point of the perpendicular wch is in least motion. For that point will be the center of the curvity of the curve at that point, upon wch the perpendicular stands.

In another leaf of the same waste-book[19] the same method is set down in other words, and fluxions applied to their fluents are represented by pricked letters.[20] And this paper is dated May 20, 1665. In another leaf[21]...the method of fluxions is described without pricked letters, & dated May 16, 1666.

---

there is yet another method better and much more compendious than both, knowne long ago to Slusius [&] Hudden...in which you are only to looke for the Aequation which expresses the nature of the Line, for which there may be presently and without any trouble derived another Aequation that gives the Construction of the Tangent, and in regard Slusius in one of his Letters sayes thus, vis ut verbo dicam, Monachos, tangens, Maxima et Minima, unum idemque sunt, it gives just doubt whether the Derivative Aequation for the tangent be not the same with that for the Limits, which you were pleased to send me saying it was couched in Huddens doctrine, this I believe is better knowne to you than to those above mentioned.' (*James Gregory Tercentenary Memorial Volume* (ed. H. W. Turnbull, London, 1939): 246–7. For Newton's exposition of his tangent-method see 4, §3 below.

(13) See 4, §2/§3.2 below. (14) Compare 1, 3, §3 above.

(15) This method first appears in manuscript in Newton's *De Analysi per Æquationes Numero Terminorum Infinitas*, apparently not composed till late in 1668. Newton would seem a little confused in his dating.

(16) Compare 5, §4 below.

(17) ULC. Add. 4004: $57^r/57^v = 6$, §3 below.

(18) Extracted from a stray fragment (101.H.1: 112) now in Lord Macclesfield's possession. (A full version of this paper was printed by S. P. Rigaud in his *Historical Essay on the first publication of Sir Isaac Newton's Principia* (Oxford, 1838): *Appendix:* 20–4.)

(19) ULC. Add. 4004: $47^r/47^v = 4$, §3.1 below.

(20) This statement is not untrue but yet neatly manages to leave the impression—as Newton would seem to have intended—that this 1665 paper already employed the standard Newtonian dot-notation for fluxions. The 'pricked letters' of this early paper, however, are a disguised partial-derivative notation which has little to do with Newton's later use of $\dot{x}$ to represent dx/dt. (This latter notation does not occur before the autumn of 1691.) For example, by writing $\ddot{a} = x(da/dx)$, $\ddot{c} = x(dc/dx)$ and $\ddot{e} = x(de/dx)$, he was able to express the incremented form of $a+cy+ey^2 = 0$ as $a+cz+ez^2+(o/x)(\ddot{a}+\ddot{c}z+\ddot{e}z^2) = 0$ and so find its total derivative in the form

$$\lim_{o\to\text{zero}}\left[\frac{c(z-y)+e(z^2-y^2)}{o}+\frac{\ddot{a}+\ddot{c}z+\ddot{e}z^2}{x}\right] = 0,$$

where a, c and e are functions of x alone, $o = dx$ and $z = y+o(dy/dx)$ (the incremented form of y).

(21) ULC. Add. 4004: $51^r/51^v = 6$, §4 below. Newton was more ready with details of its content on other occasions, writing for example 'In alio Manuscripto 16 Maij 1666 composito, methodum solvendi problemata per motum, complexus sum Propositionibus septem, quarum ultima est Regula jam descripta eliciendi velocit[at]es crescendi vel decrescendi ex æquatione quantitates crescentes vel decrescentes involvente' (ULC. Add. 3968.6: 44^r).

In a small tract,[22] dated in October 1666, the same method is again set down without pricked letters, & how to proceed in equations involving facts[23] or surds, & several rules are given for returning back from the fluents to the fluxions; & in doing this the areas of curves are represented by prefixing rectangles to the ordinates....After this, the method is applied in this Tract to the solving of problems concerning tangents, curvatures of curves, the greatest or least curvatures, the squaring of curvilinear figures, comparing their areas wth the areas of simpler curves, finding the lengths of curves, finding such curves whose lengths may be defined by equations or by the lengths of other curves, & about the centers of gravity of figures.

[D][24] ...in this smaller Tract tho I generally put letters for fluxions..., yet in giving a general Rule for finding the Curvature of Curves, I put the letter *x* with one prick for first fluxions drawn into their fluents & with two pricks for second fluxions drawn into the square of their fluents....

[E][25] All this was in the two plague years of 1665 & 1666.[26] For in those days I was in the prime of my age for invention & minded Mathematicks & Philosophy more then at any time since.

APPENDIX 2

SCHEMATIC RECONSTRUCTION OF THE GROWTH OF NEWTON'S IDEAS ON FLUXIONS

[1664–1666]

We have already insisted on several inadequacies in Newton's own chronology (as reproduced in the preceding appendix) of his early discoveries in calculus. These weaknesses—the result, for the most part, of lapses in memory and of an inadequate later appreciation on Newton's part of his own originality rather than the product of deliberate distortion—stand out when we come to analyse

(22) ULC. Add. 3958.3: 48^v–63^v = section 7 below.

(23) That is, products.

(24) ULC. Add. 3968.31: 448^r (with a first draft on 449^v).

(25) ULC. Add. 3968.41: 85^r, first printed in the *Catalogue of the Portsmouth Collection of Books and Papers* (Cambridge, 1888): xviii, but more accurately by Rouse Ball in his *An Essay on Newton's 'Principia'* (London, 1893): 7.

(26) The fragment from which the present extract is taken relates particularly to Newton's previous assertion in 1704 in introduction to his tract on the quadrature of curves that 'has motuum vel incrementorum velocitates nominando *Fluxiones* & quantitates genitus nominando *Fluentes*, incidi paulatim *Annis* 1665 & 1666 in Methodum Fluxionum qua hic usus sum in Quadratura Curvarum' (*Tractatus de Quadratura Curvarum* (London, 1704) = *Opticks* (note (8)

the existing written record of those researches. In an attempt to restore the balance we have drawn up the following reconstruction of the main lines of growth in Newton's ideas. It must firmly be stressed that the scheme depends essentially on a subjective choice of detail, but we hope that, in the early stages at least of future evaluation of the texts, it may be a rough guide to the undoubtedly complex pattern of Newton's thought, showing in particular the way in which external influences are ultimately joined with Newton's native genius in the first systematic presentation of his fluxional method in October 1666.

Little need be said on the structure of the chart. Continuous arrows suggest direct, positive growths and modifications of ideas, while the two occurrences of broken lines show probable influences: in the one case, the conjectured rôle of Barrow's lectures on motion in the formation of Newton's concept of a fluxion depends essentially for its justification on a stray remark written down by Newton in old age; the second, hazarding the impact of Newton's first researches in fluxional equations, cannot adequately be substantiated because the corresponding entry in the October 1666 tract is incomplete. The numbers in square brackets, finally, refer to the position of corresponding texts in the present volume: thus 2, 7 identifies the October 1666 tract, here ordered as section 7 of Part 2.

above): 166). The assertion was savagely attacked the following year in an anonymous review in the *Acta Eruditorum* and was one of the points at issue in the fluxion priority dispute. In the present instance, having quoted extracts from letters by John Collins and Isaac Barrow in defence of his priority, Newton proceeded to introduce the familiar passage from which we have made our extract with the declaration that 'the testimony of these two ancient, knowing & credible witnesses may suffice to excuse me for saying in the Introduction to the book of Principles [sic!] that I found the Method by degrees in the years 1665 & 1666'. (Newton makes correct reference to 'the Book of Quadratures' in his revised version of this passage, meant for insertion in a letter to DesMaizeaux in the summer of 1718, at ULC. Add. 3968.27: 390^v.)

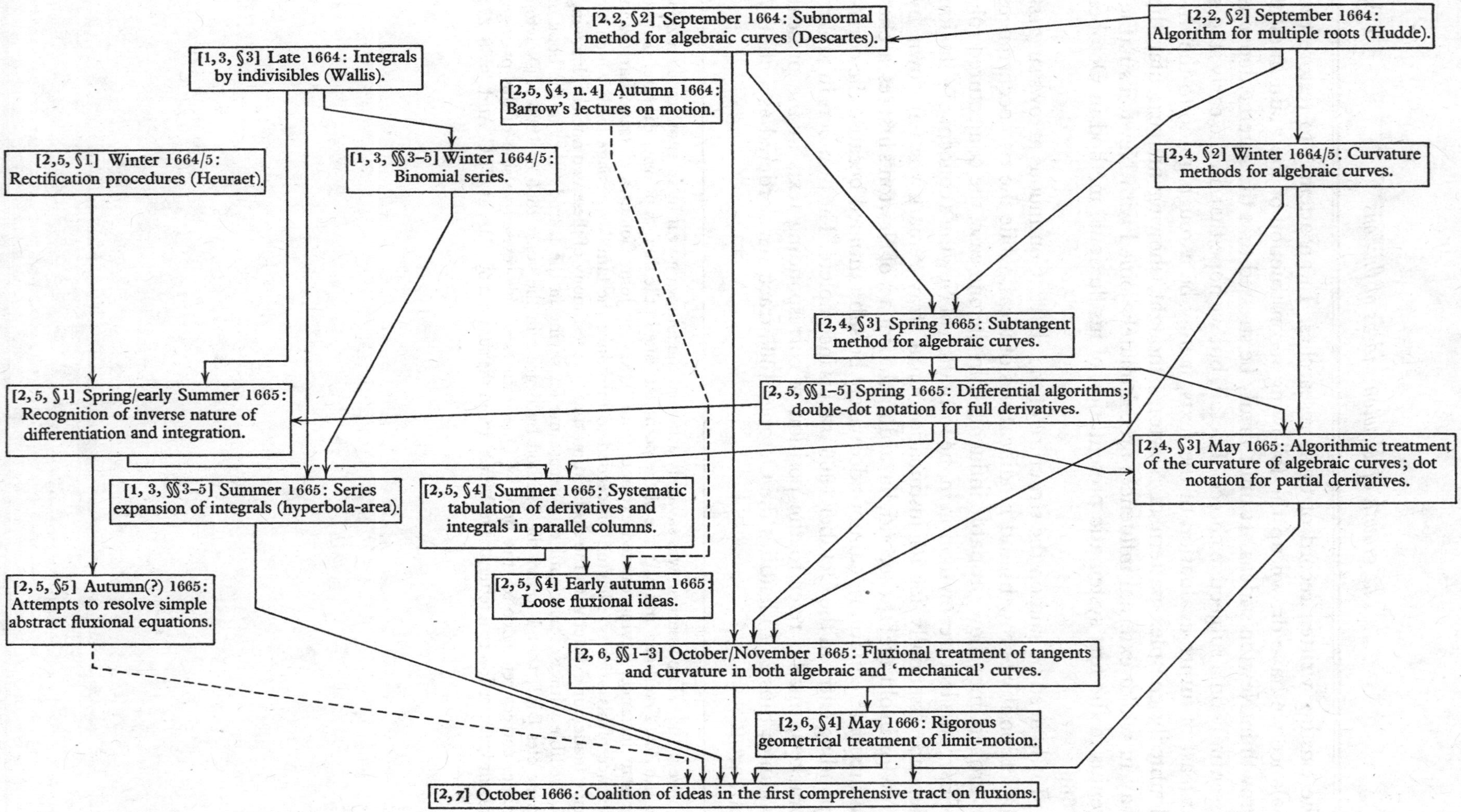

[2,2, §2] September 1664: Subnormal method for algebraic curves (Descartes).
[2,2, §2] September 1664: Algorithm for multiple roots (Hudde).
[1,3, §3] Late 1664: Integrals by indivisibles (Wallis).
[2,5, §4, n. 4] Autumn 1664: Barrow's lectures on motion.
[2,5, §1] Winter 1664/5: Rectification procedures (Heuraet).
[1, 3, §§3–5] Winter 1664/5: Binomial series.
[2,4, §2] Winter 1664/5: Curvature methods for algebraic curves.
[2,4, §3] Spring 1665: Subtangent method for algebraic curves.
[2, 5, §1] Spring/early Summer 1665: Recognition of inverse nature of differentiation and integration.
[2, 5, §§1–5] Spring 1665: Differential algorithms; double-dot notation for full derivatives.
[2,4, §3] May 1665: Algorithmic treatment of the curvature of algebraic curves; dot notation for partial derivatives.
[1, 3, §§3–5] Summer 1665: Series expansion of integrals (hyperbola-area).
[2,5, §4] Summer 1665: Systematic tabulation of derivatives and integrals in parallel columns.
[2, 5, §5] Autumn(?) 1665: Attempts to resolve simple abstract fluxional equations.
[2, 5, §4] Early autumn 1665: Loose fluxional ideas.
[2, 6, §§1–3] October/November 1665: Fluxional treatment of tangents and curvature in both algebraic and 'mechanical' curves.
[2, 6, §4] May 1666: Rigorous geometrical treatment of limit-motion.
[2, 7] October 1666: Coalition of ideas in the first comprehensive tract on fluxions.

1

EARLY NOTES ON ANALYTICAL GEOMETRY

[Autumn 1664]

From the original manuscripts in the Waste Book and an undergraduate pocket-book in the University Library, Cambridge

§1 PRELIMINARY CALCULATIONS ON TRANSFORMATION OF AXES(1)

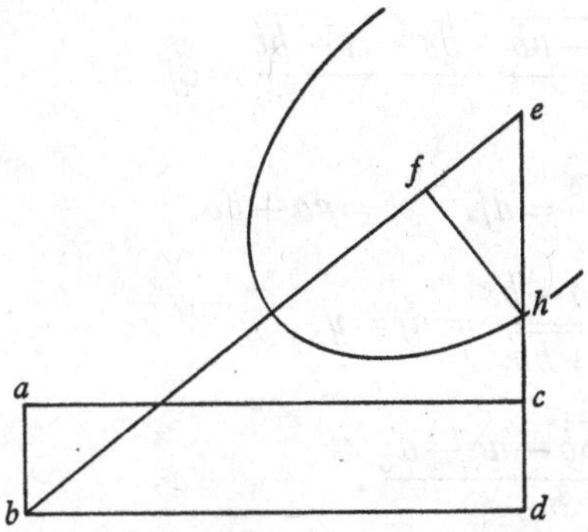

[1]

$$ab=a.\ bd:be::b:c.\ \frac{cx}{b}=be.\ \frac{c\gamma}{b}=eh.$$

$$fe^2=\frac{cc\gamma\gamma-bb\gamma\gamma}{bb}.\ fe=\frac{cx-bz}{b}=\frac{\gamma\sqrt{cc-bb}}{b}.$$

$$\frac{bz+\gamma\sqrt{cc-bb}}{c}=x.\ ed=\frac{x\sqrt{cc-bb}}{b}.$$

$$ed=\frac{\gamma cc-\gamma bb+bz\sqrt{cc-bb}}{bc}.$$

$$ch=\frac{-\gamma b+z\sqrt{cc-bb}-ca}{c}.$$ (2) $$-xx+dy+yy=0.$$

$$\frac{2b\gamma z\sqrt{cc-b[b]}}{cc}+\frac{d\gamma b}{c}+\frac{2ab\gamma}{c}+\frac{2b\gamma z\sqrt{cc-b[b]}}{cc}:-\frac{d}{2}$$ (3) $$=+a.\ b=0.\ c=$$ any finite line.

as d. [then] $x=\gamma.\ y=\frac{-1}{2}d+z.\ \gamma^2+\frac{1}{4}dd\mp dz-zz[=0.]$(4)

---

(1) Add. 4004: 6^v, 7^v–8^v, written very probably in September 1664. (f. 1^v bears the date '[S]ept 1664', and f. 8^v 'September 1664'.)

(2) Where $ac = x$, $ch = y$; and $bf = z$, $fh = \gamma$ are two perpendicular sets of co-ordinates which define the same point h, Newton finds the relation which exists between them: that is

$$\left\{\begin{aligned} x &= \frac{bz+\gamma\sqrt{(c^2-b^2)}}{c} \\ y = ch &= \frac{-\gamma b+z\sqrt{(c^2-b^2)}-ca}{c} \end{aligned}\right\}.$$

(3) Read '$+\frac{d}{2}$'.

(4) Newton uses his axis transform to simplify the defining equation of the rectangular hyperbola $-x^2+y^2+dy = 0$. First, substituting for x and y their corresponding values in z

$$y^3=axx+aax.\quad y+a=\mathfrak{z}.\quad \mathfrak{z}^3-3a\mathfrak{z}^2+3aa\mathfrak{z}-axx+aax$$ [5] $$-a^3=0.\ \text{or this}\ axx=y^3.$$

$$y=x+z.\quad -axx+x^3+3x^2z+3xz^2+z^3=0.\quad \sqrt{c}\,[:]\,\overline{axx}-x=z.$$

$$z^3+3xz^2+3xxz+x^3-dxx=0.$$ [6]

---

[2]

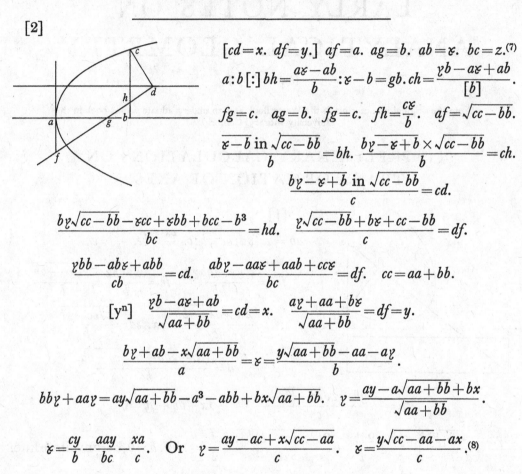

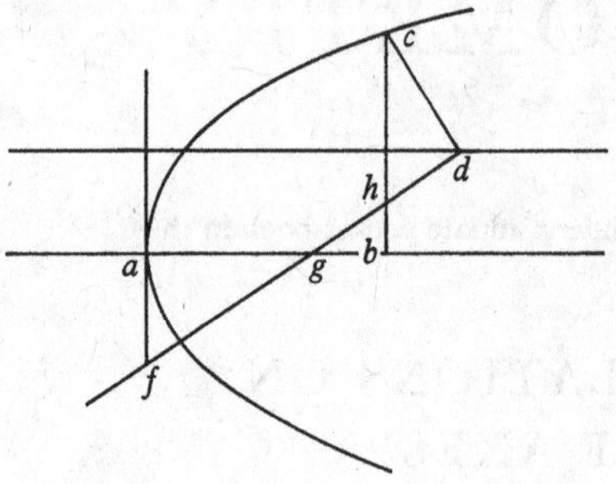

$$[cd=x.\ df=y.]\ af=a.\ ag=b.\ ab=\mathfrak{z}.\ bc=z.$$ [7]

$$a:b\,[:]\,bh=\frac{a\mathfrak{z}-ab}{b}:\mathfrak{z}-b=gb.\,ch=\frac{\mathfrak{y}b-a\mathfrak{z}+ab}{[b]}.$$

$$fg=c.\quad ag=b.\quad fg=c.\quad fh=\frac{c\mathfrak{z}}{b}.\quad af=\sqrt{cc-bb}.$$

$$\frac{\overline{\mathfrak{z}-b}\ \text{in}\ \sqrt{cc-bb}}{b}=bh.\quad \frac{\overline{b\mathfrak{y}-\mathfrak{z}+b}\times\sqrt{cc-bb}}{b}=ch.$$

$$\frac{\overline{b\mathfrak{y}-\mathfrak{z}+b}\ \text{in}\ \sqrt{cc-bb}}{c}=cd.$$

$$\frac{b\mathfrak{y}\sqrt{cc-bb}-\mathfrak{z}cc+\mathfrak{z}bb+bcc-b^3}{bc}=hd.\quad \frac{\mathfrak{y}\sqrt{cc-bb}+b\mathfrak{z}+cc-bb}{c}=df.$$

$$\frac{\mathfrak{y}bb-ab\mathfrak{z}+abb}{cb}=cd.\quad \frac{ab\mathfrak{y}-aa\mathfrak{z}+aab+cc\mathfrak{z}}{bc}=df.\quad cc=aa+bb.$$

$$[\mathrm{y}^{\mathrm{n}}]\quad \frac{\mathfrak{y}b-a\mathfrak{z}+ab}{\sqrt{aa+bb}}=cd=x.\quad \frac{a\mathfrak{y}+aa+b\mathfrak{z}}{\sqrt{aa+bb}}=df=y.$$

$$\frac{b\mathfrak{y}+ab-x\sqrt{aa+bb}}{a}=\mathfrak{z}=\frac{y\sqrt{aa+bb}-aa-a\mathfrak{y}}{b}.$$

$$bb\mathfrak{y}+aa\mathfrak{y}=ay\sqrt{aa+bb}-a^3-abb+bx\sqrt{aa+bb}.\quad \mathfrak{y}=\frac{ay-a\sqrt{aa+bb}+bx}{\sqrt{aa+bb}}.$$

$$\mathfrak{z}=\frac{cy}{b}-\frac{aay}{bc}-\frac{xa}{c}.\quad \text{Or}\quad \mathfrak{y}=\frac{ay-ac+x\sqrt{cc-aa}}{c}.\quad \mathfrak{z}=\frac{y\sqrt{cc-aa}-ax}{c}.$$ [8]

---

and $\mathfrak{y}$, he abandons his calculations when they prove too cumbersome. He then simplifies his transformation by taking $a=\frac{1}{2}d$, $b=0$, c arbitrary, so that

$$\left\{\begin{matrix} x=\mathfrak{y} \\ y=z-\frac{1}{2}d \end{matrix}\right\},\quad \text{or}\quad -\mathfrak{y}^2+z^2-\tfrac{1}{4}d^2=0.$$

(Note that the term '$\mp dz$' in the final line is Newton's contraction for '$-dz+dz$', or a vanishing term.)

(5) Read '$-aax$'.

(6) Newton adds two more simple examples, applying respectively the translation

$$\left\{\begin{matrix} x=x \\ y=\mathfrak{z}-a \end{matrix}\right\},$$

[3]

$$\mathfrak{y}=\frac{ay-ac+x\sqrt{cc-aa}}{c}.\quad \mathfrak{x}=\frac{y\sqrt{cc-aa}-ax}{c}.$$

$$\mathfrak{y}\mathfrak{y}=\frac{aayy-2aacy+aacc+ccxx-aaxx+2ayx\sqrt{cc-aa}-2acx\sqrt{cc-aa}}{[cc]}.$$

$$\mathfrak{x}\mathfrak{x}=\frac{ccy^2-aayy-2axy\sqrt{cc-aa}+aaxx}{[cc]}.\quad \mathfrak{x}\mathfrak{x}\mathfrak{y}\mathfrak{y}-\mathfrak{x}^2g^2=b^4.$$

$$\begin{array}{llllll}
aaccx^4+4a^3x^3y\sqrt{cc-aa} & +6a^4xxyy & +6a^3cxyy\sqrt{\quad} & & +aaccy^4 & \\
-a^4x^4-2a^3cx^3\sqrt{cc-aa} & -6aaccxxyy & -4a^3xy^3\sqrt{\quad} & & -a^4y^4 & [=0.]^{(9)} \\
-2aaccx^3y\sqrt{cc-aa} & +4aac^3xxy & -2a^3ccxy\sqrt{\quad} & & +2a^4cy^3 & \\
 & -6a^4cxxy & +2accxy^3\sqrt{\quad} & & -2aac^3y^3 & \\
 & +a^4ccxx & -2ac^3xyy\sqrt{\quad} & & +aac^4yy & \\
 & +c^4xxyy & & & -a^4ccyy & \\
 & [-a^2c^2y^2xx] & [+2ac^2g^2xy\sqrt{\quad}] & & [a^2c^2g^2yy] & \\
 & & & & [-c^4g^2yy] & \\
 & & & & [-c^4b^4] &
\end{array}$$

[. . . . .][(10)]

---

and the algebraic transform $\begin{Bmatrix} x=x \\ y=x+z \end{Bmatrix}$ to his chosen equations. (The latter without modification does not, of course, represent an axis transform.)

(7) Read '$bc=\mathfrak{y}$'.

(8) Where $af=a$, $ag=b$ (and so $fg=\sqrt{(a^2+b^2)}=c$) with the point c determined by the two perpendicular co-ordinate systems $fd=y$, $dc=x$; $ab=\mathfrak{x}$, $bc=\mathfrak{y}$, Newton gives the two sets of equations which define the axis-transform from one set into the other. Specifically, where $\widehat{agf}=\theta$ (or $a/c=\sin\theta$, $b/c=\cos\theta$)

$$\begin{Bmatrix} \mathfrak{y}=(y-c)\sin\theta+x\cos\theta \\ \mathfrak{x}-b=(y-c)\cos\theta-x\sin\theta \end{Bmatrix} \text{ and conversely } \begin{Bmatrix} x=\mathfrak{y}\cos\theta-(\mathfrak{x}-b)\sin\theta \\ y-c=\mathfrak{y}\sin\theta+(\mathfrak{x}-b)\cos\theta \end{Bmatrix}.$$

(9) A page later Newton takes up his calculations from the preceding, using his equations to transform to new axes x, y the curve whose defining equation is $\mathfrak{x}^2\mathfrak{y}^2-\mathfrak{x}^2g^2=b^4$. With his calculations still not quite complete, however, Newton breaks off. (Note that '$\sqrt{\quad}$' here is Newton's abbreviation of '$\sqrt{cc-aa}$'.)

(10) In these omitted calculations Newton particularizes his axis-transform relations and finds the condition for a simple rotation of axes. With his text Newton has entered no diagram,

[4]

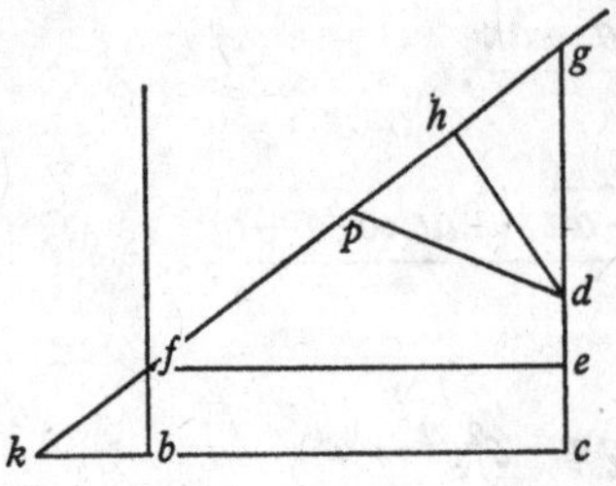

$bc=x.\ cd=y.\ bf=c.\ [f]p=\mathfrak{x}.\ pd=\mathfrak{y}.\ fe:fg::d:e.$

$fe:eg::d:f.\ hp=[t].\ fg=\dfrac{ex}{d}.\ eg=\dfrac{fx}{d}.\ pd:dh::r:s.$

$dh=\dfrac{s\mathfrak{y}}{r}.\ r:t::pd:ph=\dfrac{t\mathfrak{y}}{r}.\ d:e::dh:dg=\dfrac{es\mathfrak{y}}{dr}.\ gh=\dfrac{fs\mathfrak{y}}{dr}.$

$$fg=\mathfrak{x}+\frac{t\mathfrak{y}}{r}+\frac{fs\mathfrak{y}}{dr}=\frac{ex}{d}.\qquad \frac{dr\mathfrak{x}+dt\mathfrak{y}+fs\mathfrak{y}}{er}=x.$$

$$\frac{rfx+cdr-es\mathfrak{y}}{dr}=y.\quad \frac{fdr\mathfrak{x}+dtf\mathfrak{y}+ffs\mathfrak{y}+cder-ees\mathfrak{y}}{der}=y=\frac{fr\mathfrak{x}+tf\mathfrak{y}-ds\mathfrak{y}+cer}{er}.^{(11)}$$

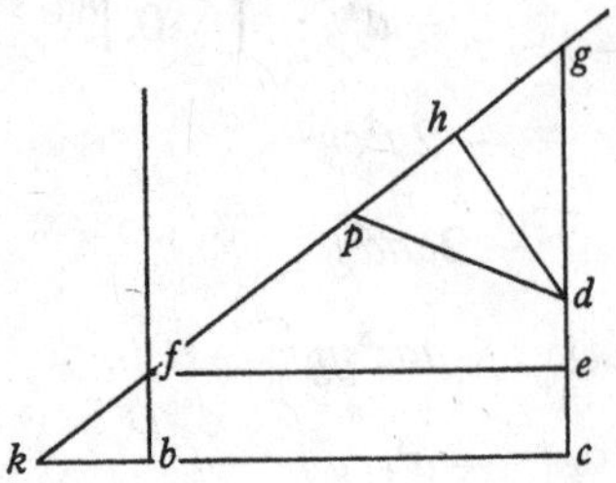

$ec=c.\ bc=x.\ dc=y.\ pd=\mathfrak{y}.\ pf=\mathfrak{x}.\ fg=\dfrac{ex}{d}.\ eg=\dfrac{fx}{d}.$

$pg=\dfrac{s\mathfrak{y}}{r}.\ gd=\dfrac{t\mathfrak{y}}{r}.\ fg=\dfrac{\mathfrak{x}r+s\mathfrak{y}}{r}=\dfrac{ex}{d}.\ x=\dfrac{dr\mathfrak{x}+ds\mathfrak{y}}{re}.$

$$eg+ec-gd=y=\frac{fr\mathfrak{x}+fs\mathfrak{y}+rec-te\mathfrak{y}}{re}.^{(12)}$$

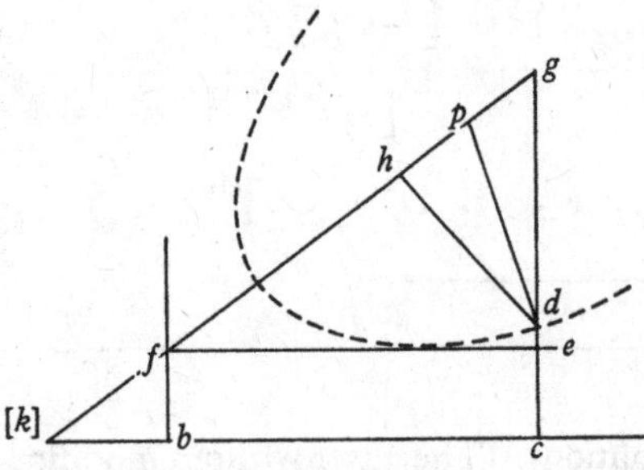

[5] $fe=bc=x.\ cd=y.\ fp=\mathfrak{x}.\ pd=\mathfrak{y}.$

$fe:fg::d:e::dh:dg.\ fe:eg::d:f::hd:hg.$

$pd[:]dh::r:s.\ pd:ph::r:t.\ \dfrac{dr\mathfrak{x}-dt\mathfrak{y}+fs\mathfrak{y}}{er}=x.$

$\dfrac{fr\mathfrak{x}-ft\mathfrak{y}-ds\mathfrak{y}+cer}{er}=y.\ d=r.\ f=\sqrt{ee-dd}.$

$s=\sqrt{rr-tt}.$ [or] $s=\sqrt{dd-tt}.$

---

but in the restored figure supposing $f\hat{a}g=\theta$ with the point c determined by the two perpendicular axis-systems $ab=x$, $bc=y$; and $ad=\mathfrak{x}$, $dc=\mathfrak{y}$, we may write Newton's result as the familiar transform

$$\begin{cases} x=\mathfrak{x}\cos\theta+\mathfrak{y}\sin\theta \\ y=\mathfrak{x}\sin\theta-\mathfrak{y}\cos\theta \end{cases}.$$

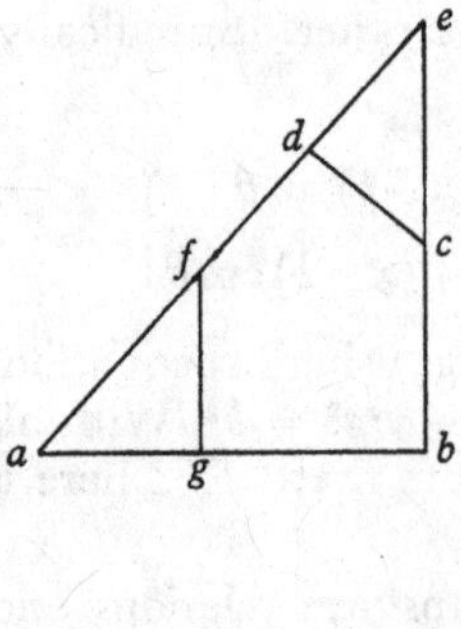

Newton himself, where we set

$$\cos\theta=\frac{ag}{af}=\frac{d}{e}\left(\text{or }\sin\theta=\frac{\sqrt{e^2-d^2}}{e}\right),$$

writes this as

$$\begin{cases} x=\dfrac{d\mathfrak{x}+\mathfrak{y}\sqrt{ee-dd}}{e} \\ y=\dfrac{-d\mathfrak{y}+\mathfrak{x}\sqrt{ee-dd}}{e} \end{cases},$$

and proceeds to evaluate the powers $e^{r+s}x^ry^s=(d\mathfrak{x}+\mathfrak{y}\sqrt{ee-dd})^r\times(-d\mathfrak{y}+\mathfrak{x}\sqrt{ee-dd})^s$ for

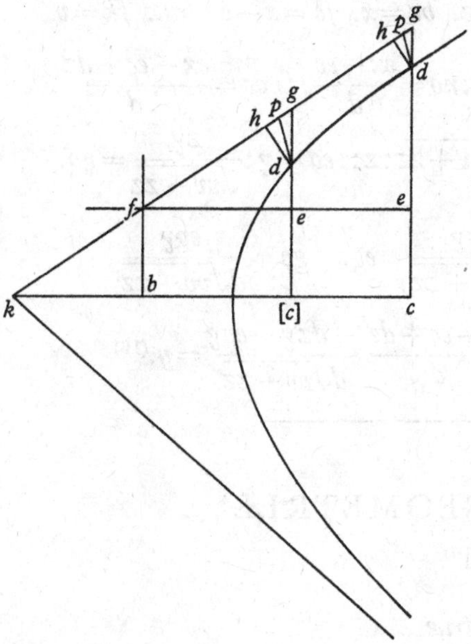

$$\frac{dd\varepsilon - dt\mathfrak{z} + \mathfrak{z}\sqrt{eedd - eett + ddtt - d^4}}{ed} = x.$$

$$\frac{d\varepsilon\sqrt{ee-dd} - t\mathfrak{z}\sqrt{ee-dd} - d\mathfrak{z}\sqrt{dd-tt} + ced}{ed} = y.$$

Lastly $dp = -dt + \sqrt{eedd - eett + ddtt - d^4}$.

$n = \sqrt{ee-dd}$. $dq = t\sqrt{ee-dd} + d\sqrt{dd-tt}$.

& Therefore

$$x = \frac{d\varepsilon + p\mathfrak{z}}{e}. \quad \& \quad y = \frac{n\varepsilon - q\mathfrak{z} + ce}{e}.$$ (13)

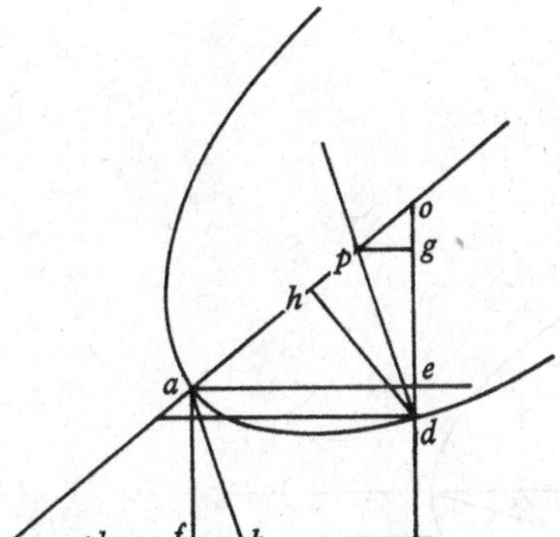

[6] $bf = c$. $fa = z$. $fk = v$. $bc = x$. $cd = y$. $ap = \varepsilon$.

$$dp = \mathfrak{z}.\ vv + zz : vv :: \mathfrak{z}\mathfrak{z} : \frac{vv\mathfrak{z}\mathfrak{z}}{vv+zz} = pg^2.$$ (14)

$$d:e:[:]\frac{v\mathfrak{z}}{\sqrt{vv+zz}} : \frac{ev\mathfrak{z}}{d\sqrt{vv+zz}} = po.$$

$$ae = x - [c] : ao = \frac{ex - e[c]}{d} :: d:e.$$

$$\frac{ex-e[c]}{d}\ \frac{-ev\mathfrak{z}}{d\sqrt{vv+zz}} = \varepsilon.\quad d\varepsilon\sqrt{vv+zz}\ [\quad]$$ (15)

$r = 0, s = 1, 2, 3, 4, 5$; $r = 1, s = 0, 1, 2, 3, 4, 5$; $r = 2, s = 0, 1, 2, 3, 4$; $r = 3, s = 0, 1, 2$; $r = 4, s = 0$; and $r = 5, s = 0$.

(11) Newton generalizes his previous treatment, now for the first time considering a transformation from perpendicular to general oblique axes. Note that $cd = y = eg + ce - dg$ with $e^2 - f^2 = d^2$.

(12) In redraft some measure of simplification is achieved.

(13) Further simplification of [4] by introduction of new constants p, q, n yields a manageable set of equations defining the general transformation from perpendicular to general oblique axes.

(14) That is, $ak:kf = dp:pg$, or ak is taken parallel to pd.

(15) Newton breaks off his calculation and starts afresh.

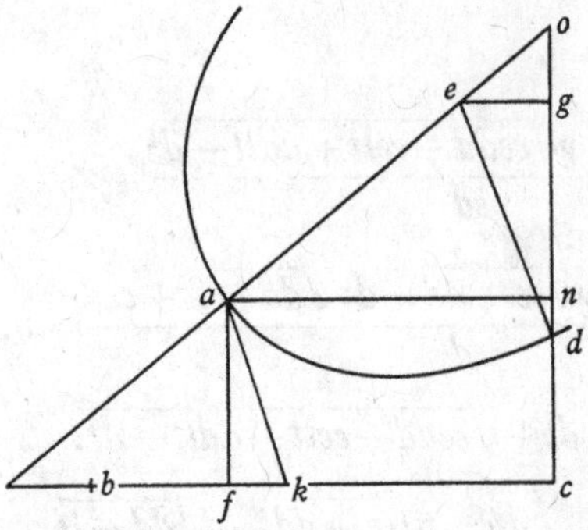

$bf=c.\ af=z.\ bc=x.\ fc=x-c=an.\ fk=v.$

$$d:e::an:no=\frac{ex-ec}{d}.\quad oc=\frac{ex-ec+dz}{d}.$$

$$ak=\sqrt{vv+zz}:z::ed=\gamma:\frac{z\gamma}{\sqrt{vv+zz}}=gd.$$

$$\frac{v\gamma}{\sqrt{vv+zz}}=eg.\quad go=\frac{ev\gamma}{d\sqrt{vv+zz}}.$$

$$\frac{ex-ec+dz}{d}-\frac{dz\gamma-ev\gamma}{d\sqrt{vv+zz}}=y.$$ (16)

§2 THE 'EPITOME GEOMETRIÆ'

[September? 1664](1)

[1](2) *Epitome Geometriæ*

$y=db.\ x=ba.\ aay=x^3.\ b+z=y.\ z=bc.\ [cd=b.]\ aab+aaz=x^3.\ b-z=y.$
$z=bf.\ [fd=b.]\ aab-aaz=x^3.\ z-b=y.\ z=bg.\ [dg=b.]\ aaz-aab=x^3.$

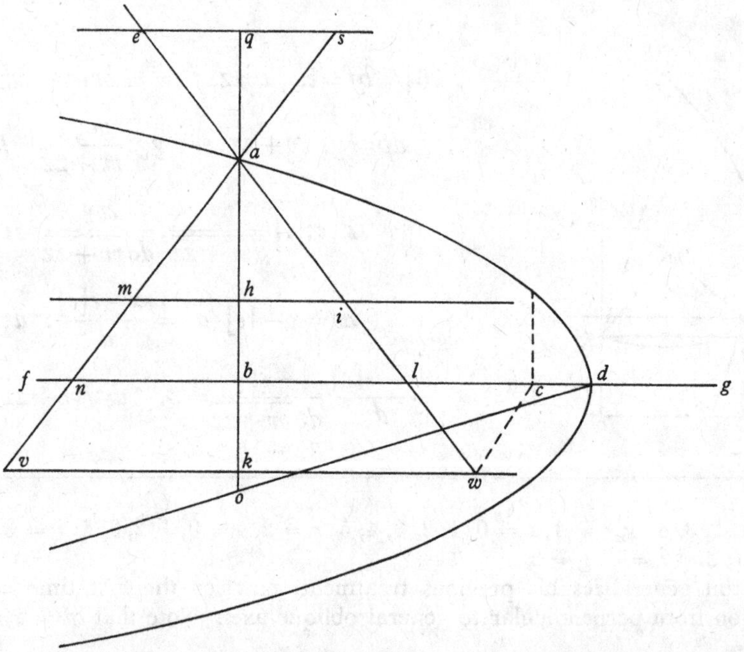

(16) In attempted further simplification Newton tries a modified approach. The line *ak*, taken parallel to *ed*, defines $\widehat{eak}=\widehat{oed}$ as the angle between the transformed axes $ae=\mathfrak{x}$, $ed=\gamma$ and on that basis Newton finds the relation connecting x, y and γ, but breaks off yet again before calculating the similar relation between x, y and $\mathfrak{x}$. (Note that $cd=y=oc-og-gd$, that $\widehat{bfa}$, $\widehat{bcd}$ and $\widehat{ego}$ are each right and that the ratio '$d:e$' is used inconsistently.)

$$x=d+\xi.\quad \xi=ah.\quad aay-d^3-3dd\xi-3d\xi\xi-\xi^3=0.$$

$$\left.\begin{array}{l} aab+aaz \\ aab-aaz \\ aaz-aab \end{array}\right\}=d^3+3dd\xi+3d\xi\xi+\xi^3.$$

$$x=[d]-\xi.\quad aq=\xi.\quad [bq=d.]$$

$$\left.\begin{array}{l} aay \\ aab\pm aaz \\ aaz-aab \end{array}\right\}=d^3-3dd\xi+3d\xi\xi-\xi^3.$$

$$x=\xi-[d].\quad ak=[\xi].$$

$$\left.\begin{array}{l} aay \\ aab\pm aaz \\ aaz-aab \end{array}\right\}=\xi^3-3d\xi\xi+3dd\xi-d^3.\text{(3)}$$

$$na=\xi.\ nd=z.\ ab:an::a:b.\ \&\ ab:nb::a:c.\ \text{then } \xi^3=\frac{b^3}{a}z-\frac{bbc}{a}\xi.\text{(4)}$$

$$\xi=d+\zeta.\ \zeta=mn.\ d^3+3d^2\zeta+3d\zeta\zeta+\zeta^3-\frac{b^3}{a}z+\frac{bbcd}{a}[=0].\text{(5)}$$
$$+\frac{bbc}{[a]}\zeta$$

---

(1) Add. 4000: 19^v-22^r (with the heading, cancelled, on 19^r). The date is confirmed by an analysis of the handwriting and by the evident connection of the piece with the preceding. The whole is cancelled and overwritten with the hyperbola calculations which Newton added about the summer of 1665 to his Wallis annotations (**1**, 3, §3), and this provides a firm post-date.

(2) The first tentative draft. In the accompanying figure the curve (*a*), whose defining equation $a^2y=x^3$, where $db=y$, $ba=x$, shows it to be the simple cubic parabola, is wrongly drawn. (It should be symmetrical round the vertex *d* as centre, passing to infinity in opposite directions.) It is tempting to assume that Newton took this example together with its incorrect shape from his reading of Wallis' *Sectiones Conicæ*, where it appears in the appended propositions XLV–XLVII as *Paraboloidis Cubicalis* (*De Sectionibus Conicis, Nova Methodo Expositis, Tractatus* (Oxford, 1655 [1656]): 105–10). It is clear that he has not appreciated the corrections communicated by Brouncker and discussed at length by Wallis himself a year later (*Adversus Marci Meibomii, De Proportionibus Dialogum, Tractatus Elencticus* (Oxford, 1657): *Dedicatio:* 3–17, 30–50, especially 35–39).

(3) Newton tabulates the effects of a translation of axes $\left\{\begin{array}{l} x=\xi+d \\ y=z+b \end{array}\right\}$ on the Wallis parabola $x^3=a^2y$.

(4) More generally, he evaluates the effect of changing to oblique axes in the simple case $\left\{\begin{array}{l} x=(a/b)\,\xi \\ y=z-(c/b)\,\xi \end{array}\right\}$.

(5) The translation $\left\{\begin{array}{l} \xi=d+\zeta \\ z=z \end{array}\right\}$ further transforms the equation. (The entry '$\zeta=mx$' implies that $ma=d$.)

[2]

$db=x.\ ba=y.\ aax=y^3.\ x=b+z.\ z=bc.$

$$\left.\begin{array}{ll} aab+aaz = & \xi^3+3c\xi^2+3cc\xi+[c]^3.\quad y=c+\xi.\quad \xi=ah. \\ (x=b-z.\ z=bf.) & \\ aab-aaz = & c^3-3cc\xi+3c\xi^2-\xi^3.\quad y=c-\xi.\quad \xi=aq. \\ (x=z-b.\ z=bg.) & \\ aaz-aab = & \xi^3-3c\xi^2+3\xi c^2-c^3.\quad y=\xi-c.\quad \xi=ak. \end{array}\right.$$

$na=y.\ dn=x.\ ab:an::a:b.\ ab:nb::a:c.$ & $y^3=\frac{b^3}{a}x-\frac{bbc}{a}y.$

Or $d^2x=eey+y^3.$(6)

$$\left.\begin{array}{ll} (x=z+o.\ z=nc.) & eey+y^3. \\ d^2z+d^2o = & e^2n+ee\xi+n^3+3nn\xi+3n\xi^2+\xi^3.\quad (y=n+\xi.\ \xi=ma.) \\ (x=z-o.\ z=gn.) & \\ ddz-ddo = & een-ee\xi+n^3-3nn\xi+3n\xi^2-\xi^3.\quad (y=n-\xi.\ \xi=as.) \\ (x=o-z.\ z=nf.) & \\ ddo-ddz = & ee\xi-een+\xi^3-3n\xi^2+3\xi n^2-n^3.\quad (y=\xi-n.\ \xi=av.) \end{array}\right.$$(7)

$al=y.\ dl=x.\ ab:al::a:b.$ & $al:bl::b:c.$ whence $y^3=\frac{xb^3+yb^2c}{a}$. Or $y^3-eey=ddx$. &c as before onely varying ye signes at *een* & *eef*.(8)

$ao=y.\ do=x.\ a:b::bd:do.\ b:c::do:ob.$ &

$$\frac{a^3x}{b}=y^3-\frac{3c}{b}xyy+\frac{3cc}{bb}xxy-\frac{c^3x^3}{[b^3]}.$$(9)

---

(6) That is, taking $d^2=b^3/a$, $e^2=b^2c/a$.

(7) Newton transforms the equation $a^2x=y^3$ much as before, first by the translation $\begin{Bmatrix} x=z+b \\ y=\xi+c \end{Bmatrix}$; and then more generally by $\begin{Bmatrix} x=x'-(c/b)\,y' \\ y=(a/b)\,y' \end{Bmatrix}$: the latter finally is transformed by the second translation $\begin{Bmatrix} x'=z+o \\ y'=\xi+n \end{Bmatrix}$.

(8) Newton considers the equivalent transform $\begin{Bmatrix} x=x'+(c/b)\,y' \\ y=(a/b)\,y' \end{Bmatrix}$, which requires only a couple of sign changes in the preceding discussion.

(9) The converse transform $\begin{Bmatrix} x=(a/b)\,x' \\ y=y'-(c/b)\,x' \end{Bmatrix}$ is applied to $a^2x=y^3$.

§3 FIRST NOTES ON FINDING THE AXES OF CURVES

[October? 1664]

[1](1)

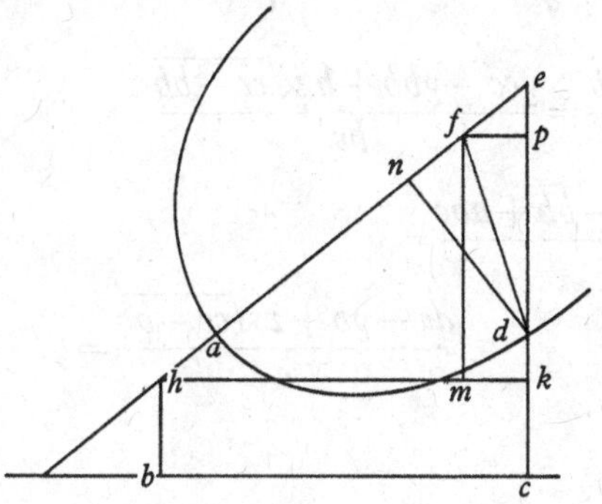

$[bc=x.\ dc=y.\ bh=c.]\ [fd=\gamma.]\ fd:fp::f:g.$

$\frac{\gamma g}{f}=fp.\ hm=\mathfrak{z}.\ \frac{f\mathfrak{z}+\gamma g}{f}=x.\ fp:pe::d:e.$

$pd=\frac{\gamma\sqrt{ff-gg}}{f}.\ fm=\frac{e\mathfrak{z}}{d}.\ \frac{e\mathfrak{z}+cd}{d}-\frac{\gamma\sqrt{ff-gg}}{f}=y.$

$f=d.$ &c. $\frac{d\mathfrak{z}+g\gamma}{d}=x.\ \frac{e\mathfrak{z}+cd-\gamma\sqrt{dd-gg}}{d}=y.$(2)

$$dg-eg-e\sqrt{dd-gg}+d\sqrt{dd-gg}=0.$$

$g=\sqrt{dd-gg}.$(3) $dd=2gg.$ or $d=e.$

$ag-2cg-2c\sqrt{dd-gg}=0.\quad \frac{ag}{2g+2\sqrt{dd-gg}}=c.$(4)

---

(1) Add. 4004: 9v.

(2) Newton constructs the transforming equations between the point d defined by Cartesian co-ordinates $bc=x$, $cd=y$ and parametrically in terms of $ef=\gamma$, $hm=\mathfrak{z}$, where the inclination of fe to hm is given by $fe:fp=f:g$ constant. (The ratio $fp:pe=g:\sqrt{[f^2-g^2]}$ is denoted by that of d (or f) to e, so that $d^2=f^2=g\sqrt{[e^2+f^2]}$.)

(3) Read '$g=-\sqrt{dd-gg}$'.

(4) These calculations are added in the margin. Newton seeks the condition for the ellipse $x^2-2xy+2y^2+ax+k=0$ to have an axis $\gamma=0$ by applying the transform

$$\begin{cases} dx = d\mathfrak{z}+g\gamma \\ dy = e\mathfrak{z}+cd-\gamma\sqrt{[d^2-g^2]} \end{cases}$$

found above and reducing the resulting equation to the form $A\mathfrak{z}^2+B\mathfrak{z}+C+\gamma^2=0$ (or $\gamma=\pm\sqrt{[-(A\mathfrak{z}^2+B\mathfrak{z}+C)]}$). Specifically the coefficient

$$(2/d^2)(d-e)(g+\sqrt{[d^2-g^2]})\mathfrak{z}+(1/d)(ag-2cg-2c\sqrt{[d^2-g^2]})$$

of γ is equated identically to zero; that is,

$$0=(d-e)(g+\sqrt{[d^2-g^2]})\quad\text{and}\quad 0=ag-2cg-2c\sqrt{[d^2-g^2]}.$$

Reversing the transform, we have $\gamma=\frac{d(ex-dy+cd)}{eg+d\sqrt{[d^2-g^2]}}$,

so that both $d=e$ and $g+\sqrt{[d^2-g^2]}=0$ cannot hold together; so that $d=e$,

$$2c(g+\sqrt{[d^2-g^2]})=ag$$

and the axis $0=\gamma$ is given as

$$x-y+\frac{ag}{2(g+\sqrt{[d^2-g^2]})}=0.$$

[2][5]

$ac=x.\ ch=y.\ bf=z.\ fh=\gamma.\ cd=a.\ db:be::fh:eh::b:c.\ \dfrac{cx}{b}=be.\ \dfrac{c\gamma}{b}=eh.$

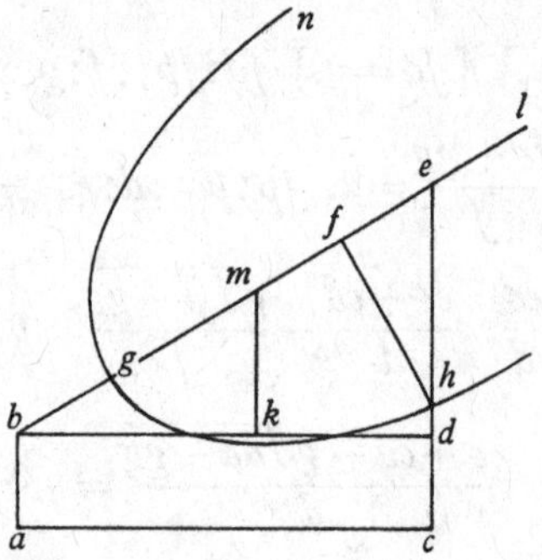

$$\frac{\gamma\sqrt{cc-bb}}{b}=fe=\frac{cx-bz}{b}.^{(6)}\quad \frac{\gamma\sqrt{cc-bb}+bz}{c}=x.$$

$$\frac{ed=x\sqrt{cc-bb}}{b}=\frac{\gamma cc-\gamma bb+bz\sqrt{cc-bb}}{bc}.$$

$$hc=\frac{-\gamma bb+bz\sqrt{cc-bb}+abc}{bc}=$$

$$\frac{ac-\gamma b+z\sqrt{cc-bb}}{c}=y.^{(7)}$$

	x	$=$	$\dfrac{+\gamma\sqrt{cc-bb}}{c}$
	xx	$=$	$\dfrac{2b\gamma z\sqrt{cc-bb}}{cc}$
for	x^3	write	$\dfrac{\gamma^3cc\sqrt{cc-bb}}{c^3}-\dfrac{\gamma^3bb\sqrt{cc-bb}}{c^3}+\dfrac{3bbzz\gamma\sqrt{cc-bb}}{c^3}$
	x^4	$=$	$\dfrac{4\gamma^3cc\sqrt{cc-bb}\times bz-4[\gamma]^3b^3z\sqrt{cc-bb}+4\gamma b^3z^3\sqrt{cc-bb}}{c^4}$
	y	$=$	$\dfrac{-\gamma b}{c}$
	yy	$=$	$\dfrac{-2\gamma abc-2\gamma bz\sqrt{cc-bb}}{cc}$
	y^3	$=$	$\dfrac{-\gamma^3b^3-3\gamma baacc-3\gamma bcczz+3\gamma b^3zz-6\gamma abcz\sqrt{cc-bb}}{[c^3]}$
	y^4	$=$	$\dfrac{-4\gamma^3b^3ac-4\gamma^3b^3z\sqrt{cc-bb}-4\gamma a^3bc^3-12\gamma aabccz\sqrt{cc-bb}-12\gamma abc^3zz+12\gamma ab^3cz^2\begin{matrix}+\gamma z^3b^3\\-\gamma z^3bcc\end{matrix}\sqrt{cc-bb}}{c^4}$
	xy	$=$	$\dfrac{ac\gamma\sqrt{cc-bb}+\gamma zcc-2\gamma zbb}{cc}$

(5) Add. 4004: 9r.

(6) That is, $be-bf$.

(7) Newton constructs the transform between the two sets of perpendicular Cartesian co-ordinates $ac = x.\ ch = y$ and $bf = z, fh = \gamma$; that is,

$$\left\{\begin{matrix} cx = \gamma\sqrt{[c^2-b^2]}+bz \\ cy = ac+z\sqrt{[c^2-b^2]}-\gamma b \end{matrix}\right\}.$$

$$xyy = \frac{-4\gamma abbcz - 3\gamma bbzz\sqrt{cc-bb} \begin{matrix}+aacc\\+zzcc\end{matrix}\gamma\sqrt{cc-bb} + 2ac^3z\gamma + b^2\gamma^3\sqrt{cc-bb}}{c^3}$$

$$xy^3 = \frac{3bbcc\gamma^3z - 4b^4z\gamma^3 + 3abbc\gamma^3\sqrt{cc-bb} + z^3c^4\gamma + 4z^3b^4\gamma - 5z^3bbcc\gamma \begin{matrix}-9abbc\\+3ac^3\end{matrix} zz\gamma\sqrt{cc-bb} \begin{matrix}-aabbcc\\+3aac^4\end{matrix} z\gamma + a^3c^3\gamma\sqrt{cc-bb}}{c^4}$$

$$xxyy = \frac{2ab^3c\gamma^3 - 2abc^3\gamma^3 \begin{matrix}+4b^3\\-2bcc\end{matrix} z\gamma^3\sqrt{cc-bb} \begin{matrix}+2bcc\\-4b^3\end{matrix} z^3\gamma\sqrt{cc-bb} + 2aabbccz\gamma\sqrt{cc-bb} + 4abc^3zz\gamma - 6ab^3czz\gamma}{c^4}$$

$$xxy = \frac{b^3\gamma^3 - bcc\gamma^3 + 2bcczz\gamma - 3b^3zz\gamma + 2abcz\gamma\sqrt{cc-bb}}{c^3}$$

$$x^3y = \frac{4b^4z\gamma^3 - 5bbccz\gamma^3 + c^4z\gamma^3 \begin{matrix}+ac^3\\-abbc\end{matrix}\gamma^3\sqrt{cc-bb} + 3bcczz\gamma - 4b^3zz\gamma + 3abcz\gamma\sqrt{cc-bb}}{c^4}$$

Haveing therefore an equation expressing yᵉ nature of a crooked line. To find its axis.

Supposeing c = some quantity most frequent in yᵉ equation subrogate $\dfrac{bz+\gamma\sqrt{cc-bb}}{c}$ into yᵉ roome of x; & $\dfrac{ac-\gamma b+z\sqrt{cc-bb}}{c}$ into yᵉ roome of y: Order yᵉ equation according to γ, make every terme = 0, in wᶜʰ γ is of one(8) dimensiō. Order every terme in this 2ᵈᵃʳʸ equation according to yᵉ dimensions of z. & supposeing every terme of each of $y^m = 0$, by yᵉ helpe of these equations (in wᶜʰ is neither x, y, z or γ) may be found yᵉ valors of a & b. Then perpendicular to ac from yᵉ point a draw $ab = a$. & from yᵉ point b draw $bk = b$, & parallell to ac. from yᵉ point k draw $mk = \sqrt{cc-bb}$, & perpendicular to bk. & through yᵉ points b, m draw bl yᵉ axis of yᵉ line hgn. & yᵗ yᵉ relation twixt

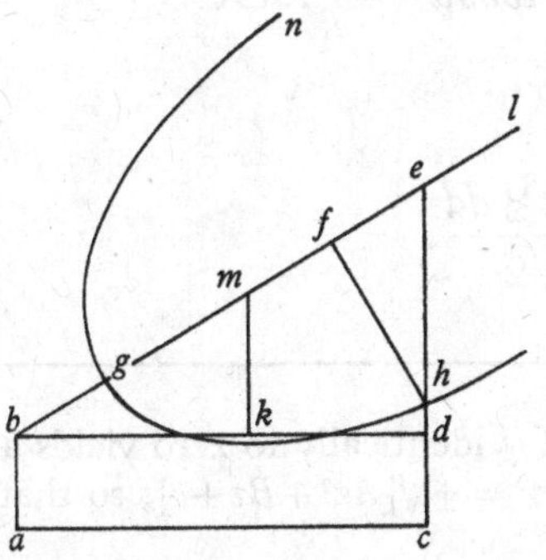

He then uses this in the sequel to tabulate a set of values of $x^r y^s$, but ignoring all but odd powers of γ 'since [in finding the condition for diameters] there is noe use of those termes in wᶜʰ γ is of eaven dimensions' (f. 17ᵛ below).

(8) Newton has cancelled '*od*'.

$bf=z$. & $fh=\gamma$ may bee had, write y[e] valors of a, b, c now found in their stead in y[e] 2[dary] equation.(9)

Example. $dd+dy+xy-yy=0$.(10) Then makeing $d=c$ I write $\dfrac{bz+\gamma\sqrt{cc-bb}}{c}$, or $\dfrac{bz+\gamma\sqrt{dd-bb}}{d}$ for x & its square for xx &c. & $\dfrac{ad-b\gamma+z\sqrt{dd-bb}}{d}$ for y, & its square for yy. & soe I have this equation,

$$0=dd+ad-b\gamma+z\sqrt{dd-bb}\,\frac{\begin{array}{rll} & -b\gamma\gamma & \\ & +bzz & \\ +adbz-2bb\gamma z & +ad\gamma & \sqrt{dd-bb}+\gamma zdd-aadd+2adb\gamma \\ & -2azd & \qquad -bb\gamma\gamma-ddzz+bbzz \\ & +2b\gamma z & \end{array}}{dd}.$$

or by ordering it according to γ,

$$\begin{array}{lllll} & & & -d^4[-ad^3] & \\ bb\gamma\gamma+b\gamma\gamma\sqrt{dd-bb} & +ddb\gamma & -da\gamma\sqrt{dd-bb} & -abdz & +2adz\sqrt{dd-bb}=0. \\ & +2bbz\gamma & -2bz\gamma\sqrt{dd-bb} & +aadd & -bzz \\ & -ddz & & +ddzz & -ddz \\ & -2abd & & -bbzz & \end{array}$$

Then by makeing those quantitys in y[e] last terme save one(11) $=0$ I have this equation $2bbz-ddz-2bz\sqrt{dd-bb}+ddb-2abd-da\sqrt{dd-bb}$. Which I divide into 2 ꝑts makeing those termes $=0$ in w[ch] z is not, & those $=0$ in w[ch] z is of one dimension. & then I have these 2 equations $2bb-dd-2b\sqrt{dd-bb}=0$. & $db-2ab-a\sqrt{dd-bb}=0$. by y[e] first $4b^4-4bbdd+d^4=4bbdd-4b^4$. Or

$$8b^4-8bbdd+d^4=0.$$

That is $$bb=\frac{dd}{2}\infty\sqrt{\frac{d^4}{8}}=\frac{dd}{2}\infty\frac{dd}{2\sqrt{2}}=\frac{dd\sqrt{2}\infty dd}{2\sqrt{2}}.$$

(9) Newton's technique of equating (here) the coefficient of γ identically to zero yields a transformed equation for the hyperbola $\gamma^2 = Az^2+Bz+c$, or $\gamma = \pm\sqrt{[Az^2+Bz+c]}$; so that the line $\gamma = 0$ which bisects the ordinate lines is a diameter.

(10) A simple (Apollonian) hyperbola.

(11) That is, the coefficient of γ.

(12) Read 'bb'.

(13) Resolving $a^2(3\sqrt{2}\pm5)-4ad(\sqrt{2}\pm1)+d^2(\sqrt{2}\pm1) = 0$ as a quadratic in a,

$$\frac{a}{d}=\frac{2\sqrt{2}\pm2\pm\sqrt{[4(\sqrt{2}\pm1)^2-(3\sqrt{2}\pm5)(\sqrt{2}\pm1)]}}{3\sqrt{2}\pm5}=\frac{(2\sqrt{2}\pm2)\pm1}{3\sqrt{2}\pm5}.$$

By the 2[d] equation I find $ddbb-4adbb+4aabb=aadd-aabb$. or $5aabb \; -aadd=0$
$\qquad -4adbb[+ddbb]$

& by writeing y[e] valor of *dd*(12) w[ch] was found before I have

$$\frac{5aadd\sqrt{2} \text{ ♉ } 5aadd-4ad^3\sqrt{2} \text{ ☊ } 4ad^3+d^4\sqrt{2} \text{ ♉ } d^4-2aadd\sqrt{2}}{2\sqrt{2}}=0.$$

Or $\underset{\text{♉ } 5aa}{3aa\sqrt{2}}-\underset{\text{☊ } 4ad}{4ad\sqrt{2}}+\underset{\text{♉ } dd}{dd\sqrt{2}}=0$. Or $aa=\dfrac{4ad\sqrt{2} \text{ ☊ } 4ad-dd\sqrt{2} \text{ ♉ } dd}{3\sqrt{2} \text{ ☊ } 5}$. &

$$a=\frac{d\sqrt{8} \text{ ☊ } 2d}{\sqrt{18} \text{ ☊ } 5} \text{ ♉ } \sqrt{\frac{8dd \text{ ☊ } 4dd\sqrt{8}+4dd-5dd \atop -dd\sqrt{36} \text{ ♉ } 5dd\sqrt{2} \text{ ♉ } dd\sqrt{18}}{18 \text{ ☊ } 5\sqrt{18}+25}}.$$

$$a=\frac{d\sqrt{8} \text{ ☊ } 2d}{\sqrt{18} \text{ ☊ } 5} \text{ ♉ } \sqrt{\frac{dd}{18 \text{ ☊ } 5\sqrt{18}+25}}. \text{ or } \frac{2d\sqrt{2}-2d}{3\sqrt{2}-5} \text{ ♉ } \sqrt{\frac{dd}{43-15\sqrt{2}}}=a.$$

& $\dfrac{dd\sqrt{2} \text{ ♉ } dd}{2\sqrt{2}}=bb$. $a=\dfrac{2d\sqrt{2} \text{ ♉ } 2d \text{ ♉ } d}{3\sqrt{2} \text{ ♉ } 5}$.(13)

---

§4(1)

[October 1664]

How to find y[e] axes[,] vertices[,] Diamiters, Centers, or Asymptotes of any Crooked Line supposeing it have them.

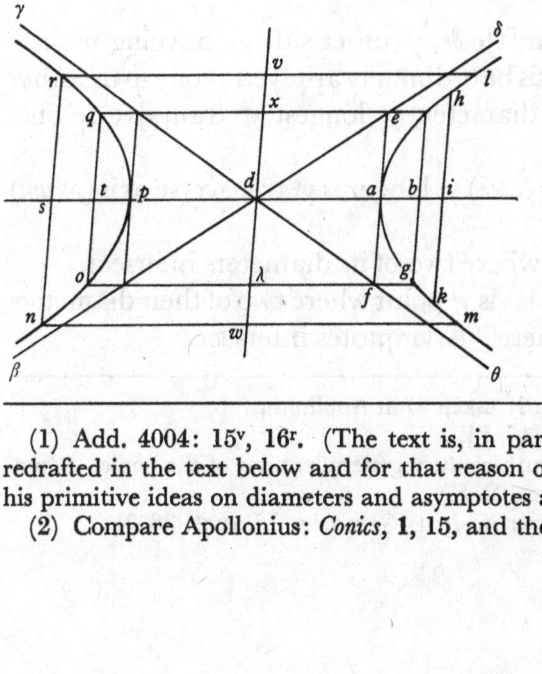

Definitions. 1 If all y[e] parallell lines w[ch] are terminated by the same or by 2 divers figures, bee bisected by a streight line; y[t] bisecting line is a diameter; & those parallel lines, are lines ordinately applyed to y[t] diameter.(2)

2 If those parallell lines intersect y[e] diameter at right angles y[t] diameter is an axis.

---

(1) Add. 4004: 15[v], 16[r]. (The text is, in part, heavily cancelled but these portions are redrafted in the text below and for that reason omitted.) Newton here begins to generalize his primitive ideas on diameters and asymptotes and set them out systematically.

(2) Compare Apollonius: *Conics*, 1, 15, and the preliminary definitions.

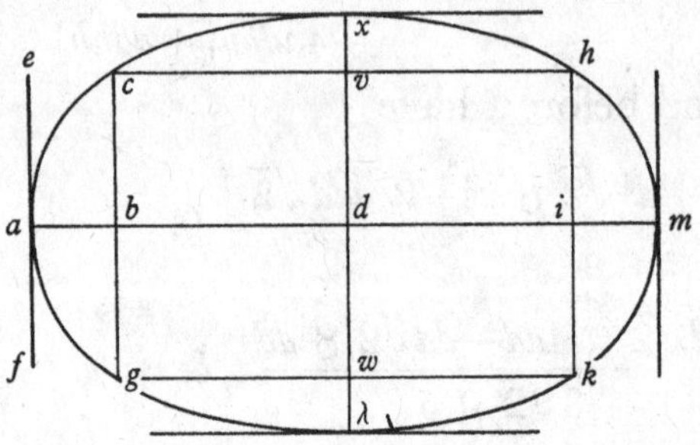

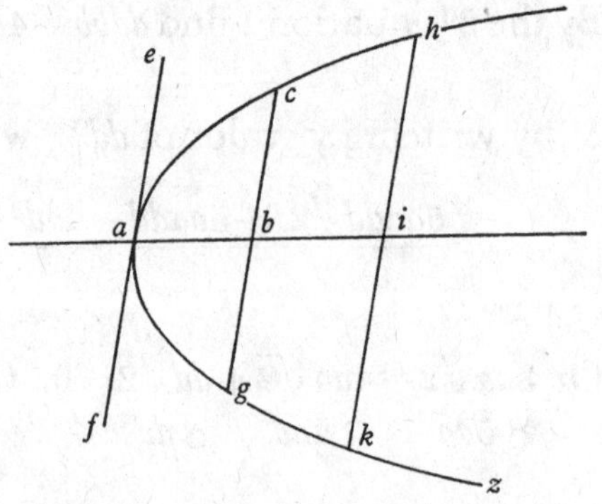

3 The Vertex of a crooked line is y^{t} point where y^{e} crooked line intersects the diameter or axis as at (*a*).[3]

4 The Asymptote[s] of crooked lines are such lines wch being produced both ways infinitely have noe least distance twixt y^{m} & y^{e} crooked line & yet doe noe where intersect it or touch it. as *dδ*, *dθ*.[4]

5 Those lines wch are limited on all sides as *acxkλ* are Ellipses of y^{e} first, 2^{d}, 3^{d}, 4th kind &c.

6 Those wch are not ellipses & have noe As$\bar{y}$ptotes are Parabolas of y^{e} first, 2^{d}, 3^{d}, 4th kind &c. as *zkah*.

7 Those wch have Asymptotes, are Hyperbolas of y^{e} 1st, 2^{d}, 3^{d}[,] 4th kind, &c: as (*rpon*) whose asymptotes are *βδ*, *γθ*.[5]

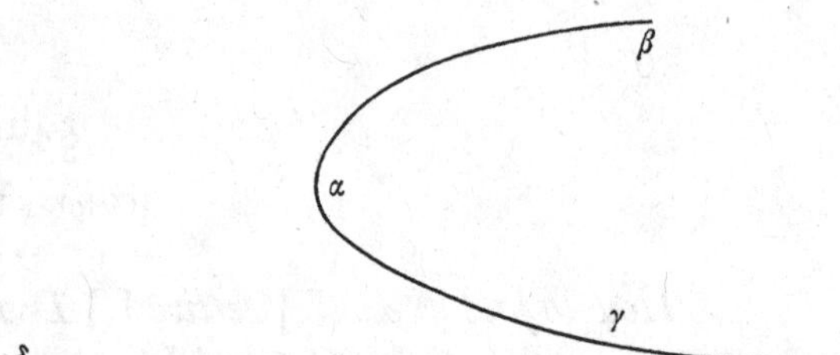

8 There are some lines of a middle nature twixt a Parab: & hyperb: haveing an Asymptote for one of its sides but none for y^{e} other as *βαγε*, one side *αγ* haveing y^{e} asympto[t]e *δε*, ye other side *αβ* haveing none.[6]

9 If two diameters of ye same Ellipsis be ordinately applyed ye one to the other ye shortest of them is called ye right diameter, ye longest ye tran[s]verse one. (as *am* & *xλ*).

10 If an Ellipsis have 2 axes (as *am* & *xλ*) ye longer is ye transverse axis (as *am*) ye shorter is ye right axis (as *xλ*).

The center of an Ellipsis is y^{t} point where two of its diameters intersect.

The center of two opposite Hyperbolas is y^{t} point where two of their diameters intersect one another or else where there[7] Asymptotes intersect.

(3) These two last definitions are apparently taken from Apollonius.

(4) Compare Apollonius: *Conics*, **2**, 1, 15, 17, 21.

(5) Definitions 5–7 are standard seventeenth-century generalizations of the conic. (Compare Wallis' *De Sectionibus Conicis:* Appendix: 101–8.)

(6) Newton probably has in mind Descartes' trident (*Geometria.* Liber II: 36–7).

Propositions. [1st] The lines ordinately applied to ye axis of a crooked line are parallell to ye tangent of ye crooked line at its vertex.

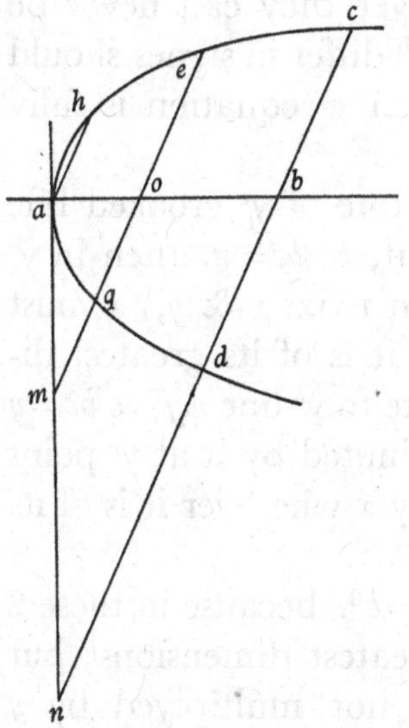

Demonstr. Suppose *chad* a Parab & *dc* (being ordinately applied to ye axis *ab*) not parallell to ye tangent *an* but to some other line as *ah*. if *dc* bee understood to move towards *a*[,] *db* continually decreaseth untill it vanish into nothing at ye conju[n]ction of ye points *a* & *d*. & since *cb* must be equall to *ah* at ye conjun[c]tion of ye point[s] *a* & *d*. it followeth yt *cb* cannot decrease so as to vanish into nothing at ye same time wch *bd* doth & therefore cannot allways be =(8) to *bd*.

Otherwise. if *dc* is not parallell to ye tangent *an* but to some other line as *ah*. Then *ab* doth not bisect all ye parallell lines (as *oe*) wch are terminated by ye crooked line *cad*. & therefore cannot bee its diam:(9)

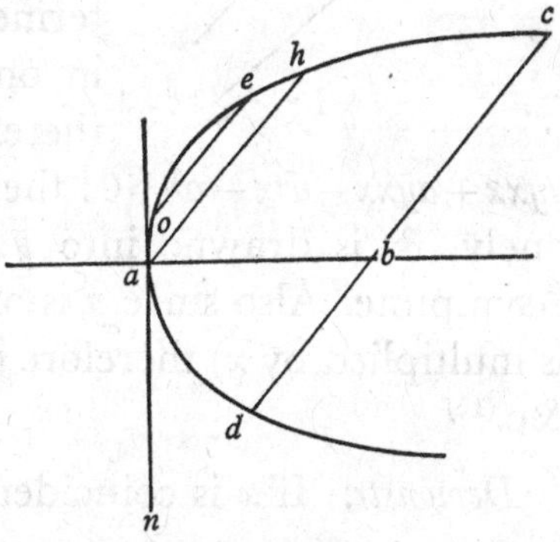

2dly. If *ad* is ye axis of a crooked line & $cb=y$, is ordinately applied to *ad*. yt is if $bc=ce=y$. Then y must be found noe where of odd dimensions in ye equation expressing ye nature of ye line *cad*. For (supposeing $y=bc=ce$ to be ye unknowne quantity) y hath 2 valors *bc* & *cd* equall to one another excepting yt ye one *bc* is affirmative, ye other *ce* is negative. wch two valors canot bee exprest by an equation in wch y is of od dimensions. for suppose $yy=aa$. yn is $\sqrt{aa}=a$, since $a\times a=aa$. & $\sqrt{aa}=-a$, since $-a\times -a=aa$. & $y=\sqrt{aa}$. therefore is $y=+a$, or $y=-a$. Soe if $y^4=a^4$. yn is $yy=aa$, & $y=a$ or $y=-a$.(10) but if $y^3=a^3$. yn $y=\sqrt{c:a^3}=a$. but not $y=\sqrt{c:-a^3}=-a$. soe if $y^5=a^5$. yn $y=a=\sqrt{qc:a^5}$

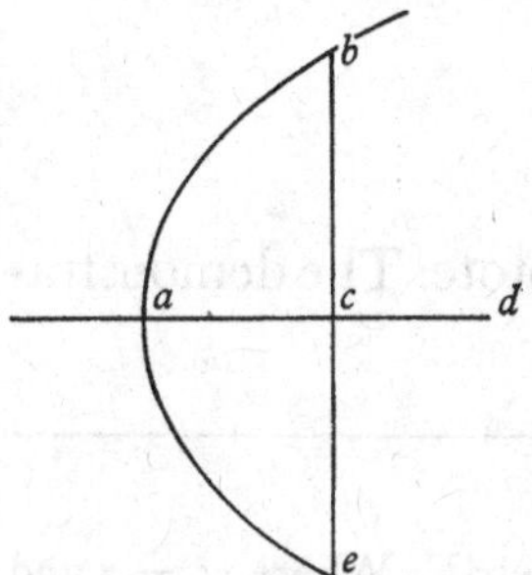

(7) Read 'their'. (8) Read 'equall'.
(9) Compare Apollonius: *Conics*, **1**, 17, 32.
(10) Neglecting the two imaginary roots $y=\pm\sqrt{-a^2}$.

but not $y=-a=\sqrt{qc:-a^5}$. The same reason is cogent in compound equations. as if $yy-2xy+xx=ax$. Then, $y=x$ ȣ $\sqrt{ax}$. where though y^e^ root a[11]$+\sqrt{ax}$ is affirmative & y^e^ roote a[11]$-\sqrt{ax}$ may bee negative yet they can never be equall in length, & though y^e^ 2 roots of an equation wch differ in signes should bee equally long yet y^t^ is when y^e^ equation is fully determined.[12]

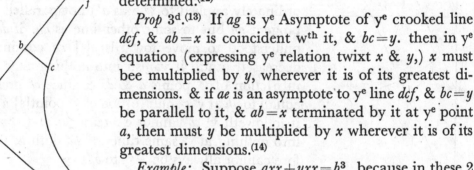

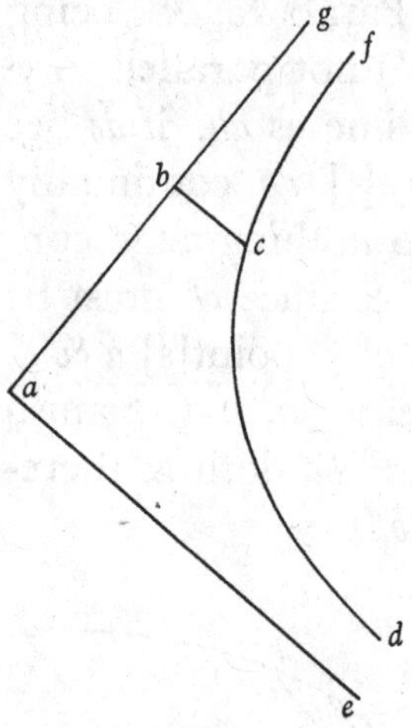

Prop 3^d^.[13] If ag is y^e^ Asymptote of y^e^ crooked line dcf, & $ab=x$ is coincident wth it, & $bc=y$. then in y^e^ equation (expressing y^e^ relation twixt x & y,) x must bee multiplied by y, wherever it is of its greatest dimensions. & if ae is an asymptote to y^e^ line dcf, & $bc=y$ be parallell to it, & $ab=x$ terminated by it at y^e^ point a, then must y be multiplied by x wherever it is of its greatest dimensions.[14]

Example: Suppose $axx+yxx=b^3$. because in these 2 termes $axx+yxx$, x is of its greatest dimensions; but in one of y^m^ (viz: axx) it is not multiplyed by y therefore x is not coincident wth ag y^e^ asymptote. If $yyxx+ayxx-a^3x-a^4=0$: then since x is of its greatest dimens: in $yyxx$ & $ayxx$ onely, & is drawne into y in both of y^m^ therefore x is coincident wth y^e^ Asymptote. Also since y is of its greatest dimensions in $xxyy$ onely, (wch terme is multiplied by x) therefore y is parallel to & x is terminated by an Asymptote &c.[15]

Demonstr: If x is coincident wth y^e^ Asymptote y^n^ $\frac{ee}{0}=x$ when $0=y$. i:e: x is infinite when y vannesheth.[16] Now suppose $yyxx+ayxx$[17]$=a^4$. y^n^ if $y=0$, it is $x=\frac{aa}{\sqrt{00+a0}}$. i:e: x is infinite. but if $yyxx+aax^2=a^4$. y^n^ if $y=0$, it is

$$x=\frac{aa}{\sqrt{00+aa}}=\text{ȣ}\, a.$$

soe y^t^ x is finite & therefore is no[t] coincident wth y^e^ asymptote. The demonstrati[on] of y^e^ other parte is likewise to be h[ad].

(11) Read 'x'.

(12) Newton first wrote 'when there is but one quantity considere[d]'. Where $ac = x$ and $cb = y$ are the Cartesian co-ordinates of the curve (b), the condition for $y = 0$ to be a diameter is that $y = \pm\sqrt{[f(x)]}$ or $y^2 = f(x)$, and so, by rationalizing the coefficients, $g(x, y^2) = 0$ (which can contain no odd powers of y).

(13) This is set below Prop: 4th in the text, but has been reordered to accord with Newton's numbering of his propositions.

(14) This condition for $y = 0$ to be an asymptote is sufficient only for conics and in general not necessary.

Prop: 4th. If x is of more dimensions in a quantity not multiplied by y then in one multiplied by it (as in $xx+xy=aa$.) y^n x is not parallell to y^e Asymptote *ag*. et e contr[a]. 2dly If y is of more dimensions in a quantity not multiplied by x y^n in one multiplied by it (as in $y^2=xy+aa$.) y^n y is not parallel to one of y^e lines Asymptotes. & e contra. Otherwise x & y are paralell to y^e Asymptotes of y^e line. et e contra.(18)

---

§5(1)

[October 1664]

October 1664 *Haveing the nature of a crooked line expresed in algebraicall termes to find its axes if it have any.*

Draw a line infinitely both ways [&] fix upon some point (as b,) for ye begining of ye unknowne quantitys (wch I call x). Then' reduce ye Equation to such an order (if it bee not already so) y^t x may be always found in ye line bc. wth one end fixed at b, & haveing y making(2) right angles wth it at y^e other end: y^t end of y wch is remote from x, describing y^e crooked line(3) wch may bee always done wthout any great difficulty. As may be perceived by these examples.

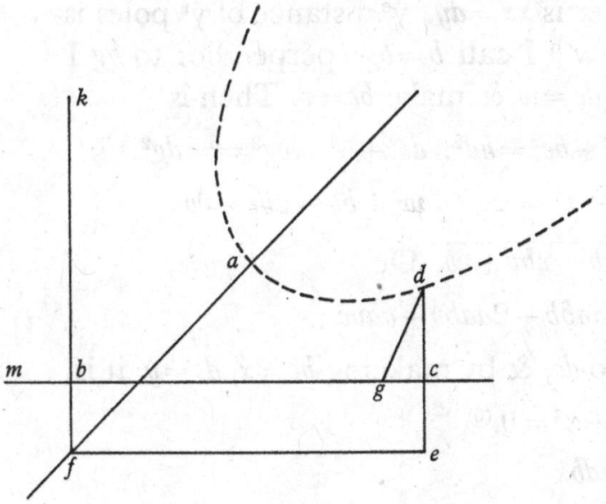

[*Example* 1st.] Suppose y^e given equation was $y^3=axx$. soe y^t $bg=x$. $gd=y$. dc being perpendicular to bc, & y^e angle dgc being given, y^e proportion twixt dg & gc is given, wch I supose as d to e. y^n is $d:e::y:dc=\frac{ey}{d}=w$. & $\frac{dw}{e}=y$.

$$gc^2 \begin{matrix} =gd^2-dc^2 \\ =yy-ww \end{matrix} = \frac{ddww-eeww}{ee}. \quad \text{or} \quad gc=\frac{w}{e}\sqrt{dd-ee}.$$

---

(15) In fact $y^2x^2+ayx^2-a^3x-a^4 = 0$ is the hyperbola-pair $(xy-a^2)(xy+a^2+ax) = 0$ with respective asymptotes $x = 0, y = 0$; $x = 0, y = -a$.

(16) Read 'vanisheth'.

(17) Newton has chosen to cancel the term $-a^3x$.

(18) The converse of Prop. 3.

(1) Add. 4004: 16v–18v, 20r. Newton generalizes the problem of finding axes to co-ordinate systems other than Cartesian.

(2) Newton has cancelled 'moving'.

(3) That is, the curve's co-ordinates are to be transformed into a perpendicular pair if they are not already so.

& $bc = v = x + \frac{w\sqrt{dd-ee}}{e}$. Or $\frac{ev - w\sqrt{dd-ee}}{e} = x$.

Therefore I write $\frac{d^3w^3}{e^3}$ for y^3. & $\frac{eevv - 2evw\sqrt{dd-ee} + ddww - eeww}{ee}$ for xx in y^e^ equation $y^3 = axx$, & soe I have this equation

$$\frac{d^3w^3}{e^3} = \frac{aeevv - 2aevw\sqrt{dd-ee} + addw^2 - aeew^2}{ee}.$$

wch expresseth y^e^ relation twixt w & v, y^t^ is twixt dc & bc, writeing therefore y for dc, & x for bc. I have this equation

$$0 = d^3y^3 + ae^3yy - aeddyy + 2aeexy\sqrt{dd-ee} - ae^3xx.^{(4)}$$

[*Example* 2^d^.] Soe if $x = bd$ turned about y^e^ pole b & $y = dg$ about y^e^ pole g describing y^e^ crooked line ad by y^r^[(5)] conjunction at y^r^ extremitys. & y^e^ equation expresing y^e^ relation wch they beare to one another is $xx = ay$. y^e^ distance of y^e^ poles is given wch I call $b = bg$. perpendic: to bg I draw $dc = w$ & make $bc = v$. Then is

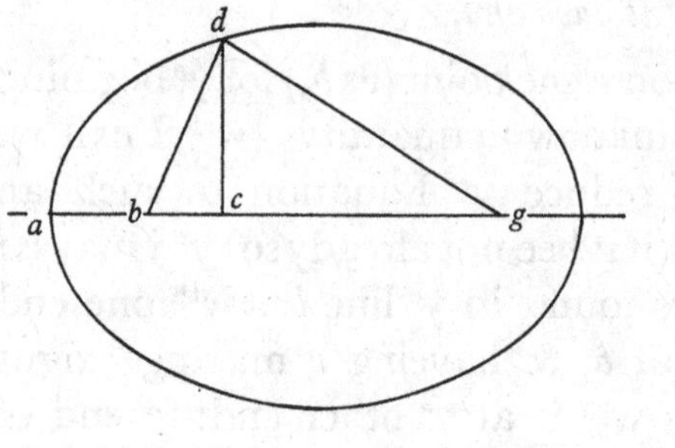

$$dc^2 + bc^2 = bd^2. \quad dc^2 + cg^2 = dg^2.$$

$$w^2 + v^2 = xx. \quad w^2 + bb - 2bv + vv = yy.$$

soe y^t^ for $xx = ay$ I write $w^2 + v^2 = a\sqrt{w^2 + bb - 2bv + vv}$. Or

$$w^4 + 2w^2v^2 + v^4 = aaw^2 + aabb - 2aabv + aavv.$$

wch expresseth y^e^ relation wch bc beareth to dc, & by makeing $bc = x$, $dc = y$, it is,

$$\begin{aligned} y^4 + 2xxyy \quad &+ x^4 = 0.^{(6)} \\ -aayy &+ 2adbx \\ &- aaxx \\ &- aabb \end{aligned}$$

(4) Newton constructs the transform from oblique Cartesian axes $bg = x$, $gd = y$ to the perpendicular pair $bc = v$, $cd = w$. Specifically, where $d\hat{g}c = \theta$ say, $\begin{Bmatrix} x = v - w\cot\theta \\ y = w\,\mathrm{cosec}\,\theta \end{Bmatrix}$, and so the cusped cubic $y^3 = ax^2$ becomes $w^3\,\mathrm{cosec}^3\,\theta = a(v - w\cot\theta)^2$.

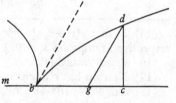

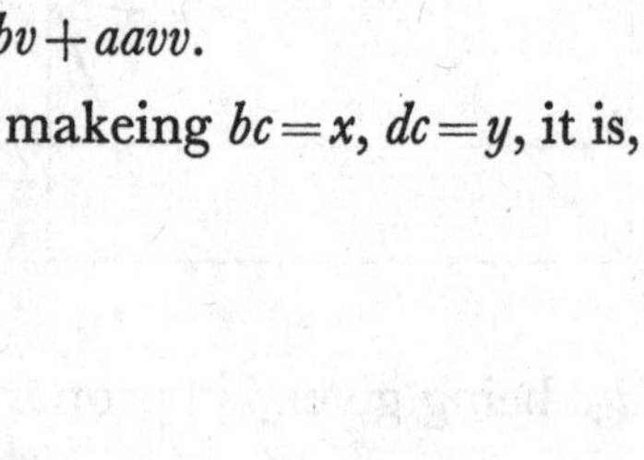

(5) Read 'their'.

(6) Newton reduces the bipolar co-ordinates $bd = x$, $gd = y$ (with $bg = b$) to the perpendicular Cartesian co-ordinates $bc = v$, $dc = w$: specifically

$$\begin{Bmatrix} x = \sqrt{[v^2 + w^2]} \\ y = \sqrt{[(b-v)^2 + w^2]} \end{Bmatrix}.$$

(Compare Descartes' treatment of his Ovals in *Geometria:* Liber II: 42 ff.)

Examp 3^{d}. If $bg=x$ be always in y^{e} line bc. & fd turning about y^{e} pole f & passing by y^{e} end of $bg=x$ wth its other end d describes y^{e} crooked line bdh, soe y^{t} calling gd y. $x=y$. y^{n} drawing ef & dc perpendicular to bh. $be=a$, & $ef=b$ are given. & I make $bc=v$. $dc=w$. y^{n} is

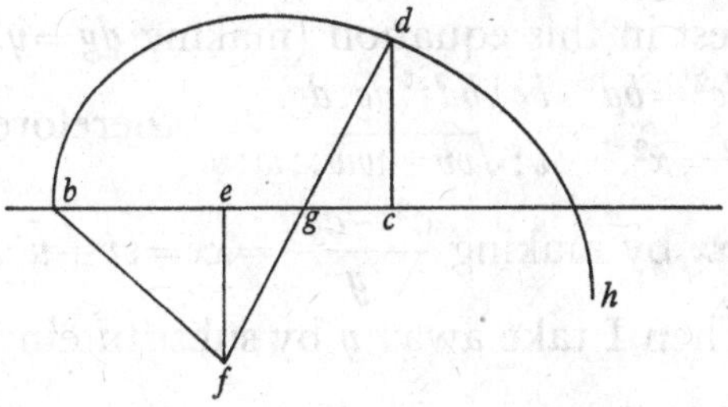

$eg=x-a$. $gc=v-x$. $\begin{matrix} ef: & eg:: & gc:cd.^{(7)} \\ b:x-a::v-x:w. \end{matrix}$

$bw=vx-av+ax-xx$. or by extracting y^{e} roote $x=\dfrac{+a+v}{2}$ ♉ $\sqrt{\dfrac{aa+2av+vv-4bw}{4}}$. againe $\begin{matrix} gc^2 & +dc^2=gd^2 \\ vv-2vx+xx+w^2=yy=xx. \end{matrix}$

Or $\dfrac{vv+w^2}{2v}=x=\dfrac{a+v}{2}$ ♉ $\sqrt{\dfrac{aa+2av+vv-4bw}{4}}$. & by transposeing $\dfrac{a+v}{2}$ to y^{e} other side & so squareing both ꝑts, $w^4-2avw^2=2av^3+v^4-4bwvz$. wch equation expresseth y^{e} relation twixt $w=dc$, & $v=bc$. & so by calling dc y, & bc x, it is, $y^4-2axyy+4bxxy-x^4-2ax^3=0$.$^{(8)}$

Example 4th. if $bd=x$ turnes about y^{e} pole b, & gd (a given line $=a$) slides upon bg wth one end & intersecting bd at right angles at y^{e} other end describes y^{e} crooked line bde by its intersection wth bd. then makeing $bc=v$, $dc=w$.

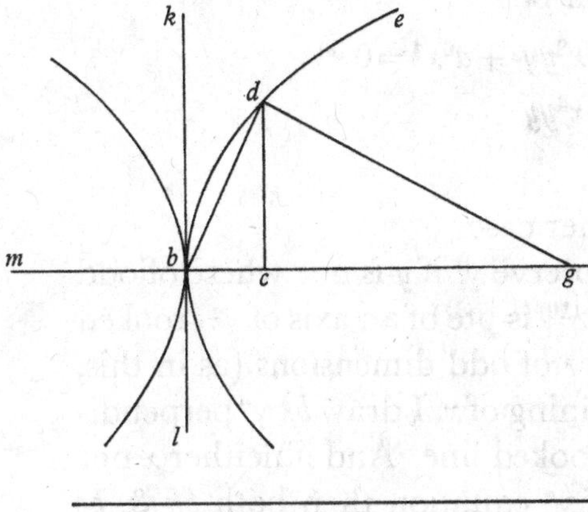

$bc^2+dc^2=bd^2$. $bc:bd::dc:dg$.

$v^2\ +w^2=x^2$. $v\ :x\ ::w\ :a$.

& $\dfrac{av}{w}=x$. therefore $vv+ww=\dfrac{aavv}{ww}$. or

(7) Read '$ef{:}fg{::}cd{:}gc$'. (The mistake vitiates Newton's further argument.)

(8) Correcting Newton's argument, we transform Newton's 'polar' co-ordinates to standard Cartesian ones by

$$\left\{\begin{matrix} x = v-mw \\ y = \sqrt{[1+m^2]}\,w \end{matrix}\right\},$$

where $\quad m = \cot \widehat{dgc} = \dfrac{x-a}{b} = \dfrac{v-a}{w+b}$.

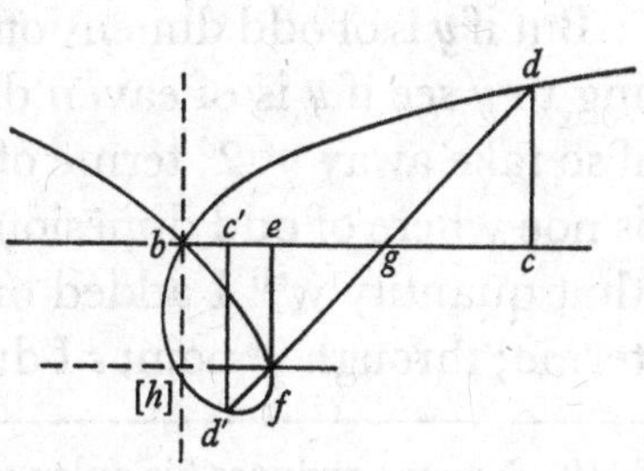

In Newton's example where $x = y$, then

$$\frac{x-a}{b} = \frac{v-x}{w} \quad \text{or} \quad x = \frac{bv+aw}{b+w},$$

with $\quad x^2 = y^2 = (v-x)^2+w^2, \quad$ or $\quad x = \dfrac{v^2+w^2}{2v}$.

Eliminating x and simplifying, we may deduce that $w^3+v^2w+bw^2-2avw-bv^2 = 0$, a nodal cubic with a double point at the origin b. (The general line $v-a = m(w+b)$ through the pole $f \equiv (a, -b)$ meets the cubic such that $(w+b)\,((1+m^2)\,w^2-(a+mb)^2) = 0$, that is, such that $w = -b$ (at f), or $\sqrt{[1+m^2]}\,w$ or $gd = y = d'g = bg = x$.)

$w^4+vvww=aavv$. & so by writeing x for v & y for w, I have y^e^ relation twixt $x=bc$ & $y=dc$ exprest in this equation $y^4+xxyy-aaxx=0$.

Or if y^e^ relation twixt bd & dg was exprest in this equation (making $dg=y$. $bd=x$.) $xxy+ayy=a^3$. then as before $\dfrac{bc^2+dc^2=bd^2.\ bc:bd::dc:dg.}{v^2+w^2=x^2.\quad v:\sqrt{vv+ww}::w:y.}$ therefore $y=\dfrac{w\sqrt{vv+ww}}{v}$. first therefore I take away xx by making $\dfrac{a^3-ayy}{y}=xx=vv+w^2$. or by ordering it $a^3-ayy-vvy-wwy=0$. Then I take away y by substituteing it[s] valor $\dfrac{w\sqrt{vv+w^2}}{v}$ into its roome & it will be

$$a^3\frac{-aw^2v^2-aw^4}{vv}=\frac{vvw\sqrt{vv+w^2}+w^3\sqrt{vv+w^2}}{v}.$$

& by □ing both ℘ts.

$$\begin{array}{l}a^6v^4-2a^4v^4w^2-2a^4vvw^4+aav^4w^4+2aavvw^6+aaw^8-v^8ww-3v^6w^4=0.\\ \qquad\qquad\qquad\qquad\qquad\qquad\qquad\qquad\qquad -3v^4w^6-vvw^8\end{array}$$

& by writeing x for v & y for w y^e^ equation will be

$$\begin{array}{lllll}aay^8+2aaxxy^6 & +aax^4y^4 & & -x^8yy & +a^6x^4=0.^{(9)}\\ -xxy^8 & -3x^4y^6 & -3x^6y^4-2a^4x^4yy & & \\ & -2a^4xxy^4 & & & \end{array}$$

The like may as easily be performed in any other case.

After y^e^ Equation is brought to this order Observe y^t^ if y is noe where of odd dimensions y^n^ y^e^ line bc (wch is coincident wth x)(10) is ℘te of an axis of y^e^ crooked line, as in y^e^ 2^d^ Example. And if x is noe where of odd dimensions (as in this, $a^4+yyxx=aax^2$.). Then from y^e^ point b at y^e^ begining of x, I draw bk y^e^ perpendicular to bc, wch is coincident wth y^e^ axis of y^e^ crooked line. And if neither x nor y bee of unequall dimensions in any terme of y^e^ equation then both bk & bc may bee taken for axes of y^e^ crooked line or lines whose natures are expressed by ye equation. As in ye 4th Example.

But if y is of odd dimensions in ye Equation then ordering ye equation according to y see if y is of eaven dimensions in ye first terme & x not found in ye 2^d^, if so take away y^e^ 2^d^ terme of y^e^ equation, & if there result an equation in wch y is noe where of odd dimēsions. Then I draw ce perpendicular to bc, & equall to that quantity wch I added or substracted fr[om] y, y^t^ I might take away y^e^ 2^d^ terme; through y^e^ point e I draw fe paralell to $b[c,]$ wch shall bee y^e^ axis of y^e^ line.

(9) Newton reduces his polar system by

$$\left\{\begin{array}{l}x = \sqrt{[v^2+w^2]}\\ y = (w/v)\sqrt{[v^2+w^2]}\end{array}\right\}$$

and applies the transform to the curves defined by $y = a$, and $y(x^2+ay) = a^3$.

(10) That is, $y = 0$.

[1st,] As in this Example, $yy+2ay-\frac{x^3}{a}=0$. Then to take away ye 2d terme

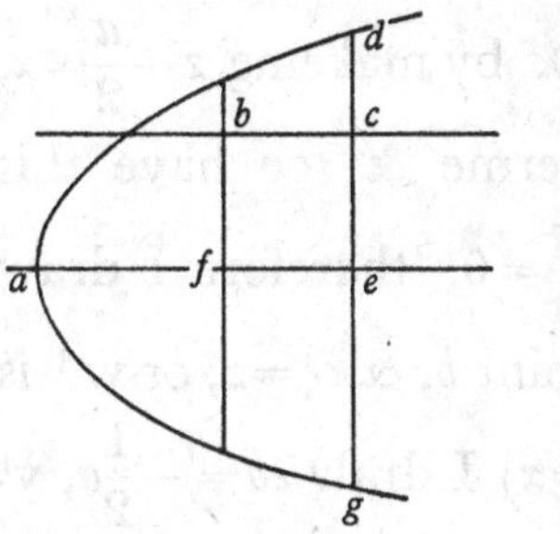

I make $z-a=y$. & soe I have, $zz * -aa\frac{-x^3}{a}=0$. in wch z is not of odd dimensions. Then drawing *bc* for *x*, *dc* for *y*, & *de* for *z*: or wch is ye same (since $z-a=y$) I make $ce=-a$ that is I draw *cd* & *ce* on 2 contrary sides of ye line *bc*. & then through ye point *e* I draw *ae* parallell to *bc* & make it ye axis of ye line *dag*.(11)

Example ye 2d.

$y^4-8ayyy+24aay^2-3axy^2-32a^3y$
$+12aaxy+16a^4=0.$

Then by makeing $z+2a=y$ I take away the 2d terme & ye Equation [is] $z^4 * -3axzz * +12a^3x=0$. in wch z is onely of eaven dimensions. Then I draw *bc* for *x*. *dc* for *y*. *de* for [*z*], or wch is ye same (sinc $z+2a=y$) I make $ec=2a$, yt is I draw *ec* & *dc* on ye same side of *bc* then through ye point *e* parallell to *bc* I draw *ea* for ye axis of ye lines *dac*[,] *dkhg*.(12)

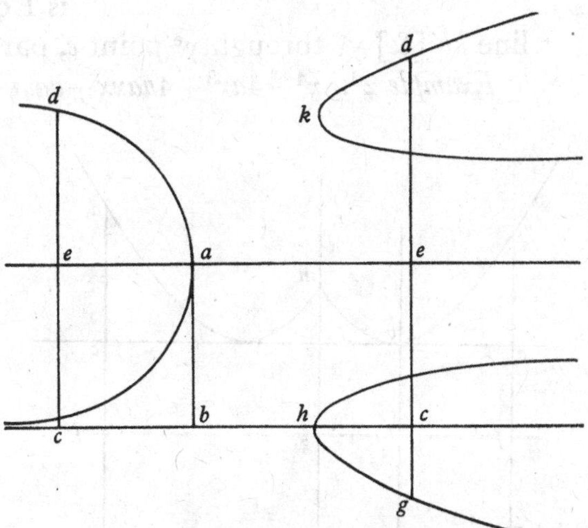

(11) The curve $ay^2+2a^2y-x^3=0$ is a pure (acnodal) cubic and the branches *d* and *g* should ultimately pass to infinity in the direction normal to the axis.

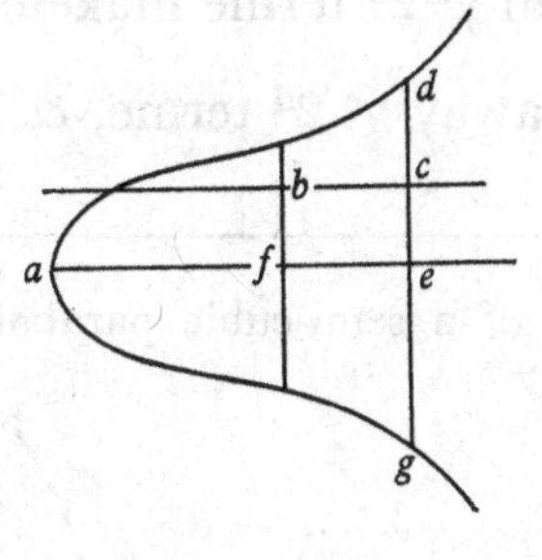

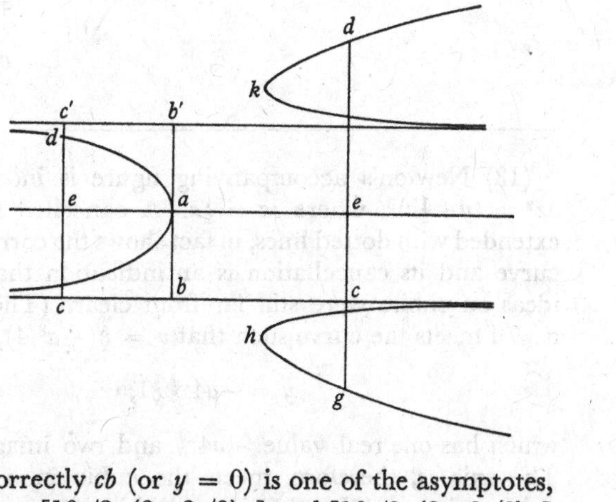

(12) Newton's figure is wrongly drawn. Correctly *cb* (or $y=0$) is one of the asymptotes, along with *c'b'* (or $y=4a$). The vertices *k*, *h* are $[16a/3, (2+2\surd 2)\,a]$ and $[16a/3, (2-2\surd 2)\,a]$ respectively. At one point the nearness of *c* to the branch *da* has misled him.

In like manner, if x is of odd dimensions in some terme of ye Equation, ye Axis bk perpendicular to $x=bc$ may bee found. As for Example.

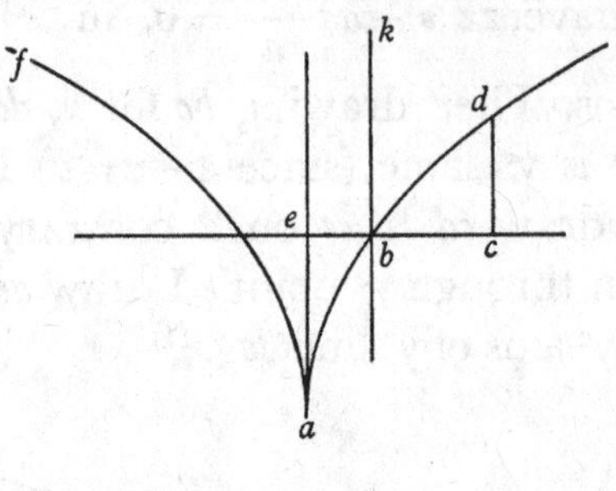

[1st] $xx+ax-\frac{y^3}{a}=0$. by makeing $z-\frac{a}{2}=x$, I take away ye 2d terme & soe have this equation $zz*\frac{-y^3}{a}-\frac{aa}{4}=0$. therefore I draw $bc=x$ from ye fixed point b, & $ce=z$, or wch is ye same (since $z-\frac{a}{2}=x$) I draw $eb=-\frac{1}{2}a$, yt is I draw eb & bc on two contrary sides of ye line kb [&] yn through ye point e, parallell to bk I draw ea ye axis of ye line.(13)

Example 2d. $x^4-4ax^3+4aaxx-aayy-aaby=0$. by makeing $x=z+a$ I have this eq: $z^4-2aazz-aayy-aaby+a^4=0$. In wch z is noe where of odd dimensions. therefore assumeing b for ye begining of x & makeing $bc=x$, & $ce=z$, or wch is ye same [if] I make $be=+a$, since $x=z+a$; that is if bc is affirmative I take be & bc on ye same side of ye line kb, otherwise I describe ym on contrary sides of it. then through ye point e parallell to bk I draw eg an axis of ye lines $dbmd$, & nhr. Againe I order ye Equation according to y & it is $a^2yy+aaby+\begin{array}{c}-z^4\\2aazz\\-a^4\end{array}=0$. & soe since x(14) is not in ye 2d terme makeing $v-\frac{b}{2}=y$, I take away ye 2d terme, & it

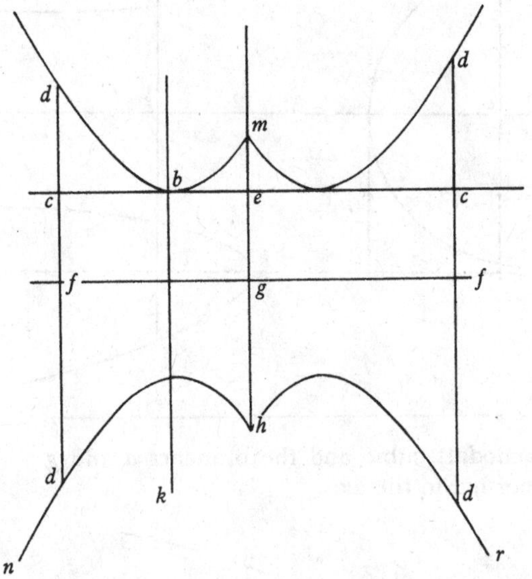

(13) Newton's accompanying figure is incorrect, being that of a semi-cubic parabola $az^2=(y+\frac{1}{2}a)^3$, where $ea=\frac{1}{2}a$. A cancelled figure, here extended with dotted lines, in fact shows the correctly drawn curve and its cancellation is an indication that Newton's ideas on cubics were still far from clear. (The axis ea or $z=0$ meets the curve such that $y=\sqrt[3]{(-a^3/4)}$, or

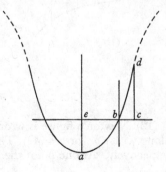

$$y=-a4^{-\frac{1}{3}}\sqrt[3]{1},$$

which has one real value $-a4^{-\frac{1}{3}}$ and two imaginary ones. The axis ea, therefore, meets the cubic in a single real point and not, as Newton 'corrects' his text, in a triple (real) point.)

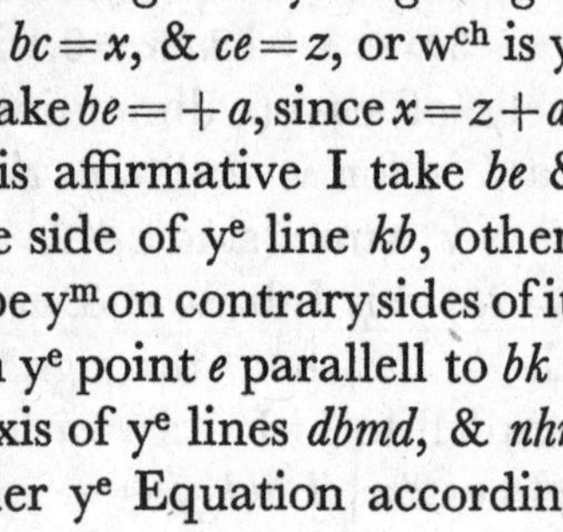

is $aavv * \begin{matrix} -z^4 \\ \dfrac{-aabb}{4} \\ -a^4 \end{matrix} + 2aazz = 0$. Therefore I draw $ec = z$, $dc = y$, & $df = v$. or wch differs not $\left(\text{since } v - \frac{b}{2} = y\right)$ I make $cf = -\frac{b}{2}$ & through y^e^ point f parallell to ce I draw fgf for another axis of y^e^ lines $dbmd$, & $ndhdr$.(15)

But if y^e^ unknowne quantity (x or y) is of odd dimensions in y^e^ first terme or if both y^e^ unknowne quantitys are in y^e^ 2^d^ terme, or if by this meanes y^e^ equation is irreducible to such a forme y^t^ x, or y, or both of y^m^ bee of odd dimensions noe where in y^e^ equat: Then try to find y^e^ axes by y^e^ following method. Observing by y^e^ way y^t^ If $+x$ begins at y^e^ one point b & extends towards c in y^e^ line sr y^n^ $-x$ is taken y^e^ contrary way towards s, & all y^e^ affirmative lines parallell to sr are drawne y^e^ same way w^ch^ $+x$ is but y^e^ negative lines parallell to sr are drawn y^e^ same way w^th^ $-x$. as if from y^e^ point m I must draw a line $= a$, I draw it towards n but if from y^e^ s[a]me point m I must draw a line $= -a$ I draw it towards l. soe if from y^e^ point d I must draw $d\lambda = +b$, y^n^ I draw it towards θ, but if $d\lambda = -b$ then I draw it towards γ. Againe if $+y$ is drawne toward p from y^e^ line sr, y^n^ $-y$ is drawne from y^e^ same line sr y^e^ contrary way towards o, & those lines wch are affected wth an affirmative signe & are paralell to y they are drawne y^e^ same way wch y is but those lines wch are negative are drawn y^e^ contrary way. as if $a = a\pi$ then I draw $a\pi$ towards e but if $-a = a\pi$ y^n^ I draw it towards t. soe if $v\mu = +d$ y^n^ I draw it from v towards δ, if $v\mu = -d$, I draw it towards w.

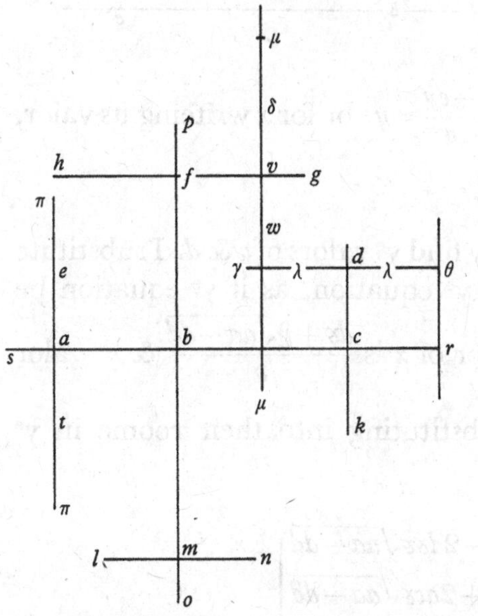

A generall rule to find y^e^ axes of any line.

Suppose $bc = x$. $cd = y$. & kg to be y^e^ axis. y^n^ paralell to y from y^e^ point b to y^e^ axis kg draw $bf = c$. from d y^e^ end of y, perpendicular to kg draw $dh = \gamma$. & make

(14) Read 'z'.

(15) In Newton's argument $bc = x$ and $cd = y$, with b the origin. The vertices m and h are the points $(a, \pm\sqrt{[a^2 + \frac{1}{4}b^2]} - \frac{1}{2}b)$ and cbc (or $y = 0$) is tangent at b.

$fh = \varkappa$. & suppose $fe:fg::d:e$. yⁿ is $fg = \frac{ex}{d}$.

$dg^2 - dh^2 = gh^2$.

& $d:e::dh = \gamma : \frac{e\gamma}{d} = dg$. $\frac{ee\gamma\gamma}{dd} - \gamma\gamma = gh^2$.

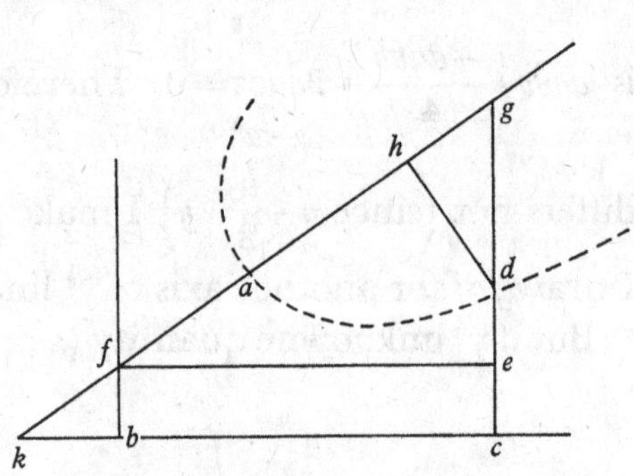

$gh = \frac{\gamma\sqrt{ee-dd}}{d}$. $gh = \frac{-fh+fg}{-\varkappa + \frac{ex}{d}} = \frac{\gamma\sqrt{ee-dd}}{d}$.

therefore $\frac{d\varkappa + \gamma\sqrt{ee-dd}}{e} = x$.

Againe $ge + ec - dg = dc$, that is $\sqrt{\frac{eexx}{d[d]} - xx} + c - \frac{e\gamma}{d} = y$. or for x writeing its valor,

$y = \frac{ce - d\gamma + \varkappa\sqrt{ee-dd}}{e}$.(16)

Now assumeing any quantity for e, y^t I may find y^e valors of c & d. I substitute these valors of x & y into theire roome in y^e equation. as if y^e equation be $x^2 - 2xy + ay + yy = 0$. by making $e = a$ y^e valor of x is $\frac{d\varkappa + \gamma\sqrt{aa-dd}}{a}$ & y^e valor of y is $\frac{ac - d\gamma + \varkappa\sqrt{aa-dd}}{a}$. wch 2 valors substituting into their roome in y^e equation, their(17) results

$$\left.\begin{array}{lr} aa\gamma\gamma & +4dd\varkappa\gamma - 2d\varkappa\varkappa\sqrt{aa-dd} \\ +2d\gamma\gamma\sqrt{a^2-d^2} - 2aa\varkappa\gamma & +2ac\varkappa\sqrt{aa-dd} \\ -a^2d\gamma & +aa\varkappa\sqrt{aa-dd} \\ -2acd\gamma & +aa\varkappa\varkappa \\ -2ac\gamma\sqrt{a^2-d^2} & -2dac\varkappa \\ & +a^3c \\ & +aacc \end{array}\right\} = 0.$$

Prop 2^d. Now y^t I may have an equation in wch γ is of eaven dimensions onely I suppose y^e 2^d terme $= 0$ & soe have this equation

$$4dd\varkappa\gamma - 2aa\varkappa\gamma - aad\gamma - 2acd\gamma - 2ac\gamma\sqrt{aa-dd} = 0.$$

(16) Much as before, Newton constructs the transform between the two sets of perpendicular Cartesian co-ordinates $bc = x$, $cd = y$ and $fh = \varkappa$, $hd = \gamma$: specifically, where

$$g\hat{f}e = \theta\left(\text{that is, } \frac{d}{e} = \cos\theta\right), \quad \left\{\begin{array}{l} x = \varkappa\cos\theta + \gamma\sin\theta \\ y - c = \varkappa\sin\theta - \gamma\cos\theta \end{array}\right\}.$$

(17) Read 'there'. (18) Read 'termes'.

(19) That is, the coefficients of the two powers of $\varkappa$, $\varkappa^1 = \varkappa$ and $\varkappa^0 = 1$.

& y[t] y[e] teres[18] in this feigned equation may destroy one another I order it according to χ & soe suppose each terme[19] $=0$. & so I have these equations

$$4dd\chi\gamma-2aa\chi\gamma=0,$$

& $-aad\gamma-2acd\gamma-2ac\gamma\sqrt{aa-dd}=0$. by y[e] first I find $2dd=aa$, or $d=\frac{a}{\sqrt{2}}$. by y[e] 2[d], $aad+2acd+2ac\sqrt{aa-dd}=0$. & by substituteing y[e] valor of d into its roome I find $\frac{a^3+4aac}{\sqrt{2}}=0$. or $c=\frac{-a}{4}$. therefore from b perpendic: to bc I draw $bf=\frac{-a}{4}$. through y[e] point f paralell to bc I draw $fe=\frac{a}{\sqrt{2}}$ & since $f[g]=a$ therefore I draw

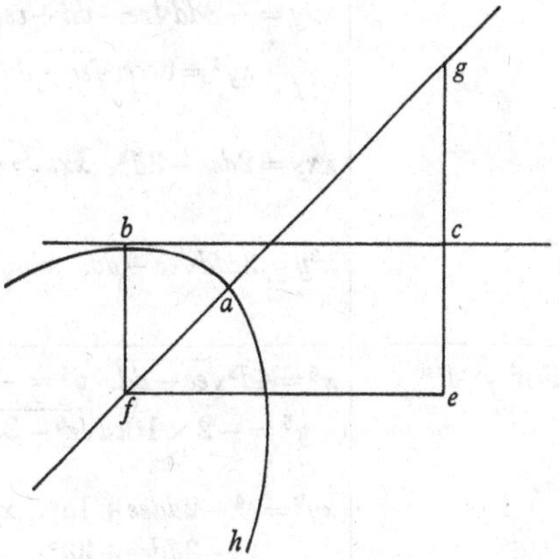

$$ge=\sqrt{aa-\frac{aa}{2}}=\frac{a}{\sqrt{2}}.$$

& lastly throu[g]h y[e] points f & g I draw fg y[e] axis of y[e] crooked line bah.

But since there is noe use of those termes in w[ch] γ is of eaven dimensions y[e] Calculation will bee much abreviated by this following table.[20]

For y[e] first equation of y[e] first sort	$x=\sqrt{ee-dd}$. $y=-d$. $yy=-2cd$. $y^3=-3ccd$. $y^4=-4c^3d$. $y^5=-5c^4d$. $y^6=-6c^5d$. &c. $xy=c\sqrt{ee-dd}$. $xyy=cc\sqrt{ee-dd}$. $xy^3=c^3\sqrt{ee-dd}$. $xy^4=c^4\sqrt{ee-dd}$. $xy^5=c^5\sqrt{ee-dd}$. &c.
For y[e] 2[d]	$xx=2d\sqrt{ee-dd}$. $yy=\left.\begin{matrix}-2\\-1\times2\end{matrix}\right\}d\sqrt{ee-dd}$. $y^3=\left.\begin{matrix}-6\\-2\times3\end{matrix}\right\}cd\sqrt{ee-dd}$. $y^4=\left.\begin{matrix}-12\\-3\times4\end{matrix}\right\}ccd\sqrt{ee-dd}$. $y^5=\left.\begin{matrix}-20\\-4\times5\end{matrix}\right\}c^3d\sqrt{ee-dd}$. $y^6=\left.\begin{matrix}-30\\-5\times6\end{matrix}\right\}c^4d\sqrt{ee-dd}$. $xy=-2dd+ee$. $xyy=-4cdd+2cee$. $xy^3=-6ccdd+3ccee$. $xy^4=-8c^3dd+4c^3ee$. $xy^5=5c^4ee-10[e]^4dd$. &c. $xxy=2cd\sqrt{ee-dd}$. $xxyy=2ccd\sqrt{ee-dd}$. $xxy^3=2c^3d\sqrt{ee-dd}$. $xxy^4=2c^4d\sqrt{ee-dd}$. $xxy^5=2c^5d\sqrt{ee-dd}$. &c.

(20) Newton tabulates the coefficients of γ and γ^3 (his 'first' and 'second sorts') in the transformed equation, where

$$\left\{\begin{matrix} x = d(\chi/e)+\sqrt{[e^2-d^2]}\,(\gamma/e) \\ y = c-d(\gamma/e)+\sqrt{[e^2-d^2]}\,(\chi/e)\end{matrix}\right\}.$$

He begins by calculating straightforwardly the coefficients of γ in x^iy^j, $i, j = 0, 1, 2, \ldots$ (here omitted), but quickly cancels his tabulation for the less cumbrous one which follows, where the kth equations of the first and second sorts evaluate the coefficients in x^iy^j of $(\chi^{k-1}\gamma)/e^k$ and $(\chi^{k-1}\gamma^3)/e^{k+2}$ respectively.

For y^e^ 3^d^

$$x^3=3dd\sqrt{ee-dd}.\quad y^3=-3dee+3d^3.\quad y^4=2\times 6cd^3-2\times 6cdee.$$
$$y^5=3\times 10ccd^3-3\times 10ccdee.\quad y^6=4\times 15c^3d^3-4\times 15c^3dee.$$
$$xyy=-3dd\sqrt{ee-dd}+ee\sqrt{ee-dd}.\quad xy^3=3cee\sqrt{ee-dd}-9cdd\sqrt{ee-dd}.$$
$$xy^4=6ccee\sqrt{ee-dd}-18ccdd\sqrt{ee-dd}.\quad xy^5=\begin{array}{l}10c^3ee\\-30c^3dd\end{array}\sqrt{ee-dd}.\ \text{\&c.}$$
$$xxy=2dee-3d^3.\quad xxyy=4cdee-6cd^3.\quad xxy^3=6ccdee-9ccd^3.$$
$$xxy^4=8c^3dee-12c^3d^3.$$
$$x^3y=3cdd\sqrt{ee-dd}.\quad x^3yy=3ccdd\sqrt{ee-dd}.\quad x^3y^3=3c^3d^2\sqrt{ee-dd}.$$
$$x^3y^4=3c^4dd\sqrt{ee-dd}.$$

For y^e^ 4th

$$x^4=4d^3\sqrt{ee-dd}.\quad y^4=-1\times 4d\sqrt{e^6-3e^4dd+3eed^4-d^6}.$$
$$y^5=-2\times 10cd\sqrt{e^6-3e^4dd+3eed^4-d^6}.\quad y^6=-3\times 20ccd\sqrt{e^6\ \text{\&c:}}.$$
$$y^7=-4\times 35c^3d\sqrt{e^6\ \text{\&c.}}$$
$$xy^3=\begin{array}{l}e^4-2ddee+1d^4\\ \quad -3ddee+3d^4\end{array}.\quad xy^4=\begin{array}{l}4ce^4-8cddee+4cd^4\\ \qquad\quad -12\qquad +12\end{array}.$$
$$xy^5=\begin{array}{l}10cce^4-20ccddee+10ccd^4\\ \qquad\qquad -30\qquad\quad +30\end{array}.\quad xy^6=\begin{array}{l}20c^3e^4-40c^3ddee+20c^3d^4\\ \qquad\qquad -60\qquad\quad +60\end{array}.\ \text{\&c.}$$
$$xxyy=\begin{array}{l}1\times 2dee\\ \quad -4d^3\end{array}\sqrt{ee-dd}.\quad xxy^3=\begin{array}{l}2\times 3cdee\\ \quad -12cd^3\end{array}\sqrt{ee-dd}.\quad xxy^4=\begin{array}{l}+12ccdee\\ \quad -24ccd^3\end{array}\sqrt{ee-dd}.$$
$$xxy^5=\begin{array}{l}+20c^3dee\\ \quad -40c^3d^3\end{array}\sqrt{ee-dd}.$$
$$x^3y=3ddee-4d^4.\quad x^3yy=6cddee-8cd^4.\quad x^3y^3=9ccddee-12ccd^4.$$
$$x^3y^4=12c^3ddee-16c^3d^4.$$
$$x^4y=4cd^3\sqrt{ee-dd}.\quad x^4yy=4ccd^3\sqrt{ee-dd}.\quad x^4y^3=4c^3d^3\sqrt{ee-dd}.\quad x^4y^4=4c^4d^3\sqrt{ee-dd}.$$

For y^e^ 5^t^

$$x^5=5d^4\sqrt{ee-dd}.\ y^5=-1\times 5de^4+10d^3ee-5d^5.\ y^6=-2\times 15cde^4+60cd^3ee-30cd^5.$$
$$y^7=-3\times 35ccde^4+210ccd^3ee-105ccd^5.\quad y^8=-4\times 70c^3de^4+560c^3d^3ee-280c^3d^5.\ \text{\&c.}$$
$$xy^4=\begin{array}{l}e^4-2ddee+1d^4\\ \quad -4ddee+4d^4\end{array}\ \text{in}\ \sqrt{ee-dd}.\quad xy^5=\begin{array}{l}5ce^4\ -10cddee+5cd^4\\ \qquad -20cddee+20cd^4\end{array}\ \text{in}\ \sqrt{ee-dd}.$$
$$xy^6=\begin{array}{l}15cce^4-30ccddee+15ccd^4\\ \qquad -60ccddee+60ccd^4\end{array}\ \text{in}\ \sqrt{ee-dd}.$$
$$xy^7=\begin{array}{l}35c^3e^4\ -70c^3ddee+35c^3d^4\\ \qquad -140c^3ddee+140c^3d^4\end{array}\ \text{in}\ \sqrt{ee-dd}.$$
$$xxy^3=\begin{array}{l}2de^4-4d^3ee+2d^5\\ \quad -3d^3ee+3d^5\end{array}.\quad xxy^4=\begin{array}{l}2\times 4cde^4-2\times 8cd^3ee+8cd^5\\ \qquad -12cd^3ee+12cd^5\end{array}.$$
$$xxy^5=\begin{array}{l}20ccde^4-40ccd^3ee+2\times 10ccd^5\\ \qquad -30ccd^3ee+30ccd^5\end{array}.\quad xxy^6=\begin{array}{l}40c^3de^4-80c^3d^3ee+2\times 20c^3d^5\\ \qquad -60c^3d^3ee+60c^3d^5\end{array}.$$
$$x^3yy=\begin{array}{l}1\times 3ddee\\ \quad -5d^4\end{array}\sqrt{ee-dd}.\quad x^3y^3=\begin{array}{l}3\times 3ddcee\\ \quad -15cd^4\end{array}\sqrt{ee-dd}.\quad x^3y^4=\begin{array}{l}6\times 3ccddee\\ \quad -6\times 5ccd^4\end{array}\sqrt{ee-dd}.$$
$$x^3y^5=\begin{array}{l}10\times 3c^3ddee\\ \quad -50c^3d^4\end{array}\sqrt{ee-dd}.\quad x^3y^6=\begin{array}{l}15\times 3c^4ddee\\ \quad -75c^4d^4\end{array}\sqrt{ee-dd}.$$
$$x^4y=1\times 4d^3ee-5d^5.\quad x^4yy=2\times 4cd^3ee-10cd^5.\quad x^4y^3=3\times 4ccd^3ee-15ccd^3ee.$$
$$x^4y^4=\begin{array}{l}4\times 4c^3d^3ee\\ \quad -20c^3d^5\end{array}.\quad x^4y^5=\begin{array}{l}5\times 4c^4d^3ee\\ \quad -25c^4d^5\end{array}.$$
$$x^5y=5cd^4\sqrt{ee-dd}.\quad x^5yy=5ccd^4\sqrt{ee-dd}.\quad x^5y^3=5c^3d^4\sqrt{ee-dd}.$$

For yᵉ 6ᵗ

$y^6 = -6\times 1d\sqrt{e^{10}-5e^8dd+10e^6d^4-10e^4d^6+5eed^8-d^{10}}$.

$y^7=-6\times 7dc\sqrt{e^{10}\ \&c:}$. $y^8=-6\times 28ccd\sqrt{e^{10}\ \&c:}$. $y^9=-6\times 84c^3d\sqrt{e^{10}\ \&c:}$.

$xy^5=1\times 1e^6-3dde^4+3d^4ee-d^6$. $xy^6=1\times 6e^6-18cdde^4+18cd^4ee-6cd^6$.

$-5dde^4+10d^4ee-5d^6$ $\qquad -6\times 5cdde^4+60cd^4ee-30cd^6$

$xy^7=1\times 21cce^6-63cce^4dd+63cceed^4-21ccd^6$. $xy^8=1\times 56c^3e^6$. &c.

$-21\times 5cce^4dd+210cceed^4-105ccd^6$

$xxy^4=2\times 1de^4-4d^3ee+2d^5\sqrt{ee-dd}$. $xxy^5=2\times 5cde^4-20cd^3ee+10cd^5$ in $\sqrt{ee-dd}$.

$-4d^3ee+4d^5$ $\qquad -5\times 4[c]d^3ee+20cd^5$

$xxy^6=2\times 15ccde^4-60ccd^3ee+30ccd^5 \quad \times\sqrt{ee-dd}$.

$-4\times 15ccd^3ee+60ccd^5$

$x^3y^3=3\times 1dde^4-6d^4ee+3d^6$. $x^3y^4=3\times 4cdde^4-24cd^4ee+12cd^6$.

$-3d^4ee+3d^6$ $\qquad -3\times 4cd^4ee+12cd^6$

$x^3y^5=3\times 10ccdde^4-60ccd^4ee+30ccd^6$.

$-30ccd^4ee+30ccd^6$

$x^4yy=4\times 1dddee-4d^5$ in $\sqrt{ee-dd}$. $x^4y^3=4\times 3d^3cee-12cd^5$ in $\sqrt{ee-dd}$.

$-2d^5$ $\qquad -6cd^5$

$x^4y^4=4\times 6ccd^3ee-24ccd^5\sqrt{ee-dd}$. $x^4y^5=4\times 10c^3d^3ee$ &c.

$-12ccd^5$

$x^5y=1\times 5d^4ee-5d^6$. $x^5yy=5\times 2cd^4ee-10cd^6$. $x^5y^3=5\times 3ccd^4ee-15ccd^6$.

$-1d^6$ $\qquad -2cd^6$ $\qquad -3ccd^6$

$x^5y^4=5\times 4c^3d^4ee-20ccd^6$.

$-4ccd^6$

$x^6=1\times 6d^5\sqrt{ee-dd}$. $x^6y=6cd^5\sqrt{ee-dd}$. $x^6yy=6ccd^5\sqrt{ee-dd}$.

$x^6y^3[=]6c^3d^5\sqrt{ee-dd}$.

&c. &c.

For yᵉ first equation of yᵉ seacond Sort

$y^3=-d^3$. $y^4=-4cd^3$. $y^5=-10ccd^3$. $y^6=-20c^3d^3$. $y^7=-35c^4d^3$.

$y^8=-56c^5d^3$.

$xyy=dd\sqrt{ee-dd}$. $xy^3=cdd\sqrt{ee-dd}$. $xy^4=ccdd\sqrt{ee-dd}$. $xy^5=c^3dd\sqrt{ee-dd}$.

$xxy=d^3-eed$. $xxyy=2cd^3-2ceed$. $xxy^3=3ccd^3-3cceed$.

$xxy^4=4c^3d^3-4cceed$.

$x^3=ee-dd$ in $\sqrt{ee-dd}$. $x^3y=cee-cdd\sqrt{ee-dd}$. $x^3yy=ccee-ccdd$ in $\sqrt{ee-dd}$.

$x^3y^3=c^3\sqrt{e^6-3e^4dd+3eed^4-d^6}$.

For yᵉ seacond

$y^4=-1\times 4d^3\sqrt{ee-dd}$. $y^5=-5\times 4cd^3\sqrt{ee-dd}$. $y^6=-15\times 4ccd^3\sqrt{ee-dd}$.

$y^7=-35\times 4c^3d^3\sqrt{ee-dd}$.

$xy^3=1\times 3ddee-3d^4$. $xy^4=4\times 3cddee \quad -12cd^4$. $xy^5=10\times 3ccddee-30ccd^4$.

$-1d^4$ $\qquad -4\times 1cd^4$ $\qquad -10ccd^4$

$xy^6=20\times 3c^3ddee-60c^3d^4$.

$-20c^3d^4$

$xxyy=-1\times 2dee+2d^3$ in $\sqrt{ee-dd}$. $xxy^3=-3\times 2cdee+6cd^3$ [in $\sqrt{ee-dd}$].

$+2d^3$ $\qquad +6cd^3$

$xxy^4=-6\times 2ccdee+12ccd^3$ [$\sqrt{ee-dd}$]. $x^2y^5=-10\times 2c^3dee+20c^3d^3$ [$\sqrt{ee-dd}$].

$+12ccd^3$ $\qquad +20c^3d^3$

$x^3y=1\times 1e^4-2ddee+d^4$. $x^3yy=1\times 2ce^4-4ceedd+2d^4c$.

$-3ddee+3d^4$ $\qquad -6ceedd+6cd^4$

$$x^3y^3 = 3cce^4 - 6cceedd + 3ccd^4. \quad x^3[y^4] = 4c^3e^4 \ \&c.$$
$$ - 9cceedd + 9ccd^4$$

$$x^4 = 4eed - 4d^3 \text{ in } \sqrt{ee-dd}. \quad x^4y = 4ceed - 4cd^3 \text{ in } \sqrt{ee-dd}.$$

$$x^4yy = -4ccd\sqrt{e^6 - 3e^4dd + 3eed^4 - d^6}. \quad x^4y^3 = -4c^3d\sqrt{e^6} \ \&c:.$$

For y^e^ 3^d^

$$y^5 = -1 \times 10eed^3 + 10d^5. \ y^6 = -6 \times 10ceed^3 + 60cd^5. \ y^7 = -21 \times 10cceed^3 + 210ccd^5.$$
$$y^8 = -56 \times 10c^3eed + 560c^3d^5.$$

$$xy^4 = 1 \times 6eedd - 6d^4 \text{ in } \sqrt{ee-dd}. \quad xy^5 = 5 \times 6ceedd - 30cd^4 \text{ in } \sqrt{ee-dd}.$$
$$-4d^4 \qquad\qquad -20cd^4$$

$$xy^6 = 15 \times 6cceedd - 90ccd^4 \text{ in } \sqrt{ee-dd}.$$
$$-60ccd^4$$

$$xxy^3 = -1 \times 3e^4d + 6eed^3 - 3d^5.$$
$$+6eed^3 - 6d^5 \quad xxy^4 = -4 \times 3ce^4d + 24ceed^3 - 12cd^5.$$
$$+24ceed^3 - 24cd^5$$
$$-4cd^5$$

$$xxy^5 = -10 \times 3cce^4d + 60cceed^3 - 30ccd^5$$
$$+60cceed^3 - 60ccd^5$$
$$-10ccd^5$$

$$x^3yy = e^4 - 2ddee + d^4 \text{ in } \sqrt{ee-dd}. \quad x^3y^3 = 3 \times 1ce^4 - 6cddee + 3cd^4$$
$$-6ddee + 6d^4 \qquad\qquad -18cddee + 18cd^4$$
$$+3d^4 \qquad\qquad +9cd^4$$

$$x^3y^4 = 6cce^4 - 12ccddee + 6ccd^4 \text{ in } \sqrt{ee-dd}.$$
$$-36ccddee + 36ccd^4$$
$$+18ccd^4$$

$$x^4y = 4de^4 - 8d^3ee + 4d^5. \quad x^4yy = 2 \times 4cde^4 - 16cd^3ee + 8cd^5.$$
$$-6d^3ee + [6]d^5 \qquad\qquad -12cd^3ee + 12cd^5$$

$$x^4y^3 = 3 \times 4ccde^4 - 24ccd^3ee + 12ccd^5.$$
$$-18ccd^3ee + 18ccd^5$$

$$x^5 = 10eedd - 10d^4 \text{ in } \sqrt{[ee]-dd}. \quad 10ceedd - 10cd^4 \text{ in } \sqrt{ee-dd} = x^5y.$$
$$x^5yy = 10ccdd\sqrt{e^6 - 3e^4dd + 3eed^4 - d^6}.$$

[&c] [&c]

The use of y^e^ precedent table in finding y^e^ Axes of crooked Lines, declared by Examples.

[*Example y^e^* 1st.] Suppose I had this equation given, $xx - 2xy + ay + yy[=0]$.(21) That I may find y^e^ axis of y^e^ line signified by it, first I observe of how many dimensions one of y^e^ unknowne quant[it]ies or y^e^ rectang[:] of y^m^ both is found

(21) A simple parabola, transformed into $Y^2 = (a/2\sqrt{2})X$ by

$$\left\{\begin{array}{l} x = \dfrac{1}{\sqrt{2}}\left(-X + Y + \dfrac{3a}{8\sqrt{2}}\right) \\ y = -\dfrac{1}{\sqrt{2}}\left(X + Y + \dfrac{a}{8\sqrt{2}}\right) \end{array}\right\}.$$

(22) The tabulated coefficients of $\mathfrak{x}\gamma/e^2$ in x^iy^j.

at most in y^e Equation, (as in this Example they have noe more y^n 2) then I take every quanti[t]y in wch one of y^e unknowne quantitys or y^e rectangle of y^m both is of soe many dimensions (wch in this case are $xx-2xy+yy$.) Then lookeing in y^e Table, (either amongst y^e rules of y^e first or second sort &c:) for a rule in wch y^e first quantity is of soe many dimensions I substitute y^e valors of y^e unknowne quantitys, found by y^t rule, into their place in y^e selected quantitys & supposeing y^e product$=0$, I find y^e proportion of d to e thereby, that is I find y^e angle wch y^e axis makes wth y^e unknowne quantity called x. As in this case I take y^e 2^d Rule of y^e first sort,(22) & by it I find $xx=2d\sqrt{ee-dd}$. $xy=ee-2dd$. $yy=-2d\sqrt{ee-dd}$. W^{ch} valors substituting into y^e roome of y^e unknown quantitys in these selected termes $xx-2xy+yy$. I have this equation.

$$2d\sqrt{ee-dd}-2ee+4dd-2d\sqrt{ee-dd}=0.$$

or, $2dd=ee$. & $e=d\sqrt{2}$. so y^t by assuming any quantity for e as a I have y^e valor of d, for $d=\frac{a}{\sqrt{2}}$. therefore

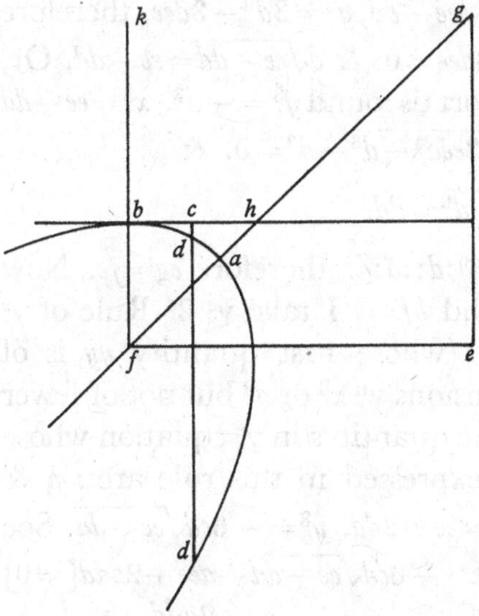

$$d:d\sqrt{2}::d:e::\frac{a}{\sqrt{2}}:a::a:a\sqrt{2}.\ \text{\&c.}$$

In y^e next place y^t I may find y^e length of y^e line $bf=c$. I take another rule whose first quantity is not of soe many nor of fewer dimensions y^n one of y^e unknowne quantitys or y^e rectangle [of] y^m both is some where in y^e equation. Then select every quantity out of y^e equation, y^e valor of whose unknowne quantity may be found by this rule, & substituting their valors, found thereby, into y^r places in these selected termes make y^e product$=0$. & find y^e valor of c thereby. As in this example I must take y^e first rule of y^e 1st sort.(23) By wch I find, $xy=c\sqrt{ee-dd}$. $y=-d$. $yy=-2dc$. but y^e valor of xx cannot be found by it.(24) therefore I onely take y^e termes $-2xy+ay+yy$, & by substituting y^e valors of y^e unknown quantitys into their roomes I have $-2c\sqrt{ee-dd}-ad-2dc=0$. Then by substituting y^e above found valors of $d=a$, & $e=\sqrt{2aa}$ into y^r places, it is ♉ $2ac+aa+2ac=0$. Or $+2ac+aa+2ac=0$. or $c=\frac{-a}{4}$. Soe y^t if I make b y^e beginning of x, & $+x$ to

(23) The tabulated coefficients of y/e in x^iy^j.

(24) $x^2 = (d(x/e)+\sqrt{[e^2-d^2]}\,(y/e))^2$ has no term y/e.

tend towards c in y[e] line bc, & $+y$ towards k perpendicularly to bc. then must I draw $bf=-\frac{a}{4}$ from y[e] point b perpendicular to bc; & $fe=a$, & parallell to bc; y[n] $eg=\sqrt{gf^2-fe^2}$ [&] $eg=a$, & parallell to bf. Lastly through y[e] points f & g draw gf y[e] axis of y[e] line sought. Otherwise it may be done thus.

$$-eg:ef::-bf:bh.$$

$$\sqrt{ee-dd}:d::-c:\frac{-cd}{\sqrt{ee-dd}}.$$

therefore I take $bh=\frac{-cd}{\sqrt{ee-dd}}=\frac{a}{4}$, & through y[e] points f & h I draw af y[e] axis sought.(25)

Example y[e] 2[d]. If y[e] Equation bee $x^3-axy+y^3=0$.(26) y[e] Rule whose first quantity is of as many dimensions as either of y[e] unknowne quantitys in this equation, is y[e] 3[d] of y[e] first sort, or y[e] first of y[e] 2[d] sort.(27) Selecting therefore onely x^3+y^3 out of y[e] equation (since in neither of these rules y[e] valor of xy is found) by y[e] 3[d] rule of y[e] first sort I find $x^3=3dd\sqrt{ee-dd}$, $y^3=3d^3-3dee$. therefore y[e] selected termes $x^3+y^3=3dd\sqrt{ee-dd}+3d^3-3dee=0$. & $d\sqrt{ee-dd}=ee-dd$. Or, $ee=2dd$. In like manner by y[e] first rule of y[e] 2[d] sort tis found $y^3=-d^3$. $x^3=ee-dd$ in $\sqrt{ee-dd}$. & therefore $x^3+y^3=\sqrt{e^6-3e^4dd+3eed^4-d^6}-d^3=0$. &

$$\sqrt{c:e^6-3e^4dd+3eed^4-d^6}=dd.$$

Or $ee=2dd$ as before. Soe y[t] $eg:fe::\sqrt{dd-ee}$(28)$:d::d:d$. therefore $eg=fe$. Now y[t] I may find $bf=c$ I take y[e] 2[d] Rule of y[e] first sort(29) (whose first quantity yy is of fewer dimensions y[n] x^3 or y^3 but not of fewer [y[n]] xy,) The quantitys in y[e] equation whose valors are expressed in this rule are xy, & y^3[,] for $xy=ee-2dd$. $y^3=-6cd\sqrt{ee-dd}$. Soe y[t] I write $-6cd\sqrt{ee-dd}-aee+2add$[$=0$] instead of y^3-axy. soe y[t] $c=\frac{2add-aee}{6d\sqrt{ee-dd}}$. Or since $2dd-ee=0$, it is $c=\frac{0\times a}{6dd}=0$. Had I taken y[e] first rule of y[e] first sort(30) I had found

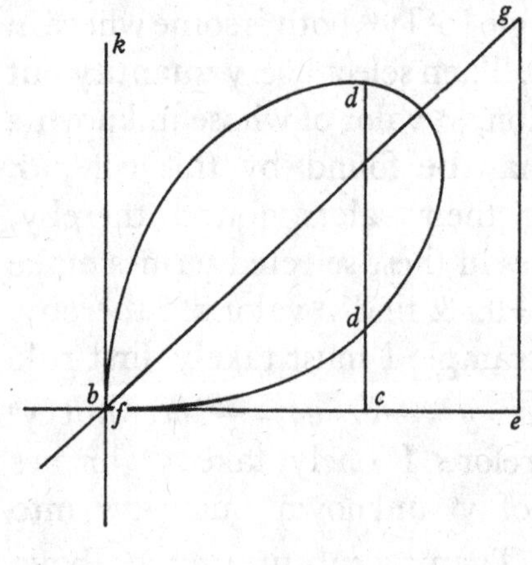

(25) Newton transforms the parabola by

$$\begin{Bmatrix} x = \frac{1}{\sqrt{2}}(\mathfrak{x}+\mathfrak{y}) \\ y = \frac{1}{\sqrt{2}}(\mathfrak{x}-\mathfrak{y})-\frac{a}{4} \end{Bmatrix}$$

into $2\mathfrak{y}^2+a\mathfrak{x}/\sqrt{2}-3a^2/16 = 0$, which has the diameter $fh(\mathfrak{y} = 0)$.

$xy=c\sqrt{ee-dd}$. & $y^3=-3ccd$. therefore $y^3-axy=-3ccd-ac\sqrt{ee-dd}=0$. wch is right since $c=0$. but by this equation c hath other valors, for $3cd+a\sqrt{ee-dd}=0$. or $3c$ ♉ $a=0$, & $c=\frac{♉ a}{3}$. &c. Whence observe y^t^ for y^e^ most ꝑte it will bee most convenient to find c by y^t^ rule whose 1st quantity hath one dimension lesse y^n^ y^e^ first qu[a]ntity of y^t^ rule by wch y^e^ proportion twixt d & e were found.(31)

*Example y^e^* 3^d^ If y^e^ equation be $\begin{matrix} xxyy+4bxyy+4bbyy \\ -2axxy-8abxy-8abby \\ -a^4 \end{matrix}=0$. $xxyy$ being of 4 dimensions I take y^e^ 4th rule of y^e^ first sort, or y^e^ 2^d^ of y^e^ 2^d^ sort.(32) By y^e^ 4th rule of y^e^ 1st sort I find $xxyy=2dee\sqrt{ee-dd}-4e^3\sqrt{ee-dd}$. & since by that rule I can find y^e^ valor of noe other quantity in y^e^ equation I make

$$xxyy=\begin{matrix} 2dee \\ -4d^3 \end{matrix}\sqrt{ee-dd}=0.$$

(26) Descartes' folium, drawn only for the first quadrant. (Compare Schooten's *Exercitationes:* 493.)

(27) The coefficients, respectively, of $\mathfrak{x}^2\mathfrak{y}/e^3$ and $\mathfrak{y}^3/e^3$ in x^iy^j.

(28) Read '$\sqrt{ee-dd}$'.

(29) The coefficients of $\mathfrak{x}\mathfrak{y}/e^2$ in x^iy^j.

(30) The coefficients of $\mathfrak{y}/e$ in x^iy^j.

(31) In effect Newton constructs the transform

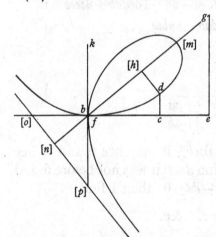

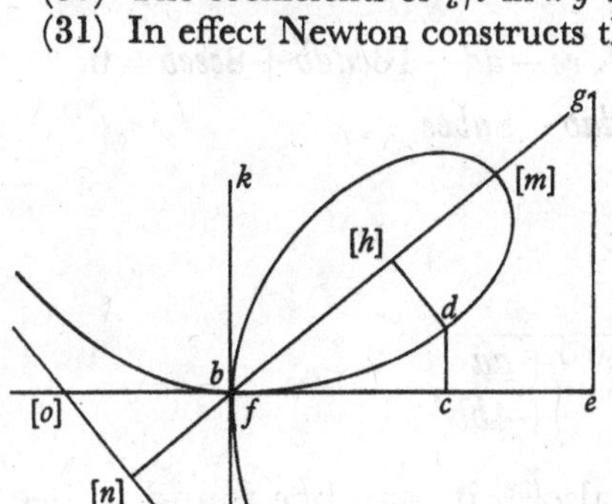

$$\left\{\begin{aligned} x &= \frac{1}{\sqrt{2}}(\mathfrak{x}+\mathfrak{y}) \\ y &= \frac{1}{\sqrt{2}}(\mathfrak{x}-\mathfrak{y}) \end{aligned}\right\}$$

between the two sets of perpendicular Cartesian coordinates $bc = x$, $cd = y$ and $bh = \mathfrak{x}$, $hd = \mathfrak{y}$, which reduces the defining equation $x^3-axy+y^3 = 0$ of the folium to $0 = \frac{1}{2}[\sqrt{2}\,\mathfrak{x}(\mathfrak{x}^2+3\mathfrak{y}^2)-a(\mathfrak{x}^2-\mathfrak{y}^2)]$, or

$$\mathfrak{y}^2 = \frac{a\mathfrak{x}^2-\sqrt{2}\,\mathfrak{x}^3}{3\sqrt{2}\,\mathfrak{x}+a}.$$

Clearly bg (or $\mathfrak{y} = 0$) is a diameter of the curve, with vertex m fixed by $x = y = \frac{1}{2}a$. Newton has not drawn the full curve (note (26)) and it is doubtful whether at this time he knew its true shape: in fact the line $onp(\mathfrak{x}=-a/3\sqrt{2})$ is an asymptote to the cubic and the whole curve is symmetrical round nbm. Descartes, who had originally proposed his folium in the summer of 1638 (compare his letters to Mersenne of 27 July and 23 August 1638 = *Œuvres* (ed. Adam and Tannery), **2**: nos. CXXXI and CXXXVIII, especially 274, 313–16 and Tannery's note on 341–2), was able successfully to draw the 'feuille' *bmf* but went on to write 'Soit ACKFA l'une des feuilles, qui fait partie de cete courbe' in his letter to Mersenne of 23 August. (*Œuvres*, **2**: 313.) Much later, in the winter of 1692/3, this expression was to occasion between Huygens and l'Hospital a brisk correspondence on the folium, closed only three years later by Huygens' death. (See *Œuvres complètes de Christiaan Huygens* **9**: 350–2, 390–1, 452–3, 461–2, 474–5, 565–6, 578, 580, 623–6, 713.) Compare 4, §3: note (66) below.

(32) The coefficients respectively of $\mathfrak{x}^3\mathfrak{y}/e^4$ and of $\mathfrak{x}\mathfrak{y}^3/e^4$ in x^iy^j.

Which is divisible by d, & $ee-2dd$, & by $\sqrt{ee-dd}$. therefore either $d=0$; or, $ee-2dd=0$; or, $ee-dd=0$. The operation is y[e] same if I make use of y[e] 2[d] rule of y[e] 2[d] sort.(33) Againe I take y[e] 3[d] rule of y[e] 1st sort(34) & by it I find,

$$xyy=ee\sqrt{ee-dd}-3dd\sqrt{ee-dd}. \quad \& \ xxy=2dee-3d^3. \quad \& \ xxyy=4cdee-6cd^3.$$

$$\text{therefore} \quad xxyy-2axxy+4bxyy=\frac{4cdee-4adee+4bee}{-6cd^3+\ 6ad^3-12bdd}\sqrt{ee-dd}=0.$$

Or $c=a\dfrac{+2bee\sqrt{ee-dd}-6bdd\sqrt{ee-dd}}{3d^3-2dee}$. & if $d=0$, then $c=a+\dfrac{2bee\sqrt{ee-dd}}{0}$. or c is infinitely long. but if $ee-2dd=0$, then $c=a \mho 4b \mho 6b$. & if $ee-dd=0$, then $c=a$. Againe I take y[e] first rule of y[e] 2[d] sort(35) & by it I find $xxyy=2cd^3-2cdee$. $xxy=d^3-eed$. $xyy=dd\sqrt{ee-dd}$. therefore

$$xxyy-2axxy+4bxyy=2cd^3-2cdee-2ad^3+2aeed+4bdd\sqrt{ee-dd}=0.$$

Now if $d=0$. or if $ee-dd=0$, then y[e] termes $ee-dd$ of this equat[n] destroy one another[,] soe y[t] y[e] valor of c may not be found thereby. but if $ee-2dd=0$, then I find $2cd^3-4cd^3-2ad^3+4ad^3+4bdd\sqrt{dd}=0$. Or $2cd^3=2ad^3 \mho 4bd^3$. or $c=a \mho 2b$. Againe I take y[e] 2[d] rule of y[e] 1st sort(36) & by it I find

$$\begin{aligned}&xxyy-2axxy+4bxyy= && 2ccd\sqrt{ee-dd}-4acd\sqrt{ee-dd}-16cddb+8ceeb=0.\\ &+4bbyy-8abxy && -8bbd\sqrt{ee-dd}+16ddab-8abee\end{aligned}$$

If $d=0$ then $cc=2ac\dfrac{-4ceeb}{0\times\sqrt{ee}}+4bb\dfrac{+4abee}{0\sqrt{ee}}$. or

$$c=a \mho \frac{2eb}{0} \mho \sqrt{\frac{4eebb}{0\times 0} \mho \frac{4aeb^{(37)}}{0} \begin{matrix}+aa\\ +4bb\end{matrix}}.$$

that is c is infinitely long as was found before. also(38) it may bee found to bee $8ceeb-8aeeb=0$, or $c=a$. but upon this supposition $d=0$ it was not before found $c=a$ & therefore $c=a$ is false, when $d=0$. If $ee-dd=0$. then I find

$$8abdd-8cbdd=0. \quad \text{or} \quad c=a. \quad \&\text{c.}$$

(33) Since $xxyy=-1\times 2dee+2d^3\left.\begin{matrix}\\ +2d^3\end{matrix}\right\}$ in $\sqrt{[ee-dd]}$.

(34) Which evaluates the coefficients of $\mathfrak{x}^2\mathfrak{y}/e^3$ in x^iy^j.

(35) The coefficients of $\mathfrak{y}^3/e^3$ in the transformed equation.

(36) The coefficients of $\mathfrak{x}\mathfrak{y}/e^2$ in x^iy^j.

(37) Read '$\dfrac{+4aeb}{0}$'.

(38) Taking $d=0$ in the preceding equation *before* dividing through by $2d\sqrt{[ee-dd]}$.

(39) Or finding the requisite coefficients of $\mathfrak{y}/e$ in the transformed equation.

If $ee-2dd=0$. then $cc\sqrt{dd}-4bb\sqrt{dd}-2ac\sqrt{dd}=0$. $c=a$ ♉ $\sqrt{aa-4bb}$. Which valor not being found before I conclude $ee-2dd=0$ to bee false. Lastly by useing y[e] first rule of y[e] first sort(39) I find,

$$4bcc\sqrt{ee-dd}-8abc\sqrt{ee-dd}-8cdbb+8abbd=4bxyy-8abxy+4bbyy-8abby=0.$$

& by supposeing $d=0$, I have $c=2a$.(40) & if $ee=dd$, then $c=a$. w[ch] being always found upon y[e] supposition $dd=ee$. I conclude y[e] valor of dd to be ee & of c to be a. & so draw the axis gf parallell to x & distant from it y[e] length of a. But here observe y[t] this might have beene better performed by taking away y[e] 2[d] terme of y[e] Equation $\begin{array}{l} xxyy+4bxyy+4bbyy \\ -2axxy-8abxy-8abby=0. \\ -a^4 \end{array}$ Or $xx+4bx+bb\,\dfrac{-a^4}{yy-2ay}=0$. as was observed before.(41)

---

(40) Newton first wrote in error '$2c=2a$. & $c=a$'. but has, in his text, cancelled only its first part. (Note that the value $c=2a$ further contradicts the supposition that $d=0$.)

(41) Newton's text lacks the clarification of a diagram but the accompanying figure is restored in keeping with those of his previous examples. Where $bc=x$, $cd=y$ the quartic (d) defined by $(x+2b)^2((y-a)^2-a^2)=0$ has in fact not one but two diameters, gk and gf

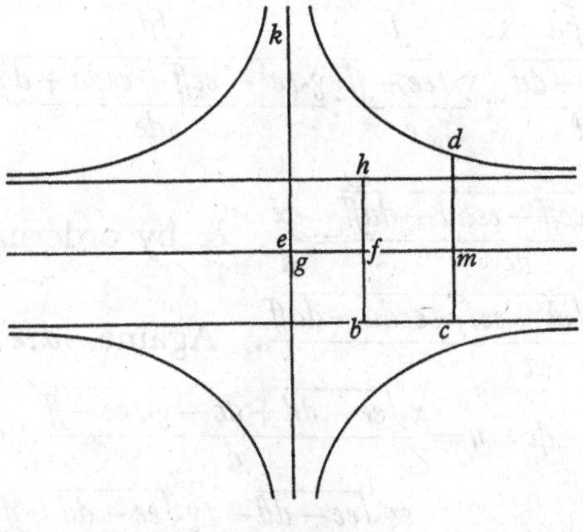

($x=-2a$ and $y=b$ respectively). (Note too that $y=0$ and $y=2a$ are asymptotes.) However, Newton's present method is limited by the nature of his general transform

$$\begin{Bmatrix} x=\mathfrak{x}\cos\theta+\mathfrak{y}\sin\theta \\ y=c+\mathfrak{x}\sin\theta-\mathfrak{y}\cos\theta \end{Bmatrix},$$

which can allow no translation of the x-co-ordinate. Here the translation $\begin{Bmatrix} x=\mathfrak{x} \\ y=a+\mathfrak{y} \end{Bmatrix}$ can reduce the quartic's equation only to $(\mathfrak{x}^2+4b\mathfrak{x}+b^2)(\mathfrak{y}^2-a^2)=0$, and the test for diameters, that only even powers of the variable exist, allows Newton to isolate $\mathfrak{y}=0$ as an axis. (Presumably Newton would urge a second application of his method with the co-ordinates x and y interchanged.)

§6[1]

[November 1664]

November 1664 *To find y^e Diameter or axis of any crooked line which hath it.*

Suppose y^e crooked line to bee (lgc), y^e diameter or Axis (kd), y^e undetermined quantitys describing y^e line to be $ab=x$, & $bc=y$. from y^e point a (y^e begining of x), perpendicular to kb draw $ah=pb=c$, cutting y^e axis kd in h. paralell to kb draw $hp=x$. & produce cb soe y^t it intersect y^e axis in y^e point d. & suppose y^t hp is to hd, as d to e: or y^t $hd=\frac{ex}{d}$. & therefore $dp=\frac{x\sqrt{ee-dd}}{d}$. let $ec=el=\mathfrak{y}$ be one of those lines wch are ordinately applied to y^e diameter $he=\mathfrak{x}$. lastly suppose y^t ec is to ef as e to f: y^n is $ef=\frac{f\mathfrak{y}}{e}$; & $fc=\frac{\mathfrak{y}\sqrt{ee-ff}}{e}$.

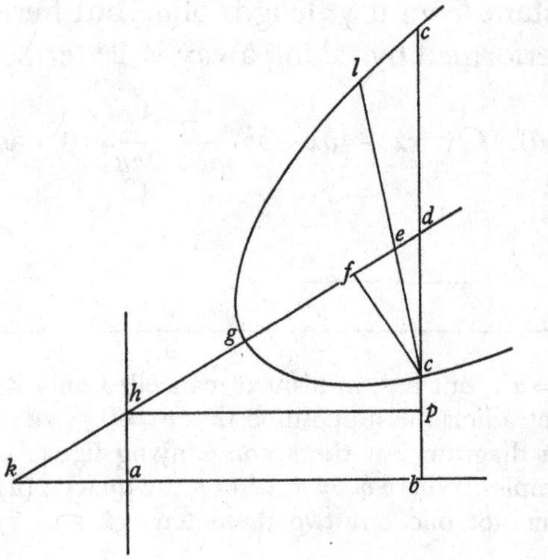

then $\quad hp : \; pd \; :: \; fc \; : \; fd.$

$$x : \frac{x\sqrt{ee-dd}}{d} :: \frac{\mathfrak{y}\sqrt{ee-ff}}{e} : \frac{\mathfrak{y}\sqrt{e^4-eeff-eedd+ddff}}{de}=fd.$$

& $hd=\mathfrak{x}-\frac{f\mathfrak{y}}{e}+\frac{\mathfrak{y}\sqrt{e^4-eeff-eedd+ddff}}{de}=\frac{ex}{d}$. & by ordering y^e equation it will bee, $x=\frac{ed\mathfrak{x}-df\mathfrak{y}+\mathfrak{y}\sqrt{e^4-eef-eedd-ddff}}{ee}$. Againe, $d:e::fc:cd=\frac{\mathfrak{y}\sqrt{ee-ff}}{d}$. & $dp=\frac{x\sqrt{ee-dd}}{d}$, $dp+pb-dc=y=\frac{x\sqrt{ee-dd}+dc-\mathfrak{y}\sqrt{ee-ff}}{d}$. or by substituteing y^e valor of x into its place it is $y=\frac{e\mathfrak{x}\sqrt{ee-dd}-f\mathfrak{y}\sqrt{ee-dd}-d\mathfrak{y}\sqrt{ee-ff}+eec}{ee}$. And y^t

(1) Add. 4004: 21^r, 21^v, an interesting first draft for §7 below. (2) Read 'lessen'.

(3) Newton constructs the transform $\left\{\begin{array}{l} x=(d/e)\,\mathfrak{x}+(s/e)\,\mathfrak{y} \\ y=(t/e)\,\mathfrak{x}-(v/e)\,\mathfrak{y}+c \end{array}\right\}$ which relates the perpendicular co-ordinates $ab=x$, $bc=y$ and the oblique set $he=\mathfrak{x}$, $ec=\mathfrak{y}$.

(4) Where $\left\{\begin{array}{l} x=(d\mathfrak{x}+s\mathfrak{y})/e \\ y=(t\mathfrak{x}-v\mathfrak{y}+ec)/e \end{array}\right\}$ he tabulates the coefficients of the various powers of $\mathfrak{x}$ and $\mathfrak{y}$ in $\mathfrak{x}^i\mathfrak{y}^j$, $i=0, j=1, 2, 3, 4, 5$; $i=1, j=0, 1, 2, 3$; $i=2, j=0, 1, 2$; $i=3, j=0, 1$; $i=4, j=0$; $i=5, j=0$. Since the tabulation is taken up again in §7 below (=Add. 4004: 23^v) we omit the present enumeration.

I may abbreviate y^e^ termes I make $\sqrt{ee-dd}=t$; & $\dfrac{+f\sqrt{ee-dd}+d\sqrt{ee-ff}}{e}=v$; & so y^e^ Equation is $y=\dfrac{t\mathfrak{x}-v\mathfrak{y}+ec}{e}$. Also by supposeing

$$s=\frac{\sqrt{e^4-eeff+ddff-eedd}-df}{e},$$

I lesson(2) the termes of y^e^ Equation $x=\dfrac{ed\mathfrak{x}-df\mathfrak{y}+\mathfrak{y}\sqrt{e^4-eeff-eedd+ddff}}{ee}$, by writeing instead of it, $\dfrac{d\mathfrak{x}+s\mathfrak{y}}{e}=x$.(3)

Now therefore by substituting these valors of x & y into y^r^ stead I take y^m^ out of y^e^ Equation expressing y^e^ rel[a]tion twixt y^m^ soe y^t^ y^n^ I have an equation expressing y^e^ relation twixt $\mathfrak{x}$ & $\mathfrak{y}$. And to that end it will bee convenient to have a table of y^e^ squares, cubes, squaresquares, square=cube, rectangles &c of y^e^ valors of x & y, After y^e^ manner of y^t^ wch follows.

[. . . . .(4)]. &c. If there bee occasion to doe these operacons in equations of 5 or 6 or more dimensions this table may be easily enlarged.

As for example. [1st] If y^e^ relation twixt x & y bee exprest in this Equation, $xx+ax-2xy+yy=0$.(5) then into y^e^ place of xx, x, xy, yy, I substitute their valors found by this table, & there results,

$$\begin{array}{lll}
+ss\mathfrak{y}\mathfrak{y} & +2ds\mathfrak{x}\,\mathfrak{y} & +dd\mathfrak{x}\mathfrak{x}=0.\\
+2sv & +aes & +ade\mathfrak{x}\\
+vv & -2ts\mathfrak{x} & -2dt\mathfrak{x}\mathfrak{x}\\
 & -2ecs & -2dec\mathfrak{x}\\
 & +2dv\mathfrak{x} & +tt\mathfrak{x}\mathfrak{x}\\
 & -2t\mathfrak{x}v & +2ect\mathfrak{x}\\
 & -2ecv & +eec[c]
\end{array}$$

Which Equation expresseth y^e^ rela[t]ion twixt $\mathfrak{x}$ & $\mathfrak{y}$. y^t^ is, twixt *ge* & *le* or *ec*. Now y^t^ $ge=\mathfrak{x}$ be y^e^ diameter & $le=ec=$ ȣ $\mathfrak{y}$ be ordinately applied to it, it is required (by Prop 2^d^) y^t^ in this equation $\mathfrak{y}$ be not of odd dimensions. & [y^t^] that may bee soe y^e^ quantitys in the 2^d^ terme (in w^ch^ $\mathfrak{y}$ is but of one dimension) must destroy one another, wch cannot be unlesse those quantitys destroy one another in wch y^e^ unknowne quantitys $\mathfrak{x}$ & $\mathfrak{y}$ are of y^e^ same dimensions. Which things being considered it will appeare y^t^ I must divide y^e^ 2^d^ terme into two ꝑts, makeing $2ds\mathfrak{x}\mathfrak{y}-2ts\mathfrak{x}\mathfrak{y}+2dv\mathfrak{x}\mathfrak{y}-2tv\mathfrak{x}\mathfrak{y}=0$; &,

$$aes\mathfrak{y}-2ecs\mathfrak{y}-2evc\mathfrak{y}=0.$$

(5) Newton repeats his first example (a parabola) from §5 above.

& divideing y^e^ first by $2\mathfrak{x}\mathfrak{y}$, & y^e^ 2^d^ by $e\mathfrak{y}$ they will be, $ds-ts-tv+dv=0$. &, $as-2cs-2cv=0$. Hitherto useing y^e^ letters s, t, & v for brevitys sake, I must now write their valors in theire stead (y^t^ I may find y^e^ length of c, & y^e^ proportion of d to e wch determine y^e^ position of y^e^ axis, & also y^e^ proportion of e to f wch determines y^e^ position of y^e^ lines applyed to y^e^ axis.) & soe instead of y^e^ Equation, $ds-ts-tv+dv=0$; there results,

$$2df\sqrt{ee-dd}=eef+ee\sqrt{ee-ff}-2dd\sqrt{ee-ff}.$$

& by squareing both ꝑts & ordering y^e^ product it is,

$$e^4-4ddee+4d^4=4ddf\sqrt{ee-ff}-2eef\sqrt{ee-ff}.$$

Which is divisible by $2dd-ee=0$, for y^e^ quote will bee $2dd-ee=2f\sqrt{ee-ff}$. & therefore $2dd=ee$. Or, $2dd=ee+2f\sqrt{ee-ff}$. Againe by inserting y^e^ valors of s & v into y^e^ Equation $as-2cs-2cv=0$, there resulteth,

$$\begin{matrix}+a\\-2c\end{matrix}\sqrt{e^4-eedd-eeff+ddff-2cf\sqrt{ee-dd}-2cd\sqrt{ee-ff}-adf+2cdf=0}.$$

& by writeing $2dd$ instead of ee & dividеing it by d there resulteth

$$\begin{matrix}a\sqrt{2dd-ff}-af=0.\\-4c\end{matrix}$$

Or $\dfrac{a}{4}\dfrac{-af}{4\sqrt{2dd-ff}}=c$. Thus haveing found y^e^ proportion of d to e, & y^e^ valor of c[,] since theire remaines noe more equations by wch I may find y^e^ proportion of e to f I concluded it to be undetermined, soe y^t^ I may assume any proportion betwixt y^m^. As if I make $f=0$. Then y^e^ angle ceh is a right one & eh y^e^ axis of y^e^ line, & $c=\frac{a}{4}=ah$. & $d:\sqrt{2dd}::d:e::hp:hd$. or $d:\sqrt{2dd-dd}::hp:dp$. that is $hp=dp$; As in y^e^ 1st figure. Or if I make $ee:ff::2:1$. that is $ee=2ff$. or $f=d$, then

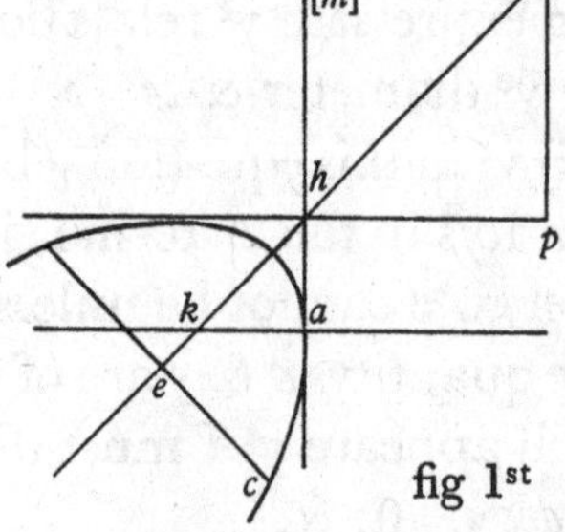

fig 1st

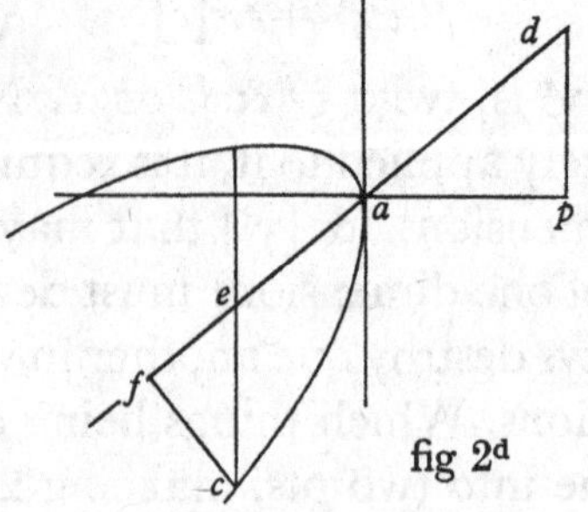

fig 2^d^

(6) Newton's general transform, where $\widehat{fec}=\theta$ and $e=\sqrt{2}\,d$ (or $hp=dp$), is

$$\left\{\begin{matrix}x=(1/\sqrt{2})\,(\mathfrak{x}-\mathfrak{y}\,(\cos\theta-\sin\theta))\\ y-c=(1/\sqrt{2})\,(\mathfrak{x}-\mathfrak{y}\,(\cos\theta+\sin\theta))\end{matrix}\right\}$$

I find y[t] $c=0$. that is that y[e] diameter *ed* intersects y[e] line *ap* at y[e] point *a* y[e] begining of x. & y[t] y[e] lines *ec* are parallell to *ma*. as in y[e] 2[d] figure &c. Soe y[t] by assumeing any proportion twixt *ec* & *ef*, that is, supposeing y[e] angle *fec* of any bigness, y[e] position of y[e] dia[m]iter *fa* may be found after y[e] same manner.(6) As if I would have y[e] angle *fec* to be an angle of 60 degrees. y[n] must $ec=\gamma$ be double to $fe=\frac{1}{2}\gamma$, & $fc=\sqrt{\frac{3}{4}\gamma\gamma}$. i.e. $e:f::2:1$. & $2dd=ee$, therefore $\frac{dd}{2}=ff$. I found before y[t] $\frac{a}{4}\frac{-af}{4\sqrt{2dd-ff}}=c$, or writeing y[e] valor of f in its roome, tis

$$\frac{a}{4}\frac{-a\sqrt{\frac{1}{2}dd}}{4\sqrt{\frac{3dd}{2}}}=c. \quad \text{that is } \frac{a}{4} \text{ ȣ } \frac{a}{4\sqrt{3}}=c.$$

Or since c must be lesse y[n] $\frac{a}{4}$(7) it must [be] $\frac{a}{4}-\frac{a}{4\sqrt{3}}=c=ah$. & $ph=dp$ since $ee=2dd$. As in y[e] 3[d] figure. But if I would make y[e] angle *ceh* of 60[degr:] then as

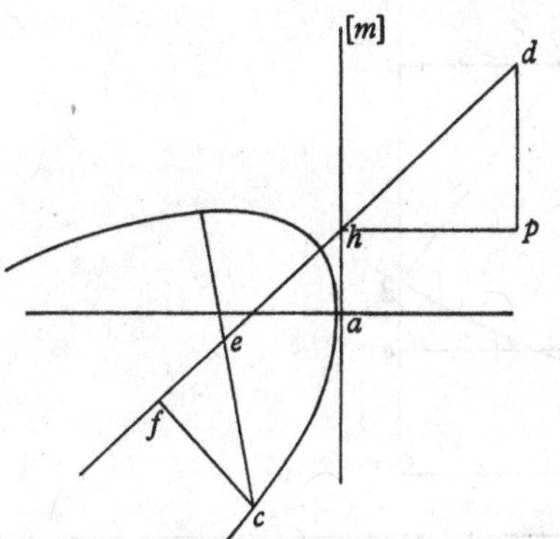

fig 3[d]

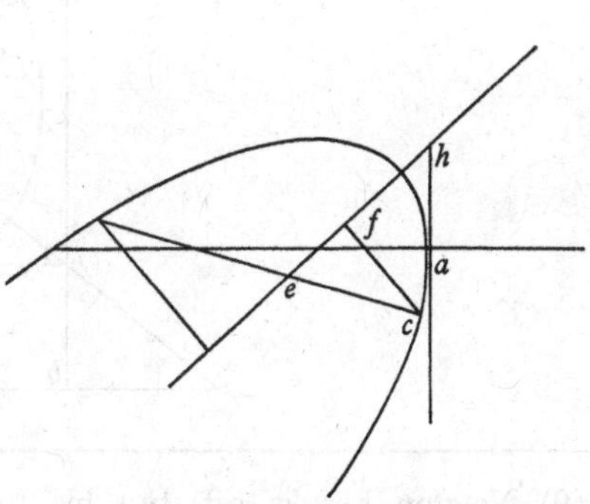

fig 4th

before $e=2f$, & $\frac{dd}{2}=ff$, & $\frac{a}{4}$ ȣ $\frac{a}{4\sqrt{3}}=c$, or since c must be greater y[n] $\frac{a}{4}$(8) tis $\frac{a}{4}+\frac{a}{4\sqrt{3}}=c$, as in y[e] 4th fig. &c.

---

with $c=\frac{1}{4}a([\sin\theta-\cos\theta]/\sin\theta)$: by its application the parabolic equation $(x-y)^2+ax=0$ is reduced to $2\gamma^2\sin^2\theta+(a/\sqrt{2})\chi+c^2=0$, which has the diameter $\gamma=0$, or *ehd*.

(7) Where $\theta=\frac{1}{3}\pi$, $c=\frac{1}{4}a(1-\cot\theta)<\frac{1}{4}a$.

(8) Where $\theta=\frac{2}{3}\pi$, $c=\frac{1}{4}a(1-\cot\theta)>\frac{1}{4}a$.

Example 2[d]. If y[e] Equation expressing y[e] nature of y[e] line be

$$x^3-3xxy+\underset{-ayy}{3xyy}-y^3=0.^{(9)}$$

---

§7(1)

[November/December 1664]

A November 1664.

To find y[e] Axis or Diameter of any Crooked Line supposeing it have y[m].

Suppose $bc=x$; $cd=y$; nad y[e] line whose axis or Diameter is sought; pk its axis or Diameter; a its vertex; $hd=hn=\mathfrak{y}=$lines ordinately applied to its Diameter; bm a perpendicular to pc drawne from y[e] point b, i.e. from y[e] begining

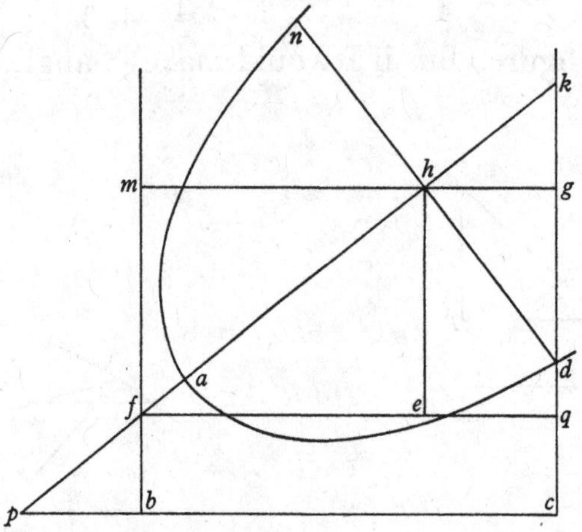

---

(9) Newton breaks off, but his further argument is easily restorable. The transform between the perpendicular co-ordinates $ab=x$, $bc=y$ and the oblique set $ae=\mathfrak{x}$, $ec=\mathfrak{y}$ given by

$$\begin{cases} x=\mathfrak{x}+(1/\sqrt{2})\,\mathfrak{y} \\ y=(1/\sqrt{2})\,\mathfrak{y} \end{cases}$$

reduces the cubic $(x-y)^3=ay^2$ to $\mathfrak{x}^3=\frac{1}{2}a\mathfrak{y}^2$ with diameter $ab\,(\mathfrak{y}=0)$ and a cusp at a.

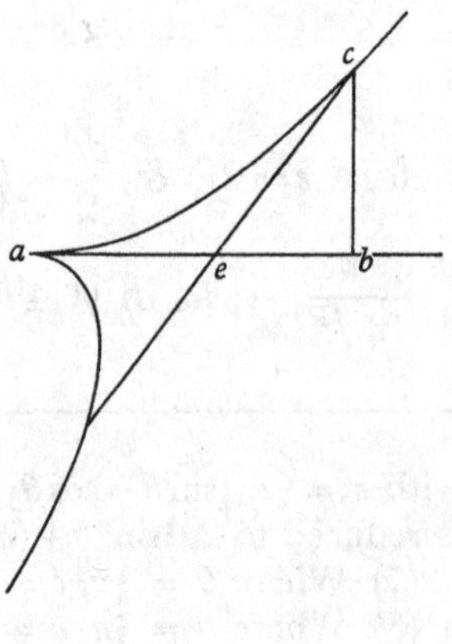

(1) Add 4004: 23[v]–23[r], 26[r], 27[r], 27[v]. (The subsectioning letters A, B, C, etc. seem to have been added later in the paper's composition.)

(2) More correctly, in Newton's first figure

$$\begin{cases} bc=mh+hg \\ cd=cq+qg-gd \end{cases} \quad\text{or}\quad \begin{cases} x=\mathfrak{x}+(f/d)\,\mathfrak{y} \\ y=c+(e/d)\,\mathfrak{x}-(\sqrt{[d^2-f^2]}/d)\,\mathfrak{y} \end{cases};$$

of x; $bf = c =$ pte of ye line bm intercepted twixt ye diameter & pc; $mh = \mathfrak{z} =$ a line parallell to bc & drawne from bm to ye intersection of fk & nd; & $he = mf$ & parallell to mf; & dgh a right angled triangle. $dh : hg :: d : f$. &,

$$mh = fe : he = mf :: d : e.$$

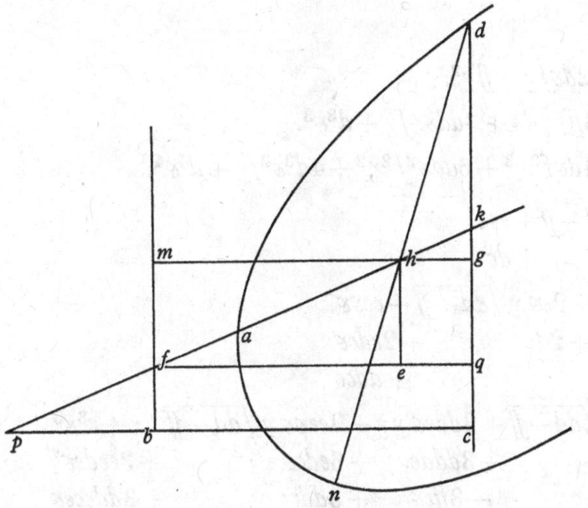

Then $d : f :: \gamma : \frac{f\gamma}{d} = hg = eq$. &, $eq + fe = \frac{d\mathfrak{z} + f\gamma}{d} = x$. Againe $gd = \sqrt{\frac{dd\gamma^2 - ff\gamma^2}{dd}}$. & $d : e :: \mathfrak{z} : he = \frac{e\mathfrak{z}}{d} = gq$. &, $gq + qc + dg = cd$; or, $\frac{e\mathfrak{z} + dc + \gamma\sqrt{dd - ff}}{d} = y$.(2) Now therefore by substituteing $\frac{d\mathfrak{z} + f\gamma}{d}$ into ye place of x, & $\frac{e\mathfrak{z} + dc[+]\gamma\sqrt{dd - ff}}{d}$

---

and in his second figure

$$\begin{Bmatrix} bc = mh + hg \\ cd = cq + qg + gd \end{Bmatrix} \quad \text{or} \quad \begin{Bmatrix} x = \mathfrak{z} + (f/d)\,\gamma \\ y = c + (e/d)\,\mathfrak{z} + (\sqrt{[d^2 - f^2]}/d)\,\gamma \end{Bmatrix}.$$

However, hd in the first figure is to be taken negative in sign, and this reconciles both forms with Newton's statement.

Though it is not immediately obvious Newton constructs the transform between the perpendicular co-ordinates $bc = x$, $cd = y$ and $fh = \mathfrak{z}'$, say, $hd = \gamma$: specifically, where

$$k\hat{h}g = \theta, \quad d\hat{h}g = \phi, \quad \begin{Bmatrix} x = \mathfrak{z}'\cos\theta + \gamma\cos\phi \\ y - c = \mathfrak{z}'\sin\theta + \gamma\sin\phi \end{Bmatrix}$$

(which may take on several variant forms according as $\phi \to -\phi$, or $\pi - \phi$). But since here we are concerned only with equating coefficients of $\mathfrak{z}'^i\gamma^j$ to zero, Newton gains simplicity in computation by expressing $\mathfrak{z}'$ parametrically in terms of $\mathfrak{z}$, the perpendicular distance of h from mf, by $\mathfrak{z} = \mathfrak{z}'\cos\theta$; so that the transform takes on the form

$$\begin{Bmatrix} x = \mathfrak{z} + \gamma\cos\phi \\ y - c = \mathfrak{z}\tan\theta + \gamma\sin\phi \end{Bmatrix}.$$

into y^e^ place of y, & theire □s & cubes &c: into y^e^ place of x^2, x^3, y^2[,] y^3 &c. I take x & y out of y^e^ Equation expressing y^e^ relation twixt y^m^ & soe have an Equation expressing y^e^ relation twixt $\mathfrak{x}$ & $\mathfrak{y}$. And to y^t^ end it will be convenient to have a table of y^e^ squares, cubes, & rectangles &c: of y^e^ valors of x & y, like y^t^ wch follows.(3)

$$dx = d\mathfrak{x} + f\mathfrak{y}.$$

$$ddxx = dd\mathfrak{x}\mathfrak{x} + 2d\mathfrak{x}f\mathfrak{y} + ff\mathfrak{y}^2.$$

$$d^3x^3 = f^3\mathfrak{y}^3 + 3ff\mathfrak{y}\mathfrak{y}d\mathfrak{x} + 3dd\mathfrak{x}\mathfrak{x}f\mathfrak{y} + d^3\mathfrak{x}^3.$$

$$d^4x^4 = f^4\mathfrak{y}^4 + 4d\mathfrak{x}f^3\mathfrak{y}^3 + 6dd\mathfrak{x}^2f^2\mathfrak{y}^2 + 4d^3\mathfrak{x}^3f\mathfrak{y} + d^4\mathfrak{x}^4.$$

$$dy = +\mathfrak{y}\sqrt{dd-ff} + \begin{array}{l} e\mathfrak{x}. \\ +dc \end{array}$$

$$ddyy = \begin{array}{l} dd \\ -ff \end{array}\mathfrak{y}\mathfrak{y} + \begin{array}{l} 2e\mathfrak{x} \\ +2dc \end{array}\mathfrak{y}\sqrt{dd-ff} + \begin{array}{l} ee\mathfrak{x}\mathfrak{x}. \\ +2edc\mathfrak{x} \\ +ddcc \end{array}$$

$$d^3y^3 = \begin{array}{l} dd \\ -ff \end{array}\mathfrak{y}^3\sqrt{dd-ff} + \begin{array}{l} 3dde\mathfrak{x} \\ +3dddc \\ -3ffe\mathfrak{x} \\ -3ffdc \end{array}\mathfrak{y}\mathfrak{y} + \begin{array}{l} 3ee\mathfrak{x}\mathfrak{x} \\ +6edc\mathfrak{x} \\ +3ddcc \end{array}\mathfrak{y}\sqrt{dd-ff} \quad \begin{array}{l} +e^3\mathfrak{x}^3. \\ +3eedc\mathfrak{x}^2 \\ +3ddcce\mathfrak{x} \\ +d^3c^3 \end{array}$$

$$d^4y^4 = \begin{array}{l} d^4 \\ -2ddff \\ +f^4 \end{array}\mathfrak{y}^4 + \begin{array}{l} 4e\mathfrak{x}dd \\ +4d^3c \\ -4e\mathfrak{x}ff \\ -4dcff \end{array}\mathfrak{y}^3\sqrt{dd-ff} + \begin{array}{l} 6ddee\mathfrak{x}\mathfrak{x} \\ -6ffee\mathfrak{x}\mathfrak{x} \\ +12d^3ec\mathfrak{x} \\ -12dffec\mathfrak{x} \\ +6d^4cc \\ -6ddffcc \end{array}\mathfrak{y}\mathfrak{y} \quad \begin{array}{l} +4e^3\mathfrak{x}^3 \\ +12dcee\mathfrak{x}\mathfrak{x} \\ +12ddcce\mathfrak{x} \\ +4d^3c^3 \end{array}\mathfrak{y}\sqrt{dd-ff} \quad \begin{array}{l} +e^4\mathfrak{x}^4. \\ +4dce^3\mathfrak{x}^3 \\ +6ddccee\mathfrak{x}\mathfrak{x} \\ +4d^3c^3e\mathfrak{x} \\ +d^4c^4 \end{array}$$

$$ddxy = +f\mathfrak{y}^2\sqrt{dd-ff} \quad \begin{array}{l} +fe\mathfrak{x}\mathfrak{y} \\ +fdc\mathfrak{y} \end{array} \begin{array}{l} +de\mathfrak{x}\mathfrak{x}. \\ +ddc\mathfrak{x} \end{array}$$
$$+d\mathfrak{x}\mathfrak{y}\sqrt{dd-ff}$$

$$d^3x^3y = +ff\mathfrak{y}^3\sqrt{dd-ff} \quad \begin{array}{l} +ffe\mathfrak{x} \\ +ffdc \end{array}\mathfrak{y}^2 \quad \begin{array}{l} +2def\mathfrak{x}^2 \\ +2ddcf\mathfrak{x} \end{array}\mathfrak{y} \begin{array}{l} +dde\mathfrak{x}^3. \\ +d^3c\mathfrak{x}\mathfrak{x} \end{array}$$
$$+2df\mathfrak{x}\sqrt{dd-ff} \quad +dd\mathfrak{x}\mathfrak{x}\sqrt{dd-ff}$$

$$d^4x^3y = +f^3\mathfrak{y}^4\sqrt{dd-ff} \quad \begin{array}{l} +ef^3\mathfrak{x} \\ +dcf^3 \end{array}\mathfrak{y}^3 \quad \begin{array}{l} +3deff\mathfrak{x}\mathfrak{x} \\ +3ddcff\mathfrak{x} \end{array}\mathfrak{y}\mathfrak{y} \quad \begin{array}{l} +3ddef\mathfrak{x}\mathfrak{x}\mathfrak{x} \\ +3cd^3f\mathfrak{x}\mathfrak{x} \end{array}\mathfrak{y} \begin{array}{l} +d^3e\mathfrak{x}^4. \\ +d^4c\mathfrak{x}^3 \end{array}$$
$$+3ffd\mathfrak{x}\sqrt{dd-ff} \quad +3ddf\mathfrak{x}\mathfrak{x}\sqrt{dd-ff} \quad +d^3\mathfrak{x}^3\sqrt{dd-ff}$$

(3) Having constructed $\left\{\begin{array}{l} dx = d\mathfrak{x}+f\mathfrak{y} \\ dy = e\mathfrak{x}+\sqrt{[d^2-f^2]}\,\mathfrak{y}+dc \end{array}\right\}$, Newton tabulates $d^{r+s}x^ry^s$ for the values $r = 0$, $s = 1, 2, 3, 4$; $r = 1$, $s = 0, 1, 2, 3$; $r = 2$, $s = 0, 1, 2$; $r = 3$, $s = 0, 1$; $r = 4$, $s = 0$.

$$\begin{array}{llllll} d^3xyy = & fdd\,\mathfrak{y}^3 & +d^3\mathfrak{z}\,\mathfrak{y}\mathfrak{y} & +eef\mathfrak{z}\mathfrak{z}\,\mathfrak{y} & +dee\mathfrak{z}^3. \\ & -f^3 & -dff\mathfrak{z} & +2cdef\mathfrak{z} & +2cdde\mathfrak{z}\mathfrak{z} \\ & & +2ef\mathfrak{z}\sqrt{dd-ff} & +ccddf & +ccd^3\mathfrak{z} \\ & & +2cdf\sqrt{dd-ff} & +2ed\mathfrak{z}^2\sqrt{dd-ff} & \\ & & & +2cdd\mathfrak{z}\sqrt{dd-ff} & \end{array}$$

$d^4xxyy =$

$d^4xy^3 =$ (4)

As for example. [1st], if ye relation twixt x & y bee exprest by,

$$xx + ax - 2xy + yy = 0.$$ (5)

then in stead of xx, x, xy, yy, writeing their valors found by this table there resulteth

$$\begin{array}{lll} dd\,\mathfrak{y}\mathfrak{y} & +2df\mathfrak{z}\mathfrak{y} & +dd\mathfrak{z}\mathfrak{z} = 0. \\ -2f\mathfrak{y}\mathfrak{y}\sqrt{dd-ff} & +adf & +add\mathfrak{z} \\ & -2ef\mathfrak{z} & -2de\mathfrak{z}\mathfrak{z} \\ & -2dcf & -2ddc\mathfrak{z} \\ & -2d\mathfrak{z}\sqrt{dd-ff} & +ee\mathfrak{z}\mathfrak{z} \\ & +2e\mathfrak{z}\sqrt{dd-ff} & +2edc\mathfrak{z} \\ & +2dc\sqrt{dd-ff} & +ddcc \end{array}$$

Which equation expresseth ye relation twixt $\mathfrak{z}$ & $\mathfrak{y}$ when any valors are assumed for c, d, e, & f. And if ye valors of c, d, e, & f bee such yt $\mathfrak{y}$ is not of odd dimensions in ye Equation (that is, yt ye 2d terme of this Equation bee wanting), then (by Prop: ye 2d)(6) ♉ $\mathfrak{y} = hn = hd$ is ordinately applyed to ye diameter pk. Now yt ye 2d terme of this Equation vanish it is necessary yt those termes destroy one

(4) Newton has left spaces to enter the values of the two last terms. Where $v = \sqrt{[d^2 - f^2]}$ they are

$$\begin{aligned} d^4x^2y^2 &= f^2v^2\mathfrak{y}^4 + 2(\mathfrak{z}(f^2ev + dfv^2) + f^2cdv)\,\mathfrak{y}^3 \\ &\quad + (\mathfrak{z}^2(e^2f^2 + 4defv + d^2v^2) + 2\mathfrak{z}(cdef^2 + 2cd^2fv) + c^2d^2f^2)\,\mathfrak{y}^2 \\ &\quad + 2(\mathfrak{z}^3(de^2f + d^2ev) + \mathfrak{z}^2(2cd^2ef + cd^3v) + c^2d^3\mathfrak{z})\,\mathfrak{y} \\ &\quad + d^2e^2\mathfrak{z}^4 + 2cd^3e\mathfrak{z}^3 + c^2d^4\mathfrak{z}^2, \end{aligned}$$

and

$$\begin{aligned} d^4xy^3 &= fv^3\mathfrak{y}^4 + (\mathfrak{z}(3efv^2 + dv^3) + 3cdfv^2)\,\mathfrak{y}^3 \\ &\quad + 3(\mathfrak{z}^2(e^2fv + dev^2) + \mathfrak{z}(2cdefv + cd^2v^2) + c^2d^2fv)\,\mathfrak{y}^2 \\ &\quad + (\mathfrak{z}^3(e^3f + 3de^2v) + 3\mathfrak{z}^2(cde^2f + 2cd^2ev) + 3\mathfrak{z}(c^2d^2ef + c^2d^3v) + c^3d^3f)\,\mathfrak{y} \\ &\quad + de^3\mathfrak{z}^4 + 3cd^2e^2\mathfrak{z}^3 + 3c^2d^3e\mathfrak{z}^2 + c^3d^4\mathfrak{z}. \end{aligned}$$

(5) The parabola of his first example in §6 above.

(6) That is, the second proposition in §4 above.

another in wch y^e^ unknowne quant[it]ys $\mathfrak{x}$ & $\mathfrak{y}$ are not diverse nor differ in dimensions. Whence it appeares y^t^ I must divide the 2^d^ terme into 2 *p*ts making

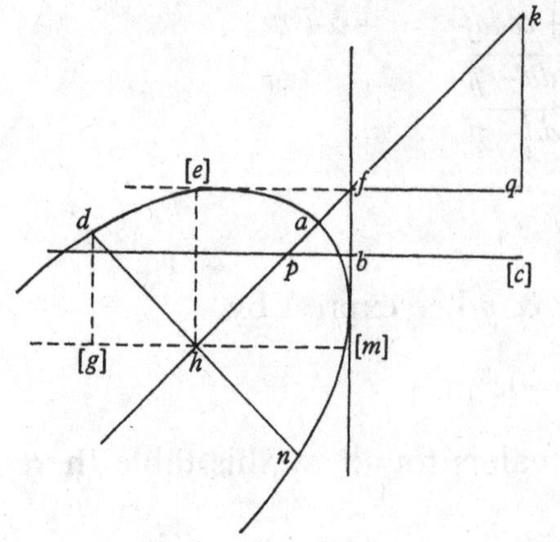

fig 1st

$$2df\mathfrak{x}\mathfrak{y}-2ef\mathfrak{x}\mathfrak{y}-2d\mathfrak{x}\mathfrak{y}\sqrt{dd-ff}$$
$$+2e\mathfrak{x}\mathfrak{y}\sqrt{dd-ff}=0.$$

& $adf\mathfrak{y}-2dcf\mathfrak{y}+2dc\mathfrak{y}\sqrt{dd-ff}=0$. Or by divideing the first of these by $2\mathfrak{x}\mathfrak{y}$, & y^e^ 2^d^ by $d\mathfrak{y}$ they are, $df-ef\,{}^{-d}_{+e}\sqrt{dd-ff}=0$, & $af-2cf+2c\sqrt{dd-ff}=0$. The first being divided by $d-e=0$. there results, $f+\sqrt{dd-ff}=0$. Therefore one or both these propositions $d=e$; $dd=2ff$, is trew. by y^e^ 2^d^ tis found y^t^ $\dfrac{af}{2f-2\sqrt{dd-ff}}=c$. Now since by assumeing some quantitys for y^e^ valors of d, c, or f I cannot find y^e^ valor of e unless by y^e^ Equation $d=e$. therefore I conclude $d=e$. whence it is not necessary y^t^ $dd=2ff$, or y^e^ proportion of d to f bee limited[,] soe y^t^ by assuming y^e^ angle *ahd* of any bigness I may find y^e^ position of y^e^ axis *ahk*. As if I suppose y^e^ angle *fhd* to be a right one (i.e. y^t^ *ah* is y^e^ axis of y^e^ line) then are y^e^ $\triangle$s *feh* & *hgd* alike, & therefore

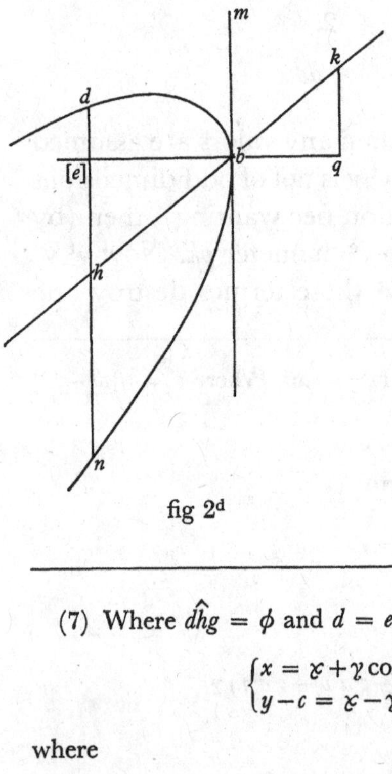

fig 2^d^

$fh:fe::dh:gh::\sqrt{dd+ee}:e::d:f$. & $\dfrac{ed}{\sqrt{dd+ee}}=f$. Or because $d=e$ therefore $+f=\dfrac{d}{\sqrt{2}}$. & $c=\dfrac{a}{4}$.

Soe y^t^ I draw $bf=\dfrac{a}{4}$. & *fq* parallell to *pb*[,] $qk=fq$ & parallell [to] *bf* & through y^e^ points *f* & *k* I draw *kh* y^e^ axis of y^e^ line *nad*, as in figure 1st. So if I would have *hd* paralell to *qk* i.e. y^e^ angle *dhf* of 45 degrees. then tis evident y^t^

(7) Where $\widehat{dhg}=\phi$ and $d=e$, the parabola $(x-y)^2+ax=0$ is reduced by the transform

$$\begin{Bmatrix} x=\mathfrak{x}+\mathfrak{y}\cos\phi \\ y-c=\mathfrak{x}-\mathfrak{y}\sin\phi \end{Bmatrix} \quad \text{to} \quad \mathfrak{y}^2(1+\sin 2\phi)+a\mathfrak{x}+c^2=0,$$

where
$$c=\frac{a\cos\phi}{2(\cos\phi+\sin\phi)}:$$

clearly the line *hk* ($\mathfrak{y}=0$) is a diameter.

$hg=0=f$. & $c=\dfrac{af}{2f \,\forall\, 2\sqrt{dd-ff}}=0$. Therefore through y^e^ point b I draw y^e^ axis kh, so y^t^ $bq=kq$, as before. &c. & note y^t^ since kh the axis is always paralell to it selfe y^e^ line dbn is a parabola.(7)

*Example y^e^ 2^d^*, $x^3+y^3=a^3$.(8) Being first to write y^e^ valors of x^3 & y^3 (found by y^e^ precedent table) into their roome, since I have noe neede of those two termes in wch γ is of eaven dimensions I leave y^m^ out, & soe for $x^3+y^3=0$ I write onely

$$\begin{matrix} f^3\gamma^3+dd\gamma^3\sqrt{dd-ff}+3ddf\varkappa\varkappa\gamma \\ -ff \end{matrix} \quad \begin{matrix} \\ +3ee\varkappa\varkappa\gamma\sqrt{dd-ff} \end{matrix} \quad +6edc\varkappa\gamma\sqrt{dd-ff}+3ddcc\gamma\sqrt{dd-ff}=0.$$

Then sorting those quantitys together in wch y^e^ unknowne quantitys are y^e^ same there [resulteth] these 4 equations (y^e^ 1st being divided by γ^3, y^e^ 2^d^ by $3[\varkappa^2\gamma]$, y^e^ 3^d^ by $6\gamma\varkappa$, y^e^ 4th by 3γ) viz: $f^3\begin{smallmatrix}+dd\\-ff\end{smallmatrix}\sqrt{dd-ff}=0$, $ddf+ee\sqrt{dd-ff}=0$;

$+cde\sqrt{dd-ff}=0$; $+ddcc\sqrt{dd-ff}=0$.

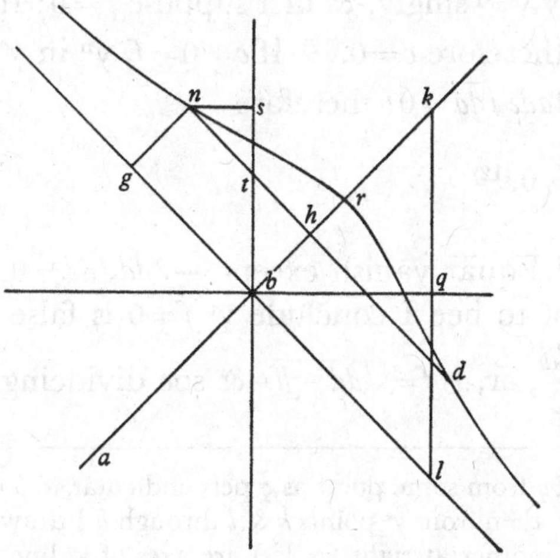

In y^e^ first equation $f^3=\begin{smallmatrix}-dd\\+ff\end{smallmatrix}\sqrt{dd-ff}$, I extract y^e^ cube roote & tis $f=-\sqrt{dd-ff}$. or $dd=2ff$. In y^e^ 2^d^ $ddf=ee\sqrt{dd-ff}$, or(9) $ddf=eef$, or $d=e$. By y^e^ 3^d^,

$$+cde\sqrt{dd-ff}=0, \quad \text{or} \quad c=\frac{+0}{edf}=0.$$

& so by y^e^ fourth. Now therefore since $c=0$. $d=e$. In y^e^ line bq from some point as q perpendicular to bq I draw

$$kq=+e=bq=d.$$

then from y^e^ point k through b I draw y^e^ line ak w^ch^ (since it cuts y^e^ line hnd applied to [it] at right angles) is an axis of y^e^ line ndr. wch appeares in y^t^ $dd=2ff$, for therefore $nt^2=2st^2=st^2+ns^2$, soe y^t^ $ns=st$ & nt [is] perpendicular to bk.(10)

(8) Compare James Stirling's treatment of this cubic in his *Lineæ Tertii Ordinis Neutonianæ* (Oxford, 1717): 126–7: *Exemplum Primum.*

(9) Using $d^2 = 2f^2$.

(10) The defining equation $x^3+y^3 = a^3$ is reduced by

$$\begin{Bmatrix} x = \varkappa+(1/\sqrt{2})\gamma \\ y = \varkappa-(1/\sqrt{2})\gamma \end{Bmatrix} \quad \text{to} \quad 2\varkappa^3+3\varkappa\gamma^2 = a^3, \quad \text{or} \quad \gamma^2 = \frac{a^3-2\varkappa^3}{3\varkappa}.$$

It follows that abk (or $\gamma = 0$) is the cubic's diameter (with gbl, or $\varkappa' = 0$, its asymptote).

Newton's text is heavily cancelled and it is clear that he found the tracing of the curve's shape difficult, at first believing it to have two branches placed symmetrically round a second

Example ye 3d. If ye nature of ye given line bee expressed in these termes $x^3-3xxy+2xyy-2ayy=0$. Then by supplanting ye valors of x & y into their roome & working as before, there will bee, $-f^3+2fdd-3ff\sqrt{dd-ff}=0$.

& 2dly $$3ddf-6def\begin{smallmatrix}-3dd\\+4ed\end{smallmatrix}\sqrt{dd-ff}+2eef=0.$$

& 3dly $$-6cddf+4cdef+4cdd\sqrt{dd-ff}-4ade\sqrt{dd-ff}=0.$$

& 4thly $2ccddf-4acdd\sqrt{dd-ff}=0$. The first of these divided by $f=0$. is
$$-ff+2dd=3f\sqrt{dd-ff}.$$

Or □ing both sides & ordering ye product tis, $10f^4-13ffdd+4d^4=0$. Which being divided by $2ff-dd=0$. there results $5ff-4dd=0$. Wherefore I conclude one of these 3 to be ye valors of f viz: $f=0[,]=\sqrt{\frac{dd}{2}}[,]=\sqrt{\frac{4dd}{5}}$. Now yt I may know wch of these is ye right valor of f I try ȳ(11) singly, & first suppose $f=0$; If so yn by ye 4th Equation $-4acdd\sqrt{dd}=0$, therefore $c=0$.(12) If $c=0=f$, yn in ye 3d Equation all ye termes vanish except $4ade\sqrt{dd}=0$: therefore
$$e=\frac{0}{4ad\sqrt{dd}}=0.^{(12)}$$

& since $c=e=f=0$, all ye termes in ye 2d Equat̄ vanish except $-3dd\sqrt{dd}=0$, therefore also $d=0$, wch since it ought not to bee I conclude yt $f=0$ is false. Therefore I passe [t]o ye 2d valor of $ff=\frac{dd}{2}$, or, ♉$f=\sqrt{dd-ff}$. & soe divideing

axis *gbl*. He first wrote in conclusion: 'In ye line *bq* from some point as *q* perpendicular to *bq* I draw $kq=+e$, & $ql=-e$, both of ym $=bq=d$. then from ye points *k* & *l* through *b* I draw ye two lines *ak* & *gl* both wch (since they cut one another at right angles) are axes of ye lines *ndr* & *nad*. wch appeares alsoe in yt $dd=2ff$....'

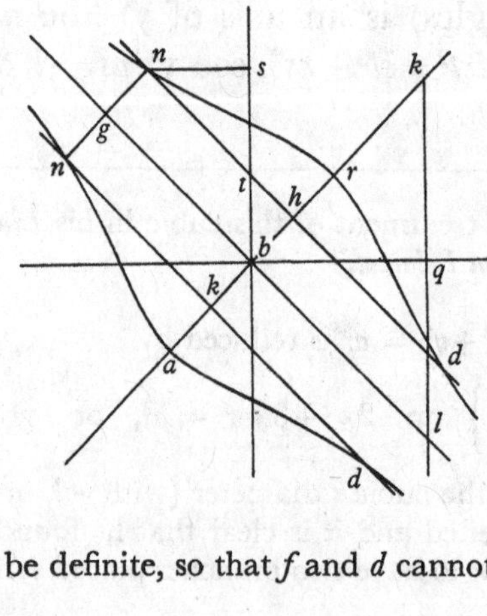

(11) Read 'ym'.
(12) The ratio $f:d$ must be definite, so that f and d cannot both be zero at the same time.

yᵉ 4th Equatⁿ by $2ddf$ t[her]e results $cc = \mho\, 2ac$: wch is divisible by $c=0$, & by $c \mho\, 2a = 0$, Now yt I may know wch is yᵉ right valor of c first I suppose $c=0$: & soe all yᵉ termes in yᵉ 3d equation vanish except, $4ade\sqrt{dd-ff}=0$. or $e = \frac{0}{4adf} = 0$. & since $e=0$, by yᵉ 2d Equation tis $3ddf - 3dd\sqrt{dd-ff} = 0$, or

$$f = \mho\sqrt{dd-ff} = \sqrt{ff} = +\sqrt{dd-ff}.$$

Which things since they agree I conclude yt $f = \sqrt{dd-ff}$, or $dd = 2ff$; $c=0$; $e=0$. Since $e=0$ the diameter must be parallel to x. & since $c=0$. it must bee coincident wth it. then in yᵉ axis bc I take some point (as h) & from it draw $[hp] = +f =$ to yᵉ perpendicular $+pq = +\sqrt{dd-ff}$ & through yᵉ points h & q I draw nd wch shall be [paral]lell to yᵉ lines ordinately applied to yᵉ diameter $bhpc$.(13)

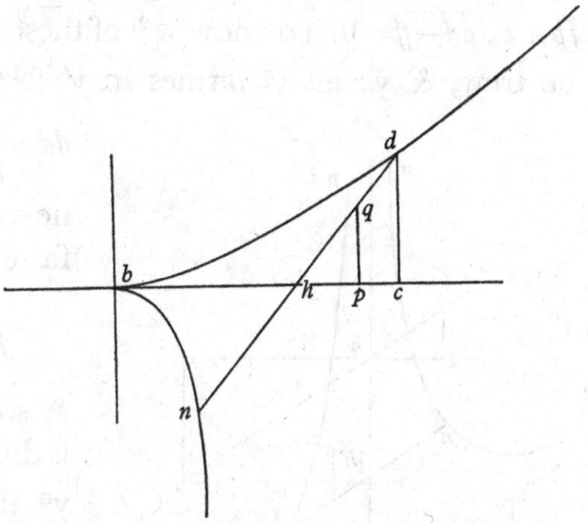

Example yᵉ 4th. If yᵉ Equation bee $bx^3 + ayxx = a^4$. by takeing only those termes (of yᵉ valors of x^3 & yxx by yᵉ precedent table) in wch y is of odd

(13) In summary, the transform $\begin{cases} x = \mathfrak{x} + (1/\sqrt{2})\mathfrak{y} \\ y = (1/\sqrt{2})\mathfrak{y} \end{cases}$ reduces $(x-y)^3 = y^2(x-y+2a)$ to $2\mathfrak{x}^3 = \mathfrak{y}^2(\mathfrak{x}+2a)$, where $bc = x$, $cd = y$ and $bh = \mathfrak{x}$, $hd = \mathfrak{y}$. The line bhc (or $\mathfrak{y} = 0$) is a diameter, meeting the cubic in the cusp b. (Note that Newton seems to know only one branch of the curve. In fact, the cubic has the three real asymptotes

$$\mathfrak{x} = -2a \quad \text{and} \quad \mathfrak{y} = \pm(\mathfrak{x} - a)\sqrt{2}$$

and so three hyperbolic branches.)

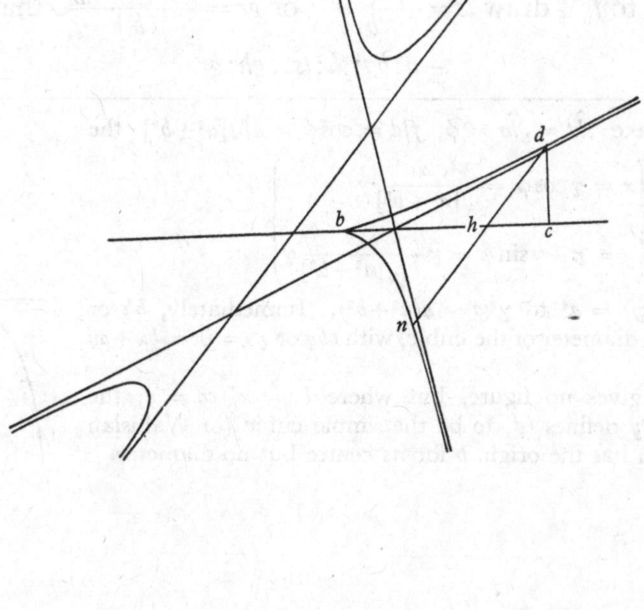

dimensions, & sorting those together in wch y^{e} unk[n]owne quantitys are y^{e} same & of y^{e} same dimensions as before. there will result these Equations. first

$$bf^3+aff\sqrt{dd-ff}=0.$$

2dly, $3ddbf+2adef+add\sqrt{dd-ff}=0$. & 3dly, $2addfc=0$. y^{e} 1st is divisible by $f=0$, $fb+a\sqrt{dd-ff}=0$. To know wch of these 2 are y^{e} valors of f first I suppose $f=0$ to be trew, & y^{n} all y^{e} termes in y^{e} 2^{d} Equation vanish except $add\sqrt{dd}=0$, or $dd=\dfrac{0}{a\sqrt{dd}}=0$. now since both d & f should never bee $=0$ therefore I conclude y^{t} $f=0$ is false & so pass to its other valor

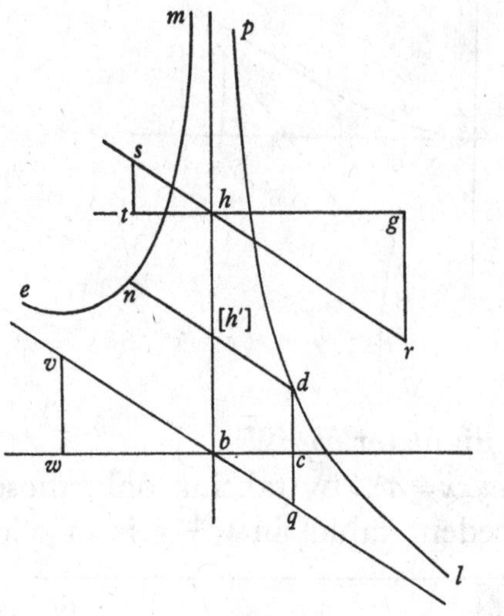

$$f=\frac{-a\sqrt{dd-ff}}{b}.\quad\text{or}\quad\frac{-bf}{a}=\sqrt{dd-ff}.$$

& soe by y^{e} 2^{d} Equation tis $ddb=-ade$. wch is divisible by $d=0$, & $db+ae=0$. If $db=-ae$ y^{n} tis $b:-a::e:d::\sqrt{dd-ff}:f$. & soe y^{e} diameter will bee parallel to y^{e} lines ordinately applied to it[,] wch cannot bee. therefore I try y^{e} other valor of $d=0$. And if $d=0$, y^{n} y^{e} 3^{d} Equation $2addfc=0$ vanisheth & soe cannot bee found & is therefore unlimited. Now since I find noe repugnancys in these Equations $f=\dfrac{-a\sqrt{dd-ff}}{b}$, & $d=0$, I conclude y^{m} trew. & since $d=0$, I draw the perpendicular to wc from b y^{e} begining of x, wch shall bee y^{e} Diameters of y^{e} lines *enm* & *dpl*. then in y^{t} diameter I take some point as b or h & from y^{t} point draw $gh=+f$ or $th=-f$, i.e. of any length, & paralell to bc. then from y^{e} points t or g perpendicular to tg I draw $ts=\dfrac{a\sqrt{dd-ff}}{b}$, or $gr=\dfrac{-a\sqrt{dd-ff}}{b}$. that is,

$$-a:b::th:ts::gh:gr.$$

(14) If we take $\hat{sht}=\hat{ghr}=\phi$, $f/d=\cos\phi=a/\sqrt{[a^2+b^2]}$ the transform

$$\begin{Bmatrix} x=\gamma\cos\phi=\dfrac{a}{\sqrt{[a^2+b^2]}}\gamma \\ y=\chi-\gamma\sin\phi=\chi-\dfrac{b}{\sqrt{[a^2+b^2]}}\gamma \end{Bmatrix}$$

reduces $x^2(bx+ay)=a^4$ to $\gamma^2\chi=a(a^2+b^2)$. Immediately, bh or $\gamma=0=x$ is the diameter of the cubic, with vbq or $\chi=0=bx+ay$ its asymptote.

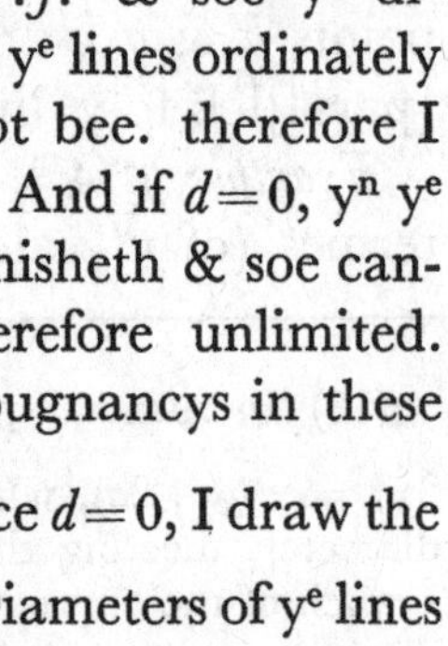

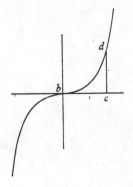

(15) Newton gives no figure, but where $bc=x$, $cd=y$, the equation $x^3=a^2y$ defines (d) to be the simple cubic (or Wallisian parabola), which has the origin b for its centre but no diameter.

& so through y^e^ points s & h or h & r I draw sr wch shall be parallel to y^e^ lines ordinately applied to y^e^ Diameter bh.(14)

Example y^e^ 5^t^. Suppose $x^3=aay$. Then by selecting those termes out y^e^ valors of x^3 & y in w^ch^ $\mathfrak{y}$ is of od dimensions, & sorting them together in wch y^e^ unknowne quantitys differ not, I have, $f^3\mathfrak{y}^3=0$; $3ddf\mathfrak{x}\mathfrak{x}\mathfrak{y}=0$; & 3dly

$$aad\mathfrak{y}\sqrt{dd-ff}=0.$$

by y^e^ first $f=0$, & therefore y^e^ 2^d^ vanisheth; & y^e^ 3^d^ divided by $aa\mathfrak{y}$ is,

$$dd\sqrt{dd-ff}=0; \quad \text{or} \quad 0=ddd.$$

Now since $d=f=0$, & y^e^ proportion of d to e & y^e^ length of c cannot bee found tis evident y^e^ line hath noe axis or diameter.(15)

B November 1664.

Observe [y^t^] y^e^ Axes, Diameters & position of y^e^ lines ordinately applied to y^m^ may bee for y^e^ most ꝑte easlier obteined by makeing

$bc=x.$ $cd=y.$ $bf=c=cq.$ $mh=\mathfrak{x}=fe.$ $hd=\mathfrak{y}=-hn.$ $hd:dg::d:f::\mathfrak{y}:\dfrac{f\mathfrak{y}}{d}=dg.$ y^e^ angles bcd, mgd, fqd, mbc, feh, right ones.

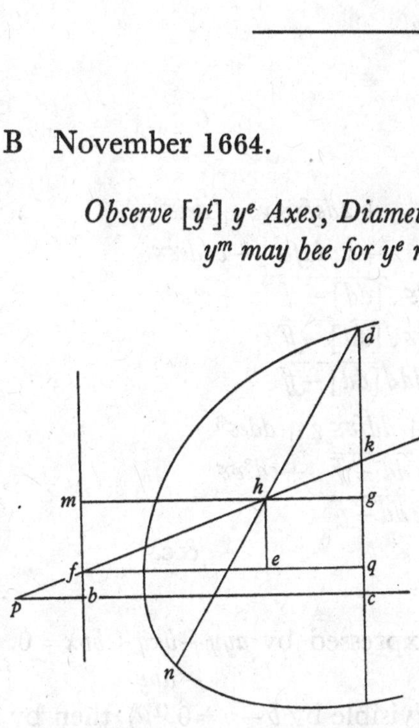

$$fe:he::d:e::\mathfrak{x}:\frac{e\mathfrak{x}}{d}=he=gq.$$

$$qc+gq+gd=cd=\frac{f\mathfrak{y}+e\mathfrak{x}+cd}{d}=y.$$

$$hg=\sqrt{\mathfrak{y}\mathfrak{y}\frac{-ff\mathfrak{y}\mathfrak{y}}{dd}}=eq.$$

$fe+eq=bc=\dfrac{d\mathfrak{x}+\mathfrak{y}\sqrt{dd-ff}}{d}=x.$ Then for readiness in these operations make a table of y^e^ □^s^, cubes, rectangles, &c of these valors of x & y. As was done before.(16)

---

(16) Where
$$\begin{cases} dx = d\mathfrak{x}+\sqrt{[d^2-f^2]}\mathfrak{y} \\ dy = f\mathfrak{y}+e\mathfrak{x}+cd \end{cases}$$

Newton tabulates $d^{r+s}x^ry^s$ for a few low values of r and s, as before, though now f and $\sqrt{[d^2-f^2]}$ are interchanged. (The transposition, yielding terms c^pf^q instead of $c^p(d^2-f^2)^{\frac{1}{2}q}$, makes for some computational simplification.)

$$dx=\mathfrak{y}\sqrt{dd-ff}+d\mathfrak{x}.$$

$$ddxx=\begin{array}{l} dd\,\mathfrak{y}\mathfrak{y}+2d\mathfrak{x}\mathfrak{y}\sqrt{dd-ff}+dd\mathfrak{x}\mathfrak{x}.\\ -ff \end{array}$$

$$d^3x^3=\begin{array}{llll} dd\,\mathfrak{y}^3\sqrt{dd-ff} & +3d^3\mathfrak{x}\mathfrak{y}\mathfrak{y} & +3dd\mathfrak{x}\mathfrak{x}\mathfrak{y}\sqrt{dd-ff} & +d^3\mathfrak{x}^3.\\ -ff & -3dff\mathfrak{x}\mathfrak{y}\mathfrak{y} & & \end{array}$$

$$dy=f\mathfrak{y}+e\mathfrak{x}+cd.$$

$$ddyy=\begin{array}{lll} ff\mathfrak{y}\mathfrak{y} & +2ef\mathfrak{x}\,\mathfrak{y} & +ee\mathfrak{x}\mathfrak{x}.\\ & +2cdf & +2cde\mathfrak{x}\\ & & +ccdd \end{array}$$

$$d^3y^3=\begin{array}{llll} f^3\mathfrak{y}^3 & +3eff\mathfrak{x}\,\mathfrak{y}\mathfrak{y} & +3eef\mathfrak{x}\mathfrak{x}\,\mathfrak{y} & +e^3\mathfrak{x}^3.\\ & +3cdff & +6cdef\mathfrak{x} & +3eecd\mathfrak{x}\mathfrak{x}\\ & & +3ccddf & +3ccdde\mathfrak{x}\\ & & & +c^3d^3 \end{array}$$

$$ddxy=\begin{array}{lll} f\mathfrak{y}\mathfrak{y}\sqrt{dd-ff} & +df\mathfrak{x}\mathfrak{y} & +de\mathfrak{x}\mathfrak{x}.\\ & +e\mathfrak{x}\mathfrak{y}\sqrt{dd-ff} & +cdd\mathfrak{x}\\ & +cd\mathfrak{y}\sqrt{dd-ff} & \end{array}$$

$$d^3xyy=\begin{array}{llll} ff\mathfrak{y}^3\sqrt{dd-ff}+ & dff\mathfrak{x}\,\mathfrak{y}\mathfrak{y} & +2def\mathfrak{x}\mathfrak{x}\,\mathfrak{y} & +dee\mathfrak{x}^3.\\ +2ef\mathfrak{x}\sqrt{dd-ff} & & +2cddf\mathfrak{x} & +cdde\mathfrak{x}\mathfrak{x}\\ +2cdf\sqrt{dd-ff} & +ee\mathfrak{x}\mathfrak{x}\sqrt{[dd]-ff} & & +ccd^3\mathfrak{x}\\ & +2cde\mathfrak{x}\sqrt{[dd]-ff} & & \\ & +ccdd\sqrt{[dd]-ff} & & \end{array}$$

$$d^3xxy=\begin{array}{llll} ddf\,\mathfrak{y}^3 & +dde\mathfrak{x}\,\mathfrak{y}\mathfrak{y} & +ddf\mathfrak{x}\mathfrak{x}\,\mathfrak{y} & +dde\mathfrak{x}^3.\\ -f^3 & -ffe\mathfrak{x} & +2de\mathfrak{x}\mathfrak{x}\mathfrak{y}\sqrt{dd-ff} & +cd^3\mathfrak{x}\mathfrak{x}\\ +2df\mathfrak{x}\sqrt{dd-ff} & & +2cdd\mathfrak{x}\mathfrak{y}\sqrt{dd-ff} & \\ & +cd^3 & & \\ & -dcff & & \end{array}$$

&c.

Example. If y[e] relation twixt bc & cd be expressed by $\begin{array}{l}ayy-bxy+bbx=0.\\ \quad\quad\quad -aby\end{array}$ (This line is a streight one y[e] equation being divisible by $b-y=0$.[(17)]) then by inserting those quanti[t]ys (of y[e] valors of x & y found by this table) in w[ch] $\mathfrak{y}$ is of odd dimensions, into [y[e]] place of yy, xy, y, x in this Equation, & supposeing those to destroy one another wch are multiplied by y[e] same unknowne quantitys there will bee these 2 Equations $2aef-bdf-be\sqrt{dd-ff}=0$, &

$$2acdf+bcd\sqrt{dd-ff}+bbd\sqrt{dd-ff}-abdf=0.$$

(17) This comment is a later marginal addition. In fact, the equation factorizes as the line-pair $(y-b)(ay-bx)=0$.

The 2ᵈ is divisible by $d=0$ & there results $2acf+bc\sqrt{dd-ff}+bb\sqrt{dd-ff}-abf=0$. Now to try wch of these two are true first I suppose $d=0$, & soe y^e^ first Equation will bee $2aef-be\sqrt{-ff}=0$. wch is impossible unlesse $f=0$, & y^n^ y^e^ valors of e & c canot bee found, Therefore $d=0$ is false. And therefore by y^e^ 2^d^ Equat^n^ $c=\dfrac{-bb\sqrt{dd-ff}+abf}{2af+b\sqrt{dd-ff}}$. & by y^e^ first $e=\dfrac{bdf}{2af-b\sqrt{dd-ff}}$. Or $2af \pm b\sqrt{dd-ff}:bf::d:e$. & $c=\dfrac{\pm bb\sqrt{dd-ff}+abf}{2af \mp b\sqrt{dd-ff}}$. Whence y^e^ proportion twixt d & f y^t^ is y^e^ angle fhn is undetermined.(18)

Endeavor not to find y^e^ quantity d in these cases, but suppose it given.(19) Or else

C ☞ For avoyding mistakes (wch might have happened in y^e^ 4th Example where I found $d=0$. & $f=\dfrac{-a\sqrt{dd-ff}}{b}$) it will not bee amisse to make

$$hd:dg::f:g::\mathfrak{y}:\frac{g\mathfrak{y}}{f}=dg. \quad \& \quad fe:he::d:e::\mathfrak{x}:\frac{e\mathfrak{x}}{d}=he.$$

& soe it will be $\dfrac{g\mathfrak{y}}{f}+\dfrac{e\mathfrak{x}}{d}+c=y$. & $\mathfrak{x}+\dfrac{\mathfrak{y}\sqrt{ff-gg}}{f}=x$. Or, $\dfrac{dg\mathfrak{y}+fe\mathfrak{x}+dfc}{df}=y$. & $\dfrac{f\mathfrak{x}+\mathfrak{y}\sqrt{ff-gg}}{f}=x$. And then observe y^t^ it can never happen y^t^ $f=0$, or $d+e=0$. Observe also y^t^ if $-d:e::g:+\sqrt{ff-gg}$. y^n^ y^e^ line fh is y^e^ Axis, otherwise y^e^ diameter of y^e^ crooked line.(20) when $d=0$ y^n^ [y^e^] axis is perpendicular to

(18) Newton adds no diagram, but in the accompanying figure (where $bc = x$, $cd = y$ and $(mh = \mathfrak{x})\, fh = \mathfrak{x}'$, $hd = \mathfrak{y}$ are the two co-ordinate systems, with $g\hat{h}k = \theta$ and $g\hat{h}d = \phi$) the transform

$$\begin{Bmatrix} x = \mathfrak{x}-\mathfrak{y}\cos\phi \\ y-c = \mathfrak{x}\tan\theta+\mathfrak{y}\sin\phi \end{Bmatrix}$$

reduces $0 = (y-b)(ay-bx)$ to

$$0 = (2a\sin\phi+b\cos\phi)^2\mathfrak{y}^2-b^2(\mathfrak{x}-a)^2$$

by taking

$$bf = c = \frac{b(a\sin\phi+b\cos\phi)}{2a\sin\phi+b\cos\phi}.$$

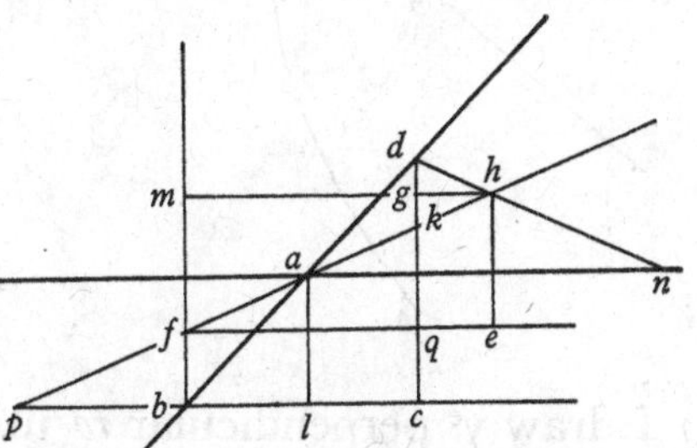

For arbitrary ϕ, pah (or $\mathfrak{y} = 0$) bisects all ordinates $\mathfrak{y}$; and similarly, though Newton does not note it, the line through the vertex $a \equiv (a, b)$ of the line-pair parallel to the ordinates $\mathfrak{y}$ (that is $\mathfrak{x}' = a\sec\theta$) bisects all lines parallel to pah.

(19) Newton added this last remark in the margin at a later date.

(20) Newton distinguishes between an 'Axis' and 'diameter': where the line pfh bisects all ordinates dn to the curve (d), then it is a diameter of the curve; and where the ordinates dhn are normal to the diameter pfh it is an axis.

x[,] as also if $c=\frac{a}{0}$:(21) And y^n it will be convenient to doe y^e worke over againe changing y^e names of x & y[,] y^t is writeing y instead of x & x instead of y.(22)

[D] December [1664].

Haveing y^e Diameter to find y^e Vertex of y^e line.

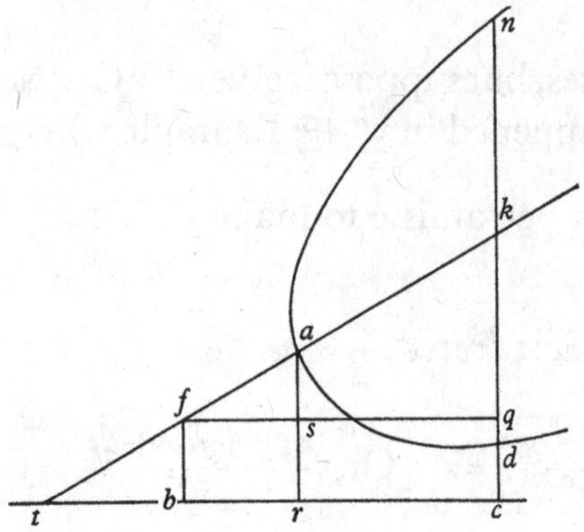

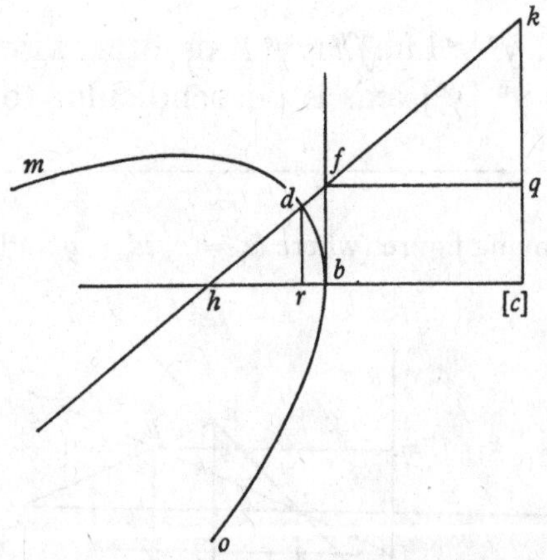

Suppose $bc=x$, or $tr=x$. cd or $ra=y$. $bf=c$. $fq:kq::d:e::fs=br=x$[:] $as=y-c$. y^t is $ex=dy-dc$. soe y^t into y^e given equation I insert this valor of $x=\frac{dy-dc}{e}$ or of $y=\frac{ex+dc}{d}$ into y^e place of x or [y] ([y^t] wch may more readily bee done).(23)

As in y^e first Example I found $d=e$, & y^e proportion twixt d & f to bee unlimited[,] so y^t if I would [have] fk to bee y^e Axis,(24) I make $-d:e::f:\sqrt{dd-ff}$. (vide C) or

$$f=-\sqrt{dd-ff}.$$

& there I found $c=\frac{af}{2f-2\sqrt{dd-ff}}$, or sinc $f=-\sqrt{dd-ff}$ it is $c=\frac{a}{4}$. As may bee seene in y^t Example. Now y^t I may find y^e vertex of y^e line [w^ch] was there exprest in these termes. $xx+ax-2xy+yy=0$, there results $ax+\frac{aa}{16}=0$. or $x=\frac{-a}{16}$. Therefore from y^e point b I draw $br=-\frac{a}{16}$. & from y^e point r I draw y^e perpendicular rd untill it cut y^e axis hd, y^t is, soe y^t $rd=hr$. & y^e point d shall bee y^e vertex of y^e Parab: mdo.(25)

---

(21) That is, when c is infinite.

(22) This interchange of x and y is needed to make Newton's transform general. Without it Newton can move his co-ordinate origin b only by a vertical translation to f parallel to the ordinate $cd = y$.

(23) Newton constructs the defining equation of the diameter fak as $dy-ex-cd = 0$, and finds its meet with the curve by eliminating between its equation and that of the curve. This meet of a curve and its diameter is, by definition, a 'vertex'.

Soe in y^e^ 2^d^ Example of y^e^ line $x^3+y^3=a^3$, it was found $d=e$. & $c=0$. & therefore $y=\dfrac{ex+dc}{d}$. or $y=x$. therefore I write x^3 for y^3 in y^e^ Equation $x^3+y^3=a^3$. & it is $2x^3=a^3$. or $x=\dfrac{a}{\sqrt{c:2}}$. therefore I take $br=\dfrac{a}{\sqrt{c:2}}$ & soe draw y^e^ perpendicular ar, which shall intersect y^e^ axis ab at y^e^ vertex of y^e^ crooked line. & y^n^ (calling $br=h$) it shall be $ar=c+\dfrac{eh}{d}[=h]$. Soe y^t^ in this case $ar=\dfrac{a}{\sqrt{[c:]2}}$.[26]

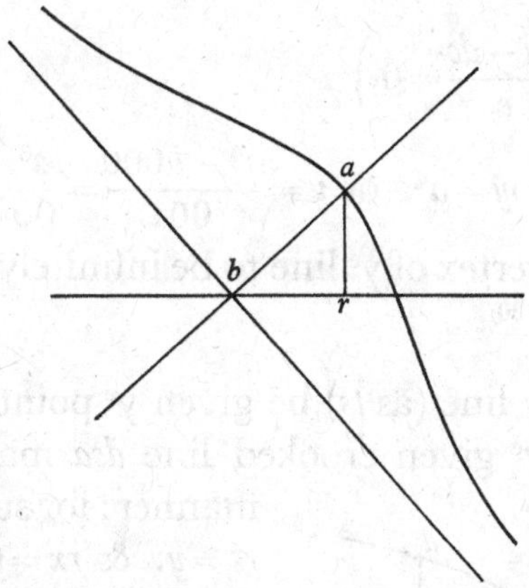

In y^e^ 3^d^ Example y^e^ Equation being $x^3-3xxy+2xyy-2ayy=0$, It was found, $c=0$. $d:e::q:0$.[27] y^t^ is $e=0$. therefore $y=\dfrac{ex+dc}{d}=0$. Therefore by writeing 0

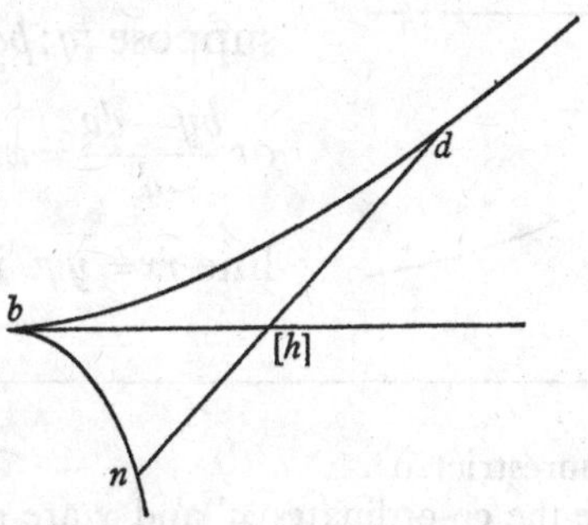

(24) Newton has cancelled 'diameter'.

(25) The condition that the diameter hfk be an axis is that $f\hat{h}d = \frac{1}{2}\pi$ or, since $f\hat{h}m = \frac{1}{4}\pi$, $\phi = d\hat{h}g = \frac{1}{4}\pi$; so that $(x-y)^2+ax = 0$ is transformed into $0 = 2\gamma^2+a\mathfrak{x}+\frac{1}{16}a^2$ by

$$\begin{Bmatrix} x = \mathfrak{x}+(1/\sqrt{2})\,\gamma \\ y-\frac{1}{4}a = \mathfrak{x}-(1/\sqrt{2})\,\gamma \end{Bmatrix},$$

whose axis is $0 = \gamma = (1/\sqrt{2})\,(x-y+\frac{1}{4}a)$. Reduction between $(x-y)^2+ax = 0$ and $x-y+\frac{1}{4}a = 0$ yields the vertex $(-\frac{1}{16}a, \frac{3}{16}a)$.

(26) Since, in Newton's figure for this example (on p. 197), $b\hat{h}d$ is right, ab is an axis of the cubic and the point a a principal vertex. The cubic should meet br normally in an inflexion.

(27) In accordance with his concluding remark in section B, Newton takes $d = q$, an arbitrary constant.

instead of y in y^e^ Equation all y^e^ termes vanish except $x^3=0$, or $x=0=br$. & $ar=c+\frac{eh}{d}=0$. soe y^t^ y^e^ vertex of y^e^ line (bdn) must bee at y^e^ point b.

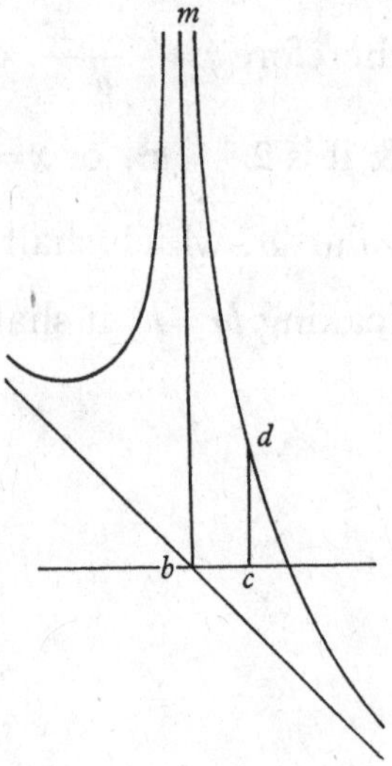

But in y^e^ 4th Example, $bx^3+axxy=a^4$. It was found $d:e::0:1$. or $d=0$. & c was unlimited,(28) I make therefore $c=0$. & since y^e^ axis(29) is perpendicular to x therefore I insert y^e^ valor of x into y^e^ equation;

$$\left(x=\frac{dy-dc}{e}=0.\right)$$

& there results $b000+a00y=a^4$. or $y=\frac{a^4-b000}{00a}=\frac{a^3}{00}$. Wherefore I conclude y^e^ vertex of y^e^ line to be infinitely distant from b towards m.(30)

[E] If y^e^ position of any line (as ts) be given y^e^ point ☞ where it intersects y^e^ given crooked line dsa may be found by y^e^ same manner; for suppose ar, or $ac=x$. cd, or $rs=y$. & $rx=yy$. $ta=a$. $tq=b$. $pq=d$. angles tqp, srt, dct right ones; y^n^, to find y^e^ point s where y^e^ crooked line dsa is intersected by y^e^ line tp, I suppose $tq:pq::tr:rs$. y^t^ is, $y=\frac{da+dx}{b}$. or $\frac{by-da}{d}=x$. & since by y^e^ nature of y^e^ line $rx=yy$. it follows y^t^ $yy=\frac{bry-dar}{d}$.

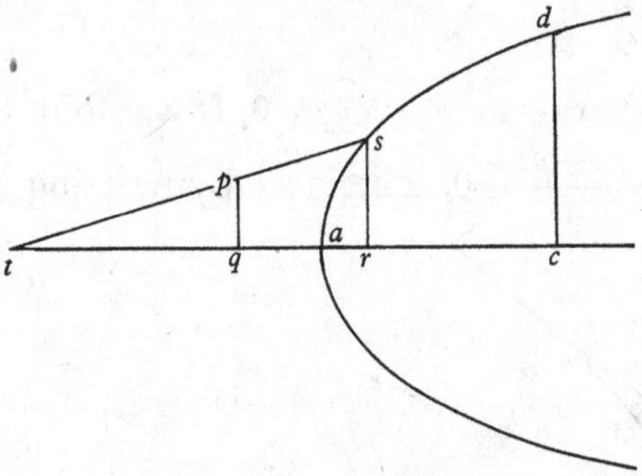

(28) Read 'indefinite' or 'unrestricted'.

(29) Read 'diameter' (since the co-ordinates $\mathfrak{x}'$ and $\mathfrak{y}$ are not inclined at right angles).

(30) That is, the vertex m is the point at infinity on the diameter.

(31) Newton finds the co-ordinates

$$\left(\frac{b^2r}{2d^2}-a\pm\frac{b}{d}\sqrt{\frac{b^2r^2}{4d^2}-ar},\ \frac{br}{2d}\pm\sqrt{\frac{b^2r^2}{4d^2}-a^2}\right)$$

the (two) meets of the line $dx-by+ad=0$ with the parabola $y^2=rx$ by eliminating y and x successively between the two equations.

(32) That is, where two curves have the Cartesian defining equations $f(x,y)=0$, $g(x,y)=0$, their meets are to be found by successive elimination of y and x between their equations. This basic axiom of Cartesian analytical geometry suggests the true theorem that, in general, two algebraic curves of respective degrees m and n meet in mn points (some or all of which may be non-real), but Newton here does not pursue the matter.

& $\dfrac{ddxx+2ddax+ddaa}{bb}=rx$. & by extracting y^e^ rootes of y^m^ both,

$$y=\frac{br}{2d}\ \text{ȣ}\ \sqrt{\frac{+bbrr-4ddar}{4dd}}.\quad \&\quad x=\frac{bbr}{2dd}-a\ \text{ȣ}\ \sqrt{\frac{b^4rr}{4d^4}-\frac{abbr}{dd}}.$$

therefore I take $ar=\dfrac{bbr}{2dd}-a\ \text{ȣ}\ \dfrac{b}{d}\sqrt{\dfrac{bbrr}{4dd}-ar}$. & $rs=\dfrac{br}{2d}\ \text{ȣ}\ \sqrt{\dfrac{bbrr}{4dd}-aa}$.(31)

By y^e^ same manner y^e^ intersection of 2 crooked lines may be found.(32)

F *Having y^e^ nature of any lines expressed in Algebraicall termes, to find its Asymptotes if [it] have any.*(33)

Suppose *rh*, & *rs* y^e^ asymptotes of y^e^ line *dtn* & *hd* parallel to *rs* drawne from y^e^ Asymptote to y^e^ line *dtn*. $bc=x$. $cd=y$. $bf=c$. $fe:he::d:e$. $hd:dg::f:g$.

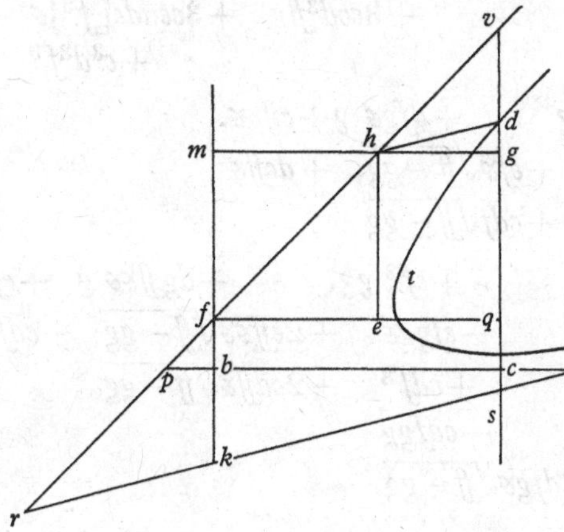

$fe=mh=\mathfrak{x}$. $hd=\mathfrak{y}$. y^e^ angles *dcb*, *cbm*, *dgm*, *bfq*, *hef*, to bee right ones. y^n^ is,

$$eh=\frac{e\mathfrak{x}}{d}.\quad dg=\frac{g\mathfrak{y}}{f}.\quad hg=\frac{\mathfrak{y}\sqrt{ff-gg}}{f}.\quad dg+gq+qc=\frac{dg\mathfrak{y}+ef\mathfrak{x}+dfc}{df}=y.$$

& $fe+eq=\dfrac{f\mathfrak{x}+\mathfrak{y}\sqrt{ff-gg}}{f}=x$. Now for readinesse in operation it [will] bee

---

(33) Much as in section A, Newton constructs the transform

$$\begin{Bmatrix} x = \mathfrak{x}+\mathfrak{y}\cos\phi \\ y-c = \mathfrak{x}\tan\theta+\mathfrak{y}\sin\phi \end{Bmatrix}$$

between the perpendicular co-ordinates $bc = x$, $cd = y$ and the oblique set $fh = \mathfrak{x}'$, $hd = \mathfrak{y}$, where $v\hat{h}g = h\hat{f}e = \theta$, $d\hat{h}g = \phi$ with $mh = \mathfrak{x} = \mathfrak{x}'\cos\theta$. He then tabulates $(fx)^r(dfy)^s$ in powers of $\mathfrak{x}$ and $\mathfrak{y}$ for $r = 0, s = 1, 2, 3$; $r = 1, s = 0, 1, 2$; $r = 2, s = 0, 1$; $r = 3, s = 0$.

convenient to have a table of these valors of x & y wch will bee y^e^ same wth y^t^ by wch y^e^ diameters of crooked lines are determined. viz.

$$fx = f\mathfrak{x} + \mathfrak{y}\sqrt{ff-gg}.$$

$$\begin{array}{llll} ffxx = & ff\,\mathfrak{y}\mathfrak{y} & +2f\mathfrak{x}\mathfrak{y}\sqrt{ff-gg} & +ff\mathfrak{x}\mathfrak{x}. \\ & -gg & & \end{array}$$

$$\begin{array}{lllll} f^3x^3 = & ff\sqrt{ff-gg}\,\mathfrak{y}^3 & +3fff\mathfrak{x}\,\mathfrak{y}^2 & +3ff\mathfrak{x}^2\mathfrak{y}\sqrt{ff-gg} & +f^3\mathfrak{x}^3. \\ & -gg & -3ggf\mathfrak{x} & & \end{array}$$

$$dfy = dg\mathfrak{y} + ef\mathfrak{x} + dfc.$$

$$\begin{array}{llll} ddffyy = & ddgg\mathfrak{y}\mathfrak{y} & +2defg\mathfrak{x}\,\mathfrak{y} & +eeff\mathfrak{x}\mathfrak{x} \\ & & +2ddcfg & +2cdeff\mathfrak{x} \\ & & & +ccddff \end{array}$$

$$\begin{array}{lllll} d^3f^3y^3 = & d^3g^3\mathfrak{y}^3 & +3ddefgg\mathfrak{x}\,\mathfrak{y}\mathfrak{y} & +3deeffg\mathfrak{x}\mathfrak{x}\,\mathfrak{y} & +e^3f^3\mathfrak{x}^3. \\ & & +3cd^3fgg & +6cddeffg\mathfrak{x} & +3cdee[f^3]\mathfrak{x}\mathfrak{x} \\ & & & +3ccd^3ffg & +3ccdde[f^3]\mathfrak{x} \\ & & & & +c^3d^3f^3 \end{array}$$

$$\begin{array}{llll} dffxy = & dg\sqrt{ff-gg}\,\mathfrak{y}\mathfrak{y} & +dfg\mathfrak{x}\,\mathfrak{y} & +eff\mathfrak{x}\mathfrak{x}. \\ & & +ef\mathfrak{x}\sqrt{ff-gg} & +dcff\mathfrak{x} \\ & & +cdf\sqrt{ff-gg} & \end{array}$$

$$\begin{array}{llllll} df^3xxy = & dffg\,\mathfrak{y}^3 & +ef^3\,\mathfrak{y}\mathfrak{y} & +dgff\mathfrak{x}\mathfrak{x}\,\mathfrak{y} & +ef^3\mathfrak{x}^3. \\ & -gggd & -efgg\mathfrak{x} & +2eff\mathfrak{x}\mathfrak{x}\sqrt{ff-gg} & +cdf^3\mathfrak{x}\mathfrak{x} \\ & & +cdf^3 & +2cdff\mathfrak{x}\sqrt{ff-gg} & \\ & & -cdfgg & & \\ & & +2dfg\mathfrak{x}\sqrt{ff-gg} & & \end{array}$$

$$\begin{array}{lllll} ddf^3xyy = & ddgg\mathfrak{y}^3\sqrt{ff-gg} & +ddfgg\mathfrak{x}\,\mathfrak{y}\mathfrak{y} & +2deffg\mathfrak{x}\mathfrak{x}\,\mathfrak{y} & +eef^3\mathfrak{x}^3. \\ & & +2defg\mathfrak{x}\sqrt{ff-gg} & +2cddffg\mathfrak{x} & +2cdef^3\mathfrak{x}\mathfrak{x} \\ & & +2cddfg\sqrt{ff-gg} & +eeff\mathfrak{x}\mathfrak{x}\sqrt{ff-gg} & +ccddf^3\mathfrak{x} \\ & & & +2cdeff\mathfrak{x}\sqrt{ff-gg} & \\ & & & +ccddff\sqrt{ff-gg} & \end{array}$$

This table may be continued when y^e^ nature of y^e^ lines are expressed by Equations of 4 or more dimensions.

This like y^e^ former rules will be better perceived by Examples y^n^ precepts. As

*Example y^e^* 1st. To find y^e^ asymptotes of y^e^ line whose nature is exprest by

(34) In fact Newton groups his substituted terms according to the scheme

$$rxx/q - yy + rx = 0.$$

$rx+\frac{rxx}{q}=yy$. first I write y^e^ valors of x, xx & yy (found by this table) into theire places in y^e^ Equation $rx+\frac{rxx}{q}-yy=0$. & there results(34)

$$\frac{rff\mathfrak{z}\mathfrak{z}-rgg\mathfrak{z}\mathfrak{z}+2rf\mathfrak{x}\mathfrak{z}\sqrt{ff-gg}+rff\mathfrak{x}\mathfrak{x}}{qff}\;\frac{\begin{array}{l}-ddgg\mathfrak{z}\mathfrak{z}-2defg\mathfrak{x}\mathfrak{z}-eeff\mathfrak{x}\mathfrak{x}\\-ccddff-2ddcfg\mathfrak{z}-2cdeff\mathfrak{x}\end{array}}{ddff}\;\frac{+rf\mathfrak{x}+r\mathfrak{z}\sqrt{ff-gg}}{f}=0.$$

or by ordering it,

$$\begin{array}{llllll}+\frac{r}{q} & \mathfrak{z}\mathfrak{z} & \frac{+2r\mathfrak{x}\sqrt{ff-gg}}{qf} & \mathfrak{z} & \frac{+r\mathfrak{x}\mathfrak{x}}{q} & =0.\\ \frac{-rgg}{qff} & & \frac{-2eg\mathfrak{x}}{df} & & \frac{-ee\mathfrak{x}\mathfrak{x}}{dd} & \\ \frac{-gg}{ff} & & \frac{-2cg}{f} & & \frac{-2ce\mathfrak{x}}{d} & \\ & & \frac{+r\sqrt{ff-gg}}{f} & & +r\mathfrak{x}-cc & \end{array}$$

Or,

$$\begin{array}{llllll}+ddffr & \mathfrak{z}\mathfrak{z} & +2ddfr\mathfrak{x}\sqrt{ff-gg} & \mathfrak{z} & +ddffr\mathfrak{x}\mathfrak{x} & =0.\\ -ddggr & & -2defgq\mathfrak{x} & & -eeffq\mathfrak{x}\mathfrak{x} & \\ -ddggq & & -2cddfgq & & +ddffqr\mathfrak{x} & \\ & & +ddfqr\sqrt{ff-gg} & & -2cdqeff\mathfrak{x} & \\ & & & & -ccddffq & \end{array}$$

Now by assumeing any valors for c; d, e; f, g. I have, by this equation, y^e^ relation w^ch^ $\mathfrak{x}$ beares to $\mathfrak{z}$, y^t^ is wch mh beares to hd. But y^t^ y^e^ valors of c, d, e, f & g, may be such y^t^ hr (to wch hd is applyed) may be one asymptote & hd paralell to y^e^ other, it is necessary (by Prop: 3^d^(35)) y^t^ neither $\mathfrak{x}$ nor $\mathfrak{z}$ bee any where of soe many or of more dimensions, y^n^ in those termes in wch they multiply one another. Therefore I consider of how many dimensions $\mathfrak{x}$ is at y^e^ most in any terme multiplied by $\mathfrak{z}$; & find y^m^ but of one. & therefore conclude y^t^ $\mathfrak{x}$ & $\mathfrak{z}$ ought to be found in noe terme in this Equation unlesse where they multiply one another.

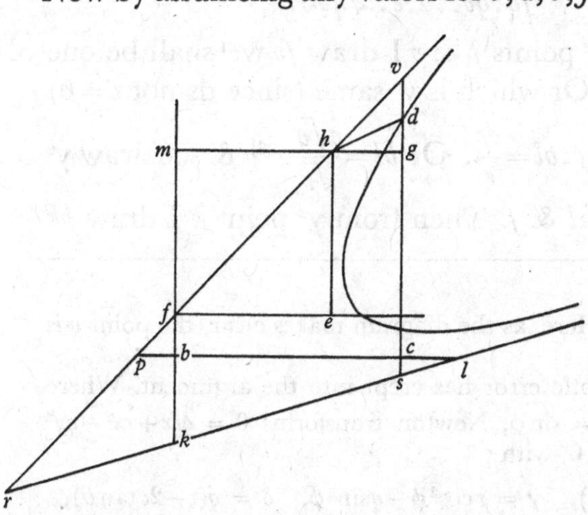

(35) That is, in §4 above.

G MOREOVER tis manifest y[t] y[e] Equation (expresing y[e] nature of y[e] given line) will ever be of one & but of one dimension more y[n] $\mathfrak{x}$ or $\mathfrak{y}$ in some termes in w[ch] they multiply one another: & therefore this may bee put for a GENERALL RULE. Viz. ALL THOSE termes must destroy one another in w[ch] there is not $\mathfrak{x}\mathfrak{y}$ & w[ch] are of as many, or want but one dimension of being as many dimensions as y[e] Equation is. Now that those termes destroy one another, tis necessary y[t] those be $=0$ in w[ch] y[e] unknowne quantitys $\mathfrak{x}$ & $\mathfrak{y}$ are y[e] same. Upon w[ch] considerations it will appeare y[t] in this Example I must make,

$$ddffr\mathfrak{y}\mathfrak{y}-ddggr\mathfrak{y}\mathfrak{y}-ddggq\mathfrak{y}\mathfrak{y}=0. \quad 2^{\text{dly}},\ +ddfqr\sqrt{ff-gg}\,\mathfrak{y}-2cddfgq\mathfrak{y}=0.$$

$$\text{thir[dl]y, } ddffr\mathfrak{x}\mathfrak{x}-eeffq\mathfrak{x}\mathfrak{x}=0. \quad 4^{\text{thly}},\ ddffqr\mathfrak{x}-2cdqeff\mathfrak{x}=0.$$

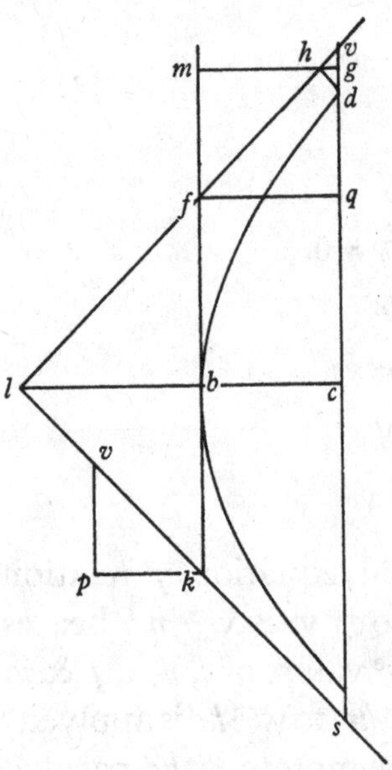

Or by dividing y[m] by those quantitys w[ch] neede not bee $=0$. they are [1st], $ddrff-ddggr-ddggq=0$. 2dly, $ddr\sqrt{ff-gg}-2cddg=0$. thirdly, $ddr-eeq=0$. 4^{thly}, $ddqr-2cdqe=0$. by y[e] 3[d], $d=e\sqrt{\dfrac{q}{r}}$. & since tis not $d=0$, by y[e] first $ff=gg+\dfrac{qgg}{r}$. by y[e] 2[d] $c=\dfrac{r\sqrt{ff-gg}}{2g}$. Or $c=\dfrac{\sqrt{rq}}{2}$. by y[e] 4th $\dfrac{dr}{2e}=c=\dfrac{\sqrt{rq}}{2}$. Therefore from y[e] point b I draw bf & $bk=c=\dfrac{\text{♉}\sqrt{rq}}{2}$. from f I draw fq paralell to bc. from q I draw qv, soe y[t]

$$fq:qv::d:e::\sqrt{q}:\sqrt{r}.$$

& through y[e] points f & v I draw fv w[ch] shall be one Asymptote. Or which is y[e] same (since tis not $c=0$) I make $e:d::c:bl=\dfrac{dc}{e}$. Or $bl=\dfrac{c\sqrt{q}}{\sqrt{r}}$.(36) & soe draw y[e] asy[m]ptote passing through y[e] points l & f. Then from y[e] point k I draw pk

(36) Or $bl = \frac{1}{2}q$.

(37) Newton's diagram has two points v. Here, as the diagram makes clear, the point v is in the asymptote lks.

(38) The construction is correct, but a subtle error has crept into the argument. Where $v\widehat{h}g = \theta$, $d\widehat{h}g = \phi$ with $e/d = \tan\theta$ and $g/f = \sin\phi$, Newton transforms $0 = qrx+rx^2-qy^2$ into the form $\alpha\mathfrak{x}^2+\beta\mathfrak{x}\mathfrak{y}+\gamma\mathfrak{y}^2+\delta\mathfrak{x}+\epsilon\mathfrak{y}+\zeta = 0$, with

$$\alpha = r-q\tan^2\theta,\quad \beta = 2(r\cos\phi-q\tan\theta\sin\phi),\quad \gamma = r\cos^2\phi-q\sin^2\phi,\quad \delta = q(r-2c\tan\theta),$$

$\epsilon = q(r\cos\phi-2c\sin\phi)$ and $\zeta = -qc^2$. He then seeks to reduce this still further to $\beta\mathfrak{x}\mathfrak{y}+\zeta = 0$ by making $\alpha = \gamma = \delta = \epsilon = 0$. However, where $fh = \mathfrak{x}' = \mathfrak{x}\sec\theta$ and $hd = \mathfrak{y}$ are not the same line, $v\widehat{h}g$ must be different from $d\widehat{h}g$ and so $\tan\theta \neq \tan\phi$. It follows that δ and ϵ cannot, as Newton would have it, be zero together. Taking $\alpha = \gamma = \delta$ we deduce easily that

paralell to *bl* & *pv*(37) paralell to *bk* soe y[t] (assumeing some other proportion twixt *d* & *e* y[n] before if there be any other) $d:e::pk:pv::-\sqrt{q}:\sqrt{r}$. & soe through y[e] points *k* & *v* I draw y[e] other Asymptote, Or since it is not $c=0$; I make $e:d::bk:\dfrac{dc}{e}=bl=\dfrac{+c\sqrt{q}}{[\sqrt{r}]}$. & soe through y[e] points *l* & *k* I draw y[e] other asymptote, w[ch] shall bee parallell to *hd*.(38)

Example y[e] 2[d]. Supose y[e] Asymptotes of $xx-yx+ay=0$ were to bee determined, Since I have noe use of y[e] termes in w[ch] is $\mathfrak{x}\mathfrak{y}$ I onely select those termes out of y[e] valors of xx, y & yx in w[ch] $\mathfrak{x}\mathfrak{y}$ is not & sorting them as was before taught I have these equations, 1st $ff\mathfrak{y}\mathfrak{y}-gg\mathfrak{y}\mathfrak{y}-g\mathfrak{y}\mathfrak{y}\sqrt{ff-gg}=0$. 2[dly] $ag\mathfrak{y}-c\mathfrak{y}\sqrt{ff-gg}=0$. 3[dly], $d\mathfrak{x}\mathfrak{x}-e\mathfrak{x}\mathfrak{x}=0$. 4[thly] $\dfrac{ae\mathfrak{x}}{d}-c\mathfrak{x}=0$. by y[e] third $d=e$. by y[e] 4th $a=c$. by y[e] 1st $g=+\sqrt{ff-gg}$. by y[e] 2[d] $c=a$.(39)

$\tan^2\theta = \tan^2\phi = r/q$, so that, say, $\tan\theta = -\tan\phi = \sqrt{[r/q]}$, with $c = r/(2\tan\theta) = \frac{1}{2}\sqrt{[qr]}$. The transform, therefore, is

$$\begin{Bmatrix} x = \mathfrak{x}+\mathfrak{y}\sqrt{[q/(q+r)]} \\ y-\frac{1}{2}\sqrt{[qr]} = \mathfrak{x}\sqrt{[r/q]}-\mathfrak{y}\sqrt{[r/(q+r)]} \end{Bmatrix}$$

and the reduced equation $0 = 2r\sqrt{[q/(q+r)]}\,(2\mathfrak{x}+q)\,\mathfrak{y}-\frac{1}{4}q^2r$. The curve (*d*) is then a hyperbola with asymptotes

$$0 = 2\mathfrak{x}+q = 1/\sqrt{r}\,(x\sqrt{r}+y\sqrt{q}+\tfrac{1}{2}q\sqrt{r}) \quad\text{and}\quad 0 = \mathfrak{y} = \tfrac{1}{2}\sqrt{[(q+r)/qr]}\,(x\sqrt{r}-y\sqrt{q}+\tfrac{1}{2}q\sqrt{r})$$

and centre *l* given by $x = -\frac{1}{2}q$, $y = 0$.

(39) Newton breaks off, leaving a space for the insertion of a diagram and details of the construction of the asymptote. Where $\theta = v\hat{h}g$ and $\phi = d\hat{h}g$ as before, the equation $x^2-xy+ay = 0$ is reduced by

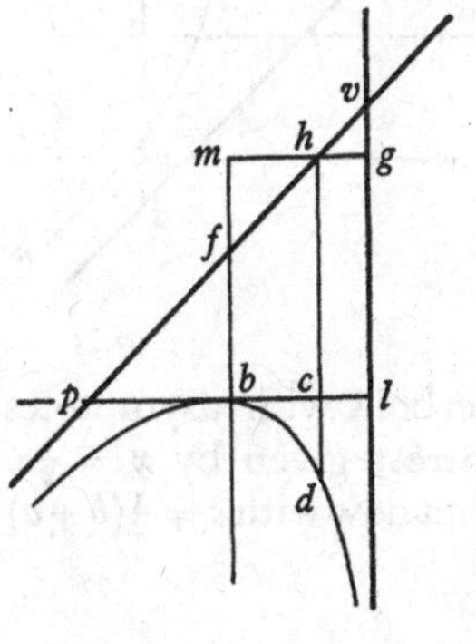

$$\begin{Bmatrix} x = \mathfrak{x}+\mathfrak{y}\cos\phi \\ y-c = \mathfrak{x}\tan\theta+\mathfrak{y}\sin\phi \end{Bmatrix}$$

to $\alpha\mathfrak{x}^2+\beta\mathfrak{x}\mathfrak{y}+\gamma\mathfrak{y}^2+\delta\mathfrak{x}+\epsilon\mathfrak{y}+\zeta = 0$, where $\alpha = 1-\tan^2\theta$,

$$\beta = 2\cos\phi-\tan\theta\cos\phi-\sin\phi, \quad \gamma = \cos^2\phi-\sin\phi\cos\phi,$$

$$\delta = a\tan\theta-c, \quad \epsilon = a\sin\phi-c\cos\phi, \quad \zeta = ac.$$

As before, δ and ϵ cannot, as Newton would have it, be zero together. However, taking $\alpha = \gamma = \delta = 0$, we may find $\theta = \frac{1}{4}\pi$, $\phi = -\frac{1}{4}\pi$ and $c = a$, with the equation reduced to

$$-(\mathfrak{x}-a)\,\mathfrak{y}+a^2 = 0 \quad\text{by}\quad \begin{Bmatrix} x = \mathfrak{x} \\ y-a = \mathfrak{x}-\mathfrak{y} \end{Bmatrix}.$$

The curve (*d*) is therefore a hyperbola with asymptotes $0 = \mathfrak{x}-a = x-a$ and

$$0 = \mathfrak{y} = x-y+a,$$

and centre at *v* given by $x = a$, $y = 2a$.

Example y^e 3^d, Suppose $xx-ax+by-yy=0$.(40) then by workeing as before I have these Equations, $ff-2gg=0$. $-a\sqrt{ff-gg}+bg-2cg=0$. $dd-ee=0$. $be-ad-2ce=0$. by y^e 4th $d=$ ♉ e. by y^e 1st $f=$ ♉ $g\sqrt{2}$ or $g=$ ♉ $\sqrt{ff-gg}$. by y^e 2^d (by suposeing $g=+\sqrt{ff-gg}$) tis $\frac{b-a}{2}=c$: & by y^e 4th (by supposeing $d=+e$) tis $\frac{b-a}{2}=c$. But by y^e 2^d (by supposing $g=-\sqrt{ff-gg}$) tis $\frac{a+b}{2}=c$. & by y^e 4th (by supposeing $d=-e$) tis $\frac{a+b}{2}=c$. Whence I conclude y^t when $c=\frac{b-a}{2}$ y^n is $d=e$, & $g=\sqrt{ff-gg}$; & when $c=\frac{b+a}{2}$ y^n is $d=-e$ & $g=-\sqrt{ff-gg}$.(41)

---

(40) Newton first took the equation '$yy = \frac{bx^2}{c}+yb$' (yet one more hyperbola) for his third example.

(41) Again Newton leaves room for future insertion of a diagram and details of the construction of the asymptotes. Where $v\widehat{h}g = \theta$, $d\widehat{h}g = \phi$, the transform

$$\begin{Bmatrix} x = \mathfrak{z}+\mathfrak{y}\cos\phi \\ y-c = \mathfrak{z}\tan\theta+\mathfrak{y}\sin\phi \end{Bmatrix}$$

reduces $0 = x^2-ax+by-y^2$ to

$$\alpha\mathfrak{z}^2+\beta\mathfrak{z}\mathfrak{y}+\gamma\mathfrak{y}^2+\delta\mathfrak{z}+\epsilon\mathfrak{y}+\zeta = 0,$$

with $\alpha = 1-\tan^2\theta$, $\beta = 2(\cos\phi-\tan\theta\sin\phi)$, $\gamma = \cos^2\phi-\sin^2\phi$, $\delta = -a+b\tan\theta-2c\tan\theta$, $\epsilon = -a\cos\phi+b\sin\phi-2c\sin\phi$, $\zeta = bc-c^2$.

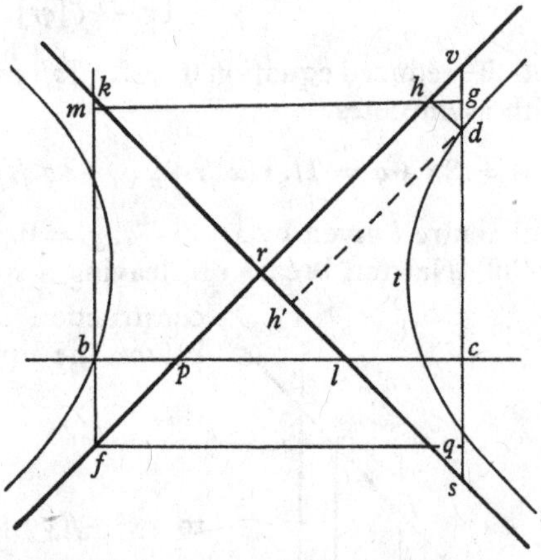

Once more Newton is wrong in thinking he can have $\delta = \epsilon = 0$. However, taking $\alpha = \gamma = \delta$, we may find $\theta = \frac{1}{4}\pi$, $\phi = -\frac{1}{4}\pi$ and $c = \frac{1}{2}(b-a)$, so that

$$\begin{Bmatrix} x = \mathfrak{z}+\mathfrak{y}/\sqrt{2} \\ y-\frac{1}{2}(b-a) = \mathfrak{z}-\mathfrak{y}/\sqrt{2} \end{Bmatrix}$$

reduces the given equation to $\sqrt{2}\mathfrak{y}(2\mathfrak{z}-a)+\frac{1}{4}(b^2-a^2) = 0$, a hyperbola with asymptotes $0 = 2\mathfrak{z}-a = x+y-\frac{1}{2}(a+b)$, $0 = \mathfrak{y} = 1/\sqrt{2}\,(x-y-\frac{1}{2}(a-b))$ and centre r given by $x = \frac{1}{2}a$, $y = \frac{1}{2}b$. The alternative values $\theta = -\frac{1}{4}\pi$, $\phi = \frac{1}{4}\pi$ yield the same result but now with $c = \frac{1}{2}(b+a)$ and the modified transform

$$\begin{Bmatrix} x = \mathfrak{z}+\mathfrak{y}/\sqrt{2} \\ y-\frac{1}{2}(a+b) = -\mathfrak{z}+\mathfrak{y}/\sqrt{2} \end{Bmatrix}$$

suitably reducing the equation.

2

WORK ON THE CARTESIAN SUBNORMAL

[Autumn 1664]

HISTORICAL NOTE

The problem of finding the tangent at a general point on a given algebraic curve exercised the minds of many of the most gifted mathematicians of the period between about 1630 and 1664, and indeed the problem was already resolved in the insights which came to both Descartes and Fermat while they pursued their 'querelle' in private correspondence in the summer of 1638. But none of this was yet published in 1664(1) and till the work of Barrow and Gregory appeared a few years afterwards the mathematical world at large had to rely on the inadequate exposition provided by Descartes in his *Geometrie*(2) and the few revealing remarks appended by Schooten in commentary.(3)

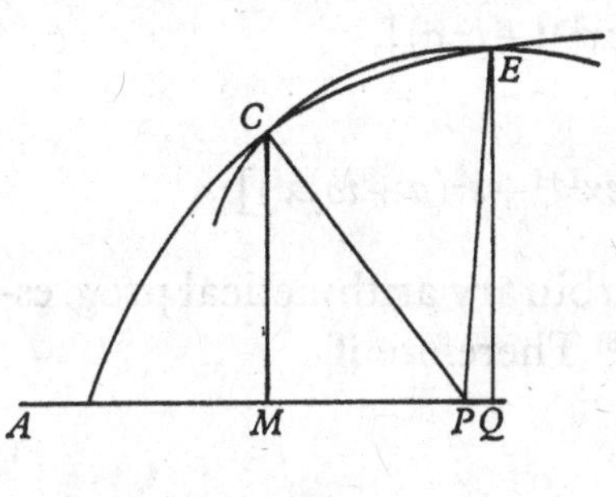

In 1637 Descartes' only method of finding the tangent at the point C on the arbitrary (smoothly continuous) algebraic curve (C) required the prior construction of the subnormal at the point, with the tangent drawn perpendicular to the subnormal through C. The subnormal from P to the curve was deduced from the condition that the circle (C) with centre P and radii $CP=EP$ should meet the curve in a double point (and so be tangent to it). This condition requires that E and C should be the same point, or, where CM and EQ are ordinates to the abscissa AP, that the segment MQ should vanish. Analytically, taking $AM=x$, $MC=y$, with the curve (C) defined by $y=f(x)$ and $AP=v$, $PC=s$, we deduce that

$$\left\{\begin{matrix} PC^2=CM^2+MP^2 \\ s^2=y^2+(v-x)^2 \end{matrix}\right\}.$$

(1) The bulk of the correspondence was first published by Clerselier in his *Lettres de M. Descartes*, Tome III (Paris, 1667) (*Où il répond a plusieurs difficultez qui luy ont esté proposées sur la Dioptrique, la Geometrie et sur plusieurs autres sujets*). See also Gaston Milhaud, *Descartes savant* (Paris, 1921): ch. VII: 149–62.

(2) See *Geometria*: Liber II: 43–50.

(3) Schooten, *Commentarii in Librum II*=*Geometria*: 246–53.

Descartes' subnormal condition, therefore, is that

$$[f(x)]^2+(v-x)^2-s^2=0$$

have a double root.

The condition that the algebraic equation $0=\phi(x)$ have a double root Descartes found by equating it term by term with the polynomial

$$0=(x-e)^2\times\sum_i(c_ix^i),$$

which clearly has the double root $x=e$. In Schooten's example of the parabola defined by $y^2=rx$[4], we have $s^2=rx+(v-x)^2$ and comparison with $(x-e)^2=0$ yields $-2e=r-2v$, or $v=e+\frac{1}{2}r$ (where e is the double root and may be interchanged with x). Thus $MP=v-x=\frac{1}{2}r$, constant.[5]

Quickly, however, computation of the subnormal by Descartes' method becomes tedious.[6] In particular a great deal of time was wasted in calculating constants c_i (in the equated polynomial $(x-e)^2\times\sum_i(c_ix^i)=0$) which were later themselves to be eliminated from the calculation. In the second Latin edition of *Geometrie* Schooten printed for the first time an algorithm for finding double roots which he had had from Hudde in private correspondence.[7]

Changing Hudde's notation slightly to accord with Descartes', we may show that, given the polynomial

$$(x-e)^2\times\sum_i(c_ix^i)=\sum_i[c_i(x^{2+i}-2ex^{1+i}+e^2x^i)],$$

the polynomial

$$\sum_i[c_i((a+(2+i)b)\,x^{2+i}-2(a+(1+i)b)\,ex^{1+i}+e^2(a+ib)x^i)]$$

derived by multiplying the terms in order by an arbitrary arithmetical progression of constant difference b has the factor $x-e$.[8] Therefore if

$$(x-e)^2\times\sum_i(c_ix^i)=0$$

(4) *Geometria*: 246.

(5) More generally and in modern terms, where $MQ=e$ and $EQ=y'=f(x+e)$, CP^2 (or $y^2+(v-x)^2$) $=EP^2$ (or $y'^2+(v-x-e)^2$), that is,

$$\frac{y'-y}{e}(y'+y)=2(v-x)-e(v-x)^2;$$

so that $$\frac{dy}{dx}=\lim_{e\to 0}\left(\frac{y'-y}{e}\right)=\frac{v-x}{y},\quad\text{or}\quad MP=v-x=y\frac{dy}{dx},$$

the required subnormal.

(6) Compare Schooten's lengthy justification (*Commentarii in Librum II*, *O = Geometria*: 249–53) of Descartes' construction for the subnormal to the conchoid.

(7) This letter of 27 January 1658 was printed a year later, in Schooten's Latin translation, as *Johannis Huddenii Epistola Secunda, de Maximis et Minimis* (= *Geometria*: 507–16).

(8) Specifically, where $A_i=a+ib$,

$$\sum_i[c_i(A_{2+i}x^{2+i}-2A_{1+i}ex^{1+i}+e^2A_ix^i)]=\sum_i[c_i(A_i(x-e)^2+2bx(x-e))\,x^i].$$

and so has the double root $x=e$, then also the derived polynomial must be zero. In other words, where $G(x)$ is derived from $F(x)$ by multiplying its terms in order by an arbitrary arithmetical progression, a general condition for $F(x)=0$ to have a double root is that $G(x)=0$.(9)

As Hudde went on to state in his letter, though without proof, the algorithm is easily extended to finding a general condition for $F(x)=0$ to have n equal roots. In particular, where $F(x)=(x-e)^3\times\sum_i(c_i x^i)$ with $G_1(x)$ derived from $F(x)$ by multiplying its terms in order by an arbitrary arithmetical progression of constant difference b and $G_2(x)$ similarly derived from $G_1(x)$ by multiplying its terms in order in the same sense by a second arbitrary arithmetical progression of the same(10) difference b, then $F(x)=0$ demands that $G_1(x)=G_2(x)=0$ likewise.(11) This algorithm for finding a triple root Newton was to make one of the foundations of his analytical investigations into the curvature of algebraic curves in the winter of 1664/5.(12)

(9) Where $F(x)$ is the polynomial $(x-e)^2\times\sum_i(c_i x^i)$, then $G(x)$ is the polynomial

$$aF(x)+bxF'(x),$$

with $F'(x)$ the derivative of $F(x)$. It follows that application of Hudde's algorithm to find the condition that $F(x)=0$ have a double root is strictly equivalent to the more familiar test $F(x)=F'(x)=0$.

(10) This restriction (not to be found in Hudde's statement in *Geometria*: 509) was in May 1665 to cause Newton some worry before he realized its necessity. (See 4, §3, note (34) below.)

(11) Where $F(x)=(x-e)^3\times\sum_i(c_i x^i)$ with $A_i=a+ib$, $B_i=\alpha+ib$, we may take

$$\begin{aligned}G_1(x) &= \sum_i[c_i(A_{3+i}x^{3+i}-3A_{2+i}ex^{2+i}+3A_{1+i}e^2x^{1+i}-A_ie^3x^i)]\\ &= \sum_i[c_i(A_i(x-e)^3+3bx(x-e)^2)\,x^i]\\ &= aF(x)+bxF'(x),\end{aligned}$$

and

$$\begin{aligned}G_2(x) &= \sum_i[c_i(A_{3+i}B_{3+i}x^{3+i}-3A_{2+i}B_{2+i}ex^{2+i}+3A_{1+i}B_{1+i}e^2x^{1+i}-A_iB_ie^3x^i)],\\ &= \sum_i[c_i(A_iB_i(x-e)^3+3(A_i+B_i+b)\,bx(x-e)^2+6b^2x^2(x-e))\,x^i]\\ &= a\alpha F(x)+(a+\alpha+1)\,bxF'(x)+b^2x^2F''(x).\end{aligned}$$

Clearly $G_1(x)$ has the factor $(x-e)^2$ and $G_2(x)$ the factor $(x-e)$, so that if $F(x)=0$, then also $G_1(x)=G_2(x)=0$ holds and so is a general condition for $F(x)=0$ to have a triple root. Further, $F(x)=G_1(x)=G_2(x)=0$ implies that $F(x)=F'(x)=F''(x)=0$, a more familiar condition for a triple root.

(12) Section 4 below, *passim*.

§1 *Early calculations using Descartes' subnormal method*

[Early Autumn 1664]

From the original in Newton's Waste Book in the University Library, Cambridge

[1](1)

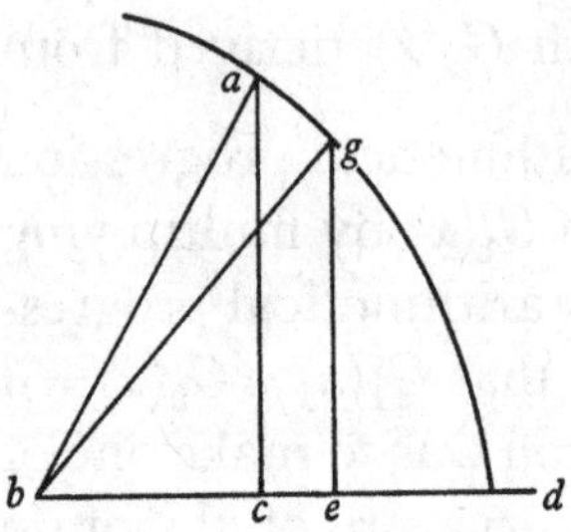

$$bd=v.^{(2)}\ cd=y.\ ac=x.\ ab=s.$$

$$\begin{array}{l}-xx+xy+ay=0.\\ \qquad +ax+yy\end{array}$$

$$\begin{array}{l}\dot{vv}+\dot{xy}+\dot{ay}=ss=\dot{vv}-2ov+\dot{xy}+xo+\dot{ax}+\dot{ay}\\ \quad +\dot{ax}+\dot{yy} \qquad\qquad\qquad\qquad +ao+\dot{yy}+2oy\end{array}$$

$$2ov=ox+ao+2oy.^{(3)}$$

$$\left[\begin{array}{l}vv+xy+ay=ss\\ \quad +ax+yy\end{array}\right]=vv+2ov+oo+rs+ra+as+ss.$$

$$v=\frac{\begin{array}{l}xy+ax+ay+yy\\ -rs-ar-as-ss\end{array}}{2o}.^{(4)}$$

$[bd=v.\ cd=y.\ ac=x.\ ab=s.]\ ed=z.\ eg=\xi.$

$$\left[\begin{array}{l}-xx+xy+ay=0.\\ \qquad +ax+yy\end{array}\right]$$

$$xy+ax+ay+2yy+vv-2vy=ss=\xi z+a\xi+az+2zz+vv-2vz.^{(5)}$$

$$\frac{xy-\xi z+ax-a\xi}{2y-2z}=b.\quad xy-\xi z-2by+2bz=0.\quad xy-2by=0.\quad x=2b.$$

$$\frac{xxyy-\xi\xi zz}{a^2x-\xi a^2}=b.^{(6)}$$

---

(1) Add. 4004: 6ᵛ, probably to be dated September 1664.

(2) In his first calculations Newton takes rather $bc = v$.

(3) Taking $bc = v$, Newton finds $s^2 = v^2+(x+y)(a+y)$ and then seeks to apply Descartes' method by considering the increment $ce = -o$: thus $v \to v-o$ and $y \to y+o$ (with o^2 neglected), but he ignores the increment of the ordinate $x(o[dx/dy])$. This seems the first use Newton made of the notation o for an increment of the basic variable (here y), though he had introduced it two pages earlier in the Waste Book in an optical problem. (See the appendix on geometrical optics in Part 3 below, especially 1, §1, note (21).) Note Newton's use of dots to cancel equal terms on either side of the equation: these should not be confused with Newton's later dot-notation for fluxions.

(4) Newton, in a second attempt, tries to allow for the increment in the ordinate by considering the limit-increases $v \to v+o$ (with the power o^2 omitted), $x \to r$ and $y \to s$ (not to

$bd = v.\ cd = x.\ ac = y.\ ab = s.$

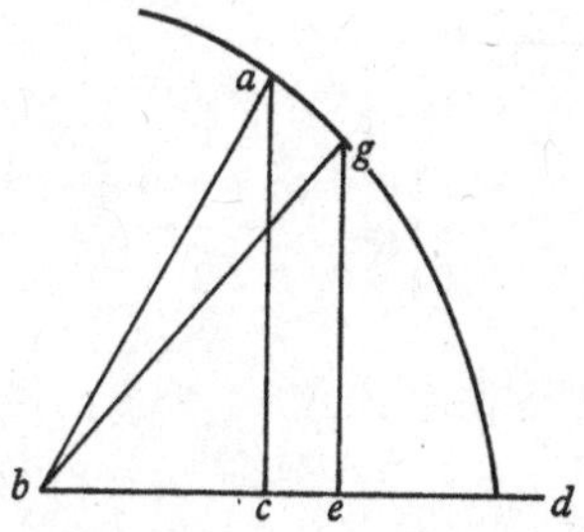

$$0 = yy - 2xy + ax - x^2.$$

$$y = x \pm \sqrt{2xx - ax} = ac.\ yy = 3xx - ax \pm 2\sqrt{2x^4 - ax^3}.$$

$$eg^2 = 3x^2 - 6ox - ax + ao$$
$$\pm [2]\sqrt{2x^4[-8ox^3] - ax^3[+3aoxx]}.$$

or $\quad eg^2 = 3z^2 - az \pm [2]\sqrt{2z^4 - az^3}.$(7)

$$[(v-x)^2 + yy = ss = (v-z)^2 + eg^2.]$$

$$\frac{\begin{array}{l}4x^2 - ax \pm 2\sqrt{2x^4 - ax^3}\\ -4z^2 + az \mp 2\sqrt{2z^4 - az^3}\end{array}}{-2z + 2x} = v.\text{(8)}$$

$$\frac{4x^2 - 4zz}{2x - 2z} = b. \qquad \begin{array}{l}2x^2 - 2z[^2] - bx + bz\,[=0.]\\ \ 2 \qquad\quad 0 \qquad\ 1 \quad\ 0\end{array} \qquad b = 4x.\text{(9)}$$

$$\left[\frac{\pm 2\sqrt{2x^4 - ax^3} \mp 2\sqrt{2z^4 - az^3}}{2x - 2z} = c.\right] \qquad \pm\sqrt{2x^4 - ax^3} \mp \sqrt{2z^4 - az^3} = cx - cz.$$

---

be confused with the denomination of *ab*). Clearly $r = -o$, but he abandons his calculation after expressing v as the limit-ratio

$$\frac{1}{2}\left(\frac{(xy - rs) + a(x-r) + a(y-s) + (y^2 - s^2)}{o}\right).$$

(5) Applying Descartes' method to $s^2 = x^2 + (v-y)^2$, or $s^2 = (x+y)(a+y) + (v-y)^2$, Newton considers the equation which results from allowing the limit-increases $x \to \xi$ and $y \to z$.

(6) From the limit-equality it follows that

$$v = \frac{(xy - \xi z) + a(x - \xi) + a(y-z) + 2(y^2 - z^2)}{2(y-z)}.$$

Newton then seeks to calculate v as $\alpha + \beta + \frac{1}{2}a + \gamma$, where

$$\alpha = \text{limit}\left(\frac{xy - \xi z}{2(y-z)}\right), \quad \beta = \text{limit}\ \tfrac{1}{2}a\left(\frac{x - \xi}{y - z}\right) \quad \text{and} \quad \gamma = \text{limit}\ (y+z) \quad \text{or} \quad 2y.$$

The last term '$\dfrac{xxyy - \xi\xi zz}{a^2x - \xi a^2}$' seems inexplicable.

(7) Newton introduces the increment $ce = -o$ in $cd = x$, but straightaway modifies his notation by taking $ed = x - o = z$.

(8) Since $v^2 - 2vx + x^2 + 3x^2 - ax \pm 2\sqrt{[2x^4 - ax^3]} = v^2 - 2vz + z^2 + 3z^2 - az \pm 2\sqrt{[2z^4 - az^3]}$.

(9) Newton attempts now to evaluate the limit-quotients in the value of v separately. First, taking $b = [2(x^2 - z^2)]/(x-z)$, it follows that $-b(x-z) + 2(x^2 - z^2) = 0$ must have a double root $x = z$ and so Newton applies Hudde's algorithm.

$$2x^4+2z^4-ax^3-az^3-2zx\sqrt{[4]x^2z^2-[2]axz^2+aazx-2azxx}$$
$$=ccxx-2c^2zx+cczz.^{(10)}$$

$$[v=]\frac{4x^2-4z^2-ax+az \mho 2\sqrt{2x^4-ax^3} \Omega 2\sqrt{2z^4-az^3}}{2x-2z}$$

$$=2x+2z-\frac{1}{2}a \mho \sqrt{2xx-ax} \mho \frac{2xx+2zx-ax}{2\sqrt{\frac{1}{2}zz-\frac{azz}{4x}}}.^{(11)}$$

$$yy-2xy+ax-xx=0.\ [y-x=\mho\sqrt{2xx-ax}.]$$

$$\frac{-2yy+ay-2xy}{-2y+2x}+x=v.\quad x+y\ \frac{+ay[-4xy]}{-2y+2x}=v.^{(12)}$$

$[2]^{(13)}$

$$ss-v^2+2vy-yy=\frac{a^4}{yy}.\quad [v=]\ \frac{-2a^4}{2y^3}+2y.^{(14)}$$

$$[0]\ [0]\ [1]\ [2]\ [-2]$$

$$ss-vv+2vx-xx=\frac{r^4}{xx}.\quad \frac{-2r^4}{2xxx}+\frac{2xx}{2x}=v.^{(15)}$$

$$0\quad 0\quad 1\quad 2\quad -2$$

---

(10) Newton tries to evaluate

$$c=\frac{\pm(\sqrt{[2x^4-ax^3]}-\sqrt{[2z^4-az^3]})}{x-z}$$

by squaring to remove the radical signs. However, this yields c as the root of a polynomial and so Newton abandons this attempt also.

(11) Read $\dfrac{\text{‘}\mho\, 2xz+2zz-az\text{’}}{2\sqrt{2xx-ax}}$ (or an equivalent), since

$$\pm z\left(\frac{\sqrt{[2x^2-ax]}-\sqrt{[2z^2-az]}}{x-z}\right)=\pm z\left(\frac{2([x^2-z^2]/[x-z])-a}{\sqrt{[2x^2-ax]}+\sqrt{[2z^2-az]}}\right),$$

with $z\to x$ in the limit.

(12) For, substituting $\mho\sqrt{[2x^2-ax]}=y-x$, with $z=x$, $v=4x-\frac{1}{2}a+(y-x)+\dfrac{4x^2-ax}{2(y-x)}$. Newton has at last succeeded in differentiating an algebraic expression involving radicals by a direct consideration of the ratios of limit-increments, though it is possible that the last two lines represent a later application of the subnormal algorithm as a check on his result that,

$$v=4x-\tfrac{1}{2}a\pm\left(\sqrt{[2x^2-ax]}+\frac{x(4x-a)}{2\sqrt{[2x^2-ax]}}\right).$$

(13) Add. 4004: 5v.

§2 *First systematic evaluation of derivatives by Descartes' subnormal method and Hudde's algorithm*

[Autumn 1664?]

From the original in a pocket-book in the University Library, Cambridge[1]

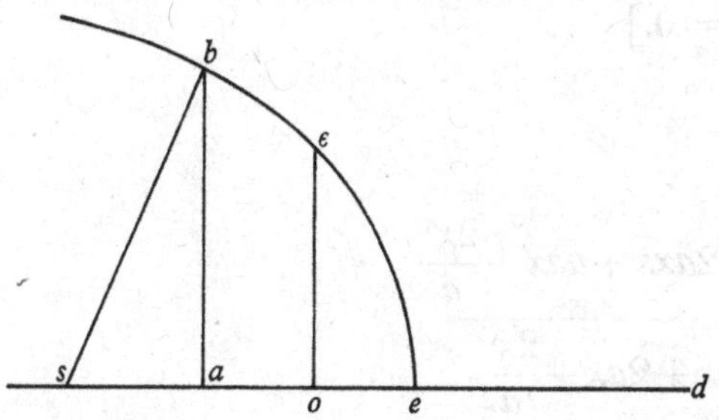

[1] $\left[y=a\sqrt{\frac{a+x}{x}}.\right]$

Let $ed=a$. $ae=x$. $ab=y$. $se=v$.

$sb=s$. $ss=y^2+vv+xx-2vx$.

$aax+a^3=yyx$. $y^2=ss-vv+2vx-xx$.

$aax+a^3=ssx-vvx+2vx^2-x^3$.

$$x^3-2vx^2+vvx+a^3=0.$$
$$\qquad\qquad -ss$$
$$\qquad\qquad +aa$$
$$2 \quad 1 \quad 0 \quad -1$$
$$2x^3 \quad -2vx^2 \quad * \quad -a^3=0. \quad \frac{2x^3-a^3}{2xx}=v.^{(2)}$$

---

[2] $\left[y=a\sqrt{\frac{x}{a+x}}.\right]$

$aax=yyx+yya$. $y^2=ss-vv+2vx-xx$.

$aax=ssx-vvx+2vxx-x^3+ssa+2avx-axx-avv$. [or]

$$x^3-2vx^2-ssx-ssa=0.$$
$$\quad +a \quad +vv \quad +avv$$
$$\qquad -2av$$
$$\qquad +aa$$

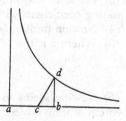

(14) Read '$\frac{-2a^4}{2y^3}+y$'. Newton finds by Descartes' method and Hudde's algorithm the subnormal $ac = v = y+x(dx/dy)$ to the hyperbola (d) defined by $xy = a^2$, where $ab = y$ and $bd = x$.

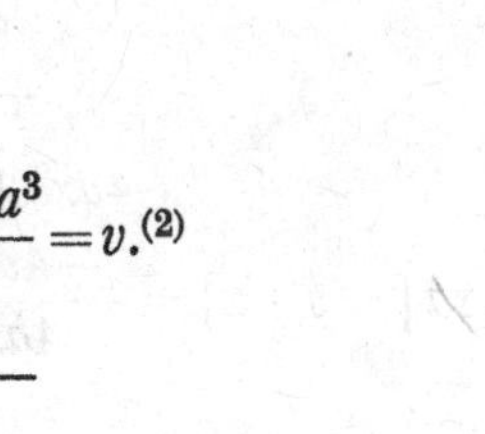

(15) Similarly, Newton finds the subnormal $v = x+y(dy/dx)$ to the hyperbola whose defining equation is $xy = r^2$.

(1) Add. 4000: 93v and the recto sides of 94–113, 115, 116.

(2) Newton uses Hudde's algorithm to calculate the subnormal sa to the curve (b) whose defining equation is $xy^2 = a^2(x+a)$, and so he would have immediately that $y(dy/dx) = sa = v-x = -(a^3/2x^2)$.

$$x^2-2ex+ee\times x+f[=]x^3-2ex^2+eex+eef=0.^{(3)}$$
$$\qquad\qquad\qquad\qquad +f \quad -2ef$$

$$-2v+a=f-2e. \quad [\text{or}]\ f=a+2e-2v. \quad eef=avv-ass.$$

$$\frac{-eea-2e^3+2eev+avv}{a}=ss.$$

$$\left.\begin{matrix} vvx-2vxx\ +x^3-a \\ vva-2vax+ax^2-x \\ +aax \end{matrix}\right\} \frac{eev+avv-2e^3-eea}{a} \quad [=0.]$$

$$[\text{or}] \quad \begin{matrix} avv & -2vx^2 & +x^3=0. \quad [e=x.] \\ x & -2vax & +ax^2 \\ -a & -2vee & +aax \\ -x-\dfrac{2x}{a} & eev & +2e^3 \\ & & +eea \\ & & +\dfrac{2x}{a}e^3 \\ & & +xee \end{matrix} \qquad v=\frac{4x^3+2axx+aax+\dfrac{2x^4}{a}}{4xx+2ax+\dfrac{2x^3}{a}}.$$

$$[\text{or}]\ v=\frac{4ax^2+2aax+a^3+2x^3}{4ax+2aa+2xx}$$

$$sa\,[=v-x]=\frac{\begin{matrix}4ax^2+2x^3+2aax+a^3\\ -4ax^2-2x^3-2aax\end{matrix}}{4ax+2aa+2xx}. \quad sa=\frac{a^3}{2x^2+4ax+2aa}.^{(4)}$$

$$o\epsilon=p. \quad [oe=x.] \quad \sqrt{\frac{aax}{a+x}}:\frac{a^3}{2x^2+4ax+2aa}::p:z.$$

$$\frac{z^2aax}{a+x}=\frac{a^6pp}{\begin{matrix}4x^4+16ax^3+8aaxx+16a^3x+4a^4\\ +16aa\end{matrix}}.$$

[or] $a^5pp+a^4xpp=4zzx^5+16zzax^4+24aazzx^3+16zza^3x^2+4a^4z^2x.$ divided by $x+a$ it produceth.

$$4zzx^4+12zzax^3+12zzaax^2+4zza^3x-a^4pp=0.^{(5)}$$

---

(3) For some reason Newton here returns to Descartes' method of finding a double root by equating coefficients with those of $(x-e)^2(x+f) = 0$.

(4) Newton finds $y(dy/dx) = v-x = a^3/[2(a+x)^2]$.

(5) Where z is the subnormal at the point ϵ given by $eo = x$, $o\epsilon = p$, Newton calculates

$$\frac{dp}{dx} = \frac{z}{p} = \frac{a^2x^{\frac{1}{2}}}{2(a+x)^{\frac{3}{2}}}.$$

(6) Newton seems to have calculated this directly from the inverse of the fundamental theorem of indivisibles, $d(x^n)/dx = nx^{n-1}$ (n integral). Here

$$v-x = y\frac{dy}{dx} = \frac{x^3}{a^2}\frac{d}{dx}\left(\frac{x^3}{a^2}\right) = \frac{3x^5}{a^4}.$$

[3] $\left[y=\frac{x^3}{a^2}\right]$.

$x^3=aay.\ v=\frac{3x^5}{a^4}+x.$(6)

---

[4] $\left[y=a\sqrt[3]{\frac{x}{a}}.\right]$

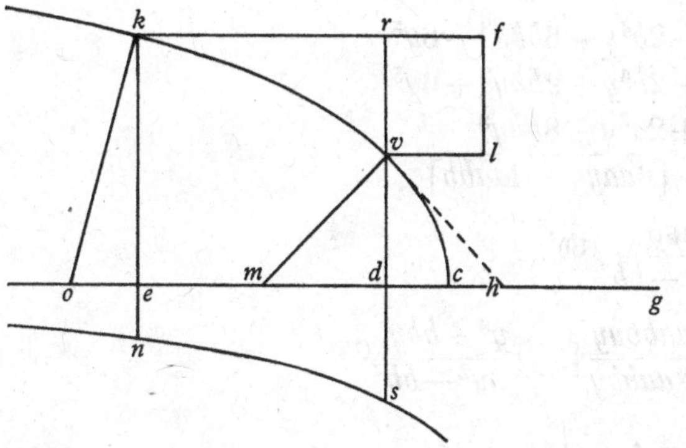

$aax=y^3.\ cd=a.\ ce=x.\ ek=y.\ oc=v.$
$ok=s.\ ss=y^2+x^2-2vx+v^2.$

$x^2=2vx+ss-y^2-v^2.\ x=v-\sqrt{+ss-y^2}.$

$aav-aa\sqrt{ss-y^2}=y^3.$

$a^4v^2-2aavy^3+y^6-a^4ss+a^4y^2=0.$
$\quad 0 \quad 3 \quad 6 \quad 0 \quad 2$

$6y^6-6aavy^3+2a^4yy=0.$

$v=\frac{a^4+3y^4}{3aay}.\quad [oe=]\,v-x=\frac{a^4+3y^4-3y^4}{3aay}.\quad v-x=\frac{a^4}{3aay}=\frac{aa}{3y}.$

$y:\frac{aa}{3y}::\frac{y^3}{aa}:\frac{aay^3}{3aayy}=\frac{y}{3}.$(7)

make $md=dv.$ $\left[\frac{aa}{3y}=y.\right]$ that is $aa=3yy.$ & $\frac{a}{\sqrt{3}}=y=md=dv.\ x=dc=\frac{a}{\sqrt{27}}.$

[make] $ds=dc=\sqrt{\frac{aa}{27}}.\ \sqrt{c:aax}:\frac{aa}{3y}::\sqrt{\frac{aa}{27}}:z=en.$ [or] $aax:\frac{a^6}{3aax}$(8)$::\frac{a^3}{27\sqrt{27}}:z^3.$

$aax^2z^3=\frac{a^7}{[729]\sqrt{27}}.\ xxz^3=\frac{a^5}{[729]\sqrt{27}}.$ wch expresseth y^e^ nature of y^e^ crooked line *ns.* & [if] $vl=ds.\ vlfr=dens.$(9)

---

(7) That is $y:y(dy/dx)=x:y/3$, or $x(dy/dx)=\frac{1}{3}y$.

(8) Read '$\frac{a^6}{27aax}$'. (The mistake is here corrected in the subsequent calculations.)

(9) Where $\alpha=a/3\sqrt{3}$, then $en=z=\alpha\frac{dy}{dx}=ds\times\frac{d(ek)}{d(ec)}$.

It follows that $(vlfr)=\int_\alpha^y \alpha.dy=\int_\alpha^x z.dx=(dens),$

since $vl=ds=\alpha$. Newton thus has a geometrical transform equating the area $\int z.dx$ under the 'derivative' curve (n) or $\phi(z,x)=0$, and the area of the rectangle $(vlfr)$ or $\alpha(y-\alpha)$. Newton will deepen his knowledge of this transform and make it the basis for his geometrical (constructive) definition of an integral. (See section 4 below.)

[5] $[y^3=b^2y+a^2x.]$

$aax+bby=y^3.\quad x=v-\sqrt{ss-yy}.\quad aav-aa\sqrt{ss-y^2}+bby=y^3.$

$$aav+bby-y^3=aa\sqrt{ss-y^2}.$$

$$\begin{array}{cccccc} a^4v^2+2aavbby & -2aavy^3 & +b^4y^2 & -2bby^4 & +y^6 & [=0.] \\ -a^4ss & & +a^4y^2 & & & \\ 0 \quad 1 & 3 & 2 & 4 & 6 & \end{array}$$

$$\begin{array}{r} -2b^4y+6bby^3-6y^5 \\ +2b^4y+2bby^3+6y^5 \end{array}$$

$$\frac{6y^5-8bby^3+2b^4y+2a^4y}{-2aabb+6aayy}=v.\quad v-x=\frac{+2a^4y-8bby^3}{[6aayy-2aabb]}.$$

$$v-x=\frac{2a^4y}{6aay^2-2aabb}.\quad [\text{or}]\ v-x=\frac{aay}{3y^2-bb}.^{(10)}$$

$$\frac{-bby+y^{3\,(11)}}{aa}:\frac{aay}{3y^2-bb}:[:]\frac{y^3-bby}{aa}:\frac{aay^4-aabbyy}{3aay^3-aabby}[=]\frac{y^3-bby}{3y^2-bb}.^{(12)}$$

[Take $md=dc$.] $\dfrac{aay}{3y^2-bb}=\dfrac{y^3-bby}{aa}.\quad a^4=3y^4-4bbyy+b^4.$

$$y^4=\frac{4bbyy-b^4+a^4}{3}.$$

[If] $a=b.\ y^2=\dfrac{4bb}{3}.\ y=\dfrac{2b}{\sqrt{3}}=dm=dv.^{(13)}$

$$\frac{8b^3}{3\sqrt{3}}-\frac{2bbb}{\sqrt{3}}-aax[=0.]\ [x=]\frac{2b}{3\sqrt{3}}=dc=ds.$$

$y:\dfrac{aay}{3y^2-bb}::\dfrac{2b}{3\sqrt{3}}:z.\ yz=\dfrac{2aaby}{9y^2\sqrt{3}-3bb\sqrt{3}}.$ [or] $9yyz-3bbz=2\dfrac{aab}{\sqrt{3}}$. An equation expressing y^e^ nature of y^e^ line *ns*.(14)

---

(10) Newton calculates $$y\frac{dy}{dx}=v-x=\frac{a^2y}{3y^2-b^2}.$$

(11) Read 'y' $\left(\text{and not } x=\dfrac{y^3-b^2y}{a^2}\right)$.

(12) That is, $x\dfrac{dy}{dx}=\dfrac{y^3-b^2y}{3y^2-b^2}$.

(13) Read '$y=\dfrac{2b}{\sqrt{3}}=dv$' (since Newton has taken $md=dc$).

(14) Much as before Newton calculates the defining equation of the 'derivative' curve (n), that is $\phi(x,z)=0$ where $z=\dfrac{2b}{3\sqrt{3}}\dfrac{dy}{dx}$.

(15) These two terms are lacking in Newton's text, and the rest of his calculation is vitiated accordingly.

(16) This is derived by an application of Hudde's algorithm, with the equation multiplied term by term by the first row of numbers and then ordered.

[6] $[y^3=x(by+a^2).]$

$$aax+byx=y^3.\quad x=v-\sqrt{ss-yy}.\quad aav+byv-y^3=\frac{aa}{by}\sqrt{ss-yy}.$$

$$\begin{array}{llllll}
a^4v^2+2aav^2by & -2aavy^3 & +bbvvy^2 & -2bvy^4 & +y^6\ [=0.] & \\
-a^4ss[-2a^2bs^2y]^{(15)} & [+2a^2by^3]^{(15)} & +a^4yy & +bb & & \\
 & & -bbss & & & \\
-2 & -1 & +1 & 0 & 2 & 4 \\
0 & 1 & 3 & 2 & 4 & 6
\end{array}$$

$$ss=\frac{2a^4vv+2aavvby+[2]aavy^3+4bvy^4-2bby^4-4y^6}{2a^4}.^{(16)}$$

$$\left.\begin{array}{l}
+8y^8bb[+]12a^4y^6-2aabbvy^5-16a^4bvy^4 \quad -12a^6vy^3+4a^8yy+4a^6vvby \\
\qquad -8b^3vy^6 \qquad\qquad\qquad +8a^4bby^4-4aavvb^3 \\
\qquad +4b^4y^6
\end{array}\right\}[=]0.$$

[or]
$$\left.\begin{array}{r} +4a^6by \\ -4aab^3y^3 \end{array}\right| vv=\frac{\begin{array}{ll} 12a^6y^3\,v & -8a^4bby^4 \\ 2aab^2y^5 & -8y^8bb \\ 16a^4by^4 & -4b^4y^6 \\ 8b^3y^6 & +12a^4y^6 \end{array}}{4a^6by-4aab^3y^3}.$$

$$v=\frac{\begin{array}{r} 6a^6y^2 \\ +a^2b^2y^4 \\ +8a^4by^3 \\ +4b^3y^5 \end{array}}{4a^6b-4aab^3y^2}\ \text{♉}\ \sqrt{\frac{\begin{array}{ll} 36a^{12}y^4 & +8aab^5y^9 \\ +12a^8b^2y^6 & +64a^8bby^6 \\ +16a^{10}by^5 & +16b^6y^{10} \\ +24a^6b^3y^7 & \\ +65a^4b^4y^8 & \end{array}}{16a^{12}b^2-32a^8b^4y^2+16a^4b^6y^4}\quad\frac{\begin{array}{r} -8a^4bby^3 \\ -8y^7bb \\ -4b^4y^5 \\ \\ +12a^4y^5 \end{array}}{+4a^6b-4aab^3y^2}}.^{(17)}$$

---

(17) The incorrect calculation is printed to show the structure of Newton's argument and as an implicit comment on Newton's arithmetical accuracy. Correctly, where $\alpha = a^2$, $\beta = by$, it follows that

$$(\alpha+\beta)\,v-y^3 = (\alpha+\beta)\sqrt{[s^2-v^2]}:$$

squaring and ordering the powers of y, we deduce

$$\begin{array}{llllll}
\alpha^2(v^2-s^2) & +2\alpha\beta(v^2-s^2) & +\beta^2(v^2-s^2) & -2\alpha y^2(vy-\beta) & -\beta y^2(2vy-\beta) & +y^6 = 0. \\
 & & +\alpha^2y^2 & & & \\
-2 & -1 & 0 & 1 & 2 & 4 \\
0 & 1 & 2 & 3 & 4 & 6
\end{array}$$

Applying Hudde's algorithm twice, we have

$$s^2\alpha(\alpha+\beta) = v^2\alpha(\alpha+\beta)+vy^3(\alpha+2\beta)-\beta(\alpha+\beta)\,y^2-2y^6,$$

and
$$s^2\beta(\alpha+\beta) = v^2\beta(\alpha+\beta)-vy^3(3\alpha+4\beta)+(\alpha+\beta)\,(\alpha+2\beta)\,y^2+3y^6,$$

[7] $\left[y=a\sqrt[4]{\frac{x}{a}}.\right]$

$$a^3x=y^4.\quad x=v-\sqrt{ss-yy}.\quad a^3v-y^4=a^3\sqrt{ss-yy}.\quad \begin{matrix} a^6v^2-2a^3vy^4+y^8+a^6yy\ [=0]. \\ -a^6ss \\ 0\quad 4\quad 8\quad 2 \end{matrix}$$

$$v=\frac{4y^6+a^6}{4a^3y^2}.\quad v-x=\frac{4y^6+a^6-4y^6}{4a^3y^2}=\frac{a^3}{4y^2}.\quad y:\frac{a^3}{4y^2}::\frac{y^4}{a^3}:\frac{a^3y^4}{4a^3y^3}=\frac{y}{4}.$$ (18)

---

[8] $[y^3=a^2(x-y).]$

$$aax-aay=y^3.\quad aav-aay-y^3=aa\sqrt{ss-yy}.$$

$$\begin{matrix} a^4v^2-2a^4vy-2aavy^3+2a^4y^2+2aay^4+y^6\ [=0.] \\ -a^4ss \\ 0\quad 1\quad 3\quad 2\quad 4\quad 6 \end{matrix}$$

$$v=\frac{3y^5+4aay^3+2a^4y}{a^4+3aayy}.\quad v-x=\frac{3y^5+4aay^3+2a^4y-aay^3-3y^5-ya^4-3aay^3}{a^4+3aayy}.$$

[or] $v-x=\dfrac{aay}{aa+3yy}.\quad y:\dfrac{aay}{a^2+3y^2}::\dfrac{a^2y+y^3}{aa}:\dfrac{a^4y^2+a^2y^4}{a^4y+3a^2y^3}=\dfrac{aay+y^3}{aa+3yy}.$ (19)

---

[9] $[a^2x=y^2(b+y).]$

$$aax=by^2+y^3.\quad aav-by^2-y^3=aa\sqrt{ss-yy}.$$

$$\begin{matrix} a^4v^2-2aabyyv-2aavy^3+bby^4+2by^5+y^6\ [=0]. \\ -a^4ss\ +a^4[yy] \\ 0\quad 2\quad 3\quad 4\quad 5\quad 6 \end{matrix}$$

$$v=\frac{3y^4+5by^3+2bby^2+a^4yy}{2aab+3aay}.$$

$$v-x=\frac{3y^4+5by^3+2bby^2-2bby[^2]-3by^3-2by^3-3y^4+a^4}{2aab+3aay}.\quad \text{[or]}$$

$$v-x=\frac{a^4}{2aab+3aay}.$$

[. . . . .(20)]

---

two values for s^2 derived by multiplying by the first and second row of numbers respectively. Hence, by elimination of s^2,

$$vy^3(\alpha+\beta)(3\alpha+2\beta) = y^6(3\alpha+2\beta)+y^2(\alpha+\beta)^3,$$

or
$$v = \frac{y^3}{\alpha+\beta}+\frac{(\alpha+\beta)^2}{(3\alpha+2\beta)\,y} = x+\frac{(a^2+by)^2}{(3a^2+2by)\,y}.$$

(18) That is $x(dy/dx) = \frac{1}{4}y$. (19) Or $x(dy/dx)$.

[10] $[axy=b^3+y^3.]$

$$axy=y^3+b^3. \quad ayv-y^3-b^3=ay\sqrt{ss-yy}.$$

$$\begin{matrix} aavvy^2 & -2avy^4 & -2avb^3y & +y^6 & +2b^3y^3 & +b^6\ [=0.] \\ -aass & +aa & & & & \\ 0 & 2 & -1 & 4 & 1 & -2 \end{matrix}$$

$$v=\frac{4y^6+2b^3y^3+2aay^4-2b^6}{4ay^4-2ab^3y}. \quad v-x=\frac{4y^6+2b^3y^3+2aay^4 \begin{smallmatrix}-2b^6\\+2b^6\end{smallmatrix} -4y^6+2b^3y^3-4b^3y^3}{4ay^4-2ab^3y}.$$

[or] $v-x=\dfrac{ay^3}{2y^3-b^3}. \quad y:\dfrac{ay^3}{2y^3-b^3}[::]\dfrac{y^3+b^3}{ay}:\dfrac{y^4+b^4}{2y^3-b^3}.$

---

[11] $\left[y=a\sqrt{\dfrac{a}{x}}.\right]$

$a^3=xyy.$ $a=dg=dp.$[21] $go=x.$ $oa=y.$ $cg=v.$ $ca=s.$ $ss=y^2+xx-2vx+v^2.$

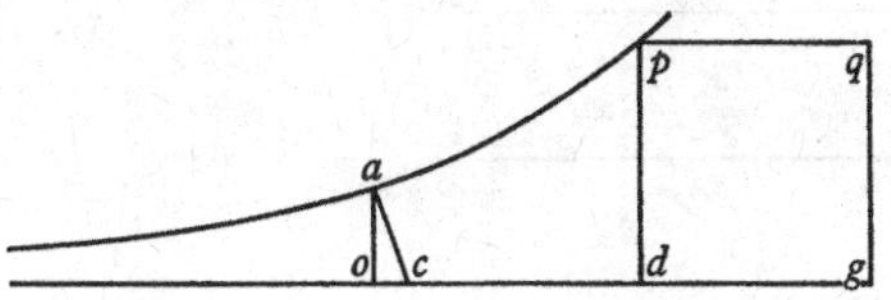

$$\begin{matrix} a^3 & -xss & +x^3 & -2vx^2 & +vvx=0. \\ -1 & 0 & 2 & 1 & 0 \end{matrix}$$

$$\frac{2x^3-a^3}{2x^2}=v. \quad x-v=\frac{a^3}{2x^2}.$$

$\sqrt{\dfrac{a^3}{x}}:\dfrac{a^3}{2xx}::p:z.$[22] $\dfrac{zza^3}{x}=\dfrac{a^6pp}{4x^4}.$ $4x^3zz=a^3pp.$

$\left[x=\dfrac{a^3}{y^2}.\right]$ $go=y.$ $oa=x.$ &c: $cg=v.$ $ca=s.$ $a^3=xyy.$ $ss=xx+yy+vv-2vy.$

$xx=ss-yy+2vy[-]vv.$

$$\begin{matrix} a^6 & -ssy^4 & +y^6 & -2vy^5 & [+]vvy^4=0. \\ -4 & 0 & 2 & 1 & 0 \end{matrix}$$

$v=\dfrac{2y^6-4a^6}{2y^5}.$ $y-v=\dfrac{2a^6}{y^5}.$[23] $\dfrac{a^3}{y^2}:\dfrac{2a^6}{y^5}::p:z.$ $z=\dfrac{2a^3p}{yyy}.$ $zy^3=2a^3p.$ wch shewes y^e^ nature of another crooked line y^t^ may be squared.[24]

---

(20) Newton's remaining calculations are faulty and not worth reproducing. Their gist is an attempt, parallel to previous cases, to calculate the defining equation $\phi(x, z) = 0$ of the curve (n), where $en = z = ds\times(oe/ek)$.

(21) The point p is therefore on the curve (a).

(22) Where $z = -p(dy/dx)$ is the ordinate of the derivative curve $\phi(x, z) = 0$.

(23) Newton calculates, conversely, $-x(dx/dy) = y-v = (2a^6/y^5)$.

(24) Specifically, where $zy^3 = 2a^3p$,

$$\int_0^y z.dy = -\int_0^x p.dx = -px.$$

[12] $[y=\sqrt[3]{ax^2}.]$

$$axx=y^3.\quad x=v-\sqrt{ss-yy}.\quad x^2=v^2-2v\sqrt{ss-y^2}+ss-y^2.$$

$$av^2+ass-ay^2-y^3=2av\sqrt{ss-yy}.$$

$$\begin{array}{llllll} a^2v^4 & -2aassy^2-2av^2y^3+aay^4-2ay^{5\,(25)}+y^6\ [=0.] \\ +aas^4 & +2vvaay^2-2assy^3 \\ -2a^2v^2ss \end{array}$$

$$+3\qquad 1\qquad 0\qquad -1\qquad -2\qquad -3$$

$$\begin{array}{l} 3a^2s^4-2aay^2ss+3a^2v^4+2v^2a^2y^2-a^2y^4+4ay^5-3y^6\ [=0.] \\ -6aavv \end{array}$$

$$ss^{(26)}=\frac{\dfrac{2yy}{3}}{\dfrac{+6vv}{3}}-\sqrt{\frac{\begin{array}{l}4y^4a^2+24aayyvv+36 \\ -9aav^4-6aayyvv+12ay^5 \\ +3aay^4+9y^6\end{array}}{9aa}}^{(27)}.$$

$$ss^{(26)}=\frac{2y^2+6v^2}{3}-\sqrt{\frac{9y^6-12ay^5+7aay^4+18aavvy^2+27aav^4}{9aa}}.^{(28)}$$

[13] $\left[y=\dfrac{a^2}{x}.\right]$

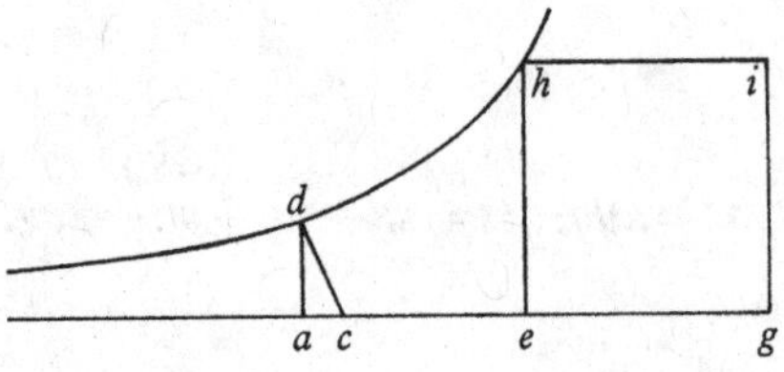

$aa=xy.$ $eg=eh=a.$[29] $ga=x.$ $ad=y.$ $dc=s.$ $cg=v.$ $y^2=ss-xx+2vx-vv.$

$$\begin{array}{l} -a^4+x^2s^2-x^4+2vx^3=0. \\ \qquad -x^2v^2 \end{array}$$

$$+2\qquad 0\qquad -2\qquad -1$$

$$\frac{2x^4-2a^4}{2x^3}=v.\quad x-v^{(30)}=\frac{a^4}{x^3}.\quad \frac{aa}{x}:\frac{a^4}{x^3}::p:z.^{(31)}\quad zxx=a^2p.$$

(25) Read '$+2ay^5$'. Newton's mistake vitiates his remaining calculations.

(26) Read '$2ss$'.

(27) Read '$-\sqrt{\dfrac{\begin{array}{l}4y^4a^2+24aayyvv+36aav^4 \\ -36aav^4-24aayyvv+48ay^5 \\ +12aay^4+36y^6\end{array}}{9aa}}$', supposing that the equation on the previous line were correct.

(28) Having made several mistakes, Newton abandons his calculation. Correctly,

$$3a^2s^4-2a^2(y^2+3v^2)\,s^2+3a^2v^4+2v^2a^2y^2-a^2y^4-4ay^5-3y^6=0,$$

or

$$s^2=v^2+\frac{y^2}{3}\pm\frac{2ay^2+3y^3}{3a}.$$

[14] $\left[y=\pm\frac{a^2}{x}.\right]$

$a^4=xxyy.$(32)

[15] $\left[x=\frac{a^4}{y^3}.\right]$

$a^4=xy^3.\quad ad=x.\quad [ga=y.\quad cg=v.\quad dc=s.]\quad -a^8+y^6ss-y^8+2vy^7-vvy^6=0.$

$\qquad\qquad\qquad\qquad\qquad\qquad\qquad\qquad +6 \quad 0 \quad -2 \quad -1 \quad 0$

$\frac{2y^8-6a^8}{2y^7}=v.\quad v-y=[-]\frac{3a^8}{2y^7}.\quad \frac{a^4}{y^3}[:]\frac{3a^8}{2y^7}::p[:]z.$(33) $\frac{za^4}{y^3}=\frac{3a^8p}{2y^7}.\quad zy^4=\frac{3a^4p}{2}.$

[16] $\left[y=a\sqrt[3]{\frac{a^2}{x^2}}.\right]$

$a^5=xxy^3.$(34)

[17] $\left[y=a\sqrt{\frac{a}{x}}.\right]$

$a^3=xyy.\quad x=v+\sqrt{ss-yy}.\quad a^3-vyy=yy\sqrt{ss-yy}.\quad a^6-2a^3vyy+vvy^4+y^6=0.$

$\qquad\qquad\qquad\qquad\qquad\qquad\qquad\qquad\qquad\qquad -ss$

$\qquad\qquad\qquad\qquad\qquad\qquad\qquad -4 \quad -2 \quad 0 \quad 2$

$[v=]\,\frac{-2y^6+4a^6}{2a^3y^2}.\quad x-v=\frac{2y^6-2a^6}{2a^3y^2}.$(35)

$2a^3y^3:2y^6-2a^6:[:]p:z.\quad 2a^3y^3z=2y^6p-2a^6p.\quad py^6-a^3y^3[z]-a^6p=0.$

Again, $s^2 = (v-x)^2+y^2$. Therefore, $2x(v-x) = y^2-x^2-(s^2-v^2)$, or, substituting for s^2-v^2 and x,

$$y\frac{dy}{dx} = v-x = \frac{a^{\frac{1}{2}}}{2y^{\frac{3}{2}}}\left[y^2\left(\frac{a-y}{a}\right)-\frac{y^2}{3}\pm\frac{2ay^2+3y^3}{3a}\right]$$

$$= \frac{2\sqrt{[ay]}}{3},$$

taking the positive value of the double sign.

(29) And so the point e is on the hyperbola.

(30) Here $-y(dy/dx)$. (31) Where $z = -p(dy/dx)$.

(32) Newton passes on without further remark, presumably because he noticed immediately that $0 = x^2y^2-a^4 = (xy+a^2)(xy-a^2)$ holds and so reduces to the previous case.

(33) Where $z = -p(dx/dy)$. The correct value of $v-y$ should be $-3a^8/y^7$.

(34) The rest of the page is left blank for future addition of the computations of

$$v-x = y(dy/dx)$$

and of the equation $\phi(x, z) = 0$ of the derivative curve, but Newton never returned to it.

(35) Read ‘$[v =]\,\frac{-2y^6+4a^6}{4a^3y^2}.\quad x-v=\frac{2y^6}{4a^3y^2}=\frac{y^4}{2a^3}$.’

This mistake vitiates the computation of the equation of the derivative curve $\phi(x, z) = 0$, which should be $py^3 = 2a^3z$. Compare [11] above, where $p^{-1}y^3 = 2a^3z^{-1}$.

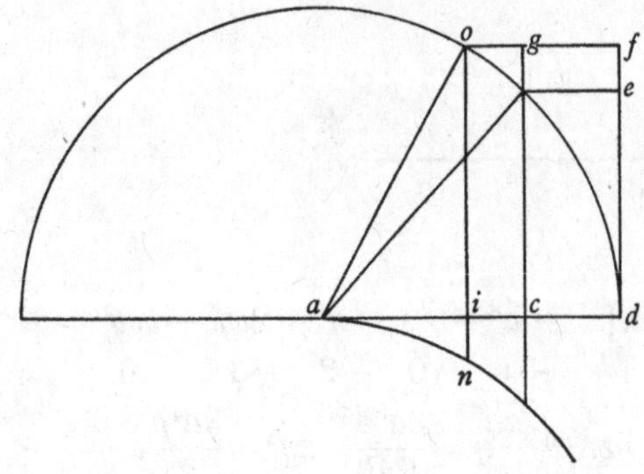

[18] $[y=\sqrt{a^2-x^2}.]$

$ao=a=ad.\quad dc=p.\quad ai=x.\quad oi=y.$
$aa-xx=y^2.\ in=z.\ xx:aa-xx::pp:zz.$
$zzxx=aapp-p^2x^2.$(36)

---

[19] $[y=\sqrt{2ax-x^2}.]$

$id=x.\quad oi[=y]=\sqrt{2ax-xx}.\quad aa-2ax+x^2:2ax-x^2::pp:zz.$

$$\begin{matrix} zzx^2-2azzx+aazz=0.^{(37)} \\ pp\quad -2app \end{matrix}$$

---

[20] $\left[y=\dfrac{a^2}{x}.\right]$

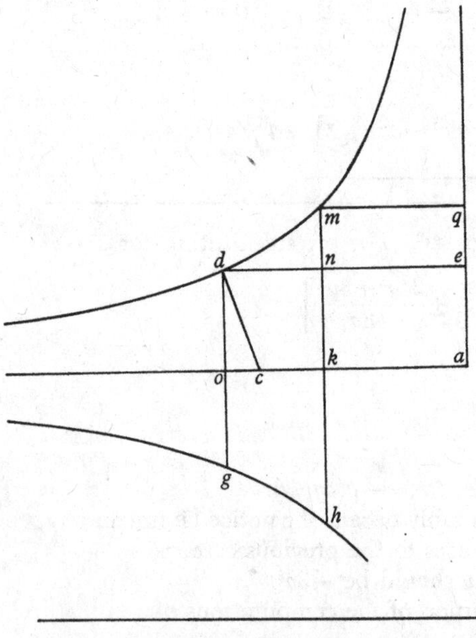

In y^e Hyperbola *dm*. suppose $ak=a=kh$. $ao=x$. $od=y$. $dc=$ a secant.(38) $ca=v$. $og=z$. $xy=aa$. $yy=ss-xx+2vx-vv$.

$$\left.\begin{matrix} -a^4+xxss-x^4+2vx^3=0. \\ -vvxx \\ +2\qquad 0\ -2\quad -1 \end{matrix}\right|\ \text{double roote equall}$$

$$\frac{x^4-a^4}{x^3}=v.\quad x-v=\frac{a^4}{x^3}=oc.$$

$od:oc::kh:og.$

$\dfrac{aa}{x}:\dfrac{a^4}{x^3}::a:z.$(39) $\dfrac{aaz}{x}=\dfrac{a^5}{x^3}.\quad zxx=a^3.$ wch equation conteines y^e nature of y^e crooked line *gh*. Now supposeing y^e line *og* always moves over y^e same superficies in y^e same time, it will increase in motion from *kh* in

---

(36) Newton has no need to calculate the subnormal $ad=v$ of the circle $x^2+y^2=a^2$ since it is the radius a. It follows that $ai=y(dy/dx)=x$, and so $z/p=\sqrt{[a^2-x^2]}/x$, the defining equation of the curve (n).

(37) The subnormal $ad=v$ of the circle $y^2=2ax-x^2$ is likewise the radius a, and so $ai=v-x=a-x=y(dy/dx)$, with $z/p=\sqrt{(2a-x^2)}/(a-x)$ the defining equation of the curve (n). Note that here, as in the previous case, Newton takes $z=p(dx/dy)$.

ye same proportion yt it decreaseth in lenght & ye line *ne* will move uniformely from (*mq*), soe yt ye space $mqen = gokh$. Suppose $ok = a$. $ao = 2a$. $od = \frac{a}{2} = nm$. & $mqen = \frac{1}{2}aa = ogkh$.(40)

---

[21] $\left[y = \frac{a^3}{x^2}.\right]$

suppose ye line last found to be *md*.(41) $ao = x$. $od = y$. $ca = v$. $yxx = a^3$.

$$yy = ss - vv + 2vx - xx.$$

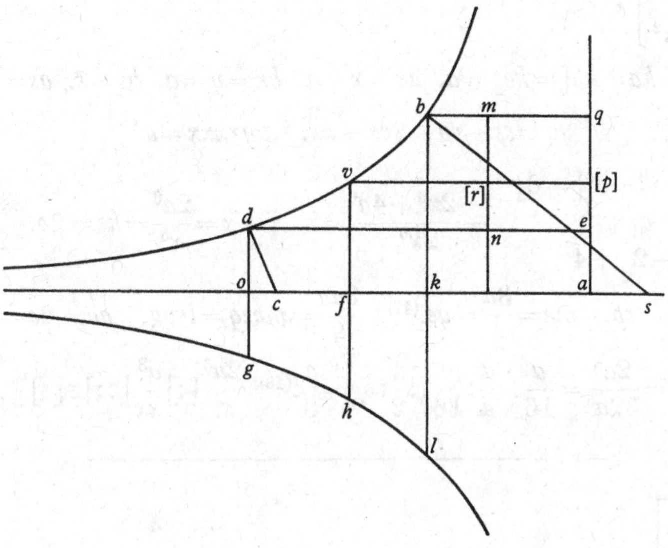

$$\begin{array}{l} ssx^4 + 2vx^5 - x^6 - a^6 = 0. \\ -vv \\ 0 \quad -1 \ -2 \ +4 \end{array} \qquad v = \frac{x^6 - 2a^6}{x^5}. \quad x - v = \frac{2a^6}{x^5}.$$

---

(38) That is, *dc* is normal to the curve.

(39) Or $z = -a(dy/dx)$, much as before.

(40) Newton begins to justify through a model of limit-motion the jump from $z = -a(dy/dx)$ to the corresponding equation of the areas or definite integrals

$$(mqen) = \int_{y=do}^{y=mk} a.dy \quad \text{and} \quad (gokh) = \int_{x=ka}^{x=oa} z.dx.$$

In particular $$\int_a^{2a} z.dx = \int_{\frac{1}{2}a}^{a} a.dy = \tfrac{1}{2}a^2, \quad \text{where} \quad z = \frac{a^3}{x^2}.$$

(41) That is, suppose the curve (*d*) has the equation $yx^2 = a^3$.

to find at w^t^ point $do = oc : \frac{a^3}{xx} = \frac{2a^6}{x^5}$. $x^3 = 2a^3$. $x = a\sqrt{c:2} = af$. $mq = fh = a$. $og = z$.

$od : oc :: fh : og$.

$\frac{a^3}{xx} : \frac{2a^6}{x^5} :: a : z$.(42) $a^3zx^5 = 2a^7xx$. $zx^3 = 2a^4$. wch shews y^e^ nature of y^e^ line (gh).

& $mneq = gofh$. or $nbpe = gokl$.(43)

suppose $ko = ka = a$.(44) $oa = 2a = x$. $od = \frac{a^3}{xx} = \frac{a^3}{4aa} = \frac{a}{4}$. bn(45) $= \frac{3a}{4}$.

$$[mq]en = \frac{3aa}{4} = gokl.^{(46)}$$

---

[22] $\left[y = \pm\frac{a^3}{x^2}\right]$

Suppose $kl = ka = kb[=ko] = a$. $ak = x = a$. $bk = y = a$. $bs = s$. $as = v$.

$$yy = ss - vv - 2vx - xx. \quad yyxxxx = a^6.$$

$$\begin{array}{l} ss\,x^4 - 2vx^5 - x^6 - a^6[=0]. \\ -vv \\ \quad 0 \quad -1 \quad -2 \quad +4 \end{array} \qquad \frac{-2x^6 + 4a^6}{2x^5} = v. \quad v + x = \frac{2a^6}{x^5} = ks = 2a. \qquad \begin{array}{l} ks : bk :: kl : fh. \\ 2a : a :: a : \frac{a}{2}. \end{array}$$

$fh = \frac{a}{2} = ne = mq = rp$. $mn = \frac{3a}{4} = qe$.(47) $\frac{3aa}{8} = mneq = lkog$. $ao = 2a = x$. $\frac{a^3}{4aa} = do$.

[or] $do = \frac{a}{4}$. $oc = \frac{2a^6}{32a^5} = \frac{a}{16}$. $\frac{a}{4} : \frac{a}{16} :: \frac{a}{2} : og = \frac{a}{8}$.(48) $\frac{2a^6}{x^5} [:] \frac{a^3}{xx} [::] z [:] \frac{a}{2}$. $a^4 = zx^3$.

---

[23] $\left[y = \frac{a^4}{x^3}.\right]$

Suppose againe y^e^ last line whose nature is compris[e]d in this equation

$yx^3 = a^4$. $ak = bk = lk = a$. $ao = x$. $ac = v$. $do = y$. $dc = s$. $og = z$.

---

(42) Or $z = -a(dy/dx)$.

(43) Read '$mneq = gokl$. or $nrpe = gofh$'.

(44) This implies that $mq = a = ka$, and so bk and mrn are coincident.

(45) That is, mn (note (44)).

(46) For $$\int_a^{2a} \frac{2a^4}{x^3} . dx = \int_a^{2a} z . dx = \int_{\frac{1}{4}a}^{a} a . dy.$$

(47) $mn = qe = bk - do = a^3/a^2 - [a^3/(2a)^2]$.

(48) Since $do : oc = fh : og$.

(49) Read 'oc'.

(50) For $mn = bk - do$.

(51) $$\int_a^{2a} z . dx = \int_{\frac{1}{4}a}^{a} a . dy, \quad \text{where} \quad zx^4 = a^5.$$

(Some unnecessary duplications in Newton's text have been omitted.)

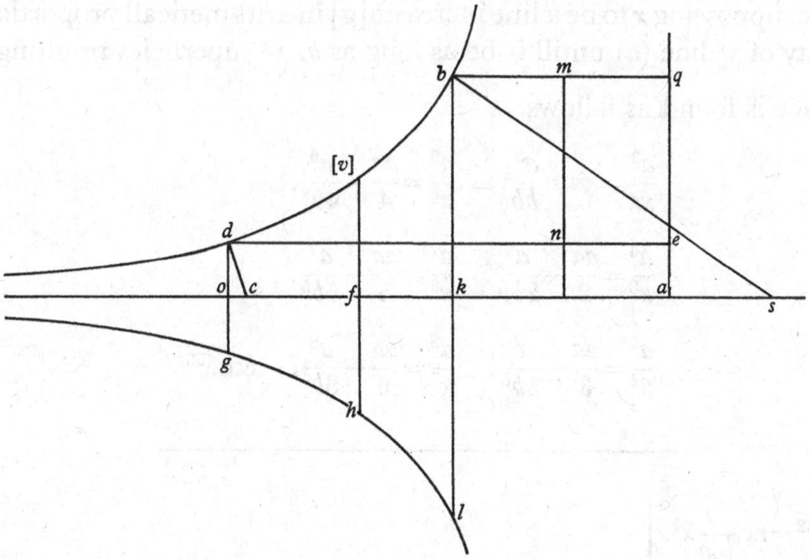

$ss\,x^6+2vx^7-x^8-a^8=0.$
$-vv$ $\qquad v=\dfrac{2x^8-6a^8}{2x^7}.\quad x-v=\dfrac{3a^8}{x^7}.$ to find where $do=dc.$(49)
$0\quad -1\quad -2\quad +6$

$\dfrac{a^4}{x^3}=\dfrac{3a^8}{x^7}.\quad a^4x^4=3a^8.\quad x^4=3a^4.\quad x=a\sqrt{qq:3}.\quad af=a\sqrt{qq:3}:$

$bk=y=a.\quad bs=s.\quad as=v.\quad ak=x=a.\quad [sk=]x+v=\dfrac{3a^8}{a^7}=3a.$

$$[sk:bk::]\,3a:a::kl(=a.):fh=\frac{a}{3}=mq=ne.$$

$oa=x=2a.\quad \dfrac{a^4}{x^3}=y=\dfrac{a^4}{8a^3}=\dfrac{a}{8}=do.\quad mn=\dfrac{7a}{8}=qe.$(50) $\quad mqen=\dfrac{7aa}{24}=lkog.$(51)

---

[24] $\left[y=\dfrac{a^5}{x^4}.\right]$

Likewise supposing ye line $yx^4=a^5.\ x-v=\dfrac{4a^{10}}{x^9}=oc.\ af=a\sqrt{qc:4}.$

$[sk=]ka(=x)+v=4a.\quad fh=\dfrac{a}{4}.\quad do=\dfrac{a}{16}.\quad mq=ne=\dfrac{a}{4}.\quad [mn=]\dfrac{15a}{16}.$

$[mqen=lkog=]\dfrac{15aa}{64}.$ &c.(52)

---

(52) af is the abscissa of the point v whose ordinate vf is equal to its subnormal (not shown in the figure); sk is the subnormal whose abscissa ka and ordinate bk are both a;

$$fh=-a(dy/dx)=a^6/x^5\quad\text{where}\quad x=af=a\sqrt[5]{4};$$

do is the ordinate corresponding to $ao=2a$; and finally $mq=en$ is made equal to fh. It follows that $mn=bk-do$, with

$$gokl=\int_a^{2a}z.dx=\int_{\frac{1}{16}a}^{a}a.dy=mqen.$$

whence supposeing x to be a line increasin[g] in arithmeticall proportion from y^e quantity of y^e line (a) untill it be as long as b. y^e superficies resulting out of $\frac{a^3}{xx}$. $\frac{a^4}{x^3}$. &c is found as follows.

$$\frac{a^3}{xx}=aa-\frac{a^3}{bb}. \quad \frac{a^6}{x^5}=\frac{aa}{4}-\frac{a^6}{4b^4}.$$

$$\frac{a^4}{x^3}=\frac{aa}{2}-\frac{a^4}{2bb}. \quad \frac{a^7}{x^6}=\frac{aa}{5}-\frac{a^7}{5b^5}.$$

$$\frac{a^5}{x^4}=\frac{aa}{3}-\frac{a^5}{3b^3}. \quad \frac{a^8}{x^7}=\frac{aa}{6}-\frac{a^8}{6b^6}. \quad \text{\&c.}^{(53)}$$

---

[25] $\left[y^2=rx+\frac{r}{q}x^2.\right]$

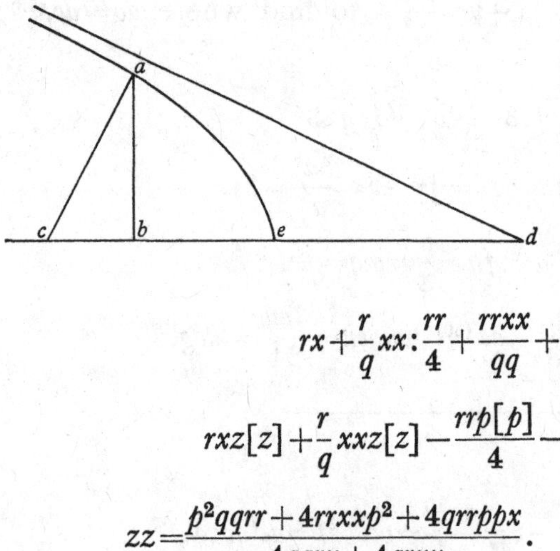

in y^e Hyperb: $ed=q$. $be=x$. $ab=y$. [$ce=v$.]

$$rx+\frac{r}{q}xx=yy=ss-xx+2vx-vv.$$

$$1 \quad 2 \qquad 0 \; -2 \; -[1] \; 0$$

$$\frac{r}{2}+\frac{r}{q}x+x=v. \quad \frac{r}{2}+\frac{r}{q}x=v-x[=cb.]$$

$$rx+\frac{r}{q}xx:\frac{rr}{4}+\frac{rrxx}{qq}+\frac{rrx}{q}::p[p]:z[z].^{(54)}$$

$$rxz[z]+\frac{r}{q}xxz[z]-\frac{rrp[p]}{4}-\frac{p[p]rrxx}{qq}-\frac{rrxp[p]}{q}[=0.]$$

$$zz=\frac{p^2qqrr+4rrxxp^2+4qrrppx}{4qqrx+4qrxx}. \quad \text{[or]} \quad zz=\frac{bbb+4cxx+4ddx}{qx+xx}.^{(55)}$$

---

(53) In generalization Newton tabulates

$$\int_a^b \frac{a^{r+1}}{x^r}.dx \quad (r = 3, 4, \ldots).$$

(54) Or $z = p(dy/dx)$.

(55) Taking $\frac{p^2qr}{4} = b^3$, $\frac{p^2r}{4q} = c$ and $\frac{p^2r}{4} = d^2$.

(56) That is, the defining equation of the derivative curve whose ordinate

$$z = b(dy/dx).$$

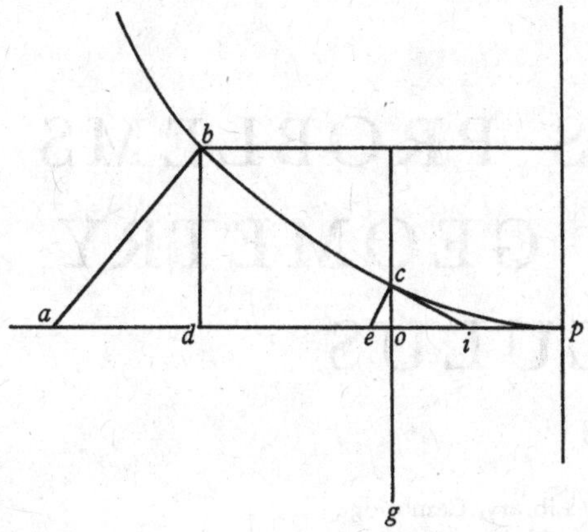

[26] $\left[y=a\sqrt{\frac{x^3}{a^3}}.\right]$

$pd=x.\ db=y.\ ayy=x^3.\ ap=v.$

$ab=s.\ op=og=b.$

$$ass-avv+2avx-axx-x^3=0.$$
$$\begin{matrix}0 & 0 & 1 & 2 & 3\end{matrix}$$

$$\frac{2axx+3x^3}{2ax}=v.\ v-x=\frac{3xx}{2a}=ad:ad^2=\frac{9x^4}{4aa}.$$

$$\frac{x^3}{a}:\frac{9x^4}{4a^2}::bb::zz.\quad \frac{z^2x^3}{a}=\frac{9x^4b^{[2]}}{4aa}.$$

$$4az^2=9b^2x.^{(56)}$$

3

MISCELLANEOUS PROBLEMS IN ANALYTICAL GEOMETRY AND CALCULUS

[Autumn 1664]

From the originals in the University Library, Cambridge

§1 THE GIRTH OF A CURVE

[September? 1664][(1)]

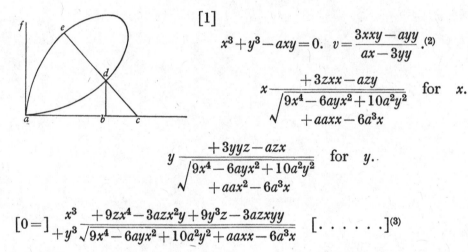

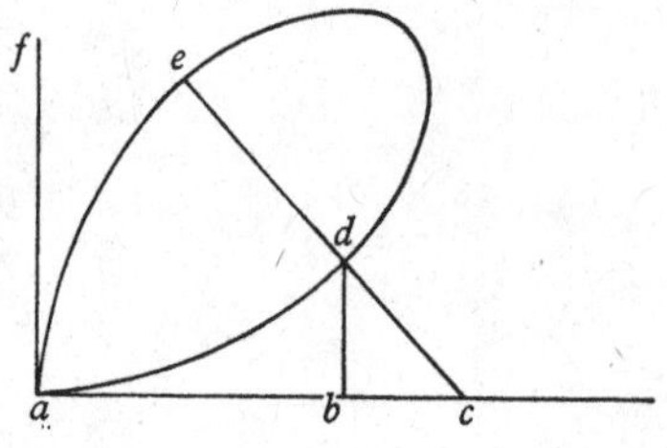

[1]

$$x^3+y^3-axy=0. \quad v=\frac{3xxy-ayy}{ax-3yy}.^{(2)}$$

$$x\frac{+3zxx-azy}{\sqrt{\begin{matrix}9x^4-6ayx^2+10a^2y^2\\+aaxx-6a^3x\end{matrix}}} \quad \text{for } x.$$

$$y\frac{+3yyz-azx}{\sqrt{\begin{matrix}9x^4-6ayx^2+10a^2y^2\\+aax^2-6a^3x\end{matrix}}} \quad \text{for } y.$$

$$[0=]\begin{matrix}x^3\\+y^3\end{matrix}\frac{+9zx^4-3azx^2y+9y^3z-3azxyy}{\sqrt{9x^4-6ayx^2+10a^2y^2+aaxx-6a^3x}} \quad [\ldots\ldots]^{(3)}$$

---

(1) These miscellaneous calculations entered in the Waste Book (Add. 4004: 5^v) are first thoughts on the problem of finding the axes of curves as diameters of maximum or minimum length. Their date is no later than that ('September 1664') of §2 below where Newton elaborates the method generally, and they form part of a series of mathematical calculations in the Waste Book of which those on f. 1^v are also dated '[S]ept 1664'. Newton seems to take his inspiration from Schooten's *Exercitationes*: 493: *Problema.... Oportet...ducere [lineam], quæ curvæ maximam latitudinem designat*; and 497–9: *Investigatio Constructionis.* There Schooten stated and proved Hudde's evaluation of the maximum girth of the folium *aed* (in the first quadrant) normal to the axis of symmetry $x = y$. (This folium is, with a changed constant, Newton's first example. Where, however, Hudde's method is highly geometrical and a straightforward reduction of the particular case of the folium to an equation which must have equal roots (and to which, therefore, he could apply his algorithm), Newton's treatment is characteristically general.)

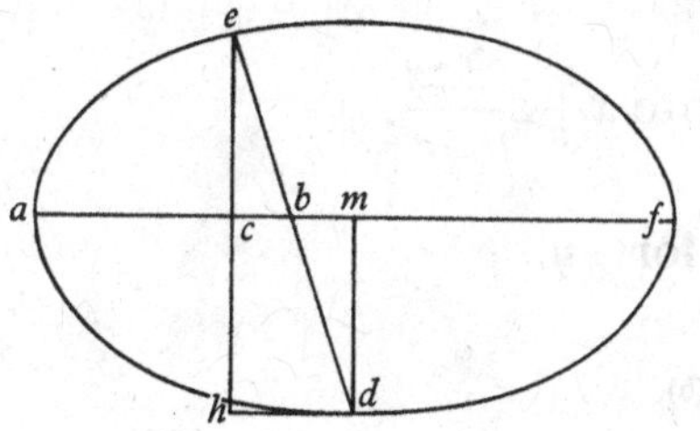

[2]

$ac = x$. $ce = y$. $ab = q$. $2qx - xx = yy$. $ed = z$.

$eb = s$. $bc = q - x$. $eb^2 = q^2$. $eb = q$. $hd = \dfrac{qz - xz}{q}$.

$am = x + z - \dfrac{zx}{q}$.[4]

---

(2) That is, the subnormal $bc = y(dy/dx)$, where in Newton's diagram $ab = x$ and $bd = y$ are the co-ordinates of the point d. (Newton's calculation of v is explained generally in §2 below.)

(3) Having found $\dfrac{v}{y} = \dfrac{3x^2 - ay}{ax - 3y^2} = \dfrac{bc}{bd}$, Newton calculates the ratios $\dfrac{dc}{bc} = \dfrac{ds}{dy} = \dfrac{\alpha}{3x^2 - ay}$ and $\dfrac{dc}{db} = \dfrac{ds}{dx} = \dfrac{\alpha}{ax - 3y^2}$, where α, incorrectly evaluated by Newton, should be $\sqrt{[(3x^2 - ay)^2 + (ax - 3y)^2]}$. Next, Newton produces the normal cd till it meets the folium in the second point e; and so finds the 'girth' $de = z$. Further,

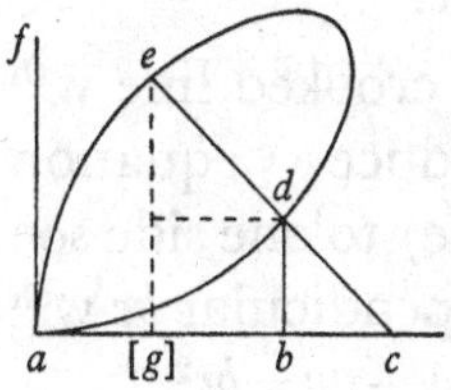

$$\left.\begin{aligned} ag &= ab - de(bc/dc) = x - z(ds/dy) = X \\ ge &= bd + de(db/dc) = y + z(ds/dx) = Y \end{aligned}\right\}$$

are the co-ordinates of a second point (X, Y) on the folium, and so $X^3 + Y^3 - aXY = 0$. Newton, finally, begins to expand this second equation in terms of x, y and z, but abandons it after a few terms.

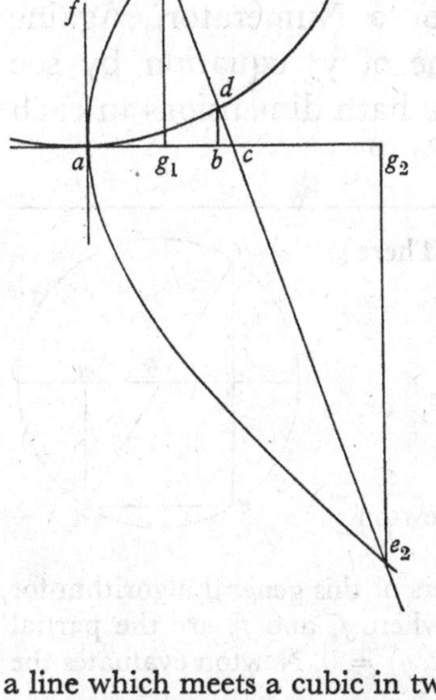

Newton's treatment in §2 below makes it clear that he now wishes to find the minimum girth $ed = z$ by eliminating one of the variables x, y between

$$x^3 + y^3 - axy = 0 = X^3 + Y^3 - aXY$$

and then evaluating z in terms of the other. The value of z will be minimum when the resulting equation,

$$\psi(x, z) = 0$$

say, has a double root, and so z will be found by Hudde's algorithm.

Or so Newton's theory would have it. In fact, cd will meet the folium in *two* other points e_1 and e_2 and the concept of a single 'girth' is possible only when one of them is at infinity (or cd is parallel to the asymptote), and we choose to consider only the finite points. It seems clear that Newton has not yet transcended the difficulties of what is visually but delusively apparent on a figure, and in particular has not yet appreciated the full implications of the truth that a line which meets a cubic in two real points must also do so in a third.

(4) Where (e) is the circle $(x - q)^2 + y^2 = q^2$, Newton calculates the subnormal cb to be $q - x$, and so (the radius) $eb = q$. Where the girth of the circle is taken to be $ed = z$, it follows

[3]

$$2rx-xx=yy.\quad v=r-x.\quad \text{♉}\,x\frac{\text{♉}\,rz\ \text{☊}\ zx}{\sqrt{rr}}.\ \text{for } x \text{ write } x+z-\frac{zx}{r}.$$

$$\text{♉}\sqrt{2rx-xx}\ \text{♉}\,z\frac{\sqrt{2rx-xx}}{r}\quad \text{for}\quad y.$$

$$\text{for } yy \text{ write } 2rx-xx-4zx\frac{+2zxx}{r}+\frac{2zzx}{r}-\frac{zzxx}{rr}.^{(5)}$$

---

§2(1)

[September 1664]

Haveing y^e nature of a crooked line expressd in Algebr: termes to find its axes, to determin it & describe it geometrically &c.

If $fd=x$. $db=y$. & y being perpendicular to x describes y^e crooked line wth one of its extreames. Then reduce y^e equation (expressing y^e nature of y^e line) to one side soe y^t it be $=0$. Then find y^e perpendicular bc wch is done by finding $dc=v$. for $vv+yy=bc^2$.

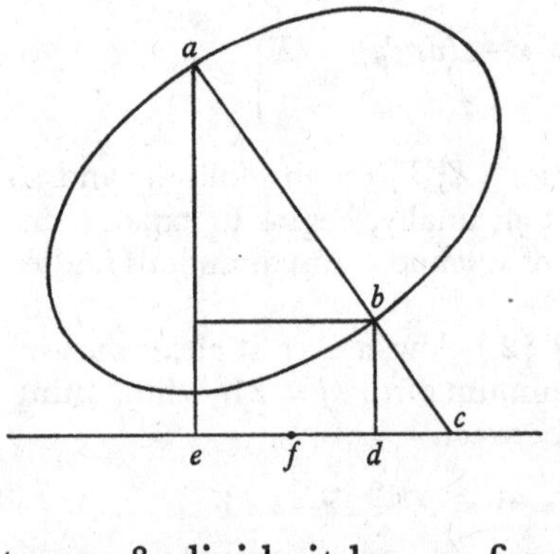

(In finding $dc=v$ observe this rule. Multiply each terme of y^e equat: by so many units as x hath dimensions in y^t terme, divide it by x & multiply it by y for a Numerator. Againe multiply each terme of y^e equation by soe ☞ many units as y hath dimensions in each terme & divide it by $-y$ for a denom: in y^e valor of v.(2)

---

that $hd=ed(cb/eb)=(z/q)(q-x)$ and so $am=x+(z/q)(q-x)$. There Newton stops, but it follows immediately that

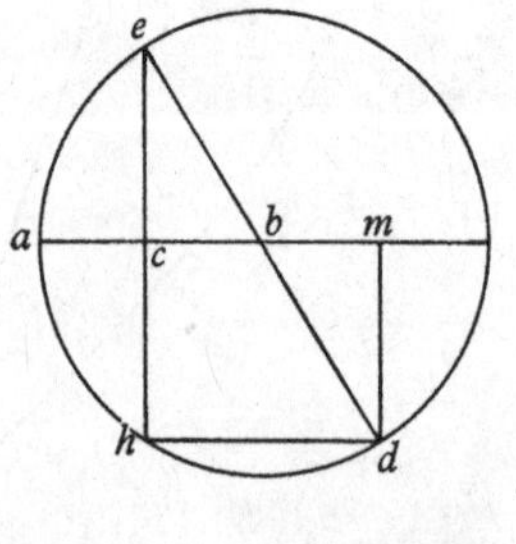

$$md=eh-ec=(z/q)y-y,$$

and so $[x+(z/q)(q-x)][2q-x-(z/q)(q-x)]=[(z/q)y-y]^2$, or, substituting $y^2=x(2q-x)$, $z=2q$.

(5) Similar calculations for the circle $(r-x)^2+y^2=r^2$.

(1) Add. 4004: 8^v. A few preliminary notes, absorbed by Newton into the present text, are omitted.

(2) This is the earliest datable statement in Newton's papers of this general algorithm for finding the subnormal to an algebraic curve. Specifically, where f_x and f_y are the partial derivatives with respect to x and y of an arbitrary equation $f(x,y)=0$, Newton evaluates the subnormal

$$dc=y\frac{dy}{dx}\quad \text{as}\quad v=\frac{(y/x)(xf_x)}{(1/-y)(yf_y)}=-y\frac{f_x}{f_y}.$$

Example, [1st.] $-rx+\frac{rxx}{q}+yy=0.$

$$\frac{-1rx+2\frac{rxx}{q}+0yy \text{ in } y}{x}=-ry+\frac{2rxy}{q}.\quad \frac{-0rx+0\frac{rxx}{q}+2yy}{-y}=-2y. \text{ therefore}$$

$$\frac{-ry+\frac{2rxy}{q}}{-2y}=v=+\frac{1}{2}r-\frac{rx}{q}.$$

[2^d.] Also if $x^3 \begin{matrix}-bxx\\+yxx\end{matrix}+yyx-y^3=0.$ then

$$\frac{\frac{3x^3\begin{matrix}-2bxx\\+2yxx\end{matrix}+yyx \text{ in } y}{x}}{\frac{-3y^3+2yyx+yxx}{-y}}=\frac{3xxy-2bxy+2xyy+y^3}{3yy-2yx+xx}=v.$$

[3^d.] And if $x^4-yyxx+aayx-y^4=0.$ then $\frac{4yx^3-2y^3x+aayy}{4y^3-aax+2yxx}=v.$ &c)

Then make $ab=z$. $fe=x\frac{+vz}{\sqrt{yy+vv}}$. $ae=y\frac{-yz}{\sqrt{yy+vv}}$. & substitute this valor of (fe) into y^e place of x & this valor of (ae) into y^e place of y in y^e equation(3) & there is a 2^d equation. then by multiplication or by some other meanes take away y^e irrationall quantity $\sqrt{yy+vv}$, & lastly take away y or x by y^e helpe of these 2 equations, soe y^t you have a 3^d equation[,] in wch multiply it according to Huddenius his Method(4) for a 4th equation & by y^e helpe of y^e 3^d & 4th equation take away y^e unknowne quantity viz: either x or y. & there will result a 5^t equation in wch is neither x nor y, & by wch y^e valor of z may be found. The greatest of whose valors signifies y^e longest, the least of y^m y^e shortest of all y^e perpendicular lines ab. & if it have other rootes they signifie other lines (ab) wch are

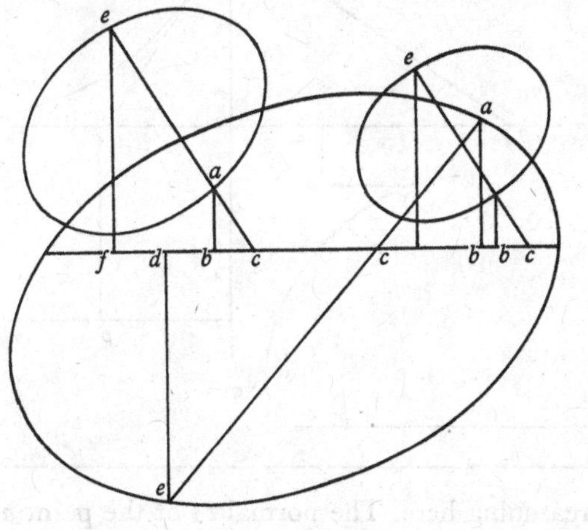

It seems probable that Newton induced this algorithm as a generalization from his researches into the Cartesian subnormal in section 2 above. (Compare 4, §3.1 below.)

(3) For the point a is also on the curve $f(x, y) = 0$, and so $f(x+z(dy/ds), y-z(dx/ds)) = 0$.

(4) To find the maximum or minimum length of $ba = z$.

perpendicular to y^e^ crooked line at both ends, *a* & *b*;(5) & some of these must signifie y^e^ axes of y^e^ line if it bee of an ellipticall nature.(6)

---

§3. *Equiponderance round an axis.*(1)

[1]

$ab=be=y.\ bd=x.\ bq=dg=b.\ nb=c.\ \dfrac{ybx}{yc}=\dfrac{bx}{c}=ef.$ Then shall *bq* &c: be y^e^ axis of gravity in *feb* &c: & *bqgd*.(2)

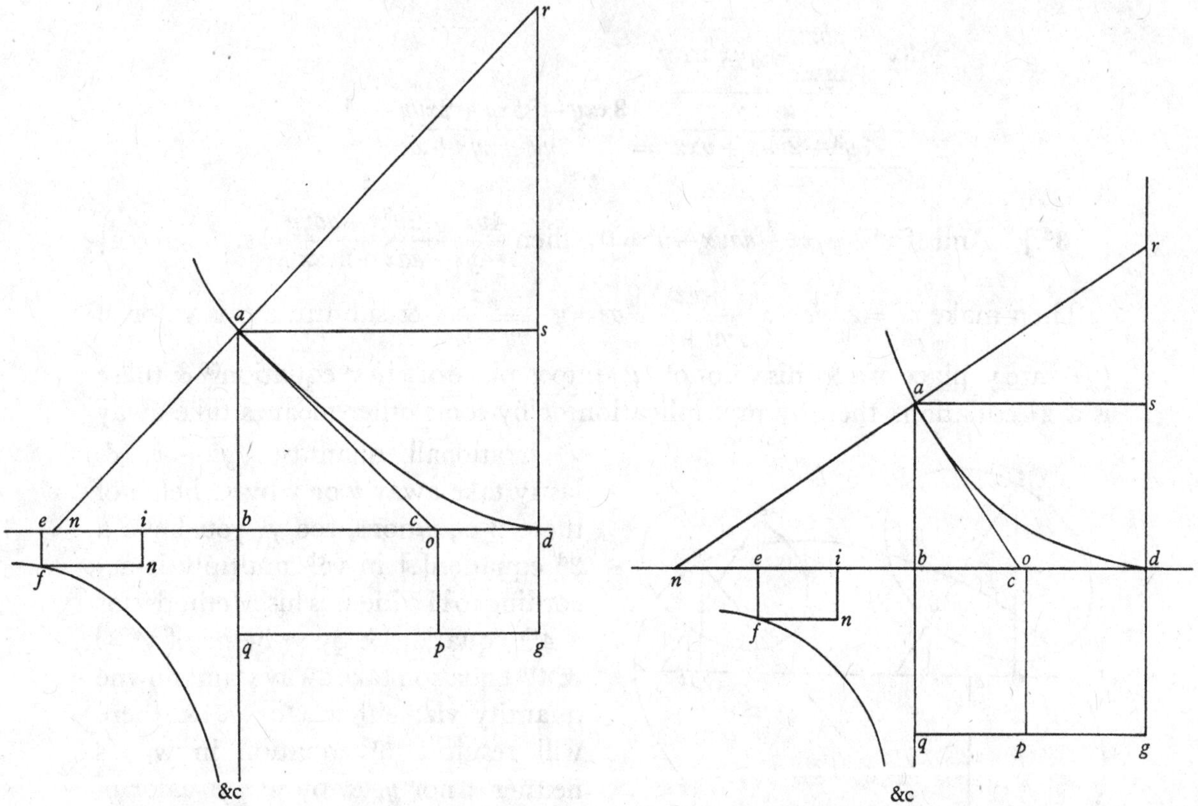

---

(5) It is a little difficult to follow Newton's reasoning here. The normal *ca* at the point *a* will not, in general, be normal at the second meet *e*. Perhaps Newton argues that, since the curve is symmetrical round an axis, with the axis given as an extreme value of *ea*, it follows that *ac* is normal at both *e* and *a* when *ea* is a minimum or maximum value of the girth.

(6) Newton's thinking is here rather wishful. Perhaps he does not realize that a general algebraic curve has no axis. Certainly the calculations even in the simplest cases of conics are formidable, and Newton in passing to the methods of finding axes already set out in section 1 above retreats to safer ground. As Newton soon found, the visually simple closed ovals (his lines 'of an ellipticall nature') are not at all tractable when they are investigated analytically.

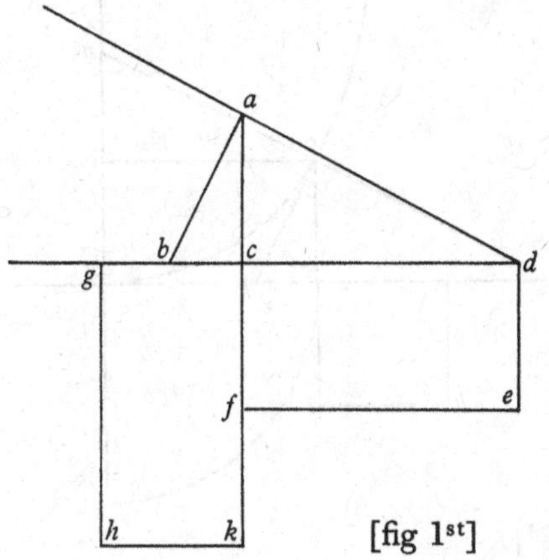

[fig 1st]

[2][(3)]

[$ca=cg=x$. $dc=z$. $cf=de=a$.]

In y^e 1st figure.

$gc:cd::cfed:ckhg=\frac{cd\times cfed}{gc}$. $ac=gc$.

$x:z::za:xy$. $\frac{zza}{x}=ckhg$. or $\frac{xxy}{z}=cdef$.

Suppose $cd:ca::ac:bc::$ y^e swiftnesse of *de* to y^e swiftnesse of *gh*.

$de\times$ its swiftnes$:gh\times$ its swiftness$::gc:cd$.

$de\times cd:gh\times ca::de\times ac:gh\times bc::gc:cd$.[(4)]

Fig 2d. 3d.

$c[k]:ca::ac:bc::nm:am::$ swiftness $de:$ swiftnesse gh.

$de\times$ its swiftnesse$:gh\times$ its swif[t]nes$::ck\times de:gh\times ac::de\times ac:gh\times bc$.

$de\times ck:gh\times ac::de\times ac:gh\times bc::de\times nm:gh\times am::gc:cd$.

In particular, the concept of the girth of a closed oval has an illusory simplicity to the eye, but mathematically is not definable except in a few simple cases, each of which requires special treatment.

(1) Add. 4000: 87v–89v, dated by an examination of Newton's writing and an assessment of the relative maturity of the mathematical content.

(2) This condition for the equiponderance of (*feb* etc.) and (*bqgd*) round the axis *abq* may be stated

$$\int ef\times eb\times\delta(eb) = \int dg\times db\times\delta(db),$$

or

$$ef = \frac{dg\times db}{eb\times[\delta(eb)]/[\delta(db)]} = \frac{bx}{c},$$

since $eb = ab = y$ with the subnormal $nb = c = y(dy/dx)$. This is closely connected with Newton's researches on the subnormal in section 2 above, and was indeed entered in the notebook Add. 4000 on the immediately preceding pages.

(3) Newton clarifies and expands his first treatment.

(4) In this first case the curves (*a*), (*h*) and so (*e*) are straight lines. As before, *bc* is the subnormal at the point *a*, with $ac = gc$, so that

$$\frac{bc}{ac} = \frac{\delta(ac)}{\delta(dc)} = \frac{\text{`swiftness' (or limit-motion) of } gh \text{ from } cfk}{\text{`swiftness' of } de \text{ from } cfk}.$$

The condition for equiponderance,

$$\int gh\times gc\times\delta(gc) = \int de\times cd\times\delta(cd),$$

is met if $gh\times gc^2 = de\times cd^2$, that is, if $gc:cd = de\times cd:gh\times gc = \square ce:\square gk$.

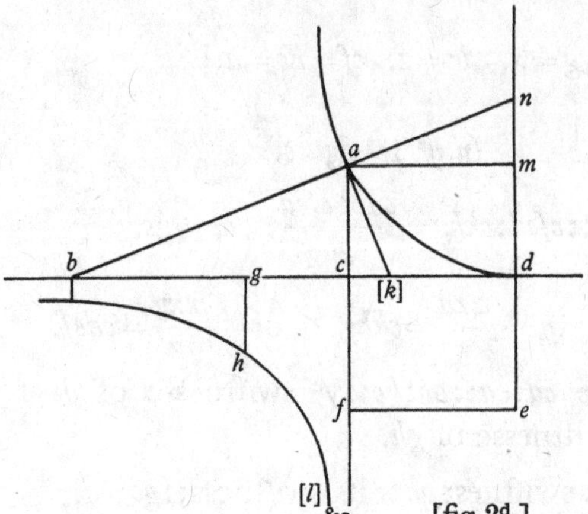

[fig 2ᵈ.]

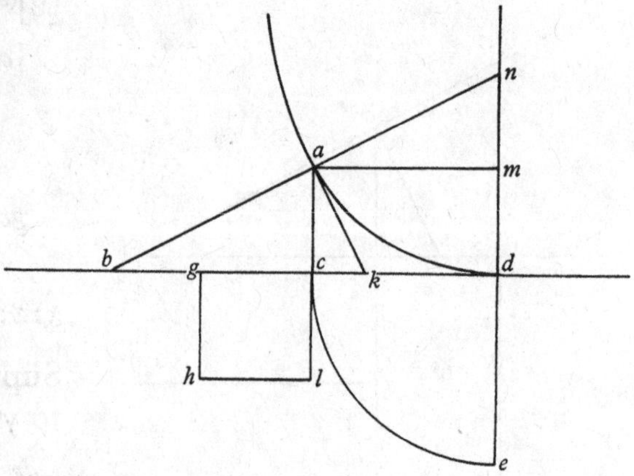

[fig 3ᵈ]

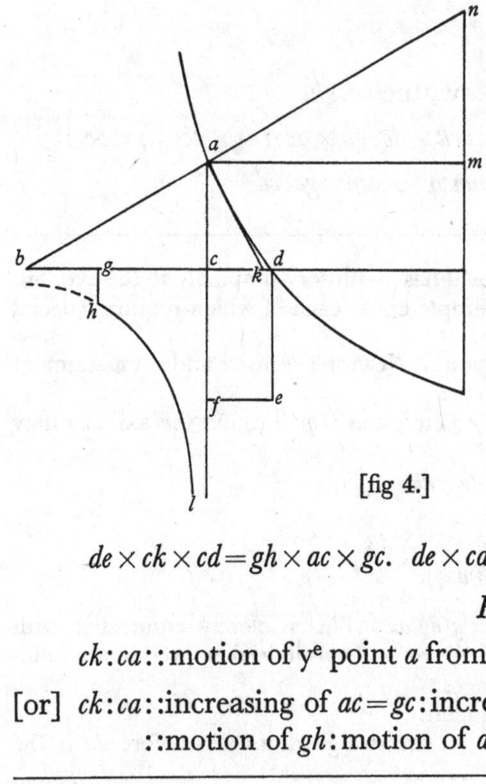

[fig 4.]

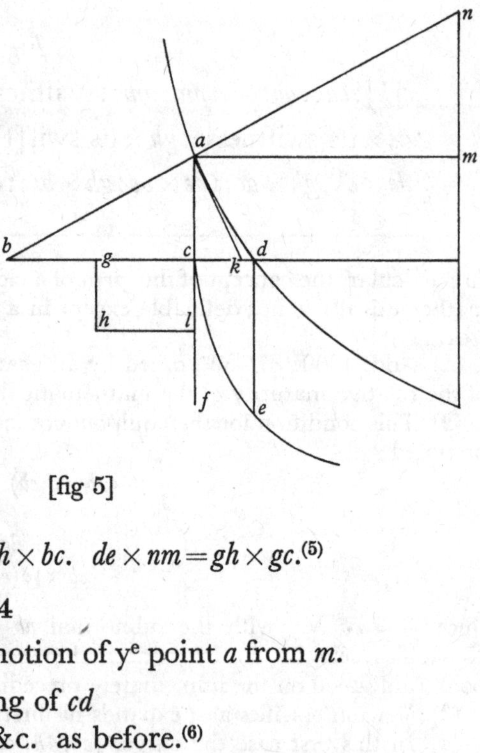

[fig 5]

$de \times ck \times cd = gh \times ac \times gc$. $de \times cd = gh \times bc$. $de \times nm = gh \times gc$.(5)

Fig. 4

$ck:ca::$motion of yᵉ point a from c:motion of yᵉ point a from m.

[or] $ck:ca::$increasing of $ac = gc$:increasing of cd

::motion of gh:motion of de. &c. as before.(6)

---

(5) More generally, where one of the curves (h) or (e) (and so (a)) is not a straight line, the equilibrium condition

$$\int gh \times gc \times \delta(gc) = \int de \times cd \times \delta(cd)$$

is met if $gh \times gc \times ac = de \times cd \times ck$, where the subnormal bc and subtangent ck are connected by

$$\frac{bc}{ac} = \frac{ac}{ck} = \frac{\text{'swiftness' of } gh \text{ from } cf}{\text{'swiftness' of } de \text{ from } cf} = \frac{\delta(gc)}{\delta(dc)},$$

since $gc = ac$.

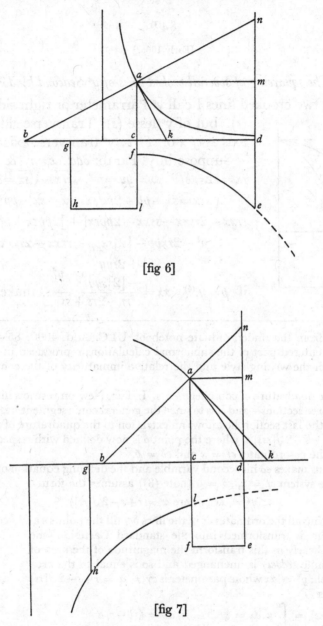

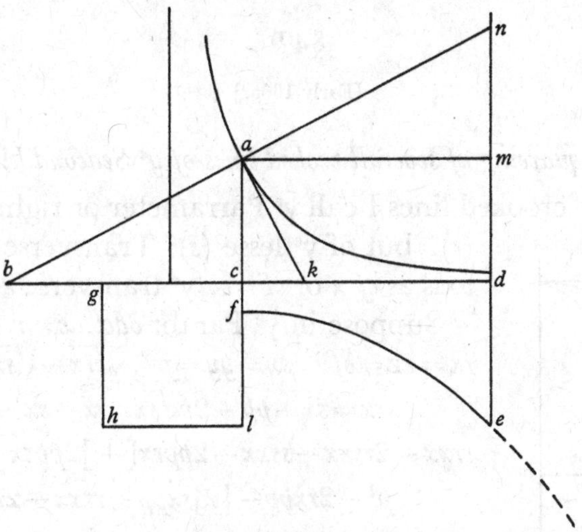

[fig 6]

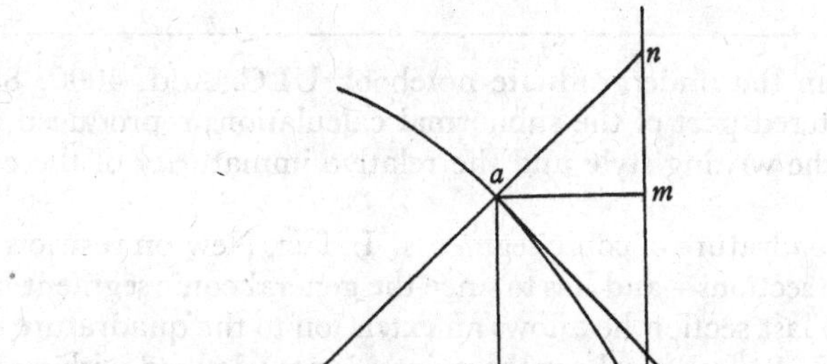

[fig 7]

These are to find such figures *cgh*[*l*], *cfed* as doe equiponderate in respect of y[e] axis *acf*[*l*].

(6) The previous remarks will hold here likewise. Figs. 5–7 are variants on the basic Fig. 4.

§4(1)

[Early 1665?]

The squareing of severall croked lines of ye Seacond kind.(2)

[1] In any two crooked lines I call ye Parrameter or right side of ye greater (r). but of ye lesse (s). Transverse side (q). ye right axis as cf x or $cf=v$. ye transverse axis as $f\epsilon$ y, or $f\delta$ z.

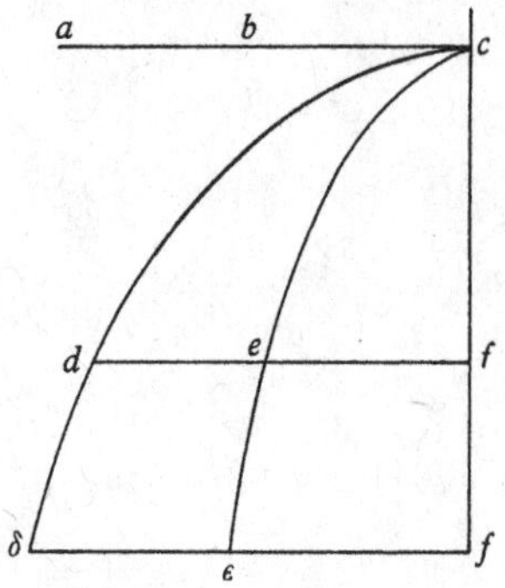

Suppose in ye Parab: δdc: $ac=r$. & [in] ϵec: $bc=s$. $rx=zz=\delta f^2$. $sx=yy=f\epsilon^2$. $\sqrt{rx}-\sqrt{sx}=\delta\epsilon=p$.(3)

$$rx=sx+pp+2p\sqrt{sx}.\ rx-sx-pp=2p\sqrt{sx}.$$

$$rrxx-2rsxx+ssxx-2pprx[+]2ppsx+p^4=4ppsx.\ \text{Or}$$

$$p^4-2rxpp-[2]sxpp+rrxx-2rsxx+ssxx=0.$$

if $p=y$.(4) $xx=\dfrac{\begin{matrix}+2ryy\\+[2]syy\end{matrix}\,x-y^4}{rr-2rs+ss}$. make $cf=a$. $f\delta=b$.

(1) Extracted from the undergraduate notebook ULC. Add. 4000: 85r–86r. Around his text Newton has entered part of the subnormal calculations reproduced in **2**, §2 above, and this, together with the writing style and the relative immaturity of the content, suggests the composition date.

(2) That is, the quadrature of conic segments. In fact, Newton restricts himself to the parabola in his first three sections—and has to since the general conic segment has no exact quadrature—though in the last section he allows an extension to the quadrature of a cubic parabola.

(3) Or $p^2 = (r+s-2\sqrt{[rs]})\,x$, where the point δ is now defined with respect to the auxiliary parabola $c\epsilon\epsilon$ by the co-ordinates $cf = x$ and $\epsilon\delta = p$.

(4) Newton thus makes $\epsilon\delta$ his second variable and the defining equation of the parabola (δ) in his co-ordinate system $cf = x$, $\epsilon\delta = y$ (note (3)) assumes the form

$$y^2 = \alpha x \text{ (where } \alpha = r+s-2\sqrt{[rs]}).$$

(5) By translating all the ordinates $\delta\epsilon$ in the lines $\delta\epsilon f$ till the points ϵ and f coincide, Newton's co-ordinate system is transformed into the standard Cartesian one $cf = x$, $f\delta = y$. Clearly in this transform the magnitude of the area of the parabolic lunule ($cd\delta\epsilon e$) is unchanged and so is equal to the area under the parabola $y^2 = \alpha x$ whose parameter is $c\gamma = \alpha = r+s-2\sqrt{[rs]}$ (note (4)), that is

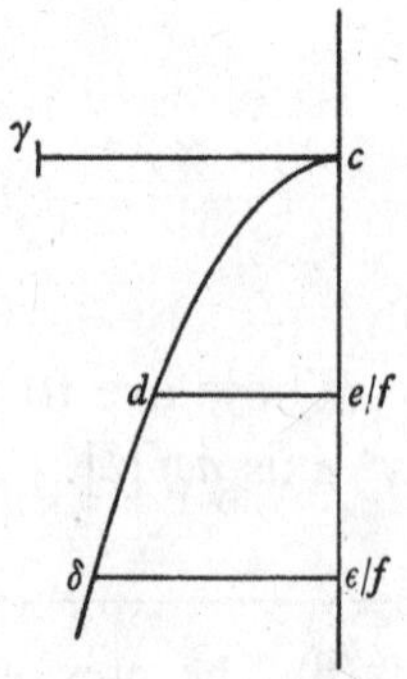

$$(cd\delta\epsilon e) = \int_0^a y\,.\,dx = \tfrac{2}{3}(\sqrt{r}-\sqrt{s})\,a^{\frac{3}{2}} = \tfrac{2}{3}(b-c)\,a$$

since $b = \sqrt{[ra]}$ and $c = \sqrt{[sa]}$.

(6) Newton substitutes qn for qi as his second variable.

(7) Much as before (note (5)) we may conclude that the area of the parabolic lunule (lqn) is unchanged in magnitude when the ordinates $qn = y$ are translated vertically downwards till the points n and i are coincident: hence

$$(lqn) = \int_0^x y\,.\,dx = k(\tfrac{1}{3}x^3-\tfrac{1}{2}bx^2),$$

where $k = (r-s)/rs$, and in particular $(lqkn) = \int_0^b y\,.\,dx = \tfrac{1}{6}kb^3$.

$f\epsilon = c$. $ce\epsilon ff = \frac{2ac}{3}$. & $cd\delta\epsilon ff = \frac{2ab}{3}$. therefore $\frac{2ab-2ac}{3} = cd\delta\epsilon e$ y^e^ square of y^e^ crooked line $cd\delta$ (when y^e^ line $ce\epsilon$ is supposed to close wth y^e^ line cf) whose nature is exprest by y^e^ foregoing Equation.(5)

2 $lk = b$. $li = x$. $qi = y$. $in = z$. $+bx - xx = ry$.

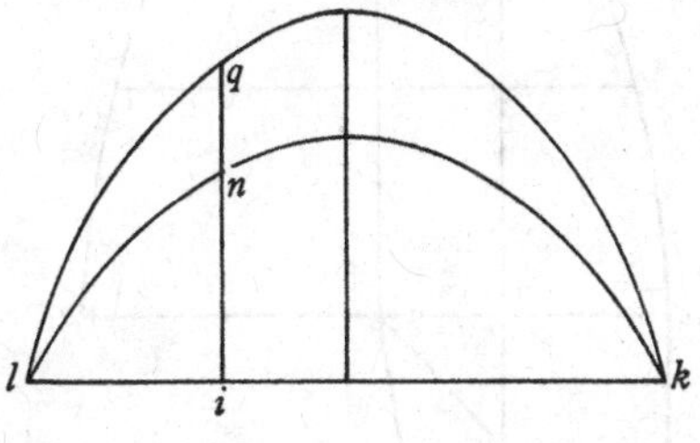

$bx - xx = sz$. $\frac{-xx+bx}{r} = y$. $\frac{-xx+bx}{s} = z$.

$$qn = \frac{-sxx+bsx+rxx-brx}{rs} = y.^{(6)}$$ or,

$$rxx - sxx + bsx - brx - rsy = 0.$$

$$xx - bx\frac{-rsy}{r[-]s} = 0.^{(7)}$$

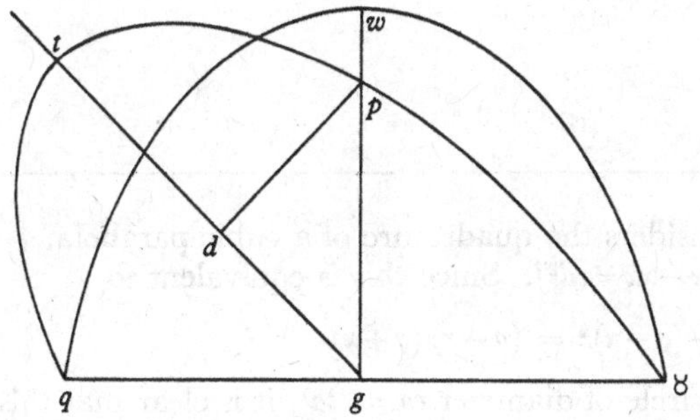

3 $tg = x$. $dg = z$. $gp = y$.

$rx - rz = dp^2$.(8) $rx - rz + zz = y^2$.(9)

$zz = rz - rx + yy$. $r : a :: rx - rz : zz$.(10)

$rzz = arx - arz$. $zz = -az + ax$.

$$z = -\frac{1}{2}a + \sqrt{\frac{1}{4}aa + ax}.$$

$ax - az = rz - rx + yy$. Or

$$yy - rx - ax = -\frac{1}{2}aa - \frac{1}{2}ar + \frac{a}{r}\sqrt{\frac{1}{4}aa + ax}.^{(11)}$$

---

(8) The curve (p) is therefore the parabola $r \times td = dp^2$.

(9) That is, $dp^2 + dg^2 = gp^2$ and the angle $p\hat{d}g$ is right.

(10) The ratio of dp to dg is therefore $\sqrt{[r/a]}$ and hence (note (9))

$$dp : dg : pg = \sqrt{r} : \sqrt{a} : \sqrt{[r+a]}.$$

(11) Since (note (10)) $y\sqrt{a/[axr]} = z = -\frac{1}{2}a + \sqrt{[\frac{1}{4}a^2 + ax]}$, this reduces to

$$y^2 - (a+r)x = \sqrt{[a(a+r)]}\,y,$$

which is the defining equation of the parabola p referred to the oblique Cartesian axes $tg = x$ and $gp = y$ (where $\tan p\hat{g}d = \sqrt{[r/a]}$). Newton, however, does not appear to see this and loses himself in the useless calculation of the quartic

$$(y^2 - (a+r)(x + \tfrac{1}{2}a))^2 = (a+r)^2(\tfrac{1}{4}a^2 + ax)$$

before breaking off. (These calculations are incorrect and not here reproduced.)

Newton's intention in this section is mysterious, and the curve qw♉ (presumably a second parabola with main axis wg) is not brought into his argument. On the lines of the surrounding sections we may perhaps assume that Newton had the quadrature of the parabolic lunules qt and w♉p in mind.

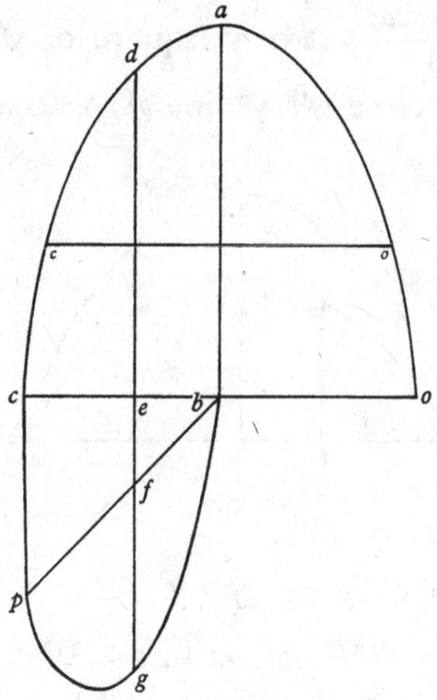

4 *In y^e Parabola.*(12)

$cb=a.\ be=x.\ [ed=y.]\ aa-xx=ry.$(13)

$\frac{aa-xx}{r}=y.\ cp=cb.$(14) $eb\times df=fg\times c=zc.$

$\frac{aax-x^3}{r}+xx=zc.\ x^3-rxx-aax+rcz=0.$(15) Since all $eb\times df$(16)$=\frac{1}{8}$ all $co^2=\frac{1}{4}ab\times ab\times r.$(17) $ab=b.$(18) all $eb^2=\frac{a^3}{3}$. therefore $[c\times]bgpf[=]\frac{1}{4}bbr+\frac{a^3}{3}$.(19)

(12) As his analysis shows, Newton here considers the quadrature of a cubic parabola.

(13) Newton first wrote '$2aa-2ax-aa+2ax-xx=ed^2$'. Since this is equivalent to

$$ed^2 \quad \text{or} \quad y^2 = 2a(a-x)-(a-x)^2 = (a-x)(a+x)$$

or $ce\times eo$ (which defines the curve (d) to be a circle of diameter $co = 2a$), it is clear that this has nothing pertinent to the following text and is here deleted. The present equation $(a-x)(a+x) = ry$ (or $ce\times eo = r\times ed$) defines the curve (d) to be a parabola of vertex a, axis ab and ordinate $cbo = 2a$.

(14) Hence $c\hat{b}p = c\hat{p}b = \frac{1}{4}\pi$ and $ef = eb = x$.

(15) This is the defining equation of a cubic parabola (g) in the oblique co-ordinate system $bf = \sqrt{2}x$, $fg = z$ with $b\hat{f}g = \frac{3}{4}\pi$. Further, when $be = x = a$ (or e and c are coincident) the corresponding value of $fg = z$ is $a^2/c \neq 0$, so that the cubic cannot, as Newton draws it, pass through the point p on bf. (The cubic, in fact, meets the line bf, or $z = 0$, such that $x = 0$ and $\frac{1}{2}r\pm\sqrt{[\frac{1}{4}r^2-cr+a^2]}$.)

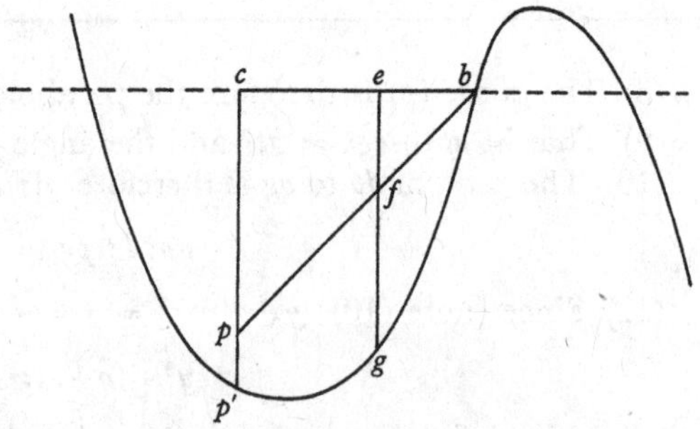

(16) Read 'de'.

(17) 'all $eb\times de$' or

$$\int_0^a xy\,.\,dx = \frac{1}{2}\int_0^b x^2.\,dy = \frac{1}{8}\int_0^b (2x)^2.\,dy \quad \left(\text{or '}\frac{1}{8}\text{ all } co^2\text{'}\right) = \frac{1}{2}\int_0^b (a^2-ry)\,.\,dy = \tfrac{1}{4}b^2r.$$

(18) That is, $b = a^2/r$.

(19) $$c\times(bgpf) = \int_0^a cz\,.\,dx = \int_0^a xy\,.\,dx+\int_0^a x^2.\,dx$$

(or 'all eb^2') $= \frac{1}{4}br^2+\frac{1}{3}a^3$ since (compare note (5)) when the ordinates fg are translated vertically upward till the points e and f coincide, the area (bgf) remains unchanged in magnitude.

4

NORMALS, CURVATURE AND THE RESOLUTION OF THE GENERAL PROBLEM OF TANGENTS

[Winter 1664–Spring 1665]

From the originals in Newton's Waste Book in the University Library, Cambridge, and in private possession

§1 ROUGH CALCULATIONS ON CURVATURE

[Late autumn 1664][(1)]

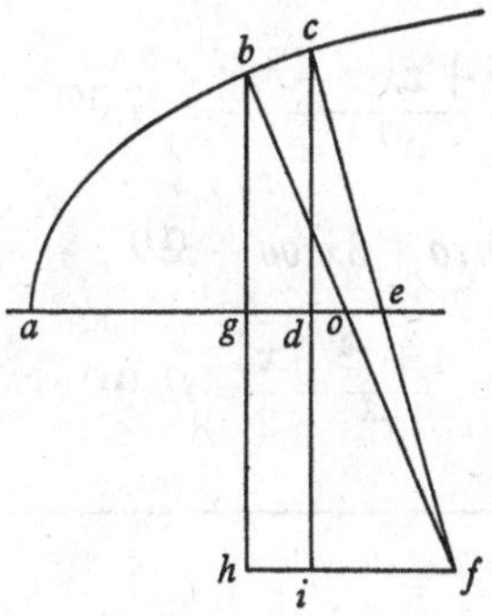

$ag=x.$ $bg=y.$ $gd=d.$ $dc=e.$ $bo=s.$ $de=w.$ $gh=b.$ $go=v.$ $if=c.$

$$e:w::e+b:\frac{ew+bw}{e}=if.$$

$$y:v::y+b:\frac{yv+bv}{y}=hf.$$

$$de+hi+\frac{gh\times de}{dc}[=hf]=go+\frac{gh\times go}{bg}.^{(2)}$$

[*Example* 1st. $y^2=rx$.]

$ag=x.$ $bg=y.$ $rx=yy.$ $go=de=\frac{1}{2}r.$[(3)] $gd=o.$ $dc=\sqrt{rx+ro}.$ $gh=z.$

---

(1) Add. 4004: 6r. The date of composition is hazarded on the basis of the position of these calculations in the Waste Book, and because they can be little earlier than their redrafted forms in §2 below (dated 'December 1664.').

(2) Where the indefinitely near points b, c on the curve are defined by $ag = x, gb = y = f(x)$, say, and $ad = x+d$, $dc = e = f(x+d)$, the centre of curvature f is found as the limit-meet (when d vanishes) of the corresponding normals bo and ce. Similarly, the subnormal $de = w = e(de/dx)$ is the increment as $x \rightarrow x+d$ of the subnormal $go = v = y(dy/dx)$. To fix f in the normal bo its projection $bh = y+b$ (or $gh = b$) is made a third variable, later to be evaluated in terms of x and y. The limit-equation Newton constructs is not very tractable and he will soon abandon it for the more usable $hi = d = (v/y)(y+b)-(w/e)(e+b)$.

(3) As Newton well knew (compare section 2 above and Schooten's *Commentarii in Librum II*, N= *Geometria*: 246) the parabola has a constant subnormal equal to half its latus rectum.

$$o+\frac{rz}{2\sqrt{rx+ro}}=\frac{rz}{2\sqrt{rx}}.$$[4] $$4o\sqrt{rrxx+rrox}+2rz\sqrt{rx}=2rz\sqrt{rx+ro}.$$

$$4oz\sqrt{rx^3+roxx}=4zzro.$$[5] $$x^3=zzr.\quad z=\sqrt{\frac{x^3}{r}}.\quad \sqrt{rx}:\frac{r}{2}::\sqrt{\frac{x^3}{r}}:\frac{x\sqrt{rx}}{2\sqrt{rx}}=\frac{x}{2}.$$[6]

$$\frac{x+r}{2}=hf.\quad \frac{x^3}{r}+\frac{9xx}{4}+\frac{3rx}{2}+\frac{rr}{4}=bf^2.$$

[*Example* 2^d. $xy=r^2$.]

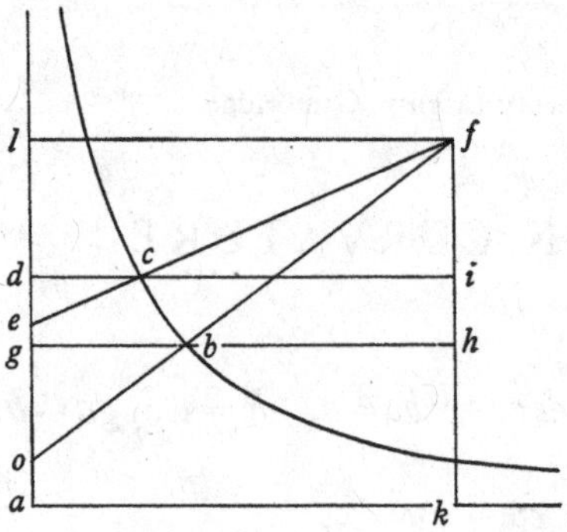

$$[ag=x.\ gb=y.]\ xy=rr.\ v=\frac{-yy}{x}=\frac{-y^3}{rr}=\frac{-r^4}{x^3}.$$

[1] $gh=z$.

$$de=\frac{+r^4}{x^3+3oxx}$$[7] $$:dc=\frac{rr}{x+o}::fi:\frac{zx-rr+zo}{x+o}.$$[8]

$$\frac{+rrzx-r^4+rrzo}{x^3+3xxo}=fi.\quad \frac{-r^4}{x^3}:\frac{rr}{x}::fh:\frac{zx-rr}{x}.$$[9]

$$\left[fh=\frac{rr(zx-rr)}{x^3}.\right]\quad \frac{-r^4+zxrr-ox^3}{x^3}=fi.$$[10]

$$-rrzx^4+r^4x^3-rrzox^3=r^4x^3-zrrx^4+ox^6+3r^4xxo-3zx^3rro+3x^5oo-.$$[11]

$$+2rrzx-x^4-3r^4=0.\quad z=\frac{3rr}{2x}+\frac{x^3}{2rr}=gh.\quad bh=\frac{rr}{2x}+\frac{x^3}{2rr}.\quad \frac{+r^4}{x^3}+\frac{x}{2}=fh.$$[12]

---

(4) Since $go = de$, $$hi+\frac{gh\times de}{dc}=\frac{gh\times go}{bg}.$$

(5) A numerical error is here introduced and the rest of the calculation is vitiated. Read correctly
'$4oz\sqrt{rx^3+roxx}=zzro.\ 16x^3=zzr.\ z=4\sqrt{\frac{x^3}{r}}$',
and so $hf = 2x+\frac{1}{2}r$, and $bh = \sqrt{(rx)}+4\sqrt{(x^3/r)}$.

(6) Since $y:v = z:(hf-go)$.

(7) That is, $\left|\frac{-r^4}{(x+o)^3}\right|$ with powers of o higher than the first omitted.

(8) $ic = z-rr/(x+o)$. (9) $v:y = fh:bh$.

(10) $fi = fh-ih$.

(11) Equating the two values of fi, cross-multiplying and ignoring powers of o higher than the first (apart from the stray '$+3x^5oo$').

(12) A numerical error enters Newton's argument and vitiates his further computation of bf^2 (here omitted). Read correctly '$\frac{+r^4}{2x^3}+\frac{x}{2}=fh$'.

[2] [$fi=z$.]

$go:gb::fh+go:hg.\ de:dc::fi+de:di.$ [or]

$$\frac{dc\times fi+dc\times de}{de}=di[=gh]=\frac{gb\times fi+gb\times dg+gb\times go}{go}.^{(13)}$$

$$\frac{\frac{rrz}{x+o}+\frac{r^6}{x^4+4x^3o}}{\frac{r^4}{x^3+3x^2o}}=\frac{\frac{rrz}{x}+\frac{r^6}{x^4}+\frac{rro}{x}}{\frac{r^4}{x^3}}.^{(14)}\quad \frac{rrz}{x+o}+\frac{r^6}{x^4+4ox^3}\text{ in }\frac{r^4}{x^3}=$$

$$\frac{rrz}{x}+\frac{rro}{x}+\frac{r^6}{x^4}\text{ in }\frac{r^4}{x^3+3ox^2}.\quad \frac{z}{xx+ox}\frac{+r^4}{x^5+4ox^4}=\frac{z+o}{xx+3ox}\frac{+r^4}{x^5+3ox^4}.$$

$$\frac{zx^3+3zoxx+r^4}{x^5+5ox^4}{}^{(15)}=\frac{zx^4+ox^4+3ozx^3+xr^4+3or^4}{x^6+6ox^5}.^{(16)}$$

$$\begin{array}{llll} zx^5+3zox^4+r^4xx & & =zx^5+ox^5+3ozx^4+xxr^4+3oxr^4. \\ +6zox^4 & +6r^4xo & +4ozx^4 \quad +4oxr^4 \end{array}$$

$$9zx^4+6xr^4=x^5+7zx^4+7xr^4.\quad 2zx^4-xr^4-x^5=0.\quad z=\frac{r^4}{2x^3}+\frac{x}{2}.$$

In general, where $ag = x$, $gb = f(x)$, we have $go = f(x)f'(x)$ and so $fh = f'(x)(z-f(x))$. Similarly, where $ad = x+o$, $dc = f(x+o)$, it follows that $fi = f'(x+o)(z-f(x+o))$ and so

$$ih = o = f'(x)(z-f(x))-f'(x+o)(z-f(x+o)),$$

or
$$z\times \lim_{o\to\text{zero}}\left[\frac{f'(x)-f'(x+o)}{o}\right]=\lim_{o\to\text{zero}}\left[\frac{f(x)f'(x)-f(x+o)f'(x+o)}{o}\right]+1,$$

with $f(x+o) < f(x)$. Hence

$$zf''(x) = f(x)f''(x)+[f'(x)]^2+1,\quad\text{or}\quad bh = z-f(x) = \frac{1+[f'(x)]^2}{f''(x)}.$$

(13) That is, where now $fi = z$ (or $fh = z+o$),

$$\frac{dc}{de}(fi+de) = \frac{gb}{go}(fi+dg+go)\quad\text{or}\quad \frac{z+f(x+o)f'(x+o)}{f'(x+o)} = \frac{z+o+f(x)f'(x)}{f'(x)}$$

is the new limit-equation.

(14) Newton substitutes $gb = f(x)$, $dc = f(x+o)$; and

$$go = f(x)f'(x),\quad de = f(x+o)f'(x+o),$$

where $f(x) = r^2/x$.

(15) Since
$$\frac{x^5+4ox^4}{x^2+ox} = x^3+3ox^2,$$

where powers of o higher than the first are omitted.

(16) Newton fails to notice that the two terms on the right side of the previous equation share a common factor $(x+3o)$ in their denominators. The result may be obtained more shortly by setting

$$\frac{z}{x^3}(x-o)+\frac{r^4}{x^6}(x-4o) = \frac{x^3(z+o)+r^4}{x^6}(x-3o).$$

$$\frac{r^4}{x^3}:\frac{rr}{x}::\frac{r^4+x^4}{2x^3}:hb. \quad \text{[or]} \quad rr:xx::\frac{r^4+x^4}{2x^3}[:]\frac{r^4+x^4}{2rrx}=bh.^{(17)}$$

$$\frac{r^8+2r^4x^4+x^8}{4r^4xx}\ \frac{+r^8+2r^4x^4+x^8}{4x^6}=\frac{3r^8x^4+3r^4x^8+x^{12}+r^{12}}{4r^4x^6}=bf^2.$$

$$\begin{bmatrix} bf^2\times 4r^4x^6=x^{12}+3r^4x^8+3r^8x^4+r^{12} \\ 0 \quad\ 6 \quad\ 2 \quad -2 \quad -6 \end{bmatrix}^{(18)} \quad 6x^{12}+6r^4x^8-6r^8x^4-6r^{12}=0.^{(19)}$$

$x^8-r^8=0.$ $x^4-r^4=0.$ $xx-rr=0.$ $x=r.$ therefore take $ag=gb=r$, & y^e^ greatest crookednes of y^e^ line cb[20] will be found at b. [&] $bh=r=fh.$ $bf=r\surd 2.$

[3] $lf=p.$ $al=q.$ $gb=y.$ $go=v.$ $ag=x.$

$$v:y::q+v-x:\frac{qy+yv-vx}{v}=p.^{(21)} \quad \left[v=\frac{y^3}{r^2}.\right] \quad yq+\frac{y^4}{rr}-rr=\frac{py^3}{rr}.$$

$$\left[gh=p=\frac{3y}{2}+\frac{r^4}{2y^3}.\right] \quad \frac{y^4}{rr}+3rr-2yq=0. \quad q=\frac{y^3}{2rr}+\frac{3rr}{2y}.$$

as before. or $q=\dfrac{yy}{2x}+\dfrac{3x}{2}=\dfrac{yy+3xx}{2x}$. $gl=\dfrac{yy+xx}{2x}$.[22]

---

§2 DETAILED RESEARCH INTO CURVATURE

[December 1664–May 1665][1]

To find y^e^ Quantity of crookednesse in lines.

[1] December 1664.

Suppose $ab=x.$ $be=y.$ $bc=o=gh.$ $bg=c.$ ed & df secants[2] to y^e^ crooked line intersecting at d. y^e^ angles abe, acf, egd right ones. & let $rx=yy$, be y^e^ relation twixt x & y. soe y^t^ aef is a Parab.

---

(17) As before. In general the limit-equation (note (12)) yields

$$z\times \lim_{o\to \text{zero}}\left[\frac{f'(x)-f'(x+o)}{o}\right]=\lim_{o\to \text{zero}}\left[f'(x+o)\left(1+f'(x)\,\frac{f(x)-f(x+o)}{o}\right)\right],$$

or $$zf''(x)=f'(x)\,(1+[f'(x)]^2) \quad \text{and} \quad z=f'(x)\times\frac{1+[f'(x)]^2}{f''(x)},$$

since here $f(x) > f(x+o)$.

(18) Applying Hudde's algorithm to find the extreme values of bf^2.

(19) That is, $0 = 6(x^4+r^4)(x^8-r^8)$, where $0 = x^4+r^4$ has no real root.

(20) That is, where the radius bf of curvature reaches a minimum.

(21) Since $go:gb = lo:lf$.

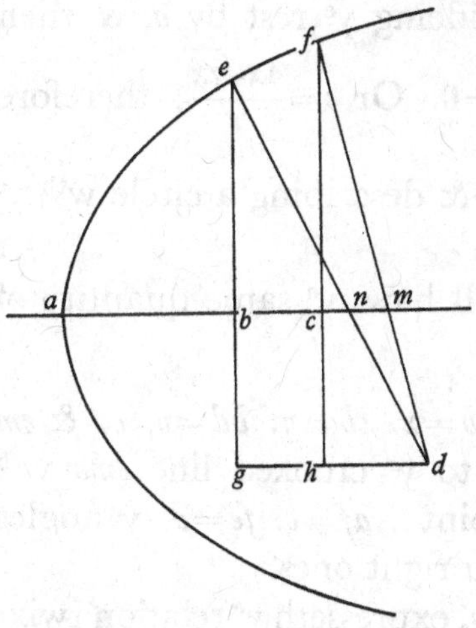

Then $be=\sqrt{rx}$. $bn=v=\frac{r}{2}$.[3]

$$eb:bn::eg \quad :gd$$

$$\sqrt{rx}:\frac{r}{2}::c+\sqrt{rx}:\frac{cr+r\sqrt{rx}}{2\sqrt{rx}}=gd=\frac{r}{2}+\frac{cr}{2\sqrt{rx}}.$$

$$cf=\sqrt{rx+ro}.$$

$$cf:cm:: \quad fh \quad :hd$$

$$\sqrt{rx+ro}:\frac{r}{2}::c+\sqrt{rx+ro}:\frac{cr+r\sqrt{rx+ro}}{2\sqrt{rx+ro}}$$

$$=hd=\frac{r}{2}-o+\frac{cr}{2\sqrt{rx}}.^{(4)}$$

That is

$$2cr\sqrt{rx}+2r\sqrt{rrxx+rrox}=2r\sqrt{rrxx+rrox}-4o\sqrt{rrxx+rrox}+2cr\sqrt{rx+ro}.^{(5)}$$

Or, sq[u]areing both sides,

$$4ccr^3x=16oorrxx+16o^3rrx\begin{matrix}-16crrxo\\-16crroo\end{matrix}\sqrt{rx}+4ccr^3x+4ccr^3o.$$

---

(22) Newton seems to intend the evaluation of the defining equation of the evolute (f) or $\phi(p, q)=0$, where $al=q$, $lf=p$ are the co-ordinates of the centre of curvature f. To do so, indeed, we need only eliminate y between

$$p=\frac{3y}{2}+\frac{r^4}{2y^3} \quad \text{and} \quad q=\frac{y^3}{2rr}+\frac{3rr}{2y},$$

their parametric equations.

In further inconsequential calculations on the same page (Add. 4004: 6r) Newton tries to extend his methods to the general conic, but with little success. After a few lines he abandons them and seems to have passed immediately to the expansion of his thoughts in §2, which follows.

(1) Add. 4004: 30v–33v. (2) That is, normals.

(3) As before, the parabola $y^2=rx$ has the constant subnormal $v=y(dy/dx)=\frac{1}{2}r$.

(4) Where $ab=x$, $ac=x+o$; $be=y=f(x)$, $cf=f(x+o)$; and $bn=f(x)f'(x)$,

$$cm=f(x+o)f'(x+o),$$

Newton constructs the limit equation $(cm/cf)fh=hd=(bn/be)\,eg-gh$, or by substitution and reordering

$$0=\lim_{o\to\text{zero}}\frac{(c+f(x+o))f'(x+o)-(c+f(x))f'(x)+o}{o},$$

that is, $0=(c+f(x))f''(x)+[f'(x)]^2+1$, and so

$$eg=|c+f(x)|=\frac{1+[f'(x)]^2}{f''(x)}.$$

Newton proceeds to work out the calculation for $f(x)=\sqrt{(rx)}$.

(5) Newton does not notice that $bn=cm=\frac{1}{2}r$ can be cancelled on both sides of the equation, but multiplies out straightforwardly by $4\sqrt{[rx]}\sqrt{[rx+ro]}$ before cancelling.

that is (by blotting out $4ccr^3x$ on both sides, divideing y^e^ rest by o, & then supposeing $o=bc$ to vanish.) $-16crrx\sqrt{rx}+4ccr^3=0$. Or $c=\frac{4x\sqrt{rx}}{r}$. therefore makeing $ab=x$. $bg=\frac{4x\sqrt{rx}}{r}$. $gd=\frac{bn\times eg}{eb}=\frac{1}{2}r+2x$. & describing a circle wth y^e^ Rad[:] $de=\sqrt{\frac{16x^3}{r}+12xx+3rx+\frac{rr}{4}}$,(6) y^e^ circle shall have y^e^ same quantity of crookednesse wch y^e^ Parabola hath at y^e^ point e.

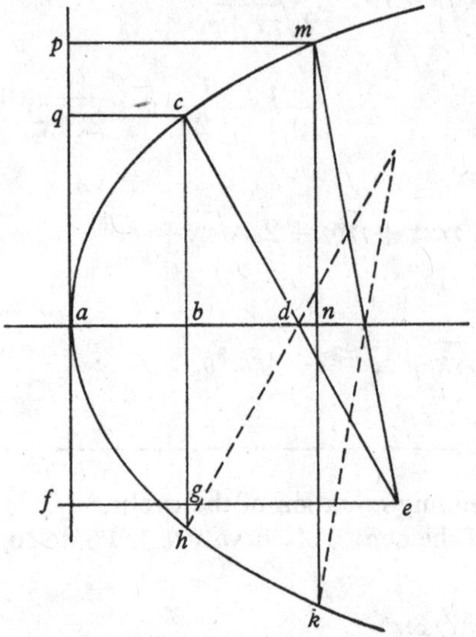

Or thus. If $ab=x$. $cb=y$. $bd=v$. cd & em perpendiculars to y^e^ crooked line cma wch intersect at y^e^ point e. $af=c$. $fe=d$. y^e^ angles abc, baf, afe, mna right ones.

Supose $rx=yy$, expresseth y^e^ relation twixt ab & bc. First I find y^e^ length of $bd=v$. (see fol: 8th hujus,(7) or Des-Cartes his Geom: pag 40(8)) wch is $v=\frac{r}{2}$.

$cb:bd::\ gc\ :\ ge.$

$y:v\ ::c+y:\frac{cv+vy}{y}$. $ab+ge$

$=x+v+\frac{cv}{y}[=fe]=d.$

Or $-dy+cv+vy+xy=0$.(9) Out of these termes first I take away v by writeing its valor in its rome(10) wch in this case is $\frac{1}{2}r$ & there results, $\frac{cr}{2}+\frac{yr}{2}+xy-dy=0$. then if I take away either x or y ([y^t^] wch may bee easliest done) by y^e^ helpe of the equation expressing y^e^ nature of the line wch is now $rx=yy$. or $\frac{yy}{r}=x$. And there results. $\frac{cr}{2}+\frac{yr}{2}+\frac{y^3}{r}-dy=0$. Now tis evident y^t(11)^ when y^e^ lines em & ce are coincident y^t^ ce is y^e^ radius of a circle wch hath y^e^ same quantity of crookednesse wch y^e^ Parabola mca hath at y^e^ point c. Wherefore I suppose cb & nm[,] 2 of y^e^ rootes of y^e^ equation $2y^3+rry-2dry+crr=0$, to be equall to one another.(12) & so by Huddenius

(6) That is, $\sqrt{[eg^2+gd^2]}$.

(7) Add. 4004: 8^v^, printed as 3, §2 above.

(8) *Geometria*: Liber II: 40ff.

(9) This equation gives the condition for normals ec to be drawn to the curve from the point e taken arbitrarily in fe (itself fixed by $af = c$). Substituting $y = f(x)$ and $v = f(x)f'(x)$, the condition becomes $f(x)\times[-d+cf'(x)+f(x)f'(x)+x] = 0$.

(10) Read 'roome'.

(11) This is, presumably, to be omitted.

his method I multiply it $\underset{3}{2y^3}+\underset{1}{rry}-\underset{1}{2dry}+\underset{0}{crr}=0$. & there results $\frac{6yy-rr}{2r}$(13) $=d$.

againe otherwise $\underset{2}{2y^3}+\underset{0}{rry}-\underset{0}{2dry}+\underset{-1}{crr}=0$, & there results $\frac{4y^3}{rr}=c$. Soe y^t if $cb=qa=y$. y^n $ab=\frac{yy}{r}$. $af=\frac{4y^3}{rr}$. $fe=\frac{6yy-rr}{2r}$.(13) then y^e circle described by y^e radius ec shall be as crooked as y^e Parabola at y^e point c.

Or Better thus. Make $ab=x$. $bc=y$. $cg=c$. $fe=d$. $bd=v$.

Then $y:v::c:\frac{cv}{y}=ge$. $cv+xy-dy=0$.

Or thus. Make $ab=x$. $bc=y$. $bd=v$. $bg=e$. $ge=f$. & $fy=vy+ev$. Thus in y^e former example $rx=yy$. $v=\frac{r}{2}$. $fy-\frac{1}{2}ry-\frac{1}{2}re[=0]$.(14)

Theorema. The crookednesse of equall portions of circles are as their diameters reciprocally.(15)

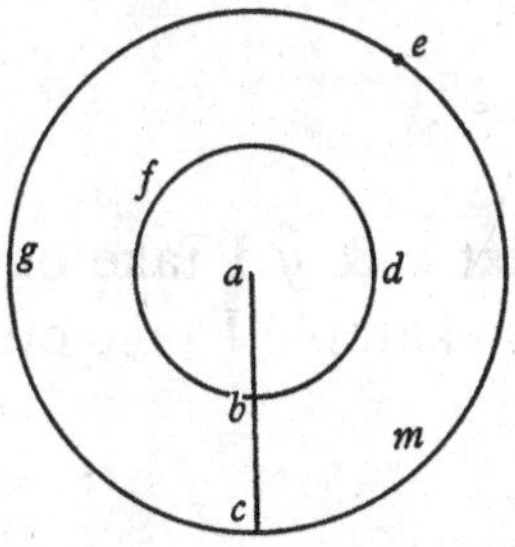

Demonstr. The crookednesse of any whole circle (*bfd*, *gcme*) amounts to 4 right angles, therefore there is as much crookednesse in y^e circle *bfd*, as in *cmeg*. Now supposing y^e perimeter *fdbf* is equall to y^e arch *cme*, Then as y^e arch *emc*=*fdbf* is to y^e circumference *cmegc*, soe is y^e crookednesse of y^e arch *cme* to y^e crookednesse of y^e perimeter *cmegc*, or of *bdfb*, soe is *ab* to *ac*.

---

(12) Newton proceeds to evaluate the double root condition by Hudde's algorithm, but in general and more modern fashion we may find the condition by equating the derivative to zero. Thus, ignoring the extraneous factor $f(x)$, the limit-equation (note (9)) produces

$$0 = (d/dx)(-d+cf'(x)+f(x)f'(x)+x) = cf''(x)+f(x)f''(x)+[f'(x)]^2+1,$$

or
$$cg = |c+f(x)| = \frac{1+[f'(x)]^2}{f''(x)},$$

as before.

(13) Read '$\frac{6yy+rr}{2r}$'.

(14) These improvements are slight and bring with them no gain in computational facility. Newton began to write a further sentence 'Or if out of y^e equa[tion]' but broke off to cancel the whole paragraph.

(15) A difference in ink shows that this was a later addition. It fills in a small gap at the bottom of f. 30^v, and seems to have been inserted there merely so that it would be with Newton's other notes on curvature. Logically, of course, it is a lemma which introduces the concept of 'crookednesse' of curves as the inverse of the radius of curvature at the point (so that the straight line shall have zero curvature) and is so placed in the October 1666 tract (Section 7: 'Prob 2^d' on page 419 below).

To find y^e Quantity of crookednesse in lines.

[2] December 1664.

Suppose ndf & efm perpendicular to y^e crooked line $adeo$, wch intersect one another at f. $ac=x$. $ce=y$. $cm=v$. $ag=ch=c$. $gf=d$. & y^e angles abd, ace, mag, agf right ones. Then,

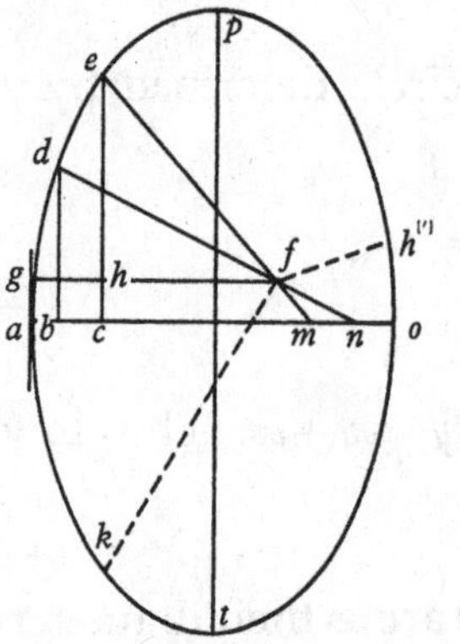

$$ec=y:cm=v::eh=y-c:hf=\frac{vy-vc}{y}.$$

$$gf=gh+hf=x+\frac{vy-vc}{y}=d. \quad \text{Or} \quad dy-vy+vc-xy=0.$$

Haveing therefore y^e relation twixt x & y (as if it bee $rx-\frac{r}{q}xx=yy$) first I find y^e valor of v (see Cartes Geom: pag 40th. or fol: 8th of this(16)) (as in this example tis $\frac{1}{2}r-\frac{rx}{q}=v$(17)) by wch I take v out of y^e equatiō $dy-xy-vy+vc=0$, (& in this case there results

$$dy-xy-\frac{1}{2}ry\frac{+rxy}{q}+\frac{1}{2}rc-\frac{rxc}{q}=0.)$$

then by meanes of y^e equation expressing y^e relation twixt x & y I take out either x or y, [y^t] wch may easliest bee done.(18) (as in this example I take out y by writeing $\sqrt{rx-\frac{rxx}{q}}$ in its stead & there results

$$\left.\begin{array}{l}+d\\ -x\\ -\frac{r}{2}\\ +\frac{rx}{q}\end{array}\right\}\sqrt{rx-\frac{rxx}{q}}=\frac{rcx}{q}-\frac{1}{2}rc. \quad \text{Or} \quad \left.\begin{array}{l}2dq\\ -2xq\\ -rq\\ +2rx\end{array}\right\}\sqrt{rxq-rxx}=2rcx\sqrt{q}-rcq\sqrt{q}.$$

(16) See notes (7) and (8) above. (17) 3, §2, Example 1 above.

(18) Substitution of $v=-y(f_x/f_y)$ in $y(d-x)-v(y-c)=0$ yields $(d-x)f_y+(y-c)f_x=0$. Elimination with $f(x,y)=0$ then produces, say, $g_1(x)=0=g_2(y)$, either of which determines the normals which can be drawn to the given curve from f.

(19) For lack of space Newton in his text added this line of Huddenian multipliers above the equation, but they have been resited here to conform with Newton's ordinary practice. The equation itself is more simply written as

$$(q(2d-r)-2x(q-r))^2rx(q-x)=c^2qr^2(q-2x)^2.$$

(20) That is, 'fix'. (21) Read 'd & e'.

(22) The equations $g_1(x)=g_2(y)=0$ (note (18)) yield the lengths $ac=x$, $ce=y$ which define the points d, e, h', k where the normals from f meet the curve; and the condition that f be the centre of curvature or the limit-meet of two indefinitely near normals is that each of

& by squareing both ꝑts

$$\begin{array}{lllllll}
+4ddq^3r\,x & -4ddqqr\,xx & +8dqqr\,x^3 & -4qqrx^4 & =q^3rrcc & -4qqrrccx & +4rrqccxx. \\
-4dq^3rr & -8dq^3r & -8dqrr & +8qrr & & & \\
+q^3r^3 & +4dqqrr & +4q^3r & -4r^3 & & & \\
 & +8dqqrr & -4qqrr & & & & \\
 & +4q^3rr & -8qqrr & & & & \\
 & -qqr^3 & +4qr^3 & & & & \\
 & -4qqr^3 & +4qr^3 & & & & \\
-1 & 0 & 1 & 2 & -2 & -1 & 0^{(19)})
\end{array}$$

Then if I assume any valors for *c* & *d* yt is if I determine(20) ye point *f*, I have an equation by wch I can find all ye perpendiculars to ye crooked line, drawne from ye point *f*. for if I tooke *x* out of ye equation, ye rootes of ye equation will bee all such lines as are drawn from ye points of intersection *d*, *e*, *k*, *h* to ye line *ao* (as *db*, *ec*, &c). but if I tooke *y* out of ye equation yn ye roots of ye equation will bee those lines drawne from *a* to ye perpendiculars (as *ab*, *ac*, &c). Now by how much ye nigher ye points *d* & *f*(21) are to one another, soe much ye lesse difference there will bee twixt ye crookednesse of ye ꝑte of ye line *de*, & a circle described by ye radius *df* or *ef*. And should ye line *df* be understood to move untill it bee coincident wth *ef*, taking *f* for ye point where they ceased to intersect at theire coincidence, ye circle described by ye radius *ef*, & ye crooked line at ye point *e*, would bee alike crooked. And when ye 2 lines *df* & *ef* are coincident 2 of ye rootes of ye equation (viz[:] *db* & *ec*, *ab* & *ac*) shall bee equall to one another; Wherefor to find ye crookednesse of ye line at ye point *e* I supose ye equation to have 2 equall rootes & so ordering it According D: Cartes or Huddenius his Method, ye valor of any of these 3 $\left.\begin{matrix}x\\y\end{matrix}\right\}$[,] *c*[,] *d* being given, ye valor of ye other 2 may be found.(22)

---

these equations has a double root. Newton applies Hudde's algorithm to derive the double root of $g_1(x) = 0$, but in more modern terms and more generally we may argue: the condition for

$$(d-x)f_y+(y-c)f_x = 0, \quad \text{where} \quad \frac{d-x}{y-c} = \frac{dy}{dx} = -\frac{f_x}{f_y},$$

to have a double root in x is that

$$0 = (d/dx)\,[(d-x)f_y+(y-c)f_x] = (d-x)\,(f_{xy}+(dy/dx)f_{yy})-f_y$$
$$+(y-c)\,(f_{xx}+(dy/dx)f_{xy})+f_x(dy/dx);$$

or, eliminating one of $d-x$, $y-c$ in terms of the other, substituting $(dy/dx) = -f_x/f_y$ and reordering

$$\frac{d-x}{f_x} = \frac{y-c}{f_y} = \frac{f_x^2+f_y^2}{f_{xx}f_y^2-2f_xf_yf_{xy}+f_{yy}f_x^2},$$

a result which Newton will state in all its generality in §3 below. (As we shall see in the next volume, Newton a few years later was able to prove that, where $f(x, y) = 0$ is of degree n, both $g_1(x) = 0$ and $g_2(y) = 0$ are of degree n^2.)

(as in this example y^e^ valor of x being given I multiply y^e^ equation to Huddenius Method & it is

$$\frac{\begin{array}{lll} -4ddq^3\,x & +8dqq\,x^3 & +8qq\;x^4 \\ +4drq^3 & -8dqr & +16qr \\ -rrq^3 & +4q^3 & -8rr \\ & -12qqr & \\ & +8qrr & \end{array}}{4qqrx-2q^3r} = cc =[23] \frac{\begin{array}{llll} -4qq\,x^4 & +8dqq\;x^3 & -4ddqq\;xx & +4ddq^3\,x \\ +8qr & -8dqr & -8dq^3 & -4dq^3r \\ -4rr & +4q^3 & +12dqqr & +q^3rr \\ & -12qqr & +4q^3r & \\ & +8qrr & -5qqrr & \end{array}}{4qrxx-4qqrx+q^3r}.$$

Then by divideing both y^e^ numerators by x & y^e^ denominators by $2qrx-qqr$, & soe multiplying y^m^ in crucem[24] & ordering y^e^ product it is.

$$\begin{array}{llll} 4ddq^4 & -4q^4r & d & +q^4rr \\ & -16q^4x & & +8q^4rx \\ & +16q^3rx & & -8q^3rrx \\ & +24q^3xx & & +12q^4xx \\ & -24qqrxx & & -36q^3rxx \\ & +16qqx^3 & & +24qqrrxx \\ & +16qrx^3 & & -24q^3x^3 \\ & & & +56qqrx^3 \\ & & & -32qrrx^3 \\ & & & +16qqx^4 \\ & & & -32qrx^4 \\ & & & +16rrx^4 \end{array} = 0.[25]$$

Now considering, y^t^ if q, r, & x bee knowne, y^t^ is, if y^e^ ellipsis *eak* be determined, & y^e^ line *ac* given, there are onely two points in y^e^ line (viz: e & k) to be

(23) That is, evaluating c^2 from the original equation.
(24) 'Crosswise'.
(25) That is, $[(2d-r)\,q-2(q-r)\,x]\,[(2d-r)\,q^3-6q^2(q-r)\,x+12q(q-r)\,x^2-8(q-r)\,x^3] = 0$.
(26) Newton's point is that, since each value of $ac = x$ determines *two* points e and k on the ellipse, the condition that $g_1(x) = 0$ have a double root is satisfied by two different situations. First, the condition can determine the point m which is the limit-meet of two normals to the ellipse, one indefinitely near to ef and the other indefinitely close to kq (where, as before, the points e and k share the same abscissa $ac = x$). Alternatively, the condition can determine the limit-meet (f and q respectively) of all normals indefinitely near to ef or to kq. As Newton goes on to remark, the problem of finding the centre of curvature at a point is concerned only with the second interpretation, and it is his immediate worry to eliminate cases of the first.
(27) That is, $ac+cm$, where $cm = \frac{1}{2}r-rx/q$ is the subnormal at e.

considered. And y^e^ valors of (d) are (gf, am, sq) [y^t^ is] such lines as are drawne from y^e^ line *gas* to y^e^ points where y^e^ perpendiculars *efm*[,] *kqm* intersect (as m) or to such points where two perpendiculars (as *ef* & *df*) ceased to intersect at theire coincidence into one (as f & q).[26] Therefore since I seke not y^e^ first sort of roots I get y^e^ valor of y^e^ line $am=x+\frac{r}{2}-\frac{rx}{q}=d$.[27] & divide this equation by it, y^t^ is, by $d-x+\frac{r}{2}-\frac{rx}{q}=0$. Or $2dq-2qx+qr-2rx=0$;[28] And there results

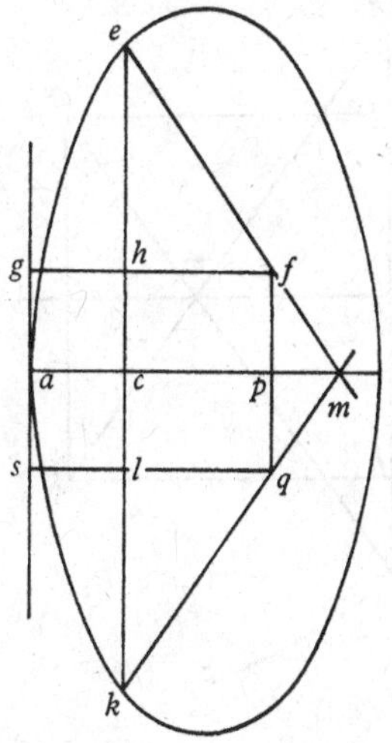

$$+2dq^3-q^3r-6q^3x+6qqrx+12qqxx-12qrxx$$
$$-8qx^3+8rx^3=0.$$

That is dividing it by $2q^3$;

$$d=\frac{r}{2}+3x\frac{-3rx-6xx}{q}\frac{+6rxx+4x^3}{qq}-\frac{4rx^3}{q^3}.$$

Which Equation expresseth y^e^ length of y^e^ lines $(qs=d$, & $gf=d)$ wch are drawne from y^e^ line *sag* to y^e^ points q & f at wch y^e^ coincident perpendiculars last intersected one another before theire coincidence. Now haveing y^e^ length of gf or sq it will not be difficult to find, $c=ag=ch$, or, $as=cl=c$; for it was found before y^t^ $dy-vy-xy+vc=0$. Or $c=\frac{vy+xy-dy}{v}$. Likewise it will not be difficult

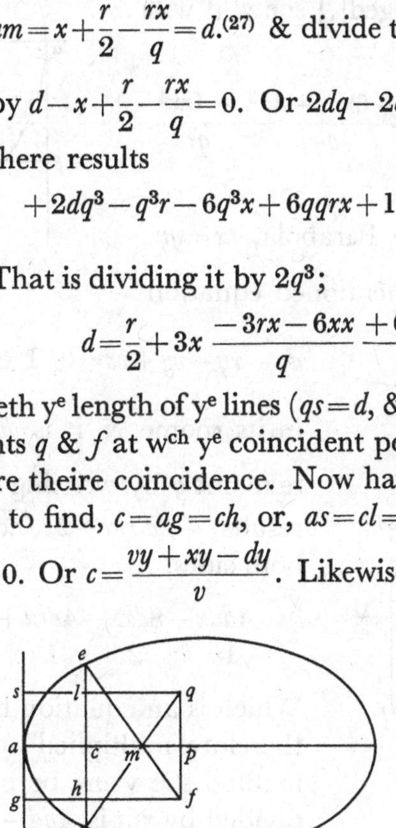

to find *ef* or *kq*, for (supposeing $lq=d-x$; $hf=d-x$; $lc=+c$; $hc=-c$; $ec=+y$; $ck=-y$. $ef=e$. or $kq=e$.) it is $yy-2cy+cc+xx-2dx+dd=ee=\begin{cases}kq\times kq\\ ef\times ef\end{cases}$, Lastly y^e^ circle described wth y^e^ radius *ef* shall have y^e^ same quantity of crookedness wch y^e^ Ellipsis hath at y^e^ point e.

Example y^e^ 2^d^. Were I to find y^e^ quantity of crookedness at some given point of y^e^ line[29] exprest by $rx+\frac{rxx}{q}=yy$; I might consider y^t^ it differs from y^e^ former

(28) Read '$d-x-\frac{r}{2}+\frac{rx}{q}=0$. Or $2dq-2qx-qr+2rx=0$.'

(29) A hyperbola. (Newton draws only one branch of it in his figure.)

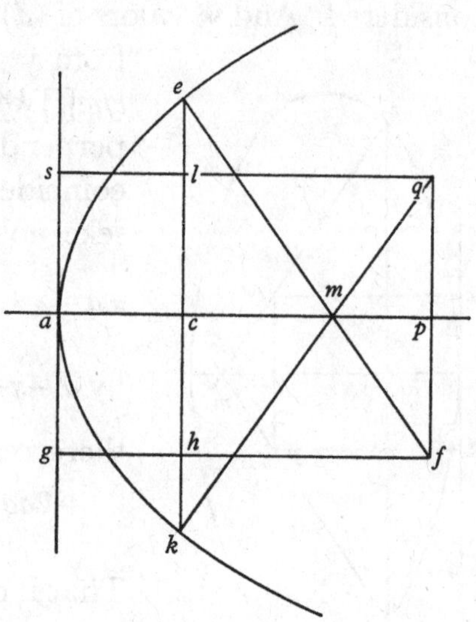

Example onely in y^t there I have $\frac{-rxx}{q}$ or $\frac{rxx}{-q}$, here $\frac{rxx}{q}$, y^t is in y^e former q was negative [&] in this tis affirmative. Soe y^t this operation will bee y^e same wth y^e former y^e signe of q being changed[,] soe y^t it will be found gf or $sq=d=$

$$\frac{r}{2}+3x\,\frac{+3rx+6xx}{q}\,\frac{+6rxx+4x^3}{qq}\,\frac{+4rx^3}{q^3}.$$

&c as before.

Example y^e 3^d. In y^e Parabola, $rx=yy$. & $v=\frac{r}{2}$. In y^e above mentioned equation

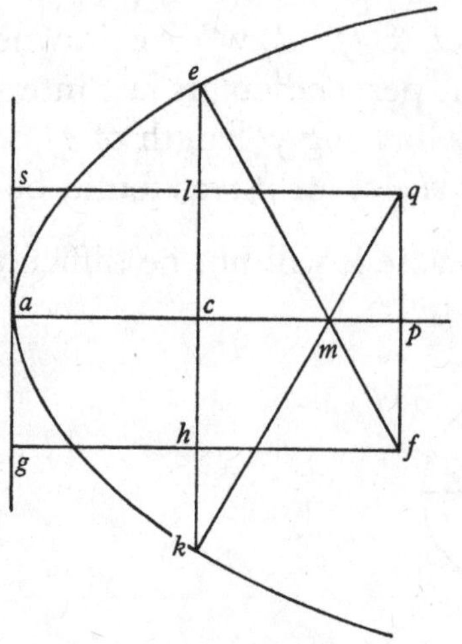

$dy-xy-vy+vc=0$ I take out v by writeing $\frac{r}{2}$ in its roome & it is $dy-xy\,\frac{-ry+rc}{2}=0$. y^n I take out y by writeing $\sqrt{rx}$ in its stead & there results $-2d\sqrt{rx}+2x\sqrt{rx}+r\sqrt{rx}=rc$. & by □ing both sides,

$$\begin{array}{ccccccc}4ddx- & 8dxx- & 4drx+ & 4x^3+ & 4rxx+ & rrx= & rcc.\\ 1 & 2 & 1 & 3 & 2 & 1 & 0\end{array}$$

Which is an equation haveing 2 equall roots & therefore multiplied according Huddenius his method soe y^t rcc be blotted out, & the result divided by x it is, $\begin{array}{l}4dd-16dx+12xx=0.\\ \quad\;\;\, -4dr\;\; +8rx\\ \quad\;\;\;\;\;\;\;\;\;\;\;\;\;\; +rr\end{array}$ Now tis evident y^t $x=ac$ being determined. there are 2 points (viz: e & k) from wch perpendiculars being drawne they intersect one another in y^e axis at m, wherefore $am=x+\frac{r}{2}$ [$=d$] is one of y^e rootes of y^e equation & therefore it being divided by $d-x\frac{-r}{2}=0$, or by $2d-2x-r=0$ there results $2d-6x-r=0$. Or $d=\frac{1}{2}r+3x=sq=gf$. Then into y^e above found equation $\frac{2x\sqrt{rx}-2d\sqrt{rx}+r\sqrt{rx}}{r}=c$,

(30) That is, $(r+4x)^{\frac{3}{2}}/2\sqrt{r}$.
(31) That is, of Hudde.
(32) Ignoring $y=0$, the equation of the axis am.

I substitute this valor of d & there results $\frac{-4x\sqrt{rx}}{r}=c=ag=ch=pf$. Soe y[t] I have $eh=\sqrt{rx}+\frac{4x\sqrt{rx}}{r}$. And $hf=\frac{1}{2}r+2x$. & therefore $ef=\sqrt{\frac{1}{4}rr+3rx+12xx+\frac{16x^3}{r}}$ $=\frac{r+4x}{2r}\sqrt{rr+4rx}$.(30) shall be y[e] Radius of a circle w[ch] is as crooked as y[e] Parabola at y[e] point e.

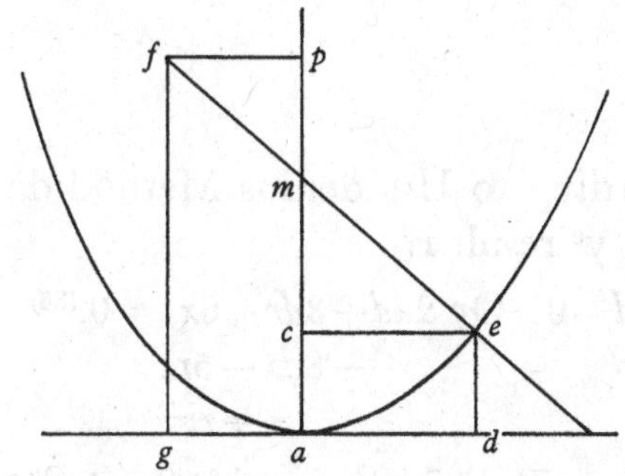

Or it might have beene done thus, haveing y[e] equation $dy-xy\frac{-ry+rc}{2}=0$, I might have writ $\frac{yy}{r}$ in stead of x, & soe have had $dy\frac{-y^3}{r}\frac{-ry}{2}\frac{+rc}{2}=0$. w[ch] must have 2

1 3 1 0

equall roots & therefore by y[e] Method de max: et min:(31) I blot out $\frac{rc}{2}$ & there results, $dy-\frac{3y^3}{r}\frac{-ry}{2}=0$. Or, $d=\frac{r}{2}+\frac{3yy}{2}$.(32) makeing $ad=y$, $de=x$. $cm=v$. $gf=d$. $ag=fp=c$. now if $ad=y$ bee determined it is manifest y[t] there is but one point of y[e] Parab: (viz: e) to bee considered[,] from w[ch] y[e] perpendiculars w[ch] are drawne doe noe where intersect one another & therefore this equation hath noe superfluous rootes like y[e] former.

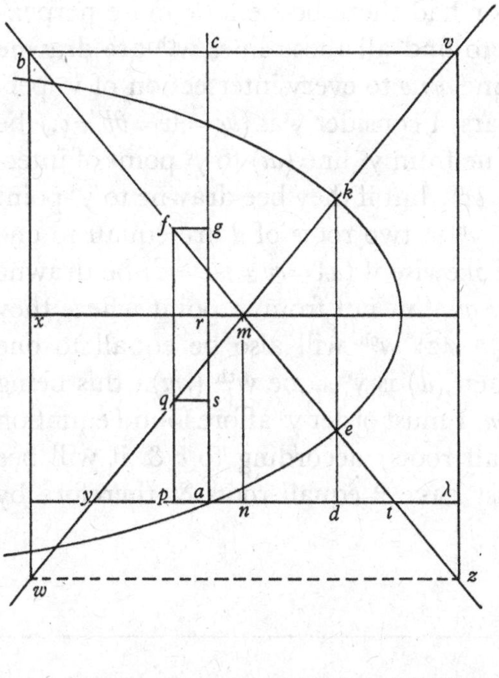

Example y[e] 4th.(33) If it bee supposed y[t] y[e] nature of y[e] line is conteined in $ry-yy-rx=0$. & if tis $ad=x$. $y=\begin{cases}ed\\ek^{(34)}\end{cases}$. $v=\begin{cases}dt\\dy\end{cases}$. fe, & qk 2 perpendiculars to y[e] crooked line, f, & q two points where y[e] coincident perpendiculars last intersected. $d=\begin{cases}ap=qs\\fg\end{cases}$. $c=\begin{cases}pq\\pf\end{cases}$. Then is $v=\frac{ry}{r-2y}$.(35) by w[ch] I take v out of y[e] above named equation $dy-xy-vy+vc=0$. & y[e] result being

(33) A second parabola $(y-\frac{1}{2}r)^2=r(\frac{1}{4}r-x)$. (34) Read '$dk$'.
(35) Compare the general subnormal rule given by Newton in section 3, §2 above.

divided by y, it is, $ry+2dy-2xy-dr+rx-cr\,[=0]$. Or $y=\dfrac{dr-rx+cr}{r+2d-2x}$. Then I substitute this valor of y into its place in [y^e] equation $rx+yy-ry=0$ & there results, $rx+\dfrac{ddrr-2drrx+rrxx+2dcrr-2crrx+ccrr}{rr-4rx+4xx+4dd+4dr-8dx}\;\dfrac{+rrx-drr-crr}{r+2d-2x}=0$. Or by ordering it, it will bee

$$\begin{array}{llll} 4x^3 & -8dxx & +4dd\,x & -ddr=0. \\ & -5r & +6dr & +ccr \\ & & +2rr & -drr \\ & & & -crr \\ 3 & 2 & 1 & 0 \end{array}$$

Which Equation must have two equall roots & therefore ordering it according to Huddenius Method de Maximis & Minimis I blot out y^e last term & y^e result is

$$\begin{array}{lll} 6xx & -8dx & +2dd=0. \\ & -5rx & +3dr \\ & & +rr \end{array} \quad \text{Or} \quad \begin{array}{lll} 2dd & +3dr & +6xx=0.^{(36)} \\ & -8dx & -5rx \\ & & +rr \end{array}$$

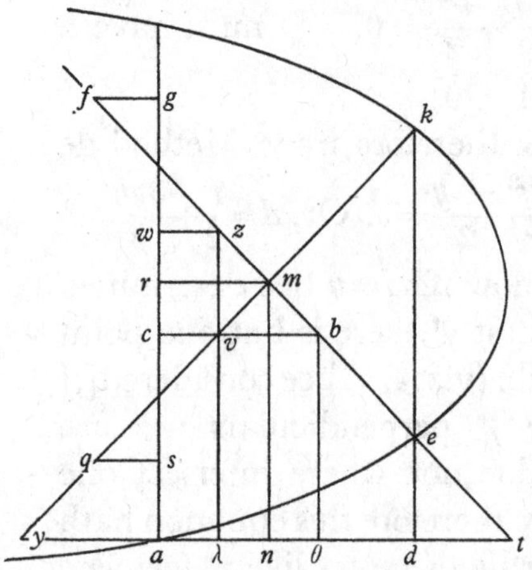

By what was said before tis evident y^t the perpendicular (rm) drawne from y^e line asg where y^e two perpendiculars [ef, kq] intersect is one of y^e rootes of this equation.(37)

And y^t I may have a generall rule to find y^e line rm (or had there beene 3 or more perpendiculars, to find all those lines wch are drawne from y^e line $acrw$ to every intersection of y^e perpendiculars) I consider y^t if $(ac=\lambda v=\theta b=c.)$ be not drawne from y^e line (at) to y^e point of intersection (m), y^n d hath two valors as $(vc$ & $bc)$. but if they bee drawne to y^e point (m), y^t is, if they be coincident wth nm; y^n y^e two roots of d are equall to one another, being y^e same wth y^e line rm. Likewise if $(a\lambda=cv=wz=d)$ be drawne from y^e line aw to y^e perpendiculars fe, qk, but not from y^e point where they intersect; then hath (c) two roots (as λv, λz) wch will also be equall to one another & coincident wth y^e line mn, when (d) is y^e same w^th (rm). this being considered; if I would have y^e valor of nm, I must order y^e affore found equation (in wch x was supposed to have 2 equall roots) according to c & it will bee

$$\begin{array}{lllll} rcc & -rrc & -drr & +6drx & -8dxx=0. \\ & & -ddr & +4ddx & +4x^3 \\ & & +2rrx & -5rxx & \\ 2 & 1 & 0 & 0 & 0 \end{array}$$

wch must have 2 equall roots & therefore by

(36) Or, $(2d-2x+r)(d-3x+r)=0$.

(37) In fact, as Newton will show generally, since $dt=v=y\,dy/dx$ and $nm=ar=\frac{1}{2}r$, then $nt=\frac{1}{2}r(r/[r-2y])$; so that $d=rm=ad-nt+dt=x-\frac{1}{2}r$. It follows that in seeking q and f we need consider only $d-3x+r=0$.

Huden^s^ Meth: de Max: & Min: I take away y^e^ last terme & soe I have $2rcc-rrc=0$, or, $c=\frac{r}{2}=nm$: But if I would have y^e^ valor of rm I order y^e^ equation according to y^e^ letter d & it is,

$$\begin{array}{ccc} 4ddx-drr & +ccr & =0. \\ -ddr+6drx & -crr & \\ -8dxx & +2rrx & \\ & -5rxx & \\ & +4x^3 & \\ 2 & 1 & 0 \end{array}$$

W^ch^ equation must likewise have two equall roots & therefore takeing away y^e^ last terme by Hud: meth: de Max: et min: there resulteth this,

$$\begin{array}{c} 8ddx-drr=0.^{(38)} \\ -2ddr+6drx \\ -8dxx \end{array}$$

Or $d=x-\frac{r}{2}=rm$. & this $\left(x-\frac{1}{2}r\right)$ is one of y^e^ rootes of y^e^ equation

$$\begin{array}{c} 2dd+3dr+6xx=0, \\ -8dx-5rx \\ +rr \end{array}$$

wch was required, therefore I must divide this equation by $d-x+\frac{r}{2}=0$. y^t^ is by $2d-2x+r=0$, & there will result, $d-3x+r=0$. That is $d=3x-r=fg=qs$. Whence it will not be difficult to find y^e^ points q & f & consequently y^e^ lines qk, fe wch shall be y^e^ radij of circles wch have y^e^ same quantity of crookednesse y^e^ line (aek) hath at y^e^ points e & k. Makeing $c=\begin{cases} as \\ ag \end{cases}$.

Note y^t^ these equations have not (rm) or (ar) for one of theire rootes unlesse when y^e^ axis of y^e^ line is parallel to x (for y^n^ onely a circle whose center is at y^e^ intersection (m) can touch y^e^ crooked line in both k & e together.) & then perhaps they may easlyer bee found y^n^ by y^e^ foregoing rule.(39)

---

[3] Feb 1664.(40)

The Crookednesse in lines may bee otherwise found as in y^e^ following Examples.

In the Parabola aeg suppose e y^e^ point where the crookednesse is sought for, & y^t^ f is the center & fe y^e^ Radius of a Circle equally crooked wth y^e^ Parabola

---

(38) Or $d(2x-r-2d)(4x-r)=0$.

(39) This, though true in the conic (the only curve here considered), does not hold in general for higher curves and is rightly cancelled by Newton.

(40) That is, February 1664/5.

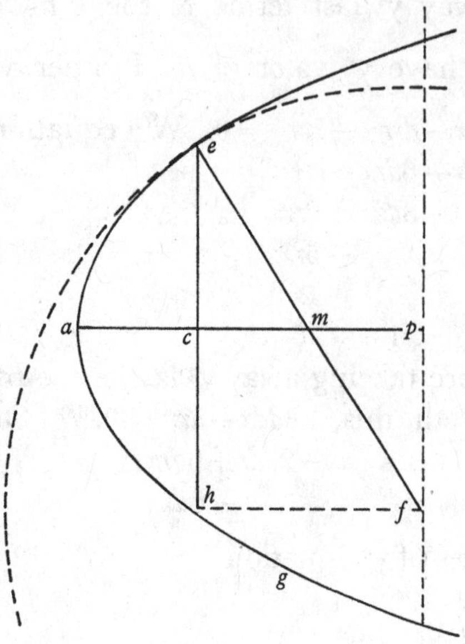

at e. Then naming y^e^ quantitys $ce=y$. $ap=d$. $pf=-c$. $ef=s$. B[y] y^e^ nature of y^e^ line $ac=\frac{yy}{r}$. $ce+pf=eh=y-c$. $cp=\frac{yy}{r}-d=hf$. $eh^2+hf^2=ef^2$. That is,

$$\frac{y^4}{rr} * \frac{-2dyy}{r} + yy - 2cy \begin{matrix}+cc\\+dd\end{matrix} = ss.^{(41)}$$

$$4 \qquad 2. \qquad 2 \quad 1 \quad 0 \quad 0$$

W^ch^ equation must have 2 equall rootes that ef may be ⊥ to y^e^ Parab: & therefore multiplyed according to Hudden's Method it produceth $\frac{2y^3}{rr} \frac{-2dy}{r} + y - c = 0$. Which

$$3 \qquad 1 \qquad 1 \quad 0$$

equation hath soe many rootes as there can be drawne perpendiculars to y^e^ Parab: from the determined point f.(42) And two of these rootes must become equall, y^t^ f may bee the center of y^e^ required Circle, therefore this equation is to bee multiplyed again,(43) & it will produce $\frac{6y^2}{rr} - \frac{2d}{r} + 1 = 0$. that is $\frac{3yy}{r} + \frac{r}{2} = d$. Or $3x + \frac{r}{2} = d$; As was found in y^e^ 3^d^ precedent example.(44)

Here observe y^t^ in y^e^ 1st of these 3 equations y hath 4 values gl, ec, hs & kv. see fig: 2^d^. when d, c, & s are determined. But d, c, & $y=ec$ being determined s hath but one valor$=ef$. And if d, s, & $y=ec$ bee determined y^n^ c hath 2 valors pf & pm. And c, s, & $y=ec$ being determined d hath 2 valors an & ap. as that first equation denotes by y^e^ dimensions of y^e^ quantitys in it.

By the 2^d^ of these equations 2 of y^e^ valors of (y) are united by y^e^ increasing or diminishing y^e^ valor of $d=[ap]$ &c. first suppose y^e^ circle soe little as noe where to intersect y^e^ Parabola, it being increased gradually will first touch y^e^ Parab: at r (fig 3^d^)[,] then ceasing to touch it intersect it in 2 points g & k (fig 2^d^)[,] wch two points grow more distant untill it touch y^e^ Parab: in t (fig: 3^d^)[,] wch being divided into two intersection points e & h (fig 2^d^) the points g & e draw neerer untill they conjoyne in y^e^ touch point w & soe y^e^ circle ceaseth (by still increasing) to touch ye Parab: or intersect it unless in h & k. Whence from one point f may

(41) Where $ac = x$ say, Newton constructs the circle $s^2 = (d-x)^2+(y-c)^2$ of fixed centre f and radius $ef = s$. He then eliminates the variable x between this and the defining equation $y^2 = rx$ of the parabola, and so finds the meets of the circle and parabola.

(42) This gives the condition that the circle $s^2 = (d-x)^2+(y-c)^2$ be tangent to the parabola, or that ef through the fixed point f be normal to the curve.

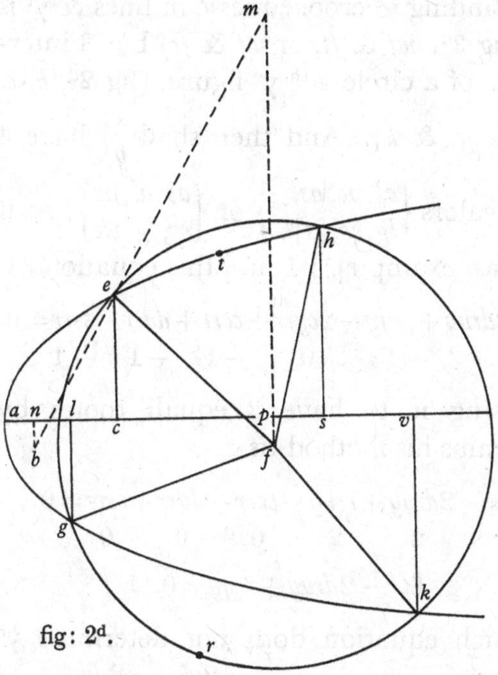

fig: 2d.

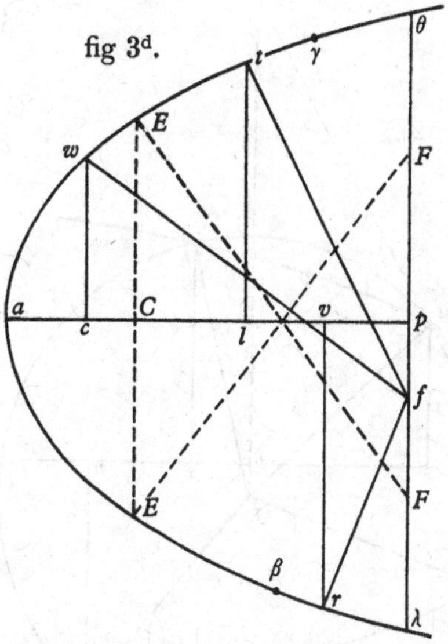

fig 3d.

be drawne 3 perpendiculars *fr*, *fw*, *ft* to ye Parabola *twar*.(45) And therefore in this 2d equation (y) must have 3 valors *wc*, *tl*, & *vr*, when $ap=d$, & $pf=-c$, are determined [&] then also hath (s) three valors *fr*, *fw*, *ft*.

By the 3d equation Two of the valors of (y) in ye 2d equation are united by increasing or diminishing ye length of $pf=-c$. For begining at the point p (from wch ye 3 perpendiculars fall upon γ, α, & β) if ye point f doth gradually move from p, the perpendicular $\begin{Bmatrix} ft \\ fw \\ fr \end{Bmatrix}$ moves from $\begin{Bmatrix} \gamma \\ a \\ \beta \end{Bmatrix}$ towards $\begin{Bmatrix} a \\ \gamma \\ \lambda \end{Bmatrix}$ Soe yt ye two perpendiculars *wf* & *tf* will at last conjoyne into one *EF*, which shall be ye Rad: of a Circle as crooked as ye Parab: at *E*.

This 3d operation might have beene done by making *pf* determined & by increasing or diminishing $ap=d$. That is by destroying ye terme $\frac{-2dy}{r}$ in stead of $-c$ in ye 2d equation. And so might ye 2d Operacō [have] beene done otherwise by determining ye circle *egk*, Or taking c or d out of ye 1st equation instead of *ss*.

(43) This further restriction is now one on the position of f: namely, it fixes f as a point from which two coincident normals may be drawn to the parabola, and so as the limit-meet of normals indefinitely near to *ef*, or as the centre of curvature at the point e.

(44) See [2] above.

(45) The fourth normal being the diameter through f, perpendicular to the parabola at the point at infinity.

Another way. There is another way of finding y^e crookednesse in lines & y^t is not by supposing two perpendiculars (fig 3^d. *wf* & *ft*, or *wf* & *fr*) but 3 intersections of a circle wth y^e figure, (fig 2^d, *h*, *e*, *g*; or *e*, *g*, & *k*). And then shall $\left.\begin{matrix}x\\y\end{matrix}\right\}$ have 3 equall valors $\left\{\begin{matrix}al, ac, as.\\lg, ec, [s]h.\end{matrix}\right\}$ or $\left\{\begin{matrix}ac, al, av\\ce, lg, vk\end{matrix}\right\}$. As if (in y^e last example[)] I had this equation

fig: 2^d.

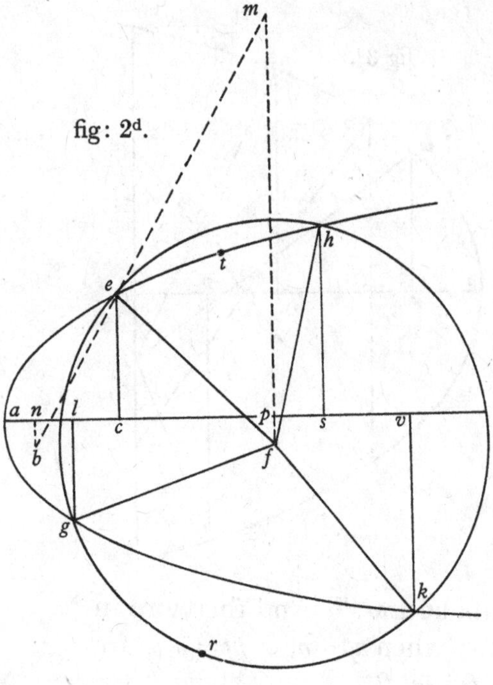

$$\begin{matrix} y^4 * -2dryy+rryy-2cyrr+ccrr+ddrr-ssrr=0.\\ 3 \quad 1 \quad 1 \quad 0 \quad -1 \quad -1 \quad -1\end{matrix}$$

Supposing it to have 3 equall rootes by Huddenius his method tis

$$\begin{matrix} 3y^4 * -2dryy+rryy-ccrr-ddrr+ssrr=0.\\ 4 \quad 2 \quad 2 \quad 0 \quad 0 \quad 0\end{matrix}$$

$$6y^4-2dryy+rryy=0.$$

(Which equation doth not determine y^e perpendiculars to (*eag*) as $\frac{2y^3}{rr} * \frac{-2dy}{r}+y=c$ doth for by this I can find y^e valor of *c* (*y* being determined) but by it I can neither find y^e valor of *c* nor [*d*] u[n]till one of y^m is taken out of y^e equation.(46) That equation multiplyed according to y^e dimensions of *y* produceth

$$6y^2-2dr+rr[=0].^{(47)} \quad \text{Or} \quad \frac{3yy}{r}+\frac{r}{2}=3x+\frac{r}{2}=d.$$

---

(46) Newton is misled when he considers this method intrinsically different from the previous one, for definitions of *f* as the centre of a circle meeting the curve in three coincident points or as the limit-meet of the normals to the curve at points indefinitely near to *e* are strictly equivalent (as the similar application of Hudde's algorithm for triple roots reveals). The differing final equations result not from a difference in method but merely from a different choice of the multiplying Huddenian numbers in the two cases.

(47) That is, ignoring the factor y^2.

(48) Since they vanish in the limit as $o \to$ zero even when first divided by o.

(49) A simple variant on Newton's first arguments. Where $x = f(y)$, and since $an = d$ (or $cn = qt = d-x$), then $bq = f'(y)\,(d-f(y))$, and so

$$bt^2 = [f'(y)\,(d-f(y))]^2+[d-f(y)]^2$$

and its incremented value $pt^2 = [f'(y+o)\,(d-f(y+o))]^2+[d-f(y+o)]^2$. Thereupon Newton introduces the centre of curvature t as the limit-meet of the (equal) normals bt, pt and so derives

$$0 = \lim_{o\to\text{zero}} [(1/o)\{(1+[f'(y+o)]^2)\,(d-f(y+o))^2-(1+[f'(y)]^2)\,(d-f(y))^2\}],$$

The same way may be done thus. If a circle touch a crooked line in one point & intersect it [in] another[,] when those two points come together y^t circle is as crooked as y^e line at y^e touch or concourse point.

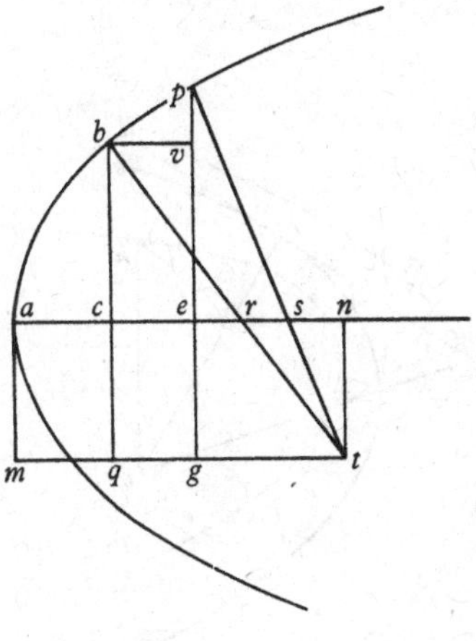

As if $bc=y.\ ac=\frac{yy}{r}.\ an=mt=d.\ pv=o.\ cr=es=\frac{r}{2}.$

$$cn=d-\frac{yy}{r}.\ \frac{r}{2}:y::\frac{dr-yy}{r}:\frac{2dry-2y^3}{rr}=bq.\ \frac{r}{2}:o+y$$

$$::gt=d\frac{-yy-2oy}{r}:\frac{2dry+2dro-2y^3-6oyy}{rr}=pg.$$

(o^2, o^3, o^4 &c: may ever bee omitted.(48)) Then since $bq^2+qt^2=bt^2=pt^2=pg^2+gt^2$. Therefore

$$6y^4o-8dryyo+rryyo+2ddrro-dr^3o=0.$$

Or, $dd=\frac{8dryy}{2rr}+\frac{dr}{2}[-]\frac{yy}{2}-\frac{3y^4}{rr}$. or

$$d=\frac{2yy}{r}+\frac{r}{4}\ \mho\ \sqrt{\frac{y^4}{rr}+\frac{yy}{2}+\frac{rr}{16}}=\frac{3yy}{r}+\frac{r}{2},$$

as before. Or else $d=\frac{yy}{r}$ wch cannot bee.(49) This is rather done by y^e convening of two perpendiculars but it might have beene done by supposeing y^e circle described by y[^e] Rad: bt to intersect y^e crooked line in p, & y^n y^e lines pt & bt to convene.

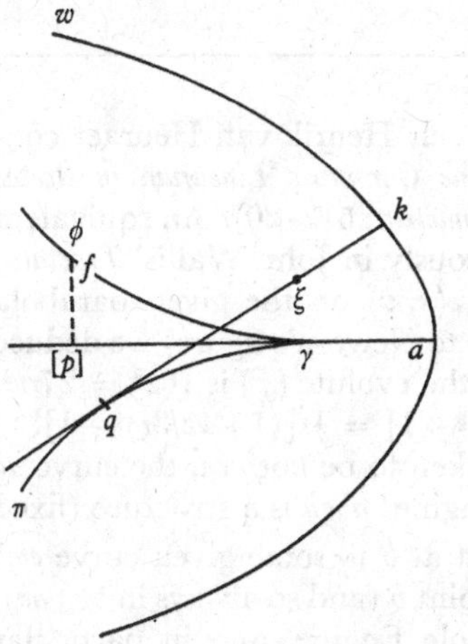

[4] May 1665. [*Notes on the evolute.*]

1. Note that y^e crooked line $\phi f\gamma q\pi$ (described by y^e points q & f,) is always touched by the (perpendicular) line kq; & that in such sort as to bee measured by it. they applying themselves the one to the other, point by point; soe y^t if $a\gamma=k\xi$ the shortest of all y^e lines qk, be substracted from qk, there remaines $q\xi=q\gamma$.(50) By this meanes y^e length of as many crooked lines may bee found as is desired.

2. Also if y^e line qk is applyed to y^e crooked line $q\gamma$ point by point, every point of y^e line qk (as k) shall describe lines (as akw) to w^ch $q\xi k$ is perpendicular.

---

or $\quad 0 = (d/dy)\{(1+[f'(y)]^2)(d-f(y))^2\}$

$$= -2(d-f(y))f'(y)(1+[f'(y)]^2)+(d-f(y))^2.2f''(y)f'(y),$$

and finally $\quad qt = d-f(y) = \frac{1+[f'(y)]^2}{f''(y)},$

as before but with x and y interchanged.

(50) For, since $q\xi$ is tangent to the arc $\pi q\gamma$ at q, the limit-increments of $q\xi$ and $q\gamma$ will be equal, with both zero at γ. (See section 7, Prob. 9 below.)

3. The line $q\pi\gamma f\phi$ is y^e^ same (if *wka* be a Parab:) wth y^t^ whose lenght Heuraet found.(51)

4. If *bc* is a perpendicular to the line *efb*, & *abd* a tangent & y^e^ position & distance of the point *a*, or *h* &c. in respect of y^e^ tangent *ab* & perpendicular *bc* bee given. that is if those points move wth y^e^ tangent (as if they were inherent in y^e^ same body) while y^e^ tangent glides over y^e^ crooked line *ebf*, soe y^t^ y^e^ point $\left.\begin{matrix} a \\ h \end{matrix}\right\}$ describe the line $\left.\begin{matrix} dag \\ hrs \end{matrix}\right\}$ then from y^e^ point *a* (or [*h*], [)] draw perpendiculars to y^e^ line *dag* (or *s*[*h*]*r*). then shall the point *c*, where those perpendiculars intersect, be y^e^ point of y^e^ Engine *abdgch* wch is in least motion, & y^e^ lines $\left\{\begin{matrix} ac \\ bc \\ [h]c \end{matrix}\right\}$ shall bee y^e^ radij of circles wch are as crooked as $\left\{\begin{matrix} dag \\ feb \\ shr \end{matrix}\right\}$ at y^e^ points $\left\{\begin{matrix} a \\ b \\ h \end{matrix}\right\}$. Soe y^t^ the point *c* determines y^e^ perpendiculars to & crookedness of an infinite company of given lines *adg*, *ebf*, *rhs*, &c.(52)

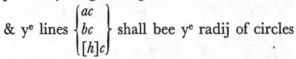

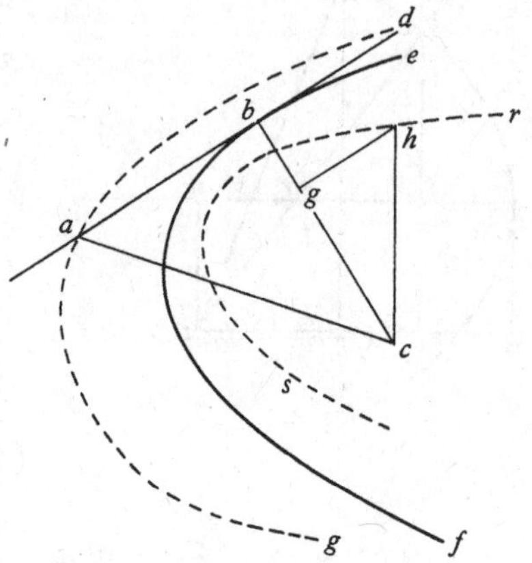

(51) Newton refers to the semicubical parabola $ky^2 = x^3$ which Henrik van Heuraet considered as the prime example of his *Epistola de Transmutatione Curvarum Linearum in Rectas* (published by Schooten from his letter of 13 January 1659 in *Geometria*: 517–20). An equivalent rectification of this cubic by William Neil appeared simultaneously in John Wallis' *Tractatus Duo* (Oxford, 1659: 90). Where f is the curvature centre of $e(x, y)$ on the given parabola $y^2 = rx$, on making $\gamma p = z$ and $pf = v$ (here added in broken line to Newton's figure) we deduce from the preceding pages that $z = 3x$ and $v = 4y^3/r^2$, so that the evolute (f) is $16z^3 = 27rv^2$. Evidently the general arc $f\gamma$ is equal to $ef - \gamma p = \frac{1}{2}r[(1+4x/r)^{\frac{3}{2}} - 1] = \frac{1}{2}r[(1+4z/3r)^{\frac{3}{2}} - 1]$.

(52) Newton's figure has two points g while the point d is taken to be both on the curve ag and the tangent ab, but there should be no confusion. The 'Engine' $abgh$ is a structure (fixed in its parts) which moves so that ab is instantaneously tangent at b to some given curve $\widehat{ebf}$. Newton argues that, where c is the centre of curvature at the point b (and so always in $bg \perp ab$), the point c is the instantaneous centre of motion of the whole Engine, and in particular therefore of any point (as a or h) in it: hence, that c is the centre of curvature at the points a and h on $\widehat{dag}$ and $\widehat{rhs}$ respectively. The argument is subtly fallacious. In fact, c is not momentarily at rest in general but will pass into c' in the indefinitely small interval of time in which points a, b, g and h pass into a', b', g' and h'. It follows that, where $cC = cc' \sin g\hat{c}h$ and $cC' = cc' \sin b\hat{c}a$ are the projections of the increment cc' in the directions of the curves $\widehat{rhs}$ and $\widehat{dag}$ at the points h and a, then the centres of curvature γ, γ' at h and a are fixed in the normals hc, ac by the proportions $h\gamma : c\gamma = hh' : cC$ and $a\gamma' : \gamma' c = aa' : cC'$. Clearly the curvature centre

[5] December 1664. *Haveing found (by ye former rule*(53)*) an equation by which ye quantity of crookednesse in any line may bee found to find ye greatest or least crookednes of that line.*

In ye 4th Example(54) I had found $gf = qs = d = 3x - r$. And by a rule there shewed viz[:] $\left.\begin{matrix} as \\ ag \end{matrix}\right\} = c = \dfrac{xy + vy - dy}{v}$: It was there found

$$\begin{array}{llll} 4x^3 & -8dxx & +4ddx & -ddr = 0. \\ & -5r & +6dr & +ccr \\ & & +2rr & -drr \\ & & & -crr \end{array}$$

γ coincides with c in the limit as $c' \to c$ only when $cC = 0$, that is when $gh = 0$ and the point h coincides with g. In that case h will develop one of the family of involutes to the evolute of the curve $\widehat{ebf}$ (which is itself the member of the family for which $bg = 0$).

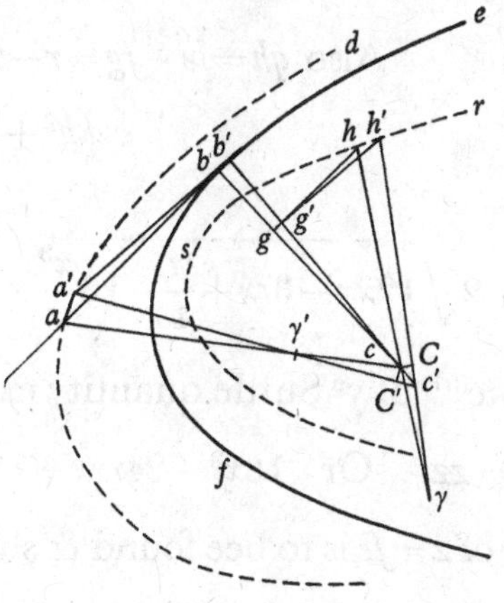

The curves (h) produced by the Engine seem not to occur in subsequent mathematical history. Analytically, if we suppose $bg = a$, $gh = b$ and determine the point $b(x, y)$ to be in the known curve $\widehat{ebf}$ by the rectangular Cartesian defining equation $f(x, y) = 0$, the Engine constructs for each point b a corresponding point $h(X, Y)$ of the curve $F(X, Y) = 0$, where $X - x = a(dy/ds) + b(dx/ds)$ and $Y - y = -a(dx/ds) + b(dy/ds)$: we may easily show that the three equations $f(x, y) = 0$, $(X-x)^2 + (Y-y)^2 = a^2 + b^2$ and

$$(X-x)(a - b(dy/dx)) + (Y-y)(a(dy/dx) + b) = 0$$

define $F(X, Y) = 0$ parametrically in terms of x and y. Further, on taking the element of arc $bb' = ds$ and the increment $cc' = d\rho$ (of the curvature-radius $bc = \rho$), it follows that γ is fixed in the normal hc by

$$h\gamma = \frac{(\rho - a)\sec\alpha}{1 - \frac{1}{2}(d\rho/ds)\sin 2\alpha},$$

where $\widehat{gch} = \alpha = \tan^{-1} b/(\rho - a)$. Newton's accompanying figure appears to illustrate the case of the parabola $y^2 = rx$, in which (h) is a double-looped 12-degree curve sharing a parabolic point with (b).

(53) That is, in [2] above.

(54) That is, the parabola $rx = y(r-y)$ in [2] above.

Now by writeing $3x-r$ in stead of d & ordering y^e^ product according to y^e^ letter c it is $ccr - crr + 16x^3 = 0$. Or extracting y^e^ roote it is

$$\begin{array}{l} ccr - crr + 16x^3 = 0 \\ \quad\;\; -12rxx \\ \quad\;\; +3rrx \end{array}$$

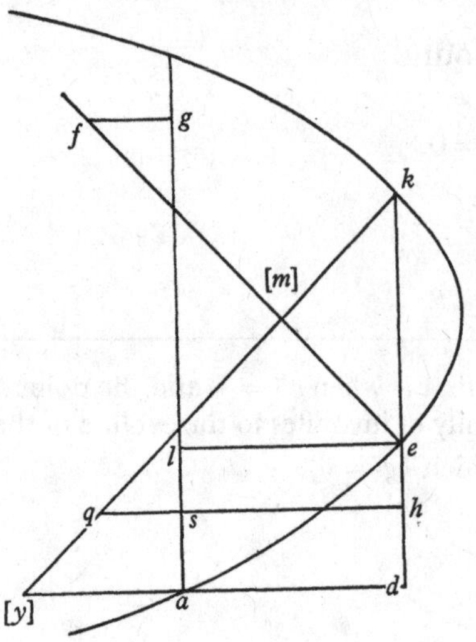

$$c = \frac{r}{2} \pm \sqrt{12xx - 3rx + \frac{rr}{4} - 16\frac{x^3}{r}}\,^{(55)} = \begin{cases} as. \\ ag. \end{cases}$$

Also by y^e^ nature of y^e^ line

$$\left.\begin{matrix} kd \\ ed \end{matrix}\right\} = y = \frac{r}{2} \pm \sqrt{\frac{rr}{4} - rx}.$$

Therefore

$$kh = gl = \begin{Bmatrix} kd - sa \\ ga - ed \end{Bmatrix}$$

$$= \sqrt{\frac{rr}{4} - rx} + \sqrt{12xx - 3rx + \frac{rr}{4} - \frac{16x^3}{r}}.^{(56)}$$

Also $qh = le - fg = r - 2x$. And since

$$kh^2 + qh^2 = qk^2$$

Therefore

$$\frac{3rr}{2} - 8rx + 16xx - \frac{16x^3}{r} + 2\sqrt{12xx - 3rx + \frac{rr}{4} - 16\frac{x^3}{r}} \times \sqrt{\frac{rr}{4} - rx} = qk \times qk = zz;$$

Supposing $qk = z$. The roote(57) of y^e^ Surde quantity extracted[,] the equation is $-\frac{16x^3}{r} + 24xx - 12rx + 2rr = zz$. Or $16x^3 - 24rxx + 12rrx - 2r^3 + rzz = 0$.(58) In wch equation y^e^ least valor of $z = fe$ is to bee found & should happen when x hath

(55) Or $c = \frac{1}{2}(r \pm \sqrt{[(r-4x)^3/r]})$.
(56) Which reduces to $(r-2x)\sqrt{[(r-4x)/r]}$.
(57) $\frac{1}{2}(r-4x)^2$. (58) Or $rz^2 = 16(\frac{1}{2}r - x)^3$.
(59) See note (65) below.
(60) In fact, $x = \frac{1}{2}r - (\frac{1}{16}rz^2)^{\frac{1}{3}}\omega$, where ω is one of the cube roots of unity.
(61) Correctly $\frac{1}{4}(r^2 + (r-2y)^2)^3 = (r^2z)^2$. (62) Read '$-20ry^4$'.
(63) This should be $\frac{1}{4}(2y-r)(r^2 + (r-2y)^2)^2 = 0$. (Newton has divided through by $12y$.)
(64) Read '$4y^4 - 8ry^3 + 8rryy - 4r^3y + r^4 = 0$'. Newton's result, $4r^2x^2 + 4r^3x - r^4 = 0$, follows by differentiating $16x^3 + 24rx^2 - 12r^2x - 2r^3 + rz^2 = 0$ and omitting the factor $-r\,dx/dy = 2y - r$.
(65) With typical acuteness Newton picks up an apparent failure of Hudde's rule for maxima and minima, but he fails to trace the error to its true source. The application of Hudde's algorithm is strictly equivalent to differentiating an algebraic function and equating the derivative to zero, and the condition that $f'(a) = 0$ does indeed demand that $f(x)$ have the factor $(x-a)^2$. Further, at a true extreme value $f(x) = 0$ must have two coincident roots, despite Newton's assertion to the contrary, and Hudde's algorithm will find all double roots of $f(x) = 0$. In his tract, however, Hudde tacitly admits that his rule yields the condition for $f(x) = 0$ to have a multiple root of any order (equivalently, $f'(a) = 0$ is the condition for

2 equall valors or rootes.(59) But because $fe = z$ being determined x can have but one valor $= ad$ y^e^ other 2 rootes being imaginary[,](60) tis impossible y^t^ it should have 2 equall rootes: Therefore I take away x out of y^e^ equation by substituting its valor $\frac{ry-yy}{r}$ in its stead & there results

$$\underset{6}{16y^6} - \underset{5}{48ry^5} + \underset{4}{24rry^4} + \underset{3}{32r^3y^3} - \underset{2}{36r^4yy} + \underset{1}{12r^5y} - \underset{0}{2r^6} + \underset{0}{r^4zz} = 0.^{(61)}$$

In wch equation z or $ef = qk$ being determined y hath 2 valors de & dk[,] y^e^ other foure being imaginary. & when ef is the longest or shortest that may bee then these two valors become one & then is y^e^ line aek most or least crooked. If therefore (y^t^ y's valors become equall) this equation is multiplyed according to its dimensions there will result $8y^5 - 20y^4$(62) $+ 8rry^3 + 8r^3yy - 6r^4y + r^5 = 0.$(63) wch is divisible by $y - \frac{r}{2} = 0$, or by $2y - r = 0$ (for there results

$$4y^4 - 8ry^3 + 4r^3y - r^4 = 0^{(64)}).$$

And if $y = \frac{r}{2}$, y^n^ is $x = \frac{r}{4}$. Therefore I take $ad = \frac{r}{4}$ & $de = \frac{r}{2}$ & at y^e^ point e shall bee y^e^ least crookednesse.

Here may bee noted Huddenius his mistake, y^t^ if some quantity in an equation designe a maximum or minimū y^t^ equation hath two equall rootes wch is false in y^e^ equation $16x^3 - 24rxx + 12rrx - 2r^3 + rzz = 0$. & in all other equations which have but one roote.(65)

$f(x) = 0$ to have a factor $(x-a)^p, p \geqslant 2$) and his only way of deciding the form of the multiple root is by further application of his rule (or, equivalently, by considering the pth order derivatives successively, $p = 2, 3, \ldots$). Thus, if we apply the algorithm to Newton's first equation above (note (58)) we find equivalently that

$$(d/dx)\,[rz^2 - 16(\tfrac{1}{2}r - x)^3] = 48(\tfrac{1}{2}r - x)^2 = 0 \quad \text{or} \quad x = \tfrac{1}{2}r$$

is the condition for the expression to have the factor $(\frac{1}{2}r - x)^2$: substituting, we may find $z = 0$ and so the condition yields a *triple* root $x = \frac{1}{2}r$. The error, then, lies not in application of the algorithm to the finding of maxima or minima, but rather—as Newton seems not to realize—in the illusory nature of the 'extreme' values of the radii of curvature which he seeks.

Implicitly, Newton allows into his argument radii of curvature only at real points of his curve, but his analysis covers all radii of curvature defined parametrically in terms of the co-ordinates x or y. In the present example, the radii z do not reach an extreme value when the ordinates of the parabola become imaginary, and, accordingly, Hudde's rule does not reveal one. If, however, we restrict our attention to real points on the parabola both

$$rz^2 = 16(\tfrac{1}{2}r - x)^3 \quad \text{and} \quad r^4z^2 = \tfrac{1}{4}(r^2 + (r-2y)^2)^3$$

yield the same apparently extreme values for z. In effect, since Newton demands that we take a positive value for the radius of curvature, then $z \geqslant 0$; further, real points on the parabola $rx = y(r-y)$ are given only for $x \leqslant \frac{1}{4}r$. Geometrically, if we represent both equations as

May 1665. *Another way.*

Or because y^e lines *fn* & *qn* described by y^e points *f* & *q* doe touch one another in y^e points *n* from wch points onely lines drawne perpendicular to y^e croked line *kea* will bee perpendicular to y^e point of greatest or least crookednesse: And

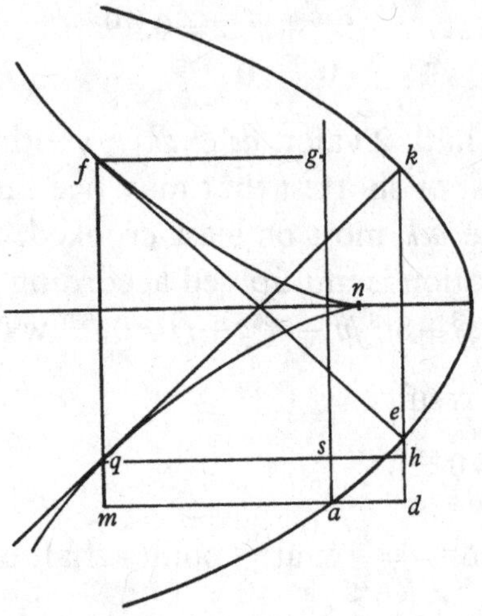

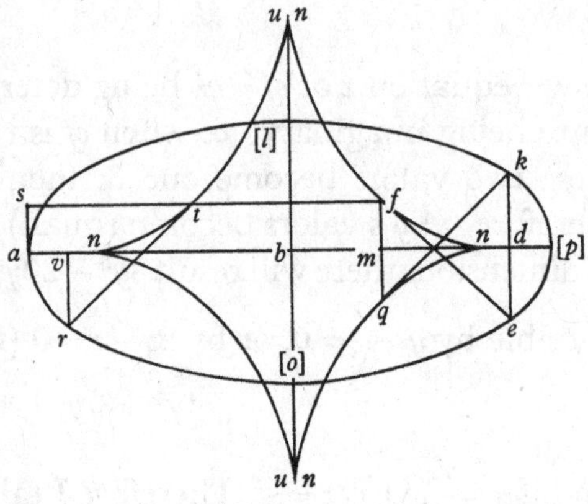

also since all those are points of greatest or least crookedness to wch such perpendiculars are drawne: The difficulty will be to find y^e point *n*. Now

Cartesian curves we restrict ourselves only to one quadrant of the plane. Specifically, $rz^2 = 16(\frac{1}{2}r - x)^3$ is the defining equation of a semicubical parabola, where $ab = x$, $bc = z$; and in the quadrant $z \geqslant 0$, $x \leqslant \frac{1}{4}r$ bc reaches an apparent minimal value at $BC = \frac{1}{2}r$

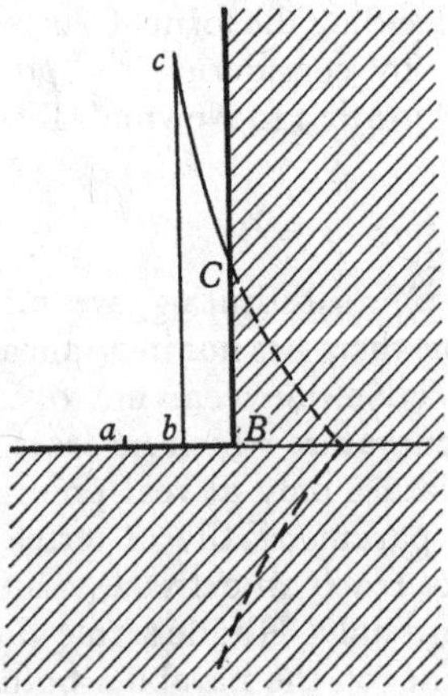

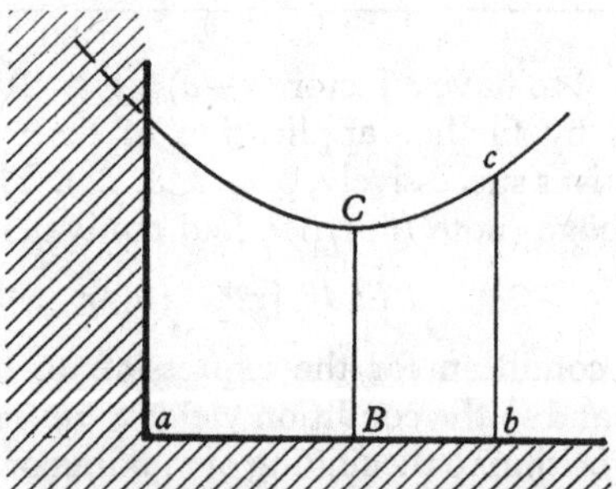

(or $aB = \frac{1}{4}r$). Likewise, where $ab = y$, $bc = z$, then $r^4z^2 = \frac{1}{4}(r^2 + (r-2y)^2)^3$ is the defining equation of a sextic; and here it is visually obvious that z takes on a true 'Huddenian' minimal value at $BC = \frac{1}{2}r$ (or $aB = \frac{1}{2}r$) in the permissible quadrant, and therefore that this value may be evaluated by application of Hudde's algorithm. (Compare note (70) below.)

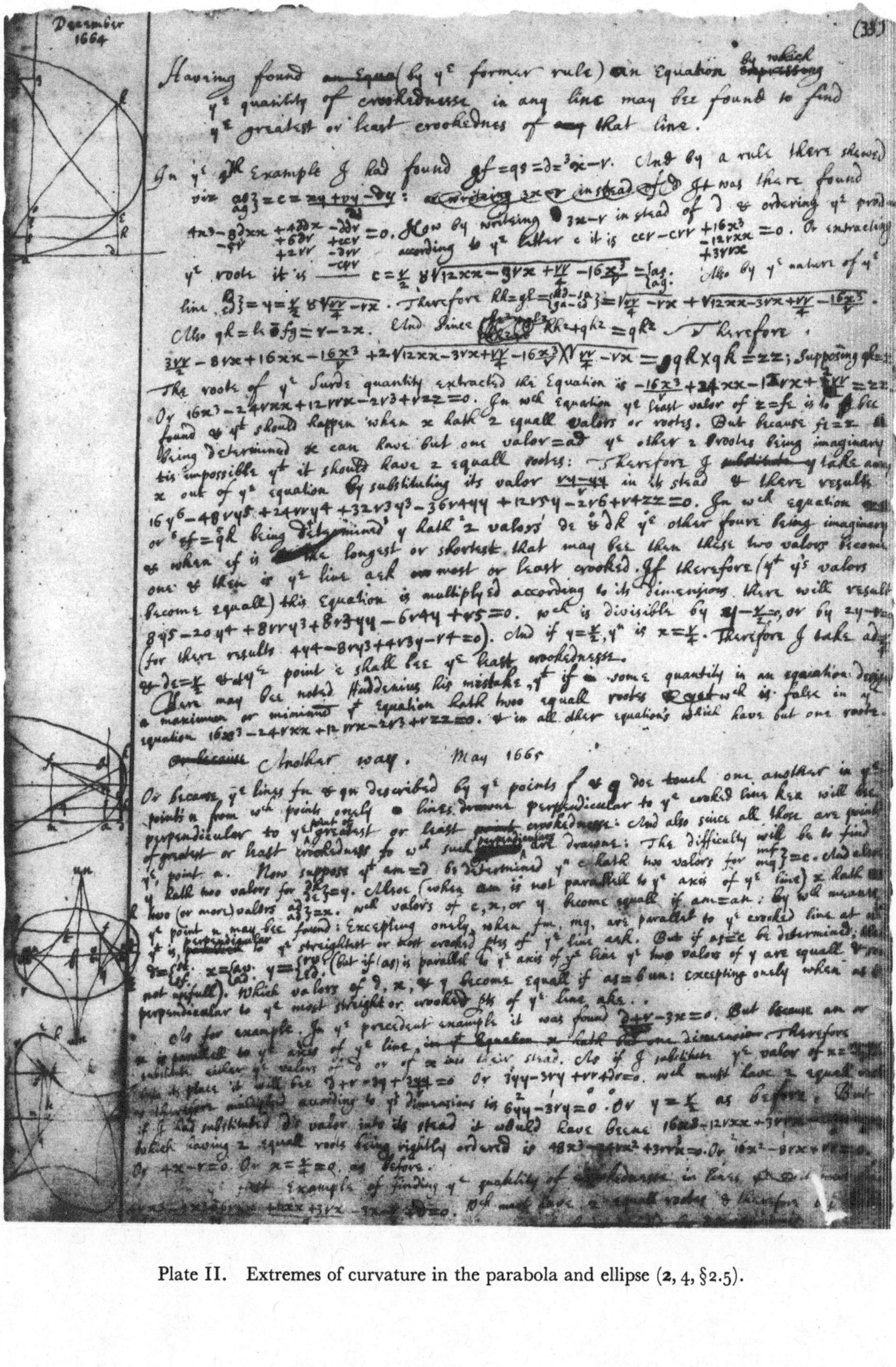

December 1664

Having found ~~an Equa~~ (by ye former rule) an Equation ~~expressing~~ by which ye quantity of crookednesse in any line may bee found to find ye greatest or least crookednes of ~~any~~ that line.

In ye 9th Example I had found $gf = qs = d = 3x - r$. And by a rule there shewed viz $\frac{ag}{ag}\} = c = \frac{xy + ry - dy}{d}$: ~~by writing 3x−r in stead of d~~ It was there found $4x^3 - 8dxx \ldots = 0$. Now by writing $3x - r$ in stead of d & ordering ye product according to ye latter c it is $ccr - crr + 16x^3 - 12rxx + 3rrx = 0$. Or extracting ye roote it is $c = \frac{r}{2} \pm \sqrt{12xx - 9rx + \frac{rr}{4} - \frac{16x^3}{r}} = \{as, ag\}$. Also by ye nature of ye line $\{kg, ed\} = y = \frac{r}{2} \pm \sqrt{\frac{rr}{4} - rx}$. Therefore $kh = gl = \{sd - sa, ga - ed\} = \sqrt{\frac{rr}{4} - rx} + \sqrt{12xx - 3rx + \frac{rr}{4} - \frac{16x^3}{r}}$. Also $qk = kd - fg = r - 2x$. And since $kh^2 + qk^2 = qk^2$. Therefore

$\frac{3rr}{2} - 8rx + 16xx - \frac{16x^3}{r} + 2\sqrt{12xx - 3rx + \frac{rr}{4} - \frac{16x^3}{r}} \times \sqrt{\frac{rr}{4} - rx} = qk \times qk = zz$; Supposing $qk = z$.

The roote of ye surde quantity extracted the Equation is $-\frac{16x^3}{r} + 24xx - 12rx + 2rr = zz$. Or $16x^3 - 24rxx + 12rrx - 2r^3 + rzz = 0$. In wch Equation ye least valor of $z = fe$ is to bee found & yt should happen when x hath 2 equall valors or rootes. But because $fe = z$ being determined x can have but one valor $= ad$ ye other 2 rootes being imaginary tis impossible yt it should have 2 equall rootes: Therefore I ~~substitute~~ take away x out of ye equation by substituting its valor $\frac{ry - yy}{r}$ in its stead & there results $16y^6 - 48ry^5 + 24rry^4 + 32r^3y^3 - 36r^4yy + 12r^5y - 2r^6 + r^4zz = 0$. In wch equation $ef = qk$ being determined y hath 2 valors de & dk ye other foure being imaginary & when ef is the longest or shortest that may bee then these two valors become one & then is ye line ark most or least crooked. If therefore (yt y's valors become equall) this Equation is multiplyed according to its dimensions there will result $8y^5 - 20ry^4 + 8rry^3 + 8r^3yy - 6r^4y + r^5 = 0$. wch is divisible by $2y - r = 0$, or by $2y - r$ (for there results $4y^4 - 8ry^3 + 4r^3y - r^4 = 0$). And if $y = \frac{r}{2}$, yn is $x = \frac{r}{4}$. Therefore I take $ad = \frac{r}{4}$ & $de = \frac{r}{2}$ & ye point e shall bee ye least crookednesse.

Here may bee noted Huddenius his mistake, yt if some quantity in an equation designe a maximum or minimum yt Equation hath two equall roots wch is false in ye equation $16x^3 - 24rxx + 12rrx - 2r^3 + rzz = 0$. & in all other equations which have but one roote.

~~Or because~~ Another way. May 1665

Or because ye lines fn & qn described by ye points f & q doe touch one another in ye point n from wch point onely lines drawne perpendicular to ye crooked line kek will bee perpendicular to ye point of greatest or least crookednesse. And also since all those are points of greatest or least crookednesse to wch such perpendiculars are drawne: The difficulty will bee to find ye point n. Now suppose yt $am = d$ be determined yn c hath two valors for $\{mf, mq\} = c$. And also y hath two valors for $\{dk, de\} = y$. Also (when am is not parallell to ye axis of ye line) x hath two (or more) valors $\{ad, ad\} = x$. wch valors of c, x, or y become equall if $am = an$: by wch meanes ye point n may bee found: Excepting onely when fn, nq, are parallel to ye crooked line at ye straightest or most crooked pts of ye line ark. But if $as = c$ be determined, ... $d = \{sk, sf\}$. $x = \{av, ad\}$. $y = \{rv, ed\}$. (but if (as) is parallel to ye axis of ye line ye two valors of y are equall & ... not usefull). Which valors of d, x, & y become equall if $as = bn$: excepting onely when as is perpendicular to ye most straight or crooked pts of ye line ark.

As for example. In ye precedent example it was found $d + r - 3x = 0$. But because an or as is parallel to ye axis of ye line, ~~in yt Equation x hath but one dimension~~ Therefore substitute either ye valors of d or of x into their stead. As if I substitute ye valor of x ... in its place it will bee $d + r - 3y + \frac{3yy}{r} = 0$ Or $3yy - 3ry + rr + dr = 0$. wch must have 2 equall roots therefore multiplyed according to yt dimensions is $6yy - 3ry = 0$. Or $y = \frac{r}{2}$ as before. But if I had substituted d's valor into its stead it would have beene $16x^3 - 12rxx + 3rrx \ldots$ which having 2 equall roots being rightly ordered is $48x^3 - 24rx^2 + 3rrx = 0$. Or $16x^2 - 8rx + rr = 0$. Or $4x - r = 0$. Or $x = \frac{r}{4} = ad$ as before.

... first Example of finding ye quality of crookednesse in lines ... $= 0$. wch must have 2 equall rootes & therefore ...

Plate II. Extremes of curvature in the parabola and ellipse (**2**, 4, §2.5).

suppose y^t^ $am=d$ be determined y^n^ c hath two valors[,] for $\left.\begin{matrix} mf \\ mq \end{matrix}\right\}=c$. And alsoe y hath two valors[,] for $\left.\begin{matrix} dk \\ de \end{matrix}\right\}=y$. Alsoe (when am is not parallell to y^e^ axis of y^e^ line x hath two (or more) valors $\left.\begin{matrix} ad \\ ad \end{matrix}\right\}=x$. wch valors of c, x, or y become equall if $am=an$: by wch meanes y^e^ point n may bee found: Excepting onely when fm, mq,

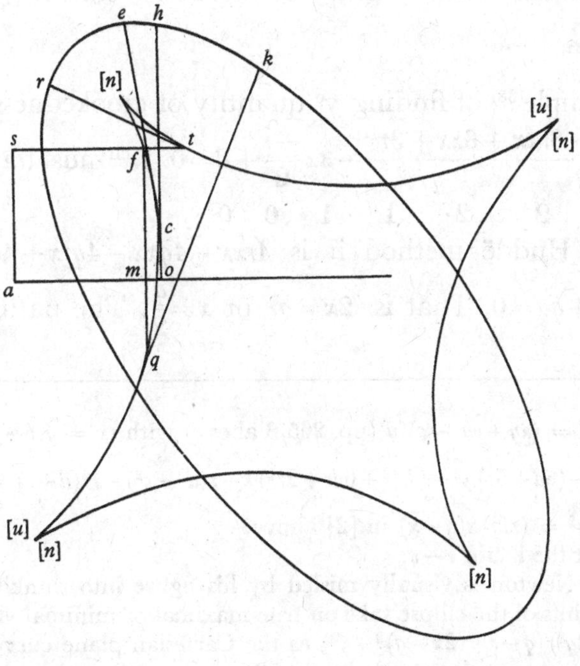

are parallel to y^e^ crooked line at n ([or] un), y^t^ is, perpendicular to y^e^ streightest or most crooked ꝑtes of y^e^ line aek. But if $as=c$ be determined, then

$$d=\begin{cases} st. \\ sf. \end{cases} \quad x=\begin{cases} av. \\ ad. \end{cases} \quad y=\begin{cases} rv. \\ ed. \end{cases}$$

(but if (as) is parallel to y^e^ axis of y^e^ line y^e^ two valors of y are equall & soe not usefull). Which valors of d, x, & y become equall if $as=bun$: excepting onely when as is perpendicular to y^e^ most streight or crooked ꝑts of y^e^ line ake.

As for example. In y^e^ precedent example it was found $d+r-3x=0$. But because am or x is parallell to y^e^ axis of y^e^ line.(66) Therefore substitute either y^e^ valors of d or of x into their stead. As if I substitute y^e^ valor of $x=\frac{ry-yy}{r}$ into its

(66) Newton has cancelled 'in y^t^ equation x hath but one dimension'.

place it will bee $d+r-3y+\dfrac{3yy}{r}=0$, Or $\underset{2}{3yy}\underset{1}{-3ry}\underset{0}{+rr}\underset{0}{+dr}=0$. w[ch] must have 2

equall roots & therefore multiplyed according to y's dimensions tis $6yy-3ry=0$. Or $y=\dfrac{r}{2}$ as before. But if I had substituted d's valor into its stead[(67)] it would have beene $\underset{3}{16x^3}\underset{2}{-12rxx}\underset{1}{+3rrx}\underset{0}{-crr}\underset{0}{+ccr}=0$. Which having 2 equall roots being rightly ordered is $48x^3-24rx^2+3rrx=0$. Or $16x^2-8rx+rr=0$. Or $4x-r=0$. Or $x=\dfrac{r}{4}$ as before.

In y[e] first example[(68)] of finding y[e] quantity of crookednesse in lines it was found $\underset{3}{\dfrac{4rx^3}{q^3}}\underset{3}{-\dfrac{4x^3}{qq}}\underset{2}{-\dfrac{6rxx}{qq}}\underset{2}{+\dfrac{6xx}{q}}\underset{1}{+\dfrac{3rx}{q}}\underset{1}{-3x}\underset{0}{\dfrac{-r}{2}}\underset{0}{+d}=0$. w[ch] must have 2 equall rootes & therefore by Huddē method it is $4rxx-4qxx-4qrx+4qqx+qqr-q^3=0$. Or,[(69)] $4xx-4qx+qq=0$. That is, $2x=q$. or $x=\dfrac{q}{2}$. The nature of y[e] line was

(67) That is, in $d=(xy+vy-vc)/y$ (pp. 265/6 above) with $rx=y(r-y)$; or in

$$4x^3-(8d+5r)\,x^2+(4d^2+6dr+2r^2)\,x-r(d^2-c^2)-r^2(d+c)=0.$$

(68) The ellipse $y^2=(r/q)\,x(q-x)$ in [2] above.

(69) Dividing out the factor $r-q$.

(70) Once again Newton is visually misled by his figure into thinking that the radii of curvature at real points of the ellipse take on true maximal or minimal values. Representing $d=f(x)=\frac{1}{2}r+(1/2q^3)\,(q-r)\,((2x-q)^3+q^3)$ as the Cartesian plane curve (c), where $ab=x$, $bc=d$, we must restrict ourselves to the region $0\leqslant x\leqslant q$, the only values of x which yield real points of the ellipse. An application of Hudde's algorithm gives a double-root condition

$$\left[\frac{df}{dx}=\right]0=\frac{3(q-r)}{q^3}\,(2x-q)^2,\quad\text{or}\quad x=\tfrac{1}{2}q.$$

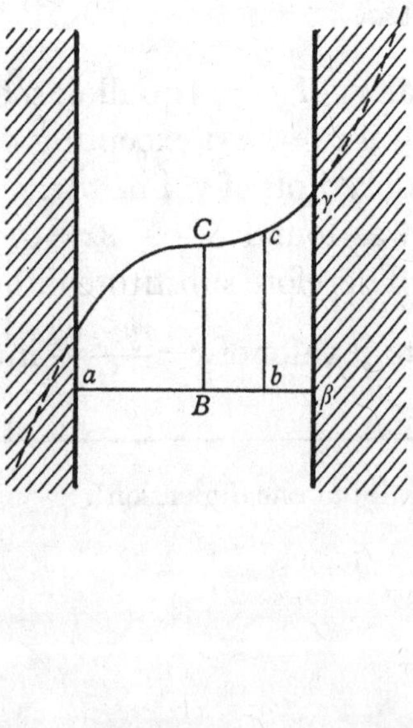

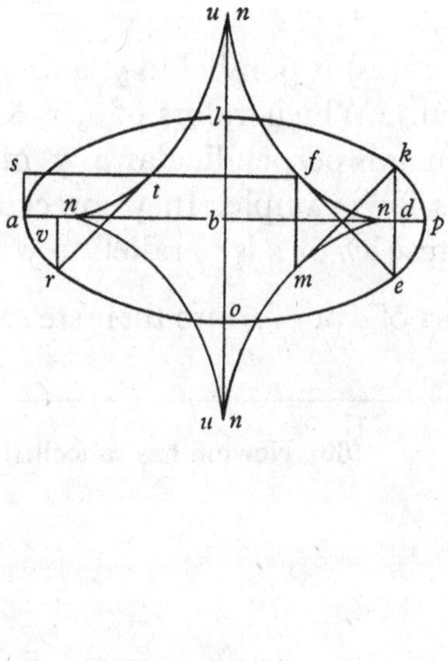

$\sqrt{rx-\frac{rxx}{q}}=y$. Soe y[t] $\sqrt{r\frac{q}{2}-\frac{rqq}{4q}}=\frac{\sqrt{rq}}{2}=y$. Soe y[t], If I take $as=\frac{q}{2}$. & $\left.\begin{matrix} st \\ sf \end{matrix}\right\}=\frac{\sqrt{rq}}{2}=y$. at y[e] points f, t shall bee y[e] greatest or least crookednesse of y[e] line *apeo*.(70) But there are two other points a & p of either greatest or least crookednesse w[ch] were not found by this method.(71)

These Equations may have superfluous rootes soe often as any of y[e] perpendiculars (as *ho*) are parallell to either x or y.(72) But there is a way pag: 48(73) w[ch] concludes, w[th]out any superfluous rootes.

The points of greatest or least crookednesse may bee yet otherwise found by an equation of 4 equall rootes, As in y[e] example of y[e] 2[d] way of finding y[e] quantity of crookednesse in lines(74) it was found

$$y^4-2dryy+rryy-2crry+ccrr+ddrr-ssrr=0.$$

W[ch] being compared wth an equation like it $y^4-4ey^3+6eeyy-4e^3y+e^4=0$. by y[e] 2[d] terme tis $4e^3y=0$, or $y=0$, & $\frac{yy}{r}=\frac{00}{r}=0=x$. Soe y[t] y[e] Parab[:] at y[e] begining is most crooked (at a).(75)

For this value, however, $d^2f/dx^2=0$ also and so this yields not an extreme value for d but an inflexion point (at C, where $aB=BC=\frac{1}{2}q$). On the ellipse itself Newton's calculations, rather than maximizing or minimizing $sf=d$, find the points $\overline{un}$, given by $ab=\frac{1}{2}q$ and $b\overline{un}=\frac{1}{2}\sqrt{(rq)}$, at which $d=f(x)$ has a turning value. It is not at all relevant that the points $\overline{un}$ are (apparently) the centres of minimum curvature at the corresponding points o and l.

(71) These apparent extreme values of z (here minimal) Newton seems to wish to calculate in similar fashion by 'minimizing' $as=c$ (expressed parametrically in terms of y) by an application of Hudde's algorithm. Such an application would, in fact, yield the points n, but as turning values of $c=g(y)$ and not as points of maximal curvature corresponding to the points a and p on the ellipse.

(72) Thus, in Newton's third figure, $as=c$ would have a double root where oc touches the evolute, but this would yield none of the points n or $\overline{un}$ of apparently extreme curvature which Newton seeks.

(73) Add. 4004: [48[r]], printed in §3.2 below.

(74) That is, in [3] above (where $y^2=rx$ is the defining equation of the parabola considered).

(75) Newton applies Descartes' technique for finding quadruple roots by comparison with $(y-e)^4=0$ as a variant on the application of Hudde's algorithm. (See *Geometria*: Liber II: 45 ff., and compare Schooten's *Commentarii in Librum II*, *N*=*Geometria*: 246 ff.) It is interesting to compare Newton's present researches in the theory of conic evolutes with the wholly independent and rather earlier investigations of Christiaan Huygens into the same topic from about 1659 (*Œuvres complètes*, **14** (The Hague, 1920): 387–406, especially 391–6). Huygens' approach is essentially that of Newton (see note (4) above). In his *Avertissement* Huygens' editor Korteweg traces his inspiration to his 1658 researches on the pendulum clock, but this cannot have been Newton's impulse: perhaps we may find that in his work on the theory of refractive surfaces (**3**, Appendix 2: especially note (25)).

§3 THE GENERAL PROBLEMS OF TANGENTS AND CURVATURE RESOLVED FOR ALGEBRAIC CURVES

[May 1665]

From the originals in the University Library, Cambridge, and in private possession(1)

[1] May 20th 1665. *A Method for finding theorems concerning Quæstions de Maximis et Minimis.*(2)

And 1st Concerning ye invention of Tangents to crooked lines.

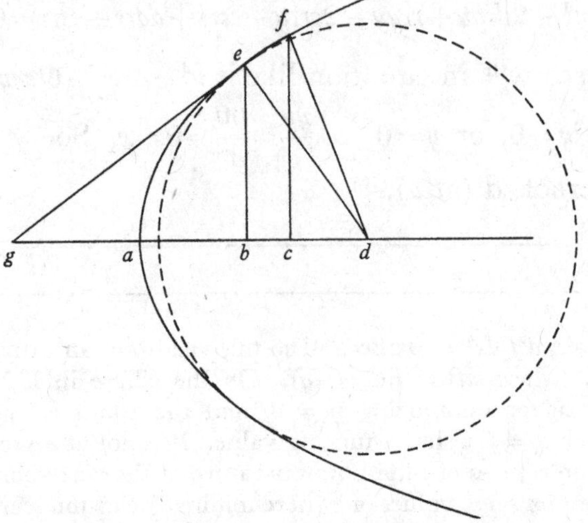

Suppose $ab=x$. $eb=y$. $bd=v$. $bc=o$. $cf=z$.(3) & $ed=df$. ye nature of ye line $ax+xx=yy$.(4) Then is $ac=x+o$. $ax+ao+xx+2ox+oo=zz$.

$$vv+yy=ed^2=fd^2=zz+vv-2ov+oo.$$

Or $yy=oo-2ov+zz$. Or $yy=oo-2ov+ax+ao+xx+2ox+oo$. & since

$$ax=yy-xx.$$

(1) As Newton wrote them, these papers formed a continuous sequence in Newton's Waste Book (Add. 4004): ff. 47r – 50r, but the two sheets ff. 48 and 49 were detached about the time of Newton's death and passed into William Jones' possession.

(2) The similarity between this and the title Hudde gave to his tract (*Geometria*: 507–16: *Epistola...de Maximis et Minimis*) can hardly be accidental.

(3) The incremented value of $eb = y$.

(4) The hyperbola $y^2-(x+\frac{1}{2}a)^2 = -\frac{1}{4}a^2$.

(5) Newton applies Descartes' method for finding the subnormal to a canonical example. (The notation of o for the increment bc is introduced for Fermat's E, or rather e as Schooten expounded Fermat's method in his *Commentarii in Librum II*, $O =$ *Geometria*: 253–5.) In general, where $y=f(x)$ Newton equates $[f(x)]^2+v^2 = [f(x+o)]^2+(v-o)^2$, using the geometrical

Therefore, $0=2oo-2vo+ao+2xo$. Or $2\times o-2v+a+2x=0$. Now y[t] *ed* may bee perpendicular to y[e] line tis required y[t] y[e] points *e*, & *f* conjoyne, w[ch] will hapen when $bc=o$, vanisheth into nothing. Therefore in the equation $2\times o-2v+a+2x=0$, Or $v=o\frac{+a}{2}+x$, those termes in w[ch] (*o*) is must be blotted out, & there remaines $v=x+\frac{a}{2}=bd$. w[ch] determines y[e] perpendicular *ed*.(5)

Observation 1st.(6) Hence it appeares y[t] in such like operations those termes may be ever blotted out in w[ch] $(o=bc)$ is of more y[n] one dimension.(7)

As if y[e] nature of y[e] line was $x^3+xxy+xyy=ayy$. Then since $ac=o+x$ it is $x^3+3x^2o+3xoo+o^3+xxz+2xoz+ooz+xzz+ozz=azz$. That is(8)

$$x^3+3x^2o+xxz+2xoz+xz^2+oz^2=az^2.$$

Also $vv+yy=vv-2vo+oo+zz$. or(8) $yy+2vo=zz$. Therefore

$$ayy-xyy-xxy(=x^3)+3xxo+xxz+2xoz+ozz(+xzz-azz=)$$
$$+xyy+2vox-ayy-2voa=0.$$

That is $xxy-3xxo-2xoz-ozz+2voa-2vox=xx\sqrt{yy+2vo}$.(9) That is (both pts □ed(10) & those terms left out in w[ch] *o* is of more y[n] one dimension)

$$x^4yy+2xxy \text{ in } \overline{2voa-2vox-3xxo-2xoz-ozz}=x^4yy+2vox^4.$$

Or *y* in $\overline{2voa-2vox-3xxo-2xoz-ozz}=voxx$. That is

$$-3xxy-2xzy-zzy=vxx+2vxy-2vay.$$

Now if $bc=o$ vanisheth y[n] is $z=y$. And consequently

$$\frac{-3xxy-2xyy-y^3}{xx+2xy-2ay}=v=\frac{3xxy+2xyy+y^3}{2ay-2xy-xx}.$$

Observacon 2[d]. Hence I observe y[t] if in y[e] valor of *y* there be divers termes in w[ch] *x* is then in y[e] valor of *z* there are those same termes & also those termes each of y[m] multiplyed by so many units as *x* hath dimensions in y[t] terme &

---

equalities $ed^2=eb^2+bd^2$, $fd^2=fc^2+cd^2$ with $ed=fd$ (since they are radii of the circle whose centre *d* is in the abscissa *ab*). The condition that *ed* be normal to the curve at *e* is that the points *e*, *f* coincide (that is, that the increment $bc=o$ vanishes, or that the circle be tangent to the curve at *e*). Analytically, after division by *o*, the condition is that

$$0=\lim_{o\to\text{zero}}[(1/o)([f(x+o)]^2-[f(x)]^2+(v-o)^2-v^2)]=2(f(x)f'(x)-v).$$

It follows immediately that $v=f(x)f'(x)=y(dy/dx)$.

(6) Newton has cancelled 'Theoreme 1st'.

(7) For even after division by *o*, $o^p (p\geqslant 2)$ remains indefinitely small and so vanishes in the limit as $o\to$ zero.

(8) By 'Observation 1st'.

(9) Substituting $z=\sqrt{[y^2+2vo]}$.

(10) Read 'squared'.

againe multiplyed by o & divided by x. As if $x^3+xxy+xyy-ayy=0$, Then, $x^3+xxz+xzz-azz\dfrac{+3x^3o+2xxoz+xozz}{x}=0$. Which operacon may bee conveniently symbolized by (ordering y^e^ equation according to y^e^ dimensions of y &) making some letter (as $a.\ e.\ m.\ n.\ p.$) to signifie a terme, & y^e^ same letter wth some marke (as $\ddot{a}$, $\ddot{c}$, $\ddot{e}$, $\ddot{g}$, $\ddot{m}$, $\ddot{n}$[,] $\ddot{p}$ &c) to signifie y^e^ same terme multiplyed according to y^e^ dimensions of x in it. as in y^e^ former example (suposing $x-a=m$. $xx=n$. $x^3=p$.) The nature of y^e^ line is

in letters $xyy-ayy+xxy+x^3=0$. $\quad xzz-az^2+x^2z+x^3\dfrac{+xozz+2xxoz+3x^3o}{x}=0$.
&
in their symbols $myy+ny+p=0$. $\quad mzz+nz+p\dfrac{+\ddot{m}zzo+\ddot{n}zo+\ddot{p}o}{x}=0$.

Soe if $\underbrace{a^4+ax^3+bbx^2-abbx}_{m}=y^4$. $\quad =y^4$. Then $\underbrace{3aox^2+2bbox-abbo}_{\frac{\ddot{m}o}{x}}+\underbrace{ax^3+bbx^2-abbx+a^4}_{m}=z^4$. $\quad + \quad =z^4$.

[y^t^ is] $\dfrac{\ddot{m}o}{x}+m=z^4$.(11)

And as any particular Equation may be thus symbolized so divers equations may bee represented by y^e^ same caracters. as, $0=a+cy+yye$ may represent all equations in wch y is of one & two dimensions.(12)

Now if a generall Theoreme be required for drawing tangents to such lines, it may bee thus found. $eb=y$, $bd=v$, $ab=x$, $bc=o$, $fc=z$, by supposition, & $a+cy+eyy=0$. Then by observation y^e^ 2^d^, $a+cz+ezz\dfrac{+\ddot{a}o+\ddot{c}oz+\ddot{e}ozz}{x}=0$. Or, $-cyx-eyyx(=xa)+czx+ezzx+\ddot{a}o+\ddot{c}oz+\ddot{e}ozz=0$. Againe $eb^2+bd^2=cf^2+cd^2$.

(11) Where $0=f(x,y)=\sum_i(a_i(x)\,y^i)$ with the a_i functions of x, Newton considers its limit-increase (as $x\to x+o$, $y\to z$), that is, $0=\sum_i(a_i(x+o)\,z^i)$. Expanding the incremented function $a_i(x+o)$ as a series in o and omitting powers of o (by 'Observation 1st') he concludes that

$$0=\sum_i(a_i(x)\,z^i)+\sum_i\left(\frac{\ddot{a}_i(x)}{x}z^io\right).$$

In effect, Newton makes use of a first mean-value theorem $a_i(x+o)\approx a_i(x)+o(\ddot{a}_i(x))/x$, where $\ddot{a}_i(x)=x(d/dx)\,(a_i(x))$ is an homogenized form of the derivative.

(12) The 'caracters' a, c and e are each, of course, functions of x.

(13) Omitting the term 'oo' by the first Observation.

(14) Cancelling the term $c^2x^2y^2$ and dividing through by $2cxo$.

(15) That is, in the limit as $o\to$ zero.

(16) More generally, where $0=\sum_i(a_iy^i)$ passes into

$$0=\sum_i(a_iz^i)+o\sum_i\left(\frac{\ddot{a}_i}{x}z^i\right),$$

it follows that

$$0=\lim_{o\to\text{zero}}\left\{\sum_i\left[a_i\left(\frac{z^i-y^i}{o}\right)\right]+\sum_i\left(\frac{\ddot{a}_i}{x}z^i\right)\right\},$$

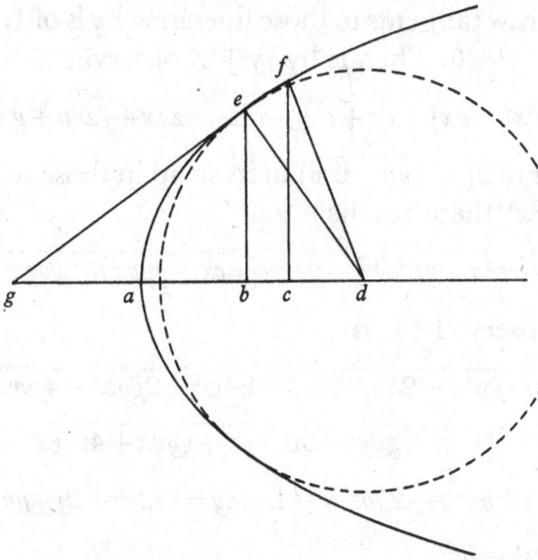

that is,(13) $yy+2ov=zz$. Which valor of zz put into its stead in y^e^ termes $ezzx$ & czx in y^e^ former equation the result is

$$+cyx-\ddot{a}o-\breve{c}oz-\ddot{e}ozz-2eovx=cx\sqrt{yy+2ov}.$$

And both ꝑts squared it is (by y^e^ first Observācon)

$$ccyyx^2-2cyx\ddot{a}o-2cxy\breve{c}oz-2cxy\ddot{e}ozz-4cxyeovx=c^2xxyy+2ccxxov.$$

W^ch^ rightly ordered is $-\ddot{a}y-\breve{c}yz-\ddot{e}yzz=2exyv+cxv$.(14) And since y^e^ points e & f conjoine to make ed a perpendicular therefore is $z=y$.(15) & consequently $\dfrac{-\ddot{a}y-\breve{c}yy-\ddot{e}y^3}{cx+2exy}=v$. W^ch^ is the Theorem sought for.(16) As for example were it required to draw a perpendicular to y^e^ line whose nature is

$$\left[\begin{cases} x^3+xxy\begin{matrix}+x\\-a\end{matrix}yy=0.\\ a\ +cy\ \ +eyy=0.\end{cases}\right] \quad \text{Then is}\quad \frac{-\ddot{a}y-\breve{c}yy-\ddot{e}y^3}{cx+2exy}=\frac{-3x^3y-2xxyy-xy^3}{x^3+2xxy-2axy}=v.$$

or $v=\dfrac{3xxy+2xyy+y^3}{2ay-2xy-xx}$.(17)

---

and since $z=\sqrt{[y^2+2ov]}=y+vo/y$ (ignoring powers of o), then $(z^i-y^i)/o=iy^{i-2}v$; so that

$$0=\sum_i(a_i iy^{i-2})\,v+\sum_i\left(\frac{\ddot{a}_i}{x}y^i\right),$$

and finally
$$v=\frac{-y\sum_i(\ddot{a}_i y^i)}{x\sum_i(a_i iy^{i-1})}.$$

(In its modern equivalent, $y(dy/dx)=-y(f_x/f_y)$.)

(17) The conclusion—that the tangent is to be drawn perpendicular to the normal to the curve at the same point, where the subnormal v may now be constructed and so also the normal—is delayed for a few paragraphs while Newton expands his argument.

In like manner to draw tangents to those lines in wch y is of 1, 2 & 3 dimensions, suppose $a+cy+eyy+gy^3=0$. Then is by [y^e^] 2^d^ observacon

$$-cyx-eyyx-gy^3x(=ax)+\ddot{a}o+czx+\ddot{c}oz+ezzx+\ddot{e}zzo+gz^3x+\ddot{g}z^3o=0.$$

& by writeing y^e^ valor of $z(=\sqrt{yy+2vo})$ in its stead in those termes in wch o is not (viz[:] $czx+ezzx+gz^3x$) there results

$$\overline{cyx+gy^3x-\ddot{a}o-\ddot{c}zo-ez^2o-gz^3o^{(18)}-2exvo}=\overline{cx+gyyx+2vogx} \text{ in } \sqrt{yy+2vo}.$$

& both pts □^ed^, by observ: 1st it is

$$\overline{cyx+gy^3x} \text{ in } \overline{cyx+gy^3x-2\ddot{a}o-2\ddot{c}oz-2\ddot{e}oz^2-2\ddot{g}oz^3-4eovx}$$
$$=\overline{cx+gyyx} \text{ in } \overline{cx+gyyx+4vogx} \text{ in } \overline{yy+2ov}. \text{ Or}$$

$$-2\ddot{a}oy-2\ddot{c}ozy-2\ddot{e}ozzy-2\ddot{g}oz^3y-4eovxy=2cxov+2gyyovx+4vogxyy.$$

That is $\dfrac{-\ddot{a}y-\ddot{c}yy-\ddot{e}y^3-\ddot{g}y^4}{cx+2exy+3gxyy}=v.$

By y^e^ same proceeding were y of 1, 2, 3, 4 dimensions as in

$$a+cy+eyy+gy^3+my^4=0.$$

it would be found $\dfrac{-\ddot{a}y-\ddot{c}yy-\ddot{e}y^3-\ddot{g}y^4-\ddot{m}y^5}{cx+2exy+3gxyy+4mxy^3}=v.$ &c.(19) Hence (calling y^t^ line x wch is perpendicular to it.(20) & y^t^ line v wch wth the perpendicular to y^e^ crooked line &c maketh a right angled triangle) this Theoreme may bee pronou[nced trew of a]ll lines in generall. That

An universall theorem for tangents to crooked lines when $y \perp x$.

Having y^e^ nature of a crooked line expressed in Algebraicall termes wch are not put one pte equall to another but all of y^m^ equall to nothing,(21) if each of the termes be multiplyed by soe many units as x hath dimensions in them. & then multiplyed by y & divided by x[,] they shall be a numerator: Also if the signes be changed & each terme be multiplyed by soe many units as y hath dimensions in y^t^ terme & y^n^ divided by y they shall bee a denominator in y^e^ valor of v.(22)

(18) Read '$-\ddot{e}z^2o-\ddot{g}z^3o$'.
(19) Compare note (16) above.
(20) That is, to the ordinate y. (The emphasis on perpendicularity is essential.)
(21) That is, when the algebraical form $f(x, y)$ is equated to zero as $0 = f(x, y)$.
(22) Newton evaluates the subnormal to $f(x, y) = 0$ as

$$v = \frac{(y/x)\,(xf_x)}{(-1/y)\,(yf_y)}.$$

(Compare 3, §2, note (2) above.)

Example 1st, If $rx+\frac{rxx}{q}-yy=0,$(23) Then

$$\frac{\overset{1}{rx}+\overset{2}{\frac{r}{q}xx}-\overset{0}{yy}\text{ in }\frac{y}{x}}{\underset{0}{-rx}-\underset{0}{\frac{r}{q}xx}+\underset{2}{yy}\text{ in }\frac{1}{y}}=\frac{ry+2\frac{r}{q}xy}{2y}=\frac{r}{2}+\frac{r}{q}x=v.$$

Example 2d, If $x^3+xxy+xyy-ayy=0$. Then

$$\frac{3x^3+2xxy+xyy\text{ in }\frac{y}{x}}{-xxy-2xyy+2ay^2\text{ in }\frac{1}{y}}=v=\frac{3xxy+2xyy+y^3}{2ay-2xy-xx}.$$

Exam: 3d. If $x^3-bxx-cdx+bcd+dxy=0$. Then

$$\frac{3xxy-2bxy-cdy+dyy}{-dx}=\frac{2by}{d}-\frac{3xy}{d}+\frac{cy}{x}-\frac{yy}{x}=v.$$

And by taking y out of ye valor of v yn,

$$v=\frac{2x^3}{dd}-\frac{3bxx}{dd}+\frac{bbx}{dd}-\frac{2cx}{d}+\frac{2bc}{d}+\frac{bcc}{xx}-\frac{bbcc}{x^3}.$$

See Des Cartes his Geometry, booke 2d, pag 42, 46, 47.(24)

---

(23) The hyperbola $y^2 = (r/q)\,x(q+x)$.

(24) That is, *Geometria*: Liber II: 42, 46, 47. The curve $(x-b)(x^2-cd) = -dxy$ is, in fact, the parabolic cubic first constructed on *Geometria*: Liber II: 36–7 as the meet of a rotating line and a continuously translated simple parabola. (Newton himself a little later when he came to enumerate cubics was to name it for all time 'Descartes' trident' from its visual form of a three-pronged fork. In fact, he called the whole species $\alpha x^3+\beta x^2+\gamma x+\delta = \epsilon xy$ the 'trident', but no new curves are introduced in the extension since, by the simple translation

$$y \to Y = y-\frac{\gamma}{\epsilon}+\frac{\alpha\delta}{\beta\epsilon},$$

the species reduces to the form

$$\left(x+\frac{\beta}{\alpha}\right)\left(x^2+\frac{\delta}{\beta}\right) = \frac{\epsilon}{\alpha}xY.)$$

The present parametric expression $v = k(x)$ for the subnormal (with x and y interchanged) appears on *Geometria*: 42, where Descartes finds the condition for the 'subnormal' circle to meet the trident as a sextic in y (with v included in its coefficients). The condition for the circle to touch the trident is then had by equating this sextic with

$$(x-e)^2 \sum_{1\leqslant i\leqslant 4}(c_i x^i) = 0,$$

from which Descartes derives the result here stated by Newton. (Newton's v is, however, Descartes' $v-y$.)

Or thus, $x^3 - bxx - cdx + dyx + bcd$ [$=0.$]
$\quad 2 \quad 1 \quad 0 \quad 0 \quad -1$

$$\frac{2xxy - bxy}{-dx}\frac{+bcdy}{dxx} = v. \quad \text{And} \quad \frac{bcy}{xx} + \frac{by}{d} - \frac{2xy}{d} = v.^{(25)}$$

Note. That haveing $\begin{Bmatrix} x \\ y \end{Bmatrix}$ given, it will be often more convenient to find $\begin{Bmatrix} y \\ x \end{Bmatrix}$ by y^e^ equation expressing y^e^ nature of y^e^ line & y^n^ having x & y to find v by them both, Then(26) to take $\begin{Bmatrix} y \\ x \end{Bmatrix}$ out of v's valor & soe to find it by $\begin{Bmatrix} x \\ y \end{Bmatrix}$ alone.

The Perpendiculars to crooked lines & also y^e^ Theorems for finding them may otherwis more conveniently be found thus.

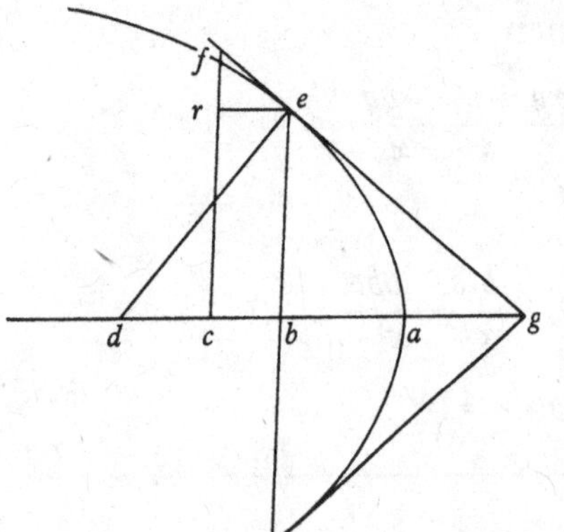

Supposing $ab = x$; $cb = o$, $db = v$, $eb = y$, $cf = z$,(27) And if y^e^ distance twixt fc & eb bee imagined to bee infinitely little, y^t^ is if y^e^ triangle efr is supposed to bee infinitely little[,] then $be:bd::bg:be::re:fr::y:v::o:z-y$. That is $yz - yy = vo$. Or $z = y + \frac{vo}{y}$.(28)

Now suppose y^e^ nature of y^e^ line bee $rx\frac{-rxx}{q} - yy = 0$.(29) Then is

$$rx + ro\frac{-rxx - 2rox - roo}{[q]} - z^2 = 0.$$

In wch equation instead of $rx\left(=\frac{rxx}{q} + yy\right)$ & $zz\left(=yy + 2vo\frac{+vvoo}{yy}\right)$ write their valo[r]s & y^e^ result is $ro\frac{-2rox - 2voq - roo}{q} - \frac{vvoo}{yy} = 0$. Or

$$r\frac{-2rx - ro}{q} - 2v - \frac{vvo}{yy} = 0.$$

---

(25) Newton uses a set of Huddenian multipliers to derive an equivalent form for the subnormal v. (This is a later marginal addition.)

(26) That is, '(more convenient) than'.

(27) It remains essential that the co-ordinates $ab = x$ and $be = y$ be perpendicular, and Newton's figure (which represents the more general case below where the co-ordinates may be oblique) should be slightly redrawn to make this clear.

(28) Newton calculates the basic relation $z = y + vo/y$ in a more direct form than his previous equivalent one, $z = \sqrt{[y^2 + 2ov]}$.

but these two termes ro, $\frac{vvo}{yy}$ are infinitely little, y[t] is if compared to finite termes they vanish[,] therefore I blot y[m] out & there rests $\frac{r}{2}-\frac{rx}{q}=v=db$.[(30)]

The two former observations may be made in this Operacon, viz: 1st to blot out those termes in w[ch] o is of more y[n] one dimension, & 2[dly] y[t] y[n] y[e] termes in y[e] valor of z are y[e] same wth the rectangle of x & the termes in y[e] valor of y & o multiplying those same termes multiplyed by soe many units as x hath dimensions in each terme. By these observations y[e] former theorems might be thus found.

Suppose y[e] nature of y[e] line be $p+qy+ryy=0$. Then (by observation y[e] 2[d]) it is $-qyx-ryyx(=px)+\breve{p}o+qzx+\ddot{q}zo+rzzx+\ddot{r}zzo=0$. Then writeing y[e] valor of $z\left(=y+\frac{vo}{y}\right)$ in its stead in these termes $qzx+rzzx$, There results

$$\breve{p}o+\frac{qxvo}{y}+\ddot{q}zo+2rxvo+\ddot{r}zzo=0.$$

Or because y[e] difference twixt z & y is infinitely little it is $\frac{\breve{p}y+\ddot{q}yy+\ddot{r}y^3}{-qx-2rxy}=v$.

An universall theorem for drawing tangents to crooked lines when x & y intersect at any determined angle.[(31)]

And though y[e] angle ebg made by y[e] intersection of x & y is not determind whither it [bee] acute[,] obtuse or a right one, yet may y[e] line bg bee found after y[e] same manner[,] w[ch] determines y[e] position of y[e] tang[nt] eg. For suppose $bg=t$. $cb=o$. $eb=y$, $fc=z$, $ab=x$, & y[t] $z\|y$. Then (supposing y[e] distance of fc & eb to be infinitely little) it is, $t:y::t+o:y+\frac{oy}{t}=z$. Now if y[e] nature of y[e] line is $p+qy+ry^2+sy^3=0$. Then is

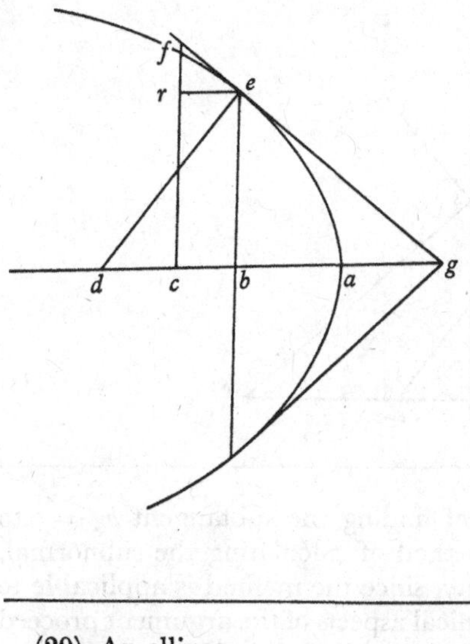

$$-qyx-ry^2x-sy^3x(=px)+\breve{p}o+qxz$$
$$+\ddot{q}oz+rxz^2+\ddot{r}oz^2+sxz^3+\dddot{s}oz^3=0.$$

And by putting y[e] valor of z into its stead in those termes in w[ch] o is not, there results

$$\breve{p}o+\frac{qxoy}{t}+\ddot{q}oz+\frac{2rxoy^2}{t}+\ddot{r}oz^2\frac{+3sxoy^3}{t}+\dddot{s}oz^3$$
$$=0.$$

(29) An ellipse.

(30) The general argument is the same as before (note (16) above).

(31) That is, the angle between the co-ordinates may now be oblique. (Compare note (27) above.)

Or $t = \dfrac{-qxy - 2rxyy - 3sxy^3}{\breve{p} + \breve{q}y + \breve{r}y^2 + \breve{s}y^3} = bg$. Soe y[t] y[e] variation of y[e] angle *ebg* makes noe variation in y[e] length of y[e] line *bg*.(32)

Note y[t] y[e] foundacon of this operacon & of y[t] by w[ch] Florimond de Beaune (in his notes on Cartes pag 131) found tangents are almost y[e] same.(33)

But since an equation is the same when multiplyed by soe many units as the unknowne quantity hath dimensions, y[t] it would bee if multiplyed by any other Arithmeticall progression (as if $x^3 + xxy + aax + a^3 = 0$. Then

$$3x^3 + 2xxy + aax = 2x^3 + xxy - a^3 = x^3 - aax - 2a^3. \quad \&c:)$$

this Theoreme may [bee] better pronounced thus—. Multiply the termes of y[e] equation ordered according to y[e] dimensions of y, by any Arithmeticall progression, w[ch] shall bee a Numerator; Againe change y[e] signes of y[e] equation & ordering it according t[o] x, multiply y[e] termes by any Arithmeticall progression & the product divided by x shall bee y[e] denominator of y[e] valor of v.(34)

---

[2]

May 21st 1665. *The Invention of Theorems for finding y[e] crookednesse in lines.*

Suppose $ac = x$. $cd = o$. $ce = v$. $df = w$. $ch = y$. $ds = z$. $pm = bc = d$. $pc = bm = c$. And if $ay = xx$. Then, $v = ce = \dfrac{2xy}{a}$. $az = xx + 2xo$. $w = \dfrac{2xz + 2zo}{a}$, by y[e] precedent

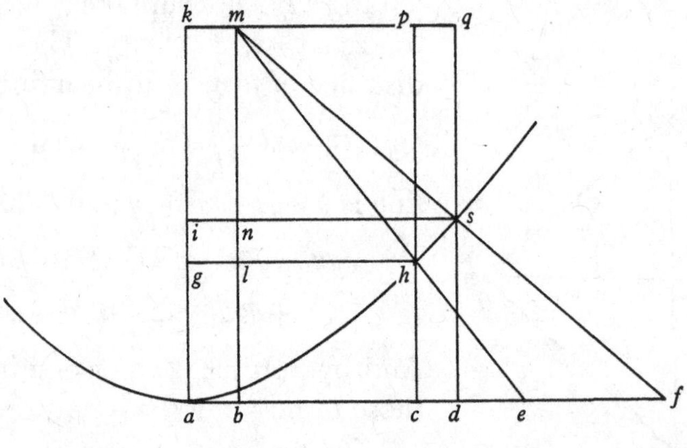

(32) Newton at last strikes on the superior method of finding the subtangent $bg = t$ to determine the tangent rather than using the devious method of calculating the subnormal. Further, he notes correctly the immense gain in generality, since the method is applicable to a general oblique Cartesian co-ordinate system. The technical aspects of the argument proceed much as before: substitution of $z = y + (oy/t)$ in the fundamental limit-relation

$$0 = \lim_{o \to \text{zero}} \left\{ \sum_i \left[a_i \left(\frac{z^i - y^i}{o} \right) \right] + \sum_i \left(\frac{\ddot{a}_i}{x} z^i \right) \right\},$$

theorems.[35] Supposing (m) to bee y^e^ point where y^e^ two perpendiculars *mf* & *mhe* intersect when their distance is infinitely little. Then is $y:v::c-y:d$. & $z:w::c-z:d+o$. that is $y:\frac{2xy}{a}::a:2x::c-y:\frac{2cx-2yx}{a}=d$. And,

$$z:\frac{2xz+2zo}{a}::a:2x+2o::c-z:[d+o=]\frac{2cx-2yx+ao}{a}.^{(36)}$$

Whence $2cx-2yx+ao=2cx+2co-2zx-2zo$.[37] Or $ao-2yx=2co-2zx-2zo$. In the former operacons it was found $z=y+\frac{vo}{y}$. That is in this example $z=y+\frac{2xo}{a}$.[38] Which valor w[r]itten in stead of z in y^e^ last equation it produceth $ao-2yx=2co-2xy-\frac{4xxo}{a}-2zo$. That is, $a=2c-\frac{4xx}{a}-2z$. Or[39] $c=\frac{a}{2}+\frac{4xx}{a}+2y=\frac{1}{2}a+3y$.

---

where $z^i-y^i = oy^i(i/t)$, with powers of o deleted, yields

$$0=\sum_i(a_i iy^i)\frac{1}{t}+\sum_i\left(\frac{\ddot{a}}{x}y^i\right),\quad\text{or}\quad t=\frac{-x\sum_i(a_i iy^i)}{\sum_i(\ddot{a}_i y^i)}.$$

Equivalently and in modern terms $y(dx/dy) = -y(f_y/f_x)$. (In the particular case where $ab\perp be$, $v:y = y:t$, or $v = y^2/t$ in agreement with note (16) above.)

(33) See *Florimondi de Beaune in Geometriam Renati des Cartes Notæ Breves* (= *Geometria* 1: 107–42): 130–3: *Ad [Geometriæ] paginam 40 et sequentes, de Modo Inveniendi Contingentes Linearum Curvarum*. De Beaune, however, there restricts himself to the particular case where the coordinate axes are perpendicular, using the proportion $t:y = y:v$.

(34) Where $$0=f(x,y)\equiv\sum_i(a_i y^i)\equiv\sum_j(b_j x^j)$$

with the a_i, b_j functions of x and y respectively, then, since

$$\sum_j(jb_jx^j)=xf_x=xf_x+\lambda f=\sum_j((j+\lambda)b_jx^j),$$

and similarly $$\sum_i(ia_iy^i)=yf_y=\sum_i((i+\mu)a_iy^i),$$

it follows that $$t=y\frac{dx}{dy}=-y\frac{f_y}{f_x}=\frac{-\sum_i(ia_iy^i)x}{\sum_j(jb_jx^j)}=\frac{-\sum_i((\lambda+\pi i)a_iy^i)x}{\sum_j((\mu+\pi j)b_jx^j)}.$$

Newton therefore has omitted the essential restricting condition that the two arithmetical progressions $(\lambda+\pi i, \mu+\pi j)$ have the same constant difference π and increase in the same sense. In practice, however, he always observes this limitation and indeed will note it specifically a little later. (See note (86) below.)

(35) That is, observations 1 and 2 on p. 273 above (which find $v/y = dy/dx$).

(36) Incrementing the former equation by $x\to x+o(d\to d+o)$ and $y\to z$, and then substituting the previous value of d.

(37) That is, $(c-z)(2x+2o)$.

(38) Since $v/y = 2x/a$.

(39) For $\lim_{o\to\text{zero}}(z) = y$. The following should read '$c=\frac{a}{2}+\frac{2xx}{a}+y=\ldots$'.

That this manner of proceeding may bee reduced to y[e] finding of Theorems it will bee convenient to let this marke (∴) signifie y[t] each terme must bee multiplyed after this manner $-1\times 0.\ \ 0\times 1.\ \ 1\times 2.\ \ 2\times 3.\ \ 3\times 4.$ That is $\frac{x}{x}.\ \ x.\ \ xx.\ \ x^3.\ \ x^4.$ &c.

$0.\ 0.\ 2.\ 6.\ 12.\ 20.\ 30.\ 42.$

$\frac{x}{x}.\ x.\ xx.\ x^3.\ x^4.\ x^5.\ x^6.\ x^7.$ &c. as if (p) signifies $a^4-3abbx+5bbxx+ax^3-2x^4$,

Then $(\dddot{p})$ signifies $\overset{0.}{a^4}.\ \overset{0.}{-3abbx}.\ \overset{2.}{+5bbxx}.\ \overset{6.}{+ax^3}.\ \overset{12}{-2x^4}=10bbxx+6ax^3-24x^4$. let also (..) signifie each terme of y[t] quantity to bee multiplied by soe many units as x hath dimensions in y[t] terme.[(40)]

Then suppose y[t] $ac=x.\ cd=o.\ ch=y.\ ds=z.\ c[f]=v.\ d[b]=w.$ Then (if hs & cd have an infinitely little distance[,] otherwise not) $z=y+\frac{ov}{y}.\ zz=yy+2ov.\ z^3=y^3+3yov.$ &c: For $v:y::y:\frac{yy}{v}=ct.$ &

$$ct:ch::td:ds=z=\frac{y\times o+\frac{yy}{v}}{\frac{yy}{v}}=y+\frac{ov}{y}.$$

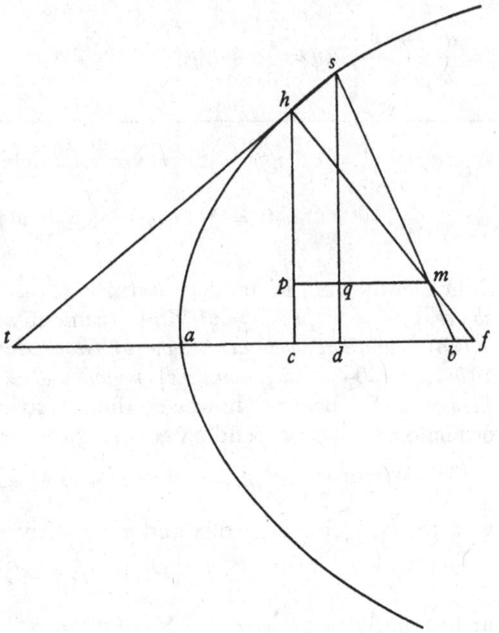

(w[ch] operacon cannot in this case bee understood to bee good unlesse infinite littlenesse may bee considered geometrically). Now if y[e] nature of y[e] line is $p+qy=0$. Then is $\frac{\ddot{p}y+\ddot{q}yy}{-qx}=v.$ &, $p+\frac{\ddot{p}o}{x}+qz+\frac{\ddot{q}oz}{x}=0.$[(41)] by y[e] seacond observation. Also $\frac{\ddot{p}z+\frac{\dddot{p}oz}{x}+\ddot{q}zz+\dddot{q}\frac{ozz}{x}}{-qx-\ddot{q}o}=w$[(42)]$=\frac{\ddot{p}xz+\ddot{q}xzz+\dddot{p}oz+\dddot{q}ozz}{-qxx-\ddot{q}ox}$. by the Theorems of Tangents. Againe $y:v::y-c:d=\frac{vy-vc}{y}$. And

$$z:w::z-c:\frac{vy-vc}{z}(=d)-o.\quad \text{Therefore}\quad \frac{wz-cw}{z}=\frac{vy-vc-oy}{y}.$$

(40) In general, if p is a given function of x, $\ddot{p}$ and $\dddot{p}$ are the homogenized first and second derivatives $x(dp/dx)$ and $x^2(d^2p/dx^2)$ respectively.

(41) The incremented form of $p+qy=0$ with powers of o ignored.

And by writeing y[e] valors of v & w in their stead in this Equation there resulteth

$$\frac{\dot{p}xz+\ddot{q}xzz+\hat{p}oz+\hat{q}ozz-\dot{p}xc-\ddot{q}xzc-\hat{p}oc-\hat{q}ozc}{+qx+\ddot{q}o}=\frac{\dot{p}y+\ddot{q}yy-\dot{p}c-\ddot{q}yc+oqx}{q},$$

That is[,] multiplyed in crucē[,][43]

$$\dot{p}qxz+\ddot{q}qxzz+\hat{p}qzo+\hat{q}qozz-c\ddot{q}xzq-\hat{p}oqc-\hat{q}qozc=$$
$$\dot{p}yqx+\ddot{q}yyqx-\ddot{q}ycqx+q^2ox^2+\dot{p}y\ddot{q}o+\ddot{q}yy\ddot{q}o-\dot{p}c\ddot{q}o-\ddot{q}yc\ddot{q}o.$$

& by substituting y[e] above found valor of $z\left(=y+\dfrac{ov}{y}\right)$ into its place in those termes in w[ch] (o) is not there results, (those terms which destroy each other being neglected)

$$\dot{p}qx\frac{ov}{y}+2\ddot{q}qxov+\hat{p}qzo+\hat{q}qzzo-\frac{\ddot{q}qcxov}{y}-\hat{p}qco-\hat{q}qcoz$$
$$=qqoxx+\dot{p}\ddot{q}yo+\ddot{q}\ddot{q}yyo-\dot{p}\ddot{q}co-\ddot{q}\ddot{q}coy.$$

Now writeing y instead of z because they are equall if compared to finite quantitys, also writeing y[e] valor of $v\left(=\dfrac{\dot{p}y+\ddot{q}yy}{-qx}\right)$ in its stead, & dividing y[e] whole equation by o, & ordering it, it is

$$\frac{+\dot{p}\dot{p}+4\dot{p}\ddot{q}y+3\ddot{q}\ddot{q}yy+qqxx-\hat{p}qy-\hat{q}qyy}{2\dot{p}\ddot{q}+2\ddot{q}\ddot{q}y-\hat{p}q-\hat{q}qy}=c=\frac{\overline{\dot{p}+\ddot{q}y}\times\overline{\dot{p}+3\ddot{q}y}\ \overline{-\hat{p}-\hat{q}y}\times\overline{qy}+qqxx}{\overline{\dot{p}+\ddot{q}y}\times\overline{2\ddot{q}}\ \overline{-\hat{p}-\hat{q}y}\times q}.$$

That is $c=y\dfrac{+\dot{p}\dot{p}+2\dot{p}\ddot{q}y+\ddot{q}\ddot{q}yy+qqxx}{2\dot{p}\ddot{q}+2\ddot{q}\ddot{q}y-\hat{p}q-\hat{q}qy}$. And

$$-hp=c-y=\frac{\dot{p}\dot{p}+2\dot{p}\ddot{q}y+\ddot{q}\ddot{q}yy+qqxx}{2\dot{p}\ddot{q}+2\ddot{q}\ddot{q}y-\hat{p}q-\hat{q}qy}.^{(44)}$$

The same may bee done more conveniently thus. supposeing as before, y[t], $ac=x$. $cd=o$. $ch=y$. $ds=z$. $ce=v$. $df=w$. $hp=c$. $pm=[d]$. (By how much y[t] lesse y[e] line hb is, soe much y[e] greater is y[e] disproportion twixt y[e] lines (ob) & (os), soe y[t] (os) is infinitely lesse y[n] (ob) when ob & hb are infinitely little (unlesse when ob

(42) The incremented form $w = z(dz/dx)$ of $v = y(dy/dx)$.

(43) And cancelling the term $-\dot{p}qcx$ on each side.

(44) That is, $$c-y=-y\frac{(\dot{p}+\ddot{q}y)^2+(qx)^2}{(\hat{p}+\hat{q}y)\,qy-2(\dot{p}+\ddot{q}y)\,\ddot{q}y}.$$

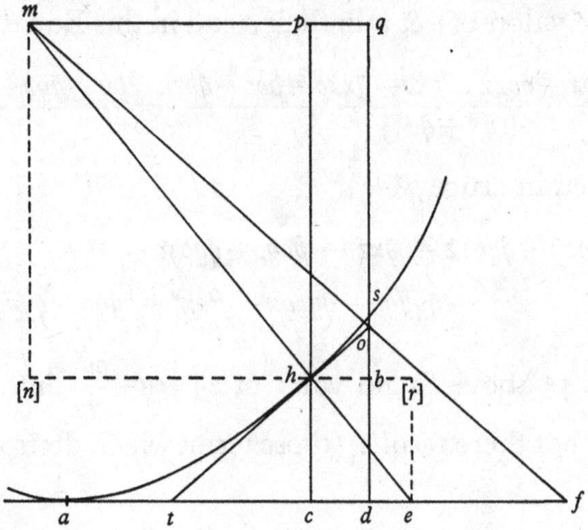

toucheth y^e^ crooked line *ahs*). y^t^ is if $hb=o$, y^n^ $ob=\frac{ao}{b}$, & $os=\frac{oo}{a}$, &c:(45) soe y^t^ *os* vanisheth if compared to *ob* when *ob* is infinitely little, therefore if I consider

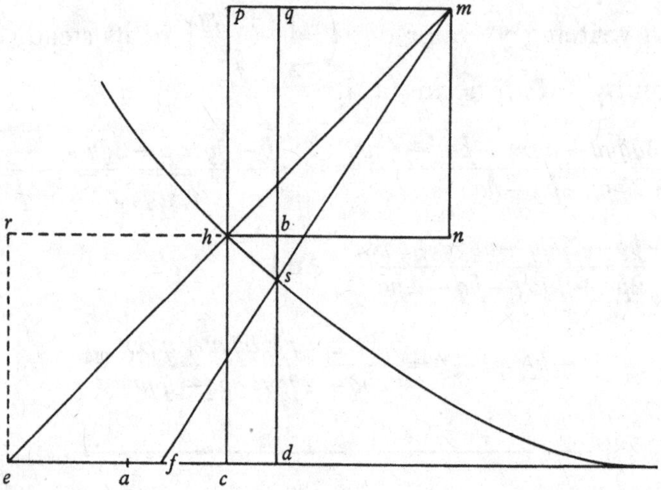

hc & *od* to differ & y^t^ but infinitely little, I may consider *do* & *ds* as equall & having noe difference at all. therefore) $y:-v::o:\frac{ov}{-y}=ob=bs.$ $sq=c-\frac{ov}{y}$.

(45) Where the curve (h) is defined by $y=f(x)$, then, since

$$ad=x+o,\quad sd=f(x+o)=f(x)+of'(x)+\tfrac{1}{2}o^2f''(x)+O(o^3).$$

However, $bd=hc=y=f(x)$, $ob=hb\,(ob/hb)=of'(x)$ and so $sd-(ob+bd)=so=\frac{1}{2}o^2f''(x)$ when o is taken indefinitely small. Newton takes therefore $a/b=f'(x)$ with $a=2/(f''(x))$. (Compare Newton's *Philosophiæ Naturalis Principia Mathematica*, London, $_1$1687: Liber II, Prop. X.)

$sd=z=y+\frac{ov}{y}$. $z^2=y^2+2ov$. $z^3=y^3+3ovy$, &c. Also $-y:v::c:d=\frac{vc}{-y}$. And $-z:w::c-\frac{ov}{y}:d-o$.[46] [soe y^t] $cvz+oyz=cyw-vow$. Or, $-vz+yw=\frac{oyz+ovw}{c}$.

Now suppose y^e nature of y^e line is $p+qy=0$. Then is $px+\ddot{p}o+qxz+\ddot{q}oz=0$.[47] And $\frac{\ddot{p}y+\ddot{q}yy}{-qx}=v$. And $\frac{\ddot{p}xz+\mathring{p}oz+\ddot{q}xzz+\mathring{q}ozz}{-qxx-\ddot{q}ox}=w$.[48] W^{ch} valors of v & w substitute into theire places in y^e termes $(-vz+yw)$ of y^e equation $-vz+yw=\frac{oyz+ovw}{c}$, And the result is $\frac{\ddot{p}+\ddot{q}y}{+q}[+]\frac{+\ddot{p}x+\ddot{q}xz+\mathring{p}o+\mathring{q}oz}{-qx-\ddot{q}o}=\frac{oyzx+ovwx}{czy}$. Then put y^e valor of $z\left(=y+\frac{ov}{y}\right)$ in its stead in those termes in which there is not o, And reduce the quantitys on y^e same side [of] y^e equation to a comon denominator & y^e result is $\frac{\ddot{p}\ddot{q}+\ddot{q}\ddot{q}y-q[\ddot{q}]\frac{xv}{y}-\mathring{p}q-\mathring{q}qy}{+qqxx}=\frac{yy+vv}{cyy}$.[49] For now v & w, & also y & z may bee made equall after y^e equation is divided by o. Againe substituting y^e valor of v into its stead & ordering y^e equation it will bee

$$2\ddot{p}\ddot{q}c+2\ddot{q}\ddot{q}cy-\mathring{p}qc-\mathring{q}qy[c]=\ddot{p}\ddot{p}+2\ddot{p}\ddot{q}y+\ddot{q}\ddot{q}yy+qqxx. \quad \text{Or}$$

$$c=\frac{\ddot{p}\ddot{p}+2\ddot{p}\ddot{q}y+\ddot{q}\ddot{q}yy+qqxx}{2\ddot{p}\ddot{q}+2\ddot{q}\ddot{q}y-\mathring{p}q-\mathring{q}qy}.\text{[50]}$$

Note y^t y^e crooked line *hs* is y^e line at wch all y^e streight lines (*hc*) or (*re*), (*ec*) or (*hr*), (*ph*) or (*mn*), *hn* or (*pm*) have their begining, soe y^t if *hp* & *hn* are affirmative then is *hc* & *hr* negative in this posit.[51]

Haveing thus found c tis easy to find,

$$d=\frac{cv}{-y}\text{[52]}=\frac{+\ddot{p}\ddot{p}\ddot{p}+3\ddot{p}\ddot{p}\ddot{q}y+3\ddot{p}\ddot{q}\ddot{q}yy+\ddot{q}\ddot{q}\ddot{q}y^3+\ddot{p}qqx^2+\ddot{q}qx^2\text{[53]}}{2\ddot{p}\ddot{q}q+2\ddot{q}\ddot{q}qy-\mathring{p}qq-\mathring{q}qqy\ [\text{in } x]}.$$

(46) The incremented form of the preceding equation.
(47) Since $p+(\ddot{p}/x)\,o+qz+(\ddot{q}/x)\,oz$ is the incremented form of the preceding.
(48) Or $\frac{\ddot{p}z+(\mathring{p}/x)\,oz+\ddot{q}zz+(\mathring{q}/x)\,ozz}{-(qx+(\ddot{q}/x)\,ox)}=w$, the incremented form.
(49) Newton has silently rounded off the denominators '$qqxx+q\ddot{q}ox$' and '$cyy+cov$'.
(50) Or $hp=-y\frac{(\ddot{p}+\ddot{q}y)^2+(qx)^2}{(\mathring{p}+\mathring{q}y)\,qy-2(\ddot{p}+\ddot{q}y)\,\ddot{q}y}$ (as before).
(51) Read 'position'.
(52) That is, $c(\ddot{p}+\ddot{q}y)/qx$.
(53) Read '$\ddot{q}q^2x^2y$'.

By y^e^ same manner of proceeding may there be found a rule for those lines in wch y is of 2 dimensions. as if, $p+qyy=0$. Then $px+\ddot{p}o+qxzz+\ddot{q}ozz=0$.(54) And $\dfrac{\ddot{p}+\ddot{q}yy}{-2qx}=v$. & $\dfrac{\ddot{p}x+\mathring{p}o+\ddot{q}xzz+\mathring{q}ozz}{-2qxx-2\ddot{q}ox}=w$,(54) by y^e^ Theorems of tangents.(55) Now subrogating these valors of v & w into theire stead in y^e^ before found equation $wy-vz=\dfrac{ovw+oyz}{c}$, The result is,

$$\frac{\ddot{p}oy+\mathring{q}ozzy+\ddot{p}xy+\ddot{q}xzzy}{-2qxx-2\ddot{q}ox}\;\frac{+\ddot{p}z+\ddot{q}yyz}{2qx}=\frac{ovw+oyz}{c}.$$

Or,

$$\mathring{p}oyq+\mathring{q}ozzyq+\ddot{p}xyq+\ddot{q}xzzyq-\ddot{p}xzq-\ddot{q}yyzxq-\ddot{p}z\ddot{q}o-\ddot{q}yyz\ddot{q}o$$
$$=\frac{-2qqxxovw-2qqxxoyz}{c}.$$

Againe by subrogate[i]ng the valor of $z=y+\dfrac{ov}{y}$ into its place in those termes in wch o is not, there results,

$$\mathring{p}oyq+\mathring{q}ozzyq+\ddot{q}qxyov-\ddot{p}qx\frac{ov}{y}-\ddot{p}z\ddot{q}o-\ddot{q}yy\ddot{q}oz=\frac{-2qqxxovw-2qqxxoyz}{[c]}.$$

Or, $\mathring{p}qyy+\mathring{q}qy^4-\ddot{p}\ddot{q}yy-\ddot{q}\ddot{q}y^4+\ddot{q}qxvy^2-\ddot{p}qxv=\dfrac{-2qqxxvvy-2qqxxy^3}{c}$. And by Subrogateing y^e^ valor of $v\left(=\dfrac{\ddot{p}+\ddot{q}yy}{-2qx}\right)$ into its place, the result is

$$\mathring{p}qyy+\mathring{q}qy^4+\frac{1}{2}\ddot{p}\ddot{p}-\ddot{p}\ddot{q}yy-\frac{3}{2}\ddot{q}\ddot{q}y^4=\frac{-\ddot{p}\ddot{p}y-2\ddot{p}\ddot{q}y^3-\ddot{q}\ddot{q}y^5}{2c}\;\frac{-2[qq]xxy^3}{c}.$$

That is, $c=\dfrac{\ddot{p}\ddot{p}y+2\ddot{p}\ddot{q}y^3+\ddot{q}\ddot{q}y^5+4qqxxy^3}{3\ddot{q}\ddot{q}y^4+2\ddot{p}\ddot{q}yy-\ddot{p}\ddot{p}-2\mathring{p}qyy-2\mathring{q}qy^4}$.(56) Haveing thus found y^e^ valor of c, y^e^ valor of d may be found by makeing

$$d=\frac{-cv^{(57)}}{y}=\frac{\ddot{p}\ddot{p}\ddot{p}+3\ddot{p}\ddot{p}\ddot{q}yy+3\ddot{p}\ddot{q}\ddot{q}y^4+\ddot{q}\ddot{q}\ddot{q}y^6+4\ddot{p}qqxxy^2+4\ddot{q}qqx^2y^4}{6\ddot{q}\ddot{q}qxy^4+4\ddot{p}\ddot{q}qxyy-2\ddot{p}\ddot{p}qx-4\mathring{p}qqxy^2-4\mathring{q}qqxy^4}.$$

---

(54) The incremented form of the preceding. (55) See pp. 279–80 above.

(56) Or $c=-y\dfrac{(\ddot{p}+\ddot{q}y^2)^2+(2qxy)^2}{(\mathring{p}+\mathring{q}y^2)\,2qy^2-2\times2\ddot{q}y^2(\ddot{p}+\ddot{q}y^2)+(\ddot{p}+\ddot{q}y^2)^2}$.

(57) That is, $c\left(\dfrac{\ddot{p}+\ddot{q}y^2}{2qx}\right)$.

(58) Or $hp=c=-y\dfrac{(\ddot{p}+\ddot{q}y^3)^2+(3qxy^2)^2}{(\mathring{p}+\mathring{q}y^3)\,3qy^3-2\times3\ddot{q}y^3(\ddot{p}+\ddot{q}y^3)+2(\ddot{p}+\ddot{q}y^3)^2}$.

(59) Since $pm=d=-\dfrac{cv}{y}=c\left(\dfrac{\ddot{p}+\ddot{q}y^3}{3qxy}\right)$.

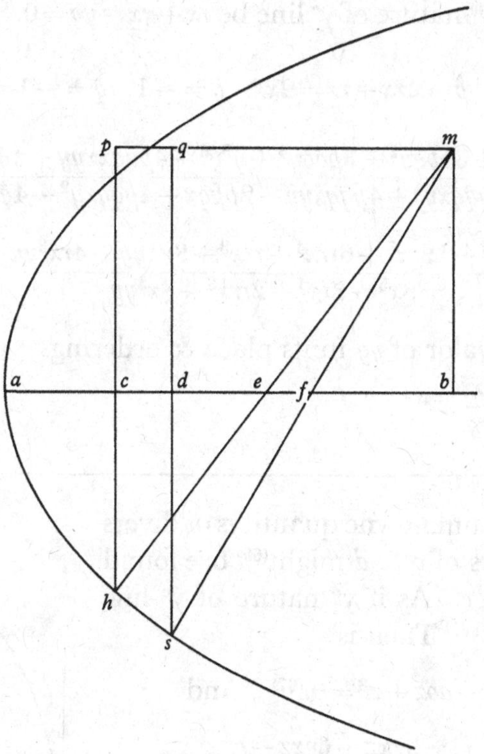

By y^e^ same meanes it may bee found y^t^ if $p+qy^3=0$. Then is

$$c=\frac{\ddot{p}\ddot{p}y+2\ddot{p}\ddot{q}y^4+\ddot{q}\ddot{q}y^7+9qqxxy^5}{4\ddot{q}\ddot{q}y^6+2\ddot{p}\ddot{q}y^3-2\ddot{p}\ddot{p}-3\mathring{p}qy^3-3\mathring{q}qy^6}.^{(58)}\quad \text{And,}^{(59)}$$

$$d=\frac{\ddot{p}\ddot{p}\ddot{p}+3\ddot{p}\ddot{p}\ddot{q}y^3+3\ddot{p}\ddot{q}\ddot{q}y^6+\ddot{q}\ddot{q}\ddot{q}y^9+9\ddot{p}qqxxy^4+9\ddot{q}qqxxy^7}{12\ddot{q}\ddot{q}qxy^7+6\ddot{p}\ddot{q}qxy^4-6\ddot{p}\ddot{p}qxy-9\mathring{p}qqxy^4-9\mathring{q}qqxy^7}.$$

Also if y^e^ nature of y^e^ line bee $p+qy^4=0$. Then is

$$c=\frac{\ddot{p}\ddot{p}y+2\ddot{p}\ddot{q}y^5+\ddot{q}\ddot{q}y^9+16qqxxy^7}{5\ddot{q}\ddot{q}y^8+2\ddot{p}\ddot{q}y^4-3\ddot{p}\ddot{p}-4\mathring{p}qy^4-4\mathring{q}qy^8}.^{(60)}\quad \text{And,}^{(61)}$$

$$d=\frac{\ddot{p}^3+3\ddot{p}\ddot{p}\ddot{q}y^4+3\ddot{p}\ddot{q}\ddot{q}y^8+\ddot{q}^3y^{12}+16\ddot{p}qqxxy^6+16\ddot{q}qqxxy^{10}}{20\ddot{q}\ddot{q}qxy^{10}+8\ddot{p}\ddot{q}qxy^6-12\ddot{p}\ddot{p}qxyy-16\mathring{p}qqxy^6-16\mathring{q}qqxy^{10}}.$$

---

(60) Or $c=-y\dfrac{(\ddot{p}+\ddot{q}y^4)^2+(4qxy^3)^2}{(\mathring{p}+\mathring{q}y^4)\,4qy^4-2\times 4\ddot{q}y^4(\ddot{p}+\ddot{q}y^4)+3(\ddot{p}+\ddot{q}y^4)^2}$.

(61) Since $d=-\dfrac{cv}{y}=c\left(\dfrac{\ddot{p}+\ddot{q}y^4}{4qxy^2}\right)$.

As for example if y^e^ nature of y^e^ line be $rx+xx-yy=0$.(62) Then is $p=rx+xx$.

$$\overset{1\ \ 2}{\breve{p}=rx+xx}=2xx+rx.\quad \overset{1\ \ 0}{\hat{p}=2xx+rx}=2xx.\quad q=-1.\quad \ddot{q}=\overset{0}{-1}=0=\hat{q}.$$

And consequently(63)

$$d=\frac{\breve{p}^3+3\breve{p}\breve{p}\ddot{q}y^2+3\breve{p}\ddot{q}\ddot{q}y^4+\ddot{q}^3y^6+4\breve{p}qqxxyy+4\ddot{q}qqxxy^4}{6\ddot{q}\ddot{q}qxy^4+4\breve{p}\ddot{q}qxyy-2\breve{p}\breve{p}qx-4\hat{p}qqxy^2-4\hat{q}qqxy^4}$$

$$=\frac{8x^6+12rx^5+6rrx^4+r^3x^3+8x^4yy+4rx^3yy}{8x^5+8rx^4+2rrx^3-8x^3yy}.$$

And by writeing y^e^ valor of yy in its place & ordering y^e^ equation the result is $d=\frac{1}{2}r+5x+\frac{12xx}{r}+\frac{8x^3}{rr}$.(64)

But were both the unknowne quantitys of divers dimensions the valors of c & d might(65) bee found after y^e^ same manner. As if y^e^ nature of y^e^ line was $x^3-axy+y^3=0$.(66) Then is

$$x^3+3oxx-axz-aoz+z^3=0.^{(67)}$$ And

$$v=\frac{3xxy-ayy}{ax-3yy}.\quad w=\frac{3xxz+6oxz-azz}{ax+ao-3zz}.^{(68)}$$

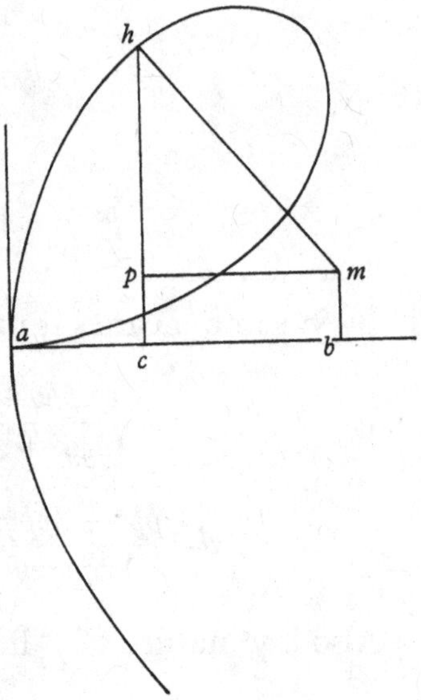

Which valors of v & w being substituted into their places in this equation $wy-vz=\frac{ovw+oyz}{c}$, in those termes $(wy-vz)$ in wch o is not there results

$$\frac{3xxzy+6oxzy-azzy}{ax+ao-3zz}\frac{-3xxzy+azxy}{ax-3yy}=\frac{ovw+oyz}{c}.$$

Or $$\frac{3xx+6ox-az}{ax+ao-3zz}\frac{-3xx+ay}{ax-3yy}=\frac{ovw+oyz}{cyz}.$$ Or

$$\frac{3aoxx-aazx-9xxyy-18oxyy+3azyy+aaxy+aaoy+9zzxx-3azzy}{aaxx+oaax-3axzz-3axyy-3aoyy+9zzyy}=\frac{ovw+oyz}{cyz}.$$

---

(62) 'Example y^e^ 2^d^' in §2.2 above with $q=r$.

(63) Since $p+qy^2=0$ $(p=x^2+rx,\ q=-1)$.

(64) For

$$8x^6+12rx^5+6r^2x^4+r^3x^3+8x^4(rx+x^2)+4rx^3(rx+x^2) = 16x^6+24rx^5+10r^2x^4+r^3x^3,$$

and

$$8x^5+8rx^4+2r^2x^3-8x^3(rx+x^2) = 2r^2x^3.$$

(65) Newton has cancelled 'as easily'.

That is(69)

$$3aoxx - 18oxyy + aaoy - aazx - 9xxyy + 3azyy + aaxy + 9xxzz - 3ayzz$$

$$= \frac{ovw + oyz}{cyz} \quad \text{in} \quad aaxx - 6axyy^{(70)} + 9zzyy.$$

Then writeing y^e^ valor of $z\left(=y+\dfrac{ov}{y}\right)$ in those places in wch o is not, there results,(71)

$$3aoxx - 18oxyy + aaoy\frac{-aaxov}{y} - 3aovy + 18xxov$$

$$= \frac{ovw + oyz}{cyz} \quad \text{in} \quad aaxx - 3axzz - 3axyy + 9zzyy.$$

And lastly writeing y^e^ valor of $v\left(=\dfrac{3xxy - ayy}{ax - 3yy}\right)$ in its stead, the result is, (o being put out(72))

$$c = \frac{a^3x^3 + 9ax^5 - 6aax^3y + a^3xyy - 9aaxxyy - 27x^4yy + 18axxy^3 - 3aay^4 + 27axy^4 - 27y^6}{2a^3xy + 54x^4y - 54axxyy + 54xy^4}.^{(73)}$$

But from these & such like consideracons may bee pronounced a generall Theoreme whereby y^e^ crookedness of any line may be readily determined.(74) To wch purpose let mee suppose $\mathfrak{X}$ to signifie all y^e^ Algebraicall termes (expressing y^e^ nature of y^e^ given line) when they are considered as equall to nothing & not some of y^m^ to others. Let $\mathfrak{X}$ signifie y^e^ same termes ordered according to y^e^ dimensions of x, & y^n^ multiplyed by any Arithmeticall progression. let $\mathfrak{X}$

---

(66) Descartes' folium. (Compare 1, §5 'Example y^e^ 2^d^' and note (31).) Note that, probably for the first time, Newton has extended it beyond the first quadrant.

(67) The incremented form of the preceding.

(68) The incremented 'valor' of v. (Compare 3, §1.1 above.)

(69) Omitting, as always, powers of o. (70) Read '$-3axzz-3axyy$'.

(71) Since $aax(z-y) = aax(ov/y)$; $9x^2(z^2-y^2) = 9x^2(2ov)$; and

$$3ay(z^2-zy) = 3ay(2ov-(ov/y)\,y) = 3ay(ov).$$

(72) And y written for z.

(73) Or

$$c = -\frac{1}{y}\,\frac{(y^2(3x^3-axy)^2 + x^2(-axy+3y^3)^2)\,(-axy+3y^3)}{(3x^3-axy)^2\,6y^3 - 2(3x^3-axy)\,(-axy+3y^3)\,(-axy) + (-axy+3y^3)^2\,6x^3}.$$

Newton will reduce this unwieldy expression in his concluding note. (Compare note (87) below.)

(74) That is, the general forms which Newton now presents are induced from the previous particular cases ($p+qy^k = 0$, $k = 1, 2, 3, 4$, and $p+qy+ry^3 = 0$ with $q = -ax$ and $r = 1$) above together with other examples, presumably worked out similarly, which Newton has not here noted.

signifie the same termes ordered according to y^e dimensions of y & then multiplyed by any Arithmeticall progression. let ${}^{\therefore}\mathfrak{X}$ signifie those termes ordered according to x & multiplyed by any two Arithmeticall progressions[,] one of them being greater then the other by a terme. Let $\mathfrak{X}^{\therefore}$ signifie those termes ordered according to y^e dimensions of y, & y^n multiplyed by any two Arithmeticall progressions differing by a terme. let ${}^{\cdot}\mathfrak{X}^{\cdot}$ signifie those termes ordered according to x & multiplyed by y^e greater of y^e progressions wch multiplyed ${}^{\therefore}\mathfrak{X}$, & then ordered according to y & multiplyed by y^e greater of y^e progressions wch multiplyed $\mathfrak{X}^{\therefore}$.(75)

Theorems whereby to find the quantity of crookedness in any parte of any given line.

Then will the Theorems bee

$$c = \frac{{}^{\cdot}\mathfrak{X}\,{}^{\cdot}\mathfrak{X}\,\mathfrak{X}^{\cdot}yy + \mathfrak{X}^{\cdot}\mathfrak{X}^{\cdot}\mathfrak{X}^{\cdot}xx}{-{}^{\cdot}\mathfrak{X}\,{}^{\cdot}\mathfrak{X}\,\mathfrak{X}^{\therefore}y + 2\,{}^{\cdot}\mathfrak{X}\,\mathfrak{X}^{\cdot}\,{}^{\cdot}\mathfrak{X}^{\cdot}y - \mathfrak{X}^{\cdot}\mathfrak{X}^{\cdot}\,{}^{\therefore}\mathfrak{X}y}. \text{ Or } c = \frac{{}^{\cdot}\mathfrak{X}\,{}^{\cdot}\mathfrak{X}yy + \mathfrak{X}^{\cdot}\mathfrak{X}^{\cdot}xx}{\dfrac{-{}^{\cdot}\mathfrak{X}\,{}^{\cdot}\mathfrak{X}\,\mathfrak{X}^{\therefore}}{\mathfrak{X}^{\cdot}}y + 2\,{}^{\cdot}\mathfrak{X}\,{}^{\cdot}\mathfrak{X}^{\cdot}y - \mathfrak{X}^{\cdot}\,{}^{\therefore}\mathfrak{X}y} = hp.$$

$$d = \frac{{}^{\cdot}\mathfrak{X}\,{}^{\cdot}\mathfrak{X}\,{}^{\cdot}\mathfrak{X}yy + {}^{\cdot}\mathfrak{X}\,\mathfrak{X}^{\cdot}\mathfrak{X}^{\cdot}xx}{-{}^{\cdot}\mathfrak{X}\,{}^{\cdot}\mathfrak{X}\,\mathfrak{X}^{\therefore}x + 2\,{}^{\cdot}\mathfrak{X}\,\mathfrak{X}^{\cdot}\,{}^{\cdot}\mathfrak{X}^{\cdot}x - \mathfrak{X}^{\cdot}\mathfrak{X}^{\cdot}\,{}^{\therefore}\mathfrak{X}x}. \text{ Or } d = \frac{{}^{\cdot}\mathfrak{X}\,{}^{\cdot}\mathfrak{X}yy + \mathfrak{X}^{\cdot}\mathfrak{X}^{\cdot}xx}{-{}^{\cdot}\mathfrak{X}\,\mathfrak{X}^{\therefore}x + 2\,\mathfrak{X}^{\cdot}\,{}^{\cdot}\mathfrak{X}^{\cdot}x \dfrac{-\mathfrak{X}^{\cdot}\mathfrak{X}^{\cdot}\,{}^{\therefore}\mathfrak{X}x}{{}^{\cdot}\mathfrak{X}}} = pm.$$

$$\sqrt{cc+dd} = \frac{{}^{\cdot}\mathfrak{X}\,{}^{\cdot}\mathfrak{X}yy + \mathfrak{X}^{\cdot}\mathfrak{X}^{\cdot}xx \text{ in } \sqrt{{}^{\cdot}\mathfrak{X}\,{}^{\cdot}\mathfrak{X}yy + \mathfrak{X}^{\cdot}\mathfrak{X}^{\cdot}xx}}{-{}^{\cdot}\mathfrak{X}\,{}^{\cdot}\mathfrak{X}\,\mathfrak{X}^{\therefore}xy + 2\,{}^{\cdot}\mathfrak{X}\,\mathfrak{X}^{\cdot}\,{}^{\cdot}\mathfrak{X}^{\cdot}xy - \mathfrak{X}^{\cdot}\mathfrak{X}^{\cdot}\,{}^{\therefore}\mathfrak{X}xy} = hm.^{(76)}$$

---

(75) In an extension of Hudde's scheme of multipliers, Newton defines operationally the homogenized first- and second-order partial derivatives of $\mathfrak{X} \equiv f(x, y) = 0$. Specifically, ${}^{\cdot}\mathfrak{X} = xf_x$, $\mathfrak{X}^{\cdot} = yf_y$; ${}^{\therefore}\mathfrak{X} = x^2f_{xx}$ ${}^{\cdot}\mathfrak{X}^{\cdot} = xyf_{xy} = xyf_{yx}$ and $\mathfrak{X}^{\therefore} = y^2f_{yy}$. (As Newton himself notes below, the arithmetical progressions must have the same constant difference.)

(76) Though Newton does not do so, these theorems are derivable by following through the structure of the previous arguments on the arbitrary algebraic function

$$\mathfrak{X} \equiv f(x, y) = 0 = \sum_i (p_i y^{\lambda_i}),$$

where the p_i are functions of x. Immediately

$${}^{\cdot}\mathfrak{X} = \sum_i (\dot{p}_i y^{\lambda_i}), \quad \mathfrak{X}^{\cdot} = \sum_i (\lambda_i p_i y^{\lambda_i}), \quad {}^{\therefore}\mathfrak{X} = \sum_i (\ddot{p}_i y^{\lambda_i}),$$

$${}^{\cdot}\mathfrak{X}^{\cdot} = \sum_i (\lambda_i \dot{p}_i y^{\lambda_i}) \quad \text{and} \quad \mathfrak{X}^{\therefore} = \sum_i (\lambda_i(\lambda_i - 1) p_i y^{\lambda_i}).$$

Also, where $\begin{Bmatrix} x \\ y \\ v \end{Bmatrix}$ increase into $\begin{Bmatrix} x+o \\ z \\ w \end{Bmatrix}$, $\mathfrak{X} = 0$ will increase into $\sum_i \left[\left(p_i + o\frac{\dot{p}_i}{x}\right) z^{\lambda_i}\right] = 0$. Then, dividing the increment of $\mathfrak{X}$ by o and considering the limit as $o \to$ zero (or in effect finding $d\mathfrak{X}/dx$),

$$0 = \sum_i \left(\frac{1}{o}\left[\left(p_i + o\frac{\dot{p}_i}{x}\right) z^{\lambda_i} - p_i y^{\lambda_i}\right]\right)$$

$$= \sum_i \left(\frac{\dot{p}_i}{x} y^{\lambda_i}\right) + \sum_i (\lambda_i p_i y^{\lambda_i - 1}) \frac{v}{y},$$

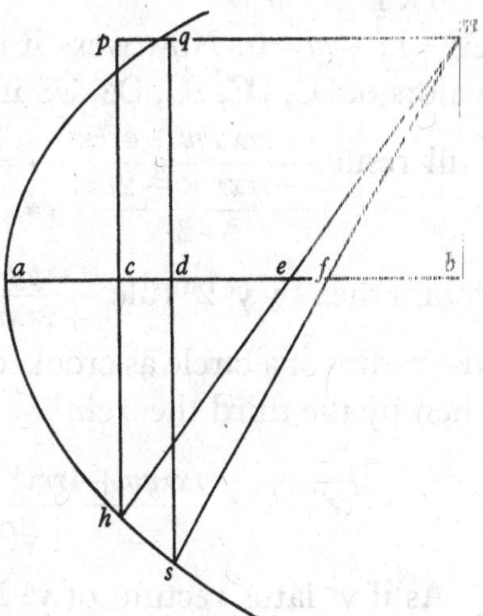

Example y^e^ first. Suppose the crookednes of some parte of y^e^ Parabola was desired, The terms expre[ss]ing its nature being $rx-yy=0$. Then by this Rule I put

$$\overset{}{\mathcal{X}}=r\overset{1}{x}-\overset{0}{yy}.\quad \mathcal{X}=rx-yy.\ \text{[or]}\ \mathcal{X}=rx.$$

$$\mathcal{X}=-\overset{2}{yy}+\overset{0}{rx}=-2yy.\ \text{Or w}^{\text{ch}}\text{ is y}^{\text{e}}\text{ same}$$

$$\mathcal{X}=-\overset{0}{yy}+\overset{-2}{rx}=-2rx.\quad \mathcal{X}=r\overset{-1\ \ 0}{x}-\overset{0\ \ -1}{yy}=0.$$

$$\mathcal{X}=-\overset{2\times1}{yy}+\overset{0\times-1}{rx}=-2yy=-2rx.$$

since
$$\lim_{o\to\text{zero}}\left(\frac{z^{\lambda_i}-y^{\lambda_i}}{o}\right)=\lim_{o\to\text{zero}}\left(\frac{z^{\lambda_i}-y^{\lambda_i}}{z-y}\right)\times\lim_{o\to\text{zero}}\left(\frac{z-y}{o}\right);$$

and so
$$\frac{v}{y}\left(=\frac{dy}{dx}\right)=-\frac{\mathcal{X}/x}{\mathcal{X}/y}.$$

'Differentiating' again we have

$$\begin{aligned}0&=\frac{d}{dx}\left(\sum_i\left[\left(p_i+o\frac{\dot p_i}{x}\right)z^{\lambda_i}\right]\right)\\&=\sum_i\left(\frac{1}{o}\left[\left(p_i+o\frac{\dot p_i}{x}+o^2\frac{\dot p_i}{x^2}\right)\left(z+\frac{ow}{z}\right)^{\lambda_i}-\left(p_i+o\frac{\dot p_i}{x}\right)z^{\lambda_i}\right]\right)\\&=\sum_i\left(\left(\frac{\dot p_i}{x}+o\frac{\dot p_i}{x^2}\right)z^{\lambda_i}+\left(p_i+o\frac{\dot p_i}{x}\right)\lambda_i z^{\lambda_i-1}\right);\end{aligned}$$

and, substituting
$$z=y+\frac{ov}{y},$$

$$\begin{aligned}0=&\left[\frac{1}{x}\sum_i(\dot p_i y^{\lambda_i})\frac{1}{x}\times\sum_i(\lambda_i\dot p_i y^{\lambda_i-1})\frac{ov}{y}+\frac{o}{x^2}\sum_i(\dot p_i y^{\lambda_i})\right]\\&+\frac{w}{z}\left[\sum_i(\lambda_i p_i y^{\lambda_i-1})+\sum_i(\lambda_i(\lambda_i-1)p_i y^{\lambda_i-2})\frac{ov}{y}+\frac{o}{x}\sum_i(\lambda_i\dot p_i y^{\lambda_i-1})\right]\\=&\left(\frac{\mathcal{X}}{x}+o\frac{v}{y}\frac{\mathcal{X}}{xy}+o\frac{\mathcal{X}}{x^2}\right)+\frac{w}{z}\left(\frac{\mathcal{X}}{y}+o\frac{v}{y}\frac{\mathcal{X}}{y^2}+o\frac{\mathcal{X}}{xy}\right).\end{aligned}$$

It follows finally, on substituting $v/y=-\mathcal{X}/x/\mathcal{X}/y$, that

$$\frac{w}{z}=-\frac{\mathcal{X}/x-o\left(\dfrac{\mathcal{X}/x}{\mathcal{X}/y}\dfrac{\mathcal{X}}{xy}-\dfrac{\mathcal{X}}{x^2}\right)}{\mathcal{X}/y-o\left(\dfrac{\mathcal{X}/x}{\mathcal{X}/y}\dfrac{\mathcal{X}}{2}-\dfrac{\mathcal{X}}{xy}\right)}.$$

Where now $ac=x$, $ch=y$, $ce=v$, $pc=c$ with the incremented forms $ad=x+o$, $ds=z$,

$\quad 0\times 1 \quad 2\times 0$
$\mathfrak{X}=rx-yy=0$. Now were it required y^t^ $c=hp$ be found Then substitute these valors of $\mathfrak{X}$, $\mathfrak{X}$, $\mathfrak{X}$, $\mathfrak{X}$, $\mathfrak{X}$ into theire places in y^e^ precedent rule And there will result $\dfrac{rrxxyy+4r^2x^4}{\dfrac{-rrxx\times-2rxy\quad **}{-2rx}}=-y-\dfrac{4xx}{y}=-y-4x\sqrt{rx}$[77] $=c$. But if d is to be found then by y^e^ 2^d^ rule $\dfrac{rrxxyy+4rrx^4}{+2rxrxx\ **}=\dfrac{yy}{2x}+2x=\dfrac{1}{2}r+2x=d$. Or if, $hm=\sqrt{cc+dd}$ the radius of a circle as crooked as y^e^ given parte (h) of y^e^ parabola bee required then by the third theorem

$$\sqrt{cc+dd}=\frac{rrxxyy+4rrx^4 \text{ in } \sqrt{rrxxyy+4rrx^4}}{+2rrxxxy^3\ **}=\frac{r+4x}{2r}\sqrt{rr+4rx}=hm.^{(78)}$$

As if y^e^ latus rectum of y^e^ Parabola (r) is an inch;[79] &, $ac=x$ two inches; then is $y=ch=\sqrt{2}$ inches $=\sqrt{rx}$. $hm=\dfrac{r+4x}{2r}\sqrt{rr+4rx}=\dfrac{27}{2}$ inches, that circle therefore whose radius is $13\frac{1}{2}$ inches is as crooked as y^e^ Parabola at h. but

---

$df=w$, Newton's first limit-relation (as $cd=o$ becomes indefinitely small) fixes the centre m of curvature at h in the normal hm by

$$\frac{w}{z}(z-c)=\frac{v}{y}(y-c)-o.$$

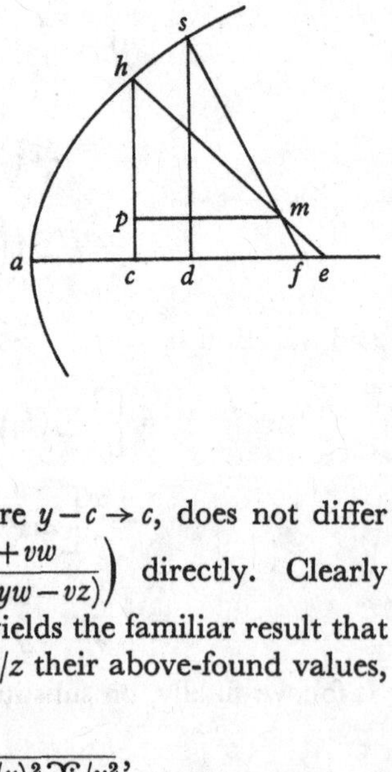

(Equivalently this may be obtained by noting that m is instantaneously at rest, or $(d/dx)\,(x+(v/y)\,(y-c))=0$.) Substituting $z=y+o(v/y)$ in this and rearranging, we have

$$0=\lim_{o\to\text{zero}}\left[\frac{1}{o}\left(\frac{w}{z}-\frac{v}{y}\right)(y-c)+\left(1+\frac{vw}{yz}\right)\right],$$

or

$$hp=y-c=\lim_{o\to\text{zero}}\left(\frac{1+\dfrac{vw}{yz}}{\dfrac{1}{o}\left(\dfrac{w}{z}-\dfrac{v}{y}\right)}\right).$$

(Newton's alternative limit-relation $c(yw-vz)=o(yz+vw)$, where $y-c\to c$, does not differ essentially from his first, but merely finds $c=\lim\limits_{o\to\text{zero}}\left(\dfrac{yz+vw}{(1/o)\,(yw-vz)}\right)$ directly. Clearly where the defining equation is given explicitly by $y=f(x)$, this yields the familiar result that $hp=(1+[f'(x)]^2)/f''(x)$.) Finally, on substituting for v/y and w/z their above-found values, Newton's basic result,

$$hp=\frac{\mathfrak{X}/y((\mathfrak{X}/x)^2+(\mathfrak{X}/y)^2)}{(\mathfrak{X}/y)^2\,\mathfrak{X}/x^2-2(\mathfrak{X}/x)\,(\mathfrak{X}/y)\,\mathfrak{X}/xy+(\mathfrak{X}/x)^2\,\mathfrak{X}/y^2},$$

follows immediately, with $\quad pm=hp\left(\dfrac{v}{y}\right)=hp\left(\dfrac{-\mathfrak{X}/x}{\mathfrak{X}/y}\right),$

and the curvature radius $hm=\sqrt{[hp^2+pm^2]}$.

would I find y^e crookednesse of y^e parabola at its vertex [a] then is $x=0$. $\dfrac{r+4x}{2r}\sqrt{rr+4rx}=\dfrac{r}{2}=hm=\dfrac{1}{2}$ an inch, soe y^t y^e Parabola is 27 times more crooked at its vertex y^n at y^e point (h), when $x=2r$.[80]

Example y^e 2^d^. Were y^e nature of y^e given line $x^3-axy+y^3=0=\mathfrak{X}$.[81] Then is

$$\begin{array}{l}\quad\;\;3\quad\;\;1\quad\;\;0\\ {\cdot}\mathfrak{X}=x^3-ayx+y^3=3x^3-ayx.\end{array}\qquad \begin{array}{l}\quad\;\;3\quad\;\;1\quad\;\;0\\ \dot{\mathfrak{X}}=y^3-ayx+x^3=3y^3-ayx.\end{array}$$

$$\begin{array}{lllll} & 3\times 2 & 1\times 0 & 0\times -1 & \\ {:}\mathfrak{X}= & x^3 & -\;ayx\;+ & y^3 & =6x^3.\end{array}\qquad \begin{array}{lllll} & 3\times 2 & 1\times 0 & 0\times -1 & \\ \ddot{\mathfrak{X}}= & y^3 & -\;ayx\;+ & x^3 & =6y^3.\end{array}$$

$$\begin{array}{lllll} & 3\times 0 & 1\times 1 & 0\times 3 & \\ {\cdot}\dot{\mathfrak{X}}= & x^3 & -\;ayx\;+ & y^3 & =-ayx.\end{array}$$

Then to find the valor of c, by y^e first theoreme

$$c=\frac{9x^6yy-6ay^3x^4+aay^4xx+9y^6xx-6ax^3y^4+aax^4yy}{\dfrac{-9x^6+6ay^3x^4-aay^4x^{(82)}}{3y^3-ayx}\text{ in }6y^4,\;-6ay^2x^4+2aay^3xx-18x^3y^4+6aayyx^4}.$$

$$\text{Or. } c=\frac{27x^4yy-18axxy^3+3aay^4+27y^6-27axy^4+9aaxxyy-9ax^5-a^3xyy+6aayx^3-a^5x}{54axxyy-54x^4y-54xy^4-2a^3xy}$$

Example y^e 3^d^. Were y^e nature of y^e line $x^3-axx+yxx-aay=0^{(83)}=\mathfrak{X}$. Then

$$\text{is } \begin{array}{l}\quad 1\qquad\;\; 0\qquad\; 0\qquad -2\\ {\cdot}\mathfrak{X}=x^3-axx+yxx-aay.\end{array}\;\; {\cdot}\mathfrak{X}=x^3+2aay.$$

$$\begin{array}{lllllll} & 1 & 0 & & 0 & -1 & \\ \dot{\mathfrak{X}}= & -aay & +x^3 & =xxy-aay= & xxy & +x^3 & .\;\; \dot{\mathfrak{X}}=axx-x^3=xxy-aay.\\ & +xxy & -axx & & -aay & -axx & \end{array}$$

$$\begin{array}{llllll} & 1\times 0 & 0\times -1 & -2\times -3 & & \\ {:}\mathfrak{X}= & x^3 & -\;\;axx & -aay & =-6aay. & \\ & & +\;\;yxx & & & \end{array}\quad \begin{array}{llll} & 1\times 0 & 0\times -1 & \\ \ddot{\mathfrak{X}}= & xxy & +x^3 & =0.\\ & -aay & -axx & \end{array}$$

$$\begin{array}{llllll} & 1\times 0 & 0\times 0 & 0\times 1 & -2\times 1 & \\ {\cdot}\dot{\mathfrak{X}}= & x^3 & -axx & +yxx & -aay & =2aay.\end{array}$$

---

(77) Read '$-y-4x\sqrt{\dfrac{x}{r}}$'. (78) Since $\dfrac{y^2+4x^2}{2y^2}=\dfrac{r+4x}{2r}$.

(79) Newton has cancelled 'As if r is a yarde'.

(80) For, when $x/r=2$, then $y/r=\sqrt{2}$ and so the preceding calculations hold with $x\to x/r$, $y\to y/r$.

(81) Descartes' folium yet again. (See note (66) above.)

(82) Read $\dfrac{\text{'}-9x^6+6ayx^4-aayyxx\text{'}}{3y^3-ayx}$.

(83) This is the degenerate cubic locus $(x-a)(x^2+xy+ay)=0$, and for values of x other than a Newton's procedure finds the curvature of the hyperbola $(x+a)(a-x-y)=a^2$.

Then subrogating these valors of $\mathfrak{X}$, $\mathcal{X}$, $\mathscr{X}$, &c into theire places in y[e] 1st Theoreme the valor of c will bee found

$$c=\frac{x^6yy+4aax^3y^3+4a^4y^4+x^6yy-2aax^4yy+a^4xxyy}{4x^3aayy+8a^4y^3+6aaxxy^3-6a^4y^3}.$$

That is $$c=\frac{2x^6+4aax^3y+4a^4yy-2aax^4+a^4xx}{4aax^3+6aaxxy+2a^4y}.^{(84)}$$

Observe that I have rather choose[n] to find the valor of c in this last example then of d or $\sqrt{dd+cc}$ because I could make $\mathcal{X}=0$, by w[ch] meanes y[e] terme $\frac{-\mathfrak{X}\mathscr{X}\mathfrak{X}\mathcal{X}}{\mathcal{X}}$ would vanish, & thereby save me the labor of a superfluous multiplication, which I could not avoyde in finding y[e] valors of d, or $\sqrt{cc+dd}$.(85)

Note [1st] y[t] all y[e] arithmeticall progressions made use of in this rule must have y[e] same difference of their termes. as if they be, $b-2a$, $b-a$, b, $b+a$, $b+2a$, $b+3a$, $b+4a$, $b+5a$ &c. $c-2a$, $c-a$, c, $c+a$, $c+2a$, $c+3a$ &c. $d-2a$, $d-a$, d, $d+a$, $d+2a$, $d+3a$, &c. Or if they bee some ꝑte of this -3, -2, -1, 0, 1, 2, 3, 4, 5, &c.

Note also y[t] y[e] progressions must proceede y[e] same way (from lesse to more) in respect of y[e] dimensions of y[e] unknowne quantitys, & not some one way, some another.(86)

Note also y[t] y[e] valor of c in y[e] 2[d] Example may bee (by y[e] equation

$$x^3-axy+y^3=0.)$$

reduced to $$\frac{27xxyy-27x^3y}{2a^3}+\frac{9xyy}{2aa}+\frac{6xx-9xy}{2a}+\frac{xx}{2y}-y=c=hp.^{(87)}$$

---

(84) That is, when $x \neq a$, $-(x+a)((2x+y)^2+(x+a)^2)/2a^2$ since $y=-x^2/(x+a)$.

(85) For d and $\sqrt{[c^2+d^2]}$ have the terms $-\frac{\mathcal{X}\mathcal{X}\mathscr{X}}{\mathfrak{X}}$ and $-\mathfrak{X}\mathcal{X}\mathscr{X}$ in their denominator, each of which requires a triple multiplication.

(86) Compare note (34) above. These two complementary remarks are crucial for the successful application of Newton's algorithm.

(87) For the denominator of c in Example 2 above (compare note (73)) is

$$54xy(x^3-axy+y^3)+2a^3xy = 2a^3xy,$$

while its numerator is

$$27x^3y^3-27x^4y^2+9ax^2y^3+6a^2x^3y-9a^2x^2y^2+a^3x^3-2a^3xy^2,$$

since $$27y^6-27axy^4 = -27x^3y^3, \quad -18ax^2y^3-9ax^5 = -9ax^2y^3-9a^2x^3y$$

and $$6a^2yx^3-a^3xy^2+3a^2y^4 = 3a^2x^3y+2a^3xy^2.$$

(88) This last, mostly cancelled passage carries on the theme of [1] above, attempting to employ a general method of tangents in a bipolar co-ordinate system. This was not the first

[3] *To draw Perpendiculars to crooked lines in all other cases.*(88)

Although y^e^ unkno[w]ne quantitys x & y are not related to one another as in the present rules (that is soe y^t^ y move upon x in a given angle) yet may there be drawne tangents to them by the same method.

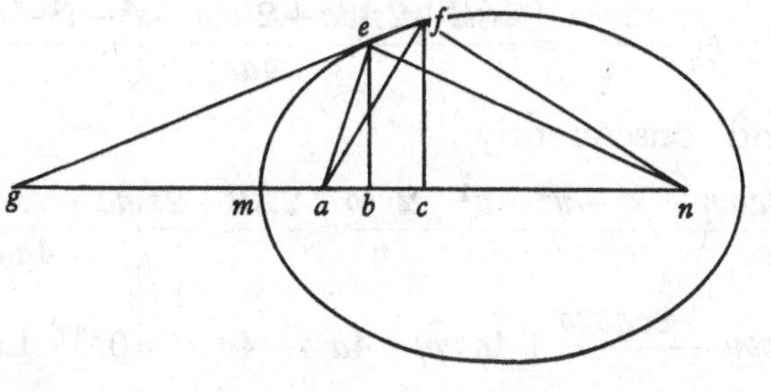

As if *efm* is an Elipsis described by y^e^ thred *aen* (as is usuall(89)) Then make $gb=v$. $ae=x$. $en=y$. $an=a$. And let y^e^ relation twixt x & y be $x+y=b$.(90) $ab=z$. Then is $bn=a-z$. $eb^2=xx-zz=yy-aa+2az-zz$. And consequently $\frac{xx-yy+aa}{2a}=z=ab$. And

$$xx\frac{-x^4+2xxyy-2xxaa-y^4+2aayy-a^4}{4aa}=eb^2=\frac{2xxyy+2xxaa+2aayy-x^4-y^4-a^4}{4aa}$$

$$=ae^2-ab^2.$$

---

time that Newton had faced the problem of limit-increments in bipolars. About September 1664 he had become immersed in the mathematical theory of the refraction of white light at a point, and inevitably came to consider the curves (Descartes' ovals, for which $x \pm (d/r)\, y$ is constant, where x and y are bipolar co-ordinates of an arbitrary point on the curve), which refract all rays within a determinable range from one fixed point to a second for a given refractive ratio. Having made a deep study of what Descartes, in *Geometria*'s Liber II, had said about them, and particularly their subnormals, he went on to widen the topic but abandoned his work when it was far from complete. (See the concluding appendix of this volume, on geometrical optics, section 1, especially notes (13) and (15).) Newton here takes the topic up again briefly, restricting his attention to the ellipse $x+y = b$ for simplicity. Later, when he returned a third time in November 1665 to the consideration of tangents in bipolar systems (6, §3: Example 2 below) he abandoned a fully analytical treatment, preferring to expound the easier semi-geometrical approach by the limit ('differential') triangle at a point.

(89) This 'gardener's' construction for the ellipse was widely known at the time, and goes back in its mathematical theory to Apollonius (*Conics*: **3**, 52) and in its explicit enunciation as a construction to Anthemius of Tralles. Newton himself had already noted it down in his earliest mathematical notes, probably from Schooten's *Exercitationes* (Liber IV, Caput IV: 325–6) though he must previously have known of Descartes' description of it in his *Dioptrique* (=*Dioptrice* (1650): Caput III: 142–3: II, *Quid sit Ellipsis, & quomodo sit describenda.*) (See 1, 1, §1.)

(90) The constancy of the sum of the focal distances (at an arbitrary point on the ellipse) expressed in bipolar co-ordinates.

Againe suppose $af=s$. $fn=t$. $bc=o$. Then is

$$ac=z+o=o+\frac{xx-yy+aa}{2a}=\frac{ss-tt+aa}{2a}.^{(91)}$$

That is $2ao+xx-yy+tt=ss$. Alsoe $v:g::v+o:g+\frac{og}{v}=fc$. And

$$fc^2=gg+\frac{2g^2o}{v}=\frac{2sstt+2ssaa+2ttaa-s^4-t^4-a^4}{4aa}.$$

Making $eb=g$. And consequently

$$\frac{2xxyy+2xxaa+2aayy-x^4-y^4-a^4}{4aa}+\frac{2g^2o}{v}=\frac{2sstt+2ssaa+2aatt-s^4-t^4-a^4}{4aa}.^{(92)}$$

Or $A=4aoxx-4aoyy\frac{+8aaggo}{v}+4aayy-4a^3o-4aatt=0.$(93) Lastly by the nature of the line $b-t=s$. And $b[b]-2bt+tt=ss=2ao+xx-yy+tt$. Or

$$\frac{-2ao-xx+yy+bb}{2b}=t.^{(94)}$$

---

(91) The increment, as $\begin{Bmatrix} x\to s \\ y\to t \\ z\to z+o\end{Bmatrix}$ of $z=\dfrac{xx-yy+aa}{2a}$.

(92) The value of g^2 is not substituted in the term $2g^2o/v$ since it is already multiplied by o and therefore its increment will not affect the calculation (by the first observation on p. 273 above).

(93) That is, substituting for s^2 its equal $t^2+x^2-y^2+2ao$, with

$$s^4-x^4 = 2x^2(t^2-y^2)+(t^2-y^2)^2+4ao(x^2+t^2-y^2)+4a^2o^2,$$
$$-2(s^2t^2-x^2y^2) = -2t^2(t^2-y^2)-2x^2(t^2-y^2)-4aot^2$$

and

$$-2a^2(s^2-x^2) = -2a^2(t^2-y^2)-4a^3o.$$

It follows that

$$(s^4+t^4+a^4-2s^2t^2-2s^2a^2-2a^2t^2)-(x^4+y^4+a^4-2x^2y^2-2x^2a^2-2a^2y^2)$$
$$= -4ao(y^2-x^2+a^2)+4a^2(y^2-t^2) = 4a^2o\left(\frac{y^2-t^2}{o}-2(a-z)\right),$$

where the term $4a^2o^2$ containing a power of o is 'blotted out'.

(94) That is, since $-x^2+(b-y)^2=0$, $t=y-(a/b)\,o$. Newton, not realizing this, falters in his further calculations, then flounders with an ugly equation and so cancels the whole paragraph.

(95) Newton here cancels his whole calculation, though little is in fact needed to complete it. Since $t=y-(a/b)\,o$ (note (94)),

$$\frac{t^2-y^2}{o}=-2\frac{a}{b}y:\ \text{further,}\ 8a^2o\left(\frac{g^2}{v}\right)=4a^2(t^2-y^2)+4ao(y^2-x^2+a^2),$$

or

$$\frac{g^2}{v}=\tfrac{1}{2}\lim_{o\to\text{zero}}\left(\frac{t^2-y^2}{o}\right)+\frac{y^2-x^2+a^2}{2a}=-\frac{a}{b}y+a-z=+\frac{a}{b}x-z,$$

and finally

$$v=\frac{g^2}{(a/b)\,x-z}.$$

And $4tt = \dfrac{4aoxx - 4aoy^2 - 4abbo + Q : -xx + yy + bb :}{bb}$. W[ch] valor substituted into y[e] Equation A, the result is

$$\frac{8aaggo}{v} \frac{+4a^3bbo - 4a^5o - x^4a^2 + 2xxyyaa + 2aabbxx - aay^4 + 2aabbyy - b^4a^2}{bb} [=0.]^{(95)}$$

As if[(96)]

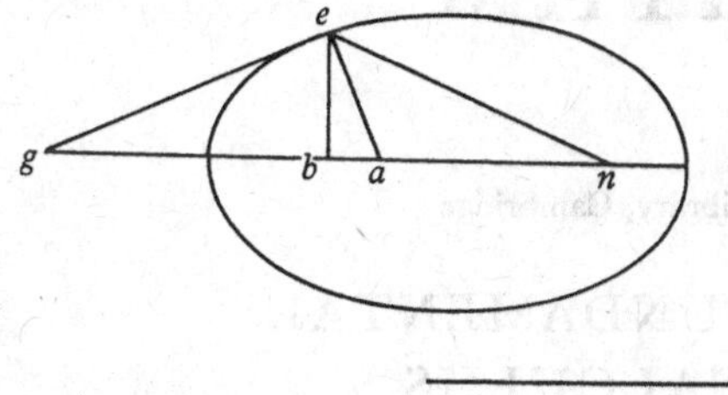

(96) Newton begins to redraft his previous calculation, but immediately leaves off. It seems clear that in his scheme *eg* and *ea* are respectively tangent and normal to the curve at *e*, with $eb \perp gban$ and *n* the further focus. Adding therefore the second focus *m*, we may easily

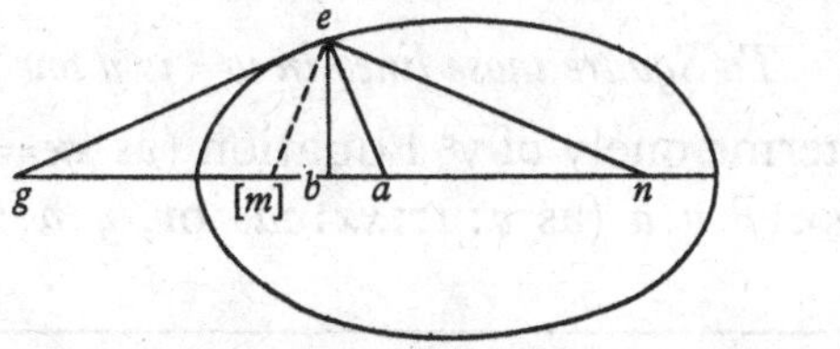

outline how Newton could have completed his general investigation of the tangent and normal in a bipolar co-ordinate system in the following way. Suppose, much as before, that $me = x$, $ne = y$, $mn = a$ and $mb = z$ with s, t and $z+o$ the increments of x, y and z respectively. Following Newton's previous argument almost through, we may show that

$$z = \frac{x^2 - y^2 + a^2}{2a} \quad \left(\text{or} \quad a - z = \frac{y^2 - x^2 + a^2}{2a}\right)$$

and, where $eb = g$ with subtangent $gb = v$ and subnormal $ba = u$,

$$g^2 = vu = x^2 - \left(\frac{y^2 - x^2 + a^2}{2a}\right)^2;$$

so that ultimately

$$\frac{g^2}{v} = u = \tfrac{1}{2} \lim_{o \to \text{zero}} \left(\frac{t^2 - y^2}{o}\right) + a - z, \quad \text{or} \quad an = a - (u + z) = -y \lim_{o \to \text{zero}} \left(\frac{t^2 - y^2}{o}\right) = -y\frac{dy}{dz}.$$

Finally, since

$$z + o = \frac{s^2 - t^2 + a^2}{2a}, \quad a = \tfrac{1}{2}\left[\lim_{o \to \text{zero}} \left(\frac{s^2 - x^2}{o}\right) - \lim_{o \to \text{zero}} \left(\frac{t^2 - y^2}{o}\right)\right] = x\frac{dx}{dz} - y\frac{dy}{dz},$$

or on substitution $ma = u + z = x(dx/dz)$. Alternatively, $a = (dx/dz)(x - y(dy/dx))$ and so, with this substituted in the preceding, $ma = u + z = ax/[x - y(dy/dx)]$. (Compare the concluding appendix on geometrical optics, section 1, note (15).)

5

THE CALCULUS BECOMES AN ALGORITHM

[Middle? 1665][1]

From the originals in the University Library, Cambridge

§1 APPROACH TO THE FUNDAMENTAL THEOREM OF THE CALCULUS

[1][2] An equation given; if both x, & y, [be of] divers dimensions, try if y[e] roote of one of ȳ[3] may be extracted:[4] & If a quantity wherein y is not is divided by x in y[e] line equall to x.[5] y[t] crooked [line] cannot be squared.[6]

To Square those lines in w[ch] is y onely.

If y is in but one terme onely of y[e] Equation (as $xx = ay$. or, $a^3 = xxy$) resolve y[e] Eq: into y[e] proport[7] $y:a$ (as $y:a::xx:aa$. or, $y:a::aa:xx$.) If y[e] line hath Assymptotes[8]

---

(1) The dating is hazarded on the basis of an examination of Newton's writing. His testimony, published and unpublished, at the time of the fluxion priority dispute half a century afterwards affords loose confirmation. (The strong basis of Newton's testimony about his early papers was his own reappraisal of the documents here printed, backed by an often blurred memory of his early years of discovery. We need not stand by it rigidly.)

(2) Fragments from Add. 4000: 92[v], 94[v] respectively.

(3) Read 'y[m]'.

(4) Newton first continued with 'whereby y[e] $=_n$ may often be reduced to fewer termes & sometimes to fewer dimensions: if it cannot y[e] line [? may not be squared]'. (As Newton writes elsewhere on the page '[For] $=_n$ [read] equation'.)

(5) Read 'y'.

(6) In general $\int x^p . dx = \frac{1}{p+1} x^{p+1}$ but, for $p = -1$, $\int \frac{1}{x} . dx = \log x$ is not quadrable. It follows that, where $y = \sum_i (a_i x^i)$ has the non-zero term $\frac{a_{-1}}{x}$, then $\int y . dx$ 'cannot be squared'.

(7) 'Proportion'.

(8) Newton breaks off in mid-sentence. A 'line hath Assymptotes' when x, y are not simultaneously infinite in its defining equation, and for the simple lines here considered this condition selects the general hyperbolas $x^p y^q = a^{p+q}$, with

$$\int y . dx = a \int x(p/q) . dx = \frac{aq}{p+q} x^{(p+q)/q}.$$

[2][9] *The line cdf is a Parab.*

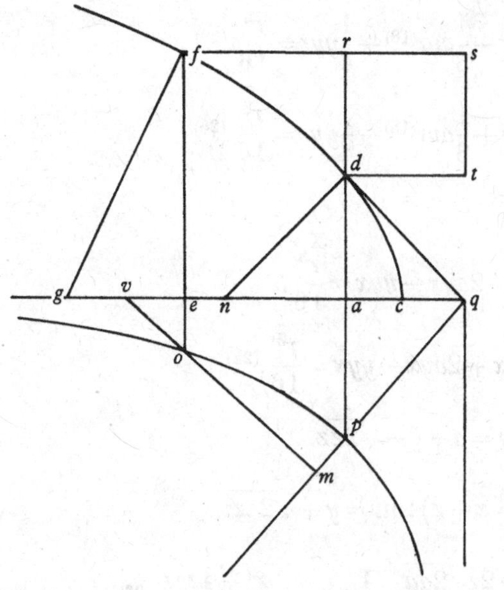

$4ac = 2ad = r = 2na = 2ge.$[10] $ce = x.$

$ef = y.$ $rx = yy.$ $ge : ef :: \frac{1}{2}r : \sqrt{rx} :: eo : ap.$

$eo = z.$ $ap = a.$ $\frac{1}{2}ra = z\sqrt{rx}.$ $\frac{1}{4}rraa = rzzx.$

$\frac{raa}{4} = zzx.$[11] or supposeing $eq = y.$[12]

$x = y + \frac{1}{4}r.$[13] & $\frac{raa}{4} = zzy + \frac{1}{4}zzr.$ wch shews y^e^ nature of y^e^ crooked line *po*. now if $dt = ap$. y^n^ $drst = eoap$. for supposeing *eo* moves uniformely from *ap*, & *rs* moves from *dt* wth motion decreaseing in y^e^ same ꝑportiō[14] y^t^ y^e^ line *eo* doth shorten.[15]

Suppos $aq = ap = \frac{r}{2} = a.$[16] & $eq = y.$

$x = y - \frac{1}{4}r.$ then $\frac{r^3}{16} = zzy - \frac{1}{4}zzr.$[17]

suppose $z = a + y$. y^n^ $\frac{r^3}{16} = aax + 2ayx + yyx.$

$$\text{Or } aax + 2ayx + yyx + \frac{1}{4}aar + \frac{1}{8}ayr^{(18)} + \frac{1}{4}yyr = \frac{r^3}{16}.^{(19)}$$

$$\text{Or } aax + 2ayx + yyx - \frac{1}{4}aar - \frac{1}{8}ayr^{(18)} - \frac{1}{4}yyr = \frac{r^3}{16}.^{(20)}$$

(9) Add. 4000: 93^r^, 94^r^.

(10) The lines *na* and *ge* respectively are the subnormals (of constant length $\frac{1}{2}r$) at the points *d* and *f* on the parabola. Since $ac = \frac{1}{4}r$, then $ad = \frac{1}{2}r$ also.

(11) That is, where $ce = x$ and $eo = z$, the defining equation of (*o*).

(12) A new variable, not to be confused with $ef = y$ above (or $eq = y$ below).

(13) Since $ec - ea = ac = \frac{1}{4}ra^2(1/a^2) = \frac{1}{4}r$.

(14) Read 'proportion'.

(15) Since $z/a = dy/dx$ (and so, where *t* is an independent variable of 'time', the 'motion' $a(dy/dt) = z(dx/dt)$), therefore

$$(eopa) = \int_{\frac{1}{4}r}^{x} z\,.\,dx = \int_{\frac{1}{2}r}^{y} a\,.\,dy = (rstd).$$

(16) Read 'Suppose $aq\left(=\frac{r}{2}\right) = ap\ (=a)$'. (The subtangent to the parabola $y^2 = rx$ is of length $2x$, with *aq* subtangent at *d* where $ac = x = \frac{1}{4}r$.)

(17) Substituting $x = y - \frac{1}{4}r$ in $\frac{1}{4}ra^2 = \frac{1}{16}r^3 = z^2x$.

(18) Read '$\frac{1}{2}ayr$'.

(19) Where $x \to x + \frac{1}{4}r$.

(20) Where $x \to x - \frac{1}{4}r$.

Or suppose $z=y-a$. y^n $\frac{r^3}{16}=aax-2ayx+yyx$.

$$\text{Or } aax-2ayx+yyx+\frac{1}{4}aar-\frac{1}{8}ayr^{(18)}+\frac{1}{4}yyr=\frac{r^3}{16}.^{(19)}$$

$$\text{Or } aax-2ayx+yyx-\frac{1}{4}aar+\frac{1}{8}ayr^{(18)}-\frac{1}{4}yyr=\frac{r^3}{16}.^{(20)}$$

Or, if $x=a-x$.(21) [y^n] $\frac{r^3}{16}=zza-zzx$.

$$\& \; a^3+2aay+ayy-aax-2ayx-yyx=\frac{r^3}{16}.$$

$$\text{Or, } a^3-2aay+ayy-aax+2ayx-yyx=\frac{r^3}{16}.^{(22)}$$

$ov^2=2z^2$.(23) $mp=\zeta$. $mq=a+\zeta$.(24) $mo=y=a+\zeta-\sqrt{2zz}$.

$$pq(=a):aq\left(=\frac{1}{2}r\right)::vq(=x+z):mv=y+\sqrt{2zz}.$$

$$ay+a\sqrt{2zz}=\frac{1}{2}rx+\frac{1}{2}rz.^{(25)} \quad \frac{2ay+2z\sqrt{2aa}}{r}-\frac{1}{2}z=x=\frac{r^3+4zzr}{16zz}.^{(26)}$$

$32zzay+32z^3a\sqrt{2}-8z^3r-r^4-4zzrr=0$.(27)

$[mp=\zeta.]$ $mv^{(28)}=\xi=a+\zeta-z\sqrt{2}$. $\frac{\xi-a-\zeta}{\sqrt{2}}=z$.(29)

[. . . . . . .](30)

---

[3](31)

By ye Squares of ye simplest lines to square lines more compound.

1s[t] those whe[re]in is y. find ye valor of y. If ye number of ye termes in ye denom: thereof be neither 1. 3. 6. 10. 15. 21. 28. &c. ye line cannot be squared.(32)

---

(18) Read '$\frac{1}{2}ayr$'.

(19) Where $x \to x+\frac{1}{4}r$.

(20) Where $x \to x-\frac{1}{4}r$.

(21) That is, if $x \to a-x$.

(22) Using the substitutions $z = a+y$, $z = a-y$ respectively.

(23) That is, $ve = eo$ and so, since $ap = aq = \frac{1}{2}r$, $v\hat{o}e = \frac{1}{4}\pi = v\hat{q}m$.

(24) Newton now takes $pq = a = \frac{1}{2}r\sqrt{2}$ or $r/\sqrt{2}$.

(25) Or, since $r = a\sqrt{2}$ (note (24)), $y\sqrt{2}+z = x$.

(26) Read correctly '$\frac{2ay+2z\sqrt{2aa}}{r}-z=x=\frac{r^3+4zzr}{16zz}$', that is, $16z^2(y\sqrt{2}+z) = r(r^2+4z^2)$.

(27) Correctly deduced from the previous line and so vitiated. Read

'$16z^3+16\sqrt{2}\,yzz-4rzz-r^3=0$'.

(28) Read 'mo'.

If it have but one terme tis squared by finding y^e square of each particular terme in y^e valor of y & y^n adding all those squares together. Example 1st. $3x^4+a^4=yaxx$. & $y=\frac{3x^4+a^4}{axx}$. Then makeing y equall to each particular terme. $\frac{3xx}{a}=y$. [&] $\frac{a^3}{xx}=y$. or $3xx=ay$. whose □ is $\frac{x^3}{a}$. & $a^3=xxy$. whose □ is $\frac{a^3}{x}$. Ad those two □s together & they $\left(\text{viz: } \frac{x^4+a^4}{ax}\right)$ are the □ of y^e line $3x^4+a^4=ayxx$. Againe $2a^7-2bx^6+x^7=a^3x^3y$. Or $y=\frac{2a^7-2bx^6+x^7}{a^3x^3}$. y^n disjoynting y^e valor of y. $y=\frac{2a^4}{x^3}$. $y=\frac{x^4}{a^3}$. $y=\frac{-2bx^3}{aaa}$. Or $x^3y=2a^4$, whose □ is $\frac{a^4}{xx}$. $ya^3=x^4$, whose □ $\frac{x^5}{5a^3}$. $ya^3=-2bx^3$, whose □ $\frac{-x^4b}{2a^3}$. which 3 □'s

$$\left(\text{viz[:]}\ \frac{10a^7+2x^7-5bx^6}{10a^3xx}\right)$$

taken together are y^e square sought for. And these lines may be ever squared unless in y^e valor of y there bee found $\frac{aa}{x}$, $\frac{ab}{x}$, $\frac{cc+de}{x}$, &c. for y^e squareing of y^t line depends on y^e squareing of y^e Hyperbola.(33) As in y^e line

$$axxy=x^4+a^3x+a^4.^{(34)}$$

---

(29) Newton changes his scheme slightly, replacing y by ξ.

(30) The remaining calculations, lengthy and erroneous, are omitted. Essentially Newton, having calculated z^2 and z^3, eliminates z from his cubic

$$(32a\sqrt{2}-8r)\,z^3+(32ay-4r^2)\,z^2-r^4 = 0$$

in favour of ζ and ξ. (The resulting cubic, if his transformation scheme were correct, would have been the transformed defining equation of (o).) Newton makes heavy weather of what should have, by now, been a simple reduction of $z^2(x-\frac{1}{2}r) = \frac{1}{16}r^3$ by the transform

$$\left\{\begin{array}{l} x = \frac{1}{2}r+\frac{1}{\sqrt{2}}(\zeta+\xi) \\ z = \frac{1}{2}r+\frac{1}{\sqrt{2}}(\zeta-\xi) \end{array}\right\} \text{ with } a = \frac{r}{\sqrt{2}}.$$

(31) Add. 4000: 95v, 96r.

(32) Newton's criterion is thus that $y = [f(x)]/[g(x)]$ is squareable only if $g(x)$ has, in general, $\frac{1}{2}n(n+1)$ terms, $n = 1, 2, 3, \ldots$. What he means is not clear. Perhaps he wishes to say that a minimum requirement for $y = [f(x)]/[g(x)]$ is that $g(x)$ be expressible as the square $(A+Bx+Cx^2+Dx^3+\ldots)^2$. At any rate Newton's criterion excludes, for example, the quadrable $y = (1+x)^{-k}$ $(k = 3, 4, 6, \ldots)$.

(33) That is, defined in Cartesian co-ordinates by $xy = k$. See also note (37).

(34) Or $y = x^2/a+a^2/x+a^3/x^2$, where the term a^2/x is not squareable.

2dly. If it have 3(35) termes. See if it may be reduced to $\left\{\begin{matrix}\text{one or}\\ \text{fewer}\end{matrix}\right\}$ dimensions by adding or subtracting a knowne quantity to or from x.

Example. $2bax+axx=bby+2bxy+xxy$.(36) w^{ch} (makeing $x+b=z$) is thus reduced $zzy=-bba+azz$. Or $\dfrac{-bba+azz}{zz}=y$.(37)

---

[4](38)

A Method whereby to square those crooked lines wch may bee squared.

That a line may be squared Geometrically tis required y^t its area may be expressed in generall by some equation in wch there is an unknowne quantity, so y^t this quantity being determined y^e area thereof (comprehended by ye crooked line, ye two lines to wch all ye points in ye crooked line are referred [& ye base](39)) is limited(40) & may bee found by ye same equation. Also every such equation must be of two dimensions because it expresseth ye quantity of a superficies.(41)

That an equation(42) expresse ye area of a crooked(43) line tis required y^t ye superficies increase in an unequall proportion, when ye line (considered as unknowne) increaseth in arithmeticall proportion,(44) wherefore (supposing x always to signifie ye unknowne quantity: a, b, c, &c; to signifie y^e quan[t]itys given) ax, or xx either alone or added to any other supperficies, serve not to find y^e area of any crooked line wch may not be found w^{th}out y^m.

Prop:(45)

Haveing an equation of 2 dimensions to find w^t crooke line it is whose area it doth expresse. suppose y^e equation is $\dfrac{x^3}{a}$. nameing y^e quantitys, $a=dh=kl$. $bg=y$. $db=mk=x=gp$. y^e superficies $dbg=\dfrac{x^3}{a}$. supose y^e square $dkhl$ is equall

---

(35) Newton has cancelled '6, 10, 15, 21,', and so has already rejected his criterion for quadrability (note (32)).

(36) Or $y=\dfrac{a[(b+x)^2-b^2]}{(b+x)^2}$.

(37) Newton begins to generalize on the particular derivatives he obtained (in section 2 above) by repeated application of Cartesian subnormal techniques. He shows both a willingness to formulate hypotheses about quadrable algebraic integrals and a readiness to reject the hypothesis when it fails. He remains, however, more than a little shaky in his control of signs and continues to accept that the 'square' of a^{r+1}/x^r ($r = 2, 3, 4, \ldots$) is

$$+\frac{1}{r-1}\frac{a^{r+1}}{x^{r-1}}.$$

(38) Add 4000: 120^r-133^v.

(39) Newton first wrote '[y^e two lines] whose intersections describe y^e crooked line, & another streight line wch guides these two lines in their motion w^n they describe it'.

to y[e] superficies *gbd*; y[n] $dk = z = bm = lh = \frac{x^3}{aa}$, & $aaz = x^3$. w[ch] is an equation expressing y[e] nature of y[e] line *fmd*.

Next makeing $nm = s$ a line w[ch] cutteth *dmf* at right angles. $nd = v$.

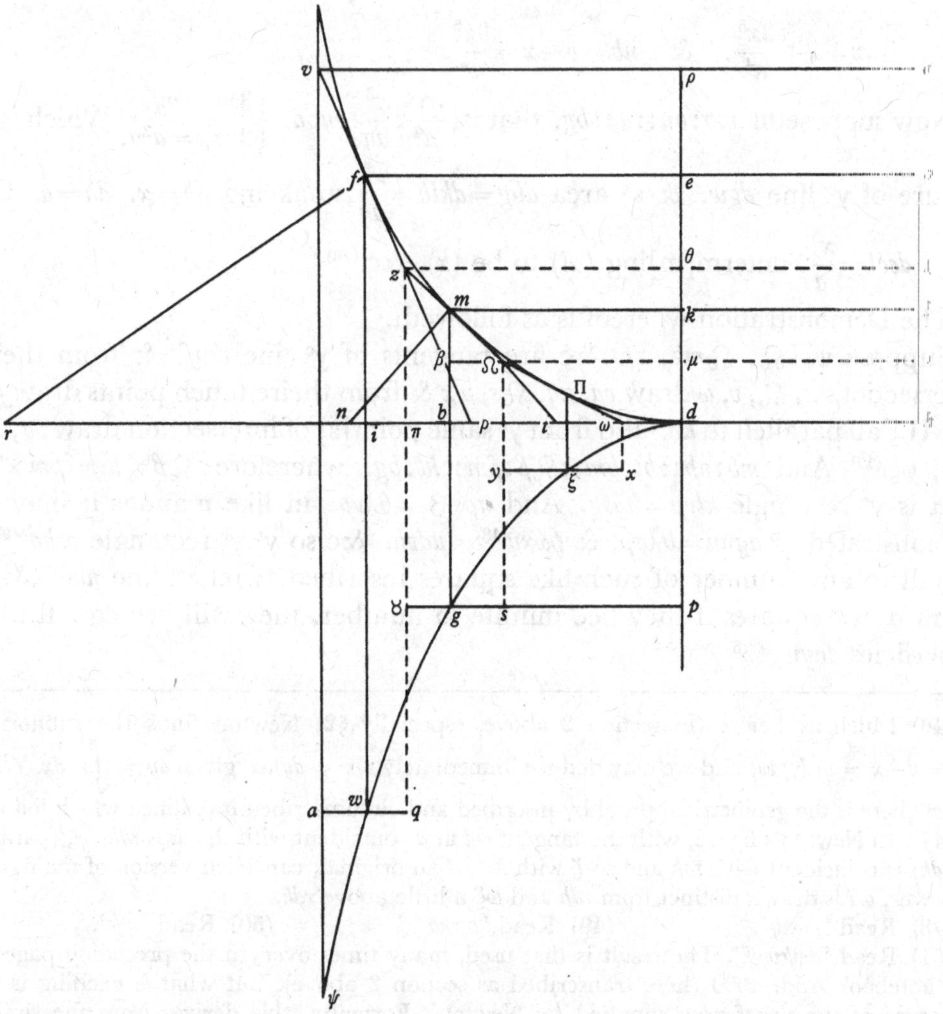

(40) That is, determined.

(41) Newton, of course, presupposes that the defining equation of his curve is homogeneous, that is, of the form

$$0 = \sum_{i,j} (\alpha_{i,j} x^i y^j),$$

where the coefficients $\alpha_{i,j}$ are of the order $n-(i+j)$.

(42) Newton has cancelled 'super[ficies]'.

(43) Newton first wrote 'streight'. (44) Newton has cancelled 'equally'.

(45) In the accompanying figure Newton has two points ρ, while the point n denotes the meets of both $va\psi$ and the normal to the curve at m with the base *rdh*. There should be no confusion, however, in identifying the required points as occasion demands.

$ss - vv + 2vx - xx = \frac{x^6}{a^4} = mb$ squared.

0. 0. 1. 2. 6.

$2vx = 2xx + \frac{6x^6}{a^4}$.

$v = x + \frac{3x^5}{a^4}$. & $nb = v - x = \frac{3x^5}{a^4}$.

wch is an equation haveing 2 equall rootes & therefore multiplyed according to Huddenius his method, produceth another.

Now supposeing $mb:bn::dh:bg$. that is, $\frac{3x^5}{a^4}:\frac{x^3}{aa}::y:a$. $\begin{cases} 3xx = ay. \\ 3xxa = a^2y. \end{cases}$ Which is y^e^ nature of y^e^ line *dgw*. & y^e^ area $dbg = dklh = \frac{x^3}{aa}$, makeing $db = x$. $dh = a$. Or. $diw = deoh = \frac{x^3}{a}$, determ[in]ing (di) to be (x), &c.(46)

The Demonstration whereof is as followeth.

Suppose $w\text{II}\Omega$, Ωmz, zfv &c are tangents of y^e^ line *dmf*. & from theire intersections z, Ω, v, ω draw va, zq, Ωs, wx & from theire touch points draw fw, mg, II ξ all parallell to kp. also from y^e^ same point[s] of intersection draw $v\sigma$, $z\lambda$, $\Omega\nu$, $\omega\zeta$.(47) And $mb:nb::bt:bm::\Omega\beta:\beta m::kl:bg$. wherefore $\Omega\beta \times bg = \beta m \times kl$. that is y^e^ rectangle $klv\mu = b\rho sg$. And $\pi\rho s ♉ = \theta\lambda v\mu$. in like manner it may be demonstrated y^t^ $aq\pi n = \theta\lambda\sigma\rho$, & $\rho\omega xy$(48) $= \mu dvh$. &c so y^t^ y^e^ rectangle ρshd(49) is equall to any number of such-like squares inscribed twixt y^e^ line ny(50) & y^e^ point d, wch squares if they bee infinite in number, they will bee equall to y^e^ superficies $dnywg\xi$.(51)

(46) Much as before (in section 2 above, especially §2) Newton finds the subnormal $nb = v - x = y(dy/dx)$ and we may deduce immediately $z/a = dy/dx$, given $ay = \int z.dx$. What is new here is the geometrical proof by inscribed and circumscribed mixtilinea which follows.

(47) In Newton's figure, with the tangent ωd at d coincident with the axis rdh, $\omega\zeta$ (parallel to rdh) is coincident with rdh and so ζ with h. In an original, cancelled version of the figure, however, ωd is drawn distinct from rdh and $\omega\zeta$ a little above rdh.

(48) Read '$\rho\omega xy$'. (49) Read '$\rho\sigma hd$'. (50) Read '$n\psi$'.

(51) Read '$dn\psi wg\xi$'. The result is that used, many times over, in the preceding pages of the notebook Add. 4000 (here transcribed as section 2 above), but what is exciting is the structure of the proof now supplied by Newton. Formally, this derives from the similar structure used by Heuraet in his letter, of 13 January 1659 to Schooten, on rectification (printed almost immediately by the latter as *Epistola de Transmutatione Curvarum Linearum in Rectas* = *Geometria*: 517–20) and Newton's figure is essentially a copy of Heuraet's. Where, however, Heuraet had used the structure in his development of a general rectification procedure, constructing geometrically

$$\int \sqrt{\left[1 + \left(\frac{dy^2}{dx}\right)\right]} dx$$

and considering as examples the general parabolas $ay^r = x^{r+1}$ $(r = 2, 3, \ldots)$, Newton here applies it to the allied quadrature problem, constructing

$$\frac{1}{a}\int z.dx \quad \text{where} \quad z = a\frac{dy}{dx} = af(x).$$

This being demonstrated that I may shunne confusion in squareing y^e lines of every sort I shall use this method in distinguishing y^m. viz:

first. such lines whose area is exprest by equations in wch y^e unknowne quantity is numerator, & y^t 1st all y^e sines(52) being affirmative, 2dly mixed.

2dly. lines whose area is exprest by quantitys in wch ye unknowne quantity is divisor, & those 1st under affirmative sines,(52) 2^d under mixed ones.

3. lines squared by equations mixt of y^e 2 former kinds, whose quantitys are all 1^s[t] affirmative[,] 2dly mixt.

The squareing of those lines whose area is exprest by affirmative quantitys in wch y^e unknowne quantity is numerat^r.(53)

The equations expressing the nature of y^e lines.		Theire square.
$3xx = ay$. Parab:	————	$\frac{x^3}{a}$.
$4x^3 = aay$.	————	$\frac{x^4}{aa}$.
$5x^4 = ya^3$.	————	$\frac{x^5}{a^3}$.
$6x^5 = ya^4$.	————	$\frac{x^6}{a^4}$.
$7x^6 = ya^5$.	————	$\frac{x^7}{a^5}$.

We may, if we wish, see either as a geometrical proof of the fundamental theorem of the calculus: that, where $F'(x) = G(x)$, then $\int G(x).dx = F(x)$. (Indeed James Gregory's reformulation of Neil's equivalent rectification procedure in proposition 6 of his *Geometriæ Pars Universalis* (Padua, 1668) has in recent years been presented in a similar capacity, and H. W. Turnbull in his printing of a revised form of the present proposition ([5] below) in his *Correspondence of Isaac Newton*, **2** (1960): 164–7, especially note (2), strongly emphasized this aspect.) Perhaps to our modern taste certain inadequacies mar Newton's proof, but they may be remedied by applying rigorously a full-blown exhaustion procedure. To that end it is necessary that the areas $(\rho\sigma hd) = (\omega naw q \,♉\, syx)$ be bounded by those of the inscribed and circumscribed rectilinea; that any interval $n\pi$, $\pi\rho$, $\rho\omega$, ωd be made arbitrarily small (by suitable subdivision of the curve $fm\mathrm{II}d$); and finally that the difference between the inscribed and circumscribed rectilinea (less than $n\psi \times \pi\rho$, where $\pi\rho$ is the greatest of the subintervals of nd) also be made arbitrarily small. Clearly all these conditions may be met on the present model. (Compare D. T. Whiteside, 'Patterns of Mathematical Thought in the later Seventeenth Century', *Archive for History of Exact Sciences*, **1** (1961): ch. IX: 331–48.)

(52) Read 'signes'.

(53) For the first time Newton begins to draft a set of tables of standard forms of integrals (or rather derivatives, since the integrals are defined merely as the inverse of the derivatives and the operation of differentiating, in the modified Cartesian form which requires prior finding of the subnormal, is clearly the more fundamental).

& soe infinitely.

$4x^3+3bx^2=bay.$ ———— $\frac{x^3}{a}+\frac{x^4}{ab}.$

$5x^4+3b^2x^2=bbay.$ ———— $\frac{x^3}{a}+\frac{x^5}{abb}.$

$6x^5+3b^3x^2=b^3ay.$ ———— $\frac{x^3}{a}+\frac{x^6}{ab^3}.$

$7x^6+3b^4x^2=b^4ay.$ ———— $\frac{x^3}{a}+\frac{x^7}{ab^4}.$

Soe y[t] y[e] nature of every crooked line, whose area is compounded of y[e] area of 2 or more of y[e] former lines, or of y[e] difference of y[e] area of 2 or more of y[e] former lines, is exprest by an equation compounded of y[e] equations expresing y[e] nature of those lines.

$4x^3-3bx^2=bay.$ ———— $\frac{x^4-x^3b}{ab}.$

$5x^4-3bbx^2=bbay.$ ———— $\frac{x^5-x^3bb}{abb}.$

$6x^5-3b^3x^2=b^3ay.$ ———— $\frac{x^6}{ab^3}-\frac{x^3}{a}.$

The squareing those lines whose area is exprest by an equation in w[ch] y[e] unknown quantity is denominator.

The equations expressing ye nature of y[e] line. The square whereof.(54)

$xxy=a^3.$ ———— $\frac{a^3}{x}.$

$x^3y=2a^4.$ ———— $\frac{a^4}{xx}.$

$x^4y=3a^5.$ ———— $\frac{a^5}{x^3}.$

(54) Newton has cancelled 'when x [is] greater then a. x lesse then a.' It seems that he intended to distinguish these two cases in parallel integral columns, but soon changed his mind. There seems little advantage in so doing.

(55) $$\int r\frac{a^{r+2}}{x^{r+1}}.dx = -\frac{a^{r+2}}{x^r} \quad (r = 1, 2, \ldots).$$

Here, as regularly below, Newton omits the necessary change of sign, and indeed it is difficult to control the sign of the derivative when one has each time first to calculate the subnormal.

$$x^5 y = 4a^6. \qquad \underline{\hspace{6cm}} \qquad \frac{a^6}{x^4}.$$

$$x^6 y = 5a^7. \qquad \underline{\hspace{6cm}} \qquad \frac{a^7}{x^5}. \text{(55)}$$

$$4yy = 9ax. \quad \text{Parab:} \quad \underline{\hspace{4cm}} \qquad x\sqrt{ax}.$$

$$4ayy = 25x^3. \qquad \underline{\hspace{4.5cm}} \qquad \frac{xx}{a}\sqrt{ax}.$$

$$4a^3 yy = 49x^5. \qquad \underline{\hspace{4.5cm}} \qquad \frac{x^3}{aa}\sqrt{ax}.$$

$$4a^5 yy = 81x^7. \qquad \underline{\hspace{4.5cm}} \qquad \frac{x^4}{a^3}\sqrt{ax}.$$

$$4a^7 yy = 121x^9. \qquad \underline{\hspace{4.5cm}} \qquad \frac{x^5}{a^4}\sqrt{ax}. \text{(56)}$$

$$4xyy = a^3. \qquad \underline{\hspace{4.5cm}} \qquad a\sqrt{ax}.$$

$$4x^3 y^2 = a^5. \qquad \underline{\hspace{4.5cm}} \qquad \frac{aa}{x}\sqrt{ax}.$$

$$4x^5 y^2 = 9a^7. \qquad \underline{\hspace{4.5cm}} \qquad \frac{a^3}{xx}\sqrt{ax}.$$

$$4x^7 y^2 = 25a^9. \qquad \underline{\hspace{4.5cm}} \qquad \frac{a^4}{x^3}\sqrt{ax}.$$

$$4x^9 y^2 = 49a^{11}. \qquad \underline{\hspace{4.5cm}} \qquad \frac{a^5}{x^4}\sqrt{ax}. \text{(57)}$$

$$9ax^2 + 12aax + 4a^3 = 4yyx + 4ayy. \qquad x\sqrt{ax+aa}.$$

$$25x^4 + 40ax^3 + 16aax^2 = 4axy^2 + 4aay^2. \qquad \frac{x^2}{a}\sqrt{ax+aa}.$$

$$49x^6 + 84ax^5 + 36a^2 x^4 = 4a^3 xy^2 + 4a^4 y^2. \qquad \frac{x^3}{aa}\sqrt{ax+aa}.$$

$$81x^8 + 144ax^7 + 64aax^6 = 4a^5 xy^2 + 4a^6 y^2. \qquad \frac{x^4}{a^3}\sqrt{ax+aa}. \text{(58)}$$

---

(56) $\qquad \int \left(\frac{2r+1}{2}\right) x \left(\frac{x}{a}\right)^{\frac{1}{2}(2r-3)} . dx = \frac{x^r}{a^{r-1}} \sqrt{(ax)} \quad (r = 1, 2, \ldots).$

(57) $\qquad -\int \left(\frac{2r-3}{2}\right) a \left(\frac{a}{x}\right)^{\frac{1}{2}(2r-1)} . dx = \frac{a^r}{x^{r-1}} \sqrt{(ax)} \quad (r = 1, 2, \ldots).$

(Newton artfully bypasses the difficulty of the sign change by squaring the derivative.)

(58) $\int y . dx = \frac{x^r}{a^{r-1}} \sqrt{[ax+a^2]} \quad$ where $\quad 2a^{r-\frac{3}{2}} y \sqrt{[x+a]} = x^{r-1}[(2r+1)x + 2ra].$

$$a^3=4xy^2+4ayy. \qquad a\sqrt{ax+aa}.$$

$$a^5x^2+4a^6x+4a^7=4x^5yy+4ax^4y^2. \qquad \frac{aa}{x}\sqrt{ax+aa}.$$

$$9a^7x^2+24a^8x+16a^9=4x^7y^2+4ax^6y^2. \qquad \frac{a^3}{xx}\sqrt{ax+aa}.$$

$$25a^9x^2+60a^{10}x+36a^{11}=4x^9y^2+4ax^8y^2. \qquad \frac{a^4}{x^3}\sqrt{ax+aa}.^{(59)}$$

$$9ax^2-12aax+4a^3=4xy^2-4ayy. \qquad x\sqrt{ax-aa}.$$

$$25x^4-40ax^3+16aax^2=4axy^2-4aayy. \qquad \frac{x^2}{a}\sqrt{ax-aa}.$$

$$a^3=4xy^2-4ay^2. \qquad a\sqrt{ax-aa}.$$

$$a^5x^2-4a^6x+4a^7=4x^5yy-4ax^4yy. \qquad \frac{aa}{x}\sqrt{ax-aa}.$$

$$9ax^2-12aax+4a^3=4ayy-4xyy. \qquad x\sqrt{aa-ax}.$$

$$25x^4-40ax^3+16aax^2=4a^2yy-4axyy. \qquad \frac{x^2}{a}\sqrt{aa-ax}.$$

$$a^3=4ay^2-4xy^2. \qquad a\sqrt{aa-ax}.$$

$$a^5x^2-4a^6x+4a^7=4ax^4yy-4x^5yy. \qquad \frac{aa}{x}\sqrt{aa-ax}.^{(60)}$$

Note y[t] the lines whose nature is exprest by y[e] 4 latter sets of equations are y[e] same wth the lines of y[e] 2 former sorts. Doubtfull.(61)

$$9aax+24axx+16x^3=4xyy+4ayy. \qquad x\sqrt{ax+xx}.$$

$$25aax^3+60ax^4+36x^5=4aaxyy+4a^3yy. \qquad \frac{xx}{a}\sqrt{ax+xx}.$$

$$49aax^5+112ax^6+64x^7=4a^4xyy+4a^5yy. \qquad \frac{x^3}{aa}\sqrt{ax+xx}.$$

$$81aax^7+180ax^8+100x^9=4a^6xyy+4a^7yy. \qquad \frac{x^4}{a^3}\sqrt{ax+xx}.$$

---

(59) $-\int y\,.\,dx = \frac{a^r}{x^{r-1}}\sqrt{[ax+a^2]}$ where $2x^ry\sqrt{[x+a]} = a^{r+\frac{1}{2}}[(2r-3)\,x+(2r-2)\,a]$.

(60) In general,

$$\int y\,.\,dx = \frac{x^r}{a^{r-1}}\sqrt{[\pm(ax-a^2)]} \quad \text{where} \quad 2a^{r-\frac{3}{2}}y\sqrt{[\pm(x-a)]} = \pm x^{r-1}[(2r+1)\,x-2ra],$$

and

$$-\int y\,.\,dx = \frac{a^r}{x^{r-1}}\sqrt{[\pm(ax-a^2)]} \quad \text{where} \quad 2x^ry\sqrt{[\pm(x-a)]} = \pm a^{r+\frac{1}{2}}[(2r-3)\,x-(2r-2)\,a].$$

(61) In fact, with a few sign changes the derivatives of $\frac{x^r}{a^{r-1}}\sqrt{[\pm(ax\pm a^2)]}$ are 'y[e] same', and so for $\frac{a^r}{x^{r-1}}\sqrt{[\pm(ax\pm a^2)]}$. They are, however, far from being identical.

$a^4+4a^3x+4aaxx=4axyy+4xxyy.$	$a\sqrt{ax+xx}.$
$a^6=4ax^3yy+4x^4yy.$	$\frac{aa}{x}\sqrt{ax+xx}.$
$9a^8+12a^7x+4a^6x^2=4ax^5yy+4x^6y^2.$	$\frac{a^3}{xx}\sqrt{ax+xx}.$
$25a^{10}+40a^9x+16a^8x^2=4ax^7yy+4x^8yy.$	$\frac{a^4}{x^3}\sqrt{ax+xx}.$
$9aax-24axx+16x^3=4ayy-4xyy.$	$x\sqrt{ax-xx}.$(62)
$4x^4+4aaxx+a^4=xxyy+aayy.$	$x\sqrt{aa+xx}.$
$9x^6+12aax^4+4a^4x^2=aaxxy^2+a^4y^2.$	$\frac{xx}{a}\sqrt{aa+xx}.$
$16x^8+24aax^6+9a^4x^4=a^4x^2y^2+a^6y^2.$	$\frac{x^3}{a^2}\sqrt{aa+xx}.$
$25x^{10}+40aax^8+16a^4x^6=a^6x^2y^2+a^8y^2.$	$\frac{x^4}{a^3}\sqrt{aa+xx}.$
$aaxx=aayy+xxyy.$	$a\sqrt{aa+xx}.$
$a^8=aax^4yy+x^6yy.$	$\frac{aa}{x}\sqrt{aa+xx}.$
$4a^{10}+4a^8xx+a^6x^4=aax^6yy+x^8yy.$	$\frac{a^3}{xx}\sqrt{aa+xx}.$
$9a^{12}+12a^{10}xx+4a^8x^4=aax^8y^2+x^{10}yy.$	$\frac{a^4}{x^3}\sqrt{aa+xx}.$
$4x^4-4aaxx+a^4=aayy-xxyy.$	$x\sqrt{aa-xx}.$
$aaxx=aayy-xxyy.$	$a\sqrt{aa-xx}.$
$a^8=aax^4yy-x^6yy.$	$\frac{aa\sqrt{aa-xx}}{x}.$
	$a\sqrt{\frac{a^3-axx}{x}}.$(63)

(62) $\int y.dx = \frac{x^r}{a^{r-1}}\sqrt{[ax\pm x^2]}$ where $2a^{r-1}y\sqrt{[ax\pm x^2]} = [(2r+1)\,ax\pm(2r+2)\,x^2]\,x^{r-1}$

and $-\int y.dx = \frac{a^r}{x^{r-1}}\sqrt{[ax+x^2]}$ where $2x^ry\sqrt{[ax+x^2]} = a^r[(2r-3)\,ax+2(r-2)\,x^2]$.

(63) $\int y.dx = \frac{x^r}{a^{r-1}}\sqrt{[a^2\pm x^2]}$ where $a^{r-1}y\sqrt{[a^2\pm x^2]} = x^{r-1}[ra^2\pm(r+1)\,x^2]$,

and $-\int y.dx = \frac{a^r}{x^{r-1}}\sqrt{[a^2\pm x^2]}$ where $x^ry\sqrt{[a^2\pm x^2]} = a^r[(r-1)\,a^2\pm(r-2)\,x^2]$.

(Note that $a\sqrt{\left[\frac{a^3-ax^2}{x}\right]} = \frac{a^{\frac{3}{2}}}{x^{\frac{1}{2}}}\sqrt{[a^2-x^2]}$.)

$$a^4+3a^2bx+4aax^2+\frac{9}{4}bbx^2+6bx^3+4x^4=\begin{matrix}a^2y^2.\\+bx\\+xx\end{matrix} \qquad x\sqrt{aa+bx+xx}.$$

$$aabb+4aabx+4aaxx=4ccyy+4bxy^2+4xxy^2. \qquad a\sqrt{cc+bx+xx}.$$

$$\sqrt{a^3+x^4}.$$

$$\sqrt{a^4+ax^3}.$$

$$9ax^4+6a^3x^2+a^5=4aaxyy+4x^3yy. \qquad \sqrt{a^3x+ax^3}.$$

$$\sqrt{a^4+x^4}.$$

$$\frac{a^3}{a+x}.$$

$$\frac{a^2x}{a+x}.$$

$$\frac{ax^2}{a+x}.$$

$$\frac{x^3}{a+x}.$$

$$\frac{a^3}{a-x}.$$

$$\frac{a^3}{x-a}.$$ (64)

$$a^3=bby+2bxy+xxy. \quad \text{———————} \quad \frac{a^3}{b+x}.$$

$$2b^4x+b^4a=x^4y+2ax^3y+aaxxy. \quad \text{———————} \quad \frac{b^4}{ax+x^2}.$$

$$3b^5x+2b^5a=x^5y+2ax^4y+a^2x^3y. \quad \text{———————} \quad \frac{b^5}{ax^2+x^3}.$$

$$4b^6x+3b^6a=x^6y+2ax^5y+a^2x^4y. \quad \text{———————} \quad \frac{b^6}{ax^3+x^4}.$$

---

(64) The corresponding entries for the derivative have been omitted—perhaps because they are contained in the more systematically arranged tables below?

(65) $-\int y.dx = \dfrac{b^{r+2}}{x^{r-1}(a+x)}$ where $x^r y(a+x)^2 = b^{r+2}[rx+(r-1)a]$,

and $\int y.dx = \dfrac{x^{r+2}}{a^{r-1}(b+x)}$ where $a^{r-1}y(b+x)^2 = x^{r+1}[(r+1)x+(r+2)b]$.

$$\frac{x^3}{a+x}.$$

$$\frac{x^4}{aa+ax}.$$

$$\frac{x^5}{a^3+xa^2}.$$

$$\frac{x^6}{a^4+xa^3}.$$

$aab=bby+2bxy+xxy.$ ———— $\dfrac{a^2x}{b+x}.$

$2bax+axx=bby+2bxy+xxy.$ ———— $\dfrac{ax^2}{b+x}.$(65)

$2a^4x=b^4y+2bbxxy+x^4y.$ ———— $\dfrac{a^4}{bb+x^2}.$

$$\frac{a^5}{bbx+x^3}.$$

$$\frac{a^6}{bbx^2+x^4}.$$

$$\frac{a^7}{bbx^3+x^5}.$$

$$\frac{x^4}{bb+x^2}.$$

$$\frac{x^5}{b^3+x^2b}.$$

$$\frac{x^6}{b^4+bbx^2}.$$

$\dfrac{x^7}{b^5+b^{[3]}x^2}.$(66)

$$\frac{a^3x}{bb+x^2}.$$

$$\frac{aaxx}{bb+xx}.$$

$$\frac{ax^3}{bb+xx}.$$

(66) $-\int y.dx = \dfrac{a^{r+3}}{x^{r-1}(b^2+x^2)}$ where $x^ry(b^2+x^2)^2 = a^{r+3}[(r+1)x^2+(r-1)b^2]$,
and similarly (though Newton makes no entries in the derivative column)

$$\int y.dx = \frac{x^{r+3}}{a^{r-1}(b^2+x^2)} \text{ where } a^{r-1}y(b^2+x^2)^2 = x^{r+2}[(r+1)x^2+(r+3)b^2].$$

$27x^2y^3=a^5.$ ———————— $a\sqrt{c:aax}.$

$27x^5y^3=8a^8.$ ———————— $\dfrac{aa\sqrt{c:aaax}}{x}.$

$27x^8y^3=125a^{11}.$ ———————— $\dfrac{a^3}{x^2}\sqrt{c:aax}.$

$27x^{11}y^3=512a^{14}.$ ———————— $\dfrac{a^4}{x^3}\sqrt{c:aax}.$

$27xy^3=8a^4.$ ———————— $a\sqrt{c:axx}.$

$27x^4y^3=a^7.$ ———————— $\dfrac{aa}{x}\sqrt{c:axx}.$

$27x^7y^3=64a^{10}.$ ———————— $\dfrac{a^3}{xx}\sqrt{c:axx}.$

$27x^{10}y^3=343a^{13}.$ ———————— $\dfrac{a^4}{x^3}\sqrt{c:axx}.$(67)

$27y^3=64aax.$ ———————— $x\sqrt{c:aax}.$

$27ay^3=343x^4.$ ———————— $\dfrac{xx}{a}\sqrt{c:aax}.$

$27a^4y^3=1000x^7.$ ———————— $\dfrac{x^3}{aa}\sqrt{c:aax}.$

$27y^3=125ax^2.$ ———————— $x\sqrt{c:axx}.$

$27a^2y^3=512x^5.$ ———————— $\dfrac{xx}{a}\sqrt{c:axx}.$

$27a^5y^3=1331x^8.$ ———————— $\dfrac{x^3}{aa}\sqrt{c:axx}.$(68)

[. . .](69)

---

(67) $-\int y.dx = \dfrac{a^r}{x^{r-1}}(a^2x)^{\frac{1}{3}}$ where $(3y)^3x^{3r-1} = (3r-4)^3a^{3r+2}$,

and $-\int y.dx = \dfrac{a^r}{x^{r-1}}(ax^2)^{\frac{1}{3}}$ where $(3y)^3x^{3r-2} = (3r-5)^3a^{3r+1}$.

(68) $\int y.dx = \dfrac{x^r}{a^{r-1}}(a^2x)^{\frac{1}{3}}$ where $(3y)^3a^{3r-5} = (3r+1)^3x^{3r-2}$,

and $\int y.dx = \dfrac{x^r}{a^{r-1}}(ax^2)^{\frac{1}{3}}$ where $(3y)^3a^{3r-4} = (3r+2)^3x^{3r-1}$.

(69) In a similar way Newton here enters on the right the integrals $\dfrac{a^{r+4}}{x^{r-1}(b^3+x^3)}$ and $\dfrac{x^{r+4}}{a^{r-1}(b^3+x^3)}$ $(r = 1, 2, 3, 4)$ together with $\dfrac{a^sx^{5-s}}{b^3+x^3}$ $(s = 1, 2, 3, 4)$, but gives no entry in the corresponding column of derivatives.

$256x^3y^4=a^7.$	————————	$a\sqrt{qq:aaax}.$
$256x^7y^4=81a^{11}.$	————————	$\frac{aa}{x}\sqrt{qq:aaax}.$
$256x^{11}y^4=2401a^{15}.$	————————	$\frac{aaa}{[xx]}\sqrt{qq:aaax}.$
$256xy^4=81a^5.$	————————	$a\sqrt{qq:ax^3}.$
$256x^5y^4=a^9.$	————————	$\frac{aa}{x}\sqrt{qq:ax^3}.$
$256x^9y^4=625a^{13}.$	————————	$\frac{a^3}{xx}\sqrt{qq:ax^3}.$
$256x^{13}y^4=6561a^{17}.$	————————	$\frac{a^4}{x^3}\sqrt{qq:ax^3}.$(70)

---

[5](71) *A Method whereby to square such crooked lines as may be squared.*

If y^e crooked lines σha & $a\vartheta\theta$ are of such a nature that (supposeing (gh) parallell to (qa), & (bh) perpendic: to σha & (an) a given line) $gh:bg::an:ge$.(72) Then y^e area $(age^{(73)})=(qlna)$ y^e rectangle made by (an) & (gh).

Demonstration.

Suppose σi, id, de, &c; are tangents of σha,(74) from whose intersections or ends are drawne ec, df, iz, σw,(75) &c & from whose touch points are drawne $\beta\theta$, $h\vartheta$, $\lambda\mu$, &c: all parallell to av. From y^e said intersections draw sw, ik, dm, es, &c. parallel to bn. Since $gh:bg::\rho d:i\rho::an:ge$.(76) $\rho d\times ge^{(76)}=ir^{(77)}\times an$. that is $\square pkmt=\square u\tau fs$. by y^e same reason $[\square]tmso=[\square]\tau rcy$; & $[\square]upkw^{(78)}=[\square]\zeta azx$. &c: Thus also it may be prooved y^t y^e $\square vwna$ is equall to any number of such like $\square$s inscribed twixt y^e line $\zeta\omega$ & y^e point a,(79) wch if they be infinite are equall to [y^e] superficies $\zeta a\omega=vwna$. also $g\pi\mu\vartheta=ql\text{II}\Omega$. &c.

---

(70) $$-\int y.dx=\frac{a^r}{x^{r-1}}(a^3x)^{\frac{1}{4}}\quad\text{where}\quad(4y)^4x^{4r-1}=(4r-5)^4a^{4r+3},$$

and $$-\int y.dx=\frac{a^r}{x^{r-1}}(ax^3)^{\frac{1}{4}}\quad\text{where}\quad(4y)^4x^{4r-3}=(4r-7)^4a^{4r+1}.$$

(71) Add 4000: 134^v–136^r, partially printed in *Correspondence of Isaac Newton*, 2 (1960): no. 190: 164–7. (The diagrams there reproduced are, however, incorrect in several important respects, and unnecessary transliterations have been carried through in the text.)

(72) Read '$g\theta$'. (73) Read '$ag\theta$'.

(74) Here and above Newton misreads his diagram, for the point σ is not on the curve. Read correctly, perhaps, 'βha'. (The line ea is tangent to the curve βha at a.)

(75) Read '$\sigma\omega$'.

(76) Read '$g\theta$'. (77) Read '$i\rho$'. (78) Read '$vpkw$'.

(79) At infinity in Newton's second figure.

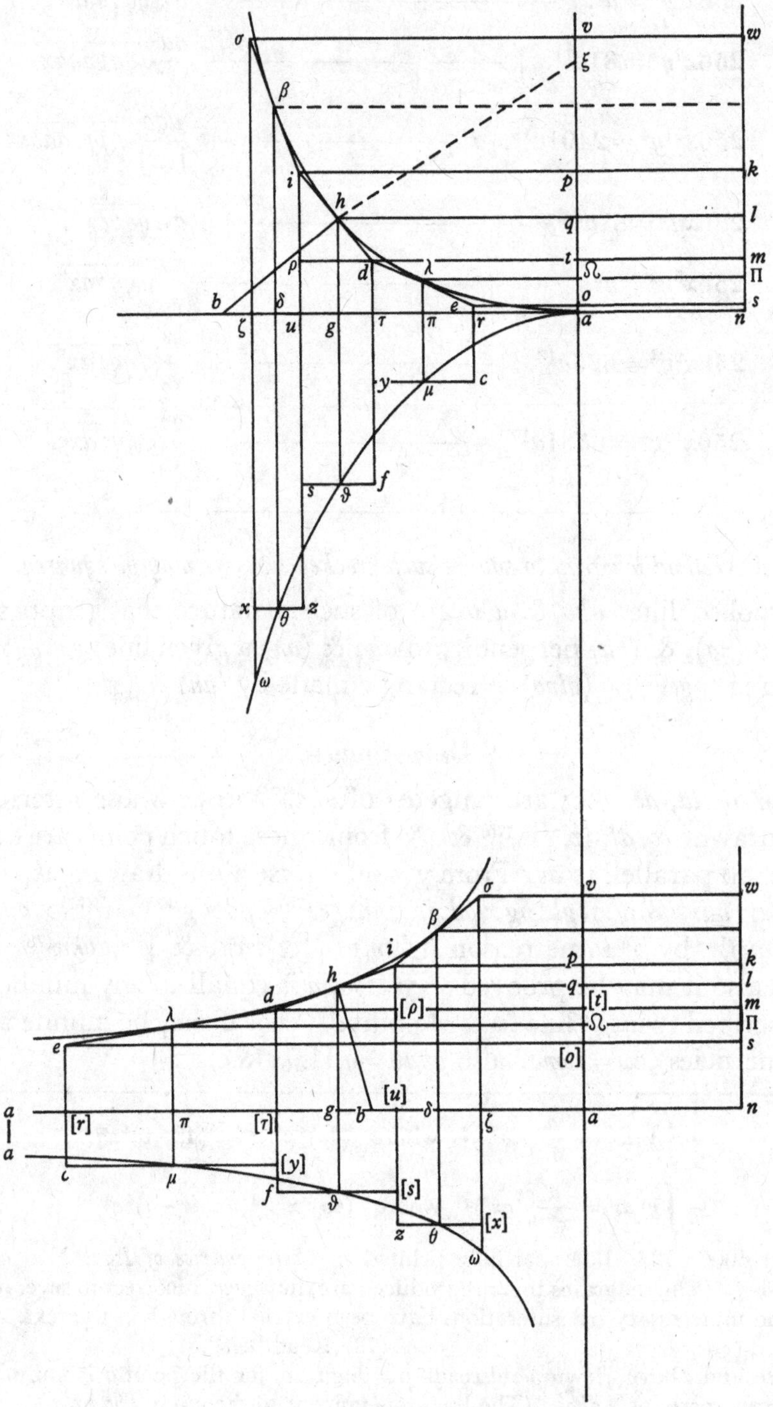
σ
v
w
ξ
β
i
p
k
h
q
l
ρ
d
λ
t
m
Ω
Π
b
e
o
s
ζ
δ
u
g
τ
π
r
a
n
y
μ
c
f
s
ϑ
x
θ
z
ω
σ
v
w
β
i
p
k
h
q
l
d
[ρ]
[t]
m
λ
Ω
Π
e
s
[o]
[u]
a
[r]
π
[τ]
g
b
δ
ζ
a
a
c
μ
[y]
f
ϑ
[s]
z
θ
[x]
ω

Prop 1[80]

To find y^e^ line whose area is exprest by any given equation. Suppose y^e^ equatiō is $\frac{x^3}{a}$.[81] nameing y^e^ quantitys $a=an$. $x=ag$. $\frac{x^3}{a}=qlna=g\vartheta a$. $gh=qa=\frac{x^3}{aa}$. $bh=s$. $ba=v$.[82]

$ss-vv+2vx-xx=\frac{x^6}{a^4}$. wch equacō hath 2 equall rootes & is therefore multiplied
0 0 1 2 6 according to Huddenius his Meth: $vx=x^3+\frac{3x^6}{a^4}$.

$gb=v-x=\frac{3x^5}{a^4}$. Wherefore if $\frac{x^3}{aa}:\frac{3x^5}{a^4}::a:\frac{3xx}{a}=g[\vartheta]$.[83] therefore $a\vartheta\omega$ is a Parab: & $ag[\vartheta]$[83] $=\frac{x^3}{a}=qlna$.

Also if y^e^ Equation be $\frac{a^3}{x}$. Then makeing $a=an$. $x=ag$. $\frac{a^3}{x}=qlna=g[\vartheta]a$.[83] $qa=\frac{aa}{x}=gh$. $bh=s$. $ba=v$.

$ss-vv+2vx-xx=\frac{a^4}{x[x]}$. wch multiplied by Huddenius his method by reasō
0 0 −1 −2 +2 of 2 equall rootes [is] $-2vx+2xx=\frac{2a^4}{xx}$. $v=x-\frac{a^4}{x^3}$.

$x-v=gb=\frac{a^4}{x^3}$. Lastly $\frac{aa}{x}:\frac{a^4}{x^3}::a:\frac{a^3}{xx}$. $\frac{a^3}{xx}$[84] $=g[\vartheta]$[83] $=y$. & $a^3=xxy$. wch last equation expresseth y^e^ nature of y^e^ line $a\vartheta o$,[85] whose surface $a[\vartheta]g$[83] $=qlan=\frac{a^3}{x}$.

Note [y^t^]. I call y^t^ line (x) to wch both y^e^ lines σha[86] & $a\vartheta\omega$ have respect as πa, ga, &c. but y^t^ line to wch but one line hath respect I call (y) as $g\vartheta$, $\pi\mu$ [&c]: or (z) as gh, $\pi\lambda$, &c.

If $ax^m=by^n$. (m & n being numbers y^t^ signifie y^e^ dimensions of x & y), then $\frac{nxy}{n+m}=ag\vartheta$, y^e^ area of y^e^ line $a\mu\vartheta$. And if $a=b\times x^m\times y^n$. y^n^ is $\frac{nxy}{n-m}=ag\vartheta$ y^e^ area of y^t^ line.[87]

---

(80) This proof, likewise modelled on that of Heuraet's rectification method (note (51)), follows much as before in [4].

(81) Which Newton will suppose equal to $ay = \int z.dx$.

(82) That is bh is the normal and $bg = v-x$ the subnormal at h.

(83) At each of these places Newton wrote 'e' instead of the correct 'ϑ'.

(84) More correctly, $-a^3/xx$.

(85) Read '$a\theta\omega$'.

(86) Read 'βha'.

(87) Where $ax^{\pm m} = by^n$, then $\int y.dx = \frac{n}{n\pm m}xy$.

The squareing of y^e simplest lines in wch y is but of one dimension.

Equations expressing y^e nature of y^e lines.		Theire squares.	Lines.		□.(88)
$3xx = ay$. Parab:	————	$\frac{x^3}{a}$.	$xxy = a^3$.	————	$\frac{a^3}{x}$.
$4x^3 = aay$.	————	$\frac{x^4}{aa}$.	$x^3y = 2a^4$.	————	$\frac{a^4}{xx}$.
$5x^4 = ya^3$.	————	$\frac{x^5}{a^3}$.	$x^4y = 3a^5$.	————	$\frac{a^5}{x^3}$.
$6x^5 = ya^4$.	————	$\frac{x^6}{a^4}$.	$x^5y = 4a^6$.	————	$\frac{a^6}{x^{[4]}}$.
		&c.			&c.(89)

The square of y^e Simplest lines in wch y is of 2 dimensions.

The lines [to be] squared.		Theire Squares.	Lines [to be] squared.		□s.
$4yy = 9ax$. Parab:	——	$x\sqrt{ax}$.	$4xyy = a^3$.	————	$a\sqrt{ax}$.
$4ayy = 25x^3$.	————	$\frac{xx}{a}\sqrt{ax}$.	$4x^3yy = a^5$.	————	$\frac{aa}{x}\sqrt{ax}$.
$4a^3yy = 49x^5$.	————	$\frac{x^3}{aa}\sqrt{ax}$.	$4x^5yy = 9a^7$.	————	$\frac{a^3}{xx}\sqrt{ax}$.
$4a^5yy = 81x^7$.	————	$\frac{x^4}{a^3}\sqrt{ax}$.	$4x^7yy = 25a^9$.	————	$\frac{a^4}{x^{[3]}}\sqrt{ax}$.
		&c.			&c.(90)

(88) Read 'Squares'.

(89) $$\int y\,.\,dx = \frac{x^{r+2}}{a^r} \quad \text{where} \quad a^r y = (r+2)\,x^{r+1},$$

and $$-\int y\,.\,dx = \frac{a^{r+2}}{x^r} \quad \text{where} \quad x^{r+1}y = ra^{r+2} \quad (r = 1, 2, 3, 4).$$

(90) $$\int y\,.\,dx = \frac{x^r}{a^{r-1}}\sqrt{[ax]} \quad \text{where} \quad 2a^{r-\frac{3}{2}}y = (2r+1)\,x^{r-\frac{1}{2}},$$

and $$-\int y\,.\,dx = \frac{a^r}{x^{r-1}}\sqrt{[ax]} \quad \text{where} \quad 2x^{r-\frac{1}{2}}y = (2r-3)\,a^{r+\frac{1}{2}} \quad (r = 1, 2, 3, 4).$$

The square of those lines where y is of 3 *dimensions onely.*

The lines squared.	Their squares.	Lines squared.	Theire Squares.
$27xy^3=8a^4$. ———	$a\sqrt{c:axx}$.	$27y^3=64aax$. ———	$x\sqrt{c:aax}$.
$27x^4y^3=a^7$. ———	$\frac{aa}{x}\sqrt{c:axx}$.	$27ay^3=343x^4$. ———	$\frac{xx}{a}\sqrt{c:aax}$.
$27x^7y^3=64a^{10}$. ———	$\frac{a^3}{xx}\sqrt{c:axx}$.	$27a^4y^3=1000x^7$. ———	$\frac{x^3}{aa}\sqrt{c:aax}$.
$27x^{10}y^3=343a^{13}$. ———	$\frac{a^4}{x^3}\sqrt{c:axx}$.	$27a^7y^3=2197x^{10}$. ———	$\frac{x^4}{a^3}\sqrt{c:aax}$.
	&c.		&c.
$27xxy^3=a^5$. ———	$a\sqrt{c:aax}$.	$27y^3=125ax^2$. ———	$x\sqrt{c:axx}$.
$27x^5y^3=8a^8$. ———	$\frac{aa}{x}\sqrt{c:aax}$.	$27aay^3=512x^5$. ———	$\frac{xx}{a}\sqrt{c:axx}$.
$27x^8y^3=125a^{11}$. ———	$\frac{a^3}{xx}\sqrt{c:aax}$.	$27a^5y^3=1331x^8$. ———	$\frac{x^3}{aa}\sqrt{c:axx}$.
$27x^{11}y^3=512a^{14}$. ———	$\frac{a^4}{x^3}\sqrt{c:aax}$.	$27a^8y^3=2744x^{11}$. ———	$\frac{x^4}{a^3}\sqrt{c:axx}$.
	&c.		&c.(91)

(91) $$-\int y.dx = \frac{a^r}{x^{r-1}}(ax^2)^{\frac{1}{3}} \quad \text{where} \quad (3y)^3x^{3r-2} = (3r-5)^3a^{3r+1},$$

$$\int y.dx = \frac{x^r}{a^{r-1}}(a^2x)^{\frac{1}{3}} \quad \text{where} \quad (3y)^3a^{3r-5} = (3r+1)^3x^{3r-2},$$

$$-\int y.dx = \frac{a^r}{x^{r-1}}(a^2x)^{\frac{1}{3}} \quad \text{where} \quad (3y)^3x^{3r-1} = (3r-4)^3a^{3r+2},$$

and finally

$$\int y.dx = \frac{x^r}{a^{r-1}}(ax^2)^{\frac{1}{3}} \quad \text{where} \quad (3y)^3a^{3r-4} = (3r+2)^3x^{3r-1} \quad (r = 1, 2, 3, 4).$$

[6][92]

A Method whereby to find ye areas of Those Lines wch can bee squared.

Prop: 1st. If $ab=x\perp y=be$. $cb=z$. $bd=v$. secant[93]$=cd$. m & n are numbers expressing ye dimensions of x, y, or z. a, b, c, d, &c. are knowne quantitys, & $\dfrac{ax^{\frac{m}{n}}}{b}=z$. yn $\dfrac{maz x^{\frac{m}{n}}}{nbx}=\dfrac{maz x^{\frac{m-n}{n}}}{nb}=v$. And in generall what ever ye relation twixt x & z bee, make all ye termes equall to nothing, multiply each terme by so many times zz as x hath dimensions in yt terme, for a Numerator: yn multiply each terme by soe many times $-x$ as z hath dimensions in yt terme for a denominator in ye valor of v.[94]

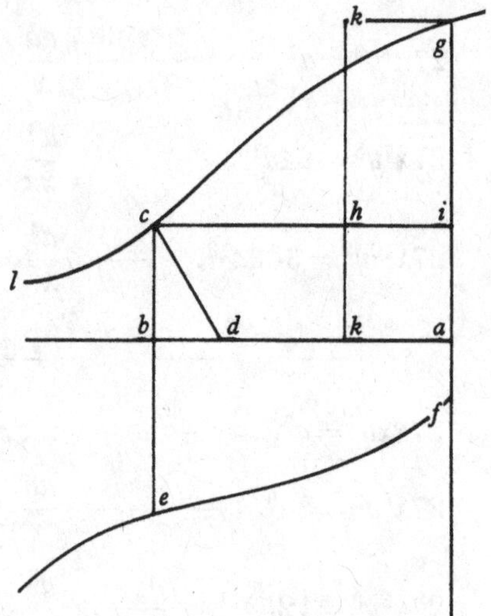

Prop: 2d. If $hi=r$. & $rv=zy$.[95] yn hi & be describe equall spaces $higk$, or $hiak$[,][96] & $abef$. that is $abef=aikh$.

Prop: 3d. If $a^n x^m=b^n y^n$. Or $\dfrac{ax^{\frac{m}{n}}}{b}=y$. yn is

$$\frac{nxy}{n+m}=\frac{n\times a\times x^{\frac{m+n}{n}}}{nb+mb}=abef^{(97)}$$

ye area of ye line $abef$. And if $\dfrac{a}{bx^{\frac{m}{n}}}=y$: yn is

$$\frac{nxy}{n-m}=\frac{na}{\overline{n-m}\times bx^{\frac{-n+m}{n}}}=abef^{(97)}$$

$$=\frac{na}{\overline{n-m}\times bx^{\frac{m-n}{n}}}.^{(98)}$$

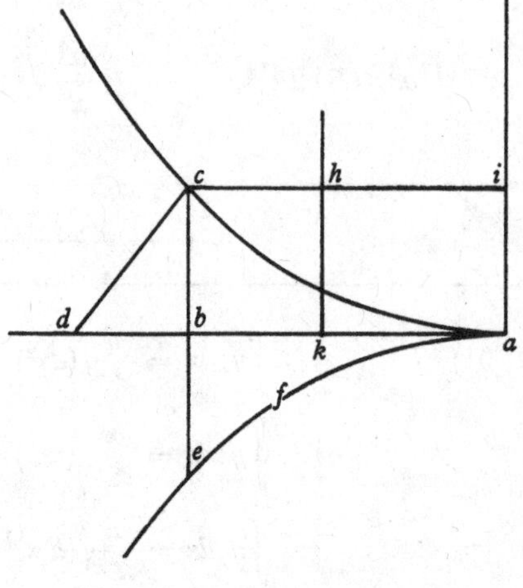

(92) Add. 4000: 147^{r}, 148^{r}, 149^{r}, printed in the *Correspondence of Isaac Newton*, **2** (1960): no. 191: 168–71. Newton revises his previous drafts and presents a finished version, in eight propositions, of his approach to the derivative by the subnormal. (Note that the first accompanying figure has two points k.)

(93) That is the normal to the curve at c.

(94) Newton inserts his general algorithm for evaluating the subnormal $v = z(dz/dx)$, at an arbitrary point of the curve whose Cartesian defining equation is

$$0=\sum_{i,j}(a_{i,j}x^i z^j),\quad \text{in the form}\quad v=\frac{\sum_{i,j}(iz^2.a_{i,j}x^i z^j)}{\sum_{i,j}(-jx.a_{i,j}x^i z^j)}.$$

(Compare 2, historical note and §2; and 4, §3.1 above.)

Demonstracon.

For Suppose *akhi* is a parallelogram & equall to $\dfrac{nax^{\frac{m+n}{n}}}{nb+mb}$. y^{n} is $\dfrac{nax^{\frac{m+n}{n}}}{nbr+mbr}=ai=z$.

& (prop 1) $\dfrac{az^2x^{\frac{m+n}{n}}}{brxz}=\dfrac{azx^{\frac{m}{n}}}{br}=v$. & (prop 2^{d}) $rv=zy$; y^{t} is $\dfrac{ax^{\frac{m}{n}}}{b}=y$.

Prop: 4th. If $y=ax^m+bx^n$. y^{n} is $\dfrac{ax^{m+1}}{m+1}+\dfrac{bx^{n+1}}{n+1}=abef$.

And in generall if y^{e} valor of y consists of severall termes so y^{t} x is not of divers dimensions in y^{e} denominator of any terme, y^{n} multiply each terme by x & divide it by y^{e} number of y^{e} dimensions of x,(99) all those products shall bee y^{e} area of y^{e} given line:(100) supposeing also y^{t} either none or all ye signes of those termes are changed by this operation. For if some bee changed & others bee not they pro[c]eed divers ways & joyne not & y^{n} y^{e} quantitys y or x must bee increased or diminished or otherwis altered.(101)

The reason of this prop: is, y^{t} y^{e} area described by y is also described by its parts[,] y^{t} is by y^{e} termes of its valor,(102) & w^{t} areas those termes describe appeares by prop 3^{d}.

(95) That is, if $be:bd = hi:bc$ or $y = r(dz/dx)$ (with $r = 1$ in Prop: 1st).

(96) In the upper and lower figures respectively.

(97) That is, $\int y.dx$.

(98) Together these two results state the fundamental integration rule that

$$\int \alpha x^p.dx = \frac{\alpha}{p+1}\,x^{p+1}$$

(with p, implicitly, not -1).

(99) That is, by the new index of x (increased by unity from the old).

(100) Or, where $\quad y = \sum_i (a_i x^i), \quad \int y.dx = \sum_i \left(\dfrac{a_i x^i.x}{i+1}\right)$.

(101) Newton here seems to require that one of the integral bounds be 0 or ∞ and the other x (freely variable). Then, since integration maintains or changes the sign of a term according as the dimension of the bound variable is positive or negative, such an expression as $\int_0^x (\alpha x^i+\beta x^j).dx$ (where i and j are of different sign) will usually be infinite: for if

$$\lim_{x\to 0}\left(\frac{1}{i+1}\,x^{i+1}\right)$$

is finite it must be zero and so $i+1$ positive, but correspondingly

$$\lim_{x\to 0}\left(\frac{1}{j+1}\,x^{j+1}\right)$$

is infinite unless $-1 < j \leqslant 0$ when $1 > i \geqslant 0$.

(102) Or $\quad \int \sum_i (a_i x^i).dx = \sum_i \left(\int a_i x^i.dx\right)$.

Prop 5[t]. The progressions in this Table may bee designed by these geomet[l]: lines. Whereby also any intermediate termes may bee found.

		a	b	b			b		b		b	
1.	1.	1.	1.	1.	1.	1.	1.	1.	1.	1.	1.	$\sim 1 = y$.
−2.	−1.	0.	1.	2.	3.	4.	5.	6.	7.	8.	9.	$\sim x = y$.
3.	1.	0.	0.	1.	3.	6.	10.	15.	21.	28.	36.	$\sim xx - x = 2y$.
−4.	−1.	0.	0.	0.	1.	4.	10.	20.	35.	56.	84.	$\sim x^3 - 3xx + 2x = 6y$.
5.	1.	0.	0.	0.	0.	1.	5.	15.	35.	70.	126.	$\sim x^4 - 6x^3 + 11xx - 6x = 24y$.
−6.	−1.	0.	0.	0.	0.	0.	1.	6.	21.	56.	126.	$\sim x^5 - 10x^4 + 35x^3 - 50xx + 24x = 120y$.
7.	1.	0.	0.	0.	0.	0.	0.	1.	7.	28.	84.	$\sim x^6 - 15x^5 + 85x^4 - 225x^3 + 274xx - 120x = 720y$.
−8.	−1.	0.	0.	0.	0.	0.	0.	0.	1.	8.	36.	$\sim x^7 - 21x^6 + 175x^5 - 735x^4 + \&c: = 5040y$.

The distance of y[e] terme *b* from y[e] terme *a* being called *x*. & the quantity of y[t] terme being *y*. & each terme being distant an unit from y[e] next.(103) The nature of w[ch] table is such y[t] y[e] sume of any figure & y[e] figure above it is equall to y[e] figure after it.(104) & y[e] nature of y[e] lines are such y[t] any figure, multiplyed by y[e] number of dimensions of *x* in y[e] first terme, being substracted from y[e] figure following it, is equall to y[e] figure under y[t] following figure.(105) And y[t] y[e] numbers of *y* may be deduced hence $1 \times 2 \times 3 \times 4 \times 5 \times 6 \times 7$ &c.(106)

Prop 6[t]. If $\frac{m}{n} = x$. This Progression

$$\frac{n \times m \times \overline{m-n} \times \overline{m-2n} \times \overline{m-3n} \times \overline{m-4n} \times \overline{m-5n}}{n \times n \times 2n \times 3n \times 4n \times 5n \times 6n} \quad \&c$$

gives all y[e] quantitys downward, in y[e] preceding table.(107) As if $m = 3$. $n = 1$. the quantitys downward are

$$\frac{1}{1}.\ \frac{m}{n}.\ \frac{m \times \overline{m-n}}{n \times 2n}.\ \frac{m \times \overline{m-n} \times \overline{m-2n}}{n \times 2n \times 3n}.\ \frac{m \times \overline{m-n} \times \overline{m-2n} \times \overline{m-3n}}{n \times 2n \times 3n \times 4n}.\ \&c.$$

y[t] is 1. 3. 3. 1. &c. so if $\frac{m}{n} = \frac{1}{2} = x$. y[n] 1. $\frac{1}{2}$. $\frac{-1}{8}$. $\frac{1}{16}$. $\frac{-5}{128}$. $\frac{7}{256}$. &c, are ye terms downward.(108)

(103) Newton tabulates $y = \binom{x}{r}$, $x = -2, -1, 0, 1, \ldots, 9$; $r = 0, 1, \ldots, 7$.

(104) Since $$\binom{x}{r} + \binom{x}{r-1} = \binom{x+1}{r}.$$

(105) Where $(x)_n = x(x-1)(x-2)\ldots(x-n+1) = \sum_{1 \leqslant i \leqslant n} (a_{n,i}x^i)$ (with the coefficients $a_{n,i}$ 'Stirling' numbers of first order), we derive the recursive scheme $(x-n)(x)_n = (x)_{n+1}$ and so, comparing coefficients, $a_{n,i-1} - na_{n,i} = a_{n+1,i}$.

(106) $(x)_n = n!y$ since $y = \binom{x}{n}$.

(107) Newton expands the general binomial coefficient $\binom{m/n}{r}$ as a product.

Prop 7th.

$$\overline{a+b}|^{\frac{m}{n}} = a^{\frac{m}{n}} + \frac{m}{n} \times \frac{b}{a} \times a^{\frac{m}{n}} + \frac{m}{n} \times \frac{m-n}{2n} \times \frac{bb}{aa} \times a^{\frac{m}{n}} + \frac{m}{n} \times \frac{m-n}{2n} \times \frac{m-2n}{3n} \times \frac{b^3}{a^3} \times a^{\frac{m}{n}}.\ \&c.$$

As may bee deduced from

$$a^{\frac{m}{n}} \times \frac{mb}{na} \times \frac{\overline{m-n}\,b}{2na} \times \frac{\overline{m-2n}\,b}{3na} \times \frac{\overline{m-3n}\,b}{4na} \times \frac{\overline{m-4n}\,b}{5na}\quad \&c.$$

The truth of this Prop: appeareth by compareing it wth ye two former as also by calculation if $\frac{m}{n}$ is a whole & affirmative number, or b lesse yn a.

Prop 8th.

$$\frac{1}{\overline{a+b}|^{\frac{m}{n}}} = \frac{1}{a^{\frac{m}{n}}} - \frac{m}{n} \times \frac{b}{a} \times \frac{1}{a^{\frac{m}{n}}} - \frac{m}{n} \times \frac{-m-n}{2n} \times \frac{bb}{aa} \times \frac{1}{a^{\frac{m}{n}}} - \frac{m}{n} \times \frac{-m-n}{2n} \times \frac{-m-2n}{3n} \times \frac{b^3}{a^3} \times \frac{1}{a^{\frac{m}{n}}}$$

&c. As may bee deduced from

$$\frac{1}{a^{\frac{m}{n}}} \times \frac{-mb}{na} \times \frac{-mb-nb}{2na} \times \frac{-mb-2nb}{3na} \times \frac{-mb-3nb}{4na}\quad \&c.$$

The truth of this appeares also by ye 5t & 6t propositions, or by calculation If $a \sqsubset b$.(109)

The truth of these two prop: is also thus demonstrated. If $\overline{a+b}|^{\frac{\bar{1}}{1}}$(110) $= \frac{1}{a+b}$ I divide 1 by $a+b$ as in decimall fractions & find ye quote $\frac{1}{a} - \frac{b}{aa} + \frac{bb}{a^3} - \frac{b^3}{a^4} + \frac{b^4}{a^5}$ &c. as appeareth also by multiplying both pts by $a+b$. So I extract ye rote of a^2+b as if they were decimall numbers & find $\sqrt{a^2+b} = a + \frac{b}{2a} - \frac{bb}{8a^3} + \frac{b^3}{16a^5}$ &c, as also may appeare by squareing both pts.(111)

---

(108) Propositions 4–6 are summarized versions of his first investigations of the binomial expansions in the winter of 1664/5. (See **1**, 3, §3.2/3.)

(109) Read 'If a is greater yn b', a crucial restriction. The symbol $\sqsubset$ is a slight modification of Oughtred's, but Newton may have become familiar with it indirectly, through his reading of Barrow's edition of Euclid, where it is extravagantly used.

(110) That is, '$\overline{a+b}|^{\frac{-1}{1}}$'.

(111) This seems Newton's first completely general statement of the binomial expansion: in modern terms

$$(a+b)^{m/n} = \lim_{j\to\infty}\left(\sum_{0\leqslant i\leqslant j}\left[a^{\frac{m}{n}}\binom{m/n}{i}\left(\frac{b}{a}\right)^i\right]\right), \quad \text{where} \quad \left|\frac{b}{a}\right| < 1.$$

§2.(1) ALGORITHMS FOR CALCULATING THE SUBNORMAL AND ATTEMPTS AT INVERTING THE PROCEDURE

[Autumn 1665?]

[1](2)

$$\frac{x^3}{aa}=y.\ \frac{x^6}{a^4}=yy.\ \frac{3x^5}{a^4}=v.$$

$$\frac{x^6}{a^4}=yy:\frac{a^4x^6+9x^{10}}{a^8}=ss::aa:\frac{a^4+9x^4}{aa}=mi^2:\text{whose } v=\frac{18x^3}{aa}.^{(3)}$$

$$\frac{mi^2}{ge^2}:aa^{(4)}::bg^2:gh^2::a^4+9x^4:a^4::xx:\frac{a^4xx}{a^4+9x^4}=q\xi^2.$$

$$\frac{a^6x+9aax^5 \text{ in } \sqrt{a^4+9x^4}}{a^8+18a^4x^4+81x^8}=q\xi.^{(5)}$$

[2](6)

$ec=x.\ de=v.\ ea=y.\ da=db=s.\ fb=y+a.$(7) [1st.] $\frac{yy}{r}=x.\ fc=\frac{yy+2ay+aa}{r}.$

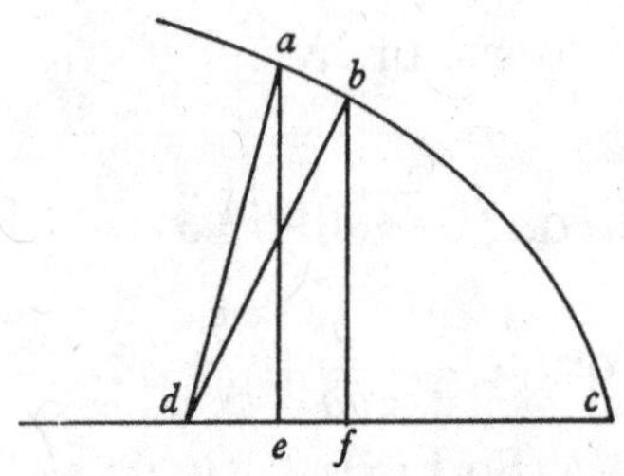

$$ef=\frac{-2ay-aa}{r}.\ df=v\frac{-2ay-aa}{r}.$$

$$vv+yy=ss=yy+2ay+aa+vv\frac{-4vay-2aav}{r}$$

$$\frac{+4aayy+4a^3y+a^4}{rr}.$$

$$a=0.\ -2ay+4\frac{avy}{r}=0.\ v=\frac{2ayr}{4ay}=\frac{r}{2}.$$

[2dly.] $\frac{r^3+y^3}{rr}=x.\ vv+yy=ss=yy+2ay$ &c: $+vv-\frac{6vyya}{rr}$ &c. $\frac{2ay}{6\frac{ayy}{rr}}=v=\frac{rr}{3y}.$(8)

(1) Add. 3960. 12: 206, 199–202, partially described by Alexander Witting in 'Zur Frage der Erfindung des Algorithmus der Newtonschen Fluxionsrechnung', *Bibliotheca Mathematica* $_3$**12** (1911–12): 56–60. On the wrapper of the manuscript Brewster in 1855 added the judicious description: 'An early paper on deducing the subnormal in a curve from equ^{ons} connecting x & y rationally, & the converse operation'.

(2) Extracts from a preliminary draft at the foot of f. 206.

(3) As before $v = y(dy/dx) = \frac{1}{2}(d/dx)(y^2)$ with $s = \sqrt{[y^2+v^2]} = y(ds/dx)$. It follows that $mi = a(s/y) = a(ds/dx)$, 'whose v' is the subnormal of the curve whose ordinate is $a(ds/dx)$, or

$$\tfrac{1}{2}(d/dx)\,(mi^2) = \tfrac{1}{2}(d/dx)\,[a^2(1+(dy/dx)^2)].$$

(4) Read '$mi^2:ge^2=aa$'.

Rule 1st. having ye valor of x in whose $\overline{\text{ratio}}$ Denom(9) y is not unless it have but one terme, to finde v: Make yy ye Numerator & ye valor of x multiplied according to ye dimens: of x in each terme, ye denom: in v's valor.(10)

As in this $\frac{yy}{r}=x.\ \frac{yy}{2\frac{yy}{r}}=v=\frac{r}{2}$.

So in this $\frac{r^3+y^3}{rr}=x.\ \frac{yy}{\frac{0r^3+3y^3}{rr}}=v=\frac{rr}{3y}$.

1st. $\frac{yy}{r+y}=x.\ ef=\frac{yy+2ay\ \&c}{r+y+a}-\frac{yy}{r+y}=\frac{+2ayr+ayy}{rr+2ry+ra+yy+ay}$.(11)

$$v=\frac{2ay}{\frac{4ayr+2ayy}{rr+2ry+yy}}{}^{(12)}=\frac{y}{\frac{2ry+yy}{rr+2ry+yy}}=\frac{rr+2ry+yy}{2r+y}.$$

Rule 2d. If y is in ye rationall denom: of x consisting of many termes, for ye Numerat: in ye valor of v multiply y by ye denom of x. squared. for ye denominatr mu[l]tiply ye Num: of x according to its dimensions & ye product by ye denom: againe multipl[y] ye Denom: according to its dim: & ye product by ye numerr & substract ye less from ye greater & divide ye diff by y.(13)

---

(5) Newton's text has no clarifying diagram and there seems no unique way of restoring one. However, the analysis is clear, requiring that
$$\frac{bg}{gh}=\frac{mi}{ge}=\frac{ds}{dx}\quad\text{with}\quad q\xi=x\frac{gh}{bg}=x\frac{dx}{ds}.$$

(6) Add. 3960. 12: 206, Newton's first extended draft.

(7) That is, Newton denotes by a the limit-increment $(bf-ae)$ of the ordinate $ae=y$. This variant on his more usual Cartesian denotation e is reminiscent of Fermat's use of E in then unprinted manuscript, but it is improbable that Newton knew that. (Apparently, Newton's only contact with Fermat's infinitesimal researches was through Schooten's inadequate exposition in *Geometria*: 253–5, where the notation e for the increment of y is used.) The use of a is a natural alternative for e and we should not make too much of it. It is possible, however, that Newton here wishes to distinguish the increments of x and y by choosing the respective denotations e and a for them.

(8) Much as before, Newton applies Descartes' method to the invention of the subnormal $v=y(dy/dx)$ in the curves $y^2=rx$ and $y^3=r^2x-r^3$.

(9) Read 'rationall Denominator'.

(10) The rule calculates $v=y(dy/dx)$ from $x=f(y)$ in the form $y^2/[yf'(y)]$, where
$$f'(y)=dx/dy.$$

(11) Neglecting powers of a. (12) Since $v=2ay/2(ef)$.

(13) More generally, where $x=f(y)/g(y)$, Newton finds the subnormal as
$$v=\frac{y.[g(y)]^2}{(1/y)\,[yf'(y).g(y)-yg'(y).f(y)]}.$$

Example. $\dfrac{yy}{r+y}=x.\quad \dfrac{y \text{ in } rr+2ry+yy}{\dfrac{2yy \text{ in } r+y,\ \begin{matrix}-1y\\-0r\end{matrix} \text{ in } yy}{y}}=v=\dfrac{rr+2ry+yy}{2r+y}.$

$$\frac{y^3}{r}=xx.\quad ec[=x]=\sqrt{\frac{y^3}{r}}.\quad fc=\sqrt{\frac{y^3+3ayy\ \&c}{r}}.\quad ef=\sqrt{\frac{y^3+3ayy\ \&c}{r}}-\sqrt{\frac{y^3}{r}}.$$

$$vv+yy=yy+2ay\ \&c+vv-2v\sqrt{\frac{y^3+3ayy}{r}}+2v\sqrt{\frac{y^3}{r}}.\quad 2ay+2v\sqrt{\frac{y^3}{r}}=2v\sqrt{\frac{y^3+3ayy}{r}}.$$

$$aayy+2avy\sqrt{\frac{y^3}{r}}+\frac{vvy^3}{r}=\frac{vvy^3}{r}+\frac{3vvayy}{r}.\quad 2avy\sqrt{\frac{y^3}{r}}=\frac{3vvayy}{r}.^{(14)}\quad \frac{2\sqrt{\frac{y^3}{r}}}{3\frac{y}{r}}=v=\frac{2}{3}\sqrt{ry}.$$

Rule 1st. If y is not in ye denom: of ye valor of xx. unlesse ye Denom: have but one terme double ye valor of x & × (15) it by y for a Numerr: Then mult: ye valor of xx according to ys dimens: in it & divide it by y for a Denom: in ye valor of v:

Exa[m]ple. $\dfrac{y^3}{r}=xx.\quad \dfrac{2y\sqrt{\frac{y^3}{r}}}{3\frac{yy}{r}}=v=\dfrac{2\sqrt{\frac{y^3}{r}}}{3\frac{y}{r}}=\dfrac{2\sqrt{ry^3}}{3y}=\dfrac{2}{3}\sqrt{ry}.\quad vv=\dfrac{2ry}{3}.^{(16)}$

$$\frac{y^3}{r+y}=xx.\quad ef=\sqrt{\frac{y^3+3ayy}{r+y+a}}-\sqrt{\frac{y^3}{r+y}}.\quad 0=2ay-2v\sqrt{\frac{y^3+3ayy}{r+y+a}}+2v\sqrt{\frac{y^3}{r+y}}.^{(17)}$$

$$2ay+2v\sqrt{\frac{y^3}{r+y}}=2v\sqrt{\frac{y^3+3ayy}{r+y+a}}.\quad aayy+2avy\sqrt{\frac{y^3}{r+y}}+vv\frac{y^3}{r+y}=v^2\left[\frac{y^3+3ayy}{r+y+a}\right].$$

(14) Ignoring the term '$aayy$'.

(15) Read 'multiply'.

(16) Read '$\dfrac{4ry}{9}$'.

(17) Since $v^2+y^2+s^2 = y^2+2ay+v^2-2vx(ef)$, with powers of a and (ef) omitted as infinitesimally small.

(18) Read '$\dfrac{\sqrt{4r^3y+12rryy+12ry^3+4y^4}}{2y+3r}$'.

(19) Read 'Denominator'.

$$2ay\sqrt{\frac{y^3}{r+y}}\frac{+vy^3}{r+y}=\frac{vy^3+3vayy}{r+y+a}.\quad 2rry+4ryy+2y^3 \text{ in } \sqrt{\frac{y^3}{r+y}}+vy^3=r+y \text{ in } 3vyy.$$

$$\frac{2r+2y \text{ in } \sqrt{ry^3+y^4}}{2yy+3ry}=v=\frac{\sqrt{4r^3y+12rryy}}{2y+3r}.^{(18)}$$

Rule 2^{d}. If y is in y^{e} Denom: of xx wch D:(19) hath many termes, Then multiply y^{e} valor of $2xy$ by y^{e} Denom of y^{e} valor of x^4 for a Num: in v. And for y^{e} denom: Multiply y^{e} numerat^{r} of xx according to its dimensions & y^{e} product by y^{e} Denom: y^{n} multiply y^{e} Denom: according to ys dimensions & y^{e} product by y^{e} num^{r}: & substract y^{e} less from y^{e} greater. & make [y^{e}] difference divided by y y^{e} Denom: in y^{e} valor of v:(20)

Example. $$\frac{y^3}{r+y}=xx.\quad \frac{rr+2ry+yy \text{ in } 2y\sqrt{\dfrac{y^3}{r+y}}}{\dfrac{3y^3r+3y^3y-1yy^3}{y}}=v$$

$$=\frac{rr+2ry+yy \text{ in } 2\sqrt{\dfrac{y}{y+r}}}{3r+2y}=\frac{2r+2y \text{ in } \sqrt{ry+yy}}{2y+3r}.$$

Or. Multiply y^{e} cube of y^{e} Den: of xx by four times y^{e} Num^{r} of xx & y^{e} product by yy for a Numerator in vv's valor; &c:(21)

$$\frac{y^4}{r}=x^3.\quad ec=\sqrt{c:\frac{y^4}{r}}:\ ef=\sqrt{c:\frac{y^4+4ay^3\ \&c}{r}}-\sqrt{c:\frac{y^4}{r}}.$$

$$0=2ay-2v\sqrt{c:\frac{y^4+4ay^3}{r}}+2v\sqrt{c:\frac{y^4}{r}}.^{(22)}$$

---

(20) Where $x^2=f(y)/g(y)$, Newton's rule evaluates the subnormal v as

$$\frac{y}{dx/dy}=\frac{2xy}{(d/dy)\,(x^2)}=\frac{2xy\,.\,[g(y)]^2}{(1/y)\,[g(y)\,.\,yf'(y)-f(y)\,.\,yg'(y)]}.$$

(The former 'Rule 1st' is the particular case where $g(y)=1$.)

(21) Alternatively, $$v^2=\frac{[g(y)]^3\,.\,4f(y)\,.\,yy}{(1/y^2)\,[g(y)\,.\,yf'(y)-f(y)\,.\,yg'(y)]^2},$$

since $x^2=\dfrac{f(y)}{g(y)}$.

(22) And so $$ay+v\sqrt[3]{\frac{y^4}{r}}=v\sqrt[3]{\frac{y^4+4ay^3+\ldots}{r}}.$$

$$+3ayvv\sqrt{c[:]\frac{y^8}{rr}}\ \&\text{c}+v^3\frac{y^4}{r}=v^3\frac{y^4}{r}+\frac{4v^3ay^3}{r}.\quad 3ayvv\sqrt{c:\frac{y^8}{rr}}=4v^3\frac{ay^3}{r}.\quad \frac{3\sqrt{c:\frac{y^8}{rr}}\text{ in }y}{4\frac{y^3}{r}}=v.$$

$$\frac{3\sqrt{c:y^8r}}{4yy}=v=\frac{3}{4}\sqrt{c:yyr}.\quad v^3=\frac{27}{64}yyr.$$

Rule 1st. If y is in ye Den[:] of x^3 unless it bee but of one terme make $3x^2y$ ye numerator & multiply ye valor of x^3 according to x's dimensions & divide it by y for a denom; of v. as

$$\frac{y^4}{r}=x^3.\quad \frac{3y\sqrt{c:\frac{y^8}{rr}}}{4\frac{y^4}{ry}}=v=\frac{3}{4}\sqrt{c:ryy}.$$

Rule 2d. I[f] y is in ye many termed Denom: of x^3 Then make $3xxy$ multipl[ye]d by ye Den: of x^6 ye Numeratr. & Multiply ye N: of x^3 by its dimensiō & ye product by ye D: of x^3; againe Mult: y[e] D: according to y's Dimensions & ye product by ye N: & substract ye lesse from ye greater & divide ye rest by y for a Denominator in ye Valor of v.(23)

$$\text{As }\frac{y^4}{r+y}=x^3.\quad \frac{rr+2ry+yy\text{ in }3y\sqrt{c:\frac{y^8}{rr+2ry+yy}}}{\frac{3y^5+4ry^4}{y}}=v=\frac{3r+3y\text{ in }\sqrt{c:ryy+y^3}}{3y+4r}.$$

---

[3](24)

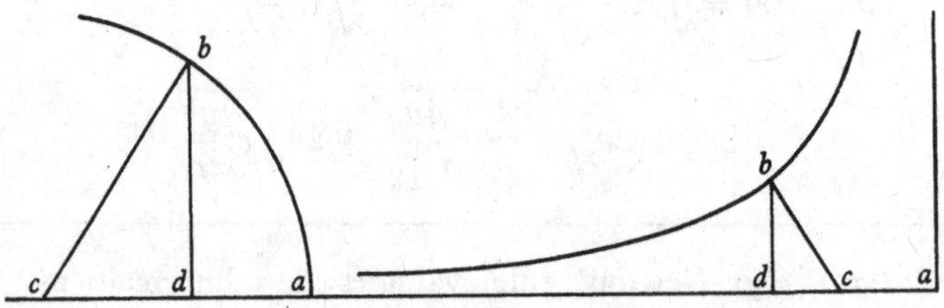

$cd=v$. $da=x$. $cb=s$. $bd=[y.]$ $\mathfrak{n}$=Numerator. D=Denominator. $\ddot{\mathfrak{n}}$, $\ddot{D}$, $\ddot{A}$, [$\ddot{B}$, $\ddot{C}$,] $\ddot{a}$, [$\ddot{b}$, $\ddot{c}$, &c sig]nifies yt every terme in ye Numerator, Denominator, or

(23) Where $x^3=f(y)/g(y)$, Newton's rule evaluates $v=y(dy/dx)$ as

$$\frac{y}{dx/dy}=\frac{3x^2y}{(d/dy)(x^3)}=\frac{[g(y)]^2.3x^2y}{(1/y)[g(y).yf'(y)-f(y).yg'(y)]}.$$

(The first rule is the particular case for which $g(y)$ is unity.)

Quantity signified by [*A*, *B*, *C*, *a*], *b*, *c* &c must bee multiplied by soe many units as y^e unknowne quantity *x*, or *y* hath dimensions in y^t [ter]me.(25) as if

$$A=\frac{x^4}{a}[+]x^3+axx-bbx+c^3+\frac{d^4}{x}-\frac{e^5}{xx} \text{ \&c then must it bee}$$

$$A=x^3+axx-bbx+c^3+\frac{d^4}{x}-\frac{e^5}{xx}+\frac{x^4}{a}=\frac{4x^4}{a}+3x^3+2axx-bbx-\frac{d^4}{x}+\frac{e^5}{xx}.^{(26)}$$

3 2 1 0 −1 −2 4

Haveing y^e rationall valor of *y*, *yy*, y^3 &c to find *v*, *vv*, v^3 &c. *When x is not found in B, unless B have but one terme.*

Rule 1st. $\frac{A}{B}=yy=C.$ $\frac{\mathfrak{C}}{2x}=$ ♉ $v.$

Example. 1st. $rx=yy.$ $\frac{1rx}{2x}=\frac{r}{2}=v.$ 2^d, $\frac{a^4}{xx}=yy.$ $\frac{-2a^4}{2x^3}=$ ♉ $v.$ $\frac{a^4}{x^3}=v.$

Rule 2^d. $\frac{A}{B}=y^4=C.$ $\frac{\frac{1}{x}\mathfrak{C}}{4yy}=$ ♉ $v.$

Example. 1st. $\frac{x^5}{r}=y^4.$ $\frac{\frac{5x^5}{rx}}{4\sqrt{\frac{x^5}{r}}}=\frac{5xx}{4\sqrt{rx}}=v.$

2^d, $\frac{a^5}{x}=y^4.$ $\frac{\frac{-1a^5}{x \text{ in } x}}{4\sqrt{\frac{a^5}{x}}}=$ ♉ $v.$ $\frac{a^3}{4x\sqrt{ax}}=v.$

Rule 3^d. $\frac{A}{B}=y^6=C.$ $\frac{\frac{1}{x}\mathfrak{C}}{6y^4}=$ ♉ $v.$

---

(24) Newton's revised draft (on f. 199) of [2] above (p. 322).

(25) Newton's earliest use of a dot-notation for the homogenized derivative: thus, where $A=f(x)$ and $a=g(x)$, then

$$A=x\frac{dA}{dx}=xf'(x) \quad \text{and} \quad \ddot{a}=x\frac{da}{dx}=xg'(x).$$

Whether this precedes his similar use of dots in May 1665 to represent the homogenized partial derivatives of the 2-valued function $\mathfrak{X}$ is not clear. (See Add. 4004: [48^v, 49^r] = 4, §3.2 above.)

(26) Read '$\frac{2e^5}{xx}$'.

Examp: 1st. $\dfrac{x^7}{r}=y^6.\quad \dfrac{\frac{7x^7}{rx}}{6\sqrt{c:\frac{x^{14}}{rr}}}=\text{♉}\,v.\quad \dfrac{7xx}{6\sqrt{c:rxx}}=v.$

2d, $\dfrac{a^8}{xx}=y^6.\quad \dfrac{\frac{-2a^8}{x^3}}{6\sqrt{c:\frac{a^{16}}{x^4}}}=\text{♉}\,v.\quad \dfrac{a^3}{3x\sqrt{c:axx}}=v.$

Rule 4th. $\dfrac{A}{B}=y^8=C.\quad \dfrac{\frac{1}{x}\mathfrak{C}}{8y^6}=\text{♉}\,v.$

Examp: 1st. $\dfrac{x^9}{r}=y^8.\quad \dfrac{\frac{9x^9}{rx}}{8\sqrt{qq:\frac{x^{27}}{r^3}}}=\text{♉}\,v.\quad \dfrac{9xx}{6\sqrt{qq:rx^3}}=v.$

2d, $\dfrac{a^9}{x}=y^8.\quad \dfrac{\frac{-1a^9}{xx}}{8\sqrt{qq:\frac{a^{27}}{x^3}}}=\text{♉}\,v.\quad \dfrac{a^3}{8x\sqrt{qq:xa^3}}=v.$

1st $\dfrac{A}{B}=yy=C;\ \dfrac{\mathfrak{C}}{2x}=\text{♉}\,v.$	2d, $\dfrac{A}{B}=y^4=C;\ \&\ \dfrac{\frac{1}{x}\mathfrak{C}}{4\sqrt{q:C}}=\text{♉}\,v.$	(27)
3d, $\dfrac{A}{B}=y^6=C.\ \&,\ \dfrac{\frac{1}{x}\mathfrak{C}}{6\sqrt{c:C^2}}=\text{♉}\,v.$	4th, $\dfrac{A}{B}=y^8=C.\ \&,\ \dfrac{\frac{1}{x}\mathfrak{C}}{8\sqrt{qq:C^3}}=\text{♉}\,v.$	&c.

Haveing ye rationall valor of y, yy, y³ &c: to find v, vv, v³ &c.

Rule 1st. If $\dfrac{A}{B}=yy=C.$ then $\dfrac{\frac{\text{♉}AB\ \text{☋}\ AB}{x}}{2B^2}=v.$ Example, $\dfrac{a^4+ax^3+x^4}{aa+xx}=yy.$

$$\dfrac{\dfrac{\text{♉}\,4x^6\ \text{♉}\,4x^4aa\ \text{♉}\,3ax^5\ \text{♉}\,3a^3x^3\ \ \text{☋}\,x^6\ \text{☋}\,1x^5a\ \text{☋}\,1xxa^4}{x}}{2x^4+4aaxx+2a^4}=v=\dfrac{3x^5+2ax^4+4a^2x^3+3a^3x^2-a^4x}{2x^4+4aaxx+2a^4}.$$

(27) In general, where $y^{2n}=C$, a rational function of x, Newton's table evaluates

$$v=y\frac{dy}{dx}\quad\text{as}\quad \frac{1}{2}\frac{d}{dx}(C^{1/n})=\frac{(1/x)\,\mathfrak{C}}{2n(C^{n-1})^{1/n}}\quad\text{since}\quad \frac{1}{x}\mathfrak{C}=\frac{dC}{dx}.$$

Rule 2^{d}. $\dfrac{A}{B}=y^4$. $\dfrac{\dfrac{\text{♉}\,\ddot{A}B\,\text{☊}\,A\ddot{B}}{x}}{4yyB^2}=v$. Ex. $\dfrac{a^5+x^5}{a-x}=y^4$.

$$\frac{\dfrac{\text{♉}\,0a^5a\,\text{☊}\,0a^5x\,\text{♉}\,5ax^5\,\text{☊}\,5x^6 \atop \text{☊}\,0aa^5\,\text{☊}\,0ax^5\,\text{♉}\,1xa^5\,\text{♉}\,1x^6}{x}}{4aa-8ax+4xx \text{ in } \sqrt{\dfrac{a^5+x^5}{a-x}}}=v=\frac{4x^5-5ax^4-a^5}{4a-4x \text{ in } \sqrt{a^6+ax^5-a^5x-x^6}}.$$

Rule 3^{d}. $\dfrac{A}{B}=y^6$. $\dfrac{\dfrac{\text{♉}\,\ddot{A}B\,\text{☊}\,A\ddot{B}}{x}}{6y^4B^2}=v$.

Rule 4th. $\dfrac{A}{B}=y^8$. $\dfrac{\dfrac{\text{♉}\,\ddot{A}B\,\text{☊}\,A\ddot{B}}{x}}{8y^6B^2}=v$.

1st, $\dfrac{a}{b}=yy$; $\dfrac{\dfrac{\text{♉}\,\ddot{a}b\,\text{☊}\,a\ddot{b}}{x}}{2bb}=v$.	2^{d}, $\dfrac{a}{b}=y^4$; $\dfrac{\dfrac{\text{♉}\,\ddot{a}b\,\text{☊}\,a\ddot{b}}{x}}{4bb\sqrt{\dfrac{a}{b}}}=v$.	(28)
3^{d} $\dfrac{a}{b}=y^6$; $\dfrac{\dfrac{\text{♉}\,\ddot{a}b\,\text{☊}\,a\ddot{b}}{x}}{6bb\sqrt{c:\dfrac{aa}{bb}}}=v$.	4th $\dfrac{a}{b}=y^8$. $\dfrac{\dfrac{\text{♉}\,\ddot{a}b\,\text{☊}\,a\ddot{b}}{x}}{8bb\sqrt{qq:\dfrac{a^3}{b^3}}}=v$.	

Haveing y^e rationall valor of x, xx, x^3 &c: to find v, when x is not found in B, unlesse B have but one terme.

Rule 1st. If $\dfrac{A}{B}=x=C$. then $\dfrac{\text{♉}\,y}{\frac{1}{y}\mathcal{C}}=v$.

Example y^e 1st $\dfrac{yy}{r}=x$; $\dfrac{\dfrac{\text{♉}\,y}{2yy}}{ry}=v=\dfrac{r}{2}$. 2^{d}, $\dfrac{rr}{y}=x$; & $\dfrac{\dfrac{\text{♉}\,y}{-rr}}{y \text{ in } y}=v=\dfrac{y^3}{rr}$.

Rule 2^{d}. If $\dfrac{A}{B}=x^2=C$. y^n $\dfrac{\text{♉}\,2yx}{\frac{1}{y}\mathcal{C}}=v$.

(28) More generally, where $a/b = y^{2n} = C$,

$$v = \frac{(1/x)\,\mathcal{C}}{2n(C^{n-1})^{1/n}} \quad \text{with now} \quad \frac{1}{x}\,\mathcal{C} = \frac{dC}{dx} = \frac{(1/x)\,(\ddot{a}b-a\ddot{b})}{b^2}.$$

An arithmetical error in the example for 'Rule 1st' has been left uncorrected.

Exā: 1st $\frac{y^3}{r}=xx;\ \frac{\text{ȣ}\,2y\sqrt{\frac{y^3}{r}}}{\frac{3y^3}{ry}}=v=\frac{2}{3}\sqrt{ry}.$ 2d, $\frac{r^3}{y}=x^2.\ \frac{\text{ȣ}\,2y\sqrt{\frac{r^3}{y}}}{\frac{-r^3}{yy}}=v=\frac{2yy\sqrt{ry}}{rr}.$

Rule 3d. $\frac{A}{B}=x^3=C.$ & $\frac{\text{ȣ}\,3yxx}{\frac{1}{y}\dot{C}}=v.$

As if $\frac{y^4}{r}=x^3$, yn $\frac{\text{ȣ}\,3y\sqrt{c:\frac{y^8}{rr}}}{\frac{4y^4}{r\text{ in }y}}=v=\frac{3}{4}\sqrt{c:ryy}.$

Rule 4th. $\frac{A}{B}=x^4=C.$ & $\frac{\text{ȣ}\,4yx^3}{\frac{1}{y}\dot{C}}=v.$

As if $\frac{r^5}{y}=x^4$, yn $\frac{\text{ȣ}\,4y\sqrt{qq:\frac{r^{15}}{y^3}}}{-\frac{r^5}{yy}}=v=\frac{4yy\sqrt{qq:r^3y}}{rr}.$

1st $\frac{A}{B}=x=C;\ \frac{\text{ȣ}\,y}{\frac{1}{y}\dot{C}}=v.$	2d, $\frac{A}{B}=xx=C;\ \frac{\text{ȣ}\,2y\sqrt{q:C}}{\frac{1}{y}\dot{C}}=v.$
3d, $\frac{A}{B}=x^3=C.\ \frac{\text{ȣ}\,3y\sqrt{c:CC}}{\frac{1}{y}\dot{C}}=v.$	4th, $\frac{A}{B}=x^4=C.\ \frac{4y\sqrt{qq[:]C^3}}{\frac{1}{y}\dot{C}}=\text{ȣ}\,v.$ (29)

Haveing ye Rationall valor of x, xx, x³ &c: to find ye valor of v.

Rule 1st. If $\frac{A}{B}=x$, yn $\frac{yBB}{\frac{AB-A\dot{B}}{y}}=\text{ȣ}\,v.$

Example $\frac{aay+y^3}{yy-aa}=x.\ \frac{y^5-2a^2yy^2+a^4y}{\frac{3y^3y^2-3y^3a^2+aayy^2-a^4y\ -2yyy^3\ -2aayyy}{y}}=v=\frac{y^5-2aay^3+a^4y}{y^4-4aayy-a^4}.$

(29) Here $x^n = C = g(y)$, so that

$$v=\frac{y}{dx/dy}=\frac{y}{(1/n)\,C^{1/n-1}\,(dC/dy)}=\frac{nx^{n-1}y}{(1/y)\,\dot{C}}.$$

Rule 2[d]. If $\dfrac{A}{B} = xx\ y^n\ \dfrac{\dfrac{2yBBx}{\dot{A}B - A\dot{B}}}{y} = \text{ȣ}\, v.$

Rule 3[d]. $\dfrac{A}{B} = x^3.\ \dfrac{\dfrac{3yxxBB}{\dot{A}B - A\dot{B}}}{y} = \text{ȣ}\, v.$

Rule 4th. $\dfrac{A}{B} = x^4.\ \dfrac{\dfrac{4yx^3B^2}{\dot{A}B - A\dot{B}}}{y} = v.$

1st $\dfrac{a}{c} = x:\ \dfrac{ccyy}{\ddot{a}c - ac} = \text{ȣ}\, v.$	2[d], $\dfrac{a}{c} = xx$, & $\dfrac{2ccy^2\sqrt{q:\dfrac{a}{c}}}{\ddot{a}c - a\ddot{c}} = \text{ȣ}\, v.$
3[d], $\dfrac{a}{c} = x^3$; &, $\dfrac{3ccyy\sqrt{c:\dfrac{aa}{cc}}}{\ddot{a}c - a\ddot{c}} = \text{ȣ}\, v.$	4th, $\dfrac{a}{c} = x^4.\ \dfrac{4ccyy\sqrt{qq:\dfrac{a^3}{c^3}}}{\ddot{a}c - a\ddot{c}} = v.$

(30)

$\dfrac{a}{b} = yy:\ \dfrac{\ddot{a}b - a\check{b}}{2bbx} = \text{ȣ}\, v.$	$\dfrac{a}{b} = y^4:\ \dfrac{\ddot{a}b - a\check{b}}{4bx\sqrt{ab}} = \text{ȣ}\, v.$	$\dfrac{a}{b} = y^6:\ \dfrac{\ddot{a}b - a\check{b}}{6bx\sqrt{c:aab}} = v.$
$\dfrac{a}{b} = y^8:\ \dfrac{\ddot{a}b - a\check{b}}{8bx\sqrt{qq:a^3b}} = \text{ȣ}\, v.$	$\dfrac{\ddot{a}b - a\check{b}}{10bx\sqrt{qc:a^4b}} = \text{ȣ}\, v$: if $\dfrac{a}{b} = y^{10}.$	
$\dfrac{a}{b} = x[:]\ \dfrac{bbyy}{\ddot{a}b - a\check{b}} = \text{ȣ}\, v.$	$\dfrac{a}{b} = xx:\ \dfrac{2byy\sqrt{ab}}{\ddot{a}b - a\check{b}} = \text{ȣ}\, v.$	$\dfrac{a}{b} = x^3:\ \dfrac{3byy\sqrt{c:aab}}{\ddot{a}b - a\check{b}} = \text{ȣ}\, v.$
$\dfrac{a}{b} = x^4:\ \dfrac{4byy\sqrt{qq:a^3b}}{\ddot{a}b - a\check{b}} = \text{ȣ}\, v.$	$\dfrac{5byy\sqrt{qc:a^4b}}{\ddot{a}b - a\check{b}} = \text{ȣ}\, v.\ \dfrac{a}{b} = y^5.$	

(31)

(30) More generally, where

$$x^n = C = \frac{a}{c} = \frac{g(y)}{h(y)},$$

then as before $v = \dfrac{nx^{n-1}y}{(1/y)\,\dot{C}}$ with $\dfrac{1}{y}\dot{C} = \dfrac{d}{dy}\left(\dfrac{a}{c}\right) = \dfrac{(1/y)(\ddot{a}c - a\ddot{c})}{c^2}.$

(31) Newton summarizes his previous results as a single table.

[4]⁽³²⁾

Rule 1st. *Ha[veing y^e v]alor of v, when y^e D: of y^t valor hath [but] one terme, to find y^e valor of yy.*

$\dfrac{E}{G}=v[=F$. Div]id each Term in $2xF$ by so many units as x hath dimensions in y^t terme & make it$=\text{♉} yy.$⁽³³⁾

Example 1st. $\dfrac{r}{2}=v.$ $2x$ in $\dfrac{r}{2}=rx.$ $\dfrac{rx}{1}=rx=yy.$

$$2^{\text{d}},\ v=\frac{b^4}{x^3}.\quad \frac{\dfrac{2b^4}{xx}}{-2}=\frac{-b^4}{xx}=\text{♉}\,yy.\quad \frac{bb}{x}=y.$$

Rule 2^d. *Haveing y^e valor of v, to find y^e valor of yy when y^e Dēn of y^e valor of v hath many dimensions.*

If $\dfrac{E}{G}=v=F$. Find all y^e least litterall divisors of G [&] set those together w^{ch} (being divided by y^e greatest rationall numbers by wch they are divisible) are =⁽³⁴⁾ to one another: Take one divisor from every sort of more yⁿ one in wch x is found, multiply y^e rest by one another for (B) y^e denom: in y^e valor of $yy=\dfrac{A}{B}=C$. Then feine an equacōn for y^e valor of A wch must have 2 dimensions more yⁿ B, But noe termes wanting, & make $\dfrac{AB-AB}{2xBB}=v=K=\dfrac{L}{N}$: Then reduce K or F to such a forme, y^t $N=G$.⁽³⁵⁾ & by compareing y^e each terme of L wth each terme of E y^e valors of y^e quantitys in y^e feined equation are found & consequently A y^e Numerat^r in y^e valor of yy is knowne.

Exāple 1st. $\dfrac{E=8x^6-8ax^5-40aax^4-9a^5x+9a^6}{G=4x^5-4ax^4-20aax^3+4a^3xx+32a^4x+16a^5}=v.$ The least litterall divisors of G are $x-2a$. $x+a$. from each of wch a divisor being taken there

$$x-2a.\quad x+a.$$
$$x+a.$$

remaines $x-2a$. $x+a$. All wch multiplyed into one another produce

$$x+a.$$
$$x^3-3aax-2a^3=B.$$

---

(32) Add. 3960. 12: 200–2. Newton attempts to apply the rules of differentiation to the inverse problem of integration, seeking to discover the integrals of simple rational functions by inspection.

(33) Where $v = F = y(dy/dx)$, $2xF = x(d/dx)(y^2) = \ddot{a}$, where $a = y^2$, and Newton merely reverses the rule for forming $\ddot{a}$ from a.

(34) Read 'equall'.

W[ch] because it is of 3 dimensions y[e] feigned equation,

$$A=bx^5+cx^4+dx^3+ex^2+fx+g,$$

must be of 5. Then is y[e] valor of

$$\frac{\dot{A}B-A\dot{B}}{2xB^2}=\frac{\begin{matrix}2bx^6-2abx^5-10aabx^4-8aacx^3-6daax^2-4aaex-2aaf\\ \quad +c\quad\quad -ac\quad\quad -e\quad\quad +ae\quad\quad +2af\quad +3ag\\ \quad\quad\quad\quad\quad\quad\quad\quad\quad\quad\quad\quad -2f\quad\quad -3g\end{matrix}}{2x^5-2ax^4-10aax^3+2a^3xx+16a^4x+8a^5}=v=K=\frac{L}{N}$$

& y[t] $N=G$ I divide F by 2 & so compareing every terme of L wth every terme of $\frac{E}{2}=4x^6-4ax^5-20aax^4**-\frac{9}{2}a^5x+\frac{9}{2}a^6$. by y[e] 1st terme tis found, $2bx^6=4x^6$, & $b=2$. by y[e] 2[d], $-2ab+c=-4a$, or $c=0$. by y[e] 3[d], $-10aab-ac=-20a^3$, & $c=\frac{0}{a}=0$. by y[e] 4th, $-8aac-e=0$, & $e=0$. by y[e] 5[t], $-6daa+ae-2f=0$, & $3daa=-f$. by y[e] 6[t] $-4aae+2af-3g=\frac{-9}{2}a^5$, or $4af=6g-9a^5$. by y[e] last, $-2aaf+3ag=\frac{9}{2}a^6$, or, $4af=+6ag-9a^6$.(36) By y[e] 5[t], $-12da^3=4af=6g-9a^5$ by y[e] 6[t] & 7th. or $f=-3da^2$ & $g=\frac{3a^5-4da^3}{2}$; & $d=\frac{3a^5-2g}{4a^3}$: Where I find f, d & g to bee undetermined, soe y[t] by assuming any valor for d, g & f are determined or by assuming any valor for g, d & f are determined &c & may bee found. as if I make $d=0$, y[n] is $g=\frac{3a^5}{2}$ & $f=0$. Then by subrogateing these valors of b, c, d, e, f, g in $A=bx^5+cx^4+dx^3+ex^2+fx+g$, There will bee found

$$\frac{A=2x^5****+\frac{3a^5}{2}}{B=x^3-3aax-2a^3}=y^2.$$

Soe if I make $g=0$, y[n] is $d=\frac{3aa}{4}$, & $f=\frac{-9a^4}{4}$, & $\frac{2x^5*+\frac{3}{4}aax^3*\frac{-9a^4}{4}x*}{x^3-3aax-2a^3}=y^2.$(37)

Soe if $f=a^4$; y[n] is $d=-\frac{aa}{3}$, & $g=\frac{13a^5}{3}$. &, $\frac{2x^5*-\frac{aa}{3}x^3*a^4x+\frac{13}{3}a^5}{x^3-3aax-2a^3}=y^2.$(38)

---

(35) Newton first wrote 'y[t] its Denom̄ bee G'. (36) Read '$+6g-9a^5$'.

(37) That is, $$\frac{2x^5+\frac{3}{2}a^5}{x^3-3a^2x-2a^3}+\frac{3a^2}{4}.$$

(38) Or $y^2=\frac{2x^5+\frac{3}{2}a^5}{x^3-3a^2x-2a^3}-\frac{a^2}{3}$. More generally, expressing f and g in terms of a and d and substituting, we may deduce that $v=\frac{2x^5+\frac{3}{2}a^5}{x^3-3a^2x-2a^3}+d$: that is, d is the arbitrary constant of integration.

Example 2d. $\left.\dfrac{3x^6+4ax^5+4a^3bbx+2a^4bb=E}{ax^4+2aax^3+a^3xx=G}\right\}=v.$ ye least litterall divisors of G are a. x. $x+a$. And by takeing one Divisō(39) from each sort in wch x is not there x. $x+a$.
remaines a. x. $x+a$. which by multiplicacō produce $axx+aax=B$. Then feigne an equacōn $A=cx^5+dx^4+ex^3+fx^2+gx+h$, of 2 more dimensions yn B. Then supposing $\frac{A}{B}=yy$, find ye valor of

$$\frac{A\dot{B}-\dot{A}B}{2xB^2}=\frac{\begin{matrix}3cx^6+4acx^5+3adx^4+2aex^3+afx^2-2hx-ah.\\ +2d \quad +e \qquad\qquad -g\end{matrix}}{2ax^4+4aax^3+2a^3xx}=v=K=\frac{L}{N}.$$

& yt N & G bee equall I multiply F by 2 & so compareing every terme of L wth every terme of $2E=6x^6+8ax^5***+8a^3bbx+4a^4bb$. by ye 1st terme tis found $3cx^6=6x^6$, & $c=2$. by ye 2d, $4ac+2d=8a$, or $2ac-4a=[-]d=0$. by ye 3d, $3ad+e=0$, or $e=-3ad=0$. by ye 4th $2ae=0$, or $\frac{0}{2a}=e=0$. by ye 6t & 7th, $-2h=8a^3bb$, & $-ah=4a^4bb$, or $h=-4a^3bb$. by ye 5t, $af-g=0$. or $af=g$. or $f=\frac{g}{a}$. where likewise f & g are undetermined, If I therefore make $g=0$. yn is $f=\frac{g}{a}=0$. Subrogateing therefor these valors of c, d, e &c in their stead in $A=cx^5+dx^4+ex^3+fx^2+gx+h$, there results $\dfrac{A=2x^5****-4a^3bb}{B=axx+aax}=yy$. or by makeing $f=aab$ g will be equall to $af=a^3b$. & $\dfrac{2x^5**+aabxx+a^3bx-4a^3bb}{axx+aax}=yy$. &c.(40)

Note 1st yt If x is found of more or of as many dimensions in G as in E ye first terme is wanting in A.

(39) Read 'Divisor'.

(40) In general, $\dfrac{A}{B}=\dfrac{2x^5-4a^3b^2}{ax^2+a^2x}+\dfrac{f}{a}$, where $\dfrac{f}{a}=\dfrac{g}{a^2}$ is the arbitrary added constant of integration.

(41) Newton tries to find $\int \frac{f(x)}{g(x)}.dx$ by supposing that $r(x).g(x) = 2[G(x)]^2$ and equating $(1/x)\,[xF'(x).G(x)-F(x).xG'(x)]$ with $r(x).f(x)$ to determine $F(x)$ term by term. If the identity holds, it follows that

$$\frac{f(x)}{g(x)}=\frac{F'(x)\,G(x)-F(x)\,G'(x)}{2[G(x)]^2}=v=y\frac{dy}{dx},$$

and so $y^2=[F(x)/G(x)]+k$, where k is an added arbitrary constant. Newton's note (2) is crucial: the method of term by term identification of coefficients demands that $F(x)$ and $G(x)$ be rational functions of x.

2dly, If x is in some divisor of G, wch is like no other Divisor, ye valor of yy cannot bee expressed in Algebraicall termes, as also if there be contradictions in ye comparisons of ye termes of L & E, ye same quantity being found greater by one terme yn by another.(41)

Haveing ye valor of vv to find ye Valor of y^4.(42)

Suppose $\frac{E}{G}=vv=F^2$. Find all ye least litterall & affirmative divisors of G: set those together wch (being divided by their greatest numerall divisors) are = to one another, & substract 2 divisors from every sort in wch x is found, multiply all ye remain[in]g affirmative divisors by one another & call ye Product B or if there remaine but one make yt $=B$, if none $B=1$: & multiply all ye remaining negative divisors by one another & call ye product P or if there remaine but one make yt $=P$, if none make $P=1$. Againe find all ye primary or simplest litterall & numerall divisors of E, from ym take away all wch are paires, add two div: to every sort remaining & multiply ym all into one another wch product call W. And if $W=0$,(44) extract ye roote of $\frac{E}{G}=vv=F^2$, otherwise extract ye root of $\frac{EW}{GW}=vv=F^2$ (observing yt $\sqrt{EW}$ is rationall but $\sqrt{GW}$ irrationall). Then feigne an equation $=H$, whose □(43) roote may bee extracted, which wants noe termes, & wch in x must have soe many Dimensions as $\frac{By^4}{PW}$ hath, or if $W=0$,(44) as $\frac{By^4}{P}$ hath: And suppose $\frac{ȣ PHW}{B}=\frac{A}{B}=y^4=C$, or if $P=0$,(44) or $W=0$,(44) leave ym out in ye Numeratr $PHW=A$. Then if B hath but one terme make $\frac{\not{C}}{4x\sqrt{C}}=v=K=\frac{L}{N}$, otherwise make $\frac{A\not{B}-\not{A}B}{4xB\sqrt{AB}}=v=K=\frac{L}{N}$. & reduce K or F to such a forme That theire denominators N & $\sqrt{GW}$ may bee equall. And then (since ye Numeratrs L & $\sqrt{EW}$ are = to one another) by compareing yr termes ye valors of ye quantitys in H are found & consequently $\frac{A}{B}=ȣ\, y^4$ is knowne.(45)

---

(42) A first cancelled version of this proposition on f. 200, fully incorporated in the present revision, is omitted. (43) 'Square'.

(44) Read '1'. Newton presumably intended 'if there is no W (or P)'.

(45) Where $y^4 = A/B$ (or $= C$) with A, B and C rational functions of x, then

$$v = \sqrt{\frac{E}{G}} = y\frac{dy}{dx} = \frac{(1/x)(\not{A}B-A\not{B})}{4B^2\sqrt{[A/B]}}\left(\text{or} = \frac{(1/x)\not{C}}{4\sqrt{C}}\right),$$

so that, for some function W, $\sqrt{[EW]} \equiv \not{A}B - A\not{B}$ and $\sqrt{[GW]} \equiv 4x\sqrt{[AB^3]}$. The former identity demands that $W = F(x)^2 . G(x)$, where $G(x)$ is the product of the non-quadratic factors of E,

Example y^e 1st. $\dfrac{E = 72x^{16} - 48a^8x^8 + 8a^{16}}{G = aax^{12} + a^{10}x^4} = vv = F^2$. The divisors of G are

$a.\ x.\ x^8 + a^8$. Two being taken from every sort in wch x is theire remaines
$a.\ x.$
$\ x.$
$\ x.$

$a.\ x.\ -x^8 - a^8$. ye product of ym wch are affirmative is $aaxx = B$. of ym wch are
$a.\ x.$

negative it is $-x^8 - a^8 = P$. The divisors of E are $2.\ 3x^8 - a^8$. From every sort
$2.\ 3x^8 - a^8.$
$2.$

substracting ye greatest eaven number of divisors yt I can there remaine 2 to wch adding two more divisors they are 2. 2. 2. wch drawn into one another mak $8 = W$. Then I extract ye roote of $FF = \dfrac{EW}{GW} = \dfrac{576x^{16} - 384a^8x^8 + 64a^{16}}{8aax^{12} + 8a^{10}x^4}$, wch is $\dfrac{24x^8 - 8a^8}{2axx\sqrt{2x^8 + 2a^8}} = v$. & since $\dfrac{PW}{B} = \dfrac{-8x^8 - 8a^8}{aaxx}$ is of 4 dimensions therefore I feine $H = bb$, & suppose $\dfrac{♉PKW}{B} = \dfrac{A}{B} = y^4 = C = \dfrac{8bbx^8 + 8b^2a^8}{aaxx}$. Then since B hath but one terme I make $\dfrac{\mathcal{C}}{4x\sqrt{C}} = \dfrac{6bx^8 - 2ba^8}{axx\sqrt{2x^8 + 2a^8}} = v = K = \dfrac{L}{N}$. & yt $N = \sqrt{GW}$, I divide F by 2 & soe compareing ye termes in $\dfrac{\sqrt{EW}}{2} = 12x^8 - 4a^8$ wth ye termes in L I find $6bx^8 = 12x^8$, & $-2ba^8 = -4a^8$, or $b = 2$. & consequently

$$y^4 = \frac{8bbx^8 + 8bba^8}{aaxx} = \frac{32x^8 + 32a^8}{aaxx}.^{(46)}$$

Example 2d. If $\dfrac{E}{G} = \dfrac{a^3x^8 - 8a^6x^5 + 16a^9xx}{32x^9 - 96a^3x^6 + 96a^6x^3 - 32a^9} = vv.$(47) The divisors of G are
$x^3 - a^3.$
$x^3 - a^3.$ from wch two being taken there remaines onely $x^3 - a^3 = B$. & there
$x^3 - a^3.$

$a.\ x.\ x^3 - 4a^3.$
being none negative, $P = 1$. The divisors of E are $a.\ x.\ x^3 - 4a^3$. And taking
$a.$

---

and in fact, though the reason for the choice is not clear, Newton selects $W = [G(x)]^3$. (In both examples below $G(x)$ is wholly numerical and so this particular choice of W is not significant.) Again, Newton chooses B the largest cubic factor in G. (It would seem more accurate, since $AB^3 = GW/16x^2$, to choose B^3 the largest cubic factor in GW, but since W in his two examples is numerical we may there slur the distinction.) Finally, since A must have a factor, say ϕ, in common with $GW/16x^2B^3$, Newton takes $A = \phi\alpha^2$, where α is a rational function (of the right dimensions in x) whose coefficients are to be evaluated from the identity $EW \equiv (AB - A\dot{B})^2$.

away every paire of divisors y^t^ I can, there onely remaines *a*. to wch I add 2 more & y^e^ sume is *a. a. a.* wch multiplied into one another make $a^3 = W$. Then extract y^e^ roote of $FF = \frac{EW}{GW} = vv = \frac{a^6x^8 - 8a^9x^5 + 16a^{12}xx}{32a^3x^9 - 96a^6x^6 + 96a^9x^3 - 32a^{12}}$ which is $F = \frac{\sqrt{EW}}{\sqrt{GW}} = v = \frac{a^3x^4 - 4a^6x}{4x^3 - 4a^3 \text{ in } \sqrt{2a^3x^3 - 2a^6}}$. And because $\frac{By^4}{PW} = \frac{y^4x^3 - y^4a^3}{a^3}$ is of 4 dimensions, y^e^ feined equation must be so two,(48) therefore make

$$cxx + dx + e\,\square^{\text{tè}(49)} = \begin{matrix} ccx^4 + 2cdx^3 + 2cx^2e + 2dex + ee = H, \\ +ddxx \end{matrix}$$

& supposeing $\frac{\text{♉}PHW}{B} = \frac{A}{B} = y^4 = \frac{a^3ccx^4 + 2cda^3x^3 + 2cea^3x^2 + 2dea^3x + eea^3 \atop + dda^3}{x^3 - a^3}$. Then since B hath more termes y^n^ one find y^e^ valor of $\frac{AB - A\dot{B}}{4xB\sqrt{AB}}$ wch is

$$\frac{a^3cx^4 - a^3dx^3 - 3a^3ex^2 - 4a^6cx - 2a^6d}{4x^3 - 4a^3 \text{ in } \sqrt{a^3x^3 - a^6}} = v = K = \frac{L}{N}.$$

& y^t^ N may be equall to $\sqrt{GW}$ divid F by $\sqrt{2}$ & compareing y^e^ termes of $\sqrt{\frac{EW}{2}} = \frac{a^3x^4 - 4a^6x}{\sqrt{2}}$, wth y^e^ termes of L. by y^e^ 1st terme there is found $\frac{a^3}{\sqrt{2}} = a^3c$. or $c = \frac{1}{\sqrt{2}}$, & $cc = \frac{1}{2}$. by y^e^ 2^d^, $-a^3d = 0$, or $d = \frac{0}{-a^3} = 0$. by y^e^ 3^d^, $-3a^3e = 0$, or $e = 0$. by y^e^ 4th $-4a^6c = \frac{-4a^6}{\sqrt{2}}$, or $c = \frac{1}{\sqrt{2}}$. & by y^e^ last $-2abd = 0$, or $d = 0$. & by substituteing these valors of c, d, e in $y^4 = \frac{a^3ccx^4 + 2cda^3x^3 + 2cea^3x^2 \,\&c \atop + dda^3x^2}{x^3 - a^3}$ there resulteth $\frac{\frac{1}{2}a^3x^4}{x^3 - a^3} = y^4$. or $\frac{a^3x^4}{2x^3 - 2a^3} = y^4$.

Note 1st That if there bee 2 divisors of G (in wch x is found,) equall to one another, but not equall to any 3^d^ divisor, or if the comparisons of y^e^ termes of L & $\sqrt{EW}$ be contradictious, y^n^ is y^e^ valor of y^4 inexpressible in Algebraical termes.

---

(46) $E = 2[2(3x^8 - a^8)]^2$, $G = a^2x^4(x^8 + a^8)$, so that $W = 2^3 = 8$ and $\sqrt{[EW]} = 8(3x^8 - a^8)$. Then for B is chosen a^2x^2 and so, since A must have the factor $8(x^8 + a^8)$, Newton takes $A = 8(x^8 + a^8)\,\alpha^2$ and finds finally that $\alpha = 2$ satisfies the identity $EW \equiv AB - A\dot{B}$.

(47) Here $E = a[ax(x^3 - 4a^3)]^2$ and $G = 32(x^3 - a^3)^3$, so that $W = a^3$, $B = x^3 - a^3$ and $A = a^3\alpha^2$ with $\alpha = cx^2 + dx + e$ evaluated as $(1/\sqrt{2})\,x^2$ by the identity $EW \equiv AB - A\dot{B}$.

(48) Read 'too'. (49) 'Quadratè', or squared.

2dly That if these opperations bee not done in ye simplest termes yt may be, they will not always hold trew.(50)

Note 3dly yt instead of subducing every paire from ye divisors of E &c(51) blot out every sort consisting of an eaven number of divisors & substract all ye paires yt may bee from every sort remaining unless they bee litterall divisors in wch x is not found[,] yn multipliing ym all into one another call ye product W.(52)

Haveing ye valor of v^3 to find ye valor of y^6.

Suppose $\frac{E}{G}=vvv=FFF$. Find all ye least litterall divisors of G [&] set those together wch (being divided by their greatest numerall divisors) are = to one another. Take away those wch have no equalls & in wch x is found, & multiplying ym into one another call ye product P or if there bee but one odd divis: make it $=P$, if none make $P=1$. Substract 3 divisors from every sort of ye rest in wch x is found & multiply all ye rest into one another [&] make ye product $=B$; or if yr remaine but one make yt $=B$, if none make $B=1$. Againe find all ye primary or simplest litterall & numerall divisors of E, Blot out every sort consisting of 3, 6, 9, 12 divisrs [&] Take away half the remain[in]g litterall divis[:] in wch x is not found: From those sorts in wch x is take away 3 divisors as often as may bee; Then multiplying all these remaining divisors by one another call ye product W, or if there bee but one divisor make yt $=W$, if none make $W=1$. Then extract ye cube roote of $\frac{EW}{GW}=v^3=F^3$, (observing yt $\sqrt{c: EW}$ is rationall, but $\sqrt{c: GW}$ is irrationall). Then feineing an equation $=H$ whose cube roote may be extracted, wch wants noe termes, & in wch x must have soe many dimensions as $\frac{By^6}{PW}$ hath, suppose $\frac{\text{♉} PHW}{B}=y^6=\frac{A}{B}=C$. & if B hath but one

---

(50) In particular, presumably, common factors of A and B are to be cancelled so that we must suppose them co-prime.

(51) Newton has cancelled 'it should bee convenienter to'.

(52) Newton fails to mention that his method, as before, yields the general solution $y^4 = A/B+p$, where p is an arbitrary added constant of integration. Perhaps he sees the point as too trivial to note.

(53) Where $y^6 = (A/B)$ (or $= C$), then

$$v = \sqrt[3]{\frac{E}{G}} = y\frac{dy}{dx} = \frac{(1/x)(AB-A\dot{B})}{6B^2\sqrt[3]{[A^2/B^2]}} \left(\text{or} = \frac{(1/x)\dot{C}}{6\sqrt[3]{C^2}}\right),$$

so that, for some term W, $\sqrt[3]{[EW]} \equiv AB-A\dot{B}$ and $\sqrt[3]{[GW]} \equiv 6Bx\sqrt[3]{[A^2B]}$. Then W is chosen as the least rational function which makes EW an exact cube and B^4 is chosen as the product of all quartic factors of G and A is equated to $F(x).\alpha^3$, where $F(x)$ is the product of the non-cubic factors of GW and α (of suitable dimensions in x) is to be evaluated, if possible, from the identity $EW \equiv (AB-A\dot{B})^3$.

terme make $\frac{\mathfrak{C}}{6x\sqrt{c:CC}}=v=K=\frac{L}{N}$, otherwise make $\frac{AB-A\not{B}}{6Bx\sqrt{c:BAA}}=v=\frac{L}{N}=K$. & reduce K or F to such a forme y^t^ theire Denom:s N & $\sqrt{c:GW}$ may be equall & so find y^e^ valors of y^e^ quantitys in H by compareing y^e^ termes of L, & $\sqrt{c:EW}$ together.(53)

[*Haveing y^e^ valor of v to find y^e^ valor of x.*]

Rule 1st. Haveing y^e^ valor of v when y^e^ Numerat^r^ of y^t^ valor hath but one terme, to find y^e^ valor of x.

Suppose $\frac{G}{E}=v=F$. & mu[ltiply] each terme in the denom of $\frac{yy}{F}$ by so many units as y hath dimensions in y^t^ terme & make y^e^ product $=\mho x$.(54)

Example 1st. $v=\frac{1}{2}r=\frac{G}{E}=F$. $\frac{yy}{F}=\frac{yy}{\frac{1}{2}r}$. $\frac{2yy}{\frac{1}{2}r}$(55) $=x=\frac{yy}{r}$. & $rx=yy$.

Example 2^d^. $v=\frac{y^3}{rr}=F$. $\frac{yy}{F}=\frac{yy}{\frac{y^3}{rr}}=\frac{rr}{y}$. $\frac{-rr}{y}=\mho x$. $\frac{rr}{y}=x$. $rr=xy$.

*Rule y^e^* 2^d^. Haveing y^e^ valor of v to find y^e^ valor of x.

Suppose $\frac{Gy}{E}=v=F$. Find all y^e^ least litterall divisors of G. set those together w^ch^ (being divided by y^r^ greatest numerall divisors) are = to one another. Take one divisor from every sort in wch y is found, multiply y^e^ rest by one another making y^e^ product $=B$. or if there be but one make y^t^ $=B$, if none make $B=1$. Then feigne an Equation $=A$ wch must have one dimension more y^n^ B but noe termes wanting & supposeing $\frac{A}{B}=x=C$, make $\frac{BByy}{AB-A\not{B}}=\mho v=K=\frac{N}{L}$, & reduce K or F to such a forme y^t^ $N=G$. & find y^e^ valors of y^e^ quantitys in A by compareing y^e^ termes of L & E.(56)

---

(54) Where $x=C=f(y)$, then $v=\frac{y}{dx/dy}=\frac{y^2}{\mathfrak{C}}$, or $\mathfrak{C}=\frac{y^2}{v}$.

(55) Read '$\frac{yy}{2\times\frac{1}{2}r}$'.

(56) Where $x=A/B$ and so

$$v=y\frac{dy}{dx}=\frac{B^2y^2}{AB-A\not{B}}=\frac{Gy}{E},$$

Newton takes for B^2 the product of the quadratic factors in G and finds A from the identity

$$\frac{1}{y}(AB-A\not{B})=\frac{E}{G/B^2}.$$

Example 1st. $\dfrac{ayy+2a^2y+a^3}{2y^2+3ay}=v=\dfrac{Gy}{E}=\dfrac{ay^3+2aayy+a^3y}{2y^3+3ayy}$. The least litterall divisors of $G=ay^2+2a^2y+a^3$, are a. $a+y$. & one taken from every one in w^ch^ $\underset{a+y.}{}$ y is there remaines a. $a+y$. wch multiplying one another make $ay+aa=B$, & since B is of 2 dimensions I feigne y^e^ equation $A=cy^3+dy^2+ey+f$ of 3 dim: & supposeing $\dfrac{A}{B}=\dfrac{cy^3+dy^2+ey+f}{ay+aa}=x$. [I] find

$$\frac{BByy}{\dot{A}B-A\dot{B}}=\text{♉}\,v=\frac{ay^3+2aayy+a^3y}{\begin{array}{l}2cy^3+3cay^2+2day+ea\\ \qquad\quad +dy^2 \qquad\qquad\quad -f\end{array}}=K=\frac{N}{L}.$$

& since y^e^ Numerators of N & Gy are =,(57) I compare y^e^ termes of y^e^ denom: L & E, & find by y^e^ first $2cy^3=2y^3$, or $c=1$. by y^e^ 2^d^, & 3^d^, $d=0$. by y^e^ 4th $ae-f=0$, or $ae=f$, soe y^t^ if I make $ae=0=f$. y^n^ $\dfrac{y^3}{ay+aa}=x$. or if I make $e=bb$ y^n^ is $ae=ab^2=f$. & $\dfrac{y^3+bby+abb^{(58)}}{ay+aa}=x$.

Haveing y^e^ valor of vv To find y^e^ valor of xx.

Suppose $\dfrac{Gyy}{E}=vv=FF$. Find all y^e^ least litterall & affirmative divisors of G, sort those together w^ch^ (being divided by their greatest numerall divisors) are = to one another. take 2 divisors from every sort in wch y is found, multiply all y^e^ remaining affirmative divisors into one another calling y^e^ product B, or if there remaine but one make it $=B$, if none make $B=0$.(59) Multiply all y^e^ other negative divisors into one another making y^e^ product $=P$, or if there bee but one negative make it $=P$, if none make $P=1$ & if P be not affirmative chang y^e^ signes of its valor. Againe find all y^e^ simplest or primary litterall & numerall divisors of E & sorting y^m^ together wch are =, blot out every sort consisting of an eaven number of divisors[,] add 2 to every sort remaining unless to litterall divisors in wch y is not found, & multiply them all into one another calling y^e^ product W, or if there bee but one make it $=W$, if none make $W=1$. Then extract y^e^ roote of $\dfrac{WGyy}{WE}=vv=FF$, observing y^t^ $\sqrt{WE}$ is rationall, but $\sqrt{WGyy}$ irrationall. Then feigne an equation H whos □ roote may bee extracted wch wanteth noe termes & wch hath soe many dimensions as $\dfrac{Bxx}{PW}$ hath: & suppose

---

(57) Read 'equall'. (58) That is, $\dfrac{y^3}{ay+a^2}+\dfrac{b^2}{a}$. (59) Read '1'.

$\frac{PHW}{B}=xx=\frac{A}{B}=C$. Then if B hath but one terme make $\frac{2yy\sqrt{C}}{\mathfrak{C}}=v=K=\frac{N}{L}$. Otherwise make $\frac{2Byy\sqrt{AB}}{AB-A\mathfrak{B}}=v=K=\frac{N}{L}$. & reduceing K or F to such a forme y^t^ $N=\sqrt{WGyy}$, Compare y^e^ termes of L & $\sqrt{EW}$ y^t^ y^e^ valors of y^e^ quantitys in H may be found, & y^n^ is $\frac{PHW}{B}=xx$ knowne. But here observe y^t^ when y^e^ valors of F & K are one affirmative y^e^ other negative all the signes of one of y^m^ must be changed, & soe must y^e^ signes of $\frac{PHW}{B}$ when it is negative.(60)

Example. $\frac{ayy}{a+y}=vv=FF=\frac{Gyy}{E}$. All The divisors of G are a soe y^t^ $B=a$; $P=1$.

$$\begin{matrix} & & a+y. \end{matrix}$$

The divisors of E are $a+y$ to wch adding two more they are $a+y.$ wch multiplying

$$a+y.$$

one another, produce $a^3+3aay+3ayy+y^3=W$. Then I extract y^e^ roote of

$$FF=\frac{WGyy}{WE}=\frac{ay^5+3aay^4+3a^3y^3+a^4yy}{y^4+4ay^3+6aayy+4a^3y+a^4} \text{ w}^{\text{ch}} \text{ is } \frac{yy+ay \text{ in } \sqrt{ay+aa}}{yy+2ay+aa}=\frac{y\sqrt{ay+aa}}{a+y}.$$

& since $\frac{Bxx}{PW}$ hath noe dimensions I make $H=ee$. &

$$\frac{PHW}{B}=xx=\frac{A}{B}=C=\frac{eea^3+3eeaay+3eeayy+eey^3}{a}.$$

& $\frac{2yy\sqrt{C}}{\mathfrak{C}}=v=K=\frac{N}{L}=\frac{2y\sqrt{ay+aa}}{3ey+3ea}$. & y^t^ $N=\sqrt{WGyy}$, multiply F by 2 & so compare y^e^ termes of $2\sqrt{WE}=2a+2y$ wth those of $L=3ea+3ey$. & there will be found $2a=3ea$, & $2y=3ey$, or $2=3e$, & $e=\frac{2}{3}$. &

$$\frac{A}{B}=xx=\frac{2a^3+6aay+6ayy+2y^3}{3a}.^{(61)}$$

---

(60) Where $x^2=(A/B)$ (or $=C$) and so

$$v=y\sqrt{\frac{G}{E}}=y\frac{dy}{dx}=\frac{2By^2\sqrt{[AB]}}{AB-A\mathfrak{B}}\left(\text{or }=\frac{2y\sqrt{C}}{(1/y)\,\mathfrak{C}}\right),$$

Newton finds B^3 as the product of the cubic factors of G and W as the cube of the product of the non-quadratic factors of E. From these the rest follows much as before with A evaluated by comparison with $AB-A\mathfrak{B}$.

(61) Which yields the correct result (apart from an added arbitrary constant) that

$$x=\sqrt{\frac{2}{3a}}\,(a+y)^{\frac{3}{2}}.$$

§3. AN EARLY TABULATION OF HYPERBOLIC INTEGRALS(1)

If	Its area(2) is
$\frac{x^6}{a+bx}=y.$	$\frac{x^6}{6b}-\frac{ax^5}{5bb}+\frac{aax^4}{4b^3}-\frac{a^3x^3}{3b^4}+\frac{a^4xx}{2b^5}-\frac{a^5x}{b^6}+$area of $\left[\frac{a^6}{b^7x+ab^6}=z.\right]$
$\frac{x^5}{a+bx}=y.$	$\frac{x^5}{5b}-\frac{ax^4}{4bb}+\frac{aax^3}{3b^3}-\frac{a^3xx}{2b^4}+\frac{a^4x}{b^5}-\square$ of $\left[\frac{a^5}{b^6x+ab^5}=z.\right]$
$\frac{x^4}{a+bx}=y.$	$\frac{x^4}{4a}-\frac{bx^3}{3aa}+\frac{bbxx}{2a^3}-\frac{b^3x}{a^4}+\square$ of $\frac{b^4}{a^5x+a^4b}.$
$\frac{x^3}{ax+b}=y.$	$\frac{x^3}{3a}-\frac{bxx}{2aa}+\frac{bbx}{a^3}-\square$ of $\frac{b^3}{a^4x+a^3b}.$
$\frac{xx}{ax+b}=y.$	$\frac{xx}{2a}-\frac{bx}{aa}+\square$ of $\frac{bb}{a^3x+aab}.$
$\frac{x}{ax+b}=y.$	$\frac{x}{a}-\square$ of $\frac{b}{aax+ab}.$ As may appeare by Division.
$\frac{1}{ax+b}=y.$	$\square$ of $\frac{1}{ax+b}.$
$\frac{1}{axx+bx}=y.$	$\square\frac{1}{bx}-\square\frac{a}{bb+abx}.$
$\frac{1}{ax^3+cxx}=y.$	$\square\frac{1}{cxx}-\square\frac{a}{ccx}+\square\frac{aa}{c^3+accx}.$
$\frac{1}{cx^3+ax^4}=y.$	$\square\frac{1}{cx^3}-\square\frac{a}{ccxx}+\square\frac{aa}{c^3x}-\square\frac{a^3}{c^4+ac^3x}.$
$\frac{1}{cx^4+ax^5}=[y.]$	$[\square]\frac{1}{cx^4}-[\square]\frac{a}{ccx^3}+[\square]\frac{aa}{c^3xx}-[\square]\frac{a^3}{c^4x}+[\square]\frac{a^4}{c^5+ac^4x}.$(3)

(1) Add. 3958. 4: 77r/80v. This tabulation is entered on the first page of a folded folio sheet, on whose three other sides Newton revised, about the same time (late summer 1665?) the calculations of hyperbola-areas which are printed in 1, 3, §5 above.

(2) That is, $\int y.dx$.

(3) Where k is a general positive integer, Newton evaluates

$$\int\frac{x^k}{a+bx}.dx=\sum_{0\leqslant i\leqslant k-1}\left[\int(-1)^i\frac{a^ix^{k-i-1}}{b^{i+1}}.dx\right]+(-1)^k\int\frac{a^k}{b^k(a+bx)}.dx,$$

In Generall, all lines in wch one of y^e^ unknowne quantitys (y) is but of one dimension may bee squared, (some Geometrically, others by supposeing y^e^ area of y^e^ Hyperbola, to bee knowne). & y^t^ by this Method;

First. If y^e^ numerator or Denominator in y^e^ valor of (y) bee multiplyed by x, or xx, or x^3 &c: Divide y^t^ Numerat^r^ by y^t^ Denominat^r^ (as in Decimall numbers) soe y^t^ y^e^ Quotient consist of ꝑts none of wch are of y^t^ nature. (As For Example. If $\frac{ax+b}{ax^4-cx^3}$ [4] $=y$. Then by Division, $\frac{-b}{cx^3}-\frac{a}{cxx}-\frac{ab}{ccxx}-\frac{aa}{ccx}-\frac{aab}{c^3x}+\frac{a^3c+a^3b}{ac^3x-c^4}=y$. As may appeare by multiplication.)

Secondly If y^e^ Numerat^r^ or Denom[:] bee neither of y^m^ multiplyed by x, Increase or diminish x untill y^e^ last terme of y^e^ Denominator vanish. And by these two operations used successively may the valor of y^at^ last bee reduced to such simple ꝑts y^t^ each of y^m^ may bee squared or else is an Hyperbola. Yet sometimes it happens y^t^ y^e^ last terme of y^e^ Denominator cannot bee taken away.[5]

---

§4. SYSTEMATIC APPLICATION OF INTEGRATION TECHNIQUES AS AN INVERSE METHOD OF FLUXIONS[1]

[1][2] [*The introduction of fluxions*]

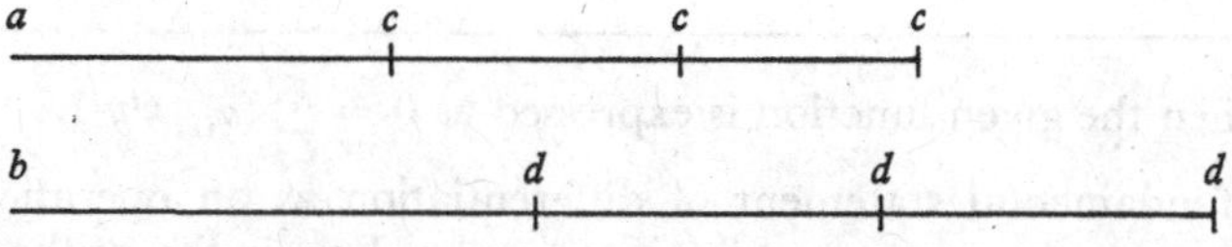

---

and $\int \frac{x^{-k}}{a+bx}.dx = \sum_{0\leqslant i\leqslant k-1}\left[\int(-1)^i\frac{a^i}{b^{i+1}x^{k-i}}.dx\right]+(-1)^k\int\frac{a^k}{b^k(a+bx)}.dx.$

Clearly $\int\frac{1}{a+bx}.dx = (1/b)\log(a+bx)$ is to be derived as the area under the hyperbola $(a+bx)y = 1$.

(4) That is, $\frac{a}{x^2(ax-c)}+\frac{b}{x^3(ax-c)}$.

(5) As the colour of the ink shows, this last sentence was added in retrospect.

(1) Add. 4000: 152^r^–163^v^. For the first time Newton introduces a fluxional notation for the derivative: specifically, where $f(x, y) = 0$ is a given (algebraic) function and t an independent variable of time, Newton denotes the fluxional 'speeds' of x and y by $p(= dx/dt)$ and $q(= dy/dt)$, so that straightforwardly $p/q = dx/dy$. He then clarifies the concept by introducing the geometrical model of points traversing lines in given periods of time. Once for all the stimulating but ultimately inadequate Cartesian method of finding the subnormal $v = y(dy/dx)$ is abandoned as a basic differentiation technique, and the true inverse nature of differentiation and integration as operations is now apparent. Much of the difficulty which Newton experienced in §2 above in integrating rational algebraic functions expressed in terms of v is banished when they are expressed in terms of $p/q = dx/dy$, and with the new freedom of an explicit notation Newton proceeds to a systematic tabulation of elementary algebraic integrals.

(2) Ff. 152^r^–153^v^.

1. If two bodys c, d describe y[e] streight lines ac, bd, in y[e] same time, (calling $ac=x$, $bd=y$, $p=$motion of c, $q=$motion of d) & if I have an equation expressing y[e] relation of $ac=x$ & $bd=y$ whose termes are all put equall to nothing. I multiply each terme of y[t] equation(3) by so many times py or $\frac{p}{x}$ as x hath dimensions in it. & also by soe many times qx or $\frac{q}{y}$ as y hath dimensions in it. the sum̄e of these products is an equation expresing y[e] relation of y[e] motions of c & d. Example if $ax^3+a^2yx-y^3x+y^4=0$. y[n] $3apxx+a^2py-py^3+aaqx-3qyyx+4qy^3=0$.(4)

2. If an equation expressing y[e] relation of their motions bee given, tis more difficult & sometimes Geometrically impossible, thereby to find y[e] relation of y[e] spaces described by these motions.(5)

If $apx^{\frac{m}{n}}=q$. then $\frac{na}{m+n}x^{\frac{m+n}{n}}=y$.(6)

As if $m=3$. $n=2$. y[n] $apx^{\frac{3}{2}}=q$, & $\frac{2a}{[5]}x^{\frac{5}{2}}=y$. Soe if $apx^{\frac{\overline{3}}{2}}=q=\frac{ap}{x^{\frac{3}{2}}}$, y[n] $m=-3$. $n=2$. & $\frac{2a}{-1}x^{\frac{-1}{2}}=\frac{-2a}{x^{\frac{1}{2}}}=y$. If y[e] valor of q consisteth of severall such termes,

---

(3) That is, when the given function is expressed as $0=\sum_{i,j}(a_{i,j}x^iy^j)$.

(4) Newton's fundamental statement of differentiation as an operation which yields a correct relation between the component fluxions p and q. Late in life, at the time of his fluxion priority dispute with Leibniz, Newton was willing to acknowledge his debt to Descartes' *Geometrie*, and Schooten's commentary upon it, for his basic ideas on indefinitely small increments (see ULC. Add. 3968.10: 131[v] and 3968.19: 290[v]), but was reticent about the origin of his clarifying concept of fluxional increase and limit-speed. In 1943 J. E. Hofmann, in commenting a later version (6, §3.1 below) of this fluxional model, made the valuable suggestion that Newton was here partially indebted to Barrow's discussion of the limit-sum of a converging geometrical progression in the third of his 1664 Lucasian lectures at Cambridge (J. E. Hofmann, *Studien zur Vorgeschichte des Prioritätstreites zwischen Leibniz und Newton um die Entdeckung der höheren Analysis*. 1. *Materialien zur ersten mathematischen Schaffensperiode Newtons* (1665–1675) = *Abhandlungen der Preussischen Akademie der Wissenschaften* (*1943*). *Math-nat. Klasse* **2** (Berlin, 1943): 115, especially note 510. Compare *Isaaci Barrow Lectiones Mathematicæ XXIII*; *In quibus Principia Matheseos generalia exponuntur*; *Habitæ Cantabrigiæ* A.D. *1664, 1665, 1666* (London, 1685): Lectio III (1664): 36–7). Newton himself wrote about 1714 in partial confirmation of the hypothesis that 'its probable that Dr Barrows Lectures might put me upon considering the generation of figures by motion, tho I not now remember it'. (ULC. Add. 3968.41: 84[v].) We must not, however, be too exact in assigning Newton's use of a geometrical model of limit-speed to a single source, and at least two other contemporaries had used the idea in print. Galileo, in the Third Day (*De Motu Locali*) of his *Discorsi e Dimostrazioni Matematiche* (Leiden, 1638, available to Newton in Salusbury's 1665 English version), had discussed uniform and uniformly accelerated motion of a point in a line, while a decade later Grégoire de Saint-Vincent gave an interesting discussion of Zeno's paradoxes of motion in much the

consider each terme severally. as if $ax+bxx=q$. y^e first terme gives $\frac{ax^2}{2}$. y^e 2^d $\frac{bx^3}{3}$. therefore $\frac{axx}{2}+\frac{bxxx}{3}=y$.

In generall multiply y^e valor of q by x & divide each terme of it by y^e logarithme(7) of x, in y^t terme: if y^t valor of q consist of simple termes.

$$\frac{-rdx^{r-1}}{ddx^{2r}+2dex^r+ee}=\frac{q}{p}. \quad \frac{2}{dx^r+e}=y.^{(8)}$$

$$\frac{\overline{m-r}\times adx^{m+r}+\overline{m-s}\times aex^{m+s}+\overline{n-r}\times bdx^{n+r}+\overline{n-s}\times bex^{n+s}}{x \text{ in } ddx^{2r}+2dex^{r+s}+eex^{2s}}=\frac{q}{p}. \quad \frac{ax^m+bx^n}{dx^r+ex^s}=y.^{(9)}$$

$$\frac{\overline{ma+\overline{3n-2m}\times bx^{n-m}}}{[2]x}\times\sqrt{ax^m+bx^n}=\frac{q}{p}. \quad \overline{a+bx^{n-m}}\times\sqrt{ax^m+bx^n}=y.^{(10)}$$

Or thus

$$\frac{\overline{ma+\overline{3n+m}\times bx^n}}{2x}\times\sqrt{ax^m+bx^{n+m}}=\frac{q}{p}. \quad \overline{a+bx^n}\sqrt{ax^m+bx^{n+m}}=y.^{(11)}$$

$$\frac{mm+8mn+15nn\times ddx^{2n}\overline{-2mn-mm}\times ee}{x^{[1-2n]}}\sqrt{ex^{[2m]}+dx^{[2]m+[2]n}}=\frac{q}{p}.$$

---

same way in his *Opus Geometricum Quadraturæ Circuli et Sectionum Coni* (Antwerp, 1647: Liber II, *De Progressionibus Geometricis*, especially Pars II: 95–106).

(5) Newton hints warily at the difficulties of the inverse operation.

(6) The fundamental integration theorem for simple powers of the variable: where

$$ax^{m/n} = \frac{q}{p} = \frac{dy}{dx}, \quad \text{then} \quad \frac{1}{(m/n)+1}ax^{(m/n)+1} = y = \int \frac{q}{p}.dx.$$

(7) That is, its index or 'dimension'.

(8) Newton for some reason has cancelled the correct '$\frac{1}{dx^r+e}=y$'. In general, where

$$y = \frac{1}{f(x)}, \quad \text{then} \quad \frac{q}{p} = \frac{dy}{dx} = \frac{-f'(x)}{[f(x)]^2}.$$

(9) More generally, where

$$y = \frac{f(x)}{g(x)}, \quad \frac{dy}{dx} = \frac{g(x).xf'(x)-f(x).xg'(x)}{x.[g(x)]^2}$$

with $f(x) = ax^m+bx^n$, $g(x) = dx^r+ex^s$ in Newton's present example. (Following his established habit he chooses to use homogenized derivatives $xf'(x)$ and $xg'(x)$.)

(10) $y = x^{-m}(ax^m+bx^n)^{\frac{3}{2}}$ and its derivative q/p.

(11) The previous example with $n \to n+m$.

And $y^n\ [x^{2n}\times]\,\overline{\overline{2m+6n}\times ddx^{2n}+2ndex^n-\overline{2m-4n}\times ee}\times\sqrt{ex^m+dx^{m+n}}=y.$

Or thus

$$\frac{\overline{3m-2n}\times maax^{m-n}+\overline{3n-2m}\times -nbbx^{n-m}}{2x}\sqrt{ax^m+bx^n}=\frac{q}{p}.$$

$$\overline{maax^{m-n}+\overline{m-n}\times ab-nb^2x^{n-m}}\sqrt{ax^m+bx^n}=y.$$ (12)

$$mac+\overline{3r-2m}\times adx^{r-m}+\overline{3m+2n}\times bcx^{m+n}+\overline{3r+2n}\times bdx^{n+r}\ \frac{\text{in }\sqrt{cx^m+dx^r}}{2x}=\frac{q}{p}.$$

$$\overline{ac+adx^{r-m}+bcx^{m+n}+bdx^{r+n}}\times\sqrt{cx^m+dx^r}=y.$$ (13)

$$\frac{\overline{3m-2n}\times\overline{2n-m}\times md^3x^{m-n}+\overline{3n-2m}\times\overline{5n-4m}\times ne^3x^{2n-2m}}{2x}\text{ in }\sqrt{dx^m+ex^n}=\frac{q}{p}.$$

$$\overline{\overline{2n-m}\times md^3x^{m-n}+\overline{2n-m}\times\overline{m-n}\times edd+\overline{n-m}\times neddx^{n-m}+\overline{3n-2m}\times ne^3x^{2n-2m}}$$
$$\times\sqrt{dx^m+ex^n}=y.$$ (14)

Or more generally,

$$\overline{3m-2n}\times mcddx^{m-n}+\overline{2m-3n}\times nceex^{n-m}+\overline{3m+2p}\times mbddx^{m+p}+\overline{3n+2p}$$
$$\times\, bedmx^{n+p}\text{ in }\sqrt{dx^m+ex^n}=\frac{q}{p}.$$

And $\overline{mddcx^{m-n}+\overline{m-n}\times cde-nceex^{n-m}+mbd^2x^{p+m}+mbdex^{n+p}}\sqrt{dx^m+ex^n}=y.$ (15)

$$\frac{\overline{5m-2n}\times\overline{2m+n}}{m}\times e^3x^{2m-n}\frac{+\overline{6n-3m}}{4n-m}\times 9nd^3x^{2n-m}\text{ in}$$
$$\sqrt{\frac{2m+n}{m}\times eex^m+\frac{9ndd}{4n-m}x^{2n-m}-3dex^n}=\frac{2qx}{p}.$$

---

(12) $y=\dfrac{(max^m-nbx^n)\,(ax^m+bx^n)^{\frac{3}{2}}}{x^{m+n}}$ and its derivative. The first form is the same apart from the substitutions $a\to d$, $b\to e$, $m\to 2m+6n$ and $n\to 2m+4n$.

(13) $y=(ax^{-m}+bx^n)\,(cx^m+dx^r)^{\frac{3}{2}}$ and its derivative.

(14) $y=\dfrac{(dx^m+ex^n)^{\frac{3}{2}}}{x^{2n+m}}\,(m(2n-m)\,d^2x^{2m}-(2n-m)\,ndex^{m+n}+(3n-2m)\,ne^2x^{2n})$

with its derivative. Newton presumably derived this rather cumbersome expression by calculating

$$F(\lambda)=\frac{d}{dx}\left(\frac{(dx^m+ex^n)^{\frac{3}{2}}}{x^{2m+n}}\,x^{\lambda}\right)=(2\lambda-m-2n)\,d.G(\lambda+m)+(2\lambda-4m+n)\,e.G(\lambda+n),$$

where $$G(\mu)=\frac{(dx^m+ex^n)^{\frac{1}{2}}}{2x^{2m+n+1}}\,x^{\mu},$$

And

$$\overline{\frac{\overline{2m+n}}{m}\times e^3x^{2m-n}\frac{\overline{+n-m}}{m}\times eedx^m\frac{+m-n}{4n-m}\times 3ddex^n+\frac{9nd^3}{4n-m}\times x^{2n-m}}$$

$$\text{in}\sqrt{\frac{2m+n}{m}\times eex^m-3dex^n+\frac{9ndd}{4n-m}x^{2n-m}}=y.^{(16)}$$

$$\frac{\overline{5n-2m}\times\overline{2m-5n}\times 3mnnb^5}{\overline{2n+m}\times\overline{16n^4-8nnmm+m^4}}\times x^{3m-2n}\frac{\overline{+2m-5n}\times 3mnb^4c}{\overline{2n+m}\times\overline{4nn-mm}}\times x^{2m-n}$$

$$\frac{\overline{+2n-2m}\times\overline{5n-2m}\times nb^3cc}{\overline{2n+m}\times\overline{4nn-mm}}x^m\frac{\overline{+2n-2m}}{2n+m}\times bbc^3x^n\frac{\overline{+m-n}}{5n-2m}\times bc^4x^{2n-m}\frac{+3m-6n}{5n-2m}$$

$$\times c^5x^{3n-2m}\ \text{in}\sqrt{\frac{\overline{5n-2m}\times nbb}{4nn-mm}x^m+bcx^n+ccx^{2n-m}}=y.$$

And

$$\frac{\overline{7m-4n}\times\overline{5n-2m}\times\overline{2m-5n}\times 3mnnb^5}{\overline{2n+m}\times\overline{16n^4-8mmnn+m^4}}\times x^{3m-2n}\frac{\overline{+4m-n}\times\overline{2m-5n}\times 3mnb^4c}{\overline{2n+m}\times\overline{4nn-mm}}x^{2m-n}$$

$$\frac{\overline{+8n-5m}\times\overline{3m-6n}}{5m-2n}\times c^5x^{3n-2m}\ \text{in}\sqrt{\frac{\overline{5n-2m}\times nbb}{4nn-mm}x^m+bcx^n+ccx^{2n-m}}=\frac{2ax}{p}.^{(17)}$$

---

and then finding the condition that $y = A.F(2m)+B.F(m+n)+C.F(2n)$ shall have the derivative $\frac{dy}{dx} = \frac{q}{p} = \alpha.G(3m)+\beta.G(3n)$. By equating the coefficients of like powers of x we derive $(3m-2n)\,Ad = \alpha$, $Ane+Bmd = 0$, $(-2m+3n)\,Be+(-m+2n)\,Cd = 0$ and $(-4m+5n)\,Ce = \beta$, or $\alpha = (3m-2n)\,(2n-m)\,md^3K$ and $\beta = (5n-4m)\,(3n-2m)\,ne^3K$, where $C/K = (3n-2m)\,ne^2$, and Newton has taken $K = 1$ for simplicity.

(15) Where $y = \left(\frac{c(m\alpha-n\beta)}{x^{m+n}}+mbdx^p\right)(\alpha+\beta)^{\frac{3}{2}}$ with $\alpha = dx^m$, $\beta = ex^n$, then

$$\frac{q}{p} = \left(\frac{c(a_1\alpha^2+a_2\beta^2)}{x^{m+n}}+mbdx^p(\alpha+\beta)\right)(\alpha+\beta)^{\frac{1}{2}},$$

in which the coefficients are determined by equating like powers of x.

(16) Where
$$y = \sum_{0\leqslant i\leqslant 3}\left(a_i\frac{\alpha^i\beta^{3-i}}{x^{m+n}}R^{\frac{3}{2}}\right)$$
with $\alpha = ex^m$, $\beta = dx^n$ and
$$R = \sqrt{\frac{A\alpha^2+B\alpha\beta+C\beta^2}{x^m}},$$
Newton chooses the coefficients a_i such that
$$\frac{dy}{dx} = \frac{q}{p} = \frac{b_1\alpha^3+b_2\beta^3}{x^{m+n}}R.$$

(17) Where
$$y = \sum_{0\leqslant i\leqslant 3}\left(a_i\frac{\alpha^i\beta^{3-i}}{x^{2(m+n)}}R^{\frac{3}{2}}\right)$$
with $\alpha = bx^m$, $\beta = cx^n$ and $R = \sqrt{[A\alpha^2+B\alpha\beta+C\beta^2]}$, Newton chooses his coefficients such that
$$\frac{dy}{dx} = \frac{q}{p} = \frac{E\alpha^5+F\alpha^4\beta+G\beta^5}{x^{2(m+n)}}R.$$

[2][18] [*Reduction of integrals by the transformation of variables.*]

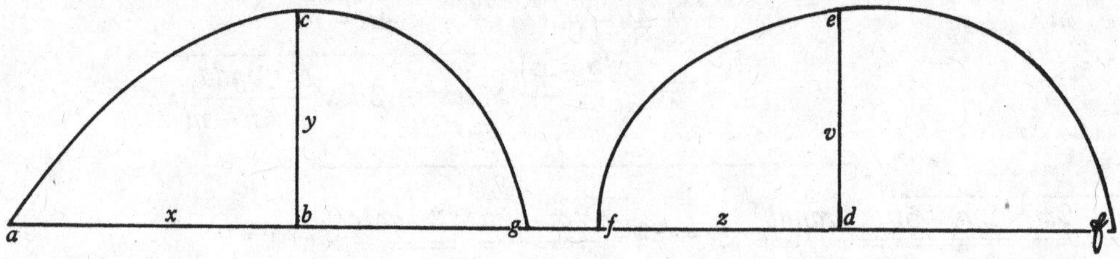

sit $ab=x$. $bc=y$. $df=z$. $de=v$.

The area *abc* of y^e^ line whose nature is	is equall to y^e^ area *fde* of y^e^ line whose nature is	supposeing y^e^ relation twixt *ab* and *fd* to bee[19]
$2xx\sqrt{c+dxx}=y.$	$\sqrt{cz+dzz}=v.$	$xx=z.$
$\sqrt{cx+dxx}=y.$	$a\sqrt{caz+daazz}=v.$	$ax=z.$[20]
$\frac{-1}{x^3}\sqrt{cx+d}=y.$	$\sqrt{cz+dzz}=v.$	$1=zx.$
$\frac{-2}{x^5}\sqrt{cxx+d}=y.$	$\sqrt{cz+dzz}=v.$	$1=zx^2.$
$\frac{-3}{x^7}\sqrt{cx^3+d}=y.$	$\sqrt{cz+dzz}=v.$	$1=zx^3.$
$3x^3\sqrt{cx+dx^4}=y.$	$\sqrt{cz+dzz}=v.$	$x^3=z.$
$4x^5\sqrt{c+dx^4}=y.$	$\sqrt{cz+dzz}=v.$	$x^4=z.$
$\frac{1}{2x}\sqrt{cx^{\frac{3}{2}}+dxx}=y.$	$\sqrt{cz+dzz}=v.$	$x=zz.$
In generall		
$nx^{n-1}\times\sqrt{cx^n+dx^{n+n}}=y.$	$\sqrt{cz+dzz}=v.$	$x^n=z.$

(18) Add. 4000: 156^r^–159^r^.

(19) Where $z=f(x)$ and $v=g(x)$, Newton's equation of the areas $(abc)=\int y.dx$ and $(fde)=\int v.dz$ implies the fundamental transforming relation $y=v(dz/dx)$.

(20) In line with the pattern of the tabulation we should have the identity $x=z$ and $\sqrt{[cz+dzz]}=v$, but Newton, apparently unwilling to use his valuable space on an example which adds nothing, slightly modifies it to yield a distinct integral.

(21) This repeated heading appears at the top of a new double page in Newton's notebook.

$2x\sqrt{c+dx^4}=y.$	$\sqrt{c+dzz}=v.$	$xx=z.$
$3xx\sqrt{c+dx^6}=y.$	$\sqrt{c+dzz}=v.$	$x^3=z.$
$-\frac{1}{x^3}\sqrt{cxx+d}=y.$	$\sqrt{c+dzz}=v.$	$1=zx.$
$-\frac{2}{x^5}\sqrt{[c]x^4+d}=y.$	$\sqrt{c+dzz}=v.$	$1=zxx.$
$\frac{1}{2x}\sqrt{cx+dxx}=y.$	$\sqrt{c+dzz}=v.$	$x=zz.$
$\frac{3}{2}\sqrt{cx+dx^4}=y.$	$\sqrt{c+dzz}=v.$	$x^3=zz.$
$-\frac{1}{2xx}\sqrt{cx+d}=y.$	$\sqrt{c+dzz}=v.$	$1=xzz.$
$\frac{-3}{2x^4}\sqrt{cx^3+d}=y.$	$\sqrt{c+dzz}=v.$	$1=x^3zz.$
In generall		
$nx^{n-1}\sqrt{c+dx^{n+n}}=y.$	$\sqrt{c+dzz}=v.$	$x^n=z.$

The area *abc* of yᵉ line	is equall to yᵉ area of yᵉ line	Supposeing y[(31)]
$2x\sqrt{c+dxx+ex^4}=y.$	$\sqrt{c+dz+ezz}=v.$	$xx=z.$
$3xx\sqrt{c+dx^3+ex^6}=y.$	$\sqrt{c+dz+ez^2}=v.$	$x^3=z.$
$4x^3\sqrt{c+dx^4+ex^8}=y.$	$\sqrt{c+dz+ezz}=v.$	$x^4=z.$
$\frac{-1}{x^3}\sqrt{cxx+dx+e}=y.$	$\sqrt{c+dz+ezz}=v.$	$1=zx.$
$\frac{-2}{x^5}\sqrt{cx^4+dxx+e}=y.$	$\sqrt{c+dz+ezz}=v.$	$1=zxx.$
$\frac{-3}{x^7}\sqrt{cx^6+dx^3+e}=y.$	$\sqrt{c+dz+ezz}=v.$	$1=zx^3.$
$\frac{1}{2x}\sqrt{cx+dx^{\frac{3}{2}}+exx}=y.$	$\sqrt{c+dz+ezz}=v.$	$x=zz.$
$\frac{-1}{2xx}\sqrt{cx+dx^{\frac{1}{2}}+e}=y.$	$\sqrt{c+dz+ezz}=v.$	$1=zzx.$
In generall.		
$nx^{n-1}\sqrt{c+dx^n+ex^{n+n}}=y.$	$\sqrt{c+dz+ezz}=v.$	$x^n=z.$

$\dfrac{b}{a+bx}=y.$	$\dfrac{1}{z}=v.$	$a+bx=z.$
$\dfrac{2bx}{a+bxx}=y.$	$\dfrac{1}{z}=v.$	$a+bx^2=z.$
$\dfrac{3bxx}{a+bx^3}=y.$	$\dfrac{1}{z}=v.$	$a+bx^3=z.$
$\dfrac{4bx^3}{a+bx^4}=y.$	$\dfrac{1}{z}=v.$	$a+bx^4=z.$
$\dfrac{-b}{axx+bx}=y.$	$\dfrac{1}{z}=v.$	$ax+b=zx.$
$\dfrac{-2b}{ax^3+bx}=y.$	$\dfrac{1}{z}=v.$	$ax^2+b=zx^2.$
$\dfrac{-3b}{ax^4+bx}=y.$	$\dfrac{1}{z}=v.$	$ax^3+b=zx^3.$
$\dfrac{b}{2a\sqrt{x}+2bx}=y.$	$\dfrac{1}{z}=v.$	$a+b\sqrt{x}=z.$
$\dfrac{3xb}{2a\sqrt{x}+2bxx}=y.$	$\dfrac{1}{z}=v.$	$a+bx^{\frac{3}{2}}=z.$
$\dfrac{-b}{[2]ax\sqrt{x}+[2]bx}=y.$	$\dfrac{1}{z}=v.$	$a+\dfrac{b}{\sqrt{x}}=z.$
$\dfrac{-3b}{[2]axx\sqrt{x}+[2]bx}=y.$	$\dfrac{1}{z}=v.$	$a+\dfrac{b}{x\sqrt{x}}=z.$

$\dfrac{a+2bx}{ax+bxx}=y.$	$\dfrac{1}{z}=v.$	$ax+bx^2=z.$
$\dfrac{a+3bxx}{ax+bx^3}=y.$	$\dfrac{1}{z}=v.$	$ax+bx^3=z.$
$\dfrac{axx-b}{ax^3+bx}=y.$	$\dfrac{1}{z}=v.$	$a[x]x+b=zx.$
$\dfrac{2a+3bx}{ax+bxx}=y.$	$\dfrac{1}{z}=v.$	$ax^2+bx^3=z.$
$\dfrac{2a+4bxx}{ax+bx^3}=y.$	$\dfrac{1}{z}=v.$	$axx+bx^4=z.$
$\dfrac{2a+5bx^3}{ax+bx^4}=y.$	$\dfrac{1}{z}=v.$	$ax^2+bx^5=z.$
$\dfrac{3a+4bx}{ax+bx^2}=y.$	$\dfrac{1}{z}=v.$	$ax^3+bx^4=z.$

$\frac{3a+5bxx}{ax+bx^3}=y.$	$\frac{1}{z}=v.$	$ax^3+bx^5=z.$
$\frac{4a+5bx}{ax+bxx}=y.$	$\frac{1}{z}=v.$	$ax^4+bx^5=z.$
$\frac{4a+6bxx}{ax+bx^3}=y.$	$\frac{1}{z}=v.$	$ax^4+bx^{[6]}=z.$
$\frac{-a+bxx}{ax+bx^3}=y.$	$\frac{1}{z}=v.$	$a+bx^2=xz.$
$\frac{-a+2bx^4}{axx+bx^5}=y.$	$\frac{1}{z}=v.$	$a+bx^3=xxz.$ (22)
$\frac{-2a-bx}{ax+bxx}=y.$	$\frac{1}{z}=v.$	$a+bx=xxz.$

In generall.

$\frac{madx^{m-1}+nbdx^{n-1}}{ax^m+bx^n}=y.$	$\frac{d}{z}=v.$	$ax^m+bx^n=z.$

Note y^t these are compounded onely of y^e first simplest Areas.

The area *abc* of y^e line	is equall to y^e area *fde* of y^e line	Supposeing that
$\frac{xx}{\sqrt{dxx-dc}}=y.$	$\sqrt{c+dzz}=v.$	$\sqrt{\frac{xx}{d}-\frac{c}{d}}=z.$
$\frac{x}{2\sqrt{dxx-dcx}}=y.$	$\sqrt{c+dzz}=v.$	$\sqrt{\frac{x-c}{d}}=z.$
$\frac{-1}{2xx\sqrt{d-cdx}}=y.$	$\sqrt{c+dzz}=v.$	$\sqrt{\frac{1-cx}{dx}}=z.$
$\frac{-1}{x^3\sqrt{d-cdxx}}=y.$	$\sqrt{c+dzz}=v.$	$\sqrt{\frac{1-cxx}{dxx}}=z.$

Or generally

$\frac{sx^{3s-1}}{\sqrt{dx^{2s}-cd}}=y.$	$\sqrt{c+dzz}=v.$	$\sqrt{\frac{x^{2s}-c}{d}}=z.$

$\frac{bcc+2bbcdx+b^3ddxx}{\sqrt{2bcx+bbdxx}}=y.$	$\sqrt{cc+dzz}=v.$	$\sqrt{2cbx+dbbxx}=z.$
$\frac{2bcc+4bbcdxx+2b^3ddx^4}{\sqrt{2bc+bbdxx}}=y.$	$\sqrt{cc+dzz}=v.$	$x\sqrt{2bc+bbdxx}=z.$

(22) Read '*xz*' and correspondingly in column 1 '$-ax$' instead of '$-a$'.

$\frac{-bccxx-2bbcdx-b^3dd}{x^3\sqrt{2bcx+bbd}}=y.$ $\sqrt{cc+dzz}=v.$ $\sqrt{2cbx+bbd}=zx.$

$\frac{-2bccx^4-4bbcdxx-2b^3dd}{x^5\sqrt{2bcxx+bbd}}=y.$ $\sqrt{cc+dzz}=v.$ $\sqrt{2bcxx+bbd}=zxx.$

In generall

$\frac{mbccx^m+2mbbcdx^{2m}+mb^3ddx^{3m}}{x\sqrt{2bcx^m+dbbx^{2m}}}=y.$ $\sqrt{cc+dzz}=v.$ $\sqrt{2bcx^m+bbdx^{2m}}=z.$

$\frac{b\sqrt{c+ad+bdx}}{2\sqrt{a+bx}}=y.$ $\sqrt{c+dzz}=v.$ $\sqrt{a+bx}=z.$

$\frac{bx\sqrt{c+ad+bdxx}}{\sqrt{a+bxx}}=y.$ $\sqrt{c+dzz}=v.$ $\sqrt{a+bxx}=z.$

$\frac{-b\sqrt{cx^2+adxx+bdx}}{2xx\sqrt{axx+bx}}=y.$ $\sqrt{c+dzz}=v.$ $\sqrt{a+\frac{b}{x}}=z.$

$\frac{-b\sqrt{cxx+adxx+bd}}{x^3\sqrt{axx+b}}=y.$ $\sqrt{c+dzz}=v.$ $\sqrt{a+\frac{b}{xx}}=z.$

In generall.

$\frac{max^m+nbx^n}{2x\sqrt{ax^m+bx^n}}\times\sqrt{c+adx^m+bdx^n}=y.$ $\sqrt{c+dzz}=v.$ $\sqrt{ax^m+bx^n}=z.$

$\frac{eb\sqrt{ac+dee+cbx}}{\overline{2a+2bx}\times\overline{a+bx}}=y.$ $\sqrt{c+dzz}=v.$ $\frac{e}{\sqrt{a+bx}}=z.$

$\frac{ebx\sqrt{ac+dee+cbxx}}{\overline{a+bxx}[\times]\overline{a+bxx}}=y.$ $\sqrt{c+dzz}=v.$ $\frac{e}{\sqrt{a+bxx}}=z.$

$\frac{-eb\sqrt{acxx+deexx+cbx}}{\overline{2ax^2+2bx}[\times]\overline{ax+b}}=y.$ $\sqrt{c+dzz}=v.$ $\frac{ex}{\sqrt{axx+bx}}=z.$

$\frac{-eb\sqrt{acxx+deexx+cb}}{\overline{ax^3+bx}[\times]\overline{axx+b}}=y.$ $\sqrt{c+dzz}=v.$ $\frac{ex}{\sqrt{axx+b}}=z.$

$\frac{4a^4cdxx+4aabcdx+bbcd}{\overline{2aaxx+2bx}[\times]\overline{4a^4ddxx+4aabddx}}=y.$(23) $\sqrt{cc+ddzz}=v.$ $\frac{cb}{2ad\sqrt{aax^2+bx}}=z.$

---

(23) Read '$\frac{-bc.\overline{4a^4cdxx+4aabcdx+bbcd}}{\overline{2aaxx+2bx}\times\sqrt{4a^4ddxx+4aabddx}}=y$'.

(24) Here also a factor $-bc$ is missing. For '$2ad$' in the denominator, read '$4aadd$'.

$$\frac{4a^4cdx^4+4aabcdxx+bbcd}{\overline{aax^3+bx}\times\overline{2adaax^4+bxx}}=y. \qquad \sqrt{cc+ddzz}=v. \qquad \frac{cb}{2adx\sqrt{a^2x^2+b}}=z.$$ [24]

$$\frac{-4a^4cd-4aabcdx-bbcdxx}{\overline{2aax+2bxx}\times\overline{2adaa+bx}}=y. \qquad \sqrt{cc+ddzz}=v. \qquad \frac{cbx}{2ad\sqrt{aa+bx}}=z.$$ [24]

$$\frac{-4a^4cdx^2-4aabcdx^4-bbcdx^6}{\overline{aa+bxx}\times\overline{2adaa+bxx}}=y. \qquad \sqrt{cc+ddzz}=v. \qquad \frac{cbxx}{2ad\sqrt{aa+bxx}}=z.$$ [24]

In generall.

$$\frac{\overline{emax^m+enbx^n}\sqrt{cax^m+cbx^n+dee}}{\overline{2ax^{m+1}+2bx^{n+1}}\times\overline{ax^m+bx^n}}=y. \qquad \sqrt{c+dzz}=v. \qquad \frac{e}{\sqrt{ax^m+bx^n}}=z.$$ [25]

[or]

$$\frac{\overline{emax^{m-1}+enbx^{n-1}}\sqrt{cax^m+cbx^n+dee}}{2aax^{2m}+4abx^{m+n}+2bbx^{2n}}=y. \qquad \sqrt{c+dzz}=v. \qquad \frac{e}{\sqrt{ax^m+bx^n}}=z.$$ [25]

$$\frac{ceb^3xx}{aa+2abxx+bbx^4}=y. \qquad \sqrt{cc\frac{-acc}{ee}zz}=v. \qquad \frac{e}{\sqrt{a+bbxx}}=z.$$ [26]

In Generall

$$\frac{nb^3cex^{\frac{3n-2}{2}}}{2aa+4abx^n+2bbx^{2n}}=y. \qquad \sqrt{cc\frac{-acc}{ee}zz}=v. \qquad \frac{e}{\sqrt{a+bbx^n}}=z.$$ [27] [25]

$$\frac{cb}{a+bx}=y. \qquad c+zv=0. \qquad \frac{1}{a+bx}=z.$$

$$\frac{2bcx}{a+bxx}=y. \qquad c+zv=0. \qquad \frac{1}{a+bxx}=z.$$

$$\frac{3bcxx}{a+bx^3}=y. \qquad c+zv=0. \qquad \frac{1}{a+bx^3}=z.$$

$$\frac{cb}{axx+bx}=y. \qquad c=zv. \qquad \frac{x}{ax+b}=z.$$

$$\frac{[4]cb}{2ax^3+2bx}=y. \qquad c=zv. \qquad \frac{xx}{ax^2+b}=z.$$

(25) Read more correctly '$-y$' here and in the preceding particular instances.

(26) Read '$aa+2abbxx+b^4x^4$'.

(27) Read '$2aa+4abbx^n+2b^4x^{2n}$'.

As before, In generall,

$$\frac{cmax^{m-1}+ncbx^{n-1}}{ax^m+bx^n}=y. \qquad c+zv=0. \qquad \frac{1}{ax^m+bx^n}=z.$$

Also

$$\frac{crdx^{r-1}+csex^{s-1}}{dx^r+ex^s}\ \frac{-cmax^{m-1}-cnbx^{n-1}}{ax^m+bx^n}=y. \qquad c=zv. \qquad \frac{dx^r+ex^s}{ax^m+bx^n}=z.$$

$$\frac{-9ac}{2bx^4+2cx}=y. \qquad a=zv. \qquad \overline{bx^3+c}\times\sqrt{bx^4+cx}=x^5z.$$

$$\frac{-3ac}{bx^3+cx}=y. \qquad a=zv. \qquad \overline{bxx+c}\sqrt{bxx+c}=x^3z.$$

$$\frac{-3ac}{2bxx+2cx}=y. \qquad a=zv. \qquad \overline{bx+c}\sqrt{bxx+cx}=xxz.$$

$$\frac{3ac}{2b+2cx}=y. \qquad a=zv. \qquad \overline{b+cx}\sqrt{b+cx}=z.$$

$$\frac{3acx}{b+cxx}=y. \qquad a=zv. \qquad \overline{b+cxx}\sqrt{b+cxx}=z.$$

As before was found,(28) In generall

$$\frac{\overline{3m+2r}\times abx^{m+r}+\overline{3n+2r}\times acx^{n+r}}{2bx^{m+r+1}+2cx^{n+r+1}}=y. \quad az=v.^{(29)} \text{ And } \overline{bx^{m+r}+cx^{n+r}}\times\sqrt{bx^m+cx^n}=z.$$

---

[3](30)

$$xx=ay. \qquad 1=v. \qquad x^3=3az.$$

$$x^3=ay. \qquad 1=v. \qquad x^4=4az.$$

$$x^4=ay. \qquad 1=v. \qquad x^5=5az.$$

$$a=xxy. \qquad 1=v. \qquad -a=xz.$$

$$a=x^3y. \qquad 1=v. \qquad -a=2xxz.$$

$$x^3=ayy. \qquad 1=v. \qquad 4x^5=25azz.$$

$$a=x^3yy. \qquad 1=v. \qquad 4a=zzx.$$

In generall

$$ax^m=y. \qquad 1=v. \qquad \frac{a}{m+1}x^{m+1}=z.$$

---

(28) Compare note (10) above.

(29) Newton intends, of course, $a = zv$, for then

$$y = v\frac{dz}{dx} = ax^{-r}(bx^m+cx^n)^{-\frac{3}{2}}.\frac{d}{dx}[x^r(bx^m+cx^n)^{\frac{3}{2}}] = \frac{a}{x}\left(\frac{3(mbx^m+ncx^n)}{2(bx^m+cx^n)}+r\right).$$

That is.

Multiply y[e] valor of y by x, & divide each terme in y[t] valor by soe many units as x hath dimensions in y[t] terme, y[e] product is y[e] area.

$\dfrac{c}{bb+2bcx+ccxx}=y.$	$1=v.$	$\dfrac{1}{b+cx}=z.$
$\dfrac{2cx}{bb+2bcxx+ccx^4}=y.$	$1=v.$	$\dfrac{1}{b+cxx}=z.$
$\dfrac{3cxx}{bb+2bcx^3+ccx^6}=y.$	$1=v.$	$\dfrac{1}{b+cx^3}=z.$
$\dfrac{-c}{bbxx+2bcx+cc}=y.$	$1=v.$	$\dfrac{x}{bx+c}=z.$
$\dfrac{-2cx}{bbx^4+2bcxx+cc}=y.$	$1=v.$	$\dfrac{xx}{bxx+c}=z.$

In generall

$$\frac{ncx^{n-1}}{bb+2bcx^n+ccx^{2n}}=y. \qquad 1=v. \qquad \frac{1}{b+cx^n}=z.^{(31)}$$

$\dfrac{b+2cx}{bbxx+2bcx^3+ccx^4}=y.$	$1=v.$	$\dfrac{1}{bx+cxx}=z.$
$\dfrac{b+3cxx}{bbxx+2bcx^4+ccx^6}=y.$	$1=v.$	$\dfrac{1}{bx+cx^3}=z.$
$\dfrac{2b+3cx}{bbx^3+2bcx^4+ccx^5}=y.$	$1=v.$	$\dfrac{1}{bxx+cx^3}=z.$

In generall

$$\frac{mbx^{m-1}+ncx^{n-1}}{bbx^{2m}+2bcx^{m+n}+ccx^{2n}}=y. \qquad 1=v. \qquad \frac{1}{bx^m+cx^n}=z.^{(32)}$$

---

(30) Add. 4000: 160r – 163v. In this final set of tables Newton considers the transform $v = \pm a$ (and mostly $v = 1$). It follows that the tabulation is straightforwardly one of integrals $z = \pm\frac{1}{a}\int y.dx$, derived from the inverse operation $y = \frac{d}{dx}(\pm az)$.

(31) That is, where $z = \frac{1}{f(x)}$ with $f(x) = b+cx^n$, then $y = \frac{dz}{dx} = -\frac{f'(x)}{[f(x)]^2}$. (Here, as in many of the preceding and following cases, Newton omits the necessary minus sign.)

(32) Here again $z = \frac{1}{f(x)}$ but now $f(x) = bx^m+cx^n$. (Compare the previous note.)

eodem modo[33]

$$\frac{-b+cxx}{bb+2bcxx+ccx^4}=y. \qquad 1=v. \qquad \frac{x}{b+cxx}=z.$$

$$\frac{-bxx-2cx}{bbxx+2bcx+cc}=y. \qquad 1=v. \qquad \frac{xx}{bx+c}=z.$$

$$\frac{-bx^4-3cxx}{bbx^4+2bcxx+cc}=y. \qquad 1=v. \qquad \frac{x^3}{bxx+c}=z.$$

$$\frac{-2bx^3-3cxx}{bbxx+2bcx+cc}=y. \qquad 1=v. \qquad \frac{x^3}{bx+c}=z.$$

$$\frac{cd-eb}{dd+2edx+eexx}=y. \qquad 1=v. \qquad \frac{b+cx}{d+ex}=z.$$

$$\frac{cd-2ebx[-ecxx]}{dd+2edxx+eex^4}=y. \qquad 1=v. \qquad \frac{b+cx}{d+exx}=z.$$

$$\frac{\overline{2cd-2eb}\times x}{dd+2edxx+eex^4}=y. \qquad 1=v. \qquad \frac{b+cxx}{d+exx}=z.$$

In generall

$$\frac{\overline{m-r}\times bdx^{m+r}+\overline{m-s}\times bex^{m+s}+\overline{n-r}\times cdx^{n+r}+\overline{n-s}\times cex^{n+s}}{ddx^{2r+1}+2edx^{r+s+1}+eex^{2s+1}}=y. \quad 1=v. \quad \frac{bx^m+cx^n}{dx^r+ex^s}=z.$$ [34]

$$\frac{-3c}{2xx\sqrt{bx^4+cx}}=y. \qquad 1=v. \qquad bx^3+c=z^2x^3.$$

$$\frac{-c}{xx\sqrt{bxx+c}}=y. \qquad 1=v. \qquad bxx+c=z^2xx.$$

$$\frac{-c}{2x\sqrt{bxx+cx}}=y. \qquad 1=v. \qquad bx+c=z^2x.$$

$$\frac{-c}{2\sqrt{b+cx}}=y. \qquad 1=v. \qquad b+cx=zz.$$

$$\frac{cx}{\sqrt{b+cxx}}=y. \qquad 1=v. \qquad b+cxx=zz.$$

---

(33) 'Similarly.'

(34) Where $f(x) = bx^m+cx^n$, $g(x) = dx^r+ex^s$ and $z = \frac{f(x)}{g(x)}$, Newton calculates

$$y = \frac{dz}{dx} = [-]\frac{g(x).xf'(x)-f(x).xg'(x)}{x.[g(x)]^2}$$

and then reverses the sequence of invention.

$\frac{3cxx}{\sqrt{b+cx^3}}=y.$	$1=v.$	$b+cx^3=zz.$
$\frac{bx^4-3c}{2xx\sqrt{bx^5+cx}}=y.$	$1=v.$	$bx^4+c=z^2x^3.$
$\frac{bx^3-2c}{2xx\sqrt{bx^3+c}}=y.$	$1=v.$	$bx^3+c=z^2xx.$
$\frac{bxx-c}{2x\sqrt{bx^3+cx}}=y.$	$1=v.$	$bxx+c=z^2x.$
$\frac{b+2cx}{2\sqrt{bx+cxx}}=y.$	$1=v.$	$bx+cxx=zz.$
$\frac{b+3cxx}{2\sqrt{bx+cx^3}}=y.$	$1=v.$	$bx+cx^3=zz.$
$\frac{b+4cx^3}{2\sqrt{bx+cx^4}}=y.$	$1=v.$	$bx+cx^4=zz.$
$\frac{2bx^3-c}{2x\sqrt{bx^4+cx}}=y.$	$1=v.$	$bx^3+c=zzx.$
$\frac{2b+3cx}{2\sqrt{b+cx}}=y.$	$1=v.$	$bxx+cx^3=zz.$
$\frac{[2]b+4cxx}{2\sqrt{b+cxx}}=y.$	$1=v.$	$bxx+cx^4=z^2.$

In generall

$\frac{mbax^m+nacx^n}{2x\sqrt{bx^m+cx^n}}=y.$	$a=v.$	$\sqrt{bx^m+cx^n}=z.$

Also more generally.

$\frac{mabx^m+nacx^n+radx^r}{2x\sqrt{bx^m+cx^n+dx^r}}=y.$	$a=v.$	$\sqrt{bx^m+cx^n+dx^r}=z.$(35)

(35) Where $z = \sqrt{[f(x)]}$ with $f(x) = bx^m+cx^n(+dx^r)$, then

$$y = a\frac{dz}{dx} = \frac{a.xf'(x)}{2x\sqrt{[f(x)]}}.$$

$\frac{-ac}{\overline{bxx+c}\sqrt{bxx+c}}=y.$	$a=v.$	$\frac{x}{\sqrt{bxx+c}}=z.$
$\frac{-ac}{\overline{2bx+2c}\sqrt{bxx+cx}}=y.$	$a=v.$	$\frac{\sqrt{x}}{\sqrt{bx+c}}=z.$
$\frac{ac}{\overline{2b+2cx}\sqrt{b+cx}}=y.$	$a=v.$	$\frac{1}{\sqrt{b+cx}}=z.$
$\frac{acx}{\overline{b+cxx}\sqrt{b+cxx}}=y.$	$a=v.$	$\frac{1}{\sqrt{b+cxx}}=z.$
$\frac{ab+2acx}{\overline{2bx+2cxx}\sqrt{bx+cxx}}=y.$	$a=v.$	$\frac{1}{\sqrt{bx+cxx}}=z.$

In generall

$\frac{mabx^{m-1}+nacx^{n-1}}{\overline{2bx^m+2cx^n}\times\sqrt{bx^m+cx^n}}=y.$	$a=v.$	$\frac{1}{\sqrt{bx^m+cx^n}}=z.$(36)

$\frac{3c}{2}\sqrt{b+cx}=y.$	$1=v.$	$\overline{b+cx}\sqrt{b+cx}=z.$
$3acx\sqrt{b+cxx}=y.$	$a=v.$	$\overline{b+cxx}\sqrt{b+cxx}=z.$
$\frac{9acxx}{2}\sqrt{b+cx^3}=y.$	$a=v.$	$\overline{b+cx^3}\times\sqrt{b+cx^3}=z.$
$\frac{-3ac}{2x^3}\sqrt{bxx+cx}=y.$	$a=v.$	$\overline{bx+c}\sqrt{bxx+cx}=zx.$(37)
$\frac{-3ac\sqrt{bxx+c}}{x^4}=y.$	$a=v.$	$\overline{bxx+c}\sqrt{bxx+c}=zxxx.$

In generall

$\frac{3nacx^{n-1}}{2}\times\sqrt{b+cx^n}=y.$	$a=v.$	$\overline{b+cx^n}\times\sqrt{b+cx^n}=z.$(38)

---

(36) Or, where $z = [f(x)]^{-\frac{1}{2}}$ with $f(x) = bx^m+cx^n$, Newton finds

$$y = a\frac{dz}{dx} = [-]\frac{a.f'(x)}{2[f(x)]^{\frac{3}{2}}}.$$

(37) Read 'zxx'.

(38) Where $z = [f(x)]^{\frac{3}{2}}$ with $f(x) = b+cx^n$, then

$$y = a\frac{dz}{dx} = \frac{3a}{2}[f(x)]^{\frac{1}{2}}f'(x).$$

(39) Where $z = x^r(bx^m+cx^n)^{\frac{3}{2}}$, then

$$y = a\frac{dz}{dx} = \frac{ax^r}{2x}(bx^m+cx^n)^{\frac{1}{2}}(3(mbx^m+ncx^n)+2r(bx^m+cx^n)).$$

$$\frac{\overline{2ba+5acx}}{2}\sqrt{b+cx}=y. \qquad a=v. \qquad \overline{bx+cxx}\sqrt{b+cx}=z.$$

$$\overline{ab+4acxx}\sqrt{b+cxx}=y. \qquad a=v. \qquad \overline{bx+cx^3}\times\sqrt{b+cxx}=z.$$

$$\frac{\overline{2abx-ac}}{2xx}\times\sqrt{bxx+cx}=y. \qquad a=v. \qquad \overline{bx+c}\times\sqrt{bx^2+cx}=z.$$

In generall

$$\frac{\overline{3m+2r}\times bx^{m+r}+\overline{3n+2r}\times cx^{n+r}}{2x}\times a\sqrt{bx^m+cx^n}=y.$$

$$a=v. \text{ and } \overline{bx^{m+r}+cx^{n+r}}\times\sqrt{bx^m+cx^n}=z.^{(39)}$$

And more generally,

$$\frac{\overline{2m+r}\times bdx^{m+r}+\overline{2m+s}\times bex^{m+s}+\overline{2n+r}\times cdx^{n+r}+\overline{2n+s}\times cex^{n+s}}{2x\sqrt{dx^r+ex^s}}\times a=y.$$

$$a=v. \quad \overline{bx^m+cx^n}\times\sqrt{dx^r+ex^s}=z.^{(40)}$$

$$\frac{3cdx}{2\sqrt{dx+e}}=y. \qquad 1=v. \qquad \overline{\frac{-2ce}{d}+cx}\times\sqrt{dx+e}=z.$$

$$\frac{3cdx^3}{\sqrt{dxx+e}}=y. \qquad 1=v. \qquad \overline{\frac{-2ce}{d}+cxx}\sqrt{dxx+e}=z.$$

$$\frac{-3cd}{2xx\sqrt{dx+exx}}=y. \qquad 1=v. \qquad \overline{\frac{-2cex}{d}+c}\times\sqrt{dx+exx}=z.$$

$$\frac{-3cd}{x^4\sqrt{d+exx}}=y. \qquad 1=v. \qquad \overline{\frac{-2cexx}{d}+c}\times\sqrt{d+exx}=z.$$

In generall

$$\frac{\overline{3m+3n}\times cdx^{3m+2n}}{2x\sqrt{dx^{3m+n}+ex^{2m}}}=y. \qquad 1=v. \qquad \overline{cx^n-\frac{2ce}{d}x^{-m}}\times\sqrt{dx^{3m+n}+ex^{2m}}=z.^{(41)}$$

---

(40) More generally, where $f(x) = bx^m+cx^n$, $g(x) = dx^r+ex^s$ and $z = f(x)\sqrt{g(x)}$, then

$$y = a\frac{dz}{dx} = \frac{a}{2x}\left(\frac{2f(x).xg'(x)+g(x).xf'(x)}{[g(x)]^{\frac{1}{2}}}\right).$$

(41) Where $R = \sqrt{[dx^{m+n}+e]}$,

$$\alpha = \frac{d}{dx}(R^3) = \frac{(m+n)\,dx^{m+n}}{2x}3R \quad\text{and}\quad \beta = \frac{d}{dx}(R) = \frac{(m+n)\,dx^{m+n}}{2x}\frac{1}{R},$$

then

$$\alpha-3e\beta = \frac{(m+n)\,dx^{m+n}}{2x}\frac{1}{R}(3R^2-3e),$$

so that $y = \frac{c}{d}(\alpha-3e\beta)$ and hence $z = \int y.dx = \frac{c}{d}(R^3-3eR)$.

$$\frac{\overline{5ab+5bbx}\times\sqrt{a+bx}}{2}=y. \qquad 1=v. \qquad \overline{aa+2abx+bbxx}\sqrt{a+bx}=z.$$

$$\overline{5abx+5bbx^3}\sqrt{a+bxx}=y. \qquad 1=v. \qquad \overline{aa+2abx^2+bbx^4}\sqrt{a+bxx}=z.$$

$$\frac{\overline{5aax+[15]abxx+[10]bbx^3}\sqrt{ax+bxx}}{2}=y. \qquad 1=v. \qquad \overline{a^2x^2+2abx^3+bbx^4}\sqrt{ax+bxx}=z.$$

$$\frac{\overline{-5abx-5bb}\sqrt{axx+bx}}{2x^{[4]}}=y. \qquad 1=v. \qquad \overline{aaxx+2abx+bb}\sqrt{axx+bx}=zx^3.$$

In generall.

$$\frac{5maax^{2m}+\overline{5m+5n}\times abx^{m+n}+5nbbx^{2n}}{2x}\sqrt{ax^m+bx^n}=y. \qquad 1=v.$$

And $\overline{aax^{2m}+2abx^{m+n}+bbx^{2n}}\times\sqrt{ax^m+bx^n}=z.$(42)

$$ax\sqrt{b+cx}=y. \qquad \frac{a}{5c}^{(43)}=v. \qquad \overline{6ccxx+2bcx-4bb}\times\sqrt{b+cx}=z.$$

$$\frac{+15aee}{x^6\sqrt{dxx+e}}=y. \qquad a=v. \qquad \overline{-3ee+4dex^2-8d^2x^4}\sqrt{dxx+e}=x^5ze.$$

$$\frac{+15aee}{2x^3\sqrt{dxx+ex}}=y. \qquad a=v. \qquad \overline{-3ee+ex^{(44)}-8d^2xx}\sqrt{dxx+ex}=x^3ze.$$

$$\frac{+15aeexx}{2\sqrt{d+ex}}=y. \qquad a+v=0. \qquad \overline{-3eex^2+4dex-8dd}\sqrt{d+ex}=ze.$$

$$\frac{+15aeex^5}{\sqrt{d+exx}}=y. \qquad a+v=0. \qquad \overline{-3eex^4+4dex^2-8d^2}\sqrt{d+exx}=ze.$$

In Generall

$$\frac{15naeex^{3n-1}}{2\sqrt{d+ex^n}}=y. \qquad a=v.$$

And $\overline{3eex^{2n}-4edx^n+8dd}\sqrt{d+ex^n}=ze.$(45)

---

(42) $z = (ax^m+bx^n)^{\frac{5}{2}}$, so that

$$y = \frac{dz}{dx} = \frac{5}{2x}(ax^m+bx^n)^{\frac{3}{2}}(max^m+nbx^n).$$

(43) Read '$\frac{a}{15cc}$', since $y = \frac{a}{c}[(b+cx)^{\frac{3}{2}}-b(b+cx)^{\frac{1}{2}}]$ and so

$$z = \int y\,.\,dx = \frac{a}{c}\left[\frac{2}{5c}(b+cx)^{\frac{5}{2}}-\frac{2b}{3c}(b+cx)^{\frac{3}{2}}\right] = \frac{2a}{15c^2}(b+cx)^{\frac{3}{2}}(3cx-2b).$$

(44) Read '$+4edx$'.

$15ddx\sqrt{dx+e}=y.$ $\quad 1=v.$ $\quad \overline{6ddxx+2dex-4ee}\sqrt{dx+e}=z.$

$60ddx^3\sqrt{dxx+e}=y.$ $\quad 1=v.$ $\quad \overline{12ddx^4+4dexx-8ee}\sqrt{dxx+e}=z.$

$\dfrac{-15dd}{x^4}\sqrt{dx+exx}=y.$ $\quad 1=v.$ $\quad \dfrac{\overline{-6dd-2dex+4eexx}\sqrt{dx+exx}}{x^3}=z.$

$\dfrac{-15dd\sqrt{d+exx}}{x^6}=y.$ $\quad 1=v.$ $\quad \dfrac{\overline{-3dd-dexx+2eex^4}\sqrt{d+exx}}{x^5}=z.$

In generall.

$\dfrac{15nddx^{2n}}{x}\sqrt{dx^n+e}=y.$ $\quad 1=v.$ $\quad \overline{6ddx^{2n}+2dex^n-4ee}\sqrt{dx^n+e}=z.$ (46)

$\dfrac{24ddxx-3ee}{x}\sqrt{dxx+ex}=y.$ $\quad 1=v.$ $\quad \left.\begin{matrix}8ddxx\\+2dex\\-6ee\end{matrix}\right\}\times\sqrt{dxx+ex}=z.$

$\overline{77ddx^4-5ee}\sqrt{dx+\dfrac{e}{x}}=y.$ $\quad 1=v.$ $\quad \left.\begin{matrix}14ddx^4\\+4dexx\\-10ee\end{matrix}\right\}\times\sqrt{dx^3+ex}=z.$

$\dfrac{8dd+eexx}{x^3}\sqrt{d+ex}=y.$ $\quad 1=v.$ $\quad \overline{-4dd-2dex+2eex^2}\sqrt{d+ex}=xxz.$

$\dfrac{\overline{45dd+3eex^4}\sqrt{dx+ex^3}}{x^6}=y.$ $\quad 1=v.$ $\quad \left.\begin{matrix}-10dd\\-8dexx\\+6eex^4\end{matrix}\right\}\sqrt{dx+ex^3}=x^5z.$

$\dfrac{32dd+4eex^4}{x^5}\sqrt{d+exx}=y.$ $\quad 1=v.$ $\quad \left.\begin{matrix}-8dd\\-4dexx\\+4eex^4\end{matrix}\right\}\sqrt{d+exx}=x^4z.$

---

(45) Where $R = \sqrt{[d+ex^n]}$,

$$\alpha = \frac{d}{dx}(R^5) = \frac{nex^{n-1}}{2R}(5d^2+10edx^n+5e^2x^{2n}),$$

$$\beta = \frac{d}{dx}(R^3) = \frac{nex^{n-1}}{2R}(3d+3ex^n) \quad \text{and} \quad \gamma = \frac{d}{dx}(R) = \frac{nex^{n-1}}{2R},$$

then $y = 3\alpha-10d\beta+15d^2\gamma$ and therefore $z = a\int y.dx = R(3R^4-10R^2d+15d^2)$.

(46) Where $R = \sqrt{[dx^n+e]}$ with

$$\alpha = \frac{d}{dx}(R^5) = \frac{5n}{2x}R(d^2x^{2n}+dex^n) \quad \text{and} \quad \beta = \frac{d}{dx}(R^3) = \frac{3n}{2x}R(dx^n),$$

then $y = 6\alpha-10e\beta$ and so

$$z = \int y.dx = R^3(6R^2-10e) = 2(3dx^n-2e)(dx^n+e)^{\frac{3}{2}}.$$

$\overline{35ddxx-8ee}\sqrt{dx+e}=y.$	$1=v.$	$\left.\begin{array}{l}10ddxx\\ +2dex\\ -6ee\end{array}\right\}\sqrt{dx+e}=z.$
$\overline{96ddx^4-12ee}\sqrt{dxx+e}=y.$	$1=v.$	$\overline{16ddx^4+4dex^2-12ee}\sqrt{dxx+e}=z.$
$\dfrac{21dd+3deex^4}{x^5}$[47] $\sqrt{dx+ex^3}=y.$	$1=v.$	$\left.\begin{array}{l}-12dd\\ -4dexx\\ +2eex^4\end{array}\right\}\sqrt{dx+ex^3}=zx^4.$
$\overline{48ddxx-15ee}\sqrt{dxx+ex}=y.$	$1=v.$	$\left.\begin{array}{l}8ddxx\\ +2dex\\ -10ee\end{array}\right\}\sqrt{dx^4+ex^3}=z.$
$\overline{117ddx^4-21ee}\sqrt{dx^3+ex}=y.$	$1=v.$	$\overline{18ddx^4+4dexx-14ee\}}\sqrt{dx^5+ex^3}=z.$
$\dfrac{12d}{x^4}\sqrt{d+exx}=y.$	$v=1.$	$\overline{-4d-4exx}\sqrt{d+exx}=zx^3.$
$\dfrac{-dd-8eexx}{xx}\sqrt{dx+exx}=y.$	$[v=1.]$	$\overline{2dd-2dex-4eex^2}\sqrt{dx+ex^2}=z.$
$\overline{63ddx^3-24eex}\sqrt{dx+e}=y.$	$v=1.$	$\left.\begin{array}{l}14ddxx\\ +2dex\\ -12ee\end{array}\right\}\sqrt{dx^5+ex^4}=z.$
$\overline{80ddx^3-35eex}\sqrt{dxx+ex}=y.$	$v=1.$	$\left.\begin{array}{l}16ddxx\\ +2dex\\ -14ee\end{array}\right\}\sqrt{dx^6+ex^5}=z.$
$\dfrac{3dd-24eexx}{x}\sqrt{dx+exx}=y.$	$v=1.$	$\overline{6ddx^5}$[48] $\overline{-2dex-8eex^2}\sqrt{dx+eex}=z.$
$\overline{99ddx^4-42eexx}\sqrt{dx+e}=y.$	$v=1.$	$\left.\begin{array}{r}18ddx^5+2dex^4\\ -16eexxx\end{array}\right\}\sqrt{dx+e}=z.$

(47) Read '$+3eex^4$'.

(48) Read '$6dd$'.

(49) Where $R=\sqrt{[dx^n+e]}$ with

$$\alpha=\frac{d}{dx}(x^{\frac{1}{2}m}R^5)=\frac{x^{\frac{1}{2}m}R}{2x}((m+5n)\,d^2x^{2n}+(2m+5n)\,dex^n+me^2)$$

and

$$\beta=\frac{d}{dx}(x^{\frac{1}{2}m}R^3)=\frac{x^{\frac{1}{2}m}R}{2x}((m+3n)\,dx^n+me),$$

then

$$y=\frac{x^{\frac{1}{2}m}R}{2x}((m+3n)\,2\alpha-(2m+5n)\,2\beta e)=[(m+3n)\,(m+5n)\,d^2x^{2n}-m(m+2n)\,e^2]\frac{x^{\frac{1}{2}m}R}{x}$$

and consequently

$$z=\int y.dx=2x^{\frac{1}{2}m}R^3((m+3n)\,(dx^n+e)-(2m+5n)\,e)$$
$$=2((m+3n)\,dx^n-(m+2n)\,e)\,x^{\frac{1}{2}m}(dx^n+e)^{\frac{3}{2}}.$$

$$\overline{8dd-35eexx}\sqrt{d+ex}=y. \qquad v=1. \qquad \overline{8ddx-2dexx-10eex}\sqrt{d+ex}=z.$$

$$\overline{120ddx^4-63eexx}\sqrt{dxx+ex}=y. \qquad v=1. \qquad \overline{20ddx^5+6dex^4-18eex^3}\sqrt{[d]xx+ex}=z.$$

$$\frac{-4dd-32eex^4}{xx}\sqrt{d+exx}=y. \qquad v=1. \qquad \overline{4dd-4dexx-16eex^4}\sqrt{d+exx}=zx.$$

$$\frac{\overline{24dd-3eexx}\sqrt{d+ex}}{x^4}=y. \qquad v=1. \qquad \overline{-8dd-2dex-6eexx}\sqrt{d+ex}=zx.$$ [3]

In Generall.

$$\overline{mm+8mn+15nn}\times ddx^{2n-1}\,\overline{-mm-2mneex^{-1}} \text{ in } \sqrt{dx^{m+n}+ex^m}=y.$$

$$1=v. \quad \&,$$

$$\overline{2m+6n}\times ddx^{2n}+2ndex^n\,\overline{-2m-4nxee} \text{ in } \sqrt{dx^{m+n}+ex^m}=z.$$ (40)

---

§5. FIRST ATTEMPTS AT THE RESOLUTION OF FLUXIONAL EQUATIONS[1]

[1][2]

[a] $[r+sy=0.]$ $\quad r+sy=0. \quad \ddot{r}+\ddot{s}y+sq=0. \quad \ddot{r}-\frac{r}{s}\ddot{s}+sq=0.$

$$\ddot{r}s-r\ddot{s}+ssq=0.$$

[b_1] $[r+sy+ty^2=0.]$ $\quad r+sy+tyy=0. \quad \ddot{r}+\ddot{s}y+\ddot{t}yy+sq+2tyq=0.$

$$\ddot{r}+sq-\frac{r\ddot{t}}{t}-\frac{s\ddot{t}}{t}+\ddot{s}y+2tqy=0.^{(3)} \qquad \frac{\ddot{r}t+sqt-r\ddot{t}}{s\ddot{t}-\ddot{s}t-2ttq}=y.$$

---

(1) Add. 3958.2: 30^v, 30^r. The writing style suggests a dating of late 1665 and the logical structure of the manuscript suggests strongly that it is later than the essay on subnormal techniques (Add. 3960.12: 199–206) printed as §3 above.

(2) F. 30^v. Newton systematically reduces $0=\sum_i(a_iy^i)$, where the a_i are functions of x, by eliminating y with the derivative $0=\sum_i(\ddot{a}_iy^i+iqa_iy^{i-1})$, where $\ddot{a}_i=da_i/dx$ and $q=dy/dx$. (Presumably, $p=dx/dt$ and $q=dy/dt$ were first chosen as the fluxions of x and y, but for simplicity Newton chose $p=1$.) The use of a double-dot superscript to denote a non-homogenized derivative is unique in Newton's early papers and is possibly a development from the homogenized notation introduced in the late spring of 1665. As we shall see, in the case of single variables Newton was soon to discard a dot-notation for the fluxions of single variables in favour of literal representations by l, m, n, p, q, r, etc. (Only in 1691 was the now-familiar dot-notation introduced by Newton.)

(3) Substituting for yy its value $-(r+sy)/t$.

$$\left.\begin{array}{l}\pm rss\ddot{t}\ddot{t}-rss\ddot{s}\ddot{t}t+r\ddot{s}\ddot{s}tt-4rstt\ddot{t}q+4r\ddot{s}tttq+4rt^4qq\\ +\ddot{r}sst\ddot{t}-\ddot{r}\ddot{s}stt+\ddot{r}\ddot{r}t^3-2\ddot{r}stttq+ssst\ddot{t}q-2sst^3qq\\ \mp rss\ddot{t}\ddot{t}+rss\ddot{s}\ddot{t}t-2\ddot{r}rtt\ddot{t}+2rstt\ddot{t}q-ss\ddot{s}ttq+sst^3qq\\ \qquad\qquad +rrt\ddot{t}\ddot{t}+2\ddot{r}stttq-2rs\ddot{t}ttq\end{array}\right\}=0.^{(4)}$$

$$\text{Or}\quad \begin{array}{c}4rt^4\\ -sst^3\end{array} qq \begin{array}{l}+s^3t\ddot{t}\\ -ss\ddot{s}tt\\ +4r\ddot{s}ttt\\ -4rstt\ddot{t}\end{array} q \begin{array}{l}+\ddot{r}sst\ddot{t}\\ -rs\ddot{s}t\ddot{t}\\ -\ddot{r}\ddot{s}stt\\ +r\ddot{s}\ddot{s}tt\\ +\ddot{r}\ddot{r}t^3\\ -2r\ddot{r}tt\ddot{t}\\ +rrt\ddot{t}\ddot{t}\end{array} =0.^{(5)}$$

$$qq\,\frac{+\ddot{s}t-s\ddot{t}}{tt}\,q\,\frac{+r\ddot{s}\ddot{s}t-\ddot{r}\ddot{s}st+\ddot{r}ss\ddot{t}-rs\ddot{s}\ddot{t}+\ddot{r}\ddot{r}tt-2r\ddot{r}\ddot{t}t+rr\ddot{t}\ddot{t}}{4rt^3-sstt}\,[=0.]$$

$[c_1]$ $[r+sy+ty^2+vy^3=0.]$

$$r+sy+tyy+vy^3=0.\qquad \begin{array}{l}\ddot{r}+\ddot{s}y+\ddot{t}yy+\ddot{v}y^3\\ \quad +sq+2tqy+3vqyy\end{array}\quad =0.$$

$$\begin{array}{l}\ddot{r}v+\ddot{s}vy+\ddot{t}vyy^{(6)}\\ +sqv+2tqvy+3vvqyy\\ -r\ddot{v}-s\ddot{v}y-t\ddot{v}yy\end{array}=0=\begin{array}{l}\ddot{r}vvy+\ddot{s}vvy^2-\ddot{t}vr-\ddot{t}vsy-\ddot{t}vtyy^{(7)}\\ -rv\ddot{v}y-sv\ddot{v}y^2+t\ddot{v}r+t\ddot{v}sy+t\ddot{v}tyy\\ +sqv^2y+2tqvvy^2-3qrvv-3vvsqy-3tvvqy^2\end{array}$$

$$\frac{r\ddot{v}+s\ddot{v}y-\ddot{r}v-\ddot{s}vy-sqv-2tqvy}{\ddot{t}v-t\ddot{v}+3vvq}=yy=\frac{t\ddot{v}r-\ddot{t}vr-3qvvr+st\ddot{v}y-s\ddot{t}vy+\ddot{r}vvy-rv\ddot{v}y-2svvqy}{sv\ddot{v}-\ddot{s}vv+\ddot{t}tv-tt\ddot{v}+tvvq}.^{(8)}$$

$$\left.\begin{array}{l}rsv\ddot{v}\ddot{v}+ssv\ddot{v}\ddot{v}y-\ddot{r}svv\ddot{v}-2\ddot{s}svv\ddot{v}y-ssvv\ddot{v}q-stvv\ddot{v}qy+r\ddot{t}vv\ddot{v}y+2s\ddot{t}vvvqy\\ -r\ddot{s}vv\ddot{v}+\ddot{r}\ddot{s}vvv+\ddot{s}\ddot{s}vvvy+\ddot{s}svvvq+\ddot{s}tvvvqy-rt\ddot{v}\ddot{v}vy-rt\ddot{v}\ddot{v}vy-3st\ddot{v}vqy\\ +r\ddot{t}tv\ddot{v}+st\ddot{t}v\ddot{v}y-\ddot{r}t\ddot{t}vv-\ddot{s}t\ddot{t}vvy-st\ddot{t}vvq-2tt\ddot{t}vvqy+3rvv\ddot{v}vqy+6svvvqqy\\ -rtt\ddot{v}\ddot{v}+stt\ddot{v}\ddot{v}y+\ddot{r}ttv\ddot{v}+\ddot{s}ttv\ddot{v}y+sttv\ddot{v}q+2tttv\ddot{v}qy-5rtvv\ddot{v}q+6r\ddot{t}vvvq\\ +rt\ddot{v}v^2q-\ddot{r}tvv^2q-stvv^2qq-2ttvvvqqy-6stvv\ddot{v}qy+5s\ddot{t}vvvqy\\ +r\ddot{t}tv\ddot{v}+r\ddot{t}\ddot{t}vv+3r\ddot{t}vvvq-s\ddot{t}tv\ddot{v}y+s\ddot{t}\ddot{t}vvy-\ddot{r}\ddot{t}vvvy\\ +rtt\ddot{v}\ddot{v}-rt\ddot{t}v\ddot{v}-2rtvv\ddot{v}q+stt\ddot{v}\ddot{v}y-s\ddot{t}tv\ddot{v}y+\ddot{r}t\ddot{v}vvy\\ +3tvv\ddot{v}rq+3r\ddot{t}vvvq+9rvvvvqq-3stv\ddot{v}vqy+3s\ddot{t}vvvqy\end{array}\right\}=0.^{(9)}$$

---

(4) Since, substituting for y in $(r+sy+tyy)\,(s\ddot{t}-\ddot{s}t-2ttq)^2 = 0$, we deduce $r\lambda^2+s\lambda\mu+tv^2 = 0$, where $\lambda = s\ddot{t}-\ddot{s}t-2ttq$ and $\mu = \ddot{r}t+sqt-r\ddot{t}$.

(5) That is, $t^2q^2+(s\ddot{t}-\ddot{s}t)\,q+\rho = 0$ where $\rho = \dfrac{\Sigma[\ddot{r}st(s\ddot{t}-\ddot{s}t)]}{t(4rt-s^2)}$.

(6) Substituting for y^3 its value $-(r+sy+tyy)/v$.

$$\left.\begin{array}{l} rsv\ddot{v}\ddot{v}-r\ddot{s}vv\ddot{v}+r\ddot{t}\ddot{t}vv-rt\ddot{t}v\ddot{v}+\ddot{r}\ddot{s}vvv-\ddot{r}svv\ddot{v}+\ddot{r}ttv\ddot{v}-\ddot{r}t\ddot{t}vv-\ddot{r}tvvvq-2rtvv\ddot{v}q \\ +9rvvvvqq-ssvv\ddot{v}q+\ddot{s}svvvq-st\ddot{t}vvq+sttv\ddot{v}q-stvvvqq-5rtvv\ddot{v}q+6r\ddot{t}vvvq \end{array}\right\}$$
$$\overline{\left.\begin{array}{l} 2\ddot{s}svv\ddot{v}-\ddot{s}\ddot{s}vvv+\ddot{s}t\ddot{t}vv-\ddot{s}ttv\ddot{v}+s\ddot{t}tv\ddot{v}-s\ddot{t}\ddot{t}vv-\ddot{s}tvvvq+2tt\ddot{t}vvq-2tttv\ddot{v}q+2ttvvvqq \\ +\ddot{r}\ddot{t}vvv-\ddot{r}tvv\ddot{v}-r\ddot{t}vv\ddot{v}+r\ddot{t}vv\ddot{v}-3rvvv\ddot{v}q+6stvv\ddot{v}q-6svvvqq-5s\ddot{t}vvvq \end{array}\right\}} = y$$

$$= y = \frac{eqq+fq+g}{hqq+kq+l}.^{(10)}$$

$[c_2]$ $[r+sy+ty^2+vy^3=0.]$

$$r+sy+tyy+vy^3=0. \qquad \frac{\ddot{r}+\ddot{s}y+\ddot{t}yy+\ddot{v}y^3}{-s-2ty-3vy^2}=q.$$

$$\begin{array}{llll} vabc & +vab\,y & +va\,yy & +vy^3=0.^{(11)} \\ & +vbc & +vb & \\ & +vac & +vc & \end{array}$$

$$\frac{\ddot{r}-a\ddot{s}+aa\ddot{t}-a^3\ddot{v}}{-s+2at-3aav}=\frac{\ddot{r}-b\ddot{s}+bb\ddot{t}-bbb\ddot{v}}{-s+2bt-3bbv}. \quad \&c^{(12)}$$

$[b_2]$ $[r+sy+ty^2=0.]$

$$r+sy+tyy=0. \qquad \frac{\ddot{r}+\ddot{s}y+\ddot{t}yy}{-s-2ty}=q=\frac{\ddot{r}+\ddot{s}a+\ddot{t}aa}{-s-2ta}=\frac{\ddot{r}+\ddot{s}b+\ddot{t}bb}{-s-2tb}.$$

$$\begin{array}{l} tab-ta[y]+t[yy]=0.^{(13)} \\ \quad -tb \end{array}$$

---

(7) Multiplying the left-hand side by vy and substituting a second time for y^3 its value derived from the original equation.

(8) Evaluating y^2 separately by the two previous equations.

(9) By cross-multiplication. On the third line the last term should be '$6svvvvqqy$'.

(10) That is, where the functions of x denoted by e, f, g, h, k and l are to be evaluated by collecting powers of q in the previous line. Substitution of this value of y in the original equation $r+sy+ty^2+vy^3 = 0$ yields a sextic in q as the eliminant. More generally, where $0 = \sum_i (a_i y^i)$ is of degree n in y, elimination of y and its powers with the derivative yields a polynomial of degree $n(n-1)$ in $q = dy/dx$.

(11) Or $v(y+a)(y+b)(y+c) = 0$, which by implicit identification with the previous equation determines $-a$, $-b$ and $-c$ as the roots of $r+sy+ty^2+vy^3 = 0$.

(12) Newton substitutes for y its values $-a$, $-b$ and $-c$. Clearly he wishes by these equations to eliminate a, b, c by using the root-properties

$$abc = r/v, \quad ab+bc+ca = s/v \quad \text{and} \quad a+b+c = t/v.$$

However, Newton sees the computational complexities involved and passes to the simpler case of $r+sy+ty^2 = 0$ in illustration.

(13) By the implicit identity of these two quadratics, Newton sets a and b as the roots of $r+sy+ty^2 = 0$.

$$[qq=]\,\frac{\begin{matrix}\ddot{r}\ddot{r}+\ddot{r}\ddot{s}b+\ddot{r}\ddot{t}bb\\+\ddot{s}\ddot{r}a+\ddot{s}\ddot{s}ab+\ddot{s}\ddot{t}abb\\+\ddot{t}\ddot{r}aa+\ddot{t}\ddot{s}aab+\ddot{t}\ddot{t}aabb\end{matrix}}{ss+2sta+2stb+4ttab}\,[=]\,\frac{\ddot{r}\ddot{r}-\frac{\ddot{r}\ddot{s}s}{t}+\frac{\ddot{s}\ddot{s}r}{t}+\frac{\ddot{r}\ddot{t}ss}{tt}-\frac{2\ddot{r}\ddot{t}r}{t}-\frac{\ddot{s}\ddot{t}rs}{tt}+\frac{\ddot{t}\ddot{t}rr}{tt}{}^{(14)}}{ss-2ss+4tr}$$

$$=\frac{\ddot{r}\ddot{r}tt-2r\ddot{r}\ddot{t}t+rr\ddot{t}\ddot{t}+r\ddot{s}\ddot{s}t-\ddot{r}\ddot{s}st-rs\ddot{s}\ddot{t}+\ddot{r}ss\ddot{t}}{4rt^3-sstt}.{}^{(15)}$$

$[d_1]$ $[s+ry^n=0.]$ $\quad 0=s+ry^n.\quad \ddot{s}+\ddot{r}y^n+nrqy^{n-1}=0.$

$$\ddot{s}-\frac{\ddot{r}s}{r}-n\frac{sq}{\sqrt{n:\frac{-s}{r}}}[=0.]\quad r\ddot{s}-\ddot{r}s-nq\sqrt{n:r^{n+1}s^{n-1}}{}^{(16)}=0.\quad \frac{r\ddot{s}-\ddot{r}s}{n\sqrt{n:r^{n+1}s^{n-1}}}=q.$$

$[b_3]$ $[s+ary+ry^2=0.]$ $\quad s+ary+ryy=0.$

$$qq=\frac{\ddot{r}\ddot{r}ss-2r\ddot{r}\ddot{s}s+rr\ddot{s}\ddot{s}}{r^4aa-4r^3s}.\quad \text{Or}\quad q=\frac{r\ddot{s}-\ddot{r}s}{r\sqrt{aarr-4rs}}.$$

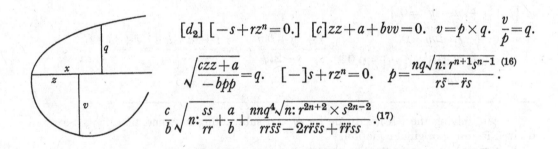

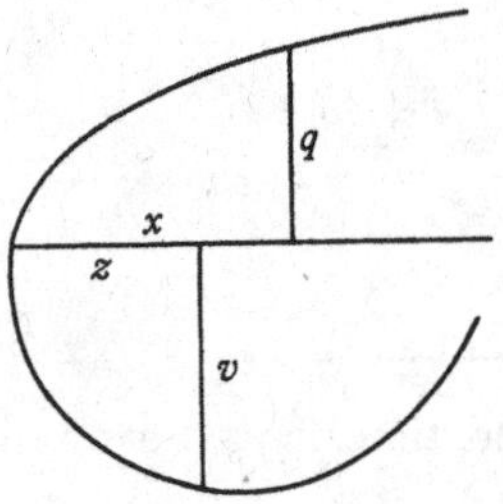

$[d_2]$ $[-s+rz^n=0.]$ $[c]zz+a+bvv=0.$ $v=p\times q.$ $\frac{v}{p}=q.$

$$\sqrt{\frac{czz+a}{-bpp}}=q.\quad [-]s+rz^n=0.\quad p=\frac{nq\sqrt{n:r^{n+1}s^{n-1}}}{r\ddot{s}-\ddot{r}s}.{}^{(16)}$$

$$\frac{c}{b}\sqrt{n:\frac{ss}{rr}}+\frac{a}{b}+\frac{nnq^4\sqrt{n:r^{2n+2}\times s^{2n-2}}}{rr\ddot{s}\ddot{s}-2r\ddot{r}\ddot{s}s+\ddot{r}\ddot{r}ss}.{}^{(17)}$$

(14) Newton eliminates the homogeneous functions of a and b through the equations $ab=r/t$, $a+b=-s/t$, so that $a^2+b^2=(s^2-2rt)/t^2$.

(15) By compounding the last equality, Newton finds the products of the roots q in $[b_1]$.

(16) Strictly '$-nq\sqrt{n:-r^{n+1}s^{n-1}}$'.

(17) The accompanying figure is a reminder that, as in §4.2 and 3 above, we are to equate $\int q.dx$ with $\int v.dz$, so that $v=q(dx/dz)$ and therefore $p=dx/dz$. Further, by $[d_1]$

$$\frac{q}{p}=\frac{r\ddot{s}-\ddot{r}s}{n\sqrt[n]{[r^{n+1}s^{n-1}]}}$$

is the derivative of $-s+rz^n=0$ (in which r and s are functions of x) or $q/p=dz/dx$, and so taking $q=1$ we have $v=dx/dz=p$. Finally, Newton eliminates z and v between

$$cz^2+a+bv^2=0,\quad -s+rz^n=0$$

and its derivative. Note that b_3 follows from b_2 by taking $r\to s$, $s\to ar$, $t\to r$.

[2][18]

Find all y^e^ divisors of y^e^ first terme in y^e^ given Equation & set y^m^ together of a sorte. from each sorte substract one divisor [&] y^e^ rectangle of y^e^ rest call A y^e^ coefficient of y^e^ first terme of y^e^ desired equation. Add y^e^ x'es greatest dimension of A to every greatest dimen: of ye other termes & also soe many dimensions more as q in y^t^ terme doth want of its greatest dimensions. & from those substract y^e^ greatest dimension of x in y^e^ first terme. y^e^ differences shall give you y^e^ greatest dimensions of x in y^e^ correspondent termes of y^e^ desired Equation.

Or thus If, $aq^n+bq^{n-1}+cq^{n-2}+dq^{n-3}+eq^{n-4}$ &c $=0$. is given &

$$fy^n+gy^{n-1}+hy^{n-[2]}+ky^{n-3}+ly^{n-4}\ \&c=0$$

is sought.[19] I first seeke f thus. viz[:] 1. find al[l] y^e^ prime divisors of a in wch x is, one of a sorte. By y^e^ rectangle of all wch divid a & y^e^ quote shall bee f. Or 2: if $b=0$, divide (a) once & againe, if it may be, (i.e. if it may bee, by y^e^ square of y^e^ divisor) by every one of y^e^ said divisors. y^e^ product call f. 3. If $b=c=0$, divide a once & again & againe by every one of y^e^ said prime divisors, calling y^e^ product f. 4. if $b=c=d=0$. divide a once & againe & againe & againe if it may be, by every one of y^e^ s^d^ Divisors &c.

☞ Where note y^t^ y^e^ Question is insoluble, if, w^n^ $\frac{b}{a}$ is reduced to its least terme some prime divisors (r) in wch x is may divide a whose square (rr) cannot divide it.

☞ Having thus found f, set downe severally y^e^ number of dimensions wch x hath in b, c, d, e, &c. to each of wch add soe many units more as y^e^ dimensions of q in y^t^ terme are lesse y^n^ its greatest dimensions n, & from each of those summes substract ye difference of ye greatest dimensions wch x hath in a & f. The remainder shall bee ye greatest number of dimensions wch x can have in ye correspondent termes g, h, k, l, &c. Therefore feigne an equation for fy^n+gy^{n-1} &c. in wch x is in each terme of noe more then y^e^ said dimensions, noe termes being wanting. Whereby get an Equation of y^e^ same forme wth aq^n+bq^{n-1} &c. & compare their termes &c:

Or better you may working by degrees 1st put $aq+b=0$ & thereby seeke $fy+g=0$. &c.[20]

(18) f. 30^r^. Newton attempts to reverse the operation of eliminating y between $0=\sum_i(a_i y^i)$ and its derivative $0=\sum_j(b_j q^j)$, where the a_i and b_j are functions of x, and $q=dy/dx$. In other words, he tries to reverse the process of finding a first-order differential equation from a given polynomial, and so find a method of resolving a general equation of this type. Quickly he finds that this is no easy task and ends in rather subdued manner by seeking a solution in the form of a polynomial (which will be infinite in general) whose terms are to be evaluated singly 'working by degrees'.

(19) Note that $q=dy/dx$ and the coefficients $a, b, c, \ldots, f, g, h, \ldots$ are functions of x.

(20) Newton has cancelled '2dly put $aqq+bq+c=0$ & thereby seeke $fyy+gy+h=0$. 3dly put $aq^3+bqq+cq+d=0$.'

Note y^t^ if 1: $b=0$. 2: $b=c=0$. 3: $b=c=d=0$. &c. Then 1: $\frac{g}{f}=0$. 2: $\frac{g}{f}$ & $\frac{h}{f}$. 3: $\frac{g}{f}$ & $\frac{h}{f}$ & $\frac{k}{f}$. [&c] is a knowne quantity. viz: in wch x is not. Perhaps they may bee put equall to 0.(21)

☞ Note also, y^t^ if y^e^ odd termes in y^e^ given equation are wanting they shall bee wanting in y^e^ sought eq.(22)

---

(21) Newton is guessing at the unknown on the analogy of the few particular results he has derived in [1].

(22) Newton offers these remarks merely as tentative suggestions and not as exact rules (which they certainly are not). As he well realizes, the difficulty in resolving the fluxional equation $f(x, q) = 0$, where $q = dy/dx$, as the polynomial $g(x, y) = 0$ is the evaluation of the term fy^n: therefore the remaining terms of $g(x, y)$ may be found by assuming for it an arbitrary polynomial which may be evaluated step by step by identifying its derivative with $f(x, q) = 0$. Six years later in 1671, when he began to collect his early calculus work in the tract *Methodus fluxionum*, Newton elaborated an improved method for the resolution in series of the general first-order linear differential equation $f(x, q) = 0$, but not for a further twenty years did he attack the general problem of the exact resolution of fluxional equations.

6

THE GENERAL PROBLEMS OF TANGENTS, CURVATURE AND LIMIT-MOTION ANALYSED BY THE METHOD OF FLUXIONS

[October 1665–May 1666]

From the originals in the University Library, Cambridge

§1.[1]

[30? October 1665].

[1] *How to draw tangents to Mechanichall lines*[2]

In the description of any Mechanicall line what ever there may be found two such motions wch compound or make up ye motion of ye point describing it

---

(1) Add. 3958.2: 34^{r}–35^{r}, 37^{v}. These mathematical fragments are extracted from two small sheets, folded in fours and pinned together at their 'spine'. Caught in with the pin, a sliver from a third sheet now torn away confirms internal evidence in the text, revealing that at least one further sheet with mathematical notes was attached. The entries on 34^{r} ([1] below) certainly antedate the more polished entry of 8 November in the Waste Book (Add. 4004: 50^{v}–51^{r}, printed as §2) and the date suggested is that which Newton set at the head of a draft, penned for his mother 'Hannah Smith', of a rent acquittal for 'The Sume of Forty pounds for one halfe yeares rent dew unto me ye 25^{t} of March past by Joseph Whiteing of Manisenderby in County Lin[colne] for certaine lands wch he holdeth by lease assigned over unto him by Mr Tho Gay'. Apart from the purely mathematical notes here transcribed the sheets contain some numerical computations, the results of Newton's experimentation with variant styles of handwriting and several anatomical drawings with descriptive text (in Newton's hand).

(2) That is, as opposed to the 'Geometricall' (or algebraic) curves to which Newton has hitherto restricted his attention. In its present form the distinction goes no further back than Descartes, who elaborated it in the introduction to the second book of his *Geometrie*. (See *Geometria*: Liber II: 17–19.) The classical cleavage between the 'geometrical' and the 'mechanical' restricted the former to the straight line and circle—a distinction still made by Viète, who in his *Apollonius Gallus* (Paris, 1600) could classify as 'mechanica' Adriaen's resolution by hyperbolas of Apollonius' problem of the construction of a circle to touch three circles given in position. Descartes' own distinction was to be reformulated a little later by Leibniz as the now familiar one between algebraic and transcendental curves.

& by those two motions may ye motion of y^{t} point bee found (*a*) whose determinacon[3] is in a tangent (*b*) to ye crooked line.[4]

As in ye Spirall line *can*[5] ye point *a* is compounded of a motion from ye center *c* & a motion in ye circumference towards *e*. wch two motions are as *ac* to *bma* (through w^{ch} *a* hath moved). therefore make *af*:*ae*::*ac*:*bma*. & complet y^{e} pgrm[6] whose diamiter *ad* shall touch y^{e} Spirall.[7] Or bisect *fe* in *r*.[8]

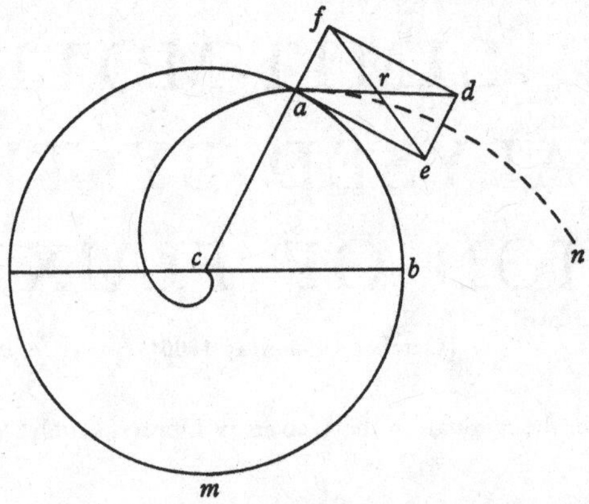

(3) That is, the 'determinacon of y^{e} motion of y^{e} point', or the instantaneous direction of the point as it describes the curve (*a*). Yet again, the terminology is Descartes' and Newton clearly absorbed it as he read through his *Principia Philosophiæ* (Amsterdam, 1644). According to Descartes the law of inertia (his *Prima lex naturæ* = XXXVII: 54–5) demands that 'omne id quod movetur, in singulis instantibus, quæ possunt designari dum movetur, determinatum esse ad motum suum continuandum versus aliquam partem, secundùm lineam rectam', and in a following example where he considered the centrifugal pull of a stone *A* swung around the head he wrote that 'eo instanti, quo est in puncto *A*, determinatus quidem est ad motum... secundùm lineam versus *C*, ita scilicet ut linea recta *AC*, sit tangens circuli'. (See *Principia Philosophiæ*: Pars secunda, XXXIX, *Altera lex naturæ*: 55–7, especially 56.) Newton himself used the concept widely in a series of notes on motion and force, based on his reading of Descartes' *Principia*, which he entered in his Waste Book some time in the autumn of 1664. (See ULC. Add. 4004: 10^{r}–15^{v}, 38^{v}.)

(4) Newton states the application of the 'composition of motions' to the tangent-problem: the tangent is to be found as the instantaneous direction of a given point in a curve suitably compounded of two generating motions. We are today, perhaps, more familiar with Roberval's development of this approach in his *Observations sur la composition des Mouvemens, et sur le moyen de trouver les Touchantes des lignes courbes*, written about 1640 but little known till it was printed in the posthumous *Divers Ouvrages de M. Personier de Roberval* (first printed in the *Memoires de l'Academie royale des Sciences* (1693), reprinted in the *Memoires de l'Academie royale des Sciences depuis* 1666 *jusqu'à* 1699, **6** (Paris, 1730): 1–478, especially 1–89). As Roberval himself—if not the modern exponents of his method—well knew, the plausible introduction of the parallelogram of forces into the approach is possible only where, as in the cycloid, the two generating motions are independent with the 'determination of the motion' of the describing point on the curve given truly as the resultant of the two component limit-motions. Where, however, the describing point is given, in some co-ordinate system (Cartesian or otherwise), as the meet of the two component co-ordinates, the parallelogram rule is inapplicable, and its use results generally (as in the quadratix below) in an erroneous construction, though by chance (as with the ellipse) it may yield a correct result. Newton, clearly ignorant of the detailed complexities of Roberval's work and

Or in y^e^ Quadratrix.(9) (if *cae* = recto) y^e^ motion of (*a*) towards (*g*) is to its motion towards (*e*) as (*bc*) is to *hap*. therefore make *af*:*ae*::*bc*:*hap*. & bisect *fe* or complete y^e^ pgrm(6) whose diagonall is tangent to y^e^ Quadratrix.(10)

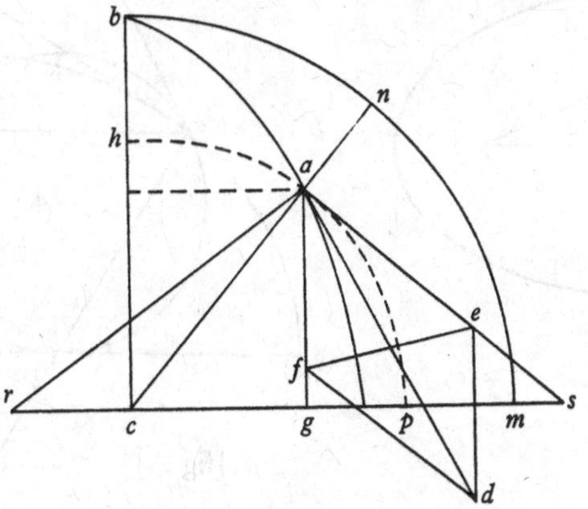

Or in y^e^ trochoides(11) (*a*) moves equally to *e* & *f* &c.(12)

still unaware of the hidden dangers in store, plunges straightaway and uncritically into the prosecution of particular examples by the sole aid of the parallelogram of limit-motions.

(5) Newton has cancelled 'Helix', proof that he derived this first example of the Archimedean spiral from Viète's *Variorum de Rebus Mathematicis Responsorum Liber VIII.* (= *Opera Mathematica* (ed. Schooten) (Leyden, 1646): Cap. XIV: 387–91), though it is referred to by Descartes in his *Geometrie* as a mechanical curve (*Geometria*: Liber II: 18). (Independently, Roberval had already considered the spiral in a similar way as his *Exemple* 8.)

(6) Read 'parallelogram'.

(7) The resulting construction and indeed the mathematics of the method are correct, but the argument by composition of motions is irrelevant. The crucial reason why the point *d* must be on the tangent at *a* is that *df* and *de* are normal to *af* and *ae*, where *af* and *ae* are proportional to the limit-increments of *ca* and $\widehat{bma}$.

(8) Noting that the diagonals *ad* and *fe* bisect each other.

(9) See Viète's *Variorum de Rebus Mathematicis Responsorum Liber VIII* (note (5)): 366: Prop. III. Descartes cites this, along with the Archimedean spiral, as an example of a mechanical curve (*Geometria*: 18), while it is also Roberval's *Exemple* 9.

(10) The construction is irredeemably false, as Newton was soon to realize. (See §2 below.) It is interesting to note that Descartes gave a similarly incorrect construction of the quadratix tangent, apparently for the same reason. (See *R. Des-Cartes Opuscula Posthuma, Physica et Mathematica*, Amsterdam, 1701: *Excerpta ex MSS. R. Des-Cartes*: *Tangens Quadratariæ*.)

(11) Roberval's *Exemple* 11. Newton seems to have taken his knowledge of the cycloid from Schooten's account of it: 'Qualem Dominus des Cartes excogitavit, atque jam pridem ejus exemplum R. P. Mersenno per literas ostendit in curva, quæ Cycloides sive Trochoides appellatur.' (Schooten, *Commentarii in Librum II*, *O*: 264–70. In particular the *Verba Authoris* on pp. 268–70 are a Latin translation of Descartes' letter to Mersenne of 23 August 1638 = *Œuvres* (ed. Adam and Tannery), **2** (Paris, 1898): 309–13). However, Descartes had previously referred to it obliquely and loosely in his *Principia Philosophiae* (Amsterdam, 1644: Pars secunda, XXXIII: 50) in connexion with his exposition of the composition of motions, a passage familiar to Newton.

(12) Here the 'determination of the motion of the point *a*', *ad*, is correctly constructed as the resultant of the two motions *ae* and *ed* (or *af* and *fd*), which are in the ratio of the limit-motions of *a* parallel to the base *bkl* and in the circle arc $\widehat{ka}$, and so equal.

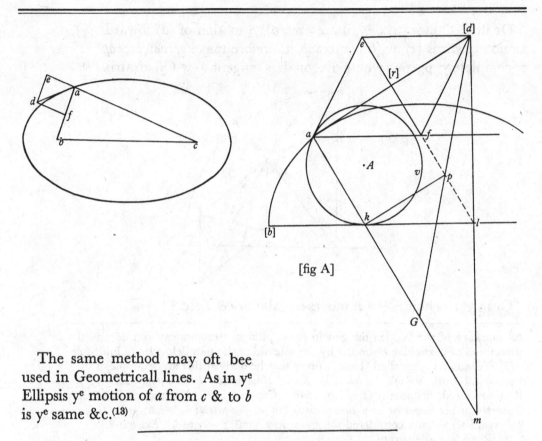

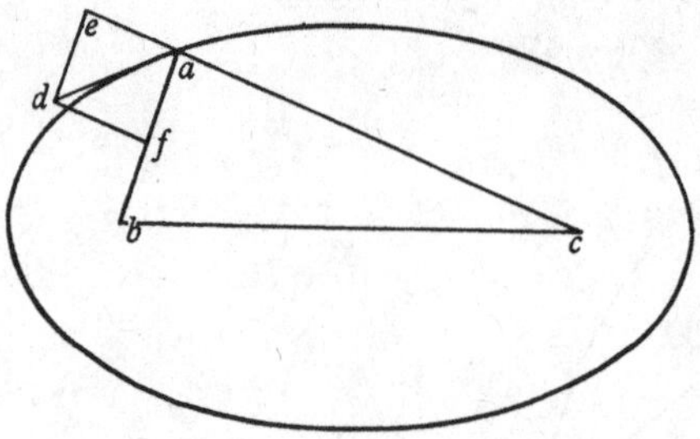

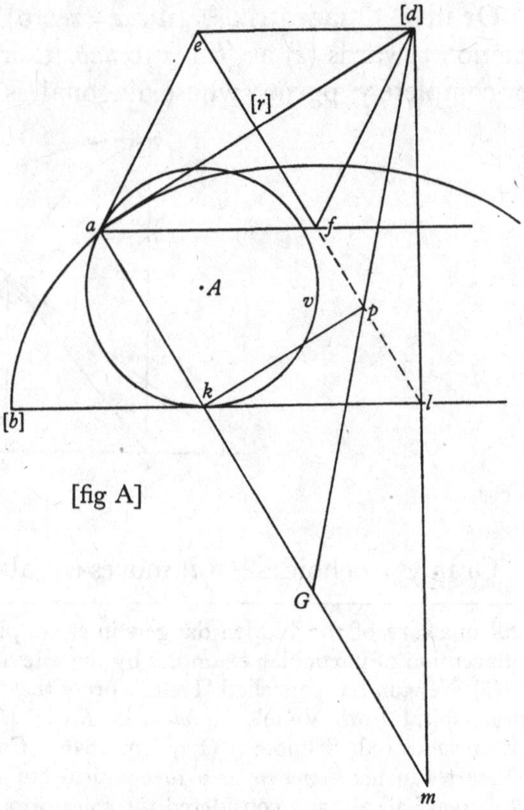

[fig A]

The same method may oft bee used in Geometricall lines. As in y^e^ Ellipsis y^e^ motion of *a* from *c* & to *b* is y^e^ same &c.(13)

The motion of y^e^ point *a* & its determinacon being thus found (& thereby y^e^ tangent [*ad*]). get y^e^ motion of some other point in y^e^ perpendicular (as of *k* to *l*.) & draw two lines through wch those two points would move in y^e^ same time (as $kl = af$. & *ac*(14). fig A(15)). Then through y^e^ second point draw a parallel & through y^e^ end [*l*] of its motion draw a perpendicular to y^e^ tangent (as *kp*, *lpe*). And through their meeting point & y^e^ end of(16) the first(17) [draw *dp*. & y^n^ *g* y^e^ meeting point of *dp* & *ak*(18) will be ye center of a circle of like crookednesse wth ye Mechanicall line at y^e^ point *a*.](19)

(13) The construction is false though, by accident, the constructed point *d* is in the tangent to the ellipse at *a*. (Correctly, where *ea* and *fa* are in the ratio of the limit-increments of *ac* and *ab*, with *fa* taken negatively as a decrement, and so (since $ba + ac$ is constant) $ea = fa$, the point *d* is to be constructed as the meet of the normals at *e* and *f* to *ea* and *fa*.)

(14) Read '*ad*', in line with his previous denotation of points. (Newton has not marked in the point *d* on his diagram, but here misreads the initial letter of the word 'complete', in his remarks on the quadratrix above, as the marking '*c*' of the point!)

(15) In Newton's diagram the letter *A* seems to have been inserted initially to denote the centre of the generating circle.

(16) Newton has cancelled 'their motion'.

(17) That is, the tangent *ad*.

(18) The point '*g*' is marked as a capital in Newton's diagram.

[As in y^e^ Trochoides.][20] Take $am^{(21)} = 2ak$, & [y^n^] $m^{(21)}$ is y^e^ center & $am^{(21)}$ y^e^ radius of a circle of like crookednesse wth y^e^ Trochoides at y^e^ point a.[22] soe y^t^ it hath no least crookednesse & at y^e^ most [is] twice as streight as y^e^ circle *akv*.[23]

---

[2][24]

mot b = mot(e) a b:mot e ad $d::b^3:abb-a^3$.
mot de e ad d:mot de c ad $a::be:ab::b:a$.[25]

---

(19) The last part of this sentence, continued by Newton on a sheet pinned to the present one and now lost (note (1)), is here conjecturally restored. Newton finds the 'center of curvity' (as he will call it) at the point a on an arbitrary curve as the limit-meet g of two indefinitely close normals to the curve, ak and $d'p'$. Having found the ratio $(ad':kl')$ of the limit-increments of a in the tangent ad and of k in kl, he finds the corresponding ratio $(kp':kl')$ of k in $kp \| ad$ and in kl, and so the ratio $(ad':kp' = ad:kp)$ of the increments of a in ad and k in $kp \| ad$. Finally, since $d'p'$ is through the centre of curvature g, so is dp, and therefore g is given as the meet of ak and dp.

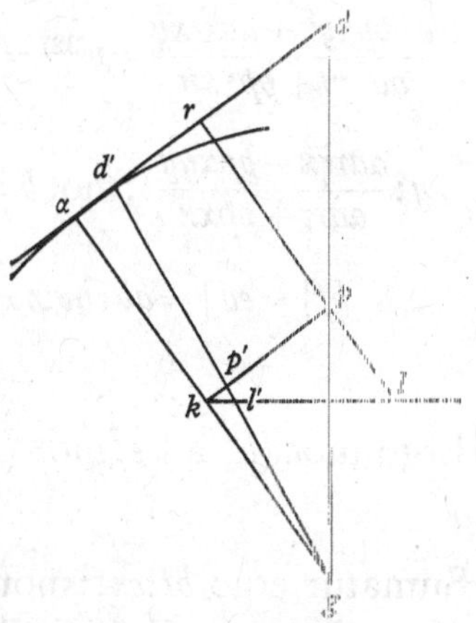

(20) In Newton's text this paragraph precedes the first. However, Newton evidently had to have the method clear in his mind before he could derive the particular result for the cycloid, and for that reason the transposition seems both justified and natural. The bracketed phrase has been added further to smooth the transition.

(21) For each occurrence of 'm' read 'g'. In Newton's diagram the point m is the meet of ak and dl, and it seems that Newton, forgetting that the increments of a and k must be in parallel lines for the equality $ad':kp' = ad:kp$ to hold, was first tempted to suppose that the ratios $ad':kl'$ and $ad:kl$ were equal (and so m truly at rest). If he did so, the absurdity of the construction for the centre of curvature at the vertex b must quickly have become apparent. (Note that two lines, undenoted but probably trial attempts at the construction of g, have been omitted from Newton's figure as here reproduced for the sake of clarity.)

(22) For $kp = ar = \frac{1}{2}ad$, and so $kg = \frac{1}{2}ag$, or $ag = 2ak$, with ag the radius of curvature at the point a on the cycloid (note (19)).

(23) The chord ak has its maximum length when it is the diameter through k perpendicular to the base bkl, and is zero when a coincides with k at the vertex b.

(24) The text begins (on f. 37^v^) in mid-sentence. A diagram, clearly concerned with the construction of tangents but otherwise seemingly inexplicable, is omitted.

(25) In translation,

'motion of b = motion of e from b: motion of e to $d::b^3:a\,(b^2-a^2)$.
motion from e to d: motion from c to $a::be:ab::b:a$'.

Newton's text lacks a diagram to which we may revert, but these fragmentary remarks clearly refer to a slightly modified form of the Archimedean spiral studied below. (Apparently Newton takes $ab = a$, $be = b$, so that 'mot de e ad d: mot de c ad a' $= be:ed$ (or ab) $= b:a$, but the first line remains mysterious.)

$ab = x$. $an = a =$ diametro[26] unius girationis.

$\frac{px}{r}$[27] $=$ peripheriæ $qsrbq$.

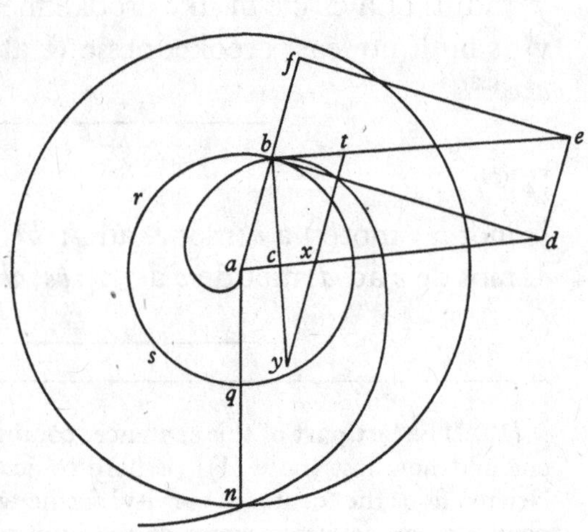

$an : ab :: a : x :: qsrbq = \frac{px}{r} : qsrb = bd = \frac{pxx}{ar}$.[28]

be[29] $= \sqrt{\frac{aarrxx + ppx^4}{aarr}} = \frac{x}{ar}\sqrt{aarr + ppxx}$.

$\frac{arx}{\sqrt{a^2r^2 + p^2x^2}} = ac = y$.[30]

$aarrxx = aarryy + ppxxyy$.[31]

$\frac{-ppxy^3 + aarrxy}{aarry + ppxxy} = v$[32] $= \frac{aarrx - ppxyy}{aarr + ppxx}$.

$y : \frac{aarrx - ppxyy}{aarr + ppxx}$ :: mot b ab a : mot c ab a :: $y : v$[33] $:: aarry + ppxxy : aarrx - ppyyx$.

$bf[= ed] = ab : be :: x : \frac{x}{ar}\sqrt{aarr + ppxx} :: ar : \sqrt{aarr + ppxx}$ :: Mot (b) ab a : Mot (b) ad $e :: y : x$.

Ergo mot (b) ad e : mot (c) ab $a :: \sqrt{aarr + ppxx} \times aarry + ppxxy : ar \times aarrx - ppyyx$ $:: aarr + ppxx : aarr - ppy^2$.

Sumatur ergo $bt : cx ::$ mot b ad t : mot c ad x[34] $:: aarr + ppxx : aarr - ppyy$.[35] duc ty et erit $by =$ radio circuli æqualis curvitatis &c.[36]

$ppxx + ppyy : \sqrt{xx - yy} :: aarr + ppxx : by :: \frac{aarrxx}{yy} : by$.[37]

(26) Read 'radio' and translate 'equal to the radius of one revolution'.

(27) That is, $\frac{\text{p(eriphery)}}{\text{r(adius)}} . x = 2\pi x$.

(28) The defining equation of the Archimedean spiral (b)—that is, where

$$q\hat{a}b = \theta\ (\widehat{qsrb} = x\theta), \quad x = a(r/p)\,\theta.$$

(29) Taking $ed = ab$ perpendicular to bd at d.

(30) $ab/(ed/eb) = ac$, since the triangles abc and ebd are similar.

(31) Newton's form of defining equation for the spiral, where $ab = x$ and $ac = y$ (the instantaneous subnormal).

(32) Where $v = y(dy/dx)$, since $2a^2r^2x^2y = (2a^2r^2 + 2p^2x^2)\,y^2(dy/dx) + 2p^2x^2y^3$ (keeping Newton's homogenized form of the derivative).

(33) Since 'motion b from a: motion c from a' $= dx/dt : dy/dt = y : v$, where t is an independent variable of time.

(34) Or 'mot b ad e: mot c ab a'.

(35) Where txy is drawn perpendicular to bd.

$$\frac{aarr+ppxx\times\sqrt{\dfrac{aarrx^2+ppx^4-aarrxx}{aarr+ppxx}}}{\dfrac{aarrppxx+p^4x^4+aarrppxx}{aarr+ppxx}}=by=\frac{aarr+ppxx\times pxx\times\sqrt{aarr+ppxx}}{2aarrppxx+p^4x^4}$$ (38)

$$=\frac{\sqrt{2:C:aarr+ppxx}}{2aarrp+p^3xx}.$$ (39)

$$by=\frac{aarr+ppxx\times\sqrt{aarr+ppxx}}{2aarrp+p^3xx}=\frac{\sqrt{a^6r^6+3a^4r^4ppxx+3aarrp^4x^4+p^6x^6}}{2aarrp+p^3xx}.$$ (39)

$$a^4r^4ppxx+aarrp^4x^4=a^6r^6.\quad x^4=-\frac{aarr}{pp}xx+\frac{a^4r^4}{p^4}.\quad xx=\frac{-aarr\pm\sqrt{5a^4r^4}}{2pp}.$$ (40)

Make m = motion of b from a. y^n^ is $\dfrac{aarrmx-ppmxyy}{aarry+ppxxy}$ = motion of c from a,(41) &, $\dfrac{m\sqrt{aarr+ppxx}}{ar}$ = motion of b to $e=\dfrac{mx}{y}$.(42) therefore, mot c [from a] : mot b [to e] :: $aarr-ppyy : aarr+ppxx$. &

$ppxx+ppyy:\sqrt{xx-yy}::aarr+ppxx$: radio quæsito.(43) [$by=z$.]

$$z:\frac{\dot{m}\dot{x}}{\dot{y}}::z-\sqrt{xx-yy}:\frac{aarr\dot{m}\dot{x}-pp\dot{m}\dot{x}yy}{aarr\dot{y}+ppxx\dot{y}}.$$ (44)

---

(36) Read 'Draw ty and then by will be equal to the radius of a circle of equal curvature [with the spiral at the point b]'. The proof is immediate by [1] above, since the ratio $bt:cx$ is the ratio of the limit-increments of b in the tangent be and of c in its parallel cd, and so the point y is instantaneously at rest in the normal bc.

(37) For $$\frac{by}{cy}=\frac{bt}{cx}=\frac{a^2r^2+p^2x^2}{a^2r^2-p^2y^2},\quad\text{or}\quad\frac{by}{bc}=\frac{a^2r^2+p^2x^2}{p^2x^2+p^2y^2},$$ where $bc=\sqrt{[x^2-y^2]}$ and $a^2r^2+p^2x^2=(arx/y)^2$.

(38) That is $$\frac{(a^2r^2+p^2x^2)\sqrt{[x^2-(arx)^2/(a^2r^2+p^2x^2)]}}{p^2[x^2+(arx)^2/(a^2r^2+p^2x^2)]}.$$

(39) Or $(a^2r^2+p^2x^2)^{\frac{3}{2}}/p(2a^2r^2+p^2x^2)$, the radius of curvature at b expressed in terms of $ab=x$ alone, with y eliminated by substitution of its value $arx/\sqrt{[a^2r^2+p^2x^2]}$.

(40) This line, squashed in at an empty space high on the right, expresses the condition that $ad=be=(1/2\pi).an$, or $x^2/y=(r/p)\,n$.

(41) Where t is an independent variable of time, Newton simplifies his previous argument by taking $m=dx/dt$ or 'the motion of b from a', so that the 'motion of c from a' or $$dy/dt=mv/y.$$

(42) Since $m=dx/dt$ is also the 'motion of e from d', with $be/ed=ab/ac=x/y$.

(43) Read 'the required radius', that is, by.

(44) Newton uses dots to cancel the factor mx/y in the denominators of the ratios. Not noticing the simplification that both ratios are equal to $\sqrt{[xx-yy]}:\dfrac{mx}{y}\left(\dfrac{aarr+ppxx}{aarry+ppxxy}\right)$, Newton proceeds straightforwardly to cross-multiply.

$$\dot{z}aarr - zppyy = \dot{z}aarr + zppx^2 - \sqrt{xx-yy} \times \overline{aarr+ppxx}.$$

$$\left[\text{or } z = \frac{aarr+ppxx \times \sqrt{xx-yy}}{ppxx+ppyy}.\right]^{(45)}$$

[3]

$ac = ab = x.$ $dc = y.$

$m^{(46)} : \frac{am}{x} :: x : a :: cd : ce.$

$vx = ay.$ or, $op = a.$(47)

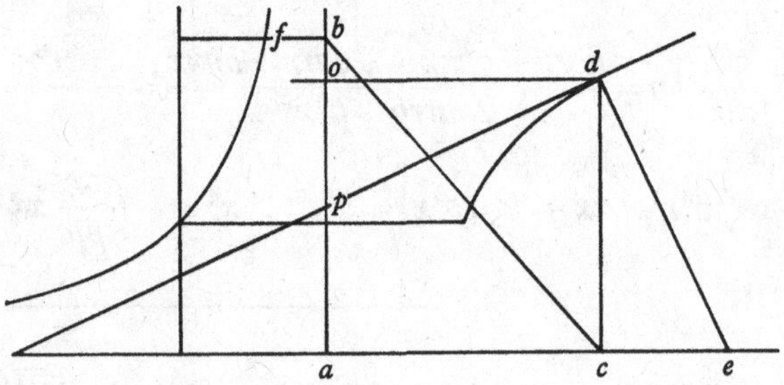

(45) As before. (Note again that the dots on the terms *zaarr* are cancellation symbols and not to be confused with a dot-notation for fluxions.)

(46) As in [2], m is dx/dt or the 'motion of b from a'.

(47) Where $ac = ab = od = x$, $dc = ao = y$ and, say, $bf = z$ with $as = sr = 1$, Newton defines the hyperbola (f) by $ab \times bf = 1$, or $xz = 1$. From this the curve (d) is defined by ao/as = Hyperbola-area $(rsbf)$ or $\frac{y}{a} = \int_1^x z.dx$. Then, where $ce = v$ is the subnormal at d, $ce/cd = dy/dx = op/od$, so that

$$op = x\frac{dy}{dx} = \frac{1}{z}\frac{dy}{dx} = a,$$

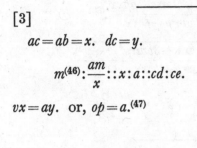

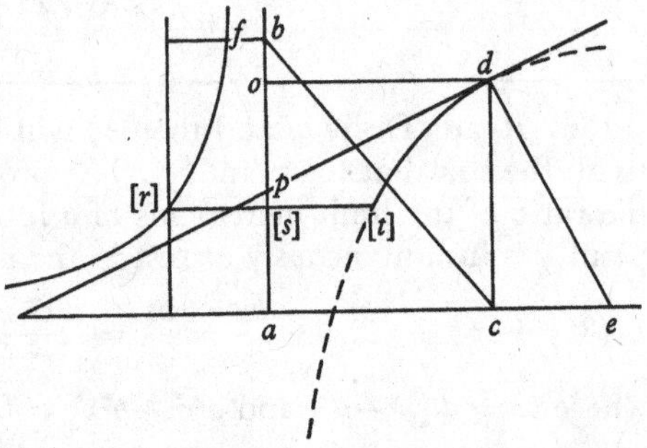

or the subtangent op is constant.

In modern terms, $y = a\log(x)$ and (d) is the logarithmic curve $x = e^{y/a}$, so that $st = e^{1/a}$. The curve had been discussed extensively by Torricelli in the 1640's, but his tract *De Hemhyperbola Logarithmica* remained unpublished till the present century. (See G. Loria: 'Le ricerche inedite di Evangelista Torricelli sopra la curva logarithmica', *Bibliotheca Mathematica* ₃1 (1900): 75–89; and E. Torricelli, *Opere*, 1, 2 (Faenza, 1919): 335–47.) Curiously, the first researches into the nature of the curve were Descartes' comments on and partial resolution of de Beaune's problem (sent to Descartes by the latter in the winter of 1638/9) of finding the curve $(y = -k\log(x-y+k)+C)$ which satisfies the relation $dy/dx = k/(y-x)$. (See Descartes' answering letter to de Beaune of 20 February 1639, first printed by Clerselier in his *Lettres de M. Descartes*, **3** (Paris, 1667): Letter 71 = *Œuvres de Descartes*, **2** (Paris, 1898): no. CLVI: especially 513–17. Compare also Christoph J. Scriba, 'Zur Lösung des 2. Debeauneschen Problems durch Descartes', *Archive for History of Exact Sciences*, **1** (1961): 407–19.) The first explicit printed account of the logarithmic curve was that inserted by James Gregory a year later in the *Proœmium* of his *Geometriæ Pars Universalis*. (It is possible that Gregory, who had spent the previous five years in Italy pursuing mathematical research under the experienced eye of Torricelli's pupil, Stefano degli Angeli, had access to a manuscript version of Torricelli's

[4]

$\frac{aayyy.\ n}{bbyy.\ D}$. $aa\Big/\frac{n}{D}=v$. $\frac{n}{y^2\times yy}$ (48)

---

§2.(1)

November 8th 1665.

How to Draw Tangents to Mechanicall Lines

Lemma

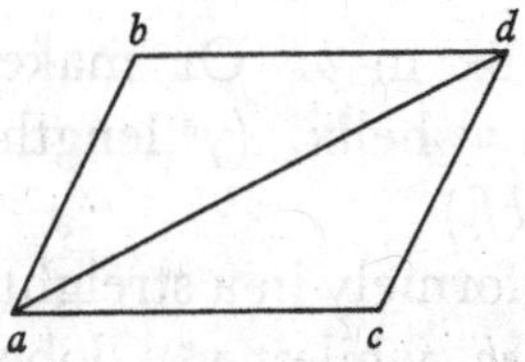

If one body move from *a* to *b* in ye same time in wch another moves from *a* to *c* & a 3d body move from *a* wth motion compounded of these two[,] it shall (completeing ye parallelogram) move to *d* in ye same time. For those motions would severally carry it ye one from *a* to *c* ye other from *c* to *d* &c.(2)

In ye description of any Mechanicall line what ever, there may bee found two such motions wch compound or make up ye motion of ye point describeing it, whose motion being by them found by ye Lemma, its determinacon shall bee in a tangent to ye mechanicall line.(3)

---

work on the curve.) Newton himself, it is clear, had no contact with any of these apart from their printed works and must be ranked as an independent discoverer. Indeed, whether or not he then fully realized its exponential nature, he had already derived the *logarithmica* some time before, probably in autumn 1664, as the 'instantaneous compound interest' curve. (See 3, 1, §3.3.)

(48) These scribbled notes (on f. 34v) seem to belong with Newton's calculations of the subnormal v in the form n/D in 5, §2 above, confirming the dating there given of autumn 1665.

(1) Add. 4004: 50v, 51r. Newton revises his rough notes (made a week earlier) in §1 above.

(2) Newton's 'Lemma' is the classical parallelogram of forces which constructs the vector *ad* as the resultant of the component vectors *ab* and *ac*. He had, some time in the autumn of 1664, already entered the axiom in his Waste Book as a paragraph 'Of compound force' (Add. 4004: 38r, especially Coroll: 1), but went on to generalize it, in a complex of early notes on motion and force written up about the same time elsewhere in the Waste Book (Add. 4004: 10r–15v), as 'Two bodys being uniformely moved in ye same plaine their center of motion will describe a streight line.... They doe ye same in divers plaines.' (13v–14v: §§27–31, especially 13v–14r: §§27, 28.) The form of Newton's presentation of the axiom at f. 38r suggests strongly that he has taken it over from his reading of Descartes' *Principia Philosophiæ* (Amsterdam, 11644: Pars secunda: 50: XXXII, *Quomodo etiam motus propriè sumptus, qui in corpore unicus est, pro pluribus sumi possit*).

Newton originally entered this 'Lemma' below the next paragraph, but in accordance with his manuscript directions their order has been reversed.

(3) This general resolution of the problem is repeated from §1, 1 above, almost word for word.

Example ye 1st. If *abe* is an helix(4) described by ye point (*b*) the line *ab* increasing uniformely whilest it also circulates uniformely about ye center *a*. Let ye radius of ye circle *dmbd* bee *ab*. & let *dmb* measure ye quantity of the giration of *ab* (viz[:] *ad* touching y^e helix at y^e center) [&] let *bf* bee a tangent to y^e circle *dmbd*. y^n is y^e motion of y^e point *b* towards *c* to its motion towards *f* as *ab* to *dmb*. therefore make

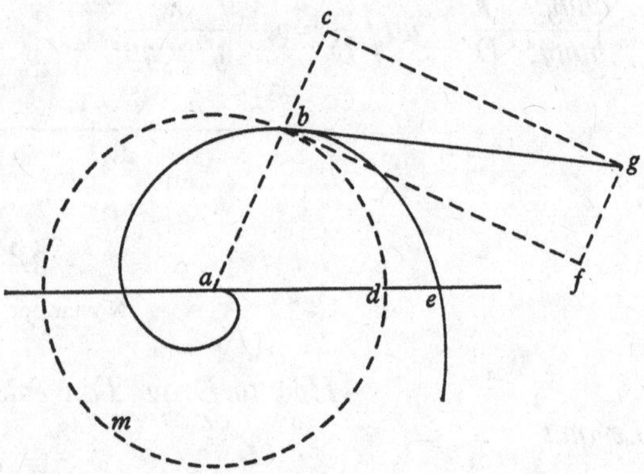

$$bc = fg : bf = cg :: ab : dmb.$$

& (by y^e Lemma(5)) y^e diagonall *bg* shall touch y^e helix in *b*. Or make $bc = fg = ab$. & $bf = cg = dmb$. the diagonall *bg* shall touch y^e helix. (y^e length of *bf* may be thus found viz; $ae : bc = ad = ab :: dmbd : dmb = bf$.)

Example y^e 2^d. If y^e center (*a*) of a globe (*bae*) moves uniformely in a streight line parallel to *eh*, whilest y^e Globe uniformely girates. Each point (*b*) in y^e Globe will describe a Trochoides:(6) to wch at y^e point *b* I thus draw a tangent. Draw y^e radis(7) *ab* & *bc* perpendicular to it[,] y^n is y^e circular motion of the point *b* determined in y^e line *bc*, & its progresive(8) in *bf*. If therefore I make $bc = fg$ to $bf = cg$ as y^e circular motion of y^e point *b* to its progressive y^e Diagonall (by y^e Lemma(9)) *bg* shall touch y^e Trochoïdes in *b*. As if y^e Globe roule(10) upon y^e plaine *eh*, & I make $bc = fg = ab$. & $bf = cg = ae$. y^n doth y^e Diagonall *bg* touch y^e Trochoides. (Or *be*, passing through y^e point in wch y^e globe & plaine touch, is a perpendicular to y^e Troch.(11))

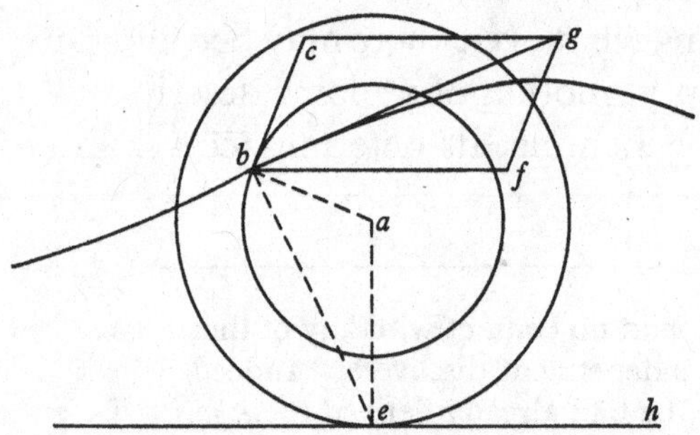

(4) That is, an Archimedean spiral. (See §1, note (5) above.)

(5) This is not true, as Newton will realize, though the resulting construction is correct. (Compare §1, note (7) above.)

(6) That is, a general cycloid (which Newton draws curtate). The example is taken from Schooten's *Commentarii in Librum II, O* (= *Geometria*: 268–70, especially 269–70). (Compare §1, note (11) above.)

(7) Read 'radius'.

(8) That is, motion.

(9) A valid application, since the component motions *bc* and *bf* are truly independent with the vector *bg* found as their resultant and so as the 'determinacon' of the motion of the point *b* in the cycloid.

(10) Read 'roll'.

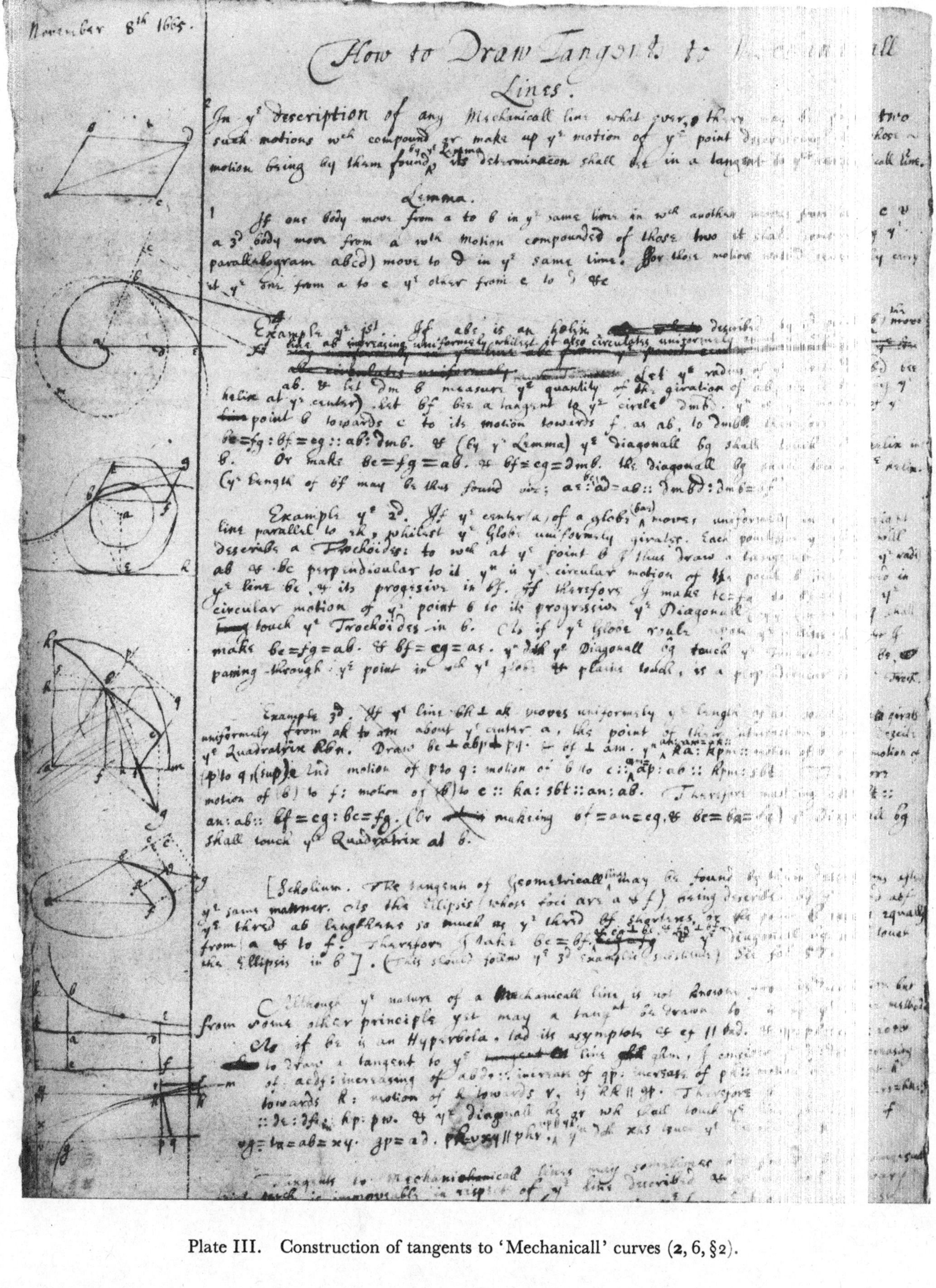
November 8th 1665.

How to Draw Tangents to [illegible] Lines.

In ye description of any Mechanicall line what ever, [illegible] two such motions wch compound or make up ye motion of ye point [illegible] motion being by them found by ye Lemma its determinacon shall be in a tangent [illegible]

Lemma.

1 If one body move from a to b in ye same time in wch another [illegible] a 3d body move from a wth motion compounded of those two it shall [illegible] ye parallelogram abcd) move to d in ye same time. For those motions [illegible] it ye one from a to c ye other from c to d &c

Example ye 1st. If abe is an helix [illegible] described by [illegible] Let ye radius ab. & let dm b measure ye quantity of the giration of [illegible] helix at ye center) let bf bee a tangent to ye circle dmb. [illegible] point b towards c to its motion towards f. as ab. to dmb [illegible] be=fg: bf=eg :: ab: dmb. & (by ye Lemma) ye Diagonall bg shall touch [illegible] b. Or make be=fg=ab. & bf=eg=dmb. the diagonall bg [illegible] (ye length of bf may be thus found viz; ae: ad=ab :: dmbd: dmb=bf)

Example ye 2d. If ye center a of a globe (bac) moves uniformly in [illegible] line parallel to ek, whilest ye Globe uniformly giratis. Each point [illegible] describes a Trochoides: to wch at ye point b I thus draw a tangent [illegible] ab & bc perpendicular to it ye n is ye circular motion of the point b [illegible] ye line bc, & its progression in bf. If therefore I make be=fg [illegible] circular motion of ye point b to its progression ye Diagonall [illegible] touch ye Trochoides in b. Also if ye Globe [illegible] make be=fg=ab. & bf=eg=ae. [illegible] ye Diagonall bg touch ye [illegible] passing through ye point in wch ye globe [illegible] plaine touch, is a perp [illegible]

Example 3d. If ye line bk ⊥ ak moves uniformly ye length [illegible] uniformly from ak to am about ye center a, the point of their [illegible] ye Quadratrix kbn. Draw be ⊥ ab ⊥ pq. & bf ⊥ am. [illegible] ka: kpm :: motion of [illegible] (p to q (sup)) ye motion of p to q: motion of b to c :: ap: ab :: kpm: sbt [illegible] motion of (b) to f: motion of (b) to c :: ka: sbt :: an: ab. Therefore making [illegible] an: ab :: bf=eg: be=fg. (Or making bf=an=eg. & be=ba=[illegible]) ye Diagonall bg shall touch ye Quadratrix at b.

[Scholium. The tangents of Geometricall lines may be found by [illegible] ye same manner. As the Ellipsis (whose foci are a & f) being described [illegible] ye thred ab lengthens so much as ye thred bf shortens, or [illegible] from a & to f. Therefore I take be=bf [illegible] ye diagonall [illegible] the Ellipsis in b]. (This should follow ye 3d example [illegible])

Although ye nature of a Mechanicall line is not knowne [illegible] from some other principle yet may a tangent be drawn to [illegible]

As if be is an Hyperbola, ad its asymptote & ef ∥ bad. [illegible] to draw a tangent to ye line [illegible] of acdf: increasing of abde :: increase of gp: increase of pk [illegible] towards k: motion of k towards v. if kk ∥ gt. Therefore [illegible] :: be: dfe :: kp: pw. & ye diagonall [illegible] shall touch ye [illegible] eg=ba=ab=xy. gp=ad. [illegible]

Tangents to Mechanicall lines may sometimes [illegible] [illegible] in respect of ye line described [illegible]

Plate III. Construction of tangents to 'Mechanicall' curves (2, 6, §2).

Example 3^{d}. If y^{e} line $bh \perp ak$ moves uniformely y^{e} length of ah whilest ab girats uniformely from ak to am about y^{e} center a, the point of their intersection (b) will desc[r]ibe y^{e} Quadratrix kbn.[12] Draw $bc\perp abp \perp pq$. & $bf \perp am$. y^{n} $ah:am=ak::ka:kpm::$ motion of (b) to f: motion of (p) to q, (sup).[13] And motion of (p) to q: motion of (b) to c::

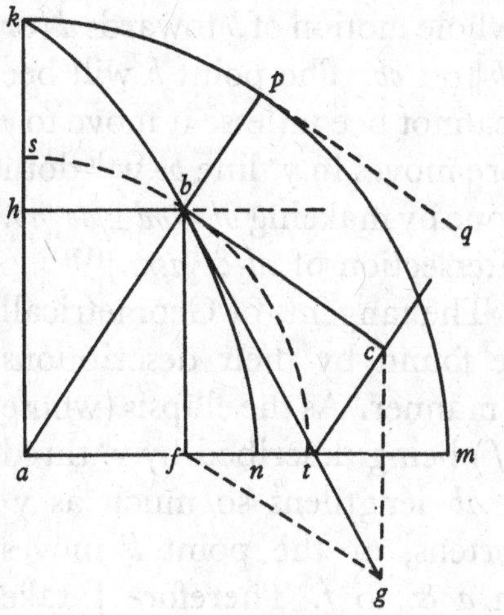

$$am = ap:ab::kpm:sbt.$$

Therefore motion of (b) to f: motion of (b) to $c::ka:sbt::an:ab$. Therefore makeing

$$ak:sbt::an:ab::bf=cg:bc=fg.$$

(Or makeing $bf=an=cg$. & $bc=ba=fg$.) y^{e} Diagonall bg shall touch y^{e} Quadratrix at b.[14]

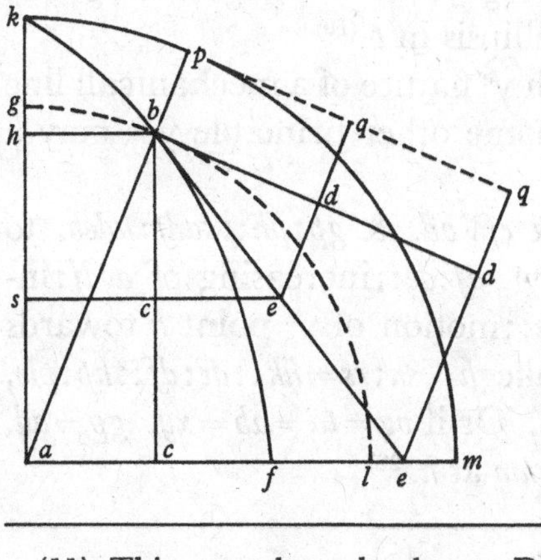

Instead of y^{e} third Example reade this.[15]

Sometimes y^{e} tangent may bee found by considering two absolute motions of y^{e} describing point (b). As if $bh \parallel am \perp ak$ moves uniformely y^{e} length of ka, whilest ab girates uniformely[16] from ak to am about y^{e} center a. The point b of their intersection will describe y^{e} Quadratrix kbf. wth y^{e} radij ag & ak draw y^{e} circles kpm & gbl, make $pq \parallel bd \perp abp \parallel ed$. &, $bc \parallel ak$. Then tis, $ka:kpm::$

(11) This second result, due to Descartes, is taken over from Schooten's *Commentarii in Librum II*, *O* (= *Geometria*: 269). In proof, $\begin{Bmatrix} bc \\ cg \end{Bmatrix}$ is equal and perpendicular to $\begin{Bmatrix} ba \\ ae \end{Bmatrix}$, so that $\triangle bcg \equiv \triangle bae$, and therefore bg is equal and perpendicular to be. Descartes' own proof, given in his letter to Mersenne of 23 August 1638 (= *Œuvres* (ed. Adam and Tannery), 2 (Paris, 1898): 308–9) and printed in Schooten's Latin translation in *Geometria* (267–8), is inadequate since it presupposes that e is the instantaneous centre of motion at the point b in the cycloid.

(12) See §1, note (9) above. (13) That is, by supposition.

(14) The construction is irredeemably false. (Compare §1, note (10) above.) Here for the first time Newton realized his error and has corrected his figure by drawing the normal at c to bc through the meet of the tangent bg and the base am. (The realization of his mistake and the dangers of an uncritical application of the parallelogram rule presumably came quickly, but not before he had drafted the similarly erroneous ellipse-tangent construction in the scholium, on the tangents of 'Geometricall lines', which follows.)

(15) This corrected version of Example 3^{d} has been transposed from the page bottom in accordance with a note added to the following scholium that 'This should follow ye 3^{d} Example's substitute'.

(16) Newton has cancelled 'the length'.

whole motion of *b* to *c* (or *ec*) [: whole motio]n of *p* to *q*. And *ap*:*ab*::*kpm*:*gbl*:: whole motion of *p* to *q*: whole motion of *b* to *d* (or to *ed*). Therefore *af*:*ab*::*ak*:*gbl*:: absolute & whole motion of *b* towards *c* (or *acf*): whole motion of *b* towards *d* (or *ed*) Soe y[t] makeing *bc*:*bd*::*af*:*ab*. *ab* ∥ *ed*⊥ *bd* & *hb* ∥ *ce*⊥ *cb*. The point *b* will bee moved to y[e] line[s] *ce* & *ed* in [y[e]] same time w[ch] cannot bee unlesse it move to *e* (their common intersection). The point *b* therefore moves in y[e] line *be* w[ch] doth therefore touch y[e] Quadratrix at *b*. (The same is done by makeing *bl*=*bd*⊥ *de* ∥*ab*. & drawing y[e] tangent *be* through y[e] common intersection of *ed* & *aem*.)(17)

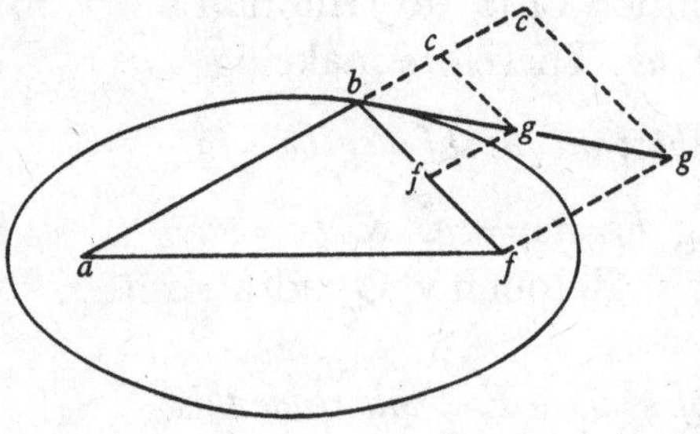

Scholium. The tangents of Geometricall lines may be found by their descriptions after y[e] same manner. As the ellipsis (whose foci are *a* & *f*) being described by y[e] thred *abf* y[e] thred *ab* lengthens so much as y[e] thred *bf* shortens, or the point *b* moves equally from *a* & to *f*. Therefore I take *bc*=*bf*=*cg*=*fg*.(18) & y[e] diagonall *bg* will touch the ellipsis in *b*.(19)

Although y[e] nature of a mechanicall line is not knowne from its description but from some other principle yet may a tang[nt] be drawne to it by y[e] same method.(20)

As if *be* is an Hyperbola, *tad* its asymptote & *cf* ∥ *ad*. & *gp*:*ph*::*cadf*:*adeb*. to draw a tangent to y[e] line *ghm*,(21) I consider y[t], *df*:*de*::increasing of *acdf*:increasing of *abde*::increase of *gp*:increase of *ph*::motion of y[e] point *h* towards *k*:motion of *h* towards *r*, if *hk* ∥ *gp*. Therefore I make *rh*=*sk*:*rs*=*hk*::*de*:*df*::*hp*:*pw*. & y[e] diagonall *hs* or *wh* shall touch y[e] line *ghm*. Or if *vg*=*ta*=*ab*=*xy*. *gp*=*ad*. *ph*=*vxy* ∥*phr*, *vp* ∥*yh*. y[n] doth *xhs* touch y[e] line *ghm* at *h*.(22)

(17) The construction is now wholly correct, and it is interesting to compare the modifications here introduced by Newton with Roberval's own implicit rejection of any application of the parallelogram rule in finding the quadratix tangent. (See G. P. de Roberval, *Observations sur la composition des Mouvemens, et sur le moyen de trouver les Touchantes des lignes courbes* (1638?): *Exemple neuviéme de la Quadratrice.* = *Memoires de l'Academie royale des Sciences Depuis* 1666 *jusqu'à* 1699, **6** (Paris, 1730): 1–89, especially 57–67.) Newton himself pondered the whole problem of limit-motion anew during the winter of 1665/6 and finally drafted the impressive paper of 14 May 1666 (§4.1 below), revising it two days later with examples (§4.2).

(18) After Newton discovered the falsity of applying the parallelogram rule in this case (compare §1, note (13) above), he emended this to read '...I take *bc*=*bf*. & *cg*⊥*bc*, & *fg*⊥*bf*.' and slightly altered his figure accordingly.

(19) Having first correctly emended this erroneous construction for the ellipse-tangent (copied from §1.1 above), Newton has cancelled the whole paragraph with the note 'See fol 57.'. In fact, on f. 57[v] of the Waste Book (printed below as §3.2) Newton entered the correct construction on 15 November.

(20) This paragraph, presumably entered before Newton discovered the falsity of his Example 3, continues to refer to a 'method' which applies the parallelogram rule uncritically.

(21) The logarithmic curve. (Compare §1.4 above.)

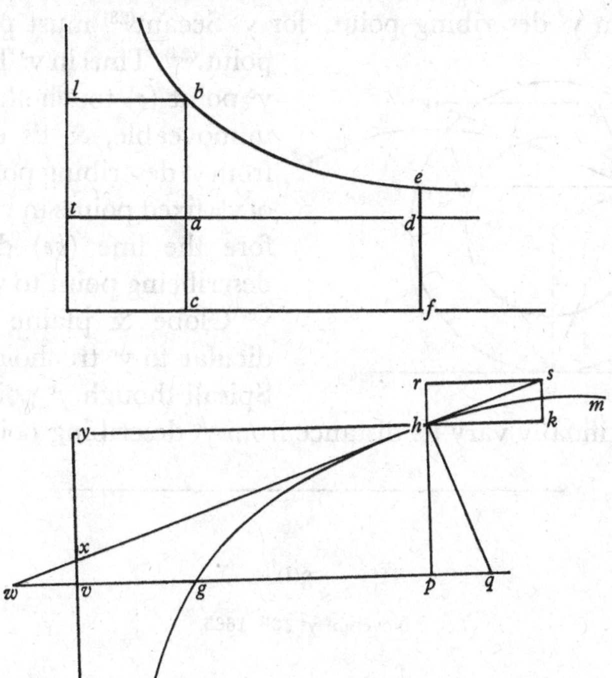

Tangents to Mechanicall lines may sometimes bee found by finding such a point wch is immoveable in respect of y^e^ line described & y^n^ also doth not vary

(22) Where $td = x$, $de = y$ and $ta = ab = \alpha$, the defining equation of the hyperbola (e) is $xy = \alpha^2$, and so, where $vp = td = x$, $ph = z$ and $vg = ta = \alpha$, $ac = df = c$, the defining equation of the (logarithmic) curve (h) is given by

$$(x-\alpha) : z = \int_\alpha^x c \,.\, dx : \int_\alpha^x y \,.\, dx,$$

or $$cz = \int_\alpha^x y \,.\, dx = \int_\alpha^x \frac{\alpha^2}{x} \,.\, dx.$$

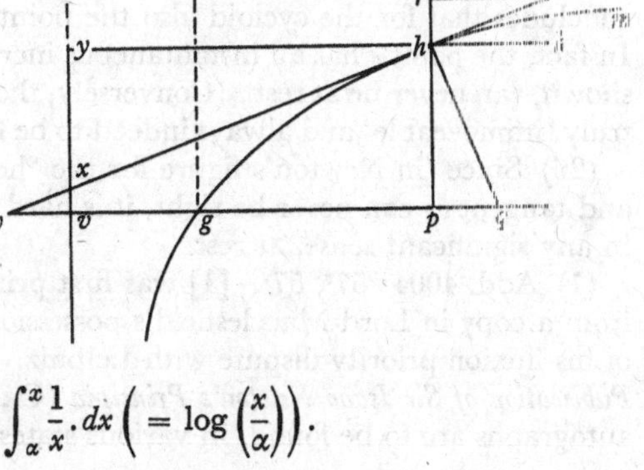

Since $hk : hr$ is the ratio of the limit-increments of $gp = x-\alpha$ and $ph = z$ $\left(\text{or } \frac{hk}{hr} = \frac{dx}{dz}\right)$, it follows that the subtangent

$$xy = hy \,.\, \frac{hr}{hk} = x\frac{dz}{dx} = \frac{\alpha^2}{c},$$

constant, the well-known defining property of the logarithmic curve. (In modern terms $cz/\alpha^2 = \log(x/\alpha)$.) In final simplification Newton takes $xy = ta = ab = \alpha$, or $c = \alpha$, so that the defining equation of (h) reduces to $\frac{z}{\alpha} = \int_\alpha^x \frac{1}{x} \,.\, dx \left(= \log\left(\frac{x}{\alpha}\right)\right)$.

in distance from y^e describing point. for y^e Secant(23) must passe through y^t point.(24). Thus in y^e Trochoides when y^e point (*e*) toucheth y^e plaine *eh*, tis immoveable, & tis ever equidistant from y^e describing point *b* (being both of y^m fixed points in y^e Globe). Therefore the line (*be*) drawne from ye describeing point to y^e touch point of y^e Globe & plaine (*eh*) is perpendicular to y^e trochoides.(25) But in y^e Spirall though y^e point *a* doth rest(26) yet it doth continually vary its distance from y^e describing point *b*.

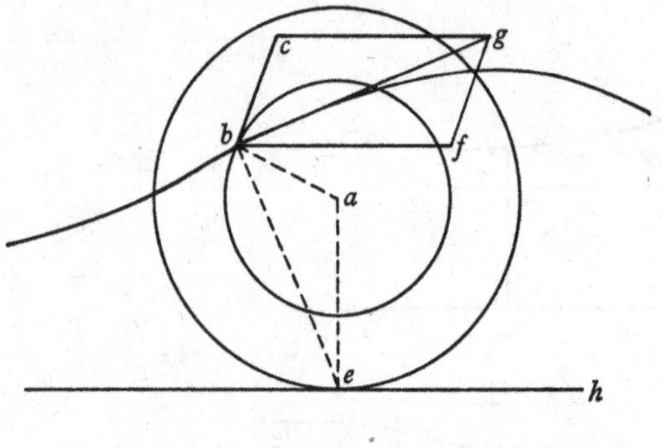

---

§3(1)

Novemb^r y^e 13th 1665.

[1]

To find y^e velocitys of bodys by y^e lines they describe.

.R.(2) An Equation being given, expressing y^e Relation of two or more lines x, y, z, &c described in ye same time by two or more moveing bodys A, B, C &c to find ye relation of their velocitys p, q, r &c:

---

(23) The normal at the point considered.

(24) Newton argues correctly that if we can find a point which is instantaneously at rest with respect to the generating point of a curve, then that point must lie on the normal to the curve (since a point instantaneously at rest as the point generates the curve must be the instantaneous centre of rotation and so the centre of curvature at the point).

(25) Newton repeats this argument of Descartes from Schooten's presentation of it (*Commentarii in Librum II = Geometria*: 269) but fails to see that, as there presented, it is fallacious. The 'determinācon' of the point *b* is correctly considered to be perpendicular to *be* (note (11) above), so that *be* is indeed normal to the cycloid. Descartes' argument, however, which considers the rolling circle as the limit of a regular polygon of an infinitely great number of sides, whose turning points are indeed in the base *eh* (compare *Geometria*: 267–8), wrongly concludes that for the cycloid also the point *e* will be a turning point and so 'immoveable'. In fact, the point *e* has an instantaneous increment horizontally and so, in the curtate cycloid shown, can never be at rest. (Conversely, the centre of curvature at the point *b*, the only point truly 'immoveable' and always indeed to be found in the normal *be*, can never coincide with *e*.)

(26) Since (in Newton's figure for the 'helix' above) the angle $a\hat{b}g$ between the radius *ab* and tangent *bg* can never be right, it is hard to see how Newton can argue that the point *a* is, in any significant sense, at rest.

(1) Add. 4004: 57^r, 57^v. [1] was first printed, inadequately and incompletely, by Rigaud from a copy in Lord Macclesfield's possession made by Newton himself at the time (*c.* 1714) of his fluxion priority dispute with Leibniz. (See S. P. Rigaud, *An Historical Essay on the First Publication of Sir Isaac Newton's Principia* (Oxford, 1838): Appendix II: 20–2. Several similar autographs are to be found, in various states of completeness, in ULC. Add. 3968.)

Resolution

Sett all y[e] termes on one side of y[e] equation y[t] they become equall to nothing.

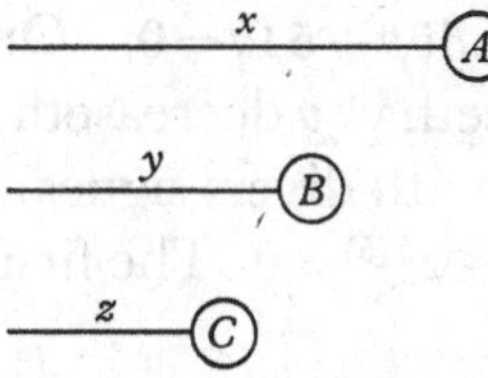

And first multiply each terme by soe many times $\frac{p}{x}$ as x hath dimensions in y[t] terme. Seacondly multiply each terme by soe many times $\frac{q}{y}$ as y hath dimensions in it. Thirdly multiply each terme by soe many times $\frac{r}{z}$ as z hath dimensions in it &c. The sum̄e of all these products shall bee equall to nothing. Which Equation gives y[e] relation of p, q, r &c.[(3)]

Or more generally thus. Order y[e] equation according to y[e] dimensions of x, & (putting a & b for any two numbers whither rationall or not) multiply y[e] termes of it by any ꝑte of this progression viz: :℈℈. $\frac{ap-3bp}{x}$. $\frac{ap-2bp}{x}$. $\frac{ap-bp}{x}$. $\frac{ap}{x}$. $\frac{ap+bp}{x}$. $\frac{ap+2bp}{x}$. $\frac{ap+3bp}{x}$. &c: Also order y[e] Equation according to y & multiply the termes of it by this progression. :℈℈ $\frac{aq-2bq}{y}$. $\frac{aq-bq}{y}$. $\frac{aq}{y}$. $\frac{aq+bq}{y}$. $\frac{aq+2bq}{y}$. $\frac{aq+3bq}{y}$. &c. Also order it according to y[e] dimentions of z & multiply its termes by this progression viz: :℈℈. $\frac{ar-3br}{z}$. $\frac{ar-2br}{z}$. $\frac{ar-br}{z}$. $\frac{ar}{z}$. $\frac{ar+br}{z}$. $\frac{ar+2br}{z}$. $\frac{ar+3br}{z}$. &c. The sume of all these products shall bee equall to nothing. Which equation gives y[e] relation of p, q, r &c.[(4)]

Example 1st. If y[e] propounded Equation bee

$$x^3-2xxy+4xx+7xyy-y^3-103=0.$$

(2) This letter was added for reference on 14 May 1666 when Newton composed the first draft (§4.1 below) of his paper on limit-motions.

(3) Thus, where $f(x, y, z) = 0 = \Sigma(a_{i,j,k}x^iy^jz^k)$, Newton's 'resolution' sets

$$0 = \Sigma\left(\frac{ip}{x}+\frac{jq}{y}+\frac{kr}{z}\right)(a_{i,j,k}x^iy^jz^k), \quad \text{or} \quad 0 = pf_x+qf_y+rf_z.$$

It follows that $p:q:r = dx:dy:dz$, and in fact, where t is an independent variable of time, Newton opts for a model in which p, q and r are the respective 'velocitys' dx/dt, dy/dt and dz/dt.

(4) This reduces to the former since, where α, β and γ are arbitrary,

$$0 = \Sigma\left(\frac{\alpha+ip}{x}+\frac{\beta+jq}{y}+\frac{\gamma+kr}{z}\right)(a_{i,j,k}x^iy^jz^k) = \left(\frac{\alpha}{x}+\frac{\beta}{y}+\frac{\gamma}{z}\right).\Sigma(a_{i,j,k}x^iy^jz^k) + \Sigma\left(\frac{ip}{x}+\frac{jq}{y}+\frac{kr}{z}\right)(a_{i,j,k}x^iy^jz^k).$$

By y^e precedent rule y^e first operation will produce $3xxp-4xyp+8xp+7yyp$. The seacond produ[c]eth $-2xxq+14xyq-3yyq$. Which two added together make $3xxp-4xyp+8xp+7yyp-2xxq+14xyq-3yyq=0$. (Now suppose a yarde to bee an unit & y^t *A* hath moved 3 yardes, y^n (by y^e 1st equation) *B* hath moved two; i,e, $x=3$. $y=2$. And at that time by y^e last Equation $55p+54q=0$. Or $55:-54::p:q::$velocity of *A*:velocity of *B*. Onely if x increaseth y^n y decreaseth, y^t is, *A* & *B* move contrary ways because p & q are affected with divers signes).

Example 2^d. If y^e Equation bee $x^3-2a^2y+z^2x-y^2x+zy^2-z^3(5)=0$. The first

$$\frac{3p}{x}.\quad *^{(6)}.\quad \frac{p}{x}.\quad 0.\quad 0.\quad 0$$

operation will produce $x^3\ *+zzx-yyx-2aay+zy^3$(7)$-z^3$. Or $3pxx+pzz-pyy$. The second produceth $-2aaq-2yxq+2zyq$. The third $+2zxr+yyr-3zzr$. The summe of wch is $3pxx+pzz-pyy-2aaq-2yxq+2zyq+2zxr+yyr-3zzr=0$. (Note y^t in this Example there being three unknowne quantitys x, y, z, There must be two of them & two velocitys supposed thereby to find y^e 3^d quantity & y^e third velocity. Or else there must be some other equation expressing y^e relation of two of these x, y, z. (as in y^e first example) whereby one quantity & one velocity being supposed y^e other quantity & velocity may bee found & y^n by this 2^d Example y^e 3^d quantity & 3^d velocity may bee found.)

Example 3^d. Of y^e more generall rule. If y^e equation bee

$$x^4-3gyxx+yyxx-ggyy-2y^4=0.$$

$$\frac{ap}{x}.\quad *.\quad \frac{ap+2bp}{x}.\quad *.\quad \frac{ap+4bp}{x}.$$

y^e first operation gives $x^4\quad *\quad \begin{matrix}+yyxx\\-3gy\end{matrix}\quad *\quad \begin{matrix}-ggyy\\-2y^4\end{matrix}$ Or

$$apx^3+apyyx-3apgyx-\frac{apggyy}{x}-\frac{2apy^4}{x}.^{(8)}$$
$$+2bpyyx-6bpgyx-\frac{4bpggyy}{x}-\frac{8bpy^4}{x}$$

the 2^d gives

$$\overline{aq-2bq}\times\overline{-2y^3}+\overline{aq}\times\overline{xxy-ggy}+\overline{aq+bq}\times\overline{-3gxx}+\overline{aq[+]2bq}\times\frac{x^4}{y}.^{(9)}$$

The sume of wch two products is equall to nothing. &c.

(5) Newton has cancelled the additional term '$+zzy$'.

(6) Read '$\frac{2p}{x}$' more correctly (though the result is unaffected).

(7) Read 'zy^2'.

(8) Which, since $(ap/x)(x^4-3gyx^2+y^2x^2-g^2y^2-2y^4)=0$, reduces to

$$(bp/x)(2y^2x^2-6gyx^2-4g^2y^2-8y^4)=(bp/x)(4x^4-6gyx^2+2y^2x^2).$$

Demonstration.[10]

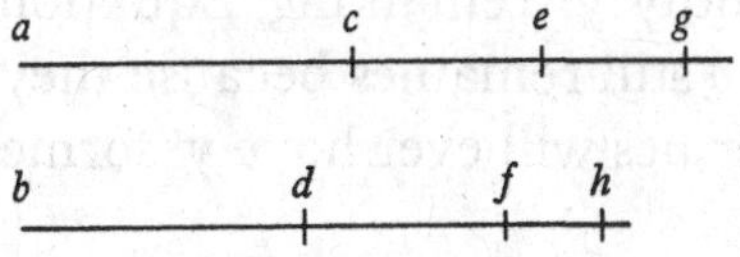

Lemma. If two bodys $\frac{A}{B}$ move uniformely ye $\frac{\text{one}}{\text{other}}$ from $\frac{a}{b}$ to $\frac{c,\, e,\, g,}{d,\, f,\, h,}$ &c in ye same time. yn are ye lines $\frac{ac}{bd}$ & $\frac{ce}{df}$ & $\frac{eg}{fh}$ &c as their velocitys $\frac{p}{q}$. And though they move not uniformly yet are ye infinitely little lines[11] wch each moment they describe as their velocitys are wch they have while they describe them. As if ye body A wth ye velocity p describe ye infinitely little line o in one moment.[12] In yt moment ye body B wth ye velocity q will describe ye line $\frac{oq}{p}$. For $p:q::o:\frac{oq}{p}$.[13] Soe yt if ye described lines be x & y in one moment, they will bee $x+o$ & $y+\frac{oq}{p}$ in ye next.[14]

Now if ye Equation expressing ye relation of ye lines x & y be $rx+xx-yy=0$. I may substitute $x+o$ & $y+\frac{qo}{p}$ into ye place of x & y because (by ye lemma) they as well as x & y doe signifie ye lines described by ye bodys A & B.[15] By doeing so there results $rx+ro+xx+2ox+oo-yy-\frac{2qoy}{p}-\frac{qqoo}{pp}=0$. But $rx+xx-yy=0$ by supposition: there remaines therefore $ro+2ox+oo-\frac{2qoy}{p}-\frac{qqoo}{pp}=0$. Or divideing it by o tis $r+2x+o-\frac{2qy}{p}-\frac{oqq}{pp}=0$. Also those termes in wch o is are infinitely lesse yn those in wch o is not therefore blotting ym out there rests $r+2x-\frac{2qy}{p}=0$. Or $pr+2px=2qy$.[16]

(9) That is, since $(aq/y)(-2y^4-g^2y^2+x^2y^2-3gx^2y+x^4) = 0$,

$$(bq/y)(4y^4-3gx^2y+2x^4) = (-bq/y)(-8y^4-2g^2y^2+2x^2y^2-3gx^2y).$$

(10) Adapted from Add. 4000: 152r (= 5, §4.1 above).

(11) Newton has cancelled 'spac[es]'.

(12) That is, since $p = dx/dt$, the segment $ce = o$ described 'in one moment', say t, is the limit-increment of $ac = x$, with $o/t = dx/dt$ (or $o = pt$).

(13) The limit-increment df of $bd = y$ is similarly $t(dy/dt) = qt = (q/p)\,o$.

(14) Newton has cancelled 'or better $p:q::po:qo$. &c:'.

(15) In general, where $0 = f(x, y)$ then also is $0 = f(x+o, y+(oq/p))$, the incremented function.

(16) More generally, $0 = f(x+o, y+(oq/p)) = f(x, y)+o(f_x+(q/p)f_y)+O(o^2)$, and so

$$0 = \lim_{o\to\text{zero}}\left(\frac{f(x+o, y+(oq/p))-f(x, y)}{o}\right) = f_x+\frac{q}{p}f_y \quad\text{or}\quad \frac{dx}{dt}f_x+\frac{dy}{dt}f_y = 0.$$

(Immediately $q/p = dy/dx = -f_x/f_y$.)

Hence may bee observed: First, y^t those termes ever vanish in wch o is not because they are y^e propounded equation. Secondly y^e remaining Equation being divided by o those termes also vanish in wch o still remaines because they are infinitely little. Thirdly y^t y^e still remaining termes will ever have y^t forme wch by ye first preceding rule they should have.(17)

The rule may bee demonstrated after ye same manner if there [bee] 3 or more unknowne quantitys x, y, z &c.(18)

[2]

By helpe of ye preceding probleme divers others may bee readyly resolved as

Of tangents to Geometricall lines. 1. *To draw tangents to crooked lines (however they bee related to streight ones).*

Resolution.

Find (by ye preceding rule) in w^t proportion those two lines to wch y^e crooked line is cheifly related doe increase or decrease; produce y^m in y^t proportion from y^e given point in y^e crooked line; at those ends draw lines in which those ends are inclined to move,(19) through whose intersection y^e tangent shall passe.

Example 1st. If $ab = id = x$. $ai = bd = y$. & $rx - \frac{rxx}{s} = yy$. Then is $p:q::2y:r - \frac{2rx}{s}$, (by y^e former rule) Therefore I draw $de:dg::p:q::2y:r - \frac{2rx}{s}$. The point g is

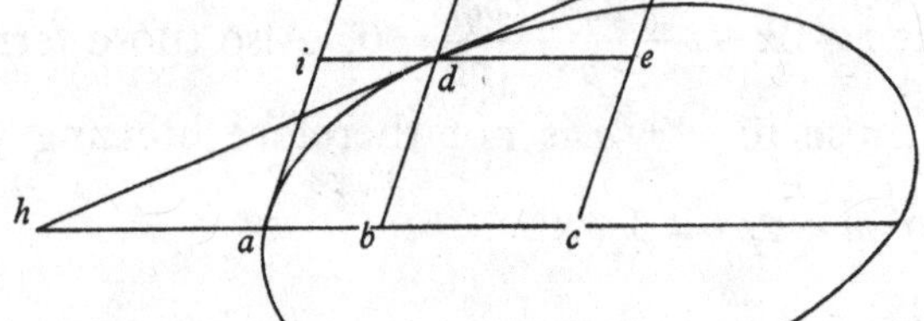

(17) Newton has cancelled 'as may partly appeare by Oughtreds Analyticall table'. (The reference is to the table of binomial coefficients ([*Tabula*] *Posterior*) given by Oughtred in his *Clavis Mathematicæ* (Oxford, ₃1652): Cap. XII, *De Genesi et Analysi Potestatum*: 34–8, especially 37.) Newton thereby seems to intend the differentiation of x^k by considering

$$\lim_{o \to \text{zero}} ([(x+o)^k - x^k]/o).$$

Oughtred's table gives the expansion of $(A+E)^k$, $k = 1, 2, \ldots, 10$ as a triangular 'Pascal' array, from which it follows that $(x+o)^k = x^k + kx^{k-1}o + O(o^2)$: immediately

$$\lim_{o \to \text{zero}} \left(\frac{(x+o)^k - x^k}{o}\right) = kx^{k-1}.$$

(18) Compare notes (15) and (16) above.

(19) Newton first wrote '...draw perpendiculars to y^m'.

inclined to move in a parallel to *abc* & y^e^ point *e* in a parallel to *aik* (for *bg* & *fe* (by supposition) move parallel to y^m^ selves y^e^ $\frac{\text{one}}{\text{other}}$ upon $\left.\frac{abc}{aik}\right)$ Therefore I draw *gf* ∥ *abc* & *ef* ∥ *aik*. & through y^r^ intersection (*f*) I draw *hdf* touching y^e^ crooked line at *d*. Soe y^t^ $gd:de::db:bh::q:p::r-\frac{2rx}{s}:2y$.

Hence may bee pronounced those theorems in Fol 47.[20]

Example y^e^ 2^d^.[21] If $ac=x$. $bc=y$, (wch move about y^e^ centers *a* & *b* as in y^e^ Hyperbola or Ellipsis by a thred) And y^e^ Equation bee $-a+x+y=0$. y^n^ is $p+q=0$.[22] or $p=-q$. therefore I make $cd:cb::p:q::1:-1$. (Note, y^t^ I draw *cd* & *cB* y^e^ one forward y^e^ other backward because *p* & *q* have contrary signes.) y^e^ points *d* & *B* are inclined to move y^e^ one in a perpendicular to *acd* y^e^ other to *bBc* (for they move in circles whose centers are *a* & *b*) therefore I draw $de\perp acd$ & $Be\perp bBc$ & y^e^ tangent *ce* through y^e^ point *e*.

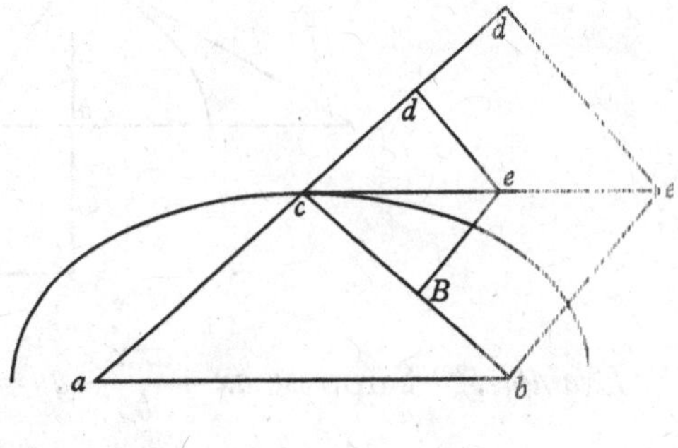

Of tangents to Mechanicall lines.

2. *Hitherto may bee reduced y^e^ manner of drawing tangents to mechanicall lines.* see Fol. 50.[23]

Of y^e^ crookednesse of Geometricall lines.

3. *To find ye quantity of crookednes in Geometricall lines.*

Resolution.

Find y^t^ point of ye perpendicular to ye crooked line wch is in least motion let y^t^ bee ye center of a circle wch passing through ye given point shall bee of equall crookednesse wth ye line at y^t^ point.

This point of least motion may bee found divers ways,[24] as

First. From any two points in ye perpendicular to ye crooked line draw 2 parallel lines in such proportion as ye perpendicular moves over y^m^; through

(20) See Newton's entry of 20 May 1665 in the Waste Book (Add. 4004: 47^r^/47^v^ ≡ 4, §3.1 above). Newton's first reference was to 'pag 47'.

(21) Corrected from the 'Scholium' on 'Geometricall lines' to §2 above. (See §2, note (19) above.)

(22) By 'ye former rule'.

(23) That is, Newton's entry of 8 November in the Waste Book (Add. 4004: 50^v^/51^r^ ≡ §2 above).

(24) See his entries in the Waste Book between December 1664 and May 1665 (Add. 4004: 30^v^ – 33^v^ ≡ 4, §2 above).

their ends draw another line wch shall intersect y^e perpendicular in y^e point required.

As if $ab=x$. $bc=y$. $ce\perp ck=$y^e tangent of y^e crooked line. $kbe=cd\parallel abef$. & as y^e motion of b from a to y^e motion of e from a so kb to ef. Then, drawing dfg through y^e points d & f, cg is y^e radius of a circle as crooked as [y^e] line acl at c.(25)

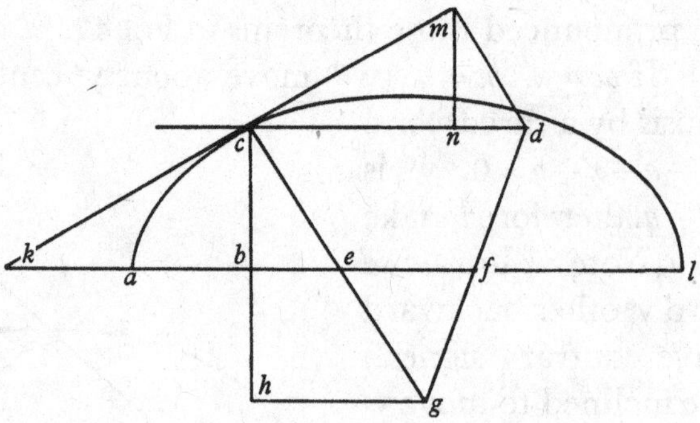

Example.(26) Suppose $ax-\frac{axx}{b}-yy=0$. y^n is $be=v=\frac{a}{2}-\frac{ax}{b}$. $kb=\frac{yy}{v}=\frac{2byy}{ab-2ax}$. And $cd=ke=\frac{aabb-4aabx+4aaxx+4bbyy}{2abb-4abx}$. $bk=p=$velocity(27) of b from a, $bc=y=$velocity of y's increase[,] $r=$velocity of v's increase.(28) [y^n]

$$2bv-ba+2ax=0.^{(29)}$$

(25) This discussion of the curvature centre as the point of instantaneous rest is elaborated from the first sketch in §1.1 above (= Add. 3958.2: 34^r).

(26) Newton first began to draft this example before the preceding paragraph was composed, but broke off after writing: 'Example. Suppose $ab=x$. $bc=y$. $x\perp y$. $ax-\frac{axx}{b}=yy$. $be=v=\frac{a}{2}-\frac{ax}{b}$. $p=$motion of b from a. $r=\frac{-ap}{b}=$ motion of e from b. & $\frac{bp-ap}{b}=$ motion of e from a. $bk=ch=\frac{yy}{v}=\frac{2bx-2xx}{b-2x}$. $nm=y$. $nd=v=\frac{ab-2ax}{2b}\equiv\frac{yy}{x}-\frac{a}{2}$. $cd=ke$.'

(27) Newton has cancelled 'motion'.

(28) Accurately the 'velocity of b from a', the 'velocity of y's increase' and the 'velocity of v's increase' are respectively the 'speeds' dx/dt, dy/dt and dv/dt (where t is an independent variable of time). However Newton represents the second by $y=y(dt/dy.dy/dt)$, so that $p=y(dt/dy.dx/dt)=y(dx/dy)$ and $r=y(dt/dy.dv/dt)=y(dv/dy)$ in his scheme.

(29) Since $v=y(dy/dx)=\frac{1}{2}(a-2ax/b)$.

(30) For $p=y(dx/dy)=y^2/v$.

(31) More generally, where the point c is defined by some given relation between

$$ab=x\perp bc=y$$

with the subtangent $kb=p=y(dx/dy)$ and subnormal $be=v=y(dy/dx)$, and where $cb=y$ is taken to represent the 'velocity of y's increase' dy/dt, then will kb represent the 'velocity of b from a' and $r=y(dv/dy)$ the 'velocity of v's increase' (or the 'velocity of e from b') and so $p+r$ the 'velocity of e from a'. However, where $cn=kb$, $nm=bc$ and md is perpendicular to

therefore $2br+2ap=0$, or $\left(\text{since } p=\dfrac{2byy}{ab-2ax}\right)$[30] tis $r=\dfrac{2yy}{-b+2x}$. Lastly $cd-ef:ce::cd:cg$ (or $v-r:y::v+p:ch$. if $ch\perp gh$) y[t] is

$$\frac{-4byy+4abx-abb-4axx}{-2bb+4bx}:y::\frac{aabb-4aabx+4aaxx+4bbyy}{2abb-4abx}:ch.$$

$$\frac{aaby+4by^3-4ay^3}{aab}=ch=\frac{4by^3-4ay^3+aaby}{aab}. \ \& \ bh=\frac{4y^3}{aa}-\frac{4y^3}{ab}.$$[31]

Hence may bee pronounced those theorems in Fol 49.[32]

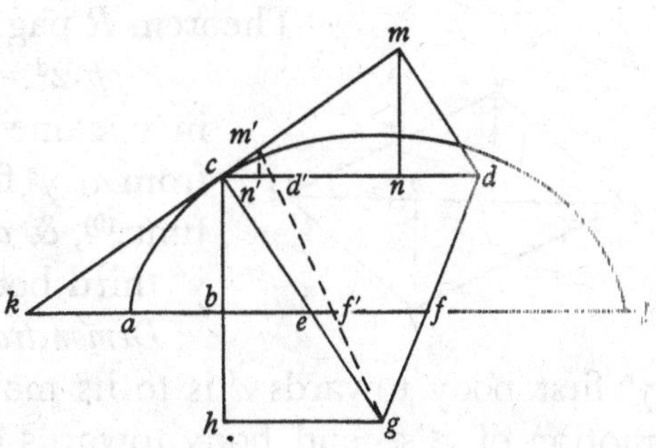

$cm = kc$, then in the same scheme $cd = p+v$ will represent the 'velocity' of the normal cg away from a in the line cd and similarly $ef = p+r$ will measure the velocity of the same normal from a (that is, the 'velocity of e from a') in the line ael. Immediately,

$$ch = cb\frac{cd}{cd-ef} = y\frac{p+v}{v-r},$$

or, since $r = y\dfrac{d}{dy}\left(y\dfrac{dy}{dx}\right) = v+y^2\dfrac{dx}{dy}\dfrac{d^2y}{dx^2}$,

$$ch = -\frac{1+(dy/dx)^2}{d^2y/dx^2}.$$

(This ingenious but rather muddy argument may become clearer if we take a normal $m'd'f'$, evidently through the centre of curvature g, indefinitely close to ce and consider the differential triangle $m'n'c$, where cn', $n'm'$, cm' and ef' are the limit-increments of ab, bc, $\widehat{ac}$ and ae respectively.)

(32) See 4, §3.2 above (=Add. 4004: [49[r]]), where Newton evaluates ch, hg and the curvature radius cg in terms of $\mathfrak{X} \equiv f(x, y) = 0$, the defining equation of (c), with $\mathfrak{X} = xf_x$, $\mathfrak{X} = yf_y$; $\mathfrak{X} = x^2f_{xx}$, $\mathfrak{X} = xyf_{xy}$ and $\mathfrak{X} = y^2f_{yy}$. Specifically, Newton notes in equivalent form that

$$ch = \frac{f_y((f_x)^2+(f_y)^2)}{(f_y)^2f_{xx}-2f_xf_yf_{xy}+(f_x)^2f_{yy}},$$

which may be derived in any of several equivalent ways from the above-found total derivative form,

$$ch = -\frac{1+(dy/dx)^2}{d^2y/dx^2}.$$

(Thus, applying the operator $d/dx = \partial/\partial x+dy/dx(\partial/\partial y)$ twice to $f(x,y)=0$, we deduce first

$$0 = f_x+\frac{dy}{dx}f_y \text{ and again}$$

$$0 = \frac{d}{dx}(f_x)+\frac{dy}{dx}\frac{d}{dx}(f_y)+\frac{d^2y}{dx^2}f_y$$

$$= f_{xx}+2\frac{dy}{dx}f_{xy}+\left(\frac{dy}{dx}\right)^2f_{yy}+\frac{d^2y}{dx^2}f_y.$$

It follows therefore that $dy/dx = -f_x/f_y$ and, with this substituted in the final line,

$$\frac{d^2y}{dx^2} = -\frac{1}{(f_y)^3}((f_y)^2f_{xx}-2f_xf_yf_{xy}+(f_x)^2f_{yy}),$$

so that the two forms of ch reduce one to the other.)

§4. REVISED THOUGHTS ON LIMIT-MOTION

[May 1666]

[1][(1)]

May 14. 1666. *To resolve these*[(2)] *& such like Problems these following propositions may bee very usefull.*[(3)]

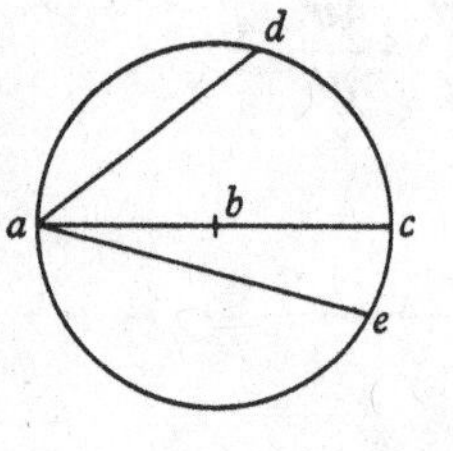

Prop 1. If y^e^ body *a* being in y^e^ circumference of y^e^ circle or sphære *adce* doth move towards its center *b* its motion or velocity[(4)] towards each point *d*, *c*, *e* of y^e^ circumference is y^n^ as y^e^ cordes *ad*, *ac*, *ae*, drawn from y^t^ body to those points are.[(5)] This may be Demonstrated by Theorem *R* pag 57.[(6)]

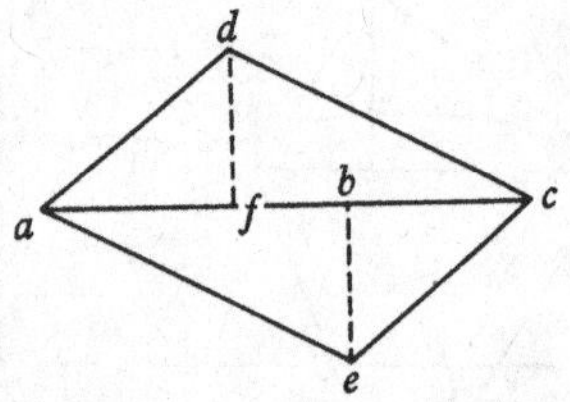

Prop 2^d^. If [$\triangle$]*adc* sim $\triangle$*aec* although they bee not in y^e^ same plane[(7)] & three bodys move uniformely from *a*. y^e^ first[(8)] to *d*, y^e^ 2^d(8)^ to *e*, y^e^ 3^d(8)^ to *c* in y^e^ same time,[(9)], & *adce* is a parallelogram then is y^e^ motion of y^e^ third body compounded of y^e^ other two.

Demonstration. For makeing $df \parallel eb \perp ac$ y^e^ motion of y^e^ first body towards *d* is to its motion towards *f* as *ad* to *af* (prop 1); & y^e^ motion of y^e^ second body towards *e* is to its motion towards *b* as *ae*[(10)] to *ba* (prop 1). But $af+ab=ac$. Therefore &c.[(11)]

Prop 3^d^. If a moving line keepe parallel to it selfe all its ꝑts have equall motion.

(1) Add. 4004: 51^r^, a first (cancelled) attempt presented more successfully in the revised version [2] below.

(2) That is, the problems of constructing tangents to 'mechanical lines' and, more general, curves of any kind, which Newton had discussed in the immediately preceding entry of 8 November 1665 (Add. 4004: 50^v^/51^r^=§2 above).

(3) Having pondered the general problem of limit-motion over the winter and early spring of 1665/6 and seeking reasons for the failure of his first fluxional constructions of the tangents to quadratrix and ellipse (in §1.1), Newton now gathers his thoughts in orderly fashion.

(4) Newton has cancelled 'acceleration' (a Galileian rather than Cartesian word).

(5) This first proposition (used here mainly in the proof of Prop. 2^d^) seems drawn in inspiration from Newton's reading of Galileo's *Discorsi e Dimonstrazioni Matematiche, intorno à due nuoue scienze Attenenti alla Mecanica & i Movimenti Locali.* (This work, first printed in the original Italian at Leyden in 1638, had in 1665 appeared in Thomas Salusbury's translation as *Mathematical Discourses and Demonstrations touching Two New Sciences pertaining to Mechanicks and Local Motion*, one of the tracts in *Mathematical Collections and Translations Englished from the originall Latine and Italian*, **2**, 1 (1662?) [London, 1665]. One of the rare copies of this second volume, apparently issued only to subscribers—the bulk of the copies were consumed in the Great Fire in 1666—was available to Newton in the library of Trinity College, Cambridge. See Stillman Drake, 'Galileo Gleanings II. A Kind Word for Salusbury' *Isis*, **49** (1958): 26–33.) In the Third Day (*De Motu Locali*) of his work Galileo began, with the *Scholium* to Theorem II, to consider the motion of bodies falling from rest under the accelerating force of

Prop 4th. If a line move in plano, so y[t] all its points keepe equidistant from some common center the motions of those points are as their distances from y[t] center.(12)

Prop 5t. If y[e] motion of a line in plano bee mixed of parallell & circular motion, y[e] motion of all its points are compound (see prop 2) of that motion which they would have, had y[e] line onely its parallel motion, & of y[t] w[ch] they would have, had y[e] line onely its circular motion.

Schol: All motion in plano is reducible to one of these three cases, & in y[e] 3[d] case any point in y[t] plaine may bee taken for a center to y[e] circular motion.(13)

Prop 6t. If y[e] streight line *ea* doth rest & *da* doth move: soe y[t] y[e] point *a* fixed in y[e] line *da* moveth towards *b*: Then from y[e] moveing line *da* drawing *de* || *ab*, & y[e] same way w[ch] y[e] point *a* moveth; These motions, viz[:] of y[e] fixed point *a* towards *b*, of y[e] intersection point *a* in y[e] line *ad* towards *d*, & of y[e] intersection point *a* in y[e] line *ae* towards *e*, shall bee one to another, as their correspondent lines *de*, *ad*, & *ae* are.

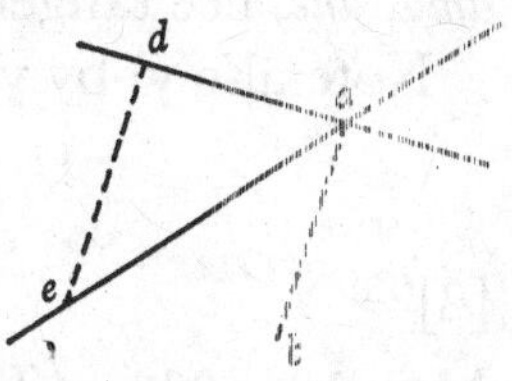

Prop 7th. If y[e] streight lines *adm*, *ane*, doe move, soe y[t] y[e] point *a* fixed in y[e] line *amd* moveth towards *b*, & y[e] point *a* fixed in y[e] line *ae* moveth towards *c*:

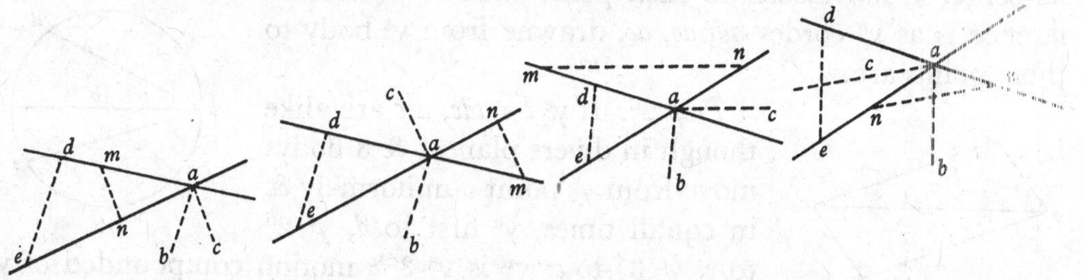

simple gravity and along planes of differing inclinations. Thus, where the diameter *ac* in Newton's figure is made vertical, Galileo was able to show in his Theorem VI that the times of fall along any chord *ae* or *ad* was equal to that of free fall down the diameter *ac*. In the limit where the circle becomes indefinitely small, this is Newton's present theorem in equivalent form.

(6) That is, Theorem '.R.' on Add. 4004: 57r (=§3.1 above: compare §3, note (2)). Presumably Newton intends this as a theorem in limit-motion (when all velocities may be considered uniform). Since then, by 'Galileo's' theorem (note (5) above) the body *a* would be moved in equal times along the chords *ae* and *ad* by a force acting in the direction *ac*, by Newton's Theorem *R* the velocities of *a* along *ad* and *ae* will be proportional to their respective lengths. (Compare §7, note (48) below.)

(7) Newton first wrote simply 'If *adce* is a parallelogram'.

(8) Newton has cancelled 'from *a*' (three times).

(9) Newton first wrote 'their motions being each to other as y[e] directing lines *ad*, *ae*, *ac* are'.

(10) Newton has cancelled '=*dc*'.

(11) A neat proof of the parallelogram law, apparently original with Newton.

(12) Since they share a common angular velocity.

(13) That is, all movement in a (Euclidean) plane may be represented as the product of a translation and a rotation.

Then[14] from the line *amd*, draw *de* || *ab* & ye same way; & from ye line *ae* draw *nm* || *ac*, & ye contrary way, to make up ye Trapezium *denm*. And if any two of these foure lines *de, mn, md, ne*, bee to any correspondent two of these foure motions, viz: of ye point *a* (fixed in ye line *dma*) towards *b*, of ye point *a* (fixed in ye line *ane*) towards *c*, of ye intersection point *a* moveing in ye line *dma* according to ye order of ye letters *m, d*, & of ye intersection point *a* in ye line *ane* according to ye order of ye letters *n, e*: Also all ye foure lines shall be one to another as those foure motions are.[15]

Note y^t in ye two last propositions ye moveing lines may bee crooked so y^t *amd, ane*, bee tangents to them in ye point *a*.[16]

Note also y^t by ye place of a body is meant its center of gravity.[17]

---

[2][18]

May 16. 1666. *To resolve Problems by motion ye 6*[19] *following prop: are necessary & sufficient.*

Prop 1. If y^e body *a* in y^e perimeter of y^e circle or sphære *adce* moveth towards its center *b*, its velocity to each point *d.c.e.* of y^t circumference is as y^e cordes *ad, ac, ae*, drawne from y^t body to those points are.

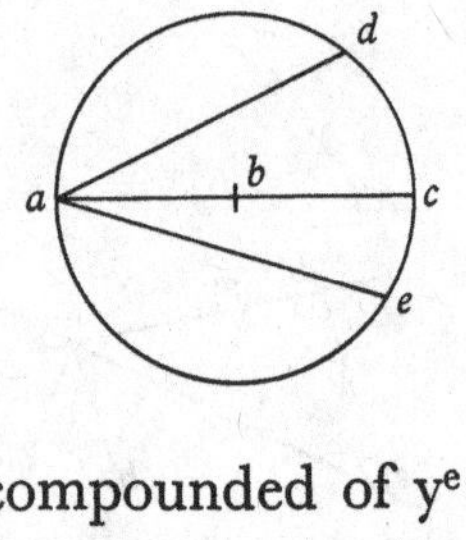

Prop 2^d. If y^e $\triangle$s *adc, aec* are alike though in divers planes; & 3 bodys move from y^e point *a* uniformely & in equall times, y^e first to *d*, y^e 2^d to *e*, y^e 3^d to *c*: y^n is y^e $3^{d's}$ motion compounded of y^e motion of y^e 1st & 2^d.[20]

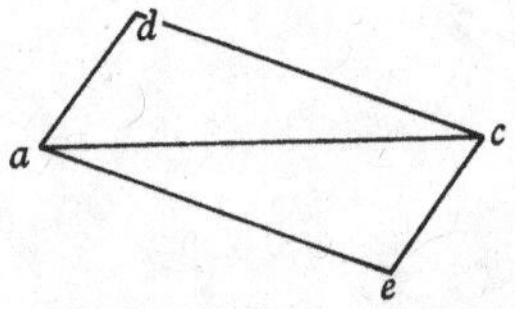

Note y^t by a body is meant its center of gravity.

Prop. 3. All y^e points of a body keeping parallel to it selfe are in equall motion.

Prop. 4. If a body onely move circularly about some axis, ye motion[s] of its points are as their distance from that axis.

Call these 2 simple motions.

Prop. 5. If ye motion of a body is considered as mixed of simple motions; ye motions of all its points are compounded of their simple motions, so as ye motion towards *c* (in prop 2^d) is compounded of ye motion towards *d* & *e*.

---

(14) Newton first wrote 'from each line to ye other draw two lines *de, mn* parallel to the mo'.

(15) This proposition (of which Prop. 6^t is the particular case where *ea* is at rest) is crucial in the application of the method of fluxions to the construction of tangents, inflexion points and extreme values of a function in general.

(16) Newton will develop this aspect in [2] below.

(17) This seems rather out of place in what is otherwise a severely mathematical piece.

(18) Add. 4004: $51^r/51^v$, partially printed in *Correspondence of Isaac Newton*, **3** (1961): no. 348.

Note y[t] all motion is reducible to one of these 3 cases: & in y[e] 3[d] case any line may bee taken for the axis (or if a line or superficies move in plano any point of y[t] plaine may bee taken for y[e] center) of motion.

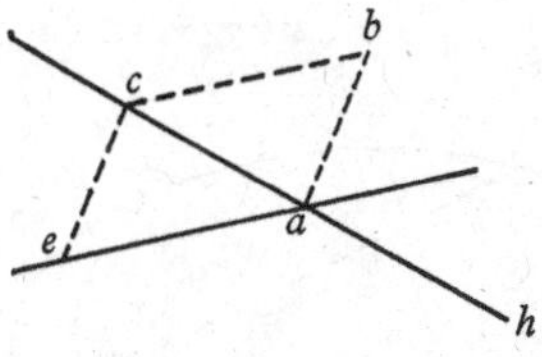

fig 3.

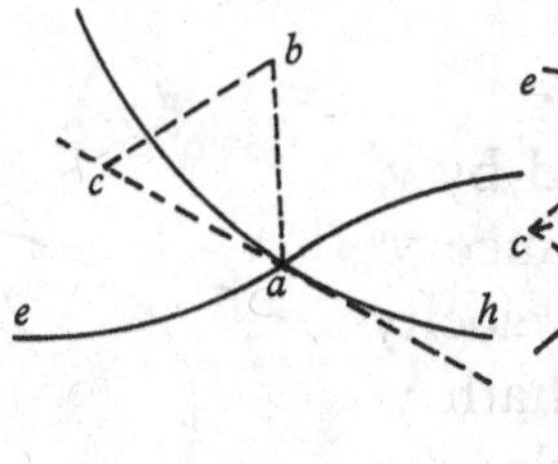

Fig 4

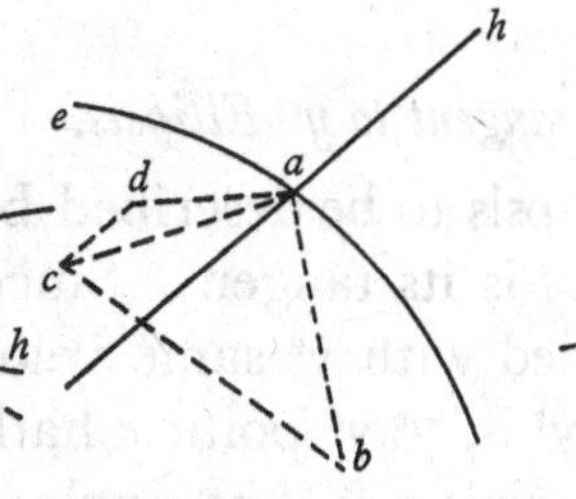

Fig 5

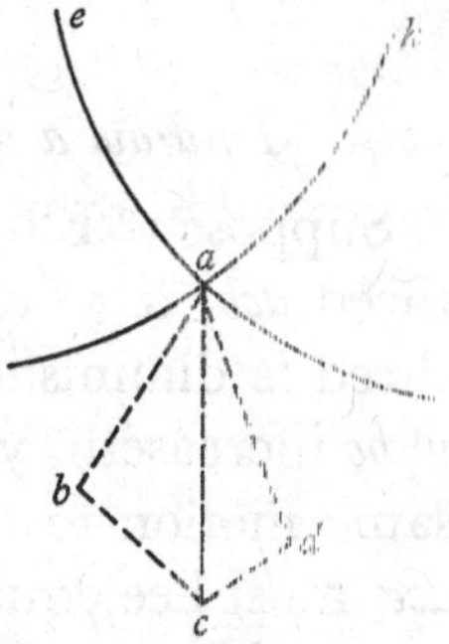

Fig 6.

Prop. 6. If y[e] lines *ae*, *ah* being moved doe continually intersect; I describe y[e] Trapezium *abcd* & its diagonall *ac*; & say y[t], y[e] proportion & position of these five lines *ab*, *ad*, *ae*, *cb*, *cd*, being determined by requisite data;[(21)] they shall designe y[e] proportion & position of these 5 motions; namely, of y[e] point *a* fixed in y[e] line *ae* & moveing towards *b*; of y[e] point *a* fixed in y[e] line *ah* & moveing towards *d*; of y[e] intersection point *a* moveing in y[e] plaine *abcd* towards *c* (for those 5 lines are ever in y[e] same plaine, though *ae* & *ah* may onely touch y[t] plaine in their intersection point); of y[e] intersection point *a* moveing in y[e] line *ae* parallely to *cb* & according to y[e] order of the letters *c*, *b*; & of y[e] intersection point *a* moveing in y[e] line *ah* parallely to *cd* & according to y[e] order of those letters.[(22)]

Note y[t] a streight line is said to designe y[e] position of curved motion in any point either when it toucheth y[e] line described by y[t] motion in y[t] point, (as *ab*, *ad*, *ac*), or when tis parallell to such a tang[nt] (as *bc*, *cd*).

Note also y[t] one line *ah* resting, (as in Fig 3 & 4) y[e] points *d* & *a* are coincident & y[e] point *c* shall bee in y[e] line *ah* if it bee streight, (fig 3), otherwise in its tangent *ac* (fig 4).

Prop. 7. Haveing an Equation expressing y[e] relation of two lines *x* & *y* described by two bodys *A* & *B* whose motions are *p* & *q*; Translate all y[e] termes to one side & multiply y[m], being ordered according to *x*, by this progression $\frac{3p}{x}.\ \frac{2p}{x}.\ \frac{p}{x}.\ 0.\ \frac{-p}{x}.\ \frac{-2p}{x}.$ &c: & being ordered by y[e] dimensions of *y*, multiply

(19) There are, in fact, seven! Prop. 7 was presumably added as an afterthought.

(20) Note that Newton has omitted the points *b* and *f* from his accompanying figure since the argument for which they are needed in [1] is abandoned.

(21) The lines *bc* and *dc* must be taken parallel to the tangents at *a* to the curves *ae* and *ah* respectively.

(22) Thus, *ab* and *ad* 'designe' the limit-motions of the curves *ah* and *ae* respectively, while *bc* and *dc* represent the 'determination' of the motions of *a* in these two curves, so that *ac* represents the composition of both sorts of motion.

them by this $\frac{3q}{y}.\ \frac{2q}{y}.\ \frac{q}{y}.\ 0.\ \frac{-q}{y}.$ &c. y^e^ summe of these products shall bee an equation expressing y^e^ relation of their motions p & q.(23)

To draw a tangent to y^e^ Ellipsis.

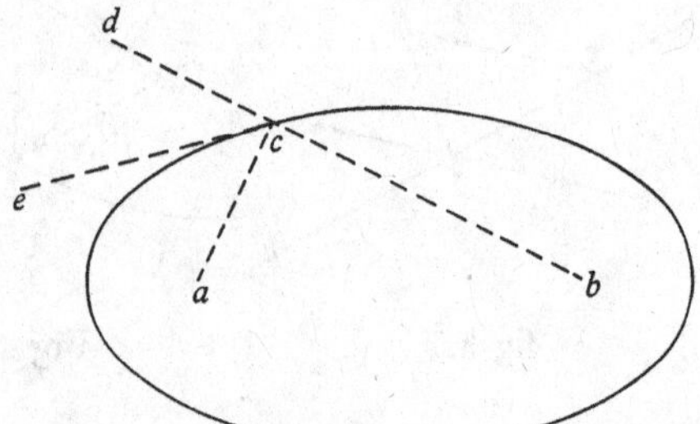

Suppose y^e^ Ellipsis to be described by y^e^ thred *acb*, & y^t^ *ce* is its tangent. Since y^e^ thred is diminished with y^e^ same velocity y^t^ *bc* increaseth, y^t^ is, y^t^ y^e^ point *c* hath y^e^ same motion towards *a* & *d*, y^e^ angles *dce*, *ace*, must bee equall, by prop 1. And so of y^e^ othe[r] conicks.(24)

To draw a Tangent to y^e^ Concha.

Suppose y^t^ *gae*, *glc*, *alf* are y^e^ rulers by wch y^e^ concha is usually described, & y^t^ $gt \parallel af \perp cb = \parallel mn$, & $ng = cl \perp tn \parallel rl$. And (since equality is more simple y^n^ proportionality) suppose y^t^ $cb = nm$ is y^e^ velocity of y^e^ point *c* towards *b*, or of *n* towards *m*. Then is *nt* y^e^ circular motion of y^e^ point *n* about *g* (prop 1), & *lr* y^e^ circular motion of y^e^ point *l* fixed in y^e^ ruler *ng*, (prop 4). And *lg* is y^e^ motion of y^e^ intersection point *l* (y^t^ is, y^e^ velocity of y^e^ point *c*) moveing in y^e^ line *glnc* from *g* (prop 6). Now since a twofold velocity of y^e^ point *c* is known namely *cb* toward *b* & *lg* towards *d*, make $fd \perp dc = lg$; & y^e^ motion of y^e^ point *c* shall bee in y^e^ line *fc* y^e^ diameter of y^e^ circle passing through y^e^ points *bcdf* (prop 1(25)) & therefore tang^nt^ to y^e^ Concha.(26)

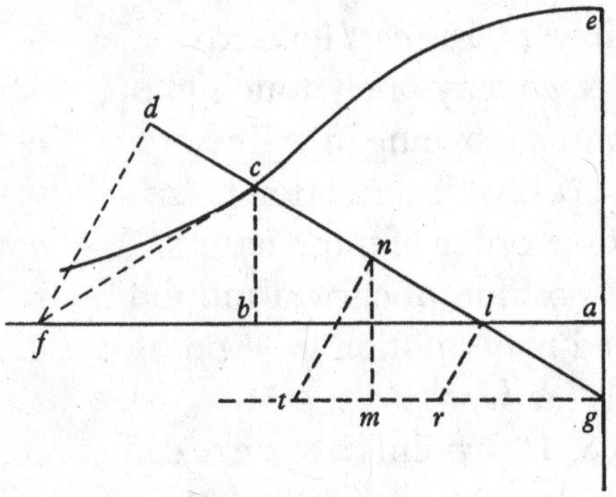

(23) Adapted from Add. 4004: 57^r^ = §3.1 above.

(24) See Add. 4004: 57^v^ = §3.2.1: Example 2. Though Newton does not say so explicitly, where the vectors *ca*, *cd*, *ce* represent the limit-motions of *a* in *ac*, *bc* and the ellipse, then $ca = cd$ and the angles $c\hat{a}e$, $c\hat{d}e$ are right.

(25) More correctly, by its converse. Newton first wrote '(prop 6)', and indeed Newton's first proposition is the particular case of his sixth where the angles $a\hat{b}c$ and $a\hat{d}c$ are right.

(26) More shortly, since $\begin{Bmatrix} cb \\ cd \end{Bmatrix}$ represents the limit-motions of

$c\begin{Bmatrix} \text{moving to } alf \text{ in rotation round } c \\ \text{in translation along } glc \text{ from } c \end{Bmatrix}$ and $\begin{Bmatrix} bf \\ df \end{Bmatrix}$ its instantaneous direction,

To find y^e point c wch distinguisheth twixt y^e concave & convex portions of y^e Concha.

Those things in y^e former prop: being supposed, make $\triangle gfh$ like *gnt* or *lbc*;(27) & $df \perp fr \parallel = hk = 2gl \perp kp$,(28) & draw *kf*. Now had y^e line *fd* onely parallel motion directed by *gd* or *rf*, (since $dc = lg$) y^e motion of all its points would bee *fr*,(29)

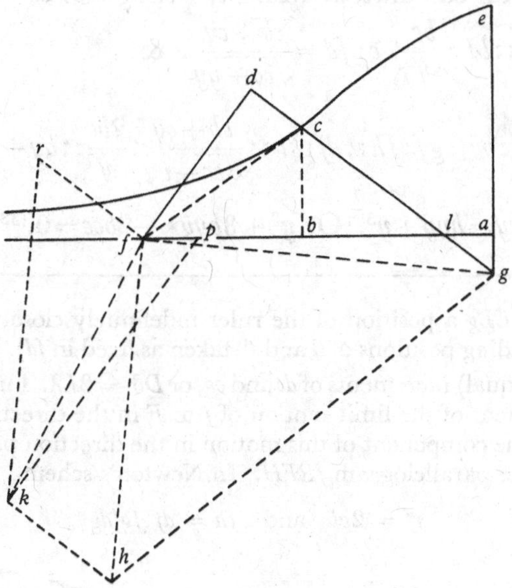

(prop 3): & if it had onely circular motion about *g*, y^e motion of y^e point *f* fixed in y^t line *df* would be *fh* (prop 4): But y^e motion of y^e point *f* is compounded of those two simple motions, & is therefore *fk* (prop 5 & 2); & ye motion of y^e

therefore *cf* represents in direction and relative magnitude the limit-motion of *c* in the conchoid *ec* and so is tangent at *c*.

Alternatively, where *CLg* is a position of the 'ruler' indefinitely close to *clg*, then the tangent *cf* at *c* is the limit-position of the chord *cC*. Then, where *fd* and *rl* are drawn normal to *clg* from *f* and to *l* respectively (meeting *CLg* in δ and λ as shown), it follows, since $cl = CL$, that the increments $C\gamma$ of *cg* and $L\lambda$ of *lg* are equal. Hence

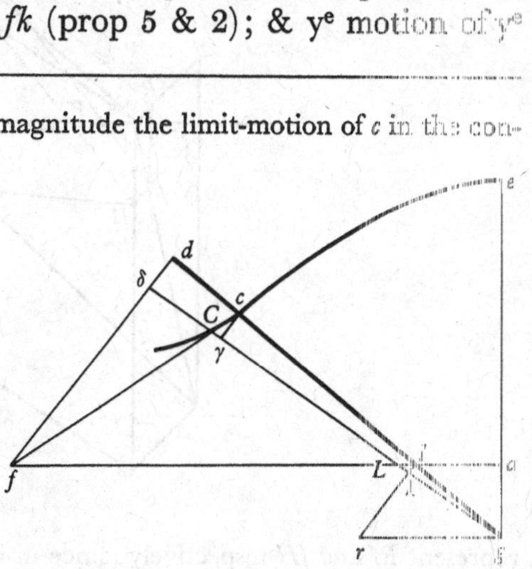

$$\frac{fd}{dl} = \frac{rl}{lg} = \frac{l\lambda}{L\lambda} = \frac{lg.c\gamma}{cg.C\gamma} = \frac{lg.fd}{cg.dc}$$

and so $\dfrac{dc}{dl} = \dfrac{lg}{cg}$, or $dc = lg$.

(27) Or, equivalently, where $\widehat{hgf} = \widehat{fgd}$ the point *h* in *hg* is constructed by taking $hf \perp fg$.
(28) Newton intends '... $hk\ (=2gl) \perp kp$'.
(29) That is, $dc + lg$ (which 'designes' the limit-motion of $dc + cg$).

intersection point f made by y^e^ lines af [&] df, & moveing in af, shall bee fp (prop 6).(30) Now if y^e^ line cf touch y^e^ concha in y^e^ required point, tis easily conceived y^t^ y^e^ motion of y^e^ intersection point f is infinitely little,(31) & therefore y^t^ y^e^ points p & f are coincident, df & fk being one streight line, & y^e^ triãgles gdf, fkh being alike.(32)

Which may bee thus calculated. Make $cl=c$. $ag=b$. $cb=y$. y^n^ is $bl=\sqrt{cc-yy}$. $2gl=\frac{2bc}{y}=hk$. $bl:bc::ld=\frac{bc}{y}+c:fd=\frac{cb+cy}{\sqrt{cc-yy}}$. &

$$\sqrt{cc-yy}:y::bl:bc::gf:fh::df:kh::\frac{cb+cy}{\sqrt{cc-yy}}:\frac{2bc}{y}::by+yy:2b\sqrt{cc-yy}.$$

Therefore $2bcc-2byy=byy+y^3$. Or $y^3+3byy*-2bcc=0$.(33)

(30) As before, take CLg a position of the ruler indefinitely close to clg with D and F (in $FD\perp Dg$) the corresponding positions of d and f (taken as fixed in fd). Then $D\delta$, the increment of dg, is the sum of the (equal) increments of dc and cg, or $D\delta = 2L\lambda$. Further, where $f\hat{g}H = d\hat{g}D$, $fH = RF$ is the component of the limit-motion of f to F in the direction normal to fg. Similarly, $D\delta = Rf\perp fd$ is the component of this motion in the direction parallel to dg. Hence F is found by completing the parallelogram $fRFH$. In Newton's scheme

$$rf = 2gl \quad \text{and} \quad fh = df.fg/dg$$

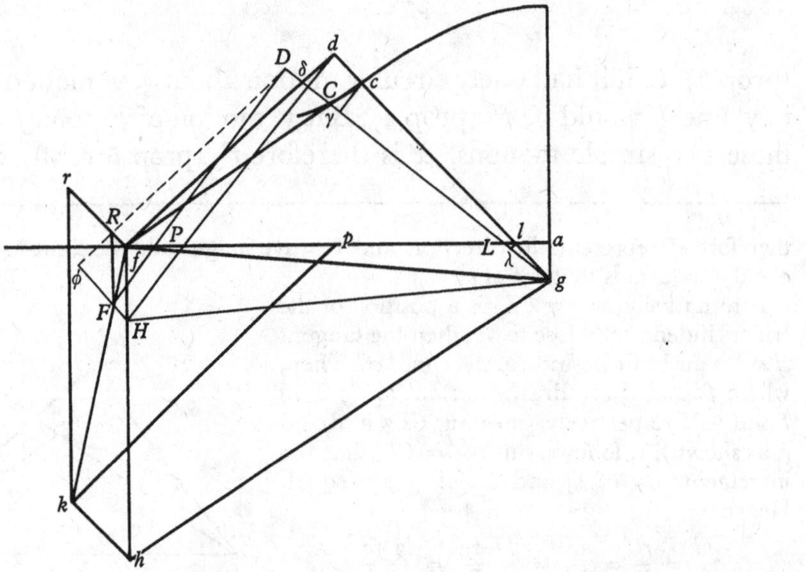

represent Rf and fH respectively (since in the same representation df and cf represent $c\gamma$ and cC), so that fk represents fF and therefore the limit-motion of f, fixed in df, in direction; finally, fp represents fP, the limit-motion of the meet of df with af. (Note that, where ϕ is the meet of DR and FH, $D\phi$ will be the parallel displacement of df outwards from g as d passes to D.)

In stead of y[e] ordinary method de Maximis et Minimis,[(34)] it will be as convenient (& perhaps more naturall) to use

This; Namely. To find y[e] motion of y[t] line or quantity & suppose it equall to nothing, or infinitely small.

But y[n] y[e] motion to w[ch] tis compared must bee finite.[(35)] That is, y[e] unknown quantitys ought not to bee at their greatest or least, both at once.[(36)]

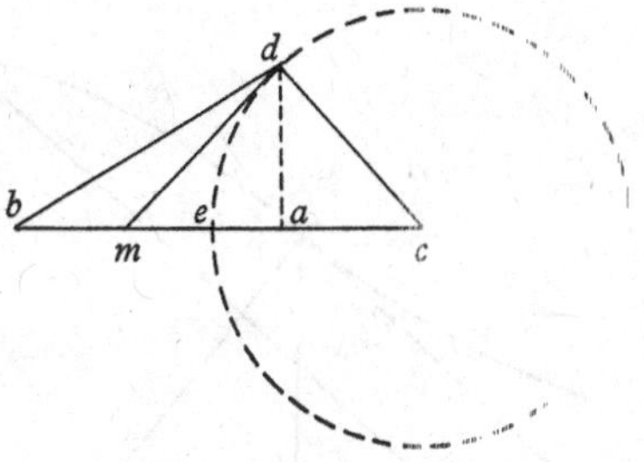

Example [1st]. In y[e] Triangle *bcd*, y[e] side *bc* being given & fixed, y[e] side *dc* being given & circulating about y[e] center *c*, I would know when *bd* is y[e] shortest it may bee. I call $bd=x$. $da=y$. $bc=a$. $dc=b$. y[n] is $\sqrt{xx-yy}=ab$. & $\sqrt{bb-yy}=ac$. & $\sqrt{xx-yy}+\sqrt{bb-yy}=a$. Or $xx-aa-bb=-2a\sqrt{bb-yy}$. &

$$\begin{array}{l} x^4-2aaxx+a^4+b^4=0. \\ \quad\;\; -2bb \quad -2aabb \\ \qquad\qquad\quad +4aayy \end{array}$$

$$\begin{array}{l} 4px^3-4aapx+8aaqy=0. \\ \qquad -4bb \end{array}$$ (prop 7).[(37)] And makeing $p=0$, tis $8aaqy=0$ Or $\frac{0}{8aaq}=y=0$. (For q signifieing y[e] motion of d towards bc may bee finite though p,

(31) For if c is an inflexion point, all tangents to the conchoid at points indefinitely close to c will coincide with cf and so, as the point c describes the conchoid, the tangent cf at c will be instantaneously at rest.

(32) It follows that the increment of P must be zero and so correspondingly must fP. Hence $dfFk$ is a straight line and $f\hat{k}h$ right, and therefore, since $g\hat{f}h$ is right (by construction), $d\hat{f}g = k\hat{h}f$ and the triangles dfg and khf are similar.

(33) This general result, first proved by Huygens on 25 September 1653 (Huygens, *Œuvres*, **12** (1910): xx: 83–6), was taken by Newton from Schooten's first printed analytical exposition of it in his commentary on Descartes (*Commentarii in Librum II*, *O = Geometria*: 252–9, especially 258–9). A fuller history of the problem is given in part 3 below (**3**, 3, §3, notes (3/4)).

(34) That is, Hudde's as expounded in his tract *de Maximis et Minimis*. (= *Geometria*: 507–16. Compare **2**, Historical note, above.)

(35) Newton states Hudde's rule in equivalent form: that is, where $f(x, y) = 0$, a *necessary* condition for y to attain an extreme value is that $dy/dx = 0$. He then deduces correctly that a *sufficient* condition for this to happen is that 'y[e] motion' (dy/dt) of y be instantaneously zero provided that the corresponding motion dx/dt of x be not also simultaneously zero. (In other words, where p and q are the fluxions of x and y, a sufficient condition for $dy/dx = q/p$ to be zero is that $q = 0$ with $p \neq 0$.)

(36) A curious turn of phrase. Newton argues that, since $q = 0$ determines an extreme value of y, then the indeterminate case $q = p = 0$ determines that y and x reach an extreme value together.

(37) That is, where p and q are the fluxions (dx/dt and dy/dt respectively) of x and y.

its motion towards b doth perish). Wherefore $xx=aa \between 2ab+bb$. or $x=a-b$. $x=b-a$. $x=b+a$. $x=-b-a$. are y^e^ greatest & least valors of y^e^ line bd.(38)

Should I have taken $ba=y$, instead of $da=y$, The effect would not have followed because both y^e^ motions p & q would have vanished at once(39) in y^e^ point e. But I might have taken y^e^ tangent dm for y, or any other line wch would not coincidere(40) wth bc at its being greatest or least.(41)

Example 2^d^. If ne(42) is y^e^ Conchoïd ($ga=b$. $ae=c=nl$.(42) nb(42)$=y$. $ab=x$.) fn(42) its tang^nt^ [&] em parallell to it. Then is $\sqrt{cc-yy}=bl:c=cl::dl=cg=c+\frac{bc}{y}$ (vide supra)$:fl=\frac{ccy+bcc}{y\sqrt{cc-yy}}$. & $fb=\frac{bcc+y^3}{y\sqrt{cc-yy}}$.(43) & $y=cb:fb::ea=c:am=\frac{bc^3+cy^3}{yy\sqrt{cc-yy}}=z$. et

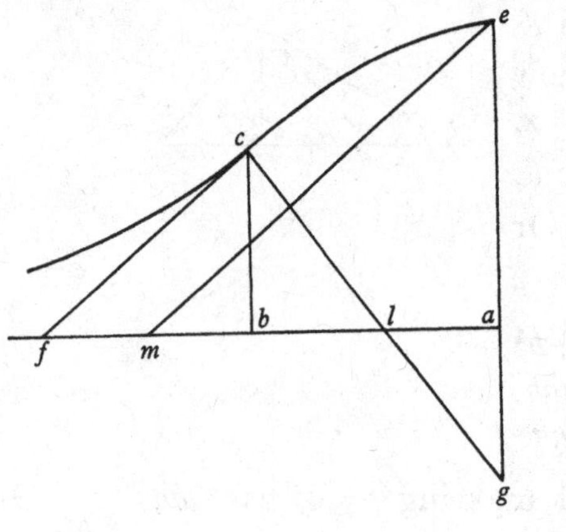

$$bbc^6+2bc^4y^3+ccy^6-ccy^4zz[+]y^6zz=0.$$

ponatur p esse motus puncti m, & q esse motus puncti c versus b. Erit(44) (prop 7)

$$6bc^4yyq+6ccy^5q-4ccy^3zzq[+]6y^5qzz$$
$$-2ccpy^4z[+]2py^6z=0.$$

(38) Since bd must have positive length, the two negative values are not representible (and appear, in fact, as extreme values of the more general analytical form

$$\pm\sqrt{[x^2-y^2]}\pm\sqrt{[b^2-y^2]}=a).$$

(39) At the point e both be and $bd=x$ reach minimum values together, and so if be is taken for y, at that point both x and y have zero increments, or $dx/dt=dy/dt=0$.

(40) Read 'coincide'. (Newton absent-mindedly uses the Latin equivalent.)

(41) Specifically, where $bd=x$ and $dm=y$, then, since

$$\cos b\hat{c}d=\frac{a^2+b^2-x^2}{2ab},\quad ac=\frac{a^2+b^2-x^2}{2a}\quad\text{and so}\quad da^2=b^2-\left(\frac{a^2+b^2-x^2}{2a}\right)^2.$$

Hence

$$dm^2=y^2=b^2.\frac{da^2}{ac^2}=b^2\left(\frac{(2ab)^2-(a^2+b^2-x^2)^2}{(a^2+b^2-x^2)^2}\right)\quad\text{or}\quad(b^2+y^2)(a^2+b^2-x^2)^2=(2ab^2)^2,$$

whose derivative is $\quad 2qy(a^2+b^2-x^2)^2-2xp(b^2+y^2)(a^2+b^2-x^2)=0,$

and so $\quad \frac{p}{q}=\frac{(a^2+b^2-x^2)y}{x(b^2+y^2)},\quad$ since $x^2\neq a^2+b^2$.

The condition $p=0$, $q\neq 0$ for an extreme value of x (which is obviously satisfied) yields $y=0$ and so $a^2+b^2-x^2=\pm 2ab$, as before.

(42) For each occurrence of 'n' read 'c'. Newton first denoted the point c by n in his accompanying diagram, but halfway through his textual argument altered the denotation to correspond with his previous figure (and Schooten's in *Geometria*: 252–62) but has not completely altered his text correspondingly.

supposeing $[p]=0$ (for when *am* is y^e^ least y^t^ it may bee, y^e^ point *c* is y^t^ wch distinguisheth twixt y^e^ concave & convex portion of y^e^ Conchoid,(45) & y^n^ y^e^ motion $[p]$ vanisheth.) it will bee.

$$\frac{3bc^4+3ccy^3}{2ccy-3y^3}=zz=\left.\overline{\frac{bc^3+cy^3}{yy\sqrt{cc-yy}}}\right\}^2.^{(46)}\qquad \frac{3c}{2cc-3yy}=\frac{bc^3+cy^3}{ccy^3-y^5}.$$

& $ccy^3=2bc^4-3bccyy$. Or $y^3=-3byy+2bcc$.(47)

---

(43) See his preceding calculations on the conchoid's inflexion point above, and note that $fb = fl-bl$.

(44) 'Make p the motion of the point m, and q the motion of c towards b. Then it will be....'

(45) Since *ae* is constant and ae/am is the slope of the tangent *cf*.

(46) Read '$\left(\frac{bc^3+cy^3}{yy\sqrt{cc-yy}}\right)^2$'.

(47) Where c is an inflexion point, the corresponding tangent *cf* and so its parallel *en* has maximum slope and *am* is a minimum. Newton finds the relation $f(y, z) = 0$ connecting $cg = y$ and $am = z$ and hence its derivative in the form $0 = qf_y+rf_z$, with $q = dy/dt$ and $r = dz/dt$. It follows that $dz/dy = -f_y/f_z = r/q$ and the condition $r = 0$ with $q \neq 0$ yields $0 = r = f_y$ as a sufficient condition for $am = z$ to attain an extreme value (here a minimum). Newton's previously found condition for an inflexion point (a cubic in y with one real root) then follows by eliminating z between $f(y, z) = 0$ and $f_y = 0$.

Unknown to Newton this method of reducing the problem of finding inflexion points to maximizing or minimizing the tangent slope had in essence been suggested by Pierre Fermat a quarter of a century before and by him written up about 1640 as an application of his general method *de maximis et minimis*. (See Fermat's *Œuvres*, **1** (Paris, 1891): 1re Partie, *Maxima et Minima*, §VI: 166–7, first printed in his posthumous *Varia Opera Mathematica* (Toulouse, 1679).) Apparently, Fermat had succeeded in constructing the general conchoid tangent about 1632, but it was Roberval who, some time before 11 October 1636 when he told Fermat that he also was able to construct tangents to the conchoid 'comme étant déterminations d'équations quarré quarrées' (Fermat's *Œuvres*, **2** (1894): 82), first noticed that the conchoid had two (finite) inflexion points and pressed Fermat to apply his method *de maximis et minimis* to their determination. (Compare Jean Itard, 'Fermat précurseur du Calcul différentiel', *Archives Internationales d'Histoire des Sciences*, **1** (1947/8): 589–610, especially 606–8: *Les points d'inflexion*.)

7

THE OCTOBER 1666 TRACT ON FLUXIONS

From the original(1) in the University Library, Cambridge

[1](2)

October 1666.(3) *To resolve Problems by Motion these following Propositions are sufficient.*(4)

1 If the body *a* in the Perimeter of y^e cirkle or sphære *adc* moveth towards its center *b*, its velocity to each point (*d*, *c*, *e*) of y^t circumference is as y^e chords (*ad*, *ac*, *ae*) drawne from that body to those points are.

fig 1

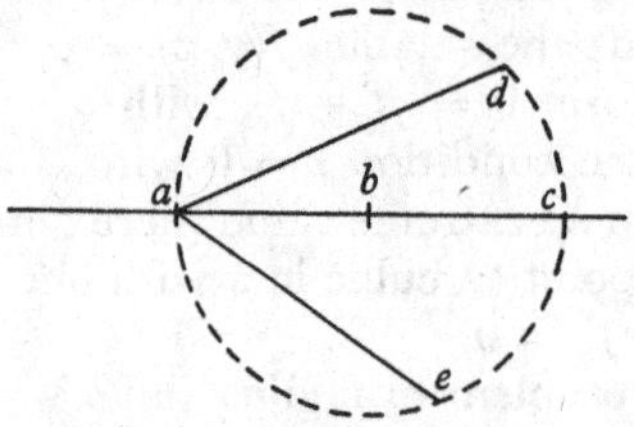

fig 2

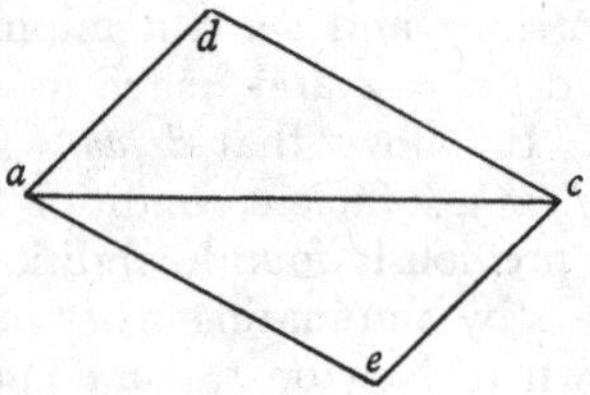

2 If y^e $\triangle$s *adc*, *aec*, are alike viz: *ad*=*ec* &c (though in divers plaines) & 3 bodys move from the point *a* uniformely & in equall times, y^e first to *d*, the 2^d to *e*, y^e 3^d to *c*; Then is the thirds motion compounded of y^e motion of the first & second.

3 All the points of a Body keeping Parallel to it selfe are in equall velocity(5).

(1) Add. 3958.3: 48^v–63^v, first printed in A. R. Hall's and Marie Boas Hall's *Unpublished Scientific Papers of Sir Isaac Newton* (Cambridge, 1962): I, 1: 15–64. In this tract Newton gathers, sifts and revises his calculus researches pursued over the two previous years, autumn 1664–May 1666. During Newton's life and especially at the time of his death the manuscript, in one or other of several variant copies, had a limited circulation among English mathematicians. There exists, in private possession, an early contemporary copy in the hand of Newton's roommate and amanuensis, Wickins, which was possibly destined for John Collins though we have no evidence to show that it ever passed out of Newton's possession before his death. With this copy are some extra sheets of notes in the hand of William Jones (and especially on Prob. 9): we may conjecture that these are first drafts for the rearranged copy in his hand in the University Library, Cambridge (Add. 3960.1: 1–50). (An eighteenth-century entry, in the hand of James Wilson, affirms correctly of the latter that 'The Transcriber has here put $\dot{x}$, $\dot{y}$ and $\dot{z}$ for *p*, *q* and *r* of the Original.... Here seems to be some transpositions and interpolations, as Mr Jones was wont to make in those papers of Sr Isaac Newton, which he distributed to his scholars, that none might make a perfect book out of them.') Compare Wilson's *Mathematical Tracts of the late Benjamin Robins, Esq*; **2** (London, 1761): Appendix: 351–6, which prints extracts from Wickins' copy of this 'small tract of fluxions', then in the hands of Newton's executor Pellet.

4 If a body move onely $\left\{\begin{matrix}\text{angularly}\\ \text{circularly}\end{matrix}\right\}$ about some axis, y^e velocity[5] of its points are as their distance from that axis.[6]

5 The motions of all bodys are either parallel or angular, or mixed of y^m both after y^e same manner, y^t the motion towards *c* (Prop 2) is compounded of those towards *d* & *e*. And in mixed motion any line may bee taken for y^t axis (or if a line or superficies move in plano, any point in y^t plane may bee taken for the center) of ye angular motion.

6 If ye lines *ae*, *ah* being moved doe continually intersect; I describe ye trapezium *abcd*, & its diagonall *ac*; & say y^t, ye proportion & position of these five lines *ab*, *ad*, *ac*, *cb*, *cd*, being determined by requisite data; shall designe y^e proportion & position of these five motions; viz: of y^e point *a* fixed in y^e line *ae* & moveing towards *b*, of y^e point *a* fixed in y^e line *ah* & moveing towards *d*; of y^e

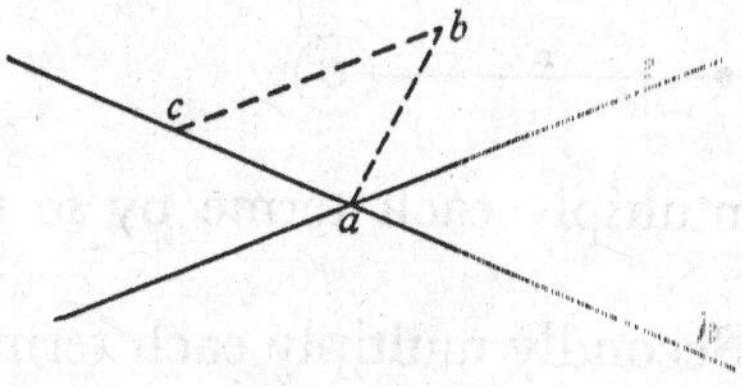

fig 3

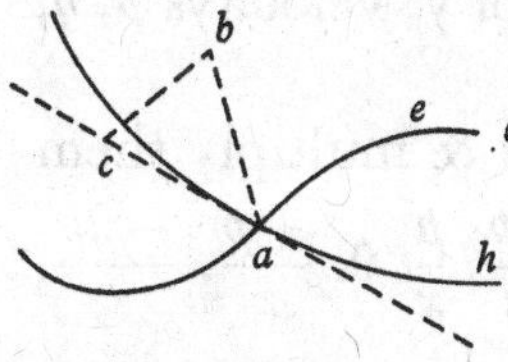

fig 4

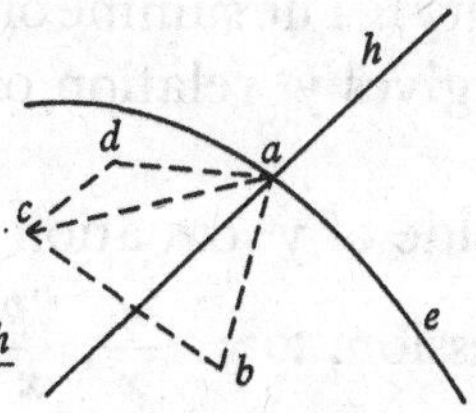

fig 5

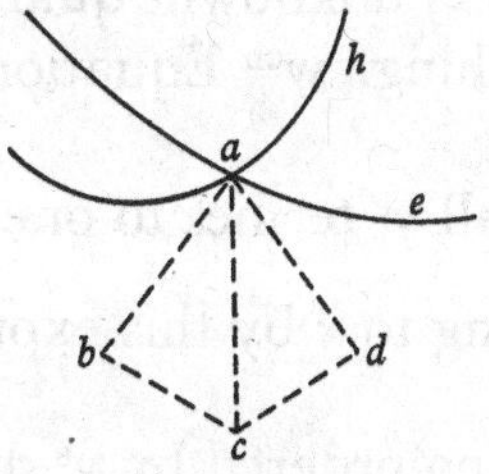

fig 6

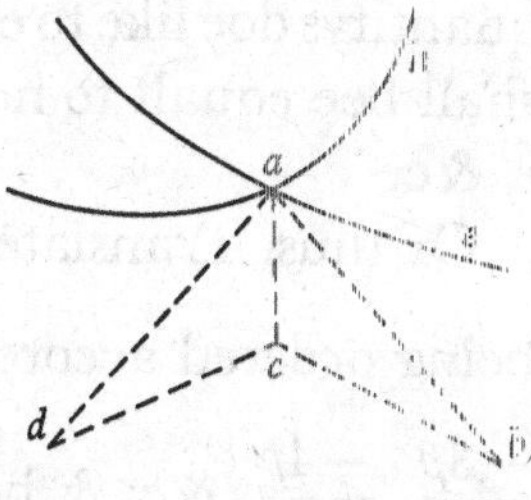

fig 7

intersection point *a* moveing in y^e plaine *abcd* towards *c*, (for those five lines are ever in y^e same plaine, though *ae* & *ah* may chanch[8] onely to touch that plaine in their intersection point *a*); of ye intersection point *a* moveing in ye line *ae* parallely to *cb* & according to ye order of ye letters *c*, *b*; & of ye intersection point *a* moving in ye line *ah* parallely to *cd* & according to ye order of the letters *c*, *d*.

(2) Ff. 48^v–52^v, the first of two parts into which the tract naturally divides itself. The analytical theory here introduced is, in the second part ([2] below), applied to the resolution of various problems.

(3) This date was squeezed in at the head of the manuscript at a later stage in its composition.

(4) A revision which compresses his previous entries on 13 November 1665 (§3.1) and 16 May 1666 (§4.2).

(5) Newton has cancelled 'motion'.

(6) Newton then copied but immediately cancelled 'Call these 2 simple motions, parallel & angular'.

(7) In each of these figures *cb* and *cd* are parallel to the tangents at *a* to the curves *ae* and *ah* respectively.

(8) 'chance'.

Note yt, one of ye lines as *ah* (fig 3d & 4th) resting, ye points *d* & *a* are coincident, & ye point *c* shall bee in ye line *ah* if it bee streight (fig 3), otherwise in its tangent (fig 4th).

7 Haveing an Equation expressing ye relation twixt two or more lines x, y, z &c: described in ye same time by two or more moveing bodys A, B, C, &c: the relation of their velocitys p, q, r, &c may bee thus found, viz:

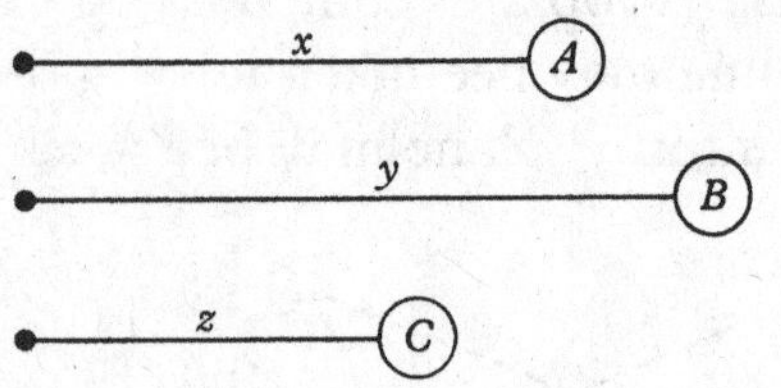

Set all ye termes on one side of ye Equation that they become equall to nothing. And first multiply each terme by so many times $\frac{p}{x}$ as x hath dimensions in yt terme. Secondly multiply each terme by so many times $\frac{q}{y}$ as y hath dimensions in it. Thirdly (if there be 3 unknowne quantitys) multiply each terme by so many times $\frac{r}{z}$ as z hath dimensions in yt terme, (& if there bee still more unknowne quantitys doe like to every unknowne quantity). The summe of all these products shall bee equall to nothing. wch Equation gives ye relation of ye velocitys p, q, r, &c.

Or thus. Translate all ye termes to one side of ye equation & multiply them being ordered according to x by this expression, :ɔꝫ. $\frac{3p}{x}.\frac{2p}{x}.\frac{p}{x}.0.\frac{-p}{x}.\frac{-2p}{x}.\frac{-3p}{x}.\frac{-4p}{x}.$ &c: & being ordered by ye dimensions of y multiply them by this, :ɔꝫ. $\frac{3q}{y}.\frac{2q}{y}.\frac{q}{y}.0.\frac{-q}{y}.\frac{-2q}{y}.$ &c. The sume of these products shall bee equall to nothing, which equation gives ye relation of their velocitys p, q, &c.

Or more Generally ye Equation may bee multiplyed by ye terme[s] of these progressions $\frac{ap+4bp}{x}.\frac{ap+3bp}{x}.\frac{ap+2bp}{x}.\frac{ap+bp}{x}.\frac{ap}{x}.\frac{ap-bp}{x}.\frac{ap-2bp}{x}.$ &c. And $\frac{aq+2bq}{y}.\frac{aq+bq}{y}.\frac{aq}{y}.\frac{aq-bq}{y}$ &c. (a & b signifying any two numbers whither rationall or irrationall).

(9) Compare 5, §4, note (7) above. (This last sentence was added after the following paragraph was begun, for it is crowded into a double space on the right-hand side of Newton's page.)

(10) For $$y = \int_0^X \frac{a}{x}.dx = a\log X - a\log(0) = a\log X + \frac{a}{0}.$$

(11) In Descartes' sense of not by an admissibly accurate mathematical procedure, and so by implication here to any desired degree of numerical approximation. (Compare 6, §1, note (2) above.)

8 If two Bodys A & B, by their velocitys p & q describe y^e lines x & y. & an Equation bee given expressing y^e relation twixt one of y^e lines x, & y^e ratio $\frac{q}{p}$ of their motions q & p; To find y^e other line y.

Could this ever bee done all problems whatever might bee resolved. But by y^e following rules it may bee very often done. (Note y^t $\pm m$ & $\pm n$ are logarithmes or numbers signifying y^e dimensions of x.)(9)

First get y^e valor of $\frac{q}{p}$. Which if it bee rationall & its Denominator consist of but one terme: Multiply y^t valor by x & divide each terme of it by y^e logarithme of x in y^t terme y^e quote shall bee y^e valor of y. As if $ax^{\frac{m}{n}}=\frac{q}{p}$. Then is $\frac{na}{m+n}x^{\frac{m+n}{n}}=y$. Or if $ax^{\frac{m}{n}}=\frac{q}{p}$. Then is $\frac{na}{n+m}x^{\frac{n+m}{n}}=y$. (Soe if $\frac{a}{x}=ax^{\bar{1}}=\frac{q}{p}$. Then is $\frac{a}{0}x^0=y$, soe y^t y is infinite. But note y^t in this case x & y increase in y^e same proportion y^t numbers & their logarithmes doe, y being like a logarithme added to an infinite number $\frac{a}{0}$.(10) But if x bee diminished by c, as if $\frac{a}{c+x}=\frac{q}{p}$, y is also diminished by y^e infinite number $\frac{a}{0}c^0$ & becomes finite like a logarithme of y^e number x. & so x being given, y may bee mechanichally(11) found by a Table of logarithmes, as shall bee hereafter showne.(12)

Secondly. But if y^e denominator of y^e valor of $\frac{q}{p}$ consist of more termes y^n one, it may bee reduced to such a forme y^t y^e denominator of each ꝑte of it shall have but one terme, unlesse y^t ꝑte bee $\frac{a}{c+x}$: Soe y^t y may bee y^n found by y^e precedent rule. Which reduction is thus performed, viz: 1st, If the denominator bee not $a+bx$, nor all its termes multiplyed by x or xx, or x^3, &c: Increase or diminish x untill y^e last terme of y^e Denominator vanish. 2dly, And when all y^e termes in y^e Denominator are multiplyed by x, xx, or x^3 &c: Divide y^e numerator by y^e Denom^r (as in Decimall numbers) untill y^e Quotient consist of such ꝑts none of whose Denominato^rs are so multiplyed by x, x^2 &c: & begin y^e Division in these termes in wch x is of its fewest dimensions unlesse y^e Denominator be $a+bx$.(13)

(12) Newton returns to this point briefly at the end of this first section (compare note (34)), but it seems possible that he intended to add, on the lines of his later *De Analysi per Æquationes Numero Terminorum Infinitas* (1668?), an extended discussion of power-series in revision of his researches into the Binomial Theorem (**1**, 3, §3 above). The basis of Newton's argument may perhaps be clearer in the modern form

$$y = \int_0^X \frac{a}{c+x}.dx = \lim_{\epsilon\to 0}\left[\int_0^X \frac{a}{(c+x)^{1-\epsilon}}.dx\right] = \lim_{\epsilon\to 0}\left[\frac{a}{\epsilon}((c+X)^\epsilon - c^\epsilon)\right] = a\log\left(\frac{c+X}{c}\right).$$

(13) Compare 5, §3 above.

If yⁿ yᵉ termes in yᵉ valor of $\frac{q}{p}$ bee such as was before required, yᵉ valor of y may bee found by yᵉ first pᵗᵉ of this Prop: onely it must bee so much diminished or increased as it was before increased or diminished by increasing or diminishing x. But if the denominator of any terme consist of more termes yⁿ one, unlesse yᵗ terme bee $\frac{a}{c+x}$. First find those pts of y's valor wᶜʰ correspond to yᵉ other pts of $\frac{q}{p}$ its valor. & yⁿ by yᵉ preceding rules &c: seeke yᵉ pte of y's valor answering to this pte of $\frac{q}{p}$ its valor.

Example 1. If $\frac{xx}{ax+b}=\frac{q}{p}$. Then by Division tis $\frac{x}{a}-\frac{b}{aa}\frac{+bb}{a^3x+aab}=\frac{xx}{ax+b}=\frac{q}{p}$. (as may appeare by multiplication.) Therefore (by 1ˢᵗ pte of this Prop:) tis $\frac{xx}{2a}\frac{-bx}{aa}+\square^{(14)}\frac{bb}{a^3x+aab}=y$. $\left(\square\frac{bb}{a^3x+aab}\right.$ signifys yᵗ pte of yᵉ valor of y wᶜʰ is correspondent to yᵉ terme $\frac{bb}{a^3x+aab}$ of yᵉ valor of $\frac{q}{p}$, wᶜʰ may bee found by a Table of logarithmes as may hereafter appeare.[(15)]$\left.\right)$

Example 2ᵈ.[(16)] If $\frac{x^3}{aa-xx}=\frac{q}{p}$. I suppose $x=z-a$. Or $\frac{z^3-3azz+3aaz-a^3}{2az-zz}=\frac{q}{p}$. And by Division $\frac{-aa}{2z}-z+a\frac{+aa}{4a-2z}=\frac{q}{p}$, (as may appeare by multiplication.) And substituteing $x+a$ into yᵉ place of z, tis $[-]x\frac{-aa}{2x+2a}\frac{+aa}{2a-2x}=\frac{q}{p}=\frac{[-]x^3}{aa-xx}$. And therefore (by pte 1ˢᵗ of Prop 8) $[-]\frac{xx}{2}+\square\frac{-aa}{2x+2a}+\square\frac{aa}{2a-2x}=y$.

But sometimes The last terme of yᵉ Denominator cannot bee taken away, (as if yᵉ Denominʳ bee $aa+xx$. or a^4+x^4. or $a^4+bbxx+x^4$. &c) And then it will

(14) Read 'yᵉ area of'. (See, for example, 5, §3: passim.)

(15) See note (12) above.

(16) Newton has cancelled a first choice, 'If $\frac{b^3}{-aa+xx}=\frac{q}{p}$. I suppose $x=a+z$, & consequently $\frac{b^3}{2az+zz}=\frac{q}{p}$. & by division, $\frac{b^3}{2az}\frac{-b^3}{4aa+2az}=\frac{q}{p}$.'

(17) The list of integrals which follows, though loosely based on his previous lists in the undergraduate notebook ULC. Add. 4000 (printed in 5, §§1/4), is for the most part freshly calculated and the pattern of the tabulation is more pronounced.

(18) Since
$$\frac{d}{dx}\left(\frac{cx}{a+bx^2}\right)=\frac{c}{a+bx^2}-\frac{2bcx^2}{(a+bx^2)^2},$$
then
$$\int\frac{c}{a+bx^2}.dx=\frac{cx}{a+bx^2}+2\int\frac{bcx^2}{(a+bx^2)^2}.dx,$$

bee necessary to have in readinesse some examples wth such Denominators to wch all other cases of like denomination may bee by Division reduced. As if[17]

$$\frac{cx}{a+bxx}=\frac{q}{p}.\ \text{Make } bxx=z, \text{ Then is } \square\ \frac{c}{2ab+2bz}=y.$$

$$\frac{cxx}{a+bx^3}=\frac{q}{p}.\ \text{Make } bx^3=z, \text{ Then is } \square\ \frac{c}{3ba+3bz}=y.$$

$$\frac{cx^3}{a+bx^4}=\frac{q}{p}.\ \text{Make } bx^4=z, \text{ Then is } \square\ \frac{c}{4ba+4bz}=y.\ \ \&c.$$

In Generall if

$$\frac{cx^{n-1}}{a+bx^n}=\frac{q}{p}.\ \text{Make } bx^n=z, \text{ \& y}^n \text{ is } \square\ \frac{c}{nba+nbz}=y.$$

Also if

$$\frac{c}{a+bxx}=\frac{q}{p}.\ \text{Make } \sqrt{\frac{c}{a}}-\sqrt{\frac{c}{a+bxx}}=z \text{ \& y}^n \text{ is } \frac{cx}{a+bxx}+\square\ 2\sqrt{\frac{2z\sqrt{ac}}{b}-\frac{azz}{b}}=y.^{(18)}$$

That is, if[19]

$$\sqrt{\frac{c}{bx}-\frac{a}{b}}=\frac{q}{p};\ \text{I make } x=zz, \text{ \& } \square 2\sqrt{\frac{c}{b}-\frac{a}{b}zz}=y.$$

Or if

$$\frac{c}{a+bxx}=\frac{q}{p}.\ \text{Make } \sqrt{\frac{c}{a+bxx}}=z=CB,^{(20)}$$

$$2\sqrt{\frac{c}{b}-\frac{a}{b}zz}=y=BD.^{(20)}\ \ \&\ \square=CDV=y.$$

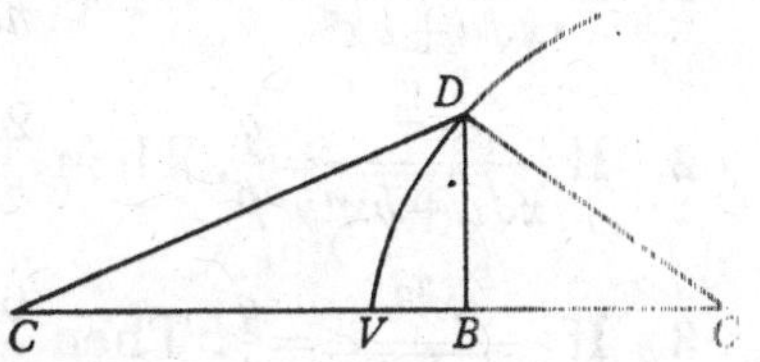

Thirdly If y^e^ valor of $\frac{q}{p}$ is irrationall being a square roote, The simplest cases may bee reduced to these following examples.

---

so that if we take $z = k+\frac{l}{(a+bx^2)^{\frac{1}{2}}}$ and so $\frac{dz}{dx} = -\frac{lbx}{(a+bx^2)^{\frac{3}{2}}}$

and equate $\int \frac{bcx^2}{(a+bx^2)^2}.dx = \int v.dz,$

we deduce that $v^2 = \frac{c^2x^2}{l^2(a+bx^2)} = \frac{c^2}{l^4b}(l^2-a(z-k)^2)$:

Newton's result follows by taking $k = \sqrt{[c/a]}$ and $l = -\sqrt{c}$.

(19) An unexpected conjunction, since the two integrals are not narrowly connected.

(20) Read '$z = VB$' and '$v = BD$'. It follows that

$$y = \int z^2.dx = z^2x-\int 2zx.dz = \tfrac{1}{2}vx-\int v.dz,$$

since $v = 2zx$, and therefore that $y = \triangle CDB-$area (VDB) = area (CDV).

1. If $\frac{cx^n}{x}\sqrt{a+bx^n}=\frac{q}{p}$. Then $\frac{2ac+2bcx^n}{3nb}\sqrt{a+bx^n}=y$.

2. If $\frac{cx^{2n}}{x}\sqrt{a+bx^n}=\frac{q}{p}$. Then $\frac{6bbcx^{2n}+2abcx^n-4aac}{15nbb}\sqrt{a+bx^n}=y$.

3. If $\frac{cx^{3n}}{x}\sqrt{a+bx^n}=\frac{q}{p}$. Then $\frac{30b^3cx^{3n}+6abbcx^{2n}-8aabcx^n+16a^3c}{105nb^3}\sqrt{a+bx^n}=y$.

4. If $\frac{cx^{4n}}{x}\sqrt{a+bx^n}=\frac{q}{p}$. Then

$$\frac{210b^4cx^{4n}+30ab^3cx^{3n}-36aabbx^{2n}+48a^3bcx^n-96a^4c}{945nb^4}\sqrt{a+bx^n}=y.$$

5. If $\frac{cx^{5n}}{x}\sqrt{a+bx^n}=\frac{q}{p}$. Then

$$\frac{1890b^5cx^{5n}+210ab^4cx^{4n}-240aab^3cx^{3n}+288a^3bbcx^{2n}-384a^4bcx^n+768a^5c}{10395nb^5}\sqrt{a+bx^n}=y.$$ (21)

---

1. If $\frac{cx^n}{x\sqrt{a+bx^n}}=\frac{q}{p}$. Then $\frac{2c}{nb}\sqrt{a+bx^n}=y$.

2. If $\frac{cx^{2n}}{x\sqrt{a+bx^n}}=\frac{q}{p}$. Then $\frac{2bcx^n-4ac}{3nbb}\sqrt{a+bx^n}=y$.

3. If $\frac{cx^{3n}}{x\sqrt{a+bx^n}}=\frac{q}{p}$. Then $\frac{6bbcx^{2n}-8abcx^n+16aac}{15nb^3}\sqrt{a+bx^n}=y$.

4. If $\frac{cx^{4n}}{x\sqrt{a+bx^n}}=\frac{q}{p}$. Then is

$$\frac{30b^3cx^{3n}-36abbcx^{2n}+48aabcx^n-96a^3c}{105nb^4}\sqrt{a+bx^n}=y.$$

---

(21) Where $f(k) = \frac{c}{n}x^{kn}(a+bx^n)^{\frac{3}{2}}$ and $F(k) = \int cx^{kn-1}(a+bx^n)^{\frac{1}{2}}.dx$,

then $$F(k+1) = \frac{2f(k)-2kaF(k)}{(2k+3)\,b}$$

and Newton uses this recursive rule to evaluate $F(k)$ in terms of $F(1) = 2f(0)/3b$.

(22) Where $f(k) = \frac{c}{n}x^{kn}(a+bx^n)^{\frac{1}{2}}$ and $F(k) = \int cx^{kn-1}(a+bx^n)^{-\frac{1}{2}}.dx$,

then $$F(k+1) = \frac{2f(k)-2kaF(k)}{(2k+1)\,b},$$

by which $F(k)$ may be expressed in terms of $F(1) = 2f(0)/b$.

5. If $\frac{cx^{5n}}{x\sqrt{a+bx^n}}=\frac{q}{p}$. Then

$$\frac{210b^4cx^{4n}-240ab^3cx^{3n}+288aabbcx^{2n}-384a^3bcx^n+768a^4c}{945nb^5}\sqrt{a+bx^n}=y.^{(22)}$$

---

1. If $\overline{\frac{b}{x}\;\frac{-a}{2x^{n+1}}}\sqrt{ax^n+bx^{2n}}=\frac{q}{p}$. Then $\frac{a+bx^n}{nx^n}\sqrt{ax^n+bx^{2n}}=y.^{(23)}$

---

1. If $\frac{3ax^n+6bx^{2n}}{x}\sqrt{ax^n+bx^{2n}}=\frac{q}{p}$. Then is $\frac{2ax^n+2bx^{2n}}{n}\sqrt{ax^n+bx^{2n}}=y$.

2. If $\frac{-15aax^n+48bbx^{3n}}{x}\sqrt{ax^n+bx^{2n}}=\frac{q}{p}$. Then is

$$\frac{-10a+12bx^n}{n}\times\overline{ax^n+bx^{2n}}\times\sqrt{ax^n+bx^{2n}}=y.$$

3. If $\frac{105a^3x^n+480b^3x^{4n}}{x}\sqrt{ax^n+bx^{2n}}=\frac{q}{p}$. Then

$$\frac{70aa-84abx^n+96bbx^{2n}}{n}\times\overline{ax^n+bx^{2n}}\times\sqrt{ax^n+bx^{2n}}=y.$$

4. [If] $\frac{-945a^4x^n+5760b^4x^{5n}}{x}\sqrt{ax^n+bx^{2n}}=\frac{q}{p}$. [Then] $-630a^{3\,(24)}$

---

(23) Where
$$f(k) = \frac{1}{n}x^{kn}(ax^n+bx^{2n})^{\frac{3}{2}} \quad\text{and}\quad F(k) = \int x^{kn-1}(ax^n+bx^{2n})^{\frac{1}{2}}.dx,$$
then
$$2f(k) = (2k+6)\,bF(k+2)+(2k+3)\,aF(k+1),$$
so that $f(-2) = bF(0)-\frac{1}{2}aF(-1)$.

(24) As in the previous note, where
$$f(k) = \frac{1}{n}x^{kn}(ax^n+bx^{2n})^{\frac{3}{2}} \quad\text{and}\quad F(k) = \int x^{kn-1}(ax^n+bx^{2n})^{\frac{1}{2}}.dx,$$
then
$$F(k+2) = \frac{2f(k)-(2k+3)\,aF(k+1)}{(2k+6)\,b}$$
and Newton uses this recursion to eliminate all $F(i)$ intervening between $F(k)$ and $F(1)$. (Example 4 is left unfinished and should read

'Then $\frac{-630a^3+756aabx^n-864abbx^{2n}+480b^3x^{3n}}{n}\times\overline{ax^n+bx^{2n}}\times\sqrt{ax^n+bx^{2n}} = y$.')

1. If $\frac{ax^n+2bx^{2n}}{x\sqrt{ax^n+bx^{2n}}}=\frac{q}{p}$. Then is $\frac{2}{n}\sqrt{ax^n+bx^{2n}}=y$.

2. If $\frac{3ax^{2n}+4bx^{3n}}{x\sqrt{ax^n+bx^{2n}}}=\frac{q}{p}$. Then is $\frac{2x^n}{n}\sqrt{ax^n+bx^{2n}}=y$.

3. If $\frac{15aax^{2n}-24bbx^{4n}}{x\sqrt{ax^n+bx^{2n}}}=\frac{q}{p}$. Then is $\frac{10ax^n-8bx^{2n}}{n}\sqrt{ax^n+bx^{2n}}=y$.

4. If $\frac{105a^3x^{2n}-192b^3x^{5n}}{x\sqrt{ax^n+bx^{2n}}}=\frac{q}{p}$. Then $\frac{70aax^n-56abx^{2n}+48bbx^{3n}}{n}\sqrt{ax^n+bx^{2n}}=y$.(25)

---

1. If $\frac{c}{xx^n}\sqrt{ax^n+bx^{2n}}=\frac{q}{p}$. Make $x^n=zz$. y^n is,

$$\frac{-2ac-2bcx^n}{nax^n}\sqrt{ax^n+bx^{2n}}+\square\frac{4bc}{na}\sqrt{a+bzz}=y.$$

2. If, $\frac{c}{x}\sqrt{ax^n+bx^{2n}}=\frac{q}{p}$. Make $x^n=zz$. And y^n is $\square\frac{2c}{n}\sqrt{a+bzz}=y$.(26)

3. If, $\frac{cx^n}{x}\sqrt{ax^n+bx^{2n}}=\frac{q}{p}$. Make $x^n=z$. Then is $\square\frac{c}{n}\sqrt{az+bzz}=y$.

---

(25) Where
$$f(k) = \frac{1}{n}x^{kn}(ax^n+bx^{2n})^{\frac{3}{2}} \quad\text{and}\quad F(k) = \int x^{kn-1}(ax^n+bx^{2n})^{-\frac{1}{2}}.dx,$$
then
$$2f(k) = (2k+1)\,aF(k+1)+(2k+2)\,bF(k+2)$$
and Newton's results follow by suitable elimination of all but two of the terms $F(k)$.

(26) Where $x^n = z^2$, $\int cx^{kn-1}\sqrt{[ax^n+bx^{2n}]}.dx = \frac{2c}{n}\int z^k\sqrt{[a+bz^2]}.dz$,

and in particular $$\int cx^{-1}\sqrt{[ax^n+bx^{2n}]}.dx = \frac{2c}{n}\int\sqrt{[a+bz^2]}.dz.$$

Further, where $f(k) = z^{2k+1}(a+bz^2)^{\frac{3}{2}}$ and $F(k) = \int z^{2k}(a+bz^2)^{\frac{1}{2}}.dz$,

then $f(k) = (2k+1)\,aF(k)+(2k+4)\,bF(k+1)$, so that $F(-1) = \frac{-f(-1)+2bF(0)}{a}$.

(27) Where $x^n = z$, $cx^{kn-1}\sqrt{[ax^n+bx^{2n}]}.dx = \frac{c}{n}\int z^{k-1}\sqrt{[az+bz^2]}.dz$,

and so in particular $$\int cx^{n-1}\sqrt{[ax^n+bx^{2n}]}.dx = \frac{c}{n}\int\sqrt{[az+bz^2]}.dz.$$

Where again $f(k) = \frac{c}{n}z^k(az+bz^2)^{\frac{3}{2}}$ and $F(k) = \frac{c}{n}\int z^k(az+bz^2)^{\frac{1}{2}}.dz$,

then $$F(k+1) = \frac{2f(k)-(2k+3)\,aF(k)}{(2k+6)\,b}$$

and it may therefore be evaluated in terms of $F(1)$ and the various $f(i)$.

4. If $\frac{cx^{2n}}{x}\sqrt{ax^n+bx^{2n}}=\frac{q}{p}$. Make $x^n=z$. Then

$$\frac{2acx^n+2bcx^{2n}}{6nb}\sqrt{ax^n+bx^{2n}}-\square\frac{ac}{2nb}\sqrt{az+bzz}=y.$$

5. If $\frac{cx^{3n}}{x}\sqrt{ax^n+bx^{2n}}=\frac{q}{p}$. Make $x^n=z$. Y^n is

$$\frac{12bcx^n-10ac}{48nbb}\times\overline{ax^n+bx^{2n}}\times\sqrt{ax^n+bx^{2n}}+\square\frac{5aac}{16nbb}\sqrt{az+bzz}=y.$$ (27)

---

1. If $\frac{cx^n}{x\sqrt{ax^n\pm bx^{2n}}}=\frac{q}{p}$. Make $\sqrt{a\pm bx^n}=z$. Y^n is,

$$\frac{2c}{na}\sqrt{ax^n\pm bx^{2n}}\mp\square\frac{4c}{na}\sqrt{\frac{zz-a}{\pm b}}=y.$$

2. If $\frac{cx^{2n}}{x\sqrt{ax^n+bx^{2n}}}=\frac{q}{p}$. Make $\sqrt{a+bx^n}=z$. y^n is $\square\frac{2c}{nb}\sqrt{\frac{zz-a}{b}}=y$.

3. If $\frac{cx^{3n}}{x\sqrt{ax^n+bx^{2n}}}=\frac{q}{p}$. Make $\sqrt{a+bx^n}=z$. y^n is

$$\frac{cx^n}{2nb}\sqrt{ax^n+bx^{2n}}-\square\frac{3ac}{2nbb}\sqrt{\frac{zz-a}{b}}=y.$$ (28)

If $\frac{cx^n}{x}\sqrt{a+bx^n+dx^{2n}}=\frac{q}{p}$. Make $x^n=z$. Then is $\square\frac{c}{n}\sqrt{a+bz+dz^2}=y$.

If $\frac{cx^n\sqrt{d+ex^n}}{x\sqrt{a+bx^n}}=\frac{q}{p}$. Make $\sqrt{a+bx^n}=z$. Then is $\square\frac{2c\sqrt{db-ae+ezz}}{nb\sqrt{b}}=y.$ (29)

---

(28) Where

$$f(k) = \frac{c}{n}x^{kn}(ax^n+bx^{2n})^{\frac{1}{2}} \quad\text{and}\quad F(k) = \int cx^{kn-1}(ax^n+bx^{2n})^{-\frac{1}{2}}.dx,$$

then

$$2f(k) = (2k+1)\,aF(k+1)+(2k+2)\,bF(k+2).$$

Hence

$$F(1) = \frac{2f(0)-2bF(2)}{a} \quad\text{and}\quad F(3) = \frac{2f(1)-3aF(2)}{4b},$$

where, if $z = \sqrt{[a+bx^n]}$ $\left(\text{or } x^n = \frac{z^2-a}{b}\right)$, we have

$$F(2) = \int cx^{2n-1}(ax^n+bx^{2n})^{-\frac{1}{2}}.dx = \frac{2c}{nb}\int\left(\frac{z^2-a}{b}\right)^{\frac{1}{2}}.dz.$$

(29) Where $\sqrt{[a+bx^n]} = z \left(\text{or } x^n = \frac{z^2-a}{b} \text{ and } nx^{n-1} = \frac{2z}{b}\frac{dz}{dx}\right)$,

$$\int cx^{n-1}\sqrt{\frac{d+ex^n}{a+bx^n}}.dx = \frac{2c}{nb}\int\sqrt{\left[d+e\left(\frac{z^2-a}{b}\right)\right]}.dz.$$

If $\dfrac{-cx^{2n}\sqrt{a+bx^n}}{x\sqrt{a-3bx^n}}=\dfrac{q}{p}$. Then is $\dfrac{ac+bcx^n}{6nbb}\sqrt{aa-2abx^n-3bbx^{2n}}=y$.(30)

If $\dfrac{cx^{3n}\sqrt{3a+bx^n}}{x\sqrt{-a+bx^n}}=\dfrac{q}{p}$. Make $\overline{\dfrac{2a}{b}+x^n\sqrt{a+bx^n}}=z$. Then is $\Box\,\dfrac{2c\sqrt{bbzz-4a^3}}{3nbb}=y$.(31)

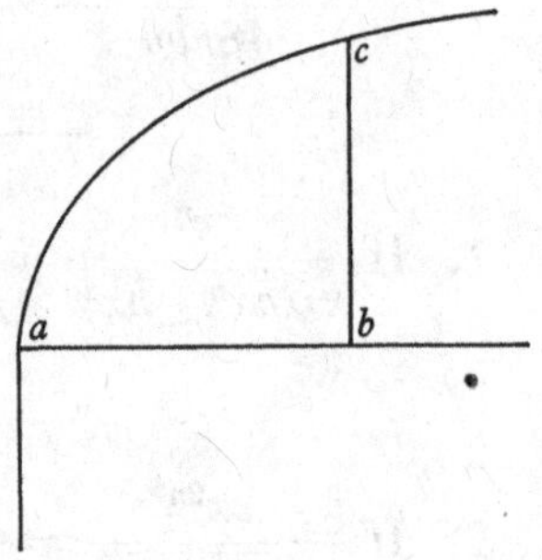

Note 1. $\dfrac{a}{b+cz}$: $\quad\Box\,\dfrac{a}{b+cz}$::

2. y[t] $\sqrt{az+bzz}$: is to $\Box\,\sqrt{az+bzz}$:: as

3. $\sqrt{a+bzz}$: $\quad\Box\,\sqrt{a+bzz}$::

y[e] ordinately applyed line bc in some of y[e] Conick sections: is to its corresponding superfices abc,(32) y[e] axis ab being in like manner related to z. But all those areas (& consequently $\Box\,\dfrac{a}{b+cz}$, $\Box\,\sqrt{az+bzz}$, $\Box\,\sqrt{a+bzz}$) may bee Mechanichally(33) found either by a Table of logarithmes or signes & Tangents.(34) And I have beene therefore hitherto content to suppose y[m] knowne, as y[e] basis of most of y[e] precedent propositions.

Note also y[t] if y[e] Valor of $\dfrac{q}{p}$ consists of severall ꝑts each ꝑt must bee considered

---

(30) Where

$$f(k) = \frac{c}{n}x^{kn}\sqrt{[a^2-2abx^n-3b^2x^{2n}]} \quad\text{and}\quad F(k) = \int\frac{cx^{kn-1}}{\sqrt{[a^2-2abx^n-3b^2x^{2n}]}}.dx,$$

then $$f(k) = ka^2F(k)-(2k+1)\,ab\,F(k+1)-(3k+3)\,b^2F(k+2).$$

Hence $$f(0) = -abF(1)-3b^2F(2),\quad f(1) = a^2F(1)-3abF(2)-6b^2F(3)$$

and so $$-\int\frac{cx^{2n-1}(a+bx^n)}{\sqrt{[a^2-2abx^n-3b^2x^{2n}]}}.dx = -aF(2)-bF(3) = \frac{1}{6b^2}(af(0)+bf(1)).$$

(31) Read correctly 'Make $\overline{\dfrac{2a}{b}+x^n\sqrt{-a+bx^n}} = z$. Then is $\Box\,\dfrac{2c\sqrt{bbzz+4a^3}}{3nbb} = y$.' (In proof, where $x^n = v$,

$$\int cx^{3n-1}\sqrt{\frac{3a+bx^n}{-a+bx^n}}.dx = \int\frac{c}{n}v^2\sqrt{\frac{3a+bv}{-a+bv}}.dv,$$

which on taking

$$bz = (2a+bv)\sqrt{[-a+bv]} \quad\text{becomes equal to}\quad \frac{2c}{3nb^2}\int\sqrt{[b^2z^2+4a^3]}.dz.)$$

Where Newton has been unable to give an exact quadrature, his general principle has been to attempt the reduction of the integral to quadrable functions and to an integral which may be interpreted as the area under a central conic, that is, which is a circular or hyperbolic function. (Compare Newton's first note immediately following.) After this set of integrals Newton left two blank pages and it would appear that he intended to extend the tabulation considerably.

severally, as if $\frac{ax^3-bbxx}{c^4}=\frac{q}{p}$. Then is $\square\frac{ax^3}{c^4}=\frac{ax^4}{4c^4}$. & $\square\frac{bbxx}{c^4}=\frac{bbx^3}{3c^4}$. Therefore $\square\frac{ax^3-bbxx}{c^4}=\frac{ax^4}{4c^4}-\frac{bbx^3}{3c^4}=y$.

Note also y[t] if y[e] denominator of y[e] valor of $\frac{q}{p}$ consist of both rationall & surde quantitys or of two or more surde quantitys First take those surde quantitys out of y[e] denominator & y[n] seeke (y) by y[e] precedent theoremes.(35)

☞(36) Note y[t] if there happen to bee in any Equation either a fraction or surde quantity or a Mechanichall one (i:e: w[ch] cannot bee Geometrically computed, but is expressed by y[e] area or length or gravity or content of some curve line or sollid, &c) To find in what proportion the unknowne quantitys increase or decrease doe thus. 1 Take two letters y[e] one (as ξ) to signify y[t] quantity, y[e] other (as π) its motion of increase or decrease:(37) And making an Equation betwixt y[e] letter (ξ) & y[e] quantity signifyed by it, find thereby (by prop 7 if y[e] quantity bee Geometricall, or by some other meanes if it bee mechanicall) y[e] valor of y[e] other letter (π). 2 Then substituting y[e] letter (ξ) signifying y[t] quantity, into its place in y[e] maine Equation esteeme y[t] letter (ξ) as an unknowne quantity & performe y[e] worke of [y[e]] seaventh proposition; & into y[e] resulting Equation instead of those letters ξ & π substitute theire valors. And soe you have y[e] Equation required.

Example 1. To find p & q y[e] motions of x & y whose relation is, $yy=x\sqrt{aa-xx}$. first suppose $\xi=\sqrt{aa-xx}$, Or $\xi\xi+xx-aa=0$. & thereby find π y[e] motion of ξ, viz: (by prop 7) $2\pi\xi+2px=0$. Or $\frac{-px}{\xi}=\pi=\frac{-px}{\sqrt{aa-xx}}$. Secondly in y[e] Equation $yy=x\sqrt{aa-xx}$, writing ξ in stead of $\sqrt{aa-xx}$, the result is $yy=x\xi$, whereby find y[e] relation of y[e] motions p, q, & π: viz (by prop 7) $2qy=p\xi+x\pi$. In w[ch] Equation instead of ξ & π writing theire valors, y[e] result is, $2qy=p\sqrt{aa-xx}\frac{-pxx}{\sqrt{aa-xx}}$. w[ch] was required.

(w[ch] equation multiplyed by $\sqrt{aa-xx}$, is $2qy\sqrt{aa-xx}=paa-2pxx$, & in stead

(32) An entirely unconventional use of the symbolism of arithmetical proportions to express a relational identity between analytical and geometrical models for the integral.

(33) See note (11) above.

(34) In the cases where comparison is made with the area under a hyperbola, and with that under a circle respectively.

(35) Newton has cancelled 'rul[es]'. More generally, as the ink and quality of the writing show, this note was added in afterthought after the following paragraph had been inserted.

(36) These two following notes were inserted, at a fairly late stage in composition, on the blank sheet (48[v]) facing the opening page of the tract. Though Newton has given no specific instructions for their reordering, their present position seems natural and justifiable by the similarity of mathematical content.

(37) That is, Newton takes π as the fluxion $d\xi/dt$ of ξ.

of $\sqrt{aa-xx}$ writing its valor $\frac{yy}{x}$, it is $\frac{2qy^3}{x}=paa-2pxx$. Or $2qy^3=paax-2pxxx$. Which conclusion will also bee found by taking y^e^ surde quantity out of y^e^ given Equation for both parts being squared it is $y^4=aaxx-x^4$. & therefore (by prop 7) $4py^3=2qaax-4x^3$,(38) as before.)

☞ Note also y^t^ it may bee more convenient (setting all y^e^ termes on one side of y^e^ Equation) to put every fractionall, irrationall, & mechanicall terme, as also y^e^ summe of y^e^ rationall termes, equall severally to some letter: & then to find y^e^ motions corresponding to each of those letters y^e^ sume of wch motions is y^e^ Equation required.

Example y^e^ 2^d^. If $x^3-ayy+\frac{by^3}{a+y}-xx\sqrt{ay+xx}=0$ is y^e^ relation twixt x & y, whose motions p & q are required. I make $x^3-ayy=\tau$; $\frac{by^3}{a+y}=\phi$; &

$$-xx\sqrt{ay+xx}=\xi.$$

& y^e^ motions of τ, ϕ, & ξ being called β, γ, & δ;(39) y^e^ first Equation $x^3-ayy=\tau$, gives (by prop 7) $3pxx-2qay=\beta$. y^e^ second $by^3=a\phi+y\phi$, gives

$$3qbyy=a\gamma+y\gamma+q\phi;$$

Or $\frac{3qbyy-q\phi}{a+y}=\gamma=\frac{3qabyy+2qby^3}{aa+2ay+yy}$. & y^e^ Third $ayx^4+x^6=\xi\xi$, gives

$$qax^4+4payx^3+6px^5=2\delta\xi;\quad \text{Or}\quad \frac{-qaxx-4payx-6px^3}{2\sqrt{ay+xx}}=\delta.$$

Lastly $\beta+\gamma+\delta$(40)$=3pxx-2qay\ \frac{+3qabyy+2qby^3}{aa+2ay+yy}\ \frac{-qaxx-4payx-6px^3}{2\sqrt{ay+xx}}=0$, is y^e^ Equation sought.

Example 3^d^. If $x=ab\perp bc=\sqrt{ax-xx}$. $be=y$. & y^e^ superficies $abc=z$ suppose y^t^ $zz+axz-y^4=0$, is y^e^ relation twixt x, y & z, whose motions are p, q, & r: & y^t^ p & q are desired. The Equation $zz+axz-y^4=0$ gives (by prop 7)

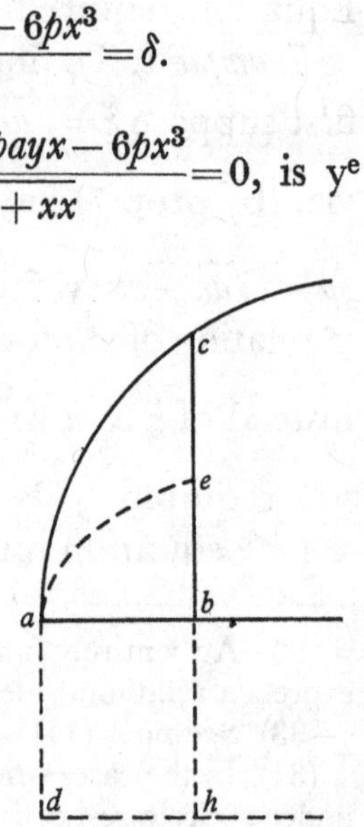

$$2rz+rax+paz-4qy^3=0.$$

Now drawing $dh \parallel ab\perp ad=1=bh$. I consider y^e^ superficies $abhd=ab\times bh=x\times 1=x$, & $abc=z$ doe increase in y^e^ proportion of bh to bc: y^t^ is, $1:\sqrt{ax-xx}::p:r$. Or $r=p\sqrt{ax-xx}$. Which valor of r being substituted into y^e^ Equation $2rz+rax+paz-4qy^3=0$, gives

$$\overline{2pz+pax}\times\sqrt{ax-xx}+paz-4qy^3=0.$$

wch was required.

(38) Read '$4qy^3=2paax-4px^3$'.

(39) That is, β, γ and δ are the fluxions $d\tau/dt$, $d\phi/dt$ and $d\xi/dt$ respectively.

How to proceede in other cases (as when there are cube rootes, surde denominators, rootes within rootes (as $\sqrt{ax+\sqrt{aa-xx}}$) &c: in the equation) may bee easily deduced from what [ha]th bee[n] already said.

But this eighth Proposition may bee ever thus resolved mechanichally, viz: Seeke y^e^ Valor of $\frac{q}{p}$ as if you were resolving y^e^ equation in Decimall numbers either by Division or extraction of rootes or Vieta's Analyticall resolution of powers;(41) This operation may bee continued at pleasure, y^e^ farther the better. & from each terme ariseing from this operation may bee deduced a ꝑte of y^e^ valor of y, (by ꝑte y^e^ 1st of this prop).

Example 1. If $\frac{a}{b+cx}=\frac{q}{p}$. Then by division is

$$\frac{q}{p}=\frac{a}{b}-\frac{acx}{bb}+\frac{accxx}{b^3}-\frac{ac^3x^3}{b^4}+\frac{ac^4x^4}{b^5}-\frac{ac^5x^5}{b^6}+\frac{ac^6x^6}{b^7}\ \&c.$$

And consequently $y=\frac{ax}{b}-\frac{acxx}{2bb}+\frac{accx^3}{3b^3}-\frac{ac^3x^4}{4b^4}+\frac{ac^4x^5}{5b^5}$ &c.(42)

Example 2. If $\sqrt{aa-xx}=\frac{q}{p}$. Extract y^e^ roote & tis

$$\frac{q}{p}=a-\frac{xx}{2a}-\frac{x^4}{8a^3}-\frac{x^6}{16a^5}-\frac{5x^8}{128a^7}-\frac{7x^{10}}{256a^9}-\frac{21x^{12}}{1024a^{11}}\ \&c$$

(as may appeare by squareing both ꝑts). Therefore (by 1st ꝑte of Prop 8)

$$y=ax-\frac{x^3}{6a}\,\frac{-x^5}{40a^3}\,\frac{-x^7}{112a^5}\,\frac{-5x^9}{1152a^7}\,\frac{-7x^{11}}{2816a^9}\ \&c.^{(43)}$$

---

(40) That is, the fluxion $(d/dt)\,(\tau+\phi+\xi)$.

(41) See his extensive notes on Viète's *De Numerosa Potestatum ad Exegesim Resolutione* (Paris, 1600 = Schooten's *Francisci Vietæ Opera Mathematica*, Leyden, 1646: 162–228) in 1, 2, §1 above.

(42) An example of resolution 'by Division':

$$\int_0^X \frac{a}{b+cx}.dx\ \left(\text{or } \frac{a}{c}\log\left(\frac{b+cX}{b}\right)\right) = \lim_{n\to\infty}\left[\int_0^X \sum_{0\leqslant i\leqslant n}\left[(-1)^i\frac{a}{b}\left(\frac{cx}{b}\right)^i\right].dx\right],$$

with implicitly $\left|\frac{cX}{b}\right| < 1$ for convergence.

(43) An example of resolution 'by extraction of rootes', which yields the binomial expansion

$$\int_0^X \sqrt{[a^2-x^2]}.dx\ \left(\text{or } a^2(\tfrac{1}{2}X\sqrt{[1-X^2]}+\tfrac{1}{2}\sin^{-1}X)\right) = \lim_{n\to\infty}\left[a\int_0^X \sum_{0\leqslant i\leqslant n}\left[\binom{\frac{1}{2}}{i}\left(\frac{x^2}{a^2}\right)^i\right].dx\right],$$

with implicitly $\left|\frac{X}{a}\right| < 1$.

Example 3. If $\frac{q^3}{p^3} * -ax\frac{q}{p} - x^3 = 0.$(44)

But y^e^ Demonstrac$\overline{\text{ons}}$ of w^t^ hath beene said must not bee wholly omitted.

Prop 7 *Demonstrated.*(45)

Lemma. If two bodys A, B, move uniformely y^e^ one/other from a/b to $c, d, e, f,$/$g, h, k, l,$ &c: in y^e^ same time. Then are y^e^ lines $ac,$/$bg,$ & $cd,$/$gh,$ & $de,$/$hk,$ & $ef,$/$kl,$ &c: as their velocitys $\frac{p}{q}$. And though they move not uniformely yet are y^e^ infinitely little lines wch each moment they describe, as their velocitys wch they have while they describe y^m^. As if y^e^ body A wth y^e^ velocity p describe y^e^ infinitely little line $(cd=)p\times o$ in one moment, in y^t^ moment y^e^ body B wth y^e^ velocity q will describe y^e^ line $(gh=)q\times o$. For $p:q::po:qo$. Soe y^t^ if y^e^ described lines bee $(ac=)x$, & $(bg=)y$, in one moment, they will bee $(ad=)x+po$, & $(bh=)y+qo$ in y^e^ next.

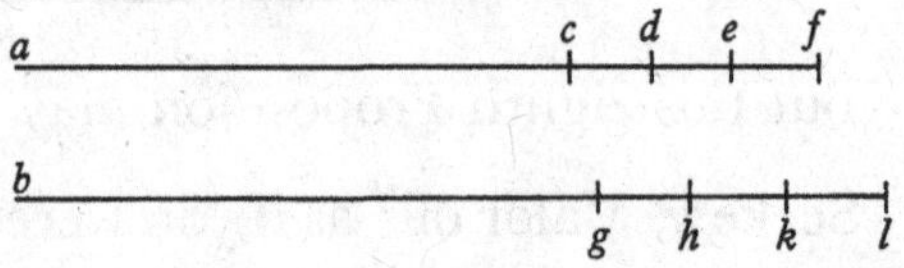

Demonstr: Now if y^e^ equation expressing y^e^ relation twixt y^e^ lines x & y bee $x^3-abx+a^3-dyy=0$. I may substitute $x+po$ & $y+qo$ into y^e^ place of x & y; because (by y^e^ lemma) they as well as x & y, doe signify y^e^ lines described by y^e^ bodys A & B. By doeing so there results

$$\begin{array}{l} x^3+3poxx+3ppoox+p^3o^3-dyy-2dqoy-dqqoo=0. \\ \quad\;\; -abx \qquad -abpo \\ \quad\;\; +a^3 \end{array}$$

But $x^3-abx+a^3-dyy=0$ (by supp). Therefore there remaines onely

$$\begin{array}{l} 3poxx+3ppoox+p^3o^3-2dqoy-dqqoo=0. \\ \qquad\qquad -abpo \end{array}$$

Or dividing it by o tis $3px^2+3ppox+p^3oo-2dqy-dqqo=0$. Also those termes are $-abp$

(44) Newton left the remainder of the page blank for the completion of this third example of resolution 'by Vieta's Analyticall resolution of powers', but never returned to it. Where $z = q/p = dy/dx$, the root z of $z^3-axz-x^3 = 0$ is to be extracted as an infinite series in powers of x, say $z = \phi(x)$: it then follows that

$$y = \int z.dx = \int \phi(x).dx.$$

As Newton soon realized there are some difficulties in applying the method, but five years later he wrote up a lengthy account of it in his 1671 tract, the *Methodus Fluxionum et Serierum Infinitarum*, which will be printed in the third volume.

(45) A slight adaptation of the Lemma in 6, §3.1 above, written on 13 November 1665.

(46) Newton has cancelled 'y^n^ those in wch tis not'.

infinitely little in wch o is.(46) Therefore omitting them there rests

$$3pxx - abp - 2dqy = 0.$$

The like may bee done in all other equations.

Hence I observe. First y^t^ those termes ever vanish wch are not multiplyed by o, they being y^e^ propounded equation. Secondly those termes also vanish in wch o is of more y^n^ one dimension, because they are infinitely lesse y^n^ those in wch o is but of one dimension. Thirdly ye still remaining termes, being divided by o will have y^t^ form w^ch^, by ye 1st rule in Prop 7th, they should have (as may ꝑtly appeare by ye second termes of Mr Oughtreds latter Analyticall table(47)).

After y^e^ same manner may this 7th Prop: bee demonstr: there being 3 or more unknowne quantitys x, y, z, &c:

Prop 8th is y^e^ Converse of this 7th Prop. & may bee therefore Analytically demonstrated by it.

Prop 1st *Demonstrated.* If some body A move in y^e^ right line $gafc$ from g towards c. From any point d draw $df \perp ac$. & call, $df = a$. $fg = x$. $dg = y$. Then is $aa + xx - yy = 0$. Now by Prop 7th, may y^e^ proportion of (p) y^e^ velocity of y^t^ body towards f, to (q) its velocity towards d bee found viz[:] $2xp - 2yq = 0$. Or $x:y::q:p$. That is $gf:gd::$ its velocity to d: its velocity towards f or c. & when y^e^ body A is at a, y^t^ is when y^e^ points g & a are coincident then is $ac:ad::ad:af::$ velocity to c: velocity to d.(48)

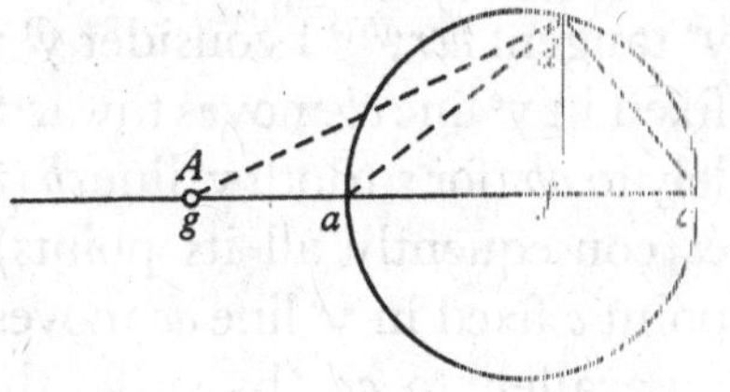

Prop 2^d^, *Demonstrated.* From y^e^ points d & e draw $df \perp ac \perp ge$. And let y^e^ first bodys velocity(49) to d bee called ad, y^e^ seconds to e bee ae, & y^e^ 3^ds^ toward c bee ac. Then shall y^e^ firsts velocity towards c bee af (by Prop 1): & The seconds towards c is ag, (prop 1). but $af = gc$ (for $\triangle adc =$(50)$\triangle aec$, & $\triangle adf = \triangle gec$, by sup). Therefore $ac = ag + gc = ag + af$. That is y^e^ velocity of y^e^ third body towards c is equall to y^e^ summ of the velocitys of y^e^ first & second body towards c.(51)

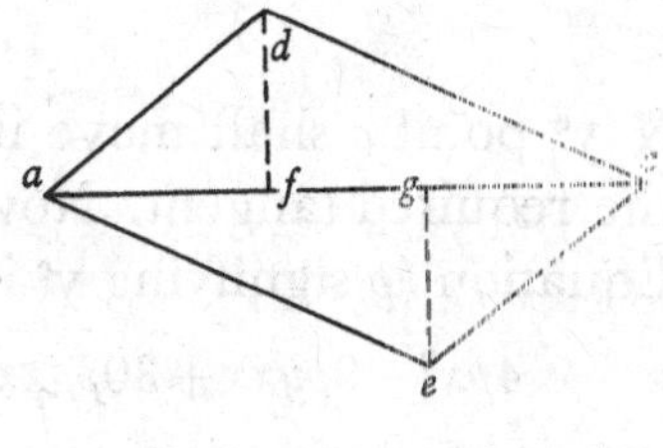

(47) That is, the second terms $(kx^{k-1}o)$ in the binomial expansion of $(x+o)^k$, which in a different notation is tabulated for a few low positive values in Oughtred's *Clavis Mathematicæ*: Cap. XII = ₃1652: 37: [*Tabula*] *Posterior.* (See 6, §3: note (17).)

(48) This seems to be the proof hinted at in the concluding sentence of Prop. 1 in his first draft of the theorem on 14 May 1666 (ULC. Add. 4004: 51^r^ = 6, §4.1 above: see especially note (6)).

(49) Newton has cancelled 'it moveing'.

(50) Newton first wrote 'similis'.

(51) The proof is repeated from 6, §4.1: Prop 2^d^. This concludes the first part of the present tract except for a blank page (53^r^) apparently left for possible future additions.

[2](52) *The former Theorems Applyed to Resolving of Problems.*

Prob. 1. *To draw Tangents to crooked lines.*(53)

Seeke (by prop 7th; or 3d, 4th & 2d, &c(54)) ye motions of those streight lines to wch ye crooked line is cheifly referred, & wth what velocity they increase or decrease. & they shall give (by prop 6t, or 1st or 2d) ye motion of ye point describing ye crooked line; wch motion is in its tangent.

Tangents to Geometricall lines.

Example 1.(55) If ye crooked line *fac* is described by ye intersection of two lines *cb* & *dc* ye one moveing parallely, viz: $cb \parallel ad$, & [ye other] $dc \parallel ab$;(56) soe yt if $ab = x$, & $bc = y = ad$, Their relation is

$$\begin{array}{l} x^4 - 3yx^3 + ayxx - 2y^3x + a^4 = 0. \\ \quad + 10a \qquad\qquad\quad - y^4 \end{array}$$

To draw ye tangent *hcr*;(57) I consider yt ye point *c* fixed in ye line *cb* moves towards *e* parallely to *ab* (for so doth ye line *cb* (by supp:) & consequently all its points): also ye point *c* fixed in ye line *dc* moves towards *g* parallely to *ad* (by sup): therefore I

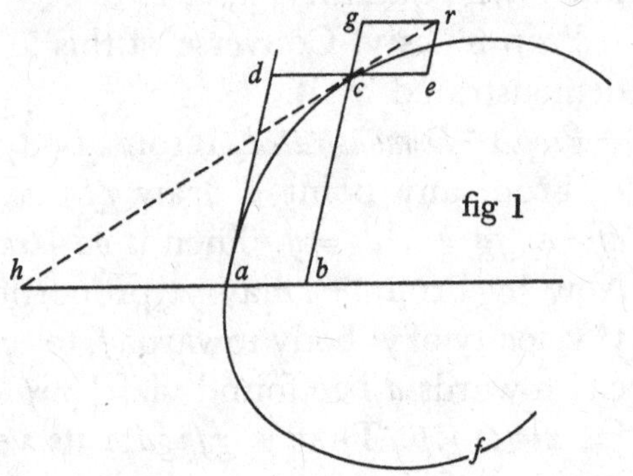

fig 1

draw $ce \parallel ab$ & $cg \parallel ad$, & in such proportion as ye motions they designe & so draw $er \parallel cb$, & $gr \parallel dc$, & ye diagonall *cr*, (by Prop 6), yt is, if ye velocity of ye line *cb*, (yt is ye celerity of ye increasing of *ab* or *dc*; or ye velocity of ye point *c* from *d*) bee called *p*; & ye velocity of ye line *cd* bee called *q*; I make

$$ce : gc :: p : q \; (:: ce : er :: hb : cb)$$

& ye point *c* shall move in the diagonall line *cr* (by prop 6) wch is therefore the required tangent. Now ye relation of *p* & *q* may bee found by ye foregoing Equation (*p* signifying ye increase of *x*, & *q* of *y*) to bee

$$4px^3 - 9pyxx + 30paxx + 2payx - 2py^3 - 3qx^3 + qaxx - 6qyyx - 4qy^3 = 0.$$

(by Prop 7) And therefore $hb = \frac{py}{q}$(58) $= \frac{3yx^3 - ayxx + 6y^3x + 4y^4}{4x^3 - 9yxx + 30axx + 2ayx - 2y^3}$. wch determines ye tangent *hc*.

(52) Add. 3958.3: 53v–62r. Newton applies the general theorems in [1] to the resolution of a set of geometrical problems.

(53) A revised version of 6, §2 and §3.2: Prob. 1 and 2, with a generalized introduction based on 6, §4.1/2.

(54) That is, analytically or geometrically.

(55) Adapted from 6, §3.2: Problem 1: Example 1st.

(56) Newton first wrote '...viz: $cb \parallel ad$ & insisting upon *ab*, ye other $dc \parallel ab$ insisting [ ]'.

☞ Hence may bee observed this Generall Theorem(59) for drawing Tang^{ts} to crooked lines thus referred to streight ones, y^t is, to such lines in w^{ch} $y=bc$ is ordinately applyed to $x=ab$ at any given angle abc. viz: Multiply the termes of y^e Equation ordered according to y^e dimensions of y, by any Ari[th]meticall progressiō wch product shall bee y^e Numerator: Againe change y^e signes of y^e Equation & ordering it according to x, multiply y^e termes by any Arithmeticall progression(60) & y^e product divided by x shall be y^e Denominator of y^e valor of hb, y^t is, of x produced from y to y^e tangent hc.

As if $rx-\dfrac{rxx}{q}=yy$. Then first $\overset{2.}{yy}\quad\overset{1.}{*}+\overset{0.}{\dfrac{rxx}{q}}=0$, produceth $2yy$, or, $2rx-2\dfrac{r}{q}xx$.
$$-rx$$

Secondly $\overset{2.}{\dfrac{-r}{q}xx}\overset{1.}{+rx}\overset{0.}{-yy}\overset{0.\,-1.\,-2.}{[=0]}$, produceth $rx-\dfrac{2r}{q}xx$. Therefore $\dfrac{2yy}{r-\dfrac{2r}{q}x}=bh$. Or else $\dfrac{2rx-\dfrac{2r}{q}xx}{r-\dfrac{2r}{q}x}=bh=\dfrac{2qrx-2rxx}{qr-2rx}$.

Example 2.(61) If y^e crooked line chm bee described by y^e intersection of two lines ac, bc circulating about their centers a & b, soe y^t if $ac=x$, & $bc=y$; their

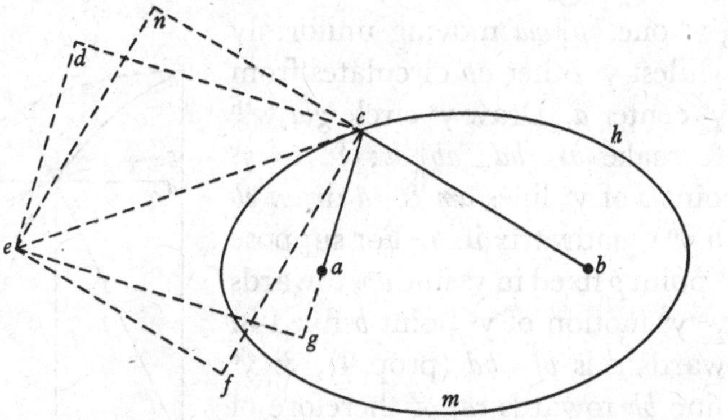

relation is $x^3-abx+cyy=0$. To draw y^e tangent ec I consider y^t y^e point c fixed in y^e line bc moves towards f in y^e line $cf\perp bc$ (for y^e tangent to a circle is perpendicular to its radius). also y^e point c fixed in y^e line ac moves towards d in y^e line

(57) A lengthy cancelled first draft, fully incorporated in the revised text, is omitted.
(58) That is, $hb = y(dx/dy)$, the subtangent.
(59) Adapted from 6, §3.1 above, but compare 4, §3.1.
(60) With necessary restriction that it have the same added difference as the former.
(61) Adapted from 6, §3.2: Problem 1: Example 2^d.

cd⊥ *ac* & from those lines *cd* & *cf* I draw two others *de*∥*cg* & *ef*∥*bc* wch must bee in such proportion one to another as ye motions represented by y^m^ (prop 6), y^t^ is (prop 6) as ye motions of y^e^ intersection point *c* moveing in y^e^ lines *ca* & *cb* to or from y^e^ centers *a* & *b*; y^t^ is (y^e^ celerity[62] of y^e^ increase of *x* being called *p*, & of *y* being *q*), *de*:*ef*::*p*:*q*. Then shall y^e^ diagonall *ce* bee y^e^ required tangent. Or wch is y^e^ same, (for △*ecg*=△*ecd*, & △*ecf*=△*ecn*), I produce *ac* & *bc* to *g* & *n*, so y^t^ *cg*:*cn*::*p*:*q*, & y^n^ draw *ne*⊥ *bn*, & *ge*⊥ *ag*; & y^e^ tangent diagonall *ce* to their intersection point *e*. Now y^e^ relation of *p* & *q* may bee found by y^e^ given Equation to bee, $3pxx-pab+2qcy=0$ (by prop 7). Or $2cy:ab-3xx::p:q::cg:cn$, wch determins y^e^ tangent *ce*.

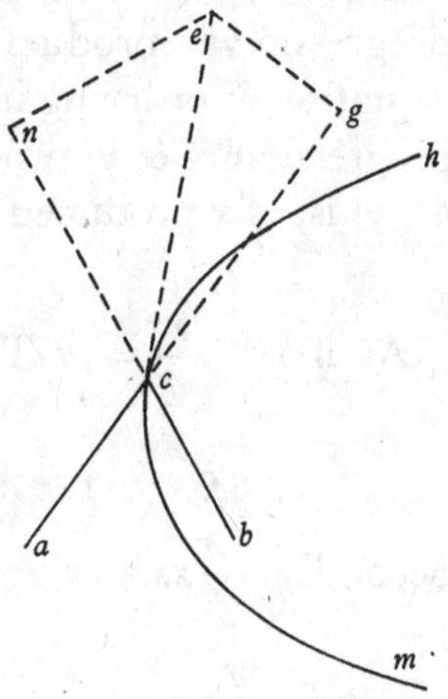

But note y^t^ if *p*, or *q* be negative *cg* or *cn* must bee drawn from *c* towards *a* or *b*, but from *a* or *b* if affirmative.

Hence tis easy to pronounce a Theorem for Tangents to such like cases & ye like may bee done in all other cases however Geometricall lines bee referred to streight ones.

Tangents to Mechanichall lines.[63]

*Example y^e^* 3^d^. If y^e^ Quadratrix *kbf* is described by y^e^ intersection of y^e^ two lines *hb* & *ap*, y^e^ one *hp*∥*ma* moving uniformly from *k* to *a*, whilest y^e^ other *ap* circulates from *k* to *m* about y^e^ center *a*. Draw y^e^ circle *gbl* wth y^e^ Rad: *ab*; & make *bl*=*bd*⊥ *ab*∥*de*; & to y^e^ intersection point *e* of y^e^ lines *am* & *ed* draw *eb* wch shall touch y^e^ Quadratrix in *b*. For suppose y^e^ motion of y^e^ point *p* fixed in y^e^ line *ap*, towards *m* to bee *pm*, y^n^ y^e^ motion of y^e^ point *b* fixed in y^e^ line *ab*, towards *d* is *bl*=*bd* (prop 4), & y^e^ motion of y^e^ line *bh* towards *ca*, & therefore of y^e^ point *b* fixed in it towards *c* (prop 3) is *ha*=*bc* (by supp). Also *ce*∥*bh* & *ed*∥*ap* (sup). Therefore (by Prop 6)[64] y^e^ intersection point *b* of these two lines *ap* & *hb*, moves in y^e^ diagonall *eb*, & consequently *eb* toucheth y^e^ Quadratrix in *b*.

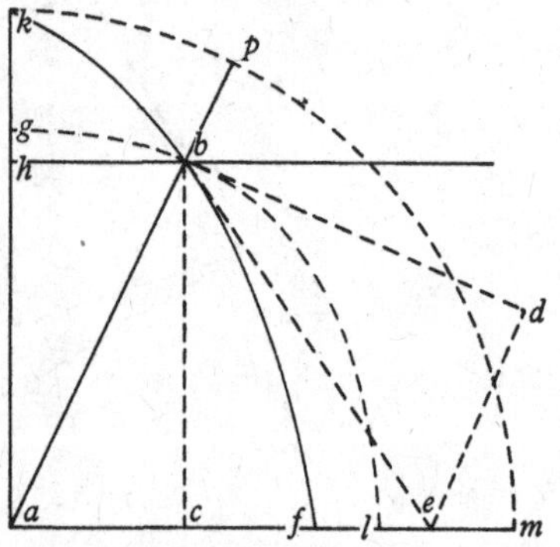

Example 4th.[65]

(62) In a cancellation Newton first wrote the more usual 'velocity' for 'celerity'.
(63) Revised from 6, §2 above.
(64) Newton has cancelled 'is y^e^ diagonall *eb* y^e^ motion of'.

Prob 2^{d}. *To find* y^e *quantity of crookednesse of lines.*[66]

Lemma.[67] The crookednesse of equall parts of circles are as their diameters reciprocally. For the crookednesse of a whole circle (*acdea*, *bfghmb*) amounts to 4 right angles. Therefore there is not more crookednesse in one whole circle *acdea* y^n in another *bfghmb*. Suppose y^e perimiter *acde* = *bfgh*. Then tis *ar*:*br*:: *bfgh* = *acde*:*bfghmb*::crookednesse of *bfgh*:crookednesse of *bfghmb* = crookednesse of *acdea*.[68]

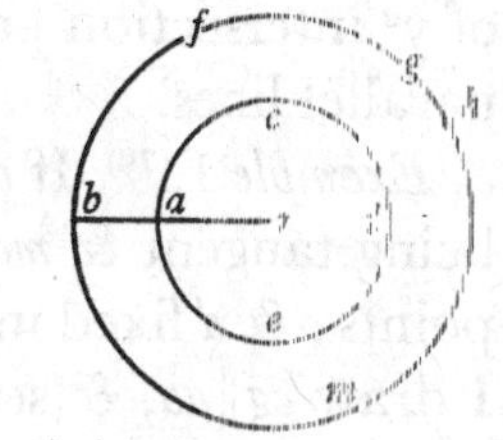

Resolution.[69] Find That point fixed in y^e crooked line's perpendicular wch is y^n in least motion, for it is y^e center of a circle wch passing through y^e given point is of equall crookednesse wth ye line at y^t given point. Now, since y^e crooked line's tangent & perpendicular &c (at y^t moment) circulate about y^t center; I observe, 1st y^t every point fixed in y^e Tang^{nt} or Perpendicular, or whose position to y^m is determined, doth

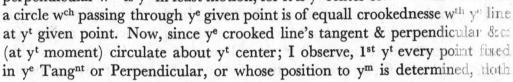

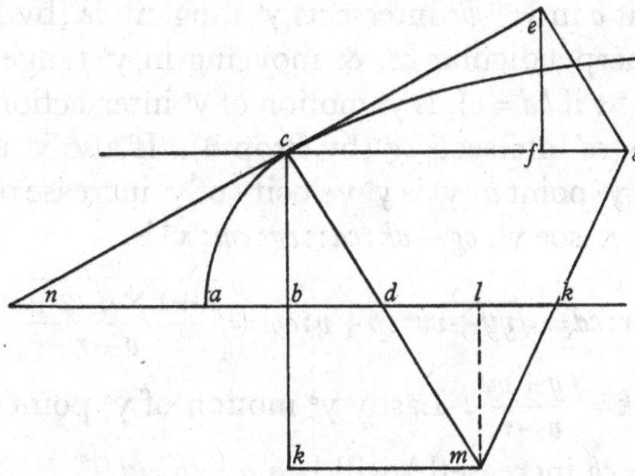

(65) A considerable space, never filled, is left on 54^r for the insertion of this second example of a mechanical curve. Presumably, on the line of his previous work, this would be either an Archimedean spiral or a general cycloid.

(66) A revised version of 4, §2.1 and 6, §3.2: Problem 3.

(67) Adapted from 4, §2.1: *Theorema.*

(68) Newton first added a second lemma as follows: 'Lemma 2. That point of a Crooked Line's Perpendicular wch is in least motion is ye center of ye circle whose crookednesse is required.

'Resolution. This may bee done by drawing perpendiculars to 2 curved lines described by ye motion of any two points fixed in ye Tangent, wch two perpendiculars shall intersect in ye center of a circle, whose crookednesse was required. One of those curve lines & perpendiculars may bee y^e given crooked line & y^e perpendicular to it at its given point.'

Immediately, however, Newton cancelled it, incorporating it in the following 'Resolution'.

(69) Adapted from Add. 4004: 57^v = 6, §3.2: Problem 3 above, where the argument is discussed in detail in the accompanying notes. See also D. T. Whiteside, 'Patterns of mathematical thought in the later seventeenth century', *Archive for History of Exact Sciences*, 1 (1961): 376–7.

describe a curve line to wch y^e^ right line drawne from y^t^ center is perpendicular, & is also y^e^ radius of a circle of equall crookedness wth it: 2dly y^t^ y^e^ motion of every such point is as its distance from y^t^ center; & so are y^e^ motions of y^e^ intersection points, in wch any radius drawn from y^t^ center intersects two parallel lines.

Example 1.(70) If $cb=y$ is ordinately applyed to $ab=x$ at a right angle abc, nc being tangent & mc perpendicular to ye curve line ac: I seeke ye motion of two points c & d fixed in ye perpendicular cd; or (wch is better & to ye same purpose) I draw $cg \| ab$, & seeke ye motions of ye two intersection points c & d in wch ye perpendicular cd intersects those fixed lines cfg, & $abdk$: & yn draw cg & dk in such proportion as those motions are, & ye line gkm drawn by their ends shall intersect ye perpendicular cd in ye required center m: mc being ye radius of a circle of equall crookednesse wth ye curve line ac at ye point c. Now, making $\triangle cegfe =$ & like $\triangle ncdbc$, suppose ye motion of ye line cb & consequently of ye points c & b fixed in it & moveing towards f & k, to bee $p=nb=cf$:(71) Then is $ce=cn$ ye motion of ye point c in wch bc intersects ye tangent ne (by Prop 6), yt is of ye point c fixed in ye perpendicular cd, & moveing in ye tangent ne: & therefore $cg=nd=p+bd$ ($=p+v$ if $bd=v$), is ye motion of ye intersection point c towards g in wch point ye perp: cd intersects cg (by Prop 6). If also ye motion of ye intersection point d from ye point b (yt is ye velocity of ye increase of v) bee called r,(72) yn is $dk=nb+r=p+r$. soe yt, $cg-dk:cd::cg:cm$; yt is,

$$v-r:cd=\sqrt{yy+vv}::p+v:cm=\frac{\overline{p+v}\times\sqrt{yy+vv}}{v-r}.$$

Also $v-r:y::p+v:ck=\dfrac{py+vy}{v-r}$. Lastly ye motion of ye point c from b, (yt is, ye velocity wth wch $y=cb$ increaseth) will bee $q=cb=y$.(73)

As if ye nature of ye crooked line bee $x^3-axy+ayy=0$. Then is

$$nb=\frac{axy-2ayy}{3xx-ay}=p \quad \text{(by examp: 1. Prob: 1.)}$$

& $\dfrac{3xxy-ayy}{ax-2ay}=v=bd$ (for $nb:bc::bc:bd$) Soe that $3xxy-ayy-axv+2ayv=0$. & therefore (by Prop 7) $6pxy-pav-2qay+3qxx+2qav-rax+2ray=0$. & substituting $\dfrac{axy-2ayy}{3xx-ay}$, & $\dfrac{3xxy-ayy}{ax-2ay}$, & y, into ye places of p, v & q in this Equation The product will bee

$$\frac{6axxyy-12axy^3}{3xx-ay}-3ayy+3yxx\frac{+6xxyy-2ay^3}{x-2y}-rax+2ray=0.$$

(70) Newton's accompanying diagram has two points k, but there is no confusion in leaving the text without emendation.

(71) That is, $p = y(dx/dy)$ represents the fluxion dx/dt.

Or $\dfrac{9yx^5-6ayyx^3-12ay^3xx+24ay^4x+3aay^3x-4a^2y^4}{3ax^4-12ayx^3-a^2yxx+12ayyxx+4a^2yyx-4a^2y^3}=r$. And therefore

$$\frac{-18yyx^4+24ay^3xx-24ay^4x-2aay^3x+2aay^4}{3ax^4-12ayx^3-aayxx+12ayyxx+4aayyx-4aay^3}$$
$$=v-r=\frac{2aay^4-2aay^3x-24ay^4x+24ay^3xx-18yyx^4}{3x^3-6yxx-ayx+2ayy \text{ in } ax-2ay}.\text{ Also}$$

$$p+v=\frac{axy-2ayy}{3xx-ay}+\frac{3xxy-ayy}{ax-2ay}=\frac{9x^4y-6axxyy+5aay^3-4aaxyy+aaxxy}{3xx-ay\times ax-2ay}.\text{ Soe y}^{\text{t}}$$

$$ck=\frac{py+vy}{v-r}=\frac{9x^4y-6axxyy+5aay^3-4aaxyy+aaxxy \text{ in } 3x^3-6xxy-axy+2ayy}{2aay^3-2aaxyy-24axy^3+24axxyy-18x^4y \text{ in } 3xx-ay}.$$

That is $ck=\dfrac{9x^5-6ax^3y+aax^3-6aaxxy+13aaxyy+12axxyy-10aay^3-18x^4y}{2aayy-2aaxy-24axyy+24axxy-18x^4}$.
Which Equation gives y[e] point k & consequently y[e] point m. for $km\,\|\,abk$.

☞ But in such cases where y is ordinately applyed to x at right angles, From y[e] consideration of y[e] Equation $\dfrac{py+vy}{v-r}=ck$; Or rather $\dfrac{py+vy}{r-v}=ck$: may y[e] following Theoreme bee pronounced.(74) To w[ch] purpose let $\mathfrak{X}$ signify the given Equation, y[t] is, all y[e] algebraicall termes (expressing y[e] nature of y[e] given line) considered as equall to nothing & not some of y[m] to others. Let ${}^{\cdot}\mathfrak{X}$ signify those termes ordered according to y[e] dimensions of x & y[n] multiplyed by any arithmeticall progressiō. Let $\mathfrak{X}^{\cdot}$ signify those termes ordered according to y[e] dimensions of y & y[n] multiplyed by any Arithmeticall progression. Let ${}^{:}\mathfrak{X}$ signify those termes ordered by x & y[n] multiplied by any two arithmeticall progressions one of y[m] being greater y[n] y[e] other by a terme. Let $\mathfrak{X}^{:}$ signify those termes ordered by y & y[n] multiplied by any two Arith: Progr: differing by a terme. Let ${}^{\cdot}\mathfrak{X}^{\cdot}$ signify those termes ordered according to x, & y[n] multiplyed by y[e] greater of y[e] progressions w[ch] multiplyed ${}^{\cdot}\mathfrak{X}$; & y[n] ordered by y & multiplyed by y[e] greater progression w[ch] multiplyed $\mathfrak{X}^{\cdot}$. Then (observing y[t] all these progressions have the same difference & proceede y[e] same way in respect of y[e] dimensions of x & y;) will y[e] 3 Theorems bee

$$1^{\text{st}}.\ \frac{{}^{\cdot}\mathfrak{X}\,{}^{\cdot}\mathfrak{X}\,\mathfrak{X}^{:}yy+\mathfrak{X}^{\cdot}\mathfrak{X}^{\cdot}\,{}^{:}\mathfrak{X}xx}{-{}^{\cdot}\mathfrak{X}\,{}^{\cdot}\mathfrak{X}\,\mathfrak{X}^{:}y+2\,{}^{\cdot}\mathfrak{X}\,\mathfrak{X}^{\cdot}\,{}^{\cdot}\mathfrak{X}^{\cdot}y-\mathfrak{X}^{\cdot}\mathfrak{X}^{\cdot}\,{}^{:}\mathfrak{X}y}=ck=\frac{{}^{\cdot}\mathfrak{X}\,{}^{\cdot}\mathfrak{X}yy+\mathfrak{X}^{\cdot}\,{}^{\cdot}\mathfrak{X}^{\cdot}xx}{\dfrac{-{}^{\cdot}\mathfrak{X}\,{}^{\cdot}\mathfrak{X}\,\mathfrak{X}^{:}y}{\mathfrak{X}^{\cdot}}+2\,{}^{\cdot}\mathfrak{X}\,{}^{\cdot}\mathfrak{X}^{\cdot}y-\mathfrak{X}^{\cdot}\,{}^{:}\mathfrak{X}y}.$$

(72) Similarly, where $v=y(dy/dx)$, $r=y(dv/dy)$ represents the fluxion dv/dt in the same scheme as the previous note.

(73) Since $q=y$ represents the fluxion dy/dt.

(74) Summarized from Add. 4004: [48[v]] = 4, §3.2 above. Note that Newton slightly changes his notation, using ${}^{:}\mathfrak{X}$ for x^2f_{xx} and $\mathfrak{X}^{:}$ for y^2f_{yy}, where $\mathfrak{X}\equiv f(x,y)=0$.

2^{d}. $\dfrac{{\cdot}\mathfrak{X}{\cdot}\mathfrak{X}{\cdot}\mathfrak{X}yy+{\cdot}\mathfrak{X}\mathfrak{X}{\cdot}\mathfrak{X}{\cdot}xx}{-{\cdot}\mathfrak{X}{\cdot}\mathfrak{X}\mathfrak{X}{:}x+2{\cdot}\mathfrak{X}\mathfrak{X}{\cdot}{\cdot}\mathfrak{X}{\cdot}x-\mathfrak{X}{\cdot}\mathfrak{X}{\cdot}{:}\mathfrak{X}x}=km=bl=\dfrac{{\cdot}\mathfrak{X}{\cdot}\mathfrak{X}yy+\mathfrak{X}{\cdot}\mathfrak{X}{\cdot}xx}{-{\cdot}\mathfrak{X}\mathfrak{X}{:}x+2\mathfrak{X}{\cdot}{\cdot}\mathfrak{X}{\cdot}x\dfrac{-\mathfrak{X}{\cdot}\mathfrak{X}{\cdot}{:}\mathfrak{X}x}{{\cdot}\mathfrak{X}}}$.

3^{d}. $\dfrac{{\cdot}\mathfrak{X}{\cdot}\mathfrak{X}yy+\mathfrak{X}{\cdot}\mathfrak{X}{\cdot}xx \text{ in } \sqrt{{\cdot}\mathfrak{X}{\cdot}\mathfrak{X}yy+\mathfrak{X}{\cdot}\mathfrak{X}{\cdot}xx}}{-{\cdot}\mathfrak{X}{\cdot}\mathfrak{X}\mathfrak{X}{:}yx+2{\cdot}\mathfrak{X}\mathfrak{X}{\cdot}{\cdot}\mathfrak{X}{\cdot}yx-\mathfrak{X}{\cdot}\mathfrak{X}{\cdot}{:}\mathfrak{X}yx}=cm=$ radio circuli æqualis curvitatis cum curva *ac* in puncto *c*.(75)

As if y^{e} line bee $x^3-axy+ayy=0$. Then is ${\cdot}\mathfrak{X}=3x^3-axy$. $\mathfrak{X}{\cdot}=-axy+2ayy$.

$3\times 2.\quad 1\times 0.\quad 0\times -1.\qquad\qquad 0\times -1.\quad 1\times 0.\quad 2\times 1$

${:}\mathfrak{X}=6x^3=\quad x^3\quad -\quad axy\quad +\quad ayy.$ $\mathfrak{X}{:}=2ayy=\quad x^3\quad -\quad axy\quad +ayy.$ &

$3\times 0.\quad 1\times 1.\quad 0\times 2.$

${\cdot}\mathfrak{X}{\cdot}=-axy=\quad x^3\quad -\quad axy\quad +\quad ayy.$

Which valors of ${:}\mathfrak{X}$, $\mathfrak{X}{:}$ &c being substituted into their places in y^{e} first rule, y^{e} result is; $\dfrac{9x^6yy-6ax^4y^3+5aaxxy^4+aax^4yy-4aax^3y^3}{\dfrac{-18ax^6y^3+12aax^4y^4-2a^3xxy^5}{2ayy-axy}+2aaxxy^3-12ax^3y^3}=ck$. Which being conveniently reduced is

$$\frac{9x^5-6ax^3y+aax^3-6aaxxy+13aaxyy+12axxyy-10aay^3-18x^4y}{18x^4-24axxy+24axyy+2aaxy-2aayy}=ck.$$

As was found before.(76) [y^{t} is] $ck=\frac{1}{2}a+3y-\dfrac{3ay}{2x}+\dfrac{ay-3xx}{2a-6x}$.(77)

Or suppose y^{e} line is a Conick section whose nature $rx+\dfrac{rxx}{q}=yy$.(78) Then is

$\qquad 2\quad 1\quad 0\qquad\qquad\qquad 0\quad 0\ 1\quad 2$

${\cdot}\mathfrak{X}=rx+\dfrac{2r}{q}x^2=\dfrac{r}{q}xx+rx-yy.$ $\mathfrak{X}{\cdot}=-2yy=\dfrac{r}{q}xx+rx*-yy.$

$2\times 1.\quad 1\times 0.\quad 0\times -1.\qquad\qquad 0\times -1.\quad 0\times -1.\ 1\times 0.\ 2\times 1.$

${:}\mathfrak{X}=\dfrac{2r}{q}xx=\dfrac{r}{q}\quad xx\quad +\quad rx\quad -\quad yy.$ $\mathfrak{X}{:}=-2yy=\dfrac{r}{q}\quad xx\quad +\quad rx\quad *\quad -yy$

---

(75) 'y^{e} radius of a circle of equall crookednesse wth y^{e} curve line *ac* at y^{e} point *c*.'

(76) In the immediately preceding example.

(77) For, since

$$(13ax+12x^2)\,ay^2=-13ax^4-12x^5+13a^2x^2y+12ax^3y \quad\text{and}\quad -10a^2y^3=10ax^4+10ax^3y+10a^2x^2y,$$

therefore

$$9x^5-6ax^3y+a^2x^3-6a^2x^2y+13a^2xy^2+12ax^2y^2-10a^2y^3-18x^4y = (ax^3+6x^3y-3ayx^2)(a-3x)+x^3(ay-3x^2)$$

and

$$18x^4-24ax^2y+24axy^2+2a^2xy-2a^2y^2=-2x^3(a-3x).$$

$$\mathfrak{X}=0=\overset{2\times 0}{\frac{r}{q}xx}+\overset{1\times 0.}{rx}-\overset{0\times 2}{yy}$$. Which valors of $\mathfrak{X}$, $\mathfrak{X}$,(79) $\mathfrak{X}$, &c being substituted into their places in ye first Theoreme, give

$$\frac{rrxxyy+4\frac{rr}{q}x^3yy+\frac{4rr}{qq}x^4yy+4xxy^4}{-rrxxy-\frac{4rr}{q}x^3y-\frac{4rr}{qq}x^4y+\frac{4r}{q}xxy^3}=ck=\frac{qqrry+4qrrxy+4rrxxy+4qqy^3}{4qryy-qqrr-4qrrx-4rrxx}.$$

Or (since $4qryy-4qrrx-4rrxx=0$) tis,

$$-ck=y+\frac{4xy}{q}+\frac{4xxy}{qq}+\frac{4y^3}{rr}=y+\frac{4y^3}{qr}+\frac{4y^3}{rr}=y+\frac{4ry^3+4qy^3}{qrr}.$$

So by ye second Theorem tis

$$\frac{rrxxyy+\frac{4r}{q}xxy^4+4xxy^4}{2rx^2yy+\frac{4r}{q}x^3yy\frac{\frac{8r}{q}y^4x^3}{rx+\frac{2r}{q}xx}}=km=bl=\frac{r}{2}+\frac{2yy}{q}+\frac{2yy}{r}+\frac{rx}{q}+\frac{4xyy}{qr}+\frac{4xyy}{qq}.^{(80)}$$

Or $bl=\frac{r}{2}+2x+\frac{3rx}{q}+\frac{6xx}{q}+\frac{6rxx}{qq}+\frac{4x^3}{qq}+\frac{4rx^3}{q^3}$.(81) And so by ye 3d Theoreme, tis

$$cm=\frac{rrxxyy+\frac{4rr}{q}x^3yy+\frac{4rr}{qq}x^4yy+4xxy^4}{2rrx^3y^3+8\frac{rr}{q}x^4y^3+\frac{8rr}{qq}x^5y^3-8\frac{r}{q}x^3y^5}\times\sqrt{rrxxyy+\frac{4rr}{q}x^3yy+\frac{4rr}{qq}x^4yy+4xxy^4}.$$

Or $cm=\dfrac{qqrr+4qrrx+4rrxx+4qqyy \text{ in } \sqrt{qqrr+4qrrx+[4]rrxx+4qqyy}}{2q^3rr}$. Or(82)

$$cm=\frac{qrr+4qyy+4ryy}{2qqrr}\sqrt{qqrr+4qqyy+4qryy}.^{(83)}$$

(78) Compare 4, §2.2: Example 2d. The 'Conick section' is here a hyperbola.
(79) Read '$\mathfrak{X}$'.
(80) Eliminating x^2 by substituting its value $-qx+(q/r)y^2$.
(81) Where correspondingly y^2 is eliminated by substituting its value $rx+(r/q)x^2$.
(82) Substituting qy^2 for $qrx+rx^2$.
(83) Newton wrote immediately afterward 'This Theoreme may bee thus Demonstrated', but cancelled it.

Note y^t y^e curvity of any curve whose ordinates are inclined from right to oblique angles is as y^e curvity of a circle whose ordinates are in like manner inclined so as to make it becom an Ellipsis.(84)

Prob: *To find ye points of curves where they have a given degree of curvity.*(85)

Prob 3^d. *To find ye points distinguishing twixt ye concave & convex portions of crooked lines.*(86)

Resolution. The lines are not crooked(87) at those points: & therefore y^e radius *cm* determining y^e crookednesse at y^t point must bee infinitely greate. To wch purpose I put y^e denominator of its valor (in rule y^e 3^d)(88) to bee equall to nothing, & so have this Theoreme ${\cdot}\mathfrak{X}{\cdot}\mathfrak{X}\mathfrak{X}{:}-2{\cdot}\mathfrak{X}\mathfrak{X}{\cdot}{\cdot}\mathfrak{X}{\cdot}+\mathfrak{X}{\cdot}\mathfrak{X}{\cdot}{:}\mathfrak{X}=0$. Or better p̸haps $\dfrac{{\cdot}\mathfrak{X}\mathfrak{X}{:}}{\mathfrak{X}{\cdot}}-2{\cdot}\mathfrak{X}{\cdot}+\dfrac{\mathfrak{X}{\cdot}{:}\mathfrak{X}}{{\cdot}\mathfrak{X}}=0.$

Example, was this point to be found in y^e Concha whose nature is

$$x^4+2bx^3\begin{array}{l}+bb\\-cc\\+yy\end{array}xx-2bccx-bbcc=0.$$

Then is ${\cdot}\mathfrak{X}=2x^4+2bx^3+2bccx+2bbcc$. $\mathfrak{X}{\cdot}=2xxyy=\mathfrak{X}{:}$. ${:}\mathfrak{X}=2x^4-4bccx-6bbcc$. ${\cdot}\mathfrak{X}{\cdot}=0$. Which valors subrogated into y^e Theoreme, they produce

$$2x^4+2bx^3+2bccx+2bbcc+\frac{\overline{2x^4-4bccx-6bbcc}\times 2xxyy}{2x^4+2bx^3+2bccx+2bbcc}=0.$$

---

(84) A neatly made point. Where *m* is the centre of the circle of curvature at the point *c* on the curve (*c*) defined by a given relation between the perpendicular Cartesian axes $ab = x$ and $bc = y$, Newton considers the curve (c') defined by the same relation between the oblique Cartesian axes $ab = x$ and $bc' = y$. In this correspondence we may without loss suppose all

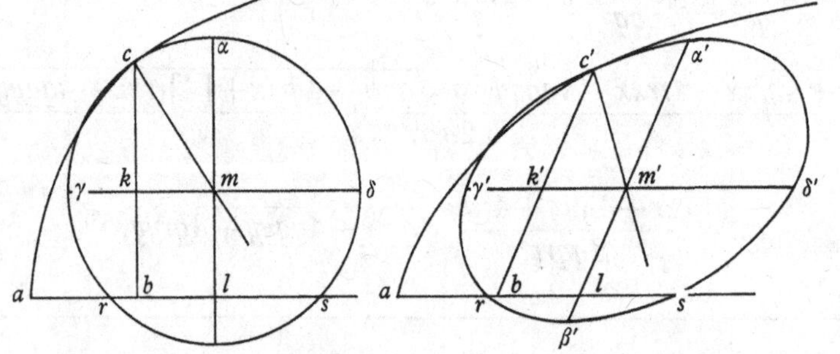

points in the axis *arbls* invariant. It follows that all perpendicular ordinates (as *bkc*, *lmα*) in the left-hand figure appear on the right unchanged in length but rotated into the given oblique angle $a\hat{b}c'$. Likewise all lines parallel to *abl* appear unchanged in length though translated in position. Hence

$$1=\frac{ck^2}{\gamma k\,.\,k\delta}=\frac{c'k'^2}{\gamma'k'\,.\,k'\delta'}$$

And subrogating $-x^4-2bx^3 \begin{smallmatrix}-bb\\+cc\end{smallmatrix} xx+2bccx+bbcc$ y[e] valor of $xxyy$ into its stead & twice reducing y[e] Equation by y[e] divisor $x+b$. Tis

$$x^3+bcc+\frac{\overline{x^4-2bccx-3bbcc}\times\overline{cc-xx}}{x^3+bcc}=0. \text{ Or } x^4+4bx^3+3bbxx-2bccx-2bbcc=0.$$

Which being againe reduced by $x+b=0$. Tis $x^3+3bxx-2bcc=0$.(89) See Geometr: Chart: pag: 259.(90)

Probl: 4. *To find y[e] points at w[ch] lines are most or least crooked.*(91)

Resol: At those points y[e] afforesaid radius cm neither increaseth nor decreaseth. So y[t] y[e] center m in y[t] moment doth absolutely rest, & therefore neither y[e] line

---

and so, as Newton correctly states, the circle of curvature (c) passes into (c') 'an ellipse of curvature' whose ordinates 'are in like manner inclined' and whose centre m' we may consider as a centre of (elliptic) curvature. (Clearly this ellipse will have a triple point in common with (c') at c' since this is true of the corresponding circle at c. Note too that, where the angle $\gamma'k'c' = abc'$ is given, the Apollonian symptom $c'k'^2 = \gamma'k'.k'\delta'$ defines a family of similar ellipses so that the osculating ellipse will be unique.)

Immediately after this note a gap follows in the manuscript. In it at some later time Newton entered the unnumbered 'Prob:' which follows.

(85) Newton restricts himself merely to the enunciation of this problem, which is indeed little more than a corollary of the preceding Prob 2[d]. Having found the radius of curvature at a general point on the given curve, we need only resolve the simultaneous equations which result (one the given defining equation of the curve, the other equating the radius of curvature to a fixed value) in order to find the values of ordinate and abscissa of the points sought.

(86) The first original section in the present tract and an ingenious corollary to the preceding problem.

(87) That is, have zero curvature.

(88) In the preceding Prob 2[d].

(89) In equivalent terms, Newton takes the defining equation of the conchoid as

$$0 = x^2y^2-(c^2-x^2)(b+x)^2 = x^2f(x,y),$$

where $$f(x,y) = x^2+2bx+(b^2-c^2+y^2)-2bc^2x^{-1}-b^2c^2x^{-2},$$

so that $$2x.f+x^2\frac{df}{dx} = 0 \quad\text{or}\quad x^2\left(f_x+f_y\frac{dy}{dx}\right) = 0,$$

with $$\mathfrak{X} = x^3f_x,\quad \mathfrak{X} = x^2yf_y,\quad \mathfrak{X} = x^4f_{xx},\quad \mathfrak{X} = x^3yf_{xy}\quad\text{and}\quad \mathfrak{X} = x^2y^2f_{yy}.$$

Hence the inflexion condition is $$\frac{xf_x(y^2f_{yy})}{yf_y}-2xyf_{xy}\frac{+yf_y(x^2f_{xx})}{xf_x} = 0.$$

(90) The reference is to Schooten's *Commentarii in Librum II*, *O*=*Geometria*: 259. (See 6, §4.2: Example 2[d] and note (33) above, and compare **3**, 3, §3.) The half page following this example is left blank for future additions.

(91) Newton finds straightforwardly the condition for the radius of curvature to attain an extreme value, namely that its instantaneous increment be zero. Further, when cm is stationary the point m is 'at rest', and therefore the increments of both al and lm will also be zero.

bk nor *al* doth increase or diminish, y[t] is, *ck* & *bl* doe soe much increase or diminish as y & x (cb & ab) doe diminish or increase. Or in a word the point m resteth. Find therefore the motion of *al* or *cm* or *lm* & suppose it nothing.

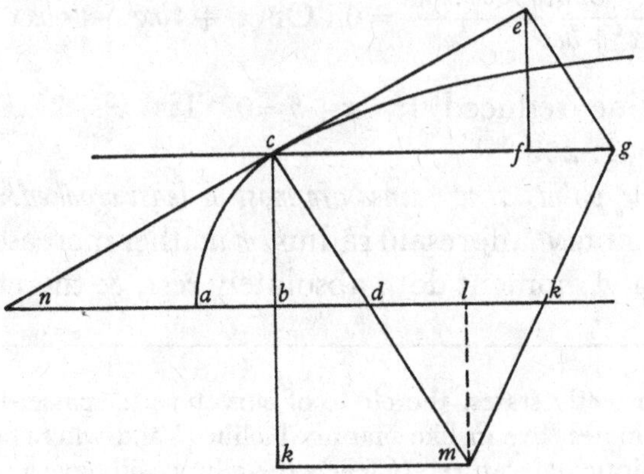

Thus(92) to find the point of least crookednesse in y[e] curve $x^3=ccy$. By the rule in prob 2 I make ${\cdot}\mathfrak{X}=3x^3$. $\mathfrak{X}{\cdot}=-ccy$. ${:}\mathfrak{X}=6x^3$. $\mathfrak{X}{:}=0$. & ${\cdot}\mathfrak{X}{\cdot}=0$ & thence obteine $ck=\frac{3}{2}y+\frac{cc}{6x}$, or $=\frac{3x^3}{2cc}+\frac{cc}{6x}$ w[ch] is least when $\frac{9x^3}{2cc}-\frac{cc}{6x}=0$ or $x=\surd 4{:}\ \frac{c^4}{27}$. And then therefore happens the greatest crookednesse.

In like manner if y[e] curve be $xxy=a^3$ The rule gives ${\cdot}\mathfrak{X}=2a^3$. $\mathfrak{X}{\cdot}=a^3$. ${:}\mathfrak{X}=-6a^3$. $\mathfrak{X}{:}=0={\cdot}\mathfrak{X}{\cdot}$.(93) And thence $\frac{2y}{3}+\frac{xx}{6y}=ck$ which is least when $y=a\surd[3{:}]\,\frac{1}{2}$.(94)

(92) With the revealing remark that 'The computation is too tedious', Newton has cancelled a first example of the conchoid $0=x^2y^2-(c^2-x^2)(b+x)^2$:

'Thus in the Concha to find the point of least crookednes beyond the point of reflection [sc. inflexion], having substituted y[e] valors of ${\cdot}\mathfrak{X}$, $\mathfrak{X}{\cdot}$ &c (exprest in the precedent problem) into this'. Clearly, it will be easiest to substitute

$$ {\cdot}\mathfrak{X} = 2(b+x)(x^3+bc^2),\quad \mathfrak{X}{\cdot} = 2x^2y^2 = 2(c^2-x^2)(b+x)^2 = \mathfrak{X}{:},\quad {:}\mathfrak{X} = 2(x^4-bc^2x-3b^2c^2) $$

and $\qquad {\cdot}\mathfrak{X}{\cdot} = 0$ in $\quad -km = \dfrac{({\cdot}\mathfrak{X}^2y^2+\mathfrak{X}{\cdot}^2x^2)\,{\cdot}\mathfrak{X}}{({\cdot}\mathfrak{X}^2\mathfrak{X}{:}-2{\cdot}\mathfrak{X}\,\mathfrak{X}{\cdot}\,{\cdot}\mathfrak{X}{\cdot}+\mathfrak{X}{\cdot}^2\,{:}\mathfrak{X})\,x}$

and we may then deduce that

$$ -km = \frac{(x^4+2bx^3+b^2c^2)(x^3+bc^2)}{x^3(x^3+3bx^2-2bc^2)}. $$

It follows that the only real points of extreme curvature are given by $x=0(y=\infty)$ and $x^3+3bx^2-2bc^2=0$, which yields the points of inflexion.

(93) Here $x^2y-a^3=x^2f(x,y)\equiv\mathfrak{X}$ with ${\cdot}\mathfrak{X}=x^3f_x$, $\mathfrak{X}{\cdot}=x^2yf_y$, ... (as in note (89)).

(94) Since $ck=\frac{2}{3}y+a^3/6y^2$ reaches an extreme value for $0=\frac{2}{3}-a^3/3y^3$.

Soe if y[e] curve bee $x^3=byy$. Then is $\mathcal{X}=3x^3$. $\mathcal{X}=-2byy$. $\mathcal{X}=6x^3$. $\mathcal{X}=-2[b]yy$. $\mathcal{X}=0$. And therefore $ck^{(95)}=3y+\dfrac{4xx}{3y}$, which hath no least nor the curve any least crookednesse.(96)

Prob 5[t]. *To find y[e] nature of y[e] crooked line whose area is expressed by any given equation.*

That is, y[e] nature of y[e] area being given to find y[e] nature of y[e] crooked line whose area it is.

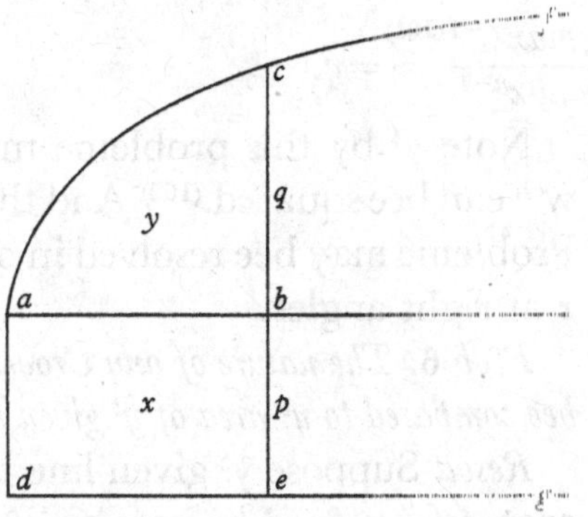

Resol. If y[e] relation of $ab=x$, & $\triangle abc=y$ bee given & y[e] relation of $ab=x$, & $bc=q$ bee required (bc being ordinately applyed at right angles to ab). Make $de \parallel ab \perp ad \parallel be=1$. & y[n] is $\square abed=x$. Now supposing y[e] line cbe by parallel motion from ad to describe y[e] two superficies $ae=x$, & $abc=y$; The velocity wth w[ch] they increase will bee, as be to bc: y[t] is, y[e] motion by w[ch] x increaseth being $be=p=1$, y[e] motion by w[ch] y increaseth will bee $bc=q$. which therefore may bee found by prop: 7th. viz: $\dfrac{-\mathcal{X}y}{\mathcal{X}x}=q=bc$.(97)

---

(95) Or $\dfrac{[(3x^3)^2y^2+(-2by^2)^2x^2].-2by^2}{-(3x^3)^2(-2y^2b)y-(-2by^2)^2(6x^3)y}$.

(96) Newton entered this in cancellation of 'which is least when $x=-\dfrac{4b}{27}$'. Where

$$ck = 3y+4x^2/3y = 3y+\tfrac{4}{3}b^{\frac{2}{3}}y^{\frac{1}{3}}, \quad \text{then} \quad d/dx(ck) = 3+4b/9x$$

which increases uniformly as x decreases and is zero for $x=-4b/27$. However Newton wishes here to consider only real points on the curve and so restricts x to the interval $[0, \infty]$. Hence this 'minimal' value for curvature defined by $x=-4b/27$ is not admissible in his scheme. More directly, we may calculate the radius of curvature to be

$$cm = \frac{(9x^2+4bx)^{\frac{3}{2}}}{6bx}, \quad \text{so that} \quad \frac{d(cm)}{dx} = \frac{\sqrt{[9x^2+4bx]}}{2bx^2}(9x^2+6bx)$$

and cm has therefore an extreme value for x in $[0, \infty]$ at $x=\infty$ (which, in fact, defines the inflexion point at infinity on the semicubic parabola $by^2=x^3$). We may also show that cm increases with x in the region of $x=0$, so that the curvature at a real point takes on an apparent maximum at $x=0$, $y=0$. (Compare the difficulties which Newton had in the winter of 1664/5 in considering the extreme values of curvature in the case of a conic. See 4, §2.5 above.)

(97) Since, where $\mathcal{X}\equiv f(x,y)=0$, we have $\mathcal{X}=xf_x$, $\mathcal{X}=yf_y$ and $f_x+f_y\dfrac{dy}{dx}=0$, therefore $q=-\dfrac{\mathcal{X}y}{\mathcal{X}x}=\dfrac{dy}{dx}$. Hence the area '$\triangle abc$' $=\int q.dx=y$, a statement of the fundamental theorem of the calculus that $\int\left(\dfrac{dy}{dx}\right).dx=y$.

Example 1. If $\frac{2x}{3}\sqrt{rx}=y$. Or $-4rx^3+9yy=0$. Then is $\frac{12rxx}{18y}=q=\sqrt{rx}$. Or, $rx=qq$ & therefore *abc* is y^e^ Parabola(98) whose area *abc* is $\frac{2x}{3}\sqrt{rx}=\frac{2qx}{3}$.

Example 2^d^. If $x^3-ay+xy=0$. Then is $\frac{3xx+y^{(99)}}{-x+a}=q$. Or $\frac{3axx-2x^3}{aa-2ax+xx}=q=bc$.

Example 3^d^. If $\frac{na}{n+m}x^{\frac{n+m}{m}}-y=0$. Then is $ax^{\frac{m}{n}}=q$. Or if $ax^m=bx^n$;(100) y^n^ is $\frac{max^{m-1}\,(101)}{nbx^{n-1}}=q$.

Note y^t^ by this probleme may bee gathered a Catalogue of all those lines wch can bee squared.(102) And therefore it will not bee necessary to shew how this Probleme may bee resolved in other cases in wch q is not ordinately applye[d] to x at right angles.

Prob 6. *The nature of any Crooked line being given, to find other lines whose areas may bee compared to y^e^ area of y^t^ given line.*(103)

Resol: Suppose y^e^ given line to be *ac* & its area $abc=s$, y^e^ sought line *df* & its area $def=t$; & y^t^ $bc=z$ is ordinately applyed to $ab=x$, & $ef=v$ to $de=y$, soe y^t^ $\angle abc=\angle def$; & y^t^ y^e^ velocitys wth w^ch^ *ab* & *de* increase (y^t^ is, y^e^ velocity of y^e^

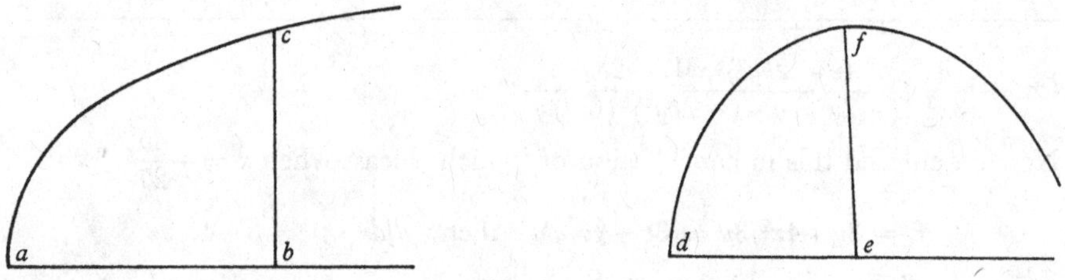

points *b* & *e*, or of y^e^ lines *bc* & *ef* moving from *a* & *d*) bee called p & q.(104) Then may y^e^ ordinately applyed lines *bc* & *ef* multiplyed by their velocitys p & q, (y^t^ is pz & qv) signify y^e^ velocitys wth wch y^e^ areas $abc=s$ & $def=t$ increase.(105) Now y^e^ relation of y^e^ areas s & t (taken at pleasure) gives y^e^ relation of y^e^ motions pz & qv describing those areas, by Proposition y^e^ 7th; Also y^e^ relation of y^e^ lines

(98) Defined by $cb^2=r.ab$.

(99) Applying $q=\frac{dy}{dx}=-\frac{\mathfrak{X}y}{\mathfrak{X}x}$.

(100) Read 'by^n'.

(101) Read '$\frac{max^{m-1}}{nby^{n-1}}$'.

(102) See Add. 4000: 152^r-163^v = 5, §4 above. Even in 1691 when he came to compose his *De Quadratura Curvarum*, Newton's only general method of algebraic integration remained a fundamental, though sophisticated, application of the inverse operation of differentiation.

$ab=x$ & $de=y$ (taken at pleasure) gives y^{e} relation of p & q, by Prop 7th: wch two equations, together wth y^{e} given Equation expressing y^{e} nature of y^{e} line ac, give y^{e} relation of $de=y$ & $ef=v$[,] y^{t} is y^{e} desired nature of y^{e} line df.

Example 1. As if $ax+bxx=zz$ is y^{e} nature of y^{e} line ac: & at pleasure I assume $s=t$ to be y^{e} relation of y^{e} areas abc & def; & $x=yy$ is to bee y^{e} relation of y^{e} lines ab & de. Then is $pz=qv$ (prop 7), & $p=2qy$ (by prop 7). Therefore $2yz=v$ (by y^{e} 2 last Equations), & $\frac{vv}{4yy}=zz=ax+bxx=ayy+by^4$. or $vv=4ay^4+4by^6$. & $v=2yy\sqrt{a+byy}$: W^{ch} is y^{e} nature of y^{e} line def whose area def is equall to y^{e} area abc, supposing $\sqrt{ab}=de=\sqrt{x}=y$.

Example 2. If $ax+bxx=zz$ as before; & I assume $as+bx=t$, & $x=yy$. Then is $apz+bp=qv$ (by prop 7), & $p=2qy$. $2azy+2by=v=2by+2ay\sqrt{ax+bxx}$. Or $v=2by+2ayy\sqrt{a+byy}$; The required nature of y^{e} line def.

Example 3^{d}. If $\frac{4c}{a}\sqrt{xx-a}=z$: & at pleasure I assume $\frac{2c}{a}\sqrt{ay+yy}-s=t$, & $xx-a=y$. Then is $4ccay+4ccyy=aass+2aast+aatt$, And (by prop 7)

$$4ccaq+8ccyq=2aaspz+2aatpz+2aasqv+2aatqv=\overline{2aapz+2aaqv}\times\frac{2c}{a}\sqrt{ay+yy}.$$

& (by prop 7) $2px=q$. Therefore $8cax+16cyx=\overline{4az+8avx}\times\sqrt{ay+yy}$. But $\frac{4c}{a}\sqrt{xx-a}=z=\frac{4c}{a}\sqrt{y}$. & $x=\sqrt{a+y}$. Therefore

$$\overline{8ca+16cy}\sqrt{a+y}=\overline{16c\sqrt{y}+8av\sqrt{a+y}}\times\sqrt{ay+yy}.$$

That is $ca+2cy=2cy+av\sqrt{ay+yy}$. Or $v=\frac{c}{\sqrt{ay+yy}}$.

Example 4th. If $ax+bxx=zz$ as before & I assume $ss=t$, & $x=yy$. Then is $2spz=qv$, & $p=2qy$ (by prop 7) Therefore

$$4syz=v=4sy\sqrt{ax+bxx}=4syy\sqrt{a+byy}.$$

(103) Generalized from the transform

$$\int v.dz = \int y.dx$$

in 5, §4.2/3.

(104) It follows that $s = \int z.dx$ and $t = \int v.dy$ with p and q the fluxions dx/dT and dy/dT respectively, where T is an independent time variable.

(105) Clearly $pz = ds/dT$ and $qv = dt/dT$.

Where note y^t^ in this case y^e^ line $v = 4syy\sqrt{a+byy}$ is a Mechanical one because s y^e^ area of y^e^ line $ax+bxx=zz$ canott bee Geometrically found. The like is to bee observed in other such like cases.(106)

Probl: 7. The Nature of any Crooked line being given to find its area, when it may bee. Or more generally, two crooked lines being given to find the relation of their areas, when it may bee.(107)

Resolution.(108) In y^e^ figure of y^e^ fift probleme Let $abc=y$ represent y^e^ area of y^e^ given line acf; $cb=q$ y^e^ motion describing y^t^ area; $abed=x$ another area wch is equall to y^e^ basis $ab=x$ of y^t^ given line acf (viz: supposing $ab \parallel de \perp be \parallel ad = 1$); & $be=p=1$ y^e^ motion describing y^t^ other area. Now haveing (by supp) y^e^ relation twixt $ab=x=abed$, & $bc=q=\frac{q}{1}=\frac{q}{p}$ given, I seeke y^e^ area $abc=y$ by y^e^ Eight proposition.

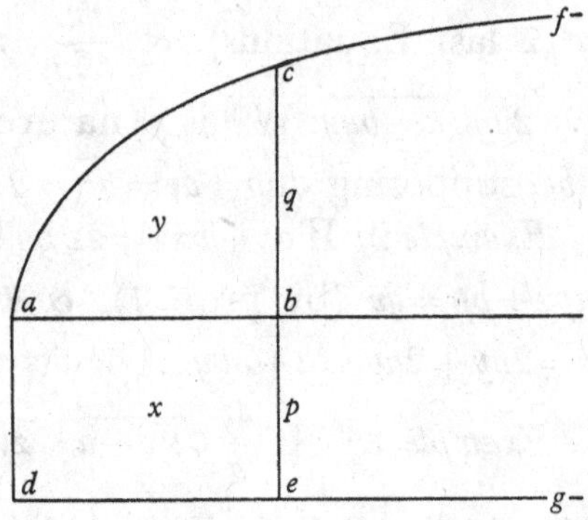

Example 1. If y^e^ nature of y^e^ line bee, $\frac{ax}{\sqrt{aa-xx}}=bc=\frac{q}{p}$. I looke in y^e^ tables of y^e^ Eight proposition for y^e^ Equation corresponding to this Equati[ō] wch I find to bee $\frac{cx^n}{x\sqrt{a+bx^n}}=\frac{q}{p}$, (For if instead of c, a, b, n I write a, aa, -1, 2, it will bee $\frac{ax}{\sqrt{aa-xx}}=\frac{q}{p}$.) And against it is y^e^ equation $\frac{2c}{nb}\sqrt{a+bx^n}=y$. And substituting a, aa, -1, 2 into y^e^ places of c, a, b, n it will bee, $-a\sqrt{aa-x^n}$(109) $=y=abc$ y^e^ required area.

Example 2. If $\sqrt{\frac{x^3}{a}}-\frac{eeb}{x\sqrt{ax-xx}}=bc=\frac{q}{p}$. Because there are two termes in y^e^ valor of bc I consider them severally & first I find y^e^ area correspondent to y^e^ terme $\sqrt{\frac{x^3}{a}}$, or $\frac{1}{a^{\frac{1}{2}}}x^{\frac{3}{2}}$; To bee $\frac{2}{5a^{\frac{1}{2}}}x^{\frac{5}{2}}$, or $\frac{2}{5}\sqrt{\frac{x^5}{a}}$. by prop: 8. part 1. Secondly to find y^e^ area corresponding to y^e^ other terme $\frac{eeb}{x\sqrt{ax-xx}}$ I looke y^e^ Equation (in

(106) As Newton's repeated use of the example $z = \sqrt{[ax+bx^2]}$ (with correspondingly $s = \int \sqrt{[ax+bx^2]}\,.dx$) shows, his main purpose here is to reduce the integral of a wide class of algebraic forms to the area under a general conic section, that is, to a circular or hyperbolic integral.

(107) The inverse of the two preceding problems.

(108) An unimportant first partial draft of this is omitted.

prop 8. part 3) corresponding to it wch is $\frac{cx^n}{x\sqrt{a+cx^n}}=\frac{q}{p}$, (for if instead of c, a, b, n, I write $eeb, -1, a, -1$ it will bee $\frac{eebx^{\bar{1}}}{x\sqrt{-1+ax^{\bar{1}}}}$, Or $\frac{eeb}{x\sqrt{-xx+ax}}$): Against which is y^e^ Equation $\frac{2c}{nb}\sqrt{a+bx^n}=y$. In which writing $eeb, -1, a, -1$, instead of c, a, b, n, the result will bee $\frac{2eeb}{-a}\sqrt{-1+ax^{\bar{1}}}$; Or, $\frac{-2eeb}{ax}\sqrt{-xx+ax}=y$ wch is the area corresponding to y^t^ other terme. Now[110] to see how these areas stand related one to another I draw y^e^ annexed scheme, in which is[111] $ab=x$. $bd=\sqrt{\frac{x^3}{a}}$. $de=\frac{eeb}{x\sqrt{ax-xx}}$. & $be=\sqrt{\frac{x^3}{a}}-\frac{eeb}{x\sqrt{ax-xx}}=\frac{q}{p}=q$. Soe that $abd=\frac{2}{5}\sqrt{\frac{x^5}{a}}$ is y^e^ superficies corresponding to y^e^ first terme bd, wch because it is affirmative must bee extended (or lye) from y^e^ line bd towards a. Also y^e^ other superficies correspondent to y^e^ 2^d^ terme de, being negative must lye on y^e^ other side bd from a, wch is therefore $gde=\frac{-2eeb}{ax}\sqrt{ax-xx}$. Lastly if $x=ab=r$. Then is $abd=\frac{2}{5}\sqrt{\frac{r^5}{a}}$. & $gde=\frac{-2bee}{ar}\sqrt{ar-rr}$. And if $aB=x=s$, y^n^ is $\frac{2}{5}\sqrt{\frac{s^5}{a}}=aBD$, And $gDE=\frac{-2bee}{as}\sqrt{as-ss}$. Soe y^t^, $\frac{2}{5}\sqrt{\frac{s^5}{a}}-\frac{2}{5}\sqrt{\frac{r^5}{a}}=bBDd$. &

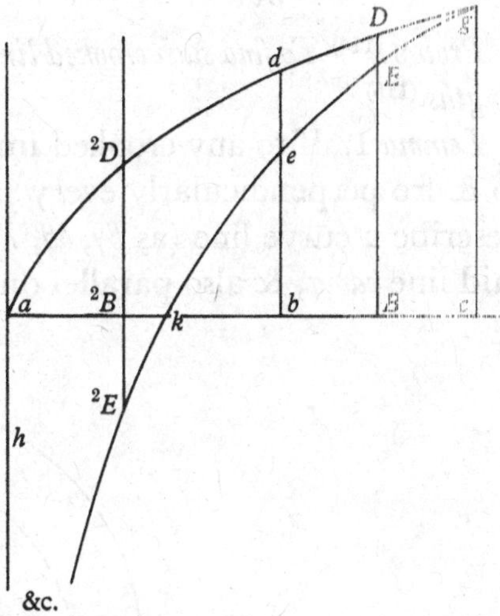

$$\frac{-2bee}{as}\sqrt{as-ss}\frac{+2bee}{ar}\sqrt{ar-rr}=DdeE.$$

& substracting $DdeE$ from $bBDd$ there remaines

$$bBEe=\frac{2\sqrt{s^5}-2\sqrt{r^5}}{5\sqrt{a}}+\frac{2bee}{as}\sqrt{as-ss}-\frac{2bee}{ar}\sqrt{ar-rr}=bBEe.^{(112)}$$

(109) Read '$-a\sqrt{aa-xx}$'.
(110) Newton has cancelled 'Lastly'.
(111) A cancelled variant of the following two lines is omitted.
(112) That is,

$$bBEe=\int_r^s\left(\sqrt{\frac{x^3}{a}}-\frac{e^2b}{x\sqrt{[ax-x^2]}}\right).dx=\left[\frac{2}{5}\sqrt{\frac{x^5}{a}}+\frac{2e^2b}{ax}\sqrt{[ax-x^2]}\right]_r^s.$$

wch is y^e^ required Area of y^e^ given line &c ${}^2EkeEg$. Where note y^t^ for y^e^ quantitys $r=ab$ & $s=aB$ taking any numbers you may thereby finde y^e^ area $bBdD$ correspond[ent] to their difference bB.

Note y^t^ sometimes one parte of y^e^ Area may bee Affirmative & y^e^ other negative. as if $a^2B=r$, & $ab=s$. Then is

$$b^2B^2Ee=kbe-k^2B^2E$$

$$=\frac{2\sqrt{s^5}-2\sqrt{r^5}}{5\sqrt{a}}+\frac{2bee}{as}\sqrt{as-ss}-\frac{2bee}{ar}\sqrt{ar-rr}=b^2B^2Ee=kbe-k^2B^2E.^{(113)}$$

Prob 9.(114) *To find such crooked lines whose lengths may bee found. & also to find theire lengths.*(115)

Lemma 1. If to any crooked immovable line acg the streight line $cdm\sigma$ moves to & fro perpendicularly every point of y^e^ said line $cdm\sigma$ (as γ, θ, σ, &c) shall describe a curve line (as $\beta\gamma$, $\delta\theta$, $\lambda\sigma$, &c) all which will bee perpendicular to y^e^ said line $cdm\sigma$, & also parallel one to another & to y^e^ line acg.(116)

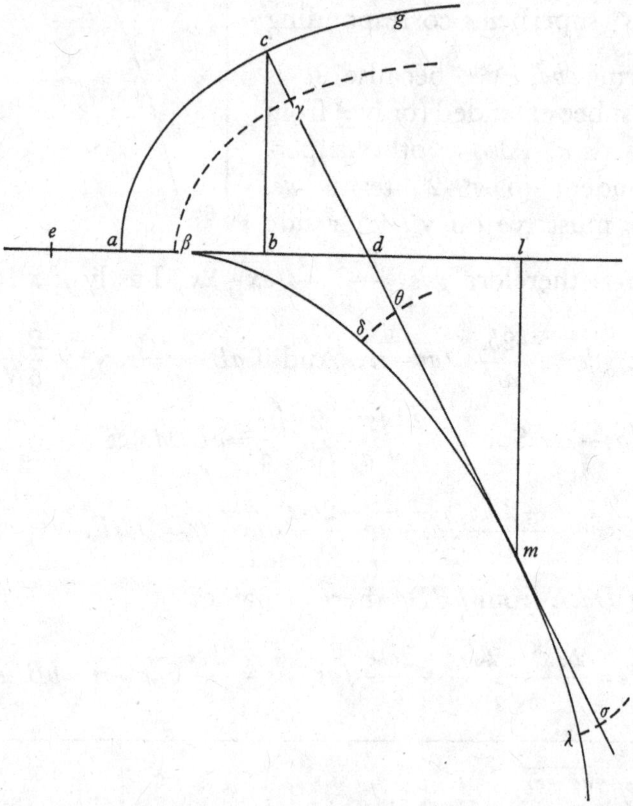

(113) The next page and a half in the manuscript is left blank. Perhaps Newton intended to add further examples, but probably the gap is left for the insertion of a Prob: 8, lacking in the tract. (The following problem was first entered as 'Prob 8' but later emended to '9'.)

2. If *acg* bee not a circle, there may bee drawne some curve line $\beta\delta m\lambda$, wth y^e^ moveing line $cdm\sigma$ will always touch in some point or other (as at m) & to wch therefore all y^e^ curve lines $\beta\gamma$, $\delta\theta$, $\lambda\sigma$, &c: are perpendiculars.

3. Soe that every point (γ, θ, m, σ, &c:) of y^e^ line $cdm\sigma$, when it begineth or ceaseth to touch y^e^ curve line $\beta\delta m\lambda$, doth y^n^ move perpendicularly to or from it: & therefore y^e^ line $\gamma dm\sigma$ doth not at all slide upon y^e^ curve line $\beta\delta m\lambda$, but exactly measure it by applying it selfe to it point by point: & therefore y^e^ correspondent parts of y^e^ said lines are equall (viz: $\beta m = \gamma m$. $\delta m = \theta m$. $\delta\lambda = \theta\sigma$. &c).

Resol. Take any Equation for y^e^ nature of y^e^ crooked line *acg*, & by y^e^ 2^d^ probleme find y^e^ center m of its crookednesse at c. That point m is a point of y^e^ required curve line $\beta\delta m\lambda$. For y^t^ point m whereat $cdm\sigma$ y^e^ perpendicular to *acg* doth touch y^e^ curve line $\beta\delta m\lambda$ is lesse moved y^n^ any other point of y^e^ said perpendicular, being as it were y^e^ hinge & center {about / upon} wch y^e^ perpendicular {moveth / turneth} at that moment.(117)

Example 1. If *acg* is a Parabola whose nature (supposing $ab = x \perp bc = y$) is $rx = yy$. By Theoreme y^e^ 1st of Problem 2^d^ I find $cb + lm = y + \frac{4y^3}{rr}$. By y^e^ 2^d^ Theoreme, $bl = 2x + \frac{1}{2}r$. And by y^e^ 3^d^ Theoreme $\frac{\overline{r+4x}\sqrt{r+4x}}{2[\sqrt{}]r} = cm$. Soe y^t^ supposing $ab^{(118)} = \frac{1}{2}r$. $bl^{(118)} = 3x = z$. $lm = \frac{4y^3}{rr} = v$. The relation twixt v & z will bee $27rvv = 16z^3$, $\left(\text{for } r^4vv = 16y^6 = 16r^3x^3 = \frac{16z^3r^3}{27}\right)$ wch is y^e^ nature of y^e^ required line $\beta\delta m\lambda$. And since $c\gamma = a\beta = \frac{1}{2}r$. Therefore

$$\gamma m = \frac{r+4x}{2r}\sqrt{rr+4xr} - \frac{1}{2}r = \frac{3r+4z}{18r}\sqrt{9rr+12rz} - \frac{r}{2}.$$

is y^e^ length of its pte $\beta\delta m$.

Example 2. Soe if $aa = xy$,(119) is y^e^ nature of y^e^ line *acg*. By y^e^ afforesaid Theoremes I find, $cb + lm = \frac{xx+yy}{2y}$; &, $bl = \frac{xx+yy}{2x}$; whereby y^e^ nature of y^e^

(114) See previous note.

(115) Generalized and adapted from his notes of May 1665 in the Waste Book (ULC. Add. 4004: $32^r = 4$, §2.2 above).

(116) In the sense that the angles $\widehat{acm}$, $\beta\widehat{\gamma m}$, $\delta\widehat{\theta m}$, $\lambda\widehat{\sigma m}$ are each right, so that the instantaneous increments of $\widehat{ac}$, $\widehat{\beta\gamma}$, $\widehat{\delta\theta}$, and $\widehat{\lambda\sigma}$ are parallel.

(117) Newton finds the evolute (m) as the locus of the instantaneous centre of rotation.

(118) Read 'β' for 'b' in each case.

(119) An Apollonian hyperbola.

curve line, $\beta\delta m\lambda$ is determined. And lastly I find $cm = \frac{xx+yy}{2aa}\sqrt{xx+yy}$ which determine[s] its length.

Prob: 10. *Any curve line being given to find other lines whose lengths may be compared to its length or to its area, & to compare ym.*

Resolution. Take any Equation for ye relation twixt $ad[=x]$ & ye perpendicular $cd=y$ (whither yt relation bee expressed by an(120) Equation or whither it bee ye same wch some streight line beares to a curved one or to its superficies &c).(121

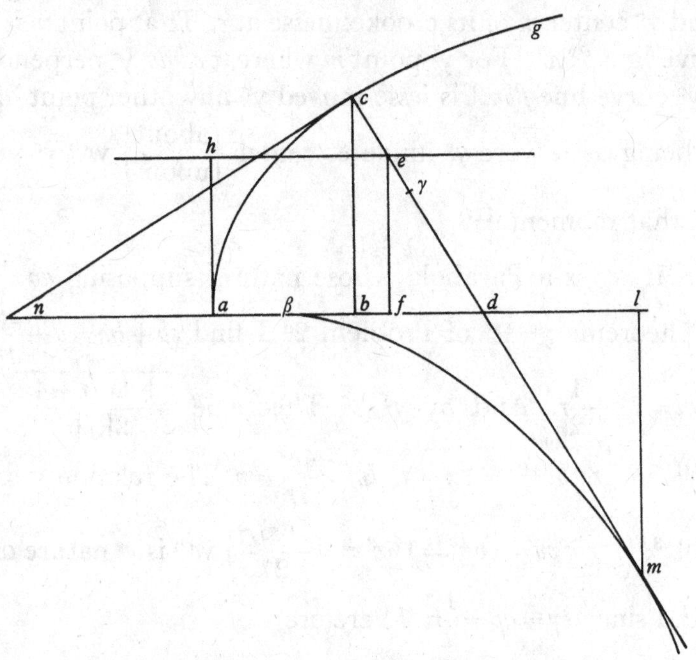

(120) Newton has cancelled 'Geometricall'.

(121) This unusual semi-intrinsic system of co-ordinates $ad = x$, $dc = y$ with dc normal to the curve (c) may be reduced to the perpendicular Cartesian one with co-ordinates

$$ab = X, \quad bc = Y \quad \text{by} \left\{ \begin{array}{l|l} X = x - y(dy/dx) & x = X + Y(dY/dX) \\ Y = y\sqrt{[1-(dy/dx)^2]} & y = Y\sqrt{[1+(dY/dX)^2]} \end{array} \right\}$$

with $$dX/dY = \sqrt{[(dx/dy)^2 - 1]}.$$

(Newton's permissible forms of defining equation,

$$f(x,y) = 0, \quad f\left(y, \int\sqrt{\left[1+\left(\frac{dy}{dx}\right)^2\right]}.dx\right) = 0$$

and more generally $f\left(y, \int\phi(x).dx\right) = 0$, seem chosen with the intention that their derivatives may be algebraic.)

Then (by prop 7) find y^e^ relation twixt y^e^ increase or decrease (p & q) of y^e^ lines $ad=x$, & $dc=y$.(122) & say (by prop 6) y^t^ $q:p::dc:dn=\frac{py}{q}$. And soe is y^e^ triangle dnc (rectangled at c) given & consequently y^e^ nature of y^e^ curve line acg to wch dc is perpendicular, & cn a tangent.

Now y^e^ center (m) of y^e^ perpendiculars motion (wch gives y^e^ nature & length of y^e^ required curve line βm) may be found as in y^e^ 2^d^ or 9th Probleme,(123) But more conveniently thus. Draw any fixed line $he \parallel ad \perp ah=a=ef \perp ad$. Also call $fd=v$. & y^e^ increase or decrease of (fd) call r. And y^e^ increase or decrease of y^e^ motions p & q, call β & γ. Now considering y^t^ y^e^ motion $(p+r)$(124) of y^e^ intersection point e in y^e^ line he is to y^e^ motion (p) of y^e^ intersection point (d) in y^e^ line (ad), as (em) is to (dm) (see prob 2); That is y^t^ y^e^ difference (r) of those motions is to y^e^ motion (p) of y^e^ point (b) soe is (ed) to (dm): First I find y^e^ valor of v. viz: $-\sqrt{pp-qq}:q::cn:cd::ef=a:fd=v=\frac{-aq}{\sqrt{pp-qq}}$. Or $aaqq+vvqq-vvpp=0$. Secondly by this Equation I find y^e^ valor of r, viz: (by Prob 7),

$$2aaq\gamma+2vvq\gamma+2rvqq-2rvpp-2vvp\beta=0.$$

Thirdly $r=\frac{aaq\gamma+vvq\gamma-vvp\beta}{ppv-qqv}:p::ed=\frac{pv}{q}:dm=\frac{p^4zz-ppqqzz}{aaqq\gamma+vvqq\gamma-vvpq\beta}$.(125) Lastly

(122) As before, p and q are the fluxions, dx/dt and dy/dt respectively, of x and y.

(123) Newton, in fact, proceeds immediately to show that

$$cm = \frac{1-(dy/dx)^2}{d^2y/dx^2},$$

which we may reduce to the familiar Cartesian form

$$cm = \frac{-(1+(dY/dX)^2)^{\frac{3}{2}}}{d^2Y/dX^2}$$

by applying the transform (note (121))

$$\left\{\begin{array}{l} x = X+Y(dY/dX) \\ y = Y\surd[1+(dY/dX)^2] \end{array}\right\}.$$

Specifically, we may prove

$$\frac{dx}{dX} = 1+\left(\frac{dY}{dX}\right)^2+Y\frac{d^2Y}{dX^2} \quad\text{and}\quad \frac{dy}{dX} = \left(1+\left(\frac{dX}{dY}\right)^2\right)^{-\frac{1}{2}}\left(1+\left(\frac{dY}{dX}\right)^2+Y\frac{d^2Y}{dX^2}\right),$$

so that
$$\frac{dy}{dx} = \left(1+\left(\frac{dX}{dY}\right)^2\right)^{-\frac{1}{2}} \quad\text{and}\quad 1-\left(\frac{dy}{dx}\right)^2 = \left(1+\left(\frac{dY}{dX}\right)^2\right)^{-1}.$$

Further, since

$$\frac{d^2X}{dY^2} = -\left(\frac{dX}{dY}\right)^3\frac{d^2Y}{dX^2}, \quad \frac{d^2y}{dx^2} = \frac{d}{dX}\left(\frac{dy}{dx}\right)\frac{dX}{dx} = \left(1+\left(\frac{dY}{dX}\right)^2\right)^{-\frac{3}{2}}\frac{d^2Y}{dX^2}\left(1+\left(\frac{dY}{dX}\right)^2+Y\frac{d^2Y}{dX^2}\right)^{-1}$$

and the reduction follows since $-cm = -[cd+(-dm)]$.

(124) That is, the fluxion of $ad-fd$.

(125) Read '$\frac{p^4vv-ppqqvv}{aaqq\gamma+vvqq\gamma-vvpq\beta}$'.

supposing y^e^ motion p to bee uniforme y^t^ its increase or decrease β may vanish, & also substituting $ppzz$(126) y^e^ valor of $aaqq+vvqq$ in its stead in y^e^ Denominator of dm, y^e^ result will bee $\frac{pp-qq}{\gamma}=dm=\frac{1-qq}{\gamma}$, if $p=1$. And to find y^e^ lines dl & ml, say $-p:q::dm:dl=\frac{-q+q^3}{\gamma}$. & $p:\sqrt{pp-qq}::dm:ml=\frac{\overline{1-qq}\sqrt{1-qq}}{\gamma}$.

Soe y^t^ by the equation expressing y^e^ relation twixt x & y first I finde q & y^n^ γ. wch two give me $\frac{1-qq}{\gamma}=dm$ &c.(127)

Example 1. If y^e^ relation twixt x & y bee supposed to bee $yy-ax=0$. Then (by Prop 7) I find, first $2qy-ap=0$, Or $2qy-a=0$. & secondly $2\gamma y+2qq=0$. And substituting these valors of $q=\frac{a}{2y}$. & $\gamma=\frac{-qq}{y}=\frac{-aa}{4y^3}$ in their stead in y^e^ Equatio $\frac{1-qq}{\gamma}=dm$, &c: The results will bee $\frac{aay-4y^3}{aa}=dm$. $\frac{-aa+4yy}{2a}=dl$. &

(126) Read '*ppvv*'.

(127) Where t is an independent variable of time, $p = dx/dt$, $q = dy/dt$ and so

$$\beta = \frac{dp}{dt} = \frac{d^2x}{dt^2}, \quad \gamma = \frac{dq}{dt} = \frac{d^2y}{dt^2} \quad \text{and} \quad r = \frac{dv}{dt},$$

where $v = aq/(\sqrt{[p^2-q^2]})$. Hence

$$r = \frac{(a^2+v^2)\,q\gamma - v^2p\beta}{(p^2-q^2)\,v} \quad \text{and so} \quad dm = \frac{p^2v}{qr} = \frac{p^2v^2(p^2-q^2)}{(a^2+v^2)\,q^2\gamma - v^2\beta pq}.$$

Newton finally takes $t = x$ (so that $p = 1$, $q = dy/dx$, $\beta = 0$, $\gamma = d^2y/dx^2$ and $r = dv/dx$) and deduces, since $p^2v^2 = (a^2+v^2)\,q^2$, that

$$dm = \frac{1-q^2}{\gamma} = \frac{1-(dy/dx)^2}{d^2y/dx^2}.$$

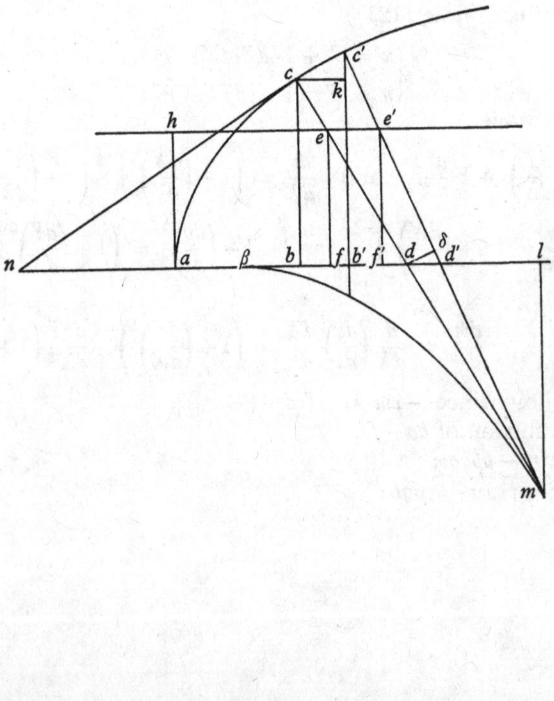

$\frac{aa-4yy}{2aa}\sqrt{4yy-aa}=lm$. (And adding $cd=-y$ to dm, & $ad=x=\frac{yy}{a}$ to dl, ye result is $cm=-\frac{4y^3}{aa}$. & $al=\frac{3yy}{a}-\frac{a}{2}$.) Wch determine ye nature & crookednesse of ye line βm. (For if $a\beta=\frac{a}{4}$. $\beta l=z$. $lm=v$. The relation twixt z & v will bee $16z^3=27vv$.[128] The length of βm being $m\gamma=\frac{4y^3}{aa}-\frac{a}{2}=\frac{3a+4z}{18a}\sqrt{9aa+12az}-\frac{1}{2}a$. as before was found).[129]

We might, alternatively, have set the argument up in terms of infinitesimal increments. Thus where $c'd'm$ is the normal at the point c' indefinitely near to c on the curve and $d\delta$, ck, $c'b'$, $e'f'$ are drawn perpendicular to $c'd'$, $c'b'$ and abd respectively, then

$$\frac{ed}{dm}=\frac{ee'\ (\text{or } ff')-dd'}{dd'}=-\frac{f'd'-fd}{dd'}$$

and the indefinitely small triangles $d\delta d'$ and ckc' are similar to cbd. Hence, where $ab = X$, $bc = Y$; $ad = x$, $dc = y$, then

$$\frac{dx}{dy}=\frac{d'd}{d'\delta}=\frac{c'c}{c'k}=\sqrt{\left[\left(\frac{dX}{dY}\right)^2+1\right]},\quad\text{or}\quad \frac{dX}{dY}=\sqrt{\left[\left(\frac{dx}{dy}\right)^2-1\right]}.$$

Further

$$fd=ef\frac{bd}{bc}=\frac{a}{\sqrt{[(dx/dy)^2-1]}},\quad\text{so that}\quad \frac{ed}{dm}=\frac{d}{dx}\left(\frac{a}{\sqrt{[(dx/dy)^2-1]}}\right)=\frac{a(d^2y/dx^2)}{(1-(dy/dx)^2)^{\frac{3}{2}}}.$$

Finally, since $ed=\frac{a}{\sqrt{[1-(dy/dx)^2]}}$, it follows that

$$dm=-\frac{1-(dy/dx)^2}{d^2y/dx^2},$$

as before.

(128) Read '$27avv$'.

(129) That is in Prob. 9, Example 1 above. Converting the given equation $y^3 = ax$ to Cartesian co-ordinates by

$$\begin{cases} X = x-y(dy/dx) \\ Y = y\sqrt{[1-(dy/dx)^2]} \end{cases}$$

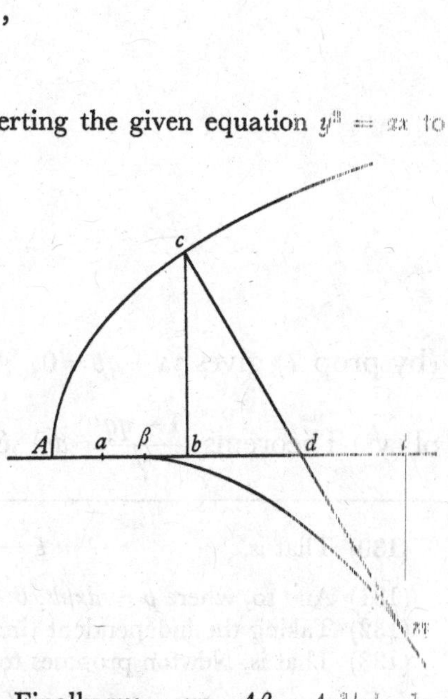

(note (121) above), we find, where $ab = X$, $bc = Y$, that $Y^2 = a(X+\frac{1}{4}a)$ is the defining equation of the curve (c)—that is, that the curve (c) is the parabola of Example 1 of the previous problem, apart from a horizontal translation through $\frac{1}{4}a$ of the origin a. It follows at once that its evolute (m) is the same semicubic parabola $z^3 = \frac{27}{16}av^2$ as before. The point β is given as the meet of (m) with the axis abd and so as the point where d and l are coincident—that is, such that

$$dl = (-a^2+4y^2)/2a = 0,\quad\text{or}\quad y = \beta A = \tfrac{1}{2}a$$

and $x = a\beta = \frac{1}{4}a$. Hence $\beta l = z = \frac{3}{4}(4y^2-a^2)/a$ and $lm = v = [(4y^2-a^2)^{\frac{3}{2}}]/2a^2$, from which the defining equation of (m) may be derived directly by eliminating y. Finally $m\gamma = cm - A\beta = 4y^3/a^2-\frac{1}{2}a$, from which Newton's alternative form is found by substituting $y = \frac{1}{6}\sqrt{[9a^2+12az]}$.

*Example y^e^ 2^d^*. If $x=ad\perp dk=\frac{aa}{x}$, is y^e^ nature of y^e^ crooked line (y^e^ Hyperbola) *gkw*: And I would find other crooked lines (βm) whose lengths may bee compared wth y^e^ area *sdkg* (calling $as=b\perp gs$) of that crooked line *gkw*: I call y^t^ area *sdkg*(130) $=\xi$, & its motion θ.(131) Now since $1:dk::p:\theta$, (see prob 5. & y^e^ Note on prop 7 in Example 3); Therefore $dk\times p=dk$(132) $=\theta=\frac{aa}{x}$. That being knowne I take at pleasure any Equation, in wch ξ is, for y^e^ valor of $cd=y$.(133)

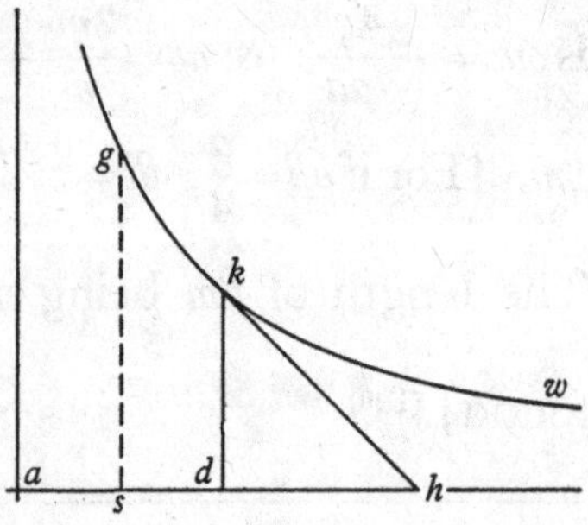

As 1st suppose $ay=\xi$,(134) That (by prop 7) gives $aq=\theta=\frac{aa}{x}$, Or $qx=a$; wch also

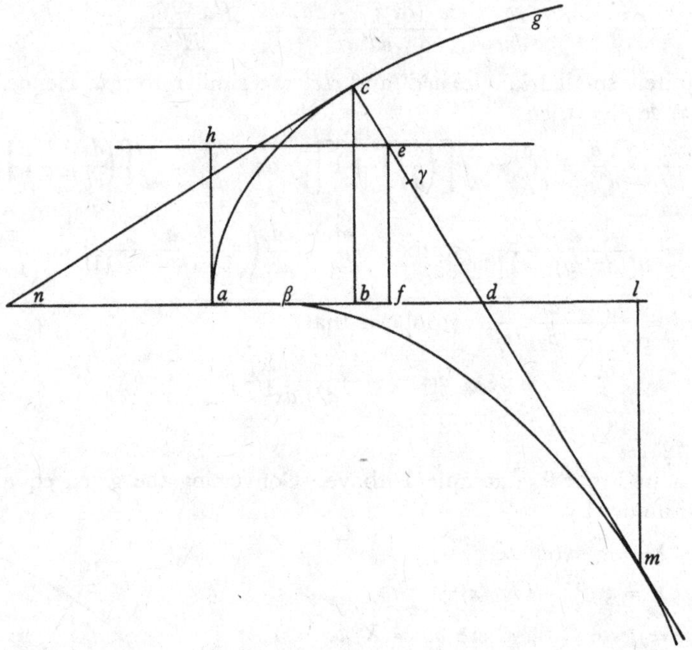

(by prop 7) gives $\gamma x+qp=0$. Which valors of $q=\frac{a}{x}$, & $\gamma=\frac{-q}{x}=\frac{-a}{xx}$; by helpe of y^e^ Theorems $\frac{1-qq}{\gamma}=dm$ &c: doe give $\frac{xx-aa}{-a}=dm$.(135) $\frac{xx-aa}{x}=dl$. &

(130) That is, $$\xi=\int_b^x \frac{a^2}{x}.dx=a^2\log\left(\frac{x}{b}\right).$$

(131) And so, where $p=dx/dt$, $\theta=d\xi/dt=a^2p/x$, or $\theta/p=d\xi/dx=a^2/x=dk$.

(132) Taking the independent time variable $t=x$, so that $p=1$ and $\theta=a^2/x$.

(133) That is, Newton proposes to consider examples of the function $$y=f\left(\int_b^x \frac{a^2}{x}.dx\right)=f\left(a^2\log\left(\frac{x}{b}\right)\right).$$

$\frac{xx-aa}{-ax}\sqrt{xx-aa}=lm.$ wch determines y^e^ nature of y^e^ required curve line βm. The length of y^t^ portion of it wch is intercepted twixt y^e^ point m & the curve line acg being $-cm=cd+dm$(135) $=\frac{\xi+xx-aa}{-a}$, Or $mc=\frac{\xi+xx-aa}{a}$.(136)

☞ (Note y^t^ in this case although y^e^ area $sdkg=\xi$ cannot bee Geometrically found & therefore y^e^ line acg is a Mechanicall one yet y^e^ desired line βm is a Geometricall one. And y^e^ like will happen in all other such like cases, when in y^e^ Equation taken at pleasure to expresse y^e^ relation twixt x, y, & ξ; neither x, y, nor ξ doe multiply or divide one another, nor it selfe, nor is in any denominator or roote, except x wch may multiply it selfe & bee in denominators & rootes, when y or ξ are not in those fractions or rootes. & herein onely doth this excell the precedent 9th probleme. Such is this Equation

$$a\xi-aby+ax^3+\frac{axx}{ax-xx}-5xx\sqrt{ab-xx}=0. \quad \&c.$$

But not this $\xi\xi=a^3y$. nor $\xi=xy$. &c.)(137)

Secondly suppose $\xi=xy$.(138) y^t^ (by prop 7) gives $\theta=y+xq=\frac{aa}{x}$; Or $xy+xxq=aa$, & y^t^ (by prop 7) gives $y+qx+2qx+\gamma xx=0$. Which two valors of $\frac{aa-xy}{xx}=q$, And $\frac{y+3qx}{-xx}=\gamma$; by meanes of y^e^ Theoremes $\frac{1-qq}{\gamma}=dm$, $\frac{q^3-q}{\gamma}=dl$, & $\frac{1-qq}{\gamma}\sqrt{1-qq}=lm$; doe determine y^e^ nature & length of y^e^ desired curve line βm.

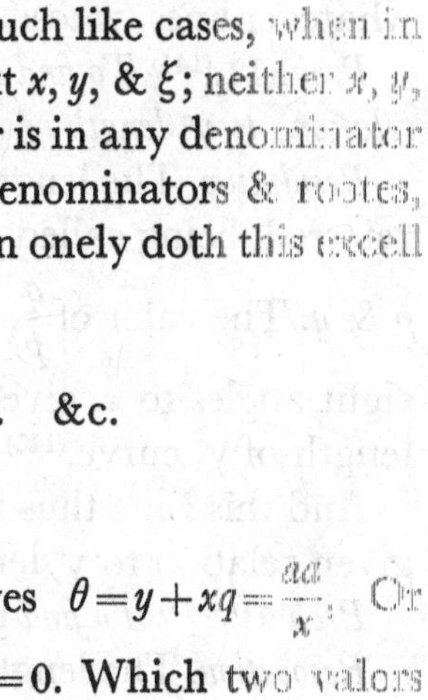

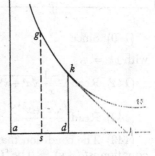

Example y^e^ 3^d^. In like manner to find curve lines whose lengths may bee compared to y^e^ length gk of y^e^ said curve line (Hyperbola) gkw. Call $gk=\xi$, & its motion θ. Now, drawing kh y^e^ tangent to gkw at k, I consider that $ad=x$ & $gk=\xi$(139) doe increase in y^e^ proportion of dk to kh;

(134) Or $y = a\log(x/b)$.
(135) Strictly, $+(-dm)$.
(136) Note that the position of β is fixed by the condition that dl be zero, or $a\beta = x = a$.
(137) Here $\theta = d\xi/dx$ is supposed algebraic (in the present example

$$\xi = a^2\log(b/x) \quad \text{so that} \quad \theta = a^2/x)$$

and the general condition for the evolute (m) to be 'Geometricall' is that dy/dx, and hence d^2y/dx^2, be algebraic.
(138) Or $y = (a^2/x)\log(x/b)$.
(139) That is $$\xi = \int_b^x \sqrt{\left[1+\left(-\frac{a^2}{x^2}\right)^2\right]}.dx.$$

y[t] is, $dh:kh::p:\theta$. Now finding (by prob 1) y[t] $dh=-x$,(140) & $kh=\frac{-a}{x}\sqrt{aa+xx}$,(141) therefore is $\frac{kh\times p}{dh}=\frac{kh}{dh}=\frac{a\sqrt{aa+xx}}{xx}$(142) $=\theta$. Which being found, I take any equation, in w[ch] ξ is, for the valor of $cd=y$. & y[n] worke as in y[e] precedent Example.

Note y[t] by this or y[e] Ninth Probleme may bee gathered a Catalogue of whatever lines, whose lengths can bee Geometrically found.(143)

Prob. 11.(144) *To find curve lines whose Areas shall bee Equall* (*or have any other given relation*) *to y[e] length of any given Curve line drawn into a given right line.*

Resolution. The length of any streight line, to w[ch] y[e] given curve line is cheifely referred, being called x, y[e] length of y[e] curve line y, & their motions of increase p & q. The valor of $\frac{q}{p}$, (found by y[e] first probleme) being ordinately applyed at right angles to x, gives y[e] nature of a curve line whose area is equall to (y) y[e] length of y[e] curve.(145)

And this Line thus found gives (by prob 6) other lines whose areas have any given relation to y[e] length (y) of y[e] given curve line.(146)

Prob 12.(147) *To find y[e] Length of any given crooked line, when it may bee.*

Resolution. The length of any streight line to w[ch] y[e] curve line is cheifly related being called (x), y[e] length of y[e] curve line (y), & theire motions (p & q) first

---

(140) Since
$$\frac{dk}{dh}=\frac{d}{dx}\left(\frac{a^2}{x}\right)=\frac{y}{-x}$$
with $dk = y$.

(141) Read '$\frac{-1}{x}\sqrt{a^4+x^4}$', since $kh = -\sqrt{[(a^2/x)^2+(-x)^2]}$.

(142) Read '$\frac{\sqrt{a^4+x^4}}{xx}$'.

(143) The condition that the length of a general arc of a curve whose Cartesian defining equation is $f(x, y) = 0$ be 'Geometrically found' is that the integral of $\sqrt{[1+(dy/dx)^2]}$ be exact and algebraic, and hence each rectification of the general arc of the evolute (m) by the technique developed in Prob. 9 may be reduced to the exact integration of a corresponding rational algebraic form. Clearly, if we were to tabulate such rectifications systematically we would derive a Catalogue of exact integrals similar to those in Prop. 8 above. Similar considerations hold for the rectification technique in the present Prob. 10.

(144) This was first entered by Newton after the following problem, but is here set before it to correspond with Newton's renumbering.

(145) Where $p = dx/dt$ and $q = dy/dt$, then the arc-length
$$y=\int\frac{q}{p}.dx,$$
where x is the length of the corresponding abscissa. Then, where $z = q/p = \phi(x)$, he finds the numerical value of the arc-length y as the area $\int z.dx$ under the curve $z = \phi(x) = dy/dx$.

(by prob 1) get an equation expressing y^e^ relation twixt x & $\frac{q}{p}$, & y^n^ seeke y^e^ valor of (y) by y^e^ Eight proposition. (Or find a curve line whose area is equall to y^e^ length of y^e^ given line, by Prob 11. And then find that area by Prob 7.)(148)

Prob 13. *To find y^e^ nature of a Crooked line whose length is expressed by any given Equation, (when it may bee done).*

Resolution. Suppose $ab=x$, $bc=y$, $ac=z$. & their motions p, q, r. And let y^e^ relation twixt x & z bee supposed given. Then (by prop 8) finding the relation twixt p & r make $\sqrt{rr-pp}=q$: (For drawing cd tangent to ac at c & $de\perp cb\perp ab$: y^e^ lines de, ec, dc, shall bee as p, q, r. but $\sqrt{dc\times dc-de\times de}=ec$, & therefore $\sqrt{rr-pp}=q$). Lastly, y^e^ ratio twixt x & $\frac{q}{p}$ being thus knowne, seeke y (by prop 8). Which relation twixt $ab=x$ & $bc=y$ determines y^e^ nature of y^e^ crooked line $ac=z$.(149)

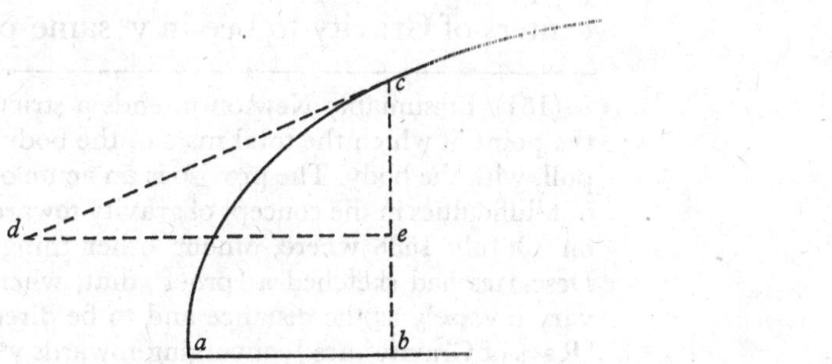

[3](150) *Of Gravity.*

Definit^n^. 1. I call y^t^ point y^e^ center of Motion in any Body, wch always rests when or howsoever y^t^ Body circulates w^th^out progressive motion. It would

(146) The remainder of the manuscript page is left blank, presumably for the future insertion of examples.

(147) Originally 'Prob 11'. (See note (144) above.)

(148) Where the ordinate z corresponds to the arc-length y and abscissa x of the curve, with their corresponding fluxions

$$r = dz/dt,\ q = dy/dt \quad\text{and}\quad p = dx/dt,$$

then $q^2 = p^2+r^2$ and so $\quad y = \int \frac{q}{p}.dx = \int \sqrt{\left[1+\frac{r^2}{p^2}\right]}.dx.$

(As before the rest of the manuscript page is left blank, probably for the insertion of examples.)

(149) The inverse of the preceding problem 12. Since the relation between x and z is given, then so is the ratio of their fluxions $p = dx/dt$ and $r = dz/dt$, so that, where $q = dy/dt$ is the fluxion of the ordinate y, $p^2+q^2 = r^2$ and finally

$$y = \int \frac{q}{p}.dx = \int \sqrt{\left[\frac{r^2}{p^2}-1\right]}.dx.$$

(Yet again the remainder of the page is left blank presumably for future insertion of examples.)

(150) Add. 3958.2: 62^r^–63^v^. This last section is virtually an appendix to the main tract, though the numbering of the problems is in unbroken sequence from [2].

always bee y^e^ same wth y^e^ center of Gravity were ye Rays of Gravity parallel & not converging towards ye center of ye Earth.(151)

Def: 2. And ye right lines passing through y^t^ point I call ye axes of Motion or Gravity.

Lemma 1. The place & distance of Bodys is determined by their centers of Gravity. Which is ye middle point of a right line[,] circle or Parallelogram:

Lemma 2. Those weights doe equiponderate whose quantitys are reciprocally proportionall to their distances from the common axis of Gravity, supposing their centers of Gravity to bee in ye same plaine wth y^t^ common axis of Gravity.(152)

(151) Presumably Newton intends a strict definition of the centre of gravity of a body as the point at which the total mass of the body 'equiponderates', under the given gravitational pull, with the body. The proviso is an acute observation, but Newton was not the first to point out difficulties in the concept of gravity towards a finite centre of force. In a letter to Mersenne on 13 July 1638 where, among other things, he discoursed on the principles of the lever, Descartes had sketched a 'proof' that, where gravity is supposed (following Beaugrand) to vary inversely as the distance and to be directed to a finite point-centre of force (so that the 'Rays of Gravity' are 'converging towards y^e^ center'), the centre of gravity of a loaded beam must move in relative position as the beam rotates round its centre. (See Descartes–Mersenne, 13 July 1638 = *Œuvres*, **2** (1898): No. CXXIX: 235 ff.: 3 *Exemple. Du Levier.*) Mersenne included extracts from the letter in Latin translation in his *Cogitata Physico-Mathematica* (Paris, 1644), but the original letter was first printed by Clerselier in his *Lettres de M. Descartes*, **1** (Paris, 1657): Lettre LXXIII, of which an English version was inserted by Thomas Salusbury in his *Mathematical Collections and Translations*, **2**, 1 (London, 1665), while the section relating to machines was republished in the posthumous *Traité de la Mechanique, composé par Monsieur Descartes* (Paris, 1668) as the *Explication des Machines et Engins, par l'ayde desquels on peut avec une petite force, lever un fardeau fort pesant.* (See especially 12–15: *Le Levier.*) Already in the middle 1650's, however, a manuscript of Descartes' text was circulating in Holland and his proof that the centre of gravity in a rotating body is not at rest had aroused a considerable degree of attention. In particular, Johann Hudde composed a valuable clarification of Descartes' remarks and this was duly printed in 1657 by Schooten with the introduction:

'Gravitatis centrum in unoquoque corpore non fixum existere, ex proprio principio ante complures annos deduxit Vir illustris Renatus des Cartes in tractatu suo de Mechanica, ubi naturam Vectis persequitur. Quoniam autem eo nondum edito centri hujus naturam cuivis ex illo perspicere haud obvium est, visum fuit hoc loco ejusdem centri explicationem, quâ natura ejus perfacilè ab unoquoque intelligatur in medium afferre; Qualem eam ingeniosissimus ac sæpiùs laudatus D. Joh. Huddenius Ger. Fil. excogitavit, mihique illius participem fecit, statuens cum D. des-Cartes gravitatem in corpore, quod grave dicitur, nil esse præter conatum aut vim descendendi versùs Terræ centrum, qui per aliorum corporum motum causetur.'

(Schooten: *Exercitationes Mathematicæ*: Liber v, Sectio XXX, *Quemadmodum centrum gravitis in magnitudinibus mobile sit intelligendum*: 515–16, especially 515.) It is difficult to believe that Newton, who as we have seen (**1**, 1, §3) had made elaborate notes on earlier *Sectiones* in the same Liber v, overlooked this provocative fragment of Descartes' thought. However, having pointed out that the centre of gravity to a finite point may not in general be identified with 'ye center of Motion' (though, implicitly, for any position of the body it is always a unique point), Newton in the following problems equates the two and hence supposes the point-source at infinity, the radiating 'Rays of gravity' to which may be considered parallel. (See also the Abbé Varignon's 'Sur le Centre de Gravité Des Corps Sphériques', *Memoires de l'Academie Royale des Sciences. Depuis 1666; jusqu'à 1699*, **10** (Paris, $_2$1730): 722–7.

Prob [14_1](153) *To find y^e center of Gravity in rectilinear plaine figures.*

1. In y^e Triangle *acd* make $ab=bc$, & $cf=fd$. & draw *db*, & *af*, their intersection point (*e*) is its center of Gravity.(154)

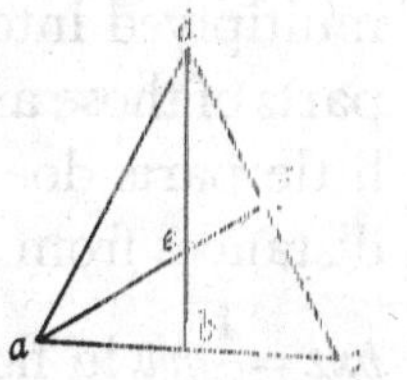

2. In y^e Trapezium *abdc*, draw *ad* & *cb*. Joyne y^e centers of Gravity *e* & *h*, *f* & *g* of y^e opposite triangles *acb* & *dcb*, *bad* & *adc* wth y^e lines *eh*, *fg*. Their intersection point *n* is y^e center of Gravity in y^e Trapezium. (And so of Pentagons, hexagons &c.)(155)

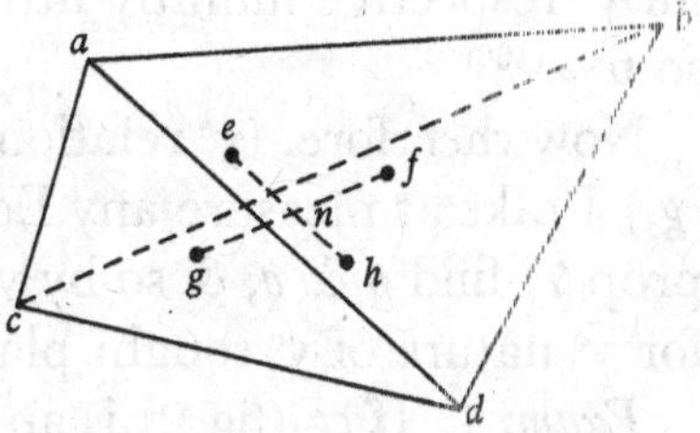

Prob: [14_2.] *To find such plaine figures wch are equiponderate to any given plaine figure in respect of an axis of Gravity in any given position.*(156)

Resol. That y^e natures & positions of y^e given curvilinear plaine (*gbc*) & sought plaine (*lde*) bee such y^t they may equiponderate in respect of y^e axis (*ak*;) I suppose $x=ab\perp bc=z$, & $y=ad\perp de=v$ to bee either perpendicular or parallel or coincident to y^e said axis *ak*: And y^e motions whereby *x* & *y* doe increase or decrease (i:e: y^e motions of *bc*, & *de* to or from y^e

fig 1.

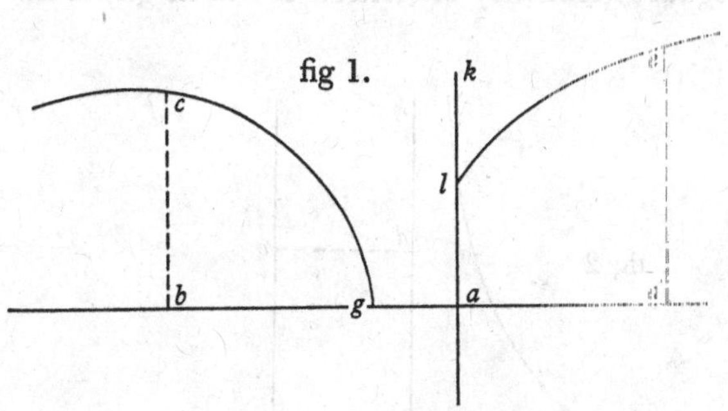

(152) These lemmas, summarizing the basis of the elementary theory of the centre of gravity derive ultimately from Archimedes' tract *On the equilibrium of planes* (see E. J. Dijksterhuis: *Archimedes* (Copenhagen, 1956): 286–313, 346–60) but more directly probably from one or other of the works relating to the finding of the centre of gravity of mathematical figures which it inspired in Europe from the middle of the sixteenth century. (We have seen that Newton had, in the winter of 1664/5, noted the sophisticated theory of the centre of gravity which Wallis made use of in his printed correspondence in 1657: see **1**, 3, §3.4.) Lemma 2, in particular, is Archimedes' celebrated postulate of equilibrium, stated most generally in the case where the two balancing point-centres are in line with the fulcrum in Proposition 7 of his Book 1.

(153) This and the following problem are left unnumbered by Newton, but room is left (between problems 13 and 15) for a single 'Prob 14' only.

(154) Proposition 13 of Archimedes' Book 1 (note (152) above) = Dijksterhuis: 309–11.

(155) A generalization of Propositions 10 and 15 of Archimedes' Book 1 (note (152) above) = Dijksterhuis (note (152) above): 308, 312/13.

(156) Compare his earlier annotations on pp. 52–6 of Wallis' *Commercium Epistolicum* (1658) = **1**, 3, §3.4.

point a) I call p & q.(157) Now y[e] ordinatly applyed lines $bc=z$, & $de=v$, multiplyed into their motions p & q (y[t] is, pz & qv) may signify y[e] infinitly little parts of those areas (acb,(158) & lde) w[ch] each moment they describe; w[ch] infinitely little parts doe equiponderate (by Lemma 1 & 2), if they multiplied by their distances from y[e] axis ak doe make equall products. (y[t] is; $pxz=qyv$, in fig 1: $pxz=\frac{1}{2}qvv$ in fig 2: $\frac{1}{2}pzz=qv\times fm$, in fig 3; supposing $dm=me$. &c.).(159) And if all y[e] respective infinitly little parts doe equiponderate y[e] superficies must do so too.(160)

Now therefore, (y[e] relation of x & z being given by y[e] nature of y[e] curve line cg,) I take at pleasure any Equation for y[e] relation twixt x & y, & thereby (by prop 7) find p & q, & so by y[e] precedent Theorem find y[e] relation twixt y & v, for y[e] nature of y[e] sought plaine(161) lde.

Exam: 1. If cg (fig 2) is an Hyperbola, soe y[t] $aa=xz$. & I suppose $2x=y$. y[n] is $2p=q$ (prop 7). & $paa=pxz=\frac{1}{2}qvv=pvv$. Or $aa=vv$. Or $a=v=de$. Soe y[t] le is a streight line, & lde a parallelogram. w[ch] equiponderates wth y[e] Hyperbola $cgkabc$ (infinitly extended towards gk) if $2ab=ad$. $al\times al=ab\times bc$.

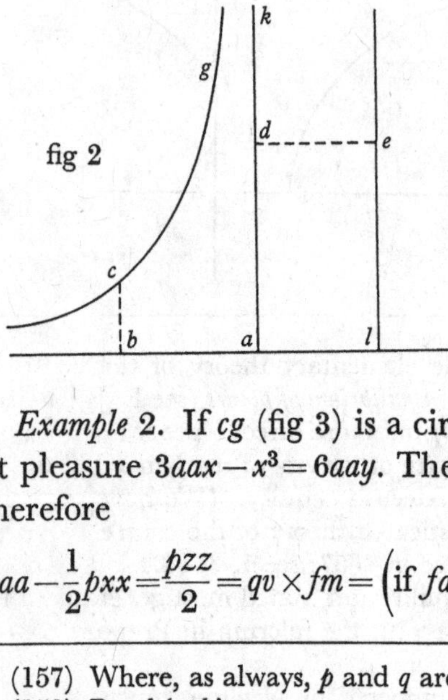

fig 2

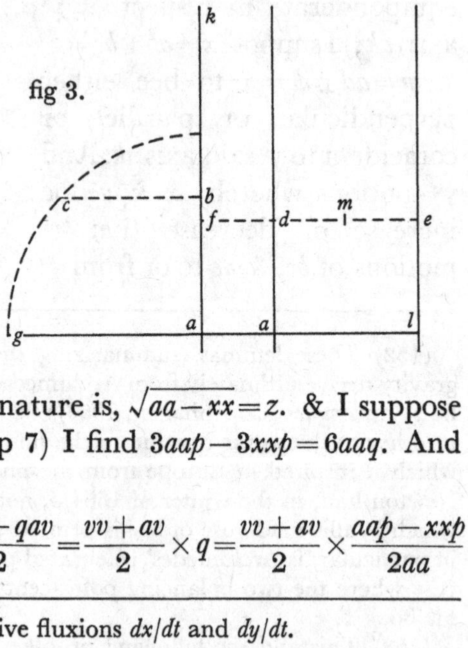

fig 3.

Example 2. If cg (fig 3) is a circle whose nature is, $\sqrt{aa-xx}=z$. & I suppose at pleasure $3aax-x^3=6aay$. Then (by prop 7) I find $3aap-3xxp=6aaq$. And therefore

$$\frac{1}{2}paa-\frac{1}{2}pxx=\frac{pzz}{2}=qv\times fm=\left(\text{if } fd=\frac{1}{2}a\right)\frac{qvv+qav}{2}=\frac{vv+av}{2}\times q=\frac{vv+av}{2}\times\frac{aap-xxp}{2aa}.$$

(157) Where, as always, p and q are the respective fluxions dx/dt and dy/dt.

(158) Read 'gcb'.

(159) These imply by the axiom of integration (note (160) below) the necessary conditions for balance

$$\int xz.dx=\int yv.dy,\quad \int xz.dx=\frac{1}{2}\int v^2.dy \quad\text{and}\quad \frac{1}{2}\int z^2.dx=\int v.fm.dy$$

respectively in figures 1, 2 and 3.

Or $aa\times\overline{paa-pxx}=\frac{vv+av}{2}\times\overline{aap-xxp}$. Or $2aa=vv+av$. Or

$$v=-\frac{1}{2}a\pm\sqrt{\frac{9aa}{4}}=a;^{(162)}$$

Soe that *le* is a streight line & *alde* is a parallelogram.

Example 3. If *abcg* is a parallelogram (fig 4) whose nature is $a=z$. & I suppose at pleasure $x=yy-b$. Then (by prop 7) tis

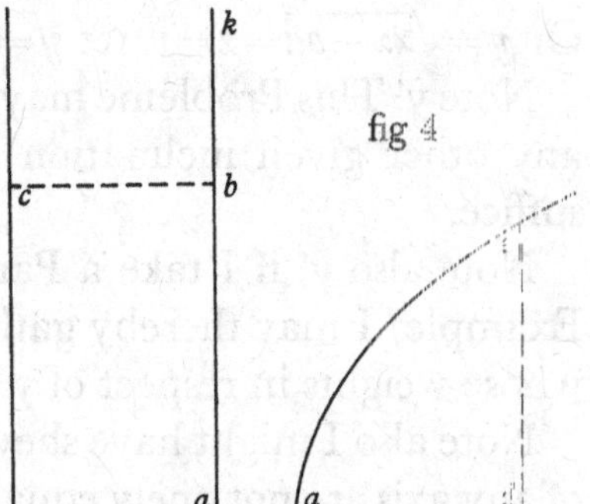

fig 4

$p=2qy$. Therefore $\frac{1}{2}aap=aaqy=\frac{1}{2}pzz=qvy$.

Or $aa=v$. Soe y[t] *aed* is a parallelogram.

Or if I suppose at pleasure, $x=y^3-b$. Then is (prop 7) $p=3qyy$. & therefore

$$\frac{1}{2}aap=\frac{3}{2}aaqyy=\frac{1}{2}pzz=qvy.$$

Or $3aay=2v$. soe that *aed* is a triangle.

Of if I suppose $x=y^4-b$. y[n] is, $p=4qy^3$ & $2aayy=v$. so that *aed* is a Parabola. (Soe if $xx=y^5$, y[n] is $2px=5qy^4$. & $\frac{5a^2qy\sqrt{y}}{4}=\frac{5aaqy^4}{4x}=\frac{1}{2}aap=\frac{1}{2}pzz=qvy$. Or

$$5aaqy\sqrt{y}=4qvy. \quad \& \quad 25a^4y=16vv.$$

soe y[t] *aed* is a Parabola.)

Example 4. If *gbc* (fig 1) is an Hyperbola whose nature is $xx-aa=zz$. & I suppose $x=y+b$. Then (by prop 7) is $p=q$. Therefore

$$px\sqrt{xx-aa}=pxz=qyv=pyv.$$

fig 1.

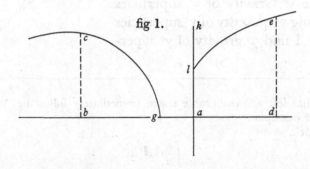

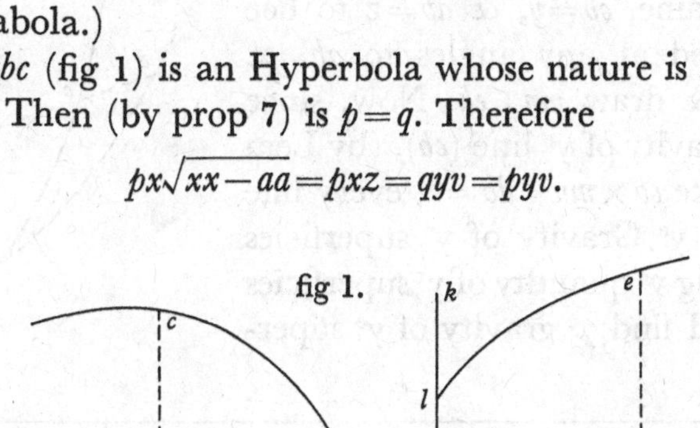

(160) A basic axiom which allows the introduction of integration into the theory. Though Newton here makes no attempt at its justification, its validity depends on the sum of the 'infinitely little [equiponderating] parts' tending to a unique limit as a definite integral.

(161) Newton has cancelled 'area'.

(162) Or $-2a$ if we were to take the sign negative.

Or $\overline{y+b}\times\sqrt{yy+2by+bb-aa}=x\sqrt{xx-aa}=yv$.

Or $\overline{yy+2by+bb}$ in $\sqrt{yy+2by+bb-aa}=yyvv$. &c.

Or if I suppose $xx=2y$. Then is $2px=2q$. Therefore $q\sqrt{xx-aa}=pxz=qyv$. Or $(xx-aa=)2y-aa=yyvv$.

Or if I suppose $xx=yy+aa$. Then (prop 7) is $2px=2qy$. Therefore

$$qyz=pxz=qyv.$$

Or $y=\sqrt{xx-aa}=z=v$. & $y=v$; so y[t] *aed* is a triangle.(163)

Note y[t] This Probleme may bee resolved although the lines x, z, y, v, & ak have any other given inclination one to another, but the p[r]cedent(164) cases may suffice.

Note also y[t] if I take a Parallelogram for y[e] knowne superficies (as in y[e] 3[d] Example) I may thereby gather a Catalogue of all such curvilinear superficies whose weights in respect of y[e] axis, may bee knowne.

Note also I might have shewn how to find such lines whose weights in respect of any axis are not onely equall but have also any other given proportion one to another. And y[n] have made two Problems instead of this, as I did in Probl: 5 & 6; 9 & 10.

Prob 15. *To find y[e] Gravity of any given plaine in respect of any given axis, given in position, when it may bee done.*

Resol: Suppose *ek* to bee y[e] Axis of Gravity, *acb* the given plaine, $cb=y$, & $db=z$ to bee ordinatly applyed at any angles to $ab=x$. Bisect *cb* at *m* & draw $mn\perp ek$. Now, since $(cb\times mn)$ is y[e] gravity of y[e] line (cb), (by Lem 1 & 2); if I make $cb\times mn=db=z$, every line *bd* shall designe y[e] Gravity of y[e] superficies *acb*. Soe y[t] finding y[e] quantity of y[t] superficies *adb* (by prob 7) I find y[e] gravity of y[e] superficies *acb*.

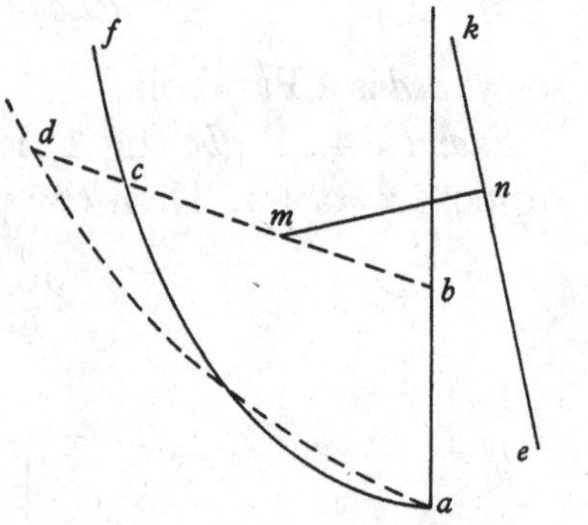

(163) Newton has left a considerable space immediately following, presumably for the insertion of further examples.

(164) 'precedent'.

(165) That is $$\int\frac{e}{d}r^{\frac{1}{2}}x^{\frac{3}{2}}.dx.$$

(166) Or $\frac{e}{d}\int z.dx$, since $dh:db = bn:ba = e:d$.

(167) Newton breaks off his exposition though a space is left for future completion of the example. The details as given are too vague for a restoration of his intended text, but presumably the defining equation of the circle will be taken in the simple form $xx+yy = 2ax$, where $ab = x$, $bc = y$ and the angle $\widehat{abc}$ must be taken right.

Example 1. If ac is a Parabola; soe y^t, $rx=yy$, & y^e axis ak is $\parallel$ to dcb. &, $nb \perp ak$, & y^t, $ab:nb::d:e$. Then is

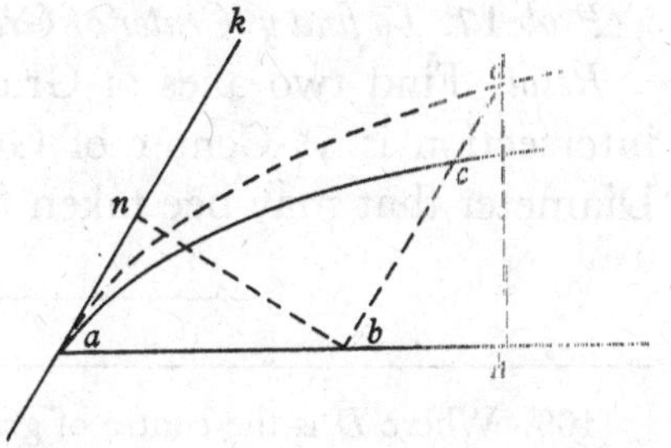

$$bc \times nb = y \times \frac{e \times ab}{d} = \frac{ex}{d}\sqrt{rx} = z.$$

Or $eerx^3 = ddzz$, is y^e nature of y^e curve line ad. whose area (were abd a right angle would be $\frac{2e}{5d}x^{\frac{5}{2}}r^{\frac{1}{2}}$(165) $= \frac{2e}{5d}\sqrt{rx^5}$. but now it) is $\frac{2ee}{5dd}\sqrt{rx^5}$,(166) (by prob 7) wch is y^e weight of y^e area acb in respect of y^e axis ak.

Examp: 2. If ac is a Circle(167)

Prob 16. *To find y^e Axes of Gravity of any Plaines.*

Resol. Find y^e quantity of y^e Plaine (by Prob 7) wch call A & y^e quantity of its gravity in respect of any axis (by prob 15) wch call B. & parallell to y^t axis draw a line whose distance from it shall bee $\frac{B}{A}$. That line shall bee an Axis of Gravity of y^e given plain.(168)

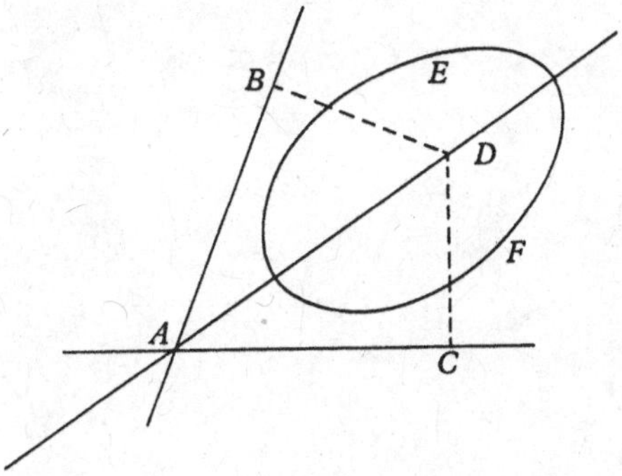

Or If you cannot find y^e quantity of the plane: Then find its gravitys in respect of two divers axes (AB & AC) wch gravitys call C & D. & through (A) y^e intersection of those axes draw a line AD wth this condition y^t y^e distances (DB; DC) of any one of its points (D) from the said axes (AB & AC), bee in such

(168) The 'gravity' of the area $A = \int y.dx$ round the ordinate ($x = 0$) at the origin is the moment of that area, $B = \int xy.dx$, round it, and the condition for $\bar{x} = x$ to be an 'axis of Gravity' of the area is that $0 = \int (x-\bar{x})\,y.dx$, or

$$\bar{x} = \frac{\int xy.dx}{\int y.dx} = \frac{B}{A}.$$

proportion as (C to D) the gravitys of the plane. That line (AD) shall bee an axis of gravity of y[e] said plane EF.(169)

Prob 17. *To find y[e] Center of Gravity of any Plaine, when it may bee.*

Resol. Find two axes of Gravity by the precedent Prop, & their common intersection is y[e] Center of Gravity desired. If y[e] figure have any knowne Diameter that may bee taken for one of it[s] axes of Gravity.(170)

---

(169) Where D is the centre of gravity of the area (EF) and C, D its respective moments round the axes AB and AC, then $(EF) = \frac{C}{BD} = \frac{D}{CD}$ or $BD:CD = C:D$, constant, and hence D will be on the line AD. Immediately, since AD is a straight line through the centre of gravity, by definition it is an axis of gravity. (Note that the 'distances' BD and CD are perpendicular to AB and AC respectively.) A small gap is left for the insertion of examples.

(170) Since a diameter bisects all ordinates at a fixed inclination to it and hence each ordinate must 'balance' on the diameter.

PART 3

MISCELLANEOUS EARLY MATHEMATICAL RESEARCHES (1664-1666)

INTRODUCTION

We need say little in general preface to the miscellany of mathematical items which form the last part of the present volume. With the exception of the pieces in the first section on trigonometry and especially of the second section in which are gathered Newton's earliest researches into the theory, resolution and geometrical construction of equations, no general themes connect them and their content is in general interesting rather than profound. Despite their chaotic nature, however, they give a vivid picture of Newton's early mathematical probings and the considerable extent to which they were still dependent for initial inspiration on the work of others, and of Descartes especially.

All these pieces here receive their first printing, in their original form at least. (It is indeed interesting to notice how many small, insignificant items—the concluding fragment on the length of 'Ellipticall lines' is a perfect example—have crept into some corner or other of Newton's later, more mature work and we can begin to appreciate that characteristic side of his personality which was unwilling to discard anything.) All were to be superseded more or less, if not by an improved full-scale later treatment (as with the researches on equations) at least by their insertion in a logically more stringent form in one or other of Newton's compendia of mathematical examples. That we have them at all is sufficient reason for inserting them in a collected edition, but in many cases their intrinsic interest is considerable. The reader, we are sure, will wish to make up his own mind from a study of the texts themselves.

1

EARLY SCRAPS IN NEWTON'S WASTE BOOK

[Late 1664?]

From the original(1) in the University Library, Cambridge

§1.(2)

[1]

[1]
[1. 1]
1. 2. 1
1. 3. 3. 1
1. 4. 6. 4. 1
1. 5. 10. 10. 5. 1
1. 6. 15. 20. 15. 6. 1
1. 7. 21. 35. 35. 21. 7. 1
1. 8. 28. 56. 70. 56. 28. 8. 1
1. 9. 36. 84. 126. 126. 84. 36. 9. 1
1. 10. 45. 120. 210. 252. 210. 120. 45. 10. 1(3)

[2]

$$\sqrt{aa}\times\sqrt{3:a^3}=\sqrt{6:[a]^{12}}=\sqrt{\quad : \quad}\ ^{(4)}$$

[3]

to find y^e^ proportion of two irrationall rootes
to free y^e^ Numerator or denom[:] from Surde quantitys

$$\frac{a}{b+\sqrt{c}}=x=\frac{a}{b+\sqrt{c}}\times\frac{b-\sqrt{c}}{b-\sqrt{c}}=\frac{ab-a\sqrt{c}}{bb-c}.^{(5)}$$

---

(1) Add. 4004. The present extracts are taken from a group of mathematical papers (mostly printed in Part 2 above) which bear dates from September to December 1664.

(2) Mathematical notes entered on the verso of the flyleaf.

(3) The 'Pascal' triangle of binomial coefficients $\binom{r}{s}$, $r = 0, 1, 2, ..., 10$. Elsewhere Newton refers to the array as 'Oughtreds Analyticall table' (**2**, 6, §3), and here also, we may presume, he has in mind the [*Tabula*] *Posterior* (Oughtred, *Clavis Mathematicæ*, ₃1652: Cap. XII, *De Genesi et Analysi Potestatum*: 37) which lays out $(A+E)^k$, $k = 1, 2, ..., 10$, as a triangular array spreading to the right.

(4) In general, '$\sqrt{aa}\times\sqrt{3:a^3}$' $= a^2 = \sqrt[n]{a^{2n}}$.

(5) Newton rationalizes the (binomial) surd denominator $b+\sqrt{c}$ of the fraction by multiplying it into the conjugate surd $b-\sqrt{c}$. (Compare §3, [2], note (18) below.)

§2.[1]

[1664/1665?]

Problems

1. To find y^e axis, diameters, centers, asymptotes & vertices of lines.[2]
2. To compare their crookednesse wth y^e crookednes of a given circle.[3]
3. To find y^e longest & shortest lines wch can be drawn w^{th}in & perpendicular to the line & to find all such lines [w^{ch}] are perpendicular at both ends to y^e given crooked line.[4]
4. To find where their greatest or least crookednesse is.[5]
5. To find y^e areas, lengths, & centers of gravity of crooked lines when it may be.[6]
6. If y (one undetermined quantity) moves perpendicularly to x (y^e other undetermined quantity[)]. if $s=$a secant[7]$=db$. $v=dc$. $y=bc$. $x=ca$. Then having y^e proportion of v to x to find y, or having y^e proportion of v to y to find x: when it may bee.[8]

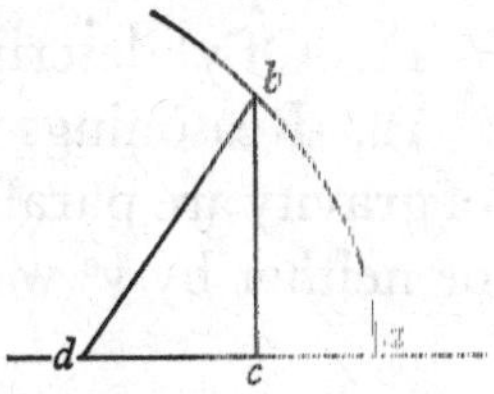

---

(1) A list of problems entered by Newton on f. 4r and probably begun in late 1664. As the variation in ink colour and writing style shows, these entries were not made at a single sitting but rather at various times over a period, and they are here broken up correspondingly into sets. What Newton's intention was in gathering and numbering these problems now seems difficult to know with any certainty. The inference, plausible at first sight, that this list is the outline of a projected general work on mathematics seems denied by the lack of order in its layout and repeated duplication in its topics. More probably, Newton is collecting for his own satisfaction and future attention brief notes on the problems that from time to time possessed his mind and lay there fermenting. However that may be, these notes offer an exciting glimpse of Newton's mind at work, and with little imagination we can see him probing, analysing and deepening the topics as they presented themselves in sequence to him. It is a satisfying comment on the completeness of our present knowledge of Newton's early mathematical researches that for almost all but the most cryptically enunciated problems we can (as we suggest in later notes) trace a corresponding place in the corpus of his extant early papers where a detailed treatment is given. (It is entirely possible, of course, that Newton never returned to those topics which do not appear in more elaborate form in his other papers.) Such a reminder that nothing, apparently, of importance in his early work on mathematics has been lost is heartening.

(2) This enunciation agrees with that on ULC. 4004: 15v ff. (=**2**, 1, §4).

(3) See ULC. Add. 4004: 30v ff. (=**2**, 4, §2.1).

(4) Compare ULC. Add. 4004: 8v (=**2**, 1, §1).

(5) ULC. Add. 4004: 33r/33v, revised in the October 1666 tract's Probl: 4. (See **2**, 4, §2.4 and **2**, 7.2 respectively.)

(6) The last clause is a late insertion. The Problem itself is very fully treated in the October 1666 tract's Prob. 5t–Prob. 17 (ULC. Add. 3958.3: 57r–63v=**2**, 7.2).

(7) That is, normal.

(8) Compare ULC. Add. 4000: 94r–116r (=**2**, 5, §1), revised as ULC. Add. 3960. 12 (=**2**, 5, §2).

7. To reduce all kinds of equations, when it may bee.(9)

8. To find tangents to any crooked lines.(10) Whither Geometricall or Mechanichall.(11)

9. To compare y^e superficies of one line wth y^e area of another & to find y^e centers of gravity twixt two lines or sollids.(12)

10. Haveing ye position(13) wch x must beare to y (as if x is always in ye same line, but y cutteth x at given angles. or if x & y wheeling about 2 poles describe ye line by theire intersection &c) to find theire position in respect of ye line soe [y^t] ye Equation expressing theire relation may bee as simple as may bee (as to find in what lines x is & w^t angles it maketh wth y; or to find ye distance of ye 2 poles & in what line they must be, soe y^t ye relation twixt x & y may bee had in as simple termes as may bee.).(14)

11. Of ye description of lines.(15)

12. Reasonings of gravity & levity upon several suppositions (as y^t ye rays of gravity are parallel or verge towards a center, y^t they are reflected, refracted, or neither by ye weighty body &c[)].(16)

13. Of ye Use of lines.(17)

14.(18) To find such lines whose areas length or centers of gravity may bee found.(19)

15. To compare ye areas, leng[t]hs, gravity of lines when it may bee & to find such lines whose lengths, areas, gravity may be compared.(20)

16. To doe y^e same to sollids in respect of theire areas, content, gravity &c wch was done to lines in respect of their lengths, areas, & gravity.(21)

---

(9) Probably by operating on the roots, as Newton explains at length on ULC. Add. 4004: 55^r–56^r (=3, §4 below).

(10) See ULC. Add. 4004: 47^r ff. (=**2**, 4, §3.1).

(11) This clause is a later addition, probably inserted in November 1665 when Newton wrote his paper on the subject (ULC. 4004: 50^v/51^r=**2**, 6, §2).

(12) Newton cancelled this problem only when he added problem 15 to his list (as a '15' written in alongside in the manuscript confirms).

(13) Newton clearly had some trouble in choosing this word, for he first wrote 'pro-[portion]' and 'respe[ct]'.

(14) This seems to correspond with Newton's treatment of co-ordinate systems in ULC. Add. 4004: 16^v ff. (=**2**, 1, §5).

(15) This enunciation is too brief to allow positive identification with any of Newton's existing papers. Perhaps he intends the problem of constructing a curve by points. (Compare ULC. Add. 4004: 2^v=§4 below.)

(16) This problem, touched on in the early Cartesian paper on mechanics, ULC. Add. 4004: 10^v–15 (not here printed) and referred to again in the opening Definitions of the concluding section 'Of Gravity' in the October 1666 tract, is not developed extensively in any of Newton's known early papers. (But compare **2**, 7.3 above.)

17. Of lines wch lye not in ye same plane as those made by ye intersection of a cone & sphære.[22]

18. Two Equations given to know whither they expresse ye same line or not.[23]

19. Of ye proportion wch ye rootes of an Equation beare to one another.[24]

20. One line being given to find other lines at pleasure of ye same length, area or gravity.[25]

21. How much doth any medium resist y^{e} motion of any given body?[26]

22. To Determin maxima & minima in Equations wch hath more then two unknowne quantitys.

23. To Determin max & min by numbers.[27]

---

(17) Perhaps in the construction of equations. (See ULC. Add. 4004: 67^{v}–69^{r} = 3, §2 below.)

(18) The remaining ten problems are entered in a smaller, blacker writing and we may set their date tentatively as mid-1665.

(19) Compare note (6) above.

(20) The generalization of the previous problem.

(21) This problem is not discussed in any extant early manuscript of Newton's and it is possible that he never persevered with it.

(22) Newton first wrote 'sphæroides' (or spheroid). No analysis of the meet of a cone and sphere appears in his early papers.

(23) That is, the general problem of linear transforms. (See ULC. Add. 4004: 6^{v}–10^{r}/16^{v}–27^{v} = 2, 1 above.)

(24) See ULC. Add. 4004: 55^{r}–56^{r} = 3, §4 below.

(25) Compare notes (6) and (19).

(26) Newton seems not to have pursued this problem in detail in his early papers, but compare ULC. Add. 4004: 1^{r} (= §3.1 below in part), 10^{v}–15^{v} (note (16)).

(27) These two concluding problems cannot closely be identified with Newton's early expositions of the theory of maxima and minima. Strictly, an equation which has more than two free variables cannot be the Cartesian defining equation of a plane curve and so cannot take on extreme values. However, Newton probably intends an equation in several variables, but only two of which are not tied by further restrictions. The determination of maxima and minima by numbers is perhaps a reference to Hudde's algorithm, widely used by Newton in his early calculus work. (See especially ULC. Add. 4004: 47^{r} ff. = 2, 4, §3.)

§3.[1]

[Late 1664?]

[1][2]

. . . .

If y^{e} body *b* moved in an Ellipsis y^{n} its force in each point (if its motion in y^{t} point bee given) will bee found by a tangent circle of equall crookednesse wth y^{t} point of y^{e} Ellipsis.[3]

. . . .

If *c* circulate in y^{e} circle *cgef*, to whose diamiter *ce* *ad* = *ab* being perpendicular y^{n} will y^{e} body *b* undulate in y^{e} same time y^{t} *c* circulates.[4]

And those bodys circulate in y^{e} same time whose lines drawne from y^{e} center *a* to y^{e} center *d* are equall.[5]

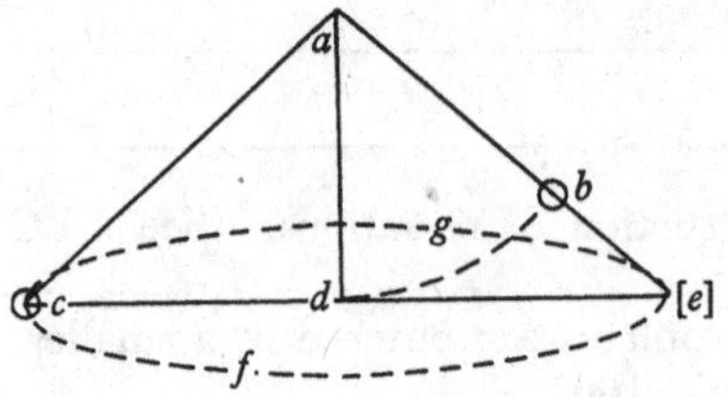

And *ad*:*dc*::force of gravity[6] to y^{e} force of *c* from its center *d*. Coroll: hence may y^{e} fo[rce o]f gravit[y] & y^{e} motion of things falling were they not hindered by y^{e} aire [bee] very exactly found (viz: *cd*:*ad*::force from *d*:force from *a*[)].[7]

---

(1) Miscellaneous short pieces from ff. 1–10.

(2) Extracts from a set of notes (on f. 1^{r}) which show Newton's early ideas on motion, force and gravity. (These notes, apparently the first to be made in the Waste Book and so presumably of late summer 1664, have been briefly described by J. W. Herivel in 'Newton's Discovery of the Law of Centrifugal Force', *Isis*, **51** (1960): 546–53). The present quotations are chosen to illustrate the close interweaving of his mathematical and physical ideas even in this first year (1664) of his creative life.

(3) Newton equates the (centripetal) force at a general point on an ellipse with that towards the same centre instantaneously at rest in the plane of the circle of curvature at the point. This insight, in its immediate extension to a point on an arbitrary curve, allowed Newton to reduce all problems of centripetal forces in arbitrary orbits to the equivalent ones which treat of centripetal attractions in circular paths. (The concept is fundamental in Newton's account of centripetal force in his *Principia*: Liber I.)

(4) That is, Newton equates the periodic time of the simple pendulum formed by the body *b* 'undulating' vertically in the arc *db* with that of the conical pendulum which is the body *c* rotating in the horizontal circle *cgef*. The equality is, of course, true for small angles $\widehat{dab}$ of vibration. (See note (7).) At a later date, perhaps about 1669, Newton entered this observation in a short paper on centrifugal forces as 'Pendulum gyrans et undulans si sint æque

[2][8]

☞ To know whither y^e^ changing of y^e^ sines[9] of an Equation change y^e^ nature of y^e^ crooked line signified by y^t^ Equation Observe y^t^

If ye sines[9] of every other terme (of y^t^ Equation ordered according to either of ye undetermined quantitys) be changed ye nature of ye line is not changed. but if some signes bee changed but not in every other terme (of it ordered according to one of ye unknowne quantitys) y^n^ [y^e^] nature of ye line is changed.[10]

Those lines may bee defined ye same whose natures may be[11] expressed by ye same equation although [y^e^] angles made by x & y are not ye same.[12]

In ye Hyperbola ye area of it beares ye same respect to its Asymptote wth a logarithme dot[h its] number.[13]

---

profunda in eodem tempore redeunt'. (ULC. Add. 3958. 5: 87^v^. This paragraph, omitted by A. R. Hall in the version of Newton's paper which he incorporated in his 'Newton on the calculation of central forces', *Annals of Science*, **13** (1957): 62–71, was first printed by H. W. Turnbull in the *Correspondence of Isaac Newton*, **1** (1959): no. 117: 299—compare also 312/13, note 13.)

(5) For in the conical pendulum the period of circulation of the body c round the perimeter $cgef$ is proportional (note (7) below) to $\sqrt{[ad]}$ and hence is independent of the radius cd of its swing.

(6) Vertically downwards and, by implication, of constant magnitude.

(7) That is, in the conical pendulum the centrifugal force, say C, of the body c outwards from the centre d of its swing is to the constant force g (acting vertically downwards) of simple gravitation as $cd:ad$. Since, by taking the speed V_c of the body c rotating in the circle $cgef$ to be expressed in suitable units, we may suppose the centrifugal force $C = V_c^2/cd$, it follows that the time of one revolution of c round d is $T_c = 2\pi.cd/V_c = 2\pi\sqrt{[cd/C]}$ and hence, by Newton's proportion, $T_c = 2\pi\sqrt{[ad/g]}$. Further, since $2\pi\sqrt{[ad/g]}$ is approximately the time of vibration T_b of the body b in the vertical plane adb through the small angle $c\hat{a}e$, Newton's assertion that $T_c = T_b$ is accurate. As Newton observes, the centrifugal force C of c from d may be directly calculated from the period T_c of its swing and thereby the force of gravity g also: specifically, $g = ad(2\pi/T_c)^2$, where ad and T_c are measured in suitable units.

(8) A set of short problems entered by Newton on f. 7^r^, apparently in late 1664 and, as the quality of the writing suggests, not at a single sitting.

(9) Read 'signes'.

(10) The condition $f(x) \equiv f(-x)$ requires that $f(x)$ contain only even powers of the variable x: $x^p = (-x)^p$ only for p an even integer. Newton has cancelled a following paragraph whose meaning is obscure: 'If y^e^ knowne quãtitys are every where divers. & one of y^m^ be bloted out y^t^ produceth a line, when one terme is already wanting.'

(11) Newton first wrote 'are'.

(12) That is, Newton considers all curves to be similar, which are defined by the same analytical equation with respect to Cartesian co-ordinates invariant in length but whose fixed angle of inclination may be taken arbitrarily.

(13) Where the Cartesian defining equation of the hyperbola is $xy = 1$, then

$$\int y.dx : \int 1.dx = \log x : x.$$

To make y[e] Equation $x^3-ax^2+abx-abc=0$. be divisible by $x-c=0$. suppose $c=x$, y[n] tis $c^3-acc+abc-a[bc=0.]$ or $c=a$. therefore write c in steade of a & it is $x^3-cx^2+cbx-bcc=0$. w[ch] is divisible by $x-c=0$.(14)

To make y[e] same Equation divisible by $xx-2ax+ac=0$ suppose it to bee divided by it & y[e] op[e]ration will bee

$$xx-2ax+ab)\underline{x^3}-\underline{ax^2}+\underline{abx}-\underline{abc}=0(x+a.$$ The quote (15) is

$$\underline{-x^3+2ax^2-abx}$$
$$0+\underline{ax^2}+0-abc$$
$$\underline{-axx+2aax-aab}$$
$$0+2aax-aab-abc$$

$2aax-abc-aab$ w[ch] [was to] have vanished therefore to make it soe suppose each terme $=0$ & there will bee $2aax=0$ & $abc+aba=0$ both w[[ch] con]clude $a=0$. Which since it cannot happen the equations cannot be divided y[e] one by y[e] othe[r].

The rootes of two divers equations may Easily be added to substracted from multiplyed & [divided] by one another while they are unknowne.(16)

That y[e] penultimate terme of y[e] Equation $x^3 * -a^2x+b^3=0$. bee wanting I multiply [y[e] rootes by $\frac{y}{d}$] & then suppose x a knowne quantity & y an unknowne one, as $x^3d^3-aadxyy+b^3y^3[=0.$ y[n]] by making $b=x$. $ad=bb$. it is

$$d^3-ayy+y^3=0.$$

By this having found y, x is $=\frac{db}{y}$.

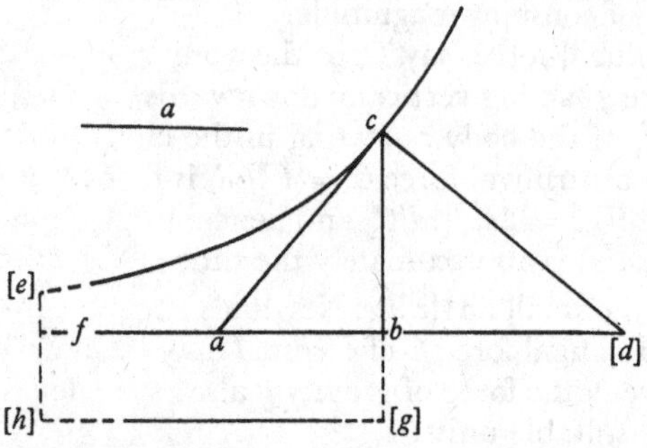

If $fb=x$. $cb=y$. ac a tangent & dc a perpendicular to y[e] crooked line [e]c, soe y[t] $ab=a$, or $bd=\left[\frac{yy}{a}.\ ef=fh=bg=b.\right]$ then $fbce:fbgh::$number$:$logarithme. but y[e] line fc is a mechanicall one.(17)

(14) In fact, $x^3-cx^2+cbx-bc^2 = (x-c)(x^2+bc)$.

(15) 'how many [are left]', that is, the remainder.

(16) Compare *Geometria*: Liber III: 71–2: *Quomodo augeri vel diminui possint Æquationis radices, ipsis non cognitis*; and 75: *Quomodo multiplicari vel dividi possint Æquationis radices, ipsis incognitis.* The topic is developed by Newton in a paper of May 1665 (ULC. Add. 4004: 55[r]–56[r] = 3, §4 below).

(17) The condition that the subtangent $ab = y(dx/dy)$ be equal to a yields, on integration

$$\int_0^x \frac{1}{a}.dx = \int_b^y \frac{1}{y}.dy \quad \text{or} \quad \frac{x}{a} = \log\left(\frac{y}{b}\right),$$

so that the curve (c) is the *logarithmica* and hence 'mechanicall' in Newton's sense. (Compare 2, 6, §1.3.) It follows readily that

$$(fbce):(fbgh) = \int_0^x y.dx\left(\text{or} \int_b^y a.dy\right): \int_0^x b.dx \doteq \left[\frac{y}{b}\right]_b^y : \left[\log\left(\frac{y}{b}\right)\right]_b^y.$$

The denominat[or] of a Fraction may bee ever freed from surde quantitys. ma[ke $b+\surd c]\times b-\surd c=[bb-c]$.[18] (just[19] as in y^e^ denom: [of a]n Equation. [ ][20]

---

[3][21]

If any crooked line be revolved about its owne axis it generates a sollid which intersected by any plaine not perpendicular to y^e^ axis produceth another line $\left\{\begin{array}{l}\text{not more compound y}^{\text{n}}\\ \text{of the same kind with}\end{array}\right\}$ y^e^ former.

But if it bee revolved about any other line it generates a Sollid which intersected by any plaine not perpendicular to y^e^ axis produceth another whose composition is not $\left\{\begin{array}{l}\text{lesse y}^{\text{n}}\text{ equall}\\ \text{more y}^{\text{n}}\text{ double}\end{array}\right\}$ to the former's.[22]

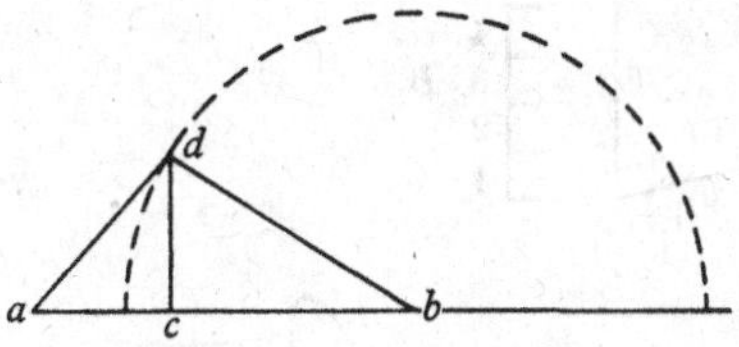

In the $\triangle adb$ if $ab=a$. & $db=b$ are definite,[23] but $ad=v$, & $bc=x$ indefined. Then y^e^ Equation is $bb-vv+aa-2ax=0$.[24] But in this case y^e^ maximū or minimum of either v or x cannot bee found according to Cartes or Huddens method, by reason y^t^ $\left\{\begin{matrix} v \\ x \end{matrix}\right\}$ hath not 2 divers valors when $\left\{\begin{matrix} x \\ v \end{matrix}\right\}$ is determined, wch become equall when $\left\{\begin{matrix} x \\ v \end{matrix}\right\}$ is y^e^ least or greatest y^t^ may be. But cd might here bee used insted of cb &c: There be other instances of this Nature against Huddenius his assertion.[25]

---

(18) Newton rationalizes the denominator of the fraction by multiplying it into the conjugate surd. (The detailed calculation is set out in 1, §1.3.)

(19) Newton has cancelled 'exactly'.

(20) The remainder of the page has crumbled away.

(21) Miscellaneous jottings on f. 9^v^.

(22) These remarks seem generalizations on Newton's observations of the properties of conoids of revolution (such as the spheroids and paraboloids). Their detailed investigation and justification would seem beyond Newton's present analytical power.

(23) That is, determined in length.

(24) Since $db^2-dc^2-cb^2 = 0$ with $dc^2 = b^2-(a-x)^2$. Compare Euclid's *Elements* II, 3.

(25) Hudde, in his *Epistola secunda de Maximis et Minimis* (=*Geometria*: 507–16) certainly claimed (p. 510) that 'hæc quidem generalis mea Methodus est', but Newton's present counter-example, though valid, is not an unqualified refutation of that claim. Implicitly, both Descartes and Hudde restricted their methods for finding maxima and minima to curves defined in a Cartesian co-ordinate system (compare 2, 2: introductory historical note), but in Newton's present example the point d is defined in a unipolar system with co-ordinates $ad = v$ and $bc = x$. The weakness of this new co-ordinate system is that it cannot distinguish between the point d in Newton's figure and its mirror-image in the line ab (both which points may be denoted by the ordered pair (v, x)) and it is this inadequacy which results in the apparent failure of a method which requires that these two points be distinct. In Newton's suggested

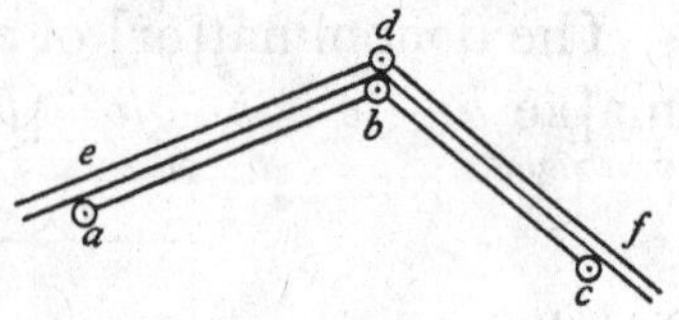

Three points *a*, *b*, *c*, being given a circle may be described (wch shall passe through them all) by an instrument whose angle $edf = \angle abc$. And soe y^e sides *ed* & *df* being moved close by y^e points *a* & *c*, y^e point (*d*) shall describe y^e arch *abc*.[26]

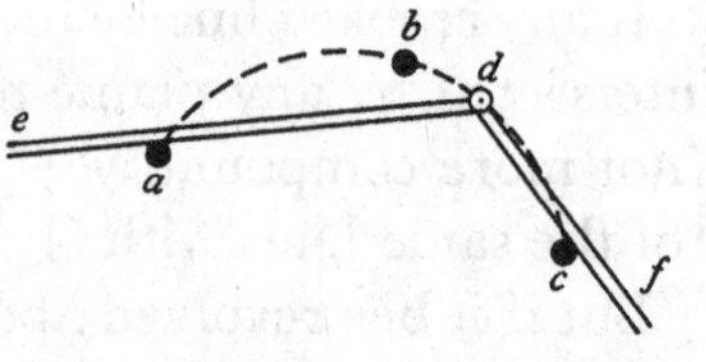

To worke mechanically & exactly by a scale it may bee better divided according to y^e fassion[27] represented by y^e figure *A*, y^n by that at *B*.[28]

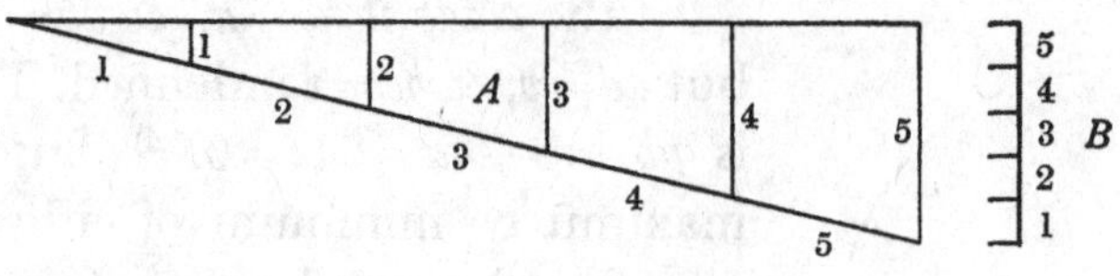

To make a plane superficies exactly: Take three plates *A*, *B*, *D*. & grind them together *A* & *B*, *A* & *D*, *B* & *D*; pressing the uppermost plate onely in the middle at *C* that it may not weare more away in the edges y^n in the midst, & move it to & fro wth but small vibrations. Soe shall the 3 fiduciall sides of y^e 3 plates bee ground exactly plane.[29]

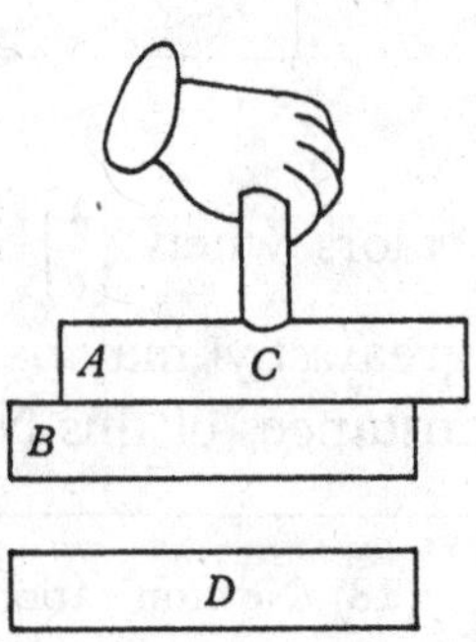

---

revision, where the point *d* is defined by the unipolar system $ad = v$, $dc = y$, we may deduce the defining equation of the circle (*d*) to be

$$\sqrt{[v^2-y^2]}+\sqrt{[b^2-y^2]} = a, \quad \text{or} \quad [(a^2+b^2-v^2)/2a]^2 = b^2-y^2:$$

hence *y* takes on the two distinct values $\pm\sqrt{[b^2-(a^2+b^2-v^2/2a)]}$ for each *v* and their coincidence (at $y = 0$) straightforwardly determines the two points at which *ad* attains its least and greatest values. (The point *d* now becomes confused with its mirror-image in the normal at *a* to *ab*, but this creates no difficulty in the present example since the implicit restrictions $ad+db \geqslant ab$ and $ab+bd \geqslant ad$ limit $ad = v$ to the interval $[a-b, a+b]$.)

On 16 May 1665 Newton inserted a revised form of this objection as an example in a paper on limit-motion (ULC. Add. 4004: 51^v = **2**, 6, §4.2), considering the varied success of Hudde's method in its application to yet further unipolar systems defined, equivalently, by $ad = v$, $ac = z$ and $ad = v$, $dm = y$, where *dm* is the length of the tangent at *d* to the circle between *d* and *ab*.

(26) A simple application of the property that the circle segment *abc* subtends a constant angle $a\hat{b}c = a\hat{d}c$.

[4][30] *Probleme. Of Usury.*

$ab=a=$ the principle. $bc=x=$ the time of y[e] money lent. $ce=\frac{dx}{e}=$ the use[31] due for y[e] principle for y[e] time bc. $de=$ the use upon use. $cd=y=$ y[e] use for y[e] principle & use during y[e] time bc. $ax:\frac{dx}{e}::a+y$ in $fc:y$.[32] df a tangent to y[e] line bd. dg a perpendicular. $cg=v$. Then as y[e] summe (ab) or principle drawne into y[e] time bc, is to ce y[e] use for it in y[t] time so is the sum̄e $ab+dc$ drawne into y[e] time fc, to y[e] use (dc) for it in y[e] time fc. therefore

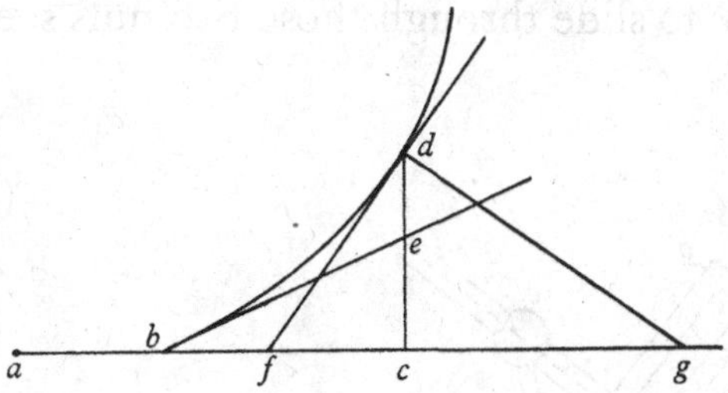

$$ax:\frac{dx}{e}::a+y \text{ in } \left\{\begin{matrix} h \\ fc \end{matrix}\right. :y. \quad axy=\frac{ahdx+yhdx}{e}. \quad \frac{eay}{ad+yd}=h:y::y:v. \quad eav=ady+dyy.$$

$$v=\frac{ady+dyy}{ea}.^{(33)} \quad d=e. \quad \frac{ay+yy}{a}=v.$$

---

(27) Read 'fashion'.

(28) A familiar use of triangulation to refine a meter-reading.

(29) This last paragraph is reproduced for completeness' sake and for its transitivity argument. Since friction between the plates acts in a direction perpendicular to that of the movement the plates will wear each other away in a horizontal plane, but to this argument, theoretically true, Newton adds some necessary practical qualifications.

(30) Extracted from f. 10[r], where it is set at the head of a page otherwise filled with notes (not here reproduced) 'Of Reflections'.

(31) That is, the interest.

(32) This proportion, repeated below, should probably be cancelled.

(33) Newton sets up a geometrical model for instantaneous compound interest. Where $bc = x$ is the period over which the principal $ab = a$ is lent and $dc = y$ the interest due at the end of that time from continuously incremented interest, then $dh = dc+ab = y+a$ is the total money returnable after the time bc and the compound interest rate, expressed as a fraction of the original capital, will be the slope of the tangent df at the point d on the curve of 'use'. It follows that, where be is tangent at b, ec is interest on the original capital and hence $dc-ec = de$ is the 'use upon use'. From these considerations Newton, in effect, derives a differential equation for the curve (d) in the form

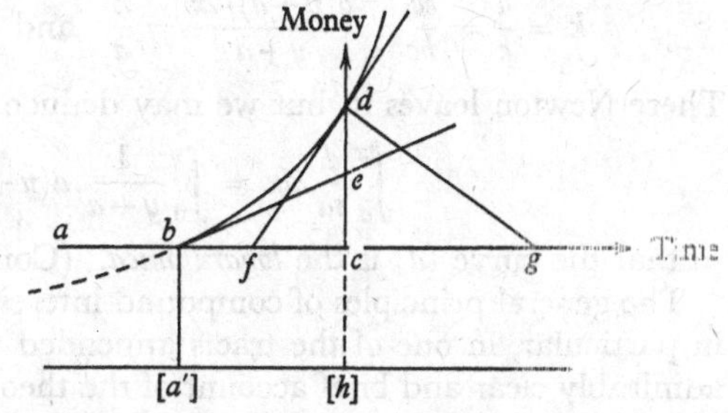

$$\frac{\text{instantaneous rate of interest } (dc/fc)}{\text{instantaneous capital } (dh)} = \text{constant} = \frac{\text{original rate of interest } (ec/bc)}{\text{original capital } (a'b = ab)}$$

[5] [*Construction of a parabolic template.*](34)

Make y^e^ line *ac* to revolve about y^e^ point *a*. on y^e^ end *c* let y^e^ Nut *c* bee fastened so as to turne about its center. make $ab = ac$ & fastend(35) a nother nut at y^e^ point *b* in y^e^ same manner. make y^e^ line *bc* to slide through those two nuts soe

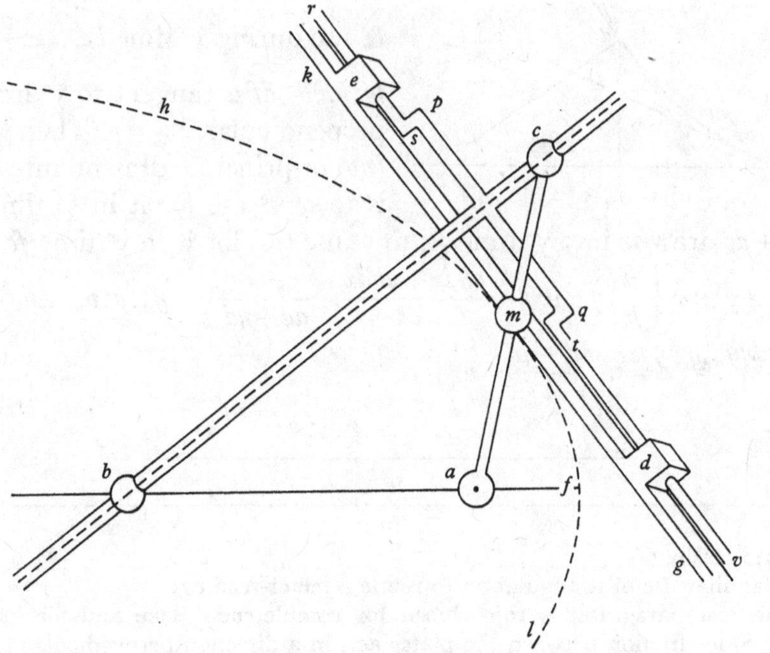

y^t^ y^e^ △*abc* will always be an isosceles. To y^e^ line *cb* fasten y^e^ line *rstv* at right angles. & make y^e^ line *kg* wth 2 nuts *e* & *d* at each end through wch y^e^ lines *rs* & *tv* must slide to keep y^e^ line *kg* perpendicular to *bc*, in y^e^ midst of *kg* fasten y^e^

---

analytically, where

$$k = \frac{d}{e} = \frac{ec}{bc}, \quad \frac{d(y+a)/dx}{y+a} = \frac{k}{a} \quad \text{and therefore} \quad cg = v = y\frac{dy}{dx} = ky\left(\frac{y+a}{a}\right).$$

There Newton leaves it, but we may deduce by straightforward integration that

$$\int_a^x \frac{d}{ea}.dx = \int_0^v \frac{1}{y+a}.d(y+a) \quad \text{or} \quad \frac{kx}{a} = \log\left(\frac{y+a}{a}\right),$$

so that the curve (d) is the *logarithmica*. (Compare note (17) above.)

The general principles of compound interest were widely known at the time and Oughtred in particular, in one of the tracts appended to the later editions of his *Clavis*, had given an admirably clear and brief account of the theory when the increments of interest are added at regular, finite intervals of time. (See his *Clavis Mathematicæ*, Oxford, $_3$1652: (Appendix): 42–4: *De Anatocismo, sive Usura Composita.*) Newton, however, in extending the theory here to interest added over indefinitely small increments of time is being highly original.

(34) Taken from f. 2^v^. The writing style of this piece and its position in the Waste Book suggests autumn 1664 as the probable date of composition.

nutt *m* so as it may turne about its center & y^t ye line *ac* may slide through it[,] then make y^t side of ye line *kg* wch is next *ab* to be a file wch must be very smooth at the point *m* but must grow rougher towards y^e ends *d* & *e*. Then by turneing y^e line *ac* to & from *l* & *h* about its center, & holding y^e file *kg* close to y^e plate *kmflab*, it shall file it into y^e shape of a Parabola.(36)

[6](37) *To describe y^e Parabola by points*

Make $ca=\frac{r}{4}$; *c* y^e vertex; *a* y^e focus; $ab=r$. y^n wth some radius as $ag=ae$, describe y^e circle *ge*: & take $bd=2ga=2ae=2de$ & y^e point *e* shall bee in y^e

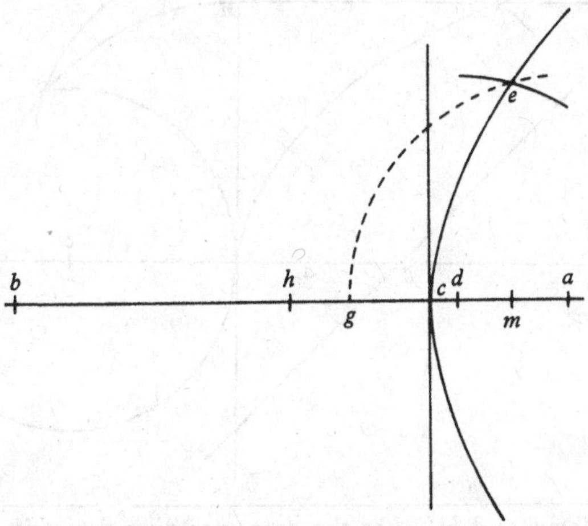

(35) Read 'fasten' simply.

(36) Newton's machine has one degree of movement, since for fixed angle $b\hat{a}c$ the construction allows motion of the line *kmg* parallel to itself. Further, since the tangent at *m* to the parabola *hmfl* (of focus *a*) bisects $a\hat{m}\mu$, where $m\mu \parallel ba$, it is parallel to the bisector of $b\hat{a}c$ and so

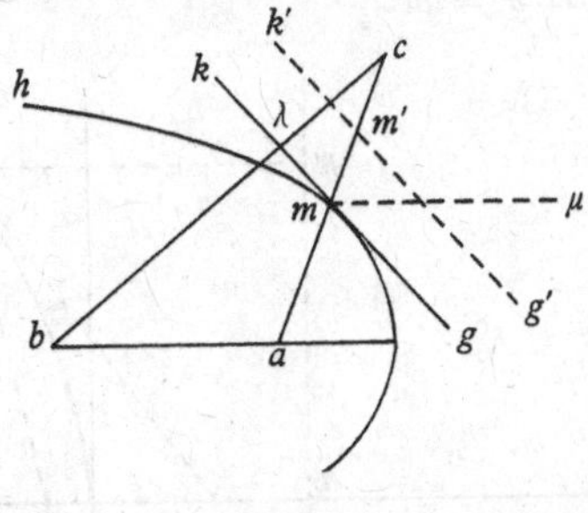

normal to *bc*. Newton's machine, therefore, fixes *kg* to be parallel to the tangent at *m* with *m* itself moving on *ac* and he requires *m* on the file *kmg* to be perfectly smooth in order that the immediate vicinity of *m* shall not be worn away. (In practice, of course, the parabolic template will already be roughly shaped and the machine is needed only to polish the cut.)

(37) Taken from f. 2^v, in immediate sequel to [5].

parabola, also if from e to g, a streight line be drawne it shall touch y^e^ Parab: in e.(38)

Or thus, take $ch=ca=\frac{r}{4}$, $hd=2gc$; or $da=2hg$: & $ga=ae=de$. &c:(39)

Or thus, take $cm=gc$, $dm=ma$; & $ga=ae=de$; &c.(39)

Or thus[,] take $cm=gc$, & raise me a perpendicular to ca, wch shall intersect y^e^ parab, & circle ge in y^e^ same point.(39)

Or thus.(40) make $ab=\frac{r}{4}$. $bd=2bc=r$. $kb=bg$. wth y^e^ Rad bc describe bed. y^e^ Circle(41)

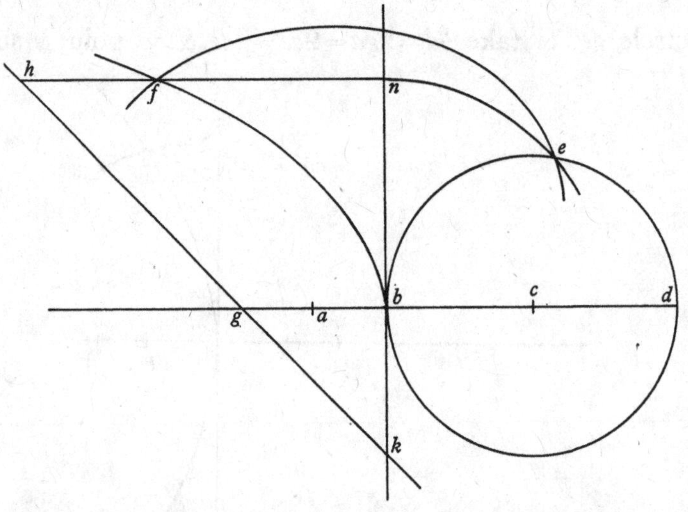

(38) Where a is its focus and $hh' \perp ba$ the corresponding directrix, this description of the parabola (e) reduces to the focus-directrix defining property. In proof, if we take

$$ae = de = ga = y,$$

then, since

$$bh = ha = \tfrac{1}{2}r \text{ and } hc = ca = \tfrac{1}{4}r, \text{ therefore } dm = ma = \tfrac{1}{2}r-y = \tfrac{1}{2}(ba-bd),$$

$$cm = ca-ma = y-\tfrac{1}{4}r = gc \quad \text{(or } ge \text{ is tangent at } e\text{)}$$

and finally $h'e = hm = ha-ma = y = ae$.

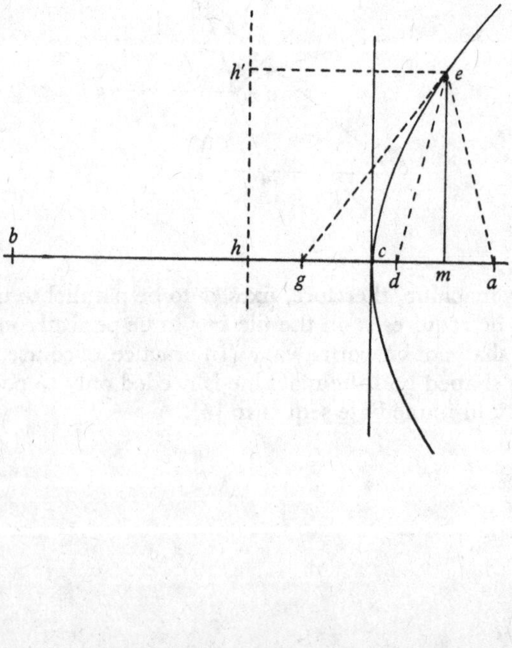

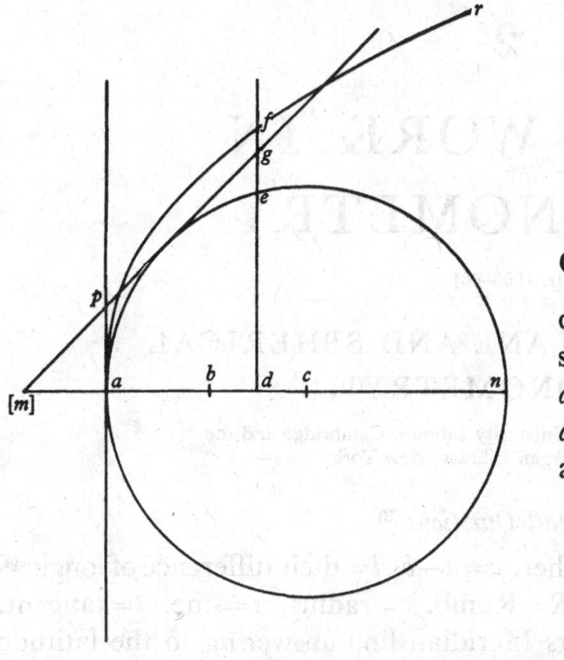

Or thus. take $\frac{r}{2} = ac = cn =$ Rad: Circle *aen*: $ab = \frac{r}{4}$. $am = ap$, & produce *mp* indefinitely. Then take some point *d* in y^e^ line *an*, & draw *dg* perpendic: to *an* y^t^ is soe y^t^ $dm = dg$, y^n^ take $df = ae$, & *f* shall be a point in y^e^ parabola *afr*.(42)

(39) These are each minor variants on the first description.

(40) Newton has cancelled 'Upon y^e^ focus or center *a* describe y^e^ center of'.

(41) The accompanying figure seems to require $ae = af$ and $be = bn$, but the text breaks off in mid-sentence. We may perhaps guess that Newton abandoned this description when he realized that the curve is not a parabola at all! In fact, if we take $bF = x$ and $Ff = y$ where *fF* and *eE* are drawn perpendicular to *ab*, then $af^2 = ae^2 = eb^2 - bE^2 + aE^2$: that is since

$$bE = \frac{be^2}{bd}, \quad y^2 + (x - \tfrac{1}{4}r)^2 = y^2 - \left(\frac{y^2}{r}\right)^2 + \left(\tfrac{1}{4}r + \frac{y^2}{r}\right)^2$$

and hence $x^2 - \frac{1}{2}rx = \frac{1}{2}y^2$, the defining equation of a hyperbola. The line *hgk* seems irrelevant.

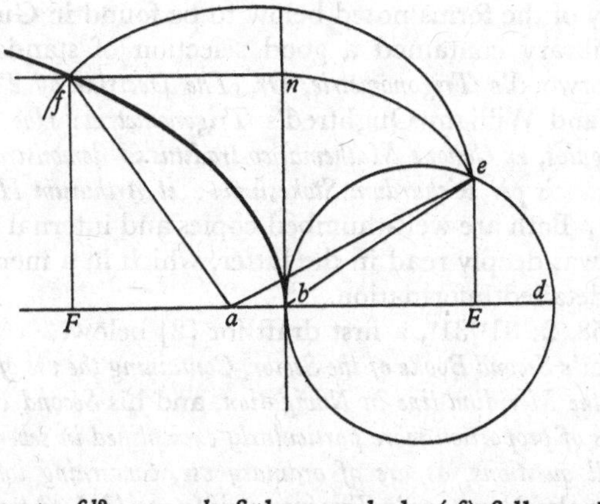

(42) The defining property $fd^2 = r.ad$ of the parabola (f) follows directly from the circle property ae^2 (or df^2) $= ad.an$. The line *mpg* is apparently superfluous.

2

EARLY WORK IN TRIGONOMETRY

[*c*. 1665–66]

§1. NOTES ON PLANE AND SPHERICAL TRIGONOMETRY(1)

From the originals in the University Library, Cambridge and the Pierpoint Morgan Library, New York

[1](2) [1665?]

Nauticall Questions.(3)

Suppose a = one latitud. b = another. $c = a - b$. l = their difference of longit:(4) d = distance(5) from y^e^ Meridian. R = Rumb. r = radius. s = sine. t = tangent. cs = cosine.(6) Ma, Mb(7) = Mercators Meridian line answering to the latitude a, b.(8)

1. Of a, b, d, R. $ra - rb(=rc) = d \times cs: R$.(9)

2. Of a, b, l, R. $r \times l = t:R \times \overline{Ma - Mb}$.(10) Or $r \times \overline{Ma - Mb} = l \times ct:R$.(11)

3. If $a, b, d : a, b, l : a, d, R : a, l, R : b, d, R : b, l, R$. bee given & $l:d:l:d:l:d$. bee sought, first find $R:R:b:b:a:a$. by y^e^ 1st : 2^d^ : 1st : 2^d^ : 1st[:] 2^d^. Proposition, & then $l:d:l:d:l:d$. by y^e^ 2^d^ : 1st : 2^d^ : 1st[:] 2^d^ : 1st. proposition.

---

(1) Newton's annotations from various authors, interlarded with his own comment, of standard formulæ in plane and spherical trigonometry, together with their application to the problem of navigating by a Mercator map. By name Newton refers for his source only to 'Gunter', that is, to Gunter's *Description and Vse of the Sector, Crosse-staffe, and other Instruments* (note (14) below), but his severe contractions for the trigonometrical operators are not Gunter's nor are many of the forms noted below to be found in Gunter's work. However, at his death Newton's library contained a good selection of standard trigonometrical texts, including Richard Norwood's *Trigonometrie, Or, The Doctrine of Triangles* (London, $_2$1645 = Trinity. NQ. 9. 33) and William Oughtred's *Trigonometria: Hoc est, Modus computandi Triangulorum Latera & Angulos, ex Canone Mathematico traditus & demonstratus. Collectus ex Chartis... Willelmi Oughtred Ætonensis pèr Richardum Stokesium...et Arthurum Haughton* (London, $_1$1657 = Trinity. NQ. 16. 182). Both are well-thumbed copies and internal evidence in the text below suggests that Newton was deeply read in the latter, which in a mere 36 pages of text presents a rich mine of exact, detailed information.

(2) ULC. Add. 3958. 2: 31^r^/31^v^, a first draft for [2] below.

(3) Compare Gunter's *Second Booke of the Sector, Containing the vse of the Circular Lines*: 103–46: chap. VI, *Of the use of the Meridian line in Navigation*, and his *Second Booke of the Crosse-staffe. Of the use of the former lines of proportion more particularly exemplified in severall kinds*; 84–99: chap. VI, *Containing such nauticall questions, as are of ordinary vse, concerning longitude, latitude, Rumb, and distance* (see note (14)); also Norwood's *Trigonometrie* (note (1)): Appendix: 119–30: *Of Sayling by Mercators Chart.* See also E. G. R. Taylor, *The Haven-finding Art* (London, 1956): 223–33.

4. If a, d, l: b, d, l: d, l, R. bee given & b, or R: a, or R: a, or b sought it cannot bee scientifically resolved.(12)

Gunter adds this $rl =$ ☋ $sR \times Ma$ ☊ $sR \times \overline{Ma\ ☊}\ d$(13) to y^e Nautical problems.(14) Tis exact enough when d is not greate.(15)

---

(4) Read 'longitude'. (5) That is, on the rhumb.

(6) This severely contracted symbolism is used, with slight modification, by both Norwood and Oughtred. (Note that $s{:}R = r\sin R$, $t{:}R = r\tan R$ and so of the rest.)

(7) This notation is apparently Newton's own invention.

(8) In fact, as Gunter noted in his *Sector*: Book 2, chap. VI, §1: 104, 'The making of this Table [of meridional parts] is, by addition of Secants': analytically,

$$Ma = \int_0^a \sec\theta . d\theta = \log(\sec a + \tan a) = \log\tan(\tfrac{1}{4}\pi + \tfrac{1}{2}a).$$

(Compare note (35) below.) The theory of the 'plain chart' (Mercator's map) is assumed.

(9) Gunter's *Sector*: Book 2: chap. VI: 122: 'As the Radius, to the sine of the complement of the Rumb from the meridian: So the distance upon the Rumb, to the difference of latitudes.'

(10) Gunter's *Sector*: Book 2, chap. VI: 127: 'As...the proper difference of latitude, is to... the difference of longitude: So...as Radius, to...the tangent of the Rumb from the Meridian.'

(11) On the analogy of his previous notation, 'ct' is a contracted notation for 'cotangent'.

(12) That is, resolved with mathematical exactness, since each of the posited reductions involves inversion of the table of meridional parts,

$$Ma = \int_0^a \sec\theta . d\theta,$$

which at this time (middle 1665?) were still calculated by Wright's primitive method of adding all the secants up to sec a at 1′ intervals. (See note (35) below.)

(13) That is, $rl = \pm s{:}R(Ma - \overline{Ma \mp d})$.

(14) See Edmund Gunter's *The Description and Vse of the Sector, Crosse-staffe, and other Instruments: With a Canon of Artificial Sines and Tangents, to a Radius of* 10000.0000. *parts, and the use thereof in Astronomie, Navigation, Dialling, and Fortification, &c. The second Edition much augmented* (London, 1636): *The Second Booke of the Sector*, chap. VI: 141: 'As...the Radius, is to...the sine of the Rumb from the meridian: So...as proper distance upon the Rumb, to...the difference of longitude.' (Or $r{:}(s{:}R) = (Ma - \overline{Ma - d}){:}l$, if we adapt Newton's notation.) Newton's copy of this work is now in Trinity College, Cambridge (NQ. 9. 160) but, as the record of purchase ('Gunters book & sector &c to Dr Fox—0.5[s].0') in the Fitzwilliam notebook confirms, Newton did not purchase it till early in 1667. Presumably at this time he was making use of Trinity's own copy in the college library.

(15) An exact observation on Newton's part. In fact, where a and b are both small, $Ma = \log(\sec a + \tan a) \approx \log(1+a)$ and correspondingly $Mb \approx \log(1+b)$. Hence

$$l = \tan R(Ma - Mb) \approx \tan R \log\left(\frac{1+a}{1+b}\right) = \tan R \log\left(\frac{1}{1-(a-b)/(1+a)}\right),$$

that is,

$$l \approx \sin R\left(\frac{1}{\cos R}\cdot\frac{a-b}{1+a}\right) \approx \sin R \log\left(\frac{1+a}{1+a-(a-b)/\cos R}\right) \approx \sin R\left(Ma - M\left(a - \frac{a-b}{\cos R}\right)\right),$$

where (by 1) $a - b = d\cos R$. (This paragraph, added in the manuscript as an afterthought at the end of his notes on spherical trigonometry below, is here reset to accord with Newton's sense.)

Of triangles right lined.

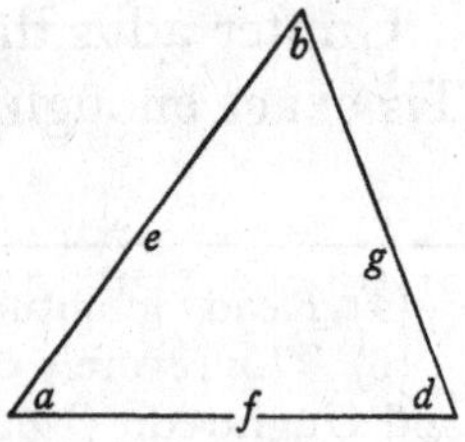

1st, of a, d, e, g. $sa\times e = sd\times g$.

2dly, of a, d, e, f. $rrf = re\times csa + e\times sa\times ctd$.

$rf\times td = re\times s{:}a + e\times td\times csa$.

3dly, Of a, e, f, g. $ree+rff = rgg + 2ef\times csa$.(16)

Speciall resolutions.

a, e, f given; d sought. $\dfrac{e-f}{e+f}\times ct{:}\dfrac{1}{2}a = ct{:}d+\dfrac{1}{2}a$. e, f, g given; a sought. if $m = \dfrac{e+f+g}{2}$. $n = m-e$. $p = m-f$. $q = m-g$. yn is $\dfrac{r\sqrt{np}}{\sqrt{ef}} = s\dfrac{1}{2}a$. $\dfrac{r\sqrt{mq}}{\sqrt{ef}} = cs\dfrac{1}{2}a$. $\dfrac{r\sqrt{np}}{\sqrt{mq}} = t\dfrac{1}{2}a$. $\dfrac{r\sqrt{mq}}{\sqrt{np}} = ct\dfrac{1}{2}a$.(17) $\sqrt{mnpq}$ = areæ trianguli.(18)

Of sphæricall triangles.

Of b, d, e, f. $sb\times se = sd\times sf$.

Speciall resolution[*s*],

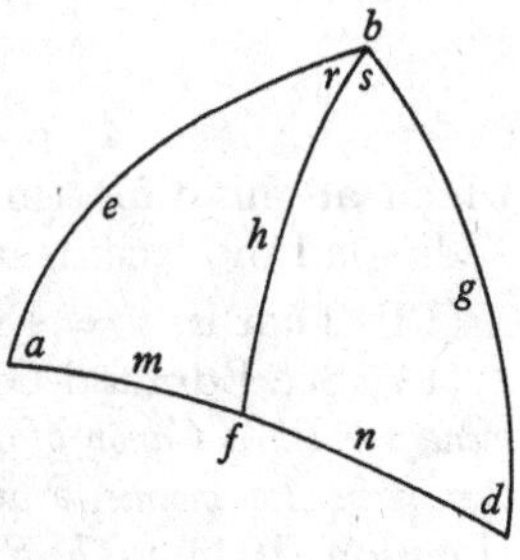

[1.] efg given[:] a sought, $\dfrac{r\times\sqrt{sp\times sn}}{\sqrt{se\times sf}} = s\dfrac{1}{2}a$.

$\dfrac{r\times\sqrt{sm\times sq}}{\sqrt{se\times sf}} = cs\dfrac{1}{2}a$. $\dfrac{r\sqrt{sn\times sp}}{\sqrt{sm\times sq}} = t\dfrac{1}{2}a$. $\dfrac{r\sqrt{sm\times sq}}{\sqrt{sn\times sp}} = ct\dfrac{1}{2}a$.

$\dfrac{\sqrt{ck{:}f\times sm\times sq}}{\sqrt{se}} = cs\dfrac{1}{2}a$.(19) &c. $\dfrac{rr\times \overline{cs{:}e-f{:}-cs{:}g}}{se\times sf} = sv{:}a$.(20)

$\dfrac{ck{:}f\times \overline{cs{:}e-f-cs{:}g}}{se} = sv{:}a$. $\dfrac{rr\times \overline{vs{:}e-f-vs[:]g}}{se\times sf} = sv[:]a$. ait Gunter.(21)

---

(16) If we denote the angles d, b and a by E, F and G, these basic rules of plane trigonometry take on the familiar forms

$$\sin G.e = \sin E.g,\quad f = e\cos G + e\sin G\cot E (= e\cos G + g\cos E)$$

and

$$e^2+f^2 = g^2+2ef\cos G.$$

(17) These standard results for the half angle are not to be found in Gunter and are probably taken from Oughtred's *Trigonometria* (note (1)).

(18) An immediate deduction from the preceding since

$$\triangle abd = \tfrac{1}{2}ef\sin a = \frac{ef}{r^2}(s{:}\tfrac{1}{2}a)(cs{:}\tfrac{1}{2}a).$$

(19) Newton quietly introduces the notation 'ck' for the cosecant: $ck{:}f = r\,\mathrm{cosec}f$.

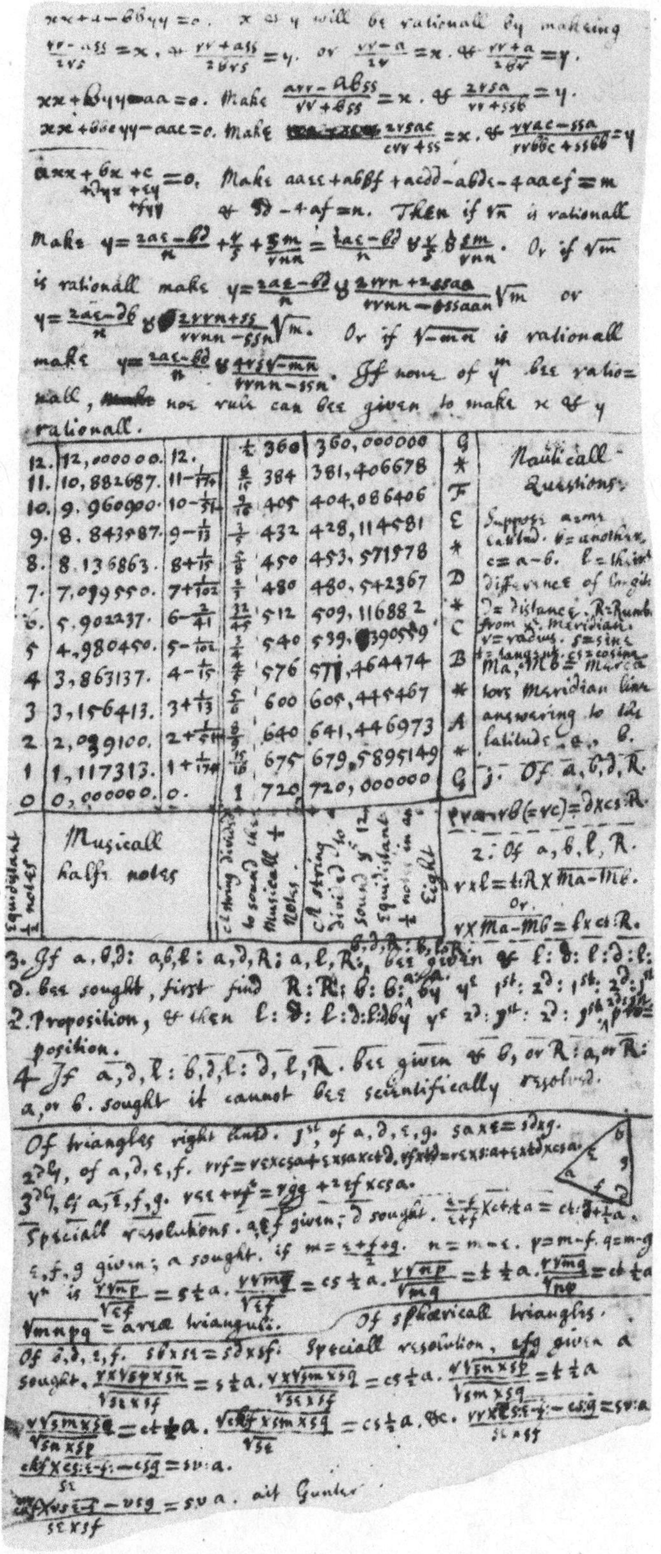

Plate IV. A 1665 scrap-sheet: notes on number theory (**3**, 4, §2), duodecimal division of the musical octave, and 'Nauticall Questions' (**3**, 2, §1).

2. a, d, f given: e, g sought.

$$s:\frac{a+d}{2}:t\frac{1}{2}f::s:\frac{a-d}{2}:t:\frac{g-e}{2}.$$

$$\text{And}\quad cs\frac{a+d}{2}:cs\frac{a-d}{2}::t\frac{1}{2}f:t\frac{g+e}{2}.$$

$$\text{And}\quad \frac{g+e}{2}+\frac{g-e}{2}=g.\quad \frac{g+e}{2}-\frac{g-e}{2}=e.$$

3. b, e, g given: a, d, sought.

$$s\frac{e+g}{2}:s\frac{e-g}{2}::ct\frac{1}{2}b:t\frac{d-a}{2}.$$

$$\text{And}\quad cs\frac{e+g}{2}:c\frac{e-g}{2}::ct:\frac{1}{2}b:t:\frac{d+a}{2}.$$

$$\frac{d+a}{2}+\frac{d-a}{2}=d.\quad \frac{d+a}{2}-\frac{d-a}{2}=a.$$

4. a, b, d, given: f sought. make $\frac{a+b+d}{2}=m.\ m-a=n.\ m-d=p.\ m-b=q.$

Then is $\frac{r\sqrt{csm\times csq}}{\sqrt{sa\times sd}}=s\frac{1}{2}f.\quad \frac{rr\times cs{:}\overline{a-d}-cs{:}b}{sa\times sd}=$ to the versed sine of y[e] complement of [f].(22) tis like y[e] 1st speciall resolution.

5. a, d, g, e given. f, or b, sought. tis y[e] converse of y[e] 2[d] & 3[d] resolution.

(20) That is, r versin $a = r(1-\cos a)$, a notation already introduced by Oughtred. (Compare F. Cajori, *A History of Mathematical Notations*, 2 (Chicago, 1929): §520.) Newton uses the variant '*vs*' on the next line.

(21) 'Or so says Gunter.' See Gunter's *Sector* (note 14): Book 2, chap. v, *Of the resolution of sphæricall Triangles*, §19, *To find an angle by knowing the three sides*: 98: 'According to Regiomontanus and others. As the sine of the lesser side next the angle required, to the difference of the versed sines of the base and difference of the sides: So is the Radius, to a fourth proportionall. Then as the sine of the greater side next the angle required, is to that fourth proportionall: So is the Radius, to the versed sine of the angle required.' (Analytically, if we suppose $e > f$, then $(s{:}f):(vs{:}g-vs{:}(e-f)) = r:\lambda$ where $(s{:}e):\lambda = r:(vs{:}a)$. The result follows easily from the preceding formulæ for the half-angle since

$$sv{:}a = r\,\text{versin}\,a = r\left(1-\left(\frac{cs{:}\frac{1}{2}a}{r}\right)^2+\left(\frac{s{:}\frac{1}{2}a}{r}\right)^2\right)$$

$$= r\left(\frac{s{:}e\times s{:}f-s{:}m\times s{:}q+s{:}p\times s{:}n}{s{:}e\times s{:}f}\right)$$

$$= \frac{1}{2}\left(\frac{[cs{:}(e-f)-cs{:}(e+f)]-[cs{:}g-cs{:}(e+f)]+[cs{:}(e-f)-cs{:}g]}{s{:}e\times s{:}f}\right).)$$

(22) That is, $r(1-\sin f)$.

6. Other cases are to bee resolved by rectanguled sphæricall $\triangle s$. To wch use may bee added these proportions.

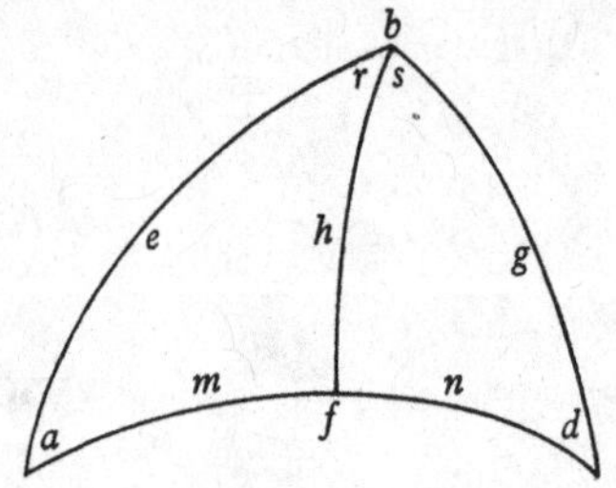

1. $cs:m:cs\,e::cs\,n:cs\,g.$

2[d], $ta:td::sn:sm.$

3[d], $cs:r:cs:s::tg:te.$

4th, $s:r:s:s::cs\,a:cs:d.$

5t(23)

---

[2](24)

Of y[e] Resolution of streight lined(25) triangles.

s=sine. t=tangent. k=secant. cs=cosine. ct=cotangent ck=cosecant. r=radius. vs=versed sine.(26)

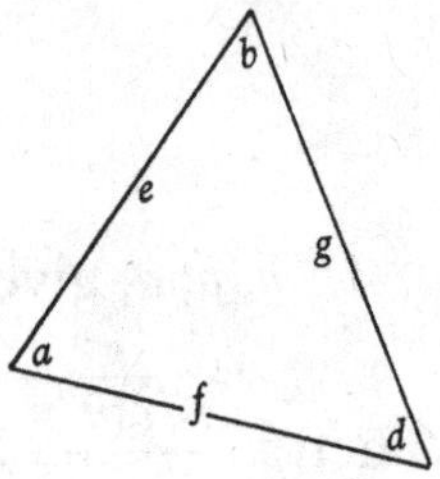

1. Of a, d, e, g. $sa\times e=sd\times g.$
2. Of a, d, e, f. $rr\times f=r\times e\times csa+e\times sa\times ctd.$
   Or, $rf\times t:d=re\times s:a+e\times t:d\times cs:a.$
3. Of a, e, f, g. $ree+rff=rgg+2ef\times csa.$

Of some speciall resolutions.

If a, e, & f are given & d sought then $\dfrac{\overline{e-f}\times ct\frac{1}{2}a}{e+f}=ct:\overline{d+\frac{a}{2}}=\dfrac{e-f}{e+f}\times ct\frac{a}{2}$. e, f, g given, & a sought. $\dfrac{r\times\sqrt{\overline{e+f+g}\times\overline{e+f-g}}}{\sqrt{4ef}}=cs:\frac{1}{2}a.$

Or, $\dfrac{r\sqrt{\overline{g+e-f}\times\overline{g+f-e}}}{\sqrt{4ef}}=s:\frac{1}{2}a.$ $\dfrac{r\sqrt{\overline{e+f+g}\times\overline{e+f-g}}}{\sqrt{\overline{g+e-f}\times\overline{g+f-e}}}=ct:\frac{1}{2}a.$

---

(23) Newton's text breaks off, but almost immediately he seems to have copied these notes into his notebook (the corresponding entry in which is here reproduced as [2]).

(24) Extracted from ff. 23[r]–25[v] of the early notebook now in the Pierpont Morgan Library, New York. (The book is well described by D. E. Smith in his *Two unpublished documents of Sir Isaac Newton* = (ed. W. J. Greenstreet) *Isaac Newton 1642–1727* (London, 1927): 16–31, especially 29.) Much of the notebook is of relatively early composition and Newton has inserted the date 1659 on the flyleaf, but the present piece is beyond doubt a revised version of [1] and by examining the handwriting we may set the same tentative date of 1665 for its composition. The major difference between the two pieces, apart from some new material (compare note (35)), is that Newton has reset the section on 'Nauticall Questions' as an appendix to the notes on trigonometry. Compare Cajori's *Notations* (note (20)) **2**: 160–1.

Or, $\frac{r\sqrt{g+e-f\times g+f-e}}{\sqrt{e+f+g\times e+f-g}}=t:\frac{1}{2}a$. (Viz: Make $\frac{e+f+g}{2}=m$. $m-e=n$. $m-f=p$. $m-g=q$. then is $\frac{\sqrt{rrnp}}{\sqrt{mq}}=t:\frac{1}{2}a$. &c. & $\sqrt{mnpq}=$ areæ $\triangle$.)

The resolution of Sphæ[r]icall triangles.

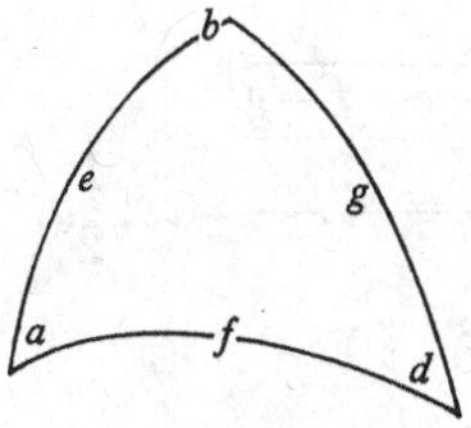

Rectanguled at b.

1.[27] Of b, e, f, g.

$r\times cs:f=cs:e\times cs:g$. $r\times cs:g=cs:f\times k:e$. $r\times cs[:]e=cs:f\times k:g$.

2. Of b, d, e, f.

$r\times s:e=s:d\times s:f$. $r\times s:f=s:e\times ck:d$. Or $r\times s:d=s:e\times ck:f$.

3. Of a, b, e, f.

$r\times t:e=t:f\times cs:a$. $r\times cs:a=t:e\times ct:f$. Or $r\times t:f=t:e\times k:a$.

4. Of a, b, e, g.

$r\times t:g=s:e\times t:a$. $r\times s:e=t:g\times ct:a$. $r\times t:a=t:g\times ck:e$.

5. Of a, b, d, e.

$r\times cs:d=s:a\times cs:e$. $r\times s:a=cs:d\times k:e$. $r\times cs:e=cs:d\times ck:a$.

6. Of a, b, d, f.

$r\times ct:a=t:d\times cs:f$. $r\times cs:f=ct:a\times ct:d$. $r\times t:d=ct:a\times k:f$.

Oblique angled triangles.

1. Of b, d, e, f. $s:b\times s:d=s:e\times s:f$.
2. Of a, e, f, g.
3. Of a, b, e, f.
4. Of a, b, d, e.

---

(25) Newton has cancelled 'Right line' (a phrase perhaps taken from Gunter).

(26) As before, these standard seventeenth-century trigonometrical functions are their modern equivalents multiplied into the radius: for example,

$$s:e = r\sin e,\quad k:\tfrac{1}{2}a = r\sec\tfrac{1}{2}a \text{ and } ct:a = r\cot a.$$

(27) This proposition, set fourth in order in Newton's text, is here reset to accord with his numbering.

The 3 last cases are ordinarily resolved by letting fall perpendiculars, viz: an angle or side being given[,] or else by y^e following speciall resolutions.

1. e, f, g given; a sought.

$$\frac{r\times\sqrt{s:\frac{e+f+g}{2}\times s:\frac{e+f-g}{2}}}{\sqrt{s:e\times s:f}}=cs:\frac{1}{2}a. \text{ Or, } \frac{r\times\sqrt{s:\frac{g+f-e}{2}\times s:\frac{g+e-f}{2}}}{\sqrt{s:e\times s:f}}=s:\frac{1}{2}a.$$

$$\frac{r\sqrt{s:\frac{e+f+g}{2}\times s:\frac{e+f-g}{2}}}{\sqrt{s:\frac{g+f-e}{2}\times s:\frac{g+e-f}{2}}}=ct:\frac{1}{2}a. \text{ Or } \frac{\sqrt{ck:f\times s:\frac{e+f+g}{2}\times s:\frac{e+f-g}{2}}}{\sqrt{s:e}}=cs:\frac{1}{2}a.$$

$$\text{Or } \frac{r\sqrt{s:\frac{g+f-e}{2}\times s:\frac{g+e-f}{2}}}{\sqrt{s:\frac{e+f+g}{2}\times s:\frac{e+f-g}{2}}}=t:\frac{1}{2}a. \text{ Or } \frac{ck:f\times\overline{-cs:g+cs:e-f}}{s:e}=vs:a.$$

$$\text{Or, } \frac{rr\times\overline{-cs:g+cs:e-f}}{s:e\times sf}=vs:a=\frac{rr\times\overline{cs:e-f-cs:g}}{se\times sf}.$$

$$\frac{\sqrt{rr\times se\times sf}}{\sqrt{s:\frac{e+f+g}{2}\times s:\frac{e+f-g}{2}}}=k:\frac{1}{2}a=\frac{\sqrt{ck:\frac{e+f+g}{2}\times se\times sf}}{\sqrt{s:\frac{e+f-g}{2}}}. \ \&c.$$

2. a, d, f given; e, g sought.

$s:\frac{a+d}{2}:t:\frac{1}{2}f::s:\frac{a-d}{2}:t:\frac{g-e}{2}$.(28) $\dfrac{cs:\frac{1}{2}a-\frac{1}{2}d\times t\frac{1}{2}f}{cs\frac{1}{2}a+\frac{1}{2}d}=t\frac{1}{2}g+\frac{1}{2}e$. The $\frac{1}{2}$ sum & $\frac{1}{2}$ diference of g & e is g &c. viz $\frac{g+e}{2}+\frac{g-e}{2}=g$. $\frac{g+e}{2}-\frac{g-e}{2}=e$.

3. b, e, g given; a, d sought.

$$\frac{s:\frac{e-g}{2}\times ct:\frac{1}{2}b}{s:\frac{e+g}{2}}=t:\frac{d-a}{2}. \ \& \ \frac{cs:\frac{e-g}{2}\times ct:\frac{1}{2}b}{cs:\frac{e+g}{2}}=t:\frac{\overline{d+a}}{2}. \ \frac{d+a}{2}+\frac{d-a}{2}=d. \ \&c.$$

tis like y^e preced^nt.

(28) Equivalently, $\dfrac{s:\frac{1}{2}(a-d)\times t:\frac{1}{2}f}{s:\frac{1}{2}(a+d)}=t:\frac{1}{2}(g-e)$.

Newton's double use of the colon is confusing.

[4.] *a*, *b*, *d* given: *f* sought.

$$\frac{r\sqrt{cs:\frac{a+b+d}{2}\times cs:\frac{a+d-b}{2}}}{\sqrt{s:a}\times\sqrt{s:b}}=s\frac{1}{2}f.$$

$$\frac{ck:d\times cs\,\overline{\overline{a-d}-csb}}{s:a}=\frac{rr\times cs:\overline{\overline{a-d}:-csb}}{sa\times sd}=vs\text{: compl of } f.^{(29)}\ \&\text{c.}$$

tis like yᵉ 1st speciall resolution.

[5.] *a*, *d*, *g*, *e*, given; *f* or *b* sought.

Tis yᵉ converse of yᵉ 2d & 3d Resolutions.

Of Questions in Navigation.

Suppose a = one latitude. b = another latitude, c = their diff[:] l = difference of longitude, d = distance, R = Rumb. r = Radius. s = sine &c: Ma, Mb = Mercators Meridian line answering to yᵉ latitudes a & b.(30)

1. Of a, b, d, R.

$$r\times\overline{a-b}\,(=r\times c)=d\times csR.\quad c\times\sec R=d\times r.$$

2. Of a, b, l, R.

$$r\times l=t{:}R\times\overline{Ma-Mb}.\quad r\times\overline{Ma-Mb}=l\times ct{:}R.$$

3. Of a, b, l, d.

$$ll\times\overline{a-b}\times\overline{a-b}=\overline{Ma-Mb}\times\overline{Ma-Mb}\times\overline{d+a-b}\times\overline{b-a+d}.^{(31)}$$

(viz $ll\times\overline{a-b}\times\overline{a-b}=\overline{Ma-Mb}\times\overline{Ma-Mb}\times dd+\overline{Ma-Mb}\times\overline{Ma-Mb}$
$\times\overline{a-b}\times\overline{b-a}$.

Or 1st. $l=\dfrac{\overline{Ma-Mb}\times\sqrt{\overline{d+a-b}\times\overline{d+b-a}}}{a-b}$.

2d, $d=\dfrac{\overline{a-b}\sqrt{ll+\overline{Ma-Mb}\times\overline{Ma-Mb}}}{Ma-Mb}$.

but if b, l, d: Or l, d, R, or a, d, l be given yⁿ a, R: or a, b: or b, R cañot be positively found.(32) The two first problems are sufficient for resolving all other nauticall problems, by yᵉ 3d, & 4th following.)

---

(29) That is, the complementary versed sine of f (in modern notation $r(1-\sin f)$).

(30) See note (8).

(31) For, by 1, $a-b=c=d\cos R$ and, by 2,

$$l=(Ma-Mb)\tan R,\quad\text{so that}\quad l^2=(Ma-Mb)^2\left(\left(\frac{d}{a-b}\right)^2-1\right).$$

(32) Compare note (12).

4. Of a, l, d, R.

$$rl+tR\times Mb=tR\times M:\frac{br+d\times csR}{r}.^{(33)}\ \text{Or}$$

$$l\times ct{:}\,R+r\times Mb=r\times M:\frac{b\times r+d\times csR}{r}.$$

5. Of b, d, l, R.[34]

Note, 1st yt these nauticall problems may be resolved in numbers if instead of $Ma-Mb$ wee write

$$\frac{\overline{\text{lt }45+\frac{1}{2}a}-\overline{\text{lt }45+\frac{1}{2}b}}{75802}\left(=\frac{\overline{\text{lt }45+\frac{1}{2}a}-\overline{\text{lt }45+\frac{1}{2}b}}{\text{lt }45.\,30'{:}\,-r}.\ \text{Or}\ \frac{\overline{\text{lt: }45+\frac{1}{2}a}-\overline{\text{lt: }45+\frac{1}{2}b}}{30\times\overline{\text{lt }45{:}\,1'.}-r}.\right)$$

supposeing lt to signifie ye logarithme of ye tangent.[35]

---

(33) From 2, $rl+t{:}R.Mb = t{:}R.Ma$ and Newton's result follows by the substitution

$$a = b+(a-b) = b+d\frac{cs{:}R}{r}\quad \text{(by 1)}.$$

(34) Newton adds no reduction but the case is almost that of 4 preceding.

(35) A most interesting paragraph which requires clarification on several counts. First, Newton's 'logarithme of ye tangent' is definable in modern terms by

$$\text{lt}{:}\phi = 10^7\log_{10}(r'\tan\phi) = 10^7\log_{10}(\tan\phi)+r,$$

where $r' = 10^{\frac{r}{10^7}}$ is the corresponding trigonometrical radius and where, following contemporary convention, the factor 10^7 is introduced so that all 7-place tabulations of $\text{lt}{:}\phi$ may be integral. Immediately

$$\text{lt}{:}(45°\ 30')-r = 10^7\log_{10}\tan(45°\ 30') \approx 75802_+,$$

from which Newton finds

$$\text{lt: }(45°\ 1')-r = 10^7\log_{10}\tan(45°\ 1')$$

by the close approximation $\frac{1}{30}(10^7\log_{10}\tan(45°\ 30'))$. (More generally, where α is small and k arbitrary, then

$$\log_k\tan(\tfrac{1}{4}\pi+n\alpha) \approx n.\log_k\tan(\tfrac{1}{4}\pi+\alpha).)$$

Finally, Newton seeks to evaluate the difference of the Mercator meridians

$$Ma-Mb = \int_b^a \sec\theta.d\theta$$

as

$$\frac{\text{lt: }(45+\frac{1}{2}a)°-\text{lt: }(45+\frac{1}{2}b)°}{\text{lt: }(45\frac{1}{2})°-r} = \frac{\log_{10}\tan(45+\frac{1}{2}a)°-\log_{10}\tan(45+\frac{1}{2}b)°}{\log_{10}\tan(45\frac{1}{2})°}$$

and hence would seem to take for $M(a°)$ the value

$$\frac{\log_{10}\tan(45+\frac{1}{2}a)°}{\log_{10}\tan(45\frac{1}{2})°} = \frac{\log\tan(45+\frac{1}{2}a)°}{\log\tan(45\frac{1}{2})°}.$$

In fact, $M(a°) = \log\tan(45+\frac{1}{2}a)°$ so that Newton has introduced the proportion factor

$$M(1°) = \log\tan(45\tfrac{1}{2})° \approx 0.0174_+.$$

2dly, y^t by the helpe of these or such like equations (viz:

$$rr=t\times ct=cs\times k=s\times ck.\quad r\times cs=s\times ct,\quad r\times s=cs\times t.\quad r\times t=s\times k,$$

$$r\times ct=cs\times ck.\quad r\times k=t\times ck,\quad r\times ck=ct\times k.\quad r=vs+cs.\quad k\times vs=r\times k\,\&\,r.$$

$$2r\times sv=ss+sv\times sv.\quad rr=ss+cc.^{(36)}\quad rr+tt=zz.^{(37)}$$

$$rrt=2rr\times t\frac{1}{2}+t\times t:\frac{1}{2}\times t:\frac{1}{2},\quad \text{or}\quad rr\times t:2=2rr\times t+t:2\times tt.$$

$r^3+rrt=rr\times k:2-tt\times k:2.$ &c) wee may get new theorems for every proposition.(38)

The integral $\int_0^a \sec\theta\,.\,d\theta$ has a long history and it is interesting that Newton is already aware of its evaluation as $\log\tan(\frac{1}{4}\pi+\frac{1}{2}a)$. This value was first publicized as a workable numerical approximation in the mid-1640's by Henry Bond, who apparently made his discovery by analysing the pattern of the 'Table for the true dividing of the meridians in the sea Chart' in Edward Wright's *Certaine Errors in Navigation,...and Tables of declination of the Sunne and fixed Starres detected and corrected* (London, 1599). Bond inserted the first printed notice of his observation in the 1645 edition of Richard Norwood's *Epitome of Navigation* (p. 14) and it quickly became common knowledge, appearing in such popular works as the 1653 (3rd) edition of Gunter's *Crosse-staffe* (Book 2: 99) and the 1659 edition of Norwood's *Epitome*. The demand for a theoretical proof of the result grew quickly, and was voiced typically by John Collins in his *Navigation by the Mariner's Plain Scale new plain'd* (London, 1659) when he wrote (p. 35) that 'True it is, this Proposition [the "perpetual Addition of Secants"] may be performed by the differences of the Logarithmicall Tangents, having a Table of them as in Norwood's *Epitome*, without the help of a Table of the Meridian-Line, but as yet we have no geometrical way known for making the Logarithmical-lines of Tangents'. Five years later, on 28 February 1664/5, Collins wrote to Wallis to the same effect (S. P. Rigaud, *Correspondence of Scientific Men of the Seventeenth Century*, **2** (Oxford, 1841): 460–2), while in 1666 Nicolaus Mercator went so far as to propose (in *Philosophical Transactions*, **1** (1665/6): 215) a sum of money for a proof (or disproof) of the result. Two years later James Gregory was the first to prove its theoretical accuracy (in his *Exercitationes Geometricæ* (London, 1668): 14 ff.), and simplified variants of his complex demonstration continued to appear over the next thirty years in the works of Isaac Barrow (1670), John Wallis (1685) and Edmund Halley (1695). (See Florian Cajori, 'On an Integration ante-dating the Integral Calculus', *Bibliotheca Mathematica* $_3$**14** (1913/14): 312–19.) Js. Lohne has recently established that the first evaluation of the integral (by stereographic projection of a spherical loxodrome from the south pole into a 'rumbus in meridiano in sphæra', or logarithmic spiral) was made by Thomas Harriot about 1594 in still unpublished researches (presumably unknown to Newton) into the theory of Mercator's projection (British Museum, Add 6789: 17r ff). Compare E. G. R. Taylor, 'The Doctrine of Nauticall Triangles Compendious. I. Thomas Hariot's Manuscript' and D. H. Sadler, 'II. Calculating the Meridional Parts', *Journal of the Institute of Navigation*, **6** (1953): 131–47.

(36) Read '$cs\times cs$'. (37) Read 'kk'.

(38) In his listing of these basic trigonometrical identities Newton has left out the variable. Thus $rr\times t:2 = 2rr\times t+t:2\times tt$ is the modern

$$r^2.(r\tan 2\theta) = 2r^2.(r\tan\theta)+(r\tan 2\theta).(r\tan\theta)^2.$$

§2. ANGULAR SECTIONS(1)

[Winter 1664/5?]

From the originals in the University Library, Cambridge

[1](2) *Of Angular sections.*

Suppose $ab = q$. $\frac{ah}{2} = r$. & $ag = x$. & y^t^ y^e^ arches hg, gb, bb(3) are equall.

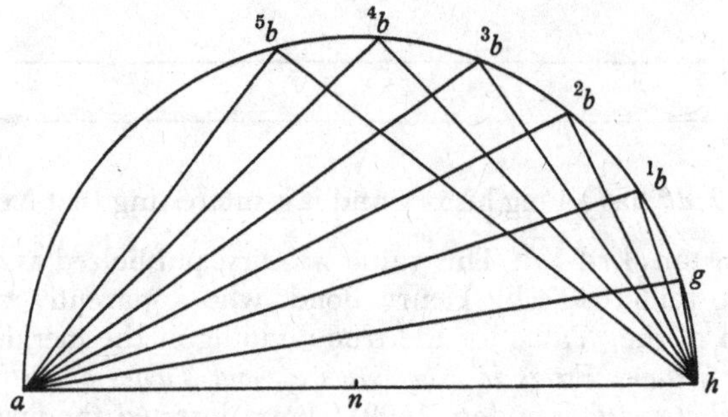

By y^e^ following Equations an angle *bah* may be divided into any number of partes.

$x = q$. unisectio.

$x^2 - 2rr = rq$. bisectio

$x^3 - 3rrx = rrq$. trisectio.

$x^4 - 4rrx^2 + 2r^4 = r^3q$. quadrisectio.

$x^5 - 5rrx^3 + 5r^4x = r^4q$. quintusectio.

$x^6 - 6rrx^4 + 9r^4xx - 2r^6 = r^5q$. sextusectio.

$x^7 - 7rrx^5 + 14r^4x^3 - 7r^6x = r^6q$. septusectio.

$x^8 - 8rrx^6 + 20r^4x^4 - 16r^6xx + 2r^8 = r^7q$.

$x^9 - 9rrx^7 + 27r^4x^5 - 30r^6x^3 + 9r^8x = r^8q$.

$x^{10} - 10rrx^8 + 35r^4x^6 - 50r^6x^4 + 25r^8xx - 2r^{10} = r^9q$.

---

(1) Newton renews his researches into the theory of multiple angles, taking as his starting-point his notes on Alexander Anderson's commentary on Viète's *Ad Angularium Sectionum Analyticen Theoremata καθολικωτερα* (= *Vieta*: 287–304). (See **1**, 2, §2.3.)

(2) Add. 4000: 81^r^, 80^v^. Compare Schooten's *Exercitationes*: Liber v, Sectio xxi: 464–75.

(3) That is, ${}^k b^{k+1} b$, $k = 1, 2, 3, \ldots$, where as in Newton's figure the circumference is divided successively by the points ${}^1b$, ${}^2b$, ${}^3b$, ${}^4b$, ….

$x^{11}-11rrx^9+44r^4x^7-77r^6x^5+55r^8x^3-11r^{10}x=r^{10}q.$

$x^{12}-12rrx^{10}+54r^4x^8-112r^6x^6+105r^8x^4-36r^{10}xx+2r^{12}=r^{11}q.$

$x^{13}-13rrx^{11}+65r^4x^9-156r^6x^7+182r^8x^5-91r^{10}x^3+13r^{12}x=r^{12}q.$

$x^{14}-14rrx^{12}+77r^4x^{10}-210r^6x^8+294r^8x^6-196r^{10}x^4+49r^{12}xx-$ &c.

$x^{15}-15rrx^{13}+90r^4x^{11}-275r^6x^9+450r^8x^7-3[7]8r^{10}x^5+140r^{12}x^3-$ &c.

$x^{16}-16rrx^{14}+104r^4x^{12}-352r^6x^{10}+660r^8x^8-672r^{10}x^6+336r^{12}x^4-$ &c.

$x^{17}-17rrx^{15}+119r^4x^{13}-442r^6x^{11}+935r^8x^9-1122r^{10}x^7+714r^{12}x^5-$ &c.

$x^{18}-18rrx^{16}+135r^4x^{14}-546r^6x^{12}+1287r^8x^{10}-1782r^{10}x^8+1386r^{12}x^6-$ &c.

$x^{19}-19rrx^{17}+152r^4x^{15}-665r^6x^{13}+1729r^8x^{11}-2717r^{10}x^9+2508r^{12}x^7-$ &c.

$x^{20}-20rrx^{18}+170r^4x^{16}-800r^6x^{14}+2275r^8x^{12}-4[0]04r^{10}x^{10}+4290r^{12}x^8-$ &c.(4)

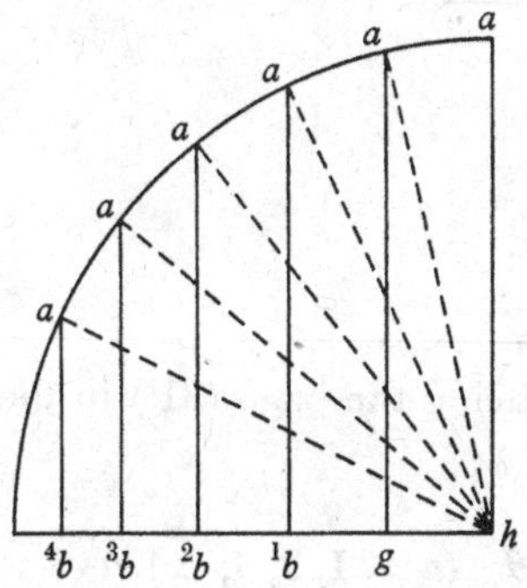

This scheame is ye former inversed.(5)

(4) Where $g\hat{a}h = \frac{1}{2}\theta$, Newton reproduces Viète's table of $CS(n\theta) = 2\cos(\frac{1}{2}n\theta)$ from his notes (1, 2, §2.3 above) on Vieta: 294–5: Theorem VI. Here, successively,

$$CS(0) = 2 = ah/r,$$
$$CS(\theta) = x/r = gh/r \quad (\text{or } gh = x),$$
$$CS(2\theta) = (x/r)^2-2 = {}^1bh/r,$$
$$CS(3\theta) = (x/r)^3-3(x/r) = {}^2bh/r,$$
$$\ldots\ldots\ldots\ldots\ldots\ldots$$
$$CS(n\theta) = \sum_{0\leqslant i\leqslant[\frac{1}{2}n]}\left((-1)^i\frac{n}{i}\binom{n-i-1}{i-1}\left(\frac{x}{r}\right)^{n-2i}\right);$$

and the basic recursive rule is

$$CS((n+1)\theta) = x/r.CS(n\theta)-CS((n-1)\theta).$$

(5) An alternative geometrical model which arranges the triangles a^kbh as a quadrant instead of a semicircle. The present figure is obtained (in mirror-image) if the triangles a^kbh in the previous scheme are rotated round h till each of the points ^kb lies in line with gh.

Suppose $gh=x$. $nh=r$. Then

$gh=x$. unisectio.

$ab\times r=2rr-xx$. bisectio.

$h^2b\times r^2=3rrx-x^3$. trisectio.

As on y^e other leafe(6) excepting some signes here changed.(7)

$a^3b\times r^3=2r^4-4rrxx+x^4$. quadrisec:

$h^4b\times r^4=5r^4x-5rrx^3+x^5$. quintusect^o.

$a^5b\times r^5=2r^6-9r^4x^2+6rrx^4-x^6$.

$h^{[6]}b\times r^6=7r^6x-14r^4x^3+7rrx^5-x^7$.

$a^{[7]}b\times r^7=2r^8-16r^6x^2+20r^4x^4-8rrx^6+x^8$.

$h^{[8]}b\times r^8=9r^8x-30r^6x^3+27r^4x^5-9rrx^7+x^9$.

$a^{[9]}b\times r^9=2r^{10}-25r^8x^2+50r^6x^4-35r^4x^6+10rrx^8-x^{10}$.

$h^{[10]}b\times r^{10}=11r^{10}x-55r^8x^3+77r^6x^5-44r^4x^7+11rrx^9-x^{11}$. [&c]

If $gh=x$. ${}^{[1]}bh=y$. Then

$y={}^{[1]}bh$. duplicatio anguli *hag*.

$yy-xx=x\times h^2b$. triplicatio anguli *hag*.

---

(6) These two concluding tabulations were entered on f. 80^v facing the page (81^r) in the notebook on which Newton wrote down his first entry.

(7) Viète's alternating table of

$$S((2n-1)\,\theta) = 2\sin(\tfrac{1}{2}(2n-1)\,\theta) \quad\text{and}\quad CS(2n\theta) = 2\cos n\theta \quad (n = 1, 2, 3, \ldots).$$

(See *Vieta*: 298–300: Theorema IX, Newton's notes on which are reproduced in **1**, 2, §2.3 above.) The kth row in the table, $f(k)$ say, is

$$\begin{cases} S(k\theta), & k \text{ odd} \\ CS(k\theta), & k \text{ even}\end{cases}$$

and Newton, taking $gh/r = x/r = CS(\theta)$, tabulates

$$\begin{aligned} f(1) &= S(\theta) = gh/r = x/r,\\ f(2) &= CS(2\theta) = {}^1ba/r = 2-(x/r)^2,\\ f(3) &= S(3\theta) = {}^2ba/r = 3(x/r)-(x/r)^3,\\ f(4) &= CS(4\theta) = {}^3ba/r = 2-4(x/r)^2+(x/r)^4,\\ &\ldots\ldots\ldots\ldots\ldots\ldots\ldots\ldots\\ f(k) &= \sum_{0\leqslant i\leqslant[\frac{1}{2}k]}\left((-1)^{[\frac{1}{2}i]}\frac{k}{i}\binom{k-i-1}{i-1}\left(\frac{x}{r}\right)^{k-2i}\right).\end{aligned}$$

As given by Viète this scheme may be derived step by step from the recursions

$$\begin{cases} CS(2n\theta) = -x/r.S((2n-1)\,\theta)+CS((2n-2)\,\theta)\\ S((2n+1)\,\theta) = x/r.CS(2n\theta)+S((2n-1)\,\theta)\end{cases},$$

but more generally we may show that

$$f(k) = (2-(x/r)^2)f(k-2)-f(k-4) \quad (\text{where } f(-1) = -x/r,\ f(0) = 2).$$

$y^3-2xxy=xx\times h^3b.$ quadruplicatio.
$y^4-3xxyy+x^4=x^3\times {}^4bh.$ quint^o^
$y^5-4xxy^3+3x^4y=x^4\times h^5b.$ sext^o^
$y^6-5xxy^4+6x^4yy-x^6=x^5\times h^{[6]}b.$ sept^o^
$y^7-6xxy^5+10x^4y^3-4x^6y\times h^{[7]}b.$ oct^o^,
$y^8-7xxy^6+15x^4y^4-10x^6yy+x^8=x^7\times h^{[8]}b.$ nonc
$y^9-8xxy^7+21x^4y^5-20x^6y^3+5x^8y=x^8\times h^{[9]}b.$ dec
$y^{10}-9xxy^8+28x^4y^6-35x^6y^4+15x^8y^2-x^{10}=x^9\times h^{[10]}b.$ und,
$y^{11}-10xxy^9+36x^4y^7-56x^6y^5+35x^8y^3-6x^{10}y=x^{10}\times h^{[11]}b.$ duod [&c][(8)]

---

[2][(9)]

versâ paginâ[(10)]

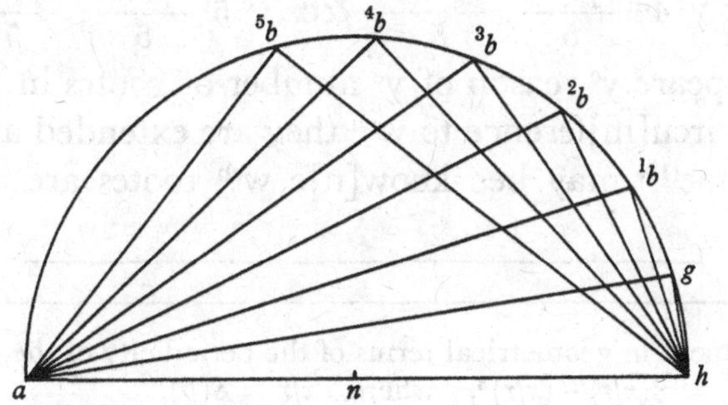

Suppose y^e^ perifery *bgh* to bee *a* & y^e^ whole perifery to bee *p*.[(11)] The line *bh* subtends these arches. *a*. *p−a*. *p+a*. 2*p−a*. 2*p+a*. 3*p−a*. 3*p+a*. 4*p−a*.

---

(8) Viète's table of $S(n\theta) = 2\sin(\frac{1}{2}n\theta)$ (= *Vieta*: 296–7: Theorema VII, noted by Newton in the annotations printed in **1**, 2, §2: Prop. 21). In Newton's scheme $ah = r$, $gh = x = rS(\theta)$ and ${}^1bh = y = rS(2\theta)$, so that, by taking $z = y/x = CS(\theta)$, we may present the tabulation as

$$[rS(\theta) = x,]$$
$$rS(2\theta) = y = z = {}^1bh,$$
$$rS(3\theta) = x(z^2-1) = {}^2bh,$$
$$rS(4\theta) = x(z^3-2z) = {}^3bh,$$
$$\cdots\cdots\cdots$$
$$rS(n\theta) = x.\sum_{0\leqslant i\leqslant[\frac{1}{2}n]}\left((-1)^i\binom{n-i}{i}z^{2(n-i)-1}\right).$$

Viète deduces this step by step from the recursion

$$S((n+1)\theta) = zS(n\theta)-S((n-1)\theta).$$

(9) Add. 4000: 81^v^, 82^v^, which continues [1].

(10) '[See] the previous page', a reminder that Newton continues to work with the diagram on f. 81^r^ (here reproduced a second time).

(11) In apparent modification of Oughtred's notation of $\frac{\pi}{\delta}$ for π, Newton uses p to represent the circumference of a circle of undefined diameter.

$4p+a$. $5p-a$. $5p+a$. $6p-a$. $6p+a$. &c:[12] All wch are bisected, trisected, quadrisected, quintusected &c after [y^e^] same manner. As for example

The rootes of y^e^ Equation $h^2b \times rr = 3rrx - x^3$[13] are 3. The first whereof subtends y^e^ arches $\frac{a}{3}$. $\frac{3p-a}{3}$. $\frac{3p+a}{3}$. $\frac{6p-a}{3}$. $\frac{6p+a}{3}$. $\frac{9p-a}{3}$ &c.[14] The second subtends y^e^ arches $\frac{p-a}{3}$. $\frac{2p+a}{3}$. $\frac{4p-a}{3}$. $\frac{5p+a}{3}$. $\frac{7p-a}{3}$. &c.[15] The 3^d^ $\frac{p+a}{3}$. $\frac{2p-a}{3}$. $\frac{4p+a}{3}$. $\frac{5p-a}{3}$. $\frac{7p+a}{3}$ &c.[16]

Soe y^e^ rootes of y^e^ Equation $h^{[4]}b \times r^4 = 5r^4x - 5rrx^3 + x^5$,[17] doe y^e^ first subtend y^e^ arches $\frac{a}{5}$. $\frac{5p-a}{5}$. $\frac{5p+a}{5}$ &c: y^e^ 2^d^ $\frac{p-a}{5}$. $\frac{4p+a}{5}$. $\frac{6p-a}{5}$. [&c:] y^e^ 3^d^ $\frac{p+a}{5}$. $\frac{4p-a}{5}$. $\frac{6p+a}{5}$ &c. y^e^ 4th $\frac{2p-a}{5}$. $\frac{3p+a}{5}$. &c. y^e^ 5^t^ $\frac{2p+a}{5}$. $\frac{3p-a}{5}$. $\frac{7p+a}{5}$. &c.[18]

Hence may appeare y^e^ reason of y^e^ number of rootes in these equations & y^t^ y^e^ points of y^e^ circu[m]ference to wch they are extended are æquidistant. & by y^e^ lower scheme[19] may bee know[n]e wch rootes are affirmative & wch negative,[20]

---

(12) A clear statement in geometrical terms of the periodicity of the general angle.

(13) $h^2b/r = S(3\theta) = 3(x/r) - (x/r)^3$, where $x/r = S(\theta)$.

(14) $n\pi \pm \frac{1}{3}a$ $(n = 0, 1, 2, \ldots)$. (15) $n\pi \pm \frac{1}{3}(\pi - a)$ $(n = 0, 1, 2, \ldots)$.

(16) $n\pi \pm \frac{1}{3}(\pi + a)$ $(n = 0, 1, 2, \ldots)$.

(17) $h^4b/r = S(5\theta) = 5(x/r) - 5(x/r)^3 + (x/r)^5$, where $x/r = S(\theta)$.

(18) Respectively

$$n\pi \pm \tfrac{1}{5}a,\ n\pi \pm \tfrac{1}{5}(\pi - a),\ n\pi \pm \tfrac{1}{5}(\pi + a),\ n\pi \pm \tfrac{1}{5}(2\pi - a) \text{ and } n\pi \pm \tfrac{1}{5}(2\pi + a) \quad (n = 0, 1, 2, \ldots).$$

(19) That is, the inverse 'scheame' (on f. 81^r^) in [1] above.

(20) Newton stresses the fundamental role which the periodicity of the trigonometrical functions $S(\theta)$ and $CS(\theta)$ (and so of $\sin\theta$ and $\cos\theta$) plays in the theory of the general n-section of angles. In particular he gives the now familiar interpretation of the angle $2k\pi + \alpha$ as being composed of k whole revolutions and the simple angle α, $0 \leqslant \alpha < 2\pi$. This, historically, seems the first coherent treatment of periodicity, though Viète in his reply to Adrianus Romanus' challenge-problem half a century before had insisted that more than one root, $S(\theta)$, of $S(45\theta) = N$ was possible. (See *Ad Problema...quod proposuit Adrianus Romanus*, especially Caput VI = *Vieta*: 310. Viète, however, never gave a clear interpretation of the 'extra' roots $S(2k\pi/45 + \theta)$ $(k = 1, 2, 3, \ldots, 44)$.)

(21) Add. 4000: 81^v^, which follows immediately on [2].

(22) That is, on ff. 81^r^, 80^v^ = [1] above.

(23) Read 'angle:angle'.

(24) Newton expands the general coefficient

$$(-1)^i.\left[\binom{n-i}{i} + \binom{n-i-1}{i-1}\right] = (-1)^i \frac{n}{i}\binom{n-i-1}{i-1} \quad (i = 0, 1, 2, \ldots),$$

as a product.

(25) Read 'progression'.

(26) See the tenth line in the first tabulation of [1].

[3](21)

The Numerall coëfficients of ye afforesaid Equations(22) may bee deduced from this progression (if $\angle : \angle$(23)::1:*n*.)

$$1\times\frac{-0+n\times-1+n}{1\times1-n}\times\frac{n-2\times n-3}{2\times2-n}\times\frac{n-4\times n-5}{3\times3-n}\times\frac{n-6\times n-7}{4\times4-n}$$

$$\times\frac{n-8\times n-9}{5\times5-n}\times\frac{n-10\times n-11}{6\times6-n}\ \&c.^{(24)}$$

As if $n=10$. ye p̄ression(25) is $1\times-10\times\frac{-7}{2}\times\frac{-10}{7}\times\frac{2}{25}\times0$. And ye coefficients 1. -10. $+35$. -50. $+25$. -2.(26)

---

[4](27)

Theorema 1.

If in the Circle *abcdeP* there be inscribed any Poligon(28) *abcde* wth an odd number of sides. & from any point in ye circumference *P* there bee drawne lines *Pe*, *Pa*, *Pb*, *Pc*, *Pd* to every corner of ye Polygon: ye summ of every other line is equall to ye summ of ye rest $Pa+Pb+Pc=Pd+Pe$. & soe are their cubes $Pa^3+Pb^3+Pc^3=Pd^3+Pe^3$. unless ye figure be a Trigon.(29)

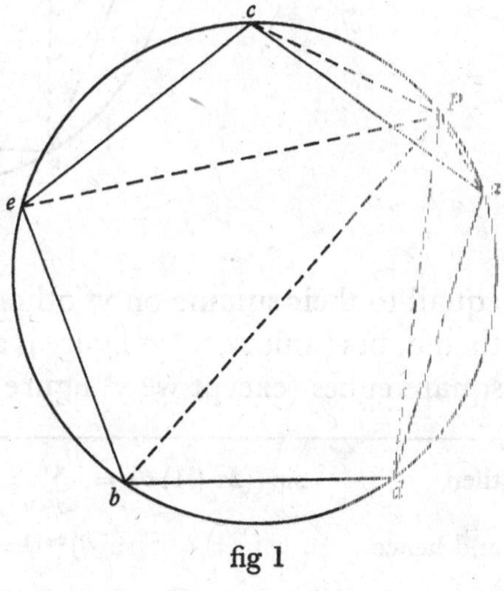

fig 1

---

(27) Add. 4000: 82v, which continues [3].

(28) Implicitly, any regular polygon.

(29) If we suppose the polygon to have $(2n+1)$ sides, then

$$a\hat{P}d = d\hat{P}b = b\hat{P}e = \ldots = \pi/(2n+1);$$

and therefore if we take $P\hat{b}a$ (subtended by the chord *Pa*) to be θ, it follows that, where the circle diameter is taken as unity,

$$Pa = \sin(\theta),\quad Pd = \sin(\theta+\pi/(2n+1)),\quad Pb = \sin(\theta+2\pi/(2n+1)),\quad \ldots,$$

and Newton's proposition is

$$0 = \sum_{0\leqslant i\leqslant 2n}\left[(-1)^i\sin\left(\theta+\frac{i\pi}{2n+1}\right)\right]^{2k+1}\quad (k \neq n,\ n = 0, 1, \ldots).$$

The proof is a neat application of the results in [1]. First, since

$$S((2n+1)\,2\theta) = 2\sin((2n+1)\,\theta)$$

can be expanded as

$$S((2n+1)\,2\theta) = \sum_{0\leqslant j\leqslant 2n}\left[(-1)^j\binom{n-j}{j}[CS(2\theta)]^{2(n-j)}\right].S(\theta),$$

Theor 2.

If from y^e points of y^e Polygon there bee drawne perpendiculars *ap*, *br*, *ct*, *ds*, *eq* to any Diameter *pt*: y^e summe of y^e Perpendiculars on one side y^e Diameter is

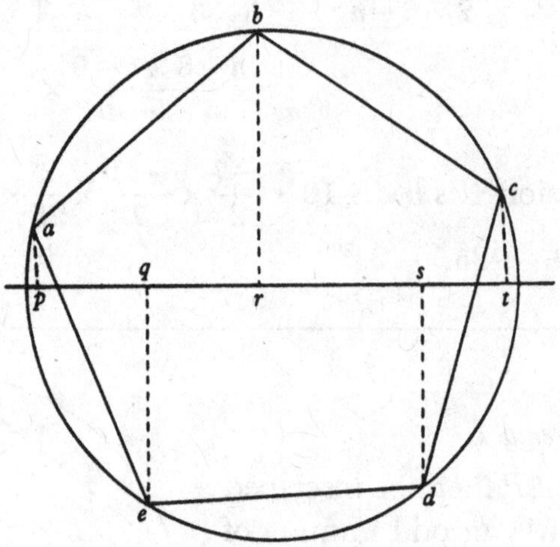

fig 2

equall to their summe on y^e other[,] $ap+br+ct=eq+ds$. & soe is y^e summe of their cubes (unlesse w^n y^e figure is a Trigon), $ap^3+br^3+ct^3=eq^3+ds^3$. & of theire square cubes (except w^n y^e figure is a Trigon or Pentagon[)]. &c.(30)

---

then $$\sin((2n+1)\theta) = \sum_{0\leqslant j\leqslant n} [a_j(\cos\theta)^{2j}].\sin\theta = \sum_{0\leqslant j\leqslant n} [b_j(\sin\theta)^{2j+1}]$$

and hence $$(\sin\theta)^{2n+1} = \sum_{0\leqslant j\leqslant n} [c_j \sin((2j+1)\theta)],$$

where the constant coefficients a_j, b_j and c_j are readily determinable. Again, if

$$\cos\left(\frac{2k+1}{2n+1}.\frac{\pi}{2}\right) \neq 0 \quad (\text{that is, } k \neq n),$$

then $$\sum_{0\leqslant i\leqslant 2n} (-1)^i \left[\sin\left(\phi+\frac{2k+1}{2n+1}i\pi\right)\right]$$

$$= \frac{1}{2\cos\left(\frac{2k+1}{2n+1}.\frac{\pi}{2}\right)} \times \sum_{0\leqslant i\leqslant 2n} (-1)^i \left[\sin\left(\phi+\frac{2k+1}{4n+2}(2i+1)\pi\right)+\sin\left(\phi+\frac{2k+1}{4n+2}(2i-1)\pi\right)\right]$$

$$= 0, \quad k \neq n.$$

From this Newton's theorem follows by expanding

$$\left[\sin\left(\theta+\frac{i\pi}{2n+1}\right)\right]^{2k+1} \quad \text{as} \quad \sum_{0\leqslant j\leqslant k} \left[c_j \sin\left((2j+1)\theta+\frac{2j+1}{2n+1}i\pi\right)\right]$$

and then taking $\phi = \theta, 3\theta, 5\theta, ..., (2n-1)\theta$ successively. (Independently, three years later James Gregory gave an ingenious geometrical proof of the first part of Newton's theorem in his *Geometriæ Pars Universalis*, Padua, 1668: Prop. 69: 128–30.)

Theor 3.

If y^e 2 circles (fig 1 & 2) be equall wth like Poligons inscribed, & *Pa* in fig 1 be assumed double to *pa* in fig 2. then are all y^e other corresponding lines in fig 1 double to those in fig 2 viz $Pb=2rb$, $Pc=2tc$, $Pd=2sd$, $Pe=2qe$.(31)

---

[5](32).

$x^3-3rrx=rrq$.(33) $[r=1.]$ $-2=q$. if $x=1$. $x=2..q=2$. $x=3$. $q=18$.

$3x-x^3=p$. $3p-p^3=q$. $9x-3x^3-27x^3+27x^5-9x^7+x^9=q$.

$x^9-9x^7+27x^5-30x^3+9x-q=0$.(34)

$3x^3-qxx=5x^3-5[x]+q$.(35) $\;2x^3|+qxx-5x+q=0$.
$\qquad -6x+q$(36)

$-11xx+2qx=-3qx+qq$.(37) $\;11xx-5qx+qq=0$.

$11qxx=121x-22q=5qqx-q^3$.(38) $\;\dfrac{22q-q^3}{121-5qq}=x$.

$-qx^4+5x^3-qxx=10x^3-10x+[2]q$.(39)

---

(30) If we take the circle radius as unity, Newton's proposition is in analytical terms

$$0=\sum_{0\leqslant i\leqslant 2n}\left[\sin\left(\theta+\frac{2i\pi}{2n+1}\right)\right]^{2k+1}\quad (k\neq n,\ n=0,1,2,\ldots):$$

which reduces to the preceding theorem since

$$\sum_{(n+1)\leqslant i\leqslant 2n}\left[\sin\left(\theta+\frac{2i\pi}{2n+1}\right)\right]^{2k+1}=\sum_{0\leqslant i\leqslant n}\left[\sin\left(\theta+\pi+\frac{(2i-1)\pi}{2n+1}\right)\right]^{2k+1}$$
$$=-\sum_{0\leqslant i\leqslant n}\left[\sin\left(\theta+\frac{(2i-1)\pi}{2n+1}\right)\right]^{2k+1}.$$

(31) This is an immediate deduction from the preceding treatment.

(32) Calculations extracted from Add. 3958. 2: 34^v. A letter draft on the same page, already quoted in 2, 6, §1, note (1), confirms a dating of late October 1665 for its composition.

(33) That is, where $x/r=S(\theta)$, $CS(3\theta)=q$.

(34) Where $r=1$ and $x/r=x=S(\theta)$, $S(3\theta)=p$ and so $S(3(3\theta))=S(9\theta)=q$. Newton proceeds to expand $S(9\theta)$ in terms of $x=S(\theta)$. (See the ninth line in the second tabulation in [1], and compare the ninth line in the first.) Note that $S(\theta)=3$ is impossible in real terms.

(35) Newton now sets $q=S(3\theta)=S(5\theta)$ where $S(\theta)=x$ and so has

$$x^2(3x-q)=[x^5=]\,5x^3-5x+q.$$

(36) Newton attempts to eliminate $2x^3$ with $3x-x^3=q$ but makes a numerical slip. Read correctly '$6x-2q$'. (The mistake is carried through the next two lines.)

(37) For $x(-11x+2q)=-qx^3=-q(3x-q)$.

(38) For $11(qx^2)=11(11x-2q)=q(11x^2)=q(5qx-q^2)$.

(39) Perhaps discovering his mistake, Newton eliminates x^5 from his first equations by

$$-x^2(2x^3+qx^2-5x+q)=0=-2(x^5-5x^3+5x-q),$$

but then breaks off his calculations.

[6][40]

Si in angulo quovis PAQ inscribantur æquales AB, BC, CD, DE, EF, FG, GH &c anguli BAP erit angulus CBQ duplus, DCP tripl, EDQ quadr[,] FEP

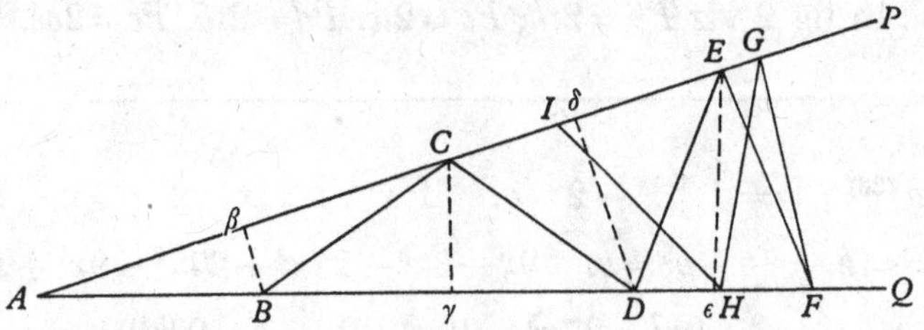

quint, GFQ sext, HGP sept. IHQ oct &c. Horū vero angulorum posito radio AB sinus erunt $B\beta$, $C\gamma$ &c cosinus $A\beta$, $B\gamma$, $C\delta$ &c. Ergo si $AB=r$ & $A\beta=x$ erit $AC=2x$. $A\gamma=\frac{2xx}{r}$. $AD=(2A\gamma-AB)=\frac{4xx-rr}{r}$. $A\delta=\frac{4x^3-rrx}{rr}$ &c.[41]

(40) A miscellaneous note (on f. 35[r]) in the undergraduate notebook Add. 4000. Its writing style and the fact of its being composed in Latin suggest a fairly late date of composition, perhaps some time in 1666.

(41) 'If in any angle PAQ there be inscribed equal [segments] AB, BC, CD, DE, EF, FG, GH &c, then will the angle CBQ be double the angle BAP, DCP its triple, EDQ four times it, FEP five times, GFQ six times, HGP seven times, IHQ eight times it &c. Further, if we take AB as radius, of these angles the sines will be $B\beta$, $C\gamma$, &c and the cosines $A\beta$, $B\gamma$, $C\delta$ &c. Hence if $AB=r$ & $A\beta=x$, then $AC=2x$. $A\gamma=\frac{2xx}{r}$. $AD(=2A\gamma-AB)=\frac{4xx-rr}{r}$. $A\delta=\frac{4x^3-rrx}{rr}$. &c.'

This scheme is a neat generalization of the technique expounded by Viète in his *Supplementum Geometriæ* (= *Vieta*: 240–57, annotated by Newton on Add. 4000: 8[r]/8[v] = **1**, 2, §2.1) and in fact is a geometrical representation of the recursion

$$\begin{Bmatrix}\cos((n+1)\theta) = 2\cos\theta.\cos(n\theta)-\cos((n-1)\theta)\\ \sin((n+1)\theta) = 2\cos\theta.\sin(n\theta)-\sin((n-1)\theta)\end{Bmatrix}.$$

In illustration Newton takes $x = r\cos\theta$ (so that

$$\cos((n+1)\theta) = 2(x/r)\cos(n\theta)-\cos((n-1)\theta))$$

and calculates successively

$$\begin{bmatrix}\cos(0) = 1,\\ \cos(\theta) = \frac{x}{r},\end{bmatrix}$$

$$\cos(2\theta) = 2\frac{x}{r}\left(\frac{x}{r}\right)-1 \quad \left(\text{or } B\gamma = r\cos(2\theta) = \frac{2x^2-r^2}{r}\right),$$

$$\cos(3\theta) = 2\frac{x}{r}\left(2\frac{x^2}{r^2}-1\right)-\frac{x}{r},$$

. . . .

§3. CONSTRUCTION OF A TRIGONOMETRICAL CANON

[Early 1665?]

From the originals in the University Library and the Fitzwilliam Museum, Cambridge

[1](1)

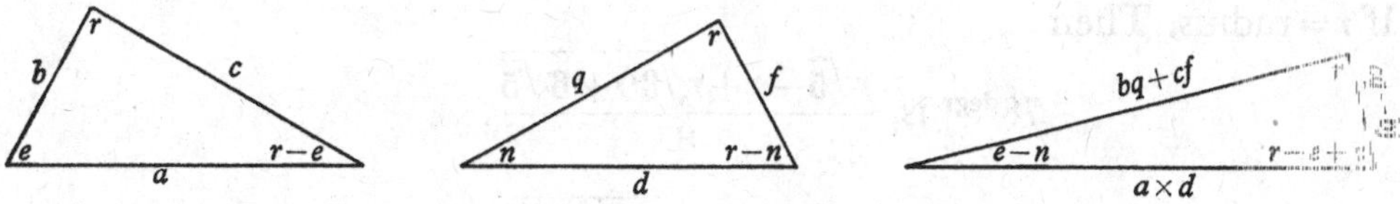

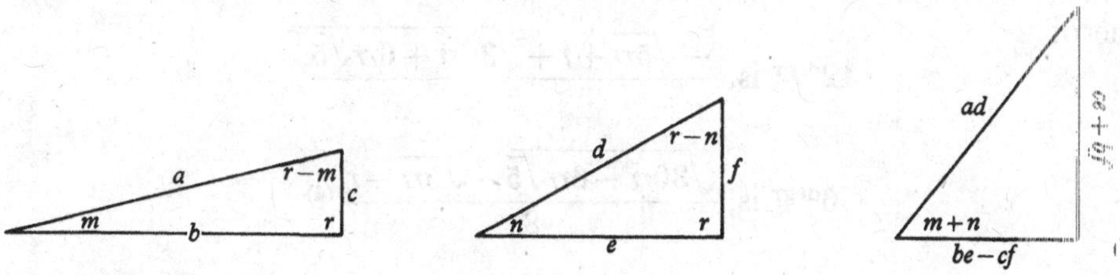

(2)

In later years this was to become Newton's favourite technique for constructing multiple angles. In an unpublished appendix to his 1670 notes on Kinckhuysen's *Algebra* he set it in revised form as one of a set of problems illustrating algebraic method (ULC. Add. 3959. 1: 21r) and five years later lectured on it in his Lucasian course on arithmetic and algebra (ULC. Dd. 9. 68: 78r/78v: Lectio 8 (1676), Prob. 15, printed as *Arithmetica Universalis* (London, 1707): 134–6: Prob. XV, renumbered as Prob. XXIX in later editions). Finally, on 24 October 1676 he communicated the scheme privately to Leibniz in his Epistola posterior (= *Correspondence of Isaac Newton*, 2 (1960): no. 188: 110–29, especially 125).

(1) Extracted from ff. 79v/80r in the undergraduate notebook Add. 4000 in the University Library, Cambridge.

(2) A pictorial representation of the addition theorems of elementary trigonometry, presumably based on Newton's previous annotations (Add. 4000: 9r, especially Propositions 14/15 = 1, 2, §2.2) of Alexander Anderson's *Ad Angularium Sectionum Analyticen Theoremata καθολικωτερα* = *Vieta*: 287–304, especially 288–90: Theoremata II/III. In the first scheme, where $a:b:c = 1:\cos e:\sin e$ and $d:q:f = 1:\cos n:\sin n$, it follows correctly that

$$1:\cos(e-n):\sin(e-n) = ad:(bq+cf):(cq-bf),$$

that is, as $1:(\cos e\cos n+\sin e\sin n):(\sin e\cos n-\cos e\sin n)$. In the latter, similarly, where $a:b:c = 1:\cos m:\sin m$ and $d:e:f = 1:\cos n:\sin n$, Newton correctly presents the ratio

$$1:\cos(m+n):\sin(m+n) \quad \text{as} \quad ad:(be-cf):(ce+bf),$$

or $$1:(\cos m\cos n-\sin m\sin n):(\sin m\cos n+\cos m\sin n).$$

Ad constructionem Canonis angularis.

$\frac{90^{gr}}{5}=18^{gr}$. $\frac{18^{gr}}{5}=3^{gr}+36'$. Et $\frac{60^{gr}}{3}=20^{gr}$. $\frac{20^{gr}}{3}=6^{gr}+40'$. $\frac{6^{gr}+40'}{2}=3^{gr}+20'$.

$3^{gr}+36'-3^{gr}-20'=16'$. $\frac{16'}{2}=8'$. $\frac{8'}{2}=4'$. $\frac{4'}{2}=2'$. $\frac{2'}{2}=1'$.(3)

If r=radius. Then

$$78^{degr}\text{ is, }\frac{r\sqrt{5}-r+r\sqrt{30+6\sqrt{5}}}{8}.$$

$$66^{degr}\text{ is, }\frac{r\sqrt{5}+r+r\sqrt{30-6\sqrt{5}}}{8}.$$

ye sine of

$$42^{degr}\text{ is, }\frac{-\sqrt{5rr}+r+\sqrt{30rr+6rr\sqrt{5}}}{8}.$$

$$6^{degr}\text{ is, }\frac{\sqrt{30rr-6rr\sqrt{5}}-\sqrt{5rr}-r}{8}.^{(4)}$$

---

[2](5)

Canones Sinuum.(6)

[1r] 0 Degr

16[']. 465 4195[0 0000].

---

(3) A neat scheme for constructing the trigonometrical functions of

$$1'=\frac{1}{2^4}\left(\frac{90^\circ}{5^2}-\frac{60^\circ}{2.3^2}\right)$$

by successive bisection, trisection and quintusection of the familiar angles 90° and 60°. The derivation is not Newton's own, however, but copied from Alexander Anderson's. (See *Vieta*: 303/4: 'Ad constructionem...primum inquiratur perpendiculum [sc. sinum] unius scrupuli, quam fieri potest accuratum, idque hac methodo...'.) The construction of the 'angular canon' is, of course, then to be performed at 1' intervals 'regrediendo ad angulos in ratione multipla' (*Vieta*: 304) with the help of the multiple-angle theorems (listed by Newton in §1 of the present section). This method of construction by subsectioning was, of course, at least as old as Ptolemy's bisectioning procedure in the *Syntaxis*, but not till Viète applied to the construction of his *Canon Mathematicus* (Paris, 1579) his new-found theorems on the multisection of angles coupled with his powerful method for the resolution *in numeris* of the resulting equations did the method become comparatively free of prohibitive back-breaking calculation. Alexander Anderson acutely observed in 1615 that 'Mathematicum igitur Canonem, secure ac fœliciter construet Analysta, & constructum examinabit, adiutus analyticis hisce principiis, & edoctus methodum resolvendi potestates quascumque sive puras, sive affectas' (= *Vieta*: 303: *Corollarium*).

[3r]		1 Gr.
	0[′].	1745 34064 3728.
[3v]		89gr.
	0′.	99984 76951 5635
[7r]		3 Gr.
	0[′].	05233 59562 4299.
[13v]		83gr.
	0′.	99254 61516 4132.
[23r]		11 Gr
	0[′].	19080 89903 76545.

(4) Newton gives no derivation of these values for what are, in modern terms, $r\sin 78°$, $r\sin 66°$, $r\sin 42°$ and $r\sin 6°$ respectively, but they follow immediately by applying the addition theorems to the primitive values $\sin 60° = \frac{\sqrt{3}}{2}$, $\sin 30° = \frac{1}{2}$, $\sin 18° = \frac{\sqrt{5}-1}{4}$ and $\sin 72° = \frac{\sqrt{[10+2\sqrt{5}]}}{4}$. (Compare *Vieta*: 303: §1.) In detail,

$$\sin 78° = \sin(60°+18°) = \sin 60° \sin 72° + \sin 30° \sin 18°,$$
$$\sin 66° = \cos 24° = 2(\sin 78°)^2 - 1,$$
$$\sin 42° = \cos 48° = 2(\sin 66°)^2 - 1,$$

and
$$\sin 6° = -\cos 96° = -2(\sin 42°)^2 + 1.$$

We might similarly derive all $\sin(6k)°$, $k = 1, 2, ..., 14$ in terms of simple surd quantities.

(5) Extracted from the final leaves of the early pocket-book in the Fitzwilliam Museum, Cambridge. Only the entries here reproduced were, in fact, made by Newton and the scheme is overwritten with the shorthand passage and elaborate list of accounts which open the reverse end of the notebook. Since the first date in the table of accounts is 23 May 1665, we may safely conclude that the present sine table (and probably also the preceding mathematical notes reproduced in **1**, 1, §1) was composed no later than the spring of 1665.

(6) The scheme for these 'Sine Tables' is entered in outline on the last 91 leaves of the pocket-book. (The pages are not numbered but the position of the following entries is shown according to their supposed pagination as 1r–91v.) The sine canon was planned to be written on both sides of each leaf (0°–45° 0′ on the recto sides, 45° 1′–90° on the verso) with two pages given over to each degree of arc and with entries for each minute: precisely, x^{gr} is entered on ff. $(2x+1)^r/(2x+2)^r$ while $(45+x)^{gr}$ is to be found on ff. $(91-2x)^v/(90-2x)^v$, except that 89^{gr} is on $3^v/4^v$. (In each case 0′–30′ are intended to be set on the first side with 31′–60′ on the second.)

[31^r]		15[gr]
	0[′].	$25881,\ 90451\ 02520\frac{444}{517}$.
[61^r]		30[gr]
	0[′].	50000, 0[0000 00000]
[61^v]		60[gr]
	0[′].	$86602,\ 54037\ 844386\Big/\frac{810}{1732}$.(7)

(7) Apparently, Newton intended to construct this canon of sines at intervals of one minute loosely on the lines of [1] and in response to Viète's expressed wish in the *Auctarium* to his *Ad Adr. Romanum Responsum.* (See *Vieta*: 323: 'Est conditus canon sinuum per singula sexagesima scrupula partium quadrantis circuli in particulis qualium totus adsumitur. Quærit aliquis summam omnium sinuum singulis scrupulis congruentium.' Compare also the *Corollarium* to Anderson's commentary on Viète's exposition of angular sections quoted in note (3) above.) The entries made in fulfilment of this intention are tantalizingly brief, but we may perhaps trace a pattern of derivation by bisection, trisection and quintusection with combination by application of the addition theorems where necessary. Thus: $\sin 30° = \frac{1}{2}$ and $\sin 60° = \frac{\sqrt{3}}{2}$ (as in [1]); $\sin 15° = \sin(\frac{1}{2}.30°)$, $\sin 3° = \sin(\frac{1}{5}.15°)$, $\sin 1° = \sin(\frac{1}{3}.3°)$; then $\sin 89° = \sin(90° - 1°)$, $\sin 83° = \sin(89° - 2.3°)$ and $\sin 11° = \sin(15° - 2^2.1°)$; finally, if we suppose $\sin 1'$ found by Anderson's scheme as noted by Newton in [1] (see note (3)), then $\sin 16' = \sin(2^4.1')$. (Note, finally, that in accordance with standard contemporary convention Newton tabulates integral values for the 'sinus' $r\sin\theta$, where r is taken variously as 10^{14}; 10^{15}; and 10^{16} according as θ takes on the values 16′, 1°, 89° and 3°; 83°, 11°, 15° and 30°; and 60°.)

3

THE THEORY AND CONSTRUCTION OF EQUATIONS

[1665–1666]

§1. THE IMPROVED RESOLUTION OF 'AFFECTED' EQUATIONS

[Early 1665?]

From the original in Newton's Waste Book[1] in the University Library, Cambridge

The resolusion of y^e affected Equation[2] $x^3+pxx+qx+r=0$. *Or* $x^3+10x^2-7x=44$.

First having found two or 3 of y^e first figures of y^e desired roote viz 2|2[3] (wch may bee done either by rationall or Logarithmical tryalls as Mr Oughtred hath tought,[4] or Geometrically by descriptions of lines,[5] or by an instrument consisting of 4 or 5 or more lines of numbers made to slide by one another wch may be oblong but better circular)[6] this knowne ꝑte of ye root I call g, ye other

---

(1) Add. 4004: 64^r. The date is suggested by the writing style of the manuscript and its position in the Waste Book.

(2) See **1**, 2, §1.2.

(3) That is, as we should now write it, '2.2'. The notation is taken from Oughtred's *Clavis* (note (4)), but Newton in the sequel seems to prefer the notation of '2,2'.

(4) See *De Æquationum Affectarum Resolutione in Numeris* = *Clavis Mathematicae*, ₃1652: 110–51, especially 113–24, where Oughtred used logarithmical methods to lighten the calculation of powers of large numbers necessary in his technique for resolving numerical equations. (Compare **1**, 2, §1, note (1).)

(5) This method is elaborated in §2 below.

(6) Newton described this machine for the numerical resolution of equations in a letter to Collins on 20 August 1672 (ULC. Add. 3977.10 = *Correspondence of Isaac Newton*, **2** (1960): 229–31, especially 230: 'The approximating y^e roots of affected Æquations by Gunters line is thus....') Essentially it is no more than an ingenious device for performing Oughtred's logarithmic calculations (note (4)) manually. Three years later, in the early summer of 1675, Collins incorporated it in a general scheme for resolving cubics which he drew up in reply to Leibniz' letter to Oldenburg of 20 May 1675 (NS) (= C. I. Gerhardt, *Der Briefwechsel von ... Leibniz mit Mathematikern*, **1** (Berlin, 1899): 122–4). In his still unpublished draft *In Answer to Mon^r Leibnitz's Letter about Solving a Cubick æquation by Plaine Geometry* (British Museum. Add. 4398: 139^r–140^r) Collins wrote (139^v): '...7 Mr Newton by helpe of the Log^mes graduated on Scales that are to lye parallel, at equall distances, or by helpe of Concentrick Circles so graduated[,] finds the rootes of æquations; 3 Rulers serve for Cubicks 4 for Biquadraticks &c, in the placing of these the respective Coefficients lye all in the same right [line], from a Point whereof, as remote from the first ruler as the graduated Scales stand from

unknowne pte I call y[,] then is $g+y=x$. Then I prosecute y^e Resolution[7] after this manner (making $x+p$ in $x=a$. $a+q$ in $x=b$. $b+r$ in $x=c$. &c.)[8]

$$\begin{array}{r|l} \times & 12=x+p \\ \hline x=2, & 24=a \end{array}\,. \quad \begin{array}{r|l} a+q=17 & \times \\ \hline b=34 & 2=x \end{array}\,. \quad r^{(9)}-b=10=h. \text{ by supposing } x=2.$$

Againe supposing $x=2\lfloor 2$.

$$\begin{array}{r|l} x+p=12{,}2 & \\ \hline 244 & 2 \\ 244\ & 2 \\ \hline 26{,}84=a & \end{array} \qquad \begin{array}{r|l} a+q=19{,}84 & \\ \hline 3968 & 2 \\ 3968\ & 2 \\ \hline 43{,}648=b & \end{array}$$

$r^{(9)}-b=00{,}352=k$. $h-k=9{,}648$. That is y^e

$$\left.\begin{array}{l} \text{latter } r-b \text{ substracted from the former } r-b \text{ there remaines} \\ \text{difference twixt this \& y}^{\text{e}} \text{ former valor of } r-b \text{ is} \end{array}\right\} 9{,}648.$$

& y^e difference twixt this & y^e former valor of x is 0,2. Therefore make

$$9{,}648:0{,}2::0{,}352:y.$$

Then is $y=\dfrac{0{,}0704}{9{,}648}=0{,}00728$ &c. the first figure of wch being added to y^e last valor of x makes $2{,}207=x$. Then wth this valor of x prosecuting y^e operacon as before tis[10]

$$\begin{array}{r|l} x+p=12{,}207 & \\ \hline 85449 & 7 \\ 2\,44140\ & 02^{(11)} \\ 24\,414\ & 2 \\ \hline 26{,}940849=a & \end{array} \qquad \begin{array}{r|l} a+q=19{,}94084 & \\ \hline 13958588 & 7 \\ 39881680\ & 02^{(11)} \\ 3988168\ & 2 \\ \hline 44{,}00943388=b & \end{array} \qquad r^{(12)}-b=-0{,}00943388.$$

---

each other, a right line is stretched over them with direction suited to the nature of the Æquation, Whereby on one of the rulers is given a pure power of the roote sought.' Some time in June Collins passed his *Answer* on to Oldenburg and it was communicated to Leibniz on the 24th of that month. (See Gerhardt's *Briefwechsel*, 1: 127–30, especially 130.) A year later, finally, during a visit to England in the middle of October 1676, Leibniz made notes on Newton's original letter of 20 August 1676 (then still in Collins' possession, presumably), and in particular, made a Latin transcript of the paragraph dealing with Newton's machine. (Leibniz' notes, now in Hanover, were first printed by J. E. Hofmann in his *Studien zur Vorgeschichte des Prioritätstreites zwischen Leibniz und Newton*... (Berlin, 1943): 80, with a photocopy inserted at p. 130.)

(7) Newton has cancelled 'Reduction'.

(8) That is, Newton develops the present cubic $x^3+px^2+qx+r=0$ as $(((x+p)\,x+q)\,x+r)=0$ for convenience of computation. In Newton's example, $p=10$, $q=-7$ and $r=-44$.

(9) Correctly, '$-r$'.

(10) Newton has cancelled a first multiplication of $12{\cdot}207\times 2{\cdot}207$ in which by error he laid out the summation rows for a multiplier 702.

w[ch] valor of $r-b$ substracted from y[e] precedent valor of $r-b$ y[e] diff: is $+0,36143388$. Also y[e] diff[:] twixt this & y[e] precedent valor of x is $0,007$. Therefore I make $0,36143388:0,007::-0,00943388:y$. That is

$$y=\frac{-5,903716}{36143388}=-0,0001633 \text{ \&c.}$$

2 figures of w[ch] (because negative) I substract from y[e] former value of x & there rests $x=2,20684$. And so might y[e] Resolution be prosecuted.(13)

(11) Read '20' correctly.

(12) Again '$-r$' correctly.

(13) Newton's argument will be clearer if we set it in modern terms. If we suppose that $y=f(x)$ is the defining relation of the curve $qA'B'C'$ with respect to the Cartesian coordinates $Op=x$ and $pq=y$ and that the curve meets the axis in D such that $OD=X$, we may consider,

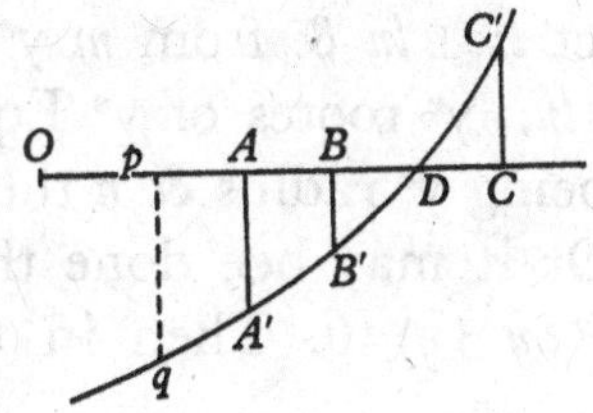

with Newton, a given set of values $OA=a$, $OB=b$, $OC=c$, ...which straddle $OD=X$ and treat the curve (q) as a straight line to a first approximation. We may deduce from these assumptions that, where A and B are on the same side of D,

$$OD \approx OB+(OB-OA)\frac{BB'}{AA'-BB'}\left(\text{or } X \approx b+(b-a)\frac{-f(b)}{-f(a)+f(b)}\right);$$

and similarly, where B and C lie on opposite sides of D,

$$OX \approx OC-(OC-OB)\frac{CC'}{CC'+BB'}\left(\text{or } X \approx c-(c-b)\frac{f(c)}{f(c)-f(b)}\right).$$

In either case, using the divided difference notation

$$f(\alpha,\beta)=\frac{f(\alpha)-f(\beta)}{\alpha-\beta} \quad\text{and}\quad f(\alpha,\beta,\gamma)=\frac{f(\alpha,\beta)-f(\beta,\gamma)}{\alpha-\gamma},$$

we may write Newton's two basic results succinctly as $0=f(a,b,X)$ and $0=f(b,X,c)$. It follows that Newton's approximation method is strictly equivalent to rounding off a divided differences interpolation of $f(x)$ in the given values $f(a), f(b), f(c)$ &c after the first two terms.

§2. THE GEOMETRICAL CONSTRUCTION OF EQUATIONS.

From the original in the Waste Book[1] in the University Library, Cambridge

May 30th 1665.

Of the construction of Problems.

The resolution of plaine problems by yᵉ Circle.

If yᵉ Equation to be resolved bee $yy \between ay - bb = 0$. Or $yy - ay + bb = 0$. in wᶜʰ yᵉ roote of yᵉ last terme (viz: b) is knowne, they may bee conveniently resolved by D. Cartes his rules.[2] Otherwise yᵉ roote of yᵗ terme must bee first extracted as in this $yy - py + q = 0$. Where I take $ln = \frac{1}{2}q$.[3] $lg = \frac{q-1}{2}$. $gs = \frac{q+1}{2}$. & soe describing yᵉ circle smf erect $lm \perp ln$ & from m yᵉ point of intersection draw $mr \parallel ln$. yᵉ rootes of yᵉ Equation shall bee mq & mr. ln being yᵉ radius & n the center of the circle.[4]

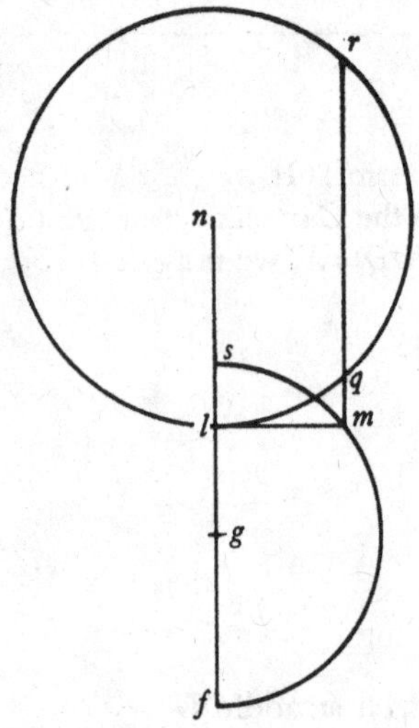

Or it may bee done thus. Let the Equation bee $yy \between py \between q = 0$. Then in the indefinite line af take $ab = \frac{\between p}{2}$.[5] erect the perpendicular $db = c$. And from

(1) Add. 4004: 67ᵛ–69ʳ, a revision with improvements and many additions of parts of Book 1 and the latter half of Book 3 of Descartes' *Geometrie*. There Descartes' aim had been to give simple geometrical constructions for the real roots of algebraic equations up to the sextic as the meets of curves suitably defined in a Cartesian co-ordinate system. For him this geometrical resolution of equations was not a mere pleasing mathematical structure but a serious attempt at solution where contemporary analytical techniques had proved inadequate, and it became important to introduce the concept of a geometrically *easiest* method of construction. This seems the reason why in Descartes' work (and, following him, in that of Newton) the circle occupied a primary position as the most easily constructible of all true curves. Similarly, we can see why the Cartesian parabola, constructible by Descartes' parabola-linkage as a simple mechanical curve though analytically cumbrous, should be his second choice in the general scheme for the resolution of quintics and sextics. (See *Geometria*: Liber III: 97–104.) In the present manuscript are to be found Newton's own first researches in the field and, once caught, his interest was held for the rest of his life. Some half dozen years later he rewrote these researches as a lengthy unpublished tract (ULC. Add. 3963. 9) on the construction of equations, a summary of which he incorporated into his Lucasian lectures on mathematics (ULC. Dd. 9. 68: Octob. 1683, Lectiones 1–10: 216–51, printed in appendix to his *Arithmetica Universalis* (London, ₁1707): 279–326: *Æquationum Constructio linearis*). (Compare C. B. Boyer, *History of Analytic Geometry* (New York, 1956): 95–9.)

(2) See *Geometria*: Liber I: 6–7, where Descartes constructs the equation $y^2 - ay + b^2 = 0$ as the meet of $x - b = 0$ and $x^2 + y^2 - ay = 0$ (which are the Cartesian representing equations of a line and circle respectively).

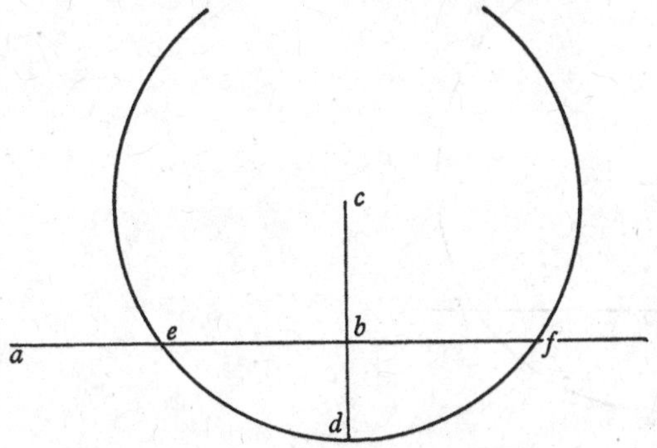

ye point d towards b draw

$$dc=\frac{pp\,\text{Ȣ}\,4q+4cc}{8c}\left(=\frac{pp}{8c}\,\text{Ȣ}\,\frac{q}{2c}+\frac{c}{2}\right)$$

wth wch radius describe ye circle *edf* & *ae*, *af* shall bee ye rootes of ye Equation.(3) Where note that any quantity may be taken for c, Soe yt ye operac̄on may thereby be made convenient, & to yt purpose the difference twixt *db* & *dc* must bee as little as may bee, (that is twixt $pp\,\text{Ȣ}\,4q$ & $4cc$). soe yt ye circle intersect not (af) over obliquely. nor ye circle be over greate.

As if I had this Equation $yy+6y-9=0$. Or $yy=-6y+9$. Then must I make

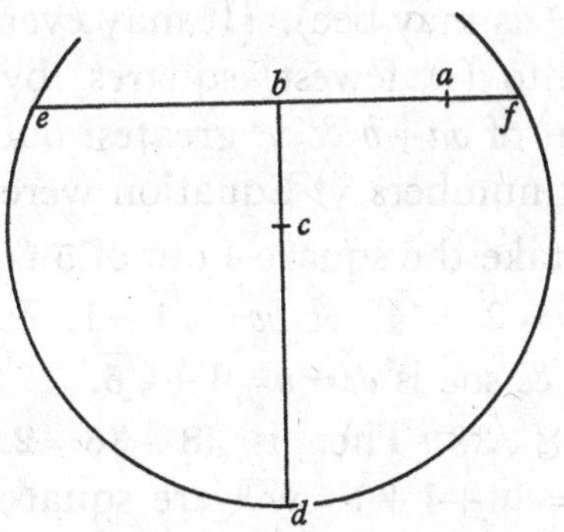

$dc=\frac{9}{c}+\frac{1}{2}c$. Then if I make $c=6$ it will bee

$d[c]=\frac{3}{2}+3=\frac{9}{2}=4\frac{1}{2}$. Therefore I take

$$ab=-\frac{1}{2}p=-3.\quad bd=c=6.\quad dc=4\tfrac{1}{2}=\frac{3c}{4}.$$

And soe describing ye circle *efc*, I have one affirmative roote *af*, another negative *ae*.(7) Or had I taken any other convenient valor for c as 1, or 3. or 4 the line *ae* & *af* would still have bene ye same.

Had I this equation $yy-8=0$. or $yy=8$. Then is $dc=\frac{4}{c}+\frac{1}{2}c$. Or makeing $c=2$, tis $dc=3$. Soe yt since p is wanting I take $ab=0$. $ad=c=2$. $dc=3$. & describing a circle ye rootes will bee *ea*, *af*.

---

(3) Read '$\frac{1}{2}p$'.

(4) Newton constructs $y^2-py+q=0$ as the meet of $x-\sqrt{q}=0$ and $x^2+y^2-py=0$ (the line *mqr* and circle *lqr* respectively in his figure).

(5) Newton went on to write 'Erect ye perpendicular $bc=\frac{pp}{8c}\,\text{Ȣ}\,\frac{q}{2c}-\frac{1}{2}c\left(bc=\frac{pp\,\text{Ȣ}\,4q-4cc}{8c}\right)$ & wth ye Radius *cd*', but quickly cancelled it.

(6) The equation $y^2\pm py\pm q=0$ is constructed as the meet of $x-c=0$ and

$$x^2+(y\pm\tfrac{1}{2}p)^2-\left(\frac{p^2\mp 4q+4c^2}{4c}\right)x=0,$$

which are the straight line *aef* and circle *edf* respectively in Newton's figure (where the general point e is defined with respect to the Cartesian co-ordinates $db=x$ and $be=\frac{1}{2}p\pm y$).

(7) In fact, $af=3(\sqrt{(2)}-1)$ and $ae=-3(\sqrt{(2)}+1)$.

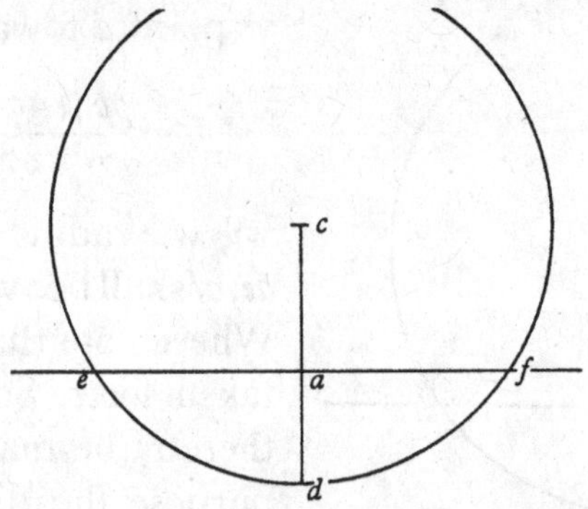

Note y[t] if dc is negative or not greater then $\frac{1}{2}ad$ y[e] circle cannot intersect y[e] line eaf & therefore y[e] rootes of y[e] equation are imaginarie.

[(8)]Or they may bee construed by drawing streight lines onely thus.

Let y[e] Equation be $yy=2ay+b$. or $y=a$ ♉ $\sqrt{aa+b}$. First I divide $aa+b$ into square numbers (as few of y[m] as may bee). (It may ever bee divided (though not) into (y[e] fewest) squares, by taking the greatest square out of $aa+b$ & y[e] greatest out of y[e] remainder &c.) as if in numbers y[e] Equation were $yy=2y+4$. Or $y=1$ ♉ $\sqrt{5}$. I take the square 4 out of 5 & there rests 1 w[ch] is also a square. Then I draw $ab=2=\sqrt{4}$. & $bc=\sqrt{1}=1$. & make $ab\perp bc$. soe is $ac=\sqrt{5}$. to w[ch] I add $ad=1$. & soe is $dc=y=1+\sqrt{5}$.

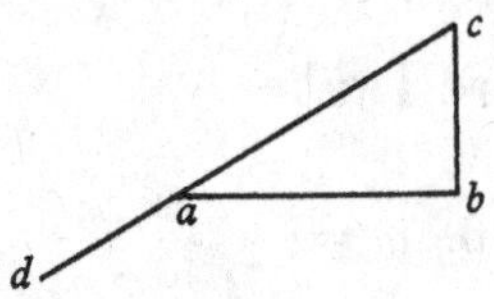

Were y[e] Equation $yy=-4y+34$. Or $y=-2$ ♉ $\sqrt{38}$. Then is $38-36=2$. $2-1=1$. & $38=36+1+1$. w[ch] are square numbers. Therefore I make $ad=\sqrt{36}=6$. $ad\perp de=\sqrt{1}=1$. & draw $ae\perp ec=\sqrt{1}=1$. & draw $ac=\sqrt{38}$. from w[ch] take $ab=2$, & there rests $bc=-2+\sqrt{38}=y$.

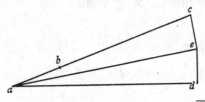

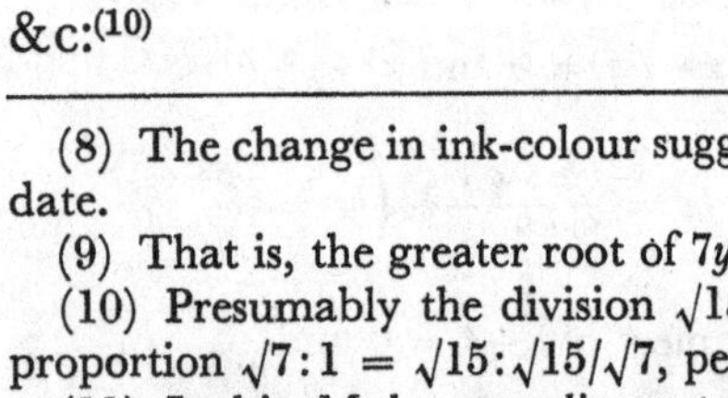

Were y[e] Equation $y=6+\sqrt{\frac{15}{7}}$.[(9)] Or $y=6+\frac{\sqrt{15}}{\sqrt{7}}$. Find $\sqrt{15}$. & $\sqrt{7}$ as before, &c:[(10)]

(8) The change in ink-colour suggests that the remainder of this section was added at a later date.

(9) That is, the greater root of $7y^2-84y+237 = 0$.

(10) Presumably the division $\sqrt{15}/\sqrt{7}$ is to be effected by constructing geometrically the proportion $\sqrt{7}:1 = \sqrt{15}:\sqrt{15}/\sqrt{7}$, perhaps by similar triangles.

(11) In his *Modus generalis construendi omnia Problemata solida, reducta ad Æquationem trium, quatuórve dimensionum* (*Geometria*: Liber III: 85–90) Descartes had shown how to reduce the general quartic, by a suitable transform of the roots, to the form $z^4-pz^2+qz-r = 0$, and this he then constructed geometrically as the meet of the circle $x^2+(z+\frac{1}{2}q)^2 = \frac{1}{4}[q^2+(p+1)^2]+r$ and the parabola $z^2=x+\frac{1}{2}(p+1)$. (There are some slight variations in his method according as the coefficients p, q and r are each positive or negative.)

If y^e^ Probleme be sollid it may bee readily resolved by the intersection of y^e^ Parabola & circle as D: Cartes hath shewed.(11) If it bee of 5 or 6 dimensions it may bee resolved by y^e^ intersection of y^e^ line $y^3-byy-cdy+bcd+dxy=0$.(12) Or $y^3-byy+bcd+dxy=0$(13) & y^e^ circle when pp ⅃(14) $4q$. & q & v affirmative. as D:C: hath explained.(15) Or it might bee done by y^e^ intersection of a circle & one of these lines, viz $y^3+byy-hx=0$ when y^e^ equation is reduced to such a forme y^t^ $pp=4q$. Or this $y^3+byy+gy-hx=0$. Or this $y^3+gy-hx=0$, s being affirmative & $p=0$. Or this $y^3+d-fyx=0$ when $p=0$, & q & v affirmative. &c.(16)

con- ction of d—& ear blems.

(12) That is, $dxy = (b-y)(y^2-cd)$, the 'trident' (as Newton was later to name the class of curves from their shape of a three-pronged fork). Descartes had given a simple mechanical description of this 'Cartesian' parabola in his *Geometrie* (=*Geometria*: Liber II: 36–7) as the meet of a pivoting line and a freely translatable parabola.

(13) Which is immediately reducible to the standard Cartesian form

$$d(x-c)y = (b-y)(y^2-cd).$$

(14) 'lesse y^n^' (Barrow's modification of Oughtred's notation).

(15) In his *Modus generalis construendi Problemata omnia, reducta ad Æquationem sex dimensiones non excedentem* (*Geometria*: Liber III: 97–104) Descartes constructed the general sextic

$$y^6-py^5+qy^4-ry^3+sy^2-ty+v = 0$$

as the meet of the trident

$$n\left(x-\frac{t}{2nv^{\frac{1}{2}}}+\frac{2v^{\frac{1}{2}}}{pn}\right)y = \left(\tfrac{1}{2}p-y\right)\left(y^2-2\frac{v^{\frac{1}{2}}}{p}\right)$$

and the circle

$$x^2+\left(y-\frac{m}{n^2}\right)^2 = \left(\frac{t}{2nv^{\frac{1}{2}}}\right)-\left(\frac{s+pv^{\frac{1}{2}}}{n^2}\right)+\left(\frac{m}{n^2}\right)^2,$$

where

$$2m = \frac{pt}{2v^{\frac{1}{2}}}+2v^{\frac{1}{2}}+r \quad\text{and}\quad n^2 = -\tfrac{1}{4}p^2+q+\frac{t}{v^{\frac{1}{2}}}.$$

The condition for real n is that $p^2 \leqslant 4(q+t/v^{\frac{1}{2}})$ but the restricted form $p^2 < 4q$ here noted by Newton is Descartes' own limitation. (Compare Claude Rabuel, *Commentaires sur la Geometrie de M. Descartes* (Lyon, 1730): Livre 3e, Partie 5e: 566–77: Section I, *Façon generale pour construire tous les Problêmes reduits à une équation, qui n'a point plus de six dimensions.*)

(16) Newton constructs the sextic

$$y^6-py^5+qy^4-ry^3+sy^2-ty+v = 0$$

under various restricting conditions:

(*a*) When $p^2 = 4q$, the sextic is given as the meet of

$$y^3-\tfrac{1}{2}py^2 = hx \quad\text{and}\quad h^2(x^2+y^2)+hrx-ty+v = 0,$$

where $h^2 = s+\frac{1}{2}pr$.

(*b*) More generally, it is given as the meet of

$$y^3-\tfrac{1}{2}py^2+gy = hx \quad\text{and}\quad (x-\alpha)^2+(\gamma-\beta)^2 = \alpha^2+\beta^2-\gamma,$$

where $g=\frac{1}{2}q-\frac{1}{8}p^2$, $\alpha h = \frac{1}{2}(-pg+r)$, $h^2 = s^2-g^2+p(\alpha h)$, $h^2\beta = \frac{1}{2}t-g(\alpha h)$ and $\gamma = v/h^2$.

(*c*) When $p = 0$, it is the meet of $y^3+\frac{1}{2}qy = hx$ and $(x-\alpha)^2+(y-\beta)^2 = \alpha^2+\beta^2-\gamma$, where

$$h^2 = s^2-\tfrac{1}{4}q^2,\quad \alpha = \frac{r}{2h},\quad \beta = \frac{t-\frac{1}{2}qr}{2(s^2-\frac{1}{4}q^2)} \quad\text{and}\quad \gamma = \frac{v}{s^2-\frac{1}{4}q^2}.$$

(*d*) Alternatively, when $p = 0$, the sextic is constructible as the meet of $y^3+v^{\frac{1}{2}} = fxy$ and $(x-\alpha)^2+(y-\beta)^2 = \alpha^2+\beta^2-\gamma$, where

$$f^2 = q-\frac{t}{v^{\frac{1}{2}}},\quad \alpha = -\frac{t}{2v^{\frac{1}{2}}f} \quad\text{and}\quad \beta = \frac{v^{\frac{1}{2}}+\frac{1}{2}r}{q-t/v^{\frac{1}{2}}}.$$

A Generall rule wherby any Probleme may bee resolved.

But all Equations in Generall may bee resolved by ye line $a^2x = y^3$,(17) after this manner. First (making $a=1$) describe ye line $x=y^3$ uppon a plate.(18) (as *cadce.* in wch $ab=x$. $bc=y$.) Then suppose ye Equation to bee resolved bee

$$y^9 * + my^7 + ny^6 + py^5 + qy^4 + ry^3 + syy + ty + v = 0.$$

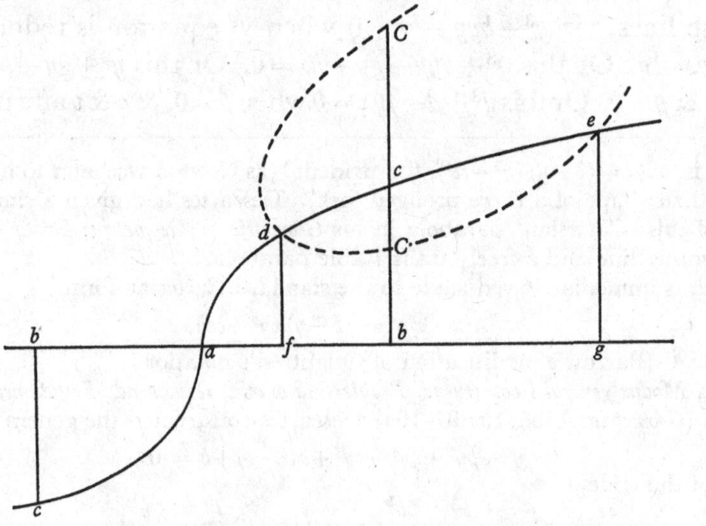

(in wch ye letters m, n, p &c: signifie ye knowne quantitys of each terme affected wth its signe + or −). I describe another line *CdCe*, whose nature (making $ab=x$, $bC=y$) is thus exprest

$$\begin{array}{llll} x^3 + my\,xx & + pyy\,x & + syy = 0. & \text{\& letting fall} \\ + n & + qy & + ty & \\ & + r & + v & \end{array}$$

perpendiculars from every point where these two lines intersect as, *df eg*, they shall bee ye rootes of ye propounded Equation.(19)

In like manner was ye Equation to bee resolved

$$y^{10} * + my^8 + ny^7 + py^6 + qy^5 + ry^4 + sy^3 + ty^2 + vy + w = 0.$$

the nature of ye line *CdCe* would bee

$$\begin{array}{llll} x^4 + my\,x^3 & + pyy\,xx & + syy\,x & + wyy = 0. \\ + n & + qy & + ty & \\ & + r & + v & \end{array}$$

Or else it might bee

$$\begin{array}{llll} y\,x^3 + myy\,xx & + qyy\,x & + tyy = 0. \\ + ny & + ry & + vy \\ + p & + s & + w \end{array}$$

Or had I this Equation

$$y^{10} + ly^9 + my^8 + ny^7 + py^6 + qy^5 + ry^4 + sy^3 + ty^2 + vy + w = 0.$$

(17) A cubic (or Wallisian) parabola.

(18) That is, a template.

(19) Newton thus constructs *df*, *eg*, etc., as the roots of the equation in y which is the meet of the cubic parabola $x = y^3$ and the cubic

$$x^3 + (my+n)\,x^2 + (py^2+qy+r)\,x + (sy^2+ty+v) = 0.$$

[t]he nature of y^e^ line *CdCe* would bee,

$$x^4+\begin{array}{l} lyy \\ +my \\ +n \end{array} x^3+\begin{array}{l} pyy \\ +qy \\ +r \end{array} xx+\begin{array}{l} syy \\ +ty \\ +v \end{array} x+wy^2=0.$$

Or,

$$\begin{array}{l} y \\ +l \end{array} x^3+\begin{array}{l} myy \\ +ny \\ +p \end{array} xx+\begin{array}{l} qyy \\ +ry \\ +s \end{array} x+\begin{array}{l} tyy \\ +vy \\ +w \end{array}=0.$$

Or it might bee,

$$\begin{array}{l} yy \\ +ly \\ +m \end{array} x^3+\begin{array}{l} nyy \\ +py \\ +q \end{array} xx+\begin{array}{l} ryy \\ +sy \\ +t \end{array} x+\begin{array}{l} vyy \\ +wy \end{array}=0.$$

If y^e^ resolved Equation have fewer dimensions[,] y^t^ is if some of y^e^ ultimate termes as, *w*, *v*, *t* &c: (or intermediate termes as *m*, *n* &c: be blotted out) Or if y^e^ Equation have more y^n^ 10 dimensions, the nature of y^e^ lines *CdCe* to bee described may be known by y^e^ same manner observing ye order of ye progression.

Tis evident alsoe y^t^ there are 3 divers lines by wch any Probl: may bee resolved unless some of them chanch to be ye same, the easiest whereof is to bee chosen.[20] It appeares also how Equations of 2 & 3 dimensions may be resolved by drawing streight lines; of 4, 5, & 6 by describing some conick section; of 7, 8, 9, by describing a line of 3 dimensions; of 10, 11, 12, by a line of 4 dimensions, &c:[21] but yet *y* is never above 2 dimensions & consequently all these lines may bee described by y^e^ rule & compasses.[22]

[23]Had I this line $y^4=x$, described on a plate & this Equation to bee resolved viz: $y^{13}+ly^{12}*+ny^{10}+py^9+qy^8+ry^7+sy^6+ty^5+vy^4+wy^3+ayy+by+c=0$. It might bee resolved by describing y^e^ line whose nature is

$$\begin{array}{l} +y \\ +l \\ +q \end{array} x^3+\begin{array}{l} nyy \\ +py \end{array} xx+\begin{array}{l} ry^3 \\ +syy \\ +ty \\ +v \end{array} x+\begin{array}{l} wy^3 \\ +ayy \\ +by \\ +c \end{array}=0.$$

A line of y^e^ 2^d^ sort.[24] Whereas by y^e^ preceding rule was required y^t^ a line of y^e^ 3^d^ sort[25] should have beene described.

(20) Where the general equation $f(y) = 0$ is to be constructed as the meet of $x = y^3$ and a secondary curve $F(x, y) = 0$, these three 'divers lines', say $F_1(x, y) = 0$, $F_2(x, y) = 0$ and $F_3(x, y) = 0$, are the result of substituting x for each y^3 in $f(y) = 0$, $yf(y) = 0$ and $y^2f(y) = 0$ respectively.

(21) For the cubic parabola $x = y^3$ meets an arbitrary curve $F(x, y) = 0$ of degree n in $3n$ points (not all of which, of course, may be real). (Compare Newton's next paragraph.)

(22) In the sense that, since the secondary curve $F(x, y) = 0$ is of the second degree only in y, it may be constructed by points merely by resolving the quadratic equation in y as

$$y = \phi(x) \pm \sqrt{[\psi(x)]}.$$

(23) Newton first began this paragraph with 'Those Equations of more then 9 dimensions may bee (though seldom soe)'.

(24) That is, a cubic curve.

(25) Or a quartic.

By what lines a Problem may bee resolved.

And here observe y^{t} taking y^{e} square number wch is next greater y^{n} y^{e} number of ye dimensions of the resolvend equation, that Equation may bee resolved by lines, ye number of whose dimensions is not greater then the roote of ye sq[u]are number. And the rectangles of those numbers wch signifie how many dimensions the lines have, may always bee greater or equall but never lesse y^{n} ye number of dimensions of ye resolvend Equations. For y^{e} number of points in wch two lines may intersect can never bee greater y^{n} y^{e} rectangle of y^{e} numbers of theire dimensions. And they always intersect in soe many points, excepting those wch are imaginarie onely. Soe that all Equations wch have noe more y^{n} $\left\{\begin{matrix}2\\4\\6\\9\\12\\16\end{matrix}\right\}$ dimensions, may always be resolved by the

In how many points two lines may intersect.

intersection of {
2 streight lines. [or]
a streight line & a conick section.
two conick sections.
a conick section & a line of 3 dimensions,
two lines of 3 dimensions.
a line of 3, & another of 4 dimensions.
two lines of 4 dimensions; or by one of 3 & another of 6 dim: &c: }

but not by any simpler lines. (From this consider[$\overline{\text{acon}}$] may Problems be distinguished into sorts.)(26) It will often bee very intricate to resolve Equations of many dimensions by the simplest lines by wch they may bee resolved & also for y^{e} most ꝑt will require a new description of two lines for every probleme. And then it may be often most convenient to use two lines whereof ye one is more compound ye other more simple & already described, As perhaps an Equation of 16 dimensions may bee more speedily resolved by two lines[,] the one of 3 the other of 6 dimensions[,] then by two lines both of wch are of 4 dimensions.

How by the Parabola plaine problems are resolved.

But it will not bee amisse to shew more particularly how these resolutions may bee performed. And that firs[t] by y^{e} parabola.

---

(26) This is apparently Descartes' aim in Book 3 of his *Geometrie*. (See especially his conclusion to the book = *Geometria*: Liber III: 105–6, where in rather confused manner Descartes lays out the basis for a general classification, 'reductis ad eandem constructionem Problematîs omnibus ejusdem generis'.)

(27) Read 'latus rectum of the parabola'.

Suppose therefore I had y^e^ parabola $x = yy$, exactly described & would resolve a plaine probleme[,] the Equation

$$yy + ky + l = 0.$$

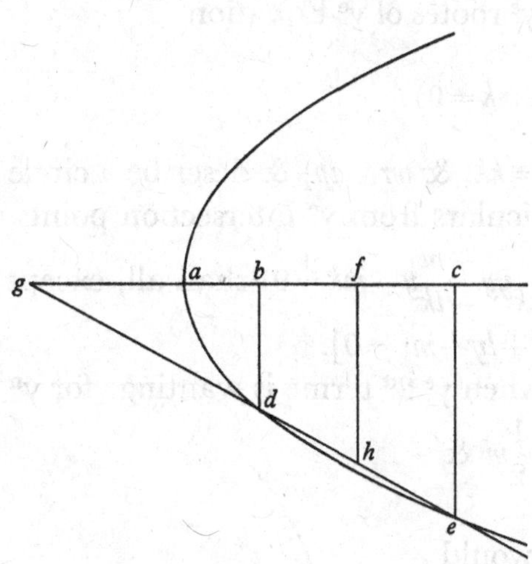

I take $ag = l$. $gf = k$. $fh = 1 =$ lateri recto Parab:(27) & so draw y^e^ line gh & from y^e^ intersection points d, e, draw db, ec perpendicular to y^e^ axis gc. wch shall bee y^e^ rootes of y^e^ Equation wch are affirmative when they fall on y^e^ contrary side to fh, but negative if on y^e^ same, as in this case.(28)

But were I to resolve a sollid problem the Equation being of 4 dimensions, I take away y^e^ 2^d^ terme, makeing it of this forme $y^4 * + lyy + my + n = 0$. Then take $ae = \frac{1}{2}$. $ep = \frac{1}{2}l$. $pq = \frac{1}{2}m$. Then perpendicular to ap draw $af = aq$. Also draw

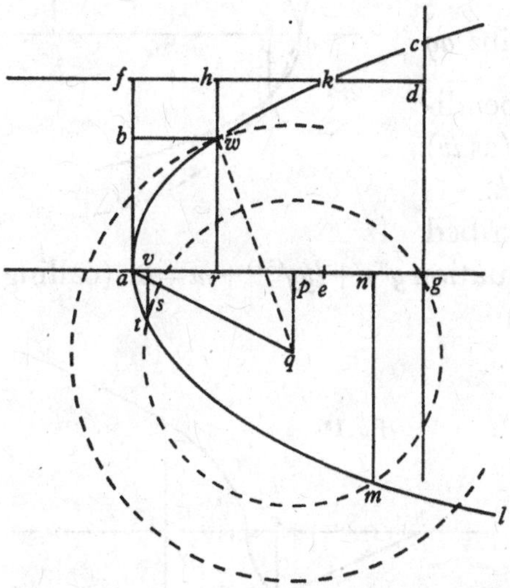

$fk \parallel ap$, & from y^e^ point of intersection k draw $kh = n$. lastly draw $hr \perp ap$, & wth the radius wr upon the center q describe y^e^ circle tsm. (or, wch is y^e^ same, take $ar = \frac{1 - 2l + ll + mm - 4n}{4}$. & soe erecting y^e^ perpendicular rw, wth y^e^ Radius rw describe y^e^ circle tsm) & from y^e^ points where it intersects y^e^ Parabola let fall perpendiculars to y^e^ axis, (tv, nm) they shall bee the rootes of y^e^ Equation[,] y^e^ affirmative ones falling on y^e^ contrary side to pq. when m is affirmative.(29)

If I would resolve y^e^ cubick Equation $y^3 + ky^2 + ly + m = 0$ (wch multiplyed by $y - k = 0$ produceth $y^4 * + lyy + my - km = 0$) I make $ae = \frac{1}{2}$. $ep = \frac{l - kk}{2}$. $pq = \frac{m - kl}{2}$.
$ - kk \quad - kl$

(28) Newton constructs the quadratic $y^2 + ky + l = 0$ as the meet of the parabola $y^2 = x$ with the straight line $x + ky + l = 0$.

(29) $y^4 + ly^2 + my + n = 0$ is constructed as the meet of $x = y^2$ with the circle

$$x^2 + y^2 + (l - 1)x + my + n = 0.$$

Newton has cancelled a further remark, 'If $n = 0$, that is if $y^3 * + ly + m = 0$, then must the circle bee described wth y^e^ Radius aq; for then is $wr = fa = aq$'.

$fk \parallel ae \perp af = aq$. $kd = km$.(30) And wth y^e radius cg upon y^e center q describe y^e circle wl. Or else doe thus (since k is one of y^e rootes of y^e Equation

$$y^4 * \begin{matrix}+l\\-kk\end{matrix} yy \begin{matrix}+m\\-lk\end{matrix} y - mk = 0)$$

make $k = ab \perp ar$ & draw $bw \parallel ae$ (or make $ar = kk$, & $wr \perp ap$) & describe a circle wth y^e radius wq. Then letting fall perpendiculars from y^e intersection points, they (being y^e rootes of y^e Equation $y^4 * \begin{matrix}+l\\-kk\end{matrix} yy \begin{matrix}+m\\-lk\end{matrix} y - mk = 0$) shall all, except $wr = k$, bee y^e rootes of y^e Equation $y^3 + kyy + ly + m[=0]$.

This operation will bee much shortened when y^e 2^d terme is wanting. for y^n since $k = 0$. it will bee $ae = \frac{1}{2}$. $ep = \frac{1}{2}l$. $pq = \frac{1}{2}m$ & aq y^e radius of y^e circle.

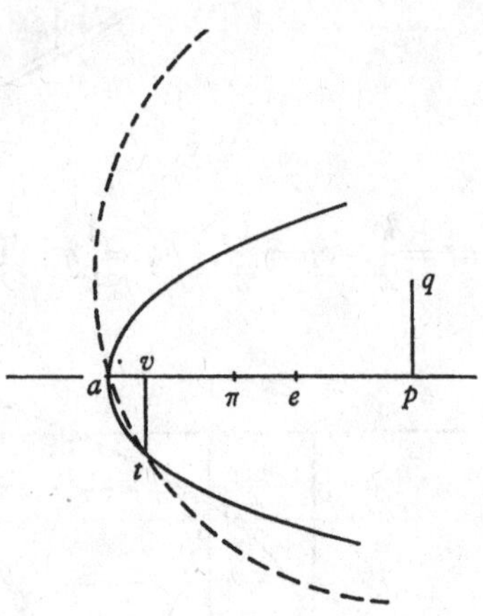

And if y^e last terme vanish that is if I would resolve this equation $yy + ky + l = 0$. by y^e intersection of a circle & parabola I must take $ae = \frac{1}{2}$. $e\pi = \frac{1}{2}l$. $\pi p = \frac{kk}{2}$. $pq = \frac{kl}{2}$. & soe wth y^e radius aq upon y^e center q describe a circle, & y^e perpendiculars from y^e intersection points to y^e axis (as tv) are y^e rootes excepting one wch is equall to k.

Constructions performed by a Parabola of y^e 2^d kind. $x = y^3$.

If I had y^e crooked line [(]fig 1st[)] described whose nature is $x = y^3$. & would resolve y^e Equation $y^3 * + lyy$(31) $+ m = 0$. (calling $ad = x$, $dg = y$; Or $a[c] = -x$. $ce = -y$) I take $ab = m$. $bd = l$. $df = 1$. & $df \perp bd$. & draw bf infinitely both ways. From y^e intersection points (as e) letting fall perpendiculars, they shall bee y^e rootes of y^e Equation $y^3 * + lyy$(31) $+ m[=0]$. as ce wch in this case is negative because on that side on wch y is negative.(32)

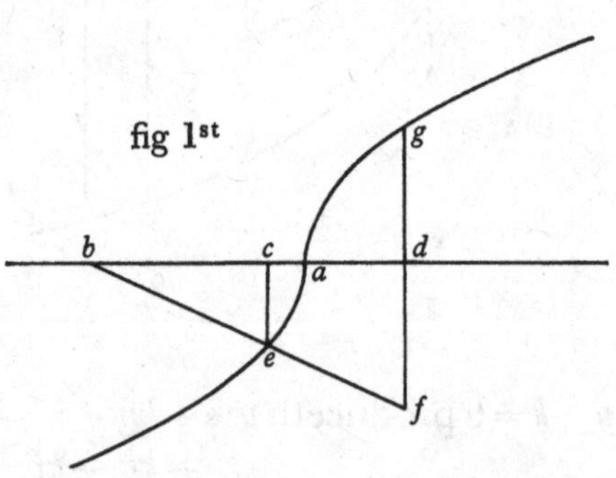

Would I resolve this equation

$$yy + ky + l = 0.$$

(wch multiplyed by $y - k$ produceth $y^3 * \begin{matrix}+l\\-kk\end{matrix} y - kl = 0$) I take $ab = kl$, (fig 2^d)

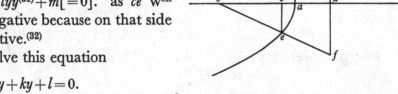

(30) That is, the coefficient km.
(31) Read '$+ly$'.
(32) Newton constructs $y^3 + ly + m = 0$ as the meet of the cubical parabola $x = y^3$ and the straight line $x + ly + m = 0$.

$b\delta = l$, $\delta d = kk$. $df = 1$, & soe through y^e points b & f draw the streight line $bf\lambda$ (Or wch is y^e same take $ab = kl$. $k = ah \perp ab$. & draw $h\lambda \parallel ab$ untill it intersect y^e crooked line in λ (i.e. untill $h\lambda = k^3$) & soe through the points λ & b draw

fig 2^d.

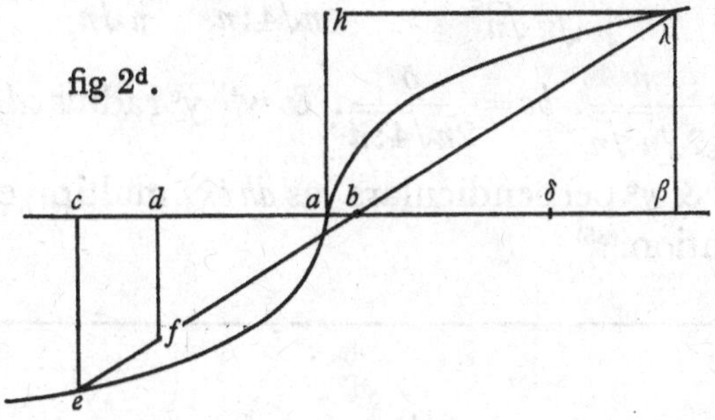

λbfe). Then from y^e intersection points to y^e axis letting fall perpendiculars they (being y^e rootes of y^e Equation $y^3 * \begin{matrix} +l \\ -kk \end{matrix} y - lk = 0$.) shall all, except $\beta\lambda = k$, be y^e rootes of y^e Equation $yy + ky + l = 0$.(33)

Would I resolve the Equation $z^4 + az^3 + bzz + cz + d = 0$. It may bee done by a circle thus. Multiply it by this Equation $zz - az + aa - b = 0$, & it will produce

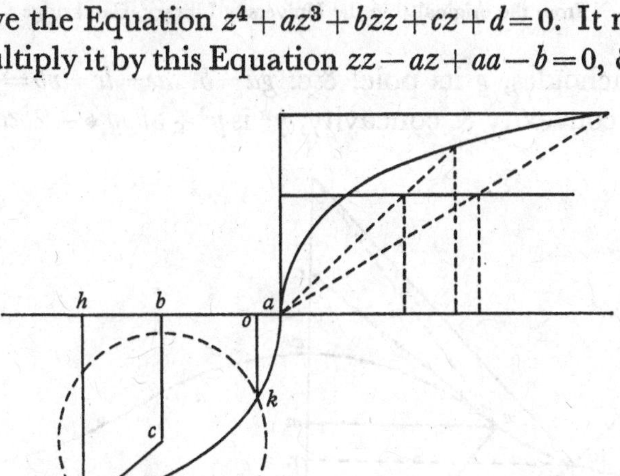

$$\begin{array}{llll} z^6 * * + c\, z^3 & + d\, zz & - ad\, z & + aad = 0, \\ \quad\;\; -2ab & -ac & +aac & -bd \\ \quad\;\; +a^3 & +aab & -bc & \\ & -bb & & \end{array}$$

Of this forme $z^6 * * + mz^3 + nzz + pz + q = 0$. In w^ch ($n$) ought to be affirmative, & if it bee not, y^n augment or diminish y^e rootes of y^e Equation $z^4 + az^3 + bzz + cz + d = 0$. & then repeate y^e operacon̄ againe

(33) $(y-k)(y^2+ky+l) = 0$ is found as the meet of $x = y^3$ and the line
$$x + (l-k^2)\, y - lk = 0.$$

untill there bee an Equation of this forme $z^6 * * + mz^3 + nzz + pz + q = 0$ in wch n is affirmative. Then (dividing this equation by $\sqrt{4:n}$(34) it is

$$z^6 * * + \frac{m}{\sqrt{n\sqrt{n}}} z^3 + zz + \frac{p}{n\sqrt{4:n}} z + \frac{q}{n\sqrt{n}} = 0$$

therefore) take $ab = \frac{m}{2\sqrt{n\sqrt{n}}}$. $bc = \frac{p}{2n\sqrt{4:n}}$. & wth y^e^ radius $cd = \sqrt{\frac{mmn + pp - 4nq}{4nn\sqrt{n}}}$, describe y^e^ circle dk & y^e^ perpendiculars (as $dh\ ok$) multiplyed by $\sqrt{4:n}$ shall bee y^e^ rootes of y^e^ Equation.(35)

§3. CONSTRUCTION OF INFLEXION POINTS IN THE CONCHOID

[Early 1665?]

From the original(1) in the University Library, Cambridge

If vce is a Conchoïdes, g its pole, &c: $ga = b$. $ae = lc = vb = c$. $ma = y$. & c y^e^ point betwixt its convexity & concavity, y^n^ is $y^3 + 3byy * - 2bcc = 0$. (see [Cartes

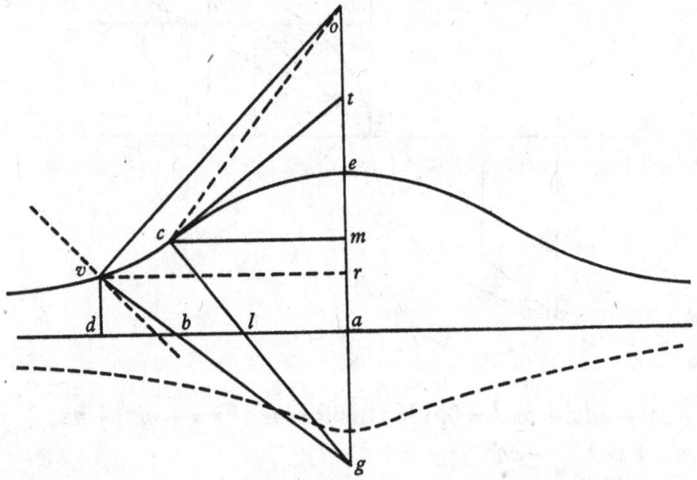

(34) That is, dividing the roots of the equation by $n^{\frac{1}{4}}$.

(35) In general, $z^6 + \alpha z^3 + z^2 + \beta z + \gamma = 0$ may be constructed as the meet of $x = z^3$ and the circle $x^2 + z^2 + \alpha x + \beta z + \gamma = 0$, and this Newton shows how to derive from the more general form $z^6 + mz^3 + nz^2 + pz + q = 0$.

(1) Add. 3958. 3: 70^r^. Newton applies the techniques of the preceding section in constructing the inflexion points in the conchoid (Descartes' 'prima Conchoïdes Veterum', as he dubs it at *Geometria*: Liber II: 49–50) as the meet of a given circle with the conchoid itself.

Geom:] pag: 259. lin:10.)(2) W[ch] Equation hath one affirmative roote (ma) referred to ye point c. & two negative rootes whereof ye greater is referred to ye lower Conch, & ye lesser (I thinke) uselesse to this question.(3)

This construction was entered by him at the head of a folded manuscript sheet (Add. 3958. 2: 70/71) which otherwise contains the revision, printed as Part I. 3, §4 above, of his early researches into the binomial theorem. The inference is that the piece was composed in the spring of 1665 and the writing style strongly confirms it.

(2) See Schooten's *Commentarii in Librum II*, O = *Geometria*: 258–9. If we take $cm = z$, we may then deduce that $z^2y^2 = (b+y)^2(c^2-y^2)$, the familiar Cartesian defining equation of the conchoid. (The result follows directly from the relations $cl:ma = lg:ag$ and $la:ag = cm:mg$.) In his commentary Schooten proceeds to find the meet of the straight line st (drawn through the fixed point t on eag) with the conchoid, where $te = v$ and the length of $es = s$ (drawn normal to eag at the vertex e) determines its slope $se/et = s/v$. Immediately the defining

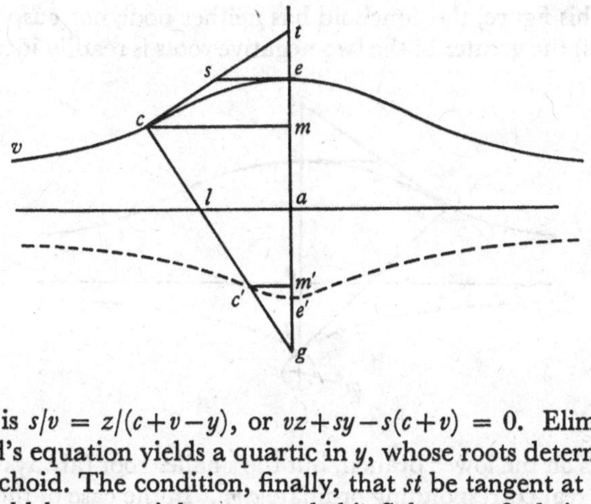

equation of the line is $s/v = z/(c+v-y)$, or $vz+sy-s(c+v) = 0$. Elimination of z between this and the conchoid's equation yields a quartic in y, whose roots determine the four meets of the line with the conchoid. The condition, finally, that st be tangent at an inflexion point c is that the quartic should have a triple root, and this Schooten finds in Cartesian fashion by identifying the quartic with $(y-e)^3(y+f) = 0$. It follows by elimination of e and f that $y^3+3by^2-2bc^2 = 0$, as Newton notes.

On p. 258 of his commentary Schooten is careful to add a reference to 'Nobilissimus D. Hugenius', who had already given an equivalent geometrical construction for the inflexion points '[in] ultimo Problematum Illustrium, quæ de Circuli magnitudine inventis adjecit'. Huygens, in fact, had already found the analytical condition on 25 September 1653 (*Œuvres complètes de C. Huygens*, **12** (La Haye, 1910): no. xx: 83–6, especially 83–4), but almost immediately reworked it into the synthetic form in which he communicated it to Schooten, without the preliminary analysis on 23 October 1653 (*Œuvres*, **1** (1888): no. 164: 245–6). A year later, likewise, when he inserted the construction in appendix to his *De Circuli Magnitudine Inventa* he gave no hint of his prior analysis nor indeed any proof at all of the result. (See *De Circuli Magnitudine Inventa*, Leyden, 1654: Appendix, *Christiani Hugenii Illustrium Quorundam Problematum Constructiones*, Probl. VIII = *Œuvres*, **12** (1910): 210–15.)

(3) Schooten (following Descartes) had in his commentary drawn only the upper shell vce of the conchoid, and it is apparently Newton's insight that the lower branch $v'c'e'$ of the conchoid has exactly the same Cartesian defining equation $z^2y^2 = (b+y)^2(c^2-y^2)$ so that its points of inflexion also are to be found by the same cubical condition $y^3+3by^2-2bc^2 = 0$. Since all points on this lower branch correspond to negative values of the co-ordinate y, Newton correctly deduces that negative roots of this cubic determine its inflexion points. In

The rootes of this Equation may bee thus found by y^e helpe of the described Conch & a circle.(4) viz: Suppose $ga=b$. $ae=lc=vb=c$. $ma=y$. (as before) $ao=s$. $oc=r$. And thereby may bee found this Equation

$$y^3\frac{\overline{+rr-ss+bb-cc}\times yy-2bccy-bbcc}{2s+2b}=0,$$

to bee compared wth y^e former,(5) (see pag 261 lin 20).(6) But their rootes cannot become equall by reason of their third termes.(7) Therefore I alter y^e rootes of y^e first equation, as, suppose I make $y=z-b$. Then is $z^3*-3bbz+2b^3-2bcc=0$, To bee compared wth y^e 2^d equation: wch cannot yet bee done wth out a contradiction, there being but two unknowne quantitys, r & s to bee found by three Equations resulting from y^e comparison of their 2^d 3^d & 4th termes. But if I

---

the case where, as in his figure, the conchoid has neither node nor cusp (and hence two pairs of real inflexion points) the greater of the two negative roots is readily identifiable with the pair

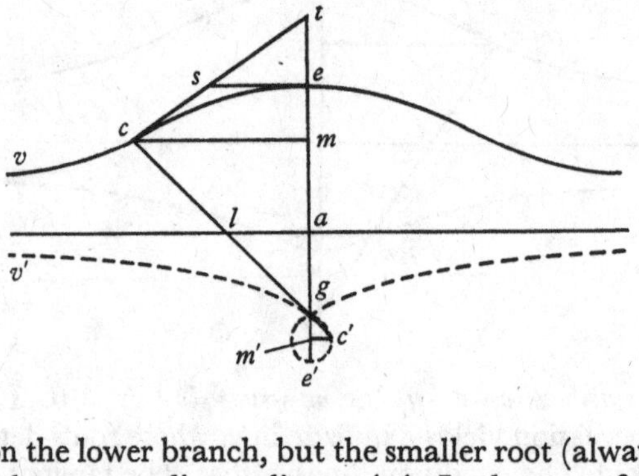

of real inflexion points on the lower branch, but the smaller root (always greater in magnitude than ag) can have no real corresponding ordinate $c'm'$. In the case of the nodal conchoid (not noticed by Newton) both negative roots are greater in magnitude than ag and hence 'uselesse

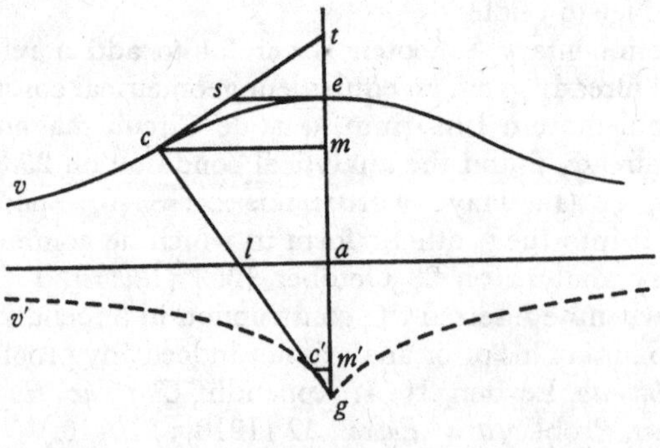

to this question'. Finally, when $b = c$ the conchoid has a cusp at g and the corresponding inflexion condition $y^3+3by^2-2b^3 = 0$ yields $y = -b$ or $-b\pm b\sqrt{3}$: the root $y = b(-1+\sqrt{3})$ determines the inflexion point c in the upper branch, $y = -b$ determines the cusp g and the lesser negative root $y = -b(+1+\sqrt{3})$ has a purely imaginary corresponding ordinate

$$z = \pm\sqrt{[-3\sqrt{3}/2]}.$$

make $z=\dfrac{4ccx-4bbx}{3bb}$, & substitute this valor into its place in y^e^ precedent Equation, the result is $x^3 * \dfrac{-27b^6x}{16c^4-32bbcc+16b^4} \dfrac{-27b^7}{32c^4-64bbcc+32b^4}=0$.(8) The termes of wch being compared with y^e^ termes of y^e^ 2^d^ Equation[,] y^e^ 3^d^ or 4th give $\dfrac{2bcc}{2s+2b}=\dfrac{27b^6}{16c^4-32bbcc+16b^4}$. Or $16c^6-32bbc^4+16b^4cc=27b^6+27b^5s$. & $s=\dfrac{16c^6}{27b^5}-\dfrac{32c^4}{27b^3}+\dfrac{16cc}{27b}-b=ao$. Their second termes give $\dfrac{rr-ss+bb-cc}{2s+2b}=0$. Or $r=\sqrt{ss+cc-bb}=ov$. THEREFORE from y^e^ intersection point v (made by y^e^ Conch & a circle whose radius is ov & center o) let fall $vd \perp ad$; & $ar=x=vd$ is y^e^ roote of y^e^ 2^d^ & last equation. Which being found make

$$y=\frac{4ccx-4bbx}{3bb}-b^{(9)}=am.$$

Which was to bee done.

---

(4) This idea of using the conchoid itself to construct the inflexion condition

$$y^3+3by^2-2bc^2 = 0$$

is due to Heuraet, and his rather inelegant solution was printed by Schooten, 'Quemadmodum id ab eruditissimo ac præstantissimo Viro-Iuvene D. Henrico van Heuraet, Harlemo-Batavo, inventum fuit, mihique ab eo communicatum', in his *Commentarii in Librum II, O* (= *Geometria*: 259–62). In September 1653 Huygens had, after deducing the cubic inflexion condition, contented himself with a straightforwardly Cartesian construction of the cubic as the meet of a circle and parabola. (Compare his *Œuvres*, **12** (1910): 84–6.) When Schooten in early 1659 published Heuraet's construction of the cubic as the meet of a circle with the conchoid itself, Huygens was quick to grasp the underlying felicity of the method and in August of that year set about cutting away the numerical complexity which encumbered Heuraet's treatment. His resulting analysis and simplified geometrical construction, forgotten till the present century when his notes were published (in his *Œuvres*, **12** (1910): 232–7), closely paralleled that of Newton, who concocted his own version wholly independently some half dozen years afterwards—indeed Huygens' approach is exactly that of the revision which Newton made a couple of years later of his present argument (as we shall see in the next volume).

(5) The circle (c) has radius $oc = r$ and centre o in ae such that $oa = s$: hence its defining equation is $r^2 = z^2+(s-y)^2$. Elimination of z^2 between this and the conchoid's equation $z^2y^2 = (b+y)^2(c^2-y^2)$ yields the present cubic (as Schooten shows in *Geometria*: 256–7).

(6) A reference to Schooten's *Commentarii in Librum II, O* = *Geometria*: 261.

(7) That is, their terms in y (which the first equation lacks).

(8) Newton chooses $z = \lambda x$ such that the equations

$$y^3+\frac{(r^2-s^2+b^2-c^2)\,y^2-2bc^2y-b^2c^2}{2(s+b)} = 0$$

and

$$z^3-3b^2z+2b(b^2-c^2) = 0$$

may be identical when $y = z$. The condition yields immediately

$$\frac{-2bc^2}{-b^2c^2} = \frac{-3b^2\lambda}{2b(b^2-c^2)} \quad \text{and so} \quad \lambda = \frac{4(c^2-b^2)}{3b^2}.$$

(9) That is, $z-b$.

§4. OPERATIONS ON THE ROOTS OF EQUATIONS

From the original[1] in Newton's Waste Book in the University Library, Cambridge

May. 1665 *Concerning Equations when the ratio of their rootes is considered.*

If to[2] of y^e rootes of an Equation are in proportion y^e one to y^e other as a to b Then multiplying y^e termes of the Equation by this progression

$$:\&c.\ \frac{a^4}{b^3}+\frac{a^3}{bb}+\frac{aa}{b}.\ \frac{a^3}{bb}+\frac{aa}{b}.\ \frac{aa}{b}.\ 0.\ -a.\ -a-b.\ -a-b-\frac{bb}{a}.\ -a-b-\frac{bb}{a}-\frac{b^3}{aa}.\ \&c.$$

(Or by y^e same progression augmented or diminished by any quantyty, as if it bee augmented by a it will bee $\frac{a^4}{b^3}+\frac{a^3}{bb}+\frac{aa}{b}+a.\ \frac{a^3}{bb}+\frac{aa}{b}+a.\ \frac{aa}{b}+a.\ a.\ 0.\ -b.\ -b-\frac{bb}{a}.\ -b-\frac{bb}{a}-\frac{b^3}{aa}.\ -b-\frac{bb}{a}-\frac{b^3}{aa}-\frac{b^4}{a^3}.$ &c. Or were it augmented by $a+b$ it would be $\frac{a^3}{bb}+\frac{aa}{b}+a+b.\ \frac{aa}{b}+a+b.\ a+b.\ b.\ 0.\ -\frac{bb}{a}.\ -\frac{bb}{a}-\frac{b^3}{aa}.\ -\frac{bb}{a}-\frac{b^3}{aa}-\frac{b^4}{a^3}.$) Then shall y^e roote wch is correspondent to (b) be a roote of y^e resulting equation: but inverting y^e order of y^e progression, that roote wch is correspondent to (a) shall bee a roote of y^e Equation resulting from such multiplication.[3]

As for example did I know y^t two of y^e rootes of y^e Equation

$$x^3-8xx+9x+18=0$$

were in proportion as 1 to 2 & would I have y^e lesser roote (viz y^t wch is correspondent to 1) I make $b=1$. $a=2$. And soe the progression will bee 28. 12. 4. 0. -2, -3. $-\frac{7}{2}$. &c. Or 30. 14. 6. 2. 0. -1. $\frac{-3}{2}$. $\frac{-7}{2}$. &c by adding 2.

(1) Add. 4004: 55^r–56^r. Newton attacks the problem of transforming the roots of equations, presumably on the style of Descartes' *Geometria*: Liber III: 69–75, especially 75: *Quomodo multiplicari vel dividi possint Æquationis radices, ipsis incognitis.*

(2) Read 'two'.

(3) Newton generalizes the rule for evaluating double roots in an equation which Johann Hudde sent to Schooten on 27 January 1658 (who printed it the following year as *Epistola secunda de Maximis et Minimis* = *Geometria*: 507–16). Indeed Newton will give below a modification of Hudde's procedure in proof of the generation of his own progressions, though these are more naturally derived in the following way: if we suppose that the equation $0 = \sum_{0\leqslant i\leqslant r}(a_i x^i)$ has two roots in the ratio a/b, then we may set $0 = \sum_{0\leqslant i\leqslant r}(a_i x^i) = (x-ak)(x-bk).f(x)$, where $f(x)$ is of degree $r-2$. Multiplying the roots by b/a, we may deduce that

$$0 = \sum_{0\leqslant i\leqslant r}\left(a_i\left(\frac{a}{b}\right)^i x^i\right) = \frac{a^2}{b^2}(x-bk)\left(x-\frac{b^2}{a}k\right).f\left(\frac{a}{b}x\right)$$

Or by adding one more it will bee 31. 15. 7. 3. 1. 0. $[-]\frac{1}{2}$. $[-]\frac{3}{4}$. $[-]\frac{7}{8}$ &c. By any of wch progressions ye Equation may bee multiplyed, as by ye 1st, $x^3-8xx+9x+18=0$. Wch produceth $7xx-24x+9=0$. Or by ye 3d

28. 12. 4. 0.

$$\begin{array}{c} x^3-8xx+9x+18=0. \\ 7. \quad 3. \quad 1. \quad 0. \end{array}$$

Wch produceth $7xx-24x+9=0$. Or by the first Otherwise, by destroying ye 1st terme, $x^3-8xx+9x+18$. Wch produceth $16xx-27x-63=0$. &c[,] the rootes

$0. \quad -2. \quad -3. \quad \frac{-7}{2}$

of wch products are, viz: of ye first $x=\frac{3}{7}$, & $x=3$. Of ye last $x=-\frac{21}{16}$, & $x=3$. Then I conclude 3 to be ye lesse, & consequently 6 the greater of those rootes of ye Equation $x^3-8xx+9x+18=0$. wch are in double proportion.(4) But was ye greater of those rootes desired yn inverting ye progression it would bee

$$\begin{array}{ccc} x^3-8xx+9x+18=0. & \text{Or} & x^3-8xx+9x+18=0. \\ 0 \quad 1 \quad 3 \quad 7 & & -\frac{7}{2} \quad -3 \quad -2 \quad 0 \end{array}$$

The first producing $8xx-27x-126=0$ whose rootes are $x=\frac{-21}{8}$, $x=6$. The 2d produceth $7x^2-48x+36=0$ whose rootes are $x=\frac{6}{7}$, $x=6$. And consequently 6 is the greater & 3 the lesse of ye rootes in duplicate proportiō.

If in ye circle *adef* *af* is ye diameter, *ah* a perpendicular to ye end of it from wch I would draw $he \| af$, wch should intersect ye circle in ye points *d* & *e* soe yt (*de*) bee triple to (*hd*), yt is *he* quadruple to *hd*. Then calling $af=g$. $ah=y$. $\left.\begin{array}{l} hd \\ he \end{array}\right\}=x$. The equation expressing ye relation twixt x & y is $yy=x(g-x)$ or

also has the root $x-bk$, and so therefore we have

$$0 = \sum_{0\leqslant i\leqslant r}\left(a_i\left[\left(\frac{a}{b}\right)^i-1\right]x^i\right) = \frac{1}{b}\left(\frac{a}{b}-1\right).\sum_{1\leqslant i\leqslant r}\left(a_i\left[\sum_{1\leqslant j\leqslant i-1} b\left(\frac{a}{b}\right)^j\right]x^i\right)$$

and more generally $\quad 0 = \lambda.\sum_{0\leqslant i\leqslant r}\left(a_i\left[\sum_{1\leqslant j\leqslant i-1} b\left(\frac{a}{b}\right)^j+\mu\right]x^i\right),$

where λ and μ are arbitrary constants. (Newton's progressions arise by choosing

$$\lambda = \left(\frac{b}{a}\right)^p, \quad p = 0 \text{ or } 1, \quad \text{and} \quad \mu = -\sum_{-s\leqslant j\leqslant 0}\left[b\left(\frac{b}{a}\right)^j\right]+qb,\ q = 0 \text{ or } 1.)$$

(4) In fact, $\quad x^3-8x^2+9x+18 = 0 = (x-3)(x-6)(x+1).$

$xx-gx+yy=0$. y[e] rootes of w[ch] equation must be quadruple y[e] one to y[e] other:

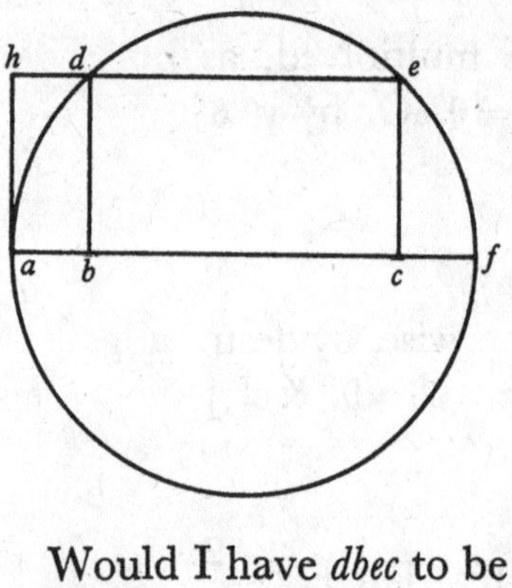

Therefore would I find (hd) y[e] lesse roote I make $b=1$. $a=4$. And y[e] progression will be, 21. 5. 1. 0. $\frac{-1}{4}$. $\frac{-5}{16}$. $-\frac{21}{32}$. &c. by w[ch] y[e] Equation being multiplyed y[e] product is $xx-gx+yy=0$. Or $x=\frac{1}{5}g$.
5 1 0

Therefore drawing $ab=\frac{1}{5}g$, Or $ac=\frac{4}{5}g$. from the point b, or c raise y[e] perpendiculars db, or ce. & soe draw hde.

Would I have $dbec$ to be a square[,] y[t] is $db=de=y$. Then to find hd I call it x & $[e]h=x+y=hd+de$. Soe y[t] y[e] lesse roote is to y[e] greater as x to $x+y$. Making therefore $b=x$, $a=x+y$, The progression will be, $\frac{+yy}{x}+3x+3y$. $2x+y$. x. 0. $\frac{-xx}{x+y}$. $\frac{-xx}{x+y}-\frac{x^3}{xx+2xy+yy}$.(5) By w[ch] the Equation $xx-gx+yy=0$, must be multiplyed. & it produceth $xx-gx+yy=0$. Or $2x^3+yxx-gxx=0$. Or
$2x+y$. x. 0.

$-2x+g=y$. & consequently $-xx+gx=yy=4xx-4gx+gg$. y[t] is $5xx=5gx-gg$. And $x=\frac{1}{2}g\text{♉}\sqrt{\frac{1}{20}gg}$. Or $x=\frac{g\sqrt{5}\,\text{♉}\,g}{2\sqrt{5}}=\frac{1}{2}g\text{♉}\frac{g}{2\sqrt{5}}$. And consequently $y=\frac{1}{\sqrt{5}}g$. Or it might have beene done thus. x substracted from y[e] precedent progression it will be, $x+y$. 0. $-x-\frac{xx}{x+y}$. &c. by w[ch] y[e] Equation being multiplyed produceth, $xx-gx+yy=0$. Or $xx+yx=yy$. And by extracting the roote,
$x+y$ 0 $-x$

$y=\frac{1}{2}x\text{♉}\sqrt{\frac{5}{4}xx}$. And therefore $gx-xx=yy=\frac{1}{4}xx\text{♉}x\sqrt{\frac{5}{4}xx}+\frac{5xx}{4}$. Or

$$4g=10x\,\text{♉}\,2x\sqrt{5}.$$

(5) Newton has cancelled a first draft of the succeeding argument, 'Or, $\frac{yy}{x}+2x+3y$. $x+y$. 0. $-x$. $-x\frac{-xx}{x+y}$. by w[ch] the Equation being multiplyed produceth $xx\underset{x+y}{}-\underset{0}{gx}+\underset{-x}{yy}=0$, Or

$x^3+yxx-yyx=0$, Or $y=\frac{x}{2}\text{♉}\sqrt{\frac{5xx}{4}}=\frac{x\,\text{♉}\,x\sqrt{5}}{2}$. And consequently

$$-xx+gx-yy=\frac{xx\,\text{♉}\,2xx\sqrt{5}+5xx}{4}.$$

Or $4g=10x\,\text{♉}\,2x\sqrt{5}$. Or $\frac{2g}{5\,\text{♉}\,\sqrt{5}}=x$.'

(6) Read '-5'.

Or $\dfrac{2g}{5\text{ȣ}\sqrt{5}}=x=\dfrac{g\sqrt{5}\text{ȣ}g}{2\sqrt{5}}$. That is $\dfrac{2g}{5+\sqrt{5}}=\dfrac{g\sqrt{5}-g}{2\sqrt{5}}=ab$. And $\dfrac{2g}{5-\sqrt{5}}=\dfrac{g\sqrt{5}+g}{2\sqrt{5}}=ac$.
And therefore

$$\frac{2g}{5-\sqrt{5}}\,\frac{-2g}{5+\sqrt{5}}=ac-ab=\frac{10g+2g\sqrt{5}-10g+2g\sqrt{5}}{25+5\sqrt{5}-5\sqrt{5}-10^{(6)}}=\frac{g}{\sqrt{5}}=ah=de=db=bc.$$

Reductions of Equations may bee perhaps(7) performed by this method. As in that problem recited by D: Cartes pag 83,(8) viz: The square *ad* & ye right

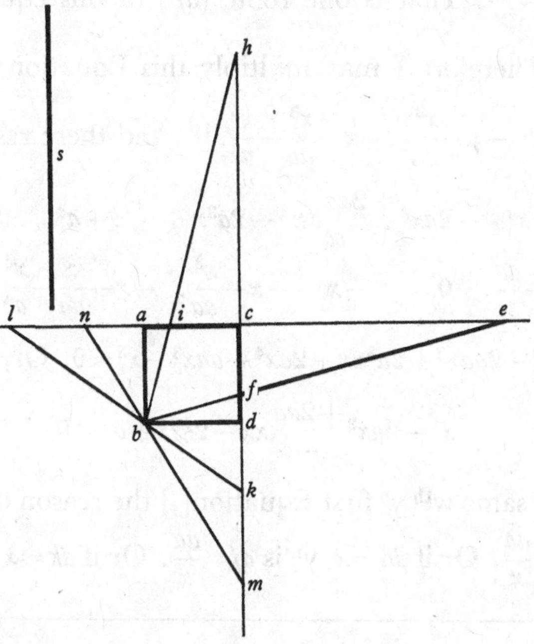

line *s* being given, to produce *ac* to *e*, soe yt *ef* drawn towards ye point *b* may bee equall to ye given line *s*.(9) Putting $df=x$ for ye unknowne quantity, $ef=c$, &

(7) Newton first wrote 'very often & readily', but has rightly cancelled it.

(8) See *Geometria*: Liber III: 82–4, but Newton takes his figure from Schooten's *Commentarii in Librum III = Geometria*: 310–15.

(9) This problem has an interesting history and, as Descartes acknowledged (*Geometria*: 83), derives from Propositions 71/72 of Book VII of Pappus' *Mathematical Collections*, which are in turn the 7th and 8th Lemmas given by him for the understanding of Proposition 8 of the first book of Apollonius' lost treatise *On Inclinations*. (See Pappus: *Mathematicæ Collectiones* (ed. Hultsch), 2 (Berlin, 1878): 782–3; also Paul ver Eecke's French translation, *La Collection Mathématique*, Paris, 1933: 2: 605–8, and compare 1: *Introduction*: lxiv–lxv.) In his Book 1, Proposition 8, Apollonius discussed the construction of a segment *EF* of given length which is required to be 'inclined' through one of the corners *B* of a rhombus *ABDC* with its end-points *E* and *F* in the opposing sides *AC* and *DC*. Propositions 71/72 of Pappus' commentary expound an ingenious construction, attributed to Heraclitus, for the present particular case where the rhombus is square: specifically the point *G* is constructed in *BD* such that *GE* is perpendicular

$bd=cd=a$. The Equation will bee $x^4-2ax^3\genfrac{}{}{0pt}{}{+2aa}{-cc}xx-2a^3[x]+a^4=0$.(10) W^ch^ having 4 rootes the Equation must have 4 divers resolutions, that is y^e^ lines *ac*, *cd*, produced both ways indefinitely, there may bee 4 divers lines drawne through the point *b*, whose parts intercepted twixt y^e^ crosse lines *lace*, *mdch*, are equall to the given line *s*: And they are *bih*[,] *bfe*, *lbk*, *nbm*. And therefore the rootes of this equation are (two affirmative) *df*, *dh*, (& two negative) *dk*, *dm*. Because $bd=ab$, $hi=fe$, Therefore $ae=dh$, $ai=fd$, $an=dk$, $al=dm$. Soe y^t^, $fd:bd::ab:ae=dh=\frac{aa}{x}$. That is one roote ($df$) of this equation is to another (dh) as x to $\frac{aa}{x}$. Therefore I may multiply this Equation by this progression $\frac{a^4}{x^3}+\frac{aa}{x}$, $\frac{aa}{x}$, 0, $-x$, $-x-\frac{x^3}{aa}$, $-x-\frac{x^3}{aa}-\frac{x^5}{a^3}$.(11) And there resulteth

$$\begin{array}{llllll} x^4 & -2ax^3\genfrac{}{}{0pt}{}{+2aa}{-cc}xx & -2a^3x & & +a^4 & =0. \\ +\frac{aa}{x}. & 0. & -x. & -x-\frac{x^3}{aa}. & -x-\frac{x^3}{aa}-\frac{x^5}{a^4} & \end{array}$$

That is $aax^3+ccx^3-2aax^3+2a^3xx+2ax^4-aax^3-x^5=0$. Or,

$$x^4-2ax^3\genfrac{}{}{0pt}{}{+2aa}{-cc}xx-2a^3x+a^4=0.$$

Which result is y^e^ same wth y^e^ first Equation[,] the reason of wch is, y^t^ if I make $df=x$ then is $dh=\frac{aa}{x}$. Or if $dh=x$, y^n^ is $df=\frac{aa}{x}$. Or if $dk=x$, y^n^ is $dm=\frac{aa}{x}$. Or if

---

to BFE. (Pappus' proof shows in essence that, where EH is drawn normal to BDG, then $BG^2 = BE^2+EG^2$ (or BF^2) and $BG.BD = BF.BE$, so that $BG.DG = BE.EF+BF^2 = BE.BF$ (or $BG.BD$) $+FE^2$, and finally $DG^2 = BG.DG-BD.DG = BD^2+FE^2$, or

$$BG = BD\pm\surd[BD^2+FE^2].)$$

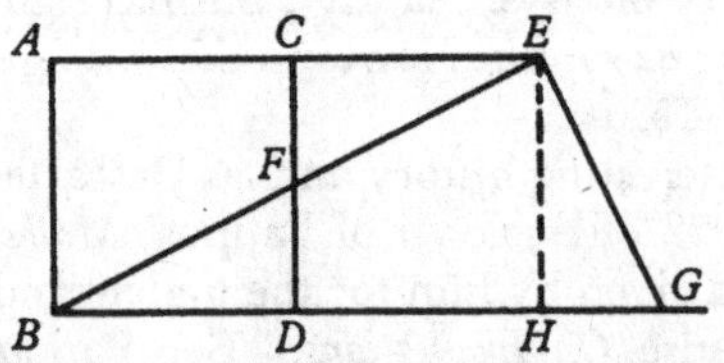

From the time of Federigo Commandino's Latin translation of Pappus' work (Pesaro, 1588), this problem of 'inclination' attracted considerable interest. Ghetaldi, for example, in his *De Resolutione et Compositione Mathematica Libri V* (Rome, 1630) and Huygens in the appendix to his *De Circuli Magnitudine Inventa* (Leyden, 1654) both considered the general problem synthetically, but it was Albert Girard who, in his *Invention nouvelle en l'Algebre* (Amsterdam, 1629), first applied analytical considerations in examining a numerical case of the particular

$dm = x$ then is $dk = \frac{aa}{x}$. Soe y[t] y[e] relation twixt all y[e] rotes(12) being reciprocally the same & not distinguishing one roote from another, tis noe wonder if they bee all indifferently expressed in y[e] resulting Equation. Otherwise y[e] reduction must have succeeded.(13)

Suppose 3 rootes of an Equation are in proportion to each other as a, b, c. Then if that roote w[ch] is correspondent to a be required, multiply y[e] termes of y[e] Equation by any of these Progressions

$$1 \quad \frac{c^3+bcc+bbc+b^3}{aa}\ \frac{+bb+bc+cc}{a}+b+c+a.\ \frac{bb+bc+cc}{aa}+b+c+a.$$

$$+b+c+a.\ +a.\ 0.\ 0.\ \frac{a^3}{bc}.\ \frac{a^3}{bc}\ \frac{+a^4b+a^4c}{bbcc}.$$

$$\frac{a^3}{bc}\ \frac{+a^4b+a^4c}{bbcc}\ \frac{+a^5bb+a^5bc+a^5cc}{b^3c^3}.\ [\&c]$$

$$2 \quad \frac{+bbc+bcc}{aa}\ \frac{+bb+bc+cc}{a}+b+c.\ \frac{bc}{a}+b+c.\ 0.\ -a.\ 0.\ \frac{+a^3+aab-aac}{bc}.$$

$$\frac{a^3+aab+aac}{bc}\ \frac{+a^3bb+a^3bc+a^3cc+a^4b+a^4c}{bbcc}.\ [\&c]$$

---

example of the square. (See his *Invention nouvelle*: F3[v]/[4][r]: *Probleme d'Inclinaison.* Girard was especially proud of his interpretation of the negative roots of the resulting quartic as allowable solutions of the problem: '& ainsi le faudra-t-il entendre de toutes solutions par moins [sc. negative solutions], qui est une chose de consequence en Geometrie, incogneue auparavant'.)

(10) For, since $cf = a-x$ and so $ce = [a(a-x)]/x$, it follows from $ef^2 = ce^2+cf^2$ that $c^2 = ([a(a-x)]/x)^2+(a-x)^2$, or $(x^2+a^2)(a-x)^2 = c^2x^2$.

(11) Read '$-\frac{x^5}{a^4}$'.

(12) An orthographic variant on 'rootes'.

(13) With a little extra work the reduction can, in fact, be made to succeed. Since the equation $(x^2+a^2)(a-x)^2 = c^2x^2$ is invariant under the transform $x \to a^2/x$, we may equate it term by term with $(x-\alpha)(x-a^2/\alpha)(x-\beta)(x-a^2/\beta) = 0$, and hence, if we define

$$f(p) = p+a^2/p$$

for arbitrary p, we may deduce that $f(\alpha)+f(\beta) = 2a$ and $c^2 = -f(\alpha)\times f(\beta)$, so that $f(\alpha)$ and $f(\beta)$ are the roots of $z^2-2az-c^2 = 0$. It follows that $\alpha+a^2/\alpha = a+\sqrt{[a^2+c^2]}$, say, and $\beta+a^2/\beta = a-\sqrt{[a^2+c^2]}$, and so, if we resolve these quadratics separately, that

$$\alpha = \tfrac{1}{2}(a+\sqrt{[a^2+c^2]}) \pm \tfrac{1}{2}\sqrt{\{c^2-2a^2+2a\sqrt{[a^2+c^2]}\}}$$

and

$$\beta = \tfrac{1}{2}(a-\sqrt{[a^2+c^2]}) \pm \tfrac{1}{2}\sqrt{\{c^2-2a^2-2a\sqrt{[a^2+c^2]}\}},$$

which are therefore the roots of $(x^2+a^2)(a-x)^2 = c^2x^2$. (Compare *Geometria*: Liber III: 32.) The two roots α are always real, yielding the line-lengths df and dh in Newton's figure; but the roots β (the lines dk, dm in the figure) are real only if $lk = nm = c \geqslant 2a\sqrt{2}$, that is, twice the diagonal of the square $abdc$ (and if real they are negative).

$$3 \quad \frac{aabb+aacc+bbcc}{a^3+aab+aac}\frac{+bc}{a}.\ 0.\ \frac{-ab-ac-bc}{a+b+c}.\ -a.\ 0.$$

$$\frac{a^3\ +bb+bc+cc}{bc\,bac+bbc+bcc}.^{(14)}\quad [\&c]^{(15)}$$

As if 3 of y^e^ rootes of this Equation $x^4 * -35ggxx+90g^3x-56g^4[=0]$ were to one another as 1. 2. 4. And I would find y^e^ roote wch is correspondent to 1. Then I make $a=1$, $b=2$, $c=4$, & soe I may have by y^e^ first progression this $+155.\ +35.\ +7.\ +1.\ 0.\ 0.\ \frac{+1}{8}.\ \frac{7}{32}.\ \frac{+70}{128}$. By y^e^ 2^d^; $+92.\ +14.\ 0.\ -1.\ 0.\ +\frac{7}{8}.\ +\frac{45}{32}$. &c. By y^e^ first of wch y^e^ Equation being multiplyed produceth

$$\begin{array}{ccccc} x^4 & * & -35ggxx & +90g^3x & -56g^4=0. \\ 35. & 7. & 1. & 0. & 0. \end{array}$$

That is $35x^4-35ggxx=0$. Or $xx=gg$. & $x=g$.(16) Or were it multiplyed by y^e^ 2^d^ progression thus $x^4 \ * \ -35ggxx+90g^3x-56g^4=0$. It would produce

$$\begin{array}{ccccc} 0. & -1. & 0. & \frac{7}{8}. & \frac{45}{32}. \end{array}$$

$$\frac{630}{8}g^3x-\frac{2520}{32}g^4=0.$$

Or $x=g$. Soe y^t^ g being the least roote, y^e^ other two rootes must be $2g$ & $4g$.(17)

If it be desired to know y^e^ length of y & z in this Equation

$$x^3-yxx+a^2x-aaz[=0.]$$

when y^e^ rootes are in proportion as 1. 2. 4. I multiply it by y^e^ precedent progressions & the results are $\begin{array}{cccc} x^3 & -yxx & +aax & -aaz=0. \\ 7 & 1 & 0 & 0 \end{array}$ Or $7x=y$. And $\begin{array}{cccc} x^3 & -yxx & +aax & -aaz=0. \\ 0 & 0 & \frac{1}{8} & \frac{7}{32} \end{array}$ Or $4x=7z$. $\begin{array}{cccc} x^3 & -yxx & +aax & -aaz=0. \\ 14 & 0 & -1 & 0 \end{array}$ Or $14xx=aa$. And consequently $7x=\frac{7a}{\sqrt{14}}=y=a\sqrt{\frac{7}{2}}.\ \frac{4x}{7}=\frac{4a}{7\sqrt{14}}=z$.(18)

Likewise were the proportion of 4 or 5 or more rootes given I might set downe progressions to find them but it will bee better to set downe y^e^ method of finding

---

(14) Read $\dfrac{`+bb+bc+cc \text{ in } a^2\text{'}}{bac+bbc+bcc}$.

(15) Where ak, bk and ck are three roots of the equation $0 = \sum_i (a_i x^i)$, it follows that ak is a root of the equations $0 = \sum_i \left(a_i \left(\frac{b}{a}\right)^i x^i\right)$ and $0 = \sum_i \left(a_i \left(\frac{c}{a}\right)^i x^i\right)$ also. From this basis we

these progressions, And it is this. Suppose two of yᵉ rootes of an Equation are as a to b. That Equation will bee of this forme $xx\begin{smallmatrix}-a\\-b\end{smallmatrix}x+ab=0$, Or of some forme compounded of it; And if a corresponds to yᵉ desired roote of yᵉ Equation this equation $xx\begin{smallmatrix}-ax\\-bx\end{smallmatrix}+ab=0$ will bee of this forme $aa\begin{smallmatrix}-a\\-b\end{smallmatrix}a+ab$. Then assuming two termes (as, *a*. 0.) & feining a third I have a progression *z. a.* 0 by which this equation being multiplyed produceth $aa\begin{smallmatrix}-a\\-b\end{smallmatrix}a+ab=0$. Or $zaa-a^3-aab=0$.

$$\begin{array}{ccc} \times & \times & \times \\ z. & a. & 0 \end{array}$$

And $z=a+b$. And soe I have 3 termes of yᵉ progression viz: $a+b$. a. 0, feigning a fourth terme [z] I againe multiply the Equation & the result is

$$aa\begin{smallmatrix}-a\\-b\end{smallmatrix}a+ab=0.$$

$$\begin{array}{ccc} \times & \times & \times \\ z. & a+b. & a. \end{array}$$

Or $aaz-a^3-2a^2b-abb+aab=0$. And $z=a+b+\dfrac{bb}{a}$. Soe yᵗ I have thus much of yᵉ progression $a+b+\dfrac{bb}{a}$. $a+b$. a. 0. And by yᵉ same proceding might continue it or get termes on yᵉ other side of yᵉ cipher. As if I multiply yᵉ Equation by this progression *a*. 0. *z* there is produced $aa\begin{smallmatrix}-a\\-b\end{smallmatrix}a+ab=0$. Or $a^3+abz=0$.

$$\begin{array}{ccc} \times & \times & \times \\ a & 0 & z \end{array}$$

And $z=\dfrac{-aa}{b}$. Againe multiplying yᵉ Equation by 0. $\dfrac{-aa}{b}$. *z*. It is

$$aa\begin{smallmatrix}-a\\-b\end{smallmatrix}a+ab=0.$$

$$\begin{array}{ccc} \times & \times & \\ 0. & \dfrac{-aa}{b}. & z. \end{array}$$

---

may deduce Newton's three progressions: for example, the general term of the first progression is

$$\frac{a^2}{b-c}\left[\frac{b}{b-a}\left(\left(\frac{b}{a}\right)^k-1\right)-\frac{c}{c-a}\left(\left(\frac{c}{a}\right)^k-1\right)\right]$$
$$=\frac{a^2b}{(b-c)(b-a)}\left(\frac{b}{a}\right)^k-\frac{a^2c}{(b-c)(c-a)}\left(\frac{c}{a}\right)^k+\frac{a^3}{(b-a)(c-a)}$$

and Newton gives the terms for $k = 4, 3, 2, 1, 0, -1, -2, -3$ and -4.

(16) Strictly $x = \pm g$.

(17) In fact, $$x^4-35g^2x^2+90g^3x-56g^4 = (x-g)(x-2g)(x-4g)(x+7g).$$

Or $\frac{a^4+a^3b}{b}+abz=0$, And $\frac{-a^3}{bb}\frac{-aa}{b}=z$. Soe that I have thus much of y^e^ progression viz: $a+b+\frac{bb}{a}.\ a+b.\ a.\ 0.\ \frac{-aa}{b}.\ \frac{-aa}{b}\frac{-a^3}{bb}$.(19)

The proceeding is [y^e^] same when y^e^ proportion of 3 rootes to one another are given, but there may bee some difference(20) when y^e^ ciphers are far distant, as if there bee three termes betwixt them, then y^e^ operation may be done thus. Let y^e^ quantitys, wch beare such proportion to one another as y^e^ rootes doe, bee, *a*, *b*, *c*. let *a* correspond to y^e^ roote wch must bee knowne And y^n^ y^t^ Equation will bee of this forme, $a^3\begin{smallmatrix}-a\\-b\\-c\end{smallmatrix}aa\begin{smallmatrix}+ab\\+ac\\+bc\end{smallmatrix}a-abc=0$. or else compounded of it. Then assuming some quantity (as *a*) for one of y^e^ termes of y^e^ progression & placing it conveniently (as(21) now equidistant(22) from y^e^ ciphers) feigne two other quantitys as *z*, *y*, for y^e^ deficient termes And the progression will bee $0.\ z.\ a.\ y.\ 0.$ By wch I multiply the Equation

$$\begin{matrix} a^3 & \begin{smallmatrix}-a\\-b\\-c\end{smallmatrix}aa & \begin{smallmatrix}+ab\\+ac\\+bc\end{smallmatrix}a & -abc=0. \\ \times & \times & \times & \times \\ 0. & z & a. & y. \end{matrix} \qquad \text{Or } \frac{-aaz-abz-acz+aab+aac+abc}{bc}=y.$$

Soe that I have the progression. $0.\ z.\ a.\ \frac{aab+aac+abc-aaz-abz-acz}{bc}.\ 0.$ by wch I againe multiply y^e^ Equation & there results

$$\begin{matrix} a^3 & \begin{smallmatrix}-a\\-b\\-c\end{smallmatrix}aa & \begin{smallmatrix}+ab\\+ac\\+bc\end{smallmatrix}a\ \ldots\ldots\ldots\ldots & -abc=0. \text{ Or} \\ \times & \times & & \times \\ z. & a. & \frac{aab+aac+abc-aaz-abz-acz}{bc}. & 0 \end{matrix}$$

(18) An incomplete first draft (where Newton considers three roots of the equation

$$x^3-ax^2+yqx-\tfrac{8}{7}a^3=0$$

also to be in proportion as 1:2:4) is here omitted.

(19) Newton generalizes Hudde's technique (note (3)) for finding double roots, the particular case when $a=b$. Since two roots of the equation are in proportion as *a* to *b* we may suppose it to be of the form

$$0=\sum_i[p_ix^i(x-ak)(x-bk)]=\sum_i[p_ix^{i+2}-(a+b)kp_ix^{i+1}+abk^2p_ix^i].$$

Hence, since *ak* is a root of the equation, it follows that, where λ, μ and ν are three adjacent terms of Newton's progression,

$$0=\lambda p_i(ak)^{i+2}-\mu(a+b)kp_i(ak)^{i+1}+\nu abk^2p_i(ak)^i$$

and so, by eliminating the common term $p_ia^ik^{i+2}$, $0=\lambda a^2-\mu(a+b)a+\nu ab$.

$z = \dfrac{aabb+aabc+aacc+abbc+abcc+bbcc}{aab+aac+bba+acc+2abc+bbc+bcc}$. wch valor of z substituted into its place in ye valor [of] y There will bee this much of ye progression

$$0.\ \frac{aabb+aabc+aacc+abbc+abcc+bbc[c]}{aab+aac+abb+acc+2abc+bbc+bcc}.\ a.$$

$$\left\{\begin{aligned} &a+\frac{aab+aac}{bc}-\frac{a^4-2a^3b-2a^3c-aabb-3aabc-aacc-abcc}{aab+aac+abb+acc+2abc+bbc+bcc} \\ &\frac{-a^4bb-a^3b^3-a^3bbc-a^4cc-a^3bcc-a^3c^3}{aabbc+aabcc+ab^3c+abc^3+2abbcc+b^3cc+bbc^3}\end{aligned}\right\}.\ 0.^{(23)}$$

The same done otherwise.[24]

Did I know yt 2 of ye rootes of this Equation $x^3 * -117x-324=0$, were in proportion as 3, -4.[25] Then I suppose one roote to be $3a$, ye other $-4a$ That is $x-3a=0$. $x+4a=0$. By one of wch I divide ye Equation as first by $x+4a=0$. And ye operation is;

$$\begin{array}{l} x+4a)\,x^3 \quad * \quad -117x-324\,(xx-4ax+16aa-117=0. \\ \quad\quad 0-4axx-117x-324 \\ \quad\quad\quad 0+16aax-324 \\ \quad\quad\quad\quad -117x \\ \quad\quad\quad\quad\quad 0-64a^3\quad[-324] \\ \quad\quad\quad\quad\quad\quad +468a \\ \quad\quad\quad\quad\quad -64a^3+468a-324=0^{(26)} \\ \quad\quad\quad\quad\quad\quad\quad 0 \end{array}$$

Againe I divide ye Quotient by ye other roote $x-3a=0$. Thus

$$\begin{array}{l} x-3a)\,xx-4ax+16aa-117=0\,(x-a=0. \\ \quad\quad 0-ax\ +16aa-117 \\ \quad\quad\quad 0\ +13aa-117=0 \\ \quad\quad\quad\quad 0 \end{array}$$

(20) Newton first wrote 'difficulty'.

(21) 'it may bee' is cancelled.

(22) Newton first wrote 'betwixt'.

(23) Much as before, we may suppose the equation with roots in the proportion $a:b:c$ to be $0 = \sum_i [p_i x^i (x-ak)(x-bk)(x-ck)]$ and so, where κ, λ, μ and ν are four successive terms of Newton's progression, derive the recursive scheme

$$0 = \kappa a^3 - \lambda(a+b+c)\,a^2 + \mu(ab+bc+ca)\,a - \nu abc.$$

(24) The writing in this final section is much larger than that of the preceding text and we may suppose it a later addition.

(25) In fact $x^3-117x-324=(x-9)(x+12)(x-3)$.

(26) Newton supposes that this remainder $(-4a)^3-117(-4a)-324$ vanishes, but does not attempt to evaluate a thereby since it is essentially his original equation.

By y[e] last division I have this equation $13aa-117=0$. Or $aa-9=0$. And $a=3$.(27) Therefore y[e] rootes of y[e] Equation $x^3-117x-324=0$ are, $3a=9=x$, $-4a=-12=x$.

If I would have y & z of such a length y[t] y[e] rootes of this equation

$$x^3-3yx+gz=0$$

be in proportion as $+1$, $+2$, -3.(28) I suppose $x-a=0$. $x-2a=0$. $x-3a=0$. And soe first divide y[e] Equation by

$$\begin{array}{l} x-a)\,x^3 \quad * \quad -3yx+gz\,(xx+ax+aa-3y=0. \\ \qquad\quad 0\ +axx-3yx+gz \\ \qquad\qquad\quad 0+aa \\ \qquad\qquad\quad -3y\ \ x+gz \\ \qquad\qquad\qquad 0+a^3-3ay+gz=0. \end{array}$$

Againe divide this product by $x-2a)\,xx+ax\ +aa-3y\,(x+3a=0$. Lastly were

$$\begin{array}{l} 0+3ax+aa-3y \\ \quad 0+7a^2-3y=0. \end{array}$$

it necessary I should have againe divided this quote $x+3a=0$ by y[e] 3[d] supposed roote of y[e] Equation (viz $x+3a=0$). By y[e] 2[d] operation I find $7a^2-3y=0$. Or $\frac{7a^2}{3}=y$. And by y[e] first $a^3-3ay+gz=0$. Or $18ay=7gz$. $18y\sqrt{\frac{3y}{7}}=7gz$. Soe y[t] If I make $\frac{18y}{[7]g}\sqrt{\frac{3y}{7}}=z$, the rootes of this Equation $x^3-3yx+gz=0$ shall bee a[s] 1. 2. -3.

But it would bee more readily done by y[e] Progressions.(29)

---

(27) Strictly $a=\pm 3$.

(28) The example is more than a little forced since, if two of the roots of $x^3-3yx+gz=0$ are a and $2a$, the third must be $-3a$. (For equation of $(x-a)(x-2a)(x-pa)=0$ with $x^3-3yx+gz=0$ yields immediately, by comparing coefficients of x^2, that $a+2a+pa=0$, or $p=-3$.)

(29) The statement seems dubious in the extreme.

(1) Two notes extracted from the loose sheet ULC. Add. 3958. 2: 31[v]. (Compare 2, §1, 1 and 4, §2 of the present part.) The writing suggests a composition date in early 1665.

(2) Newton extends Descartes' rule of signs (*Geometria*: Liber III: 70: '...tot in [unaquaque Æquatione] veras [radices] haberi posse, quot variationes reperiuntur signorum + & −; & tot falsas, quot vicibus ibidem deprehenduntur duo signa +, vel duo signa −, quæ se invicem sequuntur.') to incomplete equations, making it the true equivalent of the modern rule that an equation has (at most) as many positive roots as there are variations in the signs of its terms when they are ordered by powers of the variable. In modern terms, since Newton's example $f(x)=x^9+ax^7-bx^4-c$ has one variation and $f(-x)=-x^9-ax^7-bx^4-c$ none, then $f(x)=0$ has at most one positive root and no negative ones. (Since the degree of the equation is odd, at least one root must be real and so Newton's final statement is proved.)

§5. RESEARCHES IN THE THEORY OF EQUATIONS

[1665–1666]

From the originals in the University Library, Cambridge and in private possession

[1][1]

An Equation wants 2, 4, 6 rootes &c for every 2 4 6 termes &c wanting together. And for 1. 3. 5 termes wanting together[,] y^e next termes being affected wth y^e same signes, as $x^9 * + ax^7 * * - bx^4 * * * - c = 0$. wants 8 rootes.[2]

If $axxyy + bxxy + cxx + dxy + ex + f = 0$. Make $x = \frac{r}{z}$ & Then is,

$$arryy + brry + crr + drzy + erz + fzz = 0.^{(3)}$$

---

[2][4]

For taking of unknowne quantitys out of intricate Equations it may bee convenient to have severall formes. Now suppose x was to be taken out of y^e Equations $ax^3 + bxx + cx + d = 0$ & $fx^3 + gxx + hx + k = 0$.

I feigne y^e 3 valors of x in y^e first Equation to bee $-r$, $-s$, & $-t$. Then is[5] supposing $a = 1$, every r[6] is b. every rr[7] $= bb - 2c$. every $r^3 = b^3 - 3bc + 3d$. $rs = c$. $rrs = bc - 3d$. $r^3s = bbc - 2cc - bd$. $rrss = cc - 2bd$. $r^3ss = bcc - 2bbd - cd$.

$$r^3s^3 = c^3 - 3bcd + 3dd. \quad rst = d. \quad rrst = bd. \quad r^3st = bbd - 2cd. \quad rrsst = cd.$$

$$r^3sst = bcd - 3dd. \quad r^3s^3t = ccd - 2bdd. \quad rrsstt = dd. \quad r^3sstt = bdd.$$

$$r^3s^3tt = cdd. \quad r^3s^3t^3 = d^3.^{(8)}$$

---

(3) Newton constructs the quartic in x and y as the meet of $zx = r$ and the quadratic $ar^2y^2 + drzy + fz^2 + br^2y + erz + cr^2 = 0$.

(4) ULC. Add. 4004: 83^r, a miscellaneous note on elimination of the variable between two given equations by interpreting their coefficients as homogeneous functions of their respective roots.

(5) Newton first continued: '$r + s + t = \frac{b}{a}$. $rr + ss + tt = \frac{bb}{aa} - \frac{2c}{a}$. $r^3 + s^3 + t^3 = \frac{b^3 - 3abc + 3aad}{a^3}$. $rs + rt + st = \frac{c}{a}$. &c. that is y^e summe of y^e rootes is $\frac{b}{a}$; of their squares is $\frac{bb - 2ac}{aa}$; of their cubes is $\frac{b^3 - 3abc + 3aad}{a^3}$; of their rectangles is $\frac{c}{a}$ &c. that is.' (Here, as in Newton's succeeding discussion, there is some confusion of signs: presumably he intends $+r$, $+s$ and $+t$ to be the roots of $ax^3 - bxx + cx - d = 0$.)

(6) That is, $r + s + t$. (7) Or, $rr + ss + tt$.

(8) Before each equation after the two first read 'every'. Newton tables all homogeneous functions up to $r^3s^3t^3$ of the roots r, s and t of $x^3 - bx^2 + cx - d = 0$ in terms of the coefficients b, c and d.

Or thus, every $r^+=b$. $rs^+=c$. $rr^+=bb-2c$. $rst^+=d$. $rrs^+=bc-3d$.

$r^{3+}=b^3-3bc+3d$. $rrst^+=bd$. $rrss^+=cc-2bd$. $r^3s^+=bbc-2cc-bd$.

$r^{4+}=b^4-4bbc+4bd+2cc$. $rrsst^+=cd$. $r^3st^+=bbd-2cd$. $rrrss^+=bcc-dc-2bbd$. $rrsstt^+=dd$. $r^3sst^+=bcd-3dd$. $r^3sstt^+=bdd$. $r^3s^3t^+=ccd-2bdd$. $r^3s^3tt^+=cdd$. $r^3s^3t^{3+}=d^3$. $r^3s^{3+}=c^3-3bcd+3dd$.

Now supposing k (or any other quantity of y^e^ second Equation) to bee an unknowne quantity, it must have 3 severall valors by reason of y^e^ 3 valors of x in y^e^ first Equation, & therefore x being taken away, k will bee of three dimensions in y^e^ resulting equation.

The 3 valors of k are $+fr^3-grr+hr=k$. & $fs^3-gss+hs=k$. & $ft^3-gtt+ht=k$. Which I multiply into one another that they may produce an equation expressing y^e^ 3 fold valor of k: out of wch equation I take out r, s, t by writing b for their summe, c for y^e^ sum̄e of their rectangles $rs+rt+st$. $bc-3d$ for y^e^ summe of all theire rectangles of this forme rrs (viz: for $rrs+rrt+rss+rtt+sst+stt$). &c as in y^e^ Table. W^ch^ substitution may bee most breifly done in y^e^ said multiplication, thus; writing a(9) to make up six dimensions.

$$\begin{array}{l} k-hr+grr-fr^3 \\ \times \\ k-hs+gss-fs^3 \qquad \text{Produceth} \\ \times \\ k-ht+gtt-ft^3 \end{array}$$

$$\left.\begin{array}{l} a^3k^3-aabhkk+abbgkk-2aacgkk-b^3fkk+3abcfkk \\ -3aadfkk+aachhk-abcghk+3aadghk+bbcfhk \\ -2accfhk-abdfhk+accggk-2abdggk-bccfgk \\ +2bbdfgk+acdfgk+c^3ffk-3bcdffk+3addffk \\ -aadh^3+abdghh-bbdfhh+2acdfhh-acdggh \\ +bcdfgh-3addfgh-ccdffh+2bddffh+addg^3 \\ -bddfgg+cddffg-d^3f^3 \end{array}\right\}=0.$$

---

(9) And its powers a^2 and a^3 where necessary.

(10) ULC. Add. 4004: 84^r^. By numerical induction, it seems, Newton extends the results in [2] to the general algebraic equation. (Some ten years later, when he entered up some researches in finite differences around it, Newton cancelled the entire passage.)

(11) Here, of course, the letters y and z denote constants.

(12) Newton gives the familiar expansion of the coefficients of an equation in terms of symmetric functions of its roots. The property had already appeared in print as Definition 11 of Girard's *Invention Nouvelle en l'Algebre* (Amsterdam, 1629), but it seems unlikely that Newton, whose grasp of French was weak (compare note (1) of the introduction to the Appendix on optics), had read the work and we may assume that the expansion is his independent discovery.

(13) A first draft, unfinished and essentially the same as the text reproduced, is omitted.

[3][10]

Of the nature of Equations.

Every Equation as $x^8+px^7+qx^6+rx^5+sx^4+tx^3+vxx+yx+z=0$.[11] hath so many roots as dimensions, of wch y^e^ summ is $-p$, the summ of the rectangles of each two $+q$, of each three $-r$, of each foure $+s$: &c: & of all together ♉ z.[12]

Also y^e^ summe of their squares[,] cubes &c is as followeth.[13]

squares	$+pp-2q$.
cubes	$-p^3+3pq-3r$.
sq: squares	$+p^4-4ppq+4pr-4s+2qq$.
[sq: cubes]	$-p^5+5p^3q-5ppr+5ps-5t-5pqq+5qr$.
[cube-cubes]	$+p^6-6p^4q+6p^3r-6pps+6pt-6v+9ppqq-12pqr+6qs-2q^3$.
[sq: sq: cubes]	$-p^7+7p^5q-7p^4r+7p^3s-7ppt+7pv-7y$.
[sq: cu: cubes]	$+p^8-8p^6q+8p^5r-8p^4s+8p^3t-8ppv+8py-8z$.[14]

Or thus

If their sume is $-p=-a$. Then is y^e^ sume of their squares $ap-2q=b$. of their cubes $-pb+qa-3r=-c$. sq[u]are squares $pc-qb+ra-4s=d$. sq: cubes $-pd+qc-rb+sa-5t=-e$. cube cubes $+pe-qd+rc-sb+ta-6v$. &c[15]

Now of these rootes some are true some false & some imaginary[16]

---

(14) It seems clear that the present table is calculated from the coefficients of the octic duly expressed as symmetric functions of its roots. (Newton's table halts at the summation of the eighth powers: the tabulation alters slightly when powers higher than the equation's degree are summed.) However, it is obvious that operations on any set of symmetric functions of the same number of elements (here 8) are independent of the particular number of those elements, and the generalization is valid. There is no evidence that Newton ever thought it necessary to give the rigorous systematic derivation for an algebraic equation of arbitrary degree we now require.

(15) 'Newton's equations' for the sum of integral powers of the roots of the equation

$$x^n-px^{n-1}+qx^{n-2}-rx^{n-3}+sx^{n-4}-tx^{n-5}+vx^{n-6}\ldots=0$$

expressed in terms of its successive coefficients. In general, where s_k is the sum of the kth powers of the roots of the equation

$$x^n+\sum_{1\leqslant i\leqslant n}[(-1)^i p_i x^{n-i}]=0, \quad \text{then} \quad (-1)^r s_r=\sum_{1\leqslant i\leqslant r-1}[(-1)^{r+i+1}p_i s_{r-i}]-rp_r \quad (r=1,2,3,\ldots,n).$$

(Newton does not note, and perhaps does not yet know, the complementary form

$$(-1)^r s_r=\sum_{1\leqslant i\leqslant n}[(-1)^{r+i+1}p_i s_{r-i}]$$

for $r>n$. Compare G. Chrystal: *Algebra* (London, ${}_5$1926): 1: ch. XVIII, 436–7.)

(16) Newton breaks off but develops this theme in succeeding pages (reproduced as [5]–[8] below).

[4](17)

To find the sume of y^e squares cubes &c: of y^e rootes of an Equation.

If a, b, c, d, e, f &c be the rootes of y^e Equation

$$x^6+px^5+qx^4+rx^3+sxx+tx+v=0.^{(18)}$$

y^n is

$a+b+c+d+e+f=p.\ (=g)$

$a^2+b^2+c^2+d^2+e^2+f^2=pp-2q.\ (=pg-2q=h)$

$a^3+b^3+c^3+d^3+e^3+f^3=p^3-3pq+3r.\ (=ph-qg+3r=k)$

a^4+b^4 &c$=p^4-4ppq+4pr+2qq-4s.\ (=pk-qh+rg-4s=l)$

a^5 &c$=p^5-5p^3q+5pqq+5ppr-5ps-5qr+5t.\ (=pl-qk+rh-sg+5t=m)$

a^6 &c:$=p^6-6p^4q+9ppqq+6p^3r-12pqr-6pps+6pt-2q^3+3rr+6qs-6v.$(19)

---

[5](20)

Of Equations.

Every Equation hath soe many roots as dimensions of wch some may be true some false & some imaginary or impossible.

If there bee none imaginary then the number of true and false rootes may bee knowne by y^e signes of y^e Equations termes: Namely there are soe many true rootes as variations of signes & soe many false ones as successions of y^e same signes.(21) When any termes are wanting(22) supply their voyd places w^{th} ± 0.

But if any roots bee imaginary,(23) this rule soe far admitts of exception. Thus the signes of this Eq: $x^3-pxx+3ppx-q^3=0$. show it to have three true roots, wherefore if it bee multiplyed by $x+2a=0$ the resulting equation

$$x^4+px^3+ppxx\begin{matrix}+6p^3\\-q^3\end{matrix}-2pq^3=0$$

---

(17) ULC. Add. 4000: 79^r, a variant on [3].

(18) The signs in the table which follows require the equation to be

$$x^6-px^5+qx^4-rx^3+sx^2-tx+v=0.$$

(19) That is, $pm-ql+rk-sh+tg-6v$. In general, where s_k is the sum of the kth powers of the roots of $0=x^n+\sum_{1\leqslant i\leqslant n}[(-1)^i p_i x^{n-i}]$, it follows that

$$s_r=\sum_{1\leqslant i\leqslant r-1}[(-1)^{i+1}p_i s_{r-i}]+(-1)^{r+1}rp_r\quad (r=1, 2, 3, \ldots, n).$$

(See note (15).) Compare Girard's *Invention nouvelle en l'Algebre* (Amsterdam, 1629): F2^r.

(20) ULC. Add. 4004: 85^r–86^r.

(21) A restatement of Descartes' rule of signs (note (2)) in the restricted case where all roots are supposed real.

(22) This may contradict the previous assumption that all roots of the equation are real. (Compare [1] above.) The equation $f(x)=0$ cannot have more (non-zero) roots than the sum of the number of variations in $f(x)$ and $f(-x)$.

should have three true rootes & a false one, but the signes shew it to have three false & one true. I conclude therefore that the two roots wch in y^e one case appeare true, & in y^e other false are neither, but imaginary; & that of y^e other two roots one is true y^e other false.(24)

Hence it appeares y^t to know y^e particular constitution of any Equation it is cheifely necessary to understand w^t imaginary roots it hath. And this in some of the simplest Equations is easily discovered, thus in $xx \pm ax + bb = 0$. both roots are imaginary if $4bb \sqsubset aa$,(25) otherwis both reall. And thus in $x^3 - px - q = 0$ two roots are imaginary if $4p^3 \sqsubset 27qq$, otherwise all reall.(26) But to give accurate rules for determining the number(27) of these roots in all sorts of Equations would bee a thing not onely very difficult, but uselesse; because in Equations of many dimensions y^e rules would bee more intricate & laborious to put in practise then to solve the Equations by lines(28) or numbers.(29) Soe y^t ye accurate determination of these roots is for the most part esilyest(30) acquired by solving the Equations.

But yet because the discovery of these roots is very usefull I shall lay down rules whereby they may bee many times discovered at first sight, & almost always without much labour.

First y^n if in any three termes together ye two extreame termes having ye same signes bee neither of them as little or lesse (y^t is as little or lesse remote from nothing) then ye terme betwixt them, conclude there are two imaginary roots.(31)

---

(23) Newton first added and then cancelled '(because imaginary roots are properly neither true nor false)'.

(24) An ingenious deduction from the fact that Descartes' sign rule yields true upper bounds to the number of positive and negative roots.

(25) That is, in modern notation, $4b^2 > a^2$.

(26) This is a corollary of Cardan's resolution of the reduced cubic as expounded by Descartes (*Geometria*: Liber III: 93–5). Descartes there showed (following François Viète) that the reduced cubic could be constructed as the 'complementary subtense' equation

$$\left(\frac{x}{\sqrt{[p/3]}}\right)^3 - 3\left(\frac{x}{\sqrt{[p/3]}}\right) = \frac{3q}{p\sqrt{[p/3]}}$$

which for $4p^3 > 27q^2$ has the single real root $x = 2\sqrt{[p/3]}\cos\theta$, where $\cos(3\theta) = \frac{3q}{2p\sqrt{[p/3]}}$. Conversely, by extracting the cube roots as conjugate imaginaries we may show that, for $4p^3 \leqslant 27q^2$, Cardan's solution

$$x = \sqrt[3]{\{\tfrac{1}{2}q + \sqrt{[\tfrac{1}{4}q^2 - \tfrac{1}{27}p^3]}\}} + \sqrt[3]{\{\tfrac{1}{2}q - \sqrt{[\tfrac{1}{4}q^2 - \tfrac{1}{27}p^3]}\}}$$

yields three real roots.

(27) Newton first wrote 'species'.

(28) See §2 above.

(29) See **1**, 2, §1 and compare §1 above.

(30) That is, most easily.

(31) Thus if

$$\ldots + p_{r-1}x^{n-r+1} + p_r x^{n-r} + p_{r+1}x^{n-r-1} + \ldots$$

are 'three termes together', Newton's rule states that if $|p_r| < |p_{r-1}|$ and also $|p_r| < |p_{r+1}|$ then the equation $0 = x_n + \sum_{1 \leqslant i \leqslant n} (p_i x^{n-i})$ has (at least) two imaginary roots. Clearly, since the weaker condition $(p_r)^2 < p_{r+1} \cdot p_{r-1}$ is a consequence, the rule is a particular case of the following one.

Thus $x^3-3xx+2x-4=0$ has two roots imaginary because neither 3 nor 4 are lesse then 2. And y^e^ like of $x^3-3xx-2x-4=0$. And soe of $x^3+2x-4=0$, or $x^3+0xx+2x-4=0$, because neither 1 nor 2 is less then 0.

Secondly if uppon sight you discover three such termes together that the two extreames having the same signes their rectangle bee as greate or greater then y^e^ square of the meane terme, conclude there are two imaginary roots.(32) Thus in $x^3-pxx+3ppx-q^3=0$ are two imaginary roots because $1\times 3pp \sqsubset -p\times -p$. And soe of $x^3-pxx+ppx-2p^3=0$ because $1\times pp=-p\times -p$. or

$$-p\times -2p^3 \sqsubset pp\times pp.$$

Thirdly if(33)

First y^n^ y^e^ number of impossible roots is always eaven. If one bee impossible there must bee two, if three there must bee foure &c. And hence Equations of odd dimēsions must have one roote reall at least.

Secondly the number of reall roots of an Equation are not more then the number of its termes. Thus $x^5-3x^4+4=0$ can have but thre reall roots &(34) two must bee imaginary. Thus $x^4-2x+3=0$ cannot have all foure roots reall & therefore must have two imaginary.(35) Hence are to bee excepted Equations wch want all their odd termes as $x^6-2x^4+3xx-2[=0]$. And in this & such like cases write y for xx. And so many true roots as y^e^ product $y^3-2yy+3y-2=0$ hath, twice soe many reall roots halfe true halfe false, & foure times soe many imaginary ones y^e^ other $x^6-2x^4+3xx-2=0$ shall have.(36)

(32) That is, if the general equation $0 = x^n + \sum_{1\leqslant i\leqslant n}(p_i x^{n-i})$ be such that $p_{r-1}p_{r+1} \geqslant p_r^2$, then it has one pair of imaginary roots at least. We cannot be sure that Newton was able to prove this correct rule, but if so we may restore his argument with some confidence in the following way. Since the derivative $f'(x)$ of the (simply continuous) $f(x) = x^n + \sum_{1\leqslant i\leqslant n}(p_i x^{n-i})$ can be zero at only one point between any two adjacent real roots of the equation $f(x) = 0$, it follows that one real root at most of $f(x) = 0$ can lie between any two adjacent real roots of $f'(x) = 0$. (Geometrically, the continuous curve $y = f(x)$ meets the axis $y = 0$ in one real point at most between any two adjacent maximum and minimum values of y.) Hence if the equation $f(x) = 0$ has all its roots real so must $f'(x) = 0$, and so more generally and conversely, by repeating the argument r times, if the rth derivative $f^{[r]}(x) = 0$ has 2μ imaginary roots, then the original equation $f(x) = 0$ must have at least 2μ imaginary roots. Now the $(n-r-1)$-th derivative of $f(x)$ is

$$\frac{n!}{(r+1)!}x^{r+1} + \sum_{1\leqslant i\leqslant (r+1)}\left[\frac{(n-i)!}{(r+1-i)!}\right]p_i x^{r+1-i},$$

that is, if we substitute

$$x = \frac{1}{y}, \quad \frac{p_{r+1}(n-r-1)!}{y^{r+1}}\left(y^{r+1} + \frac{(n-r)p_r}{p_{r+1}}y^r + \frac{(n-r+1)(n-r)}{2!}\frac{p_{r-1}}{p_{r+1}}y^{r-1} + \ldots\right).$$

Hence, if all the roots of $f^{[n-r-1]}(x) = 0$ are real, it follows that all the roots of

$$y^{r+1} + \frac{(n-r)p_r}{p_{r+1}}y^r + \frac{(n-r+1)(n-r)}{2!}\frac{p_{r-1}}{p_{r+1}}y^{r-1} + \ldots = 0$$

Thirdly, if under the termes of any Equation you set a progression of fractions each having y[e] dimension of y[e] terme above it for its numerator & [y[e]] number denominating y[t] terme first second third &c[(37)] for its denominator; & y[n] if you can find any three termes together so y[t] the rectangle of the first & last multiplyed by the fraction set under the first bee greater then the square of the middle terme multipl[y]ed by y[e] fraction set under it: conclude there are two imaginary roots at least. Or if it bee equall to it in one case & not in all throughout y[e] Equation conclude also there are two imaginary roots at least.[(38)] If equall in all cases throughout the Equation, conclude that all the roots of the equation are

---

must be real and so the sum of their squares positive: that is, by the first of Newton's formulae in [3] above,

$$\left(\frac{(n-r)\,p_r}{p_{r+1}}\right)^2-2\left(\frac{(n-r+1)\,(n-r)}{2\,!}\,\frac{p_{r-1}}{p_{r+1}}\right)$$

must be positive, and so $$p_r^2>\frac{n-r+1}{n-r}\,p_{r-1}p_{r+1}>p_{r-1}p_{r+1}.$$

Conversely, if $p_{r-1}p_{r+1}\geqslant p_r^2$, then at least two roots of $f^{[n-r-1]}(x)=0$ are imaginary, and therefore *à fortiori* at least two roots of $f(x)=0$. (The result is slightly stronger than the revised rule Newton will develop below when only one pair of imaginary roots is considered.)

(33) Newton breaks off and cancels his text in favour of the revision which follows.

(34) Newton has cancelled here 'the other'.

(35) An immediate deduction from Descartes' sign-rule (note (2)) when applied to the incomplete equation.

(36) For to each real root $y=\alpha$ in the latter equation correspond the two real roots $x=\pm\sqrt{\alpha}$ in the former.

(37) That is, in the ordering of the terms of the equation by decreasing powers of the variable.

(38) Newton nowhere in his extant papers seems to have recorded the workings of his mind at this point, but we may, in fact, restore his derivation of this complex procedure on much the same lines as in note (32). Applying Hudde's method of maxima and minima (p. 214) $n-2$ times to the equation $0=x^n+\sum\limits_{1\leqslant i\leqslant n}(p_i x^{n-i})$, we may eliminate all terms except those containing p_{r-1}, p_r and p_{r+1} from its $(n-2)$-th derivative and hence evaluate it as

$$\frac{(n-r+1)\,!\,(r-1)\,!}{2\,!}\,p_{r-1}x^{n-r+1}+(n-r)\,!\,p_r x^{n-r}+\frac{(n-r-1)\,!\,(r+1)\,!}{2\,!}\,p_{r+1}x^{n-r-1}.$$

Now $f(x)=0$ has at least one pair of imaginary roots if (note (32)) $f^{[n-2]}(x)=0$ has a pair, that is $\left(\text{on dividing out the factor }\dfrac{(n-r-1)\,!\,(r-1)\,!}{2\,!}\right)$ when

$$(n-r+1)\,(n-r)\,p_{r-1}x^2+2(n-r)\,rp_r x+r(r+1)\,p_{r+1}=0$$

has only imaginary roots. Immediately, a sufficient condition for $f(x)=0$ to have imaginary roots is that

$$(n-r+1)\,(n-r)\,r(r+1)\,p_{r-1}p_{r+1}>[(n-r)\,rp_r]^2\quad\text{or}\quad\frac{n-r+1}{r}\,p_{r-1}p_{r+1}>\frac{n-r}{r+1}\,p_r^2$$

(which is Newton's condition). (Note that, since $\dfrac{(n-r)\,r}{(n-r+1)\,(r+1)}<1$, this is a weaker condition than that of note (32).)

equall.[39] If it bee greater or equall to it in two places of y^e^ Equation & not in all places betwixt conclude there are foure imaginary roots at least. If it bee greater or equall to it in three places of y^e^ Equation & not in all places betwixt, conclude there are six imaginary roots at least. And soe of the rest.[40]

Thus if the Equat^n^ be $x^3-3xx+4x-2=0$.

Then y^e^ progress: is $\frac{3}{1}.\ \frac{2}{2}.\ \frac{1}{3}.$ & because $-3x^{(41)}\times-2\times\frac{2}{2}\,(6)\sqsubset 4\times4\times\frac{1}{3}\left(\frac{16}{3}\right)$, I conclude there are two roots imaginary at least.

Thus in $x^3-6x^2+6x-2=0$. because $-6\times-2\times\frac{2}{2}\,(12)=6\times6\times\frac{1}{3}\,(12)$ but

$\frac{3}{1}.\ \ \frac{2}{2}.\ \ \frac{1}{3}.$

$1\times6\times\frac{3}{1}\,(18)\sqsupset^{(42)}-6\times-6\times\frac{2}{2}\,(36[])$ I conclude there are two imaginary roots.

Thus in $x^3-6xx+12x-8=0$ because $1\times12\times3\,(36)=-6\times-6\times1\,(36)$ and

$3.\ \ 1.\ \ \frac{1}{3}.$

also $-6\times-8\times1\,(48)=12\times12\times\frac{1}{3}\,(48)$ I conclude y^t^ all the 3 roots are equall.

---

(39) If in the equation $0 = x^n + \sum_{1\leqslant i\leqslant n}(p_i x^{n-i})$ the equality $\frac{n-i+1}{i}\,p_{i-1}p_{i+1} = \frac{n-i}{i+1}\,p_i^2$ holds for each i, $i = 1, 2, \ldots, n-1$ (with $p_0 = 1$), then, if we suppose $p_1 = nP$ $\left(\text{or } \binom{n}{1}P\right)$, it follows that $\frac{n}{1}\,p_2 = \frac{n-1}{2}\,p_1^2$ or $p_2 = \binom{n}{2}P^2$; and in general, if $p_{i-1} = \binom{n}{i-1}P^{i-1}$ and $p_i = \binom{n}{i}P^i$, we easily deduce that $p_{i+1} = \binom{n}{i+1}P^{i+1}$. The equation reduces therefore to

$$0 = \sum_{0\leqslant i\leqslant n}\left[\binom{n}{i}P^i x^{n-i}\right] = (x+P)^n.$$

(40) A first statement of Newton's 'incomplete' rule for enumerating imaginary roots. Newton's argument would here seem to be that each *separate* occurrence of adjacent coefficients p_{r-1}, p_r and p_{r+1} which satisfy $\frac{n-r+1}{r}\,p_{r-1}p_{r+1} > \frac{n-r}{r+1}\,p_r^2$ (note (38)) marks a *distinct* pair of imaginary roots in the equation. If so, we have a classic example of the productiveness of a lapse in strict logic. No matter how many separate instances occur where the test is satisfied, we may rigorously deduce no more than the primitive consequence that the equation must have at least one pair of imaginary roots. And if we wish to develop a stronger consequence we must give it independent proof. Newton's present rule is, in fact, universally valid but its rigorous justification was not forthcoming for almost exactly two centuries till Sylvester, using a result of Fourier's, produced his brilliant but technically difficult proof of Newton's more general 'complete' rule in 1865. (See J. J. Sylvester, 'On an elementary proof and generalization of Sir Isaac Newton's hitherto undemonstrated rule for the discovery of imaginary roots', *Proceedings of the London Mathematical Society*, 1 (1865/6): 1–16 = *Collected Mathematical Papers*, 2 (Cambridge, 1908): §84: 498–513. Compare also *Collected Mathematical Papers*, 2: §74: 376–479: *On Newton's Rule for the Discovery of Imaginary Roots* (where Sylvester

But in $x^3+6xx+12x-8=0$, because $1\times12\times3\,(36)=6\times6\times1$ but

$$3.\quad 1.\quad \frac{1}{3}$$

$$6\times-8\times1\,(-48)\,\urcorner\,12\times12\times\frac{1}{3}\,(+48)$$

I conclude two roots are imaginary.

In $x^4-x^3+2xx-2x+3=0$, because by y^e^ three first termes

$$\frac{4}{1}\quad\frac{3}{2}\quad\frac{2}{3}\quad\frac{1}{4}$$

$$1\times2\times\frac{4}{1}\,(8)\,\sqsubset\,-1\times-1\times\frac{3}{2}\left(\frac{3}{2}\right),$$

I conclude there are two imaginary roots at least, also by y^e^ three last termes $2\times3\times\frac{2}{3}\,(4)\,\sqsubset\,-2\times-2\times\frac{1}{4}\,(1)$ therefore all 4 roots are imaginary(43) unlesse the like happen in the three middle termes. I try therefore & find

$$-1\times-2\times\frac{3}{2}\,(3)\,\sqsubset\,2\times2\times\frac{2}{3}\left(\frac{8}{3}\right)$$

& soe can conclude but two rootes imaginary.

In $x^4-x^3+3xx-2x+3=0$ because $1\times3\times\frac{4}{1}\,(12)\,\sqsubset\,-1\times-1\times\frac{3}{2}\left(\frac{3}{2}\right)$ and

$$\frac{4}{1}.\quad\frac{3}{2}.\quad\frac{2}{3}.\quad\frac{1}{4}.$$

also $3\times3\times\frac{2}{3}\,(6)\,\sqsubset\,-2\times-2\times\frac{1}{4}\,(1)$, but not $-1\times-2\times\frac{3}{2}\,(3)\,\sqsubset\,3\times3\times\frac{2}{3}\,(6)$.

Therefore I conclude all four roots imaginary.

In $x^7\ *+x^5-2x^4+3x^3-3x^2-2x-1=0$. because the three first termes give

$$\frac{7}{1}.\frac{6}{2}.\frac{5}{3}.\quad\frac{4}{4}.\quad\frac{3}{5}.\quad\frac{2}{6}.\quad\frac{1}{7}.$$

---

proves the 'incomplete' rule for equations up to the quintic); and §108: 704–8: *On an elementary proof of Sir Isaac Newton's hitherto undemonstrated Rule*....) The proof of the restricted result given in note (38) is essentially that of George Campbell (expounded in his 'A Method for determining the Number of impossible Roots in adfected Æquations', *Philosophical Transactions*, 35: no. 404 (October 1728), II: 515–31) and was widely accepted in the eighteenth century as rigorous, though of course it can never show the existence of more than a single pair of imaginary roots. (Compare Colin Maclaurin, *A Treatise of Algebra* (London, ₁1748): Part II, chap. XI, *Of the Rules for finding the number of impossible Roots in an equation* = ₅1788: 274–85, especially 279–83.) However, at the end of the century Waring pointed out the inadequacy of such a proof: 'vulgares enim demonstrationes solummodo probant impossibiles radices in datâ æquatione contineri; non vero quod saltem tot sunt, quot invenit regula' (*Meditationes Algebraicæ. Editio tertia recensita et aucta* (Cambridge, 1782: xi)).

(41) Read '-3' simply.

(42) 'lesse y^n^'.

(43) Newton first wrote 'there are two more imaginary roots'.

$1\times1\times\frac{7}{1}(7)\sqsubset0\times0\times\frac{6}{2}(0)$ there are two imaginary roots. Also the 3[d] 4th & 5[t] terme give $1\times3\times\frac{5}{3}(5)\ [\sqsubset-2\times-2\times\frac{4}{4}(4)]$ therefore since by the 2[d] 3[d] & 4th terme tis $0\times-2\times\frac{6}{2}(0)\sqsupset1\times1\times\frac{5}{3}\left(\frac{5}{3}\right)$ I conclude there are 4 roots imaginary. Also by y[e] 4th 5[t] & 6[t] termes I find $-2\times-3\times\frac{4}{4}(6)\sqsubset3\times3\times\frac{3}{5}\left(\frac{27}{5}\right)$ but thence nothing can be concluded because those three termes are of the same condition(44) wth y[e] 3[d] 4th & 5[t] termes w[ch] immediately precede them. Lastly I find by the three last termes $-3\times-1\times\frac{2}{6}(1)\sqsubset-2\times-2\times\frac{1}{7}\left(\frac{4}{7}\right)$; And by the termes preceding them $3\times-2\times\frac{3}{5}\left(-\frac{18}{5}\right)\sqsupset-3\times-3\times\frac{2}{6}(3)$: Therefore I conclude there are two more imaginary roots; y[t] is in all 6 & but one reall.

Thus in litterall Equations, if $x^3-pxx+3ppx-q^3=0$, becausc

$$\frac{3}{1}\quad\frac{2}{2}\quad\frac{1}{3}$$

$$1\times3pp\times\frac{3}{1}(9pp)\sqsubset-p\times-p\times\frac{2}{2}(pp)$$

therefore what ever numbers are taken for p and q two roots shall bee imaginary. And soe of the rest.

This rule may be otherwise thus exprest. Over y[e] termes of y[e] Equation set a series of fractions each having y[e] dimensions of the terme under it for its numerator, & the number denominating y[e] terme first, second, third &c for its denominator.(45) Then in every three termes observe whither the square of the middle terme multiplyed by the fraction above be greater equall or lesse y[n] y[e] factus of the termes before & after it multiplyed by y[e] fraction over y[e] terme before it. If greater write y[e] signe + underneath; if equall or lesse write the signe – under-neath y[e] middle terme: & lastly set + under y[e] first terme of y[e] equation. Then observe how many changes there are from + to – & conclude that there are soe many paires of imaginary roots. Unlesse all y[e] roots bee equall.(46)

(44) Specifically, of greater inequality.

(45) That is, as before, over every term $p_r x^{n-r}$ of the equation $0 = x^n + \sum_{1\leqslant i\leqslant n}(p_i x^{n-i})$ we must set the fraction $(n-r)/(r+1)$ (and correspondingly over x^n the fraction $n/1$).

(46) Newton considers the model in which the arithmetical signs '+' and '–' denote the complementary cases where

$$\frac{n-r}{r+1}p_r^2 > \frac{n-r+1}{r}p_{r-1}p_{r+1} \quad\text{and where}\quad \frac{n-r}{r+1}p_r^2 \leqslant \frac{n-r+1}{r}p_{r-1}p_{r+1}$$

respectively.

$$\frac{3}{1}.\ \frac{2}{2}.\ \frac{1}{3} \qquad\qquad \frac{4}{1}\ \frac{3}{2}\ \frac{2}{3}\ \frac{1}{4}$$

Thus $x^3-3xx+4x-2=0$ hath 2: & $x^4-x^3+3xx-2x+3=0$ hath 4

$$+.\ +.\ -. \qquad\qquad +.\ -.\ +.\ -.$$

imaginary roots.

If you would bee more exact set downe after their signes the differences of y[e] said squares & rectangles. And then if you see three differences together wth y[e] same signe soe y[t] y[e] square of the meane diff: bee lesse then y[e] rectangle of the other two change the signe of the said meane difference.(47)

$$\frac{4}{1}.\ \frac{3}{2}.\ \frac{2}{3}.\ \frac{1}{4}.$$

Thus if $x^4-x^3+\ \ 2xx-2x+3=0$. becaus $-\frac{23}{4}{}^{(48)}\times-3\left(\frac{69}{4}\right)\sqsubset-\frac{1}{3}\times-\frac{1}{3}\left(\frac{1}{9}\right)$

$$+\frac{4}{1}.-\frac{23}{4}.^{(48)}\ -\frac{1}{3}.-3.$$

$$+.\ -.\quad +.\ -.$$

I change the signe of $\frac{-1}{3}$ & soe the signes $+.-.+.-$ shew all foure roots imaginary.

If you would bee yet more exact, augment y[e] roots of the Equatiō the more the better, & at least soe much as to make them all true. then set y[e] afforesaid differences wth their signes underneath as before. And under them the progression of fractions squared. Then if you see three differences together wth y[e] same signe soe y[t] y[e] square of the middle difference multiplyed by the fraction under it bee not greater y[n] y[e] rectangle of the other two differences multiplyed by y[e] fraction under the first: change y[e] signe of y[e] middle difference.(47)

---

[6](49)

Any Equation being propounded, set downe a series of so many fractions as y[e] Equation hath dimensions, whose numerators & denominators are a progression of units backward & forward. Divide each fraction by y[t] p'ceding it & set the quotes in order over all y[e] middle termes of the Equation.(50) Then observe of every middle terme whither its square multiplyed by y[e] fraction over it bee greater equall or lesse y[n] y[e] rectangle of y[e] two termes on either hand. If

---

(47) These observations would seem to be highly empirical, and we leave their ultimate justification or disproof for a second Sylvester.

(48) Read '$-13/2$'. Newton perhaps calculated $(-1)^2\times(3/2)^2-2\times(4/1)$ in error.

(49) Add. 4004: 86r/86v, a revised (and simplified) version of [5].

(50) That is over the term $p_r x^{n-r}$ of the equation $0 = x^n+\sum_{1\leqslant i\leqslant n}(p_i x^{n-i})$ is to be set the 'quote'

$$\frac{n-r}{r+1}\Big/\frac{n-r+1}{r} = \frac{r(n-r)}{(r+1)(n-r+1)}.$$

greater write + underneath, if equall or lesse write —.[51] Lastly set + under ye first & last terme & there shall bee soe many impossible roots as there are changes of signes. Unlesse it happen yt all ye roots are equall, for &c:[52]

Thus if $x^3-3xx+6x-4=0$. The series [of] fractions will bee $\frac{3}{1}.\,\frac{2}{2}.\,\frac{1}{3}.$ & dividing $\frac{2}{2}$ by $\frac{3}{1}$, & $\frac{1}{3}$ by $\frac{2}{2}$ their quotes will bee $\frac{1}{3}, \frac{1}{3}$, to bee set over ye middle termes of the equation thus.

$$\begin{array}{ccccc} & \frac{1}{3}. & \frac{1}{3}. & & \\ x^{[3]} & -3xx & +6x & -4 & =0. \\ + & - & - & + & \end{array}$$

Then I observe in ye 2d terme that $-3\times-3\times\frac{1}{3}\,(3)$ is lesse then 1×6 & therefore I write $-$ under it. so in ye 3d terme I find $6\times 6\times\frac{1}{3}\,(12)=-3\times-4\,(12)$, therefore I write $-$ under it. Lastly seting + under ye first & last terme I find two changes of signes & soe conclude there are [two impossible roots.]

Thus if $x^5-4x^4+4x^3-2xx-5x-4=0$. The series of fractions will bee $\frac{5}{1}.\,\frac{4}{2}.\,\frac{3}{3}.\,\frac{2}{4}.\,\frac{1}{5}.$ And dividing $\frac{4}{2}$ by $\frac{5}{1}$ & $\frac{3}{3}$ by $\frac{4}{2}$ &c there results $\frac{2}{5}.\,\frac{1}{2}.\,\frac{1}{2}.\,\frac{2}{5}$ to bee set over ye middle termes of the equation thus

$$\begin{array}{ccccccc} & \frac{2}{5}. & \frac{1}{2}. & \frac{1}{2}. & \frac{2}{5}. & & \\ x^5 & -4x^4 & +4x^3 & -2xx & -5x & -4 & =0. \\ + & + & - & + & + & + & \end{array}$$

Then in ye second terme I find $-4\times-4\times\frac{2}{5}\left(\frac{32}{5}\right)\sqsubset 1\times 4\,(4)$: therefore I set + under it. In the third $4\times 4[\times]\frac{1}{2}\,(8)=-4\times-2\,(8)$: therefore I write $-$. In the 4th $-2\times-2\times\frac{1}{2}\,(2)\sqsubset 4[\times]-5\,(-20)$ therefore I write +. In ye 5t

(51) By a slight modification of the previous rule in [5] (note (46)) the signs '+' and '−' are to be entered according as

$$\frac{r(n-r)}{(r+1)(n-r+1)}\,p_r^2 > p_{r-1}p_{r+1} \quad\text{or}\quad \frac{r(n-r)}{(r+1)(n-r+1)}\,p_r^2 \leqslant p_{r-1}p_{r+1}$$

respectively.

(52) This statement of the 'incomplete' rule for enumerating imaginary roots is essentially that expounded in [5] above, and indeed differs only in the construction of the multiplying fractions. A decade and a half later Newton inserted a brief account of the test in his Lucasian lectures (ULC. Dd. 9. 68: 155–8 = October 1681, Lectiones 6/7) together with (pp. 158/9) a sketch of the more general 'complete' rule, and this passage, duly printed in the *Arithmetica Universalis* (London, $_1$1707): 242–5 has been the sole source for all later commentary on the rule.

$-5\times-5\times\frac{2}{5}\,(10) \sqsubset -2\times-4\,(8)$ therefore I write $+$. lastly under y^e^ first & last terme I write $+$. And soe finding two changes of termes I conclude two roots to bee impossible.

$$\begin{array}{ccccc} & \frac{3}{8} & \frac{4}{9} & \frac{3}{8} & \\ x^4 & +0x^3 & -6xx & -3x & -2=0 \\ + & + & + & - & + \end{array}$$

Thus in $x^4+0x^3-6xx-3x-2=0$ two roots are impossible.

$$\begin{array}{cccc} & \frac{1}{3} & \frac{1}{3} & \\ x^3 & +0xx & +ppx & -q^3=0 \\ + & - & + & + \end{array}$$

In $x^3+0xx+ppx-q^3=0$ two roots are impossible. &c.

$$\begin{array}{cccc} & \frac{1}{3} & \frac{1}{3} & \\ x^3 & -6xx & +12x & -8=0 \\ + & - & - & + \end{array}$$

In $x^3-6xx+12x-8=0$ All y^e^ roots are equall.(53)

Sometimes there may bee impossible [roots] not by this meanes discovered, wch if you suspect, augment or diminish y^e^ roots of the Equation a little, not soe much as to make them all affirmative or all negative, or at most not much more.(54) & try the rule againe. And if there bee any impossible roots t'will rarely happen y^t^ they shall not bee discovered at two or three such tryalls. Nor can there bee an Equation whose impossible roots may not bee thus discovered.(55)

Thus if $x^3-3ppx-3p^3=0$, in wch noe impossible [roots] appeare(56) I put $x=y-p$ & the result is $y^3-3pyy-p^3=0$ in wch two appeare,(57) Or if I put $x=y-2p$ the result is $y^3-6pyy+9ppy-5p^3=0$ in wch also two appeare.(58)

(53) Since (note (39)) $6\times6\times\frac{1}{3} = 1\times12$ and $12\times12\times\frac{1}{3} = -6\times-8$.

(54) This observation, presumably wholly empirical, conflicts with his previous statement in [5] that 'If you would bee yet more exact, augment y^e^ roots of the Equatiō the more the better, & at least soe much as to make them all true'.

(55) In the restricted sense that the test will always show whether or not an equation has all its roots real this is an immediate deduction from note (38). The wider sense that the rule will, after suitable trials, invariably (at least once) indicate the full number of imaginary roots must surely be an empirical generalization.

(56)
$$\begin{array}{cccc} & \frac{1}{3} & \frac{1}{3} & \\ x^3 & * & -3ppx & -3p^3 = 0. \\ + & + & + & + \end{array}$$

(57) Since
$$\begin{array}{cccc} & \frac{1}{3} & \frac{1}{3} & \\ y^3 & -3pyy & * & -p^3 = 0. \\ + & + & - & + \end{array}$$

(58) For
$$\begin{array}{cccc} & \frac{1}{3} & \frac{1}{3} & \\ y^3 & -6pyy & +9ppy & -5p^3 = 0. \\ + & + & - & + \end{array}$$

Thus if $x^5+x^4-4x^3+5xx-2x+1=0$, I set y[e] signes + & − under it as before
+ + + + − +
and find two imaginary roots & to try if it have any more I suppose $x=y+1$ & y[e] result is $x^5+6x^4+10x^3+9xx+5x+0$.(59)

Now by this rule false roots may bee often discovered at first sight; as if you see a terme wanting twixt two others of the same signes, or if it bee greater then neither of those two or its square not greater then their rectangle; conclude there is a paire of impossible roots at least & set the signe − under y[t] terme. also set y[e] signe + on either side the term wanting.(60)

As in this $x^7+0x^6+2x^5-2x^4+3x^3-4xx+6x-2=0$. In w[ch] it appears there
+ − + − − +(61)
are 4 if not 6 impossible roots.

If there bee two or more termes wanting set signes under them successively begining wth a negative, only end wth an affirmative if the terms on either hand have contrary signes.(62) As in $x^5+ax^4 * * * +a^5=0$ so in $x^5+ax^4 * * * -a^5=0$.
− + −   − + +
The first shows 4, y[e] last two roots imaginary. Soe in

$$x^{10} * +x^8 * * -3x^5 * * * * -6=0.$$
+ − + − + + − + − + +

w[ch] hath 8 roots imaginary.

---

[7](63)

To know how many reall roots an Equation hath: Take a series of numbers gradually increasing, tis noe matter whither in any regular progression or not. Substitute them successively into y[e] Equation & set y[e] signes of the resulting quantitys in order & put + to both ends if y[e] Equation bee of eaven dimensions, otherwise put + to that end resulting frō(64) y[e] greatest quantitys substituted + & − to y[e] other.

---

(59) Read '$y^5+6y^4+10y^3+9yy+5y+2=0$.' Since

$$\tfrac{2}{5}\quad\tfrac{1}{2}\quad\tfrac{1}{2}\quad\tfrac{2}{5}$$
$$y^5+6y^4+10y^3+9yy+5y+2=0,$$
+ + − − − +

the rule again shows only one pair of imaginary roots.

(60) For if either of p_{r-1}, p_{r+1} be zero, it follows that

$$\frac{r(n-r)}{(r+1)(n-r+1)}p_r^2 > 0 = p_{r-1}p_{r+1}.$$

(61) In full the sequence of signs should be + − + − − − + +.

(62) This is required in order to make the rule consistent with Descartes' sign-rule when applied to the incomplete equation.

Example. if $x^3-9x^2+6x+1=0$. writing 1. 0. -1 for x there results -1, $+1$. -15. The signes are therefore $+, -, +, -, -$. w[ch] argue three possible roots. Thus if $x^4-4x+2=0$ writing 1. 0. for x the signes produced are $+ - + +$ w[ch] argue two possible roots, y[e] other two being imaginary because of two termes wanting together. If $x^5-3x^4+4x-1=0$. writing 2. 1. 0 for x y[e] signes will bee $+ - + - -$ w[ch] argue 3 possible roots.(65)

Note y[t] w[n] small numbers as 1 or 2 are to be substituted you need not make any exact computation but only in your mind to know w[t] the signes of the result would bee. But to substitute 3, 4, 5 &c observe this rule. If y[e] equation bee $x^3+ax^2+bx+c=0$. Then $\overline{\overline{x+a}\times x:+b}\times x:+c$ is y[e] quantity. or

$$\overline{x+a}\times x:+b+\frac{c}{x}.^{(66)}$$

Note also that by this meanes the limits of y[e] rootes are [found]. For they fall twixt those substituts w[ch] produce divers signes.(67)

To know y[e] limits of an Equation.

If $x^8+px^7+qx^6+rx^5+sx^4+tx^3+vxx+wx+y=0$. Make $p=a$. $pa-2q=b$.

$$pb-qa+3r=c. \quad pc-qb+ra-4s=d. \quad pd-qc+rb-sa+5t=e.$$

$$pe-qd+rc-sb+ta-6v=f. \quad pf-qe+rd-sc+tb-va+7w=g.$$

$$pg-qf+re-sd+tc-vb+wa-8y=h. \quad \&c.$$

Or if $p=0$. make $-2q=b$. $+3r=c$. $+2qq-4s=d$. $-5qr+5st=e$.

$$3rr-2q^3+6qs-6v=f. \quad 7qqr-7qt-7rs+7w=g. \quad [\&c.]$$

Then is a y[e] summ, b the sum̄ of y[e] squares, c y[e] summe of y[e] cubes &c: of all the roots.(68) Therefore if noe roots be imaginary $\sqrt{b}$ is greater or more distant

(63) From the recto side of a sheet, now in private possession, which once formed f. 37 of the Waste Book (ULC. Add. 4004).

(64) Read 'from'.

(65) Newton uses the continuity axiom (note (67) below) that $y=f(x)$, assumed simply continuous, has (at least) one zero in the interval $x\in[a,b]$ where $f(a)>0>f(b)$. His full rule makes implicit use of the property that, in equations of even dimension,

$$f(\infty)=f(-\infty)=+\infty$$

and correspondingly in equations of odd dimension $f(\infty)=-f(-\infty)=+\infty$ (where Newton assumes $f(x)$ of the form $x^n+\sum_{1\leqslant i\leqslant n}(p_i x^{n-i})$).

(66) Compare §1 above, especially note (8).

(67) Newton cites the continuity axiom which states, in modern terms, that any continuous function $f(x)$ has $f(a+\theta(b-a))=0$, where $f(a)>0>f(b)$ and $\theta\in[0,1]$.

(68) See section [4].

from nothing then any root or more exactly $\sqrt{\frac{d}{b}}$.(69) If but two roots be imaginary then is $\sqrt{4:d}$ y^e^ greatest limit(70) or $\sqrt{}$(71)

[8](72)

To solve numerall Equations by divisors of one dimension.

Find all the divisors of these three numbers, y^e^ aggregate of the knowne parts of the termes, y^e^ last terme & the difference of y^e^ known parts of the termes of eaven & odd dimensions. And if you see a progression of divisors differing by an unit take ye middle most & add it to x if ye first divisor is greatest, otherwis substract it & try to divide ye Equation by y^t^. If there bee noe such progression or if none of them succed, ye reduction by a divisor of one dimension is impossible.

Or thus.(73) substitute 1. 0. −1 for x. find y^e^ divisors of the results. Add & substract them from 1. 0. −[1]. respectibly(74) [&] if you see y^e^ same number in all three result[s] chang its signe adnex it to x & try the division.(75)

Example. If $x^3-xx-10x+6=0$. I find y^e^ divisors of

4.		1, 2, 4.
6.	[wch are]	1, 2, 3, 6. And seeing y^e^ progression 4. 3. 2. among them, I
14.		1, 2, 7, 14.

try the division by $x+3$.(76)

So if $x^4-15x^3+57xx-77x+70=0$. The diviso^rs^ of

36,		1, 2, 3, 4, 6, 9, 12, 18, 36.
70,	are	1, 2, 5, 7, 10, 14, 35, 70. where I see
220,		1, 2, 4, 5, 10, 11, 20, 22, 44, 55, 110, 220.

these progressions 3, 2, 1. 6, 5, 4. 9, 10, 11. And (trying y^e^ greatest number first of all becaus the roots are affirmative[) divide by $x-10$.](77)

(69) For $\sum_i(\alpha_i^2) > \alpha_p^2$ for each $p = 1, 2, \ldots, n$, where the α_i are the roots (supposed all real) of the equation and $n > 1$.

(70) Similarly, $\sum_i(\alpha_i^4) > \alpha_p^2\sum_i(\alpha_i^2)$ for each $p = 1, 2, \ldots, n$.

(71) Newton breaks off and cancels his text. Note that, much as before, $\sum_i(\alpha_i^4) > \alpha_p^4$ for each $p = 1, 2, \ldots, n$, but if two of the roots α_i are imaginary (and so conjugate complex), say $r+is$ and $r-is$ respectively, since $(r+is)^4+(r-is)^4 = r^4+s^4-6r^2s^2$, it follows that Newton's assertion is that $\sum_i(\alpha_i^4) = \Sigma$ (4th powers of real roots) $+r^4+s^4-6r^2s^2$ is greater than the fourth power of any real root.

(72) From ff. 87^v^ and 88^r^ of the Waste Book, the former of which is in private possession. Newton attacks the general problem of factorisation.

(73) This is a marginal insertion in revision of the preceding paragraph.

(74) That is, respectively!

To reduce numerall equations by divisors of 2 *dimensions.*

Substitute 2. 1. 0. −1. −2 into y^e^ place of x, & find y^e^ divisors of the resulting numbers. Adde & subduct those respectively to & from 4. 1. 0. 1. 4. And if in the results you see any Arithmeticall progression, make y^e^ middle terme $=+c$. & The difference twixt the termes of the progression $=b$. W^ch^ must bee + if the progression decreas otherwise −.[78]

Thus if $x^4-x^3-5xx+12x-6=0$ y^e^ divisors of—

$$\left\{\begin{array}{lll} 6. & 1.\ 2.\ 3.\ 6. & +4. \\ 1. & 1. & +1. \\ 6.\ \text{are} & 1.\ 2.\ 3.\ 6.\ \text{w}^{\text{ch}}\ \text{added \& substracted from} & +0. \\ 21. & 1.\ 3.\ 7.\ 21. & +1. \\ 26. & 1.\ 2.\ 13.\ 26. & +4. \end{array}\right.$$

$$\text{produ[c]e}\left\{\begin{array}{l} 5.\ 6.\ 7.\ 10.\ 3.\ 2.\ 1.\ -2. \\ 2.\ 0 \\ 1.\ 2.\ 3.\ 6.\ -1.\ -2.\ -3.\ -6. \\ 2.\ 4.\ 8.\ 22.\ 0.\ -2.\ -6.\ -20. \\ 5.\ 6.\ 17.\ 30.\ 3.\ 2.\ -9.\ -22. \end{array}\right\}\ \text{in w}^{\text{ch}}\ \text{I see these progressio}$$

2, 2, 2, 2, 2. 1, 2, 3, 4, 5. 3, 0, −3, −6, −9. −2, 0, 2, 4, 6. And so I try the division by these four divisors x^2-2. x^2+x-3. x^2-3x+3. & x^2+2x-2. & find it succeed by either of the two latter.[79]

To reduce numeral equations by divisors of 3 *dimensions.*

Get y^e^ divisors of seven such terms. Add & subduct respectively to & from 27, 8, 1, 0, −1, −8, −27, & rank them in so many series.[80] Let a b & c signify

(75) The factor $f(x) = x+k$ is isolated by deriving $f(1)-1 = f(-1)+1 = k = f(0)$.

(76) And it succeeds: $x^3-x^2-10x+6 = (x+3)(x^2-4x+2)$.

(77) These examples were cancelled apparently to avoid overloading the text and certainly not because of any supposed deficiency in them. The former, indeed, was inserted by Newton into his Lucasian lectures in 1682. (See *Arithmetica Universalis* (London, $_1$1707): 43–4.)

(78) In general, where the form $F(x)$ has the factor x^2+bx+c, then $F(n)$ has the factor n^2+bn+c, and Newton's evaluation of b and c depends on finding a sequence with

$$bn+c = (n^2+bn+c)-n^2.$$

Immediately afterward Newton has cancelled, 'And if the first terme of the progression bee greatest try the division by $xx+bx+c$ otherwise by $xx-bx+c$. If you succeed not then add 4. 1. 0. 1. 4 to the respective sorts of divisors, & find b & c as before. Then if the first terme of that progression bee greatest make $xx-bx-c$ y^e^ divisor otherwise $xx+bx-c$.'

(79) See note (77) and compare *Arithmetica Universalis* ($_1$1707): 45–6.

(80) At this point and without any preliminary comment Newton jumps back to the theme of the previous section and proceeds to construct an alternative test for quadratic factors! Presumably this is why he broke off in mid-sentence to cancel it in favour of the redraft which follows.

any three terms of 3 of those series next one another suppose of y^e^ 3 wch have fewest terms. Make $b-c=d$. $a-b=e$. $d-e=f$. & observe if you see this progression among y^e^ terms $a+e-f$. a. b. c. $c-e-2f=g$. $g-e-3f=h$. $h-e-4f$ &c.(81)

Otherwise let a b & c signify any terms of y^e^ first 4th & 7th rank. Make $\frac{b-c}{3}=m$. $\frac{a-b}{3}=n$. $d-e=f$.(82) $e-f=d$. & observe if you meet wth this progression among y^e^ terms a. $a-d-3e=g$. $g-d-2e=h$. b. $b-d=i$.

$$b^{(83)}-d+e=k. \quad c.$$

Or backward c. $c+d-2e=k$. $k+d-e=i$. b. $b+d+e=h$. $h+d+2e=g$. a. If so try the division by(84)

To reduce numerall equations by divisors of three dimensions.

Get y^e^ divisors of 6 or 7 or 8 such numbers as were described before. Add & su[b]stract them from 27. 8. 1. 0. −1. −8. −27. Take any three numbers out of the three middle ranks, r, s, t. Make $-s=c$. $\frac{r+t}{2}-s=a$.(85) $\frac{r-t}{2}=b$. Then see if you can find $4a+2b+c$ in y^e^ rank p[r]eceding them & $9a+3b+c$ in y^e^

(81) Apparently for the first time Newton uses a finite-difference scheme, testing for the quadratic factor $f(x) = \alpha x^2+\beta x+\gamma$ by the condition that, where the $f(n)$ are laid out in order for equal intervals of the argument n, then the second order differences $\Delta^2 f(n)$ must be constant. Using this as his basis, he presents the array

$$\begin{array}{ccccc} \ldots a. & b. & c. & g. & h. \ \ldots \\ \ldots e. & d = e+f. & e+2f. & e+3f. & \ldots \\ \ldots f & f & f & & \ldots. \end{array}$$

From inspection it is immediate that $c-b = d$, $b-a = e$, $d-e = f$ (and hence $f = a-2b+c$) and by substituting these values in the array we may extend it indefinitely in either direction. In modern terms, if we take $a = f(n-1)$, $b = f(n)$ and $c = f(n+1)$, we may deduce that

$$e = \Delta^1 f(n-1), \quad d = \Delta^1 f(n) \quad \text{and} \quad f = \Delta^2 f(n+k)$$

for arbitrary k and hence, by application of a Newton–Stirling central-difference interpolation formula, that

$$\begin{aligned} f(n+p) &= f(n)+\tfrac{1}{2}p[\Delta^1 f(n)+\Delta^1 f(n-1)]+(1/2!)p^2\Delta^2 f(n-1) \\ &= b+\tfrac{1}{2}p(a-c)+\tfrac{1}{2}p^2(a-2b+c). \end{aligned}$$

(82) Read '$-f$'.

(83) Read 'i'.

(84) Read '$\frac{1}{18}(a-2b+c)\,x^2+\frac{1}{6}(a-c)\,x+b$'. In this alternative discussion Newton subtabulates in his previous difference scheme, constructing the array

$$\begin{array}{ccccccc} \ldots a. & g. & h. & b. & i. & k. & c. \ldots \\ \ldots -d-3e. & -d-2e. & -d-e. & -d. & f=-d+e. & -d+2e. & \ldots \\ \ldots e. & e. & e. & e. & & e. \ldots, & \end{array}$$

rank preceding that, also $4a-2b+c$ in the rank following these & $9a-3b+c$ in the rank after that; if you can, try the division by $x^3-axx+bx-c$.[86]

Or better multiply all y^e^ numbers in y^e^ middle rank by 5 & 10. Let s, & v signify y^e^ products, [&] out of y^e^ two ranks on either side take any two eaven or two odd numbers. Let these be r & t. Then making $\frac{r+t}{2}=m$. $\frac{r-t}{2}=n$ observe if you can find $4m+2n+$ any s in y^e^ rank preceding those or $4m-2n+$ any s in y^e^

---

that is, 'backward',

$$\begin{array}{l} \ldots\ c. \quad k. \qquad\qquad i.\ \ b. \quad h. \qquad g. \qquad a.\ \ldots \\ \ldots\ d-2e.\ \ d-e=-f.\ \ d.\ \ d+e.\ \ d+2e.\ \ d+3e.\ \ldots \\ \qquad \ldots\ e. \qquad\qquad e.\ \ e. \quad e. \qquad e. \qquad e.\ \ldots. \end{array}$$

It follows that

$$b = i+d = k+2d-e = c+3d-3e \quad \text{or} \quad b-c = 3(d-e),$$

and

$$a = g+d+3e = h+2d+5e = b+3d+6e \quad \text{or} \quad a-b = 3(d+2e).$$

Hence $d = \frac{1}{9}(a+b-2c)$ and $e = \frac{1}{9}(a-2b+c)$. In the modern equivalent, where $a = f(n+3)$, $b = f(n)$ and $c = f(n-3)$, so that $d = \Delta^1 f(n-1)$, $-f = \Delta^1 f(n-2)$ and $e = \Delta^2 f(n+k)$ for arbitrary k, we may apply the Newton–Stirling central-difference formula for a difference-scheme tabulated at intervals of 3 units and deduce that

$$\begin{aligned} f(n+p) &= f(n)+\tfrac{1}{2}(p/3)\,[\Delta^1 f(n)+\Delta^1 f(n-3)]+(1/2!)\,(p/3)^2\Delta^2 f(n-3) \\ &= b+\tfrac{1}{6}p(a-c)+\tfrac{1}{18}p^2(a-2b+c). \end{aligned}$$

In a summary (ULC. Add. 3964. 2: 5^r^–6^v^: *Quomodo rationes æquationis in charta exhiberi possint*) which he composed (about 1675?), Newton omitted mention of this alternative technique of factorization by construction of a difference array, and it was this abridged version of his factorization procedure which he introduced into his Lucasian lectures (ULC. Dd. 9. 68: 173–80 = October 1682, Lectiones 1/2) and which, in slightly improved form, was eventually published to the world in his *Arithmetica Universalis* (London, $_1$1707): *De inventione Divisorum*: 42–53, especially 43–7. When Leibniz came to read the latter work early in 1708 he remained puzzled by the section on factorization and on 15 March sent a handwritten copy of it to John Bernoulli for comment. (*See G. G. Leibnitii et Johan. Bernoullij Commercium Philosophicum et Mathematicum*, 2 (Lausanne and Geneva, 1745): no. CLXXI: 182–4.) In his reply the following May (*Commercium*, 2: no. CLXXII: 185–8) Bernoulli limited himself to some pertinent comments but enclosed (*ibid.* 184–209) a long tract, *Regula Generalis inveniendi divisores rationales compositos quantitatis Algebraïcæ* $a+bx+cxx+dx^3+ex^4$, *&c.*, which had been drawn up on the subject by his young nephew Nicholas. What is, in the present context, most interesting about Nicholas' tract is that in amplification of Newton's published procedure he introduced finite-difference schemes much as Newton himself forty years before in the present manuscript. (This is, of course, yet one more example of independent rediscovery of a method initially explored by Newton but never published to the world.)

(85) Read '$-a$'.

(86) Where the factor sought by Newton is $f(x) = x^3-ax^2+bx-c$, it follows that

$$\begin{Bmatrix} r = f(1)-1^3 & = -a+b-c \\ s = f(0)-0^3 & = \quad -c \\ t = f(-1)-(-1)^3 & = -a-b-c \end{Bmatrix}: \text{ hence } \begin{Bmatrix} r-s = -a+b \\ s-t = \quad a+b \end{Bmatrix}$$

and therefore $a = -\frac{1}{2}(r+t)+s$, $b = \frac{1}{2}(r+t)$ and $c = -s$.

rank following these & y[n] $9m+3n+$any v in y[e] 2[d] rank preceding those & $9m-3n+$any v in y[e] 2[d] rank following them. If so try y[e] division by

$$x^3-m\,xx+nx-\frac{1}{5}s.^{(87)}$$
$$+\frac{1}{5}s$$

Or yet better. Do not add & subduct y[e] divisors from 9, 4, 1, 0, 1, 4, 9[(88)] but try if of those in y[e] first & last rank y[e] difference of any two eaven or two odd ones be divisible by 6. Call $\frac{1}{2}$ that difference G & $\frac{1}{2}$ y[e] summe of y[e] same termes H. Then try if in y[e] middle collumn there be any term w[ch] subducted from H or added to it produces a number divisible by 9. Call that term $+c$ if it be subducted or $-c$ if added.[(89)] & y[e] summ or difference K, & putting $\frac{1}{9}K-3=a$. & $G-I=b$.[(90)] Try if you can find $8-4a+2b-c$ in y[e] 2[d] rank above y[e] middle one or $8+4a+2b+c$ in y[e] 2[d] rank next below it. If so try y[e] division by

$$x^3-ax^2+bx-c.^{(91)}$$

Or thus against those divisors added [to] & subducted from 27. 8. 1. 0. 1. 8. 27. set $9a+[3]b+c$. $4a+2b+c$. $a+b+c$. c. $a-b+c$. $4a-2b+c$. $9a-3b+c$. then

(87) Read '$x^3\ +mxx+nx+\frac{1}{5}s$'. This cubic factor $f(x)$ is found from the tabulation of $+\frac{1}{5}s$

$\phi(x)=f(x)-x^3$ for the following values:

$$\phi(\pm 3) = 9m\pm 3n+v,$$
$$\phi(\pm 2) = 4m\pm 2n+s,$$
$$r=\phi(1) = m+n,$$
$$t=\phi(-1) = m-n,$$

and $$\phi(0)=f(0)=\tfrac{1}{10}v=\tfrac{1}{5}s.$$

(88) Read '27, 8, 1, 0, 1, 8, 27'.

(89) The terms '$+c$' and '$-c$' should be interchanged. Immediately following Newton has cancelled 'Then try if in that column next before or after y[e] middlemost w[ch] has fewest divisors there be any term w[ch] added to or subducted from $\frac{1}{3}G$ produces a number divisible by 9'.

(90) These, it appears, should read '$-\frac{1}{9}K=a$, & $\frac{1}{3}G-9=b$'.

(91) Newton constructs the cubic factor $f(x)=x^3-ax^2+bx-c$ by finding

$$G=\tfrac{1}{2}[f(3)-f(-3)]=27+3b \quad\text{and}\quad H=\tfrac{1}{2}[f(3)+f(-3)]=-9a-c.$$

Hence $H-(-c)=-9a=K$ and $\frac{1}{3}G-9=b$.

choose y^e three ranks of the fewest terms & in them three numbers one in each by wch get y^e valor of a, b, & c.(92) & those gotten will give you numbers to be sought in the other ranks, wch if you find there try the division by $x^3-axx-bx-c$. otherwise chose there(93) other numbers out of the same 3 ranks & doe so till you have gone through all variety.

Note that if y^e last term of y^e æquation(94) be p [&] y^e last but one q, Then if c & $\frac{q}{c}$ have a common divisor wch divides not q, that c is to be rejected.(95) Also if $\frac{1}{4}b\times\frac{p}{c}$ or $\frac{1}{4}a\times\frac{p}{c}$ be greater then y^e greatest term of y^e æquation y^t b or a is to be rejected.(96)

(97)Or thus best. Let y^e numbers wch arose by substituting 2. 1. 0. -1. -2 for x be G. H. I. K. L. If I end not in 5 or 0 substitute 10 & -10 for x & let y^e numbers arising be F & M. but if I end in 5 or 0 & H or K do not, increase or decrease ye root of y^e æquation by an unit. Do so also if I be an eaven number & H or K an odd one wth fewer divisors & then substitute 10 & -10 for x.(98)

---

[9](99)

How numeral æquations are to be reduced by divisors of 3 or 4 or more dimensions.

Substitute 5, 3, 2, 1, 0, -1, -2, -3, -5 & also 4 & -4 if need be, for x, & suppose ye resulting terms be F, G, H, I, K, L, M, N, O, P, Q.(100) Find all their divisors [&] set those of F & Q together by pairs whose last figures are equal or differ by 5. Gather the summs & differences of these pairs. Let [y^e] summ of any two be R, the tenth part of their difference S if their last figures be equall, otherwise if they differ by 5, let the differenc[e] of ye numbers be R & the tenth

---

(92) Three (consistent) simultaneous linear equations in a, b and c are obviously sufficient.

(93) Read 'three'.

(94) That is, the equation whose cubic factor is to be constructed.

(95) For, where π is a common factor of c and q/c and λ, μ are suitable integers, we may write $c/\lambda = q/c\mu = \pi$. It follows that $\lambda q = c^2\mu = \lambda^2\pi^2\mu$ and $\lambda\mu\pi^2 = q$, and so π^2 (and hence π) must divide q.

(96) The rule would seem to be weak and not of general application. In particular the magnitude of the term $(a+\alpha)x^4$ in $(x^3+ax^2+bx+c)(x^2+\alpha x+(p/c))$ would seem to be independent of that of $\frac{1}{4}b(p/c)$ or of $\frac{1}{4}a(p/c)$.

(97) Newton began (and quickly cancelled) a new paragraph, 'Note also y^t if y^e æquation to be reduced be of six dimensions it is not necessary both to ad & subtract y^e divisors frō 27. 8. 1. 0. -1. -8. -27.' His meaning is not clear.

(98) Newton breaks off with his intended construction unfinished.

(99) U.C. Add. 4004: $88^v/89^r$, a revision of [8].

(100) That is, where the factor $f(x)$ is to be constructed, $F=f(5)$, $G=f(4)$, $H=f(3)$, $I=f(2)$, $K=f(1)$, $L=f(0)$, $M=f(-1)$, $N=f(-2)$, $O=f(-3)$, $P=f(-4)$ and $Q=f(-5)$.

p^t(101) of their summ S.(102) And if y^e^ æquatiō be not of more y^n^ 5 dimensions so y^t^ it must be divisible (if at all) by a divisor of 2 dimensions[,] set down $xx \mp Sx \mp \frac{1}{2}R - 25$ to be tried for such a divisor. Where R & S must have y^e^ same signes if y^e^ divisor of F was greater then y^e^ divisor of Q, otherwise contrary signes.(103) Quadruple y^e^ divisors of L. & if the two last figures of any one be the same wth y^e^ two last figures of $2R$, take it from $2R$. Let y^e^ residue divided by 100, be T. Or if y^e^ two last figures of any one added to y^e^ two last figures of $2R$ make 100 add it to $2R$ & let y^e^ summ divided by 100 be T & let y^e^ number whose quadruple is added to or subducted from $2R$ be a. And if y^e^ æquation be of 6 or 7 dimensions & no more set down $x^3 + Txx + \overline{S-25}x + a$ to be tried.(104) Where note that S & T must be negative if they were found so above & a must be negative if it was added to $2R$ to make T, or els affirmative if it was subducted. & y^e^ same is to be observed of the signes in y^e^ following operations.

But if y^e^ æquation be of more then 7 dimensions then look among y^e^ divisors of K for a number wch added to or subducted from $S+T+a$ gives a number divisible by 24. This number divided by 24, call V & the divisor wch gave it call β. And set down

$$x^4 + \begin{array}{c}V\\-1\end{array}x^3 + \begin{array}{c}T\\-25\end{array}xx + \begin{array}{c}S\\-25V\\+25\end{array}x + a$$

to be tried for a divisor if the æquation be not of more then 9 dimensions. Where V must be negative(105) if it was so above.(106)

But if y^e^ æquation be of more then 9 dimensions then look among the divisors of M for a number wch added to or subducted from $-S+T+a$ gives a number

(101) 'Part'.

(102) Newton has cancelled an immediately following passage, 'But for finding this difference you must subduct y^e^ divisor of Q from y^e^ divisor of F not y^e^ [divisor] of F from y^t^ of Q[,] so y^t^ S will be negative if y^e^ divisor of Q be y^e^ greater.'

(103) Where $f(x) = x^2+px+q$ is the quadratic divisor sought, we have

$$R = f(5)+f(-5) = 50+2q \quad \text{and} \quad 10S = f(5)-f(-5) = 10p.$$

Hence $p = S$ and $q = \frac{1}{2}R-25$.

(104) Where $f(x) = x^3+px^2+qx+a$ is the required cubic divisor, we deduce that, since $L = f(0) = a$ and $R = f(5)+f(-5) = 50p+2a$, then $100T = 2R-4L = 100p$ and $p = T$; and finally that $10S = f(5)-f(-5) = 250+10q$ and $q = S-25$.

(105) Newton has cancelled 'if y^e^ number β was greater then $S+T+a$ & subducted from it, so [y^t^] $S+T+a$ increased or di[minished]'.

(106) Taking $f(x) = x^4+px^3+qx^2+rx+a$ to be the quartic factor sought, we conclude that $L = f(0) = a$, $R = f(5)+f(-5) = 1250+50q+2a$ or $q = \frac{1}{50}(R-2a)-25$,

$$10S = f(5)-f(-5) = 250p+10r \quad \text{or} \quad r = S-25p$$

and $K = f(1) = 1+p+q+r+a$: then, substituting $100T = 2R-4L$ and $24V = S+T+a-K$, we have finally $T = 25+q$ (or $q = T-25$), $24V = (25p+r)+(25+q)+a-(1+p+q+r+a)$ (or $p = V-1$) and $r = S-25(V-1)$.

divisible by 24. This number divided by 24 call W & y[e] divisor wch gave it γ. And set down

$$\begin{array}{llll} x^5+\frac{1}{2}Wx^4 & +\frac{1}{2}Vx^3 & +Txx & +Sx+a \quad \text{to be tried for a divisor if} \\ +\frac{1}{2}V & -\frac{1}{2}W & -\frac{25}{2}V & -\frac{25}{2}V \\ & -26 & -\frac{25}{2}W & +\frac{25}{2}W \\ & & & +25 \end{array}$$

y[e] æquation be not of more then 11 dimensions or supposed divisib[l]e by a Divisor of not more y[n] 5 dimensions. Atq; ita in infinitum pergitur.(107)

Now the trial of these divisors is this. Suppose y[e] divisor be

$$a+bx+cxx+dx^3+ex^4+fx^5 \text{ \&c}$$

And observe if a be among y[e] divisors of L, if this Divisor ascend but to two dimensions, & $a+b+c+1$ among y[e] divisors of K if it ascend but to two or 3 dimensions[,] & $a-b+c-1$ among the divisors of M if it ascend but to 2, 3 or 4[,] & $a+2b+4c+d+e+1$(108) among y[e] divisors of I if it ascend not to more then 5 dimensions[,] & $a-2b+4c-d+e-f+1$(109) among y[e] divisors of N if it ascend to no more then six dimensions. & $a+3b+9c+27d+81e+243f+$ &c among y[e] divisors of H if it ascend not to more then 7 dimensions, & so in infinitum. In all wch put $c=1$ & $d, e, f=0$ if y[e] divisor be of 2 dimensions or $d=1$ & $e, f=0$ if but of 3 dimensions, & so on. And when you have tried all y[e] divisors wch may be found by this rule, & rejected those wch will not hold this trial: if there remain none or if y[e] æquation be not divisible by any of those wch remaine, you may conclude y[e] æquation irreducible by any rational divisor.(110)

---

(107) 'And so on indefinitely.' Where $f(x) = x^5+px^4+qx^3+rx^2+sx+a$ is the required quintic factor, we deduce that $L = f(0) = a$, $R = f(5)+f(-5) = 1250p+50r+2a$,

$$S = \tfrac{1}{10}[f(5)-f(-5)] = 625+25q+s \quad \text{and} \quad M = f(-1) = -1+p-q+r-s+a;$$

hence $T = \frac{1}{100}(2R-4L) = 25p+r$, $V = \frac{1}{24}(S+T+a-K) = 26+p+q$ and

$W = \frac{1}{24}(-S+T+A-M) = -26+p-q$, so that finally $p = \frac{1}{2}(V+W)$, $q = \frac{1}{2}(V-W)-26$, $r = T-\frac{25}{2}(V+W)$ and $s = S-625-25[\frac{1}{2}(V-W)-26] = S-\frac{25}{2}(V-W)+25$.

(108) Read '$a+2b+4c+8d+16e+32$'.

(109) Read '$a-2b+4c-8d+16e-32f+64$'.

(110) That is, by a polynomial with rational coefficients. Newton's argument has been followed through in the restorations suggested in the preceding notes.

4

MISCELLANEOUS RESEARCHES IN ARITHMETIC, NUMBER THEORY AND GEOMETRY

[1665?]

From the originals in the University Library, Cambridge

§1. THE FUNDAMENTAL OPERATIONS OF ARITHMETIC(1)

Addition connects affirmative numbers into an affirmative sume, & negative ones into a negative one. as

$$\left.\begin{array}{r}+1352\\ \underline{+7460}\\ +8812\end{array}\right)\!\Big). \qquad \left.\begin{array}{r}-137905\\ \underline{-\ \ 68432}\\ -206337\end{array}\right)\!\Big).$$

Substracting takes y^e^ lesse Number from y^e^ Greater, the difference having the same signe præfixed wch y^e^ greater number(2) hath

$$\text{as}\quad \left.\begin{array}{r}+63579\\ \underline{-14703}\\ +48876\end{array}\right)\!\Big). \qquad \left.\begin{array}{r}-26791\\ \underline{+4503}\\ -22288\end{array}\right)\!\Big).$$

Multiplication adds one factor soe often to it selfe as there are units in y^e^ other, & if y^e^ signes of y^e^ factors bee y^e^ same y^e^ product is affirmative, if divers tis negative. As to multiply $+735$ by $+47$, doe thus

$$\begin{array}{r}5145=735\text{ in }\ 7\\ \underline{29400=735\text{ in }40}\\ +34545=735\text{ in }47.\end{array} \qquad \text{Or thus} \qquad \begin{array}{r}735\\ \underline{47}\\ 5145\\ \underline{2940\ }\\ +34545.\end{array} \qquad \text{Or thus} \qquad \begin{array}{r|l}735 & \times\\ \hline 5145 & 7\\ 29400 & 40\\ \hline +34545 & \Big)\!\Big).\end{array}$$

(1) Add. 4004: 62^v^. In these elementary notes on numerical manipulation with the operations of addition, subtraction, multiplication and division Newton is apparently interested in the logical structure rather than the technical efficiency of his exposition, but even so the topic seems trivial in comparison with the surrounding mathematical pieces in the Waste Book. Newton's source is not clear, but the use of the brace (}) and the squared-off layout derive from Oughtred's *Clavis Mathematicæ* while the description of the fundamental operations may be borrowed in part from John Wallis' *Mathesis Universalis: sive, Arithmeticum Opus Integrum, tum Philologice, tum Mathematice traditum* (printed in his *Operum Mathematicorum Pars Posterior*, Oxford, 1657).

Thus to multiply -3241 by -175 the operation will bee

3241	
16205	5
22681	7
3241	1
+567171	

Alsoe 465 multiplyed by -32

465	
930	2
1395	3

will produce -14880.

Division takes yᵉ number wᶜʰ signifies how often yᵉ divisor is conteined in & may be substracted from yᵉ divisor, the signe of wᶜʰ number or Quote is affirmative if yᵉ dividend & divisor have not divers signes, but negative if they have. For if $x=\frac{a}{b}$. then $bx=a$. Or $a-bx=0$. Suppose 34545 to be divided by 47. First gett a Table of yᵉ Divisor drawn into yᵉ 9 first units as *defg*,

d 1	2	3	4	5	6	7	8	9 *f*
e −47.	−94.	−141.	−188.	−235.	−282.	−329.	−376.	−423. *g*

cut at the bottome (*eg*) close to the figures. Then looke wᶜʰ of these 9 quantitys are most like yᵉ dividend As in this case yᵉ 7ᵗʰ 329 is. therefore substract it from yᵉ dividend 34545, & there will remaine 16.45, & then set downe its caracteristick 7 in yᵉ quote. I make a prick twixt those figures (16) wᶜʰ have or might have beene altered & those (45) wᶜʰ could not bee altered by the subtraction, & the places of yᵉ pricks will shew the places of yᵉ figures in yᵉ quotient. Againe I substract 141 from 16.4 &c: & set .3 in the quote &c.[3]

34545	+735
016.45	−000
023.5	(735
000.	

If 19489012 was to be divided by 732.

1	2	3	4	5	6	7	8	9
732.	1464.	2196.	2928.	3660.	4392.	5124.	5856.	6588.

+3.	+19489012	+30000[,]00355
−2.	−147.0988	−02009,55000
−9.	−0006.988	27990,45355.[4]
−5.	−0400.0	
−5.	−0340.0	
+3.	+0260.0	
+5.	0404.0	
+5.	0380.	

(2) This cancels Newton's first choice of 'quantity' by being written in over it.

(3) Newton has summarized the full computation in his marginal scheme of contracted division.

(4) Newton applies an adapted technique in which he allows his divisors to range from −5 to +5 (instead of from 0 to 9), which makes for some computational rapidity but requires the consideration of mixed positive and negative partial remainders.

§2. INTEGER SOLUTIONS OF QUADRATIC FUNCTIONS(1)

[1]

$xx+a-bbyy=0$. x & y will be rationall by makeing $\dfrac{rr-ass}{2rs}=x$, & $\dfrac{rr+ass}{2brs}=y$. Or $\dfrac{rr-a}{2r}=x$. & $\dfrac{rr+a}{2br}=y$.(2) $xx+byy-aa=0$. Make $\dfrac{arr-abss}{rr+bss}=x$. & $\dfrac{2rsa}{rr+ssb}=y$. $xx+bbcyy-aac=0$. Make $\dfrac{2rsac}{crr+ss}=x$. & $\dfrac{rrac-ssa}{rrbbc+ssbb}$(3) $=y$(4).

[2]

$$\begin{array}{lll} axx+bx+c & =0. \\ +dyx+ey & \\ \quad +fyy & \end{array}$$ Make $aaee+abbf+acdd-abde-4aaef=m$(5) & $dd-4af=n$.

(1) Add. 3958. 2: 31r. These loose notes are entered at the head of a sheet which contains entries on plane and spherical trigonometry (2, §1.1) and on equations (3, §5.1).

(2) The particular case of the preceding where $s = 1$.

(3) Read $\dfrac{\text{'}rrac-ssa\text{'}}{rrbc+ssb}$.

(4) These parametrizations are each derived from the fundamental identity

$$(r^2-s^2)^2+(2rs)^2 \equiv (r^2+s^2)^2,$$

which is the most general integer solution of the 'Pythagorean' function $x^2+y^2 = z^2$. (Compare Schooten's *Exercitationes Mathematicæ* (Leyden, 1657): Liber v, §§10/11: 426–32.) In particular, the first form $x^2+(\sqrt{a})^2 = (by)^2$ is parametrized as $(r^2-as^2)^2+a(2rs)^2 = (r^2+as^2)^2$ by taking $r \to r$ and $s \to s\sqrt{a}$; the second, $x^2+(y\sqrt{b})^2 = a^2$, as

$$(ar^2-abs^2)^2+b(2ars)^2 = a^2(r^2+bs^2)^2$$

by taking $r \to r\sqrt{a}$ and $s \to s\sqrt{ab}$; and the third, $x^2+(by\sqrt{c})^2 = (a\sqrt{c})^2$ or $(by/a)^2+c(x/ac)^2 = 1$, is parametrized as $(cr^2-s^2)^2+c(2rs)^2 = (cr^2+s^2)^2$ by taking $r \to r\sqrt{c}$ and $s \to s$.

It seems likely that Newton was influenced in his choice of these parametrizations by reading the researches into the Fermatian ('Pellian') equations $x^2-ny^2 = 1$ which Wallis printed in 1658 in his *Commercium Epistolicum de Quæstionibus quibusdam Mathematicis, nuper habitum*. (As we have seen in **1**, 3, §3, Newton about the same time made notes on other mathematical points in the tract.) The problem of finding integral solutions of these equations for given (non-square) integers n had been partially resolved by Bhaskara in the twelfth century (E. E. Whitford, *The Pell Equation* (New York, 1912: 31 ff.)), but in ignorance of its previous history Fermat published the problem to the world in the autumn of 1657 and on 11 September of that year Brouncker communicated it to Wallis. (*Commercium Epistolicum*: Epistola VIII, Appendix, *D. Fermatii Scriptum*: 17: 'Dato quovis numero non-quadrato, dantur infiniti quadrati qui in datum numerum ducti, adscitâ unitate, conficiant quadratum.... Canonem Generalem...inquirimus.' Compare J. E. Hofmann, *Neues über Fermats zahlentheoretische Herausforderungen von 1657* = Abh. d. Pr. Akademie der Wissenschaften (1943), no. 9 (Berlin, 1944), and see also H. Konen, *Die Geschichte der Gleichung* $t^2-Du^2 = 1$ (Leipzig, 1901). In his letter to Brouncker on 17 December 1657 Wallis, improving on

Then if $\sqrt{n}$ is rationall make $y=\dfrac{2ae-bd}{n}+\dfrac{r}{s}+\dfrac{sm}{rnn}=\dfrac{2ae-bd}{n}\text{♉}\dfrac{r}{s}\text{♉}\dfrac{sm}{rnn}$. Or if $\sqrt{m}$ is rationall make $y=\dfrac{2ae-bd}{n}\text{♉}\dfrac{2rrn+2ssaa}{rrnn-ssaan}\sqrt{m}$ or[(6)] $y=\dfrac{2ae-db}{n}\text{♉}\dfrac{2rrn+[2]ss}{rrnn-ssn}\sqrt{m}$. Or if $\sqrt{-mn}$ is rationall make $y=\dfrac{2ae-bd}{n}\text{♉}\dfrac{4rs\sqrt{-mn}}{rrnn-ssn}$. If none of y[m] bee rationall, noe rule can bee given to make x & y rationall.[(7)]

Brouncker's previous work, gave (*Commercium Epistolicum*: Epistola XVII: 56–72, especially 56 ff.) the parametrization

$$\left(\frac{r^2+ns^2}{r^2-ns^2}\right)^2-n\left(\frac{2rs}{r^2-ns^2}\right)^2=1$$

for Fermat's equation. (Immediately, if $r^2-ns^2=1$ is a first solution, then

$$(r^2+ns^2)^2-n(2rs)^2=1$$

is a second, as Bhaskara had already found.)

(5) Read '$-m$'.

(6) The particular case where $a=1$.

(7) Newton gives rational parametrizations of the general quadratic function

$$ax^2+dxy+fy^2+bx+ey+c=0.$$

Where $2ae-bd=l$, $a^2e^2+ab^2f+acd^2-abde-4a^2ef=-m$ and $d^2-4af=n$, we may write

$$\frac{1}{n}\left(\frac{ny-l}{2}\right)^2-\frac{m}{n}=(ax+\tfrac{1}{2}dy+\tfrac{1}{2}b)^2=\alpha^2,$$

say, and Newton's parametrizations follow easily. First, when $\sqrt{n}$ is rational and so

$$(\sqrt{n}\alpha)^2=\left(\frac{ny-l}{2}\right)^2-m,$$

comparison with the identity

$$\left(\frac{pr^2-qs^2}{2rs}\right)^2=\left(\frac{pr^2+qs^2}{2rs}\right)^2-pq$$

yields $\quad p=n,\quad q=m/n\quad$ and $\quad\dfrac{ny-l}{2}=\dfrac{pr^2+qs^2}{2rs},$

so that $y=\dfrac{l}{n}+\dfrac{r}{s}+\dfrac{ms}{n^2r}$. Again, when $\sqrt{m}$ is rational, comparison of

$$\left(\frac{ny-l}{2\sqrt{m}}\right)^2-1=\left(\sqrt{\frac{n}{m}}\,\alpha\right)^2\quad\text{with}\quad\left(\frac{pr^2+qs^2}{pr^2-qs^2}\right)^2-1=pq\left(\frac{2rs}{pr^2-qs^2}\right)^2$$

yields $\quad\dfrac{ny-l}{2\sqrt{m}}=\dfrac{pr^2+qs^2}{pr^2-qs^2}\quad$ and $\quad\sqrt{(pq)}=\sqrt{\dfrac{n}{m}}\,\alpha\left(\dfrac{pr^2-qs^2}{2rs}\right).$

(Since $\sqrt{m}$ is rational, we require pq/n to be square, and Newton in fact takes $p=n$ and $q=a^2$.) Finally, when $\sqrt{(-mn)}$ is rational, comparison of

$$\left(\frac{ny-l}{2\sqrt{-m}}\right)^2+1=\left(\frac{n\alpha}{\sqrt{(-mn)}}\right)^2$$

with the identity

$$\left(\frac{2\sqrt{nrs}}{nr^2-s^2}\right)^2+1=\left(\frac{nr^2+s^2}{nr^2-s^2}\right)^2\quad\text{yields}\quad\frac{ny-l}{2\sqrt{-m}}=\frac{2\sqrt{nrs}}{nr^2-s^2},$$

§3. MISCELLANEOUS THEOREMS ON THE EQUALITIES OF ANGLES(1)

Theoremata varia.

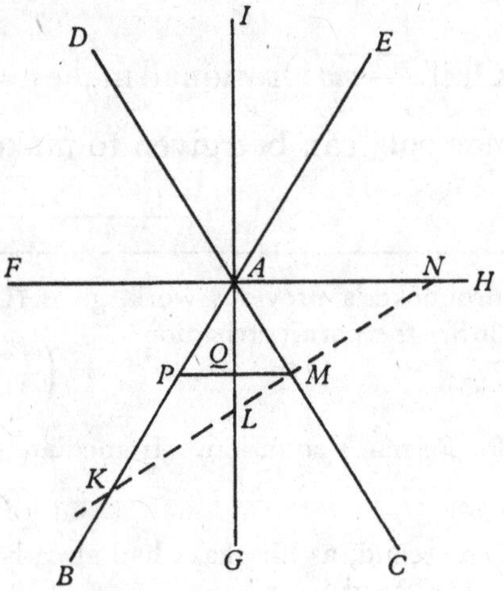

Circa angulorum æqualitates.

si ang *DAB* & *DAE* bisecentur a rectis *FH* et *IG* et ducatur quævis *KLMN*. Erit(2)

1. *AK*.*AM*::*KL*.*LM*::*KN*.*MN*.

Euclid 6 3.(3)

2. $AK \times AM = AL^q + KL \times LM = KN \times MN - AL^q$.(4)

Schootē de concinn æquat(5)

3. $AM + AK . PK :: AQ . AL$(6) posito $AP = AM$.(7)

---

and so
$$y = \frac{l}{n} + \frac{4rs\sqrt{(-mn)}}{n(nr^2-s^2)}.$$

Since these parametrizations exhaust variants on the fundamental identity
$$(\alpha^2-\beta^2)^2+(2\alpha\beta)^2 \equiv (\alpha^2+\beta^2)^2,$$
where α and β are positive integers, and since the latter is the most general integer solution of $x^2+y^2 = z^2$, the truth of Newton's last remark is assured.

(1) Add. 4000: 35r, where the present text precedes a proposition on multiple angles reproduced above as 2, §2.6.

(2) 'If the angles *DAB* & *DAE* are bisected by the straight lines *FH* and *IG* and there be drawn any [transversal] *KLMN*. Then will....'

(3) That is, by Proposition 3 of Book 6 of Euclid's *Elements*.

(4) Read 'AN^q'.

(5) The first part of this theorem, $AK.AM = AL^2+KL.LM$, is a *Theorema* given by Schooten on p. 370 of his *Tractatus de concinnandis Demonstrationibus geometricis ex Calculo Algebraïco* (= *Geometria*, 2: 341–420). In proof Schooten, taking $AM = a$, $AL = b$, $LM = c$, $AK = x$, $KL = y$ and $LQ = z$, showed that $ay = cx$ (since $MA:AK = ML:LK$),
$$2bz = b^2+c^2-a^2 \quad (\text{since } AM^2 = AL^2+ML^2+2AL.QL)$$
and $2yz = b(y-c)$ (since $[KL-KM]/KL = [KA-KM]/KA = 2QL/LA$). If we eliminate z, it follows that $c(cy+b^2) = a^2y = a(cx)$ or $cy+b^2 = ax$. The second part of the theorem, $AK.AM = KN.MN-AN^2$, is Newton's gloss and follows readily from the result
$$\begin{aligned} KN.MN-KL.LM &= KN(LN-LM)-KL(LN-MN) \\ &= (KN-KL)\,LN, \quad \text{since} \quad KL.MN = KN.LM, \\ &= LN^2 = AL^2+AN^2. \end{aligned}$$

(6) Read '*QL*'.

(7) 'Taking $AP = AM$'. Using the scheme of note (5), we deduce easily that
$$\frac{x+a}{x-a} = \frac{y+c}{y-c} = \frac{b-z}{z}, \quad \text{or} \quad \frac{AK+AM}{AK-AM} = \frac{AL-QL}{QL}.$$

§4. THE RECTIFICATION OF 'ELLIPTICALL' LINES[1]

The length of no Ellipticall line whatever of 1st 2d 3d 4th kind &c can be found. For if so the spirall lines made by them wou[l]d bee geometricall. 2dly &c.[2]

---

(1) Add. 3958. 2: 34r, a scrap extracted from the piece, apparently composed in late October 1665, which is otherwise reproduced above as 6, §1.

(2) By an 'Ellipticall' line of *n*th kind Newton seems to intend, in generalization of the (conical) ellipse of 1st kind, any simply closed portion of an algebraic curve of degree $(n+1)$. Correspondingly, the 'spirall lines made by them' seem to generalize the derivation of the Archimedean spiral from the circle. That is, given the closed curve *nc* and some fixed point *a* inside it round which rotates the line *ad*, we may describe the allied 'spirall' (*b*) by the motion of the point *b* in *ad* such that $ab:ac =$ ellipse arc $\widehat{nc}$: perimeter $\widehat{ncn}$, where *c* is the instantaneous meet of the rotor *ad* and the elliptical curve. Clearly, each time the point *c* completes a revolution round the pole *a*, so also does the point *b*, and hence (*b*) is a spiral which makes an infinite number of gyrations round *a*. Further, since an arbitrarily placed straight line would cut this spiral in an infinite number of real points, its defining equation cannot be algebraic of finite degree and therefore the spiral itself is not a 'geometricall' curve. From this transcendence of the spiral Newton would seem to argue that, likewise, the 'length' of the perimeter $\widehat{ncn}$ of the ellipse cannot algebraically be expressed in terms of any commensurable line-segment. In particular when $\widehat{ncn}$ is the perimeter of a circle whose centre is *a*, the argument would appear to demonstrate the transcendence of the constant (π) which expresses the ratio of the circumference of a circle to its diameter.

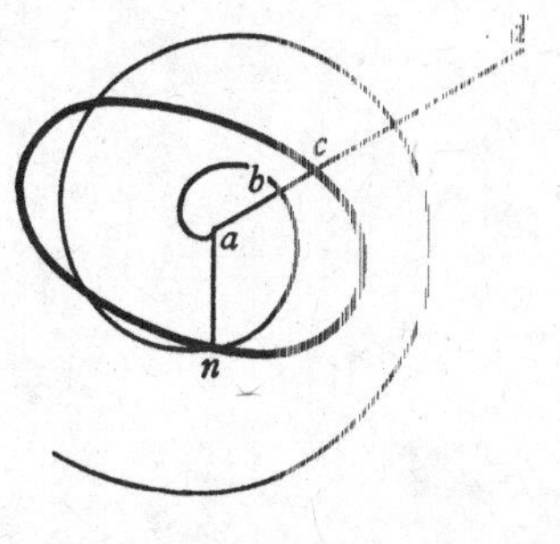

Twenty years later Newton introduced the argument into his Lucasian lectures on the motion of bodies in the autumn of 1684 (ULC. Dd. 9. 46: 96r–98r), seeking to prove the analogous proposition that 'Nulla extat Figura Ovalis cujus area rectis pro lubitu abscissa possit per æquationes numero terminorum ac dimensionum finitas generaliter inveniri', and with but slight change it appeared three years later in the *Principia* (*Philosophiæ Naturalis Principia Mathematica*: Liber I, Lemma XXVIII: 107–9). Apart from some confused criticism in the nineteenth century Newton's argument has since received little attention though its validity has been questioned on the basis of counter-examples. (Compare H. Brougham and E. J. Routh, *An Analytical View of Sir Isaac Newton's Principia* (London, 1855): 72–4, where the curve $y^m = kx^{(n-1)m}(a^n - x^n)$ is cited against the proposition.) In fact, Newton's argument founders on the rock of the periodicity of the general angle, and the infinite gyrations of the spiral merely represent an infinite number of repeated traversals of the perimeter $\widehat{ncn}$, yielding no insight into the nature of the perimeter itself (counted once). (See D. T. Whiteside, 'Patterns of mathematical thought in the later seventeenth century', *Archive for History of Exact Sciences*, 1 (1961): 179–388, especially 203–5.)

APPENDIX

EARLY NOTES ON GEOMETRICAL OPTICS (1664-1666)

HISTORICAL NOTE

Newton's interest in mathematical theories of reflection and refraction (the traditional fields, respectively, of catoptrics and dioptrics) seems first to have been aroused during his study of Descartes' scientific work in the summer of 1664. Descartes' printed contributions to geometrical optics were presented in two of the tracts (*La Dioptrique* and *La Geometrie*) appended to his *Discours de la methode* in 1637, but it is clear that Newton, whose knowledge of French was never very good,(1) read both in Latin version. The *Dioptrique*, together with the third appended tract *Les Meteores* and the *Discours* itself, was first rendered into Latin in 1644(2) and passed quickly through several further editions in the next decade and a half. The *Geometrie*, as we have seen,(3) was read by Newton in Schooten's second Latin edition of 1659. Which of these, *Dioptrice* or *Geometria*, Newton studied first we do not know, but already by the summer of 1664 he was familiar with the latter and Descartes' treatment of his 'Ovales' in its *Liber II* would lead him immediately to the other.

The next year his interest in optical theory was further stirred by the appearance of Hooke's *Micrographia*,(4) on which he made elaborate notes.(5) This new interest was more qualitative in form, relating to the theory of light, observational experiment and the practical problems of lens-grinding and the construction of telescopes and microscopes. (However, Hooke had read Descartes' published work on optics carefully and the *Micrographia* contains allusions to the mathematical topics which had already aroused Newton's interest.(6)) These non-quantitative aspects of Newton's early optical researches(7) can, unfortunately, here find no place.

---

(1) Newton wrote to Collins on 20 May 1673 that he was unwilling to express his opinion on Heuret's *Optiques*, 'not being so ready in ye French tongue myself as to reade it wthout the continuall use of a Dictionary'. (*Correspondence of Isaac Newton*, **1** (1959): no. 110: 281.) Almost twenty years later Fatio de Duillier indicated as much to Christiaan Huygens in his letter of 24 February 1689/90, announcing that Newton 'a quelque peine à entendre le François, mais il s'en tire pourtant avec un Dictionaire'. (*Œuvres complètes de Christiaan Huygens*, **9** (1901): no. 2570: 387. = *Correspondence of Isaac Newton*, **3** (1961): no. 352: 69.)

(2) *Renati des Cartes Specimina Philosophiæ: seu Dissertatio de Methodo rectè regendæ Rationis & Veritatis in Scientiis investigandæ: Dioptrice, et Meteora. Ex Gallico translata, & ab Auctore perlecta, variisque in locis emendata*, Amsterdam, 1644. The quotations in the text are made from the unchanged second printing in 1650, which is cited hereafter as *Dioptrice* (1650).

(3) See the general introduction to Part 1, especially Appendix 1, note (10).

(4) *Robert Hooke: Micrographia: or some Physiological Descriptions of Minute Bodies made by Magnifying Glasses. With Observations and Inquiries thereupon* (London, 1665 [January?]).

(5) ULC. Add. 3958. 1: 1^r–4^r: *Out of Mr Hooks Micrographia*, first printed by Geoffrey Keynes in appendix to his *Bibliography of Dr Robert Hooke* (Oxford, 1960) and reprinted in A. R. Hall and Marie Boas Hall, *Unpublished Scientific Papers of Isaac Newton* (Cambridge, 1962): Part VI: 400–13.

Descartes himself remarked in his *Dioptrique*(8) that there were many things in his work pertaining to geometry whose demonstrations he had omitted, presumably because he thought their proof an unnecessary strain on the reader's intelligence, but even in the severely mathematical *Geometrie* he had left many properties of his Ovals unexplained. (Above all, he had spurned any heuristically satisfactory deduction of their refractive properties.) Clearly it was just these difficult geometrical passages which challenged Newton's mathematical instincts, and the researches which they inspired and from which extracts are given below are a revealing account of Newton's growing analytical power.

(6) See, for example, *Micrographia*: Preface: *e*[2]r: 'But because we are certain, from the Laws of refraction (which I have experimentally found to be so...) that the [s]ines of the angles of Incidence are proportionate to the [s]ines of the angles of Refraction, therefore if Glasses could be made of those kind of Figures, or some other, such as the most incomparable Des Cartes has invented, and demonstrated in his Philosophical and Mathematical Works, we might hope for a much greater perfection of Opticks then can be rationally expected from spherical ones...'.

(7) Compare Appendix, 2, note (51) below.

(8) *Dioptrice* (1650): 196: 'Multa hîc sunt ad Geometriam spectantia quorum demonstrationes omitto'.

1. EARLY NOTES ON REFLECTION AND REFRACTION

[September 1664][1]

Extracts from Newton's Waste Book in the University Library, Cambridge

§1. THE INVENTION OF FIGURES FOR REFLECTION AND REFRACTION[2]

[S]ept 1664

[1] *The invension of Figures for reflections at right angles.*[3]

a y^e^ point reflecting. *ac* y^e^ rad: reflected to y^e^ focus *d*. *ag* y^e^ radius refle[c]ted frō y^e^ focus *b*. *aq* a perpendic: to *ed* y^e^ tangent of y^e^ crooked line sought. $ab=x$. $y=ac$, or $ag=y$. $bg=a$, or $bc=a$. $bd=v$, or $bq=v$.

fig 1st

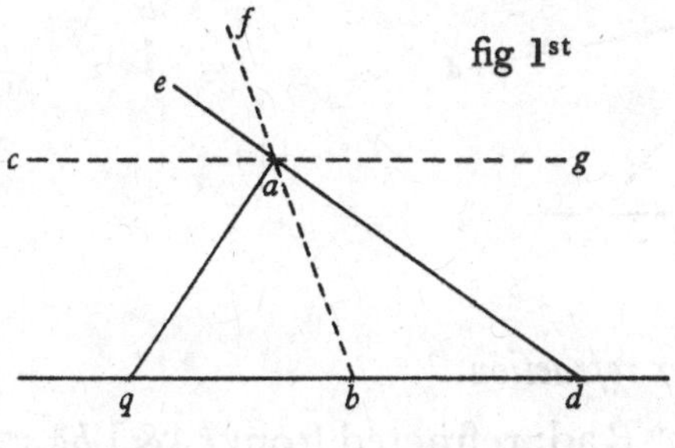

fig: 1st. $\angle eac=\angle bad=\angle adb$. Ergo, $ab=bd$, or $x=v$. & $\angle caq=qab=aqb$. Ergo $ab=qg$.[4] $x=v$.[5]

(1) The date is taken from that of §1 below.

(2) Add. 4004: 1^v^. The date is found at the head of the page, but is separated by a preliminary draft from the finished text here reproduced.

(3) That is, reflection about the normal to the reflecting curve.

(4) Read '*qb*'.

(5) Newton finds the subnormal condition for the light ray moving along *cg* parallel to the axis *qbd* to be reflected through *b*. The relation $x = v$ (or $ba = bq$) determines the curve (*a*), a parabola with focus *b* and axis *qbd*, which reflects all light rays parallel to *qbd* through *b*. (The result is classical and found in all the numerous works on 'burning mirrors' from the time of Anthemius. Compare G. L. Huxley, *Anthemius of Tralles*, Cambridge, Mass., 1959.)

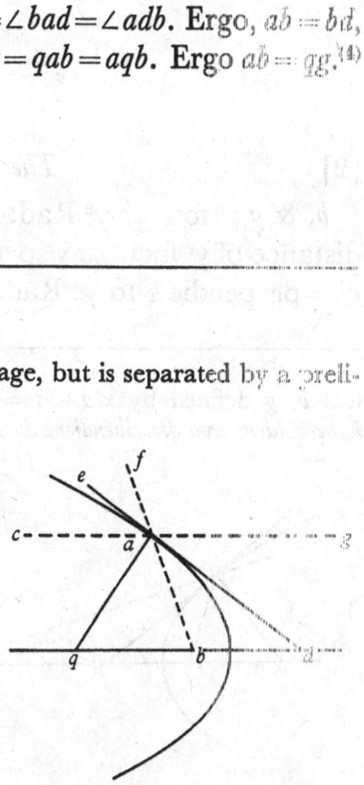

(6) Read '*dag*'.

(7) Read '*qg*'.

(8) Newton finds two equivalent subnormal conditions for a light ray *cg* travelling through *g* to be reflected to *b*. The curve (*a*) which reflects all rays through *g* to *b* is a hyperbola of

fig 2^d^. $\angle eac = \angle bad = \angle adg$.[6] Ergo, $ab:ag::bd::dg$. $ax - vx = vy$. &,

$$ab:ag::qb:bg.^{(7)}$$

Ergo $ax + vx = yv$. $v = bq$.[8]

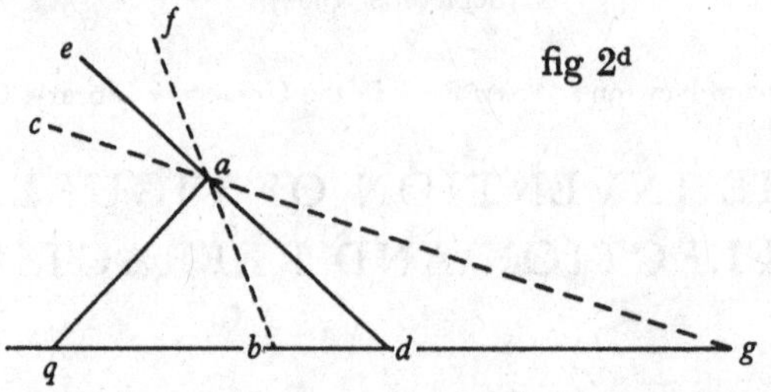

fig 2^d^

fig 3^d^. $\angle eac = \angle bad$. Ergo $\angle caq = \angle qab$. Ergo, $ca:ab::cq:qb$. & $ax - vx = vy$.[9]

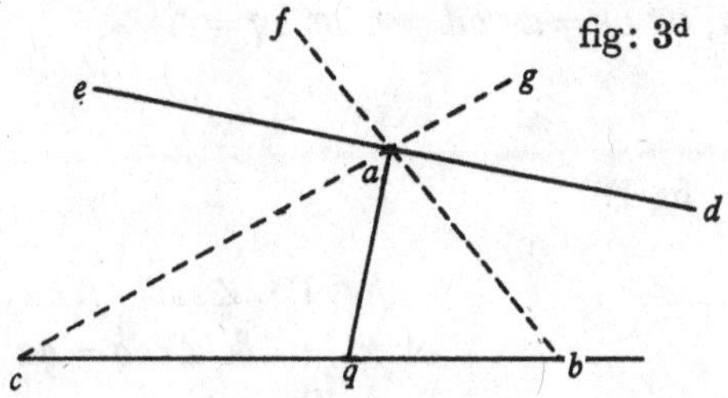

fig: 3^d^

[2] *The invention of figures for refraction.*

b, & *g* y^e^ foci, *ca* y^e^ Rad: refracted to *b*. *ga* y^e^ Rad: refracted from *b* [&] *bg* y^e^ distance of y^e^ foci. *qa* y^e^ perpend: to *de* y^e^ tangent of y^e^ crooked line sought. *qr*, *qh*=perpendic:s to y^e^ Radii *cg*, *fb*. $bg = a$. $bq = v$. $ba = x$. $ag = y$.

---

foci *b*, *g* defined by $x(a+v) = vy$. (Compare *Dioptrice* (1650): 152–3: Cap. VIII, *Quid sit hyperbola & eam describendi modus*.)

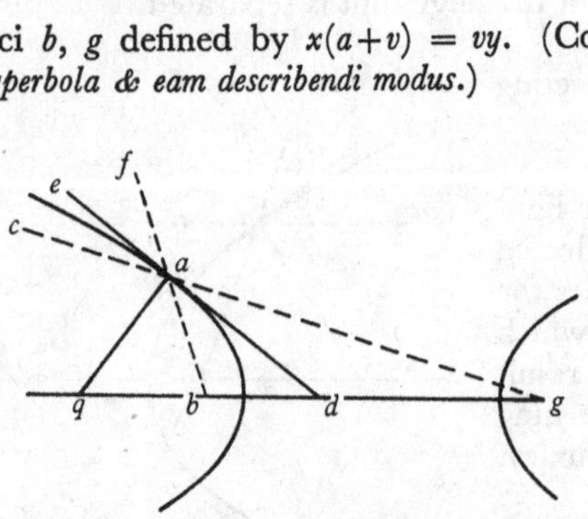

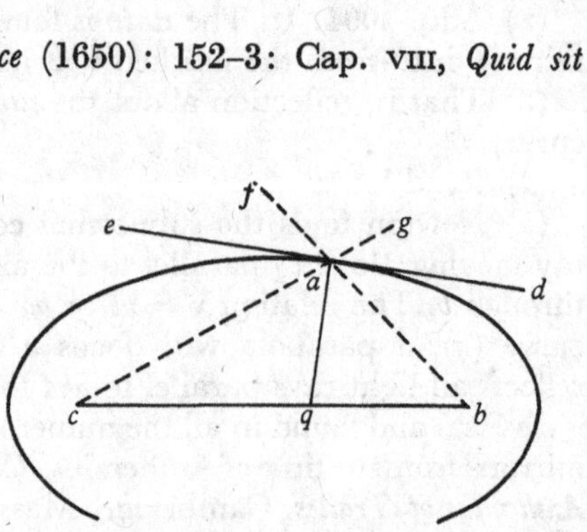

(9) The subnormal condition for the light ray *cg* travelling from *c* to be reflected back through *b*. Much as before, the curve (*a*) which reflects all rays from *c* to *b* is an ellipse of foci *c*, *b*. (Compare *Dioptrice* (1650): 142–3: Cap. VIII, *Quid sit Ellipsis, & quomodo sit describenda*.)

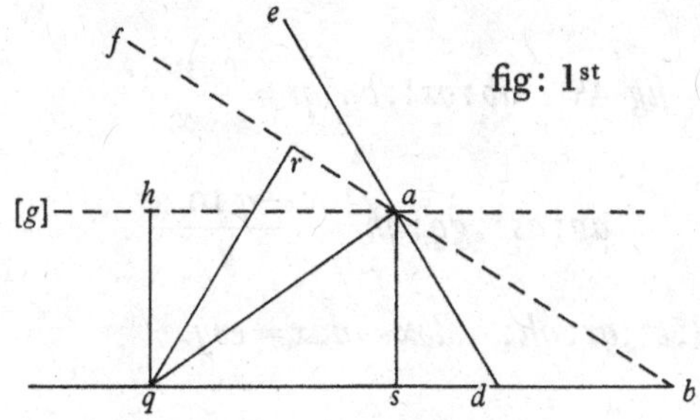

fig: 1st

fig: 1st. $ab:as::bq:qr=\frac{vy}{x}$. $d:e::qr:qh$,

Ergo. $dx=ev$.(10)

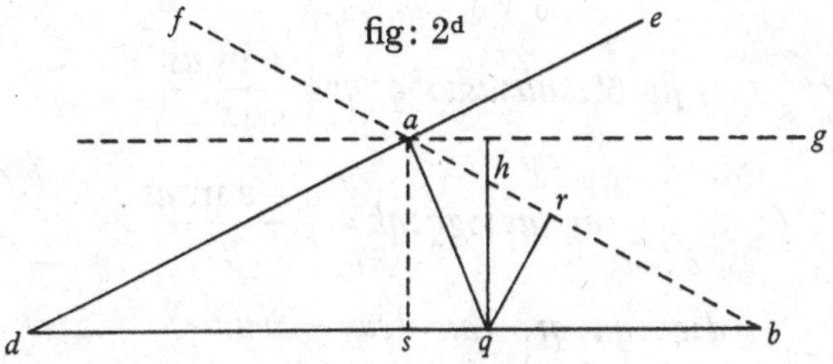

fig: 2d

fig: 2d. $ab:as::bq:qr=\frac{vy}{x}$.

$d:e::qh:qr$, Ergo $ex=dv$.(11)

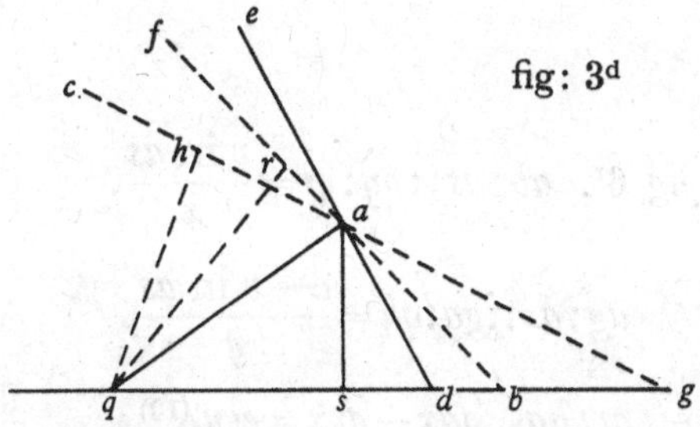

fig: 3d

fig 3d: $ab:as:[:]\,bq:qr=\frac{v \text{ in } as}{x}$.

$ag:as::gq:qh=\frac{a+v \text{ in } as}{y}$.

$d:e::qr:qh$. Ergo $dax+dvx=evy$.

(10) Newton derives the subnormal condition for the light ray moving from g parallel to the axis qdb to be refracted at a, where the ratio of the refractive indices of the two media traversed is $e:d$, $d > e$. The curve (a) which refracts all parallel light rays through the point b is a hyperbola of further focus b, where its major axis and focal distance are in the ratio $e:d$. (See *Dioptrice* (1650): Cap. VIII, *De figuris quas pellucida corpora requirunt, ad detorquendos refractione Radios omnibus modis visioni inservientibus*: 154–5: §XII, *Demonstratio proprietatis hyperbolæ quod ad refractiones* [*pertinet*].) By mistake in this and the following case Newton sets $as = y$.

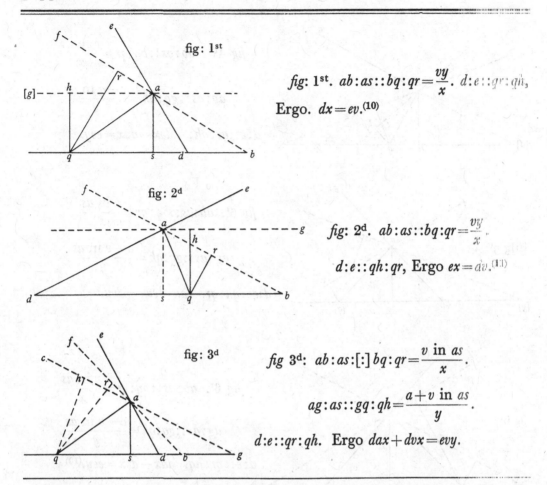

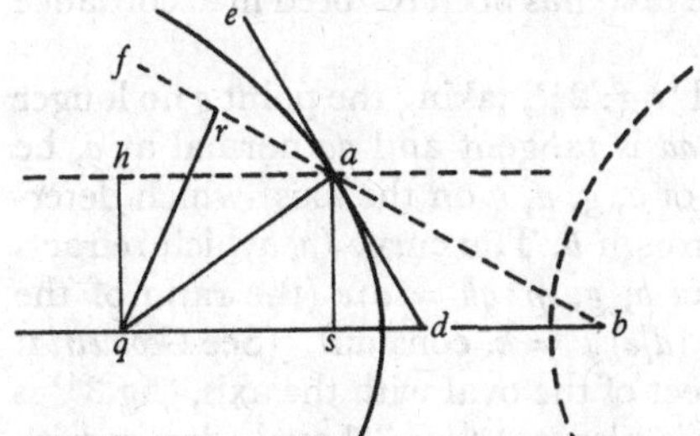

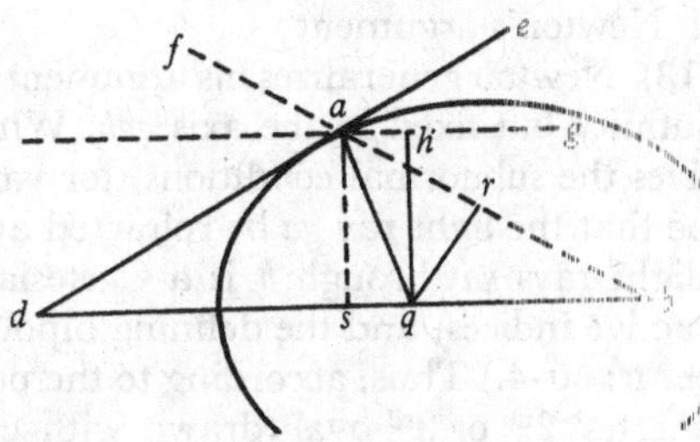

(11) The subnormal condition for the light ray moving towards g parallel to the axis to be refracted at a through b, where the ratio of the refractive indices of the media is $e:d$, $e > d$.

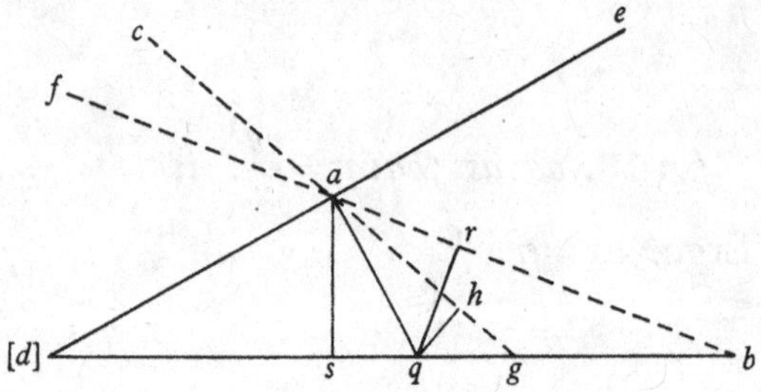

fig 4th. $ab:as::bq:qr=\frac{v \text{ in } as}{x}$.

$ag:as::gq:qh=\frac{v-a \text{ in } as}{y}$.

$d:e::qr:qh$: $dvx-dax=evy$.

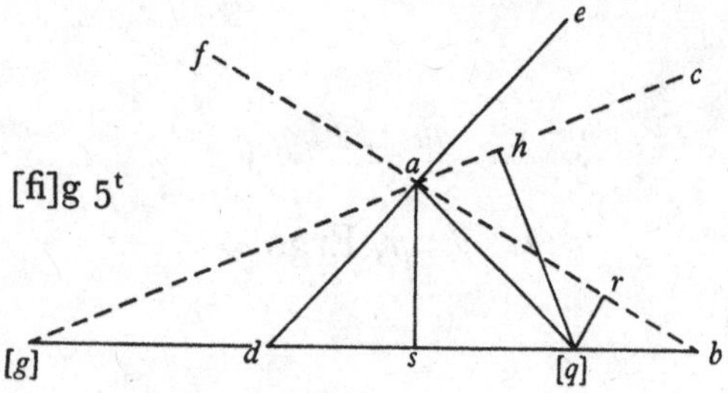

[fi]g 5^t

fig 5^t. $ab:as::bq:qr=\frac{v \text{ in } as}{x}$.

$ag:as::gq:qh=\frac{a-v \text{ in } as}{y}$.

$d:e::qh:qr$. $eax-evx=dvy$.

fig 6^t

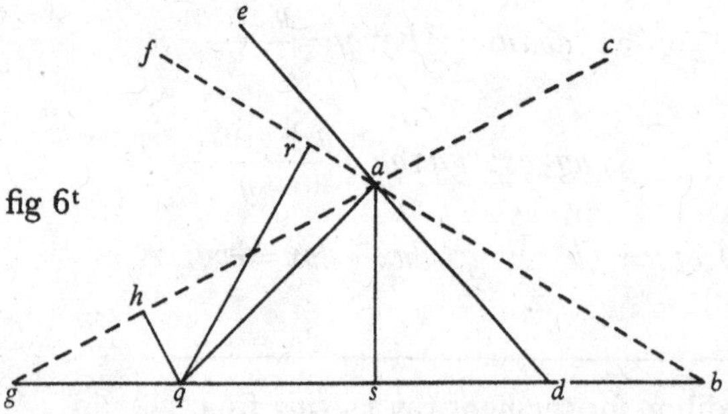

fig 6^t. $ab:as::bq:qr=\frac{v \text{ in } as}{x}$.

$ag:as::gq:qh=\frac{a-v \text{ in } as}{y}$.

$d:e::qr:hq$. $dax-dvx=evy$.(13)

The curve (*a*) which refracts all such parallel rays through *b* is now an ellipse of further focus *b*, whose major axis and focal distance are in the ratio $e:d$. (*Dioptrice* (1650): 145–6: Cap. VIII, §III, *Demonstratio proprietatis Ellipsis in refractionibus*.)

(12) This last figure, completely crumbled away in the text, has been restored in accordance with Newton's argument.

(13) Newton generalizes his argument in 'fig: 1st' and 'fig: 2d', taking the point *g* no longer at infinity but fixed in the axis *qdb*. Where, as before, *ad* is tangent and *aq* normal at *a*, he derives the subnormal conditions (for varying positions of *b*, *g*, *d*, *q* on the axis) which determine that the light ray *ga* be refracted at the point *a* through *b*. The curve (*a*) which refracts all light rays *ga* through *b* is a Cartesian oval with foci *b*, *g*, $qr:qh = d:e$ (the ratio of the refractive indices) and the defining bipolar equation $x \pm (d/e)\, y = k$, constant. (See *Geometria*: Liber II: 50–4.) Thus, according to the position of the meet of the oval with the axis, 'fig 3d' is Descartes' 2nd or 3rd oval (drawn with $d > e$); 'fig 4th' is also a 2nd or 3rd oval (drawn with $d < e$); while 'fig 5t' and 'fig 6t' are identical 1st or 4th ovals (with $d > e$ in fig 6t). (Descartes, having found the subnormal *sq*, did not in fact proceed to give Newton's subnormal conditions, though they are immediately derivable. Specifically, introducing the parametrization

§2. MISCELLANEOUS EARLY CALCULATIONS ON CARTESIAN OVALS[14]

[1]

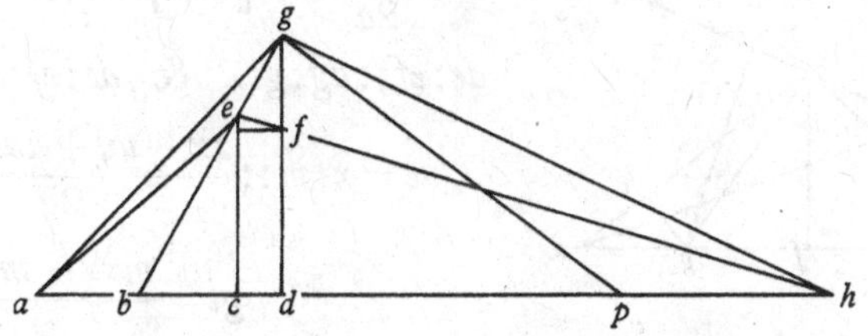

$$ag=x.\ gh=y.\ ah=c.\ dp=v.\ [gp=s.]\ ad=\frac{xx-yy+cc}{2c}.$$

$$dg^2\left[\left(\right]=\frac{2ccx^2+2xxyy+2ccyy-c^4-x^4-y^4}{[4cc]}\right)+vv=ss.^{(15)}$$

$ga = x = c+z$ and $ba = y = b-(e/d)z$, where $gc = c$ and $bs = b$ (or $b+c = a$), Descartes calculated

$$sq = \frac{bcd^2-bcde+bd^2z+ce^2z}{cd^2+bde-e^2z+d^2z}.$$

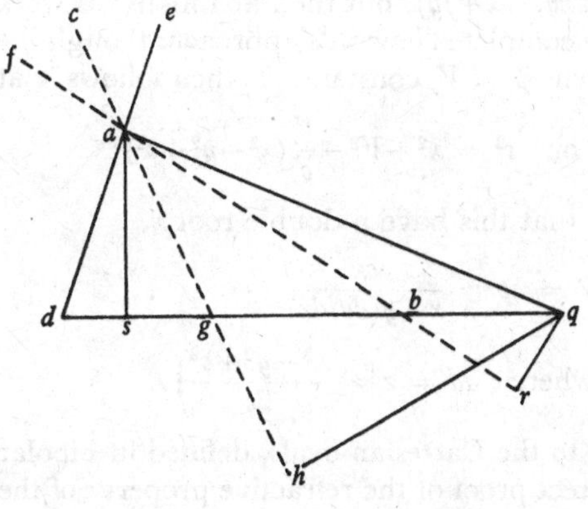

(*Geometria*: Liber II: 42–3, 48), and it remained only to show that

$$bq = bs-sq = \frac{e(b+c)(b-(e/d)z)}{d(b-(e/d)z)+e(c+z)} = \frac{eay}{dy+cx},$$

which is Newton's 'fig 6^t'.)

Note that, for symmetry, Newton would seem to need an alternative 'fig 5^t' to complement his 'fig 6^t' in the case where $d < e$. As for example:

$$ab:as::bq:qr = \frac{v \text{ in } as}{x}.$$

$$ag:as::gq:qh = \frac{a+v \text{ in } as}{y}.$$

$$d:e::qr:qh.\ dax+dvx = evy.$$

(14) Add. 4004: 1^v, 2^r.

(15) Where the bipolar co-ordinates $ag = x$, $hg = y$ (with focal distance $ah = c$) are related by some given equation $f(x, y) = 0$ in definition of the curve (g), Newton seems to be trying to improve on Descartes' method for finding the subnormal dp. Descartes himself had transformed the defining bipolar equation $x = \pm(e/d)y+k$ of his ovals into an equivalent relation in conventional Cartesian co-ordinates and so one to which his subnormal method as given in Book 2 of *Geometrie* could be applied. (See *Geometria*: Liber II: 42ff.) In contrast, Newton seeks a general subnormal method in bipolars, that is, one which does not require prior transform to Cartesian co-ordinates. He retains, however, Descartes' limit-condition for gp to be normal to the curve (g), namely that the circle whose centre p is on the axis shall have double contact with the curve in the indefinitely near points g and e, and his first step is to express ad and dg parametrically in terms of x and y. Then, taking $dp = v$ and the circle radius $gp = s$, he finds in Cartesian fashion from the relation $dg^2+v^2 = s^2$ an equation

[2]

[*a*][16] $ad=a.\ ae=x.\ ed=y.\ af=z.\ fd=a-z.\ fg=v.\ xx-zz=yy-zz+2az-aa.$

$$\frac{xx-yy+aa}{2a}[=z]=af. \quad fd=\frac{yy-xx+aa}{2a}.$$

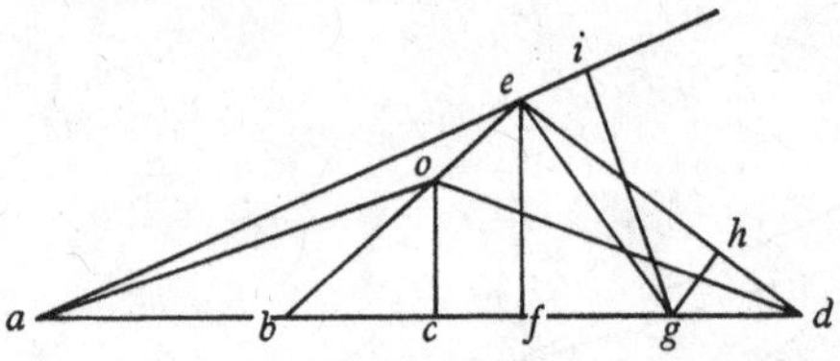

$ae:ef::ag:gi.$ & $de:ef::dg:gh.$[17] $ef=m.$

$$ae=x:m::\frac{xx-yy+aa+2av}{2a}=ag:$$

$$gi=\frac{mxx-myy+maa+2mav}{2ax}$$

$ed[=y]:m::gd:\dfrac{myy-mxx+maa-2mav}{2ay}=gh.\ gi:gh::d:e:$ therefore

$$\frac{dyy-dxx+daa-2dav}{y}=\frac{exx-eyy+eaa+2eav}{x}.^{(18)}$$

$$d=2.\ e=1.\ \begin{array}{l} y^3+2xyy-xxy\ +2aax=0.^{(19)} \\ \qquad\qquad -aay\ \ -2x^3 \\ \qquad\qquad -2avy-4axv \end{array}$$

$g(x, y, v)=0$ relating x, y and v. In further calculation (not here transcribed) Newton considers a particular defining relation $0=bx^2+c^2+dxy+ex+fy+y^2$ and eliminates y^2 from $g(x, y, v)=0$ by substituting its value $-(bx^2+c^2+dxy+ex+fy)$, but then abandons his work while it is yet unfinished. Little, in fact, is needed to complete Newton's approach, though the way would have been clearer for him if he had taken $ap=V$, constant. It then follows that

$$x^2-\left(\frac{x^2-y^2+c^2}{2c}\right)^2+\left(V-\frac{x^2-y^2+c^2}{2c}\right)^2=s^2 \quad \text{or} \quad s^2=x^2+V^2-\frac{V}{c}(x^2-y^2+c^2);$$

hence, if we apply Descartes' subnormal condition that this have a double root x,

$$0=2x-\frac{V}{c}\left(2x-2y\frac{dy}{dx}\right) \quad \text{or} \quad V=ap=\frac{cx}{x-y(dy/dx)}$$

$$\left(\text{and so} \quad ph=c-V=\frac{-cy(dy/dx)}{x-y(dy/dx)}=-y\frac{dy}{dz}, \quad \text{where} \quad ad=z=\frac{x^2-y^2+c^2}{2c}\right).$$

(16) Newton now turns his attention specifically to the Cartesian ovals, defined in bipolar co-ordinates by $x=\pm(d/e)\,y+k$. He still seeks a direct proof of the refractive property of the ovals which does not involve a reduction to Cartesian co-ordinates, and here seems to wish to derive the bipolar defining equation from the constancy of the ratio of the refractive indices $d:e=gi:gh$ (where g is the foot of the normal ge to the curve and gi, gh are drawn perpendicular to $ae=x$ and $de=y$ respectively). His calculation begins much as before by parametrizing af and fd.

(17) Newton here strayed in his calculations for several lines (not transcribed) by confusing af and ag, but quickly cancelled them when he realized his error.

(18) Where $V=ag=v+(x^2-y^2+a^2)/2a$, this reduces to

$$\frac{d(a+V)}{y}=\frac{eV}{x} \quad \text{or} \quad eVy=dax+dVx.$$

(Compare 'fig 3[d]' in §1.2 above.)

(19) To ease the burden of calculation Newton now takes the particular value 2:1 for the ratio of the refractive indices.

$$\frac{y^3+2xyy-xxy-aay+2aax-2x^3}{2ay+4ax}=v=\frac{yy-xx}{2a}+\frac{2ax-ay}{2y+4x}.$$

$$bf^{(20)}=\frac{2aaxx+2aayy+2xxyy-x^4-y^4-a^4 \text{ in } y+2x}{y^3+2xyy-xxy-aay+2aax-2x^3 \text{ in } 2a}. \quad cf=o.^{(21)}$$

$$bc=\frac{\begin{gathered}2aaxxy-x^4y-a^4y+2aay^3+2xxy^3-y^5+4a^2x^3+4aayyx+4yyx^3\\ -2y^4x-2a^4x-2x^5-2aoy^3-4aoxyy+2aoxxy+2a^3oy-4a^3ox+4aox^3\end{gathered}}{2ay^3+4axyy-2axxy-2a^3y+4a^3x-4ax^3}.^{(22)}$$

(20) That is, the subtangent to the curve and so equal to ef^2/fg, where

$$fg = v \quad \text{and} \quad ef^2 = x^2-\left(\frac{x^2-y^2+a^2}{2a}\right)^2.$$

(21) This is probably the first time Newton used his celebrated notation o for an indefinitely small increment (here a decrement of bf). Both Descartes and Schooten use the lower-case variant e of Fermat's E in this sense (*Geometria*: Liber II: 45 ff.; and *Commentarii in Librum II*, N: 249 ff.), but Newton, who has already designated e as one of his refractive constants, cannot here follow their convention. His alternative choice of o (a 'little nothing') for a vanishing increment has much to commend it visually. Indeed, the usage seems to have appealed for the same reason both earlier to Jean de Beaugrand in his account, about 1638, of Fermat's tangent-method (*Œuvres de Fermat*, 5 (Paris, 1922): 98–114) and a few years afterward to James Gregory (who introduced the notation as 'nihil, seu serum o' in his account of Fermat's method in his *Geometriæ Pars Universalis* (Padua, 1668): Prop. 7: 20).

(22) Newton expresses bc as $bf-cf$ but then abandons the calculation, presumably because of its growing complexity. Indeed, since o is the increment of bf, there seems no way of continuing short of evaluating o in terms of the increments of x and y.

(23) Newton introduces his standard subnormal condition that the ray ae be refracted at e through d. Essentially his result is that of 'fig 6t' in §1.2 above.

(24) Newton abandons bipolars, returning to Cartesian co-ordinates $af = x$, $fe = y$. As a first step he converts his previous result $d(a-v).ae = ev.de$, where ae is now $\sqrt{(x^2+y^2)}$ and $de = \sqrt{[(a-x)^2+y^2]}$. (In essence he admits defeat, for he can only modify Descartes' treatment of the subnormal if he passes to a Cartesian co-ordinate system.) This calculation also, after the last result is squared, is abandoned. Its prosecution, indeed, would require a knowledge of the techniques of differential calculus (and in particular of the theorem that

$$fg = v-x = y(dy/dx)),$$

which he did not then possess.

We may suppose that Newton was still puzzled why Descartes had chosen to define his refracting ovals in bipolar notation and how he had been able, in the first instance, to derive their defining equation from the constancy of the refractive ratio. A few years later, when the techniques of working with geometrical increments had become familiar to him, the puzzle was resolved. In 1670 in his Lucasian lectures on optics Newton made passing reference to the refractive properties of Descartes' 'Ellipses': '...Si punctorum...neutrum sit ad infinitam distantiam, Curva...erit aliqua quatuor Ellipsium quas Cartesius in hunc usum in Geometria descripsit. Sin alterutrum infinitè distet, ita ut radij punctum illud respicientes evadant paralleli, Curva erit Conica sectio, uti notum est.' (*Lectiones Opticæ*: Lectio 14 (October 1670), Prop. 34, nota 2 = ULC. Dd. 9. 67: 72, englished as *Optical Lectures read in...Cambridge, Anno Domini, 1669* (London, 1728): Part 1, Section 4, Prop. XXXIV, Note II.) However, his detailed analysis of Descartes' ovals was first published to the world in 1687 in his *Principia Philosophiæ Naturalis Principia Mathematica* (London, 1687): Liber I, Sect. XIV, Prop. XCVII (Prob.

[*b*] $ae:ef::ag:gi.\ de:ef::dg:gh.\ d:e::gi:gh.\ x:m::v:\frac{mv}{x}.\ y:m::a-v:\frac{ma-mv}{y}.$

$d:e::\frac{mv}{x}:\frac{ma-mv}{y}:$ & $adx-dvx=evy.$(23)

[*c*] $af=x.\ fe=y.\ ae=\sqrt{xx+yy}.\ ed=\sqrt{xx-2ax+aa+yy}.$

$$ad\sqrt{xx+yy}-dv\sqrt{xx+yy}=ev\sqrt{xx-2ax+aa+yy}.^{(24)}$$

§3. EARLY RESEARCHES INTO THE GENERAL PROBLEM OF TWOFOLD REFRACTION

[(25)]

XLVII): 232–3). Essentially, where o is a point on the refracting curve indefinitely close to e with $o\alpha$ and $o\beta$ drawn normal to ae and de respectively, the figures $oe\alpha\beta$ and $egih$ are similar, so that the increments $e\alpha$ of $ae = x$ and $e\beta$ of $de = y$ are related by

$$\frac{e\alpha}{e\beta} = \frac{gi}{gh} = \frac{d}{e};$$

hence, in the limit, $d(x)/d(y) = d/e$ and so

$$x = (d/e)\,y+k.$$

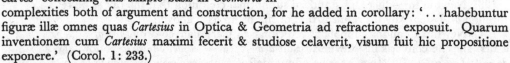

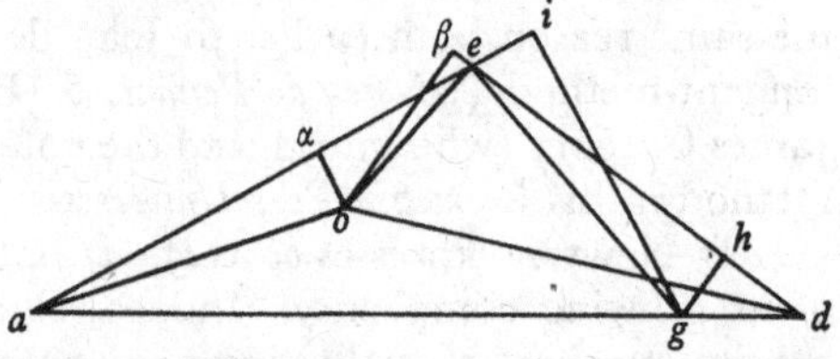

Newton seems to have felt some irritation at Descartes' concealing this simple basis in *Geometria* in complexities both of argument and construction, for he added in corollary: '...habebuntur figuræ illæ omnes quas *Cartesius* in Optica & Geometria ad refractiones exposuit. Quarum inventionem cum *Cartesius* maximi fecerit & studiose celaverit, visum fuit hic propositione exponere.' (Corol. 1: 233.)

(25) These calculations (Add. 4004: 2r, 4r, 4v) relate to Descartes' discussion of the problem of finding the surfaces of a lens which shall, by a double refraction, send all rays of white light emanating from a point through a second point. (See *Geometria*: Liber II: 60–5: *Quomodo vitrum fieri possit, cujus una superficies tam convexa aut concava sit, quàm libuerit, quod radios omnes, qui ex uno dato puncto prodeunt, colligat rursus in altero dato puncto. / Quomodo aliud fieri possit, quod idem præstet, cujusque convexitas unius superficiei datam rationem habeat ad convexitatem vel concavitatem alterius.*) Having produced an infinity of solutions by suitable pairing of his ovals Descartes added in a concluding paragraph: 'On pourroit aussy passer outre, & dire, lorsque l'vne des superficies du verre est donnée, pouruû qu'elle ne soit que toute plate, ou composée de sections coniques, ou de cercles; comment on doit faire son autre superficie, affin qu'il transmette tous les rayons d'vn point donné, a vn autre point aussy donné. Car ce n'est rien de plus difficile que ce que ie viens d'expliquer; ou plutost c'est chose beaucoup plus facile, à cause que le chemin en est ouuert. Mais i'ayme mieux, que d'autres le cherchent, affinque s'ils ont encore vn peu de peine à le trouver, cela leur face d'autant plus estimer l'inuention des choses qui sont icy demonstrées.' (*Geometrie* (1637): Livre II: 368, rendered in Latin in *Geometria*: 65). Whatever Descartes meant—and his reference to the easiness of the problem seems to envisage merely a construction by points—Newton tried to resolve the question by analytical geometry. He first (on 2r) attempted a solution where the first interface is a parabola defined in bipolar co-ordinates by $x = v$, the subnormal, and then proceeded to consider a

2. THE ESSAY 'OF REFRACTIONS'

[Winter 1665/6?][(1)]

Extracts from the original in a pocket-book[(2)] in the University Library, Cambridge and in private possession.

Of Refractions.

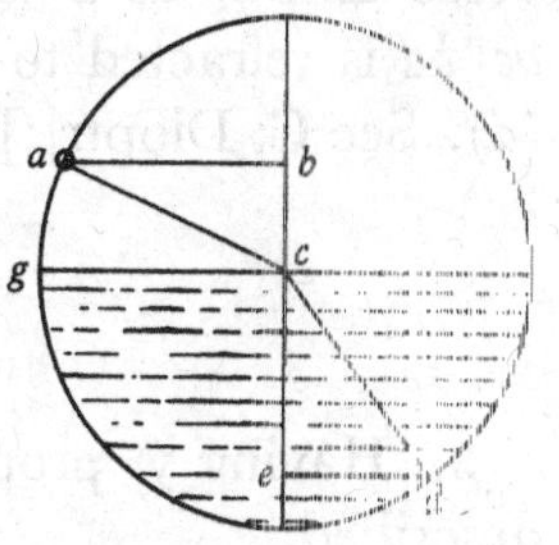

1. If y^e^ ray ac bee refracted at the center of y^e^ circle $acdg$ towards d & $ab\perp be\perp gc\parallel ed$. Then suppose

$$ab:ed::d:e.$$

See Cartes Dioptricks.[(3)]

---

hyperbola, defined in a polar co-ordinate system, as his first curve. Later (on 4^r^) he tried the general case where the first curve is defined by an arbitrary relation between Cartesian co-ordinates x and y, but in despair began (on 4^v^) to discuss the conics

$$rx = yy \quad\text{and}\quad rx \mp (r/q)\,xx = yy.$$

All proved abortive, nor had he any greater success when he again attacked the general problem in his undergraduate notebook Add. 4000 on f. 80^r^. The details of these calculations (not here reproduced) are unimportant, but there can be no doubt that the techniques he evolved in his fruitless search were soon to prove invaluable when he came to frame his general theory of spherical lenses. (See Appendix, 2 below.)

The general resolution of the problem by points is, in contrast, little more than a corollary of Descartes' researches on his ovals. Thus, where the curve (a) and the points g and p are given,

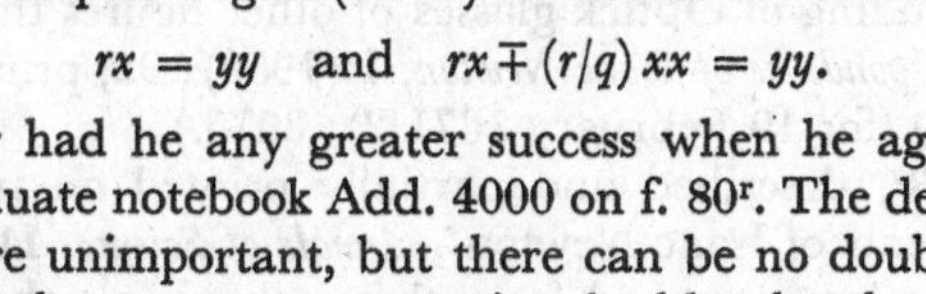

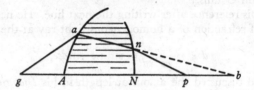

together with the constant refractive ratio $d:e$ between the media, for each point a on the curve (a) we can construct the point b on the axis through which the light ray ga from g is refracted: the point n (and its tangent) on the second curve may then be constructed as the meet with ab of a Cartesian oval of foci p, b and refractive ratio $d:e$, with the constant term determined from the implicit condition that the curve (n) be smoothly continuous. (Compare Claude Rabuel, *Commentaires sur la Geometrie de M. Descartes* (Lyon, 1730): 374–98: *La Figure qu'il faut donner aux verres, pour qu'ils rëünissent à un point donné, les rayons qui viennent d'un autre point donné*, especially §III: 390–3: *De quelques cas, que M. Descartes n'a pas resolus.*) Newton himself about 1670, when his ideas on geometrical fluxions had matured, gave an ingenious construction of the points n of the second curve by circle and straight-edge. (See ULC. Add. 4004: 90^r^, first published in 1687 in *Principia* (note (24)): Liber I, Sect. XIV, Prop. XCVIII (Prob. XLVIII): 233–4. It is perhaps referred to by Newton in his letter of 11 June 1672 to Oldenburg = *Correspondence of Isaac Newton*, 1 (1959): 173: '...for Dioptrique Telescopes I told you that the difficulty consisted not in the figure of the glasse but in y^e^ deformity of refractions. Which if it did not, I could tell you a better & more easy remedy then the use of y^e^ Conic Sections.')

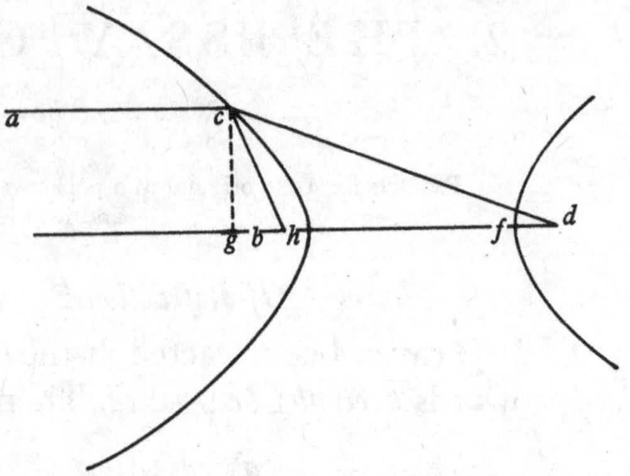

2. If there be an hyperbola the distance of whose foci are to its transverse axis *hf* as *d* to *e*. Then y^e^ ray $ac \| bd$ is refracted to y^e^ exterior focus (*d*). See C: Dioptr[:][(4)]

3. Having y^e^ proportion of *d* to *e*, or $bd:hf$. The Hyperbola may bee thus described.

(1) The date is assigned on the basis of Newton's handwriting and his later remark in his letter to Oldenburg on 6 February 1671/2 that '...in the beginning of the year 1666...I applyed myself to the grinding of Optick glasses of other figures then Spherical...'. (ULC. Add. 3970. 3: 460 = *Correspondence of Isaac Newton*, **1** (1959): 92, printed in *Philosophical Transactions*, **6** (1671/2): no. 80 (for 19 February 1671/2): 3075.)

(2) Add. 4000: 26^r^–33^v^, described and partially printed on pp. 36–43 of A. R. Hall's 'Further optical experiments of Isaac Newton', *Annals of Science*, **11** (1955): 27–43. The four concluding pages of the essay, removed from the pocket-book about the time of Newton's death and now in private possession, were unknown to Hall and their content would seem to call for some revision of his conclusion (p. 43) that 'Newton's early study of lenses was not of great profundity'. (There exists, also in private possession, a contemporary copy of the whole essay in the hand of John Collins.)

(3) Newton added this reference after writing the next line. He here expounds the fundamental (Snell) axiom of refraction of a homogeneous light ray at the meet of two media: in modern form

$$d:e = \frac{ab}{ac}:\frac{ed}{cd} = \sin \widehat{acb} : \sin \widehat{dce},$$

constant. Descartes had discussed the axiom at length in his *Dioptrique* (= *Dioptrice* (1650): Cap. II, *De Refractione*: 76–86. Newton refers especially to 82–4: § VII, *Quantum radii reflectantur a pellucidis corporibus in quæ penetrant.*). However, the notation of $d:e$ for the refractive ratio is taken over from *Geometria*: Liber II.

(4) The basic theorem which determines the refractive property of the hyperbola, and so prepares the way for Newton's subsequent work on grinding hyperbolic glasses. Descartes' proof (*Dioptrice* (1650): 154–5 = Cap. VII, § XII, *Demonstratio proprietatis hyperbolæ quod ad refractiones* [*pertinet.*]) takes $ac = cd$, and al, dk perpendicular to lck (normal to the hyperbola at c) with $bm \| lck$: then, since cd, cb are equally inclined to the normal lck and $bm \| lck$, we have $cm = cb$; so that $\sin \widehat{acl} : \sin \widehat{dck} = al:dk = ac$ (or cd) $: nd = md$ (or $cd - cb$) $: bd$, where by the focal property of the hyperbola $cd - cb = hf$. (The theorem follows more straightforwardly from the focus-directrix property of the hyperbola.)

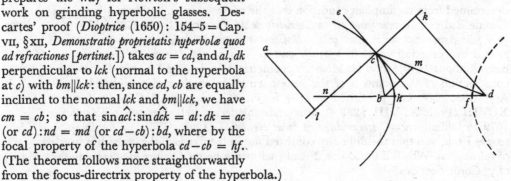

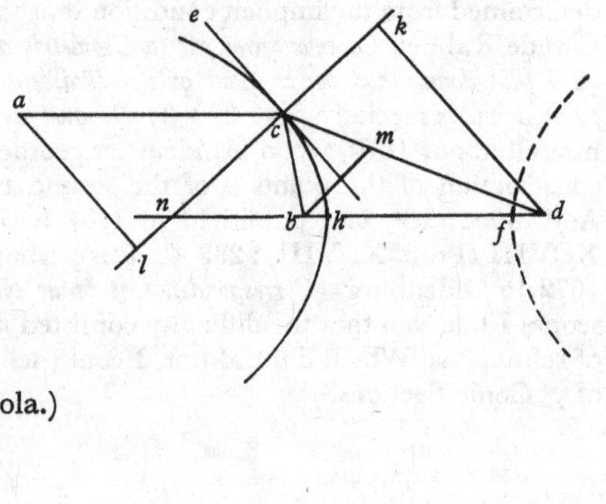

1. Upon y^e centers *a*, *b* let y^e instrument *adbtec* bee moved[,] in wch instrum^{nt} observe y^t $ad\perp de\perp cet$. & y^t the beame *cet* is not in the same plane wth *adb*: but intersects it at y^e angle *tev* soe y^t if $tv\perp ev$,(5) then $d:e::et:tv$. Or $d:e::$Rad:sine of

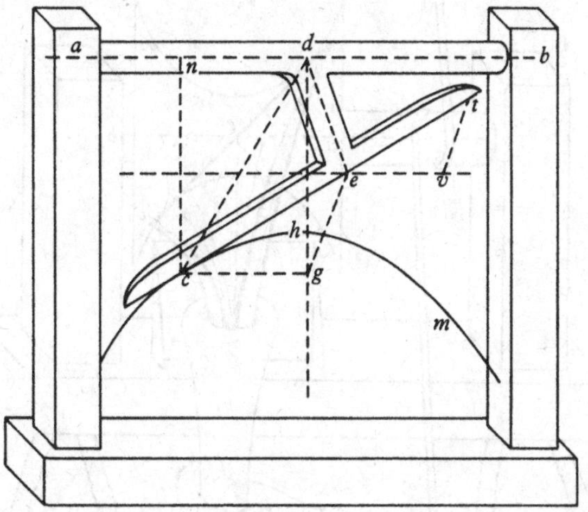

$\angle tev$. Also make $de=\frac{q}{2}$, i.e. half y^e transverse diameter. Then place the fiduciall(6) side of y^e plate *chm* in the same plaine wth *ab*. & moving y^e instrument *adbcet* to & fro its edge *cet* shall cut or weare it into y^e shape of y^e desired Parabola.(7) Or the plate *chm* may bee filed away untill y^e edg *cet* exactly touch it every where.

---

(5) *tv* is therefore parallel to *eg*.

(6) So called presumably because the template *chm* will have to be relied on when used to check the hyperbolic glass.

(7) Read 'Hyperbola'.

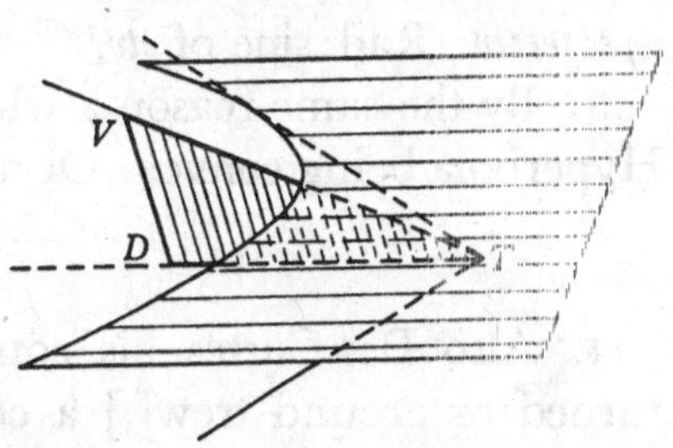

(8) Descartes' concave hyperbolical wheel (figure *A* in Newton's diagram below) was a hyperboloid of revolution fabricated in order to grind hyperbolic glasses by a suitable rotary movement. (See *Dioptrice* (1650): Cap. x, *De modo expoliendi vitra*: 190–5 = §vi, *Alia machina, quæ istius hyperbolæ figuram dat omni rei quæ ea ad vitra polienda indiget, & quomodo illâ sit utendum.*) Descartes' only method of constructing his concave wheel was by prior filing of a hyperbolic *lamen*, or template. (See *Dioptrice* (1650): 189, where the template is cut out by the edge of the triangle *VDT* rotating round the axis *DT* parallel to the plane of the template. Clearly *TV* generates a conical surface, with the hyperbola its plane section.)

(9) The central rotating bar *ab* of the lathe to which is fitted the mass *A* to be turned. Its original signification (a miner's pick) was already obsolescent at this time, but it seems that the word was revived in its present meaning by Hooke in his *Micrographia* (London, 1665: Preface: e[1]^v), where he described his own 'Engine' for grinding spherical lenses.

(10) Newton improves on Descartes by turning the hyperboloid directly with a straight-edged chisel. Interpreted mathematically, Newton's technique generates the concave hyperboloid of revolution as a ruled surface, that is, by the rotation of an obliquely set straight line.

2. By the same proceeding Des=Cartes concave Hyperbolicall wheele[8] may bee described by beeing turned wth a chissell *dtec* whose edge is a streight line

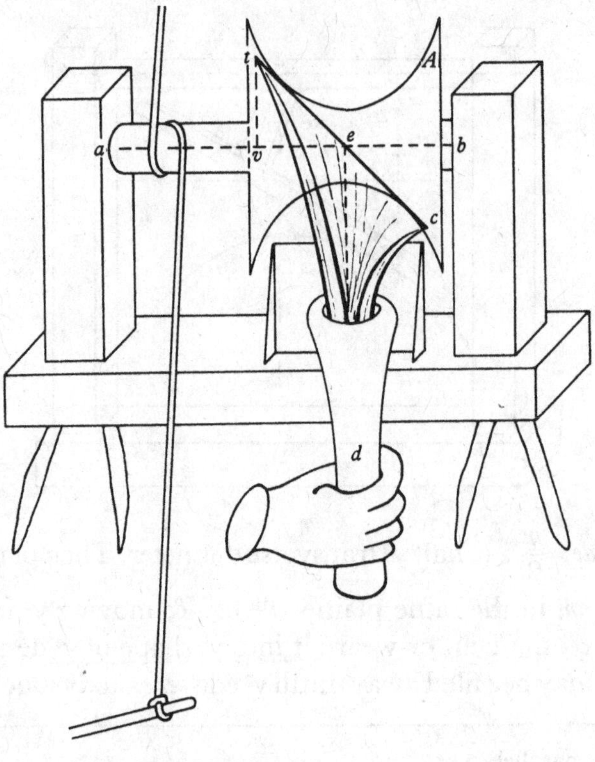

inclined to the axis of the mandrill[9] by y^e^ ∠*tev*[,] wch angle is found by making *d*:*e*::*et*:*tv*::Rad:sine of *etv*.[10]

3. By the same reason a wheele may be turned Hyperbolically concave, y^e^ Hyperbola being convex. Or a Plate may bee turned Hyperbolically concave.

4. Also Des=Cartes his convex wheele *B* may be turned or ground trew[,] a concave wheele *A* being made use of instead of a patterne.[11]

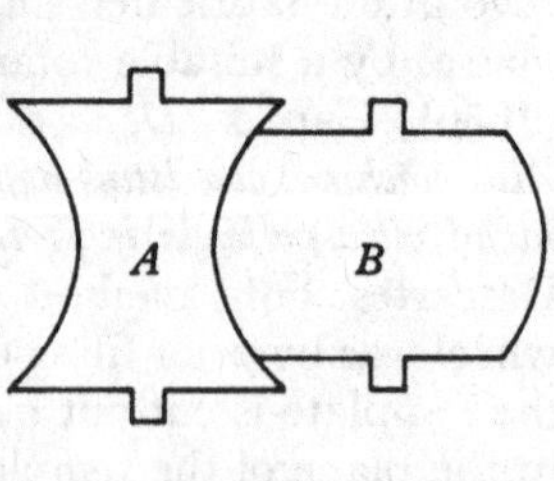

He is therefore, together with Christopher Wren, an independent discoverer of this property of the hyperboloid. (Wren published, with geometrical proof, his own variant on the lens-grinding application in 'Generatio Corporis Cylindroidis Hyperbolici, elaborandis Lentibus Hyperbolicis accommodati', *Philosophical Transactions*, **4** (1669) [London, 1670]: no. 48 (for 21 June 1669): 961–2.)

(11) That is, a template. In Newton's figure *A* and *B* are two complementary hyperboloids of revolution, concave and convex respectively.

5. . . . .$^{(12)}$

6. *The Same done by y^e helpe of a Cone.*$^{(13)}$

Draw 2 concentrick circles (*na* & *cd*) wth y^e Radij *e* & *d*. Then from y^e comon center *b* draw 2 lines *bc* & $\begin{Bmatrix} ba \\ bd \end{Bmatrix}$ at the given angle $\begin{Bmatrix} bae=abc \\ aed=cbd \end{Bmatrix}$ of $\begin{Bmatrix} \text{section}^{(14)} \\ y^e \text{ Cone} \end{Bmatrix}$.

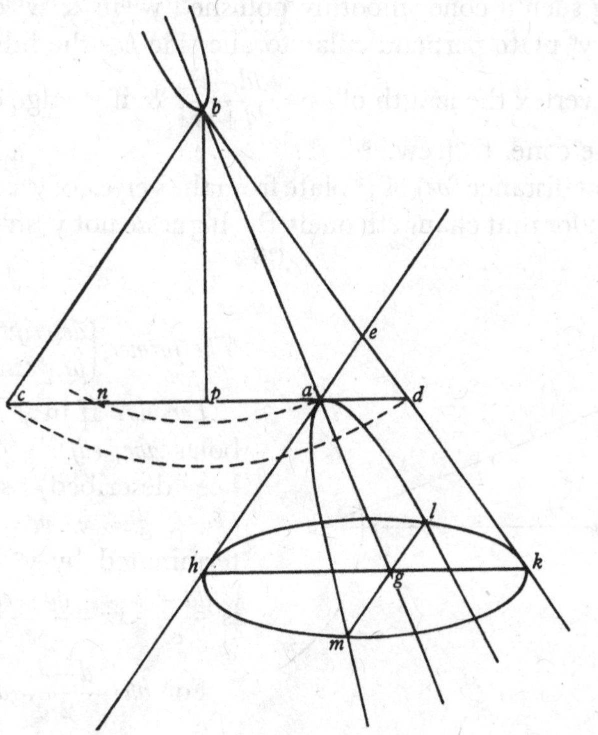

then draw a line *cad* from *c* by y^e end of y^e Rad: $\begin{Bmatrix} ba \\ bd \end{Bmatrix}$ & to y^e intersection of y^t line wth y^e circle $\begin{Bmatrix} cd \\ na \end{Bmatrix}$ draw $\begin{Bmatrix} bd \\ ba \end{Bmatrix}$ & so the ∠$^{(15)}$ of $\begin{Bmatrix} y^e \text{ cone, } hek=cbd \\ \text{section } eab=abc \end{Bmatrix}$ is found.

Or wch is the same make *ab*=*e*. *bd*=*d*. & then if y^t cone is sought the ∠*cba* being given, make *ac*=*a*. Then is $cd=\dfrac{dd-ee+aa}{a}$. & soe y^e ∠*cb*[*e*]=*aed* is knowne & also $ae=ed=\dfrac{d^3-dee}{dd-ee+aa}$, & $ad=\dfrac{dd-ee}{a}$. But if y^e ∠*bae*=*abc* of y^e section is sought[,] y^e cone being given[,] y^n *cd*=2*b*. And it will bee

$$ac=b+\sqrt{ee-dd+bb}.$$

---

(12) Paragraph 5 makes some practical remarks on grinding, and is here omitted.

(13) A first draft of this, vitiated and not here reproduced, has been struck out by Newton as 'false'.

(14) Newton has cancelled 'intersection'.   (15) Read 'angle'.

& soe $\angle abc = bae$ is given. also $ad = b - \sqrt{ee - dd + bb}$. & $ae = \dfrac{db - d\sqrt{ee - dd + bb}}{2b}$.

In generall observe y[t] in any cone cut any ways $bd = be + ea = d$, & $ba = e$.

7. Des Cartes his wheele thus described cut by any plaine produceth one of y[e] Conick Sections.(16)

9.(17) Having such a cone smoothly pollished w[th]in & w[th]out, by the helpe of a square set y[e] plate perpendicular to one side *hae* the fiduciall edge being distant from y[e] vertex the length of $ae = \dfrac{edd - e^3}{dd + ee}$. & if y[e] edge of y[e] plaine every where touch the cone, tis trew.(18)

10. The exact distance (ae) of y[e] plate from the vertex of y[e] cone neede not bee much regarded: for that changeth onely the bignesse not y[e] shape of y[e] figure.(19)

. . . .(20)

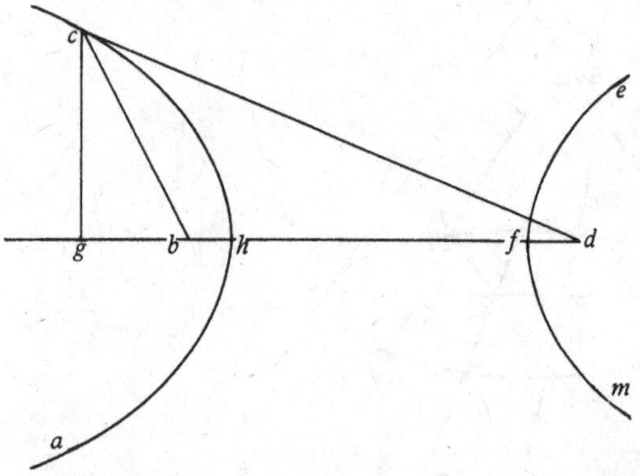

The former $\left\{\begin{matrix}\textit{descriptions}\\ \textit{propositions}\end{matrix}\right\}$ *demonstrated.*

Lemma. If in y[e] Opposite Hyperbolas *abc*, *edf*(21) (one of w[ch] are to bee described) supposing $bd = d$. $hf = e$. $gh = x$. $gc = y$. $gc \perp ghd$. & gc terminated by y[e] hyperbola. Then is $\dfrac{dd - ee}{ee} xx + \dfrac{dd - ee}{e} x = yy$.

For $bh = \dfrac{d - e}{2}$. $dh = \dfrac{d + e}{2}$.

(16) In other words, a plane section of the hyperboloid of revolution is a conic. The proposition he proved formally a few years later in his Lucasian lectures on mathematics (October 1676, Lectio 10, Prop. XIX: 80–1 = *Arithmetica Universalis* (London, 1707): 141–2, renumbered as Prop XXXIII in all subsequent editions).

(17) Newton's essay has no eighth section.

(18) This, after Newton's revision of section 6, needs some corrections. Thus, since $b\hat{a}e = a\hat{b}c$ is right, and so $ac^2 = ab^2 + bc^2$ (or bd^2), $a^2 = e^2 + d^2$, so that

$$ae = \frac{d(d^2 - e^2)}{d^2 - e^2 + a^2} = \frac{d^2 - e^2}{2d}.$$

(19) Hyperbolas of the same 'shape' (eccentricity) have the ratio d/e unchanged. Then, where the vertex distance ae' from a of a second hyperbola is fixed by the constants λd, λe,

$$ae : ae' = \frac{d^2 - e^2}{2d} : \frac{\lambda^2(d^2 - e^2)}{2\lambda d} = 1 : \lambda, \quad \text{or} \quad ae' = \lambda . ae.$$

Changing the vertex distance ae, therefore, merely alters the proportionality factor λ (where the angle $b\hat{a}e$ is kept right).

(20) Some numerical calculations on the refractive ratio between air and glass are omitted.

(21) Read '*ahc*, *efm*'.

$$bg = \frac{2x-d+e}{2}.\ gd = \frac{2x+d+e}{2}.$$

$$dc^2 = \frac{4xx+4dx+4ex+dd+2ed+ee+4yy}{4} = gd \times gd + gc \times gc.$$

$$bc^2 = \frac{4xx-4dx+4ex+dd-2ed+ee+4yy}{4} = gb^2+gc^2.^{(22)}$$

And since $dc = bc+hf$. Or $dc^2 = bc^2+2bc \times hf+hf^2$. Therefore

$$2dx+ed-ee = e\sqrt{4xx-4dx+4ex+dd-2ed+ee-4yy}:^{(23)}$$

Both ꝑts of w^{ch} $\square^{ed(24)}$ & ordered y^{e} result is

$$4ddxx-4eexx+4eddx-4e^3x-4eeyy=0.$$

That is $\frac{dd-ee}{e}x+\frac{dd-ee}{ee}xx=yy.$

For y^{e} Ellipsis $\frac{dd-ee}{d}x+\frac{ee-dd}{dd}xx=yy.^{(25)}$

---

(22) That is, $dc^2 = \left(\frac{2x+d+e}{2}\right)^2+y^2$ and $bc^2 = \left(\frac{2x-d+e}{2}\right)^2+y^2$.

(23) $dc^2-bc^2-hf^2 = hf(dc+bc-hf) = hf.2bc$, since by the focal property of the hyperbola $dc-bc = db-hb = hf$.

(24) Read 'squared'.

(25) This remark was inserted overleaf, presumably for lack of space on the manuscript page itself. The proof follows in a similar way from the modified (ellipse) property that

$$dc+bc = dh+bh = hf,$$

with $hf = e$, $bd = d$, $hg = x$, $gc = y$ and so

$$bh = \frac{e-d}{2},\quad dh = \frac{e+d}{2},\quad \text{or}\quad hf^2-dc^2+bc^2 = hf(hf-dc+bc) = hf.2bc.$$

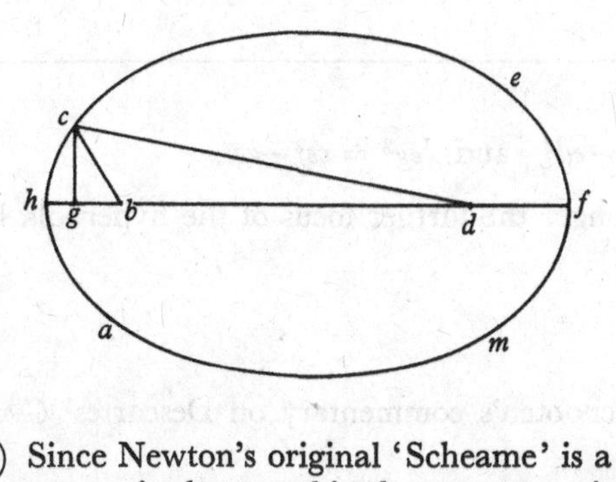

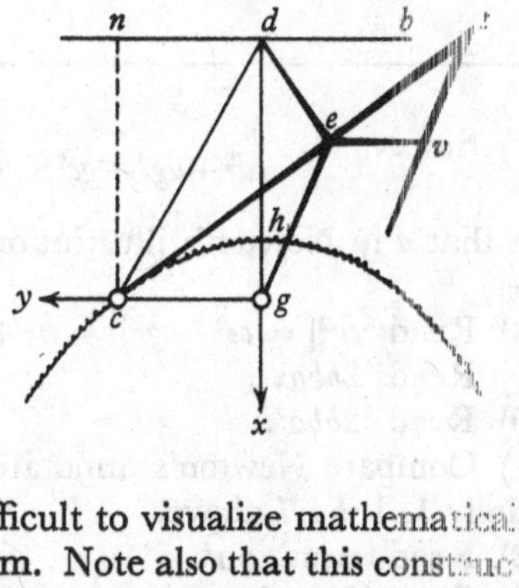

(26) Since Newton's original 'Scheame' is a little difficult to visualize mathematically, its basic structure is abstracted in the accompanying diagram. Note also that this construction is 'synthetical' in the Greek sense, that is, recomposed from a preliminary 'analytical' one (which Newton adds below).

Desc[r]iption y^e 1st demonstrated Synthetically. See y^t Scheame.(26)

Nameing y^e quantitys $ed=dh=\frac{e}{2}$. $gh=x$. $gc=y$. $dg=x+\frac{e}{2}=nc$. $cg\perp dhg$.

therefore $cd^2=x^2+ex+\frac{ee}{4}+y^2$. $ce^2=xx+ex+yy$. $eg^2=xx+ex$.(27) Also

$$d:e::et:tv::ce:eg,$$

therefore $ddxx+ddex=eex^2+e^3x+e^2yy$. That is $\frac{dd-ee}{e}x+\frac{dd-ee}{ee}xx=yy$. As in y^e lem̄a.

The same demonstrated Analytically.

Nameing y^e quantitys, $de=dh=a$. $gh=x$. $gc=y$. $dg=a+x$.

$$dc^2=aa+2ax+xx+yy. \quad ce^2=2ax+x^2+yy. \quad eg^2=xx+ex.^{(28)}$$

Suppose y^t $b:c::et:tv::ce:eg$. Then is $bbxx+bbex^{(29)}=2ccax+ccxx+ccyy$. That is $\frac{bb-cc}{cc}xx+\frac{bbe^{(30)}-2cca}{cc}x=yy$. Therefore y^e line *chm* is a Conick Section & since (bb) is greater y^n (cc) tis an Hyperbola,(31) wch y^t it may bee y^e same wth y^t in y^e lemma, Their correspond^{nt} termes are to bee compared together & soe I find y^t $\frac{bb-cc}{cc}xx=\frac{dd-ee}{ee}xx$. & $\frac{bbe^{(30)}-2cca}{cc}x=\frac{dd-ee}{e}x$. by y^e 1st $=$tion(32) $bb=\frac{ccdd}{ee}$. or $b=\frac{cd}{e}$. y^t is $b:c::d:e$. by y^e 2d $ccee-ccdd+bbee^{(33)}=2ccea$. And by substituting $\frac{ccdd}{ee}$ into the place of bb, And ordering it tis $ccee=2ccea$.(34) Or $\frac{e}{2}=a$. Therefore if I take $\frac{e}{2}=a=de$. & $d:e::b:c::et:tv$. then shall *chm* bee y^e Hyperbola desired. Q:E:D.

The 2d 3d 4th & 5th Propositions are manifest from this.

---

(27) Since $cd^2+dg^2+gc^2$, $ce^2=cd^2-ed^2$ and $eg^2=ce^2-cg^2$.

(Note that d in Newton's illustration is no longer the further focus of the hyperbola but its centre.)

(28) Read '$eg^2[=ce^2-cg^2]=xx+2ax$'.

(29) Read '$2bbax$'.

(30) Read '$2bba$'.

(31) Compare Newton's annotations of Schooten's commentary on Descartes' *Geometria*, especially 1, 1, §1.7 above.

(32) Read 'equation'.

(33) Read '$2bbae$'.

(34) Read '$e^4-e^2d^2=2ae(e^2-d^2)$'.

Description y^e 6th Demonstrated. Synthetically.

Call, $bd=d.\ ba=e.\ cp=pd=a.$

$bp=\sqrt{dd-aa}.\ ag=x.\ [gm=y.]$

$ap=\sqrt{ee-dd+aa}.\ ac=a+\sqrt{ee-dd+aa}.$

$ad=a-\sqrt{ee-dd+aa}.$

$$ba:ac::ag:gh=\frac{ax+x\sqrt{ee-dd+aa}}{e}.$$

$ba:ad::bg:gk=$

$$\frac{ea+ax}{e}\frac{-e-x}{e}\sqrt{ee-dd+aa}.$$

$gk\times gh=gm^2=y^2$. Therefore

$$\frac{dd-ee}{ee}xx+\frac{dd-ee}{e}x=yy.$$

by ordering y^e result of $gk\times gh$.

The 7th Proposition may be easyly demonstrated after the same manner.(35)

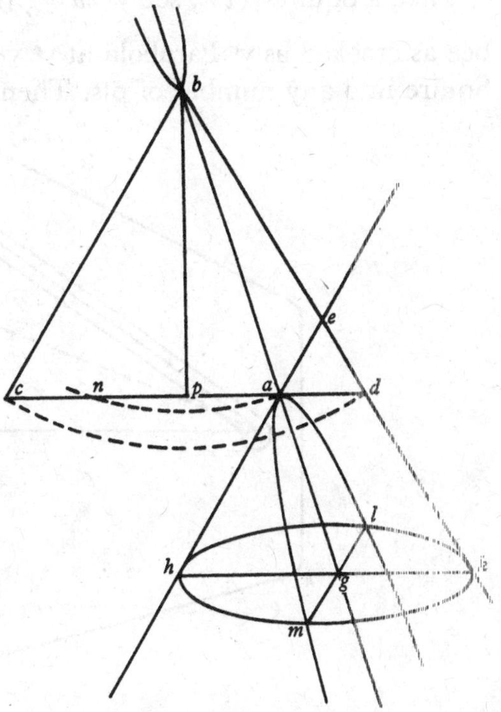

If the two equall cones *bad bcd* intersect the one the other soe y^t $ab=bc$. their intersection (bf) shall bee one of y^e Conick sections as [if] they had each beene intersected by the plane bf.(36)

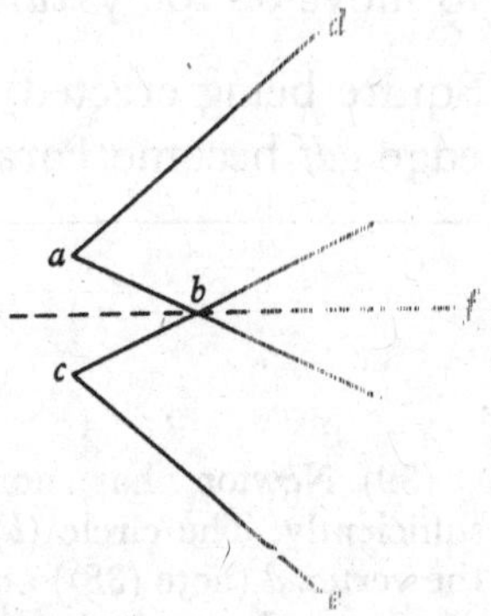

---

(35) See note (16) above.

(36) This is true only if the cones are symmetrically placed round the plane *bf*.

(37) A contemporary variant for a (carpenter's) 'square'.

(38) For, when *b* is indefinitely near to *d*, *c* will be the centre of curvature of the curve at the vertex *d* defined as the limit-meet of normals *bc*, *dc* at points indefinitely near to the vertex.

To describe ye Parabola (& other figures after ye same manner) pretty exactly.

Take a Squire[37] *cbe*, soe yt $cb=\frac{r}{2}$ (for then the circle described by (*bc*) will bee as crooked as ye Parabola at ye vertex *d*).[38] Divide ye other leg (*be*) of ye Squire into any number of ꝑts, Then get a plate of Brass &c: *lkfd* streight &

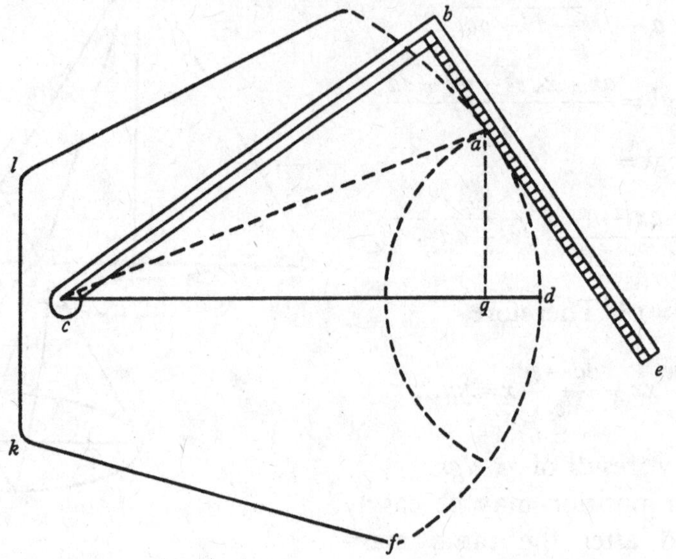

eaven And taking one point *d* for ye vertex of it & another point *c* for ye Squire to move on soe yt $cd=cb=\frac{r}{2}$, & weareing away ye edge of the plate untill (ye Squire being erected) $ab=qd$. the squire touching ye plate at *a*. thus shall ye edge *adf* become Parabolicall.[39]

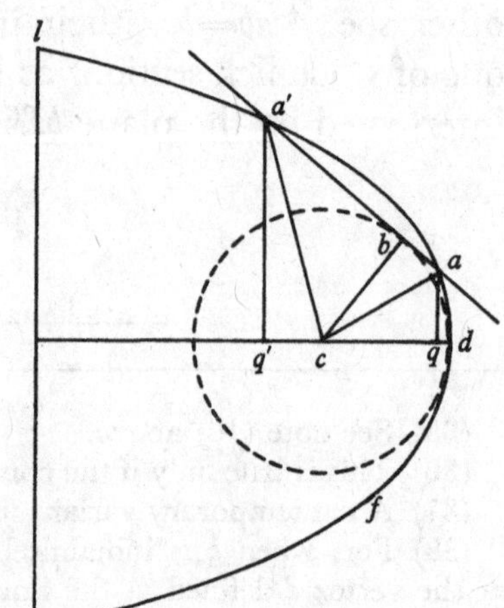

(39) Newton has not examined his construction sufficiently. The circle (*b*) is the circle of curvature at the vertex *d* (note (38)) and so will lie wholly within the parabola. In particular the 'squire' leg *ba* cannot touch it at the point *a*, but must meet it in two points *a*, *a'* where $ba = qd$ and $ba' = q'd$. The leg *ba*, therefore, cannot be used to file away the parabolic template *lkfda* but only to act as a check on the accuracy of its shape.

Demonstracon.(40)

$qd = x.\ cd = \frac{r}{2} = cb.\ cq = \frac{r}{2} - x.\ aq = \sqrt{rx} = y.\ ac^2 = \frac{rr}{4} + xx.\ ab^2 = \frac{rr}{4} + xx - \frac{rr}{4}$ &

$ab = x$. Q.E.D.(41)

Another description of ye Parabola wth ye compasses.

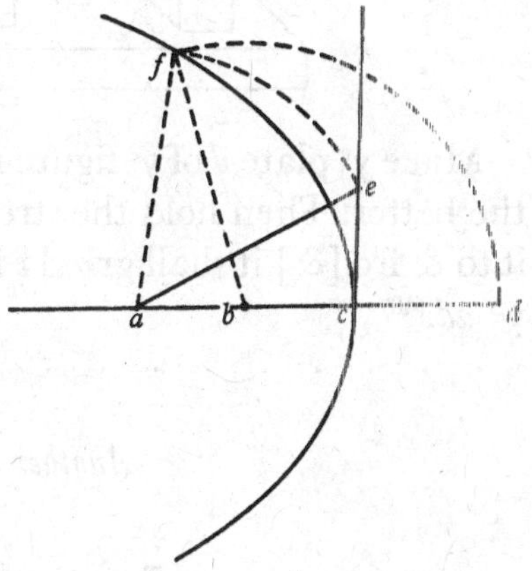

Make $ab = bc = \frac{r}{4}$. Make $ce = cd$ & $ce \perp bd$. Make $af = ae$, & $bf = bd$. then shall f be a point in ye Parabola.(42)

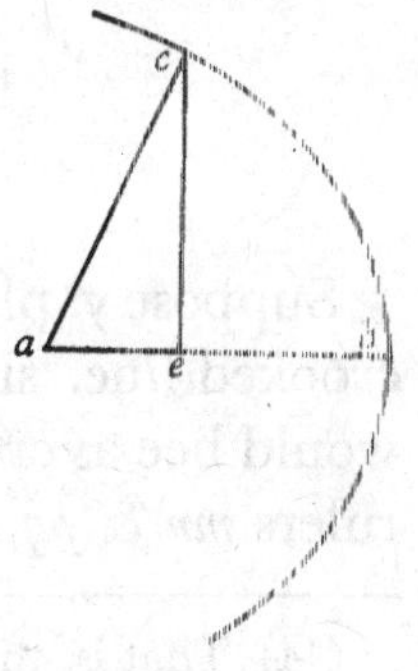

Another. Make $ab = \frac{r+x}{2} = ac$. $eb = x \perp ce$. & ye point c shall bee in ye parabola.(43) This like ye first by calculation may be made use of in other lines.

---

(40) A first 'proof' which seeks to show that $ad = ab$ is rightly cancelled. (The assumption gives, where $cb = cd = \frac{1}{2}r$ and $dq = x$, $qa = y$,

$$ab = ad = \sqrt{[x^2+y^2]}, \quad \text{or} \quad ac = \sqrt{[cb^2+ab^2]} = \sqrt{[\tfrac{1}{4}r^2+x^2+y^2]}$$

and finally $cq = \sqrt{[ac^2-aq^2]} = \sqrt{[\frac{1}{4}r^2+x^2]}$, which conflicts with $cq = \frac{1}{2}r - x$.)

(41) Newton shows that, if a is on the parabola $aq^2 = r.qd$, then $ab = qd$, the converse of which, here needed, follows easily. (A similar proof shows that, in the figure of note (39), $a'b = \sqrt{[a'c^2-cb^2]} = \sqrt{[a'q'^2-cd^2]} = q'd$, since $a'q'^2 = 2cd.q'd$ with $q'd^2 = q'c^2+cd(q'c+q'd)$.)

(42) In proof, where $cp = x \perp pf = y$,

$$af^2 = y^2+(\tfrac{1}{2}r-x)^2 = ae^2 = \tfrac{1}{4}r^2+ce^2 \quad (\text{or } cd^2)$$

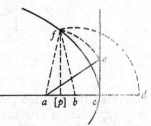

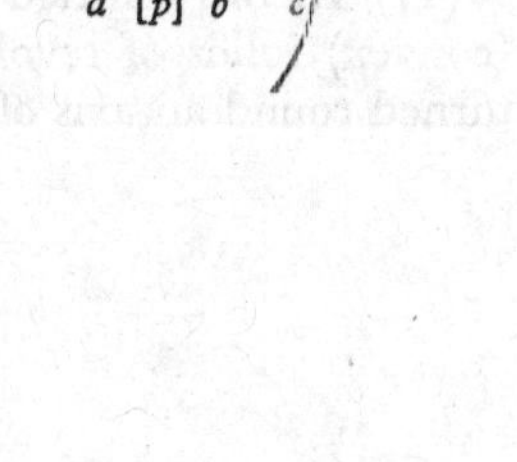

and $bf^2 = y^2+(\frac{1}{4}r-x)^2 = bd^2$, with $bd - cd = bc = \frac{1}{4}r$. Therefore $cd^2 = y^2+x^2-rx$ and $bd^2 = y^2+x^2-\frac{1}{2}rx+\frac{1}{16}r^2$, so that

$$\sqrt{[y^2+x^2-\tfrac{1}{2}rx+\tfrac{1}{16}r^2]} = \tfrac{1}{4}r+\sqrt{[y^2+x^2-rx]},$$

or $\frac{1}{2}rx = \frac{1}{2}r\sqrt{[y^2+x^2-rx]}$ and $y^2 = rx$.

(43) Where $be = x$, $ec = y$, $y^2 = \frac{1}{2}(r+x)^2 - \frac{1}{2}(r-x)^2 = rx$.

The manner whereby any kind of little lines may be described very accurately. And that the same Instrument serve for all (though never so small) differing in quantity but not in quality.(44)

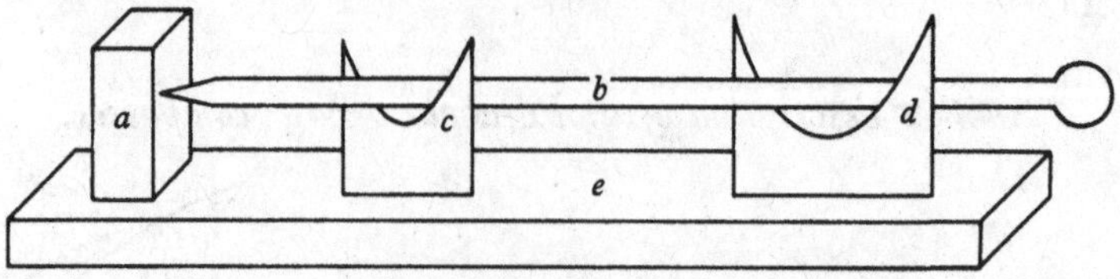

Make y^e plate *d* of y^e figure required (by some of y^e former meanes) the larger the better. Then hold the streight steele staffe *b* against the center *a* & roule(45) it to & fro [&] it shall grind *c* into y^e same figure but soe much lesse as *ac* is lesse y^n *ad*.(46)

. . . .(47)

Another way to describe lines on plates.

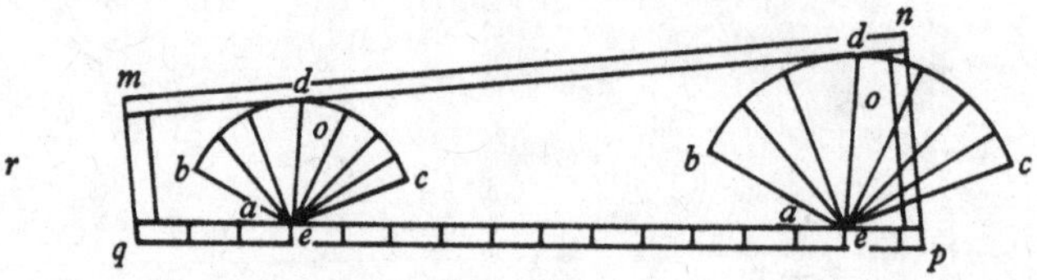

Suppose y^e plate bee *adc*, whose edg *boc* is to be made into y^e fashion of a given crooked line. suppoes (*o*) is its vertex & y^t a circle described wth y^e Radius *eo* would bee as crooked as ye given line at its vertex. Againe suppose two streight rulers *mn* & *pq* to bee very trew & steddyly fastened together wch must [be] a

(44) That is, the instrument shall construct similar figures by preserving 'shape'.
(45) Read 'roll'.
(46) The accompanying figure shows Newton's instrument in better perspective. Newton omits from his text the necessary condition that the planes *c* and *d* be parallel, but the technique remains a neat application of a homothety.

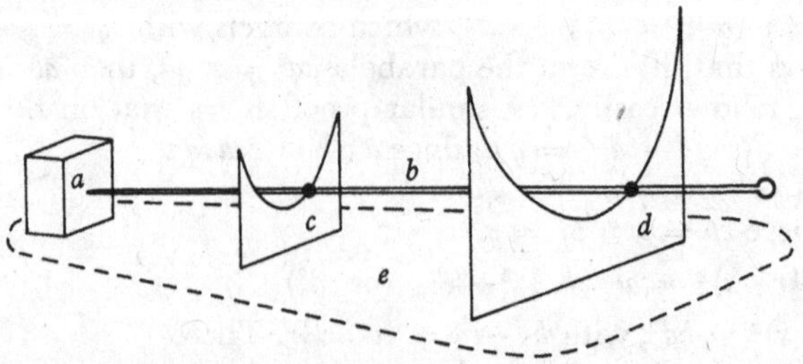

(47) In this omitted paragraph Newton applied the same technique to the grinding of (convex) solids of revolution on a lathe from the template *d* by rotating the mass *c* to be turned round an axis of symmetry.

verry little incline yᵉ one to yᵉ other, soe as that being produced they would meete at r. Then are yᵉ lines $pn=a$, & $pr=b$ given.

Suppose yⁿ yᵉ point d in yᵉ crooked line is to bee found. yⁿ is dc given by supposition, & consequently (supposing dk to bee a tangent) $dg=y$. $gc=x$. $fg=v$.[48] $fd=s$. $ec=c$. [yⁿ] $fk=v+\frac{yy}{v}$. $ef=v+x-c$. $ek=c-x+\frac{yy}{v}$. & (if $eh\perp dk\perp df$) then is

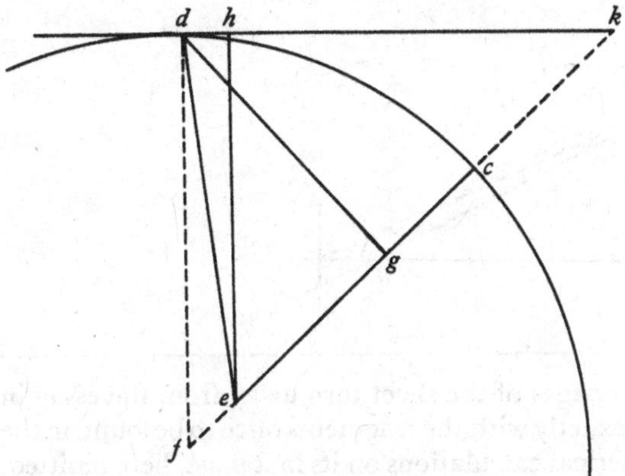

$$eh=\frac{cv-xv+yy}{\sqrt{vv+yy}}=d.^{(49)}$$

(eh) being thus found, supposing yᵗ $pn=a=ec$, then I take $re=\frac{bd}{a}$. that is $pe=b-\frac{bd}{a}$. haveing thus found yᵉ point e lay yᵉ plate twixt the two rulers so yᵗ yᵉ point of it fall upon yᵉ point e. yⁿ should yᵉ line mn touch yᵉ plate in d. But note yᵗ $pn\perp mn$.[50]

. . . .[51]

(48) The Cartesian subnormal $y(dy/dx)$.

(49) $gk = y^2/v$ with $fk = fg+gk$, $ef = fg+gc-ec$, $ek = ec-cg+gk$ and $eh = (fg.ek)/fd$.

(50) Where the given curve (d) is defined with respect to Cartesian co-ordinates $cg = x$ and $gd = y$, with vertex c the origin and e its corresponding centre of curvature, Newton derives the tangential polar of the arbitrary point $d = (x, y)$ as $eh \equiv (c-x)(dy/ds)+y(dx/ds)$. ($eh = ed$ when d and h coincide at c.) On this basis he constructs a simple gauge ($mnpqr$) with which to test the accuracy of the template $bdce$. (For simplicity, np is taken equal to ce, so that, when ec coincides with np, rmn will be the tangent at the vertex c.) Having, therefore, calculated the tangential polar $eh = d$ and, where $pn = a$ and $pr = b$, setting the point e of the template at a in pr such that $ra = bd/a$ (or $pa = b(a-d)/a$), we deduce that the template when it is rotated round e into contact should touch the base-line rmn of the gauge at its point d.

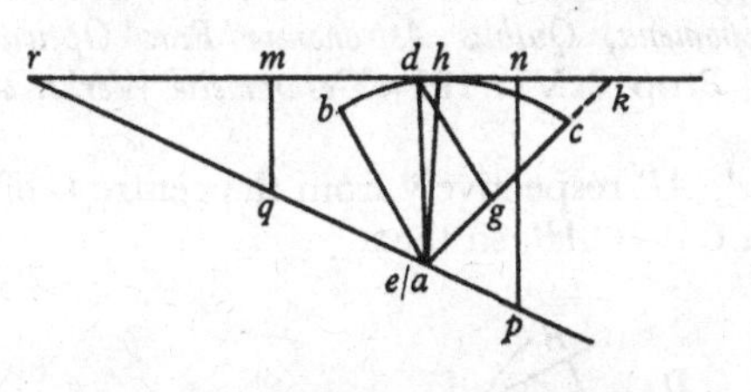

(51) In this omitted portion, printed by A. R. Hall in his *Further Optical Experiments of Isaac Newton* (note (2) above): 40–3, Newton enters some practical observations on the design of lens-systems in the telescope and microscope, and then passes on to describe a variation of his hyperboloid-instrument which shows how a 'glasse', as distinct from a template, 'may be ground Hyperbolicall'. Later, he comes to consider how 'To Grinde Sphæricall opticks Glasses', and opts for a method in which the lens is turned on a lathe while being ground by 'a circular hoope of steele'. (Compare Hooke, *Micrographia*: Preface: e[1]ᵛ–e[2]ʳ.) Finally, after a qualitative discussion of the errors which arise when telescope lenses are 'not truely ground', he tabulates the 'proportions of yᵉ motions of the Extreamely Heterogeneous Rays', when white light is refracted between air, water, glass and 'christall'. (Add. 4000: 33ᵛ, printed in *Correspondence of Isaac Newton*, **1** (1959): 103, note (6).)

[*The aberration of light refracted at a spherical surface.*](52)

[1] If $d.\,e$::maj sin. min sin of refr(53) & $d-e=h$. $OA \| CV$. And I the focus.(54) $AB=y$. $CV=r$. Then $VI=\frac{er}{h}$. $IR=\frac{ddyy}{2her}$. $IS=\frac{ddy^3}{2eerr}$. $\frac{ddhy^3}{2e^3r^3}=$ang IAR.(55)

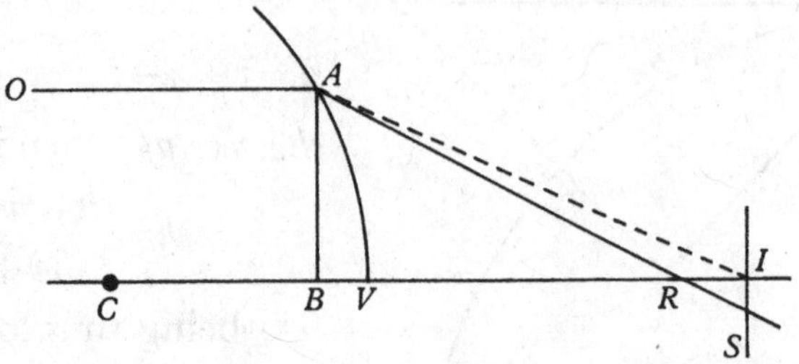

(52) Transcribed from the third and fourth pages of the sheet torn away from the essay in Add. 4000. (Two holes at the fold correspond exactly with the fragments once to be found in the spine of Newton's pocket-book, and some numerical calculations on its first page, here omitted, depend on the refraction table on 33v (note (51)). In final proof of the identity of the loose sheet, Collins' contemporary copy (note (1)) of the essay 'Of Refractions' transcribes it continuously with the preceding sections in Add. 4000.)

(53) Read 'If $d{:}e$::major sine:minor sine of refraction'.

(54) That is, the limit-meet of the axis CV with the refracted portion AR of an incident ray OA parallel and indefinitely near to CB. Later, in his *Lectiones Opticæ*, Newton introduced the notation of 'focus principalis' (principal focus) for the point I, defining it as 'Radijs in curvam quamvis superficiem quam proxime perpendiculariter incidentibus, refractorum concursu[s].' (Lectio 13, October 1670 = ULC. Dd. 9. 67: *Opticæ Pars* 1ma: Sectio 4ta, *De Refractionibus Curvarum Superficierum*, Prop. 30: 66–7, rendered in English as *Optical Lectures*... 1669 (London, 1728): Section 4, Prop. XXX: 180–1.) (Compare Kepler's 'ultimus terminus intersectionum radiorum parallelorum' in his *Ad Vitellionem Paralipomena, Quibus Astronomiæ Pars Optica traditur* (Frankfurt, 1604): Caput V, *De Modo Visionis*, Prop. XVI: 191 = *Gesammelte Werke*, 2 (Munich, 1939): 173.)

(55) Where CD, CE are drawn perpendicular to OA, AR respectively from the centre C of the spherical surface AV, then $CE = AB.CR/AR$ with $CD = AB$, so that

$$d{:}e = CE{:}CD = CR{:}AR.$$

Then, where $VB = x$, $BA = y$, the defining property of the spherical surface is $y^2 = 2rx - x^2$ and, where $CR = z$,

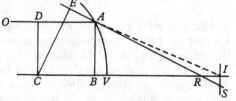

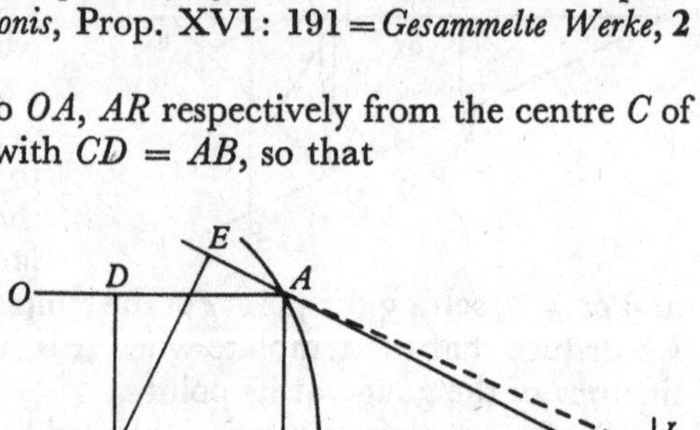

$$AR = \sqrt{[AB^2 + BR^2]} = \sqrt{[z^2 + 2xz - 2rz + r^2]},$$

so that $d{:}e = z{:}\sqrt{[z^2+2xz-2rz+r^2]}$, or

$$z^2 = \frac{d^2}{d^2-e^2}(2(r-x)z - r^2) \quad \text{and} \quad z = \frac{d}{d^2-e^2}(d(r-x) + \sqrt{[e^2r^2 - 2d^2rx + d^2x^2]})$$

$$= \frac{d}{d^2-e^2}(d\sqrt{[r^2-y^2]} + \sqrt{[e^2r^2 - d^2y^2]}).$$

Expanding the radicals in series of powers of y^2 and reducing,

$$z = \frac{dr}{d-e} - \frac{d^2y^2}{2e(d-e)r} - O(y^4);$$

[2] If $d.\ e::\text{sin. sin.}\ d+e=f.\ d-e=h.\ CV=r.\ BV=x.\ AB=y.\ FV=a.$ $FC=b.$[56] & $CR=z$. Then $ddbb$ in $rr-2rz+zz+2zx=eezz$ in $aa-2bx$.[57] & by

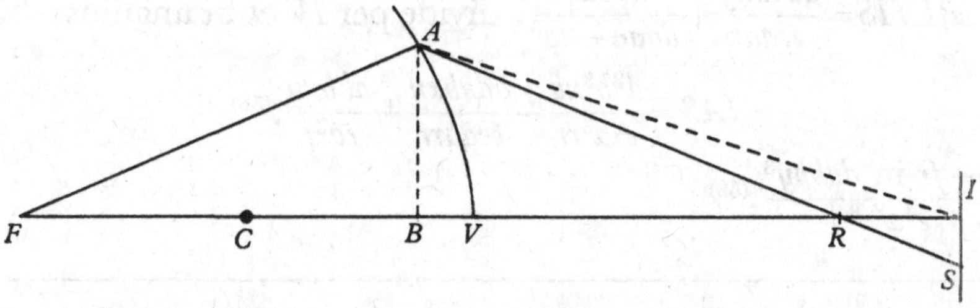

reduction $z=\dfrac{ddbbr-ddbbx+db\sqrt{eeaarr-2beerrx-2ddbbrx+ddbbxx}}{ddbb-aaee+2eebx}$. & by extraction of the root $z=\dfrac{ddbbr+dbaer-ddbbx-\dfrac{dbberx}{a}+\dfrac{d^3b^3x}{ae}}{ddbb-aaee+2eebx}$. And dividing

$$z=\frac{dbr}{db-ae}-\frac{ddbbx}{dbae-aaee}-\frac{debbrx}{ddabb-2deaab+eea^3}=CR.$$

---

but I is the limit-position of R when A, B pass into V, or $CI = \lim_{y\to 0}(z) = dr/(d-e)$, with $VI\,(=CI-r) = er/(d-e)$. The distance ('error') RI of the principal focus I from the intersection R is then

$$CI-CR = \frac{d^2y^2}{2e(d-e)\,r}+O(y^4).$$

Finally, since $BR = CR-CB = \dfrac{er}{d-e}+O(y^2)$, then $AB:BR \approx \dfrac{(d-e)\,y}{er}$,

so that $$IS = IR.\frac{AB}{BR} \approx \frac{d^2y^3}{2e^2r^2},$$

and the angle of aberration $I\hat{A}R \approx I\hat{B}S$ is measured approximately by

$$\tan(I\hat{B}S) = \frac{IS}{BI} \approx \frac{IS}{VI} \approx \frac{d^2(d-e)\,y^3}{2e^3r^3}.$$

(Compare *Lectiones Opticæ* (note (54)): Prop. 31: 67 = *Optical Lectures*: Prop. XXXI: 181–3.) Newton used this theorem to compare the errors of reflection (refraction with an index −1) and refraction at a spherical surface in his letter to Oldenburg of 11 June 1672. (*Correspondence of Isaac Newton*, **1** (1959): 172–3, and compare 189, note (6). The letter was first printed in *Philosophical Transactions*, **7**, no. 88 (for 18 November 1672): 5084–103, especially 5085.)

(56) Or $a = b+r$.

(57) Where, as before, CD, CE are drawn perpendicular to FA, AR respectively from the centre C of the spherical surface AV, then $CE = AB.(CR/AR)$ and $CD = AB.(CF/AF)$, so

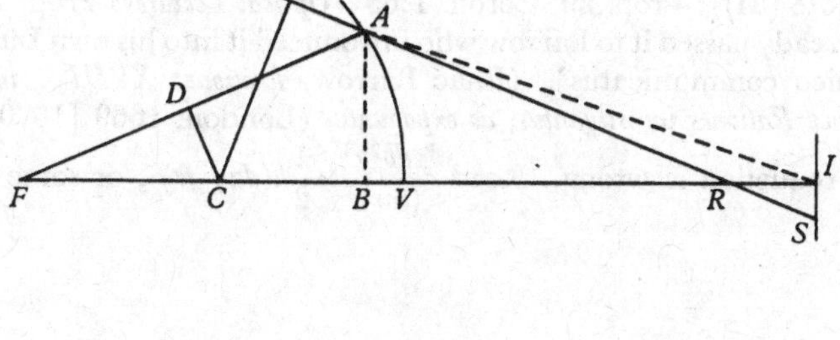

$\dfrac{dbr}{db-ae}=CI.$ $\dfrac{ear}{db-ae}=VI.$ $\dfrac{ddbbx}{deab-eeaa}+\dfrac{debbrx}{ddabb-2deabb+eea^3}=RI.$

$IV\,.\,AB::RI\,.\,IS=\dfrac{ddbbxy}{eeaar}+\dfrac{dbbxy}{dbaa-ea^3}.$ divide per IV et fit angulus

$$IAS=\frac{d^3b^3xy}{e^3a^3rr}-\frac{ddbbxy}{eeaarr}+\frac{dbbxy}{ea^3r}.^{(58)}$$

Sive $=\dfrac{ha-fr \text{ in } ddbby^3}{e^3a^3r^3}.^{(59)}$

---

that $d:e = CE:CD = CR.AF:AR.CF$. (Compare Newton's 'Fig 5t' and '6t' in 1, §1.2 above.) Then, where $VB = x$, $BA = y$, the defining property of the spherical surface is $y^2 = 2rx-x^2$ and, where $CR = z$, $FC = b$, $FV = a = b+r$, we have

$$FA = \sqrt{[(a-x)^2+y^2]} = \sqrt{[a^2-2bx]} \quad \text{and} \quad RA = \sqrt{[(z-r+x)^2+y^2]} = \sqrt{[(z-r)^2+2zx]},$$

so that $$db\sqrt{[r^2-2rz+z^2+2zx]} = ez\sqrt{[a^2-2bx]}.$$

(58) Where $\alpha = d^2b^2r$, $\beta = -d^2b^2$, $\gamma = ear$, $\delta = br(e^2r+d^2b)$, $\epsilon = db$, $A = d^2b^2-e^2a^2$ and $B = 2be^2$, by resolving the quadratic in z it follows that

$$z = \frac{\alpha-\beta x+db\sqrt{[\gamma^2-2\delta x+\epsilon^2x^2]}}{A+Bx}.$$

(Compare Newton's Lucasian lectures on mathematics, ULC. Dd. 9. 68: 79 = October 1676, Lectio 9, Prob. 17, printed as *Arithmetica Universalis* (London, 1707): 138–9: Prob. XVII, renumbered as Prob. XXXI in all subsequent editions.) Expanding the radicals, we derive

$$CR = z = \left[\alpha-\beta x+db\left(\gamma-\frac{\delta}{\gamma}x+O(x^2)\right)\right].\left[\frac{1}{A}-\frac{Bx}{A^2}+O(x^2)\right]$$

$$= \frac{\alpha+db\gamma}{A}-\frac{x}{A}\left[\frac{\beta\gamma+db\delta}{A}\right]+O(x^2) = \frac{dbr}{db-ea}-\frac{db^2}{ea}\left(\frac{d}{db-ea}+\frac{e^2r}{(db-ea)^2}\right)x+O(x^2);$$

whence $$CI = \lim_{x\to 0}(CR) = \frac{dbr}{db-ea}, \quad \text{and so} \quad VI = CI-r = \frac{ear}{db-ea}.$$

Therefore the distance ('error')

$$RI = CI-CR = \frac{db^2}{ea}\left(\frac{d}{db-ea}+\frac{e^2r}{(db-ea)^2}\right)x+O(x^2),$$

$$IS = RI.\frac{AB}{BR} \approx RI.\frac{AB}{VI} \approx \frac{db^2}{e^2a^2r}\left(d+\frac{e^2r}{db-ea}\right)xy,$$

and $$\tan(I\hat{A}R) \approx \tan(I\hat{V}S) = \frac{IS}{VI} \approx \frac{db^2}{e^3a^3r^2}(d(db-ea)+e^2r)xy^2.$$

The particular result $VI = ear/(db-ea)$ was given by Newton in modified form in his *Lectiones Opticæ* (note (54)): Prop. 29, Coroll. 1:65 = *Optical Lectures*: Prop. XXIX, Coroll. I: 174, but he had already passed it to Barrow who introduced it into his own Lucasian lectures on optics as 'ab amico communicatus'. (Isaac Barrow, *Lectiones XVIII...in quibus Opticorum Phænomenωn Genuinæ Rationes investigantur, ac exponuntur* (London, 1669 [1670]): 103.)

(59) This last reduction is wrong. Read '$\dfrac{rdb^2y^3}{(ear)^3}\times\dfrac{1}{2}h(da-fr)$', or some equivalent (since $y^2 = 2rx+O(x^2)$).

[*Notes on the theory of compound lenses.*][(60)]

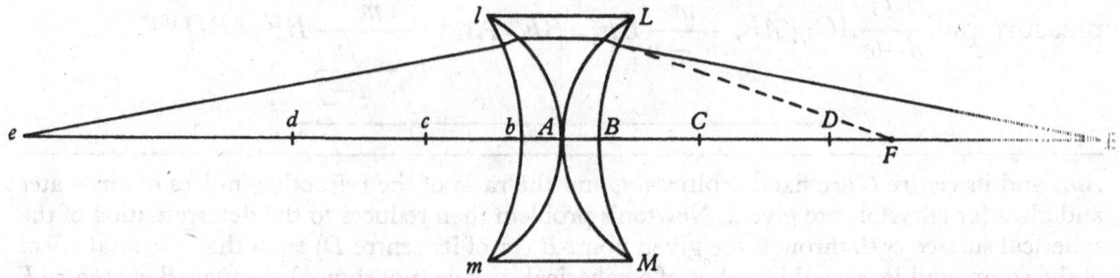

If the intervall *lAL* be water & the Lentes *lm* & *LM* glas or ch[r]istall alike figured. That the difforme rays flowing from *e* may convene in *E* [w^{ch}] points [are] æquidistant frō the lentes: Suppose *d. e. f.* the signes of refractiō in glass

(60) This seems to be the manuscript referred to by Newton in his letter of 11 June 1672 to Oldenburg (note (55)), Section 1, *Of ye Practique part of Optiques*: '...Mr Hook thinks himself concerned to reprehend me for laying aside the thoughts of improving Optiques by Refractions....what I said there [in his letter of 6 February 1671/2] was in respect of Telescopes of ye ordinary construction, signifying that their improvement is not to be expected from ye well figuring of Glasses as Opticians [Descartes and Hooke presumably] have imagined: But I despaired not of their improvement by other constructions....For although successive refractions wch are all made the same way, doe necessarily more & more augment the errors of the first refraction; yet it seemed not impossible for contrary refractions so to correct each others inequalities, as to make their difference regular, & if that could be conveniently effected, there would be no further difficulty. Now to this end I examined what may be done not onely by Glasses alone, but more especially by a complication of divers successive Mediums, as by two or more Glasses or Chrystalls with water or some other fluid between them, all wch together may perform the office of one Glasse, especially of the Object-Glasse on whose construction the perfection of the Instrument chiefely depends....' *Correspondence of Isaac Newton*, 1 (1959): 172. Compare the passage (pp. 191–2) quoted in note (20) above.) On at least two other occasions Newton was not prepared to dismiss the problem of correcting for chromatic aberration as impossible of solution. (See his *Lectiones Opticæ* (note (54)): October 1670, Lectio 15: 77 = *Optical Lectures*: 211–12; and *Philosophiæ Naturalis Principia Mathematica* (London, $_1$1687): Liber I, Sectio XIV: 234–5: *Scholium.*)

(61) In the accompanying simplification of Newton's diagram, *C* and *D* are the respective centres of the spherical surfaces $A\beta L$ and $B\gamma L$, and the light ray $\alpha\beta\gamma E$, initially parallel to the axis *AB* and travelling through water, is refracted at β into glass (or chrystal) in the direction of *F* and further refracted at γ into air through the (principal) focus *E*. The spherical surface

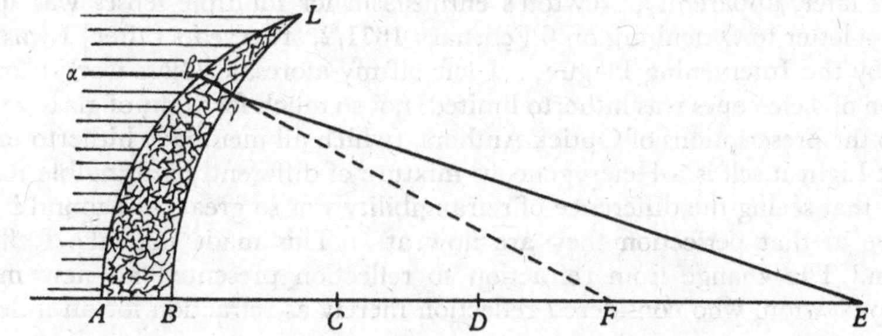

or christall, water & aire. And m to n as the meane differences of the different refractions of difform rays into air & water out of glasse. Then assuming AC at pleasure put $\frac{d}{d-e}AC=AF$. $\frac{m}{m-n}BF=BE$. And $\frac{m}{m+\frac{fn}{d-f}}BF=BD$.(61)

$A\beta L$ and its centre C are fixed arbitrarily, and the ratio of the refractive indices of air, water and glass (or chrystal) are given. Newton's problem then reduces to the determination of the spherical surface $B\gamma L$ through the given point B (or of its centre D) such that the total effect of the compound lens would be that of a spherical surface (not shown) through B, or centre F and principal focus E, which refracts light into air with a given refractive ratio m/n. Immediately, where F is the principal focus of $A\beta L$,

$$\frac{d}{d-e}AC = AF \text{ (by [1]);} \quad \frac{f.BF.BD}{f.BF-d.DF} = BE \text{ (by [2]); and} \quad \frac{m}{m-n}BF = BE \text{ (by [1]).}$$

By combining these last two results it follows that

$$\frac{m}{n} = \frac{BE}{EF} = \frac{BE}{DE}+k\frac{BD}{DE}, \quad \text{where} \quad k = \frac{f}{d-f}, \quad \text{and so} \quad \frac{m-n}{m+nk} = \frac{BD}{BE},$$

Newton's final result which fixes the required spherical centre D. Newton concludes his treatment by taking numerical examples in which the refractive ratio m/n is taken from the table on f. 33v (note (51)) as arithmetic means of the entries in the form $d-e/[k(d-f)]$, but these do not add anything to the theory and are here omitted.

It is clear that Newton in early 1666 was prepared to work at the idea of an achromatic multiple lens. However, the correction for the different refrangibilities of coloured light is not here made part of the theory of the lens, but rather, when the ratio m/n and the proportions of the lens have been derived empirically as those which substantially reduce the effects of colour, then only is the structure of the lens treated mathematically by considering a mean between the limits of dispersion of an incident ray as an approximation to its refracted path. Beyond reasonable doubt Newton had this formative idea of using a multiple lens to correct for chromaticity from Hooke. Already in 1665 in his *Micrographia* (Preface: f[2]r) the latter had found empirically that by filling the 'intermediate space' between two microscope lenses with 'very clear Water,...I could perceive an Object more bright then I could when the intermediate space was only fill'd with Air...'. A year later he improved his experimental techniques, showing that by filling the intermediate space with fluids of different kinds he could alter the focal length of the lens as a whole. (See his 'A Method, by which a Glass of a small Plano-convex Sphere may be made to refract the Rayes of light to a Focus of a far greater distance, than is usual', *Philosophical Transactions*, **1** (1665/1666): no. 12 (for 'Munday, May 7. 1666'): 202–3.) In Newton's unpublished annotations on the early numbers of the *Philosophical Transactions* Hooke's paper is duly noted as 'How Mr Hook will make any Object glasse draw any length' (ULC. Add. 3958. 1: 11v).

Two years later, apparently, Newton's enthusiasm for multiple lenses was spent. As he explained in a letter to Oldenburg on 6 February 1671/2, 'two years [after] I was forced from Cambridge by the Intervening Plague...I left off my aforesaid Glass-works; for I saw that the perfection of Telescopes was hitherto limited, not so much for want of glasses truly figured according to the prescriptions of Optick Authors, (which all men have hitherto imagined,) as because that Light it self is a Heterogeneous mixture of differently refrangible Rays....Nay, I wondered, that seeing the difference of refrangibility was so great, as I found it, Telescopes should arrive to that perfection they are now at....This made me take Reflections into consideration.' The change from refraction to reflection presented no new mathematical difficulties to Newton, who considered reflection merely as refraction for an index -1.

3. THE REFRACTION OF LIGHT AT A SPHERICAL SURFACE

[1666?]

From the originals in the University Library, Cambridge[1]

[1][2]

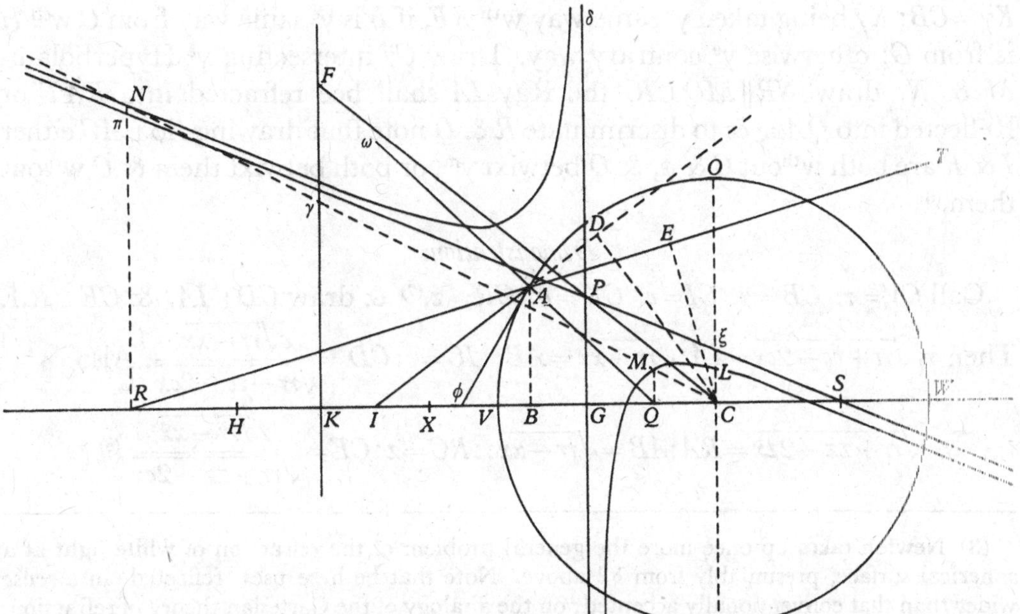

To determine how the ray IA shall bee refracted into RAT by the sphære AVW or reflected into QAω.[3]

Supposing y^e center of y^e sphæ[r]e to be C; its axis CV, intersected by all y^e rays of incidence at I. Make y^t as y^e motion of y^e $\begin{Bmatrix}\text{Refracted}\\ \text{Reflected}\end{Bmatrix}$ rays to the motion of y^e rays of Incidence$::d:e$[4]$::CI:XC::XC:CG=CS$. & the points G & I being on the same side of the center draw $\delta PG\perp CI$. Make $IC:VC::VC:IH$. & in KF

(1) Add. 3958. 6. This manuscript section contains an extensively revised worksheet and a polished final draft, both printed below, together with (at f. 93^v) an incomplete first revision. The conjectural date is offered on the basis of the handwriting style and on internal textual evidence, particularly the use of the Cartesian notation $d:e$ (rather than Barrow's $I:R$, later to be used) for the refractive ratio.

(2) The earliest draft on ff. 94^r/94^v, much cancelled and overwritten. (It seems clear that Newton destroyed almost immediately the preliminary calculations which are its foundation and will be found restored in note (10).)

drawn at pleasure perpendicular to IC[,] H falling beyond I from y[e] center[,] take $KF=GH$. Draw FC intersecting GP in P. & Draw πPS, producing it both ways. Draw $CO\perp IC$ intersecting y[e] circle in O & make $2CK:CV::CV:CL$. y[e] point L being taken on y[e] same side of C wth P if $IO \sqsubset$(5) XC. Otherwise on y[e] contrary side. And through that point L wth y[e] Assymptotes $SP\pi$, $GP\delta$, draw an Hyperbola wth its conjugate.

This Præmised, from any point of Incidence A draw $AB\perp IC$, & make $K\gamma=CB$; $K\gamma$ being taken y[e] same way wth KF, if B is y[e] same way from C w[ch] H is from G; otherwise y[e] contrary way. Draw $C\gamma$ intersecting y[e] Hyperbola in M & N. draw $NR \| MQ\perp IC$ the Ray IA shall bee refracted into RAT or Reflected into $QA\omega$. & to discriminate R & Q note that drawing $A\phi\perp AC$ either I & R are both w[th]out C & ϕ, & Q betwixt y[m], or both betwixt them & Q w[th]out them.(6)

Demonstration.

Call $CV=r$. $CB=x$. $CI=c$. $CK=a$. $CR=z$.(7) & draw $CD\perp IA$. & $CE\perp RA$.

Then is $\sqrt{rr+cc-2cx}=IA:\sqrt{rr-xx}=AB::IC=c:CD=\dfrac{c\sqrt{rr-xx}}{\sqrt{rr+cc-2cx}}$. Also

$$\sqrt{rr+zz-2zx}=RA:AB=\sqrt{rr-xx}::RC=z:CE=\frac{z\sqrt{rr-xx}}{\sqrt{rr+zz-2cx}}.^{(8)}$$

(3) Newton takes up once more the general problem of the refraction of white light at a spherical surface, presumably from §2 above. Note that he here uses 'reflected' in a sense wider than that conventionally accepted: on the analogy of the Cartesian theory of refraction, a light ray is 'reflected' at a surface between two media of 'reflective' ratio $d:e$ when the 'reflected' ray is returned into the medium of the incident ray at an angle whose sine is in proportion to that of the incident ray as $e:d$. (The only acceptable physical instance of this 'reflective' ratio has, of course, $d/e=-1$.)

(4) Since in Newton's (as in Descartes') theory of refraction $v_i \sin i = v_r \sin r$, where v_i and v_r are the speeds ('motions') of light before and after refraction with i and r the corresponding angles of incidence and refraction (here $D\hat{A}C$ and $E\hat{A}C$), it follows that

$$\frac{v_r}{v_i}=\frac{\sin i}{\sin r}=\frac{d}{e}.$$

(5) Barrow's adaptation of Oughtred's symbolism for 'greater y[n]'. Correspondingly, for '$\sqsupset$' below read 'lesse y[n]'.

(6) In the immediately following 'Demonstration' Newton gives a polished analytical justification for this synthetic construction.

(7) Newton has cancelled the next four lines, which read: '$RN=y$. And to get y[e] relation twixt z & y, Feigne an Equation $zz+mz+py=0$. for y[e] nature of y[e] Hyperbola. in w[ch] I make $+\frac{n}{m}yz+qqy$ of but one dimension because it being $\|$ to y[e] Assymptote $G\delta$ can intersect y[e] Hyperbola but in one point.'

(8) Read '$\dfrac{z\sqrt{rr-xx}}{\sqrt{rr+zz-2zx}}$', an error which is not carried through and which is independent proof that Newton is copying from a first draft.

But $d:e::CD:CE$. Therefore $\frac{ec}{\sqrt{rr+cc-2cx}}=\frac{dz}{\sqrt{rr+zz-2zx}}$.[9] That is

$$ddrrzz+ddcczz-eecczz-2ddcxzz+2eeccxz-eeccrr=0.$$

Againe (by y^e^ construct) $d:e::\overline{CI=c}:\frac{ce}{d}::\frac{ce}{d}:\frac{cee}{dd}=CG$. &

$$IC=c:VC=r::r:\frac{rr}{c}=IH.$$

Therefore $HG=\frac{rr}{c}+c-\frac{eec}{dd}=KF$. & $CK=a:KF::CG:GP=\frac{eerr}{add}+\frac{ccee}{add}-\frac{cce^4}{ad^4}$. & $\frac{1}{2}GP=\frac{ddeerr+ddeecc-e^4cc}{2ad^4}=C\xi$. Also $2CK=2a:CV=r::r:\frac{rr}{2a}=CL$. Therefore $\xi L=\frac{eerrd^2+eeccdd-e^4cc-rrd^4}{2ad^4}$ supposing $C\xi \sqsubset CL$. But

$$RG=z-\frac{eec}{dd}:GC=\frac{eec}{dd}::\pi P:P\xi::\xi L:\pi N[::]\,ddz-eec:eec,$$

by y^e^ nature of the Hyperbola, & therefore

$$\pi N=\frac{e^4crrdd+e^4c^3dd}{2d^6az-2d^4eeac}\,\frac{-e^6c^3-d^4eerrc}{2d^6az-2d^4eeac}.$$

Againe $GS=\frac{2eec}{dd}:RS=z+\frac{eec}{dd}::GP:R\pi::2eec:ddz+eec$. Therefore

$$R\pi=\frac{rrz}{2ac}+\frac{cz}{2a}-\frac{ceez}{2add}+\frac{eerr+eecc}{2add}-\frac{e^4cc}{2ad^4}.$$

& $R\pi+\pi N=RN=\frac{e^4crrdd+e^4c^3dd-}{2d^6az-2d^4eeac}$ &c. Which redu[ced] to one denomination is $\frac{ddrrzz+ddcczz-eecczz-eeccrr}{2ddacz-2eeacc}=RN$. Now $KC=a:K\gamma=x::RC=z:RN=\frac{zx}{a}$. Which two valors of RN being put equall & y^e^ Equation Reduced it is

$$ddrrzz+ddcczz-eecczz-2ddcxzz+2eeccxz-eeccrr=0,$$

As before. So y^t^ y^e^ Construction by y^e^ Hyperbola makes $z=RC$ y^e^ same Length wch y^e^ Refraction by y^e^ circle doth. Q.E.D. W^ch^ Demonstration is the same in all other cases.[10]

(9) Newton derives the fundamental condition for the light ray IA to be refracted along AT. (Compare §2, note (57) above.)

(10) This rather forbidding reduction becomes greatly clarified if we restore Newton's preliminary calculations along the lines hinted at in the cancellation recorded in note (7). Thus, taking $RN = y = zx/a$ and eliminating x between $y = zx/a$ and the refractive condition

$$\frac{ec}{\sqrt{[r^2+c^2-2cx]}}=\frac{dz}{\sqrt{[r^2+z^2-2zx]}},$$

The limits of y[e] Refracted rays are esily found by taking IV for one extreame Ray & for y[e] other make IA to touch the circle if $d \sqsubset e$ but if $e \sqsubset d$ make $C\gamma$ to touch y[e] Hyperbola & to find such rays as are refracted $\parallel$ to y[e] axis make $C\gamma \parallel PS$. In generall it may bee observed y[t] If H is scituated the same way from I w[ch] V is from C, supposing y[t] I is y[e] same way from C y[n] first if $\frac{2rr}{c} = 2IH \sqsubset KF$ all y[e] rays intersect y[e] axis towards I. secondly if $VW \sqsubset FK \sqsubset 2IH$ the[y] intersect some one way some another, 3[dly] If $FK \sqsubset VW$ they intersect all on y[e] contrary side. But supposing y[t] I is contrary to C from V Then 1st If $VW \sqsubset KF$ y[e] rays intersect some on one side some on y[e] other & 2[dly] if $KF \sqsubset VW$, they intersect on y[e] contrary side. But if y[e] scituation of H from I is contrary to y[e] scituation of V from C the rays intersect y[e] axis on y[e] same side of the tangent $A\phi$, wth I.(11)

---

we may deduce that the defining equation of (N) is

$$z^2 - \frac{2acd^2}{\lambda} zy + \frac{2ac^2e^2}{\lambda} y - \frac{c^2e^2r^2}{\lambda} = 0, \quad \text{where} \quad \lambda = d^2r^2 + c^2d^2 - c^2e^2.$$

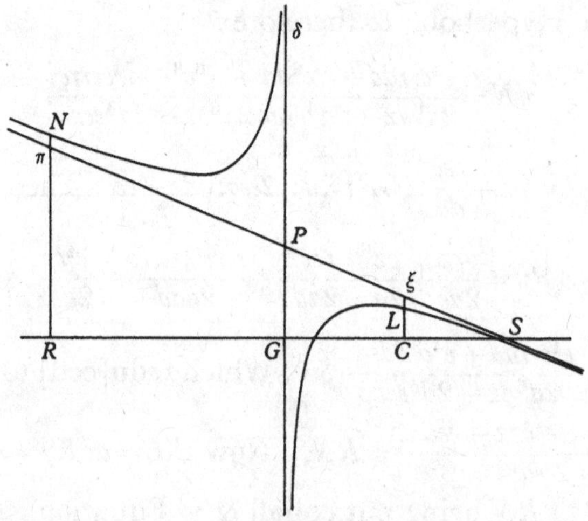

Further, if we equate this term by term with $(z+\alpha)(z+\beta y+\gamma) = \delta$, we find

$$\alpha = -\frac{ce^2}{d^2} = -\gamma, \quad \beta = -\frac{2acd^2}{\lambda} \quad \text{and} \quad \delta = c^2e^2\left(\frac{r^2}{\lambda} - \frac{e^2}{d^4}\right),$$

so that, where

$$Z = z - \frac{ce^2}{d^2}, \quad Y = y - \frac{\lambda e^2}{ad^4} \quad \text{and} \quad \mu = \frac{ce^2}{2ad^2}\left(\frac{\lambda e^2}{d^4} - r^2\right),$$

the defining equation of (N) may be set as $Z(Y - (\lambda/2acd^2)Z) = \mu$ constant. This Newton constructs by taking $CG = ce^2/d^2$ (or $Z = RG$) and $GP = \lambda e^2/ad^4$ (or $Y = RN - GP$), so that $N\pi = Y - (\lambda/2acd^2)Z$ where $GP/GS = \lambda/2acd^2$ (and hence $SG = 2ce^2/d^2 = 2CG$). Finally, where $\pi P/RG = k$, it follows that $N\pi.\pi P = k\mu$, the defining property of a hyperbola whose asymptotes δPG and $SP\pi$ are $Z = 0$, $Y = (\lambda/2acd^2)Z$ respectively. (Note that when $CR = z = 0$, then $CL = y = r^2/2a$.)

(11) These remarks are immediate deductions from the defining equation of the hyperbola (N) $\lambda z^2 - 2acd^2zy + 2ac^2e^2y - c^2e^2r^2 = 0$ (note (10)).

If $d=e$, y[e] Hyperbolas are streight lines coincident wth their Assymptotes. As also if $d:e::IC:VC$; in w[ch] case If I is beyond y[e] center from y[e] vertex y[e] Reflected Ray is given by y[e] intersection of the oblique assymptote FS, y[e] Refracted Ray by y[e] intersection of PQ. Soe y[t] y[n] y[e] point R varys not but is always coincident wth G. The contrary happens supposing I on ye other side y[e] center.(12)

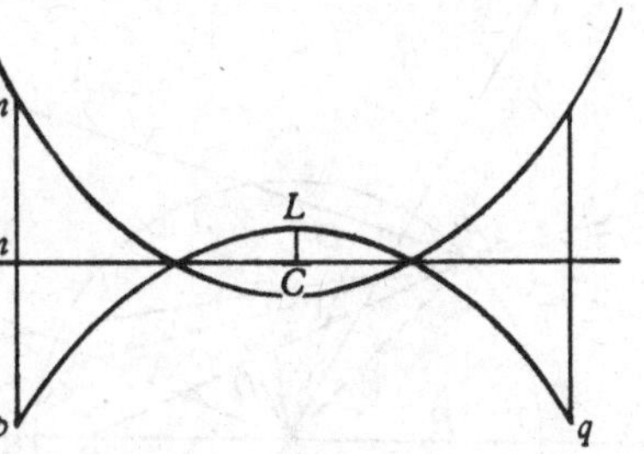

If CI is infinite y[t] is if $IA \parallel VC$ y[e] Hyperbola becomes a Parabola whose vertex is L (& making [$Cn=CV$,] $nm=2LC$ & $mp=\frac{dd}{ee}LC$ p is another point of it taken at pleasure). Or make $d-e:e::r:cn$.(13) & $CK:cn$(13)$::r:np$. then make $2CL:CV::CV:CK$. &c.(14)

---

(12) The condition for the hyperbola $Z(Y-(\lambda/2acd^2)Z) = \mu$ (note (10)) to reduce to the line-pair $(\delta PG).(SP\pi)$ is that

$$\mu = \frac{ce^2}{2ad^2}\left(\frac{\lambda e^2}{d^4}-r^2\right) = 0, \quad \text{or} \quad \lambda e^2-r^2d^4 = (d^2-e^2)(c^2e^2-d^2r^2) = 0.$$

(13) Read 'Cn'.

(14) When $CI = c$ becomes infinite, the defining equation

$$\lambda z^2-2acd^2zy+2ac^2e^2y-c^2e^2r^2 = 0$$

(note (10)) degenerates into the parabola

$$\lambda z^2 = \frac{2ae^2}{d^2-e^2}\left(\frac{r^2}{2a}-y\right),$$

which clearly has its vertex at $CL = r^2/2a$. Further, when

$$Cn = z = r, \quad np = y = \frac{r^2}{2a}\left(2-\frac{d^2}{e^2}\right);$$

and when

$$Cn = \frac{er}{d-e}, \quad \text{then} \quad np = \frac{r^2e}{a(d-e)} = Cn.\frac{CV}{CK}, \quad \text{where} \quad CK = a = \frac{CV^2}{2CL}.$$

(15) Ff. 95r/95v, a much improved version of [1]. The underlying mathematical argument is unchanged, though denotations of particular points are altered and for simplicity Newton here takes $CS = CN = ce^2/d^2 = a$. It follows that the defining equation of the hyperbola (E) may be reduced to $\lambda d^2z^2-2c^2d^2e^2zy+2c^3e^4y-c^2d^2e^2r^2 = 0$, and thus, for example, when $CG = z$ is zero, $CW = y = d^2r^2/2ce^2 = CT^2/2CN$.

[2](15)

To Determine how any Ray AB shall bee refracted into AG by the sphære ATV.

Suppose $d:e::$ as y^e^ motion of y^e^ rays w^th^in y^e^ sphære to their motion w^th^out it; y^t^ is as CD to CE y^e^ perpendiculars from y^e^ center C on those rays. Call

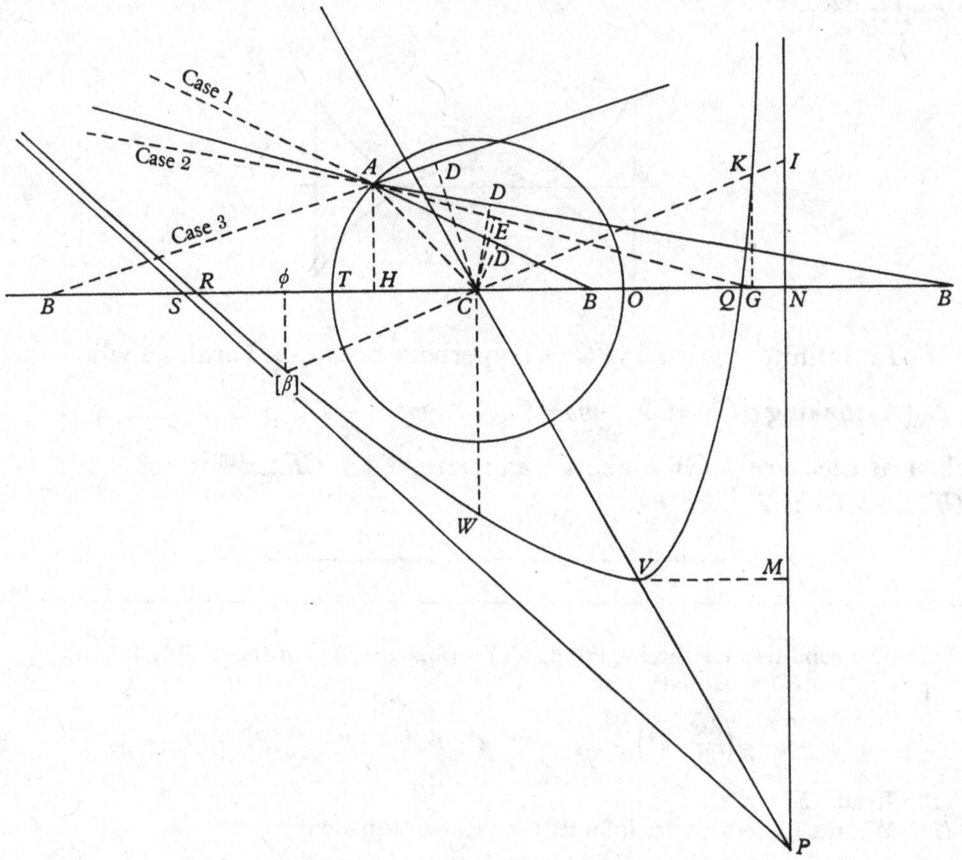

$AC=r$. & $CB=c$. Make $\frac{eec}{dd}=CS=CN\perp NP=\frac{rr}{c}+c-\frac{eec}{dd}$. & draw SP. Make $\frac{rr}{2CN}=CW\perp CB$. & through y^e^ point W about y^e^ Assymptotes NP & PS describe y^e^ Hyp^r^bola $RWQK$. $\left(\frac{er}{d}\sqrt{\frac{c}{NP}}=CR=CQ\right.$ gives also y^e^ points of y^e^ Hyperbola R & Q.$\left.\right)^{(16)}$

Then from any point of incidence A draw $AH\perp CB$, make $IN=CH$. & Draw

(16) For, when $KG = y$ is zero, then the corresponding values of $CG = z$ are found from the hyperbola's equation (note (15)) to be $CG = \pm\sqrt{[c^2e^2r^2/\lambda]}$ and Newton's reduction follows by noting that $NP = \lambda/cd^2$.

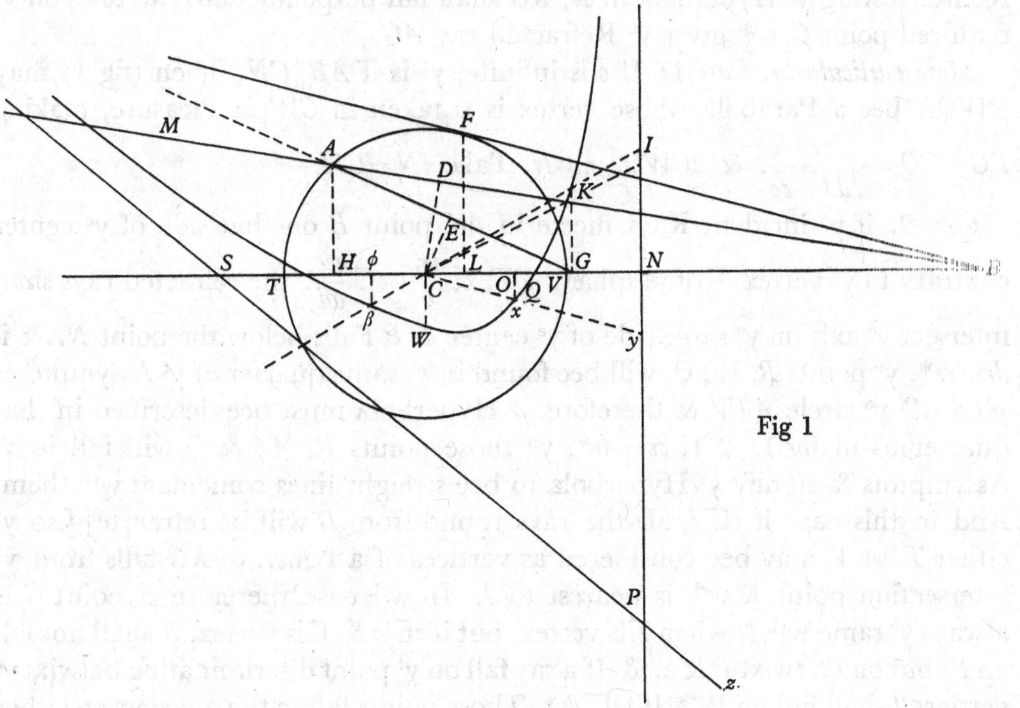

Fig 1

Fig 2.

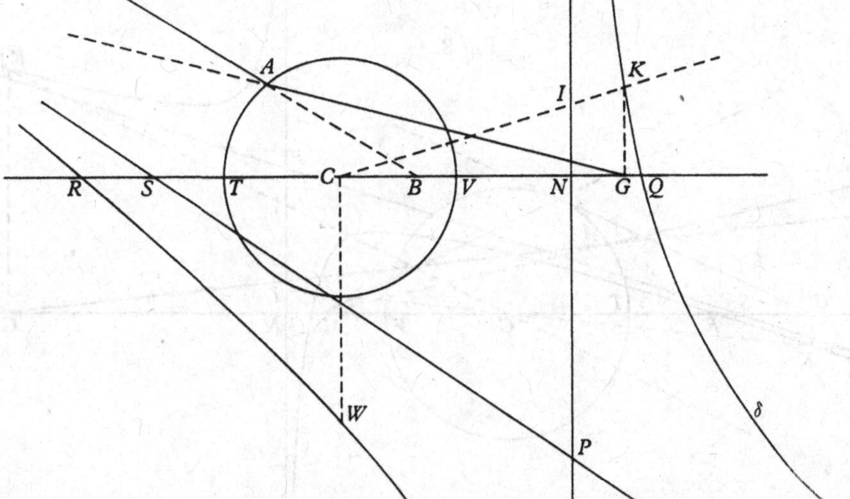

IC intersecting y^e^ Hyperbola in *K*, *KG* shall fall perpendicularly to *cb*[17] on y^e^ required point *G* wch gives y^e^ Refracted ray *AG*.

More particularly; *Case* 1: If *c* is infinite, y^t^ is if $AB \| CN$; Then (fig 1) may *RWQK* bee a Parabola whose vertex is *w* taken in *CW* at Pleasure, making $RC = CQ = \frac{re}{\sqrt{dd-ee}}$. & $2CW:r::r:CN$. Take $IN = CH$.

Case 2. if y^e^ incident Rays meete in the point *B* on that side of y^e^ center contrary to y^e^ vertex *T* of y^e^ sphære *VT*, & $\frac{rr}{c} + c \sqsubset \frac{eec}{dd}$. the refracted rays shall intersect y^e^ axis on y^e^ same side of y^e^ center & *P* Falls below the point *N*. & if $dr \sqsubset ce$*, y^e^ points *R*, *W*, *Q*, will bee found in y^e^ same quarter of y^e^ Assymptotes *NPS* wth y^e^ circle *ATV* & therefore y^e^ Hyperbola must bee described in that quarter as in fig 1. 2 If $ce = dr$*, y^n^ those points *R*, *W*, & *Q* will fall in y^e^ Assymptots & signify y^e^ Hyperbola to bee streight lines coincident wth them. And in this case if $r \sqsubset c$, all the rays round from *B* will be refrac[te]d so y^t^ either *T* or *V* may bee considered as vertices of a Lens. & *KG* falls from y^e^ intersection point *K* wch is nearest to *I*. In wch case therefore y^e^ point *G* is always y^e^ same w^th^ *N* when *T* is vertex, but if $r \sqsubset c$ & *V* is vertex, *K* shall not fall on *PI* but on *PS* twixt *P* & *S*. & If a ray fall on y^t^ point discriminating betwixt y^e^ vertices *I* shall fall on *P*. 3 If $ce \sqsubset dr$*. Those points fall in the quarters on either side of y^e^ circle in wch quarters therefore y^e^ Hyperbolas must bee described as in fig 2. & if *T* is vertex *KG* falls from y^e^ intersection point *K* of y^e^ Hyperbola *Qδ*

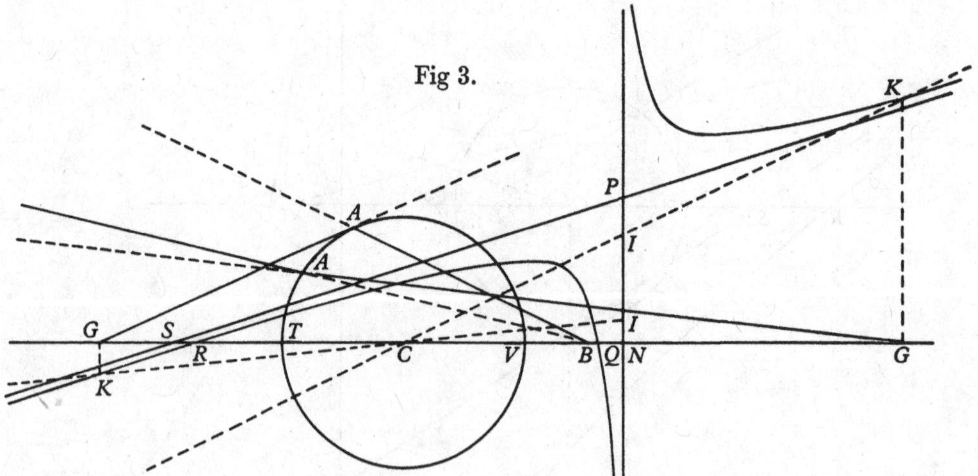

Fig 3.

(17) Read '*CB*'.

beyond I. but for the rays w[ch] respect y[e] vertex V, KG falls from that point of y[e] other Hyperbola RW w[ch] is nearest y[e] center C & remotest from I, & then CI intersects not y[e] Hyperbola δ. (*that is if $CB \sqsubset, =[,] \sqsupset CN$)

Case 3. If $\frac{rr}{c}+c+2r \sqsubset \frac{eec}{dd} \sqsubset \frac{rr}{c}+c$, y[t] is if $PN \sqsupset r$ & P falls above N (as in fig [3]) Then shall the refracted rays intersect y[e] axis some on y[e] same side some on y[e] contrary side of y[e] vertex. On y[e] same side if CI intersects y[e] Hyperbola δQ; if not y[n] on y[e] contrary side (except one ray $\parallel CN$[)] KG falling from y[e] intersection point of y[e] Hyperbola WR w[ch] is remotest from I.

Case 4. If $\frac{eec}{dd} \sqsubset \frac{rr}{c}+c+2r$. y[t] is if P falls above N & $NP \sqsubset TV$, Then all y[e] refracted rays intersect y[e] axis on y[e] contrary side of y[e] Vertex. KG falling from y[t] point of y[e] Hyperbola RW w[ch] is remotest from I.(18)

(18) Newton's discussion of the niceties of these particular cases follows straightforwardly from suitable interpretation of the hyperbola equation (note (15)).

The general intention of Newton's elaborate scheme seems to be the construction, as variants on a single basic method, of the totality of possible paths of a light ray issuing from a given point and refracted at a given spherical surface. He does not seem to have noticed that, by reversing his construction, he was now able to resolve a particular case of the generalised Alhazen problem: where the points B and G are in line with the spherical centre C, to find the points A on the sphere at which a light ray through B is refracted through G. He was, however, fully aware that his construction held true at once for both external and internal refraction and therefore that there was no need to consider such cases separately. As he wrote in a transitional draft on 93[v], 'Each operation satisfys two Quærys viz How ye rays are refracted in respect of ye Vertex W as well as V'. (The figure of [1] is understood.)

INDEX OF NAMES